STEREOCHEMISTRY
CONFORMATION AND MECHANISM

STEREOCHEMISTRY
CONFORMATION AND MECHANISM

P.S. Kalsi
Formerly Dean Colleges
Punjab Technical University, Jalandhar
Formerly Professor and Head
Department of Chemistry
College of Basic Sciences and Humanities
Punjab Agricultural University, Ludhiana

NEW AGE
TECHNO
PRESS

An Imprint of

NEW AGE INTERNATIONAL (P) LIMITED, PUBLISHERS
New Delhi • Bangalore • Chennai • Cochin • Guwahati • Hyderabad
Jalandhar • Kolkata • Lucknow • Mumbai • Ranchi
Visit us at **www.newagepublishers.com**

Published by New Age International (P) Ltd., Publishers
First Edition: 1990
Seventh Edition: 2008
Reprint: 2008

Branches:

- 36, Malikarjuna Temple Street, Opp. ICWA, Basavanagudi, **Bangalore**. ✆ (080) 26677815
- 26, Damodaran Street, T. Nagar, **Chennai**. ✆ (044) 24353401
- Hemsen Complex, Mohd. Shah Road, Paltan Bazar, Near Starline Hotel, **Guwahati**. ✆ (0361) 2543669
- No. 105, 1st Floor, Madhiray Kaveri Tower, 3-2-19, Azam Jahi Road, Nimboliadda, **Hyderabad**. ✆ (040) 24652456
- RDB Chambers (Formerly Lotus Cinema) 106A, Ist Floor, S.N. Banerjee Road, **Kolkata**. ✆ (033) 22275247
- 18, Madan Mohan Malviya Marg, **Lucknow**. ✆ (0522) 2209578
- 142C, Victor House, Ground Floor, N.M. Joshi Marg, Lower Parel, **Mumbai**. ✆ (022) 24927869
- 22, Golden House, Daryaganj, **New Delhi**. ✆ (011) 23262370, 23262368

ISBN (10): 81-224-2356-6

ISBN (13): 978-81-224-2356-3

Rs. 295.00

C-08-08-3081

Printed in India at Glorious Printers, Delhi.
Typeset at Goswami Associates, Delhi.

PUBLISHING FOR ONE WORLD
NEW AGE INTERNATIONAL (P) LIMITED, PUBLISHERS
4835/24, Ansari Road, Daryaganj, New Delhi-110002
Visit us at **www.newagepublishers.com**

UNIVERSITY OF PUNE

DEPARTMENT OF CHEMISTRY

Dr. M. S. WADIA

Emeritus Professor of Organic Chemistry

Telephone : Resi : 25534361, 25531579

Ganeshkhind,
PUNE–411 007 (INDIA)

Tel. Add. : UNIPUNA

13-2-2008

Foreword

With each edition of this extremely useful book, Prof. Kalsi has made a significant improvement in the material selected, the presentation and the problems. The main aim of this book is to make the subject interesting and to excite the students to a deeper understanding of this topic which is often not completely appreciated. The inclusion of a lucid chapter on pericyclic reactions is an important facet to enhance the utility of this book. The combination of dynamic and static ideas of stereochemistry in this edition would make this book useful not only to organic chemists but also to pharmacologists, medical chemists and biochemists. Prof. Kalsi has drawn on his long research experience in chemistry of natural products and as a teacher to nicely combine stereochemical and mechanistic aspects.

I feel confident that postgraduate students and those interested in knowing the interplay of stereochemistry conformation and mechanism will greatly appreciate this book.

M.S. WADIA

Preface to the Seventh Edition

In the seventh edition of the book Stereochemistry, Conformation and Mechanism each chapter has been reworked, updated and reorganized.

WHAT'S NEW?

- New exercises with their solutions have been added throughout the text in separate boxes. These problems provide interesting examples which illustrate the point under discussion.
- More new end-of-chapter problems have been added.
- Wealth of information in an easy-to-use format has been added in boxes throughout the text. This provides additional information on important concepts regarding the topic under discussion.
- Problem solving hints are added to emphasize the application of knowledge quickly to solve a problem. This motivates the student to active learning and its application so as to gauge his understanding of the subject.
- Extensive additional material has been added on soft descriptors *threo, erythro, syn* and *anti*. Stereodiscriptors *l*, *u* and prefixes *r*, *s* and *re*, *si* and their use has been explained by a variety of new examples.
- A new section on NMR spectroscopy and stereochemical information has been provided.
- More of the best understood and recent enantioselective reactions have been added and discussed.

It is hoped that the text in its new seventh edition with entirely new artwork, highlights, more added material and problems will now assist the students in a far better way to attain and develop knowledge in organic stereochemistry, conformation and mechanism.

Ludhiana **P.S. Kalsi**

Preface to the First Edition

The French chemist Louis Pasteur (1850) was the first to reveal that optical activity at the molecular level was due to an asymmetric placement of atoms in a molecule. He further recognized that enantiomers had equal and opposite effects on polarized light. vant Hoff and Le Bel in 1874 (Nobel Prize) proposed the tetrahedral model for the carbon as the cause of molecular dissymmetry and the resultant optical rotation. This was the beginning of stereochemistry which forms a massive and an integral part of organic chemistry.

In recent years stereochemistry has not only expanded in depth but, with the development of new kinds of nomenclature and concepts, it has also almost taken on an entirely new perspective. These developments offer knowledge of great interest to students and the present book incorporates this knowledge not only for the uninitiated student but for the expert as well.

The intention of this book, is to give equal weightage to various aspects of stereochemistry viz., chirality, stereochemical course of various reactions, molecular rearrangements and methods of assigning configurations. Though treated lightly, but several examples from the major special areas of organic chemistry: Terpenoids, steroids, alkaloids and carbohydrates have been presented to explain various stereochemical, mechanistic and conformational aspects.

The book runs into various chapters each of which is reasonably complete in itself. The first chapter describes chirality and some related aspects to provide a necessary background for understanding mechanism and conformations presented in subsequent chapters and is important in its own right. Some emphasis has also been laid on spectral methods is solving stereochemical problems.

I have tried to develop an internal order of almost all the basic concepts, to help the students to be able to apply the knowledge gained to new situations. Special attention is paid in illustrating the introduction to each chapter in some depth. The student may thus, not find himself in a situation where his understanding a concept requires knowledge that he has not gained. Special attention is paid to serve the needs of students in a way that will enhance their comprehension of the subject of stereochemistry.

It is now recognized that Hückel-Möbius approach can be gainfully and effectively used to explain pericyclic reactions, therefore, I have introduced this as the only approach to make several aspects of these reactions more understandable.

Problems and their solutions are provided at the end of each chapter and these are designed to test the students understanding of each topic up to that point. Chapter summaries are presented in the appendix.

Advances in a scientific field are often reported in somewhat condensed communications, and find elaboration in papers and books after a lapse of some years. My acknowledgement, therefore, is to many authors who have preceded me in this task of authorship and from whose efforts I have drawn innumerable facts, ideas and theories. Particularly to some of these sources, mentioned in references, I am indebted.

Ludhiana **P.S. Kalsi**

Contents

Foreword *v*
Preface to the Seventh Edition *vii*
Preface to the First Edition *ix*

1 Chirality **1**

1.1 Introduction 1
1.2 Isomers 2
1.3 Chiral Molecules-Enantiomerism (Optical Isomerism) a Summary 16
1.4 Conventions for Configurations D–L and *R*–*S* Systems 31
1.5 Interconversion Between Fischer and Three-dimensional (Wedge) Formulas With One Stereocenter 49
1.6 Stereoisomerism Resulting from More Than One Stereogenic Unit, *Threo* and *Erythro* Nomenclature 51
1.7 Some Aspects of Geometric Isomerism (*Pi* Diastereoisomerism and Torsional Chirality in Carbon Carbon Double Bonds) 71
1.8 Symmetry Elements, Operations, Point Groups and Stereochemical Properties 88
1.9 Symmetry Elements, Asymmetry, Chirality a Combined Look 93
1.10 Resolution of Racemic Mixtures 97
1.10A Measurement of Optical Activity 107
1.11 Racemic mixture and Racemization 107
1.12 Some Stereochemical Reactions Near a Stereocenter (Formation of Diastereomers) 113
1.13 Stereoselective And Stereospecific Reactions 116
1.14 Enantiomeric Excess—Optical Purity 124
1.15 Epimerization, Epimers, Anomers and Mutarotation (Asymmetric Transformations) More Methods of Total Spontaneous Resolution 125
1.16 Optical Activity-stereoisomerism in Compounds Without a Stereogenic Carbon—(Axial Chirality) 129
1.17 Optical Activity due to Stereoplane (Planar Chirality) 146
1.18 Optical Activity of Compounds due to Helicity 149
1.19 Molecular Stereoisomerism 150

1.20 Chirality in Metallic Complexes 151
1.21 Chirality Involving Atoms Other than Carbon 151
1.22 NMR Spectroscopy and Stereochemical Information 153
Problems 158
Answers to Selected Problems 164
References and Further Reading 172

2 Prochirality—Prostereoisomerism and Asymmetric Synthesis 173

2.1 Introduction—Nomenclature 173
2.2 Homotopic and Heterotopic Ligands and Faces 175
2.3 Nomenclature of Enantiotopic Ligands and Faces and Some Stereoselective Reactions 183
2.4 Asymmetric Synthesis—Stereoselective Synthesis (Stereoselectivity) 190
2.5 Stereoheterotopic Ligands and NMR Spectroscopy 218
Problems 220
Answers to Selected Problems 222
References and Further Reading 224

3 Aliphatic Nucleophilic Substitution 225

3.1 Introduction 225
3.2 S_N2 Reaction Mechanism and Evidence 226
3.3 S_N1 Reaction 242
3.4 Nucleophilic Substitution of Allylic, Systems S_N1' and S_N2' Reactions. Rearrangement in Allylic Systems 253
3.5 Nucleophilic Displacements at Allylic Halides/Tosylates 255
3.6 Nucleophilic Substitution at the Benzylic Position 256
3.7 Nucleophilic Substitution of Vinylic and Aryl Halides 257
3.5 The S_Ni mechanism 259
3.6 Mixed S_N1 and S_N2 Reactions 261
3.7 Ambident Nucleophiles–Regioselectivity 262
3.8 Set Mechanisms 263
3.9 Neighboring Group Participation—Anchimeric Assistance 264
Problems 278
Answers 282
References and Further Reading 285

4 **Conformations and Stereo-isomerism of Acyclic and Cyclic Systems** 286

4.1 Restricted Rotation about Single Bonds—Conformations of Ethane and Butane 286
4.2 Origin of Conformational Energy 292
4.3 Conformation and Chemical Reactivity in Acyclic Systems (E2 Elimination and Pinacol Rearrangement) 292
4.4 Angle and Pitzer Strain 298
4.5 Cycloalkane Rings other than Cyclohexane 298
4.6 Conformations of Cyclohexane (Conformational Analysis) 301
4.7 Equatorial and Axial Bonds in Chair form of Cyclohexane 304
4.8 Conformations of Substituted Cyclohexanes 306
4.9 Conformational Analysis of Mono and Disubstituted Cyclohexanes—Relative Stability of Conformers 312
4.10 Stereoisomerism of Disubstituted Cyclohexanes (*cis*- and *trans*-isomers and chirality) 313
4.11 Equilibria of Disubstituted Cyclohexanes and Related Systems 321
4.12 Effect of Conformation on Reactivity—Cyclic Systems—Introduction 323
4.13 Some Typical Cyclohexanes 324
4.14 Locking of Flexible Conformations 325
4.15 Conformation of Heterocycles 326
4.16 ConformatIon of Sugars and Anomeric Effect 328
(A) Conformation of Common Sugars 328
(B) Conformation of an L-Pyranose (β-L-Gulose) 329
(C) Anomeric Effect 330
4.17 Polycyclic Compounds 331
4.18 Cyclohexene and Cyclohexanones 342
Exercises and Problems 345
Answers to Selected Problems 347
References and Further Reading 350

5 **Stereochemistry of Elimination Reactions** 351

5.1 Introduction 351
5.2 Mechanisms E1, E2 and ElcB 352
5.3 E1, E2 and ElcB Variables—Their Mechanistic Spectrum 355
5.4 Competition between Elimination and Substitution 356
5.5 Stereochemistry of E2–*anti*-Elimination Reactions 359
5.6 E2-*syn*-Elimination 366

5.7 Orientation of the Double Bond—Regiochemistry of the Elimination Reactions ... 368
5.8 Pyrolytic Eliminations ... 373
5.9 Molecular Rearrangement During Elimination ... 379
5.10 Fragmentation Reactions ... 380
Problems ... 381
Answers to Selected Problems ... 382
References and Further Reading ... 384

6 Stereochemistry and Mechanism of Some Additions to Carbon–Carbon Multiple Bonds 385

6.1 Addition Reactions of Alkenes Involving Electrophiles ... 385
6.2 Addition Reactions of Alkynes Involving Electrophiles ... 402
6.3 Hydroboration/Oxidation Hydration of Alkenes: A Stereospecific Anti-Markovnikov Hydration ... 404
6.4 Hydrogenation of Alkenes and Alkynes ... 419
6.5 Hydrogenation of Aromatic Rings—Birch Reduction ... 428
6.6 Addition of Nucleophiles to Carbon-carbon Double Bond ... 429
6.7 Addition of Free Radicals to *Pi*-systems ... 433
6.8 Additions to Cyclopropane Ring Systems ... 437
6.9 Epoxidation of Alkenes ... 438
6.10 Hydroxylation of Alkenes—Diol Formation ... 443
6.11 Addition of Carbenes (Methylene) to Alkenes ... 447
Problems ... 449
Answers to Selected Problems ... 451
References and Further Reading ... 456

7 Addition to Carbon Hetero Multiple Bonds—Stereochemistry and Mechanism 457

7.1 Introduction ... 457
7.2 Some General and Stereochemical Aspects of Addition to Carbonyl Compounds ... 459
7.3 Mechanism and Stereochemistry of Metal Hydride Reduction ... 460
7.4 The Hydride Transfer Cannizzaro Reaction ... 468
7.5 Meerwein—Ponndorf Reduction the Hydride Transfer Reaction ... 468
7.6 Addition of Organometallic Compounds to Carbon—Hetero Multiple Bonds ... 470
7.7 Conjugate Addition of Organocopper Reagents to α, β-Unsaturated Carbonyl compounds–Gilman Reagents ... 477

7.8 Organozinc Compounds 478
7.9 The Wittig Reaction (Use of Carbanions in Organic Synthesis) 479
7.10 Mechanism of Some Condensation Reactions Involving Enolates 485
7.11 Hydrolysis of Esters 496
7.12 Hydrolysis of Amides 503
Problems 504
Answers to Selected Problems 506
References 507

8 Pericyclic Reactions 508

8.1 Conservation of Molecular Orbital Symmetry 511
8.2 Methods to Explain Pericyclic Reactions 511
8.3 Electrocyclic Reactions (FMO-Approach) 512
8.4 Cycloadditions (FMO-approach) 533
8.5 1,3-dipolar Cycloadditions 547
8.6 Cheletropic Reactions 548
8.7 Sigmatropic Rearrangements 550
8.8 The Ene Reaction 566
8.9 Aromatic Transition States 567
8.10 Möbius-Hückel Analysis (PMO) Approach 568
8.11 Correlation Diagram Method 574
Problems 579
Answers to Selected Problems 582
References 585

9 Aliphatic Electrophilic Substitution 586

9.1 Introduction 586
9.2 Different Mechanisms for Aliphatic Electrophilic Substitution (Displacement of One Electrophile by Another) 588
9.3 Electrophilic Substitution Accompanied by Double Bond Shift 591
9.4 Aliphatic Electrophilic Substitution in Relation to Substrate Structure, Leaving Group and Solvent Polarity 592
9.5 More Reactions Involving Aliphatic Electrophilic Substitution 593
Problems 595
Answers to Selected Problems 596
References 596

Index 597

CHAPTER 1

Chirality

1.1 INTRODUCTION

The independent work of J.H. van't Hoff first recipient of Nobel Prize in chemistry in 1901 and J.A. Le Bel provided evidence that the four bonds of the carbon atom in methane, for example, are arranged in a way, so as to point towards the corners of a regular tetrahedron (Fig. 1.1) if the carbon atom was placed at its center.

The tetrahedral geometry of carbon and a few other atoms (pyramidal) *e.g.*, N, P, Si and S and trigonal geometry of sp^2 hybrid carbon are key to the study of organic stereochemistry. The regular tetrahedron (Fig. 1.1) is highly symmetrical with four equivalent vertices. When these are occupied by four different achiral atoms or groups, symmetry disappears and a chiral complex is obtained (C_{abcd}) which is non superimposable on its mirror image (Fig. 1.1).

The tetracovalent carbon (the fifth point) placed at the center of the tetrahedron is termed a stereocenter whose presence usually generates molecular chirality. A unique feature of a stereocenter is that an exchange of any two ligands reverses its chirality to give a new stereoisomer. This *e.g.*, may be seen in a situation if all the ligands in C_{abcd} are achiral and such an exchange then gives the mirror image—the enantiomer.

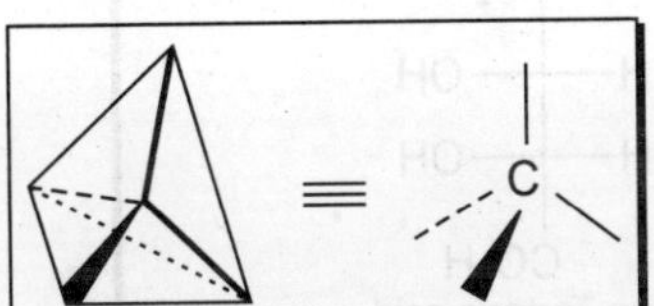

Regular tetrahedron

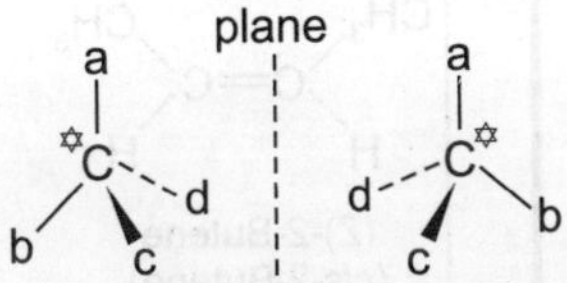

C✡ = a stereocenter

Chiral   Chiral

In tetrahedral structure of methane of the four bonds to central carbon two are in the plane of paper (continuous lines), third is below this plane (dashed wedge) and fourth is above this plane (solid wedge))

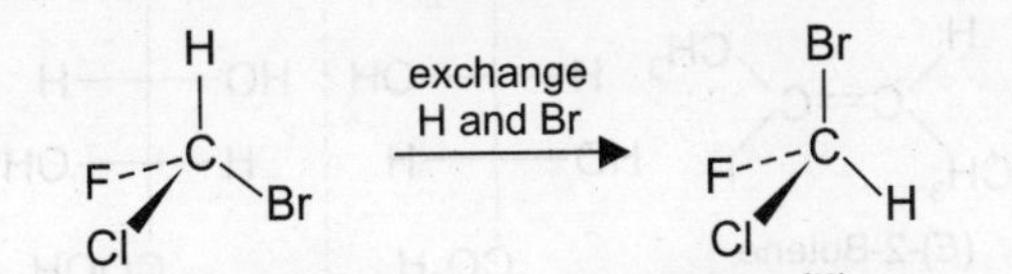

(I)
A chiral molecule

(II)
Enantiomer of (I)

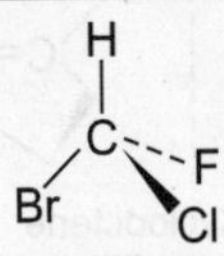

(III)
Drawan as mirror image of (I), II ≡ III.

Fig. 1.1

1.2 ISOMERS

Isomers are defined as different compounds that have the same molecular formula. These are named depending on the difference which may be structural or stereoisomeric (scheme 1.1).

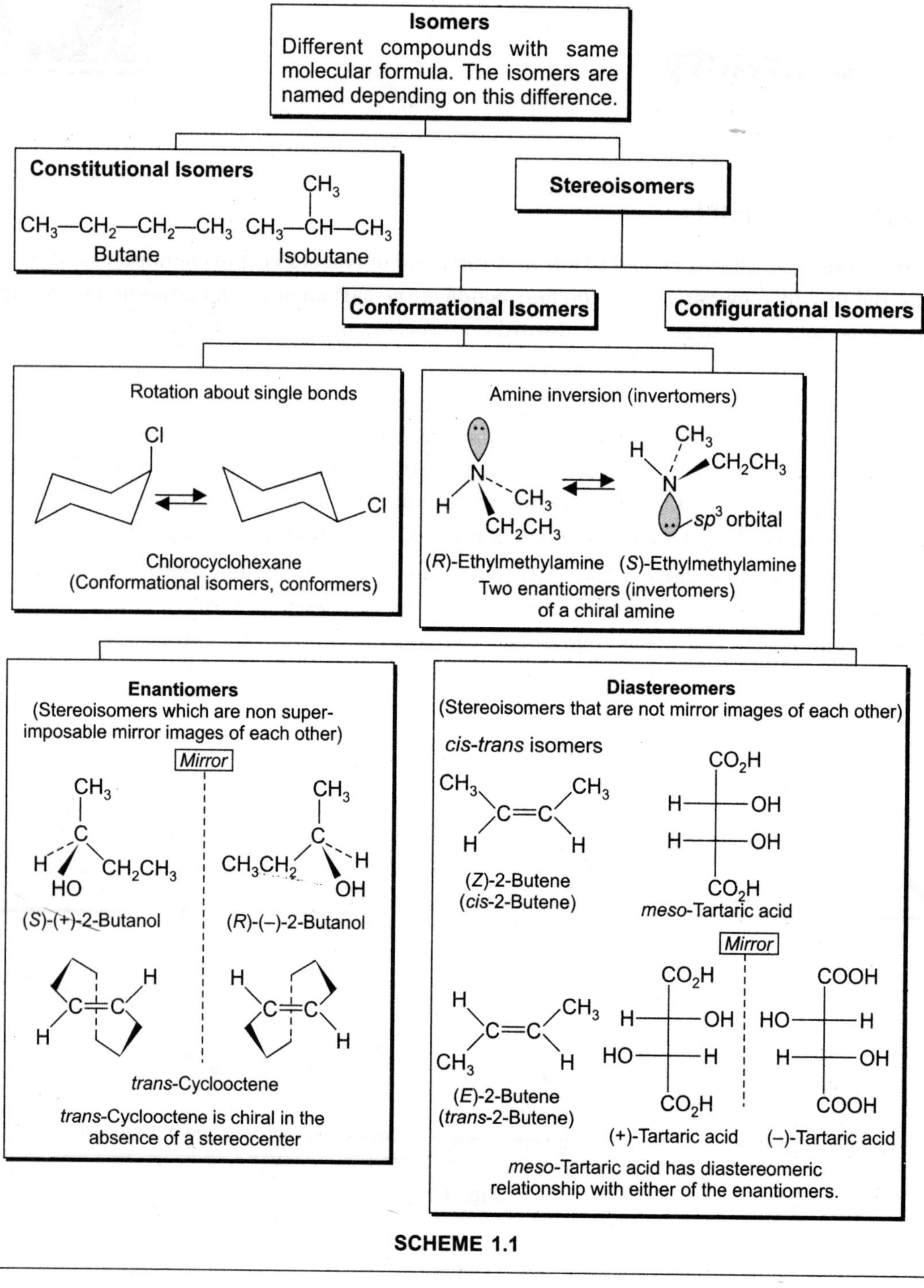

SCHEME 1.1

(A) Constitutional Isomers

Constitutional isomers have different sequence of bonding of their atoms as with the molecular formula C_2H_6O, which may represent ethanol CH_3CH_2OH or dimethyl ether CH_3OCH_3 (functional isomers) ; C_4H_{10} may be either butane or isobutane (chain isomers) and C_3H_7Cl can be 1-chloropropane or 2-chloropropane (position isomers). Glucose and fructose (scheme 1.1*a*) are constitutional isomers, glucose is an aldehyde while fructose is a ketone. Some other examples of constitutional isomers include :

- *Tautomers.* Isomers of different energies which are interconvertible *via* a low energy barrier, the isomerization involves atom or group migration.
- *Proton tautomers* (*Prototropy*). Enol-keto isomerization is an example of prototropy (*i.e.*, a change in the position of a proton) and the interconversion of the tautomers, 2-hydroxypyridine and pyridone is an example (scheme 1.1*a*) of prototropy *i.e.* change in the position of a proton and involves proton tautomers.
- *Valence isomers.* These isomers or degenerate species that are interconvertible by reorganization of some of the bonding electrons. The interconversion is accompanied by atom movement and not atom migration. Thus valence isomers are not tautomers. Valence isomers can be separately identified and in case these have the same structure (degenerate species) the individual atoms can be separately identified. The interconversion of cyclooctatetraene and its bicyclic isomer is an example of valence isomers (scheme 1.1*a*)

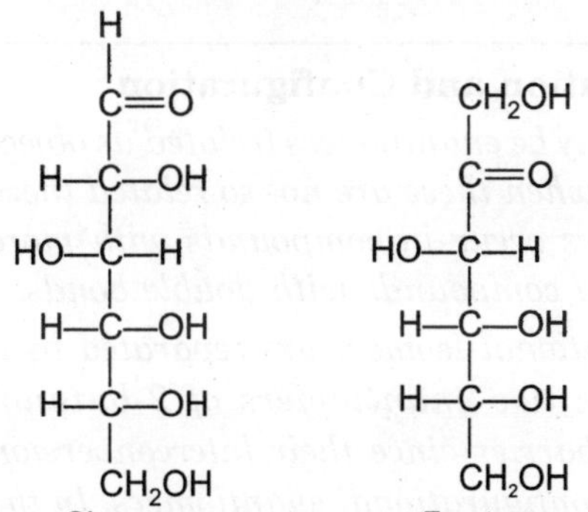

Glucose Fructose

Constitutional Isomers

> *Constitutional isomers differ in the connectivity of atoms including bond multiplicity; disregarding configuration and conformation. The equivalent older term structural isomers is obsolete and redundant. All isomers are structural isomers and structure includes constitution, configuration and conformation.*

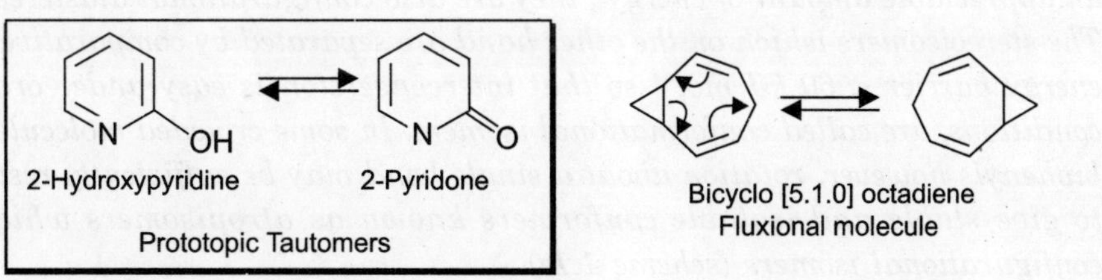

2-Hydroxypyridine 2-Pyridone

Prototopic Tautomers

Bicyclo [5.1.0] octadiene
Fluxional molecule

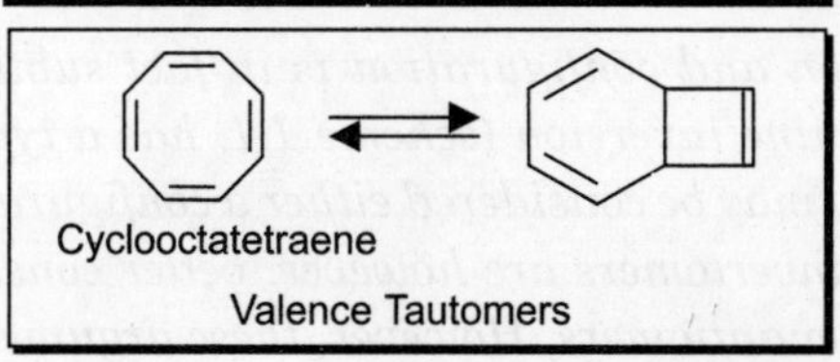

Cyclooctatetraene

Valence Tautomers

Constitutional Isomers

SCHEME 1.1a

- *Fluxional molecules.* Molecules which undergo rapid degenerate rearrangement *i.e.,* a rearrangement into indistinguishable molecules *e.g., via* bond reorganization (scheme 1.1*a*).
- *Allylic isomers*. These isomers result due to a rearrangement involving the allyl (propenyl group) and is recognized separately. These rearrangements may occur *via* the intermediate formation of a delocalized ion or a radical. The loss of X at one end and its return to the opposite end of the allyl system (scheme 1.1*b*) leads to the overall rearrangement. The allylic isomers arise due to a 1, 3-rearrangement.

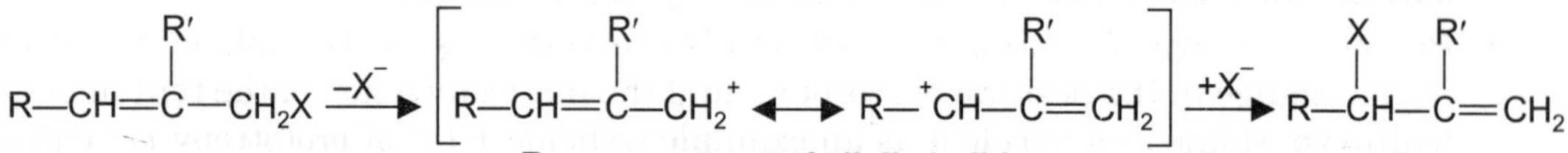

Rearrangement of allylic halides

SCHEME 1.1b

(B) Stereoisomers—An Introduction

When the isomers have the same sequence of covalent bonds, but differ in the relative disposition of their atoms in space, then the difference is stereoisomeric (scheme 1.1). Some examples are, enantiomers, diastereomers (epimers, anomers), conformational isomers (atropisomers and invertomers).

The Distinction between Conformation and Configuration

- *Based on symmetry criteria, stereoisomers may be enantiomers (related as object and nonsuperimposable mirror image) and when these are not so related these are termed diastereomers. The diastereomers occur in compounds with more than one stereocenter, cyclic compounds and compounds with double bonds.*
- *Based on energy barrier criterion, configurational isomers are separated by a high energy barrier > 100 kJ mol^{-1}. The two enantiomers of 2-butanol (scheme 1.1) are separated by high energy barrier since their interconversion involves a σ bond breaking. They represent configurational enantiomers. In the isomerisation of cis- and trans-2-butene a π bond is disrupted which requires an appreciable amount of energy ; they are also configurational diastereomers. The stereoisomers which on the other hand are separated by comparatively low energy barrier < 60 kJ mol^{-1} so that interconversion is easy under ordinary conditions, are called conformational isomers. In some crowded molecules e.g., biphenyls however, rotation about a single bond may be sufficiently restricted to give stable and isolable conformers known as atropisomers which are configurational isomers (scheme 1.1e).*

  The distinction between conformation and configuration is in fact subtle and not agreed upon universally. The amine inversion (scheme 1.1) has a typically low energy barrier (33.5 kJ / mol) and may be considered either a configurational or a conformational change. These invertomers are however, better considered as conformers or as conformational enantiomers. However, these arguments do not apply to chiral phosphines (scheme 1.3a) where inversion is associated with high energy (~ 150 kJ/mol) to result in configurational isomers.

Rotation around a single bond may be easy to give conformational isomers e.g., in n-butane (scheme 1.1c). Gauche butane is chiral. Two enantiomers interconvert by a conformational change (conformational enantiomers). Anti butane is achiral and either of the gauche butanes is its diastereomer (conformational diastereomers). Gauche butane is an example of racemization.

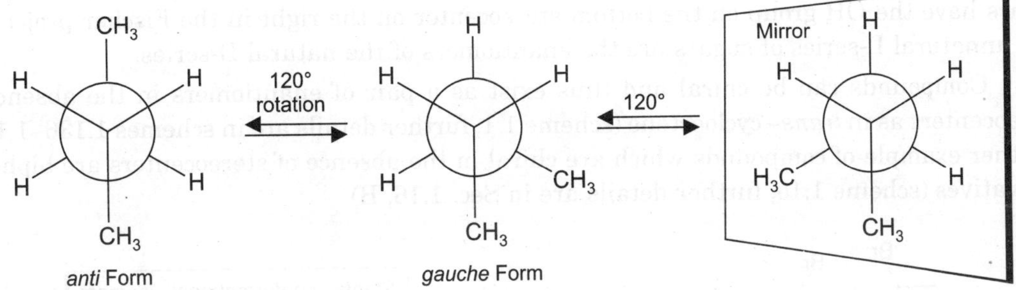

SCHEME 1.1c

Similarly chlorocyclohexane (scheme 1.1) represents a pair of conformational diastereomers. In fact chlorocyclohexane shows conformational isomerism at room temperature while configurational isomerism at – 150°C (see Fig. 4.2). Another interesting example of conformational enantiomers is in (scheme 4.33).

EXERCISE 1.1

What is the stereochemical relationship among the conformers (scheme 1.1d) ?

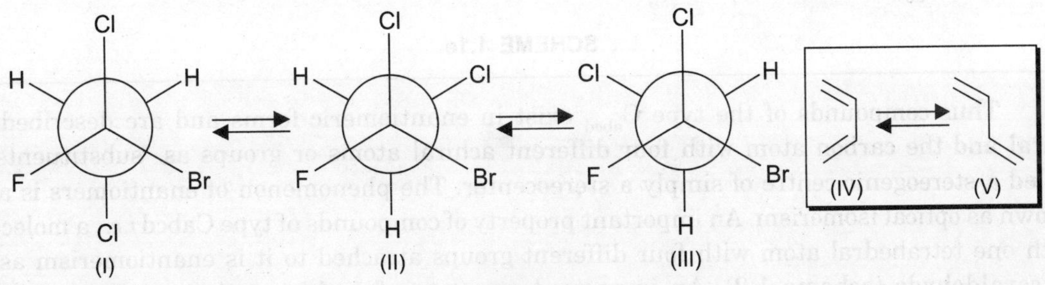

SCHEME 1.1d

ANSWER. *Compounds (I–III) are conformational diastereomers and (IV) and (V) also represent conformational diastereomers.*

1. Enantiomers—Optical Isomerism

(*a*) Simple organic molecules

Consider a simple molecule *e.g.*, a compound with an sp^3 hybridized carbon with four different substituents as in 2-butanol (scheme 1.1). The molecule cannot be superimposed on its mirror image and such molecules are said to be chiral (or handed). The pair of butanol molecules are termed enantiomers (from the Greek 'enantio' meaning opposite) which are defined as pair of molecules related as non-superimposable mirror images. The enantiomers of 2-butanol are drawn (scheme 1.1) in the three dimensional projection formulas (a procedure to draw these projections is depicted in scheme 1.15). Another example of enantiomers is in D and L-glyceraldehydes and D- and L-glucose drawn now in another projection (scheme 1.2) called a Fischer projection, (a procedure to draw these projections is detailed in scheme 1.17). The D-sugars have the OH group on the bottom stereocenter on the right in the Fischer projection. The unnatural L-series of sugars are the enantiomers of the natural D-series.

Compounds can be chiral and thus exist as a pair of enantiomers in the absence of stereocenters as in *trans*– cyclooctene (scheme 1.1, further details are in schemes 1.136–1.138). Another example of compounds which are chiral in the absence of stereocenters are biphenyl derivatives (scheme 1.1e, further details are in Sec. 1.16, B)

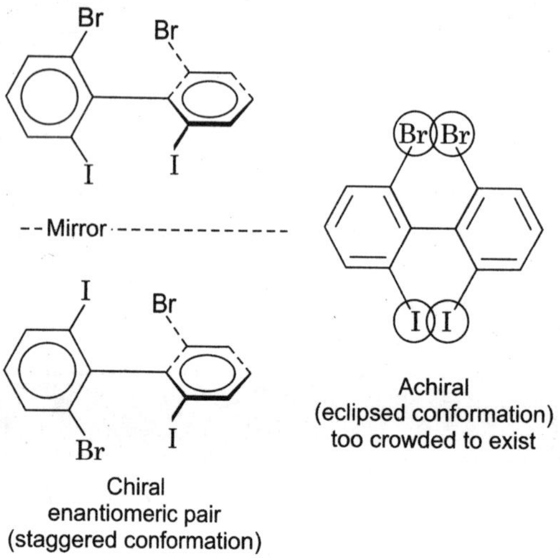

Stable conformational isomers (atropisomers) exist in compounds, e.g., biphenyls due to steric strain between the ortho substituents. These isomers become chiral when both rings are unsymmetrically substituted. The biphenyl then gets locked in one of the two chiral enantiomeric staggered conformations. This biphenyl would have been achiral if a symmetric (planar) high energy eclipsed conformation could be achieved (an impossible situation.)

SCHEME 1.1e

Thus compounds of the type C_{abcd} exist in enantiomeric forms and are described as chiral and the carbon atom with four different achiral atoms or groups as substituents is called a stereogenic centre of simply a stereocenter. The phenomenon of enantiomers is also known as optical isomerism. An important property of compounds of type Cabcd *i.e.*, a molecule with one tetrahedral atom with four different groups attached to it is enantiomerism as in glyceraldehyde (scheme 1.2). An important property of such enantiomers (*i.e.*, a chiral tetrahedral model) is that on interchanging any two groups at the stereocenter converts one enantiomer into another. In addition to the compounds of the type C_{abcd} with one stereocenter *e.g.*, glyceraldehyde (scheme 1.2) and 2-butanol (scheme 1.1) which fulfill the conditions for the occurrence of enantiomeric pairs, several other structural situations may give rise to optical isomerism. These include compounds with more than one stereocenter (as in glucose, scheme 1.2), stereocenters other than carbon (sec. B, IV) and compounds which are optically active in the absence of stereocenters (scheme 1.1*e*).

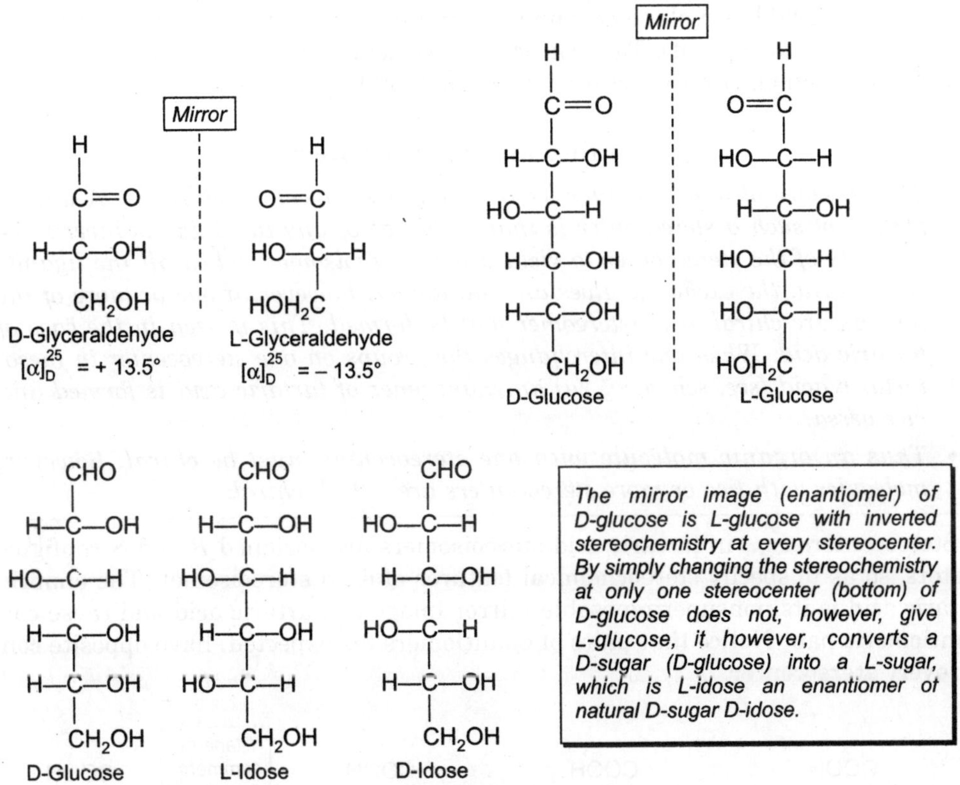

SCHEME 1.2

Chiral Organic Compounds

- *The presence of a stereocenter usually leads to molecular chirality.*
- *A tetrahedral atom or a pyramidal atom with three ligands (the lone pair of electrons serves as the fourth ligand) gives a stereocenter provided an interchange of any two ligands (this process reverses the chirality of the center) leads to a new stereoisomer.*
- *The presence of a stereocenter in an organic molecule is a sufficient condition for chirality, however it is not a necessary condition.*
- *Several molecules display chirality (optical isomerism) in the absence of stereocenters e.g., chiral biphenyls. Thus a compound is chiral if it is not superimposable on its mirror image.*
- *Organic stereochemistry is based on tetrahedral geometry of carbon which is absolutely central to its study. Study of stereochemistry is also based on atoms like N, P, Si and S and to a lesser extent on the trigonal geometry of sp^2 hybrid carbon and nitrogen.*

In tartaric acid (scheme 1.2*a*) one has two stereocenters. A molecule with two stereocenters can give rise to a maximum of four stereoisomers (2^n as also in 2-bromo-2-butanol, see scheme 1.33). However, if the two stereocenters carry an identical set of substituents, the number of stereoisomers is less than 2^n, since there will be a *meso* compound (scheme 1.2*a*).

The *meso* tartaric acid has a plane of symmetry and is achiral (it has a mirror image which is, however, superimposable on it). Tartaric acid stereoisomers are drawn in Fischer projections (The Fischer projections are the eclipsed conformations).

Chirality and Stereocenters

- *The presence of a stereocenter usually imparts molecular chirality. A unique feature of such a stereocenter is that exchange of any two ligands inverts the chirality of the stereocenter to yield a new stereoisomer. When all the ligands are achiral, the exchange gives an enantiomer, however, if one or more of the ligands are chiral, a diastereomer will be formed. This is seen in the case of tartaric acid. When one interchanges the groups on one stereocenter in meso-tartaric acid (see, scheme 1.2a) an enantiomer of tartaric acid is formed and vice versa.*
- *Thus an organic molecule with one stereocenter must be chiral, however, molecules with two or more stereocenters are not all chiral.*

Stereocenters *e.g.*, in tartaric acid stereoisomers are assigned *R* and *S* configurational descriptors, so as to specify stereochemical features of each stereoisomer. The enantiomer of (+) tartaric acid is its nonsuperimposable mirror image (–) tartaric acid and these constitute an enantiomeric pair. Notice that pairs of enantiomers (as expected) have opposite configuration at every stereocenter.

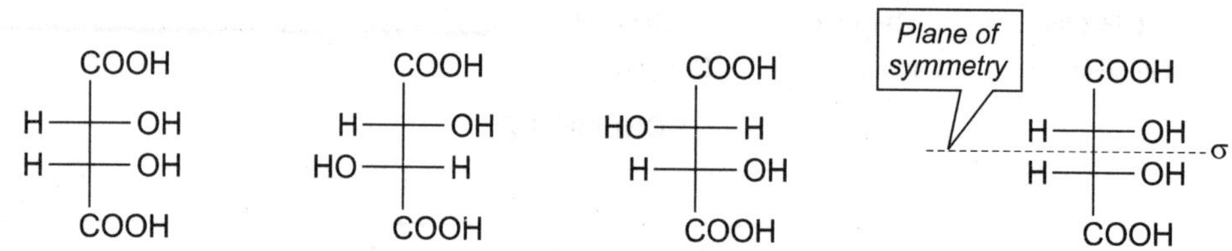

(2*R*,-3*S*)-tartaric acid (a *meso* compound) — (2*R*,-3*R*)-tartaric acid, (2*S*,-3*S*)-tartaric acid (a pair of enantiomers) — *meso*-tartaric acid (achiral)

Stereoisomers of tartaric acid in Fischer projections

Physical properties of stereoisomers of tartaric acid

	Melting point, °C	$[\alpha]_D^{25°C}$	Solubility, g/100 g H_2O at 15°C
(2R, 3R)-(+)-tartaric acid	*171*	*+ 12.7°*	*139*
(2S, 3S)-(–) tartaric acid	*171*	*– 12.7°*	*139*
(2R, 3S)-tartaric acid	*140*	*0°*	*125*
(±)-tartaric acid	*206*	*0°*	*20.6*

SCHEME 1.2a

(*b*) Complex organic molecules and biomolecules

Except for few low molecular weight organic compounds, the organic substances found in living systems both animals and plants are chiral. No doubt these molecules (with several stereocenters) can theoretically exist as a number of stereoisomers, almost invariably only one stereoisomer is found in nature. Naturally occurring alkaloid brucine has several stereocenters which are located in fused ring systems, however, nature makes only one enantiomer (–)-brucine. Naturally occurring amino acids (with the exception of achiral glycine) are chiral. There are two possible enantiomers (optical isomers) for each amino acid, but only one of them (L-form) exists in the body. Enzymes are proteins which are derived from chiral amino acids, thus an enzyme is also chiral and can exist as enantiomers, however only one enantiomer exists naturally (since an amino acid exists only as one enantiomer these will construct only one mirror image form of the enzyme). Thus enzymes provide a chiral environment.

Enzymes catalyzed reactions are stereospecific and stereoselective

- *For definition of terms, stereospecific and stereoselective (see Sec. 1.13)*
- *Enzymes are chiral and enantiomerically pure*
- *Enzymes display stereospecificity and stereoselectivity (see Sec. 2.3 D)*
- *All stereospecific reactions are necessarily stereoselective, however, the converse is not true.*

2. Properties of Enantiomers

Each enantiomer of a pair has the same physical and chemical properties in achiral environments with the important exceptions of their interactions with (*i*) plane polarized light (optical activity) and (*ii*) chiral reagents. When plane polarized light is passed through the solution of each enantiomer (in the same solvent, using the same cell and same concentration), then the plane of polarized light is rotated in opposite directions by the same amount as in glyceraldehyde enantiomers (scheme 1.2). Similarly the enantiomers of tartaric acid have *e.g.*, the same melting point (171°C), the same value of pK_a (25°C pK_1 = 2.98 ; pK_2 = 4.34), but different signs of specific rotation (+)-tartaric acid + 12.7

Plane polarized light is in fact an equal mixture of left and right circularly polarized light which propagates through space as left handed and right handed helices respectively. Due to the chirality of the circular components of the plane polarized light the two enantiomers of a compound react with it differently.

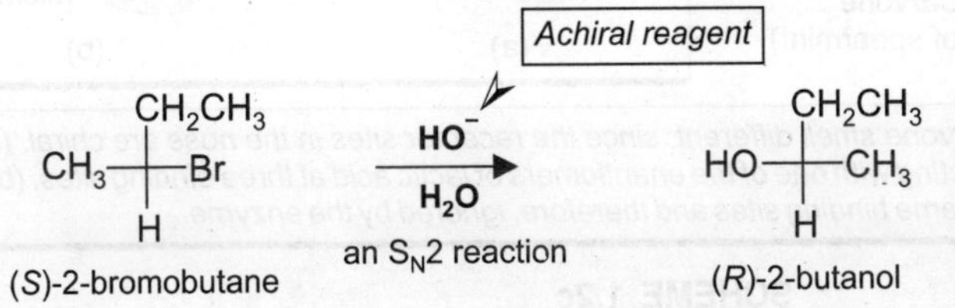

Achiral reagents e.g. hydroxide ion react with enantiomers (reaction only with (S)-enantiomer is shown) at the same rate. Chiral reagents discriminate between them.

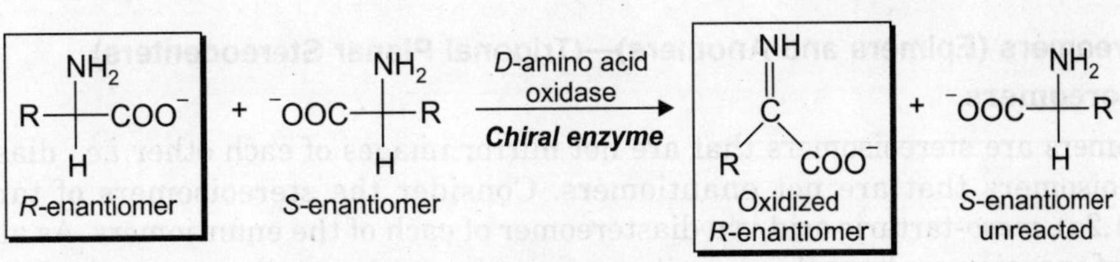

SCHEME 1.2b

while (–)-tartaric acid – 12.7 (scheme 1.2*a*). Each enantiomer shows the same chemical reactivity with achiral reagents *i.e.*, enantiomers react with achiral reagents at the same rate. Thus *e.g.* (*S*)-2-bromobutane reacts with achiral hydroxide ion to give (*R*)-2- butanol (scheme 1.2*b*) by an S_N2

mechanism. The rate of this reaction is found to be the same with the enantiomeric (*R*)-2-bromobutane with hydroxide ion to give (*S*)-2-butanol. When, however, the reagent is chiral *e.g.*, an enzyme, the two enantiomers will react at different rates. Thus the enzyme D-amino acid oxidase reacts only with one of the enantiomers–the (*R*)-enantiomer, the (*S*)-enantiomer remaining unchanged (scheme 1.2*b*). Another example is found during the kinetic resolution of amino acids (See, scheme 1.86).

This is an example of stereospecificity in general and the reaction with only one enantiomer shows that the enzyme displays total enantioselectivity.

Receptors are proteins which are chiral and thus these will bind one of the enantiomers better than the other *i.e.*, one enantiomer binds with a particular receptor whereas the other does not. Receptors located on the exterior of nerve cells in the nose are thus able to differentiate odours. The enantiomers of carvone (scheme 1.2*c*) smell different since each fits into a different receptor.

In summary chiral substances react only with substances that match their own chirality. This forms the basis for an enzyme to distinguish between two enantiomers of a compound, during enzyme catalyzed reactions. The enzyme first positions a molecule at the binding site on its surface (via, hydrogen bonds, electrostatic attractions, dispersion forces or even covalent bonds). An enzyme with specific bonding sites for three of the four groups on a stereocenter can make a distinction between a molecule and its enantiomer (or one of its disastereomers). From (scheme 1.2*c*), one can see that one enantiomer can be absorbed at its three bonding sites while the other cannot. The enzyme lactate dehydrogenase *e.g.*, brings about the conversion of only one of the enantiomers of lactic acid to pyruvic acid ($CH_3COCOOH$, scheme 1.2*c*).

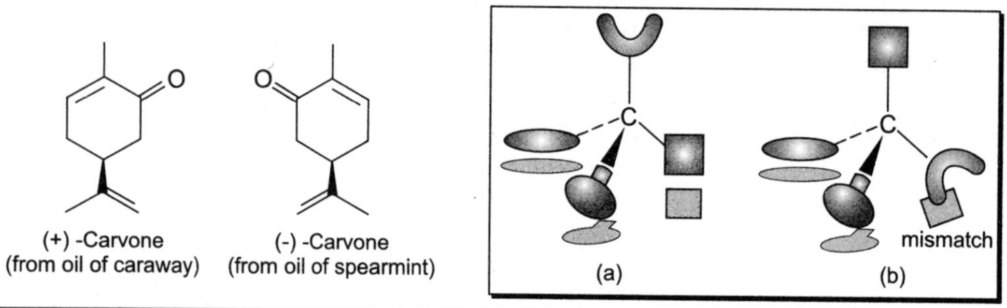

The individual enantiomers of carvone smell different, since the receptor sites in the nose are chiral. (a) An enzyme surface capable of interacting with one of the enantiomers of lactic acid at three binding sites. (b) The other enantiomer does not fit the same binding sites and therefore, ignored by the enzyme.

SCHEME 1.2c

3. Diastereomers (Epimers and Anomers)—(Trigonal Planar Stereocenters)

(*i*) Diastereomers

Diastereomers are stereoisomers that are not mirror images of each other *i.e.*, diastereomers are stereoisomers that are not enantiomers. Consider the stereoisomers of tartaric acid (scheme 1.2*a*), *meso*-tartaric acid is a diastereomer of each of the enantiomers. As already said the pairs of enantiomers have the opposite configuration at every stereocenter *i.e.*, enantiomeric tartaric acids are (2*R*, 3*R*) and (2*S*, 3*S*). While a diastereomer *i.e.*, *meso*-tartaric acid has one stereocenter of common configuration while the other of opposite configuration (2*R*, 3*S*). Another example of diastereoisomers is in aldohexoses. Diastereomers are nonenantiomeric stereoisomers which have two or more stereocenters and differ in the projection of at least one of them. Thus galactose, glucose and mannose are all diastereomers (scheme 1.2*e*).

(*ii*) Epimers

When two diastereomers differ in the stereochemistry (*i.e.* projection) at only one stereocenter then these are called epimers. The term is quite general, however, it is not used for molecules with only two stereocenters. Glucose and galactose are epimers at C4. Similarly glucose and mannose are epimers at C2.

The stereochemical relationship between the compounds (I and II, scheme 1.2 d, containing bicyclo [2.2.1]-heptane system) is epimeric. The compounds (III and IV, scheme 1.2d) containing bicyclo [2.2.2] octane system have enantiomeric relationship . In (III and IV) the two unsubstituted bridges are equivalent. Thus excluding the possibility of epimers. Rotation of e.g. (IV, scheme 1.2d) around an axis passing through C1 and C4 in an anti-clockwise fashion by 120° infact gives an orientation which can be clearly seen to have an enantiomeric relationship with (III). Study of (III) and (IV) with models will be helpful.

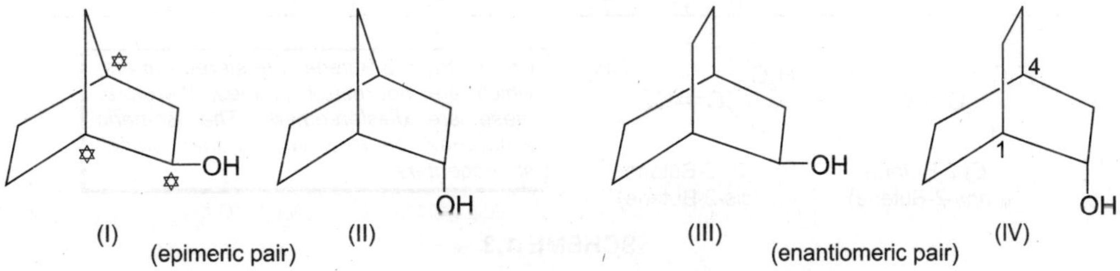

SCHEME 1.2d

(*iii*) Anomers

To give a pyranose structure, the OH group at C5 of open chain form of glucose attacks the aldehyde (carbon C1) to form a hemiacetal. A new stereocenters at C1 is generated and a pair of diastereomers is formed. These diastereomers (in the case of a monosaccharide) which differ in the configuration at C1 (called anomeric carbon) are called anomers (scheme 1.2*e*).

It is important to note that in general a molecule can have only one enantiomer, but it may have many diastereomers (*i.e.*, consider the case of cholesterol scheme 1.68*d*).

(*iv*) Trigonal planar stereocenters

A stereocenter is defined as an atom having groups of suitable nature so that an interchange of any two groups will give a stereoisomer. However, all stereocenters are not tetrahedral the unsaturated carbon atoms of *cis* and *trans*-2-butene (scheme 1.3) are examples of trigonal planar stereocenters, since an interchange of groups at these stereocenters gives a stereoisomer (a disastereomer).

cis- and *trans*-2-Butene (scheme 1.3) are not mirror images of each other, *i.e.* if a structural model of *cis*-2-butene is shown to a mirror, the arrangement which one sees in the mirror is not *trans-2*-butene. However, *cis-* and *trans*-2-butene are stereoisomers and, since they are not related to each other as an object and its mirror image, they are thus distereomers. Diastereomers are stereoisomers which are not mirror images of each other.

Although *cis-* and *trans* isomers of alkenes are diastereomers (see scheme 1.3) that are achiral, the majority of diastereomeric compounds are chiral compounds however, which have more than one stereocenter. For an example of diastereomers in alicyclic compounds (see, scheme 1.67).

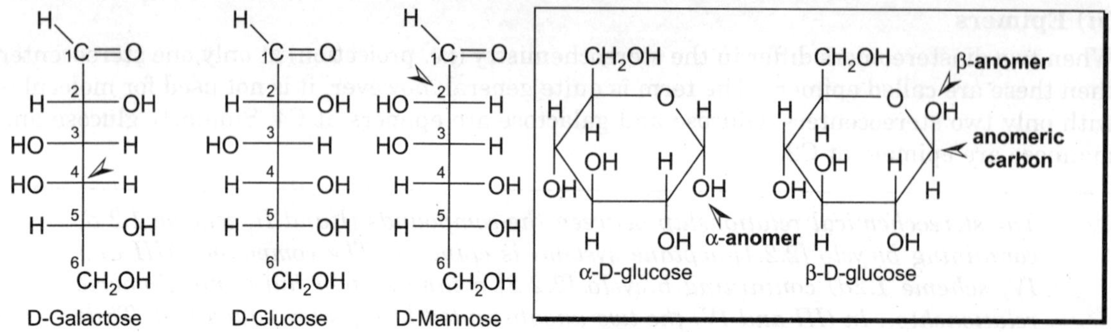

D-Galactose, D-glucose and D-mannose are diastereomers, D-galactose and D-glucose are C4 epimers while D-glucose and D-mannose are C2 epimers.α–β-Anomers of D-glucose-are also diastereomers differing in configuration at

SCHEME 1.2e

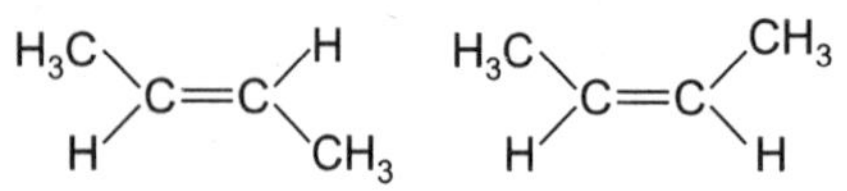

(*E*)-2-Butene (*trans*-2-Butene) (*Z*)-2-Butene (*cis*-2-Butene)

cis-and trans-2-Butenes are stereoisomers, which are not mirror images, therefore, these are diastereomers. The isomeric 2-butenes each have trigonal planar stereocenters.

SCHEME 1.3

4. Properties of Diastereomers

Diastereomers have a major advantage over enantiomers from a practical point of view. Diastereomers have different physical properties like m.p., b.p., solubility, retention times and R_f values and have different rates of reactions (chemical properties) even in achiral environments. Standard techniques like crystallization distillation or chromatography can therefore, be used to separate diastereomeric mixtures. For example, *meso* tartaric acid (specific rotation = 0) melts at 140°C as against either of the enantiomers (m.p. 171°C) and has different value of pK_a. Use of this difference in properties of diastereomers is made in the resolution of racemic mixture (See scheme 1.82). Another example is of erythroses and threoses (See scheme 1.32).

Enantiomers and Diastereomers

One may note the following points :

- *Enantiomers (mirror images of each other) are related by symmetry elements of the second kind, i.e., σ plane, i, and S_n axis while diastereomers are not related by any such symmetry element.*
- *Since a molecule (or an object) can have only one mirror image, enantiomers can exist only in pairs. On the other hand, structural conditions permitting, a molecule can have any number of diastereomers, e.g., meso tartaric acid has two diastereomers (diastereomeric relationship with either of the enantiomers).*
- *Two stereoisomers can not be enantiomers and diastereomers at the same time, i.e., enantiomeric and diastereomeric relationships are mutually exclusive.*
- *Diastereomers may be (or may not be) chiral in which case each of the diastereomer will show enantiomerism. Thus cholesterol with 8 stereocenters has 256 stereoisomers. Cholesterol is one of these and the second is its mirror image (enantiomer). Cholesterol is thus diastereomeric with 254 molecules.*

- *Diastereomers include all stereoisomers (but for enantiomers), optically active diastereomers, geometrical isomers and cis-trans isomers of classical stereochemistry.*
- *Enantiomeric relationship can only be specified by comparison with a chiral reference (plane polarized light). The diastereomeric relationship can be established without any external reference.*

5. Chiral, Achiral Compounds and Meso Compounds (An Introduction)

Consider the stereoisomers of tartaric acid (two stereocenters with identical set of substituents at each stereocenter, scheme 1.2*a*). The pair of enantiomers (which are non-superimposable mirror images) constitute chiral compounds whereas the *meso* stereoisomer is achiral (optically inactive) even though it has two stereocenters. The configuration of the *meso-* tartaric acid is 2*R*, 3*S* (see scheme 1.2*a*) if one draws its mirror image (2*S*, 3*R*) it is found to be superimposable with it. The *meso* compounds have two features in common : (*a*) a plane of symmetry and (*b*) two stereocenters with opposite stereodescriptors. *Meso* compounds are optically inactive due to self-cancelling stereocenters of opposite configuration. The optically inactive compounds with two or more stereocenters are termed *meso* and possess a plane of symmetry. *meso* Compounds also exist in cyclic structures (See scheme 1.66).

6. Invertomers—Stereogenic Nitrogen and Phosphorus—Conformational and Configurational Isomers

Acyclic amines of the type (scheme 1.3*a*) in which the three groups are different and the lone pair on nitrogen is classed as a formal substituent meets all the requirements of a stereocenter. However, no optical activity is observed in amines of this type, even though these are chiral. This is due to very rapid pyramidal inversion (energy barrier to inversion is small ~ 25 kJ mol^{-1}) which interconverts enantiomers (called invertomers), further details are in schemes 1.5 and 1.142. Inversion is slower for third-row elements. Thus, phosphines, (R_3P) high energy barrier ~ 150 kJ mol^{-1} and sulfoxides ($R_2S{=}O$) can be obtained in optically active form, and thus configurational isomers are encountered. Amine inversion is an example of conformational isomers (invertomers), while biphenyl enantiomers may be isolated due to restricted rotation. Such isomers are called atropisomers.

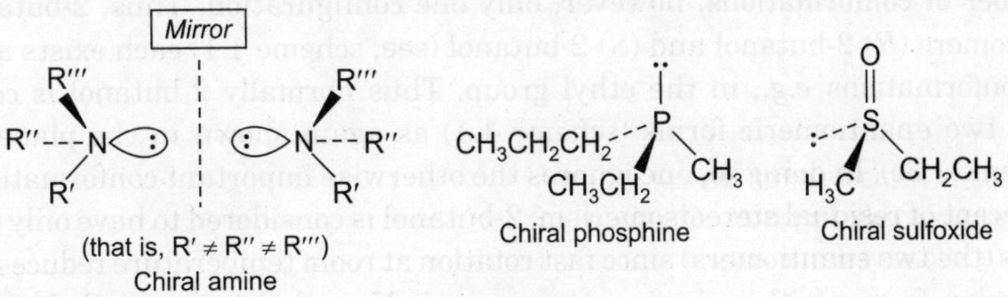

Chiral amine

Chiral phosphine

Chiral sulfoxide

SCHEME 1.3a

7. Homomers

A molecule may be written in two or more orientations which infact represent the same molecule but at first sight look different. Such different orientations of the same compound (which are superimpossable) are called homomeric. For example, for (*S*)-2-bromobutane one can write many equivalent orientations, three of which are presented (Ia-Ic, scheme 1.3*b*). These three structures are thus homomeric (one may note that one gets S-2-bromobutane by exchanging any two substituents of the (*R*)-configuration (II, scheme 1.3*b*).

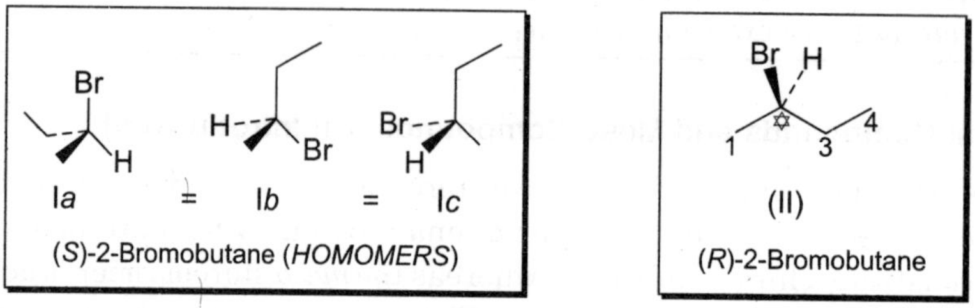

SCHEME 1.3b

8. Homochiral Molecules

When the molecules have the same sense of chirality, they are called homochiral, *e.g.*, the right hands of a group of people are homochiral (Kelvin 1904). However, recently, in violation of the original definition the term has been used to depict enantiomeric purity, when one uses it to refer to such a compound as homochiral. In the original sense *e.g.*, the hog-kidney acylase hydrolyzes the natural L-enantiomers of *N*-acylamino acids, no matter what the structure of the *R* group is, *N*-acyl-L-amino acids thus are homochiral molecules (see scheme 1.86).

9. Configuration and Conformation—Residual Stereoisomers

The actual three-dimensional arrangement of groups around a stereocenter is termed its configuration and the term configurational isomers is synonymous with stereoisomers. One way to designate the configuration of geometrical isomers is to use the terms *cis* or *trans* (or better *E*, *Z* nomenclature).

The term conformations (conformational isomers, conformers or rotamers) refers to various shapes that a molecule can adopt by rotation about single bonds *e.g.*, an eclipsed or a staggered conformation of ethane and different conformations of chlorocyclohexane (see scheme 1.1). Unless it is held rigid by a small ring or double bonds, a molecule could have an infinite number of conformations, however, only one configuration. Thus, 2-butanol has two stable enantiomers (*R*)-2-butanol and (*S*)-2-butanol (see, scheme 1.1) each exists as a dynamic mixture of conformations *e.g.*, in the ethyl group. Thus normally 2-butanol is considered to exist in only two enantiomeric forms (scheme 1.1) as again shown in the planar projection formula (scheme 1.3c). In doing so, one ignores the otherwise important conformational details. Using the concept of residual stereoisomerism, 2-butanol is considered to have only two residual stereoisomers (the two enantiomers) since fast rotation at room temperature reduces the number of isolable species. In case with amine invertomers as in N-methyl-1-phenylethyl amine (scheme 1.3c) although there are two stereocenters at carbon and nitrogen, it has only two residual stereoisomers (residual enantiomers with different configuration only at the benzylic carbon).

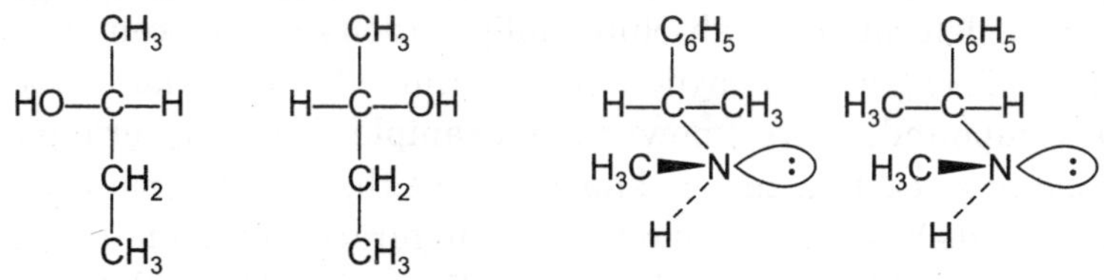

(R)-2-butanol (S)-2-butanol N-methyl-1-phenylethylamine

"Residual stereoisomers"

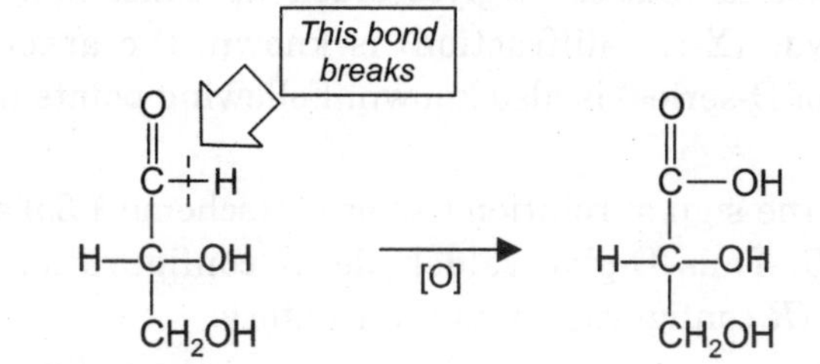

(+)-Glyceraldehyde is
D-(+)-Glyceraldehyde is
R(+)-Glyceraldehyde

D-(-)-Glyceric acid

(+)-Glyceraldehyde is converted chemically into (-)-glyceric acid with retention of configuration. Thus the two compounds have the same relative configuration. Since absolute configuration of (+)-glyceraldehyde (X-ray diffraction) is known to be (D-) as shown, (-)-glyceric acid has also the same absolute configuration and still belongs to D-series.

SCHEME 1.3c

Whether two isomeric structures constitute different configurations or merely different conformations depends on now readily they interconvert. For example the different conformations of ethane derivatives interconvert millions of times per second at 25°C, while the Z and E configurations of ethylene derivatives donot interconvert spontaneously even at 200°C. Since in principle it should be possible to stop conformational interconversion by sufficient cooling, the operational distinction between configurations and conformations usually is based on whether the structures interconvert at room temperature (25°C, for a good example see Fig. 4.2).

(*a*) Absolute configuration

There are further two terms absolute configuration and relative configuration. Absolute configuration is the spatial arrangement of the atoms in a chiral molecule which distinguishes it from its mirror image and its stereochemical description (R or S, M or P) as in the case of *e.g.*, 2-butanol enantiomers (scheme 1.3*c*).Until a special X-ray technique was developed in 1951 it was not possible to determine the absolute configuration of any compound. Although samples of one or both enantiomers of a large number of compounds were known, there was no experimental method to know (prior to 1951) if that enantiomer has R or the S-configuration *i.e.* what was its absolute configuration. This, however, did not present a major problem since an organic chemist could convert one chiral molecule into another employing reactions whose stereochemical effects were well known (*e.g.*, S_N2 reaction). Thus it was possible to relate the configuration of one compound to that of another.

(*b*) Relative and inverted configuration

(+)-glyceraldehyde was arbitrarily assigned the configuration (see, scheme 1.2). Luckily this (50-50) chance of assignment was fortunately found to be correct in 1951 using X-ray diffraction.

Chemists prior to 1951 could not determine the absolute configurations of stereocenters, instead configurations relative to that of (+)-glyceraldehyde were determined. Thus prior to 1951, only relative configurations of compounds were known. For example, if (+)-glyceraldehyde is converted to glyceric acid using a reaction that is known to put the COOH group where the CHO was, then the two compounds have the same relative configuration (scheme 1.3*c*) at their stereocenters even though one may not know their absolute configuration. The oxidation reaction (scheme 1.3*c*) does not break a bond to the stereocenter in (+)-glyceraldehyde and proceeds with retention of configuration. Thus the product glyceric acid of this conversion though levorotatory has the same relative configuration as that of (+)-glyceraldehyde. Since now the absolute configuration of D-(+)-glyceraldehyde (X-ray diffraction) is known, the absolute configuration of glyceric acid (still a member of D-series) is also known. Following points may be noted:

- There is no simple relation between the sign of rotation (+)- or (–) (scheme 1.3*c*) and the absolute configuration *R* and *S*. Thus D-glyceraldehyde (*R* configuration) is dextrorotatory while D-glyceric acid (*R* configuration) is levorotatory.
- Two closely related compounds with the same relative configuration may have different descriptors (*R* or *S*, See scheme 1.26*e*).
- In summary relative configuration is the relationship between configurations of two chiral molecules. When two such molecules are chemically interconverted without breaking any bonds to the stereocenter, these are said to have the same relative configuration independent of the direction of rotation of the plane polarized light and independent of the (*R* and *S*) designation (see, scheme 1.26*e*).
- Retention of configuration is the conversion of one molecule into another with the same relative configuration.
- Inversion of configuration is the conversion of one molecule into another which has the opposite relative configuration. Inversion of configuration does not require a change in designated absolute configuration, similarly a change in designated absolute configuration does not require an inversion of configuration. The reaction (III, scheme 1.26*e*) proceeds with inversion although (*R*) reactant gives (*R*)-product.
- Relative configuration at a stereocenter is also the relation with that of any other stereocenter in the same molecule.

1.3 CHIRAL MOLECULES-ENANTIOMERISM (OPTICAL ISOMERISM) A SUMMARY

Enantiomers occur only with those compounds whose molecules are chiral. A chiral molecule can be defined as one that is not superposable on its mirror image.

(A) The Terms Chiral, Stereogenic Center (Stereocenter)

(*i*) The Term Chiral

The word chiral (Greek word Chier, meaning hand) is used for those objects which have right-handed and left-handed forms, *i.e.*, molecules which have "handedness" and the general property of "handedness" is termed chirality. An object which is not superimposable upon its mirror

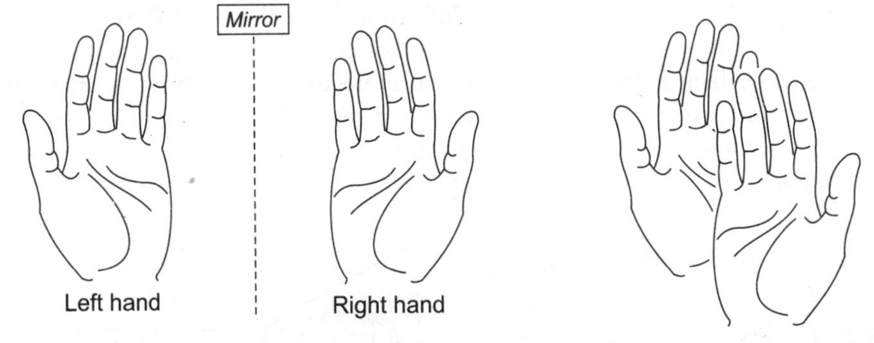

The mirror image of left hand is a right hand Left and right hands are not superimposable

Fig. 1.2

image is chiral. Enantiomers occur only with those compounds whose molecules are chiral. Thus *e.g.*, 2-butanol molecule is chiral as well as *trans*-cyclooctene (scheme 1.1). Human hands are related to each other as object and its mirror image typical of enantiomers (Fig. 1.2). Thus *e.g.*, one's right hand will fit into a right glove and not into a left glove. The term chiral is used to describe molecules of enantiomers since these are related the same way that a right hand is related to a left hand. A structurally chiral system is the human body itself, the heart lies to the left of center and the liver to the right. From an evolution point of view most people are right handed. Several plants display chirality by way they wind around supporting structures and thus represent as a left-handed or a right-handed helix. All but one of the 20 amino acids, the components of proteins are chiral and left handed. The molecules of almost all natural sugars on the other hand are right handed. The enantiomers display physiological differences, one enantiomeric form of a terpenoid limonene smells like oranges while the other enantiomer smells like lemons. Ones nose is capable of distinguishing between enantiomers, *i.e.*, the receptor sites for the sense of smell are chiral (see, scheme 1.2*c*).

Generally, otherwise, enantiomers have identical properties in a symmetrical environment, enantiomers react at the same rate with achiral compounds. Their properties may however, differ in an unsymmetrical environment, and enantiomers may react at different rates with other chiral compounds (see scheme 1.2*b*). This difference is reflected in a compound being biologically active, while its enantiomer is not.

(*ii*) The Term Achiral

Objects and molecules which are superimposable on their mirror images *e.g.* a cup (Fig. 1.2*a*) is achiral. An internal plane of symmetry is a hypothetical plane which bisects an object or a molecule into mirror–reflective halves. An object or a molecule with an internal plane of symmetry is achiral (can be superposed on its mirror image). Thus a cup is achiral since it can be divided into two equal halves by its plane of symmetry. Similar is the case with *meso*-tartaric acid (drawn as Fischer projection or perspective formula Fig. 1.2*a*). A hand is chiral since a plane cannot split it into two equal halves. Similarly either of the (+)- or (–)-enantiomers of tartaric acid is chiral.

(*iii*) The Terms Asymmetric Center and Chiral Center

Three terms are used to designate *e.g.*, a carbon atom bonded tetrahedrally to four different substituents in a chiral molecule: asymmetric atom, chiral center (*i.e.*, chiral atom) or a stereocenter. From the time of Le Bel and vant Hoff an atom with four different substituents was called an asymmetric atom. This is indeed so, since compounds with one such atom are truly asymmetric as they lack symmetry. There are molecules which have atoms with four different substituents and which also have various symmetry elements, including planes of symmetry as in *meso* tartaric acid Fig. 1.2*a*. Use of the term asymmetric atom in these type of cases may be confusing.

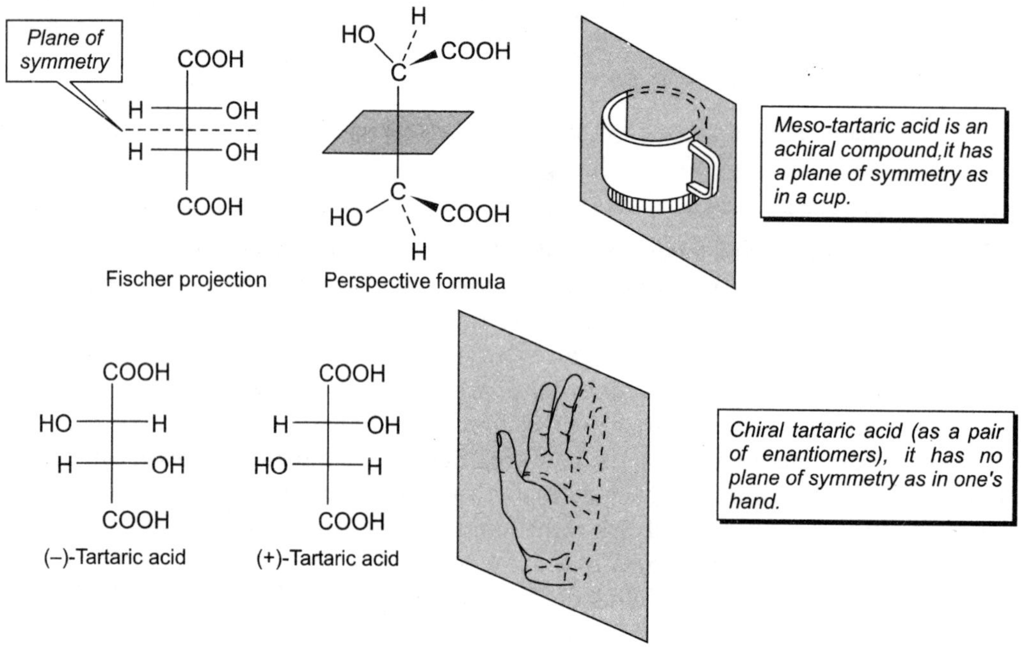

Fig. 1.2a

Similarly, like asymmetric atom the term chiral atom (used prior to 1984) may also cause a conceptual confusion, although these terms, particularly the use of term chiral center are well accepted in literature. Chirality is a geometric property which influences and affects all parts of a chiral molecule. All the atoms of a chiral molecule like 2-butanol (scheme 1.1) are placed in a chiral environment and are therefore chirotopic. Thus when, C-2 carbon atom in 2-butanol is termed a chiral center (or chiral carbon to distinguish it from chiral nitrogen chiral phosphorus etc.), it may cause a conceptual confusion. The more ideal term for such a center or atom is stereogenic center.

(*iv*) The Terms Stereogenic Center-Stereocenter

A stereogenic center or in short a stereocenter is an atom having groups of such nature that an interchange of any two groups will produce a stereoisomer. C-2 in glyceraldehyde (scheme 1.2) is an example of a tetrahedral, stereocenter. It is easy to see that interchanging *e.g.*, the OH and H in the Fischer projection of glyceraldehyde (scheme 1.2) converts one enantiomer into other. A carbon atom that is a stereocenter is also called a stereogenic carbon. All stereocenters may not be tetrahedral, thus the carbon atoms of *cis*- and *trans*-2-butene provide an example of trigonal planar stereocenters, since an interchange of groups at either atom leads to a stereoisomer (a diastereomer).

(*v*) Summary of the Terms

All three alternative terms, asymmetric center, chiral center or stereogenic center (stereocenter) are in current use for carbon atoms which have four different substituents. The term chiral is applied to the whole molecule to mean being capable of existing as nonsuperimposable object and mirror image forms. Chirality is the property of the whole molecule and cannot be localized in a particular atom or a center. Thus the more widely used term 'chiral center' like asymmetric center is strictly incorrect and should be replaced by stereogenic center or simply stereocenter *i.e.* a center giving rise to stereoisomers.

(B) The Chiral Molecules

(*i*) Nonsuperimposability of a Structure on Its Mirror Image

The ultimate criterion for chirality *i.e.* existence of enantiomers in a molecule is the nonsuperimposability of a structure and its mirror image. [*Thus enantiomers must be isomers and mirror images as well, i.e., they must not be superimposable. The adamantane derivative (See scheme 1.3e) has a mirror image however, the mirror image on rotation around an appropriate axis is superimposable, such structures are homomeric and not enantiomeric*].

(*ii*) Elements of Symmetry and Chirality—Rotation/Reflection Axis

Chirality—A Necessary and Sufficient Condition

Only those molecules are chiral which do not have an alternating axis of symmetry S_n (i.e., absence of a rotation/reflection axis). The chiral molecules can, however, have an axis of rotation (C_n).

An alternative approach to decide if or not a structure is chiral *i.e.* capable of existing in enantiomeric forms (optically active forms) is to examine the symmetry of a molecule. A structure which lacks an alternating axis of symmetry is chiral and is not superimposable on its mirror image and therefore, can exist in optically active forms. Symmetry operations are discussed in detail (see Sec. 1.8), however, here the common practice is to look for a plane of symmetry (a one fold alternating axis of symmetry S_1) and a center of symmetry which is equivalent to S_2 will be discussed while higher subscripts for S_n are rare (A molecule is said to have an *n* fold alternating axis of symmetry S_n if rotation of 360°/*n* about an axis followed by reflection in a plane perpendicular to that axis brings the molecule into a position indistinguishable form the original).

meso-Tartaric acid is achiral even though it has two stereocenters since it has a plane of symmetry (see, scheme 1.2*a* and Fig. 1.2*a*).

When one cannot detect a mirror plane of symmetry in a molecule, the molecule is not necessarily chiral. Thus the cyclobutane derivative (I, scheme 1.3*d*) has no internal mirror plane of symmetry, yet the mirror image is superposable on the original molecule (when *e.g.*, the mirror image is rotated the two structures are superposable. Unlike a three dimensional formula which does not change on rotation, the two dimensional representations of three dimensional molecules like Fischer projections can only be manipulated in specified ways). In fact (I, scheme 1.3*d*) has a center of symmetry *i.e.*, it contains an S_2 axis (an inversion center). A molecule is said to have a center of symmetry or an inversion center if all straight lines that can be drawn through the center of the molecule meet identical atoms (or points) at the same distance from the center. In other words inversion of all atoms (or points) in the molecule through the point leads to an arrangement indistinguishable from the original. There can be only one such inversion center in the molecule.

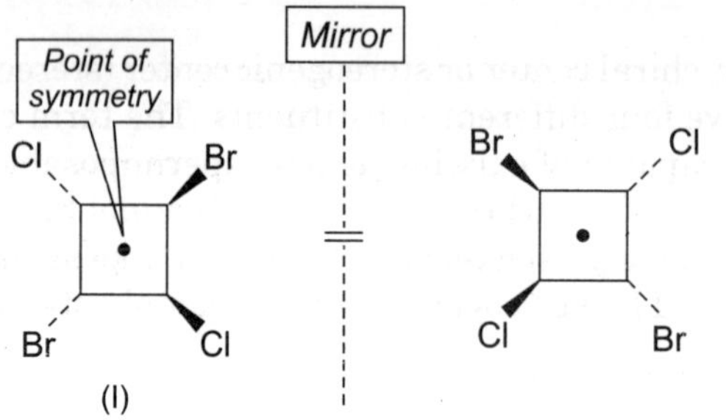

In this achiral cyclobutane derivative no plane of symmetry is detectable. Infact the compound has a center of symmetry (point of symmetry) and the two structures are superimposable on their mirror images.

SCHEME 1.3d

Molecules With a Plane or Centre of Symmetry and Axis of Rotation

The mirror plane of symmetry (S_1) is the most frequently occuring rotation reflection axis in organic molecules. Examples of further compounds which have this axis are in (I and II scheme 1.3e), cis-1, 2-dichlorocyclohexane and meso-2, 3-dichlorosuccinic acid are achiral even though each has two stereocenters. Dichlorocyclohexane dicarboxylic acid (III) on the other hand is achiral even though it has four stereocenters due to the presence of a point of symmetry—presence of S_2 axis which is somewhat rare in organic chemistry. A molecule can, however, be chiral if it has only an axis of rotation. Both trans-1, 2-dichlorocyclohexane and dichlorosuccinic acid (IV and V respectively, scheme 1.3e) have a two fold axis of rotation (C_2) as the only symmetry element and both are chiral.

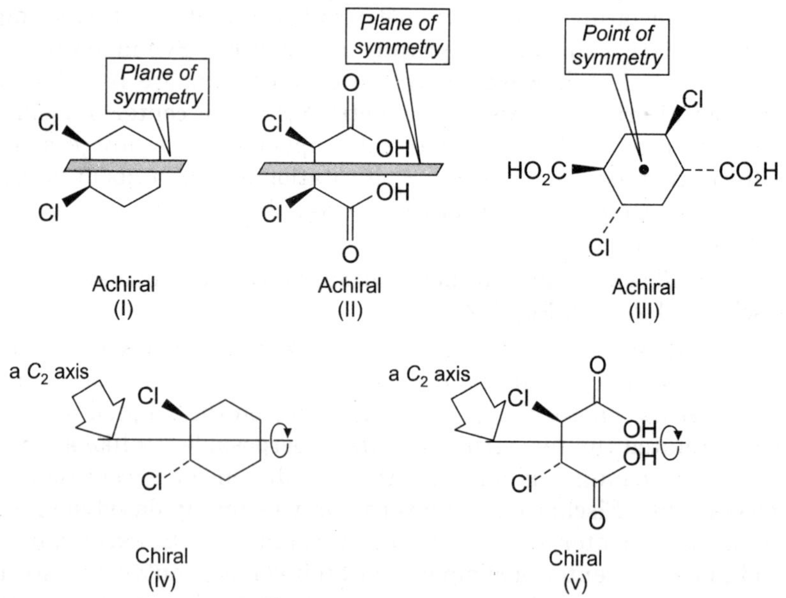

SCHEME 1.3e

EXERCISE 1.2

Pick out chiral and achiral structures from the following (scheme 1.3f)

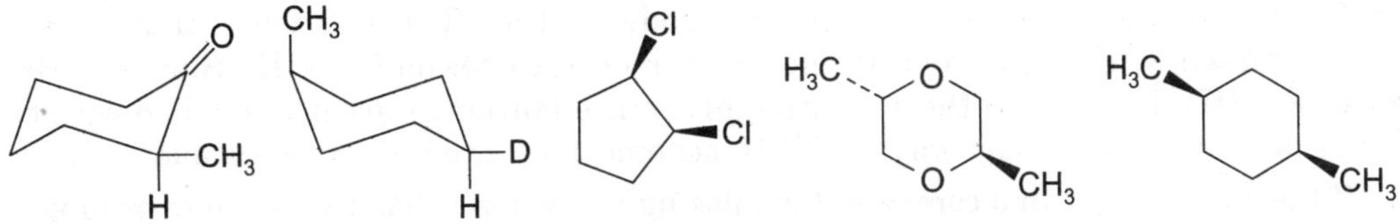

(I) (II) (III) (IV) (V)

SCHEME 1.3f

ANSWER. *(I) chiral ; (II) achiral ; (III) achiral (a plane of symmetry ; (IV) achiral has no plane of symmetry but has a center of symmetry (scheme 1.3g models will help) ; (V) achiral, has a plane of symmetry. It is always helpful to indicate H atoms in the projections.*

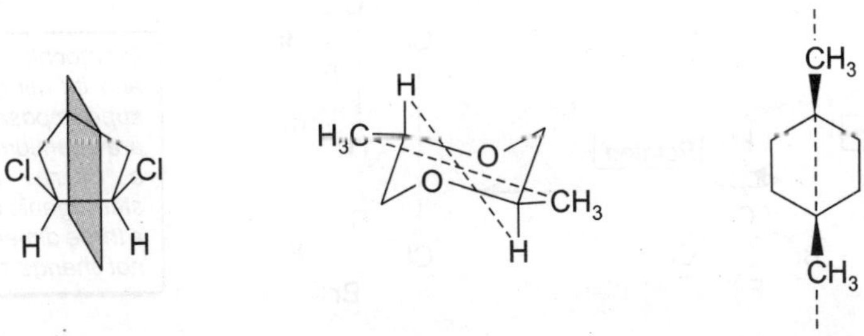

(III) (IV) (V)

SCHEME 1.3g

(iii) Presence of a Stereogenic Carbon

A molecule Cabcd is necessarily chiral when $a \neq b \neq c \neq d$ and none of these ligands are themselves chiral. In some cases a chiral ligand can generate mirror symmetry on a molecule Cabcd (*meso* compounds). The ligands can be alkyl, aryl, heteroatomic or their combination. A stereogenic carbon in a molecule is stared for emphasis.

All molecules with a single stereocenter (point group Cl) must display optical activity as in the case of glyceraldehyde. Moreover, even a minor difference among the four groups is sufficient to generate chirality as in (I, scheme 1.4) which is optically active. The chirality of this center is retained even when two substituents show as small a difference as a hydrogen atom and its isotope deuterium as in 1-butanol-1-*d* (scheme 1.4).

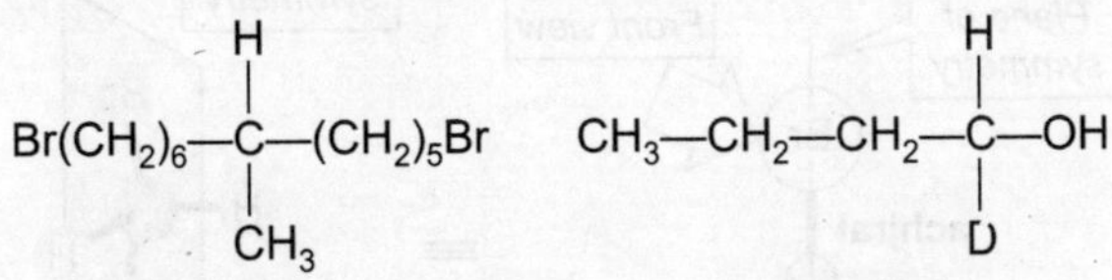

(I) 1-butanol-1-*d*

SCHEME 1.4

The two enantiomers (I and II, scheme 1.4a) of the chiral compound bromochlorofluoromethane as *e.g.*, ball and stick models are not superposable. The mirror image of (I) is rotated so that Cl in one enantiomer (I) can match with the Cl in the other enantiomer held in hand (Remember one can rotate a three dimensional formula as one likes without losing identity, however, a Fischer projection can be rotated only in specified ways). When tried to match I with II (scheme 1.4a) by superimposing one model on the other, it is found that no doubt H and CI of I match with H and Cl of (II), however, F of (I) comes on Br of (II). Similarly Br of (I) comes on F of (II). Thus two of the substituents of one enantiomer are mismatched with two on the other and one may try any way (I and II, scheme 1.4*a*) are not superposable.

If the stereocenter in a compound is missing *i.e.,* when it has two or more groups which are same, the compound becomes achiral. Since now the object and its mirror image can be superposed. This is so in bromochloromethane which has a plane of symmetry and is therefore, achiral. The plane of symmetry can be shown in different ways (scheme 1.4*b*).

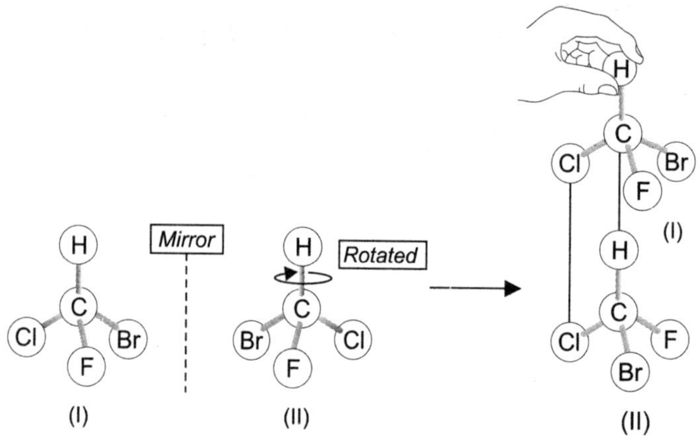

Bromochlorofluoromethane (I) and its mirror image II are not superimposable. Thus (I and II) are enantiomers. Therefore, it is a chiral molecule, with stereogenic carbon. Rotation of a three dimensional model does not change its configuration.

Demonstration of chirality

SCHEME 1.4a

As a second example consider the molecule of 2-propanol (scheme 1.4*c*) which is achiral since it is shown to have a plane of symmetry and object and mirror image are superposable.

(*iv*) Chirality in Compounds Lacking a Stereogenic Carbon Atom

(a) Compounds with tetrahedral stereocenters other than carbon

A molecule with an atom with its four bonds pointing to the corners of a tetrahedron will display chirality provided the four groups are different. Some of the tetrahedral centres are Si (Silanes) and Ge (Germanes) derivatives as well as N in quaternary salts (shown as enantiomers) and phosphine oxides (scheme 1.5).

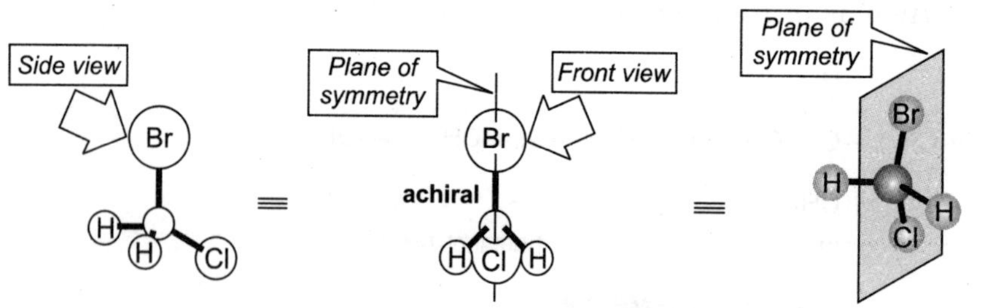

The carbon atom in bromochloromethane is not stereogenic (not a stereocenter) since two ligands (H's) are identical. The compound has a plane of symmetry and is achiral.

SCHEME 1.4b

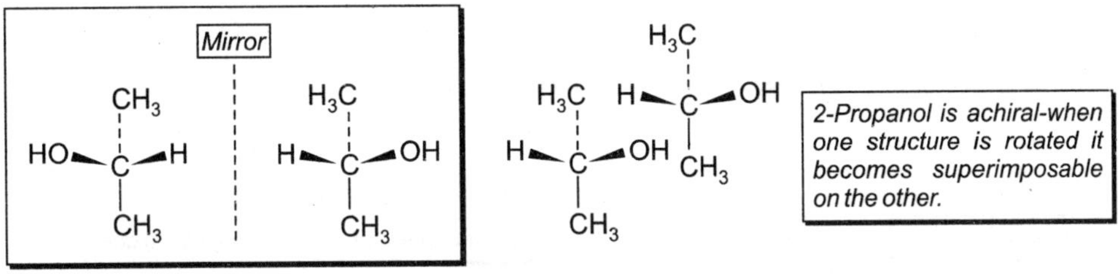

SCHEME 1.4c

In sulfones sulfur bonds tetrahedrally, however, as two of the groups have to be oxygen, normally the sulfones do not display chirality. An optically active sulfone (VI, scheme 1.5) has however, been prepared in which both the oxygens are different one being ^{16}O and the other ^{18}O. An adamantane derivative with four different bridgehead substituents becomes chiral and optically active as in (A scheme 1.5) which has been resolved. The molecule is completely asymmetric (Cl point group) and represents a kind of expanded tetrahedron. The center of chirality is represented by a dot in the unoccupied space of the adamentyl carbon framework. This is an example to show that a stereocenter in a molecule may not necessarily lie on an atom.

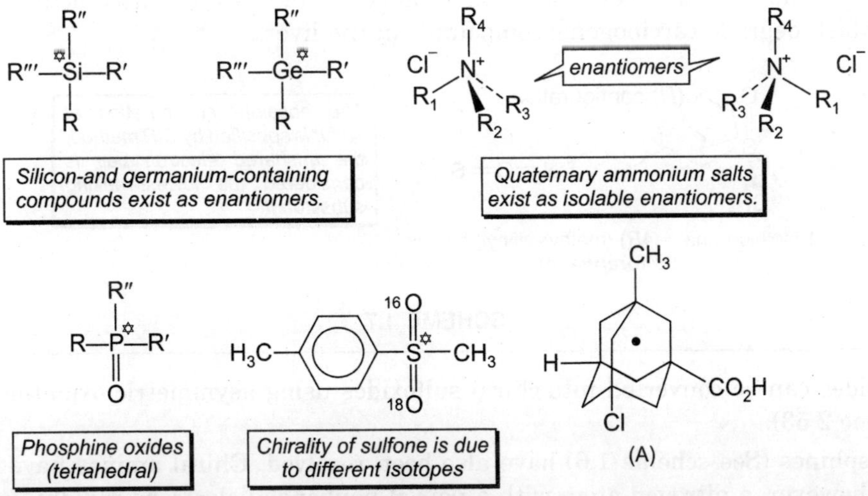

$$R^1R^2R^3R^4N^+X^- \rightleftharpoons R^1R^2R^3N + R^4X \quad (I)$$

SCHEME 1.5

(b) Amines, phosphines and sulfoxides

Atoms with pyramidal; bonding may be chiral and therefore, can give rise to optical activity provided the atom has three different groups attached to it. The unshared pair of electrons then represents the fourth group (scheme 1.6).

Sulfur exhibits pyramidal bonding in sulfoxides, sulfonium salts etc., many of which have been resolved. An interesting example of a sulfoxide which is chiral due to two alkyl groups differing only in ^{12}C versus ^{13}C is (VI, scheme 1.6). In these compounds sulfur atom maintains a rather rigid geometry and does not interconvert between the two enantiomeric forms.

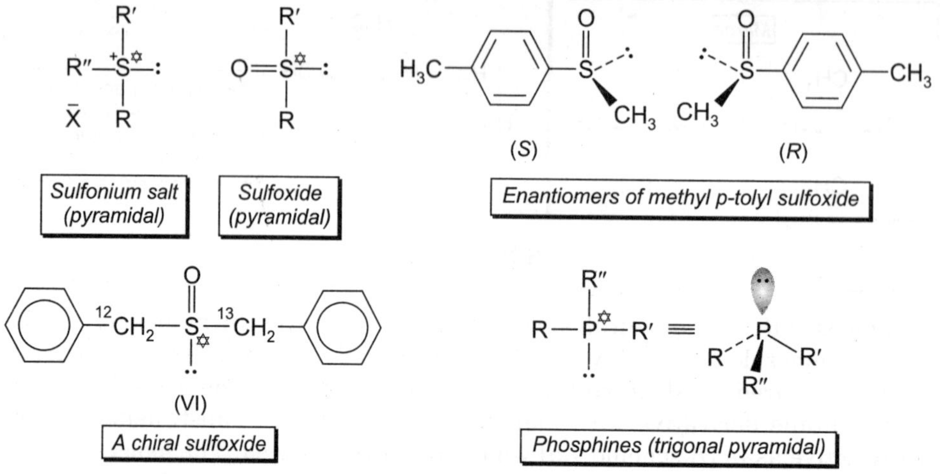

SCHEME 1.6

An interesting role is played by (*R*)–sulforaphane (scheme 1.7) which occurs in nature in broccoli (a component in green salad). This compound induces increased activity in certain enzymes which degrade carcinogenic compounds in the liver.

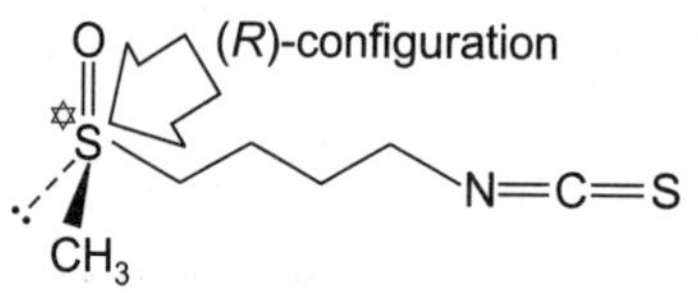

1-Isothiocyanato-(4*R*)-(methylsulfinyl) butane
(***Sulforaphane***)

The absolute configuration at sulfur is specified by CIP method, the unshared electron pair is considered the lowest-ranking substituent.

SCHEME 1.7

Sulfides can be converted into chiral sulfoxides using asymmetric oxidation condition (See, scheme 2.53).

Phosphines (See scheme 1.6) have also been resolved. Chiral amines have pyramidal geometry, however, a nitrogen atom with a pair of nonbonded electrons rapidly turns inside-out at room temperature. This is called amine inversion. The best way to picture amine inversion is to compare it to an umbrella that turns inside out in a windstorm (see scheme 1.3a) one may note that amines in which the chirality is due to the presence of a stereogenic carbon atom can however, be resolved into enantiomers (scheme 1.8).

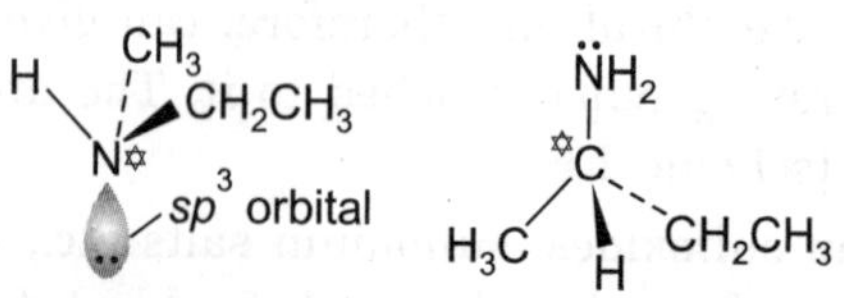

(I) (*S*)-Ethylmethylamine

(II) (*S*)-2-Butanamine

A chiral amine e.g. (I) cannot be resolved (amine inversion). The amines in which the chirality is due to a stereogenic carbon as in (II) can be resolved.

SCHEME 1.8

Due to rapid ; inversion, the individual amine isomers cannot be isolated. The two amine structures provide an interesting example of conformational isomers. The pair of nonbonded electrons are required for amine inversion, therefore, quaternary ammonium ions, with four bonds to nitrogen with no nonbonded electrons, do not invert (See scheme 1.5).

Some quaternary salts do racemize, (Eq., I, scheme 1.5) the tertiary amine so formed inverts configuration rapidly and recombination with the alkyl halide gives a racemic salt. The less nucleophilic the counterion X, the more stable will be the quaternary salt configurationally.

Phosphorus is in the same group of the periodic table as nitrogen and tricoordinate phosphorus compounds e.g., phosphines like amines are trigonal pyramidal. Interestingly phosphines undergo pyramidal inversion comparatively very slowly than amines and similar is the case with chiral tricoordinate sulfur compounds like sulfoxides (further details are in schemes 1.142–1.143).

Chiral carbanions racemize readily while carbocations and carbon free radicals being close to planarity are achiral independent of their substituents (scheme 1.9).

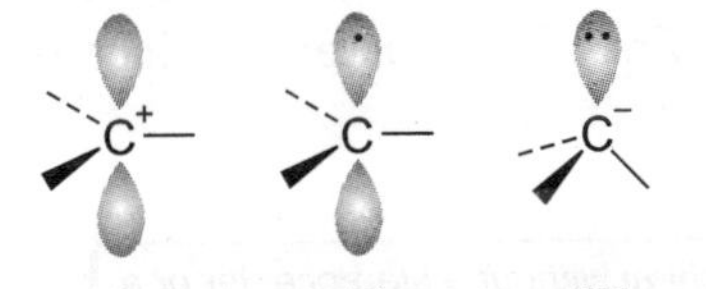

(I) Carbocation (II) Radical (III) Carbanion

The carbocation and free radicals are planar with sp^2 hybridized central carbon. A carbanion has a sp^3 hybridized central carbon and is isoelectronic with amines and undergoes inversion very easily.

SCHEME 1.9

(v) Molecules with Chiral Axis (Stereoaxis)—Conformational Enantiomers and Diastereomers

Several molecules which do not have a stereocenter are, however, capable of holding unlike pairs of groups is mutually perpendicular planes to make such molecules like allenes (VI, scheme 1.10, C_2 symmetry) and spirans (VII) chiral and to exist in nonsuperimposable mirror image forms (scheme 1.10). These molecules contain a chiral axis, which in the case of allenes (in allenes the central carbon in *sp* bonded), includes the allene system C=C=C. The geometry of the 1, 2-diene system forces the ligands attached to the termini of the allene to occupy mutually perpendicular planes. When each terminus carries a pair of unlike groups (though the sets may be identical) the compound becomes chiral. The same conditions apply to spirans.

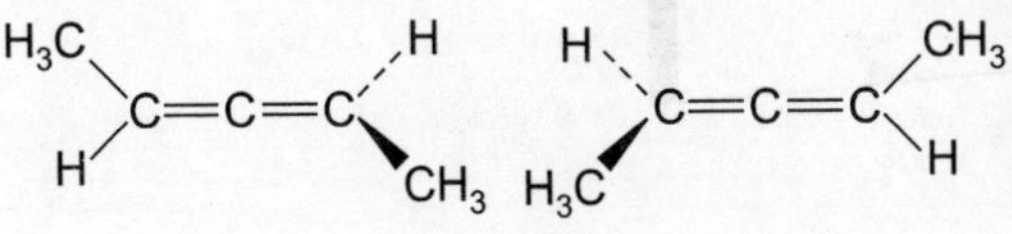

(VI)
A pair of enantiomers of 2, 3-pentadiene (a chiral molecule)

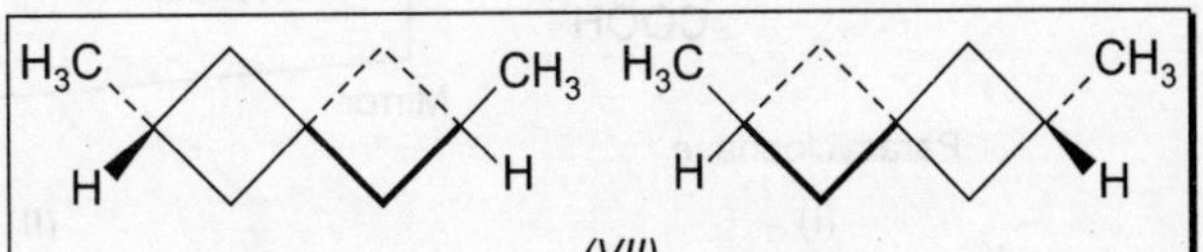

(VII)
A pair of enantiomers of 2, 6-dimethylspiro [3,3] heptane

SCHEME 1.10

Suitably substituted biphenyls also have a chiral axis. Biphenyls are twisted from the coplanar (achiral) conformation as a result of steric strain between the *ortho* substituents on adjacent phenyl rings. In case both rings are unsymmetrically substituted as in (VIII, scheme 1.11) a biphenyl has a chiral axis and exists in enantiomeric forms.

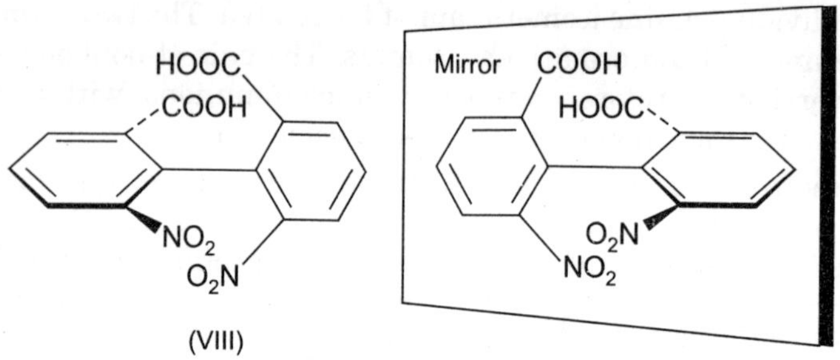

(VIII)

Correctly substituted ortho-biphenyls are chiral. Rotation about the junction bond will racemize a pure enantiomer.

SCHEME 1.11

(*vi*) Molecules with a Chiral Plane (Stereoplane)

Optically active *trans*-cyclooctene (scheme 1.12) has been prepared and characterized at room temperature. The chirality of *trans*-cyclooctene is due to the presence of a plane of chirality, since there are two asymmetric arrangements of the tetramethylene bridge either below or above the plane of the carbon-carbon double bond. *Trans*-cyclooctene is locked into one of these enantiomeric conformations.

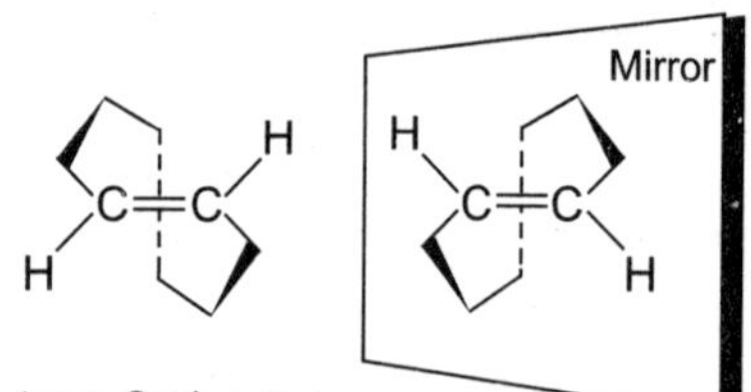

trans-Cyclooctene
(*E*)-Cyclooctene
(I) (II)

trans-Cyclooctene is chiral without a stereocenter or a stereoaxis (chiral axis) but contains a stereoplane (chiral plane). trans-Cyclooctene has either tetramethylene bridge (above one enantiomer) or below (second enantiomer). The plane of trigonal atoms (chiral plane). This chiral plane is defined by double bond and its four directly attached atoms.

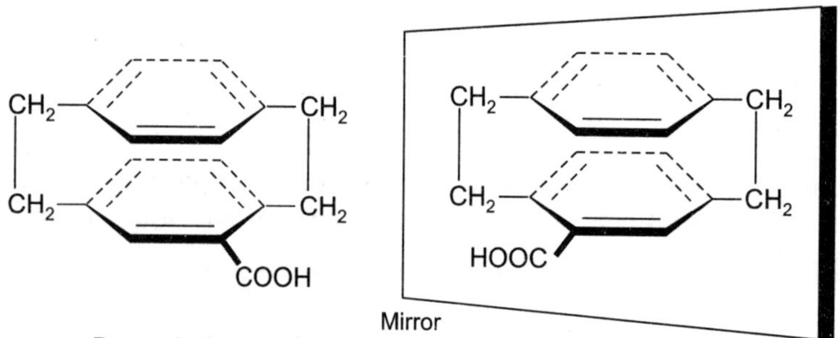

Paracyclophane
(I) (II)

A bridged paracyclophane (suitably substituted to generate chirality) is chiral due to the presence of a chiral plane.

SCHEME 1.12

Bridged aromatic rings like paracyclophanes have a plane of chirality which is specified by the substituted benzene ring and the atoms directly bonded to it.

(*vii*) Molecules with Helical Shape

Several molecules *e.g.*, hexahelicene (scheme 1.13) display chirality due to helical shape, it has no stereocenter, however.

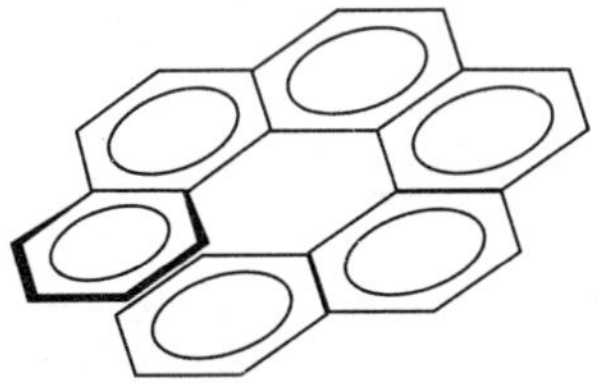

Fusing benzene rings in a linear fashion as in naphthalene, anthracene etc. does not generate chirality. However, if rings are added to produce a curve, the addition of the sixth ring produces a chiral molecule-hexahelicene (helical shape)

SCHEME 1.13

(*viii*) Chirality Due to Restricted Rotation of Other Types

Metallocenes substituted with at least two different groups on one ring become chiral and have been resolved (I, scheme 1.14). Other metallic complexes also become chiral if geometry is suitable (II, scheme 1.14). Thus fumaric acid-iron tetracarbonyl has been resolved into optically active forms.

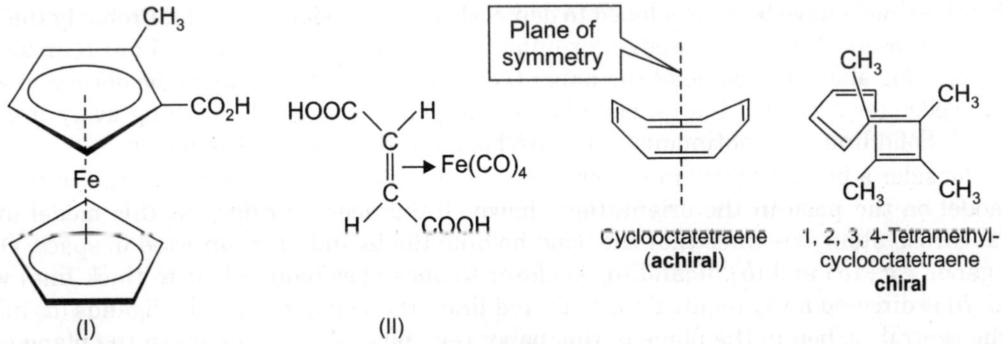

SCHEME 1.14

Cyclooctatetraene is a conjugated eight-carbon ring system (scheme 1.14) the molecule is not planar but is tub-shaped. A regular octagon has angles of 135° while sp^2 angles are most stable at 120°. In order to avoid this angle strain the molecule assumes a nonplaner conformation. 1,2,3,4-Tetramethyl-cyclooctatetraene is chiral due to its tub shape and no plane of symmetry.

Stereoisomerism–Stereogenic Units in Organic Molecules—A Summary

A stereogenic unit e.g., a stereocenter, a chiral axis, or a chiral plane when present in molecules generate chirality (scheme 1.14a). In these cases an exchange of pair of ligands around the stereocenter, chiral axis or the chiral plane leads to an enantiomer. An interesting situation is in (A, scheme 1.14a), the exchange of COOH with any of the drawn H atoms gives the enantiomeric structure (One may consider the enantiotopic relationship between the pair of hydrogens in the parent paracyclophane). The examples (B and C) represent achiral diastereomers.

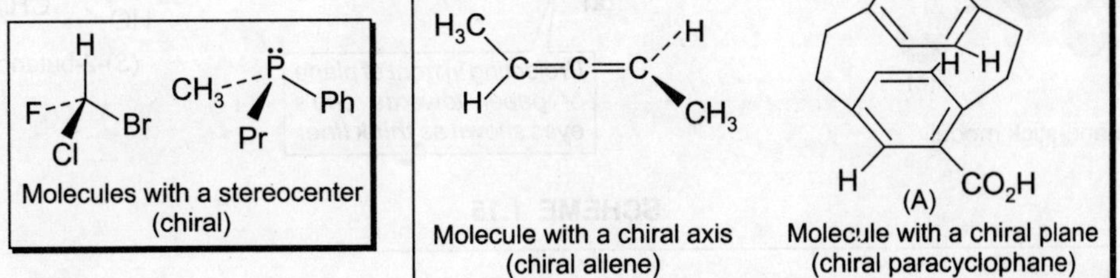

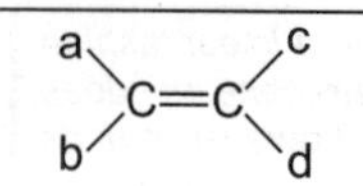

(B)
π-Diastereoisomerism
a ≠ b and c ≠ d
(achiral)

(C)
Two stereogenic centers (e.g. exchange of H with OH gives the *cis* diastereomer). Achiral due to these on a plane of symmetry.

SCHEME 1.14a

(C) Representation of Three Dimensional Molecules

(*i*) Wedge Formula

Since our major method of communication is illustrations on paper (two-dimensional) various graphical methods have been developed to depict three-dimensional objects. Probably the most common representation is the wedge formula. A solid wedge (or a heavy line) represents a bond projecting above the plane of the paper (*i.e.*, bonds pointing towards the observer) and a broken wedge (or a dashed line) a bond below the plane (*i.e.*, bonds pointing away from the observer). Solid lines (or continuous lines) are bonds in the plane of the paper.

Consider a ball and stick model of a chiral molecule (Cabcd, scheme 1.15). Firstly place the model on the page in the orientation shown. If one looks straight at this model at the central carbon atom, one finds that the bond holding the ligand (*d*) is directed in space. Of the two ligands *e.g.*, (*a*) and (*b*), ligand (*a*) is closer to ones eyes (equivalent to thick line) while ligand (*b*) is directed away (equivalent to dotted line). If one places *e.g.*, the ligands (*d*) and (*c*) and the central carbon in the plane of this paper (*i.e.*, now (*d*)—C—(*c*) are in the plane of the paper), one of the remaining C—(*a*) bonds sticks out in front of the plane towards one's eyes (equivalent to a thick line or a solid wedge), the other C—(*b*) bond projects behind the plane of the paper away from one's eyes (equivalent to a dotted line or a dashed wedge). Thus the convention for drawing and depicting the three dimensional nature of carbon compounds is to place any two C-ligand bonds on the plane of paper and indicate them by continuous lines and C-ligand bond which projects out of this plane by a thick line and the one, which projects behind this plane by a dotted line. One such operation (scheme 1.15) leads to a three dimensional sketch. Thus when the ligands are defined one depicts an enantiomer of 2-butanol as shown (scheme 1.15).

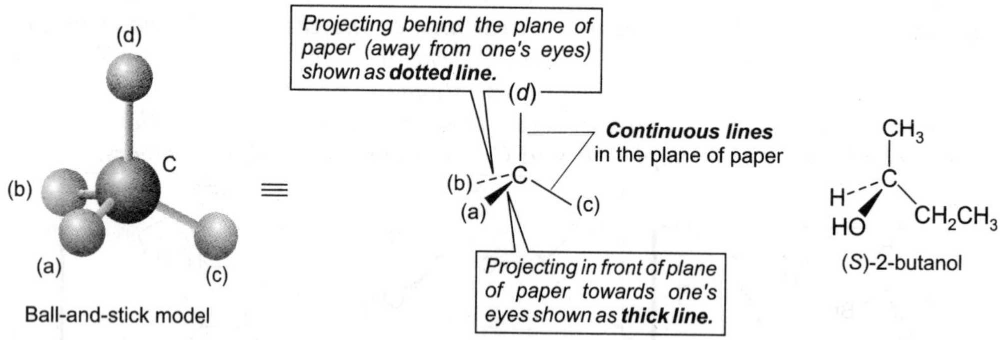

SCHEME 1.15

If now the ball and stick model (I, scheme 1.16) is picked up from the page and is held in space with *e.g.*, bond holding the ligand (—CH_2CH_3) directed perpendicular to the page one of the orientations of this ball and stick model, is depicted (II, scheme, 1.16). One finds that in this orientation the ligands (H) and (OH) are closer to one's eyes (heavy wedges) while the ligand (CH_3) is directed away (broken wedge) and this three dimensional picture (III) in which the bond holding the ligand (– CH_2CH_3) is placed perpendicular to the page is shown by a solid line is yet another useful way to depict chiral structures. These operations are equivalent to rotation of *e.g.*, your right hand (a chiral object). One of many orientations, may have thumb closer to you (equivalent to a solid wedge) or directed away from you (equivalent to a broken wedge). Whatsoever may be the orientation, the right hand remains right hand. Thus a three dimensional model of an enantiomer will retain its identity in any orientation you may depict it. In short, a right hand glove will remain right hand glove, whatever is the orientation.

A yet another orientation *e.g.*, (V, scheme 1.16) of S-2-butanol is also drawn. This is equivalent to tilting the model (II) in space to an orientation (IV) when two groups are close to ones eyes while two are away.

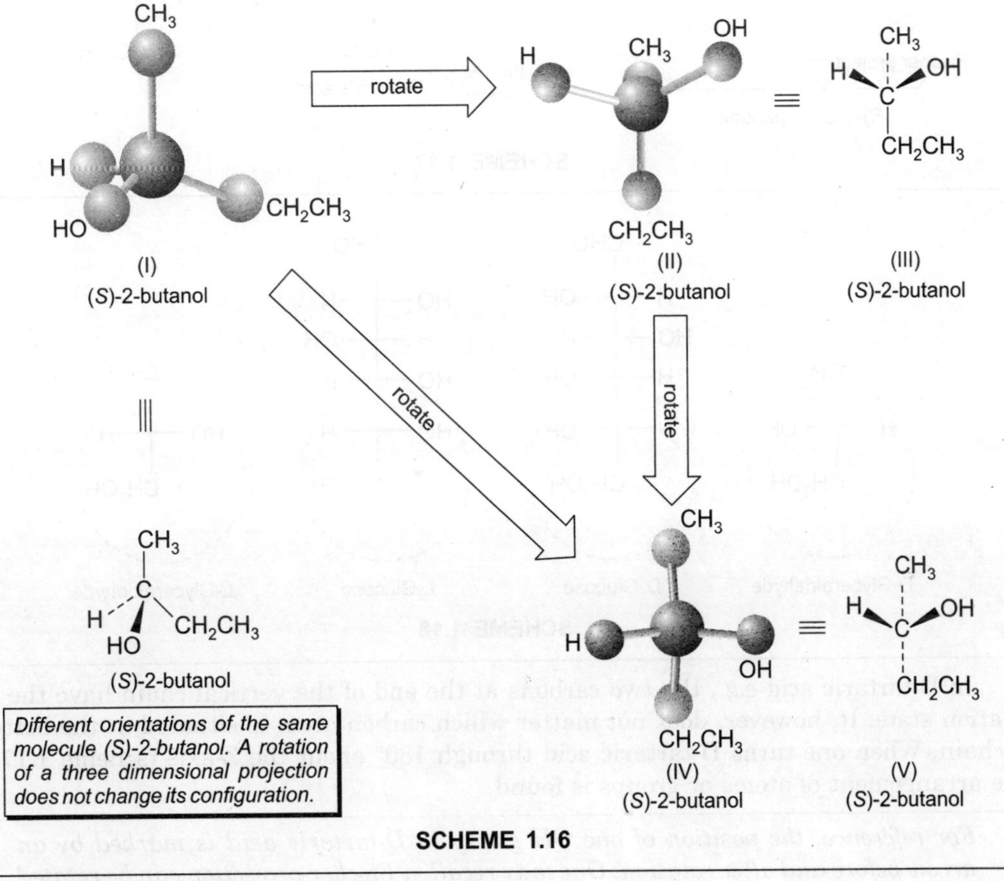

Different orientations of the same molecule (S)-2-butanol. A rotation of a three dimensional projection does not change its configuration.

SCHEME 1.16

(ii) Fischer Projections

Fischer developed a two dimensional plane projection formulas for three-dimensional molecules. These projections are particularly useful for carbohydrates and amino acids, and can also be used to represent many chiral molecules. Fischer projections are particularly useful for compounds with two or more stereocenters.

One must remember that Fischer projections represent eclipsed conformations and cannot tell anything about the preferred conformation of a molecule.

A Fischer projection uses a cross to represent the stereocenter ; the atom that is the stereocenter is often omitted in which case it is symbolized by the interaction point of the two lines of a cross. All the bonds are drawn as solid lines with the understanding that horizontal bonds point toward the observer and vertical bonds point away. The carbon chain is drawn along the vertical line of the Fischer projection usually with the most highly oxidized end carbon atom at the top as in glyceraldehyde (scheme 1.18). In (R)-1, 2-propanediol, the hydroxymethylene group being more oxidized is placed at the top (scheme 1.17). In the case of 2-butanol, neither end carbon is more oxidized than the other and in this type of cases one simply keeps the carbon chain along the vertical.

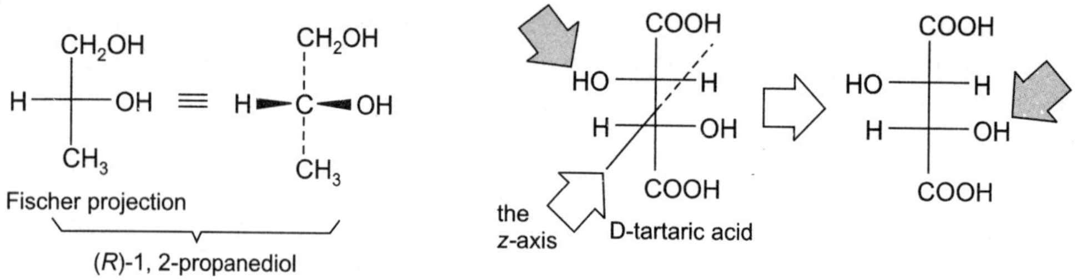

SCHEME 1.17

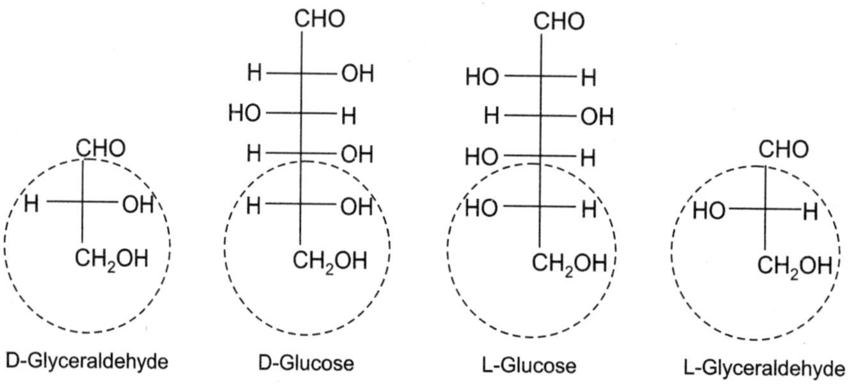

SCHEME 1.18

In D-tartaric acid *e.g.*, the two carbons at the end of the vertical chain have the same oxidation state. It, however, does not matter which carbon atom is situated at the bottom of the chain. When one turns D-tartaric acid through 180° about the Z-axis (scheme 1.17), the same arrangement of atoms or groups is found.

For reference, the position of one OH group in D-tartaric acid is marked by an arrow before and after rotation. One may recall, a Fischer projection can be rotated through 180° in the plane of the paper without creating problems. However, one cannot turn the projection through 90° or 270° which gives the mirror image. The projection cannot be taken out of the plane of the paper and rotated.

When one rotates the Fischer projection by 90°, the vertical bonds becomes horizontal and the horizontal bonds become vertical. The viewer is confused and assumes that the horizontal bonds come forward and the vertical bonds go back. One sees a different view of the molecule (actually, the enantiomer of the original molecule). A 90° rotation of the Fischer projection in the plane of the paper (gives the enantiomeric structure) is not allowed (see scheme 1.28*a*). Similarly its 180° rotation out of the plane of paper (gives the enantiomeric structure) is not allowed (see scheme 1.28*a*). A rotation of Fischer projection in the plane of paper by 180° (gives the same structure) is however, allowed (see scheme 1.28*a*).

(*iii*) Newman and sawhorse representations and interconversion of different projections

These are discussed in detail in schemes 1.29, 1.30 and 1.34c–1·38.

1.4 CONVENTIONS FOR CONFIGURATIONS D–L AND *R–S* SYSTEMS

(A) The D and L Designations—Configuration of Stereocenters

The configuration of a stereocenter is the specification of the relative spatial arrangement of the four groups attached to it. Absolute configuration specifies their order so as to distinguish the two enantiomers and thereby define their chirality *i.e.*, the specification of (R) or (S) at each stereocenter (and dissymmetric grouping) in a molecule. The term relative configuration specifies the relation of the two stereocenters to each other.

The former DL system to denote absolute configuration is not without faults and, therefore, this system is seldom used today except for some classes of compounds like carbohydrates and amino acids. The (+) enantiomer of glyceraldehyde has its OH group on the right of the Fischer projection (D-configuration) while the (–) enantiomer has it on the left (L-configuration). The D- and L- glyceraldehydes are the standard reference compounds for the DL system of absolute configuration. Thus the D-series of sugars are those with the OH group on the highest numbered stereocenter (*i.e.*, the bottom stereocenter) on the right in the Fischer projection and L-series have it on left (scheme 1.18). An aldose is degraded which results in the loss of the aldehyde carbon atom to afford a smaller sugar. Finally one obtains a triose which is then identified by comparing with glyceraldehyde enantiomers and assigned proper absolute configuration. D and L configurations are not related to the optical rotations of the sugars to which they are applied *i.e*, some sugars are D-(+)-while others are D-(–), thus D # *d* and L # *l*.

D-(+)-glucose means that dextrorotatory glucose has the same configuration at the highest number stereocenter as that of D-glyceraldehyde (the OH group on the bottom stereocenter on the right in Fischer projection) and the configuration at the other stereocenters must be memorized. Thus D—L system has the disadvantage of specifying the configuration of only one stereocenter.

Some weaknesses of D—L nomenclature. The DL nomenclature is restricted to those molecules which can be unambiguously drawn in the Fischer projection and which simultaneously obey all relevant rules. Several difficulties were faced while working with D, L nomenclature. In the case of amino acids, L-(–) serine (scheme 1.19) is taken as a reference. A further complication is introduced in assigning configuration to a molecule containing both a hydroxyl as well as an amino group. The 2-amino-2-deoxymannonic acid (A, scheme 1.19) way well be assigned D-configuration with reference to D-glyceraldehyde and L-configuration with reference to serine (C2). No doubt the complication was removed by employing subscripts '*g*' for glyceraldehyde and '*s*' for serine so that the compound could be assigned either of configurations Dg or Ls. Moreover, it was found that in several cases both the enantiomers of a compound could be chemically correlated to the same glyceraldehyde to show the ambiguity of the D, L system of nomenclature. In the first series, of reactions (scheme 1.19) one relates the levorotatry lactic acid to D-(+) glyceraldehyde, thus (–)-lactic acid has also a (D)-configuration. In the second reaction sequence one finds that the enantiomeric (+)-lactic acid is also related to D-(+)-glyceraldehyde. Both enantiomers, however, cannot have the same configuration.

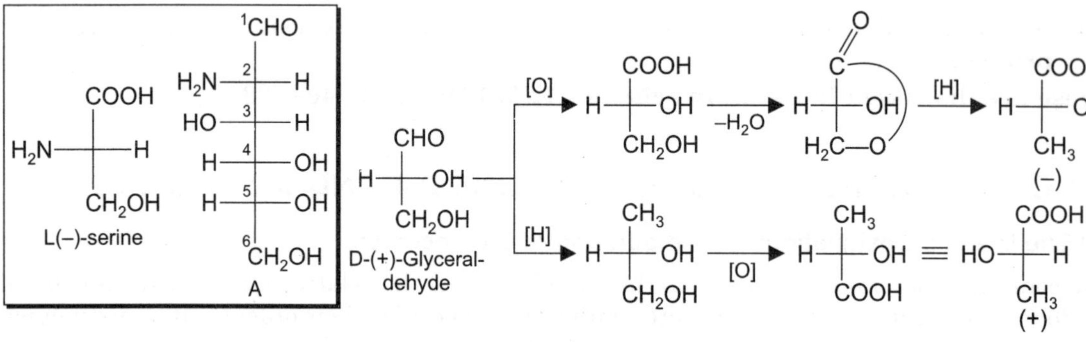

SCHEME 1.19

(B) The *R* and *S* Designations From Perspective Formulas-Three Dimensional Drawing

The system now widely used is the Cahn-Ingold-Prelog (R-S) system (which relates all compounds with only one set of rules) in which the four groups attached to the stereocenter are numbered 1, 2, 3 and 4 and are ranked according to a set of sequence rules so that they can be assigned priorities for arrangement in a sequence 1 > 2 > 3 > 4. In a simple case the priorities follow the order of decreasing atomic number of the atom that is directly attached to the stereocenter *i.e.*, Br. (35), Cl(17), S(16), O(8), N(7), C(6) and H(1). The application of the rule can be illustrated with 2-butanol enantiomers (scheme 1.20) which has a sequence OH, C_2H_5, CH_3 and H.

The, oxygen has the highest priority (1) and the hydrogen the lowest atomic number priority (4). To make the priority assignment for the methyl and ethyl groups where the atom that is directly attached to the stereocenter is a carbon atom the next sets of atoms in the groups are examined. Thus, the ethyl group takes precedence (C, H, H) as compared with methyl (H, H, H).

In table 1.2, atoms and groups are listed with increasing priority with the exception of a lone pair. H has the lowest priority. The following points may be noted:

Table 1.2. Atoms and groups with increasing priority

1—H	10—CH = CH_2	19—NH_2	28—SOR
2—D	11—$C(CH_3)_3$	20—$NHCH_3$	29—SO_2R
3—CH_3	12—C ≡ CH	21—NO	30—Cl
4—CH_2CH_3	13—C_6H_5	22—NO_2	31—Br
5—$CH_2(CH_2)NCH_3$	14—CH_2OH	23—OH	32—I
6—CH_2—CH = CH_2	15—CHO	24—OCH_3	
7—CH_2—C ≡ CH	16—COR	25—OC_6H_5	
8—$CH_2C_6H_5$	17—$CONH_2$	26—F	
9—$CH(CH_3)_2$	18—CO_2H	27—SH	

- **If the four atoms attached to the stereocenter are different, priorities depend on atomic numbers, the atom of higher atomic number is assigned higher priority. If two atoms are isotopes of the same element, the atom of higher mass number has the higher priority (I, scheme 1.19*a*)**

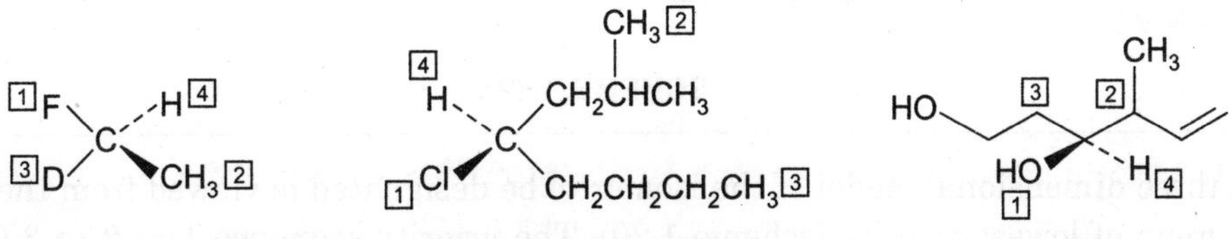

(*R*)-1-Deuterio-1-fluoroethane (I) (*R*)-4-chloro-2-methyloctane (II) (*S*) (III)

SCHEME 1.19a

- **If the relative priorities of two groups cannot be determined as above, these are decided by a comparison of priority atoms at the first point of difference (II, scheme 1.19*a*). In the case of (III scheme 1.19*a*) of the two carbon chains attached to the stereogenic carbon, the chain with a secondary carbon is of higher priority than the one with primary carbon. In the case of (I, 1.19*b*) of the four chains attached to stereogenic carbon, butyl has priority over ethyl group. In the case of "oxygen chains" the first point of difference is reached at the carbon attached to oxygen. Similarly priorities are given in compound (II scheme 1.19*b*). Often it may be necessary to proceed atom by atom till a point of difference is obtained as in compounds (III and IV scheme 1.19*b*). The cyclic compounds are treated by the same rules.**

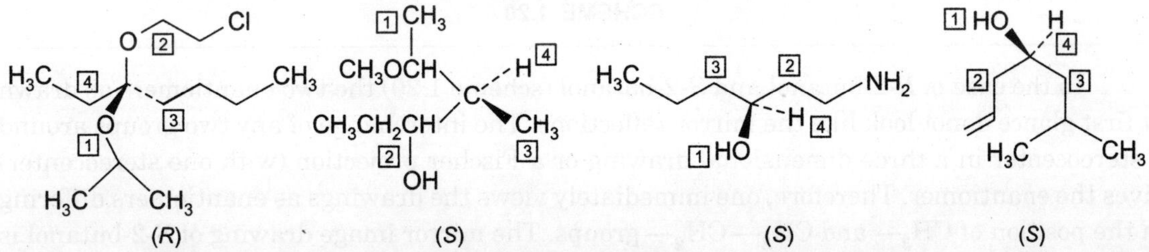

(*R*) (I) (*S*) (II) (*S*) (III) (*S*) (IV)

SCHEME 1.19b

- **In the case of double or triple bond, both atoms are considered to be duplicated or triplicated. For example, in glyceraldehyde (scheme 1.19*c*) the –OH group has the highest priority of all and the O, O, H of –CHO takes priority over the O, H, H of – CH_2OH. The complete sequence is thus –OH, –CHO, –CH_2OH, –H (scheme 1.25). In 1-amino-2-methyl-1-phenylpropane (I, scheme 1.19*c*), therefore, the C, C, C, of phenyl take priority over the C, C, H of isopropyl, but not over N, which has a higher atomic number. The sequence is thus –NH_2, C_6H_5, C_3H_7–, H.**

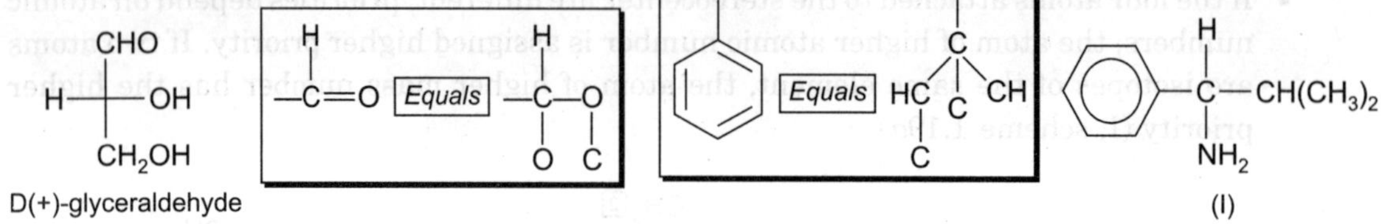

SCHEME 1.19c

A three dimensional model of the isomer to be designated is viewed from the side opposite the group of lowest priority (scheme 1.20). The priority sequence $1 \rightarrow 2 \rightarrow 3$ (decreasing priority) of the remaining three groups is determined and if it is found clockwise, the symbol (*R*) is used to designate the configuration and if the sequence is counter clockwise the symbol (*S*) is used. (*R*) and (*S*) are from Latin words *rectus* and *sinister* meaning right and left respectively.

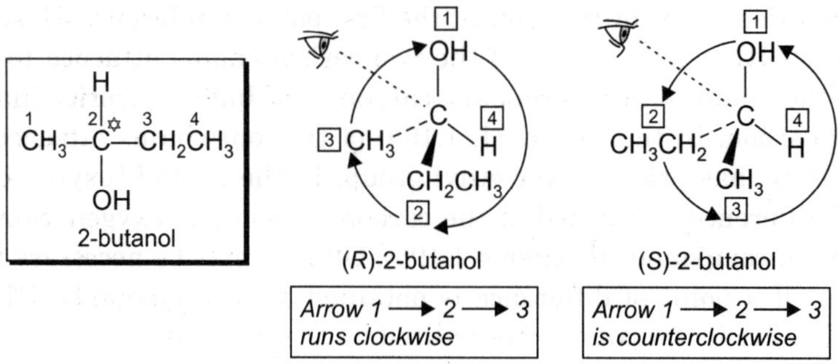

SCHEME 1.20

In the case of *R*-2-butanol and *S*-2-butanol (scheme 1.20) the two enantiomers as drawn at first glance donot look like the mirror reflections. The interchange of any two groups around a stereocenter in a three dimensional drawing or a Fischer projection (with one stereocenter) gives the enantiomer. Therefore, one immediately views the drawings as enantiomers differing in the position of CH_3— and CH_3— CH_2— groups. The mirror image drawing of *S*-2-butanol is *R*-2-butanol and is drawn suitably in (scheme 1.20*a*).

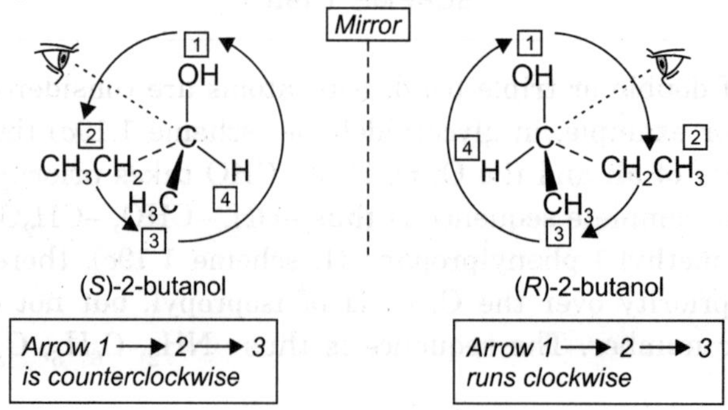

SCHEME 1.20a

One may position the same molecule, on paper or by using molecular model set to simply give different views of the same molecule. Some effort will then be needed to establish their identity. This may be done mentally or by manipulation on paper, and the later process is easy. Thus, yet another view of *R*-2-butanol is presented (scheme 1.20*b*), that this and the drawing of (*R*)-2-butanol (scheme 1.20*a*) represent the same molecule can be proved by interchanging the position of any two groups in one of these drawings twice (This process leads to retention of configuration at the stereocenter). Thus in drawing (scheme 1.20*b*) first exchange 1 with 3 and then 2 with 4 and the arrangement thus obtained is identical to the one in (scheme 1.20*a*). Identical names including *R* and *S* designation show that different orientations of a compound are infact same *i.e.* homomeric. This procedure shows that *e.g.*, different drawings of 2-butanol (schemes 1.16, 1.20-1.20*c*) represent the same enantiomer (*S*)-2-butanol.

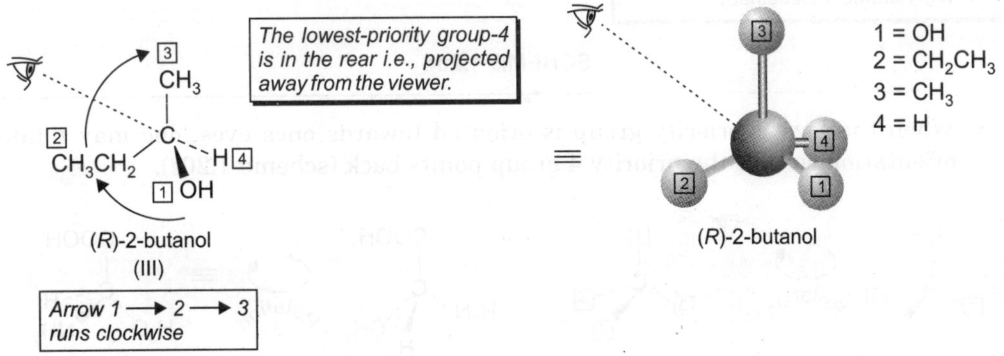

SCHEME 1.20b

To assign absolute configuration as *R* or *S* the three dimensional drawing, of the molecule has to be viewed with the lowest priority group at the back. The compounds (I-III, scheme 1.20*c*) are already drawn (oriented) so that each orientation can be viewed either from left or right with the priority group 4 already oriented toward the rear.

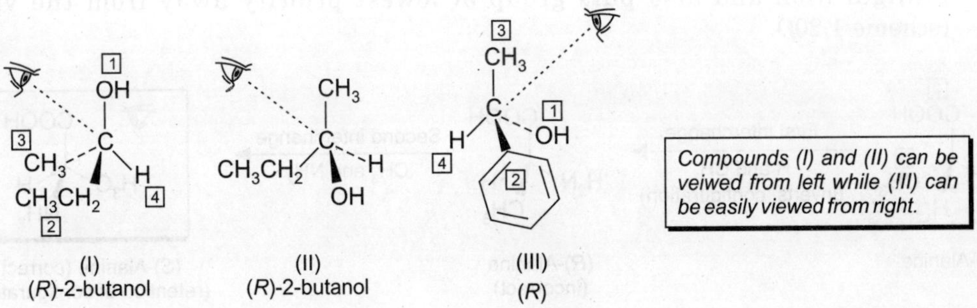

SCHEME 1.20c

One may have many options when the priority 4 group as drawn does not point away from one's eyes. One can then reorient the substituents so that the priority group 4 points away from ones eyes and the orientation of the compound can be visualized easily for the assignment to *R* and *S* descriptor to the stereocenter. Consider the following situations:

- When the group of lowest priority is drawn at the top, one can look along the C—H bond from below the orientation. Alternatively one can tilt the orientation so that it can be viewed conveniently for the assignment of configuration (scheme 1.20d).

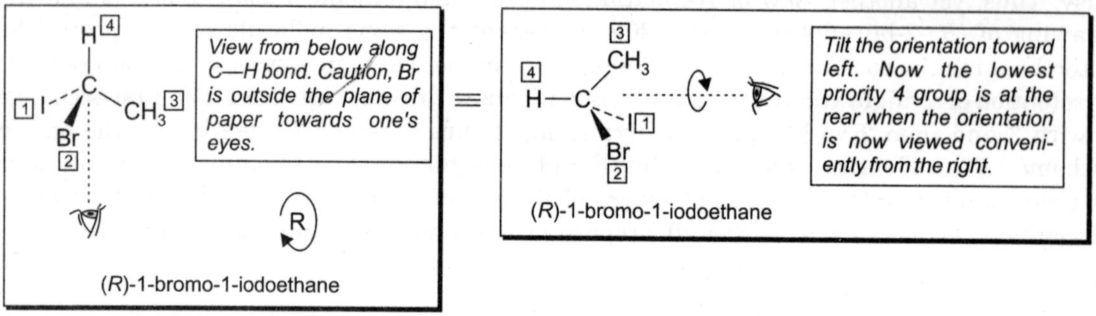

SCHEME 1.20d

- When the lowest priority group is oriented towards ones eyes, one may rotate the orientation so that the priority 4 group points back (scheme 1.20*e*).

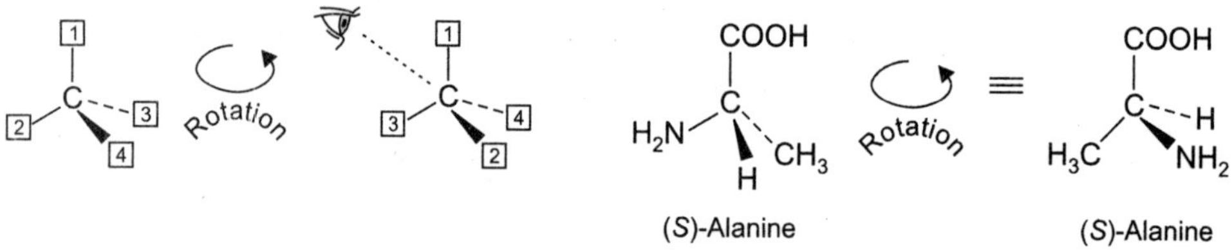

Turn the molecule to put the group of priority 4 at the rear for the assignment of configuration

SCHEME 1.20e

- When the lowest priority group is towards one eyes one may adopt another option. Interchange of the position of any two groups twice at the stereocenter retains its configuration and also puts group of lowest priority away from the viewer (scheme 1.20*f*).

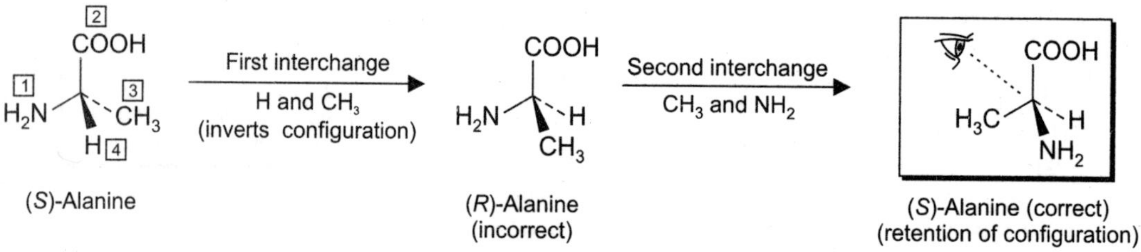

SCHEME 1.20f

One can assign R or S configuration directly from the Fischer projection by following the procedure given (See, scheme 1.26b).

Another useful exercise to show that interchanging the position of any two groups around a stereocenter retains its configuration is the relation between a-D-glucose in common Fischer projection (I, scheme 1.20g) and Haworth formula (II, scheme 1.20d). All the groups in the common Fischer projection formula which are on the right of the projection are below the plane of the hexagon in Haworth structure, but for the C5 H which is below the plane of hexagon even though it is not lying on the right of the Fischer projection (I, scheme 1.20g). In fact one would find that interchanging two groups twice (first H with CH_2OH and then H with ether link) gives the equivalent structure (modified Fischer projection III, scheme 1.20g) in which H at C5 comes to right of the projection. Six membered ring exists in (energy minimum) chair form and is not flat.

The, flat Haworth formula can be relaxed to chair by putting one carbon (C4) up and other (C1) down as far a-D-glucose (scheme 1.20g). In the chair form the substituents, maintain their identity i.e. a group closer to the observer remains closer and the one directed away remains directed away. One may remember that there are two chair forms possible interconverted by ring flip.

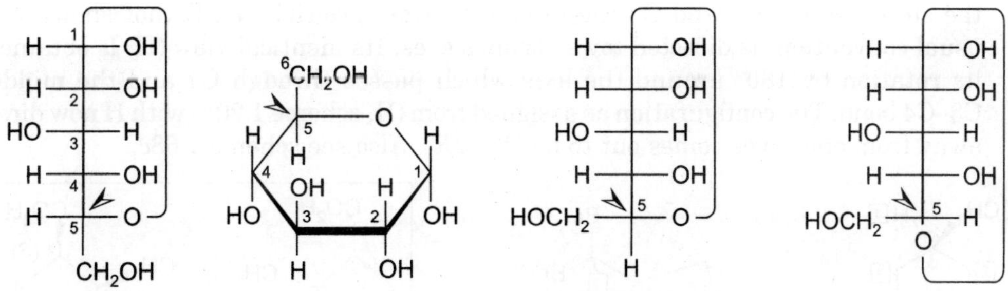

(I) α-D-Glucose (Fischer Projection)

(II) α-D-Glucose (Haworth formula)

one interchange of groups around stereogenic center inverts its configuration (not allowed)

(III) α-D-Glucose (Modified Fischer Projection) (two interchanges of groups around a stereogenic center retain its configuration)

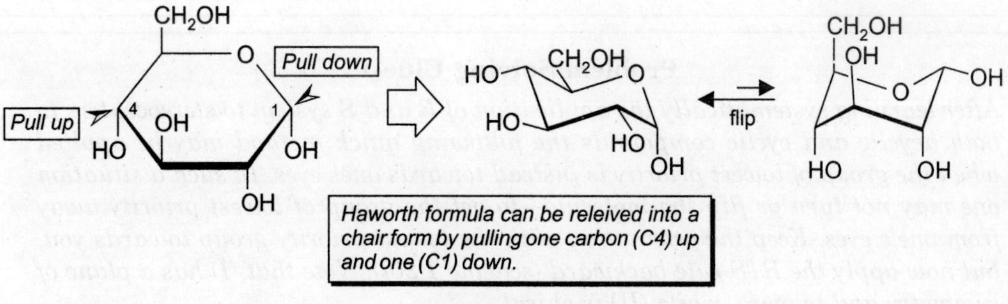

SCHEME 1.20g

The following is the procedure for converting a Haworth projection of glucose into a chair conformation.

- *Place the ring oxygen at the back right-hand corner. The primary alcohol group (—CH_2OH) is the largest of the substituents and thus would be more stable in the equatorial position (C5 position).*

- *The OH group at C4 is trans to the —CH_2OH group (see Haworth projection), this is thus placed at the equatorial position (1, 2-diequatorial substituents are trans to one another (see scheme 4.25).*
- *The C3 OH group is also trans to the C4 OH group, thus the C3 OH group is also placed in the equatorial position.*
- *Moving around the ring, thus gives the conformation of α-D-glucose.*
- *The axial positions at C5, C4, C3 and C2 are all occupied by hydrogens (not shown) which require little space and thus experience little steric strain. There is more on conformation of sugars (Sec. 4.16).*

- When the stereocenter is part of the ring, the paths around the ring are considered as independent substituents. **[*For determining the retative priority (1 to 4) see the summary of sequence rules*]**. For the assignment of (R or S) descriptors the lowest priority group has to be directed away (scheme 1.20*h*) and thus the absolute configuration of 4-methylcyclohexene (scheme 1.20h) is (*R*). Also see (Problem 4.16).
- When in a cyclic system the group of lowest priority is instead directed toward the viewer *i.e.*, it is on the thick line, the structure is rotated 180° so that it gets directed away and the usual procedure is then applied. In structure (I, scheme 1.20*h*) at both the stereocenters C1 and C2 the group of lowest priority *i.e.*, H (not shown as per usual convention) is directed towards ones eyes. Its identical view (II) is obtained by its rotation by 180° around the axis which passes through C1 and the middle of C3–C4 bond. The configuration as assigned from (II, scheme 1.20*h*) with H now directed away from ones eyes comes out to be (1*S*, 2*R*). Also see scheme 1.68*c*.

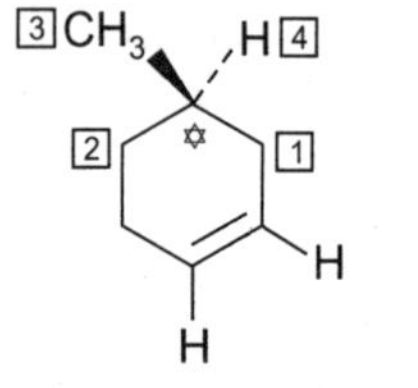

(*R*)-4-Methylcyclohexene

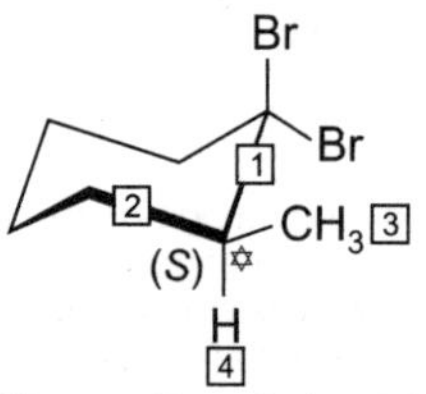

1, 1-Dibromo-2-methylcyclohexane

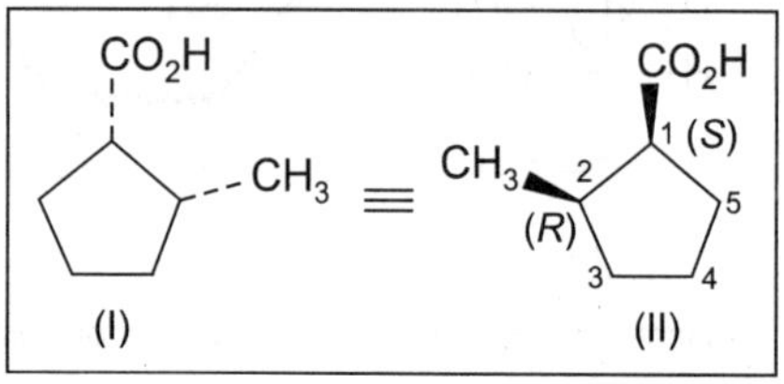

SCHEME 1.20h

Problem Solving Clue-1

After learning systematically the application of R and S system to stereocenters to both acyclic and cyclic compounds the following quick method may be applied when the group of lowest priority is instead towards ones eyes. In such a situation one may not turn or flip the molecule—to get the group of lowest priority away from one's eyes. Keep the orientation with the lowest priority group towards you, but now apply the R/S rule backward (scheme 1.20i). Note that (I) has a plane of symmetry and is meso, while (II) is chiral.

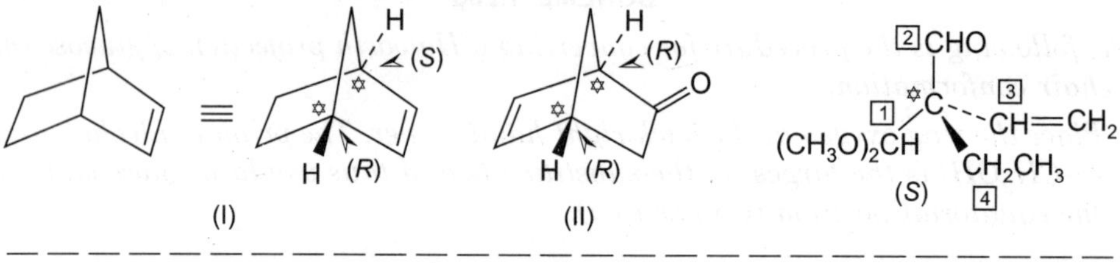

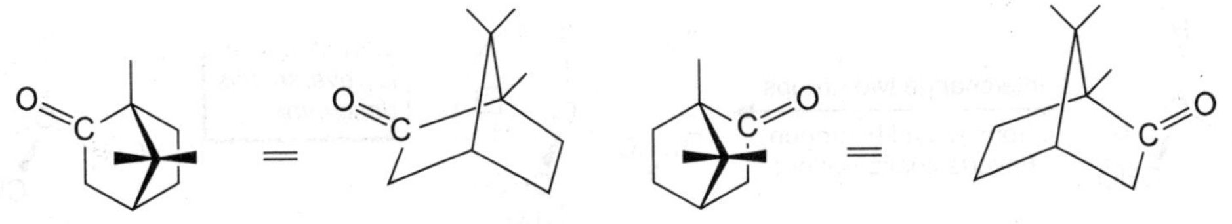

(1*R*)-(+)-Camphor (1*S*)-(–)-Camphor

SCHEME 1.20i

Thus in the molecule of camphor (two stereocenters) the methyl group containing stereocenter is already away in (I, scheme 1.20j) and when viewed from front the configuration at this stereocenter comes out to be (R). In the equivalent orientation (II, scheme 1.20j) now the methyl group of lowest priority is instead towards one's eyes. To assign configuration to this stereocenter, there is no need to turn the orientation to put CH_3 group away, but simply apply the R/S rule backwards on (II, scheme 1.20j) itself which gives R description to this stereogenic carbon.

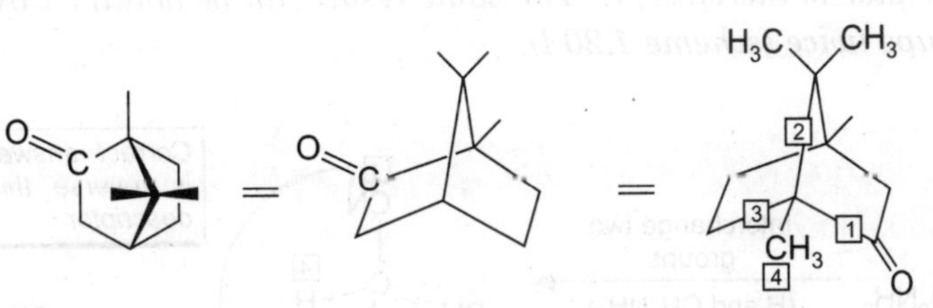

(I) (II)

(1*R*)-(+)-Camphor

SCHEME 1.20j

Problem Solving Clue–2

When one interchanges any two groups attached to a stereocenter its configuration is inverted. When e.g., there is only one stereocenter in a molecule inverting its configuration gives the enantiomer. One can utilize this option to assign configurations. Interchange (switch) the position of a group pointing away with that of the priority 4 group, assign the configuration and thus the correct configuration will be opposite to this (scheme 1.20 k). This can be applied to any of the situations already discussed and additional examples are presented.

Example 1. 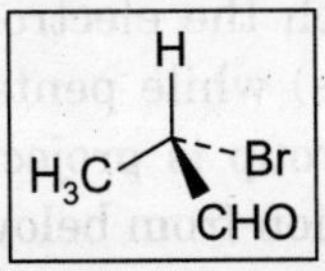

As already discussed (scheme 1.20d) either view this orientation from below along the C—H bond or tilt the orientation, so that the orientation can be instead conviently viewed from left or right. The procedure described above may provide a third simple option (scheme 1.20k).

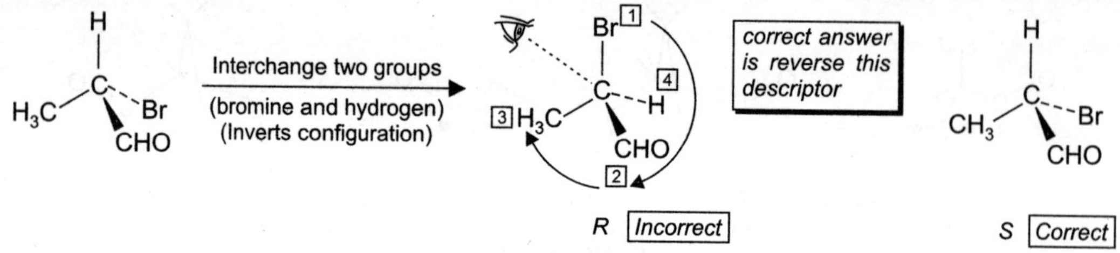

SCHEME 1.20k

Example 2. 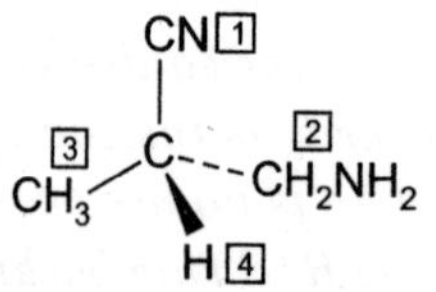 *The group of lowest priority is towards one's eyes, thus either work with problem solving clue-1 i.e., the path 1 → 2 → 3 is R the correct descriptor is therefore, S. The same result can be obtained by interchanging two groups twice (scheme 1.20 l).*

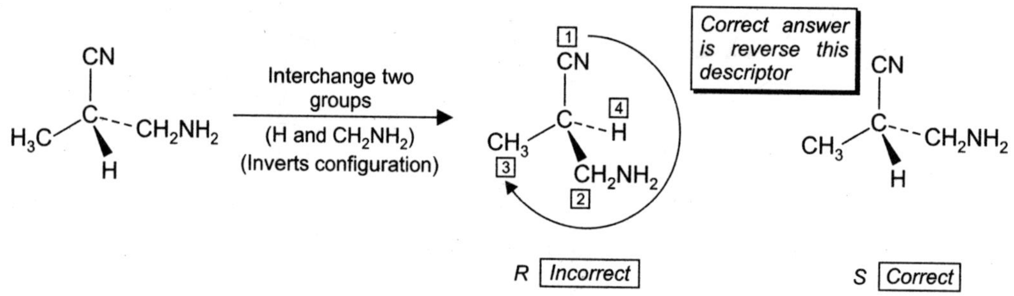

SCHEME 1.20l

(C) The *R* and *S* Designations to Compounds Containing Nitrogen Phosphorus and Sulphur

One may recall that amines and phosphines are approximately tetrahedral with the fourth ligand being the lone pair which is the lowest priority group. Phosphines, phosphine oxides, sulfoxides and amine oxides display chirality provided the ligands are non-equivalent (see scheme 1.6). These are assigned configurational descriptors as already discussed in the case of chiral compounds which contain a stereogenic carbon.

Consider the chiral phosphine (scheme 1.21) in which the electron lone pair and the C_6H_5 group are in the plane of the paper (continuous lines) while pentafluorophenyl group C_6F_5 is projected in front of the plane while the methyl group is projected away. One may assign absolute configuration either by viewing the orientation from below or tilt the orientation to conveniently view it from the side with the lowest priority 4 group (electron lone pair) directed away as done with a compound with a stereogenic carbon (see scheme 1.20*d*). It is also easy to use the problem solving clue 2 (page 39) as explained in (scheme 1.21).

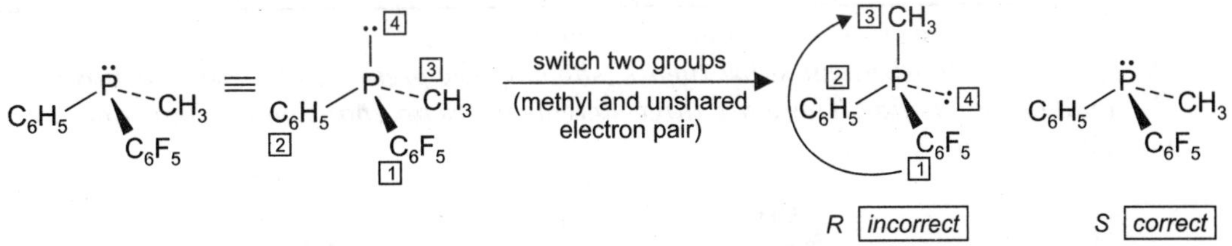

SCHEME 1.21

EXERCISE 1.3

Assign absolute configuration to the stereogenic atoms in the compounds (scheme 1.22).

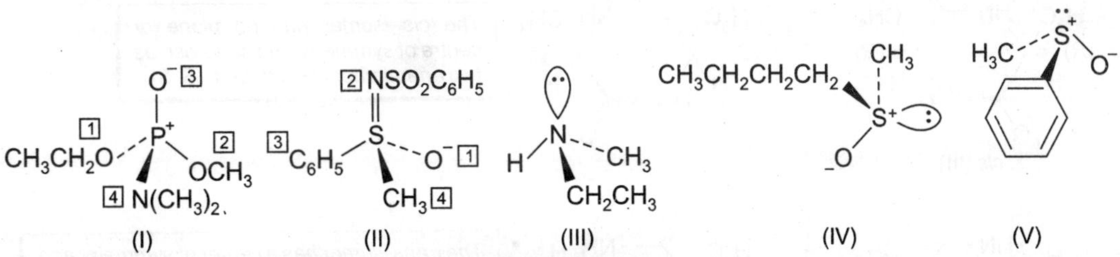

SCHEME 1.22

ANSWER. *(1) R ; (II) R ; (III) R ; (IV) S ; (V) S.*

[***Hint.*** *(Both I and II) are S as drawn since priority group 4 is coming forward so the actual configuration is R in both.*]

Relationship between possible stereoisomeric structures

Determine the R and S configuration at each stereocenter. For two stereocenters R, R and S, S are enantiomers of each other; R, S and S, R are enantiomers of each other. Any other combination is a pair of diastereomers.

EXERCISE 1.4

Discuss the stereochemical features in the compounds (scheme 1.23).

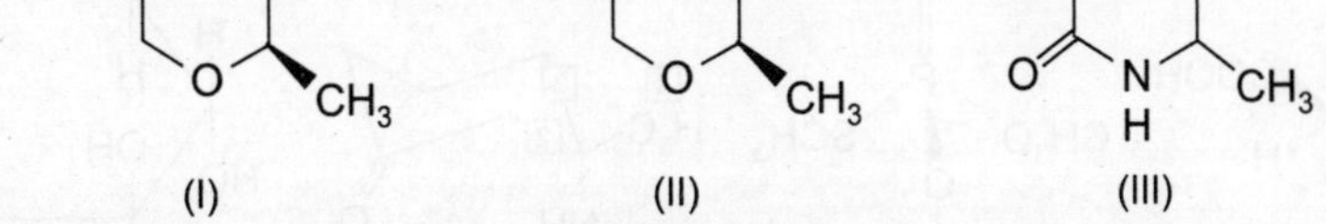

SCHEME 1.23

ANSWER. *(I) is achiral with a center of symmetry (see scheme 1.3 g) ; (II) is chiral and since at every stereocenter the group of 4 priority is below the projection, configuration is directly assigned (scheme 1.24), its enantiomer has inverted* *configuration at every stereocenter. The compound (III) has one ring and two stereocenters with identical substituents (showing the presence of meso structure), the amide group is flat. There are three stereoisomers an enantiomeric pair and a meso compound.*

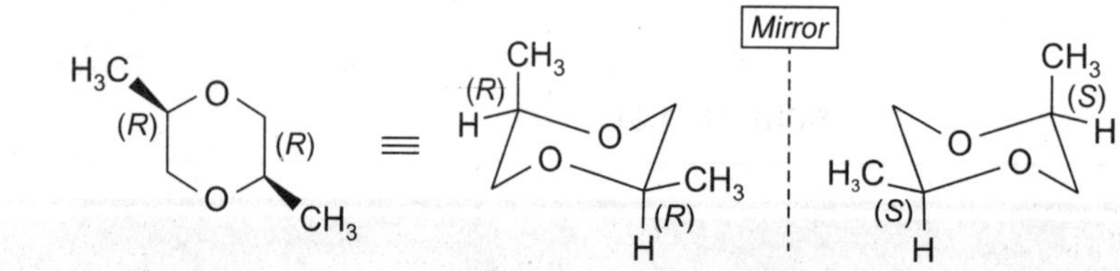

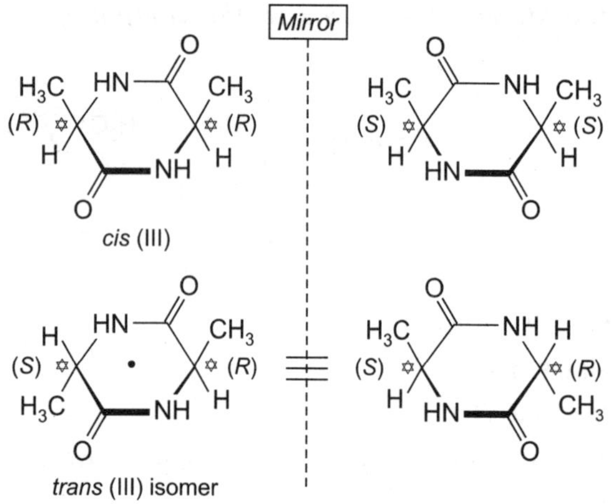

The cis isomer has no plane or centre of symmetry and can exist as two enantiomers (R, R)- and (S, S)

The trans isomer has a center of symmetry and has identical achiral meso-structures. (R, S) is diastereoisomeric with the two enantiomers (R, R) and (S, S).

SCHEME 1.24

EXERCISE 1.5

Assign configuration to the compounds (scheme 1.25).

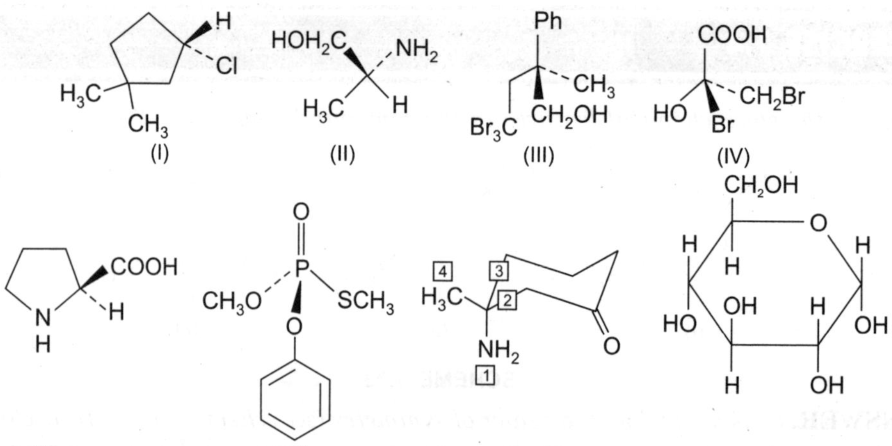

SCHEME 1.25

ANSWER. *(I) S ; (II) R ; (III) S, [hint bromines are too remote to be decisive] ; (IV) S, (V) S ; (VI) S ; (VII) R, hint give priorities NH_2 is 1, CH_3 is 4 since 4 is towards ones eyes reverse the answer obtained from $1 \rightarrow 2 \rightarrow 3$; (VIII) [hint priorities at C1 are OR > OH > C (OH) > H, the path $1 \rightarrow 2 \rightarrow 3$ gives R, however since group of lowest priority H is forward the correct configuration at C1 is S (scheme 1.26)] ; priorities at C2 are OH > C(OH)OR > C(OH)C > H and following the same procedure as for C1, configuration at C2 is R. A Caution is needed at C3 priorities are OH > CH(OH) COR > CH(OH) COH > H, group of lowest priority H is down, note that OH at C3 is not in the plane of paper but towards one's eyes so the path $1 \rightarrow 2 \rightarrow 3$ should be correctly followed. The configuration at C3 is S which is more clearly derived from the conformational formula. The configuration at other centers C4 and C5 is as indicated.*

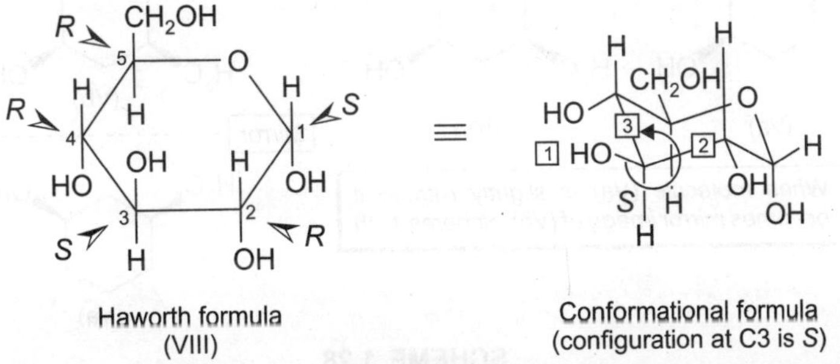

SCHEME 1.26

EXERCISE 1.6

Give the stereochemical relationship between the pair of compounds (scheme 1.27)

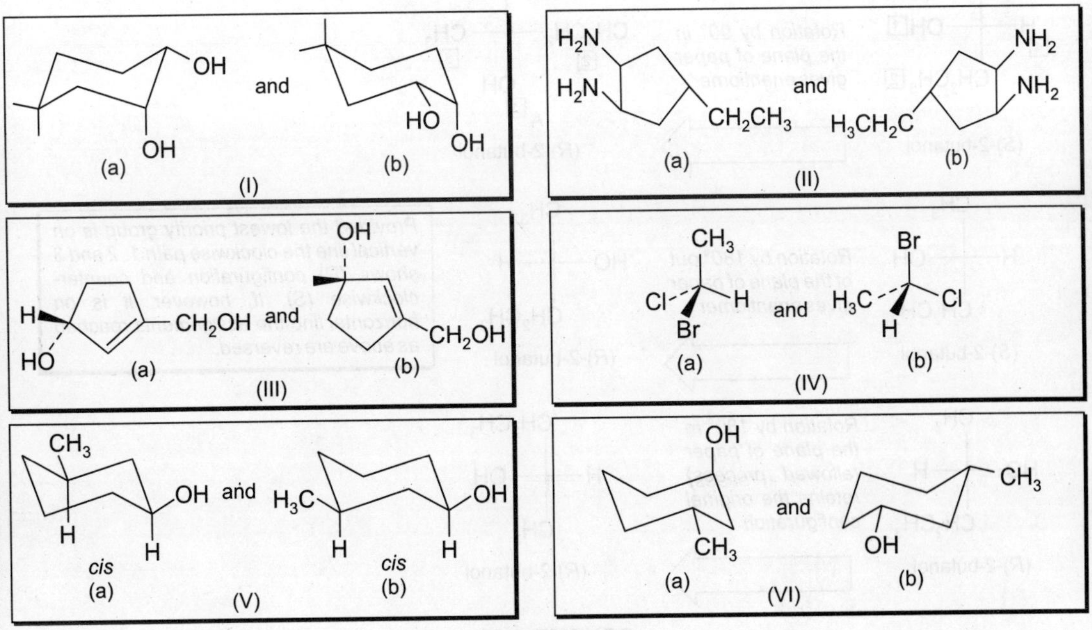

SCHEME 1.27

ANSWER. *(I) Conformational isomers (both 1 S, 2 R, configuration does not change with conformation) ; (II) same ; (III) enantiomers [turn the molecule (b) clockwise so that the two CH_2OH groups in (a) and (b) are next to each other, these come out to be non-superimposable mirror images] ; (IV) enantiomers [(a) is R while (b) is S] ; (V) enantiomers, firstly the configuration of (a) is 1S, 3R in (b) it is 1R, 3S, i.e. configuration at each center is inverted. Moreover, one may draw the planar formulas of both ignoring H (Scheme 1.28) to show that these are mirror images ; (VI) diastereomers (a) is trans while (b) is cis. For a related example see problem 1.48.*

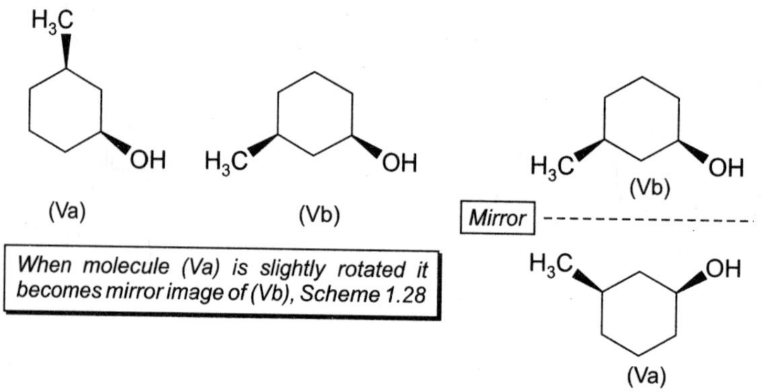

SCHEME 1.28

(D) The R and S Designations from Fischer Projections

A simple and most widely accepted procedure used now to assign R and S designations to the stereocenters when a compound is written in Fischer projection involves only two operations.

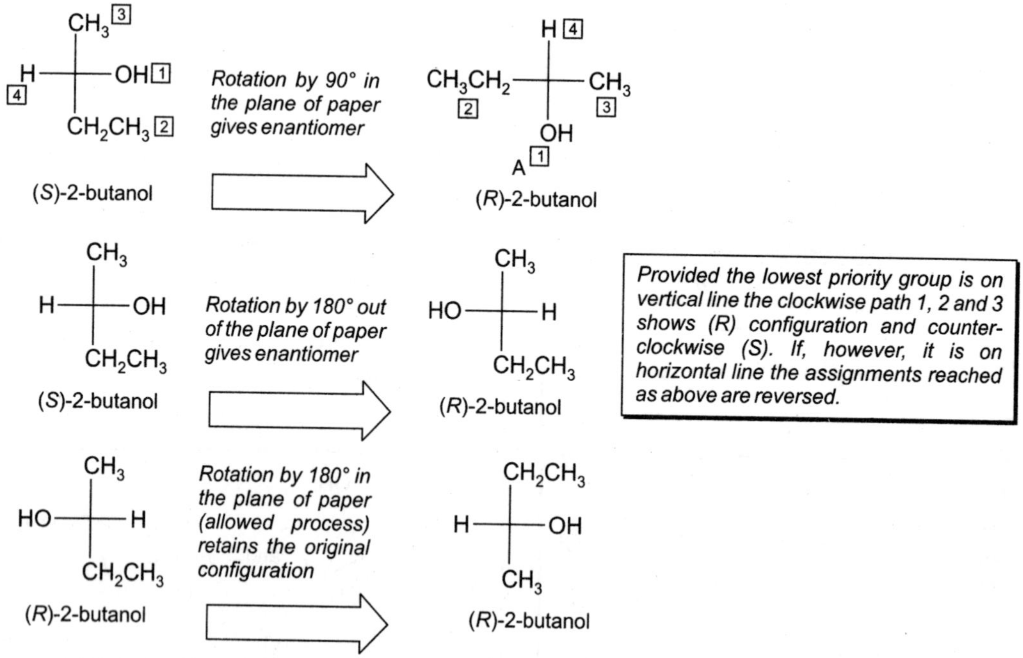

SCHEME 1.28a

Assign the priority order to the ligands and trace a semicircle joining ($1 \rightarrow 2 \rightarrow 3$ ignoring '4' the lowest priority group. When '4' is on the vertical line in the Fischer projection (no matter if it is at the top or at the bottom) the sequence gives the correct designation, if however, 4 is on the horizontal line the sequence gives the wrong answer and should be reversed. Thus, *e.g.*, in (*S*)-2-butanol (scheme 1.28*a*) H is on the horizontal line, therefore, the designation arrived at from the sequence $1 \rightarrow 2 \rightarrow 3$ is to be reversed. Rotation of a Fischer projection by 90° in the plane of paper is not allowed since it gives the enantiomer. Thus rotation of Fischer projection of (*S*)-2-butanol by 90° in the plane of paper gives (*R*)-2-butanol (A, scheme 1.28*a*) in which now H is on the vertical line and the designation arrived at from the sequence $1 \rightarrow 2 \rightarrow 3$ in (A, scheme 1.28*a*) gives the correct designation (*R*). One can confirm from proper nomenclature including (R and S) designation that rotation of a Fischer projection by 180° out of plane of paper gives the enantiomer and is not allowed, however, its rotation by 180° in the plane of paper maintains the configuration and is allowed.

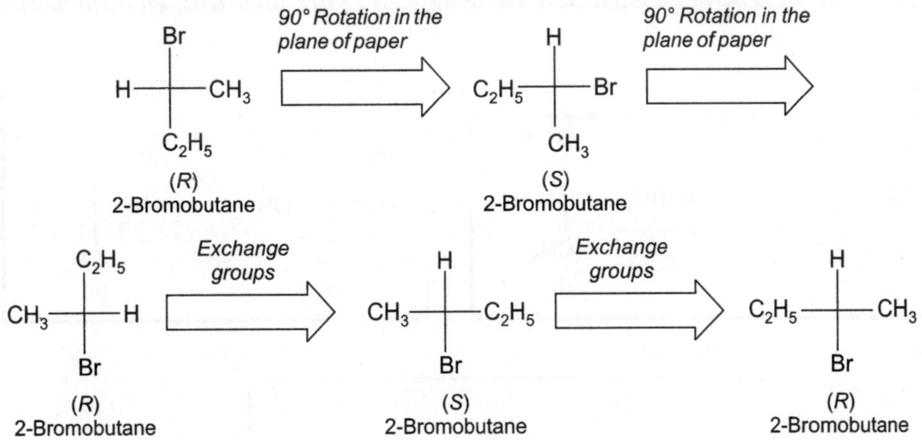

SCHEME 1.28b

This is again seen in the case of 2-bromobutane (scheme 1.28*b*). A 90° rotation of the Fischer projection in the plane of the paper gives the enantiomer and is thus not allowed. A 180° rotation (*i. e.*, two 90° rotations) in the plane of the paper retains the original configuration and is allowed. By proper nomenclature including (R and S) designation one can also prove that exchange of any two groups around a stereocenter once inverts its configuration to give its enantiomer while interchanging the position of any two groups twice retains the original configuration of the stereocenter.

Unlike a Fischer projection a three dimensional structure like a wedge projection can however, be manipulated (without loosing its identity) in any way one wishes *i.e.*, turning rotating or tumbling (see, scheme 1.20*d*). A useful check for identity of two structures (drawn as Fischer projection or three dimensional drawing) would be to name each compound including its (*R*—*S*) designation. In case the names come out to be the same then the structures are the same. It is indeed important that one should be able to assign absolute configuration to a stereocenter so as to understand the structures drawn in different ways.

Relationship between Possible Stereoisomeric Forms

- *Recall that quickest method is by determining R, S configuration at each stereocenter to reach the conclusion (See Exercise 1.6).*

- *In Fischer projections rotation through 180° in the plane of paper of a projection for comparison with others is helpful. Also placing as many similar groups as possible in identical positions vertical or horizontal in one projection compared to the other by interchanging the position of two substituents at the same stereocenter for comparison is also helpful (Practice Exercise 1.7). On the basis of R, S configuration for comparison, the same conclusion is reached, thus e.g., (Ia and Ib, scheme 1.28c) have both S configuration and are thus identical.*
- *Never operate on more than one stereocenter at a time while interconverting Fischer projections.*

EXERCISE 1.7

Show if the structures in each pair (scheme 1.28 c) are identical, enantiomers or diastereomers.

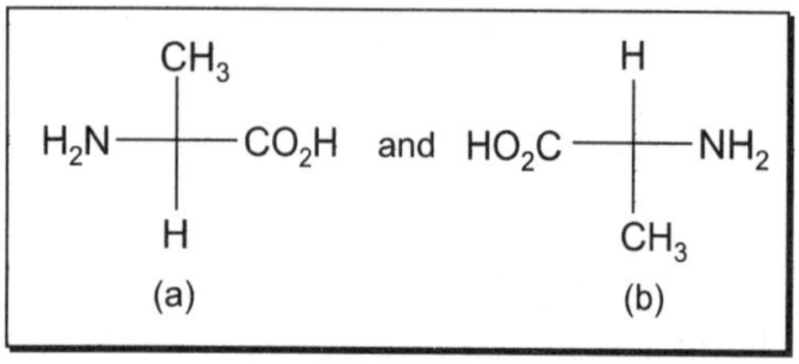

(I)

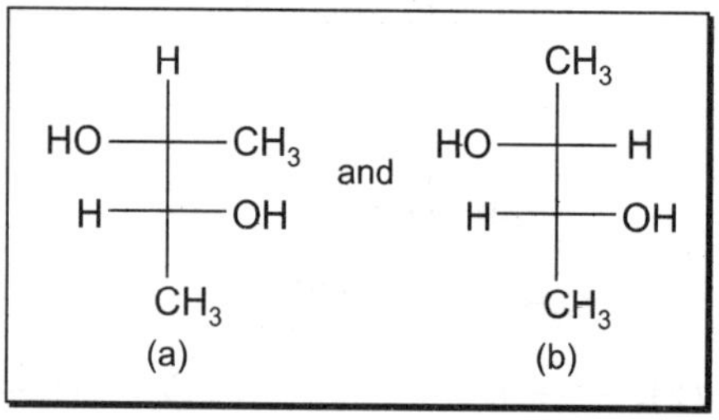

(II)

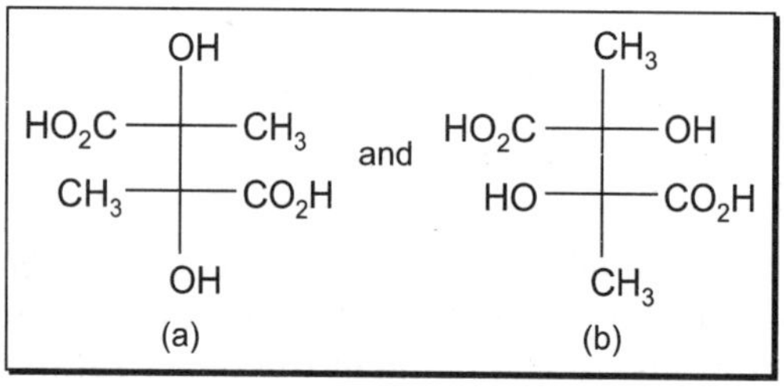

(III)

SCHEME 1.28c

ANSWER. *[I] Identical (180° rotation in the plane of paper of either structure gives the other) ; [II] diastereomers [working hint for such problems is to place as many corresponding (similar) groups as possible in identical positions and then decide relationship. Interchanging the position of two groups twice does not change the configuration of a stereocenter. In (II) at the top carbon the methyl groups are not in the same position in (a) and (b). When one does two interchanges at this stereocenter (first CH_3 with H and then H with OH without disturbing the bottom carbon the equivalent structure of (IIa) becomes (IIc) which is a meso compound. Thus (IIa = IIc) is diastereomeric with (IIb). [III] By doing a similar manipulation, the structures come out to be enantiomers [interchanging a pair of groups once on each stereocenter e.g. , in (III a) CH_3 with*

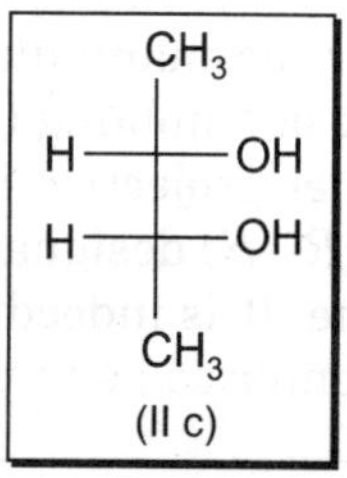

OH on top stereocenter and again OH with CH_3 at the bottom stereocenter inverts configuration at every stereocenter i.e., produces a mirror image and regenrates (III b). Thus (III a and III b) represent an enantiomeric pair.

(E) The Comparison Between *D*, *L* and *R*, *S* Conventions

An important caution about the sequence rule (*R/S*) is that chemical and biogenetic families are not necessarily correlated *e.g.*, the natural α-amino acids can be represented as structure (I, scheme 1.28*d*) in Fischer projections (the L-configuration).

On the basis of Cahn-Ingold-Prelog system the absolute configuration of the α-carbon atom in each amino acid comes out to (*S*) for most of them, however, L-cysteine has the (*R*) configuration due to higher priority given to sulfur over C, N and O. Since for both these amino acids, the side chains are similar, the L-prefix is less confusing (scheme 1.28*d*) and indicates that the absolute configurations for both these compounds are identical. This is the reason for the survival and utility of local systems *e.g.*, D and L for carbohydrates and amino acids and the α, β nomenclature for steroids.

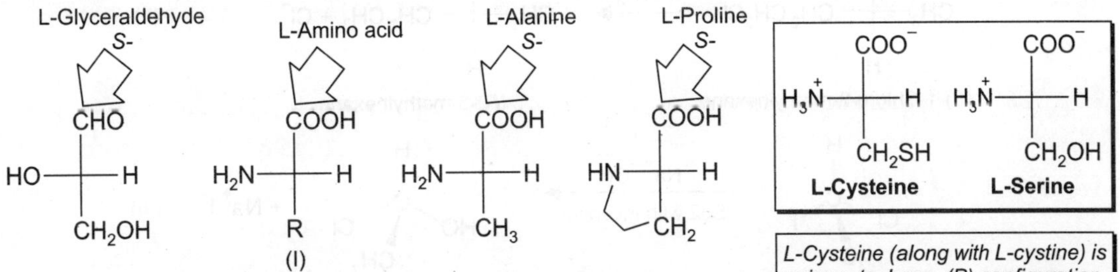

*All protein amino acids belong to **L-series**. Most of these have **S configuration**.*

L-Cysteine (along with L-cystine) is unique to have (R)-configuration. The CH_2SH group of cysteine has priority over COO^- since S has a greater atomic number than O.

SCHEME 1.28d

(F) Relationship Between Configuration and Configurational Nomenclature (*R* and *S*, D and L)

There is no necessary connection between the two systems of nomenclature. The D, L and *R*, *S* systems. The D, L system is based on chemical transformation of a compound to give either D- or L-glyceraldehyde. Thus the compounds which can be correlated with D-glyceraldehyde belong to D-series and those correlated with L-glyceraldehyde belong to L-series.

Amino acids from proteins, thus belong to L-series (I, scheme 1.28*d*) *e.g.*, L-analine and the heterocyclic amino acid L-proline. The use of D and L-system, continues since it shows the essential similarity between all the protein amino acids all belonging to L-series. However, the *R*, *S* system suffers from the disadvantage since the members of the L-series can have either *R* or *S* configuration depending on the priority of *R* ligand (I, scheme 1.28*d*) [Generally all L-amino acids have *S*-configuration. Thus *e.g.*, a change from R = CH_3 in alanine to a different alkyl group in another amino acid does not change the priorty order NH_2 > COOH > R > H and the configuration still remains *S* (I, scheme 1.28*d*). An example where two closely related compounds with the same absolute configuration have different descriptors (*R* or *S*) is of L-cysteine and L-serine (scheme 1.28*d*).

Maintaining the configuration of the molecule means that the four groups bonded to the stereocenter maintain their relative positions. It doesn't, however, necessarily mean that an *S* reactant will always give an *S* product or a *R* reactant will always give a *R* product. When in a reaction a bond to a stereocenter is broken, the resulting confguration of the stereocenter depends on the mechanism of the reaction. Thus *e.g.*, when the hydroxide ion reacts with (*S*)-2-bromobutane in a one step (S_N2) reaction (eq. I, scheme 1.28*e*), the product of this reaction (*R*)-2-butanol has the opposite relative configuration and the reaction is said to have proceeded

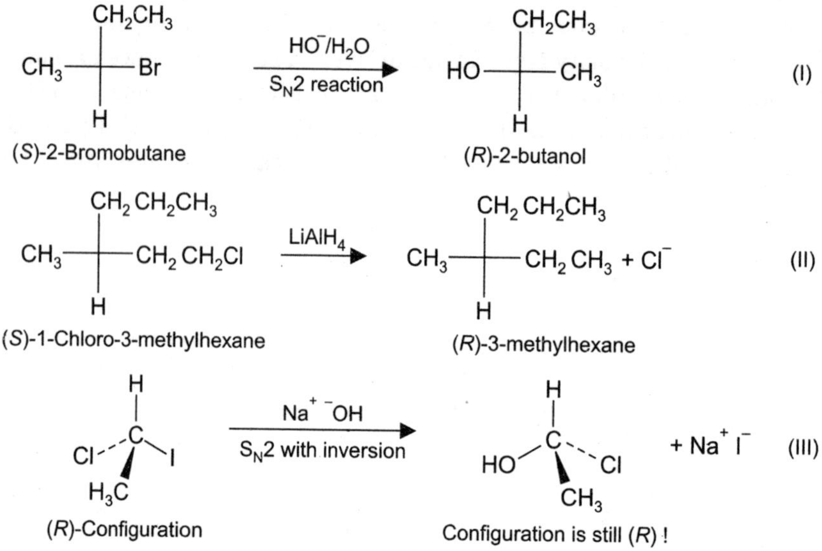

SCHEME 1.28e

by an inversion of configuration. However, inversion of configuration does not require a change in designated (*R* or *S*) absolute configuration, nor does a change in designated absolute configuration require an inversion of configuration.

In reaction (eq. II, scheme 1.28*e*) the relative position of the groups as present in the reactant is maintained in the product. However, the *S* reactant gives *R* product, the change in the priorities of the groups as defined by CIP rules and not a change in their position (inversion of configuration) is what leads an *S* reactant to become the *R* product. Again the (*R*) reactant (eq. III, scheme 1.28*e*) on S_N2 reaction gives a product with inversion of configuration which however, is still (*R*). Thus there is no guarantee that inversion will change an (*R*) stereocenter to (*S*) stereocenter. This example further shows that among the halogens an iodide ion is the best leaving group.

(G) Correlation between Configuration and the Direction of Rotation of Plane Polarized Light

The direction of rotation of plane-polarized light is often specified and incorporated with the names of optically active compounds. Each optically active compound has a characteristic specific rotation. The enantiomers of a compound (mirror image relationship) rotate the plane of polarized light to the same amount but in opposite directions (scheme 1.28*f*).

If the configuration of a chiral compound *i.e.*, an enantiomer is known to be either (*S* or *R*) it, however, does not tell the direction that the enantiomer will rotate the polarized light. Some (*R*) molecules rotate polarized light to the right and are therefore, dextrorotatory (+) and some rotate it to the left and are thus levorotatory (–). One can assign (*S* or *R*) configuration to an enantiomer of a compound by looking at the structure and applying CIP rules. The only way one can tell if a compound is (+) or (–) is by examining the compound with a polarimeter. Thus *e.g.*, (*S*)-lactic acid and (*S*)-sodium lactate have the same configuration yet one is dextrorotatory and the other is levorotatory (scheme 1.28*f*).

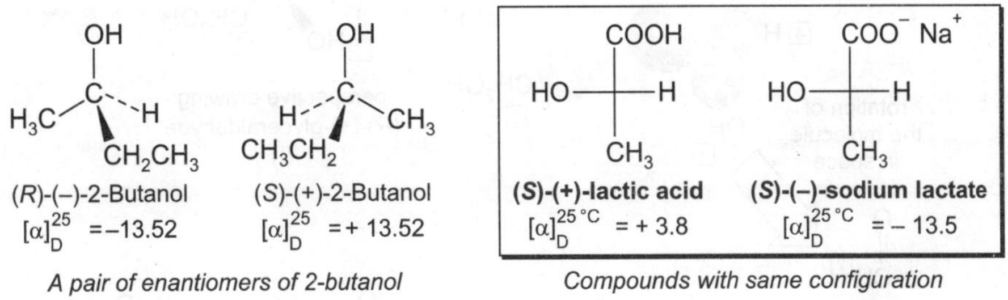

SCHEME 1.28f

The enantiomer of (*S*)-(+) lactic acid is (*R*)-(–) lactic acid. A way to determine which enantiomer has which configuration is using *x*-ray crystallography and the analysis of crystals of pure enantiomers.

In summary there is no correlation between the configurations of enantiomers and the direction in which these rotate the plane polarized light. There is also no correlation between (*R* and *S*) designation and the direction of rotation of plane polarized light.

1.5 INTERCONVERSION BETWEEN FISCHER AND THREE-DIMENSIONAL (WEDGE) FORMULAS WITH ONE STEREOCENTER

In the carbohydrate field it is customary to use the older system of configurational notation for glyceraldehyde (2,3-dihydroxypropanal). The molecule has one stereocenter and thus exists as a pair of enantiomers (see scheme 1.2). One of these enantiomers D-(+)-glyceraldehyde may be described in a manner that specifies the absolute configuration and comes out to be (R)-glyceraldehyde. It may often be necessary to rewrite a three dimensional wedge formula (perspective formula) in the Fischer projection and vice-versa.

Any two bonds at an sp^3 hybridized carbon atom are in a plane at right angles to the plane defined by the other two bonds (see II and IV scheme 1.16). The Fischer projection at a given carbon atom is just to translate this view of the molecule in two dimensions. The conversion of a three dimensional wedge formula of *e.g.*, (R)-glyceraldehyde into its Fischer Projection is shown (scheme 1.29). Consider the following points:

- Orient the wedge formula (or model) in a way so that two groups *e.g.*, H and OH are facing you and two groups CHO and CH_2OH are away from you as in (Ia and Ib scheme 1.29).
- Now simply press the molecule flat in the plane of the paper to get the Fischer projection (In the Fischer projection—thus obtained the sequence 1 → 2 → 3 is counterclockwise but; since the group of lowest priority is on horizontal line, the

designation thus arrived is reversed). Similar procedure when used for (*R*)-(–)-lactic acid is given (scheme 1.30). In the case of L-amino acid (*S*)-alanine the wedge formula is similarly translated into its Fischer projection (scheme 1.31).

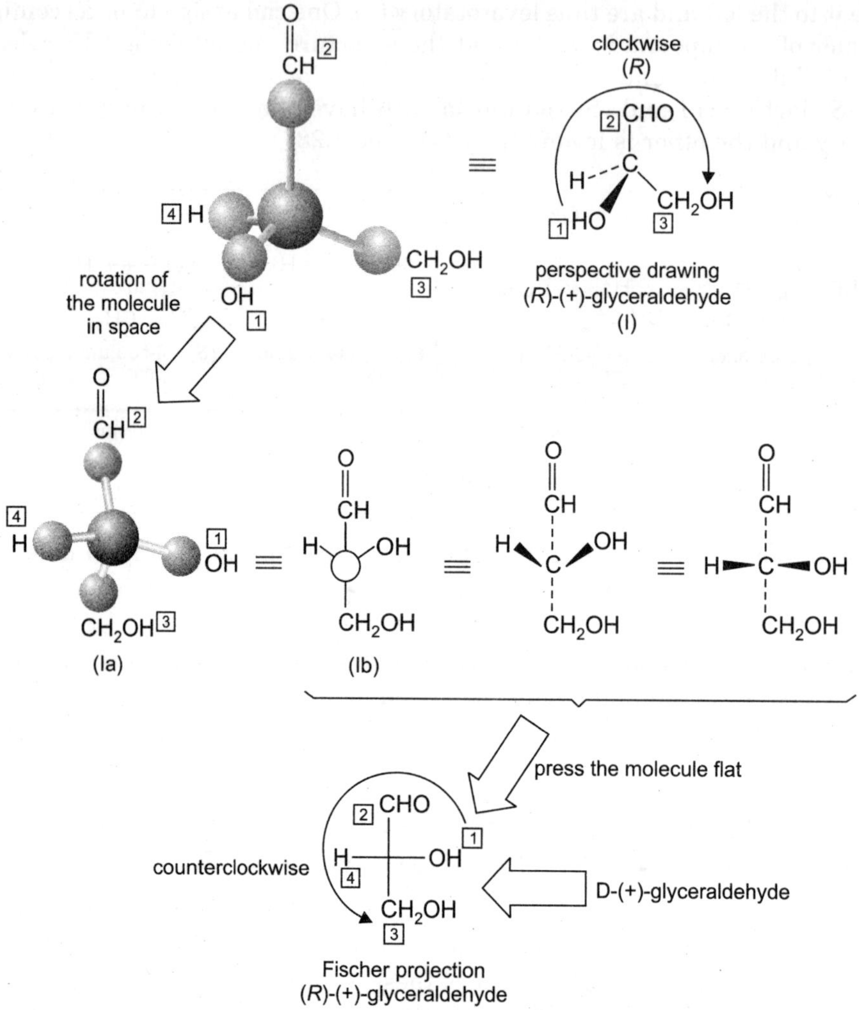

(A Simple Mental Exercise)
Conversion of Dashed-Wedge Line Structure into Fischer Projection

SCHEME 1.29

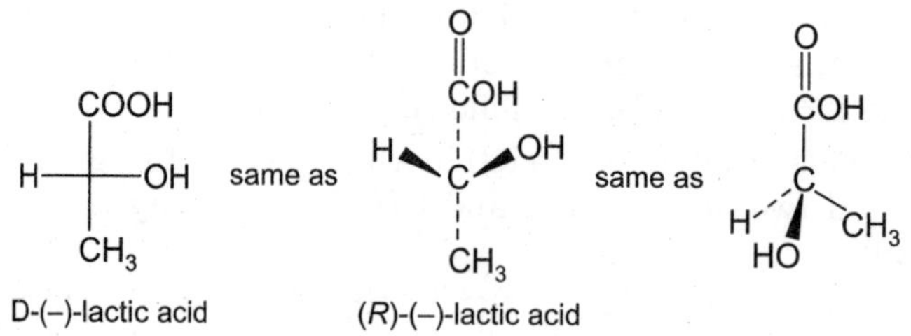

SCHEME 1.30 (Continued)

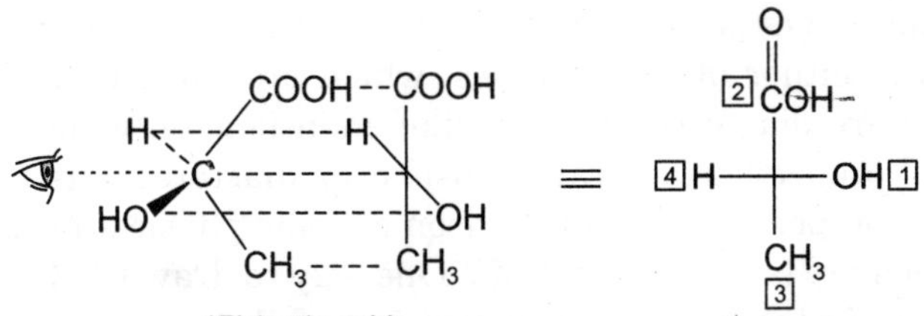

(R)-lactic acid

SCHEME 1.30

(S)-Alanine (L)-Alanine

SCHEME 1.31

1.6 STEREOISOMERISM RESULTING FROM MORE THAN ONE STEREOGENIC UNIT, *THREO* AND *ERYTHRO* NOMENCLATURE

(A) Number of Possible Stereoisomers (Acyclic systems and their Representation)

In many organic compounds, there is more than one stereocenter and such a compound can have a maximum of 2^n stereoisomers where n equals the number of stereocenters. Thus, an aldotetrose, $C_4H_8O_4$ has two stereocenters and can exist in four stereoisomeric forms (scheme 1.32) as two pairs of anantiomers.

One pair of enantiomers A second pair of enantiomers

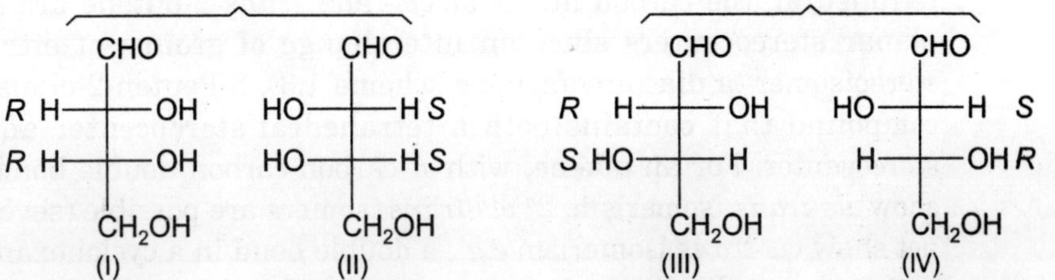

	(I) D-Erythrose	(II) L-Erythrose	(III) L-Threose	(IV) D-Threose
$[\alpha]_0$	–21.5	+ 21.5	– 29.1	+ 29.1
Melting point	(liquid)	(liquid)	130°C	130°C
Solubility in ethanol	Very soluble	Very soluble	Slightly soluble	Slightly soluble

SCHEME 1.32

An aldotetrose has two stereocenters and there are four stereoisomeric aldotetroses called erythose and threose (see, scheme 1.32). An aldohexose on the other hand has four stereocenters and a total of 2^4 = 16 stereoisomers. The names of eight of these are allose, altrose, glucose, mannose, gulose, idose, galactose and talose. The other eight are mirror images of these *e.g.*, for D-glucose, it is L-glucose (see, scheme 1.2). The following points may be noted:

- When more than one stereocenter is present in a molecule, both enantiomers and diastereomers are possible. The four stereoisomers of an aldotetrose consist of two pairs of enantiomers (scheme 1.32). Neither of the erythroses bears a mirror image relationship to either of the threoses, which are therefore, diastereomers (threose and erythose are diastereomers). Unlike enantiomers, which have similar physical and chemical properties (achiral atmosphere), diastereomers have different physical and chemical properties. For comparison of some physical properties of enantiomers and diastereomers (see, scheme 1.32). One way to draw the four stereostructures for the aldotetrose in the Fischer projection is as follows:

- First draw the stereostructure of the molecule as (I, scheme 1.32).
- Draw its mirror image (II), which is the second stereoisomer.
- Now in (II) interchange (switch) the positions of two groups of any stereocenter *e.g.* of C2. This operation gives the third stereoisomer (III).
- Draw the mirror image of (III) to give (IV) the fourth stereoisomer.
- D-glyceraldehyde and L-glyceraldehyde (see, scheme 1.2) serve as reference points for the assignment of relative configuration to all aldoses and ketoses. The reference point is the stereocenter farthest from the carbonyl group. Since this stereocenter is the next to the last carbon on the chain it is termed penultimate carbon which in the case of an aldotetrose (scheme 1.32) is C3 and D or L configuration is assigned whether the OH on C3 is on the right (D configuration) or left (L configuration).
- One may note that pairs of enantiomers have the opposite configuration at both stereocenters, while pairs of diastereomers have the opposite configuration at one stereocenter and the same configuration at the other (scheme 1.32).
- In case a compound contains both a stereocenter and a double bond, the stereocenter will have (*R* or *S*) configuration while the double bond could have either (*E* or *Z*) geometry. Thus *e.g.*, in the case of 3-penten-2-ol there are four stereoisomers (scheme 1.32*a*). 3-Penten-2-ol has a stereocenter or more precisely in this situation it will be termed a tetrahedral stereocenter. All stereocenters may however, be not tetrahedral, the carbon atoms of *cis*- and *trans*-2-butene are examples of trigonal planar stereocenters since an interchange of groups at either atom will give a stereoisomer (a diastereomer see scheme 1.3). 3-Penten-2-ol also is an example of a compound that contains both a tetrahedral stereocenter and a trigonal planar stereocenter. For an alkene, with n carbon-carbon double bonds each of which can show *cis-trans*-isomerism, 2^n *cis-trans* isomers are possible (several double bonds can not show *cis-trans* isomerism *e.g.*, a double bond in a cyclohexane ring can have only *cis*-geometry). In the case of 3-penten-2-ol (I and I*a*, scheme 1.32*a*) are enantiomers while (I) is the diastereomer of (II) and (II*a*).

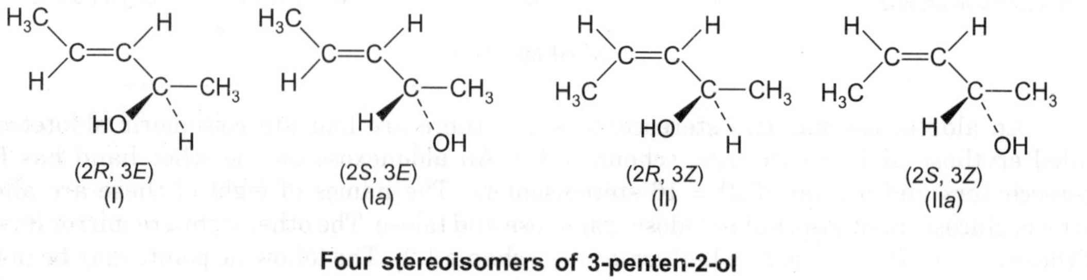

Four stereoisomers of 3-penten-2-ol

SCHEME 1.32a

(B) *Threo* and *Erythro* Nomenclature

A molecule with two adjacent stereocenters, when there are two groups which are common to each carbon while the third is different *i.e.*, $C_{abx} - C_{aby}$ gives rise to *erythro* and *threo* diastereomers. Since erythrose and threose (see scheme 1.32) are diastereomers, other diastereomeric pairs of molecules which have two adjacent stereocenters are designated as *erythro* or *threo* depending on whether similar groups are on same side (*erythro*) or on opposite sides (*threo*) of the Fischer projection respectively. Thus *e.g.*, 3-bromo-2-butanol with two stereocenters has (2^2) four stereoisomers, which are named (scheme 1.33).

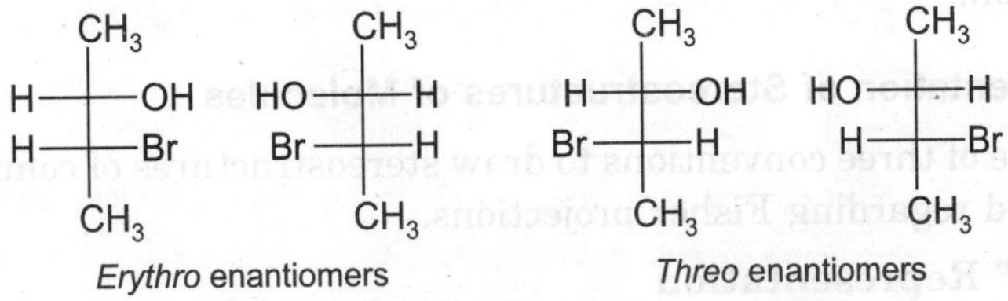

Four stereoisomers of 3-bromo-2-butanol

SCHEME 1.33

Erythro and *threo* is the short hand method employed by organic chemists to name appropriate compounds. However, if one has to refer to a particular enantiomer of the *erythro* or *threo* pair, (*R* and *S*) notation is used with the name of each. *Erythro* and *threo* diastereomers can be identified from a sawhorse formula (see, scheme 1.35).

(C) Use of Terms *Erythro, Threo, Meso and dl*

The terms *erythro* and *threo* are generally applied only to those molecules which do not have symmetric ends (scheme 1.33*a*). When however, the ends are symmetric as *e.g.* in 2, 3-dibromobutane (scheme 1.33*a*) and tartaric acid the terms *meso* and *dl* are preferred since the use of these terms shows if or not a diastereomer has an enantiomer. A use of the prefixes *erythro* and *threo* is made to name dissymmetric molecules where the ends are different. For the formation of *erythro* or *threo* products during addition to *e.g.* an olefin (see, problem 6.23 and schemes 6.51 and 6.52).

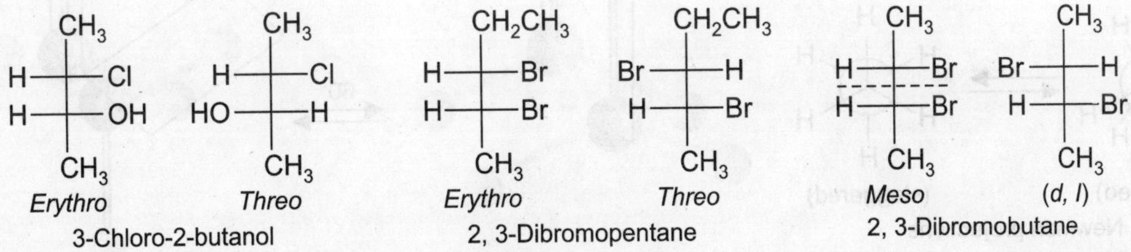

SCHEME 1.33a

(D) Disadvantages of Fischer Projection

- These projections can be turned/rotated only in certain specified ways (see scheme 1.28*a*).
- In compounds with more than one stereocenter, a Fischer projection implies an eclipsing relationship of the groups attached to the two stereocenters. One knows, however, that staggered conformations are more stable.

- Many organic reaction mechanisms (*e.g.* E2 reaction) necessitate that one should write the structures in the staggered form while only in few cases *e.g.*, pyrolytic eliminations an eclipsing relationship is required.
- In summary, Fischer projections are useful since, these allow to repesent the stereoisomers in two dimensions, unfortunately, however, these are in fact the relatively unstable eclipsed conformations.
- Perspective (three-dimensional) formulas are not only more accurate representations but at the same time these depict the molecule in its more stable staggered configuration.

(E) Realistic Representation of Stereostructures of Molecules

Mention here is made of three conventions to draw stereostructures of compounds which avoid difficulties mentioned regarding Fisher projections.

(*i*) The "Sawhorse" Representation

In the "sawhorse" representation, the bond between the two carbon atoms *e.g.* ethane (scheme 1.33*b*) is drawn diagonally and is slightly elongated. This projection shows the spatial relationship of atoms in a two-carbon system, which can be depicted, in the eclipsed or in staggered conformation as shown in ethane. It is just a simplified varison of a ball and stick model where dihedral angle has been shown.

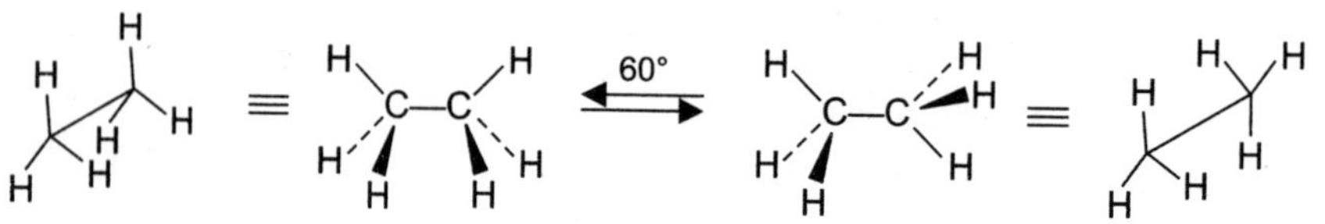

Sawhorse (eclipsed) Flying-wedge structure Sawhorse (staggered)

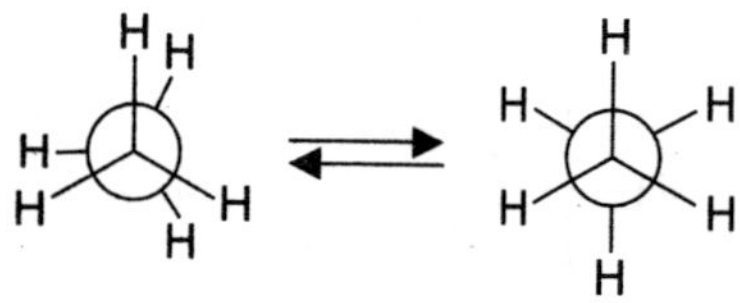

(eclipsed) (staggered)
Newman projections

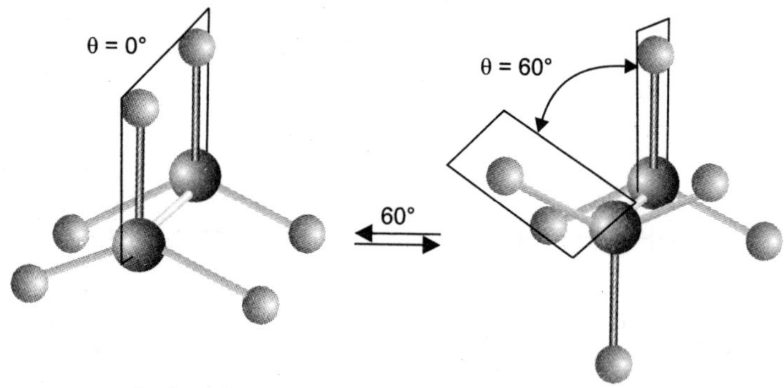

(eclipsed) (staggered)
Ball and stick model showing dihedral angle θ

In ethane a 60° rotation of one carbon converts an eclipsed conformation (least stable, substituents on adjacent carbons as close as possible) to a staggered conformation when the substituents are as far apart as possible (In a staggered conformation, one sees that each C-H bond bisects the angle between the C-H bonds on the adjacent carbon in the middle) Torsional strain is the destabilization due to increased electron-electron repulsion in eclipsed bonds.

SCHEME 1.33b

(ii) Newman Projection

In this projection the molecule is viewed along the bond joining the two carbon atoms, the front carbon atom is shown by three lines (three bonds) which come together in a 'Y' shape. The back carbon is shown by a circle with three bonds pointing out from it (scheme 1.33*b*).

(iii) The Flying Wedge Projection (Zig-zag conformation, Masamune projection)

The longest chain (backbone) is drawn as a "zig-zag" in the plane of the paper as a staggered arrangement (continuous lines) as shown for glucose (scheme 1.33*c*). The substituents (other than H) in front and behind the plane are drawn with solid wedge lines and dotted lines respectively (Masamune projection). In fact here each carbon individually has the stereostructure as discussed in scheme 1.15). A "zig zag" can also be drawn in an eclipsed form (as in scheme 1.35), to clarify some stereochemical problems.

The Fischer projection was devised long before when nothing was known about the preferred conformations of molecules. It is now usual to follow a convention for acyclic molecules devised by Masamune. Consider the following points :

- If one considers D-glucose, each stereocenter is given (*R* or *S*) descriptor along with the positional number and D-glucose will be properly designated as (2*R*, 3*S*, 4*R*, 5*R*)-2, 3, 4, 5, 6-pentahydroxyhexanal (scheme 1.33*c*). The H atoms are assumed and may not be drawn. A Fischer projection of D-(+)-glyceraldehyde is compared with the wedge formula and the wedge formula drawn as zig zag structure (scheme 1.33*c*).

For showing a relation of the Fischer projection of glucose (I, scheme 1.33c) with the "zig zag" orientation consider a pair of stereorgenic carbons at a time e.g., C2–C3 in Fischer projection. To make a start, look at C2 from the side of the arrow, OH is closer to ones eyes, thus it is placed on thick line in (II, scheme 1.33c). At C3 in (I, scheme 1.33c) OH is away, but Fisher projection is eclipsed conformation and this arrangement is shown in eclipsed sawhorse orientation (III, scheme 1.33c) . The rotation of C3 at C2-C3 bond in (III, scheme 1.33c) to get a staggered arrangement (as is so in "zig zag" orientation) will bring OH at C3 towards one's eyes and is thus placed on thick line in (II, scheme 1.33c). Similarly the relative geometry of OH groups at C4 and C5 in (I, scheme 1.33c) becomes trans (II, scheme 1.33c).

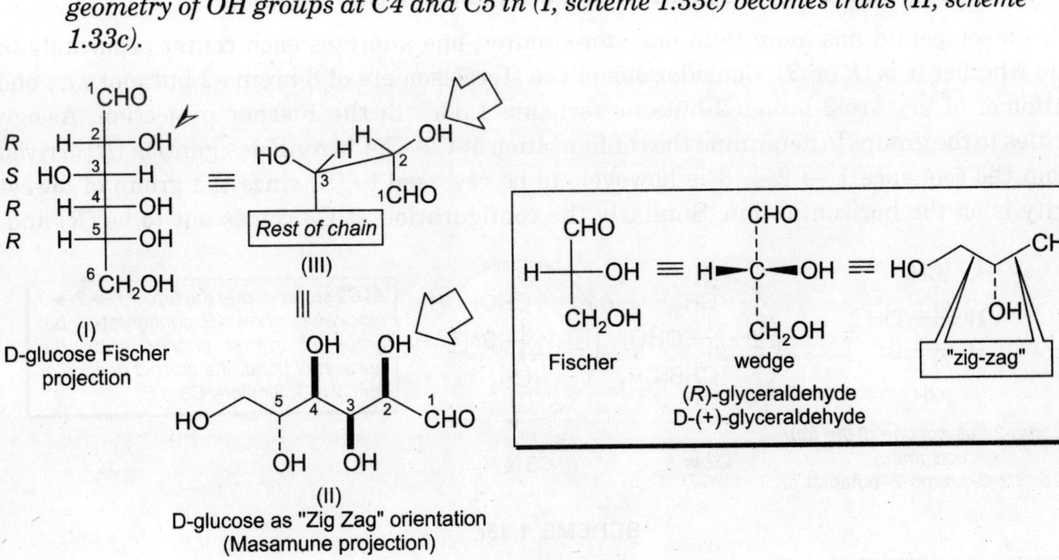

SCHEME 1.33*c*

- *Meso* stereoisomer(s) may not be easily detected while dealing with symmetric molecules. Thus in such a molecule with two stereocenters the Masamune projection will have a center of symmetry i if the stereocenters are adjacent (or separated by an even number of C atoms) but a plane of symmetry σ if they are separated by an odd number of C atoms in the main chain (scheme 1.33*d*).

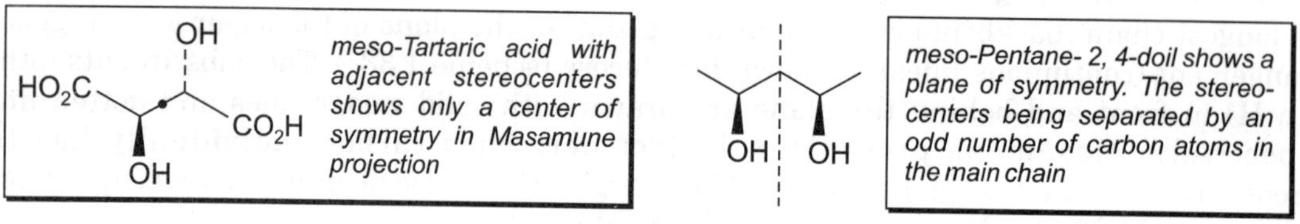

SCHEME 1.33d

- A new method for the designation of stereochemical relationship of diastereomers has been developed by aldol chemists. The carbon backbone is extended in the plane of the paper in the horizontal direction. If two substituents extend in the same direction, it is called *syn* and when these extend in opposite directions the stereochemical relationship is called *anti* (scheme 1.33*e*).

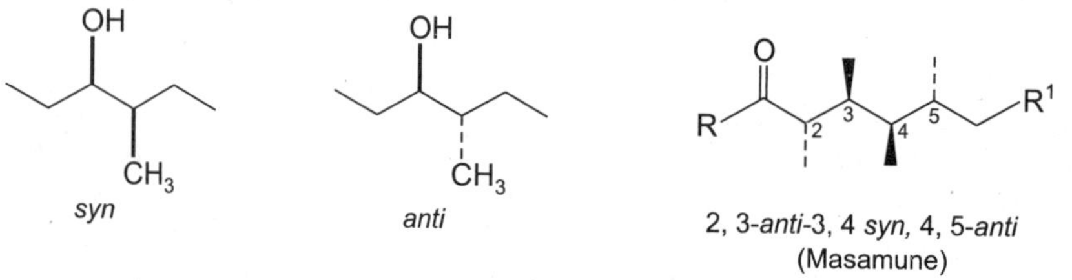

SCHEME 1.33e

(F) Naming Compounds (*R*, *S*) System with More than One Stereocenter

In case a compound has more than one stereocenter, one analyzes each center separately to decide whether it is (*R* or *S*). Consider one of the stereoisomers of 3-bromo-2-butanol *e.g.*, one enantiomer of *erythro*-3-bromo-2-butanol (scheme 1.33*f*) in the Fischer projection. Assign priorities to the groups to determine the configuration at C2. The "wrong"designation (*R*) arrived at from the sequence 1 $\rightarrow$ 2 $\rightarrow$ 3 is however, to be reversed to (*S*) since the group of lowest priority is on the horizontal line. Similarly the configuration at C3 comes out to be (*R*) and

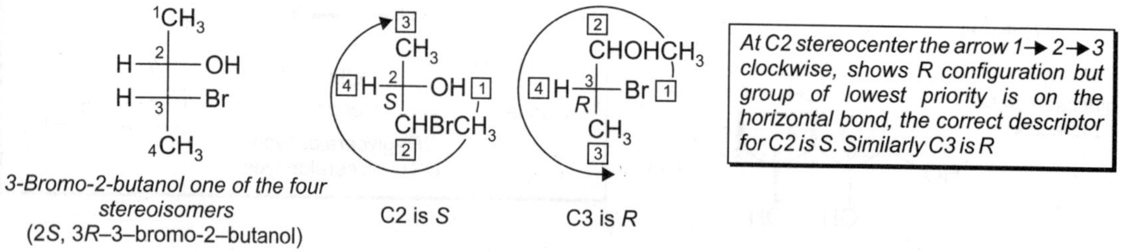

SCHEME 1.33f

the otherwise shorthand name (*erythro*) of this stereoisomer in IUPAC terms comes out to be (2*S*, 3*R*)-3-bromo-2-butanol. Similarly in a three dimensional wedge representation (perspective formula) with two stereocenters the configuration to each stereocenter is assigned separately and one at a time. The stereocenter is viewed from the side opposite to the group of lowest priority. Thus at C2 (scheme 1.33*g*) the group of lowest priority is directed away and viewing it from right hand side gives the correct descriptor *S*. At C3 the lowest priority group is instead towards ones eyes, the arrow $1 \rightarrow 2 \rightarrow 3$ is *S* as drawn, but since the priority 4 group is coming forward so the actual configuration is *R*.

The configuration at *C2* is *S* **The configuration at *C3* is *R***

SCHEME 1.33g

(G) *Meso* Compounds and Pseudoasymmetric Centers (Pseudochirality Centers)

The maximum number of stereoisomers for a compound with *n* stereocenters is 2^n. However, some compounds with *e.g.*, two stereocenters have only three stereoisomers. In such cases the four substituents bonded to one stereocenter are identical to the four substituents bonded to the other stereocenter. An example is of tataric acid which has a pair of enantiomers and one stereoisomer is the *meso* compound (scheme 1.34). *Meso* compounds contain stereocenters as well as a plane of symmetry as shown in *meso*-tartaric acid (see scheme 1.2*a*). A compound with a plane of symmetry is achiral (can be superposed on its mirror image). Thus *meso*-compounds will not be optically active even though they have stereocenters. In short tartaric acid exists as a pair of entantiomers (chiral) and an achiral *meso* diastereomer (see schemes 1.1 and 1.2*a* and 1.41). Two views of *meso*-tartaric acid are again presented (scheme 1.41). A plane of symmetry exists in the eclipsed form and a center of symmetry in the stable staggered arrangement.

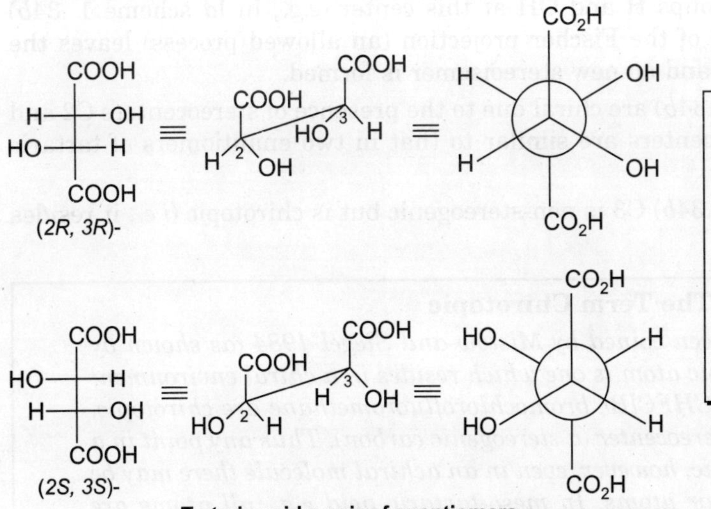

Tartaric acid a pair of enantiomers

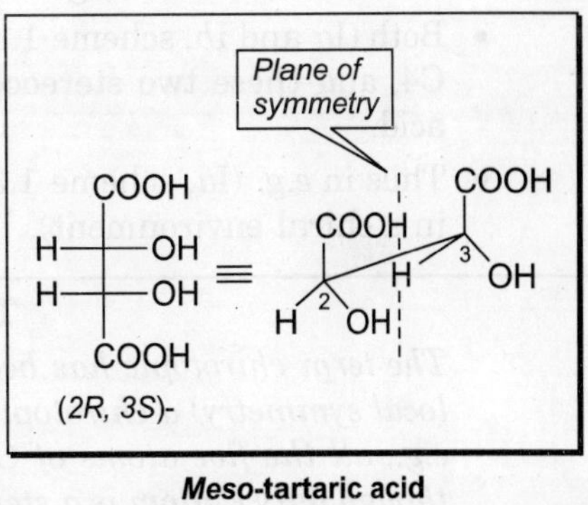

***Meso*-tartaric acid**

SCHEME 1.34

The *meso* compounds *e.g.* *meso*-tartaric acid have two features in common : (*i*) a plane of symmetry and (*ii*) two stereocenters with opposite stereo descriptors. The latter condition renders the molecule optically inactive since there is internal compensation, the rotation due to (*R*) stereocenter is cancelled by that caused by (*S*) stereocenter.

Meso compounds also exist in molecules with more than two stereocenters (see scheme 1.34*b*) and also in cyclic structures. Thus compound (I, scheme 1.34*a*) has a plane of symmetry which passes through C1, C7 and C4 and the stereocenters C2 and C6 have opposite configuration thus it represents a *meso* compound. The compound (II, scheme 1.34*a*), however, does not have a plane of symmetry and configuration is also *S* on both C2 and C6, this compound is thus not *meso*. Another example is in cyclohexane derivatives (III and IV, scheme 1.34*a*), (III) has a plane of symmetry passing through C1 and C4).

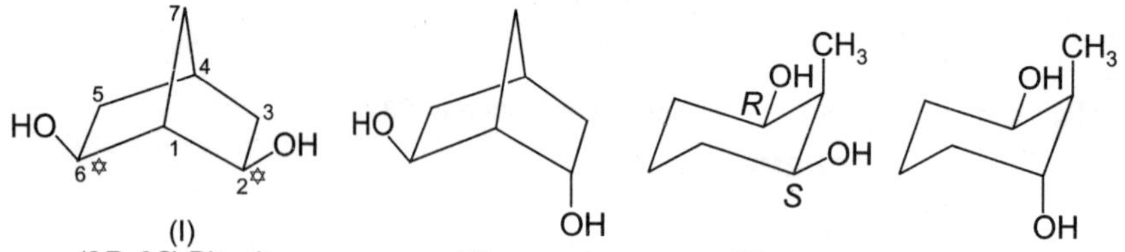

(I) (2*R*, 6*S*) Bicyclo-[2.2.1] heptane 2, 6-diol (*meso* compound)

(II) 2*S*, 6*S* (not *meso*)

(III) (*meso*)

(IV) (not *meso*)

A meso compound has a plane of symmetry and two stereocenters with opposite stereodescriptors.

SCHEME 1.34a

Consider a constitutionally symmetrical molecule, 2, 3, 4-trihydroxyglutaric acid (scheme 1.34*b*) which has three stereocenters while it exists only as four stereoisomers–a pair of enantiomers (I*a* and I*b*) and two *meso* forms, since the groups on the stereocenters are same. Consider the following points regarding the enantiomeric pair.

- The C3 center is achiral since two of the groups (*RR* in I*a* and *SS* in I*b*) attached to this center are identical.
- No configurational descriptor (*R* or *S*) can thus be given to C3.
- The C3 in either of the enantiomers is nonstereogenic (it is not a stereocenter) since an interchange of two groups H and OH at this center (*e.g.*, in I*a* scheme 1 .34*b*) followed by 180° rotation of the Fischer projection (an allowed process) leaves the structure (I*a*) unchanged and no new stereoisomer is formed.
- Both (I*a* and I*b*, scheme 1.34*b*) are chiral due to the presence of stereocenters C2 and C4, and these two stereocenters are similar to that in two enantiomers of tartaric acid.
- Thus in *e.g.* (I*a*, scheme 1.34*b*) C3 is non-stereogenic but is chirotopic (*i.e.*, it resides in a chiral environment).

The Term Chirotopic

The term chirotopic has been coined by Mislow and Siegel 1984 (as shown by local symmetry) a chirotopic atom is one which resides in a chiral environment e.g., all the five atoms of CHFClBr bromochlorofluoromethane are chirotopic, though only C atom is a stereocenter (a stereogenic carbon). Thus any point in a chiral molecule is chirotopic, however, even in an achiral molecule there may be several chirotopic points or atoms. In meso-tartaric acid e.g., all atoms are chirotopic, but for the center of symmetry (midway between C2 and C3 which is achirotopic). All points on a plane of symmetry in a molecule are achirotopic.

- Thus chirality does not reside on stereogens (*i.e.*, stereocenters) and the chiral sense of the whole molecule in that what matters.

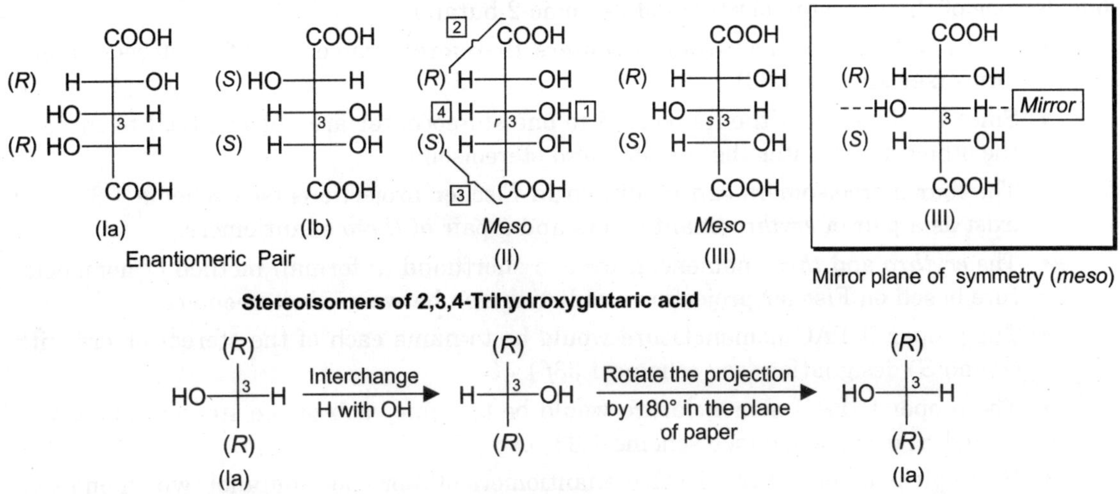

SCHEME 1.34b

Now one may consider the following points about the *meso* forms (II and III, scheme 1.34*b*). These compounds are examples of compounds with more than two stereocenters which are *meso*,

- The C3 atom in both (II and III, scheme 1.34*b*) is a stereocenter (*i.e.*, C3 is stereogenic) as seen by the interchange of ligands H with OH in either of these gives diastereomers (this process is fact converts one *meso*-stereoisomer into the other).
- In the *meso* forms (II and III, scheme 1.34*b*) C-3 is achirotopic (due to the presence of σ plane) but is stereogenic.
- A stereogenic center which is achirotopic *i.e.*, which has two identical enantiomorphic ligands and two other non-identical ligands and which lies on a plane of symmetry is termed a pseudoasymmetric center (pseudochirality center) and is designated $C_{a^+a^- bc}$ where a^+ and a^- represent enantiomorphous ligands. (For other examples of compounds having, pseudoasymmetric centers see scheme 1.68*b*).
- The sterodescriptors to these centers are given as lower case italic letters to differentiate pseudoasymmetric centres from true stereogenic centers. Since the two of the ligands on the pseudoasymmetric center are the same in terms of atomic number a further criterion is invoked an *R* stereocenter has priority over an *S* (*R* is before *S* in the alphabet). Thus C3 in *e.g.*, (II, scheme 1.34*b*) is *r* while C3 in III is *s*.

Stereogenicity and Chirotopicity

It has been suggested that stereogenicity and chirotopicity are two different and distinct characters of a stereocenter. These characters not only overlap, but are closely associated in organic chemistry (also see, scheme 1.68b). Moreover, application of CIP rules when applied to pseudoasymmetric centers become less straightforward. One has already seen cases e.g., amino acids where according to D-L system of nomenclature all naturally occuring amino acids belong to L-series while on the basis of CIP nomenclature majority of them belong to S configuration while some to R (see, scheme 1.28d).

(H) Interconversion of Fischer-Sawhorse-Newman and Flying Wedge Projections–3-Bromo-2-butanol

Consider one of the four stereoisomers of 3-bromo-2-butanol:

- 3-Bromo-2-butanol has two stereocenters, therefore, it can exist in four stereoisomeric forms (see scheme 1.33).
- Since the four substituents bonded to one stereocenter are not identical to those of the other stereocenter there is no *meso* stereoisomer.
- The four stereoisomers can be drawn in Fischer projections (see scheme 1.33) and exist as a pair of *erythro* enantiomers and a pair of *threo* enantiomers.
- The *erythro* and *threo* nomenclature is a shorthand (informal) method of nomenclature based on Fischer projections to distinguish between diastereomers.
- The proper IUPAC nomenclature would be to name each of the stereoisomers with (*R* and *S*) designation (see scheme 1.33*f*).
- The proper IUPAC nomenclature would be to name each of the stereoisomers with (*R* and *S*) designation (see scheme 1.33*f*).
- If one picks up one of the *erythro* enantiomers of 3-bromo-2-butanol (which one *e.g.*, 2*S*, 3*R* stereoisomer I, scheme 1.34*c*) it can be converted into "sawhorse", Newman or wedge representation as shown.

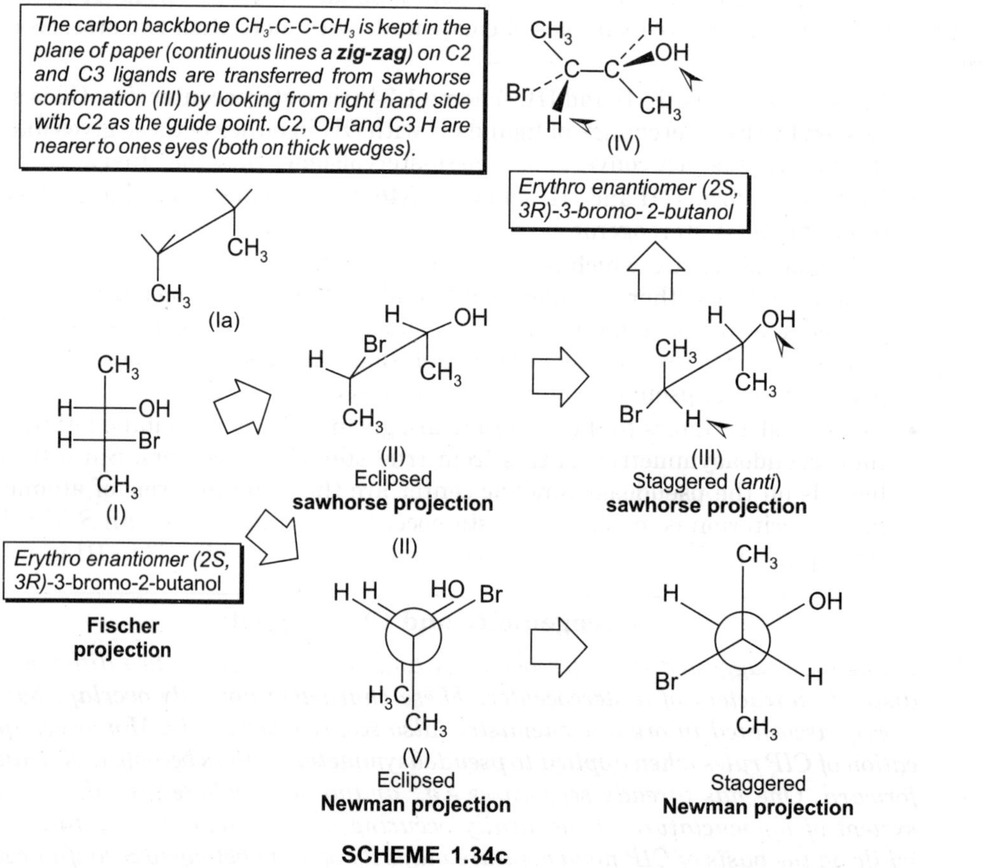

SCHEME 1.34c

- In the Fischer projection (I, scheme I.34*c*) the substituents on the two stereocenters are eclipsed. To transform it into its sawhorse or Newman projection one must write first these, projections in their eclipsed form (II and V scheme 1.34*c* respectively). For a beginner it may be helpful to do it stepwise. In the sawhorse representation in its eclipsed form, in fact each stereocenter may be looked up as the letter 'Y'. As a first step write the carbon backbone as in (Ia, scheme 1.34*c*) and put the end groups (CH_3, CH_3) on the vertical line of the Fischer projections on the "tails of each Y". Then transfer the substituents on the horizontal lines of the Fischer projection on the "head" of each 'Y' on the right hand and the left hand as they appear on the Fischer projection to give (II and V scheme 1.34*c*). Then stagger the substituents by rotation of one carbon keeping the other carbon static.
- The staggered conformation (III, scheme 1.34*c*) is more stable as compared to (II) but (III, scheme 1.34*c*) is not properly arranged to show stereochemistry according to Fischer projection method. Rotation about C-2-C-3 bond gives back the eclipsed conformation (II) and projection of this eclipsed conformation on to the page gives the correct Fischer projection (I).
- To show a relation of the wedge formula (IV staggered "zig zag" form, scheme 1.34*c*) with Fischer projection (I, scheme 1.34*c*) the relationship of model III to IV *i.e.*, identity is obvious on the C-1, C-2, C-3 and C-4 carbons ('zig-zag' drawn as continuous line). When one looks to the model (III, imagine from the side of the arrow), OH on C-2 and H on C-3 are closer to the observer (shown as, thick wedges in IV) while H on C-2 and Br on C-3 are directed away (shown as hatched wedges). To draw the perspective formula of stereoisomers, follow the method given in (Scheme 1.34*c*).

EXERCISE 1.8

Convert one of the stereostructures (a zig-zag) of tartaric acid (I, scheme 1.34d) in to its Fischer projection and assign configuration to each stereocenter.

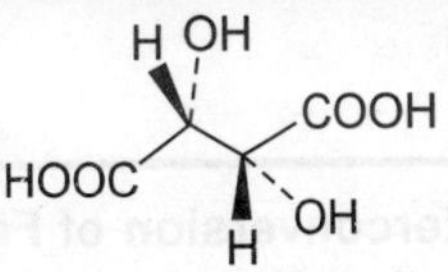

(I)

SCHEME 1.34d

ANSWER. *It is a zig zag staggered orientation (see the main carbon chain) and is to be converted into Fischer projection which is eclipsed. Conformation (scheme 1.34e).*

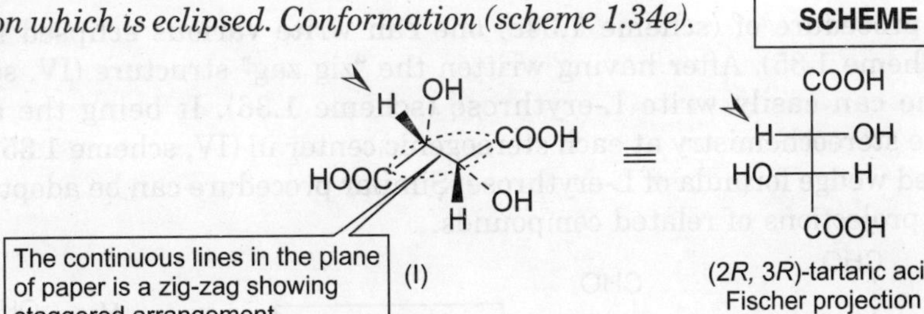

(2*R*, 3*R*)-tartaric acid Fischer projection

SCHEME 1.34e

Transfer the groups of one stereocenter shown by arrow on the Fischer projection. The H atom is forward (thick wedge) it is put on the left side of Fischer projection. On the second stereocenter H atom is again forward but it appears on the right side of Fisher projection. Since (I) is a staggered conformation one must eclipse it and the rotation of the second stereogenic carbon to do so puts H atom in the rear.

EXERCISE 1.8a

When fumarate reacts with D_2O in the presence of the enzyme fumarase only one stereoisomer of deuterated malate is formed. Explain if this a syn or anti-addition?

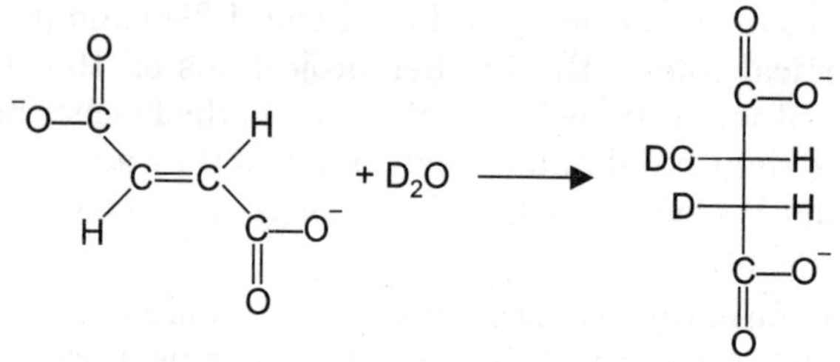

fumarate (2*S*, 3*R*)-malate-3-*d*

SCHEME 1.34f

ANSWER. *Convert the Fischer projection of the product into a three dimensional perspective drawing (first eclipsed form then stagger it so as to bring two COO⁻ groups anticoplanar i.e., these two to have 180° angle as in the starting material). Examine the relationship of the –D and –OD groups. Thus the addition is anti (Scheme 1.34g).*

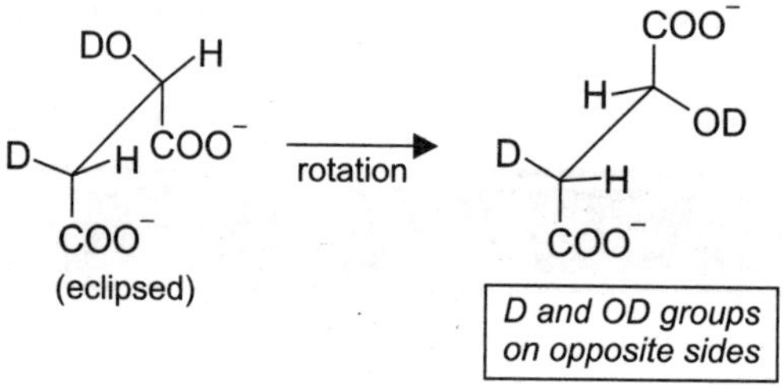

SCHEME 1.34g

(I) Interconversion of Fischer-Sawhorse Projections—Erythrose and Threose. Threo and Erythro Nomenclature from Staggered Conformations

Following the procedure of (scheme 1.34c) one can write various eclipsed forms *e.g.*, for D-erythrose (scheme 1.35). After having written the "zig zag" structure (IV, scheme 1.35) of D-erythrose one can easily write L-erythrose (scheme 1.36). It being the enantiomer of D-erythrose, the stereochemistry at each stereogenic center in (IV, scheme 1.35) is inverted to get the staggered wedge formula of L-erythrose. Similar procedure can be adopted to write the stereochemical projections of related compounds.

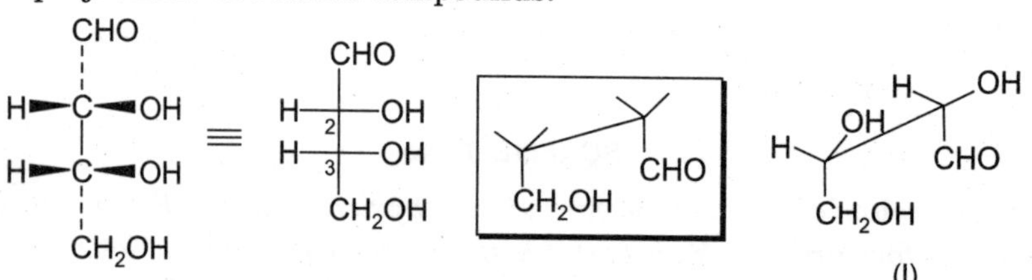

Fischer projection (eclipsed)

(I) Sawhorse projection (eclipsed)

D-Erythrose drawn in different projections

SCHEME 1.35 (Continued)

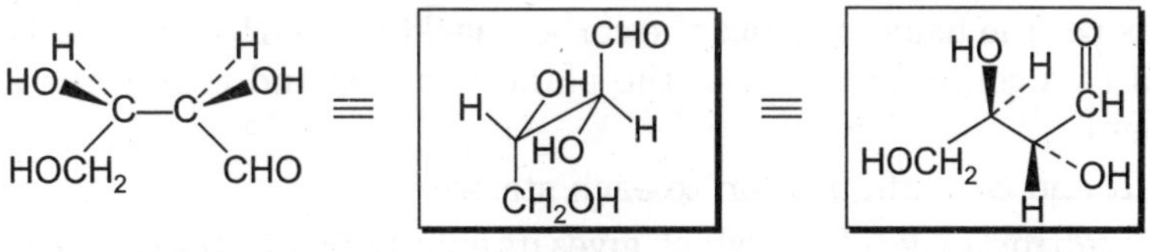

(II) Wedge formula (eclipsed) | (III) Sawhorse projection (staggered) | (IV) Wedge formula (zig-zag) (staggered)

D-Erythrose drawn in different projections

SCHEME 1.35

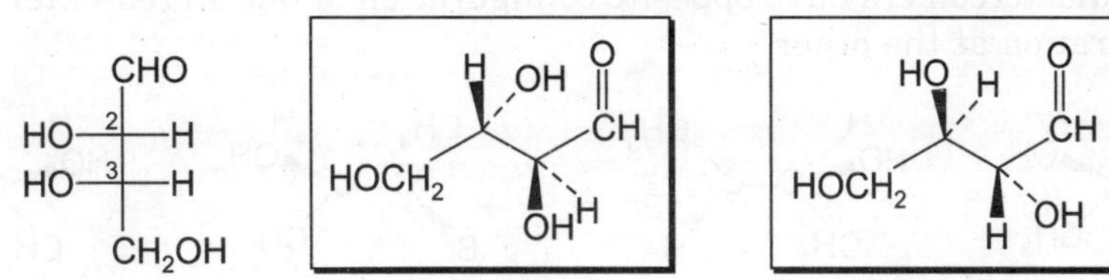

L-Erythrose Fischer projection (eclipsed) | L-Erythrose 'Zig zag' (staggered form) | **D-Erythrose** 'Zig zag' (staggered form)

SCHEME 1.36

If as in (I or II, scheme 1.35) one can eclipse every like group *i.e.*, H with H, OH with OH and the unlike groups as well CHO with CH_2OH, the configuration is *erythro*. Thus a staggered conformation like (III and IV) must be converted into eclipsed conformation to label it *erythro*. However, if one can eclipse only either any of the identical pair or unlike groups at any time then the configuration is *threo*. In the case of *threo* enantiomer (scheme 1.37), CHO and CH_2OH are eclipsed while H and H and OH and OH are not, it is therefore, *threo* configuration.

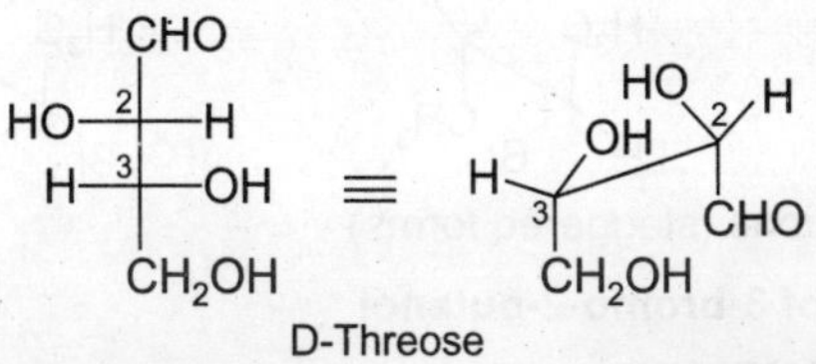

D-Threose

A threo configuration in the parent compound threose - when on two adjacent stereogenic centers the common substituents are on opposite sides of Fischer projection.

SCHEME 1.37

(J) Writing the Wedge (Perspective) Formula of Compounds with Two Stereocenters

One has already learnt to draw the Fischer projections of compounds with two stereocenters (see, schemes 1.32 and 1.33) and their conversion into sawhorse, Newman and wedge structures (see scheme 1.34*c*). Here some more examples are given to initiate the beginner, when the perspective (wedge) formulas which may be harder to visualize, are to be considered.

(*i*) 3-Bromo-2-Butanol

The compound has two stereocenters and thus has four stereoisomers. One way to write quickly the structures of these four stereoisomers involves the following steps:

- First draw a stereostructure of the molecule in any way *e.g.*, as is drawn, in (A, scheme 1.38).

In (A), as per the usual practice the carbon backbone (CH_3-C-C-CH_3 is drawn as a 'zig zag' and as a continuous line in the plane of paper and represents a staggered arrangement.

- Assign (*R* and *S*) configuration to each stereocenter.
- Draw the mirror image of (A), which gives its enantiomer as the second stereoisomer (B).
- In (A) exchange the positions of H and Br only at C-3. The configuration at C-3 is inverted and third stereoisomer (C, scheme 1.38) is produced.
- Draw the mirror image of (C) to get (D the enantiomer of C) as the fourth stereoisomer.
- A useful check is that in a pair of enantiomers there is opposite configuration at both the stereocenters (*i.e.*, A is 2*S*, 2*R*, scheme 1.38 while its enantiomer B is 2*R*, 2*S*). The pair of diastereomers have opposite configuration at one stereocenter while the same configuration at the other.

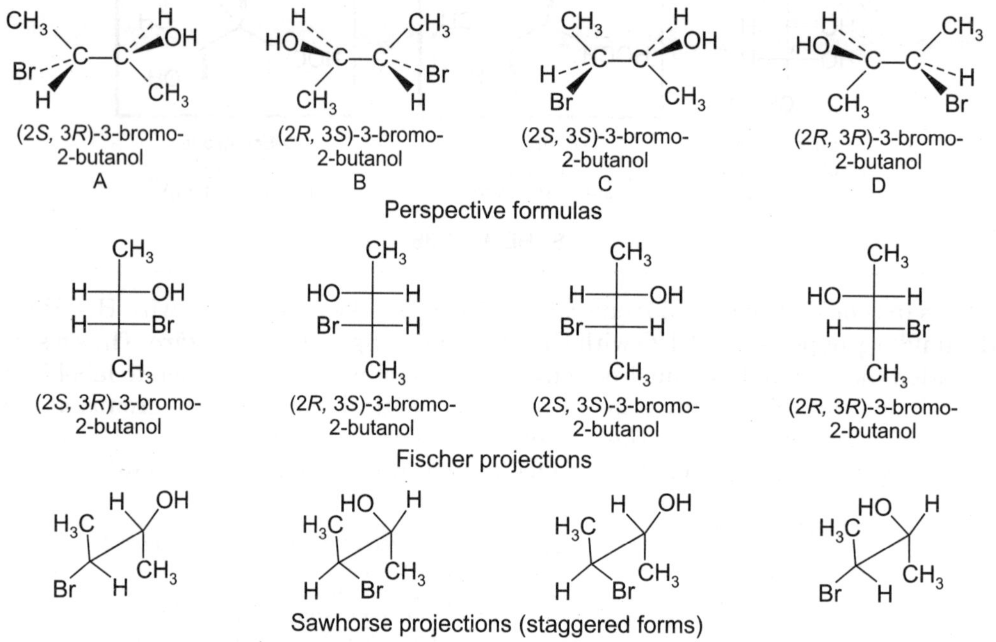

Stereoisomers of 3-bromo-2-butanol

SCHEME 1.38

(*ii*) 2, 3-Dihydroxybutanoic Acid

The compound has two stereocenters and thus like 3-bromo-2-butanol it will also have four stereoisomers. One may draw these four stereoisomers by the approach followed for 3-bromo-2-butanol. However, alternately one draws any one stereostructure *e.g.*, (I, scheme 1.39) for one of the stereoisomers and assigns (*S* and *R*) configuration to both the stereocenters and follows the following procedure :

- In (I, scheme 1.39) an interchange of the position of groups both at C-2 and C-3 gives its enantiomer (II) (here one has however, not drawn the mirror image of I). As drawn, I and II at first sight do not look like object and mirror image, this relationship can be shown by reorienting the molecules *e.g.*, when (II) is rotated 180° out of the plane of paper all groups in front go at the back and vice versa. The new orientation is indeed the mirror image of (I, scheme 1.39).

- Now in (I, scheme 1.39) an interchange of groups only at one stereocenter *e.g.*, C-3 gives the third stereoisomer (III, scheme 1.39). One can see that (III) has a diastereomeric relationship with both (I and II).
- If now in (III) one interchanges the position of groups at both C-2 and at C-3 one gets the fourth stereoisomer (IV, scheme 1.39). From configurational descriptors one can see that (III and IV) are also enantiomers.
- Thus in a molecule with two stereocenters, a stereoisomer can be converted into its enantiomer by either drawing its mirror image or simply by interchanging the position of groups at the both the stereocenters (change of configuration at both stereocenters). Interchanging the position of groups at only one stereocenter (reversing the configuration at only one stereocenter) converts it into its diastereomer.

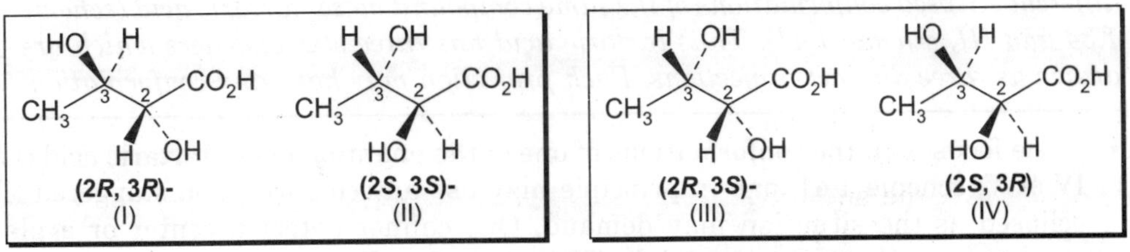

Stereoisomers of 2, 3-dihydroxybutanoic acid

SCHEME 1.39

(K) Determining Relationship Among Stereoisomers–Stereoisomeric Tartaric Acids Drawn in Different Orientations

One will see that there can be several correct ways of depicting a molecule in dashed-wedge line notation as well as Fischer projections. Firstly an example of one of the stereoisomers of a compound *meso*-tartaric acid is discussed (see, scheme 1.34). As in the case of compounds with one stereocenter (see, scheme 1.20*c*) there can be many different orientations of the same stereoisomer of a compound with two stereocenters as well. It will then be important to determine whether the given orientations represent the same or different compounds. The following points may be noted, in the case of *meso*-tartaric acid:

- *Meso*-tartaric acid is an optically inactive achiral compound and therefore, one must be able to locate a plane of symmetry and a center of symmetry in the given orientation. A plane of symmetry can be seen in an eclipsed configuration while a center of symmetry in a staggered conformation.
- When maleic acid is reacted with OsO_4 *syn*-hydroxylation occurs (both hydroxyl groups add to the double bond from the same face) to give *meso*-tartaric acid. The different orientations (I-IV scheme 1.40) of the same *meso* compound have a plane of symmetry. The planar geometry of maleic acid can be depicted using wedge and dashed bonds or in the usual way. Although orientations (I and II, scheme 1.40) are the perspective formulas *i.e.*, wedge projections and (III, Scheme 1.40) is the sawhorse projection but as shown these projections suffer from eclipsing interactions. These must be depicted properly by a σ bond rotation to get the stable staggered conformations according to already described procedures (see scheme 1.34*c*). These staggered conformations are however, not drawn here.
- The eclipsed sawhorse conformation (III, scheme 1.40) is properly arranged to show stereochemistry according to the Fischer projection (IV, scheme 1.40). Any staggered

conformation of (III) must be rotated about C2–C3 bond to get first the eclipsed conformation before it can be translated into Fischer projection (see, scheme 1.34*c*).

- When one looks to the staggered stereostructures (I and II, scheme 1.41), these are related as object and mirror image. However, since their systematic names come out to be the same (*i.e.*, both are 2*R*, 3*S* tartaric acids) these orientations represent the same *meso* stereoisomer. Recall a *meso* compound has a mirror image but it is superimposable on the parent.

Conformation and Configuration

A compound can have infinite number of conformations but only one configuration—configuration does not change with conformation. Examine . e.g., the two different eclipsed conformations of the same compound meso- tartaric acid (scheme 1.34 and III, scheme 1.41). Thus tartaric acid has three stereoisomers which are drawn as three distinct projections. Each projection may have any conformation.

- If one focuses on the conformations of one of the enantiomers of tartaric acid (2*R*, 3*R* IV or V, scheme 1.41, one in principle may use any conformation staggered (IV) or eclipsed as the situation may demand. One cannot detect a center or a plane of symmetry in (IV or V) respectively. [The projection (IV, scheme 1.41) can be obtained by switching the position of groups at one of the stereocenters, C3 in (I, Scheme 1.41). The projections (I and IV) are diastereomes, recall that diastereomers, (with two stereocenters) have same configuration on one stereocenter and inverted configuration on the second. The enantiomers have inverted stereochemistry at every stereocenter.
- The orientations (VII and VIII scheme 1.41) are identical since rotation of the structure through an horizontal axis passing through C2–C3 in (VII) by 180° out of the plane of paper gives (VIII) Recall unlike a Fischer projection which cannot be lifted out of the plane of paper a perspective drawing does not lose its identity.

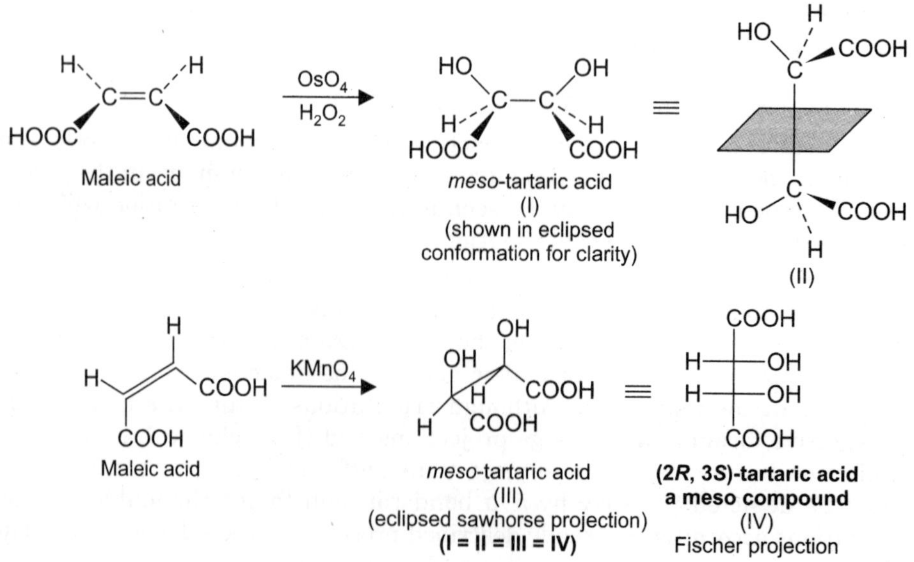

SCHEME 1.40

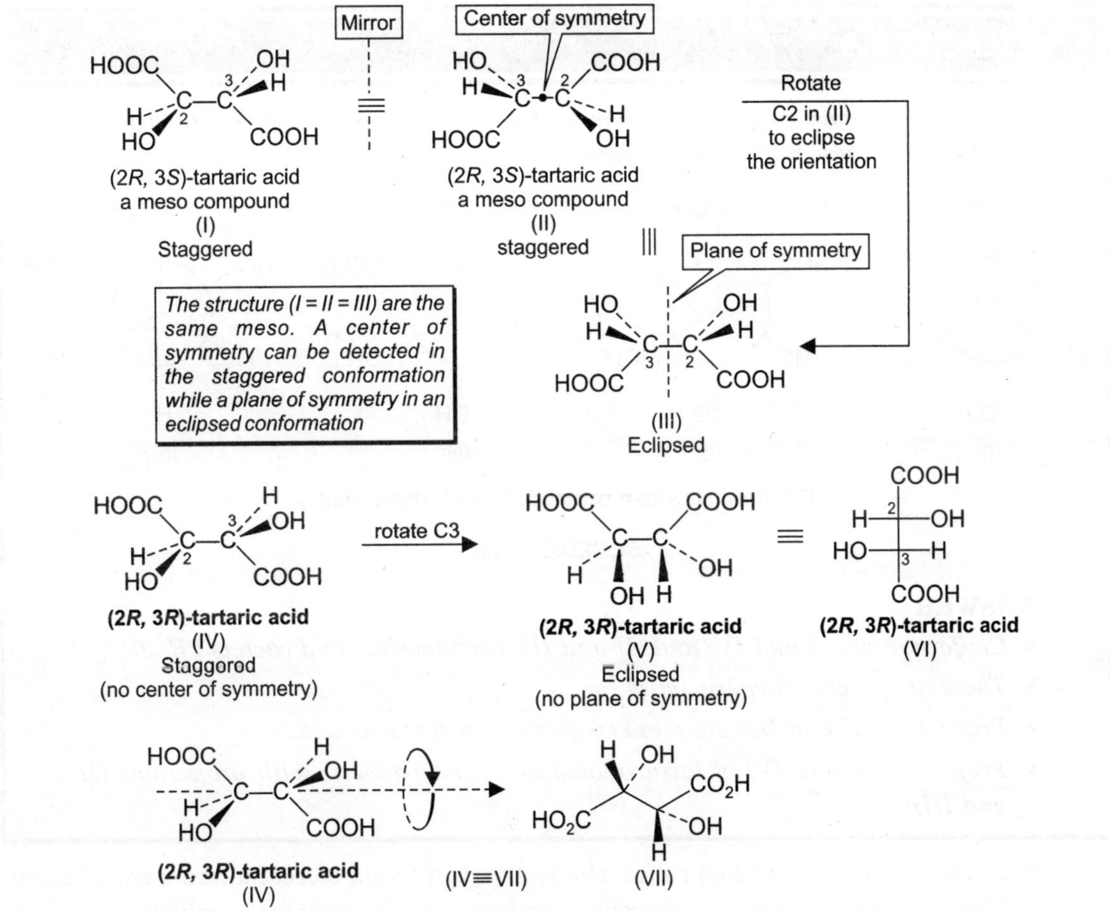

SCHEME 1.41

- One can recognize a *meso* compound from among the orientations drawn *e.g.*, in the Newman projection by rotation about the C—C bond. That stereoisomer is *meso* if one succeeds in getting an eclipsed conformation in which all the substituents are matched up (Scheme 1.42) one can also recognize a center of symmetry in the staggered conformation of *meso*-tartaric acid.

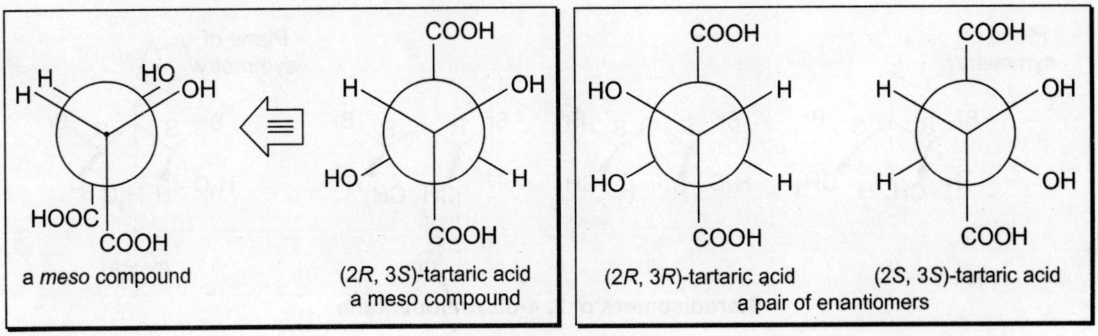

SCHEME 1.42

EXERCISE 1.9

From the Newman projections (scheme 1.42a) point out which are identical ; enantiomers ; meso ; or diastereomers.

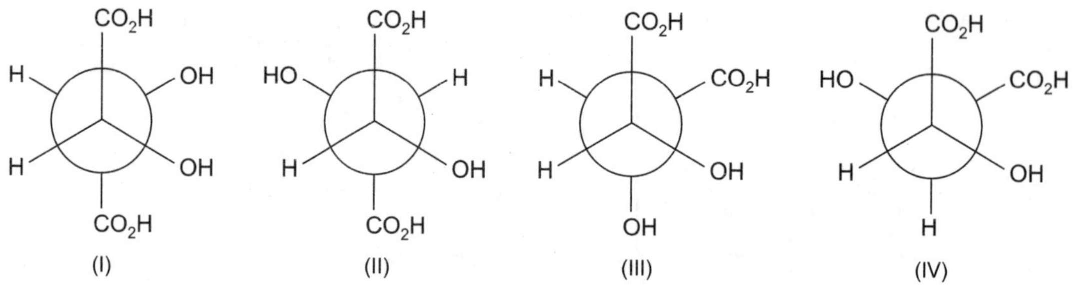

Different Newman projections of tartaric acid

SCHEME 1.42a

ANSWER

- *Conformations (I and IV) and (II and III) are identical and each is 2R, 3S.*
- *There are no enantiomeric pairs.*
- *Projections (II and III) are meso conformers of tartaric acid.*
- *Projections (I and IV) of tartaric acid are diastereomeric with projections (II and III).*

- In the compounds studied so far, the two carbon based stereocenters were adjacent. The same stereochemical relations, however, hold for systems in which centers are separated by one or more atoms. When two stereocenters are different, four stereoisomers are again possible. In case the substitution patterns of both stereocenters is the same there will be three stereoisomers, a pair of enantiomers and one *meso* compound (scheme 1.43). Structure (IV, scheme 1.43) is superimposable on (I) by 180° rotation and is therefore, the same compound (both I and IV scheme 1.43 are called homomers). The *meso* compound (I) is diastereomeric with either of the enantiomers (II or III, scheme 1.43).

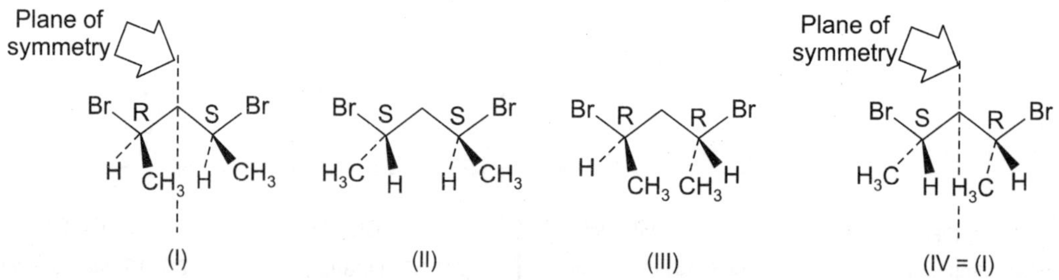

Stereoisomers of 2, 4-dibromopentane

SCHEME 1.43

EXERCISE 1.10

What is the stereochemical relationship between the configurational isomers (scheme 1.43a).

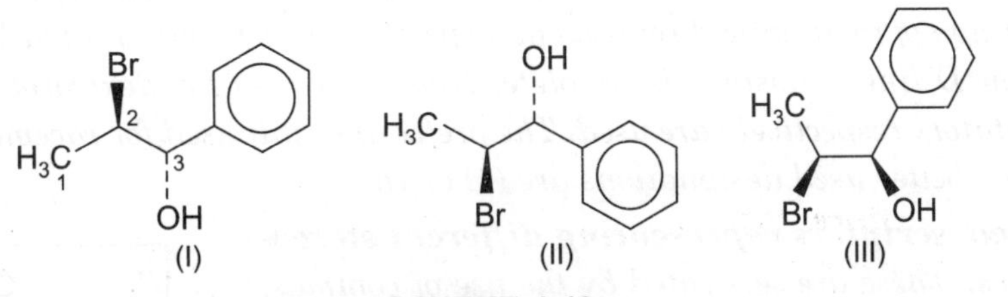

SCHEME 1.43a

ANSWER. *One may follow several approaches as already discussed*

[A] *Assign R and S descriptors to each stereocenter,*

(I) is 2R, 3S

(II) is 2S, 3R

(III) is 2S, 3R

	Priorities
C2	*Br > CH (OH) Ph > CH_3 > H*
C3	*OH > CH(Br) CH_3 > Ph > H*

Thus (I and II) are enantiomers, since at every stereocenter the configuration is inverted, while (II and III) are same.

[B] *Reorient the projections by rotation of one stereocenter or the whole projection. One may show the H atoms also for clarity (scheme 1.43 b).*

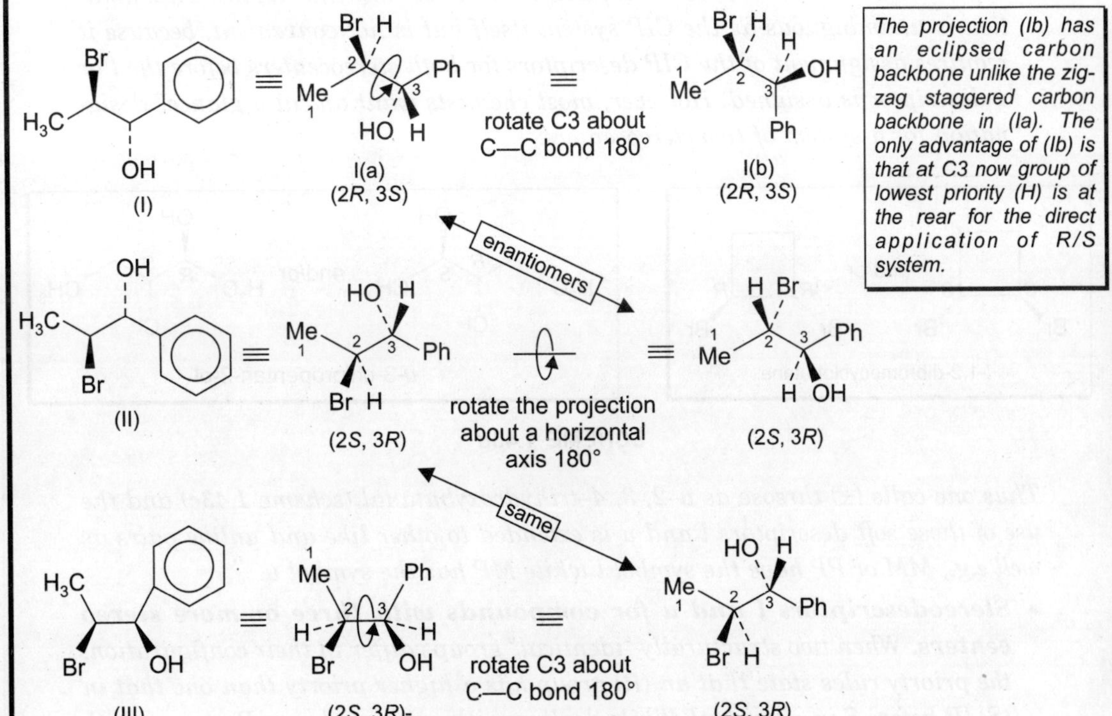

The projection (Ib) has an eclipsed carbon backbone unlike the zig-zag staggered carbon backbone in (Ia). The only advantage of (Ib) is that at C3 now group of lowest priority (H) is at the rear for the direct application of R/S system.

SCHEME 1.43b

More on Relative Configurations and Configurational Descriptors

- ***D, L are configurational descriptors for carbohydrates and amino acids.*** *Their use for other types of chiral compounds is obsolete.*
- ***d, l Characterization of enantiomers.*** *This differentiation of enantiomers by the sign of their optical rotation at a specified wavelength (normally 589 nm, sodium D line emission) is obsolete. Now (+) or (–) for dextrorotatory and levorotatory respectively are used. The prefix dl is still used for racemates, however the better used designations are (±) or rac.*
- ***Stereodescriptors representing different stereocenters.*** *These are separated by the use of commas, thus L-threose is represented as (2R, 3S)–threose along with the usual positional number (scheme 1.43c). The pair of descriptors RS is used for a racemate with one stereocenter. Two pairs of stereodescriptors imply a racemate with two like (RS, RS) as in erythrose or two unlike (RS, SR) stereocenters as in threose and in other systems (scheme 1.43c).*

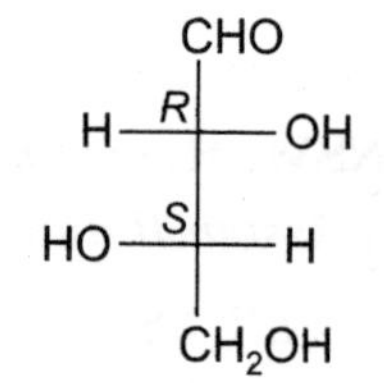

2*R*, 3*S*-threose

SCHEME 1.43c

- ***Stereodescriptors l and u.*** *These are the stereodescriptors used for compounds with two stereogenic centers. One uses the symbol l (for like) and u (for unlike) when the CIP descriptors for the two stereocentres are like (R, R or S, S or RS, RS) or unlike (R, S or S, R or RS, SR) respectively (scheme 1.43d). This notation is unambiguous as the CIP system itself but is not convenient, because it requires assignment of the CIP descriptors for both stereocenters before the l or u descriptor is assigned. However, most chemists want an "at a glance" designation for a system of two stereocenters.*

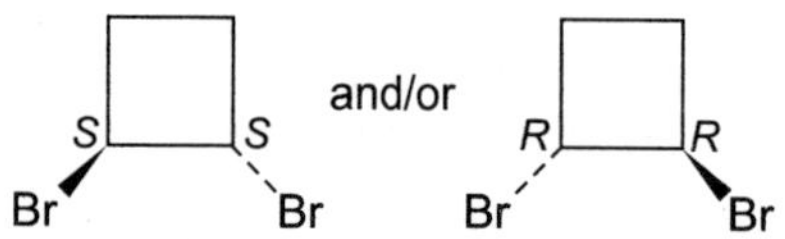

l-1,2-dibromocyclobutane

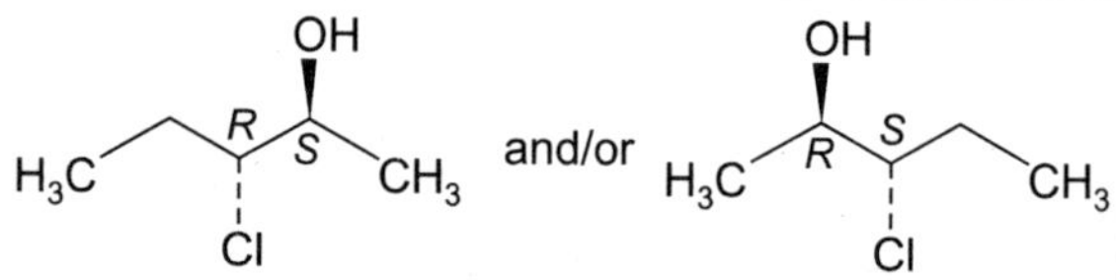

u-3-chloropentan-2-ol

SCHEME 1.43*d*

Thus one calls (±) threose as u–2, 3, 4–trihydroxybutanal (scheme 1.43c) and the use of these soft descriptors l and u is extended to other like and unlike pairs as well e.g., MM or PP have the symbol l while MP has the symbol u.

- ***Stereodescriptors l and u for compounds with three or more stereo centers.*** *When two structurally "identical" groups differ in their configuration, the priorty rules state that an (R) group has a higher priorty than one that in (S) [R before S in alphabet]. While dealing with compounds e.g. D-glucose with* *the configuration 2R, 3S, 4R, 5R (scheme 1.33c) the center with the lowest number is compared with each of the others and racemic glucose is, ull.*

- ***Priorty between groups with l an u configuration.*** *A CIP sub-rule states that of the two constitutionally similar groups the one with an l configuration e.g., R, R or S, S has pirorty over the groups with a u configuration. Thus the compound (I, scheme 1.43e) is assigned a proper stereodescriptor at each of the five stereocenters.*
- ***The prefixes threo and erythro.*** *This is an easy "at a glance" nomenclature for compounds with two stereocenters and was devised several years ago. This useful nomenclature can run into problems (II scheme 1.43e).*

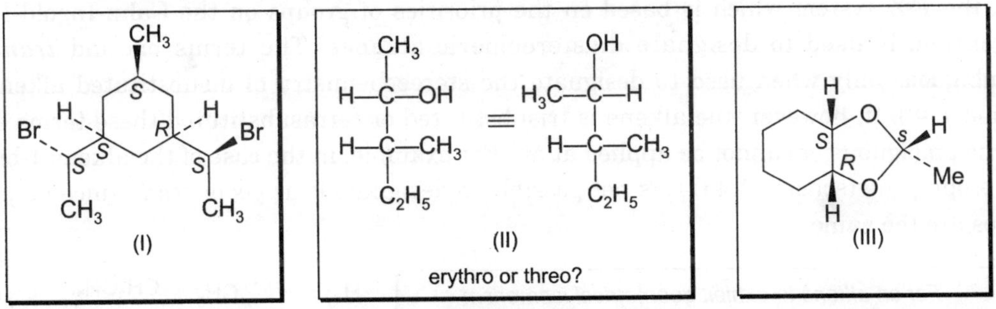

SCHEME 1.43e

- ***The prefixes r and s.*** *The stereocentres are designated by R and S and also by r and s depending on whether they are chirality or pseudochirality centres, respectively. An example of a pseudochirality center is in (III, scheme 1.43e).*
- ***Summary.*** *The soft descriptors (threo, erythro; syn, anti etc) are the shorthand methods used by organic chemists and have a danger of running into misuse. The relative configurations thus are only reliable when the orientations display 3-D relationships.*

1.7 SOME ASPECTS OF GEOMETRIC ISOMERISM (*pi* DIASTEREOISOMERISM AND TORSIONAL CHIRALITY IN CARBON CARBON DOUBLE BONDS)

In case the rotation about a bond joining two mutivalent atoms is restricted, then suitable substitution leads to isomerism. The most common method of preventing rotation about a bond is either the formation of a multiple bond, as in π-diastereoisomerism or by the incorporation of the bond in a ring system as in σ-diastereoisomerism. Isomerism arises in both cases to give isomers which represent different chemical entities known as geometrical isomers. Isomers with similar groups on the same side of the molecule are known as *cis* and those with similar groups on opposite sides are known as *trans*.

This class of diastereomers includes *cis*, *trans*, *E*, *Z* as well as *syn* and *anti*-isomers.

(A) Geometric Isomerism on Double Bonds

For an alkene to show geometrical isomerism the two groups on one end of the double bond must be different and the two groups on the other end of the double bond must also be different.

That is, in terms of the structure (scheme 1.44), A must be different from B, and D must be different from E. In such a situation both of the carbons of the double bond are said to be stereocenters (trigonal planar). A stereocenter or stereogenic atom is defined as an atom at which the interchange of two groups produces a stereoisomer. While considering in general terms the relative orientation of the groups on a double bond the use of the terms *cis* and *trans* to mean "on the same side" and "on opposite sides" respectively is fine (scheme 1.44). Thus *cis* isomer has the two ester groups on the same side of the double bond and the *trans* isomer has the ester groups on the opposite sides (I, scheme 1.44).

(B) The *E-Z* System of Nomenclature of Alkenes and Nitrogen Containing Compounds

Now, the *E-Z* system which is based on the priorities of groups on the Cahn-Ingold-Prelog convention, is used to designate diastereomeric alkenes. The terms *cis* and *trans* are unambiguous only when used to designate the stereochemistry of disubstituted alkenes (I, scheme 1.48). If, however, the alkene is trisubstituted or tetrasubstituted these terms either become ambiguous or cannot be applied at all. For example, in the case of the alkene 1-bromo-1-chloropropene (scheme 1.45) it is not possible to designate it as *cis* or *trans* since no pair of groups are the same.

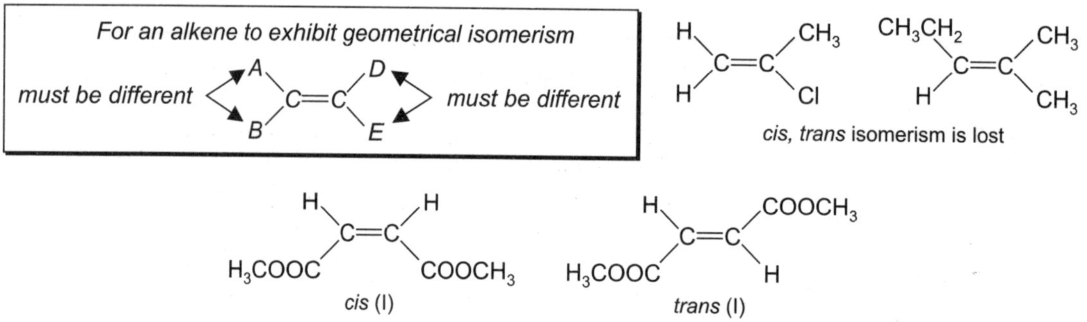

SCHEME 1.44

In the *E-Z* system the two groups attached to each carbon of the double bond are arranged in order of priority *i.e.*, Br > Cl and CH_3 > H. If the two groups of higher priority are located on the same side of the double bond the alkene is designated *Z* isomer (from the German word *Zusammen* meaning together). If on the other hand, the two groups of higher priority are on opposite sides of the double bond then the alkene is designated *E* isomer (from the German word, *Entegegen* meaning opposite). In compound (I, scheme 1.45) one carbon of the double bond has OCH_3 and COOH, OCH_3 being of higher priority than COOH. The second carbon has two ring residues, one of these residues has higher priority over the other and the compound (I, scheme 1.45) therefore, represents the *Z* isomer and similarly compound (II, scheme 1.45) is also the *Z* isomer (Cl > $CH_2CH_2CH_3$ and CH_3 > H). If an alkene has more than one double bond, the stereochemistry about each double bond is to be specified (III scheme 1.45).

While in most cases (*Z*) is equivalent to the conventional *cis* and (*E*) to *trans*, this may not always be the case. For example, compounds (I, II, scheme 1.46) are now called *E* and *Z* against the conventional nomenclature *cis* and *trans* respectively. Diastereomeric oximes and other nitrogen containing compounds are also now given (*E, Z*) nomenclature (scheme 1.46). In the case of compounds having more than one non-cumulated (belonging to different carbon atoms) double bonds the number of π-diastereomers increases and thus proper (*E* and *Z*) descriptors are to be applied to each diastereogenic unit. For an alkene with n carbon-carbon double bonds, each of which can show *cis-trans* isomerism, 2^n *cis-trans* isomers can exist. Four carbon-carbon double bonds are present in the chain of atoms attached to the substituted cyclohexene ring in vitamin A, and each has the potential for *cis-trans* isomerism, there are thus 2^4, or 16, *cis-trans* isomers possible. Vitamin A is the all-*trans* isomer (see scheme 1.52).

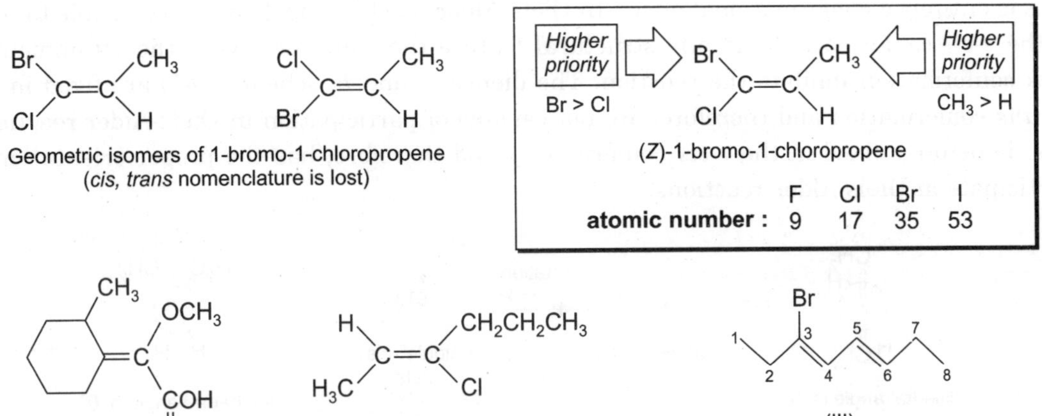

Geometric isomers of 1-bromo-1-chloropropene (*cis, trans* nomenclature is lost)

(*Z*)-1-bromo-1-chloropropene

(I) *Z*-isomer

(II) *Z*-isomer

(III) 3-bromo-(3*Z*, 5*E*) octadiene

SCHEME 1.45

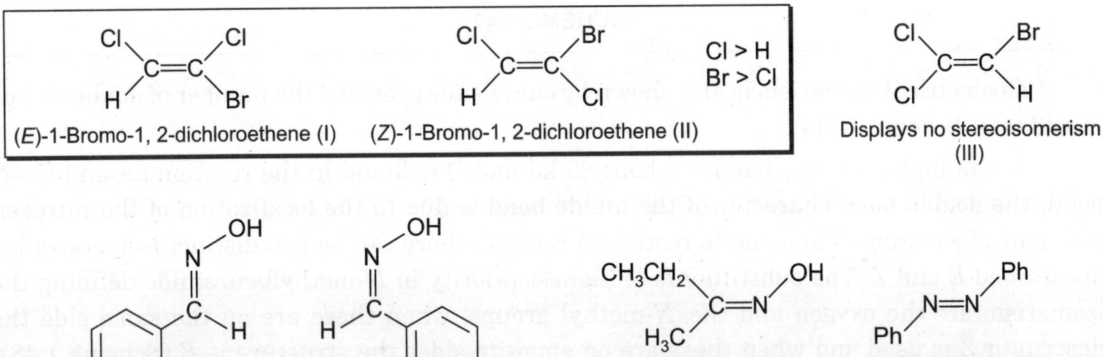

(*E*)-1-Bromo-1, 2-dichloroethene (I)

(*Z*)-1-Bromo-1, 2-dichloroethene (II)

Displays no stereoisomerism (III)

(*E*)-*anti*

(*Z*)-*syn*

Isomers of benzaldehyde oxime

(*Z*)-butanone oxime

E-azobenzene

SCHEME 1.46

A Typical Case of *E*/*Z* Isomerism

Consider 4-methylhepta-2, diene which can exist as four configurational isomers since both double bonds could be *E*, *E* or *Z*, *Z*. These can also have different configuration (scheme 1.46a) and in that case C4 would become a stereocenter. This can be assigned a stereodescriptor by CIP rules since a *Z* configured group has higher priority than *E* group.

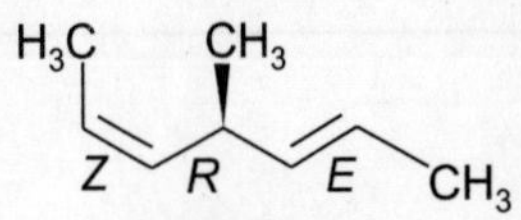

SCHEME 1.46a

In molecules of the type A=CH—CH=B, the fully eclipsed conformation with a dihedral angle of 0° is termed *s-cis*, while the conformation with a dihedral angle of 180° is called *s-trans*. In butadiene *e.g.*, (scheme 1.47), the *s* preceding the *cis* and *trans* refers to *cis* and *trans* with respect to a single bond possessing some double bond character which provides the torsional barrier nearly 25 kJ mol^{-1} (which is much higher than in ethane) to interconversion. This is entirely a conformational effect. In Diels-Alder reaction the diene must be able to react in the *s-cis* conformation and any structural feature that interferes with the attainment of s-*cis* conformation inhibits the reaction. The dienes (*a* and *b*, scheme 1.47) are fixed in the *s-trans* conformation and therefore, are not capable of participation in Diels-Alder reactions. The diene (*c*) is fixed in the *s-cis* conformation and therefore, has the proper orientation to participate in Diels-Alder reactions.

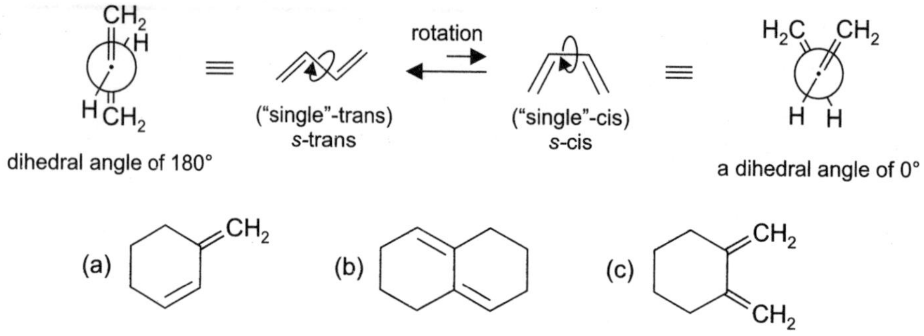

SCHEME 1.47

Geometrical isomerism is also shown by cumulenes provided the number of double bonds in odd (see, Scheme 1.114).

A still higher energy barrier (about 89 kJ mol^{-1}) is found in the rotation around C—N bond, the double bond character of the amide bond is due to the localization of the nitrogen lone pair of electrons. Thus due to restricted rotation there can be two distinct isomers which are termed *E* and Z. The substituents of highest priority in *N*-methylbenzamide defining the isomerism are the oxygen and the *N*-methyl groups, when these are on the same side the descriptor *Z* is used and when these are on opposite sides the structure is *E* (Scheme 1.48).

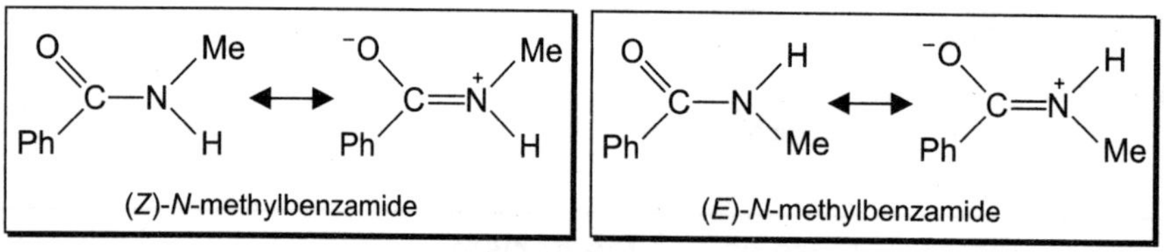

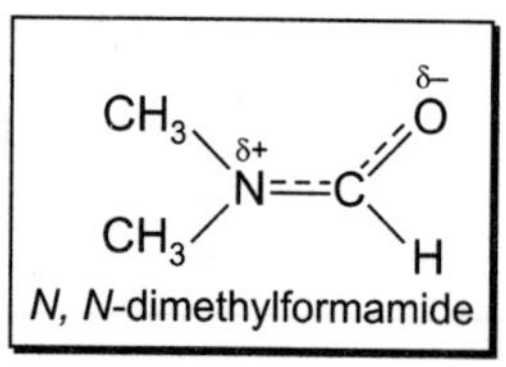

SCHEME 1.48

SCHEME 1.49

The rotation is however, too fast to allow their isolation, although these can be distinguished by low temperature NMR. Thus the absorption of protons of two methyls in *N–N*-dimethylformamide at room temperature occurs at different fields in ^{1}HNMR spectrum (a unified signal is however, observed at raised temperature when rotation around the C–N bond starts to occur scheme 1.49). Significantly the barriers in peptidylprolines as in (scheme 1.50) are so important for protein folding that to bring about such a change in proline isomers in nature there is a wide-spread enzyme (peptidylprolyl-*cis, trans*- isomerase PPI) to bring about this rotation (scheme 1.50).

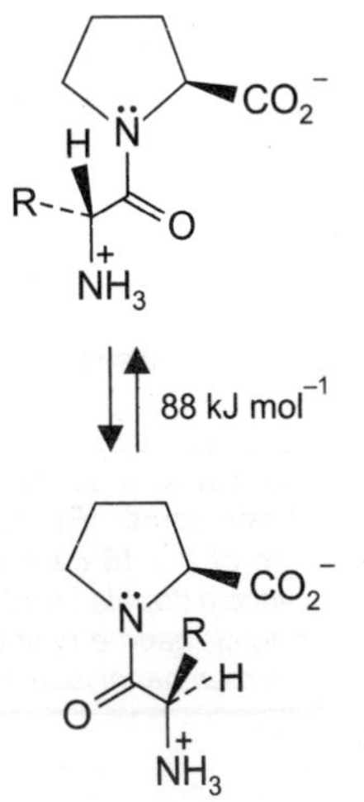

SCHEME 1.50

> *Any two species in equilibrium cannot ordinarily be separated at room temperature unless these are separatedy by a minimum energy barrier of about 100 kJ mol^{-1}.*

On the other hand there are compounds in which nearly free rotation is possible around a formal C = C double bonds. In these cases the double bonds are seriously weakened *via* resonance which decreases the double bond character to allow easier rotation. Thus *e.g.*, the compound (scheme 1.51) has a barrier to rotation of 92 kJ mol^{-1} compared to a simple alkene ~ 260 kJ mol^{-1}. Such alkenes are called "push pull" alkenes. In these push pull alkenes the barrier to rotation is close to the value found for some C—C single bonds.

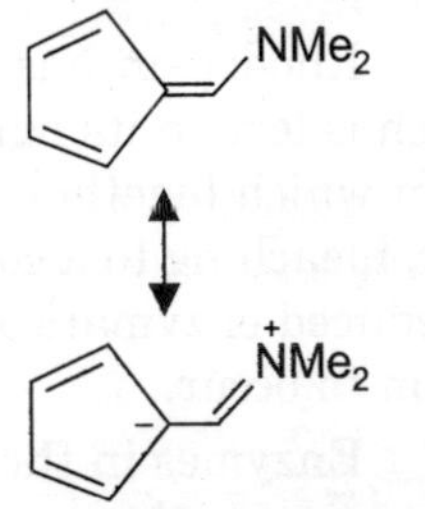

SCHEME 1.51

(C) Vision Process

Rhodopsin is formed in the retina by a condensation reaction involving 11-*cis*-retinal and a protein opsin between the aldehyde group of 11-*cis*-retinal and an amino group present on the surface of the protein. The site on the surface of the protein is such, on which the *cis*-retinal precisely fits, (scheme 1.52). The chemical changes which take place when light falls on the retina of the eye then involve its absorption by the conjugated polyene system of 11-*cis* retinal and the interconversion of *cis-trans* isomers. The visual process starts when rhodopsin absorbs a photon of light (*hv*) and two phenomena may be presented here.

> *On isomerization, 11-cis-retinal to all-trans-retinal, its shape changes and it does not fit into the pocket of opsin. Thus, the iminium ion link becomes exposed and canbe hydrolyzed. The isomerization also separates the iminium ion's positive charge from its balancing negative charge in opsin and this charge separation is one reason for the high energy content of the photoisomerization product.*

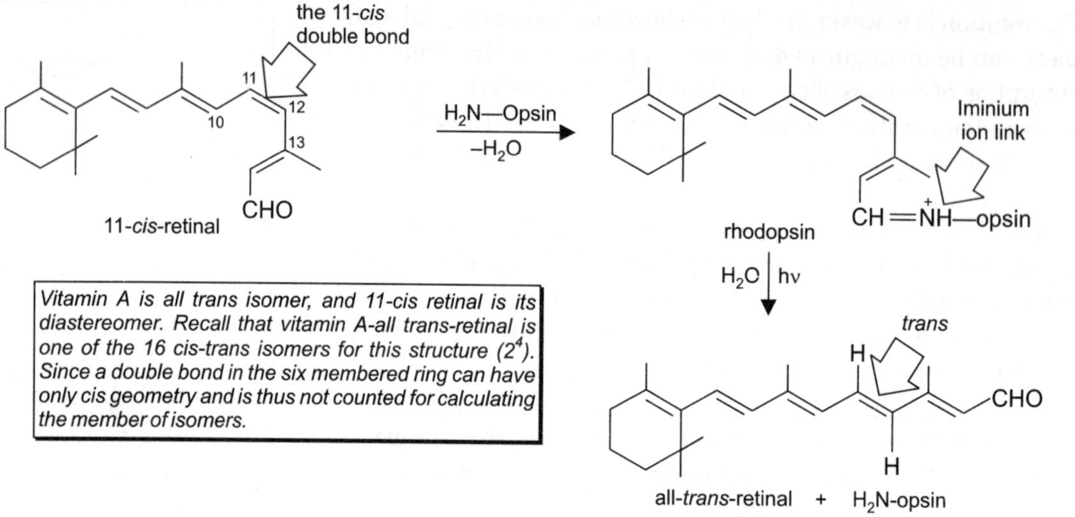

Vitamin A is all trans isomer, and 11-cis retinal is its diastereomer. Recall that vitamin A-all trans-retinal is one of the 16 cis-trans isomers for this structure (2^4). Since a double bond in the six membered ring can have only cis geometry and is thus not counted for calculating the member of isomers.

SCHEME 1.52

Rhodopsin has its absorption maximum at 500 nm which gives it its bright red colour which is lost on its light-initiated transformation to all-*trans*-retinal (vitamin A aldehyde) and opsin which together have an absorption maximum at 387 nm and are therefore yellow. Further, bleaching to a colorless form takes place when the all *trans*-retinal (vitamin A aldehyde) is reduced enzymatically to all-*trans*-vitamin A. Rhodopsin must be regenerated for sustained vision to occur.

Enzymes in the liver transform all-*trans*-vitamin A into 11-*cis*-vitamin A and the latter is returned to the eye where it is reoxidized to 11-*cis*-retinal and is utilized for the synthesis of rhodopsin.

(D) Stereoisomerism—Geometric (*cis-trans*) Isomerism in Monocylic Compounds (A Summary)

(*i*) Cis-Trans Isomerism and Chirality-Cycloalkenes

In the cycloalkenes from cyclopropene to cycloheptene the configuration about the double bond in each case is *cis* (scheme 1.53). A *trans*-double bond *e.g.*, cannot be geometrically accommodated in these *e.g.*, a six membered ring. A cyclopropane ring (bond angle 60°) is destabilized due to large deviation from the normal 109.5° angles associated with sp^3 hybridized carbon. Cyclopropene (only *cis* double bond) is however, more strained since the deviation from the normal sp^2 hybridized bond angles (120°) is even more. Interestingly however, a cyclopropene ring system is present in sterculic acid a natural product (scheme 1.53).

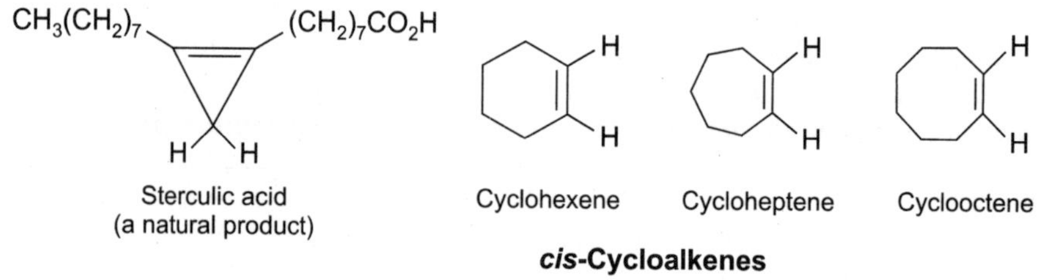

SCHEME 1.53

Cyclohexene is a *cis* alkene, and necessarily so, as the chain with four sp^3 hybrid carbons is not long enough to link C(1) and C(2) with formation of a *trans* alkene. A hypothetical *trans*-cyclohexene (I, scheme 1.54) would be too highly strained to exist at room temperature.

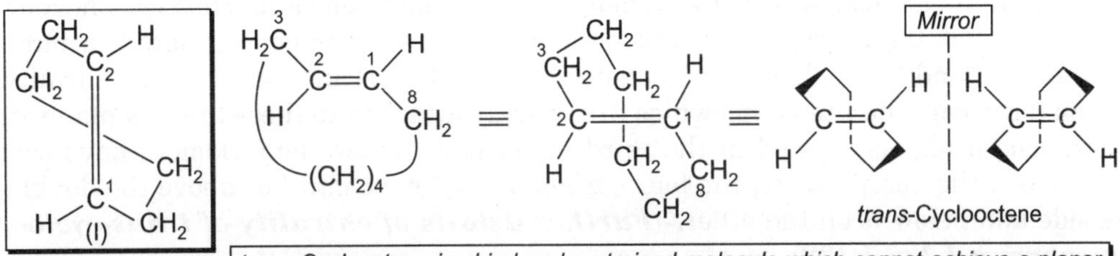

trans cyclohexene is too strained to exist

trans-Cyclooctene is chiral and a strained molecule which cannot achieve a planar symmetric conformation. It is optically active, the two enantiomers may be regarded to have helical shape.

SCHEME 1.54

EXERCISE 1.10

Assign E/Z nonenclature to the compounds (scheme 1.55)

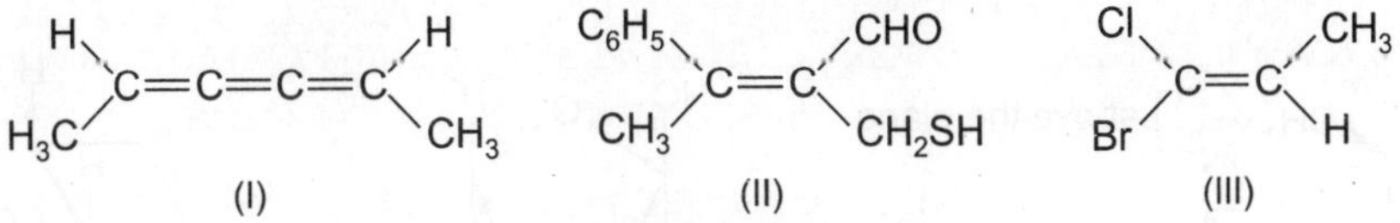

SCHEME 1.55

ANSWER. *(I), (Z)-hexa-2, 3, 4-triene ; **(II)**, E ; **(III)** E.*

EXERCISE 1.11

Assign E/Z nomeclature to the geometric isomers of compounds (scheme 1.56).

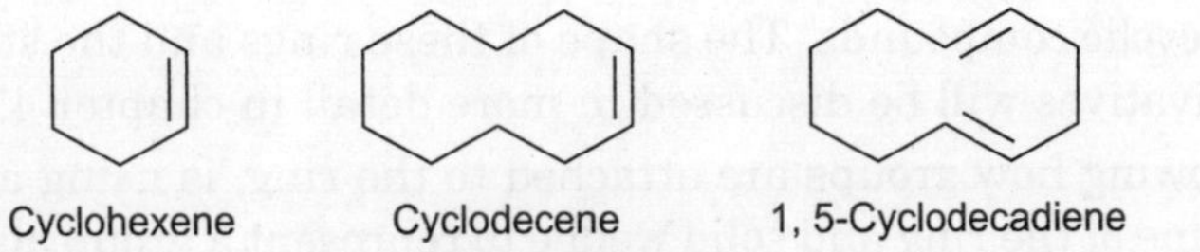

SCHEME 1.56

ANSWER. *Cyclohexene has no E isomer which would be too strained to exist. The E/Z isomers for other compounds are in (scheme 1.57).*

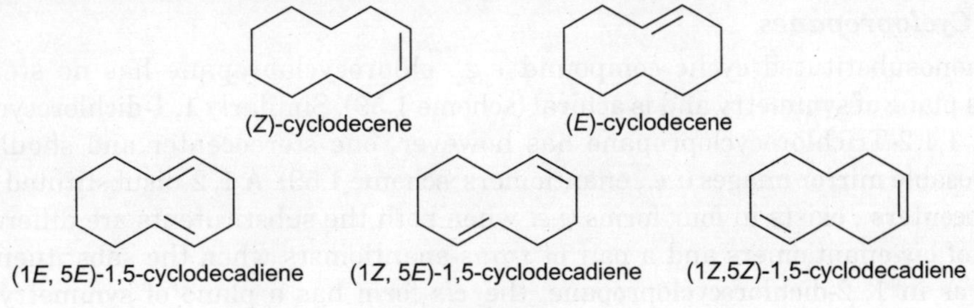

SCHEME 1.57

To date *trans*-cyclooctene is the smallest *trans* cycloalkene that has been prepared in pure form which is stable at room temperature (the *cis-trans*-cyclooctenes represent a diastereomeric pair). In *trans*-cyclooctene (scheme 1.54) there is considerable strain and it is less stable than its *cis* isomer by 9.1 kcal/mol. *Trans*-cycloheptene and *trans*-cyclohexene have been detected, but these are so unstable that these have never been isolated in pure form under ordinary conditions. *Trans*-cyclooctene is a chiral molecule with planar chirality. The two trigonal carbons and the atoms which are directly attached to these are in a plane and the polymethylene bridge is skewed in the third dimension. Its two enantiomers have been isolated. In one of the enantiomers, the four carbon methylene chain lies above the double bond on one side and below it on the other. ***(Further details of chirality of trans-cyclooctene are in schemes 1.136-1.138).***

(*ii*) Geometric and Stereoisomerism in Cycloalkanes (Alicylic Compounds)

The presence of a ring like a double bond prevents rotation and when properly substituted makes *cis-trans* isomerism possible in cycloalkanes. The substituted carbons need not be adjacent (as also is the case in acyclic compounds see scheme 1.43). Rotation around the sigma bonds forming the ring would require the attached atoms or groups to pass through the center of the ring. This process is prevented from happening unless the ring contains eight or more carbon atoms, depending on the size of the substituents. Thus *cis-trans* isomers can exist in rings and properly designated (scheme 1.58).

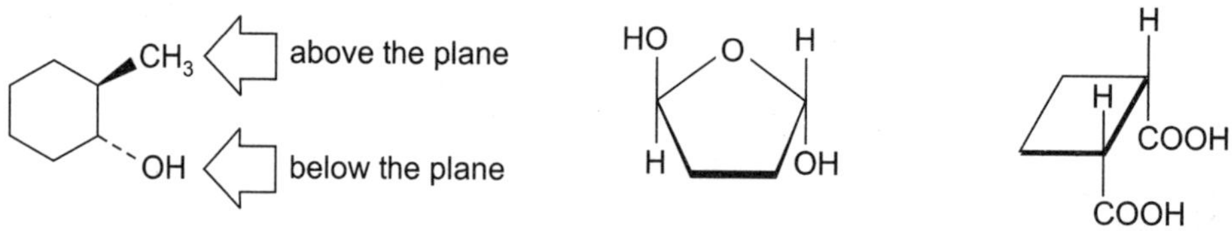

A *trans* compound

trans (OH groups on opposite sides of the ring)

cis (COOH groups on the same side of ring)

SCHEME 1.58

The stereochemistry of cyclic compounds with two or more stereocenters is basically similar to that in acyclic compounds. The shape of these rings and the stereochemical features of cyclohexane derivatives will be discussed in more detail in chapter 4.

A way of showing how groups are attached to the ring, is using a dotted line to show a group below the plane of the ring and solid wedge to represent a group above the plane (scheme 1.58). A part of the molecule when shown by thick line(s) is to orient it in a way so that this thick lined portion is projected towards ones eyes, this is often done to depict stereochemical features more clearly. The *cis* and *trans* isomers in cycloalkanes are in fact diastereomers.

(a) Cyclopropanes

A monosubstituted cyclic compound *e.g.*, chlorocyclopropane has no stereocenter, possesses a plane of symmetry and is achiral (scheme 1.59). Similarly 1, 1-dichlorocyclopropane is achiral. 1,1,2-Trichlorocyclopropane has however, one stereocenter and should exist as nonsuperposable mirror images (*i.e.*, enantiomers, scheme 1.59). A 1, 2-disubstituted compound (two stereocenters , exists in four forms (*i.e.* when both the substitutents are different): these are a pair of *cis*-enantiomers and a pair of *trans*-enantiomers when the substituents are the same *e.g.*, as in 1, 2-dichlorocyclopropane, the *cis* form has a plane of symmetry and thus represents a *meso*-type achiral compound which has superposable mirror image (scheme 1.60).

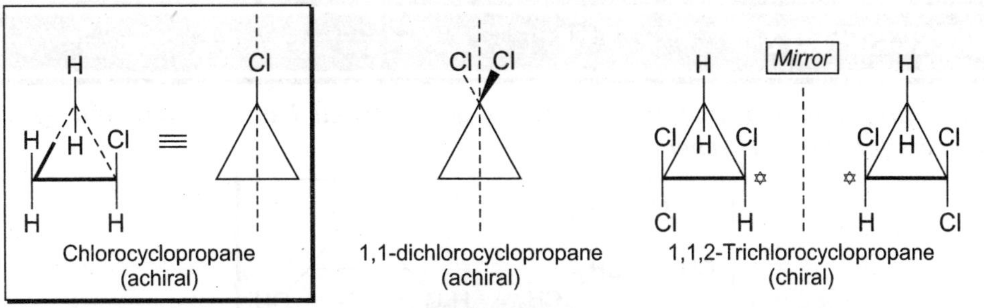

SCHEME 1.59

In the case of *trans*-1, 2-dichlorocyclopropane, the mirror image is not superposable on the original. It is, therefore, a chiral molecule and two resolvable enantiomers exist.

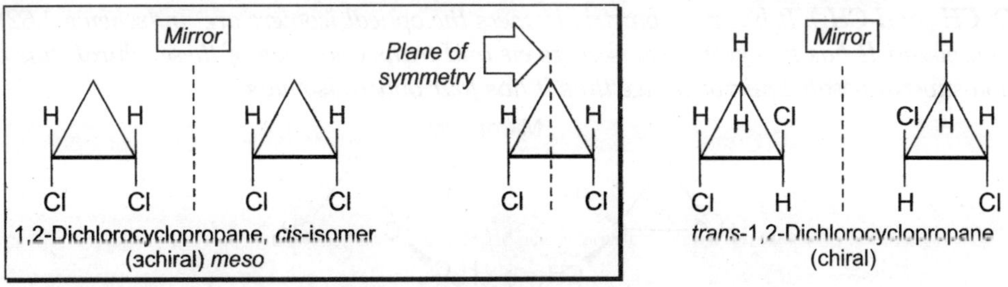

SCHEME 1.60

(b) Cyclobutanes

As with cyclopropanes, monosubstituted and 1, 1-disubstituted cyclobutanes exist in only one achiral form, whereas 1, 2-disubstituted cyclobutanes exist as a pair of enantiomers and a *meso* form (when the two substitutents are same). Both the *cis* and the *trans* 1, 3-disubstituted derivatives possess a plane of symmetry so that there are only two geometric isomers and none is optically active. (This is true of any disubstituted cycloalkane containing an even number of carbon atoms, where the substituents are on the opposite sides of the ring). One can also detect a center of symmetry in the case of *trans*-1, 3-dimethylcyclobutane which is at the center of the ring (this, however, is not shown, also see scheme 1.68*e*).

The mirror image of *e.g.*, *trans*-1, 3-dimethylcyclobutane is thus superimposable on the original scheme 1.61). This one can see by simply rotating the structures. Further proof that it is an achiral compound is due to the fact that it has a plane of symmetry. The plane of symmetry is lost in (I, scheme 1.61) which is chiral.

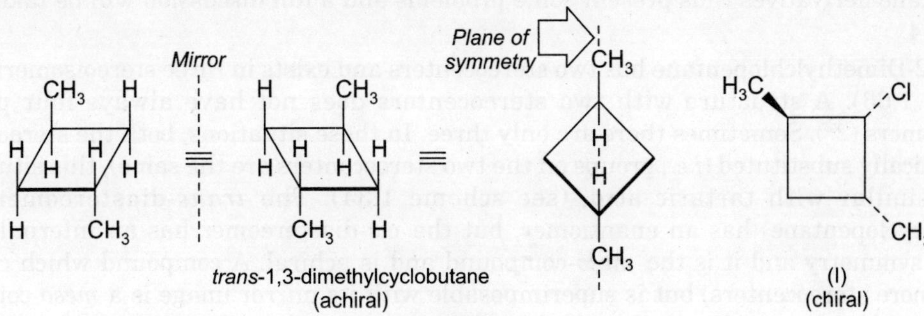

SCHEME 1.61

EXERCISE 1.12

Draw projections to show the presence of geometric cis-trans isomers and optical isomers for (I and II scheme 1.61).

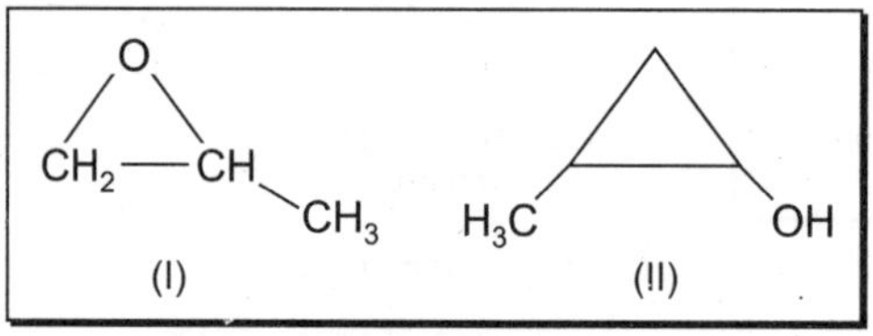

SCHEME 1.61

ANSWER. *In (I) there is one stereogenic carbon (the four different substituents being H, O, CH_2 and CH_3). It has no geometric isomers the optical isomers are in (scheme 1.62). Compound II has two geometric isomers cis and trans and each of these (chiral) has a non-superimposable mirror image thus it has four optical isomers.*

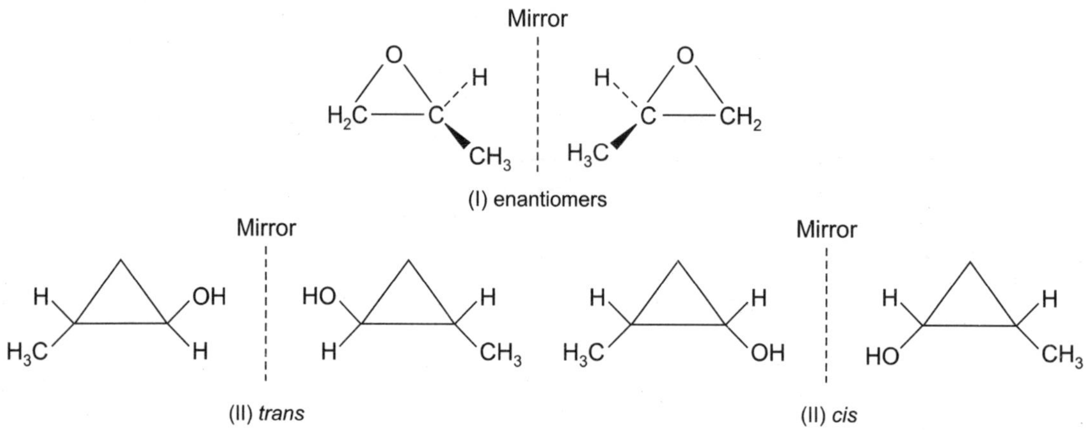

SCHEME 1.62

(c) Cyclopentanes

The discussion may be taken up with cyclopentane ring, which like cyclopropane and cyclobutane is essentially planar. The planar projection formulas of cyclohexane derivatives can obviously not be considered as entirely correct representations. The planar representations are however, useful to symbolize and to count the various possible stereoisomers. Cyclohexane derivatives thus present some problems and a full discussion will be taken up in Chapter 4.

1,2-Dimethylchlopentane has two stereocenters and exists in three stereoisomeric forms (seheme 1.63). A structure with two stereocenters does not have always four possible stereoisomers (2^2). Sometimes there are only three. In these situations, both the stereocenters are identically substituted (*i.e.*, groups on the two stereocenters are the same), this situation is indeed similar with tartaric acid, (see scheme 1.34). The *trans*-diastereomer (1, 2-dimethylcyclopentane) has an enantiomer, but the *cis*-diastereomer has an internal mirror plane of symmetry and it is the *meso*-compound and is achiral. A compound which contains two (or more stereocenters) but is superimposable with its mirror image is a *meso* compound (tartaric acid with two identically substituted stereocenters has also a *meso* compound (see scheme 1.34).

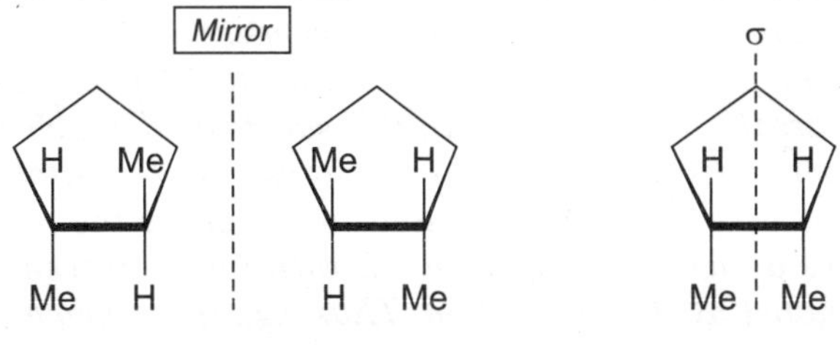

trans-1,2-Dimethylcyclopentane (chiral) *cis*-1,2-Dimethylcyclopentane *meso* (achiral)

SCHEME 1.63

(d) Cyclohexanes (an introduction)

Unlike the essentially planar cyclopentane derivatives *e.g.*, 1, 2-dimethylcyclopentane discussed in (scheme 1.63), examine the formula of 1, 2-dimethylcyclohexane which also has two stereocenters. One may expect 2^2 *i.e.* four stereoisomers. There are indeed four (one can however, isolate only three stereoisomers for further details see chapter 4). The *trans*-1, 2-dimethylcyclohexane like *trans*-1, 2-dimethylcyclopentane exists as a air of enantiomers. Significantly the *cis*-isomer (scheme 1.64) has also two enantiomers, since when this isomer is drawn in the chair form one can not detect a plane of symmetry. However, when one draws the *cis* isomer (scheme 1.64) as a hexagon it apparently looks to have a plane of symmetry and thus like *cis*-1, 2-dimethylcyclopentane one may be deceived to conclude that it should exist as a *meso* compound. It is thus important to remember that the plane of symmetry must be apparent in the case of a cyclohexane, whether the molecule is drawn as a hexagon or a chair conformation. However, it is not necessary to study the conformations to know if a cyclic compound is optically active. A cyclic compound cannot be optically active if its planar structure has a plane of symmetry (it may be meso or racemic), for further details see schemes 4.32–4.34 and 4.61.

These stereochemical complications are removed when in *cis*-1, 2-dimethylcyclohexane the ring is cleaved at the C-C bond opposite the one connecting the stereocenters (scheme 1.64). The resulting acyclic compound 3, 4-dimethylhexane, has three stereoisomers a *meso* compound and two enantiomers.

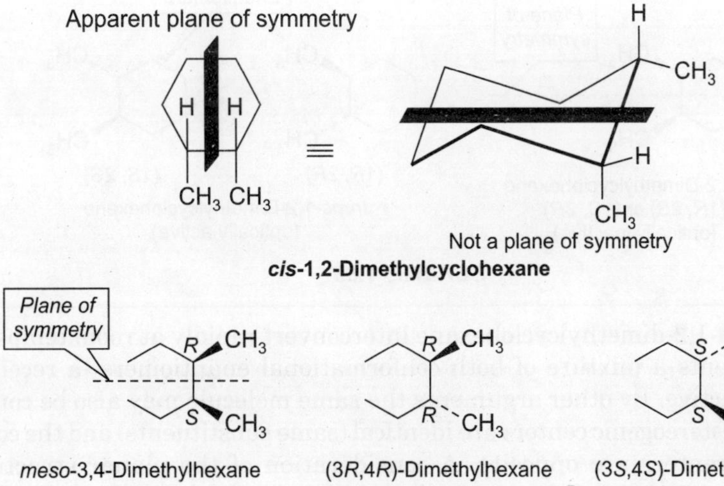

meso-3,4-Dimethylhexane (3*R*,4*R*)-Dimethylhexane (3*S*,4*S*)-Dimethylhexane

SCHEME 1.64

In summary, therefore, the cis, trans or geometric isomers are diastereomers which differ in chemical and physical properties. The structural features leading to geometric isomerism are the presence of a suitably substituted double bond (carbon-carbor, carbon-nitrogen, nitrogen-nitrogen) or the presence of a suitably substituted cyclic structure. The cis/trans notation is used to designate the configuration in carbocyclic structures as well. In addition, chirality of stereocenters has to be evaluated and designated by (R, S) notation. Thus e.g., (Ia, scheme 1.65) is trans-2-methylcyclohexanol, and it is correctly named as (1R, 2R)-2-methylcyclohexanol since its mirror image (enantiomer) (1S, 2S)-2-methyl cyclohexanol is also the trans isomer. The cis-isomer (IIa) is also chiral and has a diastereomeric relationship with the trans isomer, the two enantiomers are (IS, 2R)-2-methyl cyclohexanol and (IR, 2S)-2-methylcyclohexanol (scheme 1.65).

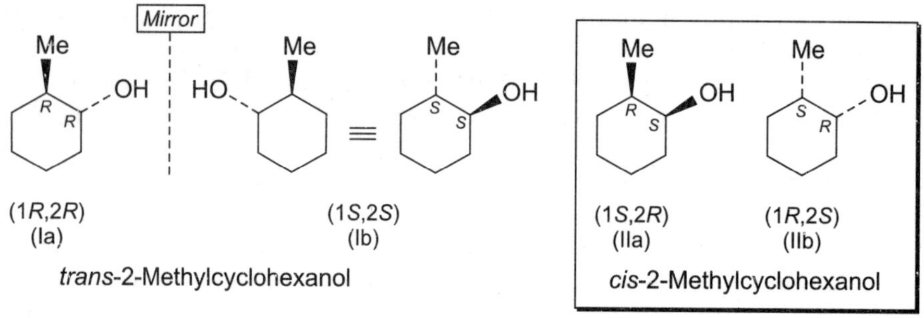

SCHEME 1.65

(e) Cyclohexanes as Planar Structures

Cyclohexane rings are chair shaped rather than being flat. Some stereochemical analysis regarding the number of stereogenic centers, detection of plane of symmetry and whether a molecule would display optical activity or not can be done with drawings using planar rings. Careful analysis of one chair conformation of *cis*-1,2-dimethylcyclohexane (see, scheme 1.64) shows that it is chiral—it is not superimposable on its mirror image. The ring flipped conformation, however, represents the enantiomer of the original conformation (see, scheme 4.33).

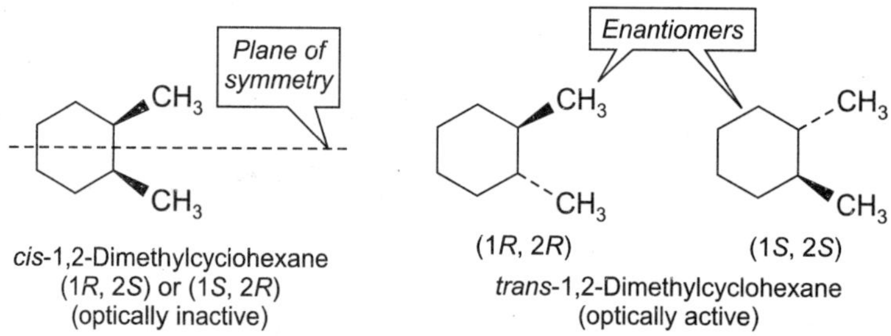

SCHEME 1.66

The conformers of *cis*-1,2-dimethylcyclohexane interconvert rapidly at room temperature, thus the molecule represents a mixture of both conformational enantiomers (a recemic mixture) and is not optically active. By other arguments the same molecule may also be considered as a *meso* compound. The stereogenic centers are identical (same substituents) and the configurations at two stereogenic centers are opposite. A consideration of the planar structure of *cis*-1, 2-dimethylcyclohexane (scheme 1.66) shows that it must not be optically active since it has a plane of symmetry. One however, cannot detect a plane of symmetry in the planar ring of *trans*-1,2-dimetylcyclohexane (scheme 1.66), it is thus expected to be chiral and optically active. The following points may be noted :

- A flat (planar) structure of *e.g., cis*-1,2-dimethylcyclohexane can be drawn in two different ways. This depends if one views the conformation from top or bottom. Thus the two structures (I and II, scheme 1.67) are equivalent. One can confirm it—if (I) is turned 180° out of the plane of paper one gets (II). One may note that during this operation a group closer to one's eyes (on thick wedge) goes away (on dotted line) and vice versa. The structures thus remain identical.

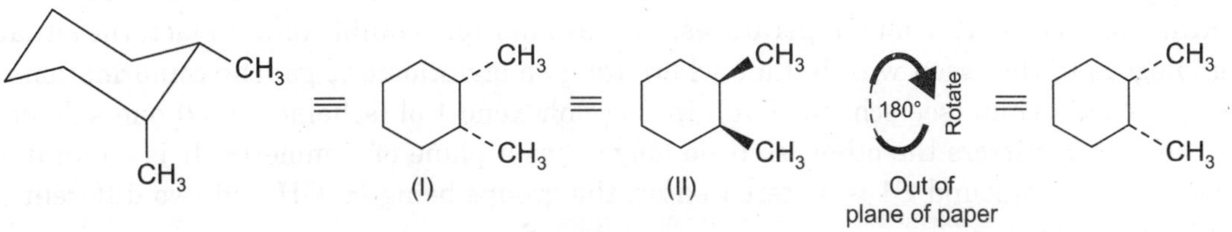

SCHEME 1.67

- If the planar structure of a cyclic compound has a plane of symnetry, the compound cannot be optically active. This is seen in *cis*-1,2-dimethylcyclohexane. In such a situation the planar structure is superimposable on its mirror image (scheme 1.67*a*).

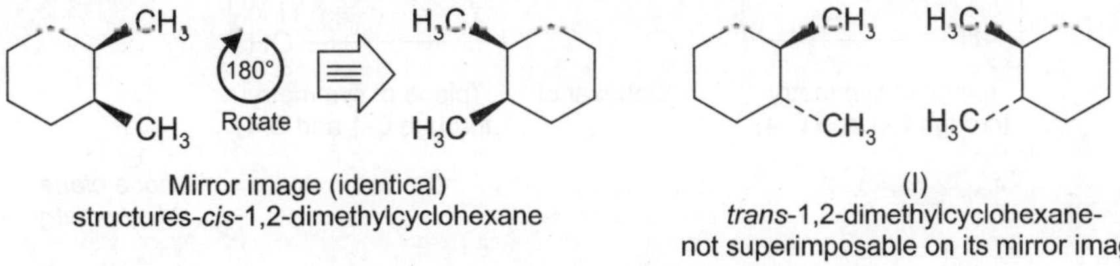

SCHEME 1.67a

- A planar structure of *trans*-1,2-dimethylcyclohexane is expected to be optically active (no plane of symmetry) since the planar structure and its mirror image are not superimposable (I, scheme 1.67*a*) (for further details see schemes 4.40 and 4.41).

EXERCISE 1.13

Draw the planar structures of all possible stereoisomers of 1, 3-dimethylcyclohexane. Show their stereochemical relationship by assigning R, S nomenclature to stereocenters.

ANSWER. *It is in scheme 1.68.*

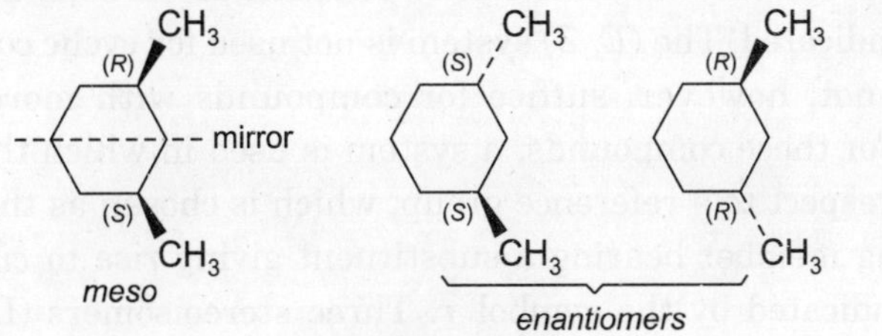

SCHEME 1.68

Cyclohexanol (like *e.g.*, a monosubstituted cyclopropane) is an achiral compound, it has a plane of symmetry which is perpendicular to the page and that passes through the hydrogen and hydroxy group attached to it, and C4. Note how one side of the ring mirrors the other side. The plane of symmetry is again detectable in both the chair conformations in one of which OH is equatorial while in other it is axial (scheme 1.68*a*). The two chair conformations of a monosubstituted cyclohexane *e.g.*, 1-clorocyclohexane actually have diastereomeric relationship, since they are stereoisomers that do not have a mirror image relationship. However, the energy barrier for ring flip of one cyclohexane chair conformation to another is of the order of 10 kcal/mole. Thus, at room temperatures, they are in rapid equilibrium. In fact, the situation is analogous to that seen with butane where the two enantiomeric gauche conformations are in rapid equilibrium (see scheme 1.1*c*). In 2-cyclohexene-1-ol (scheme 1.68*a*) one side of the ring no longer mirrors the other, so it no longer has a plane of symmetry. It is a chiral molecule. In this compound C1 is a stereocenter, the groups being H, OH and two different ring residues.

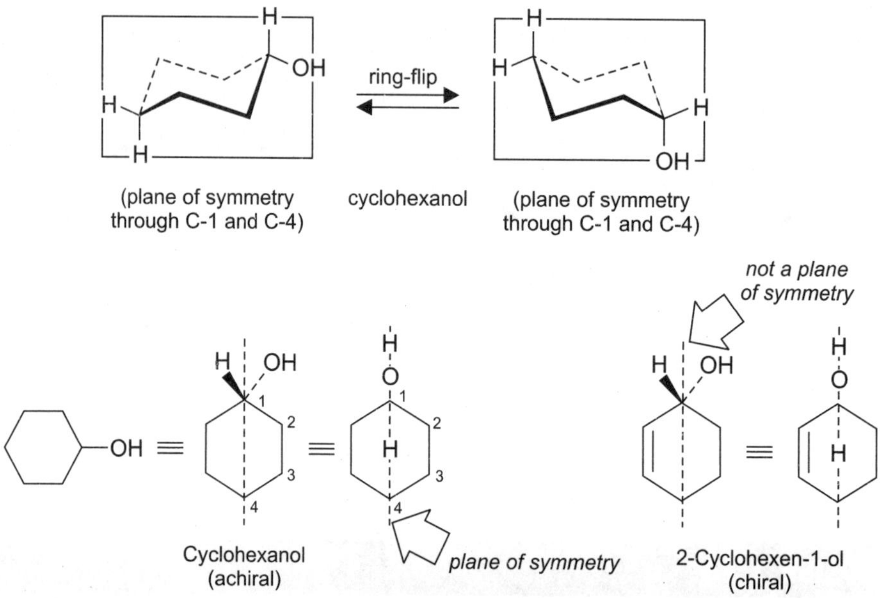

SCHEME 1.68a

(*iii*) Nomenclature of Poly-Substituted Cycloalkanes

Cyclic stereoisomers with only two differently substituted carbons are named either *cis* or *trans* as previously indicated. The (*E, Z*) system is not used for cyclic compounds. The *cis-trans* nomenclature does not, however, suffice for compounds with more than two differently substituted atoms. For these compounds, a system is used in which the configuration of each group is given with respect to a reference group, which is chosen as the group attached to the lowest numbered ring member bearing a substituent giving rise to *cis-trans* isomerism. The reference group is indicated by the symbol *r*. Three stereoisomers (I, III scheme 1.68*b*) are named according to this system.

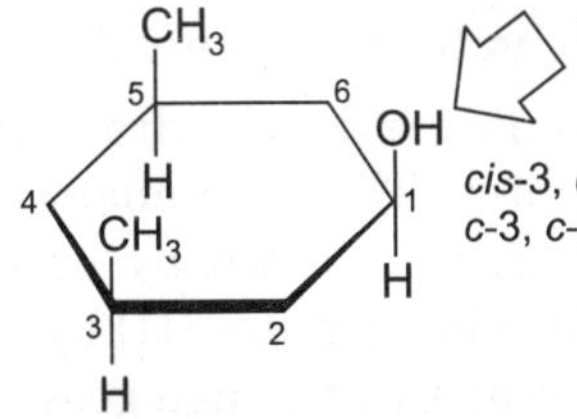

reference group is indicated by the symbol *r*

cis-3, *cis*-5-Dimethylcyclohexan-*r*-1-ol
c-3, *c*-5-dimethylcyclohexan-*r*-1-ol

meso form
(I)

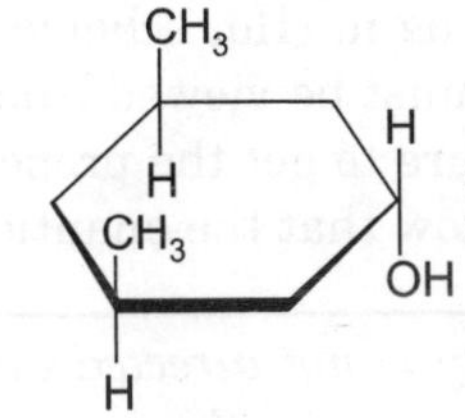

trans-3, *trans*-5-Dimethylcyclohexan-*r*-1-ol
t-3, *t*-5-dimethylcyclohexan-*r*-1-ol

meso form
(II)

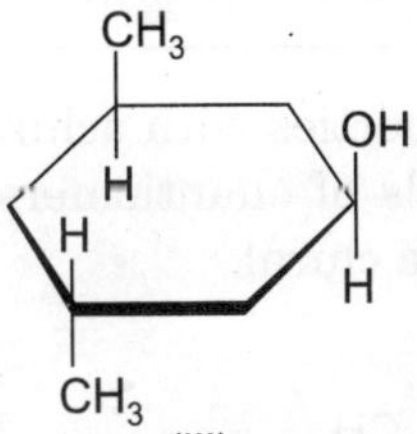

(III)

enantiomers

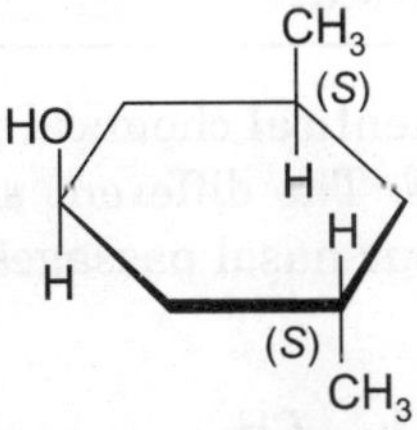

(IIIa)

cis-3, trans-5-Dimethylcyclohexan-*r*-1-ol
c-3, *t*-5-dimethylcyclohexan-*r*-1-ol

SCHEME 1.68b

When there are two otherwise equivalent ways of going around the ring. The rule is to choose the path that gives the *cis* designation to the first substituent after the reference (III, scheme 1.68*b*).

Thus, the *cis* or *trans* relationship of the methyl groups to the hydroxyl group (chosen as reference) is specified. However, until recently, there had been ambiguity in naming such compounds. For example (I, scheme 1.68*b*) could unambiguously be given the prefix *cis, cis* but (II, scheme 1.68*b*) might be *trans, trans* or perhaps *cis, trans*. However, for a chiral compound use of *RS* system to each stereocenter would help but this system cannot be applied here.

More on Nomenclature of Polysubstituted Cycloalkanes

Consider the enantiomer of (III, scheme 1.68b), the structure is chiral, however, it has S-configuration both at C3 and C5. Thus, one cannot assign a configurational descriptor to C1. As a matter of fact C1 is chirotopic (for description of the term chirotopic, see scheme 1.34a) but is non-stereogenic (since exchange of H and OH does not lead to any new stereoisomer). Interestingly in the case of (I, Scheme 1.68b) C1 is achirotopic (presence of a σ plane) but stereogenic, however, (interchange of H and OH does give another stereoisomer (II, scheme 1.68b). Similar is the situation with II and in both (I and II, scheme 1.68 b) C1 is a pseudoasymmetric center (pseudochirality center) (see also scheme 1.34b).

(*iv*) Representation and Number of Stereoisomers in Cyclic Compounds

Just like in acyclic systems, the enantiomers may be represented either by drawing a mirror image or by inverting the stereochemistry at each stereocenter. Thus the enantiomer of (I, scheme 1.68*c*) is its mirror image (II) or (II*a*) and the later is drawn by inverting the stereochemistry in (I) at the stereocenter [one may note that if one rotates (II) by 180° out of the plane of paper one gets (II*a*), the group which was shown as thick wedge goes away from ones eyes and is thus shown as hatched *i.e.*, dotted wedge] and vice versa. A three dimensional figure does not loose its identity on rotation, however, a Fischer projection can only be rotated in specified ways. Thus if an enantiomer of carvone is written as in (II*a*, scheme 1.68*c*), to assign the stereodescriptor, to the stereocenter, the compound must be viewed from the side opposite the group of least priority. One then rotates the structure to get the proper orientation (II, scheme 1.68*c*) and the path $1 \rightarrow 2 \rightarrow 3$ is clockwise to show that the enantiomer is *R*.

> *One has already learnt, that if the group of lowest priority is not directed away from you as in (IIa, Scheme 1.68c). It may not be necessary to turn the structure. Keep it as such, however apply the R / S system of nomenclature backward i.e., the descriptor thus obtained must be reversed.*

The enantiomeric compounds have identical chemical properties with achiral reagents, but behave differently with chiral reagents. The different smells of enantiomers of carvone (scheme 1.68*c*) show that receptor sites in our nasal passages are chiral.

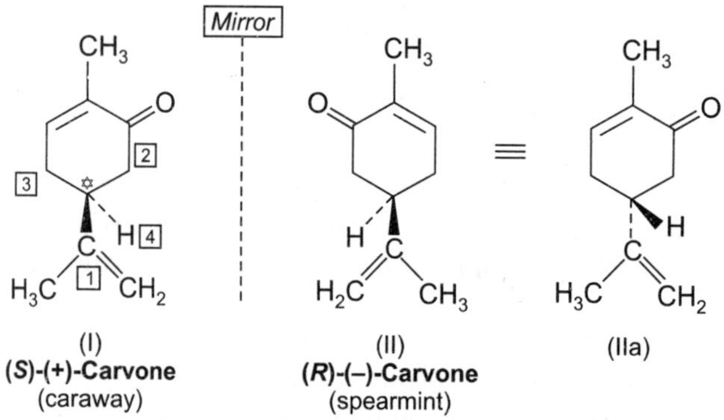

SCHEME 1.68c

Cholesterol a constitutionally unsymmetrical compound, as expected, with eight tetrahedral stereocenters can exist as $2^8 = 256$ stereoisomers. One only considers tetrahedral stereocenters although a double bond is also present in the molecule. The double bond in cholesterol being in a six-membered ring can only be *cis*, thus it does not contribute to the number of stereoisomers (compare with retinal scheme 1.52). However, the number of stereoisomers can be less than expected and the following situations may be considered :

- The existence of *meso* compounds decreases the number of possible stereoisomers than expected from the maximum number calculated from 2^n stereoisomers. Several situations in acyclic and cyclic systems have already been discussed. Another example is of 1, 2 pentanediol which can exist as three unique stereoisomers–an enantiomeric pair and a *meso* compound (scheme 1.68*d*).

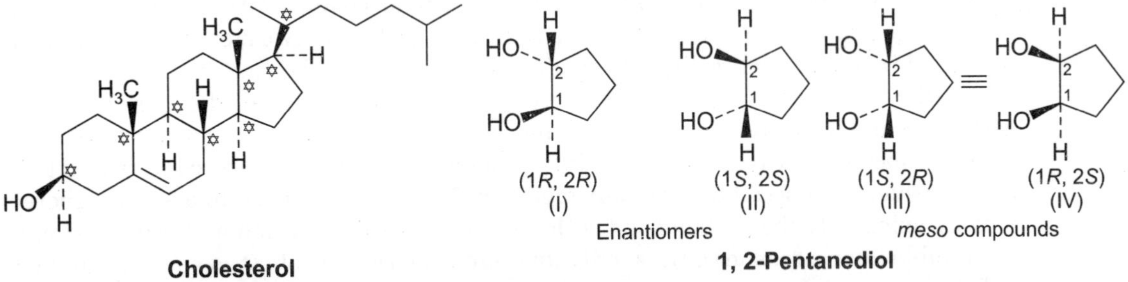

SCHEME 1.68d

- A symmetrical molecule like twistane (scheme 1.68*e*) with four equivalent stereocenters exists only in two enantiomeric forms. (One can easily detect a main C_2 axis in this chiral compound, as one shall see below, the presence of one or more C_n axes does not interfere with the chirality in a molecule). Consider 1-bromo-2-hydroxycyclobutane. The compound with two stereocenters exists as two diastereomers (cis I and trans II scheme 1.68*e*) and each is resolvable and thus one has in all expected 4 stereoisomers. However, with increased symmetry 1-bromo-3 hydroxycyclobutane with two non-equivalent stereocenters exists only in two diasteromeric *meso* forms (III and IV Scheme 1.68*e*). Recall that in 1-bromo-3-hydroxybutane there are two non-equivalent stereocenters [a stereocenter leads to a new stereoisomeric molecule on interchange of any ligands (H with OH gives the *trans*-diastereomer)]. In the case of this molecule both C1 and C3 are therefore, stereogenic but are achirotopic because of their presence on a plane of symmetry.

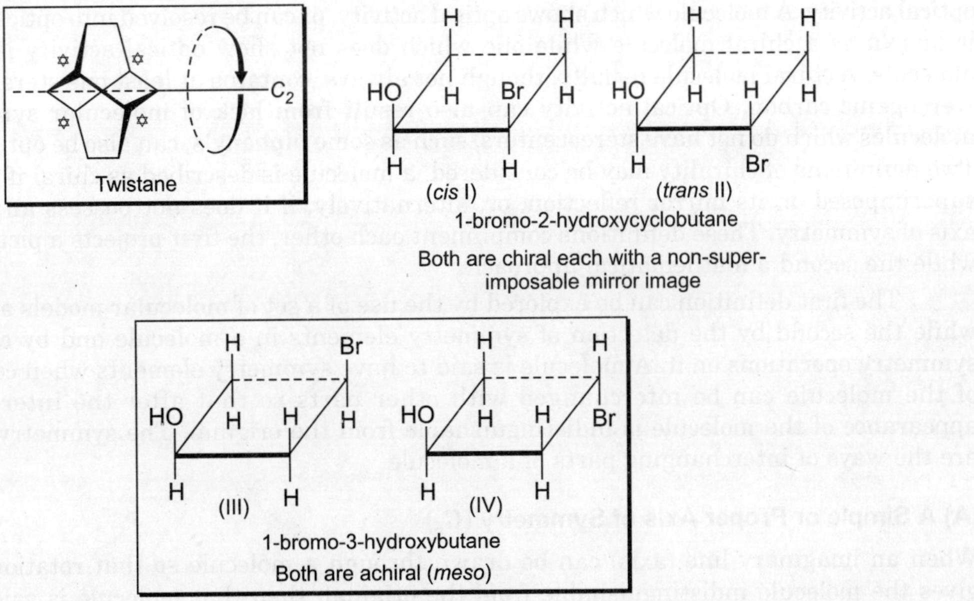

SCHEME 1.68e

- When some diastereomers cannot form due to steric reasons (configurational restriction), the number of stereoisomers is less. Thus in the case of camphor the bridge can only be *cis* and camphor (with norbornane system) with two stereocenters (two bridge head carbons) exists only as a (±) pair, and these two stereocenters behave as a single element of chirality (scheme 1.68*f*). Similarly norbornane monosubstituted in position 2 (II scheme 1.68*f*) has three stereocenters and again for steric reasons carbons 1 and 4 behave as a single element of chirality and only four stereoisomers exist for such derivatives. Similarly in α-pinene there are two stereocenters and four stereoisomers are possible. The four membered ring can be fused to a six membered ring by using only the *cis* bonds on cyclobutane in the 1,3-position and *trans*-fusion is impossible (steric reasons), thus, only an enantiomeric pair of *cis*-isomer of α-pinene is known.

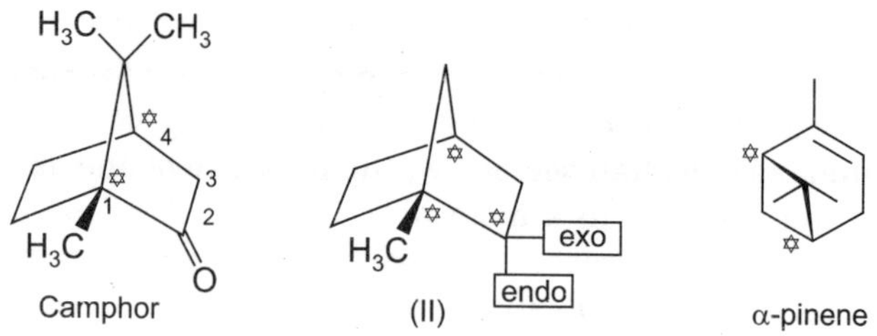

SCHEME 1.68f

1.8 SYMMETRY ELEMENTS, OPERATIONS, POINT GROUPS AND STEREOCHEMICAL PROPERTIES

The phenomenon of rotation of the plane of the polarized light by some compounds is known as optical activity. A molecule which shows optical activity, or can be resolved into optical antipodes, is known as a chiral molecule while one which does not show optical activity is an achiral molecule. A chiral molecule usually, though not always, contains at least one stereocenter *e.g.*, stereogenic carbon. Optical activity can also result from lack of molecular symmetry and molecules which do not have stereocenters, such as some biphenyls, can also be optically active. Two definitions of chirality may be considered; a molecule is described as chiral if it cannot be superimposed on its mirror reflection, or, alternatively, if it does not possess an alternating axis of symmetry. These definitions compliment each other, the first projects a pictorial aspect while the second a mathematical approach.

The first definition can be explored by the use of a set of molecular models and a mirror while the second by the detection of symmetry elements in a molecule and by carrying out symmetry operations on it. A molecule is said to have symmetry elements when certain parts of the molecule can be interchanged with other parts so that after the interchange the appearance of the molecule is indistinguishable from the original. The symmetry operations are the ways of interchanging parts of a molecule.

(A) A Simple or Proper Axis of Symmetry (C_n)

When an imaginary line (axis) can be drawn through a molecule so that rotation by 360°/*n* gives the molecule indistinguishable from the original, then that molecule is said to have a rotation axis (C_n of order *n*).

For example, (*E*)-1,2-dichloroethene has a simple axis of rotation that passes through the midpoint of the molecule and is perpendicular to the plane described by the atoms of the molecule. Rotation through 180° about the axis leads to an arrangement identical to the original.

The (E)-1, 2-dichloroethene is said to have two-fold axis of rotation. The net result of this C_2 operation is interchanging the position of two hydrogen atoms and two chlorine atoms in the (E)-1, 2-dichloroethene molecule. As the hydrogen atoms and chlorine atoms are indistinguishable so the new arrangement is indistinguishable from the old (the H and Cl which were close to one eyes initially shown by thick wedges go away from you, now shown by hatched wedges). If however, one of the hydrogen atoms was replaced by deuterium the molecule would no longer have C_2 axis.

Boron trifluoride has, in a similar way an axis passing through its midpoint, about which a 120° rotation produces an orientation indistinguishable from the original (Scheme 1.69).

The planar forms of cyclobutane and cyclopentane can be shown to have four-fold and five-fold simple axis of rotation, respectively, and benzene has a six-fold (C_6) simple axis of rotation. All linear molecules have a C_∞ axis as in O=C=O and CH=CH an equivalent arrangement is always obtained whatever be the angle of rotation.

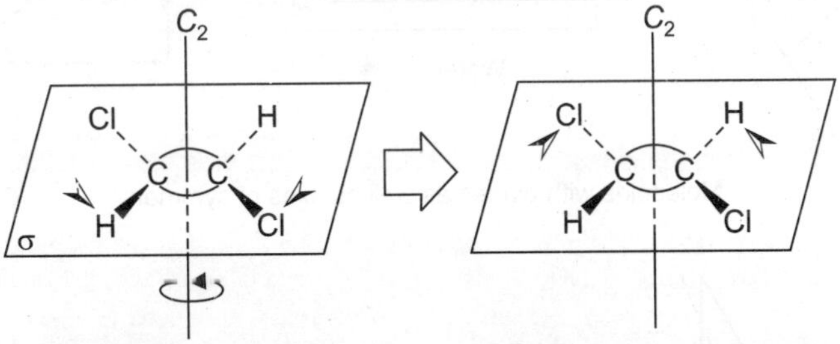

(*E*)-1, 2-Dichloroethene has a 2-Fold axis of symmetry (C_2)

If a rotation by 360°/n of a molecule through an imaginary axis-gives a molecule identical with the original it is said to have n fold simple axis of rotation (C_n).

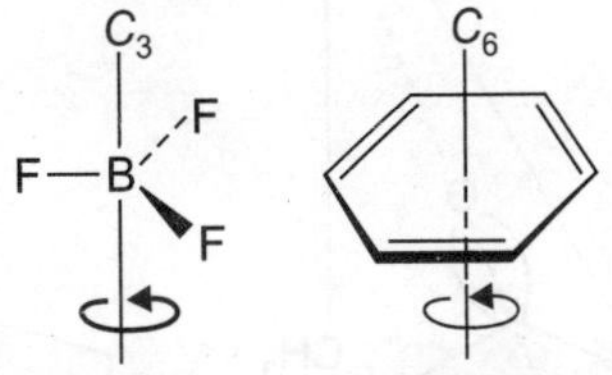

3-Fold axis 6-Fold axis

SCHEME 1.69

(B) Plane of Symmetry (σ)

A molecule possesses a plane of symmetry if:

(*i*) All the atoms of the molecule are in the same plane. For example, (E)-1, 2-dichloroethene possesses a plane of symmetry as well which is the plane of the molecule and includes all the atoms (scheme 1.69). The designation sigma (σ) for planes of symmetry comes from the German word spiegel (meaning mirror).

Thus, all planar molecules posses atleast one plane of symmetry, identical with the molecular plane. Linear molecules have an infinite number of σ planes which intersect along C_∞. The planes of symmetry which are perpendicular to the principal axis are ternmed σ_h (*h* for horizontal) whereas those containing the principal axis are marked σ_v (v for vertical).

(*ii*) A molecule has a plane of symmetry if an imaginary double-sided mirror imagined to be inserted through the molecule reflects both the halves so that the new arrangement is indistinguishable from the original molecule. In other words, if a plane (mirror-plane) can be passed through the molecule, so that it divides the molecule into two symmetrical halves, one half-reflecting the other (Scheme 1.70).

A plane of symmetry is equivalent to a one-fold alternating axis of symmetry (S_1). Some molecules have symmetry planes in addition to simple axes of rotation, while some others have a symmetry plane as their sole symmetry-element (scheme 1.70). Water has two mutually perpendicular σ-planes both containing the C_2 axis and intersecting along it. All planar molecules such as water, must contain at least one symmetry plane which is the molecular plane.

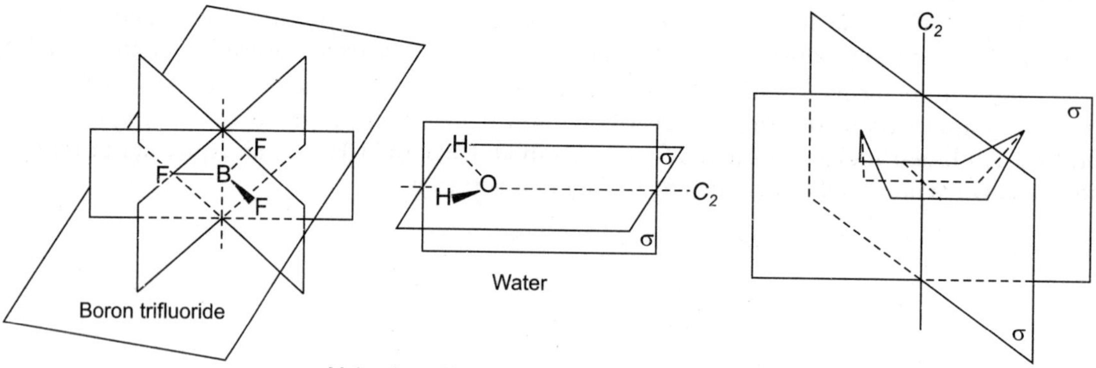

Molecules with planes as well as axes of symmetry

Cyclohexane (boat)

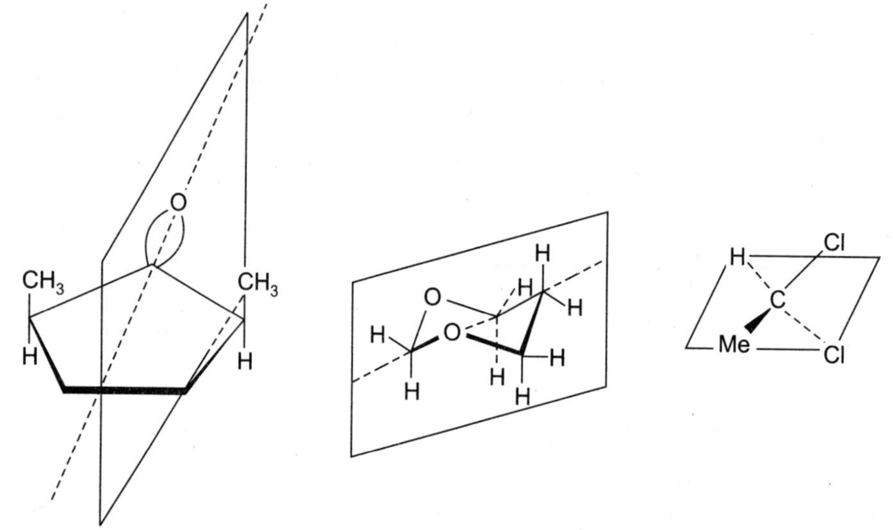

cis-2, 5-Dimethyl-cyclopentanone

1, 3-Dioxane

1, 1-Dichloroethane

Molecules with a plane of symmetry and no axis of symmetry

SCHEME 1.70

If one considers the boat conformation of cyclohexane (scheme 1.70) one can identify two perpendicular mirror planes of symmetry which intersect on the C_2 symmetry axis.

(C) Center of Symmetry (Point of Symmetry)

A molecule is said to have a center of symmetry if all straight lines that can be drawn through the center of the molecule meet identical atoms at equal distance from the center.

One has seen that in an acylic compound a plane of symmetry can be best seen in its eclipsed form, when it is written in Fischer, sawhorse, a Newmann or a wedge projection (see scheme 1.41). One can however, detect a center of symmetry (point of symmetry when such a center is present in the given compound) by writing the staggered conformation. Thus *meso*-tartaric acid is shown to have a center of symmetry in its staggered (*anti*) conformation (scheme 1.41). One has already seen a C_2 axis and a plane of symmetry (σ) in (*E*)-1, 2-dichloroethene (see scheme 1.69). A center of symmetry is equivalent to a two-fold alternating axis of symmetry S_2. The objects with a point of symmetry (center of symmetry) are termed centrosymmetric.

(D) Alternating Axis of Symmetry (Rotation Reflection Axis) S_n

A molecule has an *n*-fold alternating axis of symmetry (S_n) if the molecule when turned 360°/*n* about an axis followed by reflection in a plane perpendicular to that axis brings the molecule in a position indistinguishable from the original.

It turns out that S_1 is the same as plane of symmetry (σ), S_2 is the same as center of symmetry (*i*) and higher descriptors of S_n are rare. Thus 1, 3-dichloro-2, 4-difluoro-cyclobutane (I, scheme 1.71) has a center of symmetry as its only symmetry element. When (I, scheme 1.71) is rotated 180° about the axis passing though the center of the molecule and then reflected in a mirror perpendicular to this axis, an arrangement superimposable on the original is obtained. The multiplicity *i.e.*, the foldedness of the alternating axis is given by the extent of the rotation in Ist step *i.e.*, 360°/180° = 2; thus (I, Scheme 1.71) has a two fold axis (S_2).

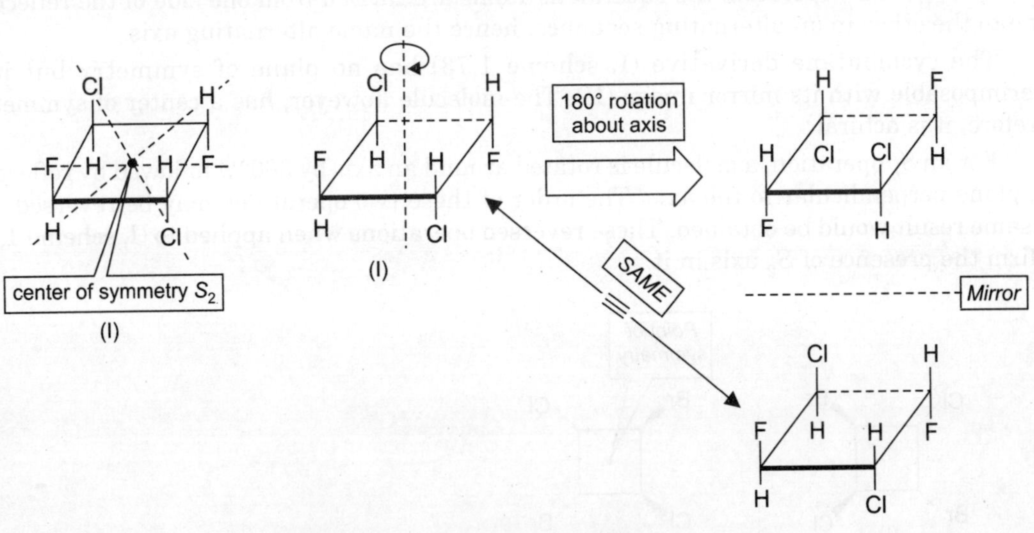

Demonstration of center of symmetry S_2 in (I)

SCHEME 1.71

For detecting an alternating axis of symmetry, the two operations may be reversed and the net result remains the same, as shown for *meso* tartaric acid drawn in sawhorse projection (I*a*, scheme 1.72). The following points may be considered:

- In the sawhorse projection drawn in the eclipsed form (I*a*, scheme 1.72, plane of symmetry is clearly visible) C3 is rotated to get the staggered form (I*b*, scheme 1.72) where now the center of symmetry is also clearly seen.

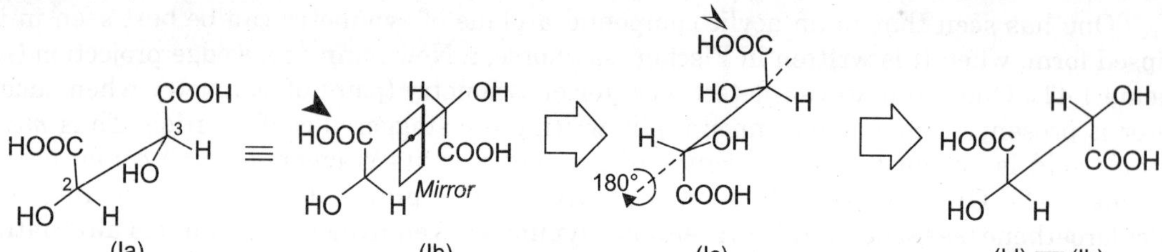

(Ia) (Ib) (Ic) (Id) ≡ (Ib)

Presence of S_2 axis in *meso*-tartaric acid (drawn as sawhorse staggered conformation)

SCHEME 1.72

- Reflect the conformation (I*b*, scheme 1.72) in a mirror placed at the center of the C2-C3 axis and at right angles to it.
- The orientation (I*c*) is thus obtained. The frontal *C*2 carbon shown by a solid thick arrow in (I*b*) will take a position in the rear shown by partly filled arrow in the reflection).
- When (I*c*, scheme 1.72) is rotated 180° around the axis perpendicular to the mirror an orientation (I*d*) is obtained which is identical (superposable) on the original.
- The presence of an S_2 axis is also seen by drawing a wedge projection of *meso*-tartaric acid (see, scheme 1.78).

Under the S_n operation the equivalent atoms are carried from one side of the reflection plane to the other in an alternating sequence, hence the name alternating axis.

The cyclobutane derivative (I, scheme 1.73) has no plane of symmetry but it is superimposable with its mirror image (II). The molecule however, has a center of symmetry, therefore, it is achiral.

For an S_n operation, a molecule is rotated around an axis by 360°/*n* followed by reflection in a plane perpendicular to the axis. The order of these two operations may be reversed and the same result would be obtained. These reversed operations when applied to (I, scheme 1.73) confirm the presence of S_2 axis in it.

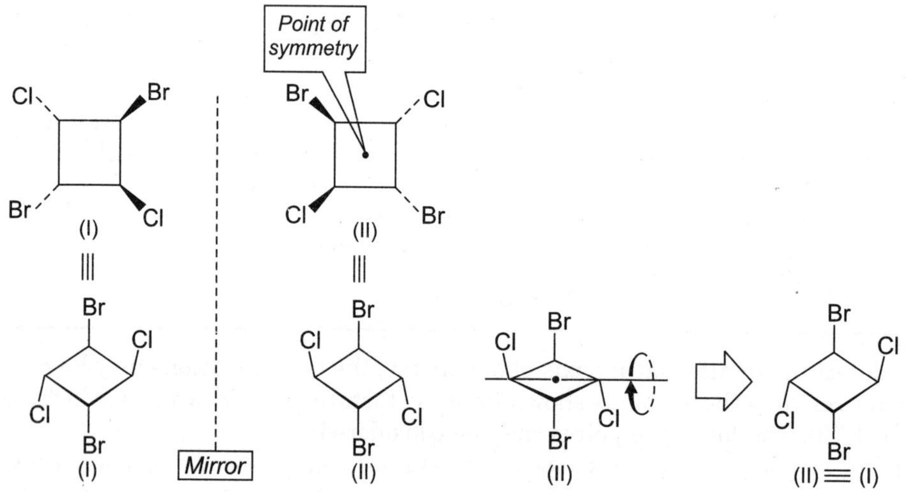

SCHEME 1.73

1.9 SYMMETRY ELEMENTS, ASYMMETRY, CHIRALITY A COMBINED LOOK

(A) Determination of Chirality

An achiral molecule is identical and can be superimposed upon its mirror image, whereas a chiral molecule is one which cannot be superimposed on its mirror image. The property of chirality can thus be described by constructing the models of two molecules under study and look for the existance of non-superimposable mirror image. One may describe the chirality in terms of symmetry elements of the molecule under study. When the molecule has a center of symmetry (C_i) or a plane of symmetry (σ) or a n-fold alternating axis of symmetry (S_n) the mirror images of the molecule are superimposable and the molecule is achiral (optically inactive).

The compound (I, scheme 1.74) is an attracant for the olive fly. The molecule has no stereocenter and no plane or center of symmetry and is chiral. This can be seen from its three dimensional model. The molecule is however, not superimposable on its mirror image. Thus the presence of a stereocenter is a sufficient but not necessary condition for chirality.

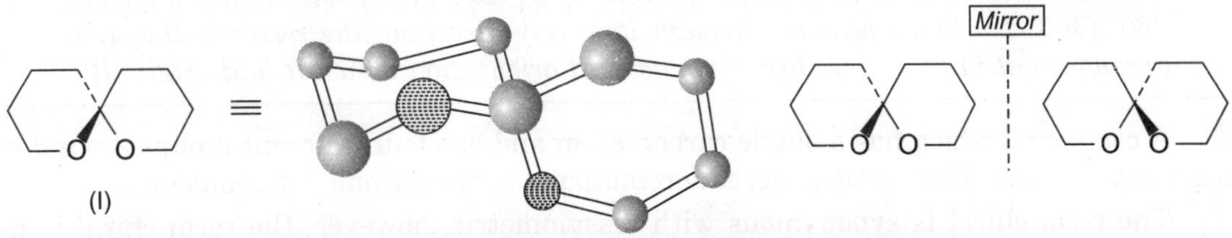

(I)

Mirror image of (I) is not superimposable on it, the molecule has no stereocenter or a center or plane of symmetry

SCHEME 1.74

Care must be used if this approach is adopted; for example, the spiro compound (a, scheme 1.75) possesses neither a center nor a plane of symmetry, but it cannot exist in enantiomeric forms because it possesses a four-fold alternating axis of symmetry (S_4). Indeed (a) is superimposable on its mirror image.

To show that (a) does contain a four-fold alternating axis of symmetry, an orientation of (a, Scheme 1.75) is rotated through 90° about the co-axis of both rings to get (b). Reflection of (b) through the central plane (*i.e.*, through the N atom) perpendicular to this axis yields a molecule identical with (a, scheme 1.75). During the process of reflection the upper half of (b) coincides with the lower half of (a) and the lower half of (b) with the upper half of (a).

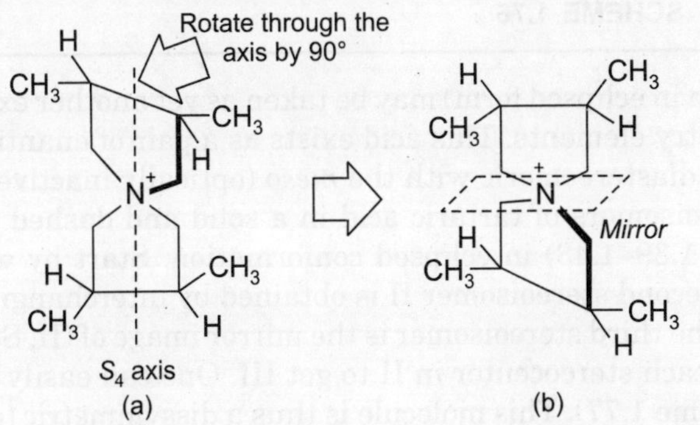

(a) (b)

The spiro structure (as a salt) has no plane (S_1) or center of symmetry (S_2) but is achiral, however. The molecule has a four fold alternating axis of symmetry (S_4)

SCHEME 1.75

It is not a necessary condition for chirality that a molecule should have no symmetry elements. A necessary and sufficient condition for a molecule to have a non-superimposable mirror image is that it must not contain a plane, a centre or a n-fold alternating axis of symmetry. However, these requirements do not preclude the presence of a simple axis of symmetry in a chiral molecule. There are several examples of such compounds *e.g.*, (I and II, Scheme 1.76) with non-superimposable mirror images; (chiral compounds), in which a symmetry element (C_2) is detectable. The allenes provide yet another example of chiral molecules with C_2 axes.

(B) The Terms Chiral, Asymmetric and Dissymmetric

One uses three terms (chiral, asymmetric and dissymmetric) for molecules which show enantiomerism (optical isomerism, *i.e.*, chirality). An asymmetric compound is that (chiral) compound which lacks all but C_1 symmetry. Thus, an asymmetric compound lacks all elements of symmetry but for C_1.

> *All molecules contain an infinite number of C_1 axes i.e., if one rotates anything 360° (360°/1 = 360°) passing through it in any direction, the result is the same arrangement in the same three dimensional orientation which it had originally.*

A compound which has a single carbon atom and has four different groups around it is always asymmetric. Thus bromochlorofluoromethane is an asymmetric compound.

The term chiral is synonymous with dissymmetric, however, the term chiral is now widely used. When the chiral compound has a simple axis of symmetry (usually a C_2 axis) and no other symmetry element is present the compound is then termed dissymmetric. Thus, the chiral molecules (I and II Scheme 1.76) are dissymmetric. The molecule (I scheme 1.76) has no stereocenter, it has no plane or center of symmetry.

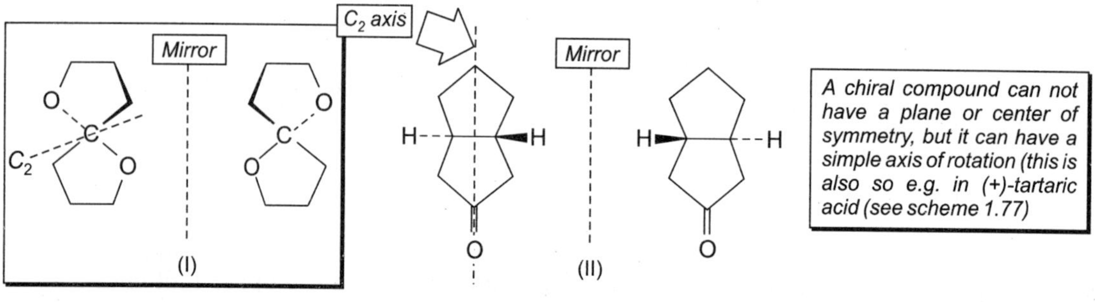

SCHEME 1.76

Tartaric acid (Scheme 1.77, drawn in eclipsed form) may be taken as yet another example to study its chirality in terms of symmetry elements. This acid exists as a pair of enantiomers (I and II Scheme 1.77) each of which is diastereomeric with the *meso* (optically inactive) form (III). It is easy to write the three stereoisomers of tartaric acid in a solid and dashed wedge fonnula (a sort of zig-zag, see Scheme 1.39–1.43) in eclipsed conformation. Start by writing *meso* isomer (plane of symmetry). The second stereoisomer II is obtained by interchanging the position of groups at one stereocenter. The third stereoisomer is the mirror image of (II, Scheme 1.77), so invert the stereochemistry at each stereocenter in II to get III. One can easily detect a C_2 axis in *e.g.*, the enantiomer (I, Scheme 1.77). This molecule is thus a dissymmetric (chiral) molecule. A plane of symmetry is easily detected in the *meso*-form (III). One can also detect a center of symmetry (C_i) in the staggered *meso* form of tartaric acid (Scheme 1.78). That a center of symmetry is equivalent, to a two fold alternating axis of symmetry (S_2) is again proved in (scheme 1.78) by following the usual operations one has already seen the presence of an S_2 axis in *meso*-tartaric acid when drawn in sawhorse projection (see, Scheme 1.72).

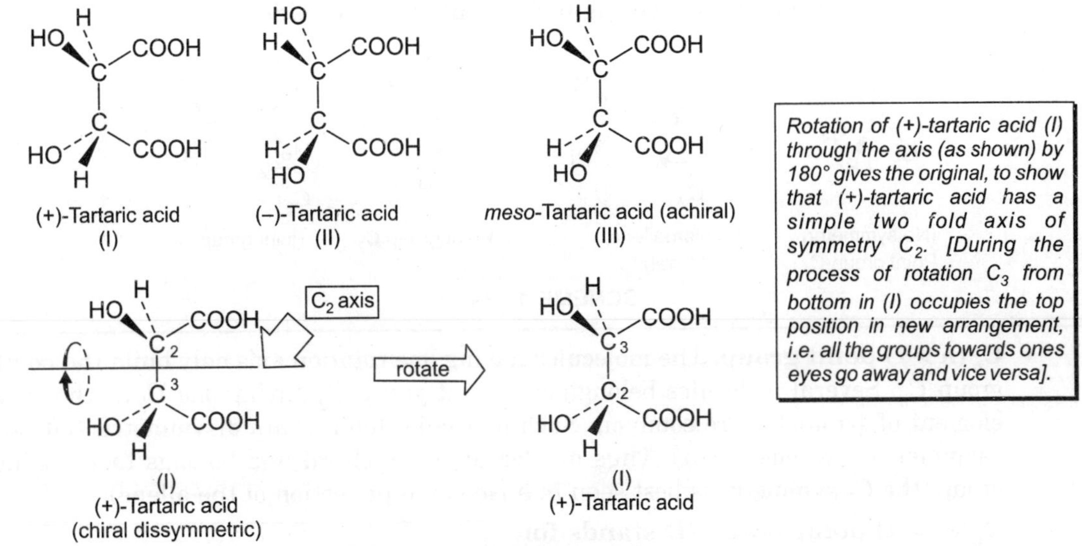

SCHEME 1.77

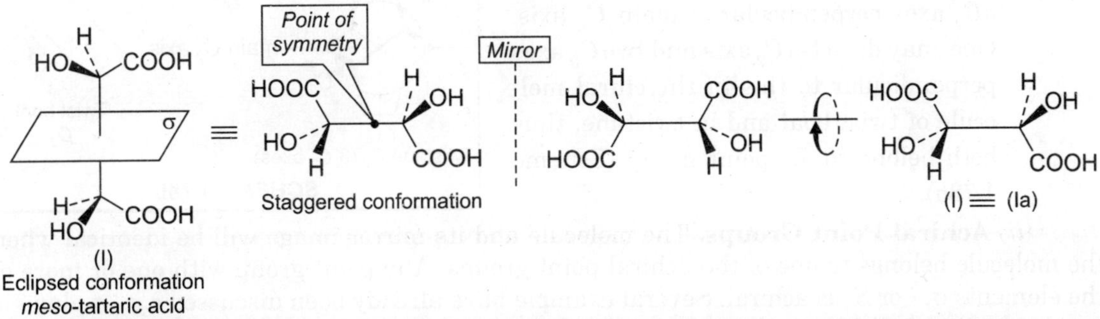

Demonstration of the presence of S_2 axis in *meso* tartaric acid drawn in wedge formula

SCHEME 1.78

(C) Point Group Classification

One may group together the molecules on the basis of the symmetry operations that can be performed on them. These operations are termed point symmetry operations, since under such symmetry operations one point, the center of mass always remains unchanged. Molecules which possess identical symmetry elements, and only these elements, belong to the same group. To be exact these belong to the same point group. The point group of a molecule "X" is the ensemble of the symmetry operations which transform X to a molecule to which it is superimposable. These symmetry operations are dependent on symmetry elements and both terms are essential to define symetry. Point groups can be further classified into two main groups (*i*) structures lacking reflection symmetry and (*ii*) structures possessing reflection symmetry.

(i) Chiral Point Groups

- **C_1 Point group.** When no symmetry element is present in a molecule (expect for C_1), the molecule is then asymmetric and belongs to C_1 point group. All organic molecules with one stereogenic carbon belong to C_1 group (scheme 1.78*a*).

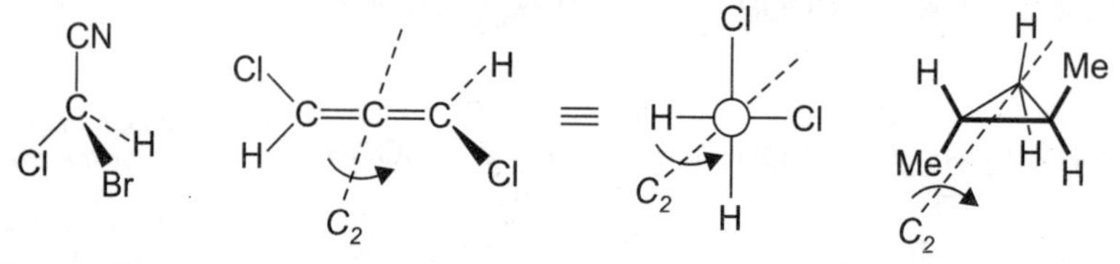

(No symmetry) Point group C_1 — Dichloroallene (chiral) — Point group C_2 — Point group C_2

SCHEME 1.78a

- **C_n (n > 1) point group.** The molecules having one rotation axis only build the point group C_n. Several molecules belonging to point group C_2 having one C_2 as the only element of symmetry are common. Such molecules (chiral) are disymmetric but not asymmetric (scheme 1.78*a*). Thus dichloroallene is chiral and belongs to C_2 point group (the C_2 symmetry is best seen in a Newman projection of the allene).
- **D_n (n > 1) point group (D stands for dihedral symmetry).** In this (chiral) point group in addition to C_n there are nC_2 axes perpendicular to main C_n axis. One may detect-a C_2 axis and two C_2 axes perpendicular to this in the chiral molecule of twist boat and in twistane, thus both belong to D_2 point group (Scheme 1.78*b*).

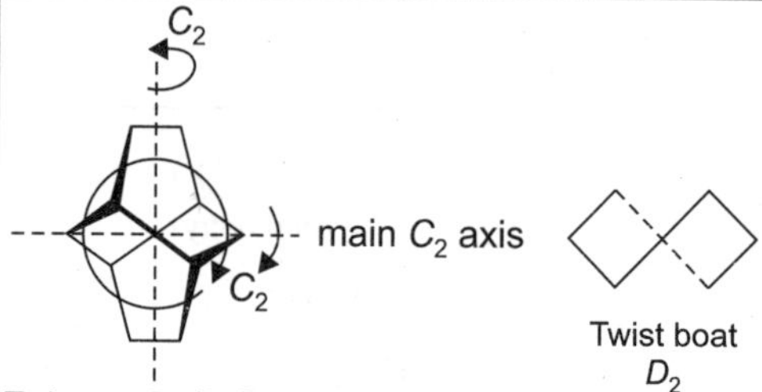

Twistane D_2 (3 C_2 axes) — Twist boat D_2

SCHEME 1.78b

(*ii*) **Achiral Point Groups.** The molecule and its mirror image will be identical when the molecule belongs to one of the achiral point groups. Any point group with one or more of the elements σ, *i* or S_n is achiral. Several example have already been discussed *e.g.*, σ plane is present in *cis*-1,2-dichlorocyclopropane (see, scheme 1.60) point group C_s; center of symmetry *i* is present in (I, scheme 1.73) point *group* C_i; S_n rotation, reflection axis S_4 is present in spiro structure (scheme 1.75) point group S_4.

In several cases (scheme 1.78*c*) achiral point groups may be built out of combination of both σ planes and C_n axes. With one C_n and $n\sigma$ planes intercepting at C_n the point groups are C_{nv} and the planes are σ_v *e.g.*, water and pyridine belongs to C_{2v} and chloroform to C_{3v} point group.

Molecules which have one C_n axis and one σ_h plane (σ_v planes being absent) belong to the group C_{nh} *e.g.*, *trans*, 1-2-dichloroethylene belong to C_{2h} group. Thus (*E*)-1,2-dichloroethene (scheme 1.69) with a C_2 axis perpendicular to the horiozontal σ plane belongs to C_{2h} point group.

When one adds σ planes to the axes present in D_n (C_n axis and nC_2 axes) one leads to D_{nd} and D_{nd} and D_{nh} point groups. When the σ planes intersect along the principal axis, these bisect pairs of C_2 axis and are termed as diagonal in D_{nd} , and an example is of allene (scheme 1.78*c*). In case there is a horizontal plane which is accompanied by *n* vertical σ planes which include rather than bisect the *n* horizontal C_2 axes, then point group is called D_{nh}.

Thus benzene belong to D_{6h} point group since it has one C_6, six C_2, six σ_v and one σ_h, similarly, acetylene has $D_{\infty h}$ symmetry *i.e.*, cylindrical symmetry.

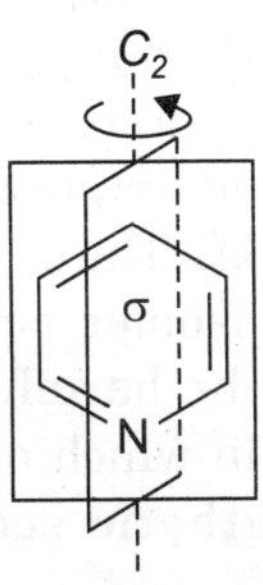

C_{2V} point group. The C_2 axis lies along the intersection of the 2 vertical σ planes.

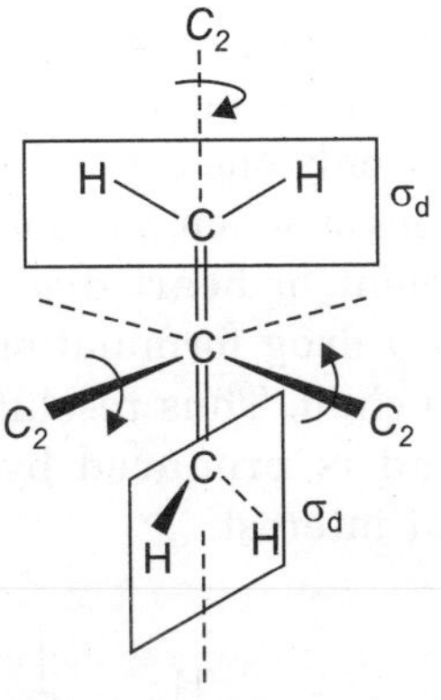

Allene D_{2d} point group. 3 C_2 axes with the 2 σ planes intersecting the principle (vertical) C_2 axis.

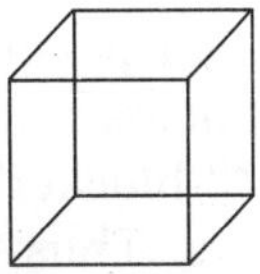

Cubane

O_h point-group. 3 C_4, 4 C_3 and 6 C_2 axes and 9 σ planes

SCHEME 1.78c

Lastly mention may be made of the point groups T_d, which is applicable to regular tetrahedral molecules like, CH_4, CCl_4 etc. and O_h, octahedral point group to which the well known compound cubane belongs (scheme 1.78*c*). The final high symmetry point group is K_h which is applicable to objects having all symmetry elements. Molecules, however, cannot have K_h symmetry and this point group is applicable only to single isolated atoms.

1.10 RESOLUTION OF RACEMIC MIXTURES

It is important not to confuse *meso* compound with racemates both of which do not show optical activity. A *meso* compound is a single achiral substance while racemate is a 50:50 combination of a pair of enantiomers with the mixture having zero optical rotation.

Enantiomers have similar physical and chemical properties in achiral environment while diastereomers display different physical and chemical behaviour even in achiral environments. In a pair of enantiomers the individual components are mirror-images, all the interatomic distances in both molecules remain the same. Physical and chemical properties depend on the atoms in a molecule and their interatomic distances. A pair of diastereomers no doubt contain identical atoms, and moreover each atom is joined to the same atoms in each molecule however, there is a difference in arrangement of the atoms in space. Consequently, diastereomers differ in energy content, and they, therefore, show different chemical and physical properties.

The energy difference between a pair of diastereomers therefore, provides the basis of a method for separating pairs of optical isomers (enantiomers) which are the constituents of a racemic mixture.

The term resolution is used to describe the separation of a racemic mixture into its enantiomeric constituents. One can obtain an optically active compound from a racemic mixture using two methods:

(*a*) via resolution of individual enantiomers from a racemic mixture and

(*b*) by carrying out an asymmetric synthesis using a chiral reagent or a catalyst.

The chiral compounds are very important substances. A large number of natural products, medicinal compounds and biomolecules exist as single optically active stereoisomers. Significantly, the opposite enantiomer or diastereomer may not have any physiological activity and may infact have a detrimental physiological effect. (+)-Morphine is a powerful pain killer (analgesic) while its enantiomer is not. Chloramphenicol (chloromycetin) has two stereocenters and of the four possible stereoisomers, only one acts as an antibiotic. Another simple drug for which enantiomers have different modes of action is propranolol (Scheme 1.78*d*). The (*S*) isomer acts as a "β-blocker" for the treatment of heart disease, while the (*R*) isomer acts as a contraceptive. Thus the presence in a drug formulation (*R*) isomer could be harmful for a patient who was trying to conceive a child. Thus resolution and reactions in which only one stereoisomeric form of a compound is produced by a particular synthetic sequence (stereoselective reactions) are of great interest.

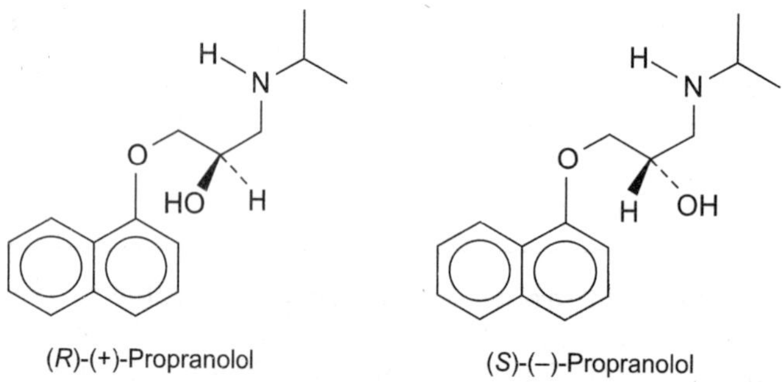

Two enantiomers have different mode of action, (S)-isomer is a β blocker for the treatment of heart disease while (R)-isomer is a contraceptive.

SCHEME 1.78d

The individual enantiomers have identical physical and chemical properties and can only be distinguished by a chiral environment. Plane polarized light is such a chiral environment and one enantiomer is dextrorotatory and one is levorotatory. The diastereomers are however, different substances and have different properties. The resolution involves the conversion of a pair of enantiomers of a racemic mixture into a pair of diastereomers with an optically active reagent. The diastereomers are then separated by making use of their different properties *i.e.*, crystallization, solubility, chromatography etc.

The optically active reagents used for resolution are mostly obtained from biological sources where these occur as only one enantiomer. Thus, though brucine has six stereocenters and can theoretically exist as $2^6 = 64$ stereoisomers nature only makes one enantiomer (–)-brucine. Instances are there when more than one stereoisomer is found but these rarely exist together in the same biological system. Fortunately nature does not waste its precious energy and resources unnecessarily, only one of the stereoisomers is produced and utilized by any given organism. The chiral substances react only with substances that match their own chirality (two enantiomers of carvone have different odors, see scheme 1.2c).

Each diastereomer is then treated separately with a second reagent to regenerate the resolving reagent and the pure enantiomer.

The following methods can be employed for the resolution of racemic mixtures.

(A) Mechanical Separation—Crystallization Method—Resolution via Conglomerates

Direct crystallization of an optically active compound from a racemate may be achieved provided the crystals of the latter form as conglomerates.

> *A conglomerate is an exactly 50 : 50 mixture of crystals of pure (+)-enantiomer with crystals of the pure (–)-enantiomer. When crystallization of a racemate leads to the formation of a conglomerate, then by definition one has achieved spontaneous resolution.*

In very rare cases the two enantiomers of a racemic mixture (racemate) will crystallize separately. The racemic sodium, ammonium salt of tartaric acid crystallized by slow evaporation of an aqueous solution is a conglomerate and Pasteur could pick out the crystals of (+)- and (–)-enantiomers, since the crystals have enantiomeric relationship (Scheme 1.79).

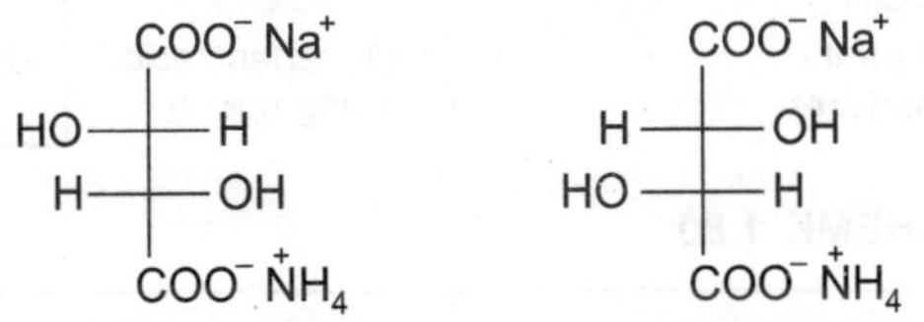

Sodium ammonium tartrate *left-handed crystals* Sodium ammonium tartrate *right-handed crystals*

Rarely, (as in the case of sodium ammonium tartrate), the two enantiomers of a racemic mixture may crystallize separately and then separated mechanically.

SCHEME 1.79

Pasteur's work was a unique combination of sensitive judgement, a firm hold on the problem, good luck and hard work. Pasteur's salt was a conglomerate only at temperatures below 25°C (work carried out in Paris).

(B) Resolution Through Preferential Crystallization

Biologically active α-methyl-L-dopa *e.g.*, (for structure, see Scheme 2.23*a*) is obtained in optically active form by circulating a supersaturated solution of the racemate through two chambers each of which contains seed crystals of the respective enantiomers. One may also bring about crystallization of one of the enantiomers of a racemate while the other is kept in a supersaturated state. If the crystal of one form is not available it is sometimes possible to induce selective crystallization by seeding with a crystal of an optically active form of another molecule.

(C) Resolution Through the Formation of Diastereomers

A racemic mixture of an alcohol *e.g.*, 2-butanol reacts with pure (+)-tartaric acid (the resolving agent as a naturally occurring pure enantiomer) to give diastereomeric esters (scheme 1.80). These esters are separated by chromatography. The resolving agent is then cleaved from the separated enantiomers of 2-butanol by hydrolysis to give (*R*)-2-butanol and (*S*)-2-butanol in their pure form. The recovered (+)-tartaric acid would probably be thrown away since it is cheap (available from any winery) and nontoxic.

Many other chiral resolving agents are expensive which can be recovered and recycled. Organic acids (*e.g.*, tartaric acid) and bases (*e.g.*, brucine) are important groups of naturally occurring compounds used as resolving agent (scheme 1.81).

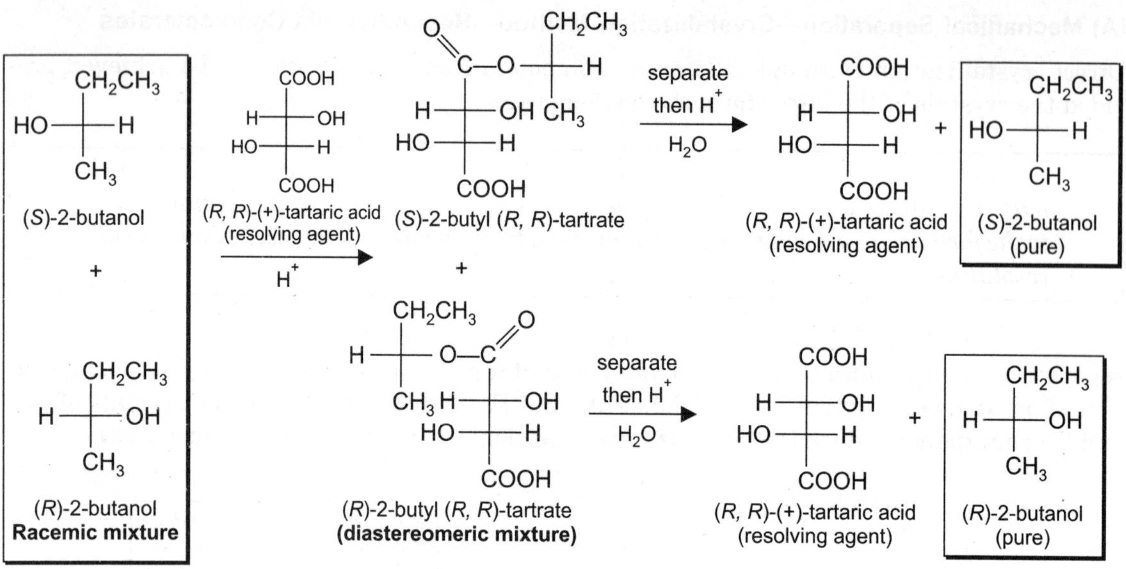

SCHEME 1.80

The most important criterion is the proper choice of the resolving agent. Thus a racemic mixture of organic acids is often separated into its pure enantiomers by using the naturally occurring (optically active—one enantiomer) alkaloid brucine (a base). The diastereomeric mixture of salts formed initially is separated by crystallization from which pure enantiomeric acids are separated by treatment with HCl (Scheme 1.82).

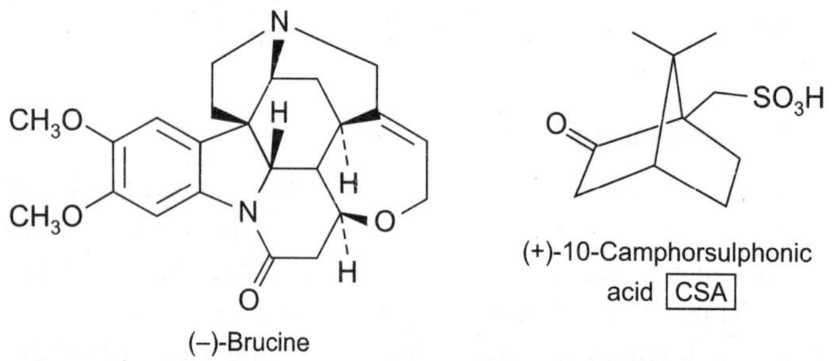

Most resolving reagents are naturally occuring. Nature makes only one enantiomer, (even though several stereocenters may be present). Racemic mixture on reaction with this pure enantiomer gives a mixture of separable diastereomers.

SCHEME 1.81

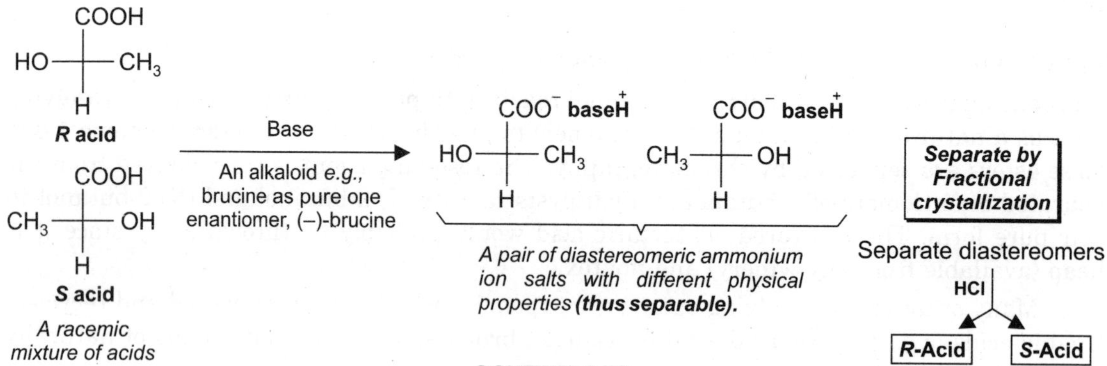

SCHEME 1.82

A racemic alcohol (neutral compound) may also be resolved by first attaching an "acidic" handle. The alcohol is first converted into ; a half ester of phthalic acid by heating with phthalic anhydride. The half ester is then typically resolved as an acid by using *e.g.*, the optically active alkaloid brucine (scheme 1.83). The optically active enantiomers of half-acid-half ester are then hydrolyzed under basic conditions to give two optically active enantiomers of the alcohol.

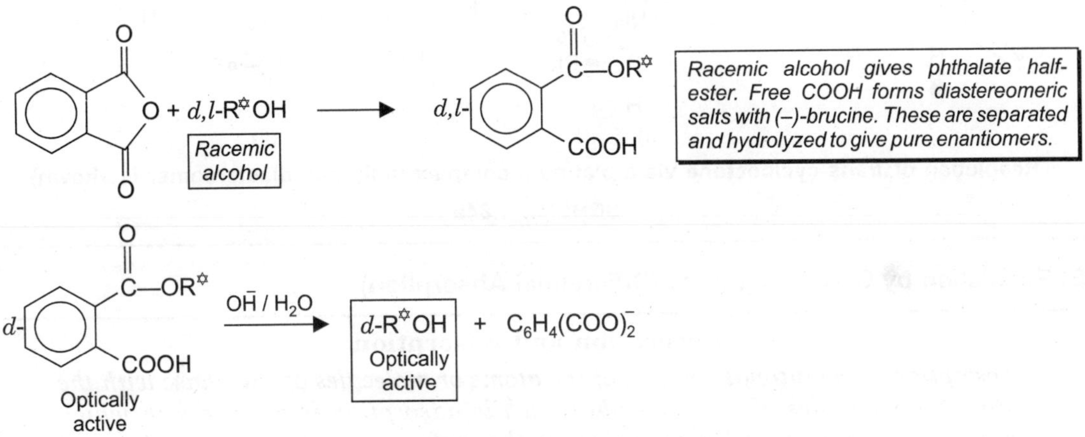

SCHEME 1.83

(D) Resolution Through the Formation of Molecular Complexes

Instead of forming stable salts or covalent compounds between the substrates (a racemic mixture) and the resolving reagents, it has now become more and more common to use weak interactions (chromatography, complexes) as a means of resolving enantiomers. In some cases it is possible to make molecular complexes. These complexes should be easy to form and easy to decompose. For example the optically active fluorenone derivative (I, scheme 1.84) TAPA has been used to resolve hexahelicene. (–)-TAPA and (–)-hexahelicene crystallize whereas the diastereomeric π-complex of (+)-hexahelicene remains in solution.

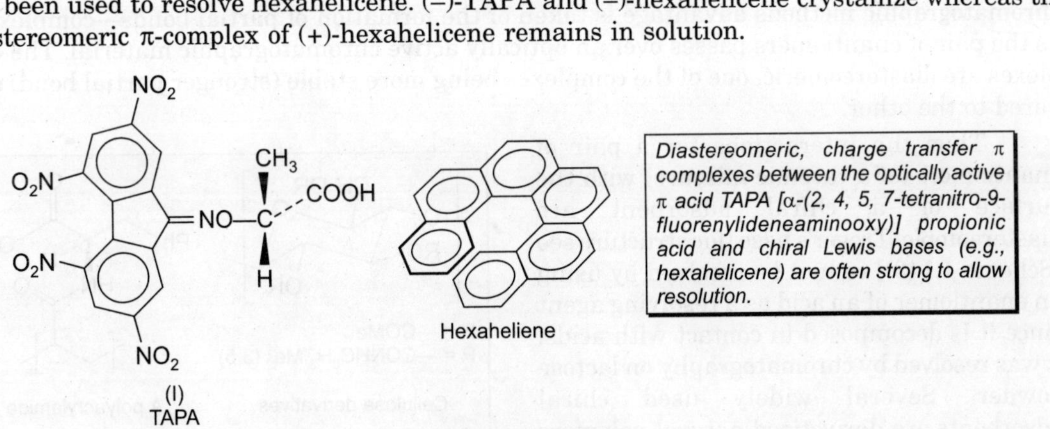

SCHEME 1.84

Compounds in which conventional functionality is absent, like chiral alkenes, arenes, sulfoxides and phosphines can be resolved by the use of a metal centered complex to give diastereomeric derivatives with *e.g.*, a recemic alkene like *trans*-cyclooctene. The coordination of the twisted Carbon-carbon double bond of *trans*-cyclooctene is more stronger than ethene with platinum in the square planar complex (I, scheme 1. 84*a*) containing a chiral unit *S*-1-phenylethyl amino moiety. After the diastereomeric complexes are separated, their treatement with KCN is attendend with the displacement of a pure enantiomer (from *e.g.*, II scheme 1.84*a*) by CN^-.

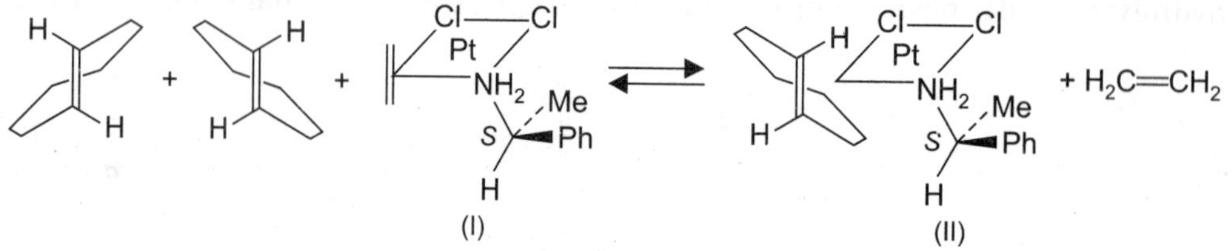

Resolution of *trans*-cyclooctene via a platinum complex (only one diastereomer is shown)

SCHEME 1.84a

(E) Resolution by Chromatography (Differential Absorption)

> **Absorption and Adsorption**
>
> *Absorption is the intimate mixing of the atoms or molecules of one phase with the atoms or molecules of a second phase, while adsorption is a process in which molecules (either gas or liquid) adhere to the surface of a solid.*

Component enantiomers of the racemic mixture can be separated by chromatography on an optically active support *e.g.* silica in which surface hydroxyl groups have been derivatized by treatment with (*R*)-phenylglycine. The diastereomeric adsorbates which are formed have different stabilities, thus one enantiomer will be held more tightly than the other and will pass through the chromatographic column more slowly than the other which would be eluted first. As a second example racemic mandelic acid is resolved by column chromatography over starch. In these chromatographic methods when the column is packed with a chiral material, the enantiomers move along the column at different rates and are separated without having to be converted to diastereomers *i.e.*, chemical bonds are not formed as in (scheme 1.82). In such chromatographic methods advantage is taken of the formation of partial bonds—complexes—as the pair of enantiomers passes over an optically active chromatographic material. The complexes are diastereomeric, one of the complexes being more stable (stronger partial bond) compared to the other.

Thus the interactions for a pair of enantiomers (of a racemic mixture) with the surface of a chiral adsorbent are diastereotopic. Troger's base [for structure see (Scheme 1.143)] cannot be resolved by using an enantiomer of an acid as a resolving agent since it is decomposed in contact with acids, it was resolved by chromatography on lactose powder. Several widely used chiral adsorbents are derivetized natural polymers *e.g.*, of cellulose (Scheme 1.84*b*) and other fully synthetic chiral polymers.

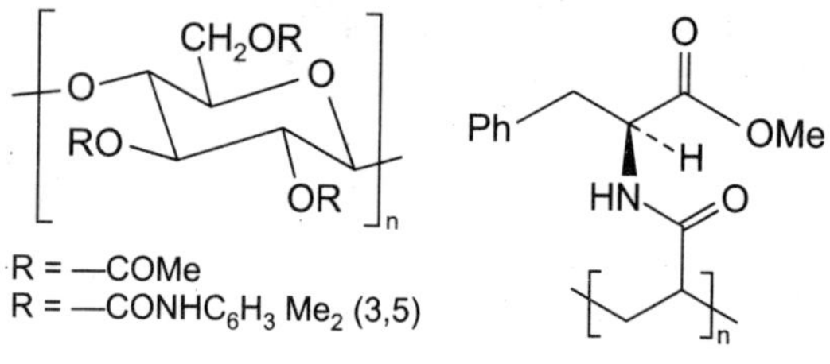

Cellulose derivatives　　A polyacrylamide

SCHEME 1.84b

(F) Biochemical Methods of Resolution—Use of Enzymes

The laboratory reduction of 2-butanone (an achiral substrate) gives only a racemic product of 2-butanol. However, the reduction with an enzyme (a chiral reagent) gives an optically pure

(R)-2-butanol (scheme 1.85, this is a typical example of an asymmetric synthesis, also see scheme 2.60a).

The mould *Penicillium glaucum*, preferentially destroys the (+) isomer of racemic ammonium tartrate while the (–) isomer is left intact. This is an example of asymmetric destruction.

One may like to learn here an example of asymmetric induction as well (this asymmetric induction (Scheme 1.84c) is infact one of the procedures to carry out asymmetric synthesis detailed in chapter 2. Thus achiral (I, Scheme 1.84c) on reaction with optically active (II, N-benzyl-3-(hydroxymethyl)-4-methyl-1, 4-dihydropyridine] gives (S)(+)-mandelic acid (97.5%) while (R) (–)-mandelic acid is produced to the extent of only 2.5%. In this process where a chiral reagent (II) has given up its chirality to another (i.e., I) while (III) becomes achiral is termed self immolative.

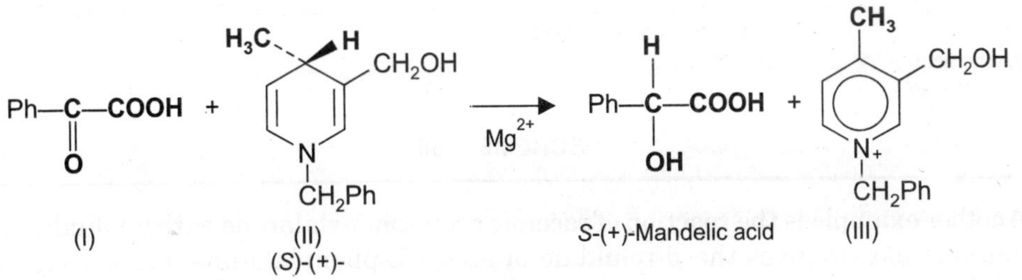

SCHEME 1.84c

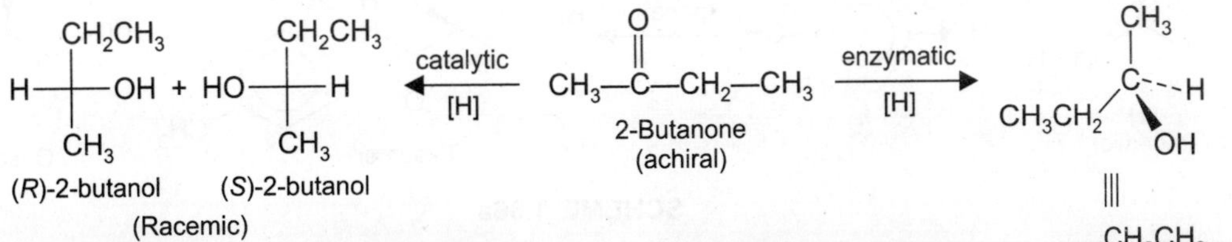

2-Butanone on hydrogenation gives chiral but racemic 2-butanol. Addition of H_2 to one face gives the (R) product, while addition to the other face gives the (S) product. In fact the two faces of a carbonyl compound R_1–CO–R_2 ($R_1 \neq R_2$) are enantiotopic. Achiral reagents donot differentiate between these faces. The chiral reagents like enzymes however, do this distinction and a prochiral ketone is thus reduced to an optically active secondary alcohol.

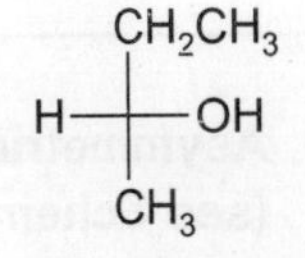

(R)-2-butanol
(pure enantiomer)

SCHEME 1.85

The use of enzymes to resolve enantiomers has become very popular only very recently. The enzymes which are chiral materials display high selectivity for one enantiomer of the racemic mixture. Enzymes are soluble in aqueous solution, thus it is practically very difficult to have reasonable concentration of organic substrates in the aqueous medium to achieve resolution at good rate. However, a variety of esterases lipases function well in organic solvents and thus the major stumbling block is removed.

It is easy to acetylate a racemic alcohol and treat the racemic esters with a lipase. One enantiomer is hydrolyzed to the alcohol while the other remains as an ester. These are then separated by chromatography. This recent method of resolution of a racemic alcohol is gaining importance and is far superior to the technique discussed under Scheme 1.83.

The biochemical methods named kinetic resolution have found important application to resolve (±)-amino acids. A (±)-amino acid is acylated and acylated racemic mixture is treated with an enzyme "acylase 1" (hog kidney acylase). This enzyme is capable of hydrolyzing amide links (– NHAc) of L-amino acids only (stereospecificity) (scheme 1.86). Thus at the end a free L-enantiomer with the D-enantiomer which is still acetylated is obtained. This mixture is now easily separated by usual methods. Because the resolution of the enantiomers depends on the difference in the rate of reaction of the enzyme with the two *N*-acetylated compounds, the method is called kinetic resolution.

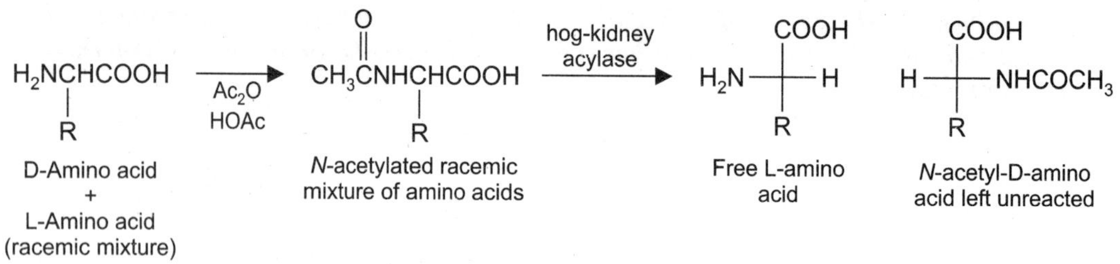

SCHEME 1.86

Another example is the reaction of racemic acetylphenylalanine with *p*-toluidine catalyzed by the enzyme papain gives the *p*-toluidide of acetyl-L-phenylalanine and leaves unchanged D-phenylalanine (scheme 1.86*a*).

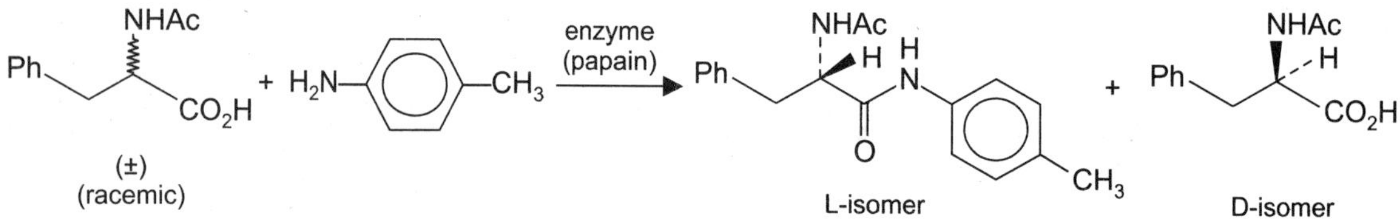

SCHEME 1.86a

(G) Asymmetric Transformation of Racemates and Total Spontaneous Resolution (see Schemes 1.109b–1.109d)

(H) Resolution via Chiral Recognition and Inclusion Compounds

(*i*) Use of chiral crown ethers

A chiral host can form diastereomeric inclusion compounds and often it is possible for the host to form an inclusion compound with one enantiomer of a racemic guest but not with the other—a process termed *chiral recognition i.e.*, when only one enantiomer fits into the chiral host cavity, while the other does not. Generally, both diastereomers may be formed, however, one is formed more readily than the other. If one removes the guest, it has already been partially resolved. Thus chiral crown ether (chirality due to restricted rotation, Scheme 1.86*a*) partially resolves *rac*-methyl phenylglycinate HPF_6. When the aqueous solution of *rac*-amine salt is mixed with a chloroform solution of optically active crown ether (Scheme 1.86*a*) followed by separation of layers, the chloroform layers contained about twice as much of the complex formed between the chiral crown ether and (*R*) amine salt. (In inclusion compounds there is no bonding between the host and the guest except van der Waals forces).

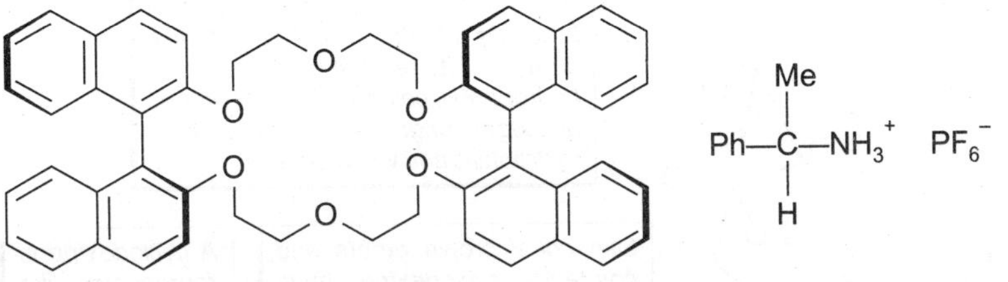

A **chiral** crown ether

rac-methyl phenyl glycinate HPF_6

Inclusion of one enantiomer may be preferential in the chiral cavity of crown ether. Diastereomeric inclusion compounds may have difference in stability or solubility.

SCHEME 1.86a

In summary the following points may be noted :

- Preferential inclusion of one of the enantiomers may be due to favourable hydrogen bonding and van der Waals attraction.
- Proper match between the size and shape of the guest molecule with that of the host cavity (*e.g.*, the size and shape of the guest molecule can vary with a conformationally flexible molecule, there may also be play in the space in the cavity).
- The diastereomeric inclusion compounds are separable due to their differential stability or solubility.

(*ii*) Use of cyclodextrins

Cyclodextrins (natural origin, chiral) are made up of six, seven or eight glucose units connected in a large ring are termed respectively α, β or γ cyclodextrins. These have a shape of a tub with primary OH groups projecting from the narrow side of the tub while the secondary OH groups from the wide side (Scheme 1.86*b*). The cyclodextrins are used for the resolution of a variety of compounds *via* inclusion complexes.

One distinguishes two broad classes of inclusion compounds. The type discussed in above two examples involving chiral crown ethers and cyclodextrins is termed 'Cavitates' in which the guest molecule i.e., the chiral substrate to be resolved is partially or entirely enclosed within a second chiral substance having a chiral cavity. The second type is termed clathrates where the guest molecules are surrounded by several molecules of the resolving agent to form a cage or channel.

(*iii*) Use of TOT in resolution

TOT (tri-*O*-thymotide) (Scheme 1.86*c*) provides a remarkable recent example of resolution by inclusion compound formation. TOT crystallizes in the absence of any guest (a solvent) as a racemic compound [(*P*)-(+)-right handed propeller and (*M*)-(–)-left handed propeller]. In the presence of a guest (*e.g.*, benzene) it however, crystallizes as a conglomerate (formation of clathrate inclusion complexes). In such a conglomerate the crystals may have TOT in either the (*P*)-(+)- or (*M*)-(–)-form and consequently the cavities become chiral. The guest containing crystals on dissolution separately at room temperature lose their optical activity due to interconversion (enantiomerization) of TOT enantiomers via bond rotations. Thus the channel clathrates between TOT and chiral guests are used in resolution.

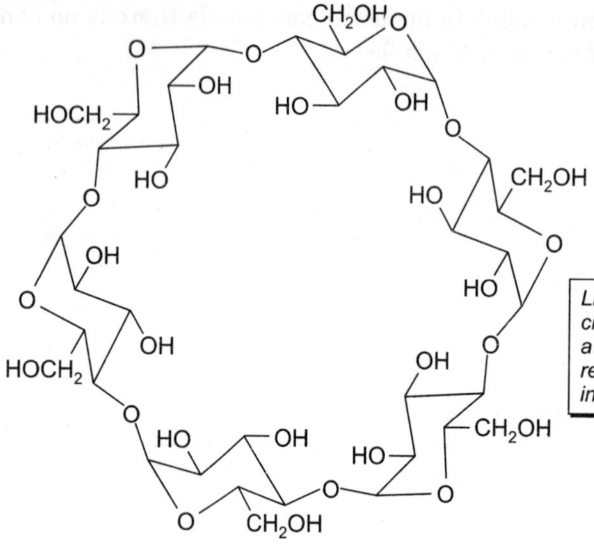

α-Cyclodextrin

> The cyclodextrins are macrocyclic glucose oligomers with six (α) seven (β) or eight (γ) glucopyranose units linked α-1,4. These are degradation products of starch with a tub like shape. These have non polar, hydrophobic cavities. The OH groups of sugar rings make the exterior hydrophilic and water soluble.

> Like chiral crown ethers and cryptands, cyclodextrins have also been employed for resolution via crystalline 1 : 1 inclusion complexes in solution

> A cyclodextrin is represented like a pale type figure

SCHEME 1.86b

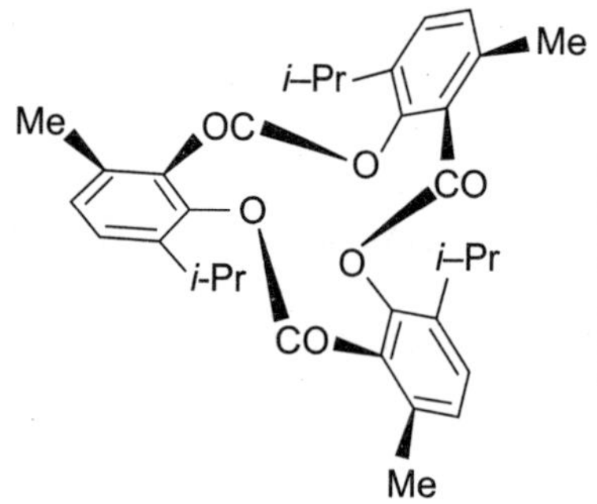

TOT (tri-*O*-thymotide)

> TOT (tri-O-thymotide) is racemic when guest free. In the presence of a guest it forms clathrate inclusion complexes which are conglomerates, with TOT as (P)-(+)-right handed propeller or (M)-(–)-left handed propeller.

SCHEME 1.86c

(*iv*) Use of other chiral hosts in resolution via inclusion compounds

The alkaloid brucine (see Scheme 1.82) has been used to resolve chiral CHBrClF via complexation (CH...N bonding is relatively stable). Both the enantiomers have been isolated via dissolution of the complex in acid.

Preresolved 2,2′-dihydroxy-1, 1′-binaphthyl (for structure see scheme 1.125) acts as an inclusion compound host in the resolution of sulphoxides, phospine oxides and phosphinates.

(l) Other Methods of Resolution

Optically active diisopinocamphenyl borane (see scheme 2.39) can be used to resolve racemic olefins. The reagent adds to one enantiomer and the other is unchanged. Chiral allylic alcohols have been resolved with chiral epoxidizing agents made from tartarate complexes of titanium (see scheme 2.52). One enantiomer is epoxidized while the other remains unchanged.

1.10A MEASUREMENT OF OPTICAL ACTIVITY

Light is a wave phenomenon in which vibrations take place at right angles to the direction in which the light travels. Infinite number of planes pass through the line of propagation and in ordinary light vibration take place in all these planes. Plane polarized light is light in which vibrations take place in only one plane, and this is realised by passing ordinary light through a polarizer which forms an important component of a polarimeter (Fig. 1.3).

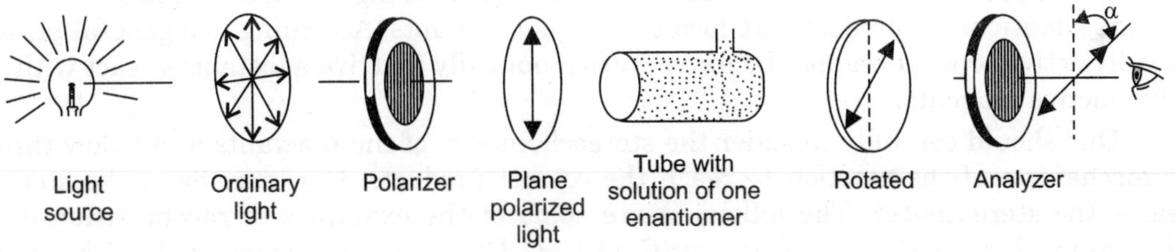

Fig. 1.3

An optically active compound is one which rotates the plane of polarized light. An optically active substance which rotates the plane-polarized light to the right (clock-wise) is said to be dextrorotatory often abbreviated by the letter *d*, and its optical rotation is given a (+) sign. A substance is said to be levorotatory (often abbreviated by the letter *l*) if it rotates plane-polarized light to the left (counter-clockwise), and its optical rotation is given a (–) sign.

Optical rotation is a function of concentration, sample thickness, temperature, wavelength of polarized light, etc. It is usually recorded in the literature in terms of specific rotation $[\alpha]_{\lambda}^{t}$.

where t = temperature measurement in °C

λ = wavelength of polarized light
(usually sodium D line, 5893 Å)

α = observed angle of rotation in degrees

l = sample thickness in decimeters

c = concentration of solution in g/100 ml

1.11 RACEMIC MIXTURE AND RACEMIZATION

Recemic mixture or a recemate is an equimolar mixture of two enantiomers. Since a recemic mixture contains equal numbers of *dextrorotating* and *levorotating* molecules, the net optical rotation is zero. A racemic mixture is often symbolized by (±) or (*dl*).

The process whereby a pure enantiomer is converted into a racemic mixture is called racemization. Racemization may be accomplished in a trivial sense by simply mixing equal amounts of two pure enantiomers. Racemization may also result from the following chemical interconversions.

(A) Formation of Enantiomers (Recemization During Reactions That Create Stereocenters)

Several organic reactions can yield a chiral product from an achiral starting material. Thus the addition of hydrogen to the carbon-oxygen double bond of 2-butanone (achiral compound) in the presence of a catalyst creates a stereocenter. The two enantiomers of 2-butanol (achiral compound) however, are produced in equal amounts—the product is racemic (see, Scheme 1.85). The hydrogen has exactly an equal chance of attacking above or below the plane of the double bond in 2-butanone molecule (see, Scheme 2.16). There is no reason why the hydrogen should prefer one approach over the other ; as long as there is nothing else that is chiral in the reaction, the enantiomeric products are formed in equal amounts. According to a general principle optically active products cannot be formed when optically inactive substances react with optically inactive reagents.

One should carefully consider the stereochemistry of the reactants and follow through the mechanism of the reaction to reach the correct products. One may focus the step that creates the stereocenter. The following are some of the examples of racemization during reactions in which neither the reactants (C=O, C=C, C^+ etc.) nor the reagents (H_2, Br_2, etc.) are chiral. Addition to either face is equivalent to produce a racemic mixture (these processes are described as having no enantioselectivity).

(*i*) Mechanism Involving a Radical Intermediate—Bromination of Butane

The radical bromination of butane (achiral molecule) at C2 yields a chiral molecule (scheme 1.87), however, the product is obtained in a racemic form. Abstraction of either methylene hydrogen at C2 by bromine gives an achiral radical, Reaction of Br_2 with this radical is equally likely at either the top or the bottom face, a situation which gives a racemic mixture of products (the C2 hydrogens are enantiotopic).

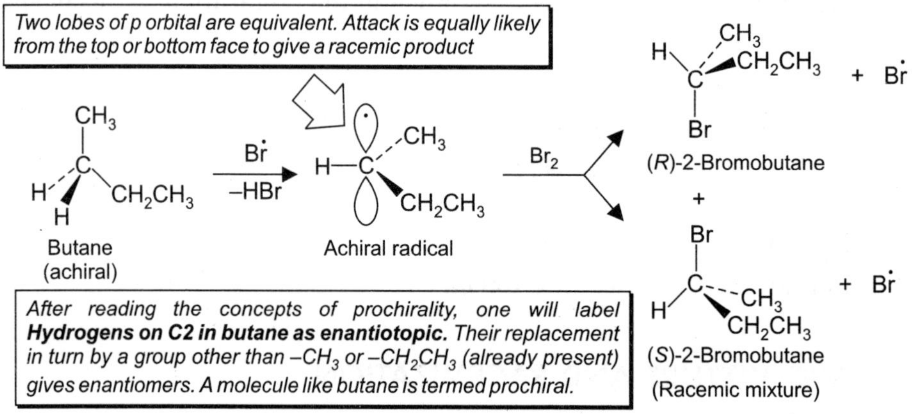

SCHEME 1.87

(*ii*) Mechanism Involving a Carbocation—Addition of HBr to 1-Butene

This addition proceeds via the intermediate formation of a planar carbocation (scheme 1.88) to give again a racemic mixture of products.

(*iii*) Mechanisms Involving Stable Cyclic Intermediates—Syn Addition to Diastereomeric Substrates

The *syn*-hydroxylation of maleic acid gives *meso* tartaric acid [see, scheme 1.40 where attack from one face is shown, the attack from the other face of the double bond is equally possible and takes place, to give again the same *meso* tartaric acid (the two faces of maleic acid are equivalent *i.e.*, homotopic)]. Infact the *syn*-hydroxylation proceeds through the cyclic osmate ester which in the case of maleic acid (scheme 1.89) is achiral having a σ plane. Attack on either face of the double bond in maleic acid gives the same achiral intermediate. Its decomposition directly leads to the product and thus the resulting tartaric acid must also be achiral *i.e., meso*. Fumaric acid on the other hand can give two enantiomeric cyclic intermediates (I and Ia, with C_2 axis scheme 1.90) in equal amounts and since the stereochemistry is again maintained in the product, the tartaric acid thus obtained is chiral but in the form of a racemic mixture of two optically active enantiomeric forms.

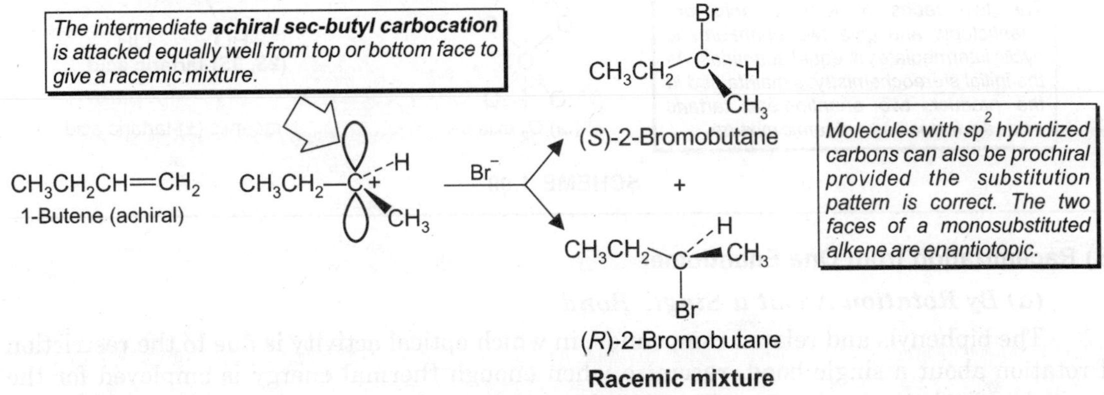

SCHEME 1.88

This result may well be compared with e.g., bromination of E-and Z-2-butenes where a Z alkene gives a racemic mixture while the E isomer gives the meso compound, this being the result of anti addition (see scheme 1.100). During syn hydroxylation, the initial geometry of the alkene is maintained in the product, during bromination it is maintained up to the formation of bromonium ion only and disturbed during its S_N2 opening of the bromonium ion.

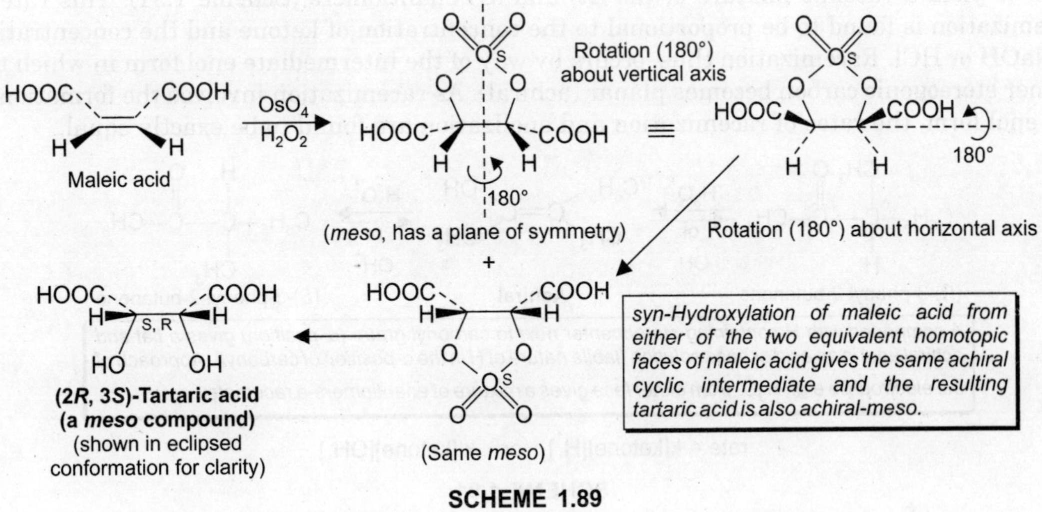

SCHEME 1.89

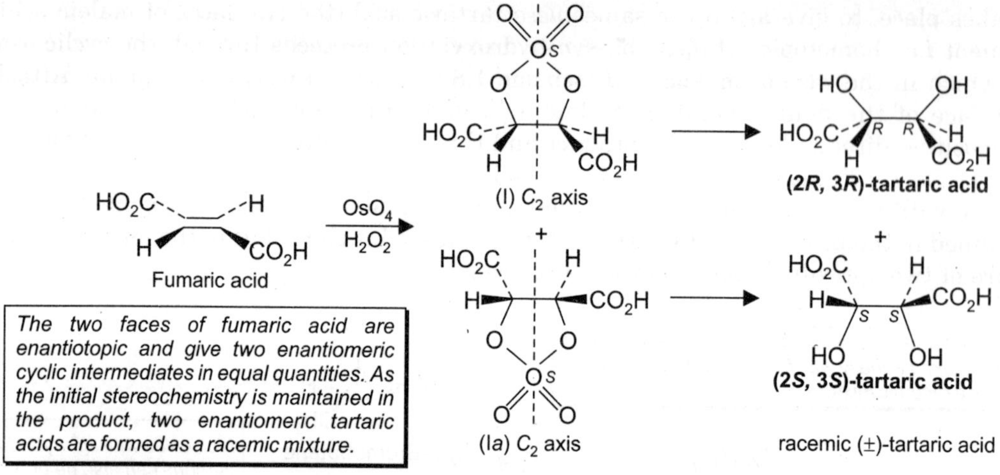

The two faces of fumaric acid are enantiotopic and give two enantiomeric cyclic intermediates in equal quantities. As the initial stereochemistry is maintained in the product, two enantiomeric tartaric acids are formed as a racemic mixture.

SCHEME 1.90

(B) Racemization from One Enantiomer

(a) By Rotation About a Single Bond

The biphenyls and related compounds, in which optical activity is due to the restriction of rotation about a single bond, racemize when enough thermal energy is employed for the energy barrier between the enantiomers to be surmounted at a practicable rate (see Scheme 1.118). Amine inversion is yet another example of recemization.

Cyclic compounds which exist in enantiomeric conformations *e.g.*, *cis*-1, 2-dimethylcychlohexane (see Scheme 4.33) undergo racemization *via* ring inversion and apparently do not involve any achiral intermediate or transition state.

(b) Via an Enol or Enolate Anion

Racemization occurs in those compounds, in which a carbonyl function is attached to a stereocenter that also carries a hydrogen. When (*R*)-3-phenyl-2-butanone is dissolved in aqueous ethanol that contains NaOH or HCl, the optical rotation of the solution gradually drops to zero, to yield a racemic mixture of the (*R*) and (*S*) enantiomers (Scheme 1.91). This rate of racemization is found to be proportional to the concentration of ketone and the concentration of NaOH or HCl. Racemization thus, occurs by way of the intermediate enol form in which the former stereogenic carbon becomes planar (achiral). As racemization involves the formation of the enol form, the rates of racemization and enolization are found to be exactly equal.

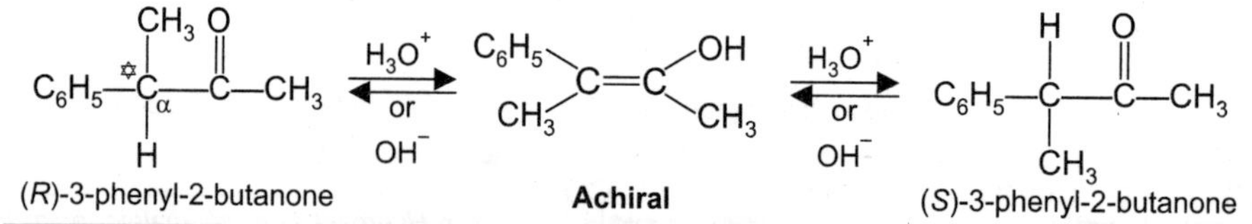

A compound with H-containing stereocenter next to carbonyl group (α-position) gives a flat and achiral enol in an acid or base solution (labile nature of H in the α-position of carbonyl). Approach of the electrophile e.g., H_3O^+ from either face gives a mixture of enantiomers-a racemate.

$$\text{rate} = k[\text{ketone}][H^+] \quad \text{or} \quad k'[\text{ketone}][OH^-]$$

SCHEME 1.91

Racemization of an optically active ketone occurs only if the stereocenter is α to the carbonyl group. If the aldehyde or ketone is chiral because of asymmetry at some other carbon, the enol form is also chiral, enolization in such a case does not lead to racemization (Scheme 1.92).

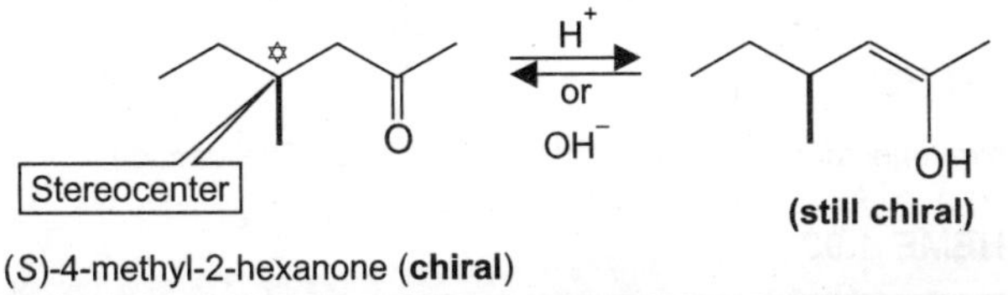

When in a chiral ketone the H-containing stereocenter is not in the α-position, racemization does not occur in acid or base solution.

SCHEME 1.92

EXERCISE 1.14

Groups others than the carbonyl group also activate the adjacent protons which are easily removed by base, examples are nitro compounds, sulfones and nitriles. Thus the optically active nitrile (I) is expected to racemize when treated with base. However, (I) undergoes deuterium exchange 4000 times faster than it racemizes on treatment with sodium methoxide in deuteriomethanol. Explain.

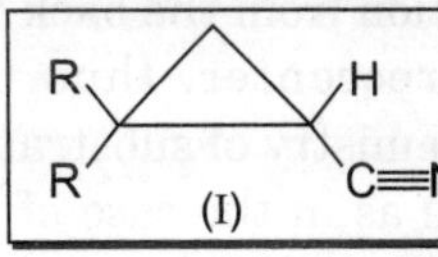

ANSWER. *The inversion of carbanion (II) formed by the abstraction of a proton from the position adjacent to the nitrile group with base is difficult since the nitrile group has to become coplanar with the cyclopropane ring and this flipping process is unfavourable since excessive strain energy will be involved. (Scheme 1.92a)*

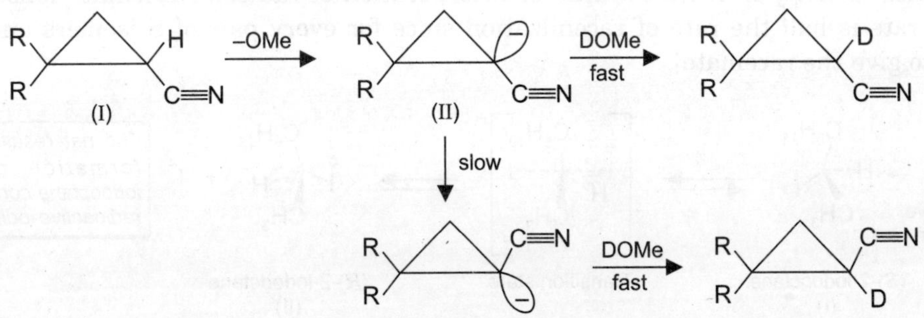

SCHEME 1.92a

Racemization is often a facile process which is troublesome as well as it has nuisance value. The presence of both the enantiomers of thalidomide in a drug formulation led to birth defects in the children of some women who took the substance during pregnancy. Even if the pure (*R*) form was given to patients, the mutagenic (*S*)-isomer was formed by racemization under physiological conditions because the stereocenter is α to a carbonyl group (scheme 1.93). Similarly amino acids can undergo racemization due to the presence of a stereocenter α to the carbonyl group in these compounds. In proteins, the peptide bonds render this structural feature prone to racemization, however, it does not occur under physiological conditions.

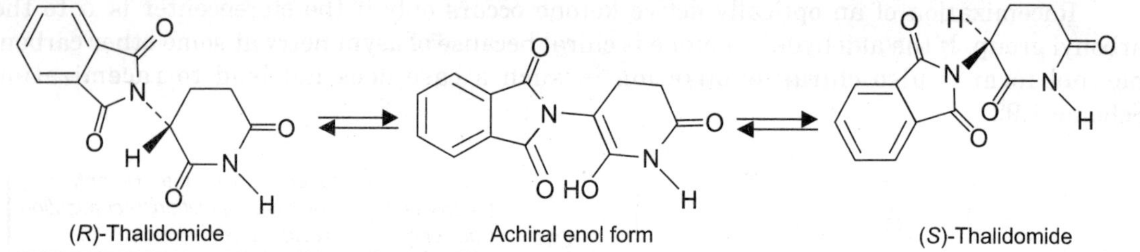

SCHEME 1.93

(c) *By Substitution Reactions*

(*i*) S_N2 reactions

Normally in an S_N2 reaction the incoming nucleophile (*e.g.*, OH^- in scheme 1.28*e*, eq. I) initiates the reaction from the back while Br^- leaves the molecule from the front in a concerted process. The stereocenter, thus undergoes an inversion of configuration with respect to the stereochemistry of substrate. In case the incoming nucleophile and the out going leaving group are same as in the case of reaction of optically active 2-iodoctane (Scheme 1.94) with sodium iodide, the reaction becomes reversible and an equilibrium is set up between the two enantiomers (I and II, Scheme 1.94) leading to racemization. Conclusive proof that inversion of configuration occurs during an S_N2 reaction was provided by studying the reaction between optically active 2-iodooctane and radioactive iodide ions. This is one of the simplest possible types of bimolecular substitution reacton, and involves replacement of iodide ions by radioactive iodide ions so that product and starting material are chemically identical. The process also involves inversion of configuration, and is thus accompanied by a loss of optical activity. It was seen that the rate of loss of optical activity is twice the rate of incorporation of radioactive iodide ions [*i.e.*, the inversion rate is half the rate of racemization since for every pair of S isomers only one is inverted to give the racemate].

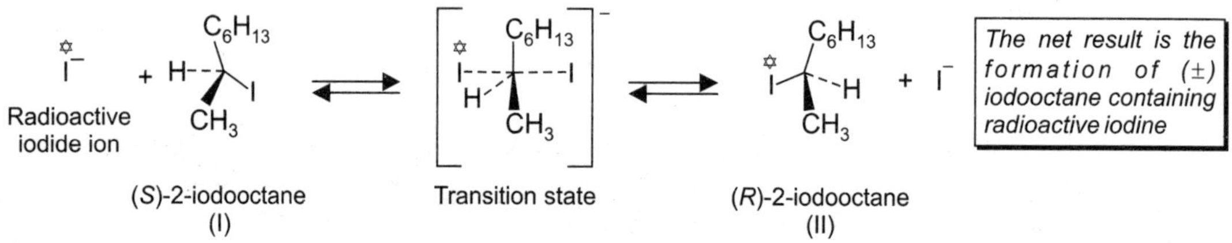

(Racemization via an S_N2 reaction)

SCHEME 1.94

In case the process involved an achiral intermediate, like a carbocation, the rate of racemization will be equal to the rate of incorporation of radioactivity. On the another hand, in an S_N2 reaction every substitution (eq. I, scheme 1.28*e*) involves inversion. Racemization is, therefore, complete when half of the material gets inverted (and has incorporated radioactivity) so that the rate of racemization is twice the rate of incorporation of radioactivity. This experiment provides the most convincing proof to date that an S_N2 reaction is accompanied by inversion of configuration.

***(ii)* S_N1 reaction**

In S_N1 reaction a planar carbocation intermediate is formed. This is attacked by the nucleophile from both sides leading to extensive racemization (Scheme 1.95).

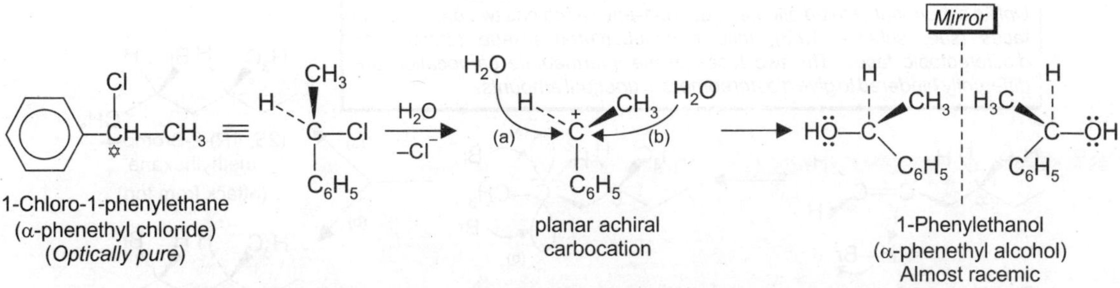

Initial ionization gives a planar achiral carbocation, when trapped by water yields racemic alcohol

SCHEME 1.95

Although many S_N1 reactions proceeds with racemization, many others result in more inversion of configuration in the product than retention. This is due to initial ion pair formation where the leaving group blocks the front side of the carbocation to favour inversion. It is only when the leaving group diffuses away leaving the carbocation in free form that the nucleophile can attack equally well from either side to result in equal amounts of inversion and retention.

***(iii)* During a molecular rearrangement**

Optically active monoterpenoid camphene undergoes racemization on treatment with acid. The reaction of (I, scheme 1.96) with a proton leads to the formation of a carbocation which undergoes a methyl shift with its bonding pair of electrons to give another carbocation, the loss of a proton gives (II) which is the mirror image of (I, Scheme 1.96).

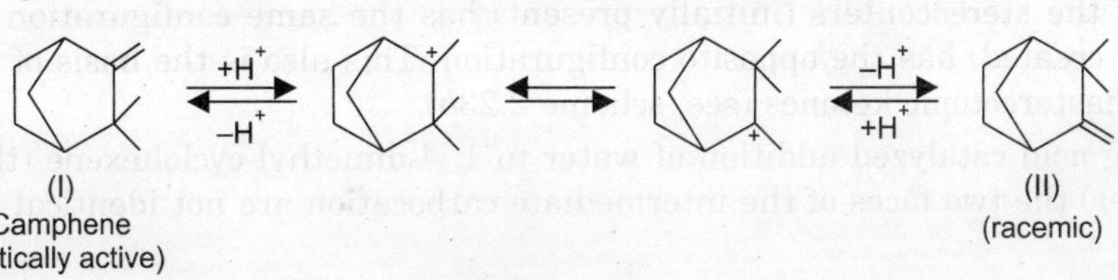

SCHEME 1.96

1.12 SOME STEREOCHEMICAL REACTIONS NEAR A STEREOCENTER (FORMATION OF DIASTEREOMERS)

Several reactions *e.g.*, electrophilic addition of HBr to monosubstituted alkene (see, scheme 1.88, achiral reactants) or the radical bromination of butane at C2 (see, scheme 1.87) introduce chirality in the molecule, however, the product obtained is racemic. In these reactions a planar sp^2 hybridized and therefore, achiral intermediates (a carbocation and a radical respectively) are formed. The two faces are equally susceptible to attack on two equivalent reaction sites to result in racemic products.

The presence of a stereocenter in the starting substrate affects the outcome of the reaction to give an unequal mixture of product diastereomers. Thus the addition of HBr to (*R*) enantiomer of 4-methyl-l-hexene proceeds through a carbocation (Scheme 1.97).

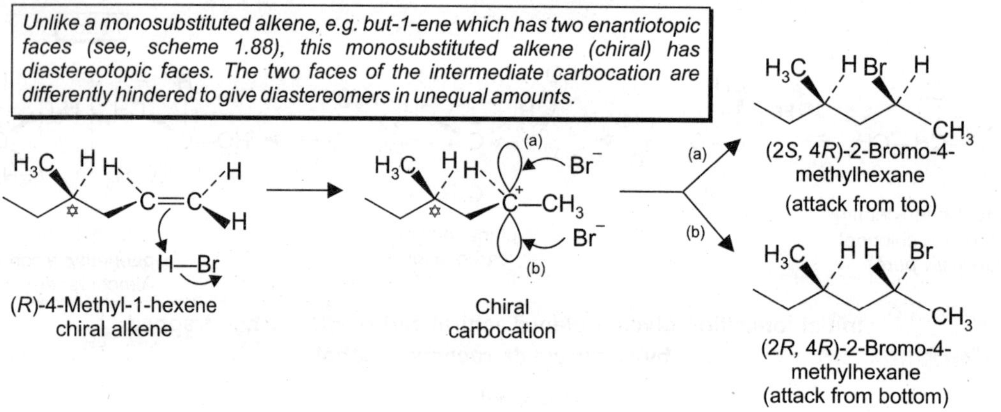

Formation of unequal mixture of diastereomers

SCHEME 1.97

Because there is a stereocenter in the carbocation intermediate, it is a chiral species and therefore, it does not have a plane of symmetry. One face of the carbocation will be more sterically hindered than the other, the incoming bromide ion will therefore, have greater access to the less sterically hindered face. As a result of this the two diastereomers will be formed in unequal amounts. The reaction is termed as stereoselective since more of one stereoisomer is formed than the other, however, precisely this is a diastereoselective reaction since the stereoisomer produced in excess is a diastereomer (for further details see under asymmetric synthesis—also see schemes (1.105–1.107). The reaction (scheme 1.97) creates a new stereocenter, thus two stereoisomers are formed. These stereoisomers are diastereoisomers since, one of the stereocenters (initially present) has the same configuration in both while other (newly created) has the opposite configuration. This also is the basis of Crams rule to addition to diastereotopic ketones (see, scheme 2.23*b*).

During acid catalyzed addition of water to 1, 4-dimethyl-cyclohexene (the presence of a stereocenter) the two faces of the intermediate carbocation are not identical (scheme 1.98).

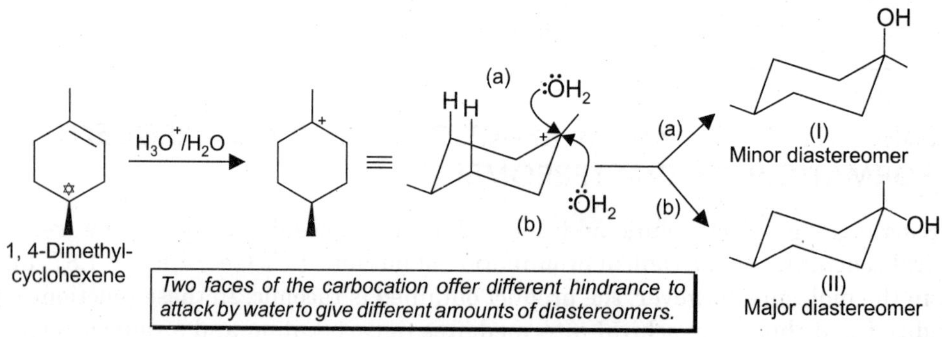

SCHEME 1.98

Attack of water from the axial direction is hindered by the two axial hydrogen atoms and is therefore, slower to give the minor diastereomer (I, scheme 1.98) and the addition of water to the equatorial face dominates and the major product is the diasteromer (II, scheme 1.98) where the hydroxyl group is *trans* to the methyl group at C4. Both the diastereoisomers are achiral (plane of symmetry).

The chlorination of (*S*)-2-bromobutane (Scheme 1.99) at C3 again gives a mixture of diastereomers in unequal amounts based on arguments similar to that for scheme 1.97 [*The methylene protons close to a stereocenter are generally diastereotopic, (see scheme 2.10)*].

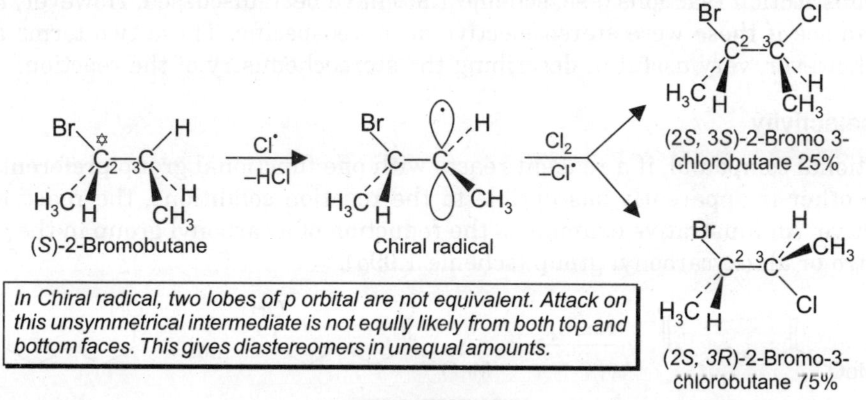

SCHEME 1.99

The radical chlorination of optically active 2-bromobutane *e.g.*, the (*S*) enantiomer at two terminal methyl groups at C1 or C4 will produce optically active products since the original stereocenter remains intact. The chlorination at C2 would lead to loss of optical activity since now the bond to stereocenter is cleaved to give a planar achiral sp^2 hybridized radical. Its chlorination will lead to racemization (compare with scheme 1.87).

Effect of a given chirality in a substrate on addition reactions (Diastereoselectivity)

- *The stereocenter already present can be R, S or both (i.e., in last possibility, the starting material is racemic.)*
- *In case the existing stereocenter has R configuration (i.e., the starting material is enantiometrically pure R stereoisomer as in scheme 1.97), the products of addition (diastereomers) will have either the R, R, or R, S configuration. Each diastereomer will be optically active since only one enantiomer is formed.*
- *In case the existing stereocenter was with S configuration the products (diastereomers) formed on addition would be S, S or S, R. These also would be optically active and infact represent the enantiomers of the diastereomeric pair formed from the starting material in which the existing stereocenter had R configuration.*
- *In case the starting material is a racemic mixture, all the four stereoisomers would be formed i.e., product will be optically inactive. The outcome of the reaction will still be two diastereomers (each is formed as a pair of enantiomers) and the same diastereoselectivity will be observed.*

In summary, therefore, during *e.g.*, addition reactions to an olefin which has a stereocenter elsewhere in the molecule, if optically active starting materials are employed the resulting diastereomers will be optically active. The racemic starting materials yield optically inactive diastereomeric mixture, in both situations the diastereomers will be formed in unequal amounts.

1.13 STEREOSELECTIVE AND STEREOSPECIFIC REACTIONS

In the previous section several stereochemical reactions *e.g.*, addition (see, scheme 1.97 and 1.98) and substitution reactions (see, scheme 1.99) have been discussed. However, no attempt was made to see if these were stereoselective or stereospecific. These two terms along with others are however, very useful in describing the stereochemistry of the reaction:

A. Chemoselectivity

In a bifunctional compound, if a reagent reacts with one functional group preferentially, even though the other is apparently susceptible to the reaction conditions, the reaction is called chemoselective. An illustrative example is the reduction of a carbonyl group in the presence of a cyano, nitro or alkoxycarbonyl group (scheme 1.99*a*).

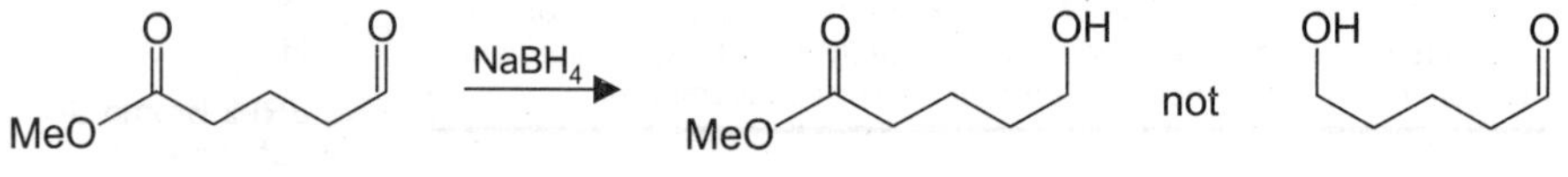

SCHEME 1.99a

B. Regioselectivity

Regioselectivity in a reaction, which proceeds without skeletal rearrangements, is seen when a molecule has two or more sites of reactivity arising from the presence of one functional group, each of which the reagent may attack, leading to the formation of constitutional isomers (scheme 1.99*b*).

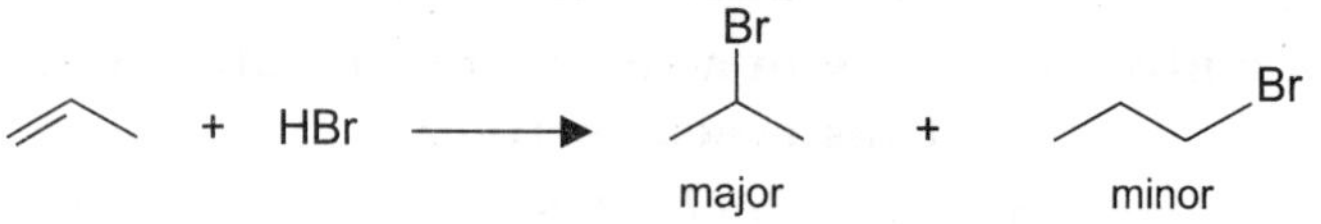

SCHEME 1.99b

C. Stereospecificity—A Common Definition

A reaction is stereospecific provided the reactant can exist as stereoisomers and each stereoisomeric reactant gives a different stereoisomeric product which may be (±) pair.

(*i*) Bromination of (E- and Z-2-Butene)—A Stereospecific *anti*-addition

Starting compounds differing only in their stereoisomerism must be converted into stereochemically different products. Accordingly one says that bromine addition to (Z) or (E)-2-butene is stereospecifically *trans*. *Addition to* (Z)-2-butene gives a racemic mixture of 2, 3-dibromobutane while addition to (E)-2-butene gives a *meso* stereoisomer. *Cis*-2-butene (Z-2-butene) adds bromine *e.g.*, from the top face (scheme 1.100) to give the intermediate bromonium ion (bridged ion) which is achiral *i.e.*, it has a plane of symmetry along with bromide ion [*reaction at the other face of the alkene bottom face not shown, however, would give the same bromonium ion, the two faces of (Z)-2-butene are* overall homotopic]. The bromide ion then attacks the bromonium ion at either carbon from bottom (S_N2) type of displacement at equal rates to yield the two enantiomers in equal amounts *i.e.*, as the racemic form.

trans-2-Butene reacts with bromine *e.g.*, from the top face to give a chiral bromonium ion with a C_2 symmetry and bromide ion [*reaction at the other face (bottom) would yield the enantiomer of the bromonium ion, this attack has not been shown however, the two faces of trans-2-butene are overall enantiotopic.*] When the bromonium ion is opened by S_N2 type displacement at either carbon, the same achiral *meso* compound is formed (see, Schemes 6.8 and 6.8*a*).

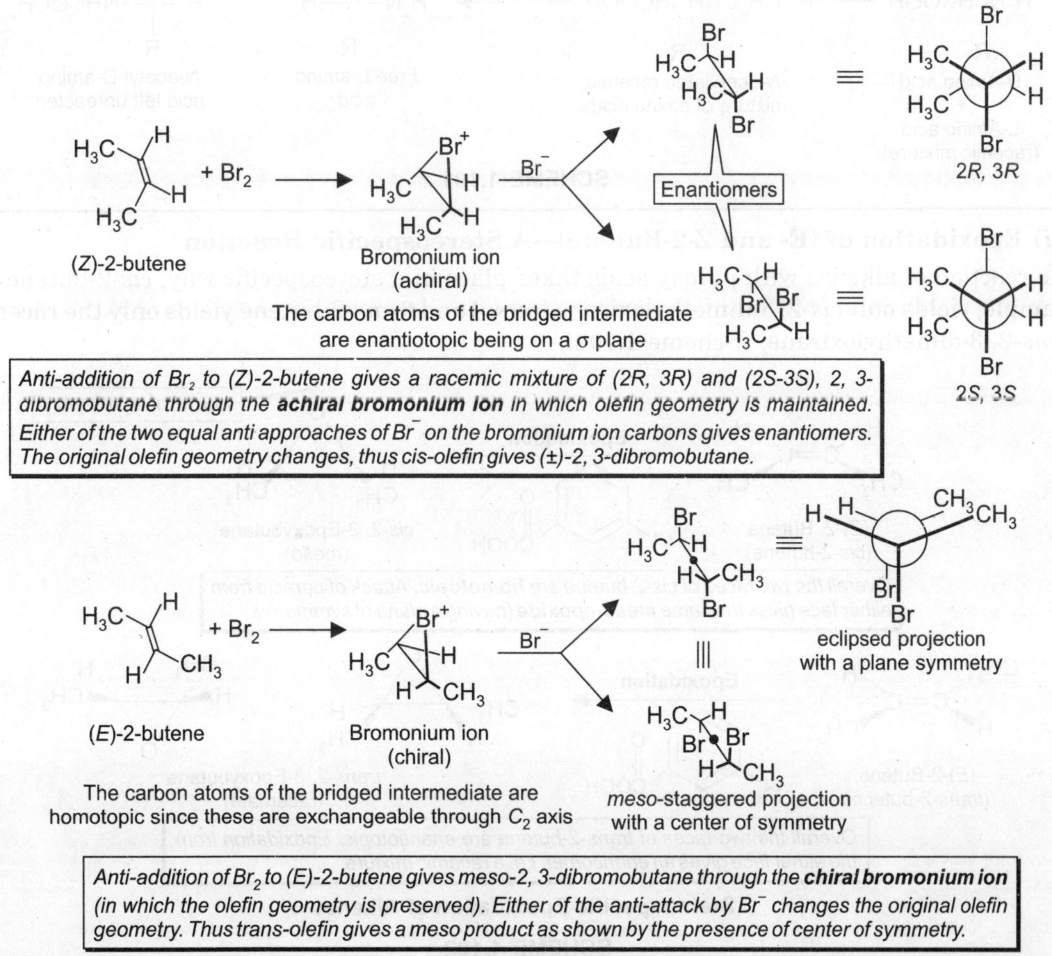

SCHEME 1.100

Thus in this stereospecific reaction two factors combine to determine as to which stereoisomers are actually formed in the reaction (*i*) The (*E*) or (*Z*)-(configuration of the starting alkene *i.e.*, which particular stereoisomer (diastereomer) is employed as the reactant, (*ii*) the *anti* stereochemistry of addition.

In summary stereospecificity is possible only if the inherent facial relationship of the olefinic bond is maintained throughout the addition process and only one bromine atom adds to each face. The bridged bromonium ion intermediate not only maintains the olefin geometry but also forces the second bromine to add from the opposite direction (*anti*, one may contrast this to an addition of HCl or H_2O where the intermediate carbocation is free to rotate so that the olefin geometry is lost and both the proton and the nucleophile can add to either face).

(*ii*) Stereospecificity of Enzymes

The enzymes react with one of the two enantiomers and are thus said to display stereospecificity. Thus during enzymatic hydrolysis of *N*-acetyl derivatives of racemic D, L-pair of amino acids only the L isomer is hydrolysed (Scheme 1.101).

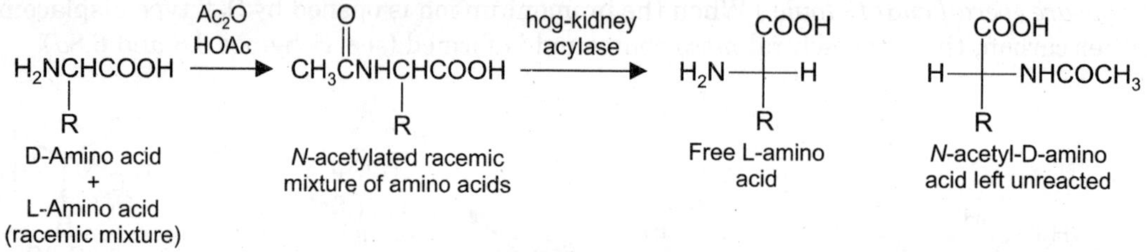

SCHEME 1.101

(*iii*) Epoxidation of (E- and Z-2-Butene)—A Stereospecific Reaction

The reaction of alkenes with peroxy acids takes place in a stereospecific way, *cis*-2-butene for example, yields only *cis*-2, 3-dimethyloxirane (*meso*) and *trans*-2-butene yields only the racemic *trans*-2, 3-dimethyloxirane, (Scheme 1.102).

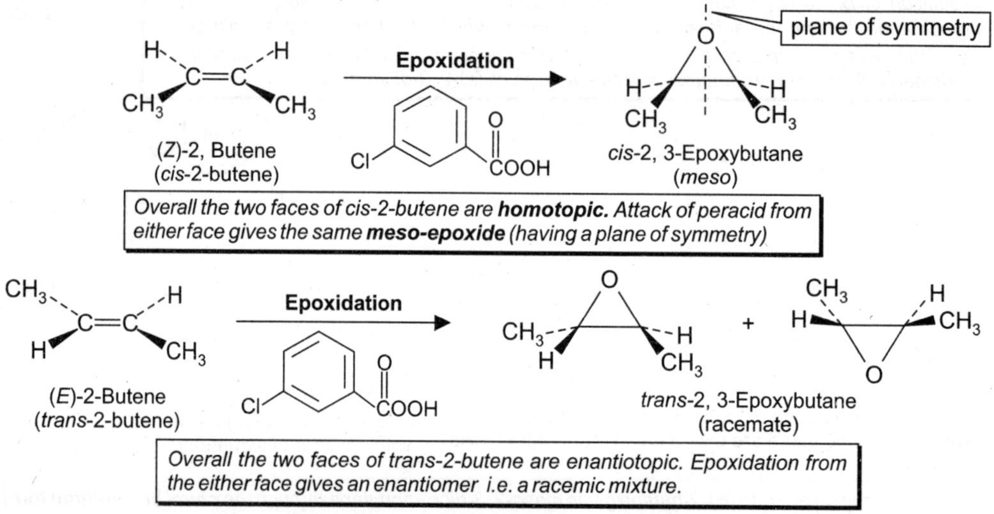

Stereospecific Epoxidation of Alkenes

SCHEME 1.102

D. Stereoselectivity—A Common Definition

A stereoselective reaction is one in which the reactant irrespective of any stereoisomerism (not necessarily chiral) produces predominantly or exclusively one stereoisomeric form of the product (or a certain subset of stereoisomers *e.g.*, (±) from among all those that are possible).

(*i*) Bromination of Cyclohexene

When cyclohexene is brominated, the product is one stereoisomeric product *trans*-1-2-dibromo-cyclohexene (scheme 1.103) no *cis*-isomer is formed. One may compare this stereoselective

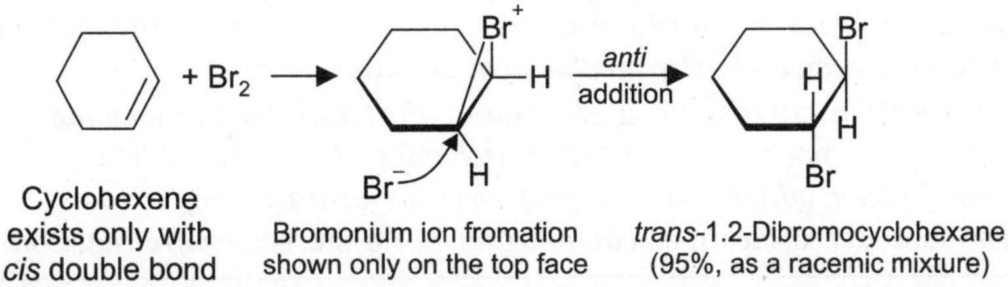

anti-Addition of bromine via bromonium ion which is attacked by Br^- from the opposite face (bottom face) (i.e., the two bromine atoms in the product are on opposite faces of the ring) gives only trans-1, 2-dibromocyclohexane as a racemic mixture. No cis-isomer is formed.

A stereoselective reaction

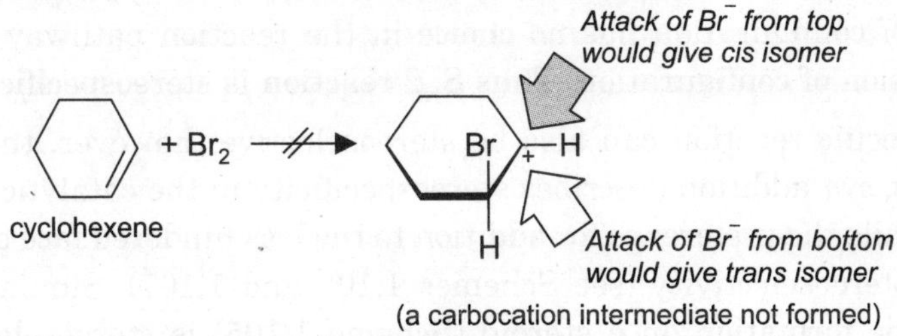

SCHEME 1.103

bromination of cyclohexene (cyclohexene has a unique structure and has no stereoisomers, it can only exist with *cis* geometry of double bond) with stereospecific bromination of 2-butene stereoisomers. Moreover, the bromination proceeds through the cyclic bromonium ion, if the intermediate were a simple carbocation both *cis* as well as *trans*-1, 2-dibromocyclohexane would have formed.

(*ii*) Stereoselectivity of Enzymes

The enzymes when in contact with prochiral molecules, react only with one of the enantiotopic ligands or faces, a property called stereoselectivity. For an example, glycerol undergoes phosphorylation exclusively at the *pro-R* hydroxymethylene group with adenosine triphosphate (ATP) in the presence of an enzyme-glycerol kinase (Scheme 1.104).

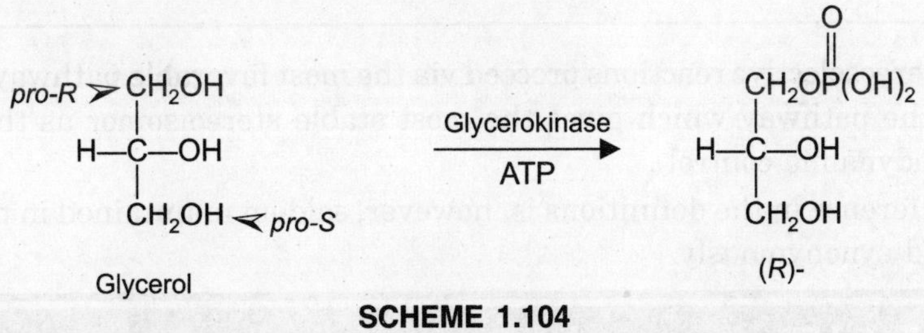

SCHEME 1.104

A Closer Look at Stereospecific and Stereoselective Reactions—A Tricky Situation

- *There appears to be confusion over the terms stereoselective and stereospecific.*
- *The common definitions given above suffer from the disadvantage e.g., a reaction which is stereospecific e.g., for two stereoisomeric alkenes must be called*

stereoselective for a small ring alkene (e.g., cyclohexene) which has no stereoisomers, however, the mechanism is same.

- *Thus a reaction carried on a compound that has no stereoisomers cannot be stereospecific, however, at most stereoselective. Thus the addition of bromine to methylacetylene which gives a preferential formation of trans-1, 2-dibromopropene is a stereoselective reaction and not a stereospecific reaction.*

More Satisfactory Definitions—Stereoselectivity and Stereospecificity

- One may look at the mechanism, when there is a choice in the stereochemical pathway *e.g.*, addition to the less hindered face of the double bond the reaction is stereoselective. When there is no choice the reaction is stereospecific *e.g.*, S_N2 displacement with inversion of configuration has no choice in the reaction pathway which must lead with inversion of configuration. Thus S_N2 reaction is stereospecific.
- A stereospecific reaction can also be stereoselective (however, the converse is not true). Thus, *syn* addition describes stereospecificity in the catalytic hydrogenation of alkenes, while the preference for addition to the less hindered face of the double bond describes stereoselectivity (see Schemes 1.106 and 1.107). Similarly one describes that epoxide formation on a steroid (Scheme 1.105) is stereoselective on the less hindered face of the C=C via a stereospecific *syn* addition. The ring opening of the epoxide by the nucleophile (Br^-) occurs stereospecifically to give overall *anti* addition.

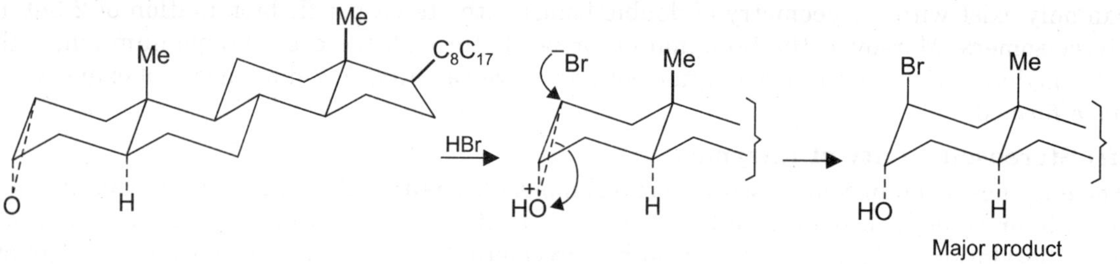

SCHEME 1.105

- Thus stereoselective reactions proceed via the most favorable pathway (kinetic control) or via the pathway which gives the most stable stereoisomer as the major product (thermodynamic control).
- This difference in the definitions is, however, seldom maintained in practice and both are used synonymously.

EXERCISE 1.15

Write the structure of the product(s) from the epoxidation of 1-butene. Assign R and S descriptors to stereocenters. Is the reaction stereoselective?

ANSWER. *The faces of a monosubstituted alkene are enantiotopic. Epoxidation of 1-butene affords a racemic mixture of epoxides (Scheme 1.105a). No.*

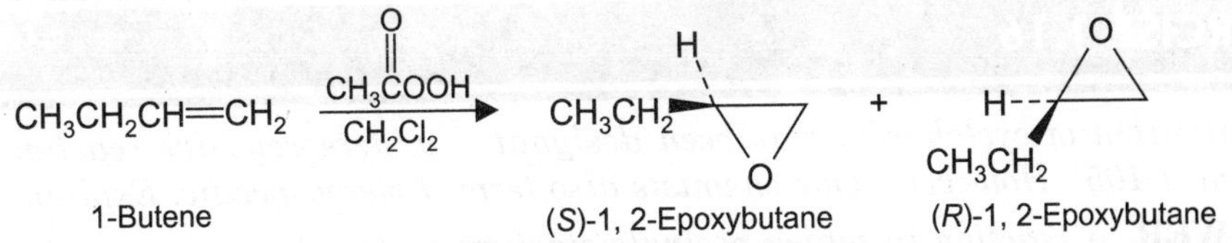

SCHEME 1.105a

(*ii*) Hydrogenation (a syn addition reaction) of a chiral alkene

A reaction which introduces a second stereocenter into a starting reactant which already has one, may not give equal quantities of two possible diastereomers (see scheme 1.97). The catalytic hydrogenation of 2-methyl (methylene) cyclohexane gives both *cis* and *trans*-1, 2-dimethylcyclohexane, but in unequal amounts (Scheme 1.106). More *cis*-1, 2-dimethylcyclohexane is formed than *trans*. The less hindered face of the double bond approaches the catalyst surface, and this is the face to which hydrogen is added (a *syn* Addition). Hydrogenation of 2-methyl (methylene) cyclohexane occurs preferentially at the side of the double bond opposite that of the methyl group to give *cis* stereoisomer as the major product. Thus it is a stereoselective reduction *i.e.*, stereoisomeric products are formed in different quantities from a single starting material.

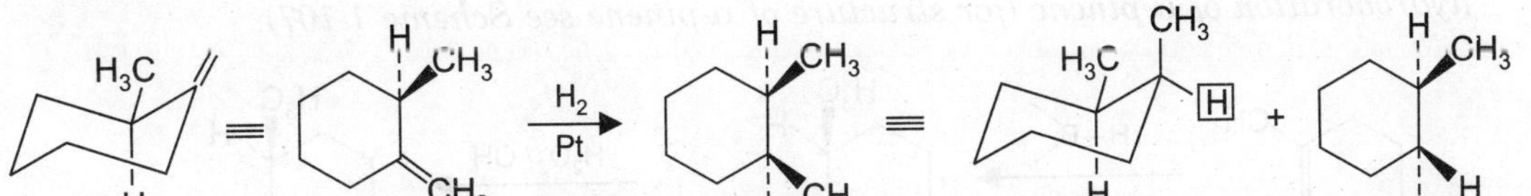

A stereoselective reaction

SCHEME 1.106

The catalytic hydrogenation of α-pinene is also a stereoselective *syn* addition of hydrogen. The observed stereoselectivity depends on the mode of alkene approach to the catalyst surface. The two hydrogens (shown by arrows) are added on the less hindered side of α-pinene. *Cis*-pinane and *trans*-pinane are common names which show that relationship between the pair of methyl groups on the bridge and the third methyl group (Scheme 1.107). (Enzyme-catalyzed reactions are also completely stereoselective as shown in Scheme 2.19).

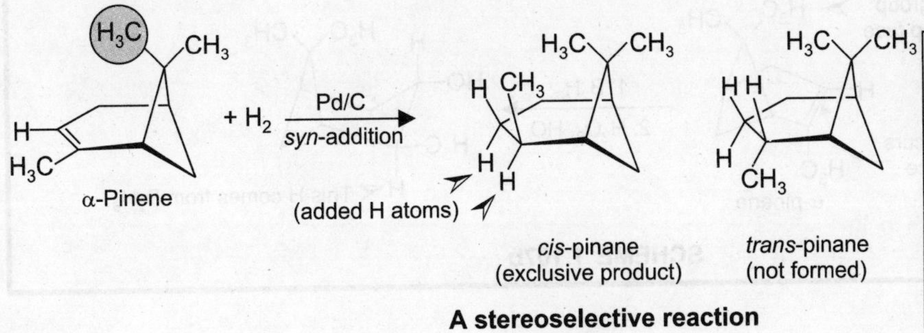

The methyl group which lies over the double bond of α-pinene shields its top face, preventing a close approach to the surface of the catalyst. Hydrogenation of α-pinene, therefore, takes place preferentially from the bottom face of the double bond

A stereoselective reaction

SCHEME 1.107

EXERCISE 1.16

Bromination of cyclohexene has been designated a stereoselective reaction (see, scheme 1.105). However, some chemists also term it stereospecific. Explain?

ANSWER. *A reaction in which a single starting material can give two or more stereoisomeric products but yields one of them in greater amounts than the other (or even to the exclusion of the other) is said to be stereoselective. However, the reaction may also be called stereospecific, since there is no choice in the stereochemical pathway. This pathway involves the formation of a bromonium ion and its S_N2 opening. Thus the reaction process is mechanistically constrained to proceed in a stereochemically defined manner just like S_N2 reaction.*

EXERCISE 1.17

The hydroboration-oxidation of an alkene brings about the net syn-addition of the elements of H—OH to the double bond. By convention this reaction is stereospecific syn addition (Scheme 1.107a). Predict the stereostructure of the product of hydroboration of α-pinene (for structure of α-pinene see Scheme 1.107).

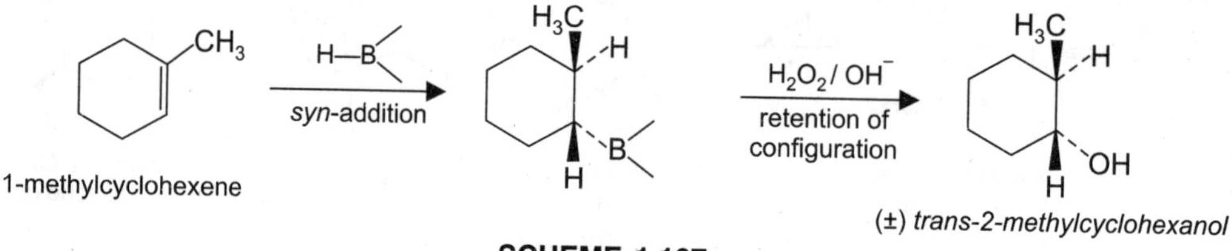

SCHEME 1.107a

ANSWER. *Recall that a stereospecific reaction can also be stereoselective. Thus, though syn addition during the catalytic hydrogenation of an alkene describes stereospecificity, while the preference for addition to the less hindered side of the double bond reflects stereoselectivity. Hydroboration-oxidation of α-pinene like hydrogenation also occurs from the less hindered bottom face and therefore, it is a stereoselective reaction (Scheme 1.107b, note that here α-pinene is drawn slightly differently than in scheme 107).*

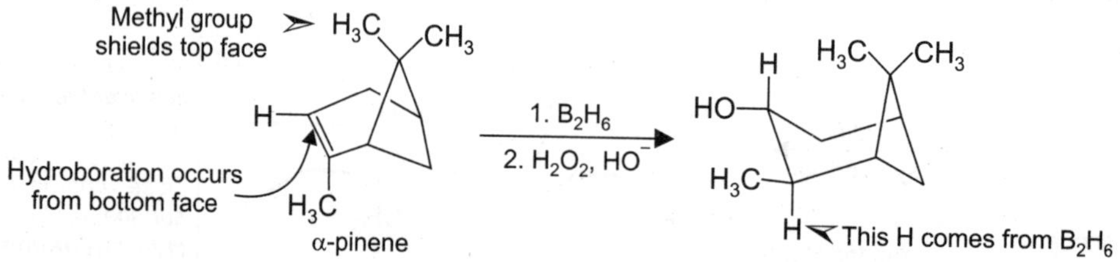

SCHEME 1.107b

EXERCISE 1.18

Ketone (I, Scheme 1.107c) on reduction with $NaBH_4$ gives a diastereomeric mixture of alcohols (II and III). Is this reaction stereospecific or stereoselective ?

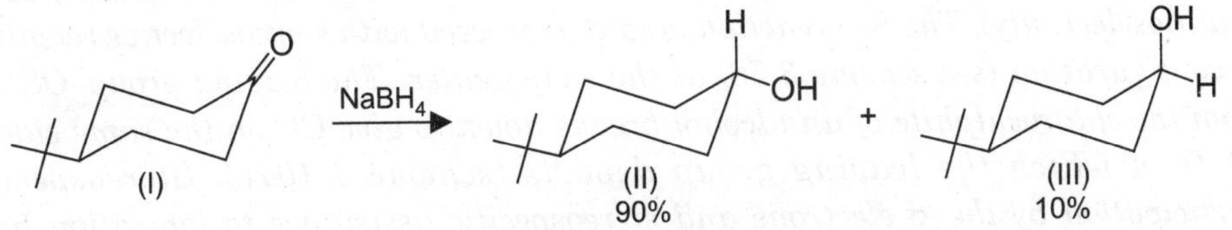

SCHEME 1.107c

ANSWER. *There is a choice in the stereochemical pathway, the hydride can be delivered from the top axial or bottom equatorial face of the carbonyl group in (I). Since addition to the top face is preferred to give thermodynamically more stable equatorial alcohol, thus the reaction is highly stereoselective* (Also see, Schemes 7.6 and 7.8).

EXERCISE 1.19

Depict each of the following reactions (I and II, scheme 1.107d) as stereospecific or stereoselective.

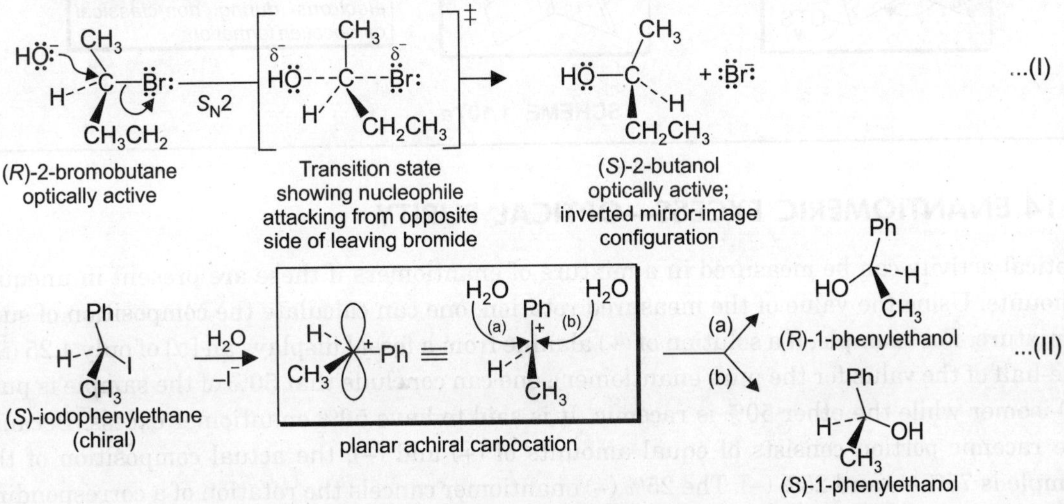

Initial ionization gives a planar achiral carbocation, when trapped by water yields racemic alcohol

SCHEME 1.107d

ANSWER. *Reaction (I) is S_N2 displacement with inversion of configuration, there being no choice in the reaction mechanism, the reaction is stereospecific. In the case of (II) a racemic mixture is formed via the intermediate formation of a carbocation. There is however, a choice in the reaction pathway, the addition could either occur from the one face to give one enantiomer or from the other face to give the other enantiomer since both enantiomers are favoured, the reaction is not stereoselective.*

More Examples of Stereoselective and Stereospecific Reactions

Often one comes across S_N1 reactions which through involve carbocations, but which favour inversion due to shielding effect of the departing anion, such reactions (see, scheme 3.39) are termed to proceed with low stereoselectivity (i.e., some enantioselectivity). The S_Ni reaction is said to proceed with stereoselective retention of configuration (see scheme 3.56) at the stereocenter. The leaving group–OSOCl from the chlorosulphite of an alcohol breaks down to give Cl^- on the same side of C^+ from which the leaving group departs (scheme 1.107e). Stereospecific participation by the σ electrons and stereospecific assistance to ionization by π electrons is observed during the non-classical carbocation formation in norbornane systems (scheme 1.107e).

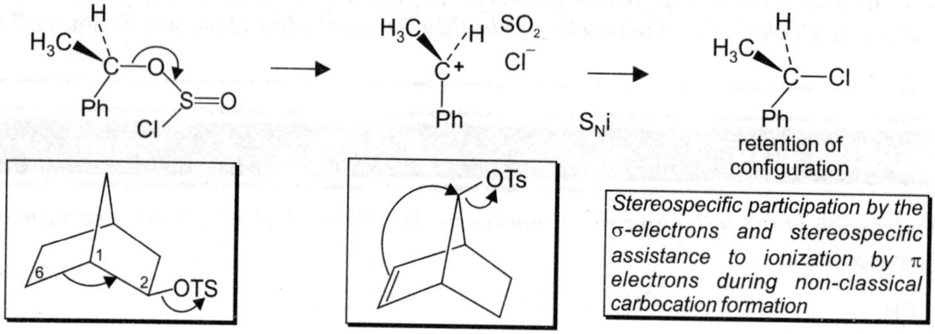

SCHEME 1.107e

1.14 ENANTIOMERIC EXCESS—OPTICAL PURITY

Optical activity can be measured in a mixture of enantiomers if these are present in unequal amounts. Using the value of the measured rotation, one can calculate the composition of such a mixture. For example, if a solution of (+) alanine from a fossal displays an [α] of only 4.25 (*i.e.* one-half of the value for the pure enantiomer), one can conclude that 50% of the sample is pure (+)-isomer while the other 50% is racemic. It is said to have 50% enantiomer excess. Because the racemic portion consists of equal amounts of (+) and (–), the actual composition of the sample is 75% (+) and 25% (–). The 25% (–) enantiomer cancels the rotation of a corresponding amount of the (+) enantiomer. This mixture is called 50% (*i.e.*, 75%–25%) optically pure. The observed optical rotation is one-half that of the pure dextrorotatory enantiomer. Optical purity can be found from the following relationship:

$$\% \text{ optical purity} = \left(\frac{[\alpha]_{\text{observed}}}{[\alpha]} \cdot 100 \right) = \text{enantiomer excess}$$

The enantiomeric excess can also be determined from NMR spectroscopy (see, Schemes 1.152 and 1.154).

1.15 EPIMERIZATION, EPIMERS, ANOMERS AND MUTAROTATION (ASYMMETRIC TRANSFORMATIONS) MORE METHODS OF TOTAL SPONTANEOUS RESOLUTION

Epimerization is a change in configuration at one stereocenter in a compound which has more than one such center. It leads to the formation of a disastereoisomer and not the enantiomer, of the starting material. The mechanisms of epimerization parallel those of racemization. Thus, glucose and mannose are C-2 epimers which are interconvertible under basic conditions (Scheme 1.108). The hydrogen atom on the carbon in the α position to the carbonyl group (open chain formula) is acidic and removed to give an enolate ion (which is protonated by water to give the intermediate 1, 2-enediol). Thus the C2 becomes trigonal planar and has lost its chirality. Protonation of the carbanion at C2 gives mannose or glucose depending on which side of the molecule the reaction takes place.

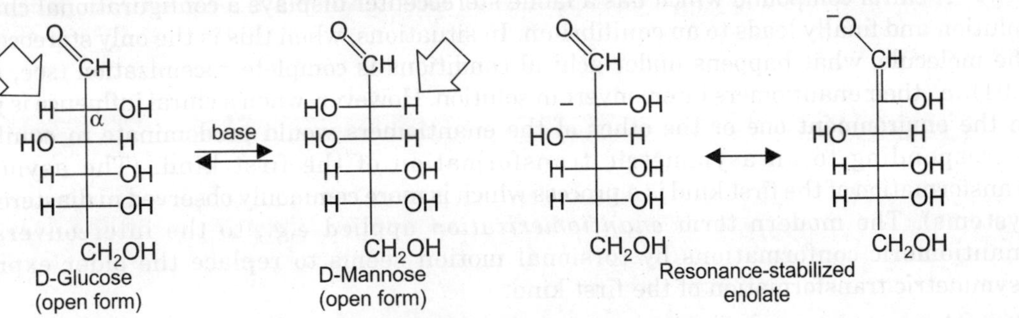

D-Glucose (open form) D-Mannose (open form) Resonance-stabilized enolate

SCHEME 1.108

Generally all the simple sugars exist in the six-membered ring form and when the hemiacetal is formed, the former aldehyde carbon becomes a stereocenter. Therefore, two cyclic forms of glucose exist; the two cyclic isomers of glucose differ only in the stereochemistry at C-1

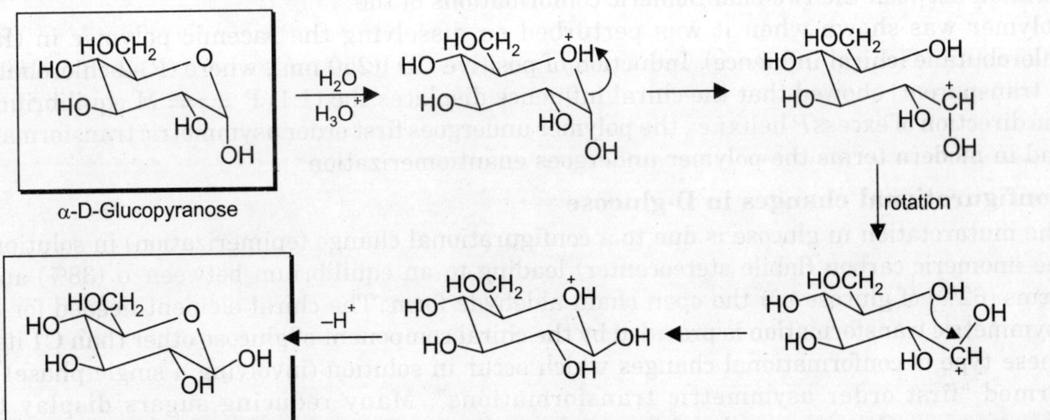

α-D-Glucopyranose

β-D-Glucopyranose

SCHEME 1.109

(anomeric carbon) and are called anomers. These anomers are differentiated by the Greek letters α and β. In the α anomer the hydroxy group on this carbon is axial and in the β anomer it is equatorial. Both α-D-glucopyranose and β-D-glucopyranose can be isolated in pure form and these being diastereomers have different physical properties. For example, the α-stereoisomer has a specific rotation of + 112.2° and that of β isomer is + 18.7°. When either of these pure stereoisomers is dissolved in water, the specific rotation undergoes a slow change to a value of + 52.7°. The process of change of optical rotation of a compound on standing in solution is called mutarotation. In aqueous solution each of these two forms is in equilibrium with the open chain form with the free aldehyde group. The mechanism of mutarotation of glucose (Scheme 1.109) involves protonation of the oxide ring, deprotonation of 1-OH by base followed by ring opening to the aldehyde form (not shown as such) and its subsequent ring closure to the original hemiacetal or its epimer under acid catalysis.

(a) First Order Asymmetric Transformation (Enantiomerization)

A chiral compound which has a labile stereocenter displays a configurational change in solution and finally leads to an equilibrium. In situations, when this is the only stereocenter in the molecule, what happens under achiral conditions is complete racemization (see, scheme 1.91) *i.e.*, their enantiomers interconvert in solution. However, when a chiral influence is present in the environment one or the other of the enantiomers would predominate in equilibrium corresponding to an asymmetric transformation of the first kind. (The asymmetric transformation of the first kind is a process which is more commonly observed in diastereomeric systems). The modern term *enantiomerization* applied *e.g.*, to the interconversion of enantiomeric conformations by torsional motion seems to replace the older expression asymmetric transformation of the first kind.

Enantiomerization of a chiral polymer

Polyisocyanates *e.g.*, poly (*n*-hexyl isocyanate Scheme 1.109*a*) do not have stereocenters and their chirality stems from helical conformations. These chiral polymers display optical activity due to the presence of both right (*P*) and left (*M*) handed helical forms. That dynamic equilibrium exists in solution between the two enantiomeric conformations of the polymer was shown when it was perturbed on dissolving the racemic polymer in (R)-2-chlorobutane (chiral influence). Induction of positive CD [(250 nm), where (R)-2-chlorobutane is transparent] showed that the chiral influence displaces the (1:1) P $\rightleftharpoons$ M equilibrium in the direction of excess *P* helix *i.e.*, the polymer undergoes first order asymmetric transformation and in modern terms the polymer undergoes enantiomerization.

$$\text{-}\!\!\left[N(n\text{-}C_6H_{13})\text{—}C(=O)\right]\!\!\text{-}_n$$

Poly (n-hexyl isocyanate)
a chiral polymer with no stereocenter

SCHEME 1.109a

Configurational changes in D-glucose

The mutarotation in glucose is due to a configurational change (epimerization) in solution at the anomeric carbon (labile stereocenter) leading to an equilibrium between α (38%) and β forms (62%) of glucose *via* the open chain aldehyde form. The chiral element needed for this asymmetric transformation is provided by the chiral component of glucose other than C1 itself. These type of conformational changes which occur in solution (involving a single phase) are termed "first order asymmetric transformations". Many reducing sugars display this phenomenon. Other example of first order asymmetric transformation accompanied by mutarotation is displayed by several biphenyl derivatives under outside chiral influence where the rotation around the pivotal single bond is slightly restricted to allow interconversion between two enantiomeric atropisomers while in solution (for details see scheme 1.122). Asymmetric transformations of first kind (epimerisation) of covalent diastereomers or diastereomeric salts are more common than with enantiomers.

(b) Second Order Asymmetric Transformations (Crystallization Induced Asymmetric Transformation)

When during an asymmetric transformation one of the enantiomers or diastereomers comes out of the solution as crystals or immiscible liquid, the transformation is then called second order asymmetric transformation. This process may often be useful in resolution when a pure enantiomer may be obtained from a racemic mixture.

Many asymmetric transformations of the first kind may be turned into second kind by changing the solvent. Thus *e.g.*, when a solution of glucose in ethanol is concentrated, the less soluble α-form cyrstallizes first and equilibrium is once again established and finally only α-form of glucose is obtained. Similarly β form is obtained by crystallization of glucose from pyridine.

(c) Asymmetric Transformations and Resolution. Total Spontaneous Resolution in Configurationally Labile Systems

The traditional older term asymmetric transformation of second order as applied to resolution when a single enantiomer may be isolated has now come to be known as "*crystallization-induced asymmetric transformation*". This term is applicable to both enantiomers and diastereomers. The term crystallization-induced asymmetric transformation when applied to a system containing enantiomeric molecules refers to a process for the conversion of a racemate in to a pure enantiomer of the substrate. In these resolutions the target molecule should have a racemizable stereocenter. A further characteristic of these resolution strategies is that the mother liquor remains racemic, since the enantiomer which comes out as crystals is replenished by a (+) $\rightleftharpoons$ (–) equilibrium (50:50) in solution. This result is therefore, crystallization of one enantiomer with simultaneous racemization of the other. In some cases of crystallization induced asymmetric transformation an enantiomer of the substrate itself provides the necessary chiral environment. In such cases one enantiomer may happen to crystallize first and provide a seed and thus the racemic form must necessarily be a conglomerate. The following are some examples of these resolutions.

(*a*) Resolution of rac-N, N, N-allyl-ethylmethylanilinium iodide. It is a configurationally labile system from which enantiopure crystals deposit on slow crystallization from chloroform (Scheme 1.109b, also see problem 1.18).

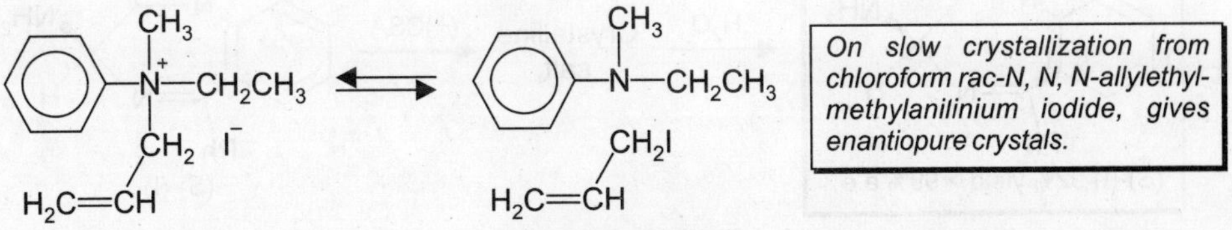

SCHEME 1.109*b*

(*b*) Resolution of rac-1, 1′-binaphthyl. In 1, 1′-binaphthyl the two rings are not coplanar in the ground state and the racemization takes place by rotation around 1, 1′-bond (Scheme 1.109*c*). This rotation is restricted by van der Waals interactions between the hydrogen atoms shown in the structures. There is crowding between these hydrogens when the two naphthyl groups become coplanar. The racemization process requires the hydrogens to move past each other (Scheme 1.109*c*). *rac*-1, 1′-Binaphthyl undergoes spontaneous resolution on crystallization either from solution or from the melt.

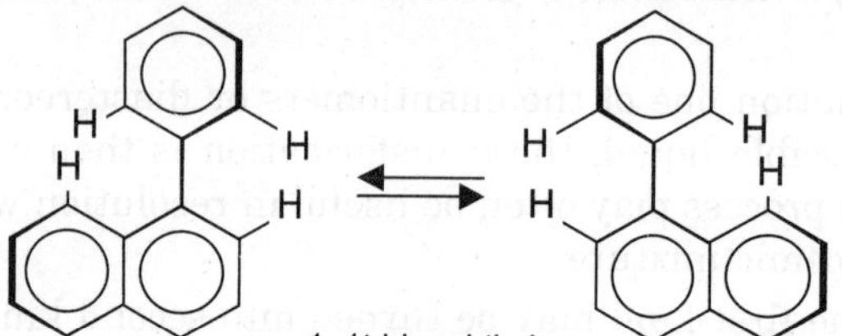

1, 1′-binaphthyl

1, 1′-Binaphthyl is resolvable via crystallization-induced-asymmetric transformation. The peri–hydrogens provide enough steric hinderance to keep the rings non-planar.

SCHEME 1.109c

(c) Resolution of chiral ketones with a stereogenic atom α- to the carbonyl group. Crystallization–induced asymmetric transformations are observed with chiral ketones which have a stereogenic atom alpha to the carbonyl group in the basic medium. The necessary condition is the presence of a hydrogen atom on the stereocenter. Thus *p*-anisyl α-methylbenzyl ketone (Scheme 1.109*d*) displays this phenomenon.

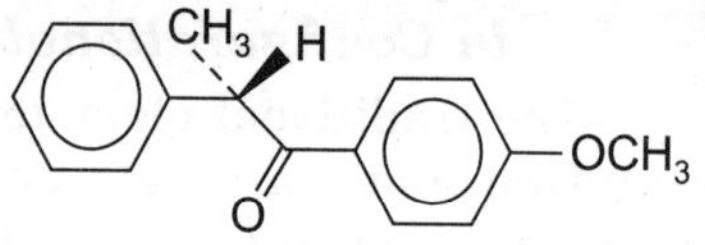

SCHEME 1.109d

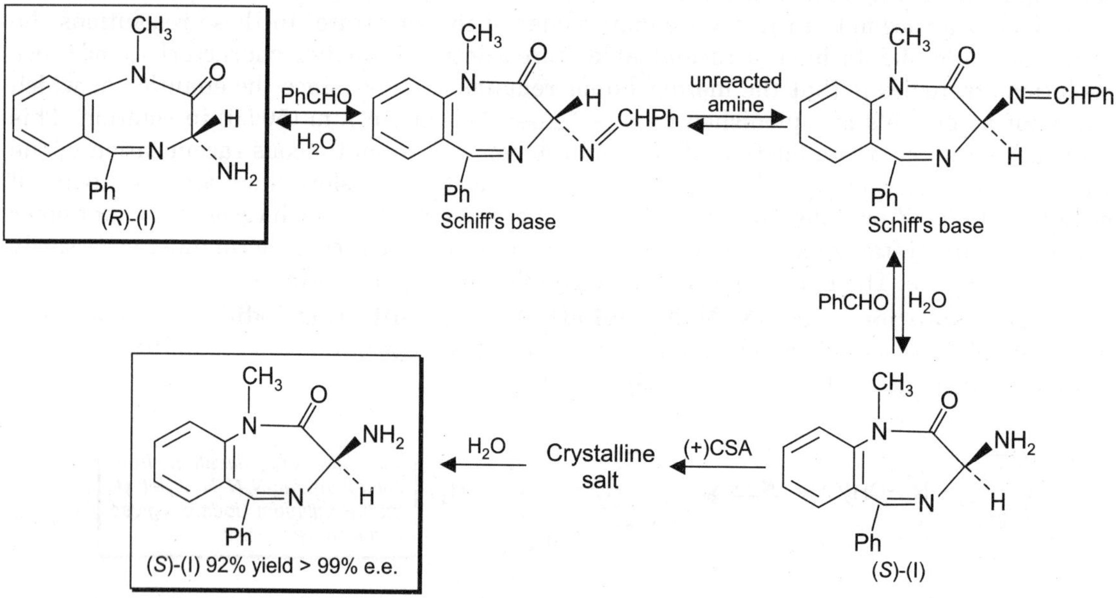

SCHEME 1.109d

The technique of "crystallization induced asymmetric transformation" was employed by Merck company to synthesize optically active 3-aminobenzodiazepinone (sleep inducer). 3-Aminobenzodiazepinone has a racemizable stereogenic carbon (I, Scheme 1.109d). The key to the resolution step was the reaction of (*rac*-I Scheme 1.109*d*) with (+)-camphorsulphonic acid [(+)-CSA]. The (*S*)-enantiomer of (I) leads to a less soluble salt with (+)-CSA compared to (*R*)-enantiomer. Thus this salt crystallizes out of the solution as the diastereomeric species.

The resolution was carried out in the presence of 3 mole per cent of benzaldehyde whose role was limited to make the proton next to the carbonyl group more acidic via Schiff's base formation. Racemization (via enolization not shown here) of the Schiff's base is brought about by the small quantity of the free amine present in the system. At the end the desired (S)-enantiomer was isolated in 92% yield (> 99% ee) *i.e.*, in essentially optically pure form.

1.16 OPTICAL ACTIVITY-STEREOISOMERISM IN COMPOUNDS WITHOUT A STEREOGENIC CARBON—(AXIAL CHIRALITY)

The chiral compounds discussed so far contains one or more stereocenters and have their chirality specified at one or more such centres. As an additional and a specific case of chirality one may consider the suitably substituted adamantanes with four different groups at the bridgehead positions which are chiral and therefore, display optical activity. The adamantane derivative (Scheme 1.110) may be regarded as the formal analogue of 2-bromopropionic acid and is completely asymmetric (belonging to C_1, point group). This has four nonequivalent stereocenters and due to highly symmetrical structure of its parent hydrocarbon admantane however, it exists only as a (±) pair. In fact, the four substituents form a regular tetrahedral arrangement and the arrangement has a center of chirality (shown by a dot) in the unoccupied space of the molecular framework of the adamantane. Two points come to light, firstly the center of chirality in a molecule may not always lie on an atom (as *e.g.*, on carbon in the case of 2-bromopropionic acid). Secondly, the number of stereoisomers may be lesser than calculated from 2^n (see Scheme 1.68*d*).

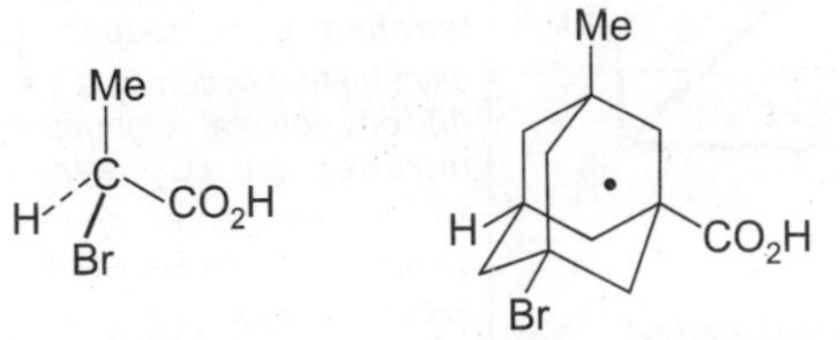

2-Bromopropionic acid 3-Methyl-5-bromoadamantane-carboxylic acid

The adamantane derivative has four bridgehead substituents and four stereocenters. It has however, only a (±) pair. A stereocenter thus may not always reside on an atom.

SCHEME 1.110

However, in some compounds with nonsuperimposable mirror images it is not possible to identify a stereocenter and it then becomes necessary to focus our attentions on a larger portion of the molecule. In the same sense as we have spoken of chirality in the case of compounds with a stereocenter, we may in the case of some other chiral compounds speak of axial chirality, planar chirality and helicity.

Whilst for models with a stereocenter all the four groups have to be different, for ones with an axis of chirality (stereoaxis) a smaller number of differences are sufficient. As one will see later on; for models with a plane of chirality (stereoplane), even one difference is enough.

Thus, the presence of a stereocenter is not a necessary and sufficient condition for molecular dissymmetry. The overall chirality of a molecule can be factorized into three elements, stereocenters, stereoaxes, and stereoplanes, while still another element of chirality is helicity.

Several important classes of chiral compounds have one or more stereocenters. Thus, for example, a stereocenter X can be detected in a molecule when the four different ligands *a*, *b*, *c* and *d* of a central atom X are located on corners of a tetrahedron (I, Scheme 1.111). In case this center is replaced by a linar grouping *e.g.*, C—C or C=C=C, the tetrahedron becomes elongated *i.e.*, extended along the axes of the grouping. On speculative elongation of this tetrahedron as in (II), the stereocenter is extended to produce a chiral axis (the stereoaxis). In this type of an extended tetrahedron on which axial chirality is based, the conditions for chirality are much less stringent when compared to a regular tetrahedron. Indeed a reference to (II) shows that an extended tetrahedron (with lesser symmetry) will be chiral if the pair of ligands at one end of the axis and the pair at other end constitute two different ligands, *i.e.*, the minimum condition for chirality is that the ligand $a \neq b$ (Scheme 1.111).

Elongated tetrahedron approach can be applied to a variety of compounds *e.g.*, allenes, spiranes, biphenyls etc. which are chiral not due to the presence of a stereocenter but a stereoaxis. The application to an allene (scheme 1.111) shows, that an allene *e.g.*, 2, 3-pentadiene will be chiral if (minimum) the two substituents at each end are different *i.e.*, ($a \neq b$, Scheme 1.111).

The allenes with three or four different substituents are also chiral, the prediction that suitably substituted allenes could be chiral was made by van't Hoff in 1875 and was verified in I935 by Maitland and Mills.

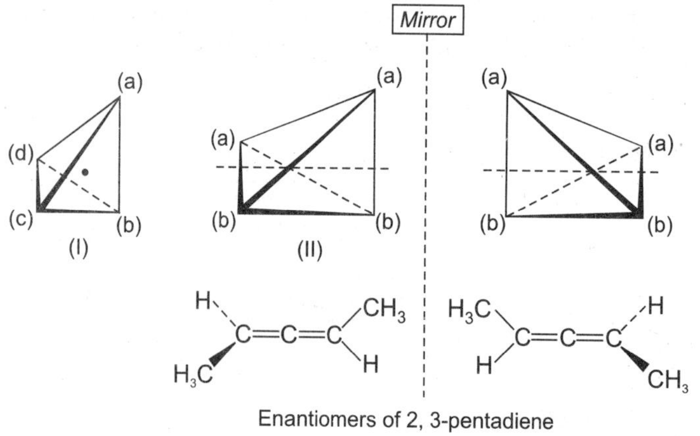

A stereocenter, Cabcd in a usual tetrahedral compound ((I) C_1 asymmetric) becomes a stereoaxis (II) on notional elongation of a tetrahedron (C_2 symmetry). Minimal conditions for chirality in compounds with stereoxis e.g., an allene is ($a \neq b$).

SCHEME 1.111

Several of these axially chiral compounds have a two fold axis of symmetry (C_2) and therefore, cannot be termed asymmetric (see sec. 1.9*b*).

One has already seen that one of the enantiomers of tartaric acid e.g. (+)-enantiomer though chiral has also a C_2 axis (see scheme 1.77a). For these reasons the term asymmetric carbon has been replaced by stereogenic carbon i.e., a stereocenter. Interestingly, however, the term asymmetric synthesis still survives. In short chirality can exist in a molecule with an axis of symmetry, but it cannot exist in a compound with a plane of symmetry, a center of symmetry or an alternating axis of symmetry.

A shrewd eye for symmetry will detect a C_2 axis *e.g.*, in the substituted allene 1, 3-dichloropropa-1, 2-diene in its Newman projection ; this C_2 axis passes through the center of the molecule and rotation of the molecule about this axis by 180° gives an identical molecule (II, Scheme 1.112). Thus it can be shown that (Ia and Ib, scheme 1.112) are enantiomers. The enantiomer (Ib, scheme 1.112) has an equivalent representation in (Ic) and the later is obtained by simply exchange of groups in (Ia) in left end carbon. Thus exchange of ligands at either of the terminal atoms across the stereoaxis reverses the chirality. The structure (Ib) when rotated 90° around C=C=C axis gives the equivalent structure (Ic, scheme 1.112).

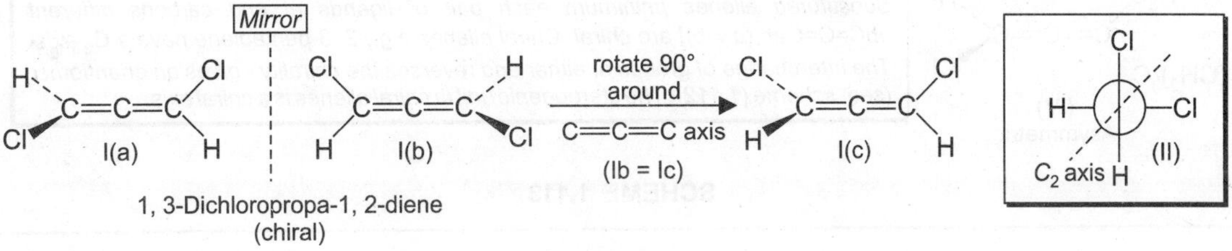

SCHEME 1.112

The allenes of the general formula Cab = C = Cab possess a C_2 axis and belong to point group C_2. In case the three or four of the end groups of an allene are different the C_2 axis disappears and then the molecule becomes totally asymmetric belonging to C_1 point group (See Scheme 1.113).

(A) Optical Isomerism of Allenes Spiranes and Related Compounds (Axial Chirality)

In the spatial arrangement of the cumulative double bonds of allene, the four substituents of the allene grouping are situated at the apexes of an imaginary tetrahedron, elongated. The following points may be noted :

- In an allene, the central carbon atom is *sp* hybridized and linear, and the two outer carbon atoms are sp^2 hybridized and trigonal. The central, *sp* hybrid carbon atom must therefore, use different *p* orbitals to form the *pi* bonds with the two outer carbon atoms. The two unhybridized *p* orbitals on a *sp* hybrid carbon atom are perpendicular, so the two *pi* bonds must also be perpendicular (scheme 1.113).
- In allene (I, 1, 2-propadiene, symmetry group D_{2d}) the C=C=C is a potential chiral axis. The planes defined by H(Cl)H and H(C3)H are mutually perpendicular.
- Allene itself is achiral since it has two planes of symmetry. In order to generate chirality, the two planes of symmetry must be eliminated. When unlike substituents are added at each end of the C=C=C unit, the substituted allene becomes chiral. This is so in 2, 3–pentadiene (Scheme 1.113) substituent at one end (H and CH_3 ; H ≠ CH_3) and the other end (H and CH_3 ; H ≠ CH_3).
- An allene with unlike substituents only at one end as in (scheme 1.113) leaves one symmetry plane and the allene is achiral.
- The allene (IV, Scheme 1.113) with all the four end groups different (this is also so when three end groups are different) loses the C_2 axis and becomes asymmetric belonging to C_1 point group.

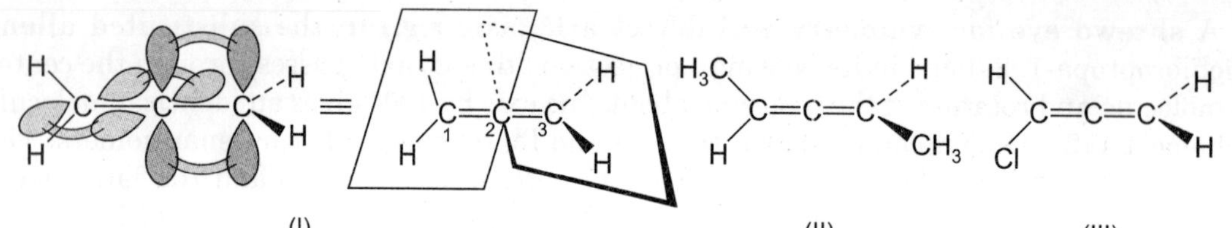

(I) Allene (propa-1, 2-diene) (achiral)

(II) 2-3-pentadiene (chiral)

(III) (achiral)

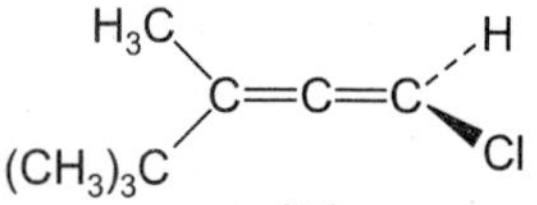

(IV) Asymmetric

Substituted allenes [minimum each pair of ligands at end carbons different abC=C=Cab (a ≠ b)] are chiral. Chiral allenes e.g., 2, 3-pentadiene have a C_2 axis. The interchange of groups at either end reverses the chirality - gives an enantiomer (see, scheme (1.112). The stereogenic unit in chiral allenes is a chiral axis.

SCHEME 1.113

The cumulated bonding systems (compounds with two or more successive double bonds) of the type (I, Scheme 1 .1 14) with an even number of double bonds do not have a plane of symmetry or a center of symmetry and therefore, must show optical isomerism and must be resolvable into enantiomers.

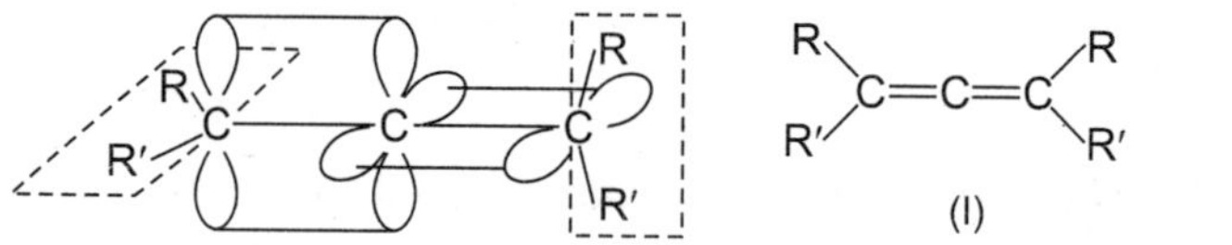

(I)

A cumulene (even number of double bonds) shows enantiomerism if R ≠ R′

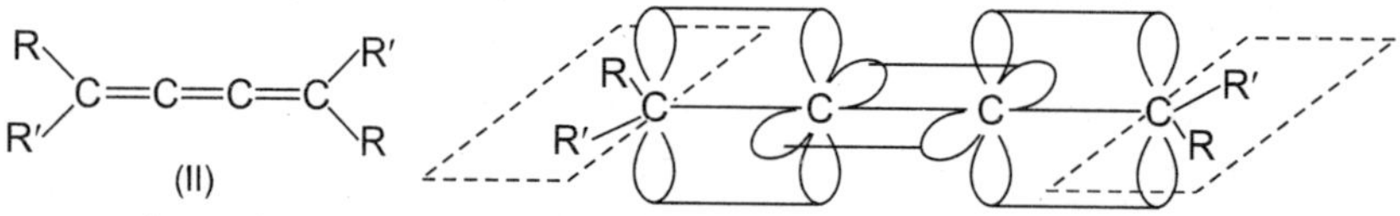

(II)

A cumulene (odd number of double bonds) shows geometrical isomerism.

SCHEME 1.114

Interestingly the compounds with odd number of cumulated double bonds intead *display Z-E* (geometrical) isomerism. When the allene chain of compound (I, Scheme 1. 114) is extended by one more double bond (introduction of another *sp*-hybridized carbon atom) one then gets a system (II, Scheme 1.114) in which the substituted groups at the two ends of the cumulated chain now lie in the same plane and geometrical isomerism is shown.

In Summary unsymmetrically substituted cumulenes with even number of double bonds are chiral while with odd number show geometric (*E-Z*) isomerism.

The replacement of one double bond in an allene by a ring gives alkylidenecycloalkanes (sometimes referred to an hemispiranes) does not alter the basic geometry of the system of allenes and suitably substituted compounds, therefore, exist in optically active form *e.g.*, (I and II Scheme 1.115). Related compounds in which sp^2 -carbon is replaced by nitrogen, *e.g.*, compound (III, Scheme 1.115) has also been obtained as enantiomers.

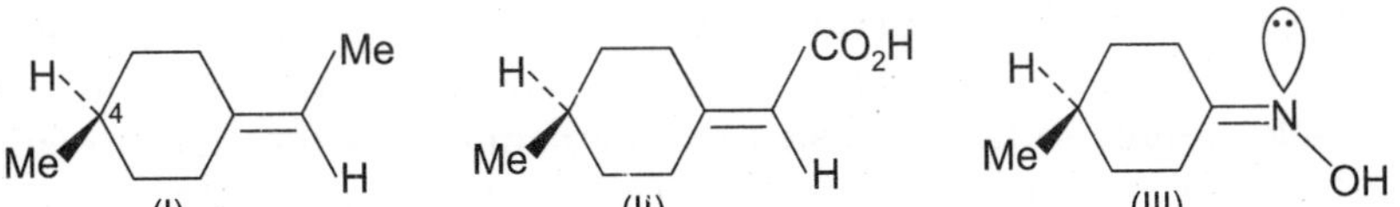

(I) (II) (III)

SCHEME 1.115

The replacement of both double bonds in an allene by ring systems gives a spiran ; appropriately substituted compounds have been obtained in optically active forms (IV and V scheme 1.115*a*).

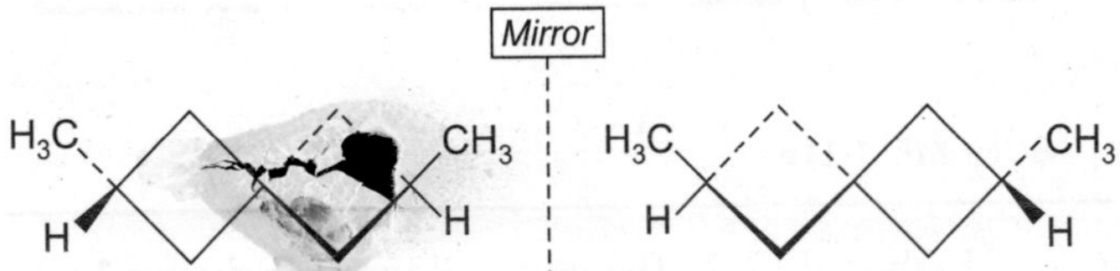

(IV)

Enantiomers of 2, 6-dimethylspiro [3, 3] heptane (a-chiral spiran)

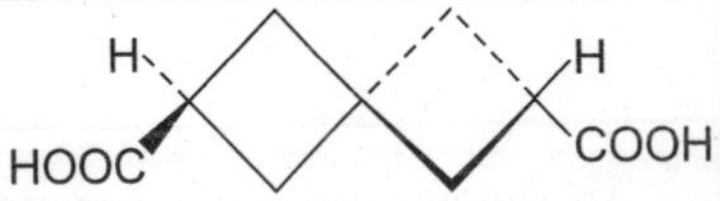

(V) chiral

SCHEME 1.115a

EXERCISE 1.20

Which of the following compounds (Scheme 1.115b) is expected to be chiral ?

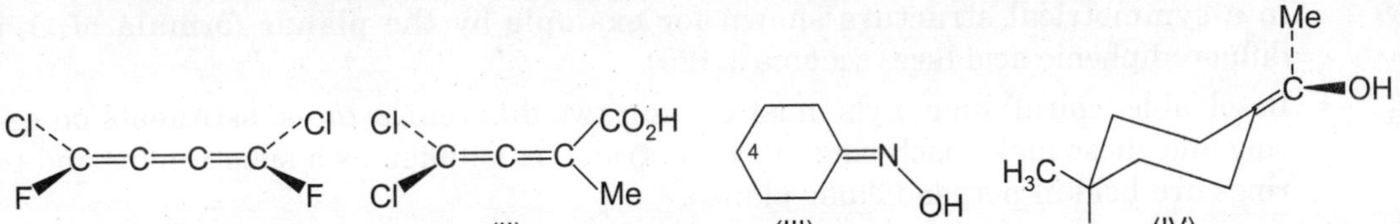

(I) (II) (III) (IV)

SCHEME 1.115b

ANSWER. *Compound (I) is achiral (the molecule contains three adjacent π bonds and thus the molecule behaves like an alkene and displays geometrical isomerizm and the configuration in (I) is Z ; (II) is achiral since in this allene one carbon carries two identical substituents ; (III) is achiral since there are no substituents at c_4 of the six-membered ring ; (IV) is chiral, however it is an enol and is convertible to more stable achiral ketone (Scheme 1.115c). The conversion of this ketone under acid or base catalysis will give back (IV) along with its enantiomer.*

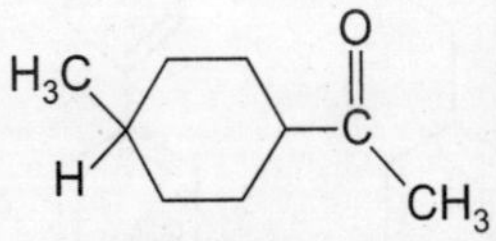

SCHEME 1.115c

(B) Optical Isomerism in Biphenyls (Atropisomerism)

(a) Optically Active Biphenyl Derivatives

The two conformations of *meso* -2, 3-dichlorobutane are nonsuperimposable mirror reflections of each other (Scheme 1.116). It is, however, not possible to separate the enantiomeric forms since rotation about the central C—C bond occurs very rapidly resulting in the interconversion of the two conformations. When the barrier to rotation about a C—C exceeds about 80kJ per mole in some suitably substituted compounds, the rotation at room temperature is slow enough to allow isolation of the two optically active isomers. This is so in the case of some biphenyls.

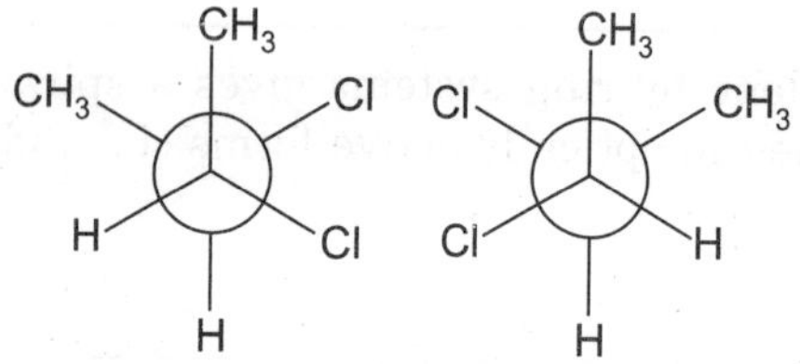

***meso-2-3-Dichlorobutane** has two conformations which are non-superimposable mirror images. The two enantiomers can not be isolated due to fast rotation about central C-C bond.*

SCHEME 1.116

In the crystal, both benzene rings of biphenyl lie in the same plane (I, Scheme 1.117). However, in solution and vapour phase the two rings are twisted with respect to each other by an angle of 45° due to steric interactions between the 2, 2′ and 6, 6′ pairs of hydrogens (II, Scheme 1.117). These interaction effects are further enhanced by *ortho* substituents larger than hydrogen so that:

- The rotation about the bond linking the two phenyl rings does not occur due to steric hindrance between the bulky *ortho* substituents.
- The two rings lie in different planes which are perpendicular, to make it impossible for the molecule to achieve a conformation in which the two aromatic rings are coplanar and this conformation has a plane of symmetry and this rules out chirality. This is so in a symmetrical structure shown for example by the planar formula of O, O′ difluorodiphenic acid (see scheme 1.120).
- Resolvable (chiral) biphenyls must contain two different *ortho* substituents on each ring and these make each ring unsymmetrical (H is taken as a substituent) and two rings are held in perpendicular planes.

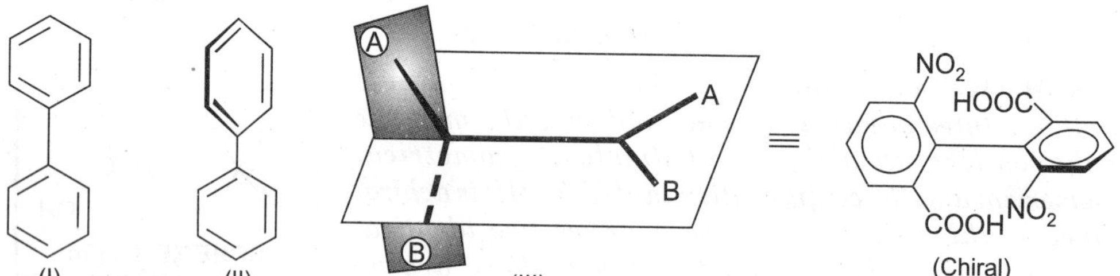

(I) Crystal (II) Solution
Bipnenyl

(III) Substituted biphenyl with different bulky *ortho* substituents A and B on each ring

(Chiral)

A = NO_2
B = COOH

SCHEME 1.117

- Each ring in a biphenyl individually represents a plane of symmetry irrespective of *ortho* substituents (and other groups). Consider the biphenyl which can be schematically represented (III Scheme 1.117) depicting the rings in perpendicular planes with two bulky *ortho* substituents (A) and (B) on each ring (A ≠ B). Such a biphenyl is chiral since neither of these planes can bisect the other into two equal halves.

- The biphenyl (III, Scheme 1.117) has a mirror image (Scheme 1.118) and can be resolved into its enantiomers and each enantiomer is stable indefinitely. The nitro and carboxylic groups are so bulky that they cannot pass by each other.

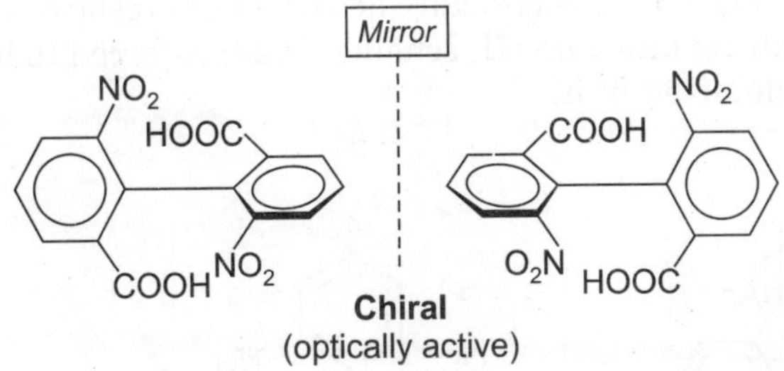

Bulky O-substituents restrict free rotation, the rings remain orthogonal (perpendicular to each other). When the O-substituents on either side of inter-ring bond are different the biphenyl becomes chiral.

SCHEME 1.118

- None of the rings should be symmetrically substituted, so that the molecule cannot have a plane of symmetry. Thus, the biphenyl (scheme 1.118) is chiral (A ≠ B in either pair, COOH ≠ NO_2 of the *ortho* substituents. The biphenyl (scheme 1.119) is, however, achiral. In this case *e.g.*, ring (II) is symmetrically substituted (A = B = COOH). A plane drawn perpendicular to ring (II, scheme 1.119) contains all the atoms and groups of ring (I) in it, hence it is a plane of symmetry and since it bisects the plane of ring B into two equal halves, thus the biphenyl (scheme 1.119) is achiral.

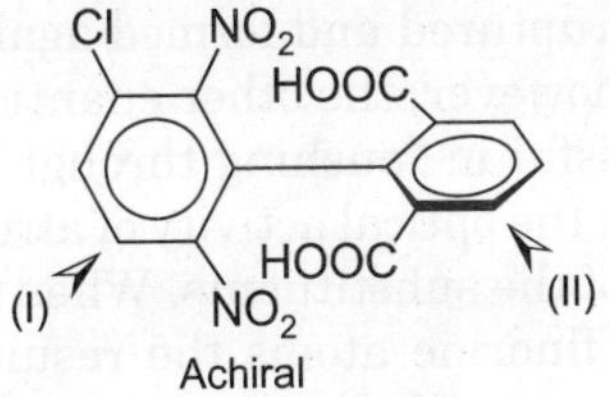

If one or both rings in a biphenyl have identical ligands, it is achiral. Each ring individually represents a plane of symmetry irrespective of O-groups. When held in mutually perpendicular planes, one plane of symmetry should not bisect the other (or vice versa) into two equal halves.

SCHEME 1.119

Atropisomerism

The chirality in biphenyls is generated by restricted rotation around a single bond provided each ring has an appropriately substituent pattern. Isolable stereoisomers resulting from restricted rotation about single bonds are called atropisomers while rotamers are stereoisomers obtained by rotation about a single bond.

The term atropisomer is derived from the Greek "without turning." Atropisomers may be configurationally stable at room temperature when the hindrance to rotation may be rather severe. Racemization can take place when the sizes of the ortho-substituents are such that these can slip past each other and the two rings become co-planar.

In order to display optical activity the substituents in the *ortho* position must be large enough to prevent the two rings from becoming copanar (the rotational energy barrier must be high enough so that interconversion of enantiomeric conformers does not occur). Thus *e.g.*, all attempts to resolve diphenic acid (scheme 1.120) have failed. The process of slipping a small hydrogen past the carboxylic acid group is very facile so that racemization of enantiomers occurs very rapidly through the planar form. In the planar from the center of symmetry is clearly seen. Interestingly the diamide (I, Scheme 1.120) is optically active and is resolvable. The activity is lost on hydrolysis since the resulting dilactam (II, scheme 1.120) is forced to be planar and a point of symmetry can be easily detected in it.

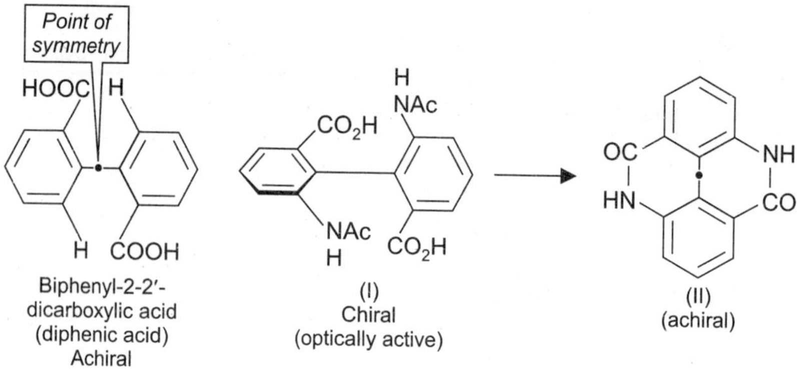

SCHEME 1.120

In the case of compounds with a stereocenter, or in the case of allenes, one enantiomer may interconvert into the other only if a chemical bond is ruptured and formed again, *i.e.*, by means of a chemical reaction. In the case of atropisomers, however, the other enantiomer may be formed if the substituent in one *ortho*-position is successful in "pushing through" past the smaller *ortho*-substituent on the second ring. The stability of the optical activity of atropisomers may thus serve as a measure of the size (effective volume) of the substituents. When the bulky nitro groups (Scheme 1.118) are replaced by the smaller fluorine atoms the resulting compound, 6, 6′-difluorobiphenyl-2, 2′-dicarboxylic acid can still display optical activity (Scheme 1.121). However, the compound racemizes readily, *i.e.*, the enantiomers are readily interconverted. The process involves squeezing fluorines past the adjacent carboxyl groups *via* the planar conformation. Once they reach the planar conformation the chirality is lost and racemization results. This transition state is congested and requires the bending of bonds. The process takes energy and is measurably slow.

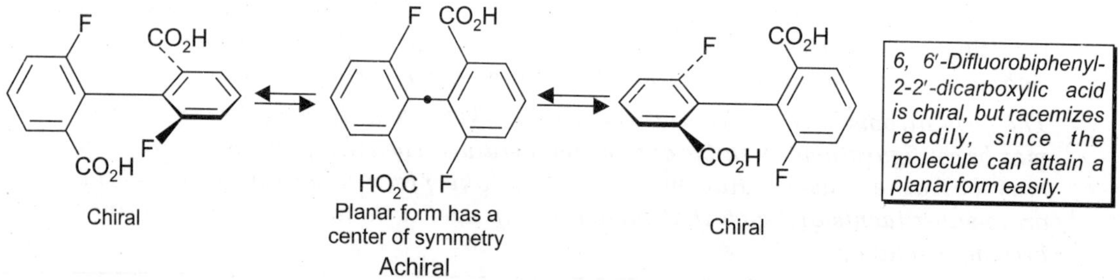

SCHEME 1.121

- Although the hydrogen atom is quite small, optically active compounds exist with two or even three *ortho*-positions of the biphenyl occupied by hydrogen. In the second case the benzene ring that is unsubstituted in the *ortho*-positions must have a substituent in a *meta*-position. The *meta*-substituent has no influence on rotation but it creates the necessary chirality of the molecule ; 3-bromobiphenyl-2′-trimethylarsonium iodide (I, Scheme 1.122) may serve as an example.
- In the case of (I, scheme 1.122) the rotation around the pivotal bond is only moderately restricted to allow interconversion between two enantiomeric atropisomers in solution and thus mono-*ortho*-substituted biphenyls in general are not resolvable, but show mutarotation when chiral additives are mixed. Thus the (+) camphorsulfonate of the arsonium salt (II, scheme 1.122) shows mutarotation (first order asymmetric transformation, see scheme 1.109) in solution. Here the chiral counterion (counter ion, an ion associated with another ion of opposite charge) (+) camphorsulfonate discriminates between the readily interconvertible atropiosomers of the biphenyl to give two unequally populated diastereomeric salts. Thus an interconversion of diastereomers occurs in solution since the two diastereomers (*i.e.*, diastereomeric salts) are not equally stable.

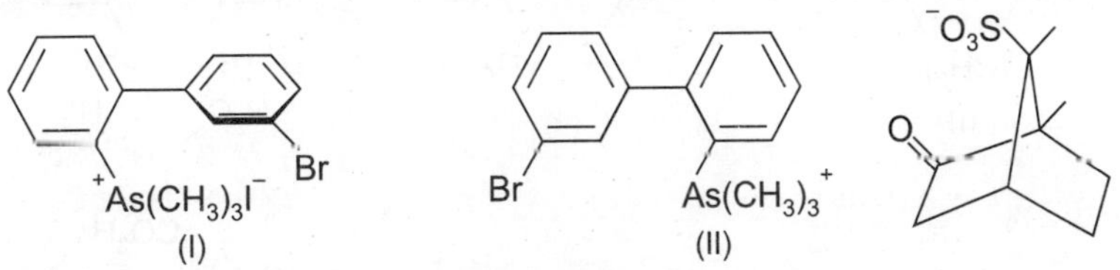

SCHEME 1.122

- In addition to the bulk of the *ortho* substituents, the substituents in the *meta*-position tend to enhance racemization barriers by what is called as "buttersing effect", *i.e.*, by preventing the outward bending of an *ortho* substituent, which would otherwise occur in the transition state (coplanar conformation) for racemization. This bending would allow the *ortho* substituents to slip past each other more readily. Thus the rate of racemization of the 3-nitroderivative (I, scheme 1.123) is much lower compared with the 5′-nitroderivative (II, scheme 1.123).

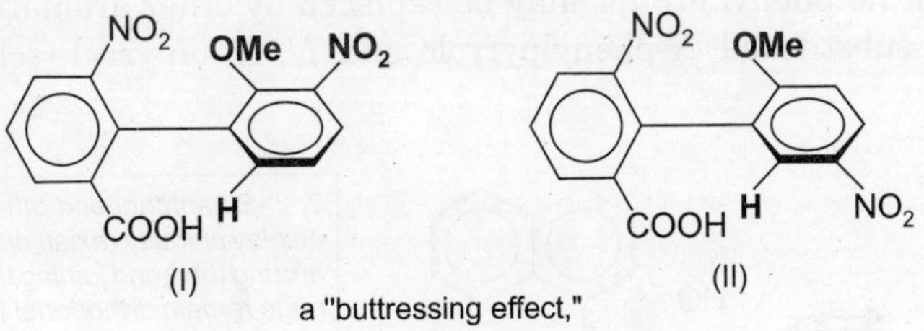

SCHEME 1.123

The apparent order of steric hindrance produced by different groups (as gauged by racemization rates of differently *ortho*-substituted biphenyls) in Br > CH_3 > Cl > NO_2 > COOH > OCH_3 > F > H. This order roughly parallels van der Waals radii of atoms and groups.

(b) Bridged Biphenyls

When a 2, 2′-bridge of the type shown in (I, scheme 1.124) with $n = 1$, one has a disubstituted fluorene, this being a planar molecule, is not resolvable. When $n = 2$, the compound is a disubstituted 9,10-dihydrophenanthrene. The nonplanar six membered ring can give rise to atropisomerism and the compound is resolvable, provided the other two *ortho*- positions are substituted with bulkyl groups *i.e.*, methyl (II, scheme 1.124), otherwise they slip through the plane readily (please consider models). In case n is larger than 2 the bridged biphenyls lead to atropisomers irrespective of the bulk of two *ortho* substituents. In these cases the non-planarity is maintained by the puckering of the rings which in planar configuration suffer from angle strain and non-bonded interactions. An example of such a compound is (III, scheme 1.124), however, these bridged biphenyls undergo easy racemization since the angle strain and steric interaction in medium rings is not very large.

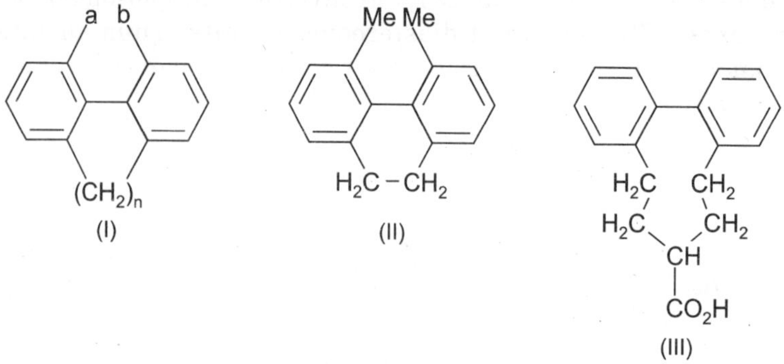

SCHEME 1.124

(c) Atropisomerism in Compounds Other Than Biphenyls

1, 1′-Binaphthyl may be looked as a di-*ortho*-substituted biphenyl (for structure see Scheme 1.109c) and has been resolved. The 2, 2′-binaphthol (scheme 1.125) is a constituent of a useful chiral reagent BINAL -H (see, scheme 2.54). The rotation around the pivotal bond is hindered due to steric factors. In 1, 1′-binaphthyl itself *e.g.*, the peri H's provide enough steric hinderance to keep the rings non-planar.

One or both of the phenyl groups may be replaced by other aromatic or heteroaromatic rings, appropriately substituted *N*-phenylpyrrole and *N, N′*-bipyrryl (scheme 1.125) are resolvable.

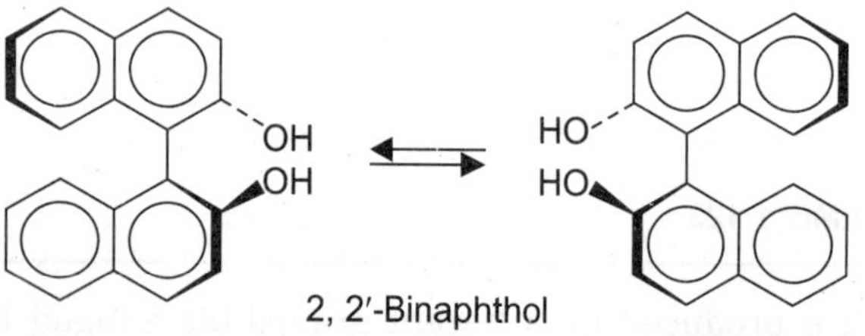

2, 2′-Binaphthol and other related compounds display chirality (axial) due to restricted rotation around the bond joining the aromatic units. The rings remain orthogonal (perpendicular) to one another and the structure is able to exist in two enantiomeric forms (atropisomerism). 1, 1′-Binaphthyl is constituent of several chiral reagents (see schemes 2.54 and 6.59b)

SCHEME 1.125

Optical isomerism also arises in suitably substituted polyphenyls, however, the stereochemistry is more complex since both *meso* from and geometrical isomerism are also possible. For example, consider the terphenyl drivatives (I and II, scheme 1.126), where restricted rotation may arise around two pivotal bonds, consequently the two terminal phenyl groups are co-axial as well as coplanar. The terphenyl compounds of the type shown (Scheme 1.126) exist in three stereoisomeric forms: a chiral *cis* isomer (exists as an enantiomeric pair) while the *trans* isomer (in which the CH_3 groups on the end rings are on opposite sides) with a center of symmetry is an optically inactive *meso* form.

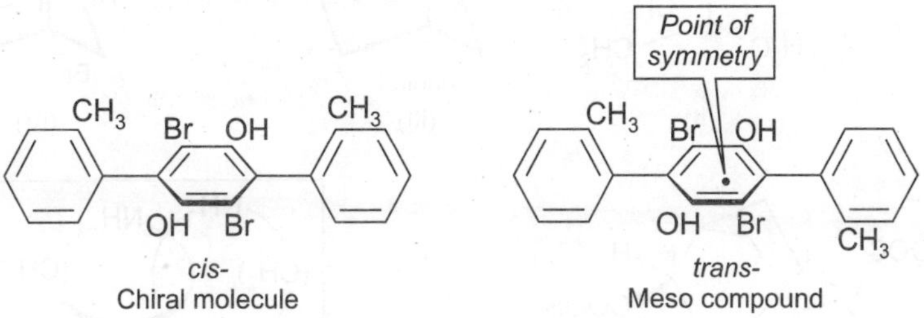

SCHEME 1.126

Atropisomers may also be displayed by compounds in which one benzene ring is replaced by a substituted ethylene group styrenes or another substituted group *i.e.* situations with restricted rotation about single bonds (Scheme, 1.127).

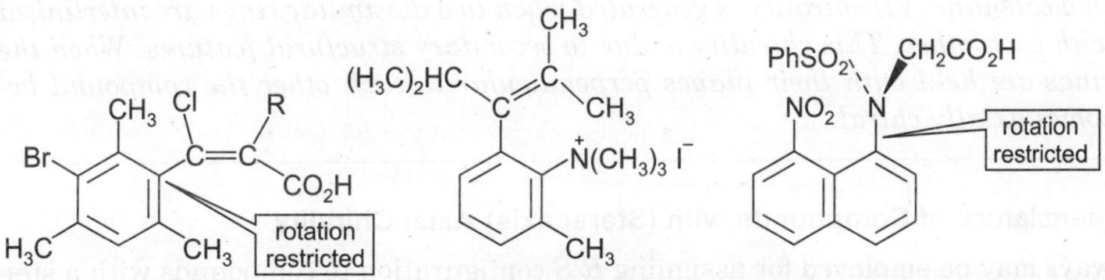

SCHEME 1.127

Optical Activity of Adamantanes and Catenanes

An example of a cyclic compound e.g., an adamantane derivative where the center of chirality is not located on any atom but in the unoccupied space inside the molecule has been presented (see, scheme 1.110). Adamantane has only two different carbons, the CH and CH_2 groups. Symmetry makes the four methine groups equivalent and the six methylene groups equivalent (scheme 1.127a). This makes (III and IV) equivalent (see the relative location of substitutents in both III and IV). The compound with two stereocentres is achiral with a plane of symmetry. One may note that dimetrically opposite methylene planes are perpendicular to one another and with proper substitution axial chirality is generated as is so in allenes. One such example of an optically active adamantoid is (V, Scheme 1.127a) which has axial chirality. In this case C2 and C6 methylenes are not planar and dissymmetrically substituted. In such chiral adamantoids one can detect an (imaginary) chiral axis which passes through the two substituted terminal carbon *atoms and the geometrical center of the adamantane ring system. One has already seen this geometrical center becoming a stereocenter when the bridgehead carbons carry different substituents. The configuration to an adamantoid e.g., (V, scheme 1.127a) is assigned as described (see problem 1.38).*

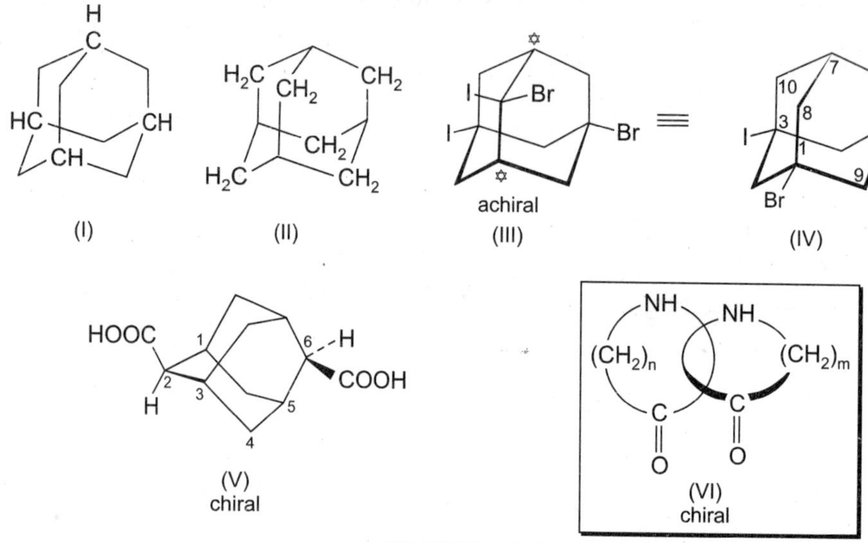

SCHEME 1.127a

In a catenane (VI) chirality is generated when two dissimilar rings are interlinked with each other. This chirality is due to secondary structural features. When the rings are held with their planes perpendicular to each other the compound becomes axially chiral.

(C) Nomenclature of Compounds with (Stereoaxis) Axial Chirality

Three ways may be employed for assigning *R*/*S* configuration to compounds with a stereoaxis. One may adopt any of these. The last third method is developed as a problem solving hint and is the easiest.

First method

(*i*) Allenes

For the assignment of configurational nomenclature to axially chiral molecules *e.g.*, allenes, the orbital picture (see scheme 1.113) is projected to a Newman formula *i.e.*, (I → Ia scheme 1.128). The following points may be noted when the procedure is applied to (*S*)-2, 3-pentadiene.

- The molecule can be viewed from either end of the stereoaxis to give the same descriptor. The near groups are given priority over far groups.
- For better visualization it is always better to put the groups nearest the view direction on a thick line (horizontal or vertical).
- Thus in the projection formula of (*S*)-2, 3-pentadiene (scheme 1.128), the horizontally placed, (nearest the view direction) CH_3 and H are to be numbered 1 and 2 and the remaining vertically placed (rear) ligands are simply assigned priorities according to sequence rules *i.e.* CH_3 and H are numbered 3 and 4 respectively. The sequence 1 → 2 → 3 gives the configurational descriptor *S* (anticlockwise).

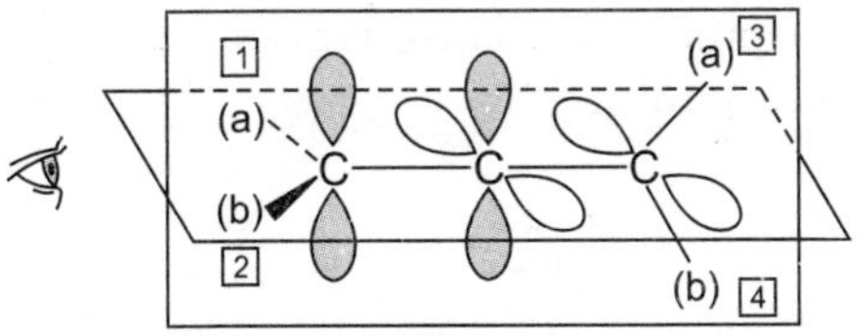

(I)
a > b

The near groups (direction of view) have priority over far groups

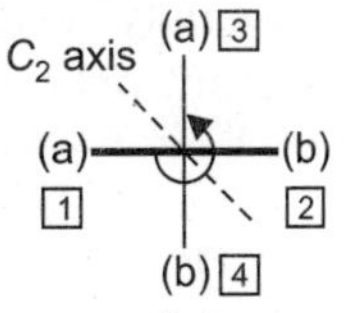

(Ia)
Priority a > b
The (*S*)-form

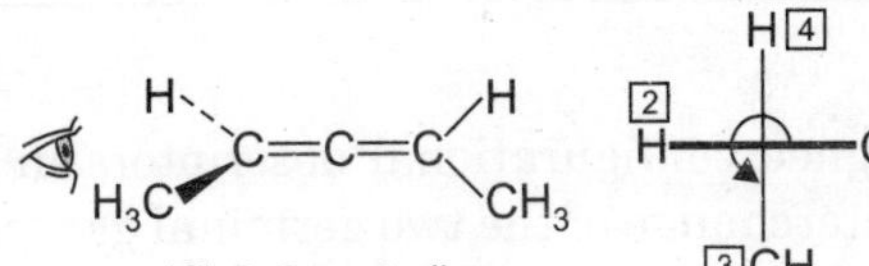

(*S*)-2, 3-pentadiene (II) (*S*)- (IIa)

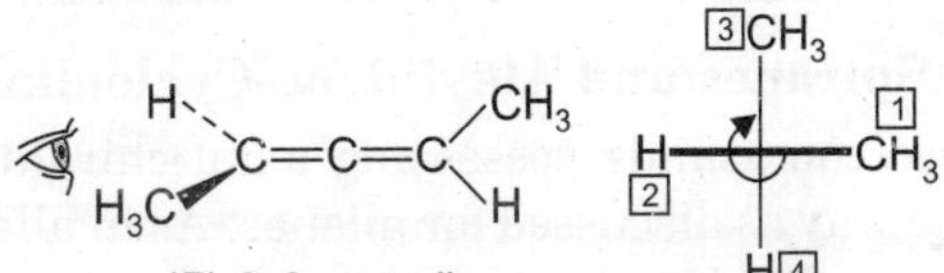

(*R*)-2, 3-pentadiene (*R*)–

SCHEME 1.128

- Interchange of the two geminal groups at one end in these molecules gives the enantiomer, *e.g.*, (*R*)-2, 3-pentadiene is obtained from the (*S*)-enantiomer (scheme 1.128) and is properly named as shown.
- As already said, it makes no difference which end the structure is viewed, the stereodescriptor will remain unchanged. Consider again (*S*)-2-,3-pentadiene now viewed from other direction (scheme 1.129) as compared to that in (scheme 1.128). Now the far groups (view, scheme 1.128) become the near groups and consequently get priority. These are put on the thick vertical line and the configuration again comes out to be (*S*).

> *A beginner may note (ignoring the priority numbers) that the position of vertical groups in Newman projection (IIa, scheme 1.128 and III, scheme 1.129 remains the same from either view direction (front or rear) however, the groups on horizontal line have interchanged positions—the H on the horizontal line in (view II, scheme 1.128) is on the left but will come on the right when viewed from the other direction.*

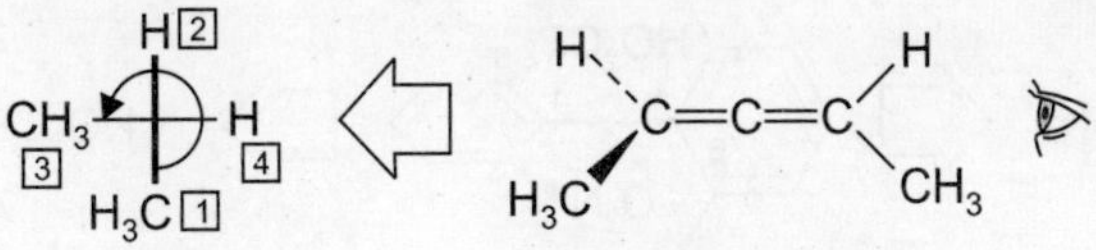

(*S*)- (III) (*S*)-2, 3-pentadiene

SCHEME 1.129

- For practice one may assign stereodescriptors to the two enantiomers of 2, 3-pentadiene drawn in different orientation (Scheme 1.130).

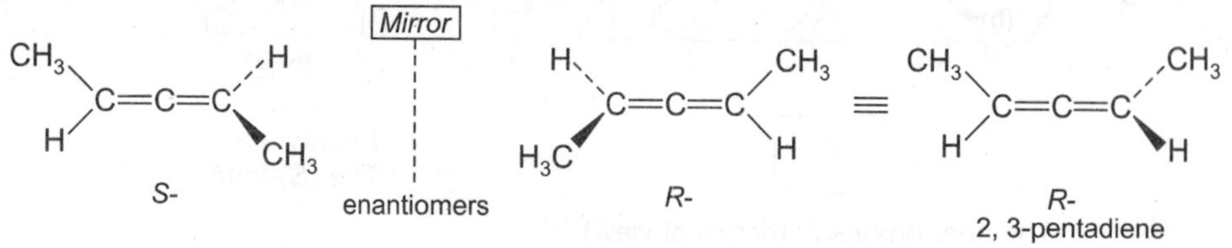

SCHEME 1.130

(*ii*) Spiranes and Alkylidene-Cycloalkanes

These compounds possessing axial chirality are assigned configurational descriptors in the same way as discussed for allenes. As in allenes, the interchange of the two germinal groups in these molecules also leads to enantiomers (II compared to I, scheme 1.131).

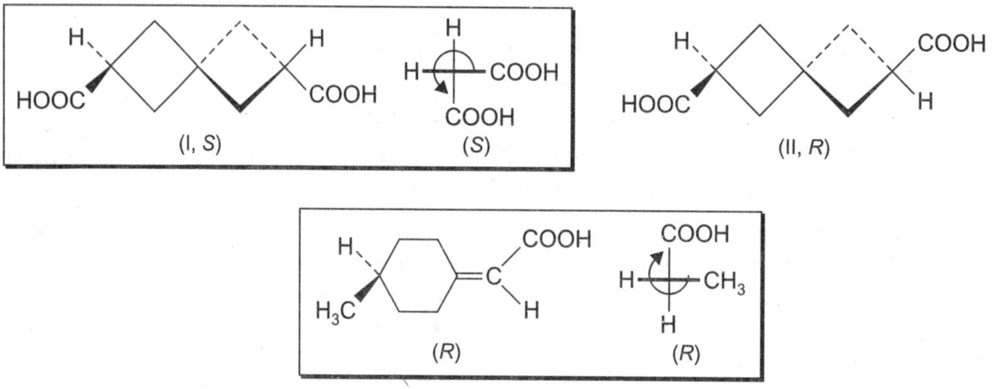

SCHEME 1.131

(*iii*) Biphenyls

In the case of biphenyls, however, it is important to note that the four *ortho*-carbons are sequenced properly according to CIP rules. Thus in the biphenyl (scheme 1.132) in the left ring the sequence is C-2 > C-6 > (OCH_3 > C-H) while in the other ring C2′ > C6′ (N > C) and therefore, configuration is S when the biphenyl (scheme 1.132) is viewed from either direction.

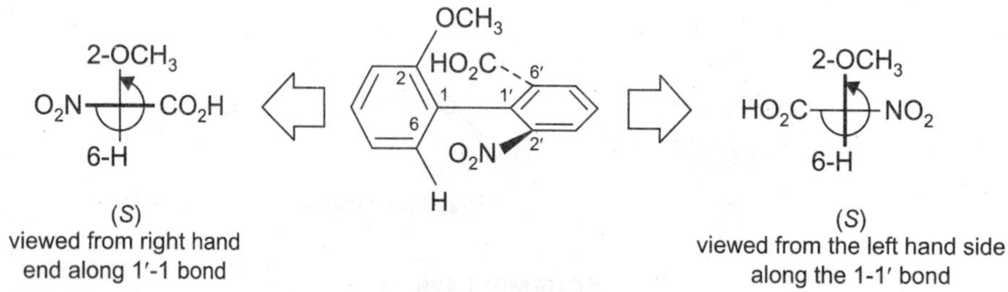

SCHEME 1.132

When in a biphenyl the C2 and C6 in a ring are attached to identical atoms, the priority order of the *ortho* carbon atoms is then determined through exploration around the ring or side chain. Thus in the biphenyl (scheme 1.133) the 2-H adjacent to 3-Br precedes 6-H.

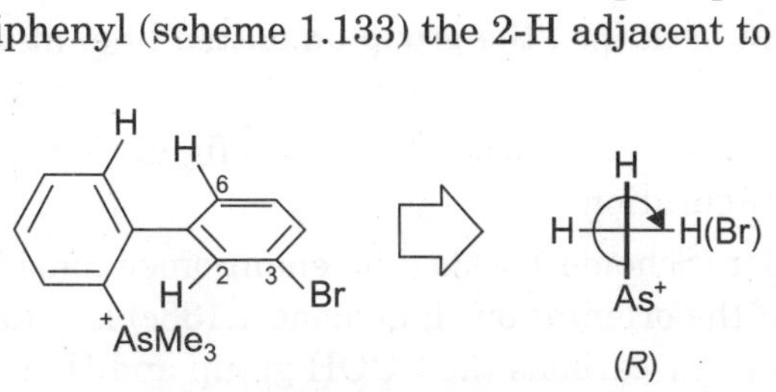

SCHEME 1.133

Second method

A simple mental exercise to assign R/S configuration to compounds with axial chirality

P.S. Kalsi: *A Simple Mental Exercise for Assigning Configurational Descriptors (R/S) to Molecules with a Stereoaxis (Chiral axis)*, Int. J. Chem. Sci. 2(3), 2004, 285–290.

This is the simplest exercise developed by author and can be used quickly to assign configuration to any compound with a stereoaxis and is used by adopting the following procedure:

- Turn the molecule to bring the groups in the plane of paper (continuous lines) at the top while the pair of groups which are perpendicular are put at the bottom (I-II, Scheme 1.133*c*).
- Always view the molecule from bottom, thus the bottom groups take precedence over the other pair at the top (II, Scheme 1.33*c*) and thus assign priorities to all the four groups.
- Draw a cross (III, Scheme 1.133*c*) and put the groups (as their priorities) on the top of (II) on the horizontal line as you see these in (II). Thus Br(3) is on the right hand side while CH_3(4) is on the left.

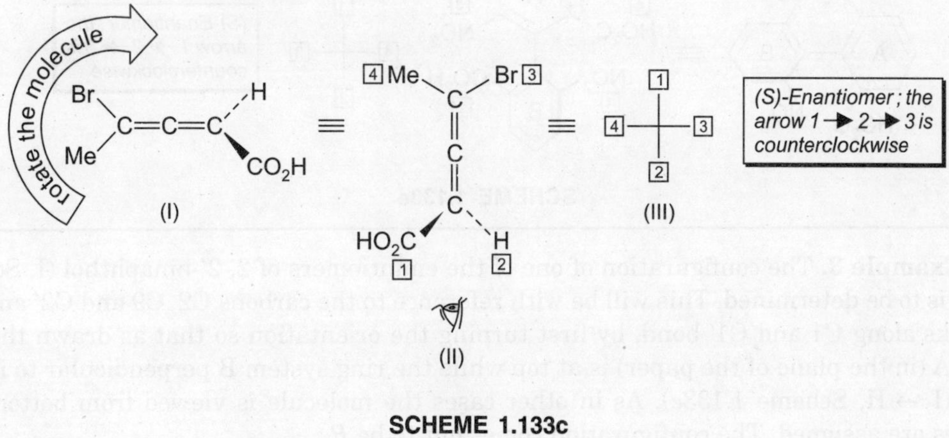

SCHEME 1.133c

- Put the groups of the bottom pair (perpendicular to the plane of paper) on the vertical line of the cross with the group on the thick line (*i.e.*, the group at the front) at the top while the group at back (*i.e.*, the group on dotted line) at the bottom of the vertical line.
- The sequence $1 \rightarrow 2 \rightarrow 3$ clockwise show *R* configuration while $1 \rightarrow 2 \rightarrow 3$ anticlockwise shows *S* configuration.

Example 1. Consider (Scheme 1.133*d*) the enantiomer considered here is the same as (I, Scheme 1.133*c*). Thus if the orientation (I, Scheme 1.133*c*) is rotated clockwise around the axis defined by the three allenic carbons the COOH group and H on right occupy the plane of paper while Br and Me become perpendicular to give the same enantiomer (Scheme 1.133*d*).

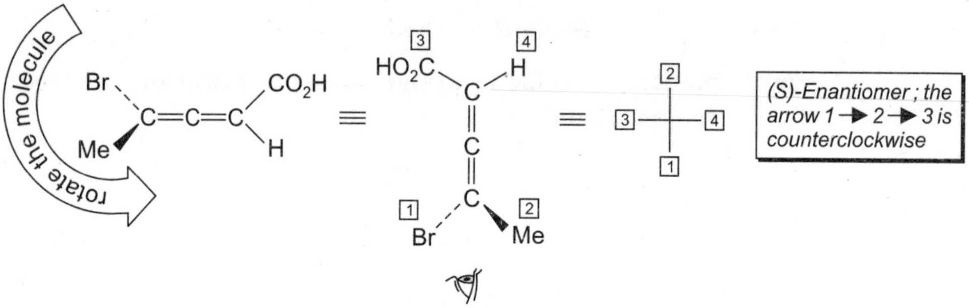

SCHEME 1.133d

Example 2. Consider one of the enantiomers of a chiral biphenyl by considering the relevant *ortho* groups (Scheme 1.133*e*). One turns the molecule so as to put the ring A (in the plane of paper) at top while ring B which is perpendicular to ring A at the bottom. Looking from bottom and assignment of priorities to the *ortho* groups shows that it is *S* enantiomer. The NO_2 group in ring B is on a thick portion of the molecule so it is at the top of vertical line of the cross.

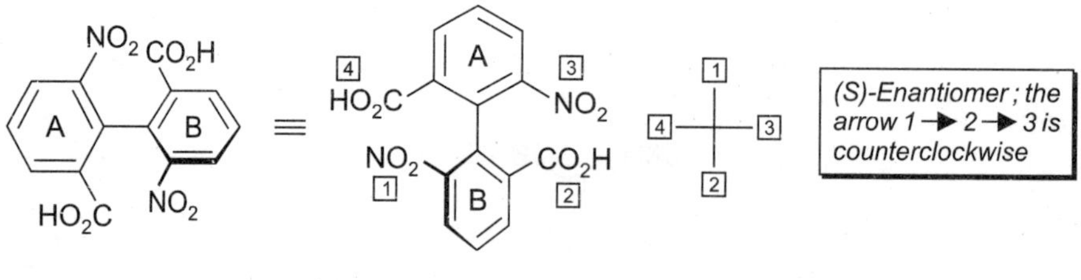

SCHEME 1.133e

Example 3. The configuration of one of the enantiomers of 2, 2′-binaphthol (I, Scheme 1.133*f*) is to be determined. This will be with reference to the carbons C2, C9 and C2′ and C9′. One looks along C1 and C1′ bond, by first turning the orientation so that as drawn the ring system A (in the plane of the paper) is at top while the ring system B perpendicular to it is at bottom (I → II, Scheme 1.133*e*). As in other cases the molecule is viewed from bottom and priorities are assigned. The configuration comes out to be *R*.

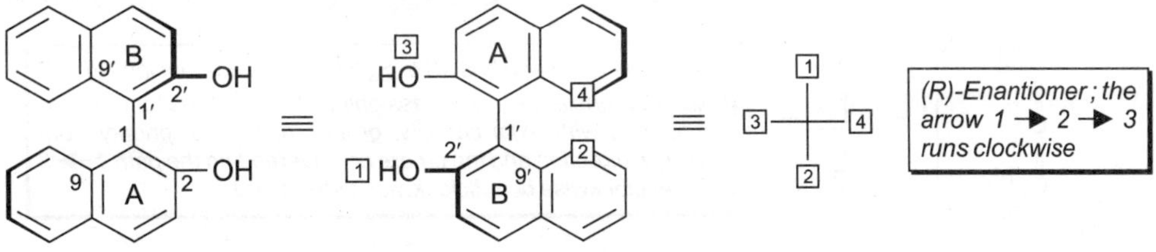

2, 2′-Binaphthol (I)

(II)

SCHEME 1.133f

Example 4. A similar procedure is applied to cyclic molecules with an exocyclic alkylidene moiety *e.g.*, (Scheme 1.133*g*). This system has two different groups at C4 of the cyclohexyl ring while two different groups are present at the π bond. The cyclohexyl "arms" (ring residues) are in the plane of the π bond and COOH and H group along with the cyclohexyl arms are in the plane of paper and are already at the top. The CH_3 and H at C4 of the cyclohexane ring are in a different plane and these constitute the bottom. The molecule when viewed from bottom with assigned priorities come out to have R configuration.

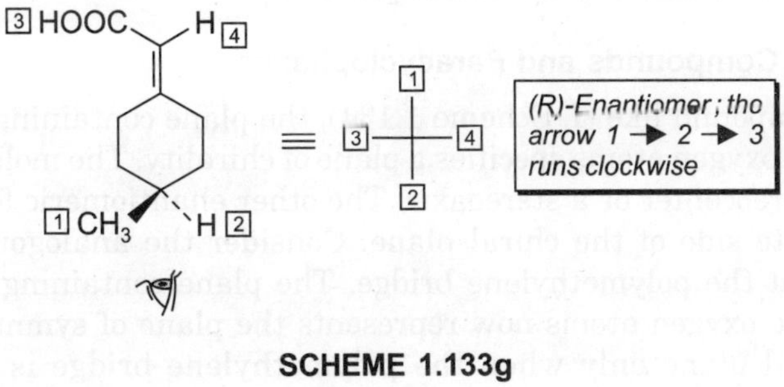

SCHEME 1.133g

(D) Helical Descriptors *M* and *P* for Molecules with Chiral Axes

The perpendicular segment of the substituents about the chiral axis (stereoaxis) means that such compounds like, allenes, biphenyls, alkylidenecycloalkanes, etc. have very short helical segments and their configuration may be denoted as *P* or *M* (as for compounds which are chiral due to helical structure). For assignment of helix descriptors only the groups of highest priority nearer to the view direction and the far are considered *e.g.*, in (II, scheme 1.128) the considered, ligands are 1 and 3. In case the turn from priority front group 1 to priority rear group 3 is clockwise. The stereodescriptor is *P*, if anticlockwise it is *M*. Thus (*S*)-2, 3-pentadiene (II, Scheme 1.128) is *aS* (chiral axis nomenclature) or *P* (helix nomenclature). The correspondence of *aR* with *M* and *aS* with *P* is general.

One has seen that the smallest carbocyclic ring that can incorporate a double bond with trans-geometry is eight membered (see schemes 1.54 and 1.137). Similarly the smallest sized ring which can contain an allene has nine carbons. One enantiomer of this compound of cyclonona-1, 2-diene (Scheme 1.133*h*) has the *S* configuration as arrived by adopting third method or *P* configuration (helix nomenclature).

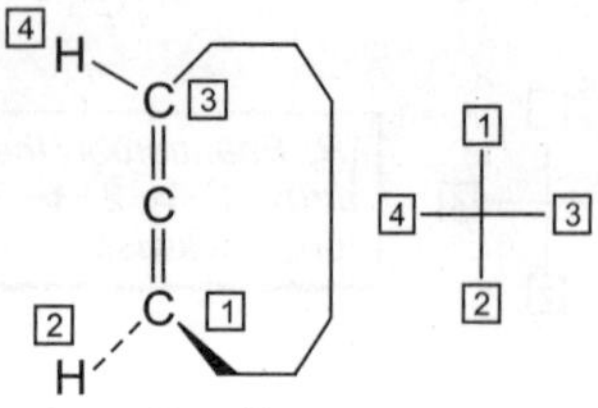

aS or P

> *Cyclonona-1, 2-diene has aS or S_a (chiral axis nomenclature) or P (helix nomenclature). For assigning M or P descriptors to compounds with axial chirality, groups of highest priority from near view direction and the far are considered and the path 1 → 3 whether clockwise or anticlockwise determined.*

There is a correlation between the two types of stereodescriptors

$R_a \Rightarrow M$; $S_a \Rightarrow P$

SCHEME 1.133h

1.17 OPTICAL ACTIVITY DUE TO STEREOPLANE (PLANAR CHIRALITY)

A chiral plane contains as many of the atoms of the molecule as possible, however, not all. The chirality of such compounds is due to the fact that normally one or more groups are not placed in the chiral plane. Thus *e.g.*, in a paracyclophane (Scheme 1.135) the more substituted benzene ring at the bottom is taken as the chiral plane.

(A) Chirality of Ansa Compounds and Paracyclophanes

If one examines a compound like (I, Scheme 1.134), the plane containing the substituted benzene ring and the two oxygen atoms specifies a plane of chirality. The molecule is chiral, though it does not have a stereocenter or a stereoaxis. The other enantiomeric form of I will have the bridge on the opposite side of the chiral plane. Consider the analogous molecule 2-bromohydroquinone without the polymethylene bridge. The plane containing the substituted benzene ring and the two oxygen atoms now represents the plane of symmetry and this attains the status of a chiral plane only when the polymethylene bridge is incorporated into 2-bromohydroquinone. Thus, the derivatives of hydroquinone, are optically active having planar chirality.

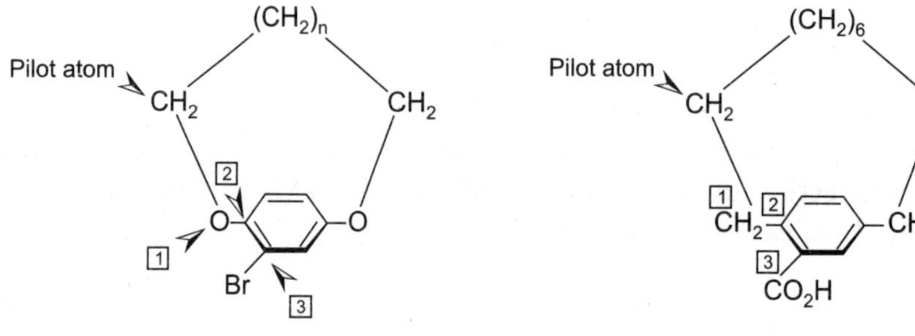

(I) (R)- (II) (R)-

SCHEME 1.134

These derivatives of hydroquinone are called *ansa* compounds (ansa-handle in *Latin,* in ansa compounds, the two *para* positions of an aromatic ring are attached to hetero-atoms which in turn are connected through a polymethylene chain). In all chiral compounds of this type the aromatic ring is dissymmetrically substituted (*i.e.*, two-dimensionally chiral) and the polymethylene bridge can be either above or below the plane of the aromatic ring to give enantiomers.

Optically active forms (enantiomers) of (I, Scheme 1.134) have been isolated if $n = 8$. The methylene ring is perpendicular to the plane of the benzene ring ; substituent – Br, prevents the rotation of the benzene nucleus inside the large ring. The compound (I, Scheme 1.134 $n = 8$) is resolvable, however when $n = 9$ the racemization is very easy since now the out of plane movement of the benzene ring is no longer restricted due to steric factor and the two enantiomeric forms are not configurationally stable.

A simple paracyclophane with one aromatic ring (II, scheme 1.134) which resembles an ansa compound has been resolved. In other paracyclophanes the two benzene rings are approximately parallel (arranged one above the other). To have enantiomerism at least one of the rings must be dissymmetrically substituted, (scheme 1.135) and when the methylene chains are small, the substituted ring cannot rotate within the cyclophane system so that interconversion of enantiomers (racemization) does not occur, and this is a form of atropisomerism. When each methylene bridge contains four methylene groups the compound cannot be resolved. Evidently, rotation of the substituted phenyl ring is now rapid for racemization.

(B) Nomenclature of Compounds with Planar Chirality

To assign configuration to compounds with chiral planes the sequence rule is extended by choosing a pilot atom. This is a priority atom directly linked to the plane but is not itself in the plane, thus it is first out of the plane atom. The pilot atom is chosen from that side of the plane which is most preferred (CIP rules). In the case of (I, scheme 1.134), the preferred side of the plane is the one which has the *ortho* bromine and therefore, the methylene group on the left hand side is the pilot atom. When both sides are equivalent, the pilot atom can be chosen from any side of the plane. Then one classifies the adjacent three atoms (again chosen by precedence if there is a choice) of the plane as these are encountered along the bonds and given numerals 1, 2 and 3. Now viewed from the pilot atom if the order $1 \rightarrow 2 \rightarrow 3$ describe a clockwise array in the chiral plane, the configuration is *R* if the array is counter-clockwise, the configuration is *S*. This procedure is amply clarified in compounds (Schemes 1.134 and 1.135).

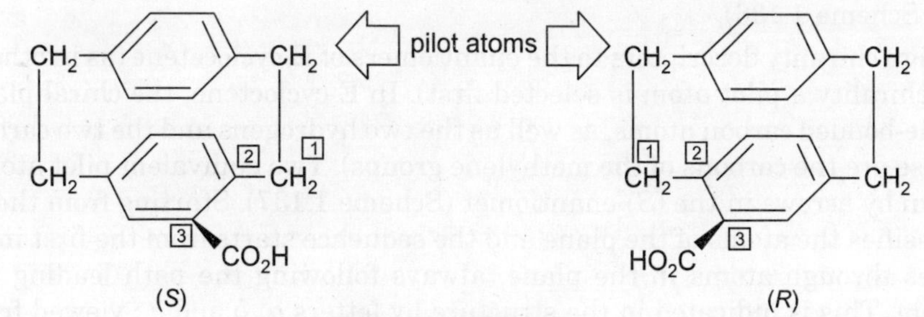

Paracyclophanes

SCHEME 1.135

In the molecule (I, Scheme 1.134) *e.g.*, this path is O—C—C (Br), which is clockwise and thus, the configuration is (*R*).

(C) Trans Cyclooctene

Another special example of planar chirality is *trans*-cyclooctene (see, Scheme 1.54). The introduction of a *trans*-double bond into rings containing six carbon atoms or less is not possible since it would introduce large strain in the molecule. *Trans*-cycloheptene has only been detected with spectrometers, it has a short lifetime and has not been isolated.

Trans-cyclooctene, on the other hand, has been isolated. The ring is large enough to accommodate the geometry required by a *trans* double bond, and thus, *cis*-and *trans*-forms of cyclooctene have independent existence (see, scheme 1.54) *trans*-Cyclooctene is chiral due to planar chirality the two double bonded carbon atoms, as well as the two hydrogens and the two carbons directly adjacent constitute the chiral plane and polymethylene bridge (4 methylene chain) is skewed in the third dimension. One may recall that *cis* and *trans* 2-butenes on the other hand have trigonal stereocenters these molecules are not chiral since each has a plane of symmetry (the plane of the molecule).

The interconversion of the two enantiomers in *E*-cycloctene (Scheme 1.136) requires the swinging of the tetramethylene chain over and below the plane of the trigonal atoms (chiral plane) and this swinging is opposed by ring strain. In the case of higher homologues the mobility of the polymethylene chain increases and consequently the rotational barrier decreases. As a result *e.g.*, in the case of *trans*-cyclononene the optically active form can be detected only at – 80°C while *trans*-cyclodecene is an extremely mobile system.

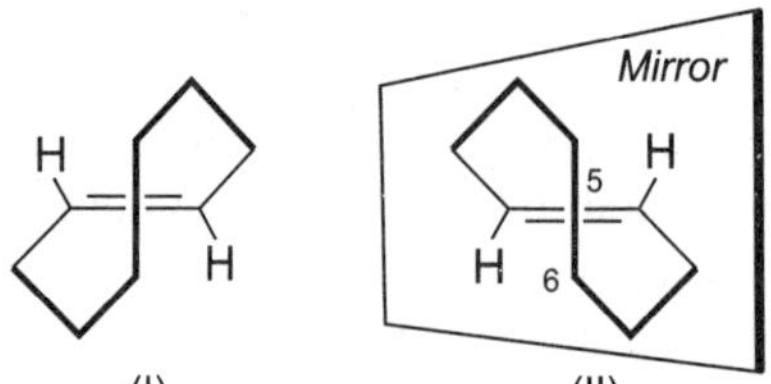

trans-Cyclooctene is chiral due to the presence of a stereoplane (planar chirality), constituted by two trigonal carbons, the two hydrogen atoms on these and two directly adjacent carbons. The four methylene bridge is skewed in the third dimension.

(I) (*R*)-Enantiomer   (II) (*S*)-Enantiomer

trans-Cyclooctene

SCHEME 1.136

The molecule of *trans*-cyclooctene (*i.e.*, E-cycloctene) has a C_2 axis which passes through the center of the double bond and bisecting C5, C6 bond and the molecule thus belongs to point group C_2 (II, Scheme 1.136).

To assign chirality descriptors to the enantiomers of E-cyclocetene (as in other molecules with planar chirality a pilot atom is selected first). In E-cycloctene, the chiral plane contains the two double-bonded carbon atoms, as well as the two hydrogens and the two carbons directly adjacent (these are the carbons of the methylene groups). Two equivalent pilot atom therefore, exist as shown by arrows in the (*S*)-enantiomer (Scheme 1.137). Starting from the either pilot atom one classifies the atoms of the plane and the sequence starts from the first in-plane atom and continues through atoms in the plane, always following the path leading to the more preferred atom. This is indicated in the structure by letters *a*, *b* and *c* ; viewed from the pilot atom if the order $a \to b \to c$ is clockwise, the configuration is (*R*) and if it is anticlockwise, the configuration is (*S*) (Scheme 1.137). The configuration comes out to be (*S*) irrespective of the chosen pilot atom since in each case the paths are C—C=C. The path $a \to b \to c$ is to be explored from the pilot atom and for better visualization, the pilot atom should be brought to the front with the double bond on the back.

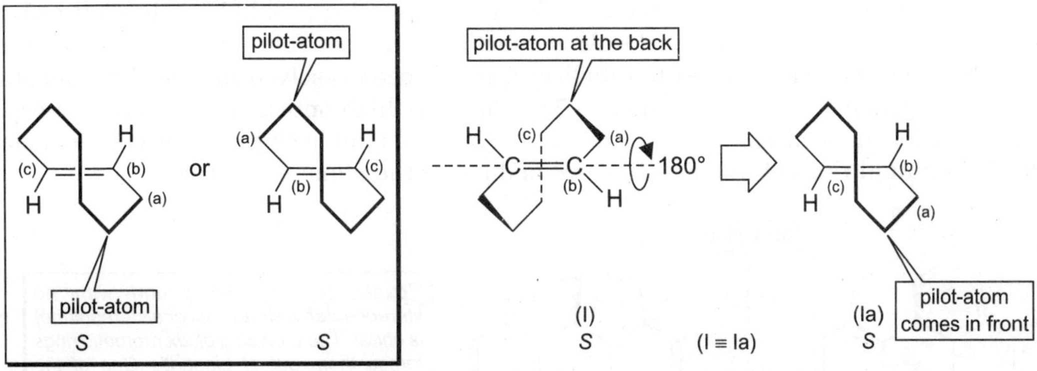

SCHEME 1.137

If an enantiomer of *trans*-cyclooctene is drawn as in (I, scheme 1.137) with the double bond in front and then the pilot atom is at the back. However, to assign configuration the molecule is to be viewed from the pilot atom. An easy procedure is to turn the molecule so that the rotated (equivalent) orientation (*Ia* scheme 1.137) can be easily viewed from the pilot atom which is now in the front. The configuration thus comes out to be *S*.

Just like axial chirality, planar chirality can be looked as a kind of helicity and specified as P or M. To assign P or M nomenclature the pilot atom is included with atoms *a*, *b* and *c*. It comes out that the *p*R enantiomers (chiral plane nomenclature) correspond to P while *p*S correspond to M *i.e.*, opposite to the correlation in axial chirality.

1.18 OPTICAL ACTIVITY OF COMPOUNDS DUE TO HELICITY

Helices (Scheme 1.138) are chiral objects, a right-handed helix (a clockwise rotation when viewed along the axis and moving from the front to the rear) is designated P (plus) while a left-handed helix is M (minus). The concept of helicity provides a simple method for designating the chirality of compounds with this feature. The terminal benzene rings in hexahelicene (Scheme 1.138), for example, cannot occupy the same plane without coming in serious conflict with one another. Therefore, the molecule is forced to adopt a nonplanar shape in which one side of the molecule must lie above the other because of crowding.

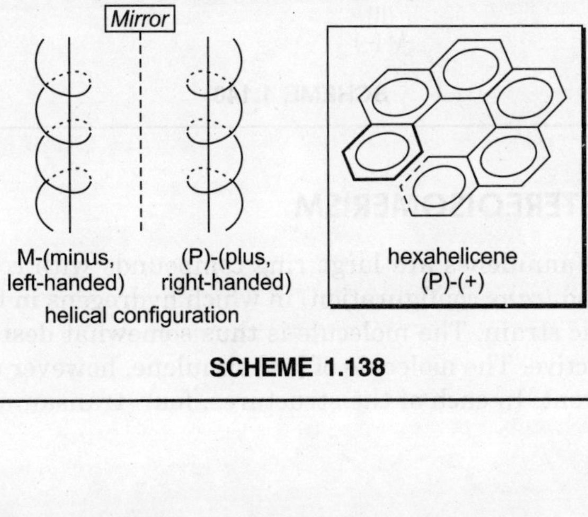

SCHEME 1.138

Hexahelicene is chiral by virtue of its helical shape which could be either left-or-right-handed in orientation. The entire molecule is infact less than one full turn of the helix, but this is enough to generate chirality in hexahelicene, it has been resolved into remarkably stable enantiomers (Scheme 1.139) which display spectacularly high optical activity and correspond to 'right' and 'left-handed' spirals. In hexahelicene and the middle rings (3 and 4) lie in a plane, while the terminal rings (1 and 6) fall above and below the plane respectively.

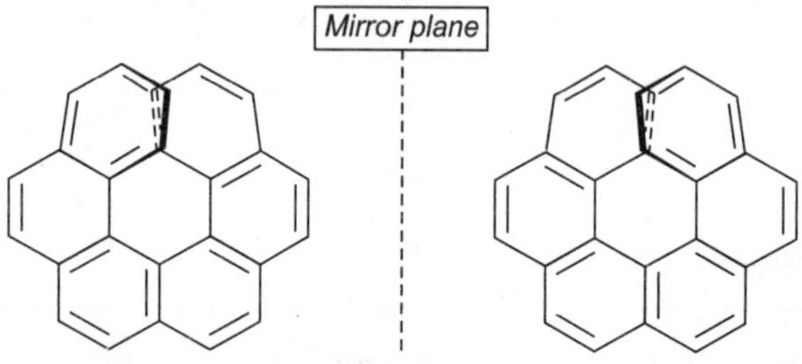

Hexahelicene

Hexahelicene (a compound with no stereocenter, a stereoaxis or a stereoplane) is chiral. The crowding of six aromatic rings forces them out of planarity. One of the terminal rings lies above (one enantiomer) or below (second enantiomer) the other so that the molecule adopts a helical shape.

SCHEME 1.139

Aromatic compounds represent another category, in which suitable substitution forces the molecule to buckle from the most favourable, planar arrangement. For example, phenanthrene, a planar molecule, when substituted in positions 4 and 5, for instance, by methyl groups becomes somewhat skewed and exists as two enantiomers. Consequently, it has been possible to isolate the optically active (I, Scheme 1.140). Conformational helicity is also encountered in benzophenanthrene in which again the terminal rings and substituents are in different planes to make the molecule exist as two helical enantiomers (II, scheme 1.140). Conformational helicity is encountered in some secondary structures of polypeptides, and many proteins have significant portions of their chains stabilized in the α-helix, which can be either left-or right-handed.

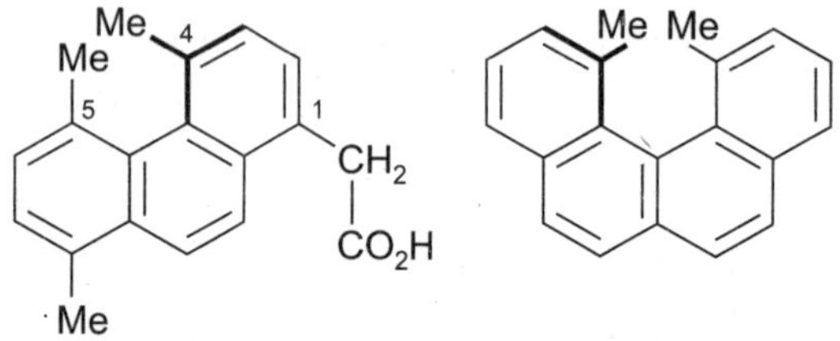

(I) P-(+)

(II) *M*-(–)

Phenanthrenes and benzophenanthrenes are normally expected to be planar. However, due to molecular overcrowding the ring structures assume a helical shape.

SCHEME 1.140

1.19 MOLECULAR STEREOISOMERISM

Consider [14] annulene (annulenes are large ring compounds with conjugated double bonds which may exist in *cis* and *trans* configuration) in which hydrogens in the ring with four *trans* double bonds cause steric strain. The molecule is thus somewhat destabilized and somewhat nonplanar and quite reactive. The molecule of [14] annulene, however does show the presence of diamagnetic ring current. In each of the structures, four transannular hydrogens overlap

Annulenes

These are monocyclic conjugated polyenes and were synthesized to test Hückel's rule. Their name uses a bracketed numerical prefix to indicate the number of carbons followed by the word annulene. [4n] Annulenes have alternating short (double) and long (single) bonds and are antiaromatic. The expected aromatic character of [4n + 2] annulenes is rendered low by angle and van der Waals strain, and is regained when the ring contains 18 or more carbons.

with each other to give two conformational diastereomers which differ in the mode of overlapping. The two forms are separable (on silica gel column impregnated with silver nitrate) but are easily interconvertible (Scheme 1.141).

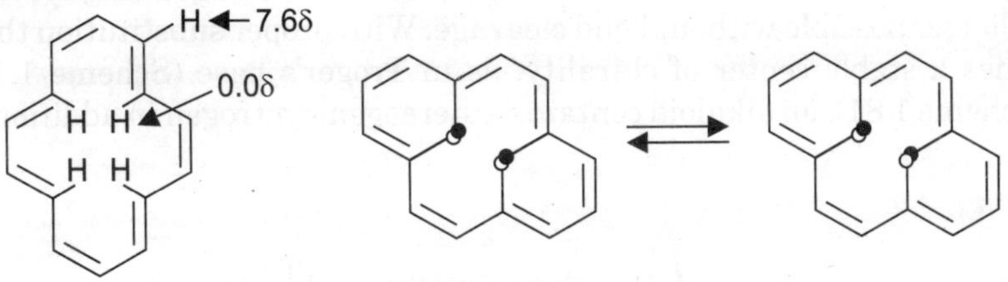

[14] Annulene

SCHEME 1.141

1.20 CHIRALITY IN METALLIC COMPLEXES

Several metallocenes substituted with atleast two different groups on one ring become chiral and have been resolved (see, Scheme 1.14). Fumaric acid-iron tetracarbonyl has also been resolved while its corresponding maleic acid compound has a plane of symmetry and is therefore achiral.

1.21 CHIRALITY INVOLVING ATOMS OTHER THAN CARBON

An atom which is an analogue of carbon in terms of a chiral centre is nitrogen. It has the same tetrahedral arrangement of electron pair as the sp^3-hybrid carbon atom. The only difference from carbon being, that one of these electron-pairs is usually a lone pair, which is not involved in bonding. Thus, nitrogen generally has three substituent groups. Tertiary amines when suitably substituted (see schemes 1.8 and 1.22) are chiral, but do not, however, display optical activity due to chirality at the nitrogen atom. The reason for this is believed to be that the groups on the nitrogen atom undergo rapid inversion as indicated in the equilibrium (Scheme 1.142). This inversion process interconverts the enantiomeric forms a amine.

In fact, no bonds are broken in the change and the configurational nomenclature is retained as the change can be described as S → R or R → S. The amine interconversion is described as an inversion (turning inside-out of an umbrella) and to avoid confusion, the enantiomers are termed invertomers. When the lone pair is donated to a substituent, giving the $[NR^1 R^2 R^3 R^4]^+$ species, this rapid inversion is prevented. The tetra-alkyl ammonium salts with four different alkyl groups diaplay optical activity (See, scheme 1.5).

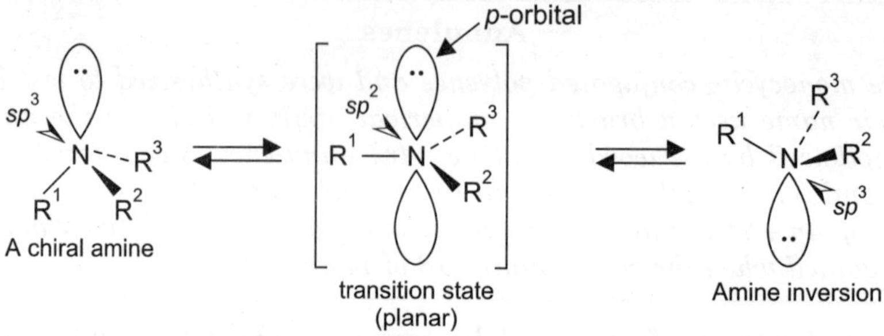

SCHEME 1.142

When the nitrogen atom is present at a ring junction in bridged ring systems, pyramidal inversion is no longer possible without bond cleavage. With proper substitution the tricoordinate nitrogen becomes a stable center of chirality, as in Troger's base (Scheme 1.143). Similarly brucine (see, Scheme 1.81), an alkaloid contains a stereogenic nitrogen in addition to stereogenic carbons.

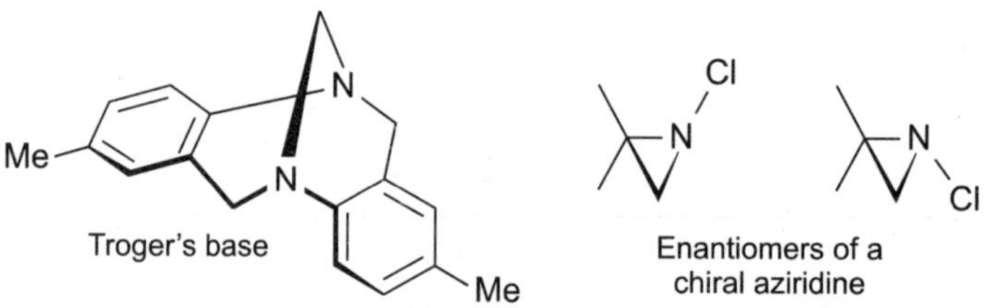

SCHEME 1.143

Optical stability is the measure of resistance of a pure enantiomer toward racemization under a given set of conditions. One type of nitrogen atom that inverts particularly slowly is a nitrogen atom in a three-membered ring. The planar transition state for flapping of some aziridines is sufficiently stained for the isolation of individual enantiomers (Scheme 1.143). In addition to nitrogen being a part of three membered ring, another factor which imparts configurational stability in amines in the presence of an adjacent atom with atleast one lone pair of electrons *e.g.* Cl, O, N is so is oxaziridine (Scheme 1.143).

Unlike the low barriers for pyramidal inversion for the first row elements, the heavier elements have much higher barriers to inversion. At the trivalent phosphorus and sulfur the preferred bonding angle is around 100°. A larger distortion is thus required to reach the planar transition state, consequently many phosphines and sulfoxides have been isolated in optically active form (see, scheme 1.6). These undergo racemization by pyramdial inversion only at high temperature.

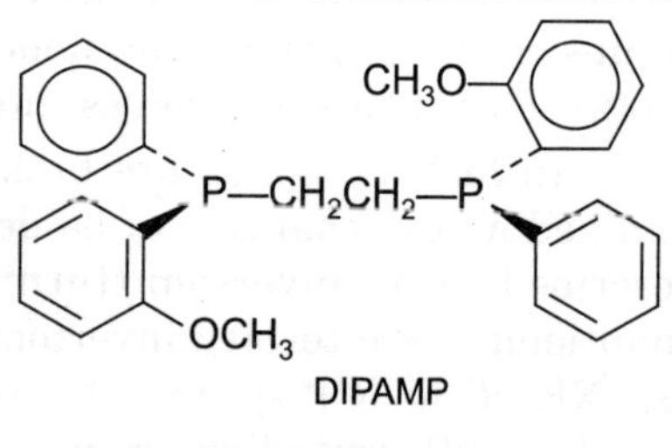

SCHEME 1.144

Several chiral phosphines are integral components of rhodium and ruthenium complexes (chiral phosphine complexes which are termed chiral Wilkinson type catalysts) and are used for enantioselective reduction. An effective chiral phosphine is DIPAMP, (Scheme 1.144 an ethane-1, 2-diphosphine) which has two stereogenic phosphorus atoms. The stereoisomeric forms are a pair of enantiomers, RR and SS and an optically inactive *meso* diastereomer (RS and SR) which are identical due to plane of symmetry. The lone pair is the least preferred ligand. For details of hydrogenation with chiral Wilkinson's type catalysts see scheme 6.59 and 6.59*a*.

Organic sulfides undergo two important reactions involving the electron pairs on sulfur. They are easily oxidized to sulfoxides and sulfones (see, schemes 1.5, 1.6 and 2.53) and they act as nucleophilic agents towards substances which undergo nucleophilic displacement readily to give sulfonium salts (Scheme 1.145). The formation of sulfonium salts from alkyl halides is reversible, and heating of the salt causes dissociation into its components. Sulfonium salts are identical with quaternary ammonium salts; sulfonium hydroxides, $R_3S^+OH^-$, like quaternary ammonium hydroxides, $R_4N^+OH^-$ are strong bases.

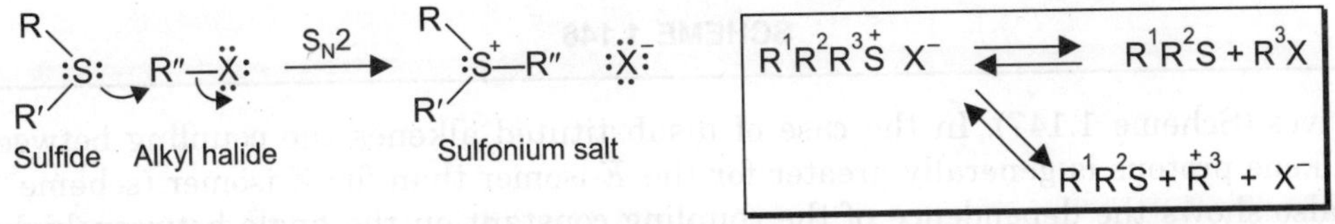

SCHEME 1.145

A special features of sulfonium ions is that when these are substituted with three different groups (Scheme 1.6), they can usually be separated into enantiomers. Thus, the reaction of methyl ethyl sulfide with bromoacetic acid gives a sulfonium ion which is separable into enantiomers (on crystallization as the salt of an optically active amine).

The chirality of these ions results from the nonplanar configuration of the bonds formed by sulfonium sulfur. The optically active forms of unsymmetrically substituted sulfonium ions are stable as compared with the low configurational stability of analogously constituted amines. A pure enantiomer of a sulfonium salt can undergo racemization through reversible dissociation similar to one suggested for tetraalkyl ammonium salts, reversible dissociation to a carbocation and sulfide and also pyramidal inversion at sulfur similar to nitrogen inversion in amines (Scheme 1.142).

1.22 NMR SPECTROSCOPY AND STEREOCHEMICAL INFORMATION

A working knowledge of NMR spectroscopy is assumed. One can get both conformational and configurational information :

- The chemical environment around a proton(s) can be determined from chemical shifts.
- The intensity of a peak points to the number of chemically equivalent protons.
- Steric disposition of neighbouring protons and their number can be determined from *J* values.
- Coupling constant *J* and chemical shift data give information regarding the relative stereochemistry of diastereomers.

(A) Coupling Constant and Relative Configuration

Karplus curve (scheme 1.146) gives relationship between the vicinal coupling constant and the dihedral angle. This information is useful to assign stereochemistry in rigid cyclohexane

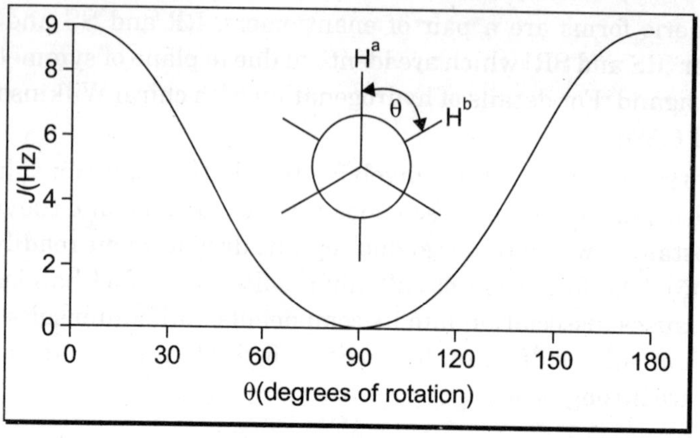

Effect of vicinal coupling constant $3J_{H^a—C—C—H^b}$ (Hz) on the dihedral angle. For θ of 0°, 60°, 90° and 180° the values of J are around 10, 2.5, 0 and 10 Hz respectively.

SCHEME 1.146

derivatives (Scheme 1.147). In the case of disubstituted alkenes the coupling between the *trans*-alkene protons is generally greater for the *E*-isomer than for *Z* isomer (scheme 1.148) which also shows the dependence of the coupling constant on the angle between hydrogens attached to sp^2 hybridized carbons. When one considers the stereostructures (I and II, scheme 1.148) the steric disposition of H_a and H_b are reflected by the values of coupling constants. Small coupling constant in (I) is indicative of dihedral angle between H_a and H_b close to around 90° while in (II) it should be around 180°.

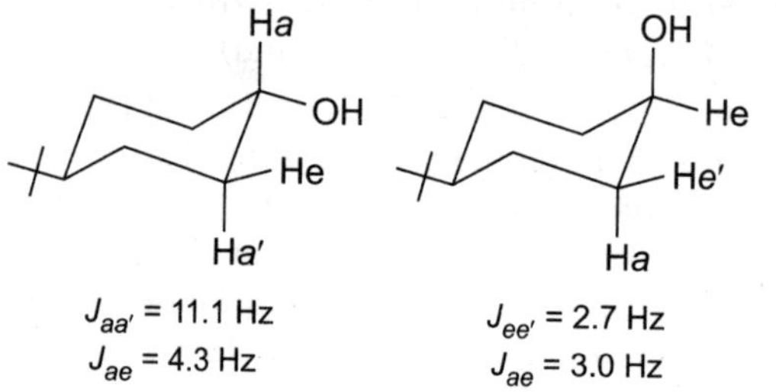

Dihedral angle (θ) between two axial hydrogens is 180° while the axial-equatorial and equatorial-equatorial dihedral angles are both 60°.

SCHEME 1.147

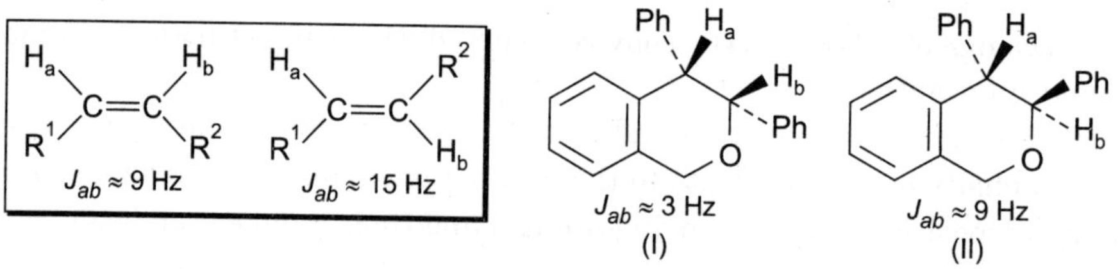

SCHEME 1.148

(B) Chemical Shift and Configuration

A proton or a group of protons which are coplanar with an aromatic ring, C=C or C=O groups are deshielded while the protons which are held above their π electron clouds are shielded as

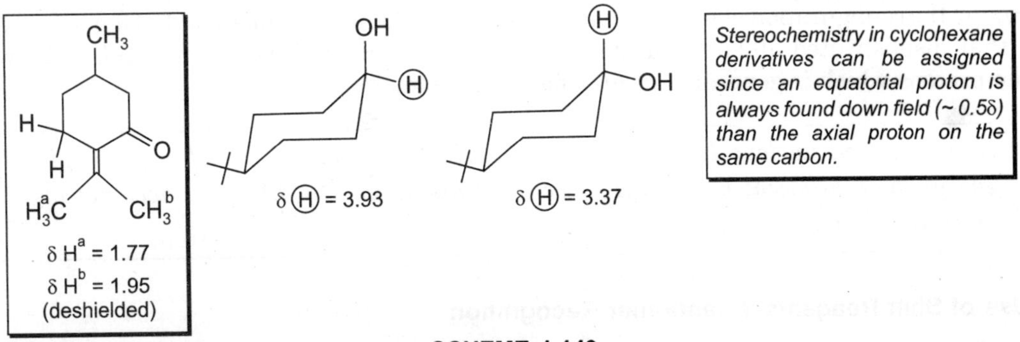

SCHEME 1.149

is so in (I, scheme 1.149). In cyclohexanes the equatorial proton is always deshielded due to C—C bond anisotropy (scheme 1.149).

EXERCISE 1.49

How from ^1H NMR chemical shift data one can assign stereostructure in the pairs (I and II, bicyclo [2.2.1] hept-2-en-7ols and α and β-D glucose III and IV Scheme 1.150).

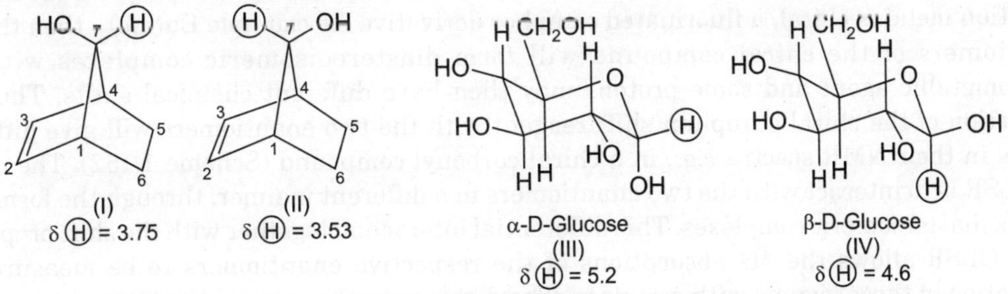

SCHEME 1.150

ANSWER. *In the anti isomer (II) the proton at C7 is held over the double bond it is therefore, shielded compared to that in (I). In the glucose molecule the proton on the anomeric carbon is the most deshielded and can be detected in the ^1H NMR spectrum. However, in (III) the proton on the anomeric carbon being equatorial is more down field compared to the same proton in (IV).*

(C) Nuclear Overhauser Effect (NOE)

When two nuclei which have different chemical shifts, but are close in space *e.g.*, H_a and H_b in (I, Scheme 1.151, but not in II) and the molecule is simultaneously irradiated with the ratio frequency v_a which is the resonance frequency of H_a during recording of NMR spectrum one observes two things. Firstly the signal corresponding to H_a disappears because of saturation while the signal of H_b is enhanced in intensity because of increased spin relaxation. One use of NOE is *e.g.*, to

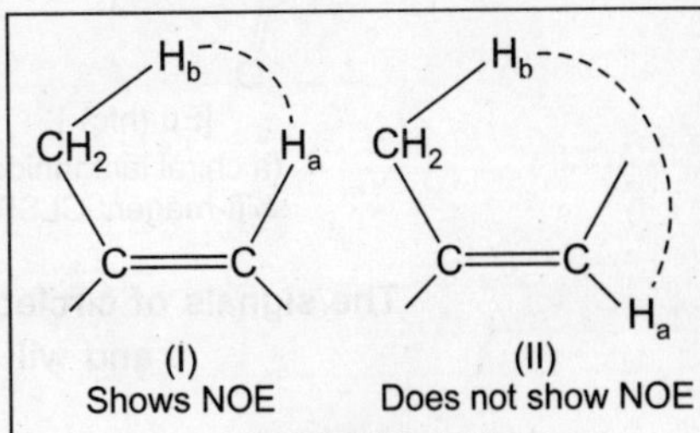

SCHEME 1.51

determine the configuration of more heavily substituted alkenes (scheme 1.151). One has already seen that disubstituted alkenes can be distinguished as *E* or *Z* based on coupling between alkene protons which is generally greater for the *E* isomer.

> *A spin excited nucleus H_a undergoes spin relaxation via the transfer of its spin energy to an adjacent nucleus H_b over short distances (2 – 4A°) and does not involve coupling.*

(D) Use of Shift Reagents–Enantiomer Recognition

Shift reagents were introduced in 1969, for spreading out NMR absorption patterns without increasing the strength of the applied magnetic field. Addition of shift reagents forms labile molecular associates with electron-donating polar groups such as OH, C=O, and NH_2 to bring about large changes in the chemical shifts of protons (or carbons in ^{13}C-NMR). The shifts known as lanthanide-induced shifts (LIS) are inversely proportional to the third power of the distance of the nuclei from the lanthanide and provide a sensitive method to know the relative distance of various groups and atoms from the complexation site. A common shift reagent, which is a paramagnetic metal [*e.g.*, an europium (III) salt] derivatized to generate solubility in an organic solvent, associates with polar functional groups in a molecule, to bring about a downfield shift of the resonance frequency of protons in the locality. If the ligand around the transition metal is chiral, a fluorinated camphor derivative for example Eu(hfc)$_3$, then the two enantiomers of the chiral compound will form diastereoisomeric complexes with the organometallic agent and some protons may then have different chemical shifts. Thus the interaction of the chiral europium shift reagent with the two enantiomers will give different signals in their NMR spectra *e.g.*, in a chiral carbonyl compound (Scheme 1.152). The role of the CLSR is to interact with the two enantiomers in a different manner, through the formation of weak diastereomeric complexes. This differential interaction together with the shift properties of the CLSR allows the ^{1}H absorptions of the respective enantiomers to be measured by integration of these signals with certainty. The diastereomeric excess (*de*) of one diastereomer X over another Y is given by the expression (scheme 1.153). Enantiomeric excess can also be measured by NMR spectroscopy and thus 90 : 10 ratio of two enantiomers gives an enantiomeric excess of 80%.

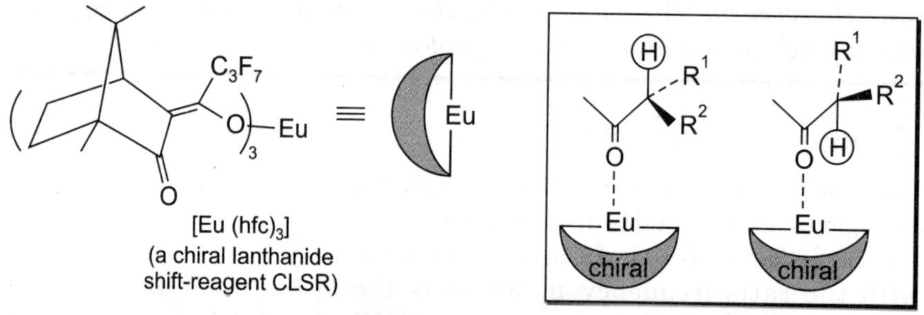

The signals of circled hydrogen atoms will be shifted down field and will have different chemical shifts

SCHEME 1.152

$$\text{d.e.} = \frac{\%\ \text{diastereoisomer X} - \%\ \text{diastereoisomer Y}}{\%\ \text{diastereoisomer X} + \%\ \text{diastereoisomer Y}}\ \%$$

$$(\text{e.e.}) = \frac{\%\ \text{enantiomer X} - \%\ \text{enantiomer Y}}{\%\ \text{enantiomer X} + \%\ \text{enantiomer Y}}\ \%$$

SCHEME 1.153

(E) Role of Mosher's Acid (MTPA)

In another method a mixture of enantiomers is reacted with an optically pure substance to give a new covalent bond and two different diastereomers and these will have different chemical shifts. The optically pure compound used for this purpose should have easily identifiable features in the NMR spectrum. One such widely used compound to determine the ratio of two enantiomeric alcohols in a mixture is Mosher's acid (3, 3, 3-trifluoro-2methoxy-2-phenylpropanoic acid MTPA). Mosher's acid gives a pair of diastereomers *e.g.*, with a racemic alcohol 1-phenylethanol (Scheme 1.154). The diastereomeric ratio can be assessed from the ^{1}H NMR spectrum by integration of the methoxy proton absorptions in the diastereomers or by integration of the CF_3 groups of the diastereomeric pair.

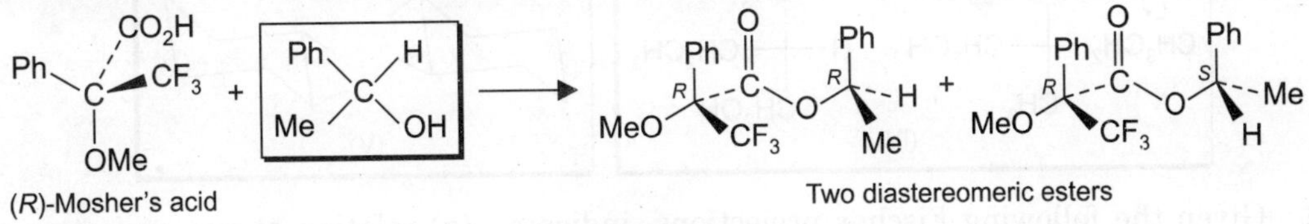

SCHEME 1.154

Thus by employing the esters of this type it is possible to find the ratio of two enantiomeric alcohols in the given mixture. The integrated intensity values of the two signals gives the ratio between the amounts of each enantiomer. If the ratio is 1 :1 one is having a racemic mixture, if the ratio is 9 :1 then the mixture consists of 90% of one enantiomer and 10% of the other and thus the *ee* value is the difference between these percentage values *i.e.*, 90 – 10% = 80% *ee*.

Asymmetric Reaction and *ee*

Earlier (around 1970's) the ee of an asymmetric reaction was determined by measuring the optical rotation of a racemic mixture (see, sec. 1.14). A useful development to find out the enantiomeric purity of a mixture is gas or liquid chromatography with chiral supports. This object can also be achieved by NMR spectroscopy when enantiomers give separate signals with a chiral lanthanide shift reagent (CLSR), or via reaction of a completely or partially racemic alcohols with Mosher's acid. The integrated intensity value of distinct signals in the diastereomes involved in either case gives the ratio between the amounts of each enantiomer.

PROBLEMS

1.1. Water has a twofold axis of symmetry (C_2), while benzene has a C_6 axis, comment on the molecules with C_∞ and C_1 axes.

1.2. Explain briefly a relation between elements of symmetry and optical activity.

1.3. Depict the symmetry plane on the structure of *cis*-1, 2-dichlorocyclopropane molecule. Does molecule of H_2O also have a symmetry plane ?

1.4. How can you demonstrate a highly symmetrical molecule like methane to have multiple S_4 axes ? Relate it with a cube.

1.5. Label the following pairs of compounds as homomers, constitutional isomers, enantiomers or diastereoisomers. Assign *R*/*S* configuration to the stereogenic centers in (I and II).

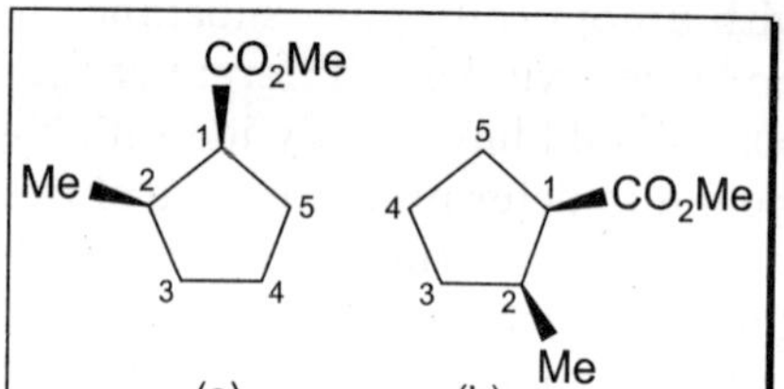
(a) (b) (I)

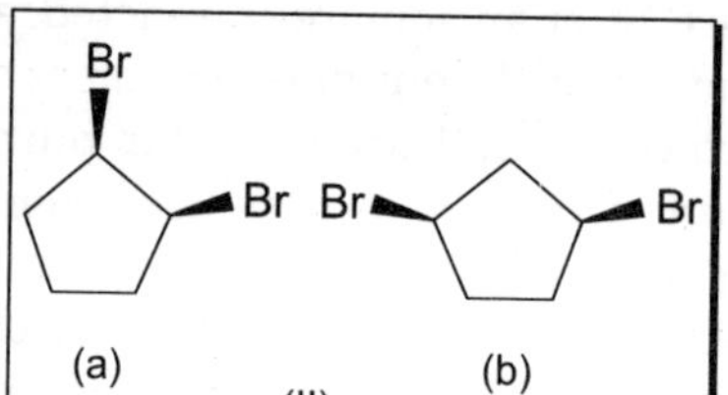
(a) (b) (II)

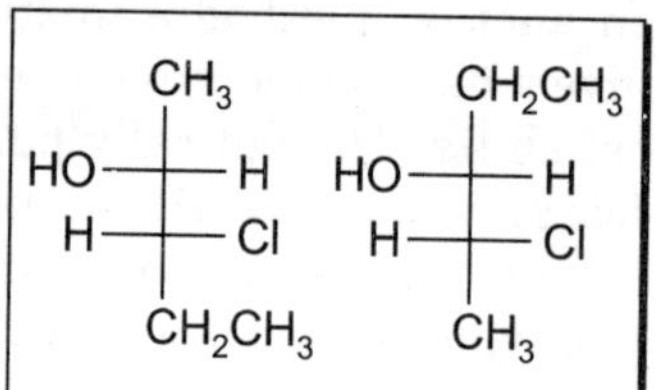
(III)

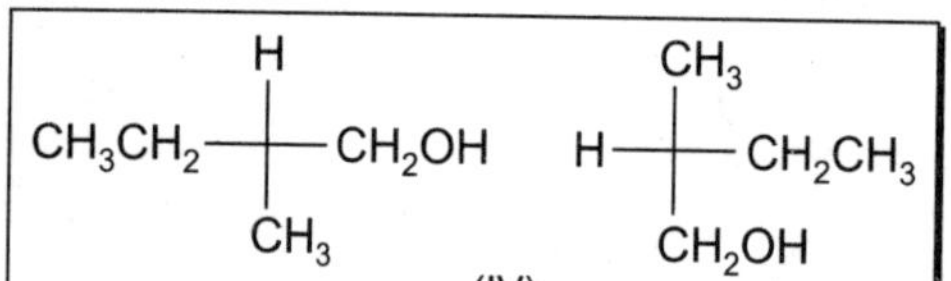
(IV)

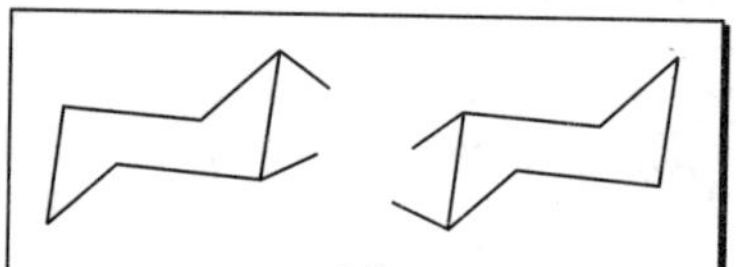
(V)

1.6. Given the following Fischer projections, indicate : (*a*) relation of one projection with other : (*b*) their names and assignment to (*R*) or (*S*).

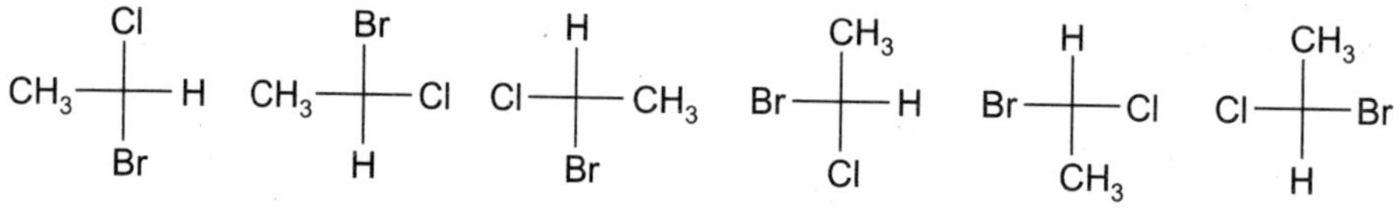

1.7. How many stereoisomers can exist for each of the compounds (I—IV)?

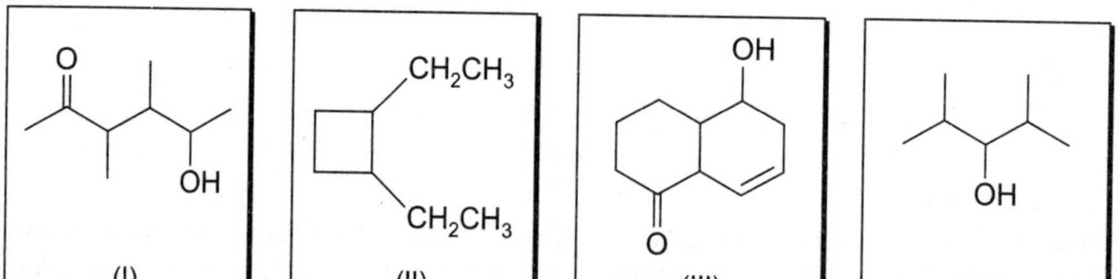
(I) (II) (III) (IV)

1.8. Write the Fischer projection for (*R*)-2-iodobutane and convert it into its "Wedge and dotted line" representation.

1.9. How are the following conformations of 2, 3-dibromobutane related with one another ?

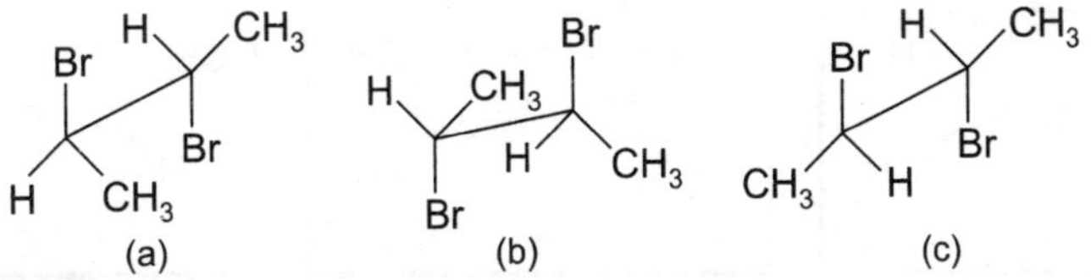

1.10. What is the stereochemical relationship, (enantiomers or diastereomers) between each of the pairs (I and II)?

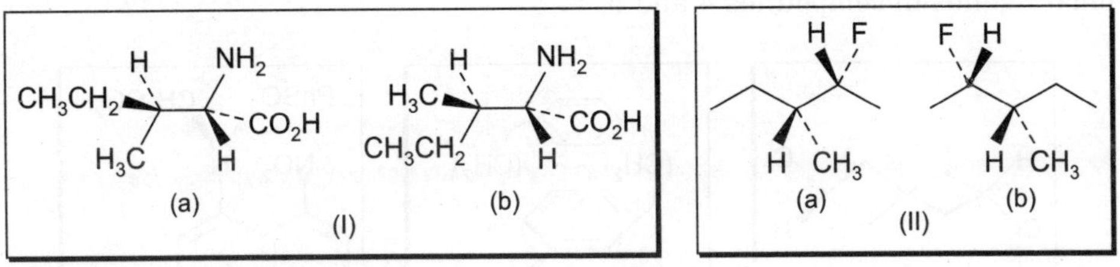

1.11. Butane on monochlorination gives a mixture of 1-chloro and 2-chlorobutane. Comment on their chirality by writing their stereostructures.

1.12. The following biphenyl is chiral. Designate its configuration.

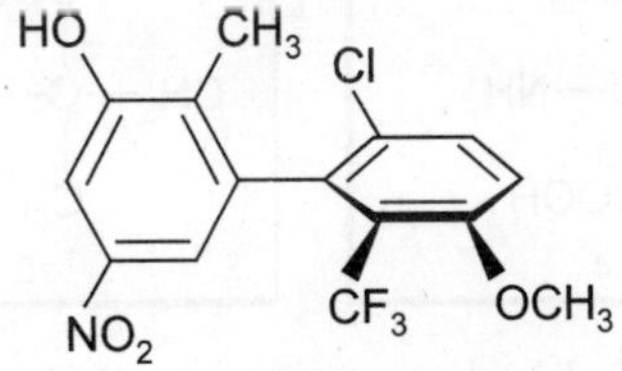

1.13. Comment on the chirality (optical isomerism) of the following biphenyls.

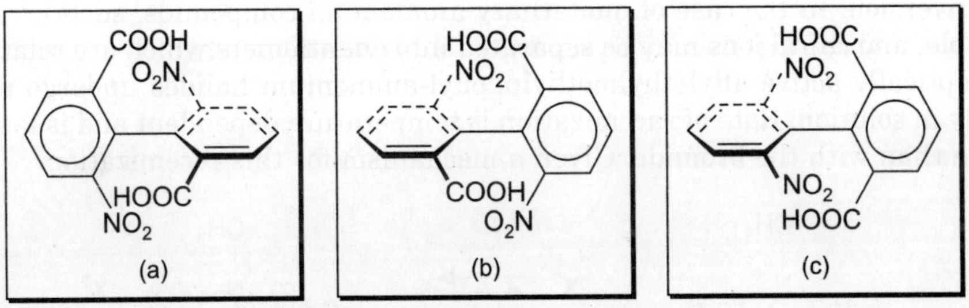

1.14. Which of the following molecules are chiral.

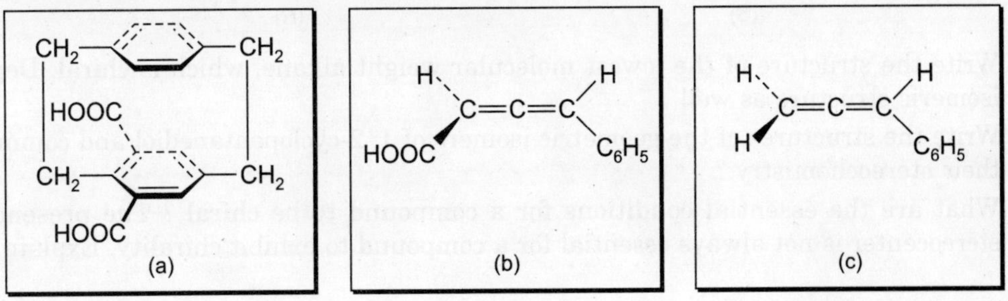

1.15. Comment on the identity and chirality of the following structure of penta-2, 3-diene.

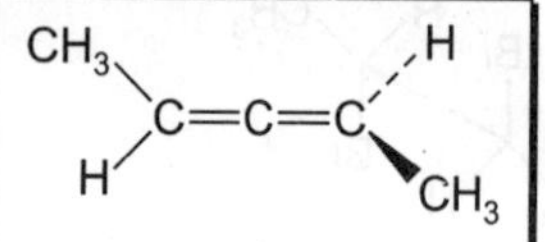

A

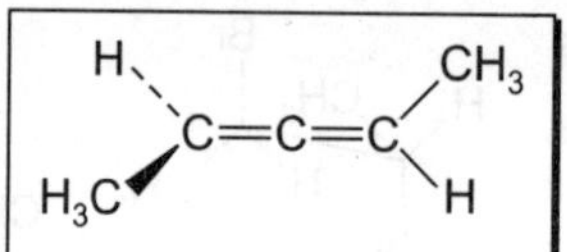

B

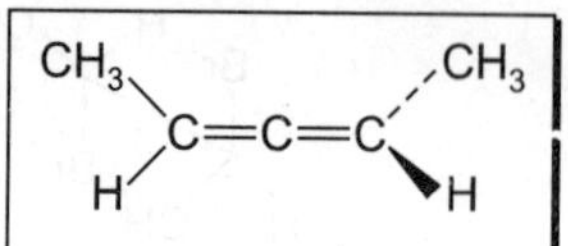

C

1.16. Label each compound as either having center, axes or planes of symmetry. How the chirality of compounds 2 depends on m, n, and x. Assign the order of priorities to individual ligands in compounds 4 and 5.

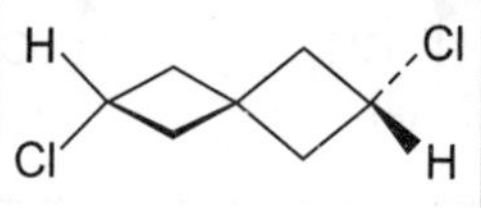

1

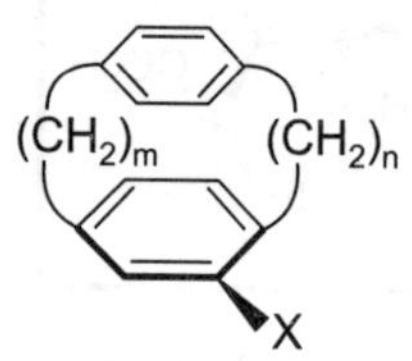

2

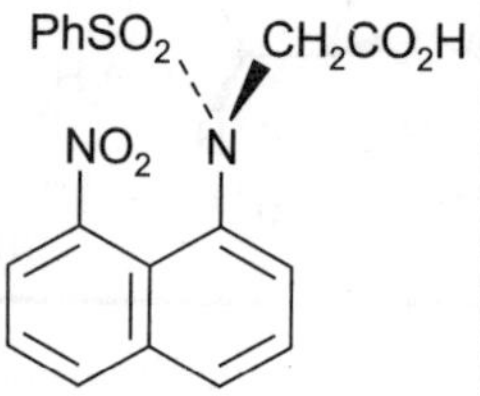

3

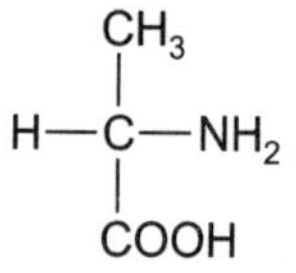

4

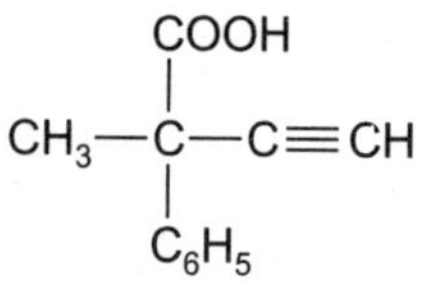

5

1.17. Give a general definition of chirality and support it with examples.

1.18. The two enantiomers of a chiral amine readily interconvert by a process known as nitrogen inversion. In the case of quaternary ammonium compounds, such inversion is not possible, and chiral ions may be separated into enantiomers which are relatively stable. The optically active allylethylmethylphenyl-ammonium halides undergo racemization slowly in solution. Rate of racemization is temperature dependent and is faster with the iodide than with the bromide. Given a mechanism for this recemization.

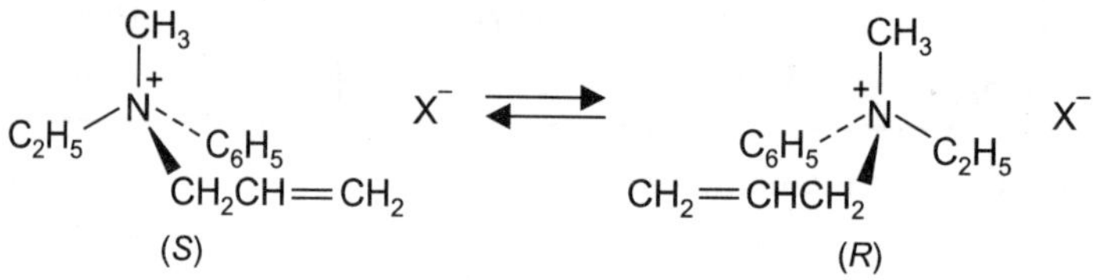

(*S*) (*R*)

1.19. Write the structure of the lowest molecular weight alkane, which is chiral. Depict its isomeric structure as well.

1.20. Write the structures of the geometric isomers of 1, 2-cyclopentanediol and comment on their stereochemistry ?

1.21. What are the essential conditions for a compound to be chiral ? The presence of a stereocenter is not always essential for a compound to exhibit chirality. Explain.

1.22. Why is it not possible under ordinary conditions to resolve amines even though three different groups are attached to nitrogen atoms ?

1.23. Comment on the chirality of unsymmetrical sulfoxides.

1.24. Draw the structures of any three dimethylcyclobutanes including the chiral stereoisomer.

1.25. How the projections (I—IV) are related with one another ?

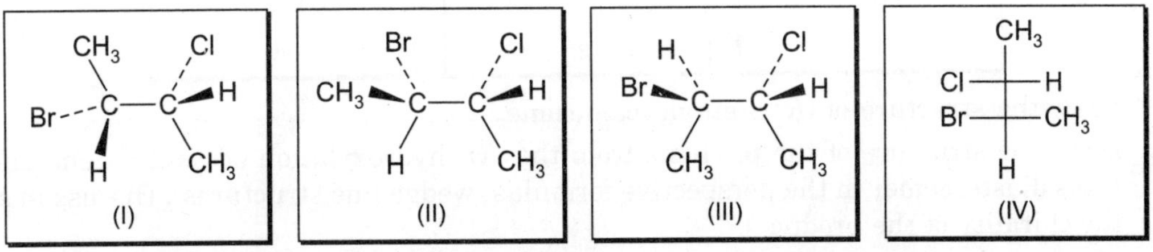

1.26. How are the projections (I, II, and III) are related to one another?

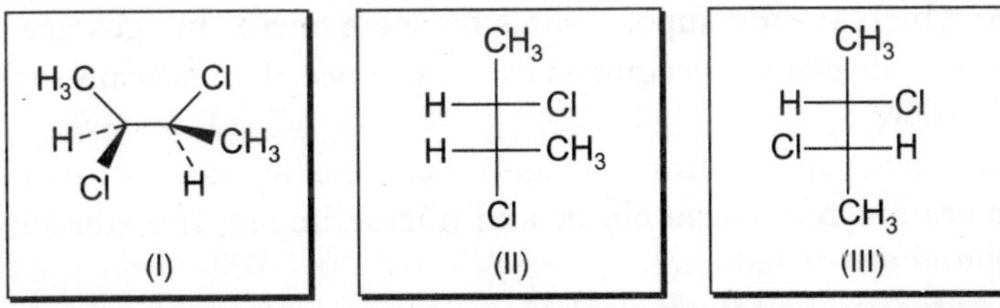

1.27. (*a*) Identify the stereochemical relationship between the pairs of compounds 1 and 2.

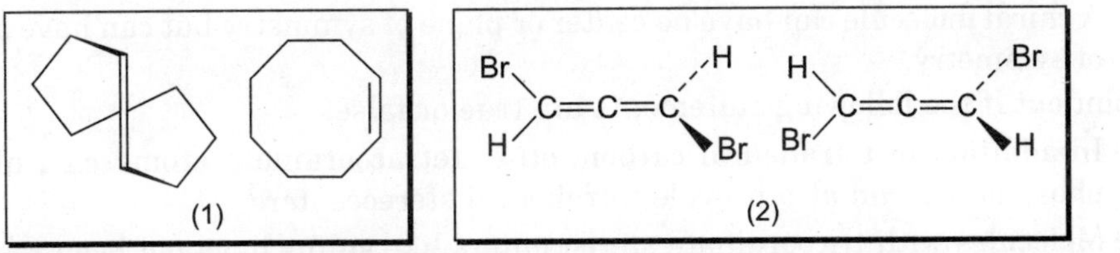

(*b*) Write the structure of (*R*)-*trans*-cyclooctene.

1.28. What is the relationship between (I and II) ? Decide by interconverting one projection into other?

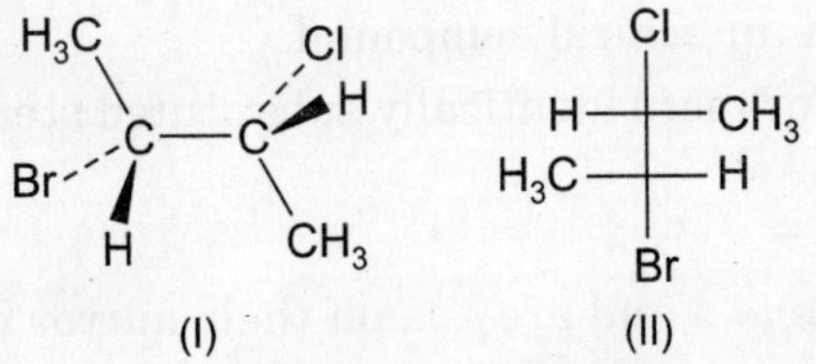

(I) (II)

1.29. Discuss the stereochemical relationship between the following two molecules, by assigning *R* and *S* nomenclature (See problem 1.31).

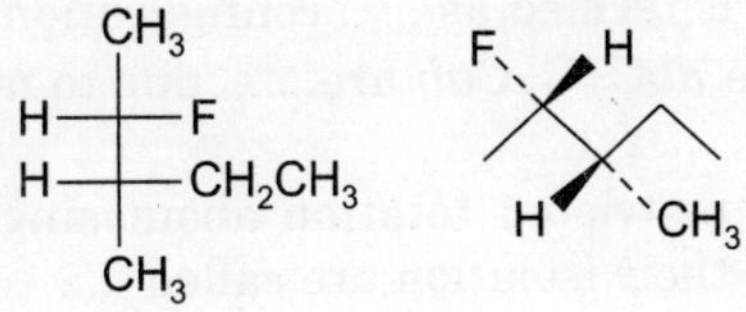

1.30. Given the priorities designate the structures below as R or S.

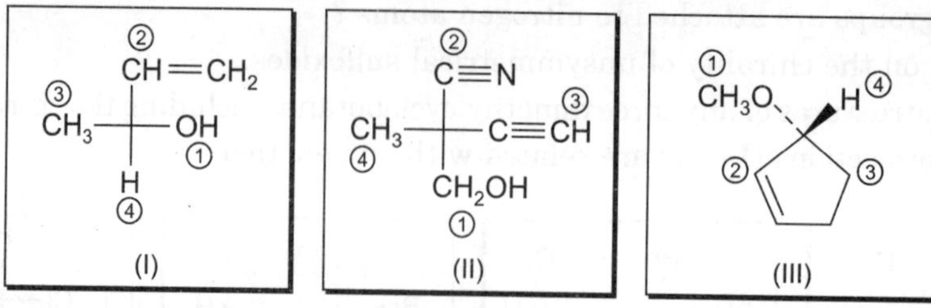

1.31. Draw the structure of (R)-3-ethylcyclohexene.

132. Draw the structure of the products from the *syn* hydroxylation of *cis*-2-butene and its *trans* diastereomer in the perspective formulas (wedge line structures). Discuss in short the chirality of the products.

1.33. Draw the Fischer projections of all the possible stereoisomers for 2, 3-butanediol.

1.34. Fill in the blanks :

(*i*) Molecules which are not superposable on their mirror images are called ....

(*ii*) A chiral molecule can be recognized by an absence of an alternating axis of symmetry S_n of ...... order.

(*iii*) Generally a molecule is chiral if it has no plane of symmetry (σ) or ...... of symmetry.

(*iv*) Enantiomers are related as object and mirror image, the stereoisomers which are not so related are called ....

(*v*) A stereocenter can be tetrahedral or ...... planar.

(*vi*) Exchange of a pair of groups around a stereocenter gives a different ....

(*vii*) A chiral molecule can have no center or plane of symmetry but can have a simple ...... of symmetry.

1.35. Point out if the following statements are true or false.

(*i*) In addition to tetrahedral carbon, other tetracoordinate atoms *e.g.*, nitrogen and phosphorous can also provide tetrahedral stereocenters.

(*ii*) Molecules with tricoordinate stereocenters like sulfur have not been resolved.

(*iii*) A molecule is achiral in the absence of a center of chirality.

(*iv*) The chiral biphenyls belong to the group of compounds which display axial chirality.

(*v*) Haxahelicene is chiral due to the presence of helical structure.

(*vi*) *trans*-Cyclooctene is an achiral compound.

(*vii*) Molecules with two or more identically substituted stereocenters may exists as *meso* stereoisomers.

1.36. Fill in the blanks :

(*i*) *Meso* compounds are ...... and are ...... on their mirror images.

(*ii*) When in a reaction no bonds to the stereocenter are broken the reaction proceeds with ...... of configuration.

(*iii*) The spatial arrangement of the atoms in a chiral molecule which distinguishes it from its mirror image is termed as ...... configuration.

(*iv*) The allenes of the type $abC{=}C{=}Cab$ are ...... due to non planar arrangement of four groups called a chiral ....

(*v*) Stereoisomers due to restricted rotation about single bonds with high rotational barriers which permit their isolation are called ....

(*vi*) A chiral molecule lacking all symmetry elements except the trivial C_1 axis is called ....

(*vii*) A chiral molecule with one or two ...... axes is called dissymmetric.

(*viii*) A center of symmetry is equal to ...... operation.

(*ix*) A plane of symmetry is equal to ...... operation.

1.37. Comment on the chirality of the naturally occurring antibiotic mycomycin.

$$HC{\equiv}C{-}C{\equiv}C{-}CH{=}C{=}CH{-}CH{=}CH{-}CH{=}CH{-}CH_2COOH$$

1.38. Give a stereodescriptor to the following chiral compounds.

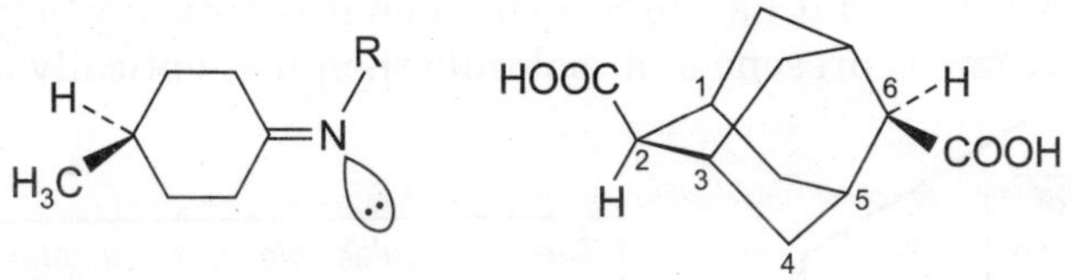

1.39. Write the structure of *meso* tartaric acid in Newman projection and translate it into Fischer projection.

1.40. Maleic acid is reacted with osmium tetroxide followed by hydrolysis. Write the stereostructure of the product, in two different ways.

1.41. Show that *meso* tartaric acid contains an S_2 axis.

1.42. Write stereostructures to show as to why the reaction of bromine with cyclohexene is stereoselective.

1.43. How many stereocenters are present in 1, 3-dimethylclobutane ? Discuss briefly the chirality of the compounds.

1.44. Draw all the possible stereoisomers of 3-pentene-2-ol.

1.45. How many stereoisomers are possible in 2, 3-dibromobutane ? How one can identify the *meso*-stereoisomer ? Explain drawing all structures in solid and dashed wedge perspective formulas ?

1.46. The molecules of water and phenanthrene belong to the same point group. Explain.

1.47. Assign configuration, if (*R*) or (*S*) to the following compounds.

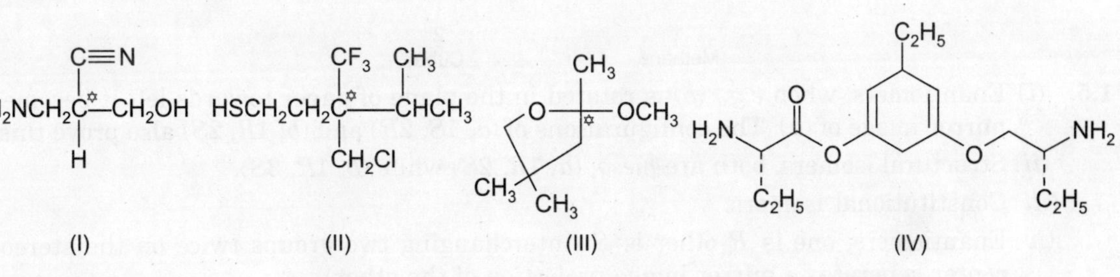

(I) (II) (III) (IV)

1.48. What is the stereochemical relationship between compounds (I and II).

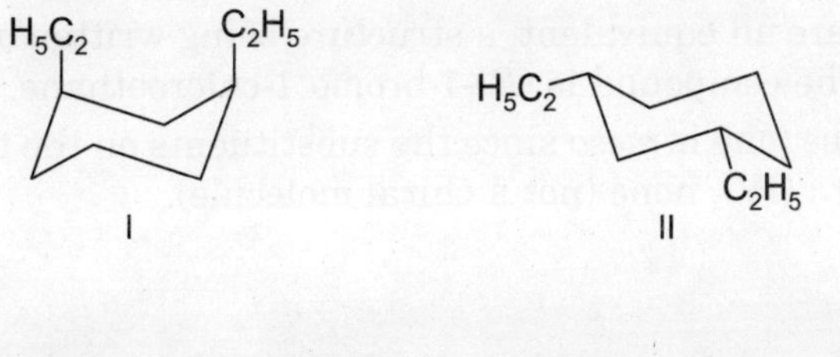

I II

ANSWERS TO SELECTED PROBLEMS

1.1. Linear molecules like acetylene have a C_∞ axis. In this case even an infinitesimal rotation (360°/∞) about this axis leads to an orientation indistinguishable from the original. The trivial onefold axis C_1 is never considered, as all molecules have an infinite number of C_1 axes.

1.2. If a molecule is not superimposable on its mirror image, then it can display optical activity. Such molecules are (*i*) asymmetric *i.e.*, these do not have any symmetry elements (except a trivial axis, C_1) or (*ii*) dissymmetric (chiral) these posses proper axes but no alternating axis of symmetry *i.e.* improper axis. Thus, if a molecule has an S_n axis, then it is not optically active. If it has no S_n axis, then it is optically active. If a C_n axis is the only symmetry element present in a molecule then it is optically active.

1.3.

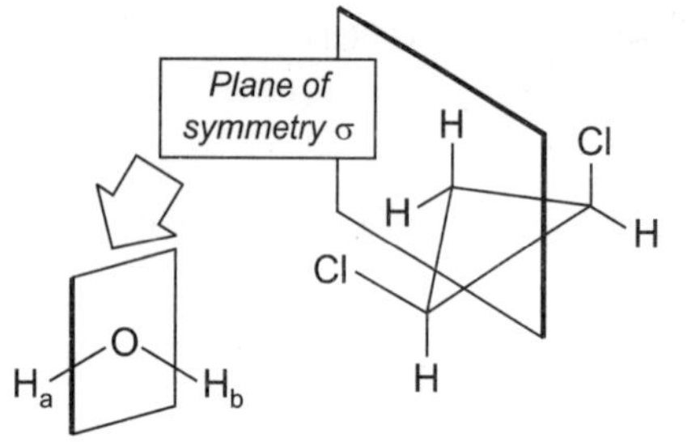

The molecular plane of a planar molecule is its plane of symmetry (e.g., in H_2O). Moreover, a plane perpendicular to the molecular plane and which includes a C_2 axis is also a mirror plane as shown by arrow in H_2O molecule.

1.4. The axis depicted in 1 passes through the middle of the molecule and bisects opposite HCH angles. If one considers the clockwise rotation of the molecule by 90° around this axis and subsequent reflection in a plane parpendicular to it, one gets an orientation of the molecule superimposable on the original. One can easily depict not only the same axis by inscribing the tetrahedral model of methane in a cube 2, but the other S_4 axes as well. The S_4 axis shown bisects the opposite sides of the cube. A cube has six equivalent faces, therefore, there are three S_4 axes. Notably each S_4 in methane molecule is also a C_2.

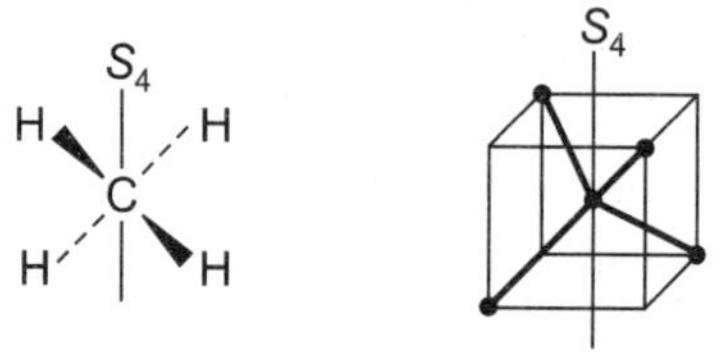

1 Methane   2 Cube

1.5. (*i*) Enantiomers, when *e.g.*, (*b*) is rotated in the plane of paper towards left it becomes mirror image of (*a*). The configurations of (*a*, 1*S*, 2*R*) and (*b*, 1*R*, 2*S*) also prove this.

(*ii*) Structural isomers, both are *meso*, (*a*, 1*R*, 2*S*) while (*b*, 1*R*, 3*S*).

(*iii*) Constitutional isomers.

(*iv*) Enantiomers, one is *R* other is *S* (interchanging two groups twice on the stereocenter generates a mirror image projection of the other).

(*v*) Enantiomers.

1.6. These structures are all equivalent, a structure being written by interchanging any two pairs of groups. The compound is (*R*)-1-bromo 1-chloroethane.

1.7. (I) Eight ; (II), three (one is *meso* since the substituents on the two stereocenters are the same) ; (III), eight ; (IV), none (not a chiral molecule).

1.8. 

Fischer projection "wedge- and-dotted line" structure

1.9. Conformations (*a*) and (*b*) are an enantiomeric pair while (*c*) is the *meso*-form. [**Hint.** The sawhorse projections are drawn in staggered conformations. Since the groups on each stereocenter are same there must be *meso* compound and a center of symmetry is seen in (*c*). When in (*c*) one carbon is rotated one gets an eclipsed conformation now with a plane of symmetry].

1.10. This type of problems can be solved in several ways and some of these have been already discussed. In the case of (I) both molecules are drawn as staggered conformations (see the zig-zag of continuous lines). These can be first converted into eclipsed conformations and then translated on to Fischer projections to see the relationship. These come out to be diastereomers. Even as such if one looks at the dashed-wedged line formulas in both (I*a* and I*b*). The configuration at the right hand carbon is same but differs only at the left hand carbon to show that these are diastereomers.

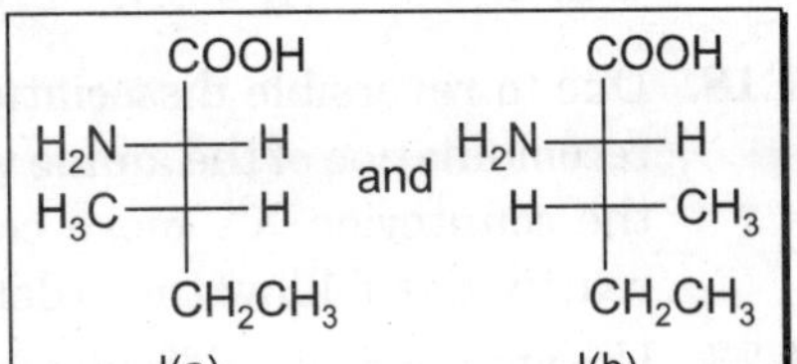

I(a) I(b)

In the case of (II*b*) its rotation out of the plane of paper (180°) gives an enantiomeric projection (inverted stereochemistry at each stereocenter). Thus II*a* and II*b* are enantiomers.

1.11. 1-Chlorobutane is achiral. The 2-chlorobutane formed in this reaction is a racemic mixture *i.e.*, an equimolar mixture of (*R*)-2-chlorobutane and (*S*)-2-chlorobutane. The reactive intermediate in the reaction leading to 2-chlorobutane is the *sec*-butyl free radical, which is almost planar and therefore, is achiral and reacts with Cl_2 on either side of the molecule. Also see (Scheme 1.87).

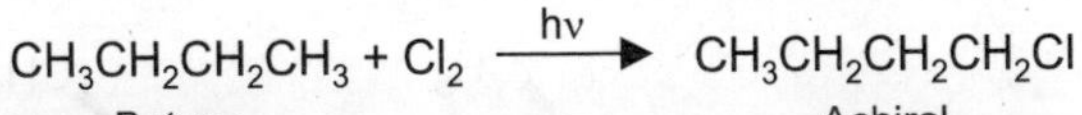

Butane Achiral

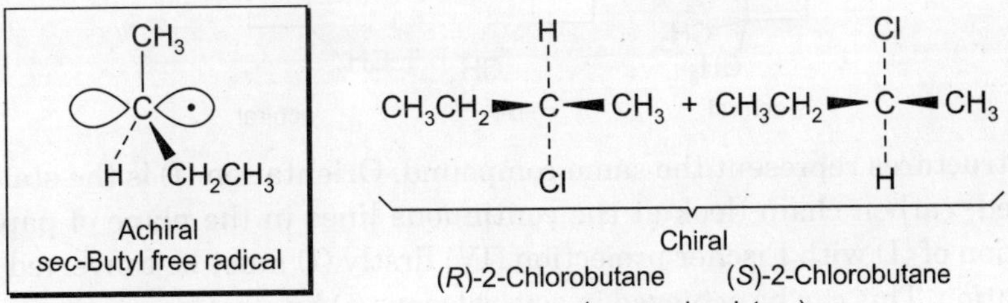

Achiral *sec*-Butyl free radical

Chiral (*R*)-2-Chlorobutane (*S*)-2-Chlorobutane (a racemic mixture)

1.12. (*R*)-Configuration.

1.13. The biphenyls (*a*) and (*b*) have the two differently substituted *ortho* positions on each ring, the groups being indeed large to pass each other on rotation about the central bond. Thus, these represent chiral molecules ; a pair of enantiomers. Compound (*c*) is a structural isomer with a plane of symmetry, therefore, it is not chiral even though in this case the rotation is equally restricted.

1.14. Only (*b*).

1.15. It is chiral allene. A and B are enantiomers and B and C are equivalent.
[**Hint :** B rotated around the axis of allenic carbons clockwise gives C].

1.16. Compounds of the type 2 (paracyclophanes) have been resolved when m and n are fairly small ($m = 3, n = 4$ and $X = CO_2H$). It is then that the benzenoid rings cannot rotate at a significant rate. When the connecting methylene chains are long, rotation is so fast that the enantiomers rapidly equilibrate.

4. NH_2 [1], COOH [2], CH_3 [3], H [4] ; 5. COOH [1], C_6H_5 [2], $C \equiv CH$ [3], CH_3 [4]

1.18. Due to reversible dissociation into an amine and an alkyl halide (see scheme 1.5). The recombination of the amine with the alkyl halide gives a racemic salt. The less nucleophilic the counterion X^-, move configurationally stable is the quaternary salt, nucleophilic reactivities folllow the order $I^- > Br^- > Cl^- > F^-$.

1.23. Unsymmetrical sulfoxides are chiral, the oxygen and lone electron pair play the role of substituents. The electron lone pair has the lowest priority while the oxygen the highest (See also Scheme 1.3*e*). For *R* and *S* designation, the molecule has to be viewed from the side opposite to the lone pair, *i.e.* from bottom one may rotate the figure along the O-S bond when the lone pair comes behind the plane of page for more easy assignment (see, scheme 1.20*d*).

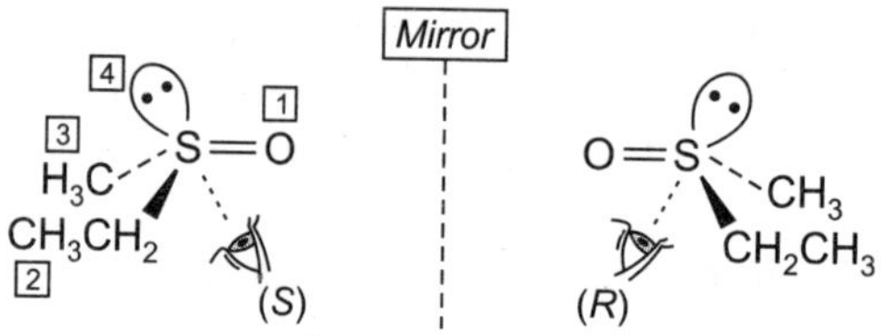

Chiral sulphoxide

1.24.

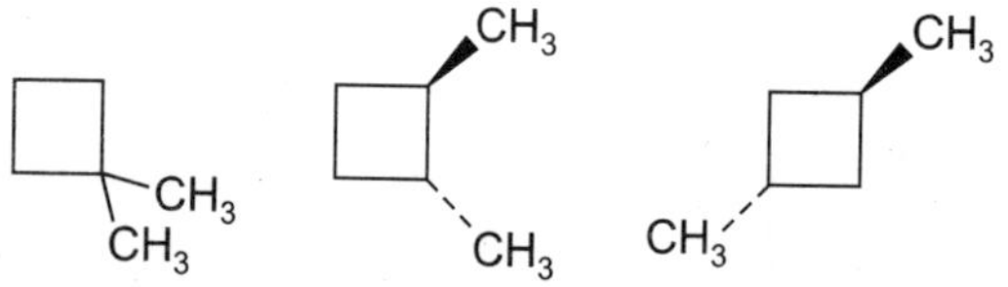

achiral chiral achiral

1.25. All the structures represent the same compound. Orientation (I) is the standard zig-zag (staggered) carbon chain (look at the continuous lines in the plane of paper). To know the relation of (I) with Fischer projection (IV) firstly (I) must be converted into eclipsed conformation. This can be achieved in several ways (*i*) by *e.g.* any rotation by 60° towards you by this procedure, the left hand methyl group will go out of the plane of page and will be projected towards you to give an orientation (II). This can be translated on an eclipsed Fischer projection. (*ii*) You may rotate the left hand carbon in (I) towards you by 180° to get another equivalent orientation (III) which can be translated on to Fischer projection (V).

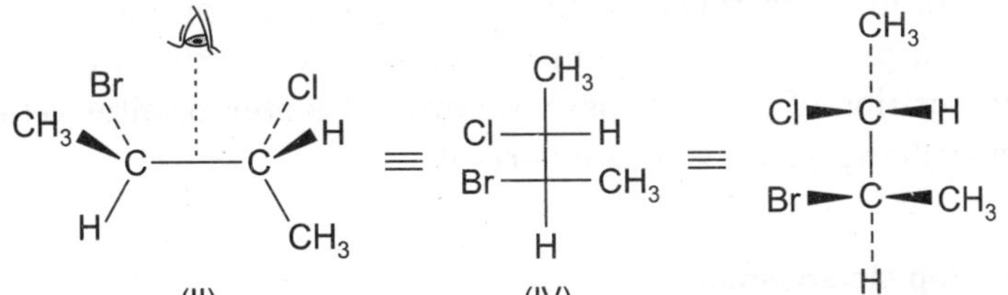

(II) (IV)

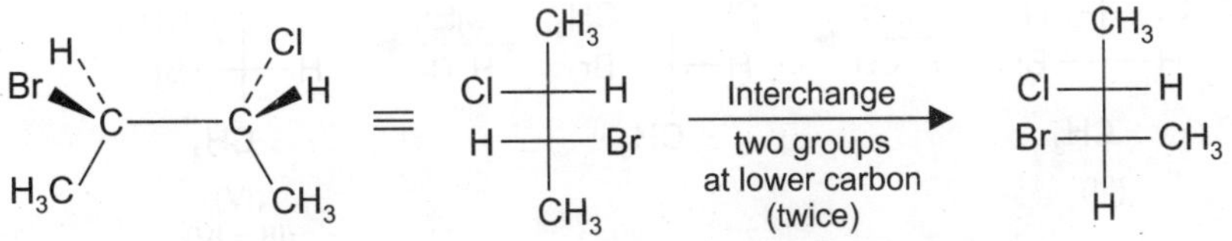

(III) (V) (V = IV) (IV)

In (V) if one switches first (CH_3, Br) and then (Br, H) on the lower stereocenter, its configuration remains unchanged and it comes out to be same as (IV).

1.26. All the orientations represent the same compound.

[**Hint :** The compound is already an eclipsed conformation which is translated on to the Fischer projection (II). The Fischer projection (III) is obtained by switching two groups twice and also provides a more conventional drawing with carbon chain vertical].

1.27. (*a*, 1) Diastereomers. (2) Same enantiomer of a chiral allene.

[**Hint :** One interchange of groups at a geminal carbon of a chiral allene gives an enantiomer, two such changes will leave the structure unchanged (see, scheme 1.128).

(*b*)

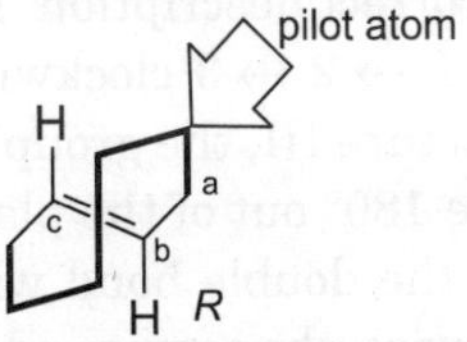

1.28. Two structure are identical. One may convert the dashed-wedge line perspective formula which is presented here in a staggered zig-zag like structure (see the continuous lines in the plane of paper in a dashed-wedged line formula whether it is staggered or eclipsed) into the Fischer projection. Since Fischer projection is eclipsed, one must first derive an eclipsed conformation from the dashed-wedge line structure by rotating either stereogenic carbon and then transferring the groups on the Fischer projection (as done in Scheme 1.34*c*).

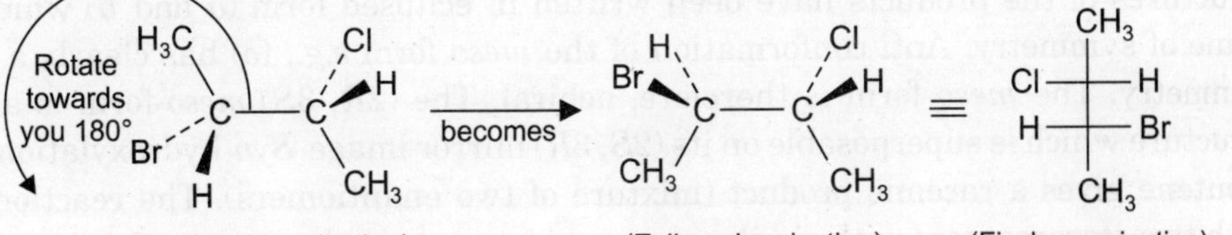

(Staggered projection) (Eclipsed projection) (Fischer projection)
(III)

Now for determining the relationship of (III) with (II, problem 1.31) place as many similar groups in projection (III) in identical positions to that in (II). Recall that interchanging the position of two groups twice around a stereocenter retains its configuration while interchanging only once inverts it.

Top stereocenter

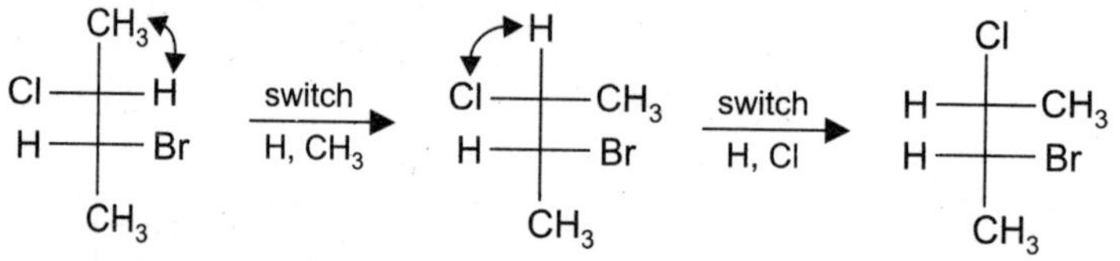

(III) (IV) (III ≡ IV)

Bottom stereocenter

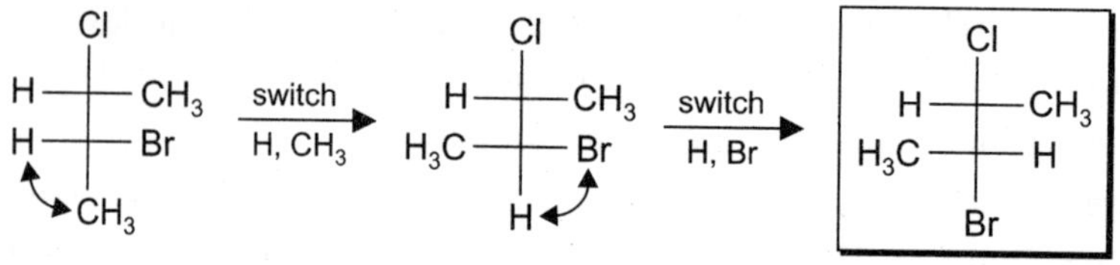

(V) (III = IV = V)

1.29. Both are identical compounds (2*S*, 3*S*)-2-fluoro-3-methylpentane.

1.30. In Fischer projection (I) the group of lowest priority is on the vertical line thus the path $1 \rightarrow 2 \rightarrow 3$ anticlockwise gives the correct description. In II, the group of lowest priority is on the horizontal line, the path $1 \rightarrow 2 \rightarrow 3$ clockwise (*R*) is to be reversed and the configuration is *S*. In the cyclic structure III, the group of lowest priority is not away. In order to do that, rotate the molecule 180° out of the plane of paper (now H will be away from you—OCH_3 towards you and the double bond will move to the right hand side), after this the path $1 \rightarrow 2 \rightarrow 3$ is clockwise, the compound is *R* [or the path in III $1 \rightarrow 2 \rightarrow 3$ *i.e.*, *S* must be reversed to get the correct cofiguration *R* since the group of lowest pirority is towards ones eyes].

1.31.

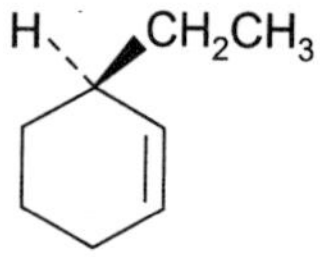

1.32. *cis*-2-butene gives only a *meso* stereoisomer by attack from either face. For clarity the structures of the products have been written in eclipsed form (*a* and *b*) which have a plane of symmetry. Anti conformation of the *meso* form *e.g.*, (*c*) has clearly a center of symmetry. The *meso* form is therefore, achiral. The (2*R*, 3*S*) *meso*-form is an achiral structure which is superposable on its (2*S*, 3*R*) mirror image. *Syn* hydroxylation of *trans*-2-butene gives a racemic product (mixture of two enantiomers). The reaction product with two stereocenters with similar groups exists only in three stereoisomeric forms an achirl *meso* compound and a chiral compound which exists as a pair of enantiomers in equal amounts.

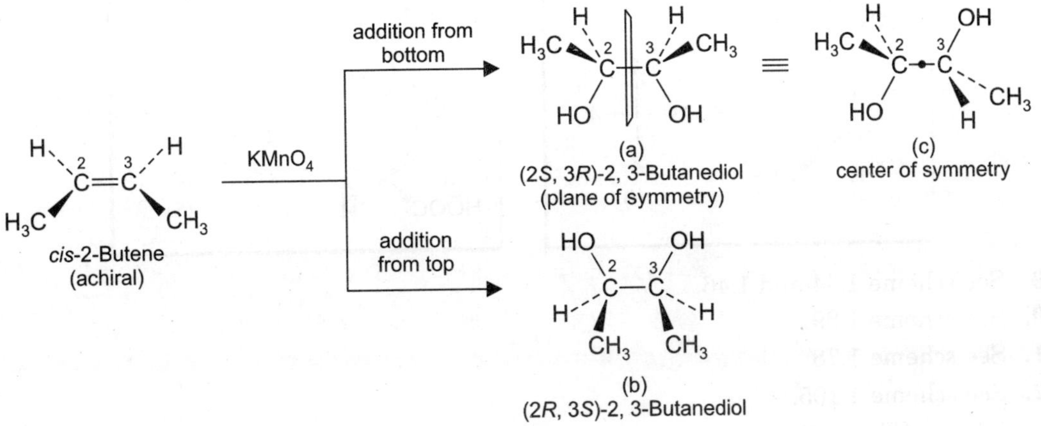

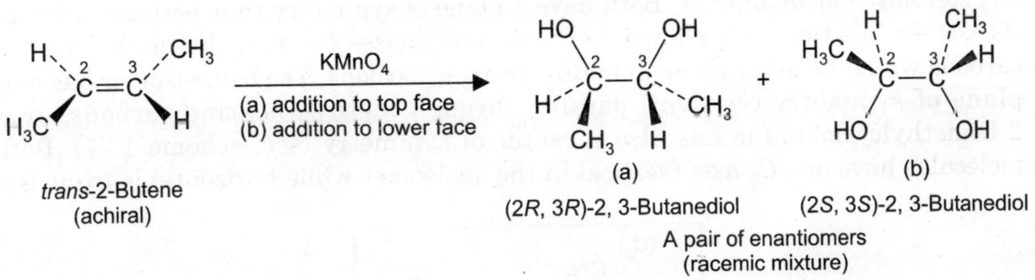

1.33.

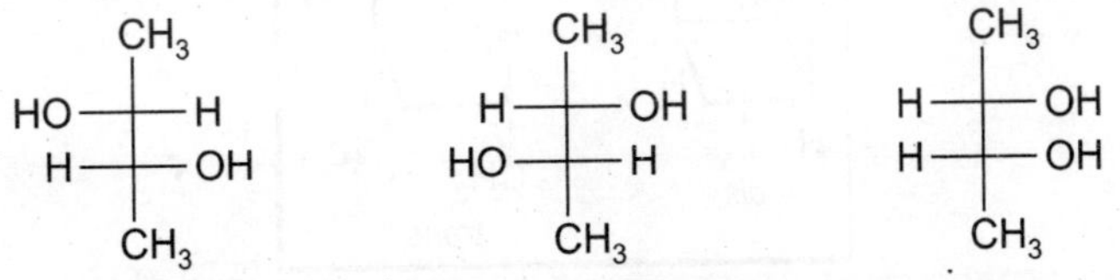

(2*R*, 3*R*)-2, 3-Butanediol (2*S*, 3*S*)-2, 3-Butanediol *meso*-2, 3-Butanediol

1.34. (*i*) Chiral ; (*ii*) any ; (*iii*) *i*, center ; (*iv*) diastereomers ; (*v*) trigonal ; (*vi*) stereoisomer ; (*vii*) axis.

1.35. (*i*) true, (*ii*) false, (*iii*) false, (*iv*) true, (*v*) true, (*vi*) false, (*vii*) true.

1.36. (*i*) achiral, superposable, (*ii*) retention, (*iii*) absolute, (*iv*) chiral, axis (*v*) atropisomers ; (*vi*) asymmetric ; (*vii*) C_n, (*viii*) S_2, (*ix*) S_1.

1.37. It is a chiral allene with the grouping of the type *ab*C=C=C*cd*. The chirality arises due to nonplanar arrangement of four groups about an axis called a chiral axis (stereoaxis).

1.38. Follow the modified simple procedure given in (sec. 1.16).

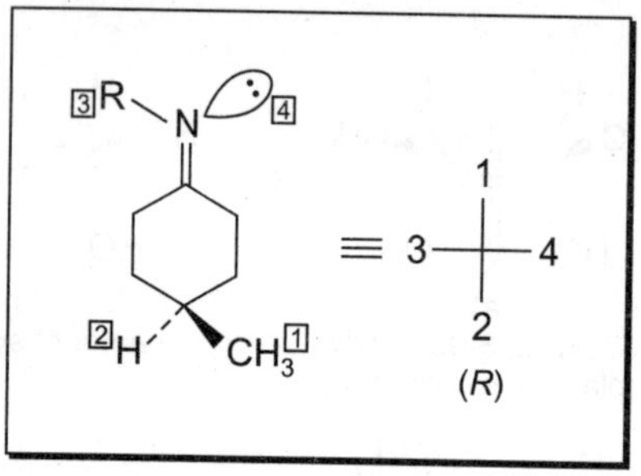

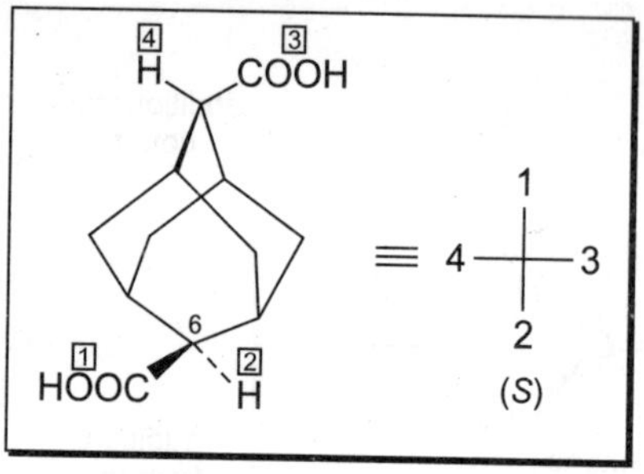

1.39. See scheme 1.44 and 1.46.

1.40. See scheme 1.89.

1.41. See scheme 1.78.

1.42. See scheme 1.105.

1.43. C1 and C3 are stereocenters, since on switching (interchanging) H and CH_3 converts one stereoisomer to another. Both have a plane of symmetry thus both are achiral. The *cis* isomer has two such planes (vertical), one passes through the methylene bearing carbons while the other through methyl bearing carbons. The *trans*-isomer has only one plane of symmetry (vertical) passing through methy-l-bearing carbons. *trans* 1, 2-Dimethylcyclobutane has also a center of symmetry (see, scheme 1.71). Both the molecules have one C_2 axis (vertical in the *cis* isomer while horizontal is *trans* isomer.

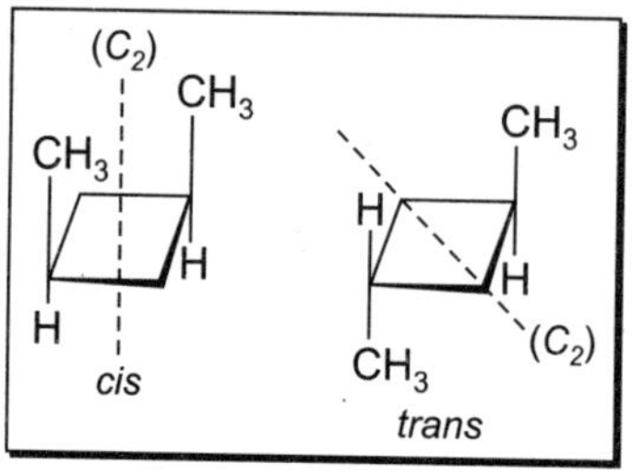

1.44.

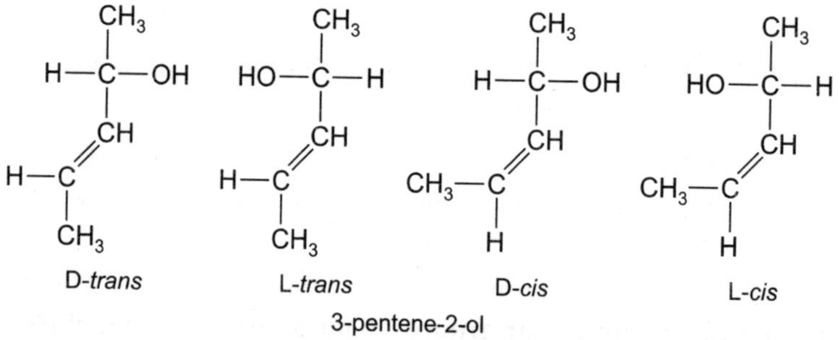

1.45. Three. Draw the eclipsed conformation so that the plane of symmetry can be drawn. This is the *meso* form (compare with tartaric acid, scheme 1.77*a*). Now convert it into one of the staggered conformations where center of symmetry is now clearly seen.

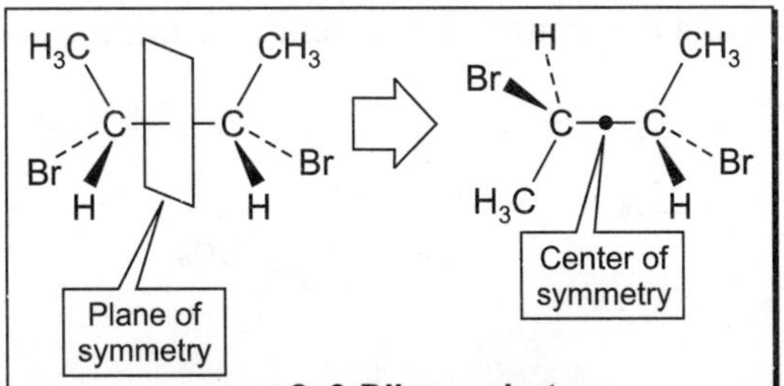

***meso*-2, 3-Dibromobutane**

1.46. Since both of these molecules have two planes of symmetry and a C_2 axis, the point group is C_{2v}, since the C_2 axis and two planes of symmetry which include the axis.

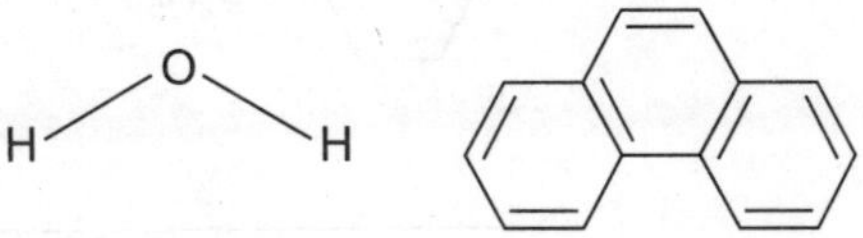

1.47. The priority of groups in the compounds (I and II) are as shown. Therefore (I) is (*S*) and (II) is also (*S*), the compound (III) is (*R*). In the case of compound (IV).

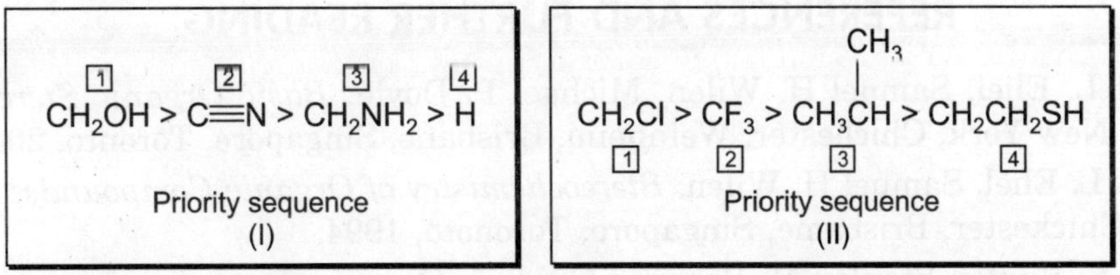

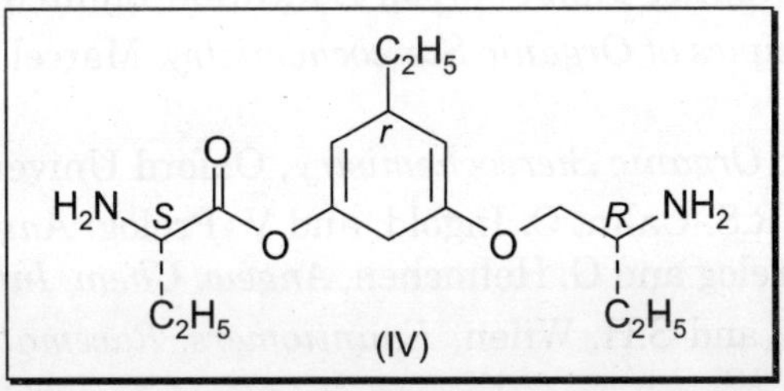

There are two stereocenters in the side chain and there is a pseudochirality center (pseudoasymmetric center) in the ring.

1.48. Both represent the same compound. For a related example see exercise 1.6.

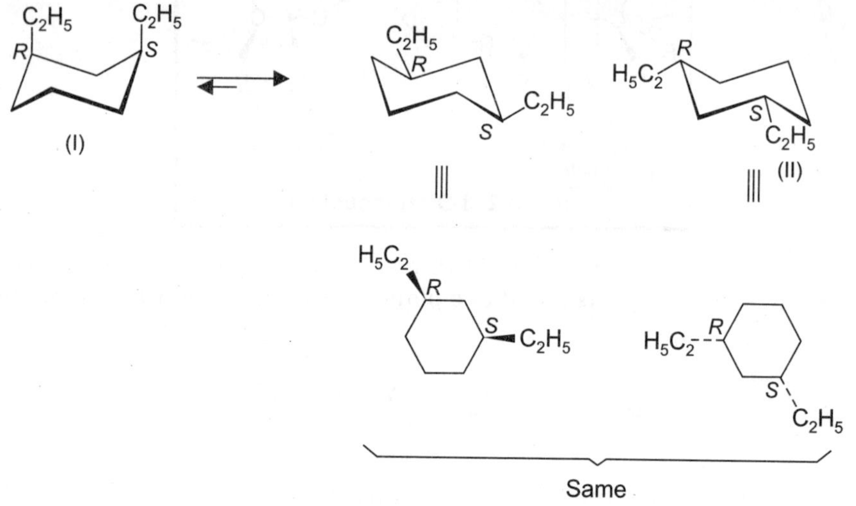

REFERENCES AND FURTHER READING

1. Ernest L. Eliel, Samuel H. Wilen, Michael P. Doyle: *Basic Organic Stereochemistry*, Wiley, New York, Chichester, Weinheim, Brisbane, Singapore, Toronto, 2001.
2. Ernest L. Eliel, Samuel H. Wilen: *Stereochemistry of Organic Compounds*, Wiley, New York, Chickester, Brisbame, Singapore, Toronoto, 1994.
3. Sheila R. Buxton Stanley M. Roberts: *Guide to Organic Stereochemistry from methane to macromolecules*, Prentice Hall/Pearson Education Limited, Harlow, 1996.
4. Bernard Testa: *Principles of Organic Stereochemistry*, Marcel Dekker, New York, Basel, 1979.
5. Michael J. Robinson: *Organic Stereochemistry*, Oxford University Press, Oxford, 2000.
6. The CIP system: (*a*) R.S. Cahn, C. Ingold, and V. Prelog: *Angew. Chem. Int. Ed. Engl.*, 1966, 5, 385; (*b*) V. Prelog and G. Helmchen, *Angew. Chem. Int. Ed. Engl., 1982, 13, 263.*
7. J. Jacques, A. Collet, and S.H. Wilen, *Enantiomers, Racemates and Resolution*, Wiley, New York, 1981.
8. *Chirality in Industry: the Commercial Manufacture and Applications of Optically Active compounds*, ed. A.N. Collins, G.N. Sheldarke, and J. Crosby, Wiley, Chichester, U.K. 1992.
9. David G. Morris: *Stereochemistry*, The Royal Society of Chemistry, Cambridge, 2001.
10. K.H. Hellwich, C.D. Siebert: *Stereochemistry Workbook*, Springer–Verlag Berlin Heidelberg, 2006.
11. P.S. Kalsi: *A Simple Mental Exercise for Assigning Configurational Descriptors (R/S) to Molecules with a Stereoaxis (Chiral axis)*, Int. J. Chem. Sci. 2(3), 2004, 285–290.

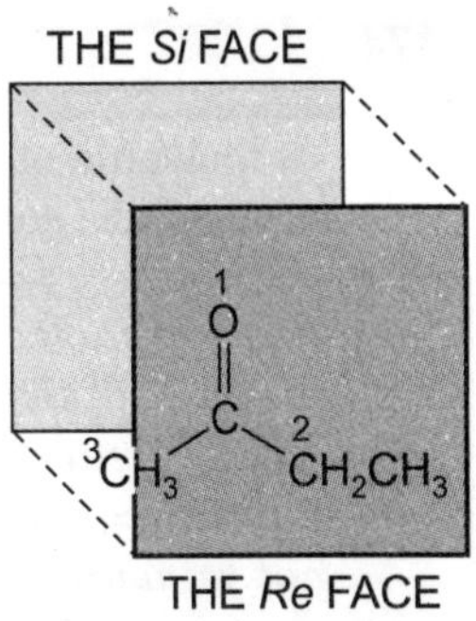

CHAPTER 2

Prochirality—Prostereoisomerism and Asymmetric Synthesis

2.1 INTRODUCTION—NOMENCLATURE

Prochirality/Prostereoisomerism

Prostereoisomerism/prochirality is the property of certain molecules due to which these can be converted into stereoisomers (enantiomers or diastereomers).

In a molecule like propionic acid (Scheme 2.1) the carbon atom marked C_α is not a stereocenter (or one may put it as C_α is a nonstereogenic carbon since a carbon atom that is a stereocenter is also called a stereogenic carbon). The carbon C_α can be made a stereocenter by replacing one or other of the two apparently identical ligands (in this case the ligands are two H atoms of the methylene group) in turn by a different ligand *e.g.*, OH to give enantiomeric products (*S*)-lactic acid and (*R*)-lactic acid. The following points may be noted :

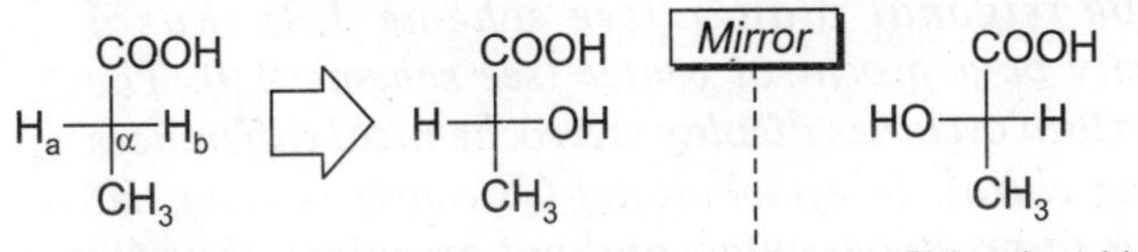

Propionic acid a prochiral molecule H_a and H_b are enantiotopic hydrogens. (*R*)-Lactic acid (*S*)-Lactic acid

When a carbon is bonded to two hydrogens and two different groups as in propionic acid the two hydrogens are termed enantiotopic hydrogens. Propionic acid is a prochiral molecule and Cα is termed a prochiral center.

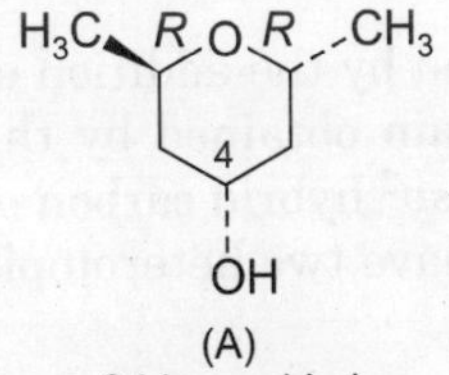

(A)
C4 is prochiral

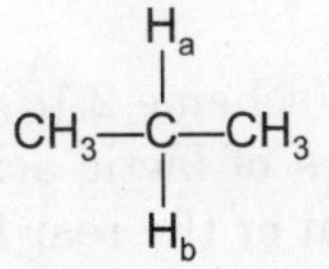

Propane
H_a and H_b are
Homotopic hydrogens

The two hydrogens H_a and H_b in propane are homotopic since the other two groups are identical. The replacement of either hydrogen by e.g., OH group gives the same compound.

SCHEME 2.1

- **Prochiral Center—Prochiral Molecules**

  A tetrahedrally bonded atom (of an achiral compound) of the general formula $Cabc_2$ (as in propionic acid) which becomes a stereocenter Cabcd and the compound becomes chiral on replacement of one of the identical groups with a different group *d* is called a prochiral center and the molecule as prochiral. This situation is available in the molecule of propionic acid. Similarly in (A, scheme 2.1) C4 is a prochiral center as the absolute configuration at two stereocenters is similar.

A chiral molecule Cabcd with four different groups attached to the carbon is said to have a stereogenic center or simply a stereocenter, the term is preferred to a "chiral center". However, a prochiral molecule (like propionic acid) is said to have a prochiral center, the term "prostereogenic center" or simply "prostereocenter" is not yet widely used.

- **Homomorphic Ligands**

  The two apparently identical ligands in propionic acid (scheme 2.1) *i.e.* H atoms of the methylene group are called "homomorphic" from Greek *homos* meaning same and *morphe* meaning form, these are indistinguishable when considered in isolation.

- **Stereoheterotopic Ligands**

  The two (apparently identical) H atoms of methylene group in propionic acid (scheme 2.1) and the methylene hydrogens of ethyl group in 2-butanol (see scheme 1.85) are called stereoheterotopic groups. Stereoheterotopic ligands may be enantiotopic (as in propionic acid), diastereotopic (as in 2-butanol) or a molecule may have only homotopic ligands (as in propane).

Prochirality and Prostereoisomerism

The molecules which have stereoheterotopic ligands e.g., propionic acid are prochiral molecules. These display prochirality so that appropriate replacement of such ligands, in turn in an achiral precursor leads to chiral products. Recall that a stereocenter may also be trigonal planar (see scheme 1.3) thus a prostereocenter may not necessarily be a prochiral center (see scheme 2.9). The cis-trans isomers of alkenes and certain cyclanes display stereochemical differences (diastereoisomerism) without being chiral. In cyclobutanol C3 center bearing H^e and H^f hydrogens is prostereogenic (a prostereocenter) and not prochiral, thus C3 in cyclobutanol exhibits prostereoisomerism and not prochirality. "Prostereoisomerism" is a more general term.

- **Heterotopic Faces**

  When the carbonyl group in pyruvic acid (scheme 2.1*a*) is reduced by the addition of hydride from *e.g.* $NaBH_4$ the enantiomers of lactic acid are again obtained by the equally feasible addition to either the front or the rear face. The *sp*^{2} hybrid carbon of carbonyl group in pyruvic acid is called prochiral and is said to have two heterotopic faces which are enantiotopic.

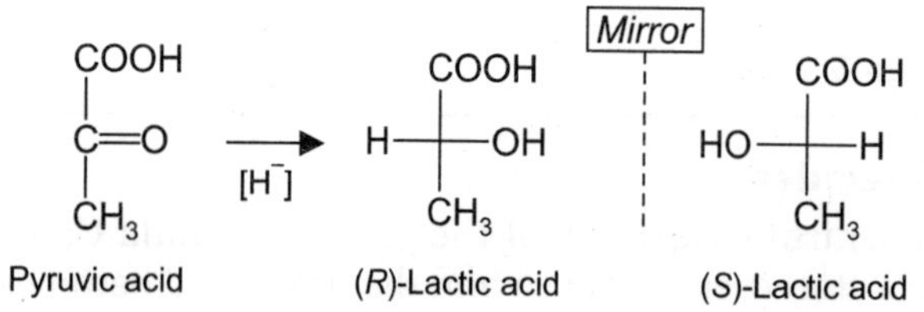

Pyruvic acid is a prochiral molecule. Chirality is introduced by nucleophilic attack, [H$^-$]. The product formed by attack on top face is the mirror image of that formed by attack on the bottom face. The two faces of the carbonyl group in pyruvic acid are termed enantiotopic.

SCHEME 2.1a

- **Homotopic Ligands and Faces**

  Unlike in propionic acid and pyruvic acid, in formaldehyde the replacement of two H atoms in turn by a test group does not generate isomers but the identical compounds to show that these two hydrogens are homotopic. Similarly there is no way to distinguish between the two faces of formaldehyde–addition of CH_3Mgl to either face gives the same compound ethanol. This shows that the two faces in formaldehyde are also homotopic.

2.2 HOMOTOPIC AND HETEROTOPIC LIGANDS AND FACES

Two criteria—substitution/addition or symmetry may be employed to determine the topic relationships of homomorphic ligands and faces and one of these tests is however, sufficient. Unfortunately the still commonly used term "equivalent" lacks precision–the two methylene hydrogen atoms of propionic acid (scheme 2.1) are equivalent when detached (*i.e.*, these are homomorphic) from the system of propionic acid. These protons are not equivalent due to their placement in propionic acid, where, these are heterotopic. Heterotopic ligands and faces may be either enantiotopic or diastereotopic.

A. Homotopic Ligands and Faces

(*i*) Substitution—Addition Criterion

Two homomorphic ligands are homotopic if substitution (replacement) of each one of them in turn by another atom or group leads to the same structure (the replacement ligand must be different not only from the original one, but also from all other ligands attached to the same carbon). Thus the three methyl hydrogen atoms in acetic acid or toluene are homotopic because replacing any one of these by *e.g.*, chlorine gives the same chloroacetic acid or benzyl chloride respectively. The two methine hydrogen atoms in (*R*, *R*)–(+)–tartaric acid (I, scheme 2.2)

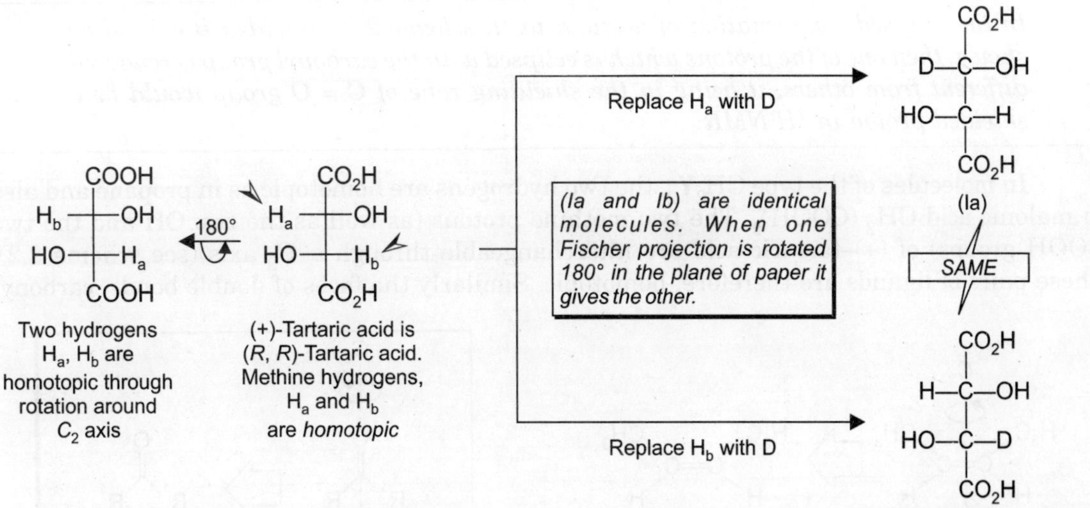

Scheme 2.2

or in its enantiomer are homotopic since replacement of either of them *e.g.*, by deuterium gives the same (2*R*, 3*R*)-tartaric-2-*d* acid. This example shows that chirality does not interfere in having homotopic ligands.

Two faces of a double bond are homotopic provided addition to either face gives the same product. Thus the two faces of *e.g.*, *cis*-2-butene are homotopic (epoxidation is used as the test reaction) since epoxidation on either face gives the same *meso* product (see, scheme 1.102). Similarly the two faces in formaldehyde (and also in acetone) are homotopic (see scheme 2.5*a*).

(*ii*) Symmetry criterion

Ligands or faces of a double bond are homotopic if they can interchange places through a rational axis. In a molecule *e.g.*, acetic acid two successive 120° rotations of a methyl group around the axis (C_3 rotational axis, scheme 2.3) allow each hydrogen to take the position of either of the other two (with no structural change). Thus in a rapidly rotating methyl group of acetic acid all the hydrogens are rendered equivalent and are thus homotopic. The methyl group is not stationary, if it was so then *e.g.*, in one of its eclipsed conformations, the hydrogens would have been rendered heterotopic which is not the case.

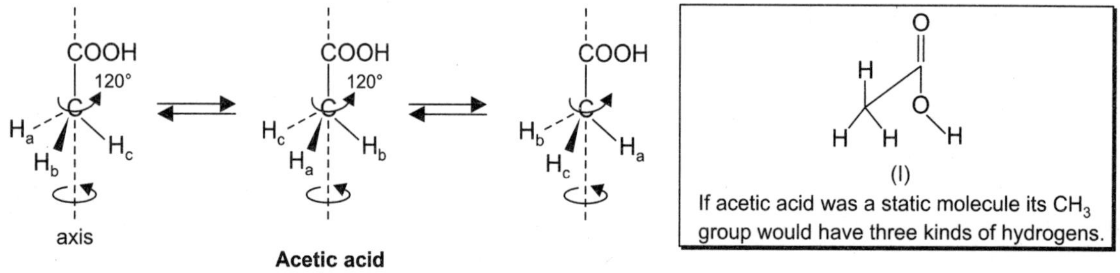

If acetic acid was a static molecule its CH_3 group would have three kinds of hydrogens.

SCHEME 2.3

> *In the eclipsed conformation of acetic acid (I, scheme 2.3) provided it existed as shown, then one of the protons which is eclipsed with the carbonyl group is rendered different from others, it being in the shielding zone of C = O group would be a shielded proton in ^1H NMR.*

In molecules of the type CH_2Y_2, the two hydrogens are homotopic as in propane and also in malonic acid $CH_2(COOH)_2$. The two methine protons (as well as the two OH and the two COOH groups) of (+)–tartaric acid are interchangeable through a C_2 axis (see scheme 2.2). These pairs of ligands are therefore, homotopic. Similarly the faces of double bonds, carbonyl

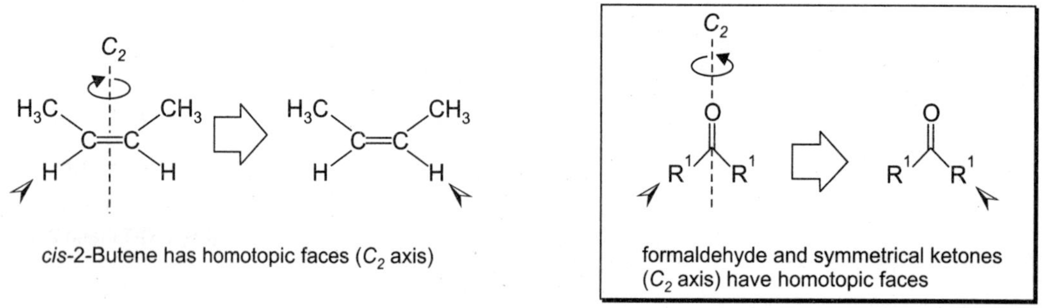

cis-2-Butene has homotopic faces (C_2 axis)

formaldehyde and symmetrical ketones (C_2 axis) have homotopic faces

SCHEME 2.4

(compounds) or carbocations are homotopic if these interchange by a C_2 axis. This is so *e.g.*, in *cis*-2-butene, formaldehyde and symmetrical ketones (scheme 2.4). The diol (I, scheme 2.5) has a two fold axis of symmetry (C_2 to show that the two hydroxyl groups are homotopic. This fact is proved experimentally since on monoacetylation the diol just gives only one product.

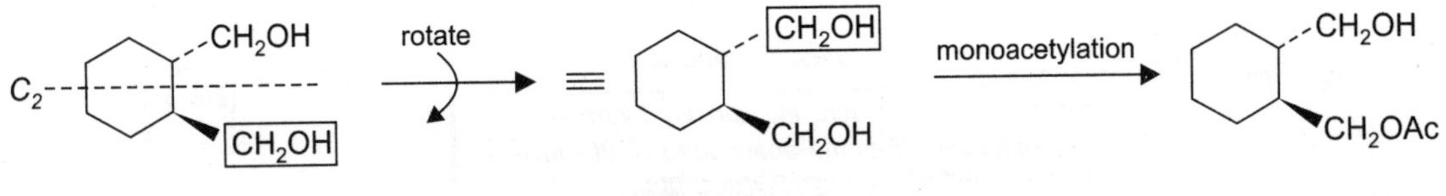

SCHEME 2.5

EXERCISE 2.1

Demonstrate that the two faces of formaldehyde (methanal) are homotopic. Are the faces in cyclopropanone and trans-2, 3-dichlorocyclopropanone homotopic ?

ANSWER. *Approach of a reagent to add a new ligand not already present in the molecule (which should not be either H or O in the case of e.g., methanal) yields the same product (scheme 2.5a). Thus approach of a reagent e.g., EtLi to either face F and F′ of methanal affords the same product. Moreover, the faces are also exchangeable through* C_2 *axis to make the faces homotopic. Based on similar arguments the faces in both cyclopropanone and trans-2, 3-dichlorocyclopropanone are also homotopic.*

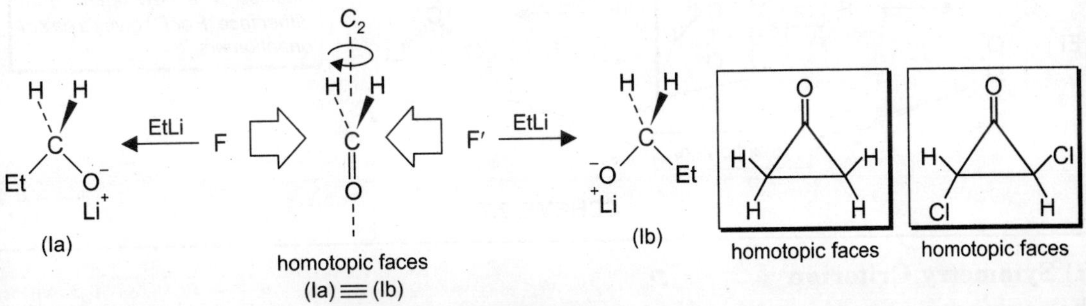

SCHEME 2.5a

B. Enantiotopic Ligands and Faces

(*i*) Substitution—Addition Criterion

Two heterotopic ligands are enantiotopic if their replacement in turn by a different achiral ligand gives rise to enantiotopic products. In the case of propionic acid (CH_3CH_2 COOH the protons of the methylene group are enantiotopic (see, scheme 2.1). On similar arguments the C2 methylene hydrogens of butane are enantiotopic (see scheme 1.87). The protons H_1 and H_2 in *meso*- tartaric acid are enantiotopic since their replacement in turn by a test group D gives rise to enantiomeric structures (scheme 2.6). Interestingly *meso*-tartaric acid is a rare example of a molecule with heterotopic ligands but no discernible prochiral atom or other element of prochirality.

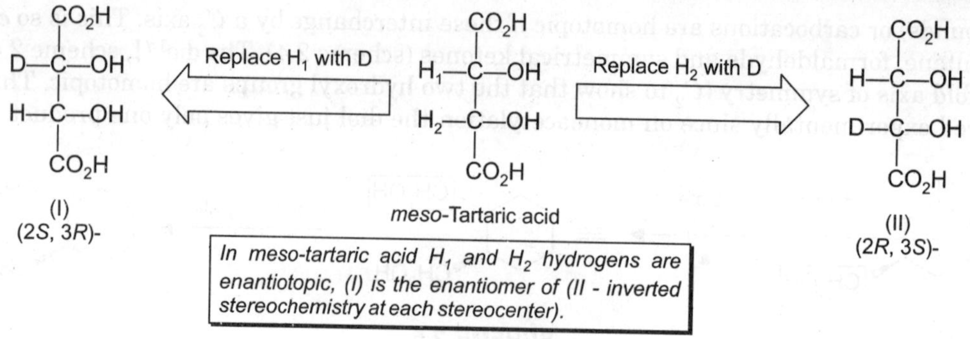

SCHEME 2.6

Based on similar criterion, the faces F and F′ of aldehydes (other than formaldehyde) *e.g.*, ethanal and unsymmetrical ketones R—CO—R′ (R ≠ R′) *e.g.*, 2-butanone, acetophenone etc. are enantiotopic (scheme 2.7).

These molecules with sp^2 hybridized carbons which becomes chiral (to give a 50 : 50 mixture of enantiomers *via* addition from either top face or bottom face) are also prochiral molecules. The two faces of the carbonyl group in these compounds are termed enantiotopic (scheme 2.7). Consider the case of *trans*-2-butene which has enantiotopic faces since epoxidation to either face gives an enantiomer (see, scheme 1.102.)

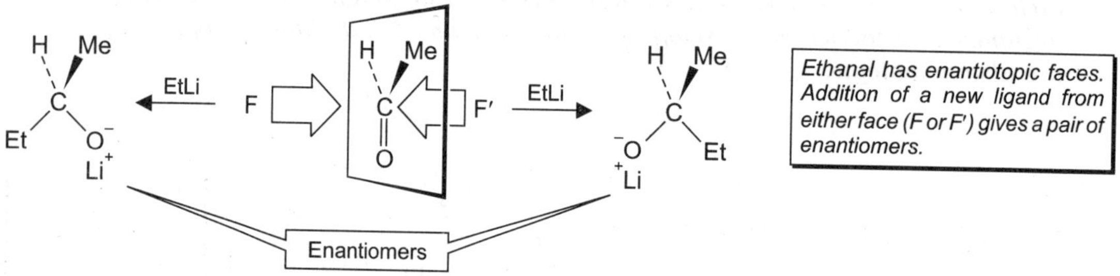

SCHEME 2.7

(*ii*) Symmetry Criterion

The enantiotopic ligands and faces cannot be interchanged by a simple axis of symmetry C_n (operation of symmetry element of first kind) but are interchangeable by σ (plane of symmetry), center of symmetry i or alternating axis of symmetry S_n (operation of a symmetry element of second kind).

> *Chiral molecules can have no enantiotopic ligands or faces since these cannot have a symmetry element of the second kind.*

The methylene protons of propionic acid CH_3CH_2COOH can exchange places through a σ plane (mirror images of each other). These ligands are therefore, enantiotopic. In *meso*-tartaric acid a σ plane of symmetry (see scheme 1.2*a*) is obvious which shows that two H atoms, two OH groups and two COOH groups are enantiotopic. Enantiotopic ligands are also interchangeable through the operation of a center of symmetry *i*. The center of symmetry is also obvious in the stable staggered conformation of *meso*-tartaric acid (see scheme 1.72). This fact again shows the enantiotopic relationship between two H atoms, OH groups and two COOH groups. In ethanal there is a σ plane (plane of the molecule) but no simple rotation axis. The faces of ethanal are therefore, enantiotopic (scheme 2.7). The enantiotopic faces of 2-butanone (see scheme 1.85) are also interchangeable through a σ plane which coincides with the plane of the molecule itself.

The two faces of *trans*-2 butene are enantiotopic (presence of plane of symmetry, and center of symmetry) in contrast to that in *cis*-2-butene (plane of symmetry, but C_2 also) where the two faces are homotopic. Consider the corresponding epoxides or analogous derivative (scheme 1.102), the epoxide form *trans*-2-butene is chiral (it has a C_2 axis only, many chiral compounds have a C_2 axis *e.g.*, (+)-tartaric acid scheme 1.77) while the one from *cis*-2-butene has a plane of symmetry and is achiral hence the faces of the butenes from which these were formed are enantiotopic and homotopic respectively. A mono substituted alkene *e.g.*, but-1-ene has two enantiotopic faces (see, scheme 1.88). Enantiotopic faces are often called prochiral faces.

EXERCISE 2.2

Identify the indicated hydrogens/ligands in (scheme 2.8) as homotopic or enantiotopic.

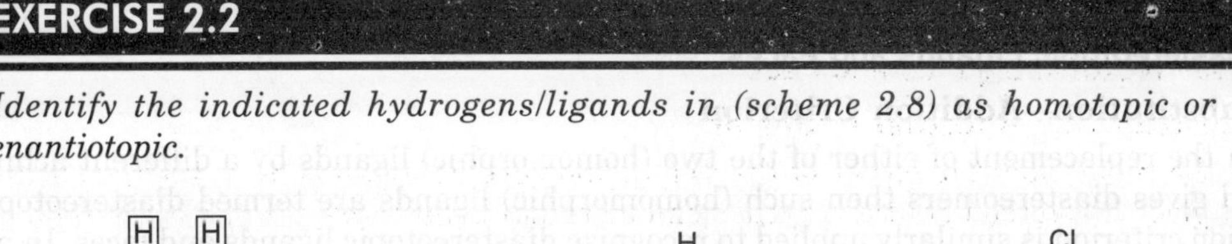

(I) *trans*-1,2-dimethylcyclopropane (II) (III)

SCHEME 2.8

ANSWER. In (I) the hydrogen atoms of the methylene group are homotopic since *C_2 symmetry axis interchanges them (IV, scheme 2.8a). Moreover, replacement*

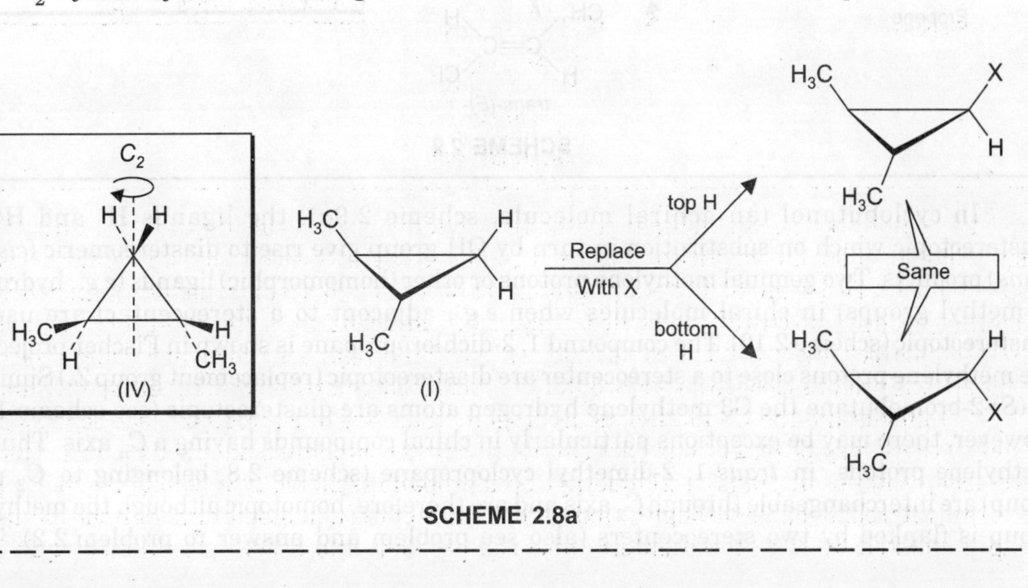

SCHEME 2.8a

with X of either of these hydrogen atoms in turn gives the same compound (turn either one over it becomes superimposable on the other). In (II) mono esterification gives an enantiomeric pair ((V) and (VI), scheme 2.8b) thus the two ligands are enantiotopic. In (III) since the protons cannot be interchanged by a simple symmetry axis but are related by a mirror plane of symmetry (VII, scheme 2.8b) these are enantiotopic.

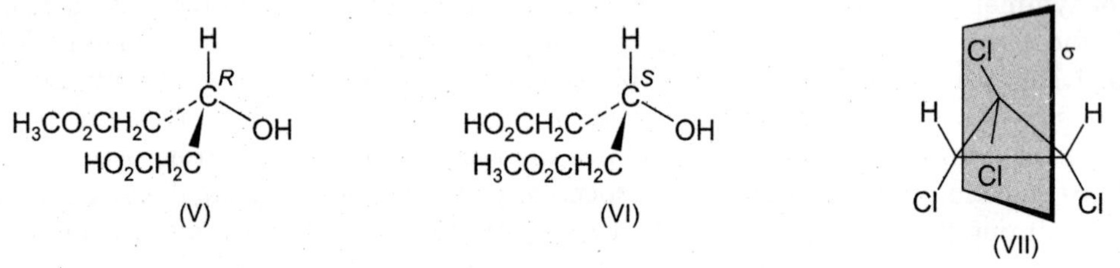

SCHEME 2.8b

(C) Diastereotopic Ligands and Faces

(*i*) Substitution—Addition Criterion

When the replacement of either of the two (homomorphic) ligands by a different achiral test ligand gives diastereomers then such (homomorphic) ligands are termed diastereotopic. The addition criterion is similarly applied to recognize diastereotopic ligands and faces. In propene (scheme 2.9) replacement of H_a and H_b in turn gives diastereomeric alkenes (*cis*- or *trans*) thus H_a and H_b are diastereotopic hydrogen atoms.

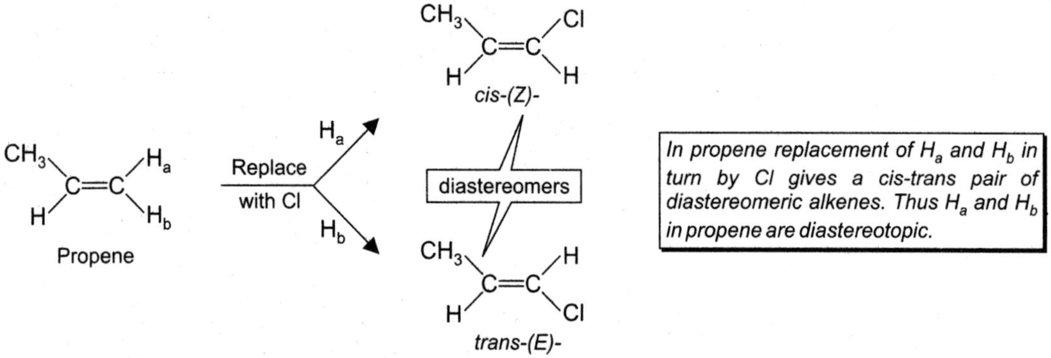

SCHEME 2.9

In cyclobutanol (an achiral molecule, scheme 2.9*a*), the ligands H^c and H^f are diastereotopic which on substitution in turn by OH group give rise to diastereomeric (*cis* and *trans*) products. Two geminal methylene protons or other (homomorphic) ligands (*e.g.*, hydrogens of methyl groups) in chiral molecules when *e.g.*, adjacent to a stereocenter) are usually diastereotopic (scheme 2.10). The compound 1, 2-dichloropropane is shown in Fischer projection, the methylene protons close to a stereocenter are diastereotopic (replacement group Z.) Similarly in (*S*)-2-bromobutane the C3 methylene hydrogen atoms are diastereotopic (see scheme 1.99). However, there may be exceptions particularly in chiral compounds having a C_n axis. Thus the methylene protons in *trans*-1, 2-dimethyl cyclopropane (scheme 2.8, belonging to C_2 point group) are interchangeable through C_2 axis and are therefore, homotopic although the methylene group is flanked by two stereocenters (also see problem and answer to problem 2.2). Thus chirality is neither a necessary nor a sufficient condition for the presence of heterotopic ligands. The achiral molecules, on the other hand may have diastereotopic ligands (as in cyclobutanol see scheme 2.9*a*).

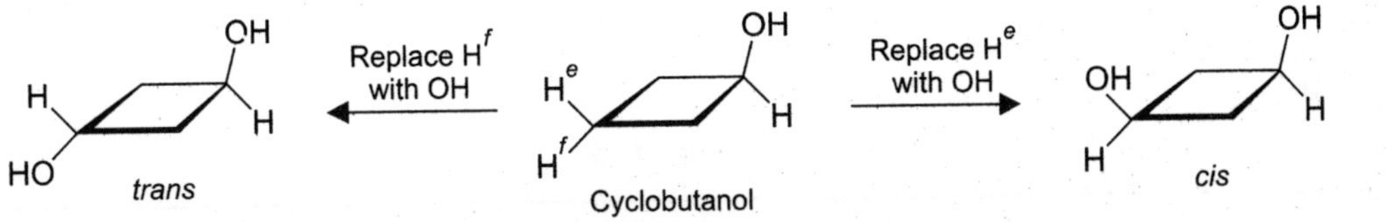

Protons H[e] and H[f] in cyclobutanol are ***diastereotopic****, their replacement in turn by "OH" gives cis - and trans - diols which are diastereomeric. Since in either cis - or trans diol chirality is not introduced at C3 (plane of symmetry in the diols)* ***C3 in Cyclobutanol is not a prochiral center but is a prostereocenter (i.e., prostereogenic center)****. Thus C3 in cyclobutanol displays prostereoisomerism and not prochirality.*

SCHEME 2.9a

The two faces of a carbonyl group close to a stereocenter are diastereotopic. Thus addition of HCN to methyl α-phenethylketone (scheme 2.12) gives diastereomers. Similarly achiral 4-*t*-butylcyclohexanone has two diastereotopic faces since addition of hydride can occur from the axial or the equatorial side to give diastereomers (scheme 2.13).

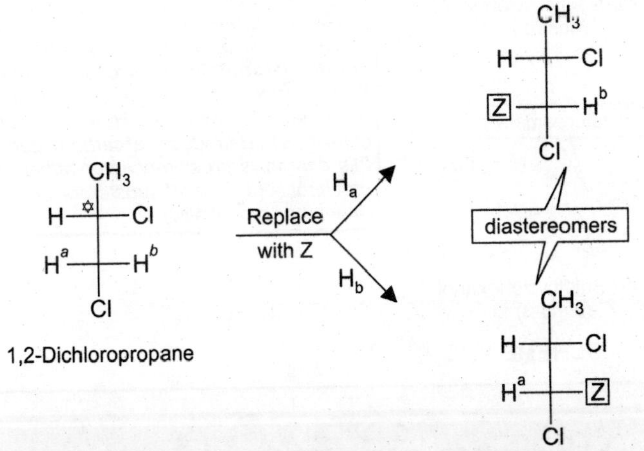

Geminal methylene protons adjacent to a stereocenter are usually diastereotopic

SCHEME 2.10

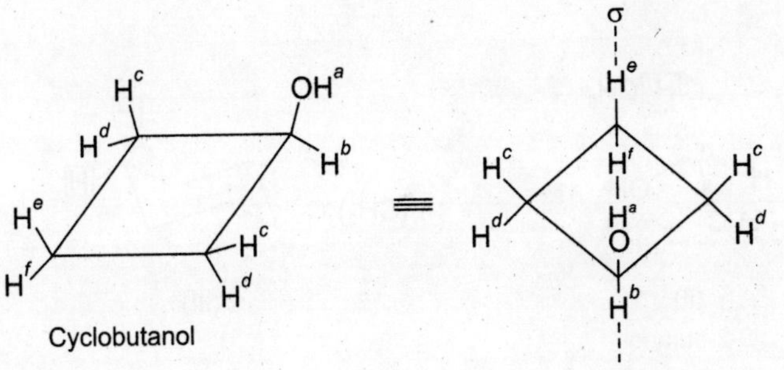

In cyclobutanol protons H[e] and H[f] are diastereotopic. Cyclobutanol has also a plane of symmetry. Protons H[c] are mirror images of each other and are enantiotopic, similarly protons H[d] are also enantiotopic. The relationship between each H[c] and H[d] pair is again diastereotopic.

SCHEME 2.11

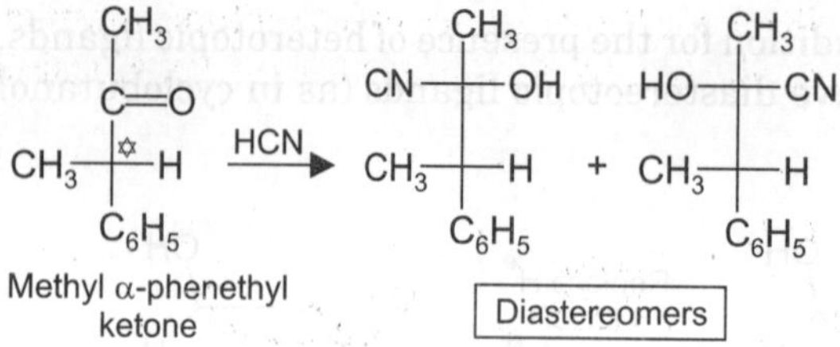

> *When in aldehydes (other than formaldehyde) and unsymmetrical ketones R-CO-R′ (R≠R′) the groups R, R′ contain one or more stereocenters, the faces of the carbonyl group become diastereotopic. Addition to one face gives a diastereomer and addition to the other face gives another diastereomer.*

SCHEME 2.12

(*ii*) Symmetry Criterion

The diastereotopic ligands or faces can not be related by any symmetry relation *i.e.*, C_n (axis of symmetry) or S_n (alternating axis of symmetry). One example is taken to explain this fact. 2-Butanone has enantiotopic faces since these faces are interchangeable through a σ plane coinciding with the plane of the molecule, however, when the C=O group is adjacent to a stereocenter as in α-phenethyl ketone (scheme 2.12), then its two faces are no longer interchangeable through a σ plane.

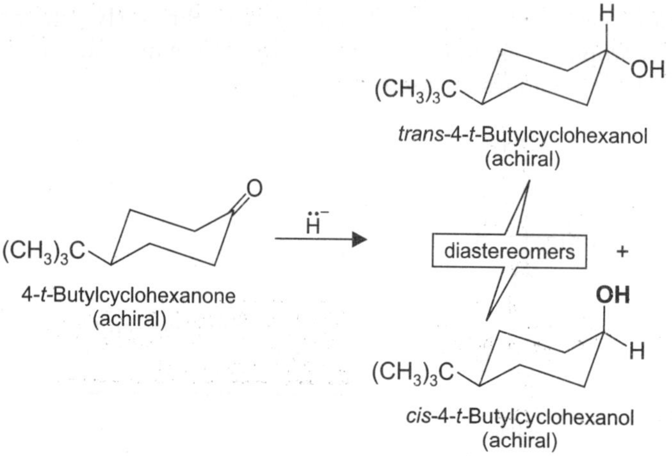

> *4-t-Butylcyclohexanone is an achiral molecule, the faces of the carbonyl group are* ***diastereotopic*** *- addition of hydride to the two faces (in turn) gives two (achiral) diastereomeric products. Here also, since chirality is not introduced at carbonyl carbon, this carbon is prostereocenter rather than prochiral - (a case of prostereoisomerism rather than prochirality).*

SCHEME 2.13

EXERCISE 2.3

Identify the indicated hydrogens (scheme 2.13a) as diastereotopic giving arguments.

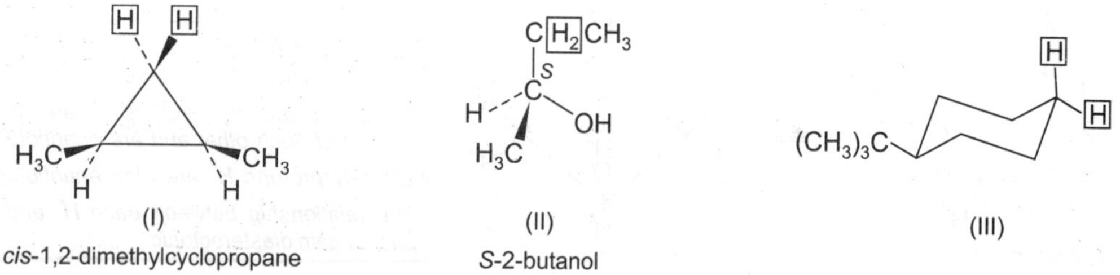

SCHEME 2.13a

ANSWER. *In (I) the methylene protons are close to a stereocenter (infact two stereocenters) and are thus diastereotopic. These are also not related by simple symmetry axis. Moreover, their replacement by a group X gives diastereomers (scheme 2.13b). In (II) one has both a stereocenter and a prochiral center. The methylene protons are diastereotopic as shown by isotopic replacement (scheme 2.13c). The indicated hydrogens in (III) are diastereopic since a replacement test gives diastereomers (scheme 2.13c).*

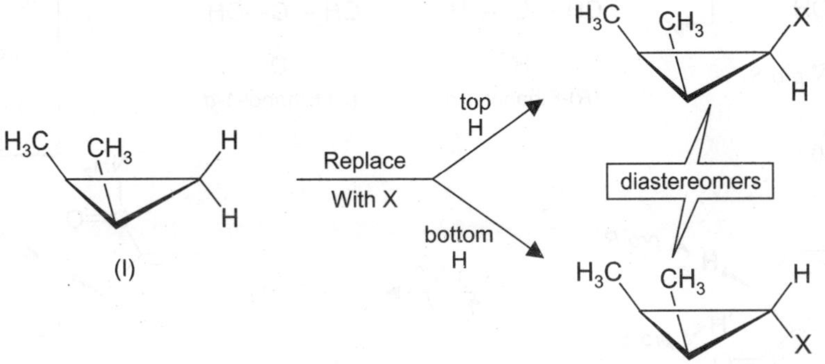

SCHEME 2.13b

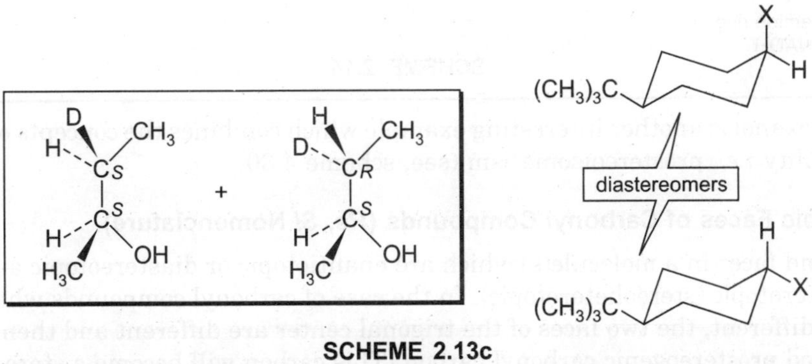

SCHEME 2.13c

2.3 NOMENCLATURE OF ENANTIOTOPIC LIGANDS AND FACES AND SOME STEREOSELECTIVE REACTIONS

A. Nomenclature of Enantiotopic Ligands

To name the enantiotopic ligands at a prochiral center *e.g.*, H_a and H_b in ethanol (scheme 2.14), the ligand to be labelled is arbitrarily assigned a higher CIP priority over the other.

If H_a is arbitrarily preferred over H_b in the sequence rule, the sequence is OH > CH_3 > H_a > H_b and the (hypothetical) configurational symbol to the prochirality center in ethanol will be *R*, thus H_a is designated *pro-R* and by default H_b becomes *pro-S*. The same answer would have come out if H_b was given precedence over H_a, in that case the sequence would have been OH > CH_3 > H_b > H_a. The hypothetical configurational symbol for ethanol is then *S*, hence H_b is *pro-S*. It might be noted that the same result would have been obtained by replacing first, one hydrogen and then the other by deuterium since deuterium has priority over hydrogen. Replacement of H_a by D gives (*R*)-ethanol-1-*d* and hence H_a is *pro-R*; similarly, replacement of H_b by deuterium gives (*S*)-ethanol-1-*d* and hence H_b is *pro-S*. Another example is found in propionic acid (I, scheme 2.14), it is always better to assign a descriptor to H_a first *i.e.* a group directed towards you then automatically H_b of lowest priority will be away from you *e.g.*, an arrangement for the direct assignment of CIP nomenclature. Then by default H_b will have the opposite descriptor. Similarly in (NADH, II, scheme 2.14) the assignment can be made easily.

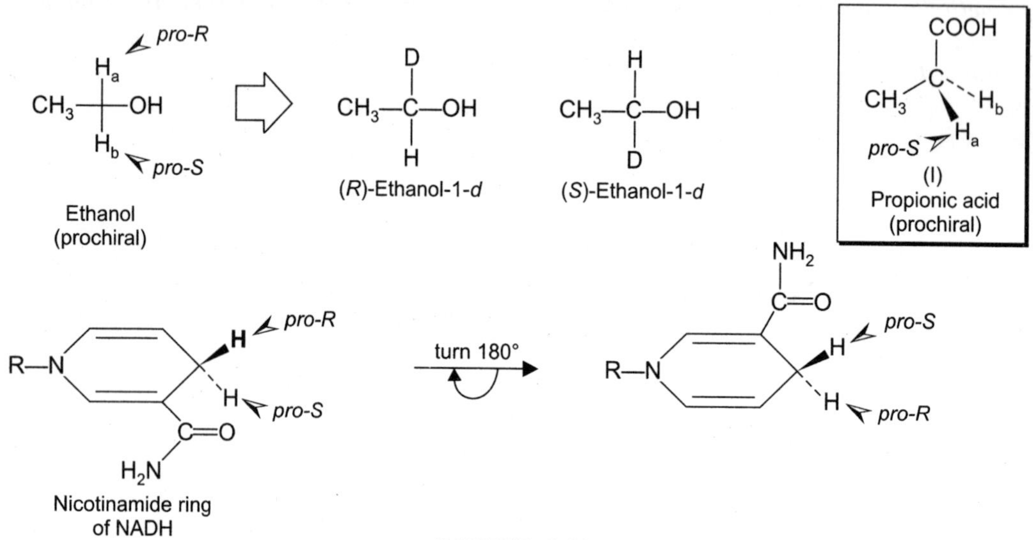

SCHEME 2.14

Cyclohexanol is another interesting example which combines the concepts of prochirality and proachirality *i.e.*, prostereoisomerism (see, scheme 4.30).

B. Enantiotopic Faces of Carbonyl Compounds (*Re*, *Si* Nomenclature)

The groups and faces in a molecule(s) which are enantiotopic or diastereotopic are collectively termed as heterotopic (stereoheterotopic). In the case of carbonyl compounds when the groups R and R′ are different, the two faces of the trigonal center are different and then the carbonyl carbon is called prostereogenic carbonyl carbon [This carbon will become a stereocenter when attacked by a group unlike any of the groups already bonded to it and the addition product will be a pair of enantiomers, see scheme (1.85)].

The faces of the carbonyl group are differentiated by the *Re–Si* nomenclature. The groups around the carbonyl group are given priorities (CIP rules for *R, S* nomenclature). If going from the group of highest priority of the group of lowest priority around the face of the carbonyl group, the path is clockwise, the face is *Re* and if it is anticlockwise, the face is *Si* (scheme 2.15).

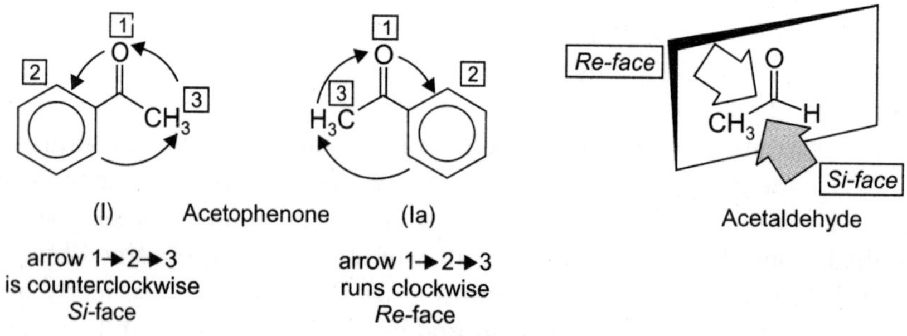

SCHEME 2.15

The closest face of the carbonyl group *e.g.*, in acetophenone (I, scheme 2.15) is *Si* face. To explore the other face one either goes behind the page or rotates the molecule (180° out of the plane of paper *e.g.*, to get the view of the *Re* face (I*a*, scheme 2.15). Both the faces may be viewed simultaneously as shown for acetaldehyde (scheme 2.15). Thus 2-butanone during its reduction with hydride (scheme 2.16) presents both the *Re* face (*i.e.* added H is closest to ones eyes and is shown by a thick wedge) to give (*S*)-2-butanol as well the *Si* face (*i.e.*, added H is away from ones eyes and is shown by a hatched wedge) to give (*R*)-2-butanol. Significantly during enzymatic reduction, 2-butanone, presents specifically only one of the faces *i.e.*, the *Si* face to give only one enantiomer *i.e.*, (*R*)-2-butanol. One may however, note that as the CIP

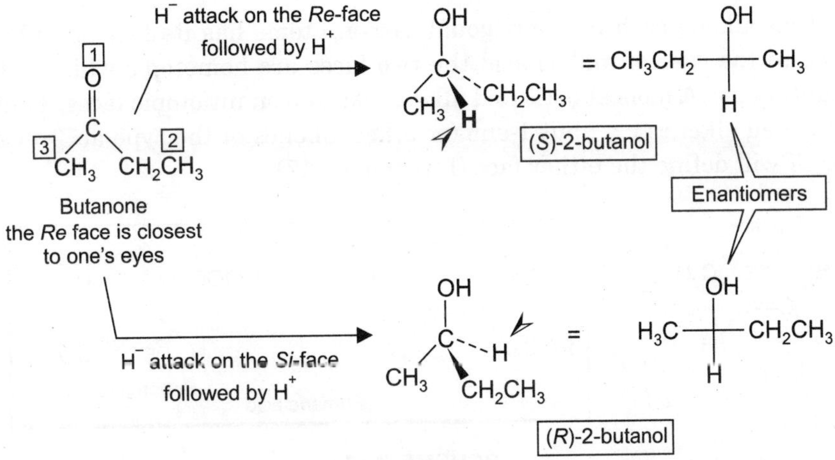

SCHEME 2.16

rules operate, there is no relationship between the face attacked by the nucleophile and the stereodesignator of the stereocenter in the product. In the preceding example (scheme 2.16) hydride addition to the *Re* face gives (*S*)-enantiomer whereas hydride addition to the *Si* face gives the (*R*)-enantiomer.

EXERCISE 2.4

Depict the top face in methyl acetate as presented (scheme 2.16a) as Re or Si.

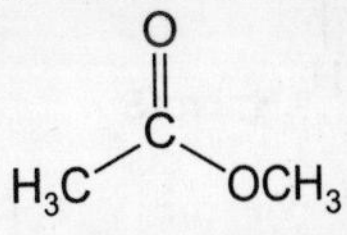

Methyl acetate

SCHEME 2.16a

ANSWER. *This is a general structure R′-CO-R² where one of the groups contains an oxygen. This situation requires the assignment of priority between the carbonyl oxygen and the R-group oxygen to assign Re, Si faces of such molecules. Considering CIP rules R′ = CH_3 has the least priority. One knows from CIP rules that double bonds are considered as two single bonds and triple bonds as three. Thus one may represent the carbonyl oxygen of methyl acetate as (I, scheme 2.16b) where the new carbon atom has no further substituent and is thus taken to be attacked to a 'phantom atom' of atomic number zero (II). Thus (I) is to be considered as (III). This shows that methoxy oxygen has priority over the carbonyl oxygen and therefore, the face presented is Si (IV, scheme 2.16b).*

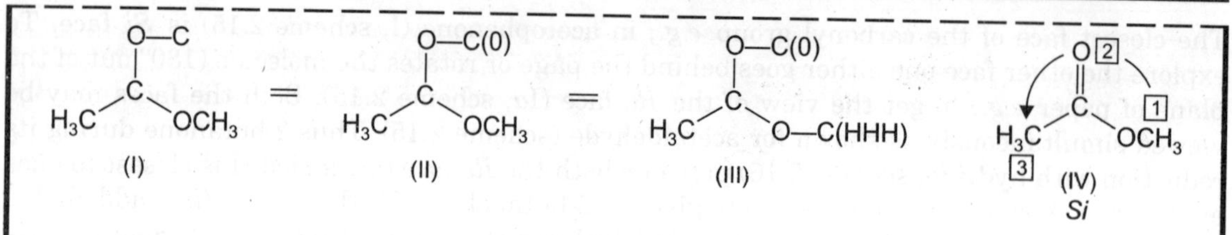

SCHEME 2.16b

C. Faces of Alkenes (*Re*, *Si* Nomenclature)

In the case of an alkene each of the trigonal carbon atoms has its face specified separately (scheme 2.17). In the case of maleic acid the two faces are homotopic while in fumaric acid these are enantiotopic. A monosubstituted alkene has two enantiotopic faces. Thus in the case of monosubstituted alkenes *e.g.*, but 1-ene (or other alkenes of the type abC = Cdd) only one symbol *Re* or *Si* will define the either face (I, scheme 2.17).

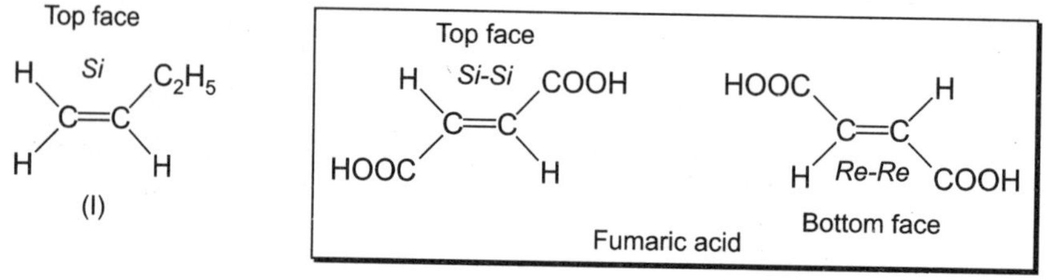

SCHEME 2.17

D. Asymmetric Enzymatic Transformations (Heterotopic Ligands and Faces and Biochemical Reactions)

Unlike the achiral chemical reagents, the chiral enzymes are capable of discriminating between two ligands and faces which are enantiotopic, as well as between two enantiomers (scheme 2.18).

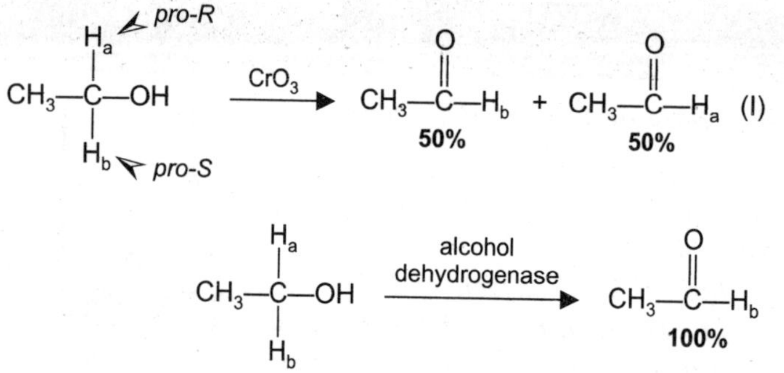

The pro-R and pro-S hydrogen can not be differentiated with achiral chemical reagents. Chiral reagents e.g., enzymes, however, distinguish between these.

SCHEME 2.18

(*i*) Enzyme Mediated Oxidation

The *pro-S* and *pro-R* hydrogens cannot be distinguished by achiral chemical reagent (scheme 2.18), however, these hydrogens can be distinguished *e.g.*, during oxidation will enzymes only *pro-R*-hydrogen being removed.

Stereospecificity and Stereoselectivity of Enzymes

The enzymes usually react with one of the two enantiomers and are then said to display stereospecificity. An example is found during enzymatic hydrolysis of the N-acetyl derivatives of racemic D, L-pair of amino acids when only the naturally occuring L-isomer is hydrolyzed (see scheme 1.86).

The enzymes when in contact with prochiral molecules, react only with one of the enantiotopic ligands or faces a property called stereoselectivity. For an example glycerol undergoes phosphorylation exclusively at the pro-R hydroxymethylene group with adenosine triphosphate (ATP) in the presence of an enzyme-glycerol-kinase.

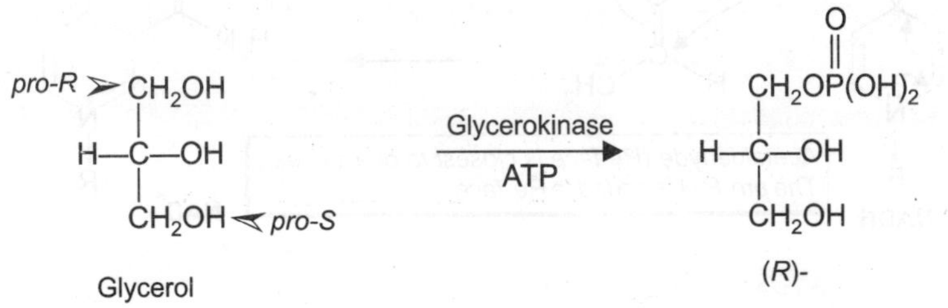

SCHEME 2.18a

(*ii*) Fumarase Catalyzed Hydration and Dehydration

The enzyme fumarase catalyzes the hydration of fumaric acid to malic acid (scheme 2.19) to yield only the *S*-enantiomer. In this reaction the hydroxyl group is added to the *Si* face of one of the carbon atoms of the double bond.This reaction is reversible with pronounced stereochemical requirements, thus neither the *cis* isomer of fumaric acid (maleic acid) nor the *R* enantiomer of malic acid can serve as substrates in this fumarase catalyzed hydration-dehydration equilibrium.

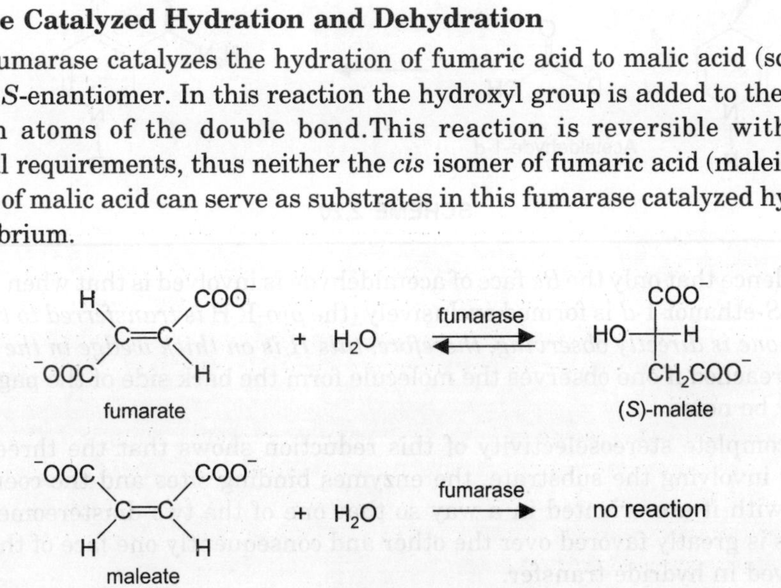

SCHEME 2.19

(*iii*) Enzyme Mediated Reduction—Bakers' Yeast

When the enzyme alcohol dehydrogenase converts acetaldehyde to ethanol using the coenzyme NADH they discriminate between the two faces of the trigonal planar carbonyl substrate so one of the two possible stereoisomeric forms of the tetrahedral product predominates. The preference of several NADH dependent enzymes for either *Re* or *Si* face of their respective substrates is known. One of the most widely used is yeast alcohol dehydrogenase (YAD). For example acetaldehyde is reduced to ethanol with yeast alcohol dehydrogenase (YAD) Bakers yeast in the presence of the hydride donating coenzyme NADH. Acetaldehyde has two enantiotopic faces (*Re* and *Si*) and NADH has two diastereotopic hydrogens (*pro R* and *pro S*. Since R attached to N is chiral). It is established that during the reduction of acetaldehyde (scheme 2.20) the *pro R* H from NADH is transferred only to the *Re* face of acetaldehyde.

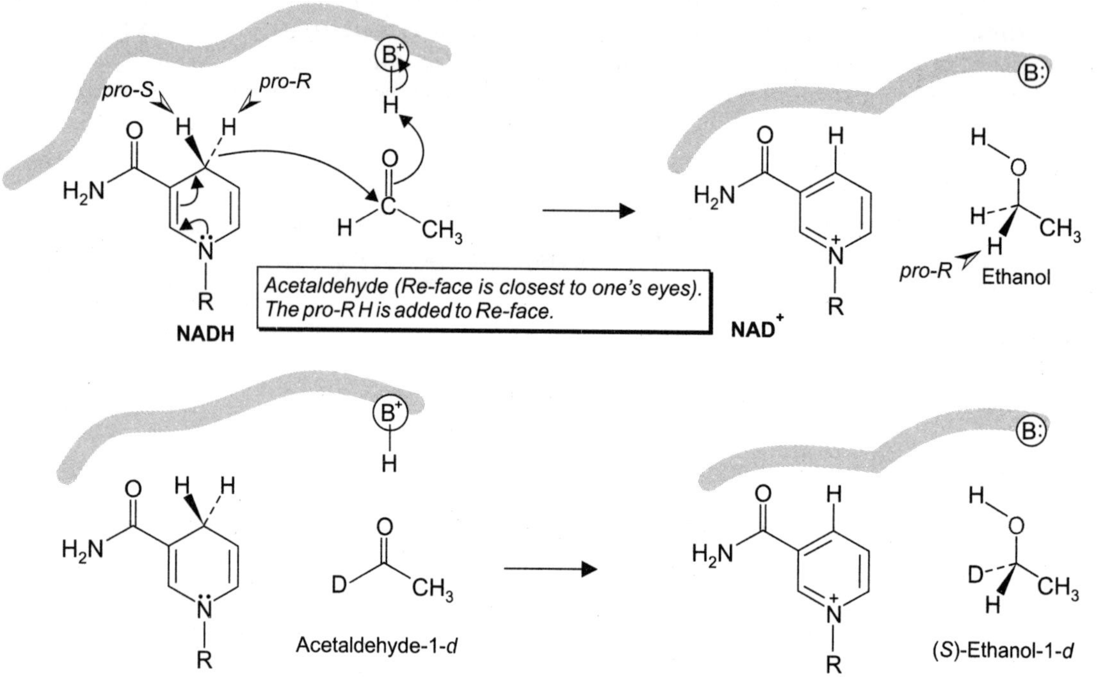

SCHEME 2.20

The evidence that only the *Re* face of acetaldehyde is involved is that when acetaldehyde-1-*d* is reduced *S*-ethanol-1-*d* is formed exclusively (the *pro*-R H *is transferred to the Re face i.e., the face which one is directly observing, therefore, this H is on thick wedge in the product).* The *Si* face can be reached if one observes the molecule form the back side of the page. The following points may be noted :

- The complete stereoselectivity of this reduction shows that the three dimensional array involving the substrate, the enzymes binding sites and the coenzyme associated with it are oriented in a way so that one of the two diastereomeric transition states is greatly favored over the other and consequently one face of the substrate is involved in hydride transfer.

  An enzyme has a specific binding site for the coenzyme and after binding, the enzyme blocks one side of the coenzyme. When the enzyme blocks the B-side of NADH, the substrate will then bind to A-side leading to the transfer of H_a hydride ion (*pro-R* hydrogen). When the A-side of the coenzyme is blocked by the enzyme, then the substrate binds to the B-side and not the H_b hydride ion (*pro-S*) will be transferred. A large number of dehydrogenases are known which can either transfer *pro-R* or *pro-S* hydrogen of NADH (scheme 2.20*a*).

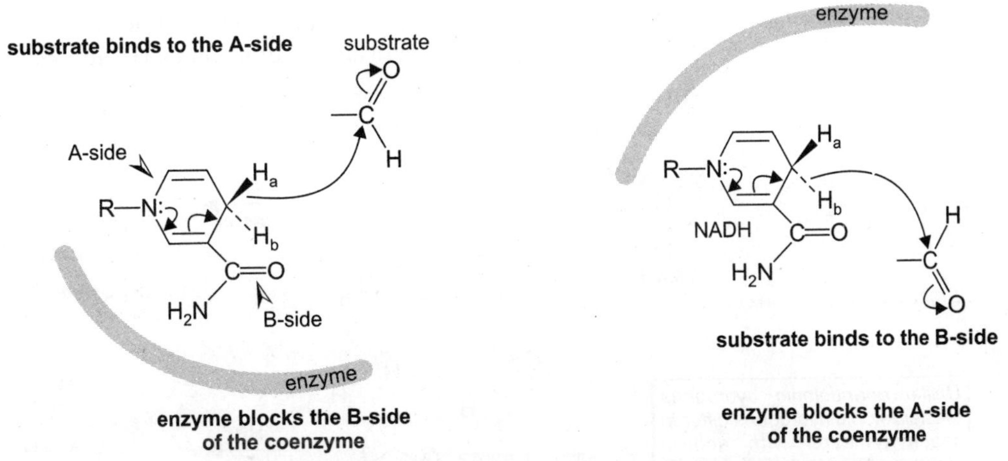

SCHEME 2.20a

- Transfer of the hydride ion concomitant with protonation of the oxygen atom form an acidic side chain of an appropriate amino acid results in reduction of the ketone or aldehyde to an alcohol. This process is similar to the laboratory method that uses a hydridic reagent followed by protic workup.
- Recall that enzymatic reduction of achiral 2-butanone gives only one enantiomer of 2-butanol (see scheme 1.85). Biological reduction of pyruvic acid catalyzed by the enzyme lactate dehydrogenase give only (*S*)-lactic acid. During biological reduction, pyruvic acid presents only one of its faces to the chiral enzyme while achiral hydride addition occurs on both faces to give a racemic mixture (see scheme 2.1*a*).

E. The *re si* Nomenclature

One uses descriptors *Re* and *Si* to define the two heterotopic faces of a trigonal planar prochirality center to which three constitutionally different groups are attached (I and II, scheme 2.20*b*). When, however, the two of the three groups are enantiomorphic, decriptors *re* and *si* are used. Thus the top face of (2*R*, 4 *S*) – 2, 4 – dibromopentan-3-one in the *re* side (*R* > *S*).

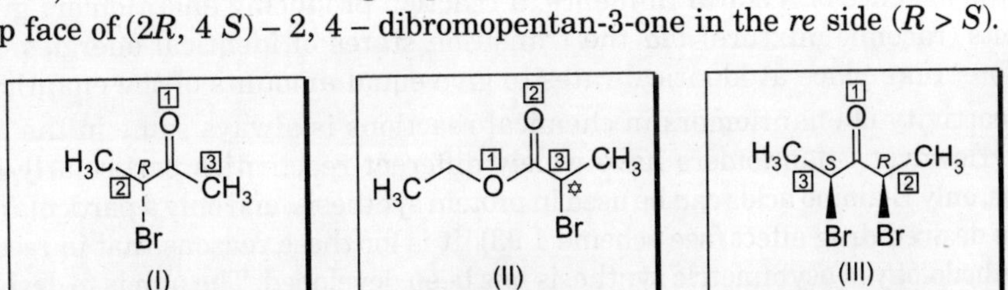

(I) top face in *Si* (II) top face in *Re* (III) top face in *re*

Scheme 2.20b

F. Reactivity of Diastereotopic Ligands with Achiral Chemical Reagents and Enzymes

The diastereotopic hydrogens however, as expected have different reactivities both with chemical reagents and enzymes. Thus for E2 elimination from 2-bromobutane (with H^a and H^b as diastereotopic protons) the substituents to be eliminated (*i.e.* H and Br) must be *anti* to each other. There are two staggered conformers (scheme 2.21) in which this arrangement is possible.

One of these conformers, (I*a*, scheme 2.21) is however, less stable due to more crowding and therefore, more of the *trans*-2-butene is formed (preferential elimination of H^b and Br).

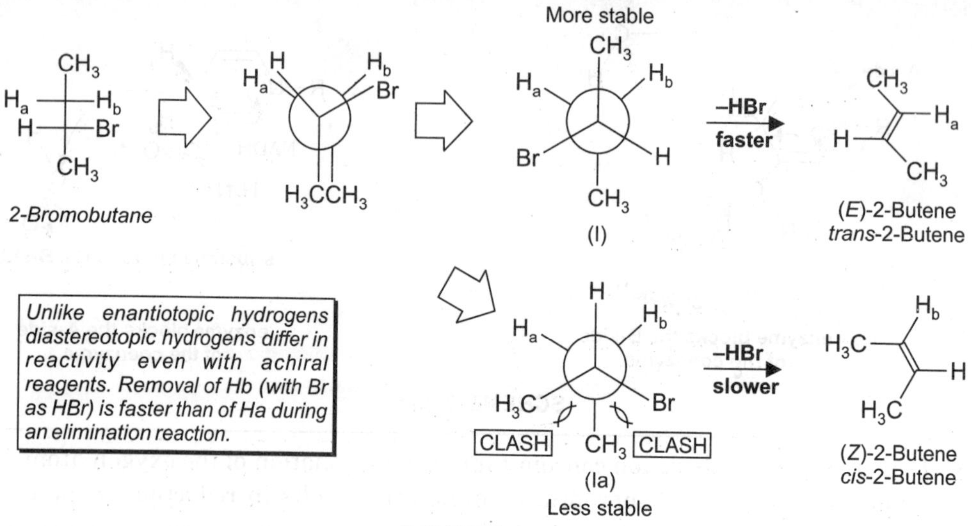

SCHEME 2.21

2.4 ASYMMETRIC SYNTHESIS—STEREOSELECTIVE SYNTHESIS (STEREOSELECTIVITY)

1. Asymmetric Synthesis—An Introduction

Optically active compounds, before 1940 could be obtained in stereoisomerically pure form only by isolation from natural sources, via the resolution of a racemic mixture, or by a few laboratory controlled enzymic reactions. In general the chemical reactions give products which contain either chiral centres, axes, or planes, however, the isolated material is only the racemic form. This is due to the fact that reactants, reagents or solvents are achiral and are themselves racemic. In the absence of a chiral influence, a reaction producing enantiomers gives them in equal amounts (racemic mixture) *via* the transition states of identical energies. These reactions, therefore, take place at identical rates to give equal amounts of the enantiomers.

The reactivity of enantiomers in chemical reactions is always same in the absence of a chiral agent. However, enantiomers have widely different reactivities particularly in biological systems. Thus, only L-amino acids can be used in protein synthesis, and only a particular enantiomer may have the desired drug effect (see scheme 1.93). It is for these reasons that in recent years an extensive methodology of asymmetric synthesis has been developed. The aim is to design synthesis in which one enantiomer, from 2^n stereoisomers is formed (stereoselective synthesis).

2. Principle of Asymmetric Synthesis

For the preferential formation of one stereoisomer (either enantiomer or diastereomer) over the other, either the reactant, or the reagent, or the solvent must be the pure enantiomeric form. The chiral agent must play an active part in the reaction and has to be integral to the transition state, so that two diastereoisomeric transition states are formed. Consequently one stereoisomer is produced more rapidly than the other. In summary an asymmetric synthesis involves competing reactions with diastereoisomeric transition states which takes place at different rates.

3. Principle Categories of Asymmetric Synthesis

(*a*) The Use of Chiral Substrates (First Generation Methods)

These methods (first generation methods) require the use of enantiomerically pure natural products *e.g.*, stereoids, terpenoids, alkaloids, amino acids etc. (Chiral pool).

Some Examples of Molecules in the Chiral Pool :

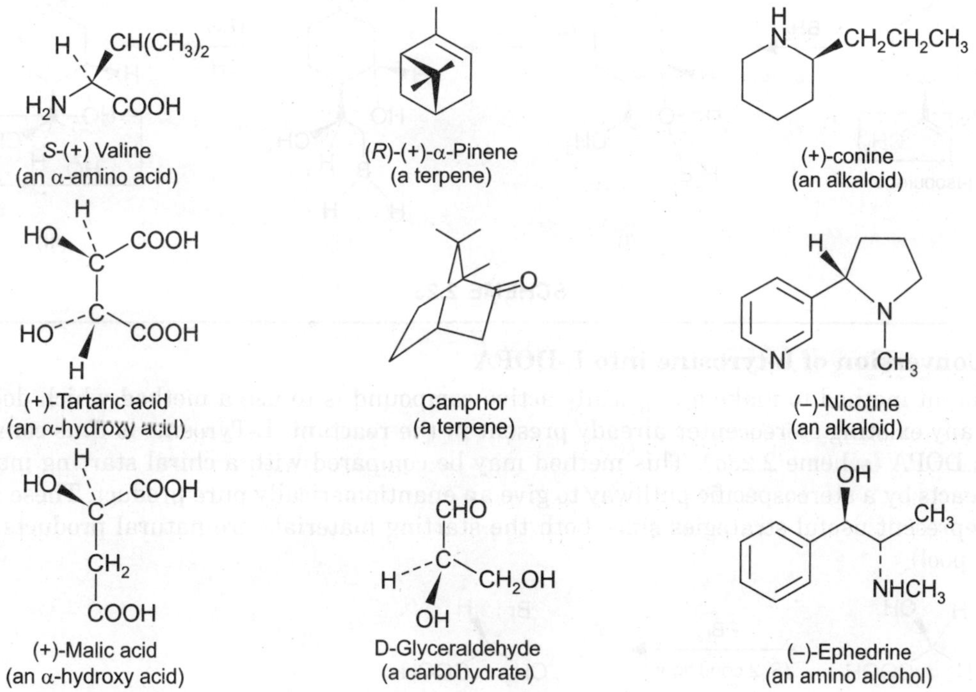

SCHEME 2.21a

(*i*) Stereoselective reduction of cholestan-3-one (Diastereoselectivity)

Cholestan-3-one on reduction with lithium aluminum hydride gives exclusively the equatorial alcohol (I, scheme 2.22) by attack of the reagent from the less hindered face (α face) of the molecule. The alcohol (I, scheme 2.22) is the more stable of these two diastereomers (I and II, scheme 2.22). The conformationally fixed steroid molecule has sterically hindered β face due to angular methyl groups. It is an example of diastereoselective reduction.

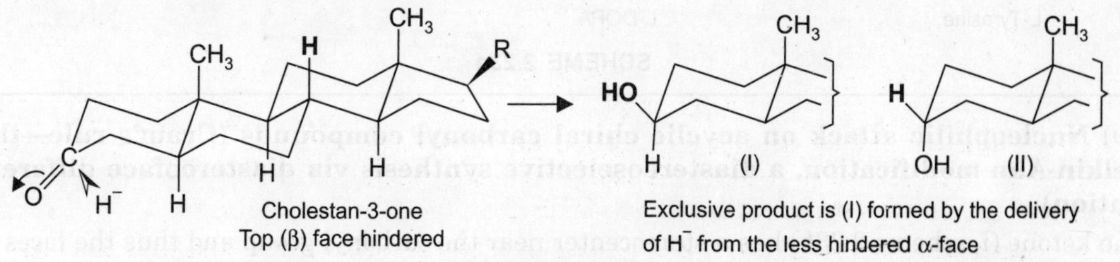

SCHEME 2.22

(ii) Hydroboration of (–)-isopulegol

In this case as well, the new stereocenter is generated under the influence of stereocenters already present and in particular the one bearing a hydroxyl group. The interaction of the hydroxyl group (nucleophilic group) with the electron deficient borane directs the hydroboration from the rear of the molecule (scheme 2.23).

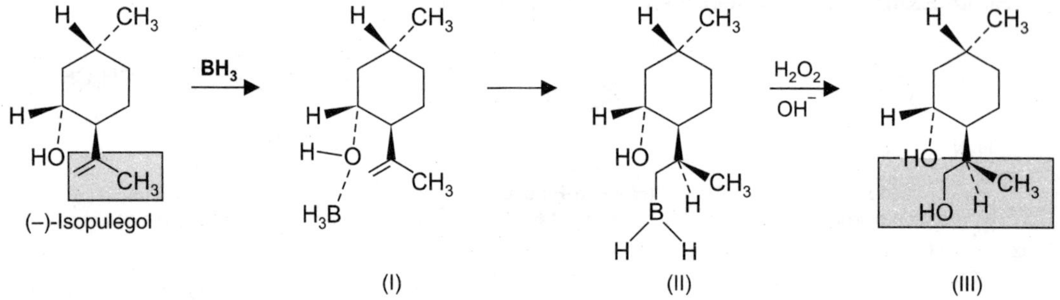

SCHEME 2.23

(iii) Conversion of L-tyrosine into L-DOPA

A different method to make an optically active compound is to use a method which does not effect any existing stereocenter already present in the reactant. L-Tyrosine is thus converted into L-DOPA (scheme 2.23*a*). This method may be compared with a chiral starting material that reacts by a stereospecific pathway to give an enantiomerically pure product. These examples represent useful strategies since both the starting materials are natural products (from chiral pool).

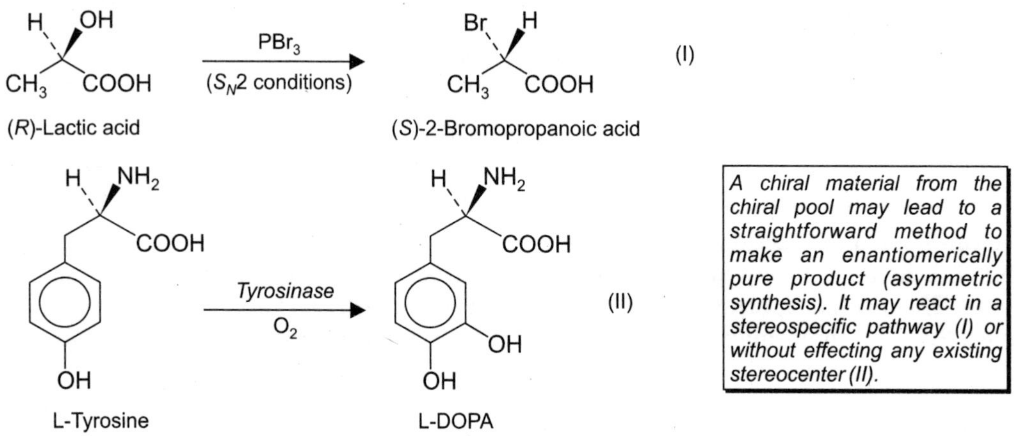

SCHEME 2.23a

(iv) Nucleophilic attack on acyclic chiral carbonyl compounds (Cram's rule—the Felkin-Ahn modification, a diastereoselective synthesis via diastereoface differentiation)

The ketone (I, scheme 2.23*b*) has a stereocenter near the carbonyl group and thus the faces of the carbonyl group are diastereotopic. This stereogenic center will influence the approach of the attacking reagent to the prochiral carbonyl. Thus the approach of the reagent to one of the faces of this type of a molecule will be easy compared to the other leading to diastereoselectivity.

A nucleophile *e.g.*, will approach (I, scheme 2.23*b*) from the face which presents less steric hindrance and this is evidently over the "hydrogen face" (*i.e.*, *Si* face) to give (II, scheme 2.23*b*). However, if the nucleophile approaches the more hindered "methyl face" (*i.e.*, *Re* face) the product would be (III). The following points may be noted :

- Provided the conformation of (I, scheme 2.23*b*) is frozen as shown the preference would be for the formation of (II).

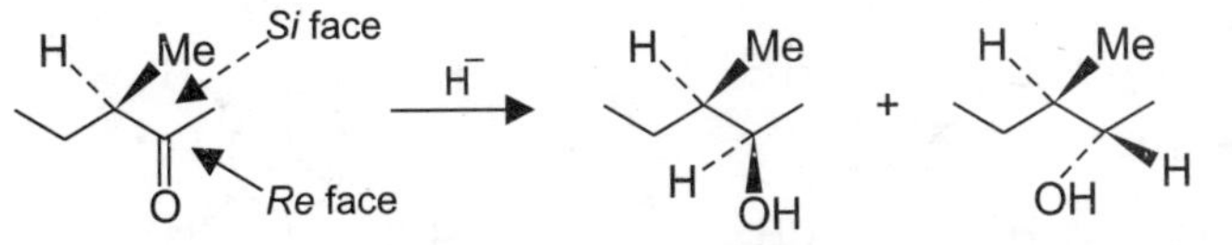

(I) (II) (major) attack from rear *Si* face (III) (minor) attack from frontai *Re* face

With the locked confirmation as in (I) the hydride would prefer to attack from less hindered rear face of C=O (side of hydrogen Si face) to be at the rear (not shown in the figure) rather than more hindered side of methyl (Re face)

SCHEME 2.23b

- The acyclic compounds of the type being discussed are, however, not locked into a particular conformation. The conformation adopted by the substrate may be locked (in the presence of the reagent *i.e.*, attacking nucleophile) in the diastereomeric transition state leading to products.
- Thus for the diastereoselectivity the facial and orientational preference has to be addressed and at the same time the conformation of the substrate in the diastereomeric transition state must be known.
- For certain additions to the carbon oxygen double bond of chiral aldehydes and ketones Cram's rule is useful to predict as to which diastereomer of the two will predominate.

The four groups attached to the stereocenter (at α carbon) are COR, S, M and L where S, M and L designate small, medium and large group respectively. The oxygen of the carbonyl orients itself so as to be between the small and the medium sized groups. A typical Cram's model is shown as its Newmann projection (scheme 2.24). Thus L being the largest group is oriented as far away as possible from the carbonyl group. In this orientation L will however, eclipse the group R. The rule is that the incoming group preferentially attacks on the side of the plane containing the small group (*i.e.*, the less hindered face). (Scheme 2.24). Based on Cram's model reaction of (I, scheme 2.25) with Grignards reagent gives (II, scheme 2.25) as the major product (also see scheme 2.18*b*).

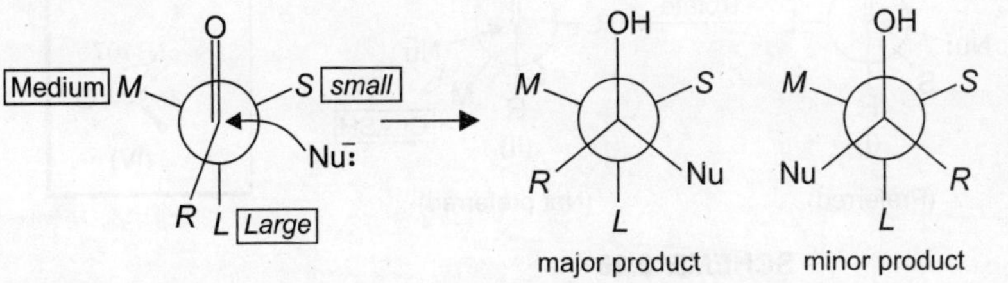

Crams rule predicts the nucleophilic attack on acyclic diastereotopic aldehydes and ketones. Group L orients trans to the carbonyl oxygen and nucleophile prefers to enter from less hindered side.

SCHEME 2.24

The Felkin—Ahn developed a variation based on theoretical calculations in terms of steric intersections and predicts more fully, the results of this chiral induction and is preferred.

The Felkin Ahn (FA) model differs from Cram's rule in the conformation adopted by the carbonyl compound. The following points may be noted:

- In the FA model the C—L bond is positioned perpendicular to the carbonyl group (scheme 2.26). This is unlike Cram's rule where L is assumed to be antiperiplanar to the C=O group. This arrangement removes unfavorable eclipsing interactions between L and R.

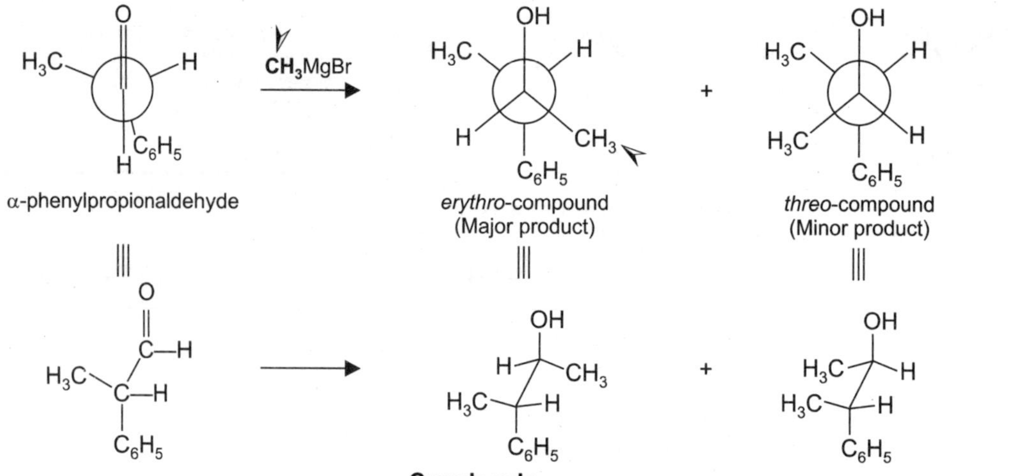

Cram's rule

SCHEME 2.25

- There are two such conformations (I and II, scheme 2.26).
- The nucleophile approaches the carbonyl carbon in a plane perpendicular to that of the –CO– fragment from the side opposite the C—L bond and at an obtuse angle with C=O which corresponds nearly to the tetrahedral angle of Nu—C—O in the product (IV, scheme 2.26).
- The consideration of the reactive conformations (I and II, scheme 2.26) shows that (I) is of lower energy due to less steric interaction between the nucleophile and the smallest group S to give (III, scheme 2.26) as the major product.

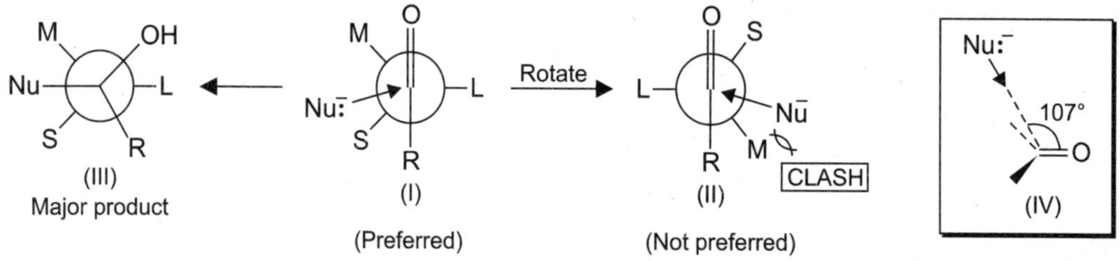

SCHEME 2.26

Thus the reduction of (*S*)-3-phenylbutanone (*S*)-PhCHMe-COMe (in which the sizes of the groups decrease in the order Ph > Me > H) with lithium aluminium hydride gives a product which predominates in the diastereomer predicted by Cram's rule (Scheme 2.27).

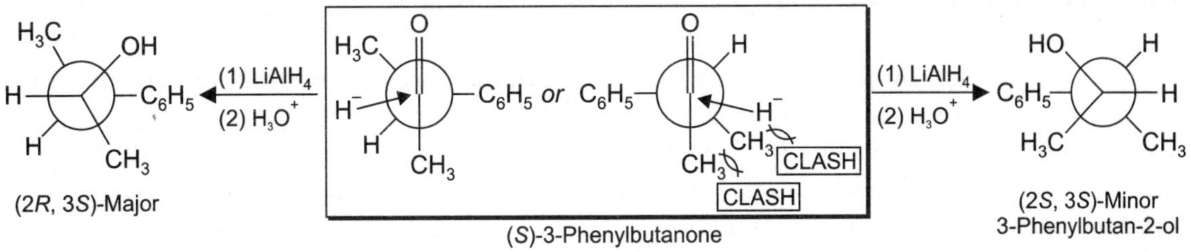

SCHEME 2.27

The preference for one product stereoisomer over the other depends on the difference in sizes between S and M. In case S and M are very similar, there is likely to be little difference in their interactions with the incoming nucleophile (as well as R) and the diastereoselectivity of such molecules is likely to be poor.

> **Summary of Felkin-Ahn model**
>
> - *Label the largest and the smallest groups.*
> - *Place the largest group perpendicular to carbonyl in Newman projection. The two remaining groups on the stereocenter can then be arranged in two ways (i) the smaller group towards or away from carbonyl oxygen.*
> - *The nucleophile adds to the carbonyl group at an optimum angle of around 107°. The attack of the nucleophile occurs from the face of the carbonyl group from the small sized group, since it will be hindered by the medium sized group..*

There are situations when Cram's rule may not be followed since the conformations of the carbonyl compounds in the transitions state are no longer dependent mainly on steric factors.

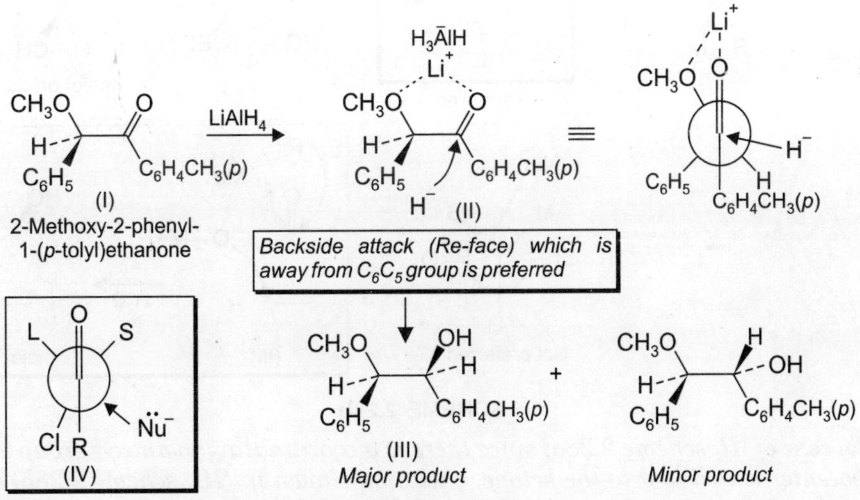

SCHEME 2.28

In such cases one of the substituents on the stereocenter is an alkoxy, hydroxyl or other complexing group. Lithium cations *e.g.*, coordinate effectively with oxygen atoms and thus reduction of ketones of the type (I, scheme 2.28) with lithium aluminium hydride proceeds through a relatively rigid chelate compound of type (II, scheme 2.28). Thus the reagent brings about the reduction by first acting as a conformational lock by coordination with both the methoxy oxygen and the ketone oxygen. A high degree of stereoselectivity is observed since the hydride attacks the chelate from the less hindered side (*i.e.* away from the phenyl group, in this case it is the *Re* face which offers the least resistance to the approach of the reagent) but not necessarily in the Cram's sense.

In cases when one of the substituents is strongly electronegative (*e.g.*, chlorine) the preferred transition state corresponds to conformation (IV, scheme 2.28 due to the tendency of the negatively polarized oxygen and chlorine atoms to be as far apart as possible. Again in such cases Cram's rule is not necessarily followed.

EXERCISE 2.5

Write the structure of major product of addition of a nucleophile ($N\bar{u}$:) to the ketones ($N\bar{u}$: to I and RMgBr to II, Scheme 2.28a).

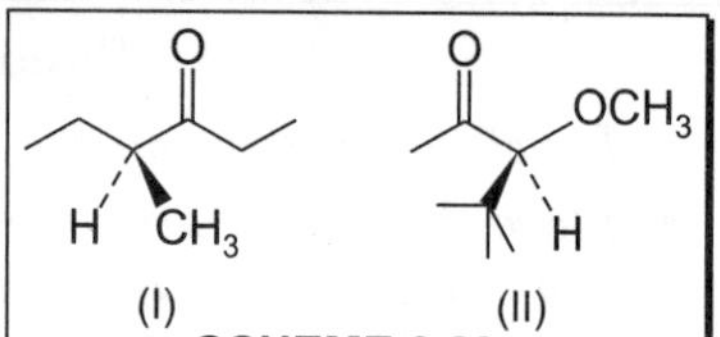

SCHEME 2.28a

ANSWER. *In the case of (I) first identify the largest, medium and smallest groups. Write the proper Felkin-Ahn model (scheme 2.28b) in the two conformations. The nucleophile will add to the carbonyl group at an obtuse angle with C=O from the small sized group.*

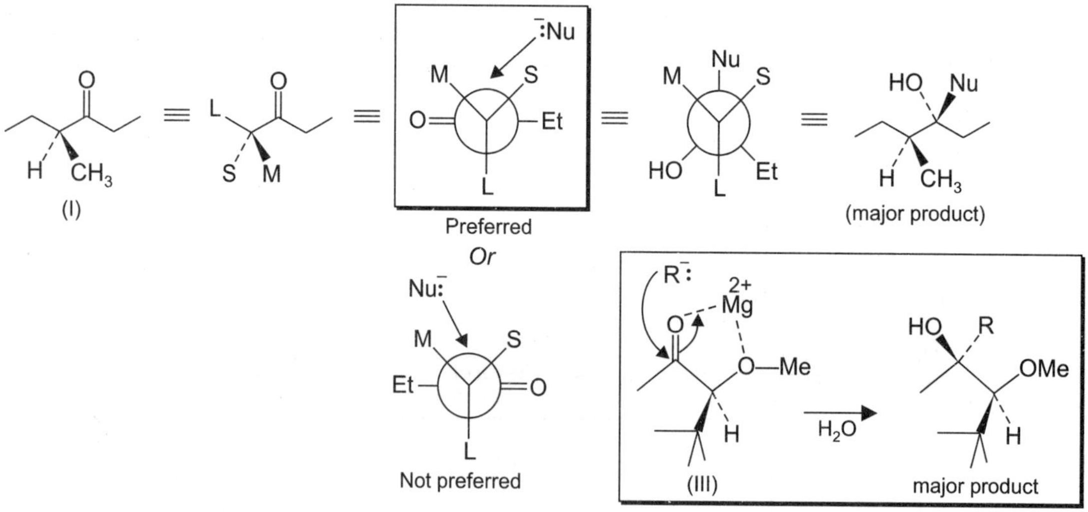

SCHEME 2.28b

In the case of (II, scheme 2.28a) since there is a coordinating methoxyl group at the stereocenter which is α to the ketone, chelation shown in (III, scheme 2.28b) locks the molecule and now the nucleophile is delivered from the less hindered side i.e., away from the bulky t-butyl group.

(b) Diastereoselectivity in Aldol Reactions (Directed Aldol Reaction)

(*i*) Introduction

The following points may be noted: (for introduction to aldol condensation see chapter 7)

- The aldol reaction creates two stereocenters from achiral starting materials and in a most general case, there are four stereoisomers of the aldol product (scheme 2.29). Thus *syn* or *anti* diastereomers are produced, each as a pair of enantiomers.
- Regarding the stereochemistry of the reaction, therefore, one has to control diastereoselection *i.e.*, whether (racemic) *syn* or (racemic) *anti* product is formed as the major product. Secondly one has also to aim at enantioselection *i.e.*, formation of one of the four possible stereoisomers.

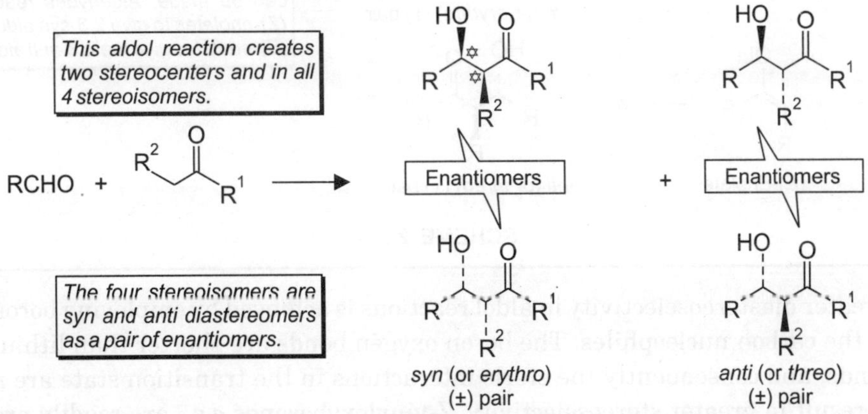

SCHEME 2.29

- Diastereoselectivity in the aldol reaction is achieved by employing the enolate of desired stereochemistry (*E* or *Z*).
- Enolates are generated *e.g.*, from a ketone and a base in the presence of chlorotrimethylsilane when the enolates are trapped as silyl enol ethers (scheme 2.30). These

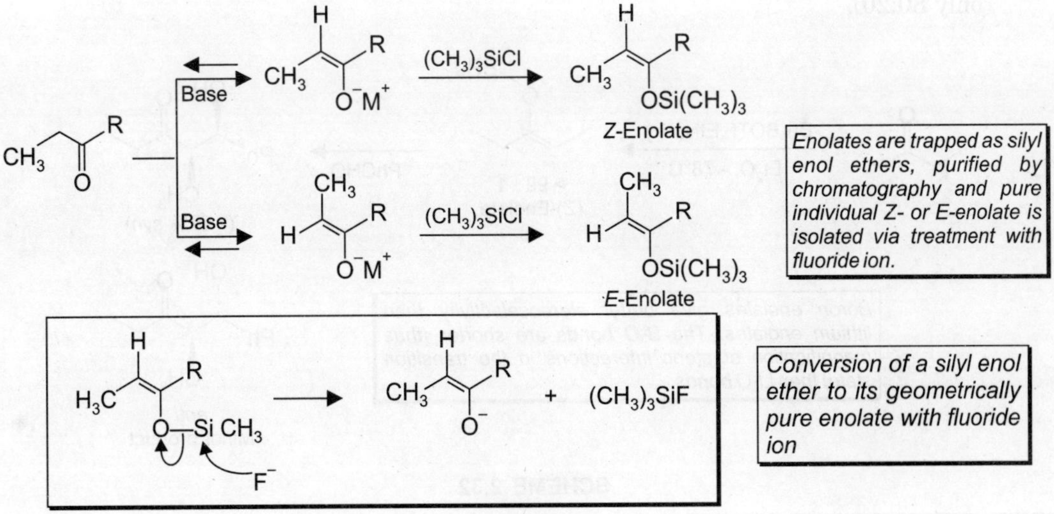

SCHEME 2.30

are separated and purified by chromatography and then converted into pure (*Z*)- or (*E*)-enolate with fluoride ion (nucleophilic substitution at the silicon atom by fluoride ion, fluoride ion has a high affinity for silicon, since Si-F bonds are very strong).

- Methods are available to produce either *E* or *Z* enolates in pure forms.
- *Z* enolates give mainly 2, 3-*syn* aldols while the *E*-enolates give the 2, 3-*anti* aldols (scheme 2.31).

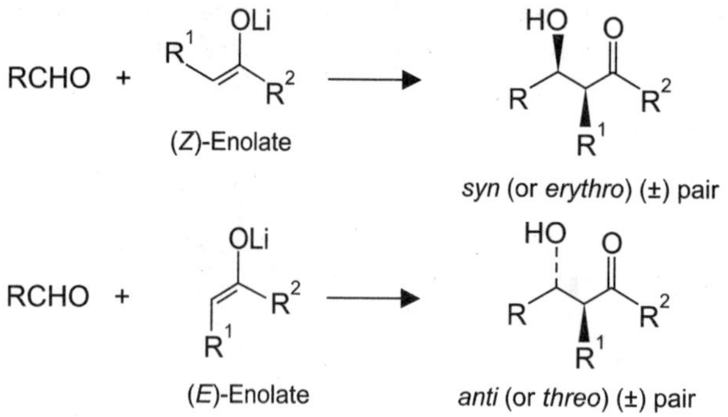

Specifically pure (E)-or (Z)-enolates can be made. Aldehydes react with (Z)-enolates to give 2,3-syn aldols and (E)-enolates give the 2,3-anti aldols.

SCHEME 2.31

- Greater diastereoselectivity in aldol reactions is achieved by employing boron enolates as the carbon nucleophiles. The boron oxygen bonds are shorter than lithium oxygen bonds, and consequently the steric interactions in the transition state are magnified to result in greater stereoselectivity. Z-vinyloxyboranes *e.g.,* are readily prepared by reacting ketones with a dialkylboron-trifluoromethanesulphonate (triflate) and a mild base diisopropylethylamine (in these enolates the boron atom is bonded to the oxygen atom of the ketone) and these react with aldehydes to give *syn* diols in high yield (scheme 2.32). 3-Pentanone by this method gives *Z* and *E* enolates in a ratio of > 99:1 and subsequent reaction with benzaldehyde give the *syn* and *anti* aldols in a ratio > 97:3 (The same condensation when carried out with lithium enolates gave a ratio of only 80:20).

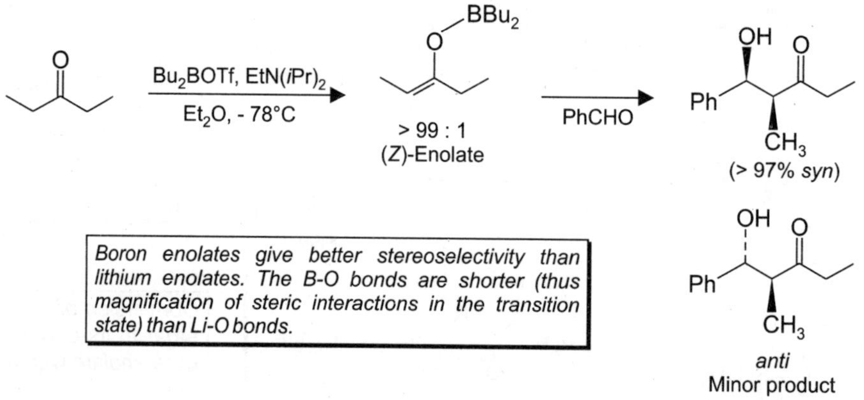

Boron enolates give better stereoselectivity than lithium enolates. The B-O bonds are shorter (thus magnification of steric interactions in the transition state) than Li-O bonds.

SCHEME 2.32

(*ii*) The diastereoselectivity of the aldol reaction—A chair like transition state

The diastereoselectivity is achieved by the reaction (*i.e.* formation of the new C—C bond) proceeding via a chair like six membered transition state in which the ligated metal atom is bonded to the oxygen atom of the aldehyde as well to that of the enolate. If the geometry of the enolate is fixed, the only variable is the orientation of the aldehyde, and therefore, one deals with transition states of different stabilities. This is so since the electron rich enol double bond and the electrophilic carbonyl carbon atom can be brought into close proximity via two transition states *e.g.*, in the case of *Z* enolate (I, scheme 2.33). One of the transition states III is disfavored due to 1, 3-non-bonded interactions between the substituents and thus the reaction takes place largely *via* transition state (II, scheme 2.33) to give *syn* aldol. Similar arguments show that the reaction with the *E*-enolate proceeds preferentially through the transition state (IV, scheme 2.33). The aldol condensation can be made enantioselective (see, scheme 2.35), by introducing chirality into one partner which can then be removed, thus aldol condensation now selects for one enantiomer. This is an example of double asymmetric synthesis (see, scheme 2.35).

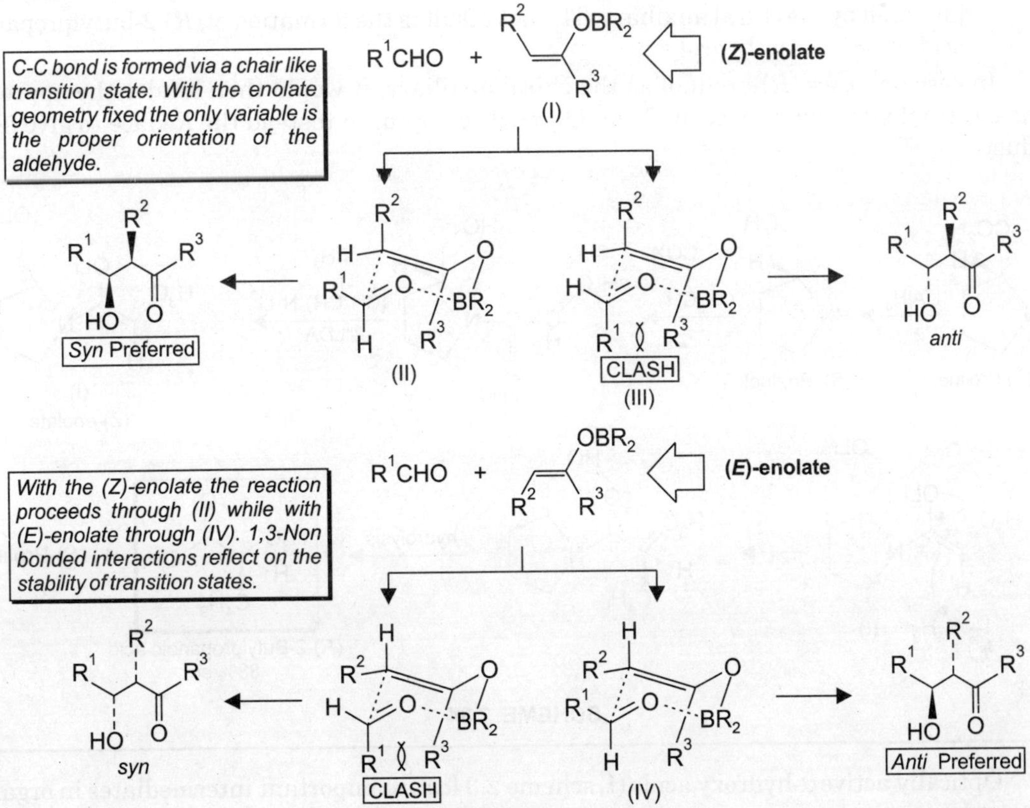

SCHEME 2.33

(c) The Use of Chiral Auxiliaries (Second Generation Methods)

During asymmetric synthesis a chiral auxiliary is attached chemically to the achiral substrate to give a chiral intermediate. This is followed by the reactions of asymmetric synthesis, during which the auxiliary dictates the preferred stereochemistry. At the end of the synthesis the chiral auxiliary is removed.

(*i*) Alkylation of chiral enolates (asymmetric α-substitution of a carboxylic acid and synthesis of optically active α-hydroxy acids)

In (scheme 2.34) alkylation of propanoic acid is taken as an example which involves the following steps:

- The chiral pyrrolidine unit is derived from *S*-(–) proline.
- The carboxylic acid substrate (activated propanoic acid) reacts with (*S*)-prolinol to give an amide.
- Deprotonation with LDA gives a chiral enolate entirely in the *Z* from and the hydroxylic hydrogen is also replaced by a lithium atom. The complex (I, scheme 2.34) thus maintains its structure by the hydroxymethyl group of the chiral auxiliary by chelating in its lithiated form with the enolate OLi group.
- The chiral enolate is alkylated with butyl iodide which mainly occurs form the top face of the molecule (the less hindered direction of approach, *i.e.* the added alkyl group C_4H_9-group is shown by a thick wedge). Thus the electrophilic attack occurs on the *Si* face of the double bond (at C2 the site of electrophilic attack), the *Re*–face being hindered by the chiral auxiliary. The net result is the formation of (*R*)-2-butylpropanoic acid in 88% *ee* on hydrolysis.

In case one uses (*R*)-prolinol as the chiral auxiliary, it would then hinder the approach of the electrophile to the *Si* face at C2 and force it to approach instead the *Re* face to give (2*S*)-product.

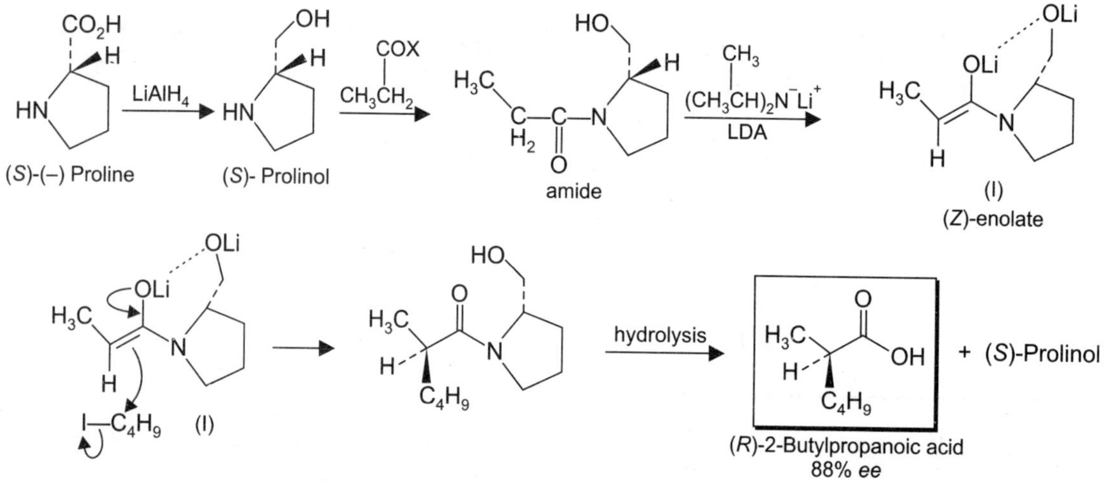

SCHEME 2.34

Optically active α-hydroxy acids (I, scheme 2.34*a*) are important intermediates in organic synthesis and occur in nature. An interesting method leading to their synthesis involves an asymmetric halolactonization (Scheme 2.34*a*). The symmetric synthesis starts with a suitable α, β–unsaturated acid which is converted into the acyl derivative (II, scheme 2.34*a*) of *S*-proline (the chiral auxiliary). Reaction with *N*–bromosuccinimide gives a bromonium ion which lactonizes to a bromolactone (III, scheme 2.34*a*). Its reduction with tributyltin hydride gives the lactone which on hydrolysis gives the optically active α-hydroxy acid. Thus tiglic acid gives (*R*-2-hydroxy-2-methylbutanoic acid by following this reaction sequence (scheme 2.34*a*).

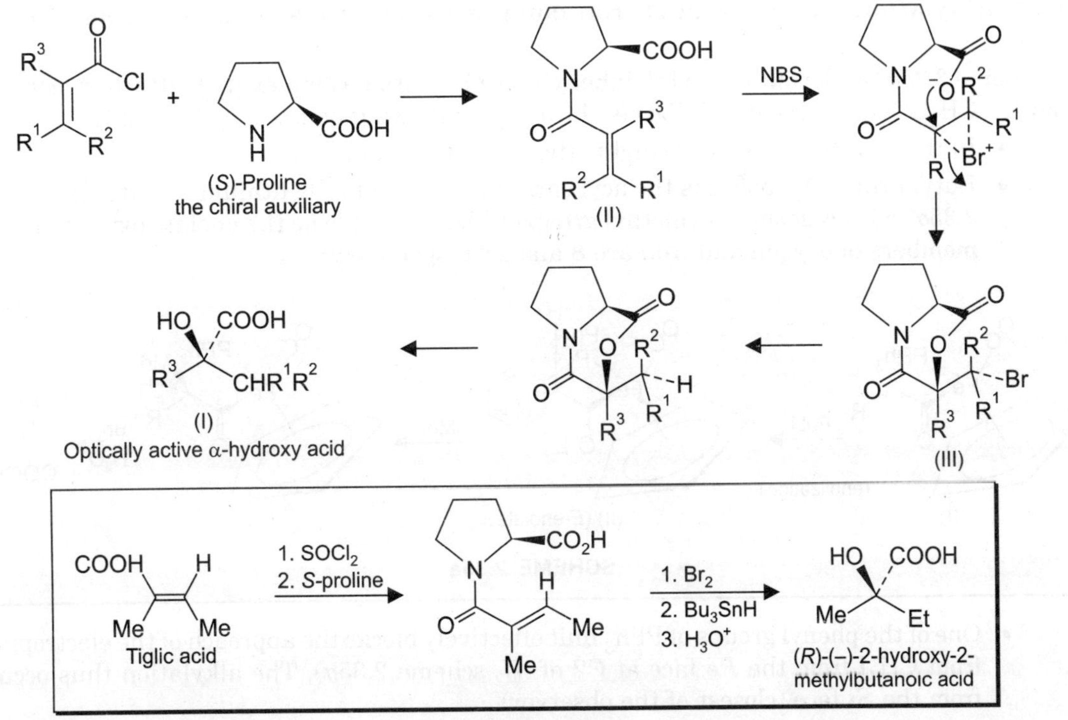

SCHEME 2.34a

(*ii*) Enantioselective aldol condensation via chiral enolates (Double asymmetric synthesis)

When one of the components in aldol condensation is optically active, the reaction may becomes highly enantioselective in which case only one of the four stereoisomers (see scheme 2.29) predominates. Boron enolates with chiral auxiliary group (chiral boron enolates) have been used with success. Thus chirality is introduced into one partner and after the reaction if is removed.

The propionamide (I, scheme 2.35) as it *Z*-boron enolate reacts with achiral aldehydes to give *syn* aldols (II, scheme 2.35) almost exclusively. On hydrolysis enantiomerically pure 3-hydroxy-2-methyl carboxylic acids (III, scheme 2.35) are formed.

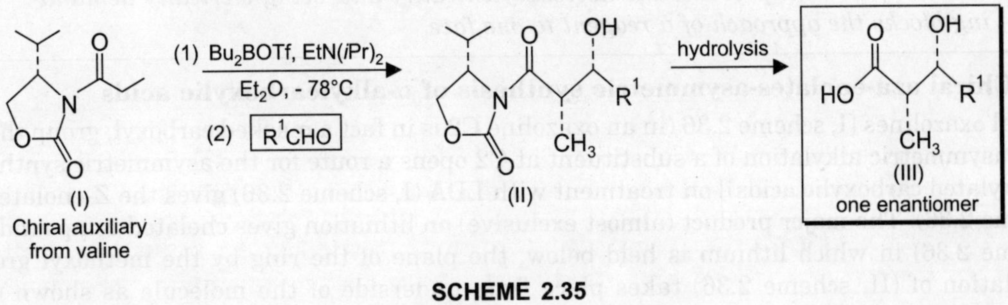

SCHEME 2.35

This type of a reaction where an optically active substrate (chiral enol derivative) reacts to generate both new stereocenters enantioselectively is called double asymmetric synthesis.

(*iii*) C-Alkylation via chiral acyl iron complexes (enantioselective synthesis of an acid)

In (scheme 2.35*a*) cyclopentadienyl (triphenylphosphine iron) complex with attached acyl unit from R—CH_2COOH (I, scheme 2.35*a*) is chiral. The following points may be noted.

- In this synthesis iron carbonyl is the chiral auxiliary.
- Butyl lithium transforms the acyl carbonyl group into the lithium enolate (II, scheme 2.35*a*) with *E* geometry (methyl group on the same side as the enolate oxygen, atomic members of oxygen and iron are 8 and 28 respectively).

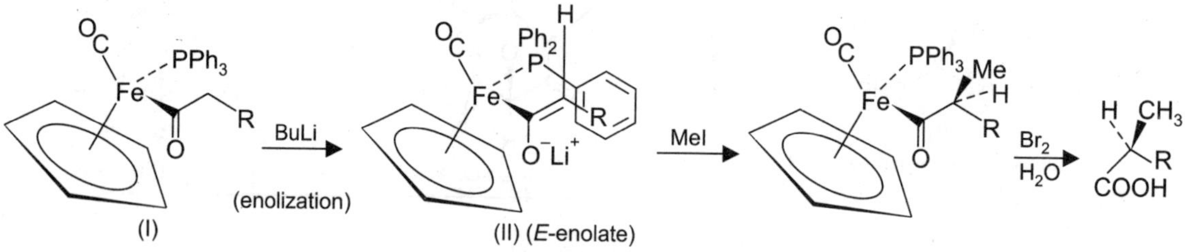

SCHEME 2.35a

- One of the phenyl groups of PPh_3 unit effectively blocks the approach of the electrophile from CH_3I from the *Re* face at C2 of (II, scheme 2.35*a*). The alkylation thus occurs from the *Si* face (closest of the observer).
- On one electron oxidation with bromine in water gives an alkylated acid in an enantiomerically pure form. Oxidation with bromine cleaves the C—Fe bond and generates the acid derivative.

> **Enantioselective Synthesis of an Acid**
>
> *The enantioselective synthesis of an acid (Scheme 2.35a) is an important application of coordination to metal centers. The enantiotopic faces can be rendered diastereotopic via ligation of the substrate to metal centers which are stereogenic. A chiral acyl iron complex can undergo C alkylation with high diastereoface selection. The outcome of observed stereochemistry of the final product (enantioselectivity) is based on the assumption that the intermediate enolate ion has E geometry, one face of which is effectively blocked by one of the phenyl groups on the iron complex. The alkylation e.g., of the enolate occurs from other side. The iron complex thus provides the necessary chirality and being sterically demanding blocks the approach of a reagent to one face.*

(*iv*) Chiral aza-enolates-asymmetric synthesis of α-alkylcarboxylic acids

Chiral oxazolines [I, scheme 2.36 (in an oxazoline C2 is in fact a masked carboxyl, group therefore, asymmetric alkylation of a substituent at C2 opens a route for the asymmetric synthesis of alkylated carboxylic acids)] on treatment with LDA (I, scheme 2.36) gives the *Z*-enolate (II, scheme 2.36). The major product (almost exclusive) on lithiation gives chelated compound (II, scheme 2.36) in which lithium is held below, the plane of the ring by the methoxyl group. Alkylation of (II, scheme 2.36) takes place from underside of the molecule as shown (III, scheme 2.36) *i.e.*, on the side opposite the bulky phenyl substituent, (this direction is provided by the lithium and further shows that both the methoxy group and phenyl group are essential for the observed optical yield).

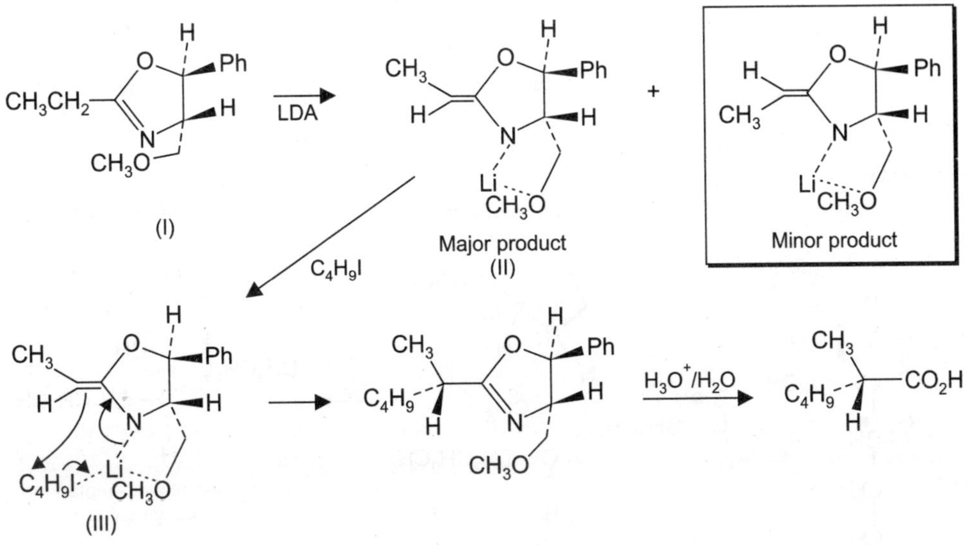

SCHEME 2.36

(*v*) Enatioselective alkylation of aldehydes and ketones via chiral hydrazones

S-1-Amino-2-methoxy-methylpyrrolidine is called SAMP and its enantiomer as RAMP and are prepared from *S*- and *R*- prolines respectively. They bear a chelating methoxy group (scheme 2.37) and are employed as chiral auxiliaries in enantioselective alkylation of aldehydes and ketones.

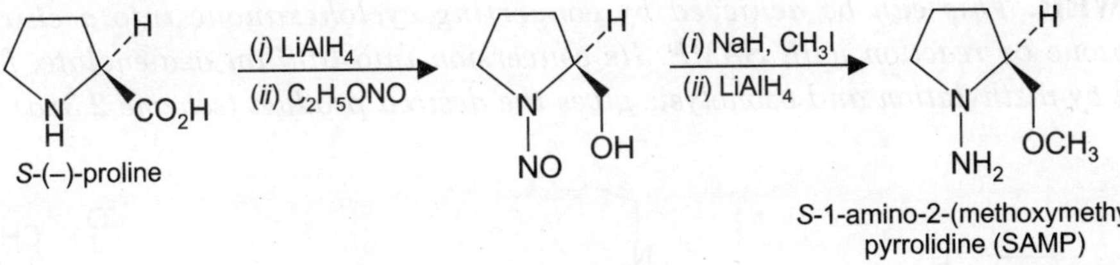

SCHEME 2.37

An easy way to bring about an asymmetric alkylation α to a carbonyl group is to just convert a ketone to a chiral imine or hydrazone followed by deprotonation with a strong base. Thus the hydrazone of 3-pentanone on treatment with LDA affords a complex (I, scheme 2.38, the α-hydrogens of hydrazones are acidic). The resulting lithium aza-enolate adopts an orientation and is stabilized by forming a complex involving the lithium complex, the α-carbanion and the chiral auxiliary (at the methoxyl group). This chiral chelated carbanion is a nucleophile which will react *e.g.*, with alkyl halides (1-iodopropane in scheme 2.38) *via* nucleophilic substitution to form exclusively only one of the two diastereomers (*S*, *S* scheme 2.38) and not (*R*, *S*). On ozonolysis the *S*, *S* diastereomer (II, scheme 2.38) afforded the *S*-enantiomer (99% of the product) of 4-methyl-3-heptanone, and the chiral auxiliary is recovered as the *N*-nitroso compound which may be reduced back to SAMP, RAMP and SAMP reacted with the same achiral carbonyl compound to permit enantioselective α-alkylation to occur in either direction.

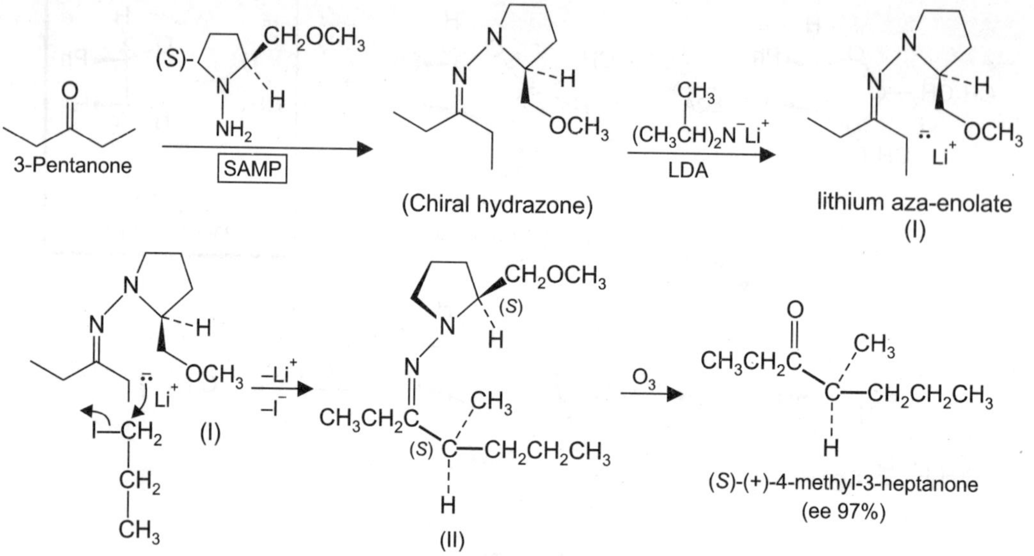

SCHEME 2.38

EXERCISE 2.6

How one can carry out asymmetric alkylation of cyclohexanone α to the carbonyl ?

ANSWER. *This can be achieved by converting cyclohexanone into a chiral hydrozone by reaction with SAMP. Its conversion into lithium aza-enolate, followed by methylation and ozonolysis gives the desired product (scheme 2.38a).*

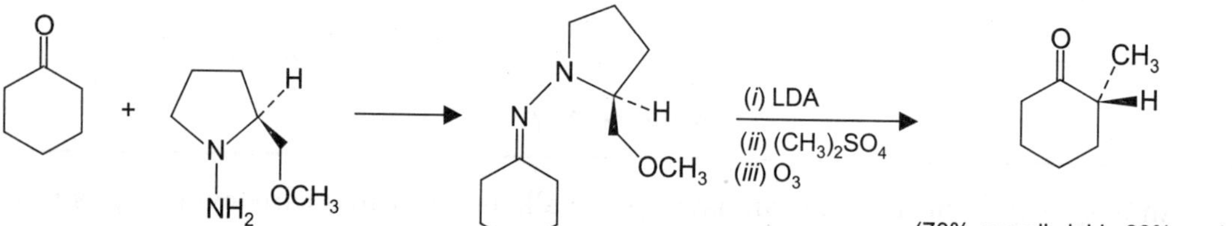

(70% overall yield ; 99% e.e.)

SCHEME 2.38a

(d) The Use of Chiral Reagents (Third-Generation Methods) and Chiral Catalysts

(*i*) Asymmetric hydroboration with diisopinocampheylboranes (an enantioselective reduction of alkenes to alcohols)

Hydroboration of a naturally occurring terpenoid hydrocarbon α-pinene (which is chiral) gives chiral alkylboranes–monoisopinocampheylborane $(Ipc)BH_2$ or diisopinocampheylborane $(Ipc)_2BH$ (scheme 2.39).

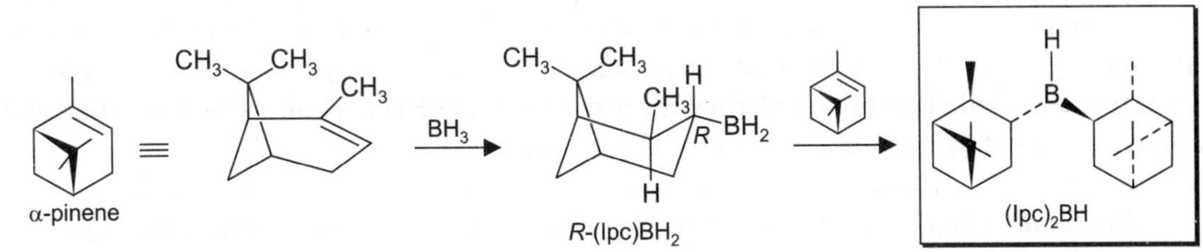

R specifies configuration at carbon atom bonded to boron

SCHEME 2.39

The use of chiral borane $(Ipc)_2BH$ in its optically pure form in the hydroboration of the prochiral (*Z*)-but-2-ene, followed by oxidation gives (*R*)-butan-2-ol with high optical purity (optical purity more than 90% scheme 2.40). Monoisopinocampheylborane (Ipc) BH_2 reacts with (*Z*)-but-2-ene to give (*S*)-butan-2-ol, and this reflects on the importance of careful reagent

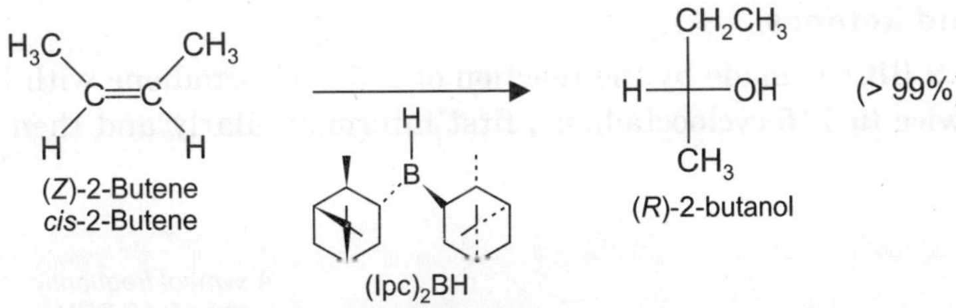

SCHEME 2.40

preparation/purification. It has been shown that $(Ipc)BH_2$ is a very good enantioselective reagent for hydroboration of *trans*-alkenes, trisubstituted alkenes and 1-substituted cycloalkenes. $(Ipc)_2BH$ on the other hand, works very well with *cis* olefins with high enantioselection. The mechanism of the reaction between (*Z*)-but-2-ene (*cis*-2-butene) with $(Ipc)_2BH$ (scheme 2.41) involves a different interaction of the two faces of the alkene with the chiral borane to give diastereomeric (triorgano) boranes (addition occurs preferentially to one face). A chiral alcohol

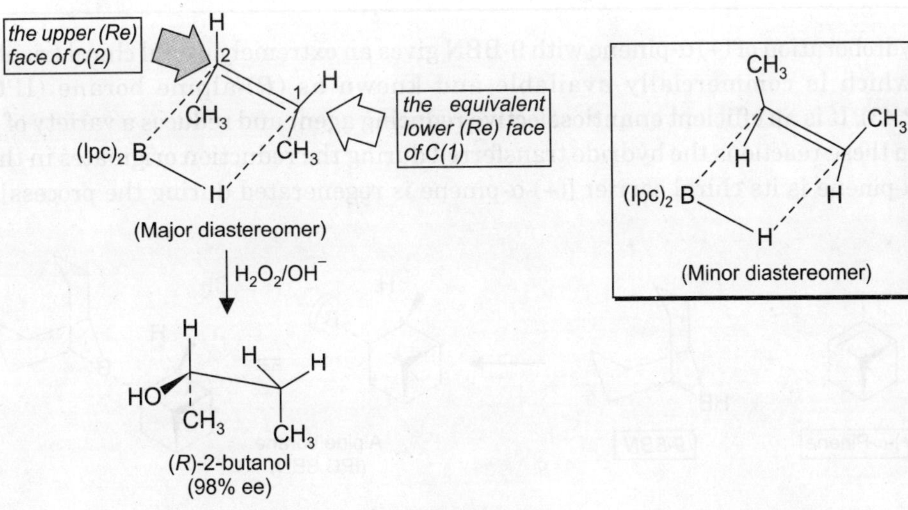

SCHEME 2.41

is then produced after oxidative hydrolysis with hydrogen peroxide and base. The transition state leading to (*R*) alcohol (scheme 2.41) is much lower in energy (one may consider the orientation of the alkyl groups on olefin with respect to $(Ipc)_2BH$ portion in the transition state leading to major diastereomer, (scheme 2.41).

> *The results of scheme 2.41 are interesting since cis-2-butene (a prochiral alkene) is devoid of any inbuilt structural features that could block the entry of a reagent to one face compared to other (camphor scheme 7.8a has, however, such inbuilt structural features). $(IPC)_2$ BH is a very bulky dialkyl chiral borane, when compared to cis-2-butene in which the comparison to resistance by attack of $(IPC)_2$ BH is H versus methyl. The alcohol is formed by the attack of bulky borane to the upper (Re) face of C2 or the equivalent lower face (Re) of C1.*

(*ii*) Asymmetric reduction using chiral trialkylboranes (enantioselective reduction of aldehydes and ketones)

A useful reagent 9-BBN is made by the reaction of 1, 5-cyclooctadiene with BH_3. In this reaction BH_3 adds twice to 1, 5-cyclooctadiene, first intermolecularly and then intramolecularly (scheme 2.42).

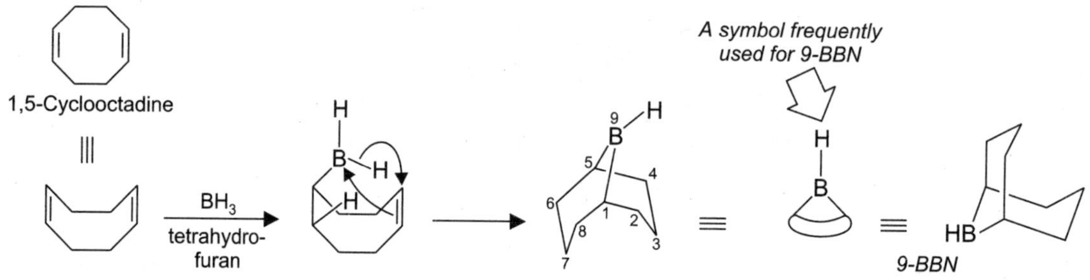

SCHEME 2.42

Hydroboration of (+) α-pinene with 9-BBN gives an extremely useful chiral trisubstituted borane which is commercially available and known as (*R*)-alpine borane (IPC. BBN, scheme 2.43). It is an efficient enantioselective reducing agent and reduces a variety of carbonyl groups. In these reactions the hydride transferred during the reduction originates in the 9-BBN and (+)-α-pinene is its chiral carrier [(+)-α-pinene is regenerated during the process].

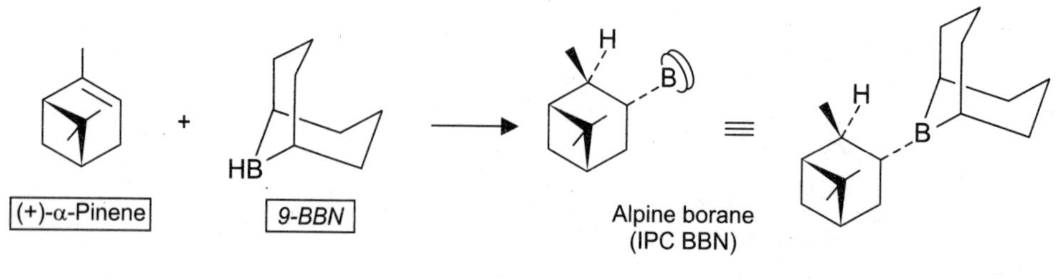

SCHEME 2.43

It is thought that the adduct transfers a hydride to the ketone via a six membered boat shaped cyclic transition state, where the larger group on the ketone (Ph of $PhCOCH_3$) preferentially lies away form the α-pinene moiety to make the steric congestion minimum (scheme 2.44). The adduct of 9-BBN can be made with both (+) and (–)-α-pinene (the later is known as *S*-Alpine Borane) to synthesize both the enantiomers individually. Thus the reduction of unsymmetrical (prochiral ketones with enantiotopic faces) ketones to optically active secondary alcohols is achieved.

Monodeuteriated and therefore, chiral primary alcohols can be made by this method by employing deuteriated analogue of Alpine Borane (scheme 2.44).

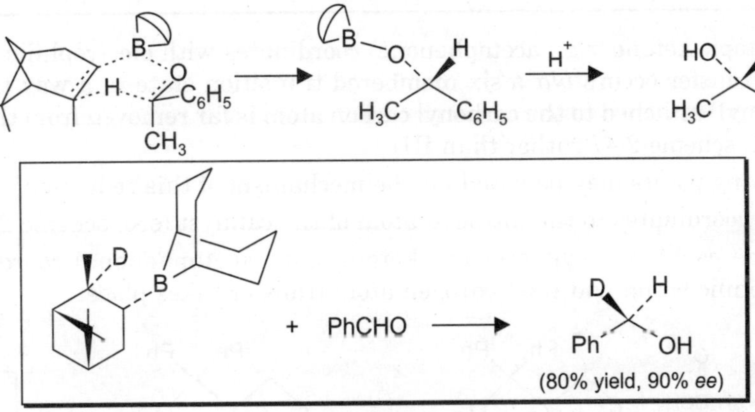

SCHEME 2.44

(*iii*) Use of chiral oxazaborolidine complex of borane (enantioselective reduction of ketones)

The borohydride (II/BH_3 scheme 2.45) derived from the reaction of oxazaborolidine (I, scheme 2.45) with diborane is an effective reducing agent and brings about an enantioselective reduction of ketones. A ketone may be reduced stoichiometrically with $(I)/BH_3$ (scheme 2.45) or the reaction can also be carried out under catalytic conditions using a small amount of (I, scheme 2.45) and excess BH_3/THF. Under stoichiometric conditions, only two of the three hydride ions attached to boron are reactive. The absolute configuration of the product can be predicted *e.g.* in the present case, the reagent which has the (*S*)-configuration at the secondary center produces the (*R*)-alcohol from the ketone, while the (*R*) reagent (prepared from D-proline) yields the (*S*)-alcohol. The reduction of acetophenone provides a specific example of this reaction (scheme 2.46).

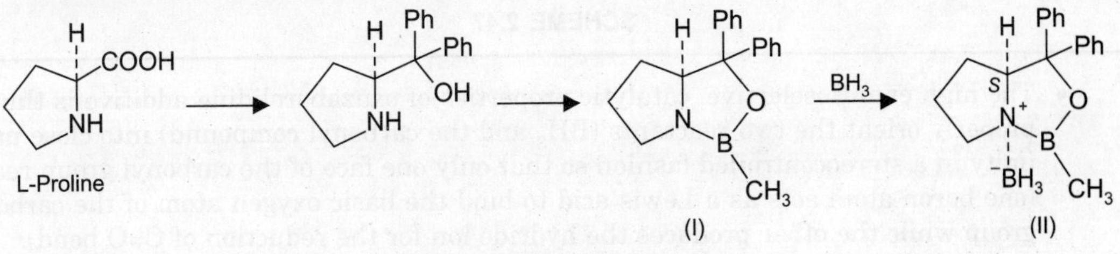

SCHEME 2.45

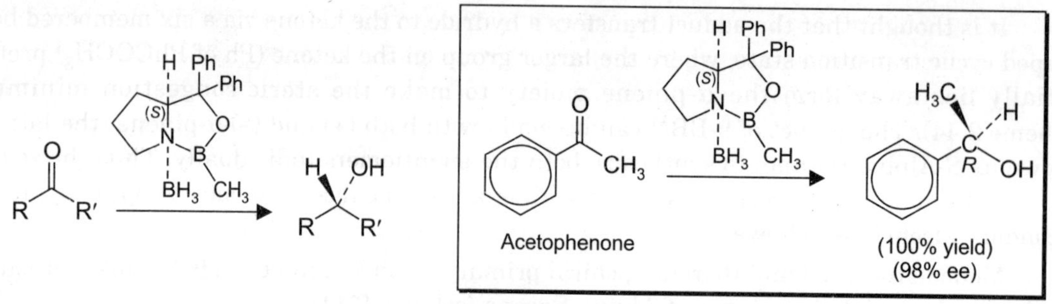

SCHEME 2.46

An enatiotopic ketone (*e.g.*, acetophenone) coordinates with electrophilic boron and the hydrogen atom transfer occurs *via* a six membered transition state in a way that the larger substituent (phenyl) attached to the carbonyl carbon atom is far removed from the heterocyclic complex as in (II, scheme 2.47 rather than III).

The following points may be noted for the mechanism of this reduction.

- Borane coordinates to the nitrogen atom of the catalyst (see, scheme 2.45).
- Next the carbonyl oxygen of the ketone (*e.g.*, acetophenone) coordinates to the electrophilic boron and the hydrogen atom transfer takes place.

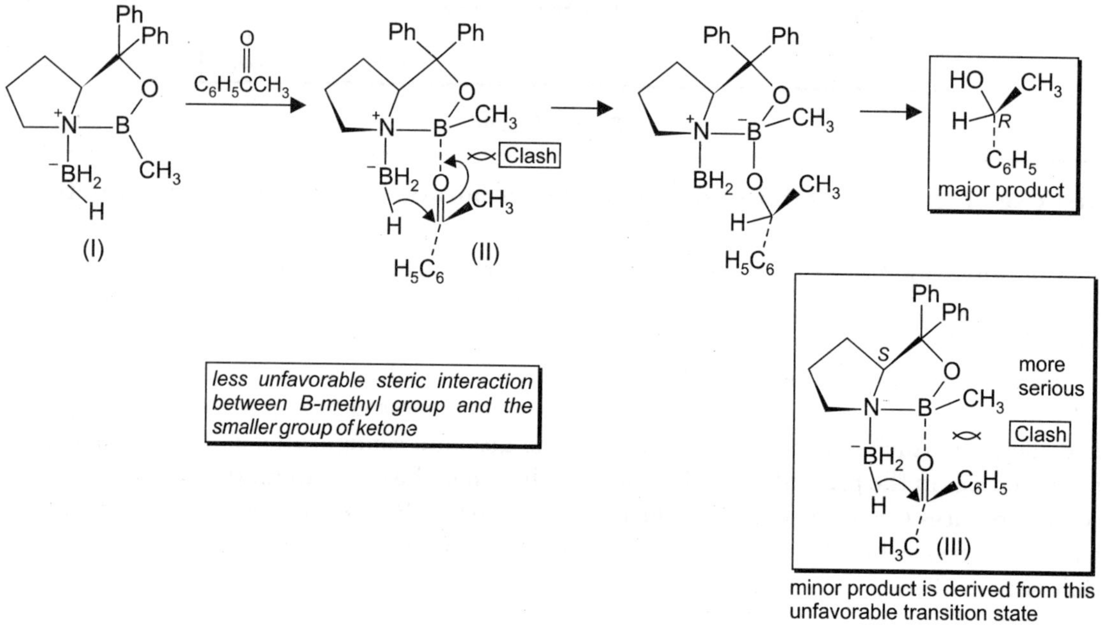

SCHEME 2.47

- The high enantioselective, catalytic properties of oxazaborolidine additive is thus to properly orient the two reactants (BH_3 and the carbonyl compound) into close proximity in a stereocontrolled fashion so that only one face of the carbonyl group reacts (one boron atom acts as a Lewis acid to bind the basic oxygen atom of the carbonyl group while the other produces the hydride ion for the reduction of C=O bond).

(iv) Sharpless enantioselective epoxidation of alkenes

The epoxidation of an alkene is achieved by reaction with perbenzoic acid or *m*-chloroperbenzoic acid. Chiral epoxides prepared this way lead only to recemates as no chiral inducing agents are present. The asymmetric synthesis of optically active epoxides is the need of the day, since these are required for the total synthesis of biologically active compounds.

Several transition metal oxidants convert alkenes to epoxides and a useful procedure involves *t*-butylhydroperoxide as the stoichiometric oxidant in combination with vanadium (scheme 2.48, epoxidation of allylic alcohols is achieved by using a vanadium compound and *t*-butylhydroperoxide the process gives a racemic mixture when the starting compound is achiral) molybdenum or titanium compounds. In the Sharpless asymmetric epoxidation (discovered 1981, Nobel prize 2001) the most useful substrate for oxidation is an allylic alcohol. The epoxidation is carried out with *t*-butylhydroperoxide and titaniumtetraisopropoxide and the procedure becomes highly enantioselective when enantiomerically pure tartrate esters are employed (scheme 2.49).

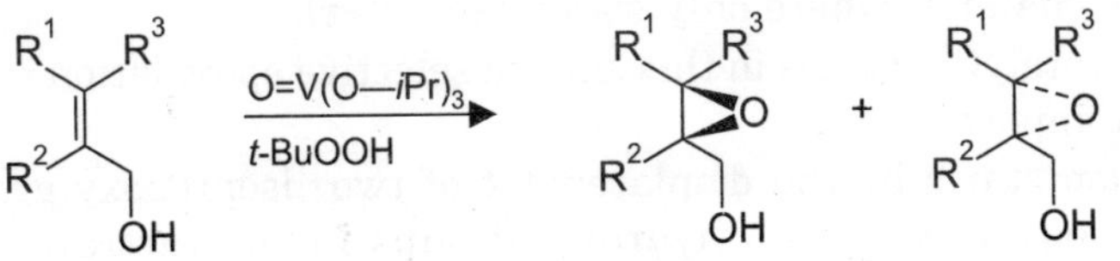

SCHEME 2.48

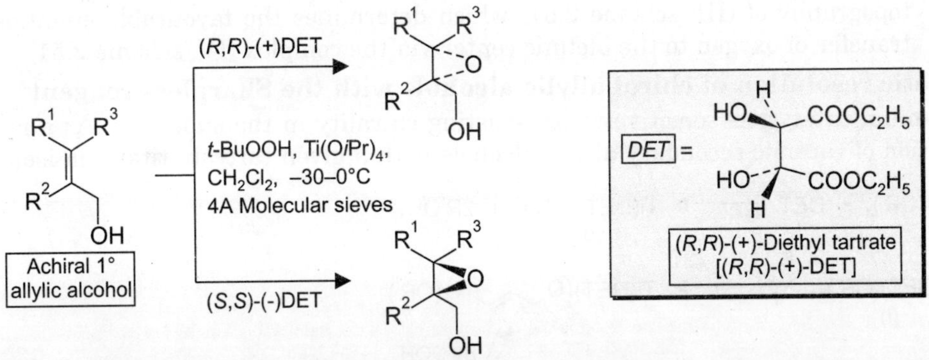

SCHEME 2.49

The technique is one of the best methods for the conversion of an achiral allylic alcohol into a chiral epoxide. The following points may be noted:

- The expoxidation is brought about by *tert*-butylhydroperoxide, catalyzed by titanium (IV) tetraisopropoxide in the presence of (+) or (–)-diethyltartrate (DET).
- The reaction creates two contiguous stereocenters with predictable stereochemistry, depending on which enantiomer of DET is used. With the use of (*R*, *R*) isomer of DET one isomer of the epoxide is obtained predominantly. By replacing with the other isomer of DET (*S*, *S*), the enantiomeric epoxide is obtained as in the case of geraniol (scheme 2.50).

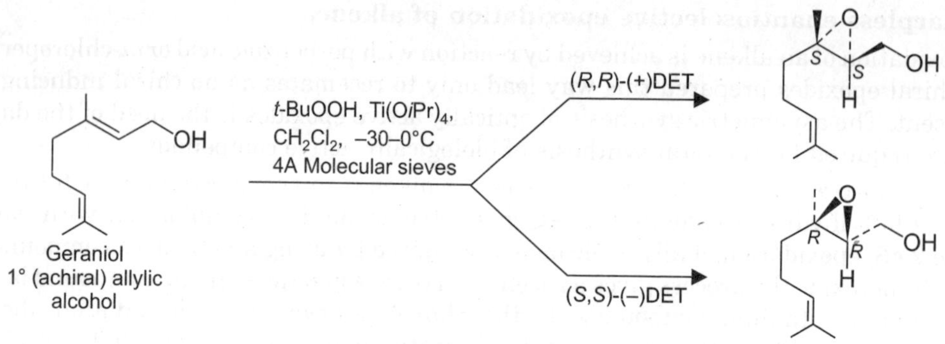

SCHEME 2.50

- The reaction is highly sensitive to the presence of water (hydrolysis of the complex) it is therefore, carried out in the presence of molecular sieves (inorganic silicates which have small channels where only water can enter).
- The major catalytic species in this enantioselective epoxidation reaction is a binuclear titanium complex.
- The reaction starts by the displacement of two isopropoxy group in one titanium tetraisopropoxide with two hydroxyl groups in the tartrate ester (DET) to give (I, scheme 2.51). Further the remaining isopropoxy groups are replaced with the hydroxyl group of allylic alcohol followed by the hydroxyl group of the peroxide. These successive displacements set up the preferred disposition of the alkene and the oxidant as in (II, scheme 2.51). The coordination activates the peroxide and it is this topography of (III, scheme 2.51) which determines the favourable enantioselective transfer of oxygen to the olefinic center via the complex (IV, scheme 2.51).

(v) Kinetic resolution of chiral allylic alcohols with the Sharpless reagent

The titanium catalyst is sensitive to pre-existing chirality in the substrate. As a result, the epoxidation of racemic secondary allylic alcohols with a given tartrate-titanium-isopropoxide

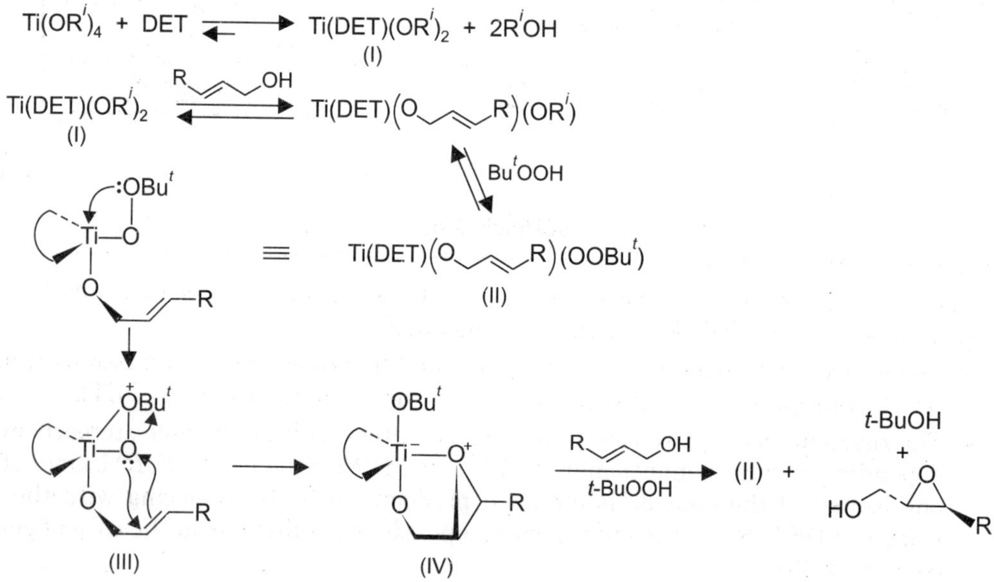

SCHEME 2.51

combination occurs rapidly only with one of the enantiomers (scheme 2.52). The other slower reacting enantiomer is left behind. This procedure thus, provides a method for obtaining two different optically active compounds a chiral epoxide and a resolved chiral allylic alcohol (a route to optically pure allylic alcohols). The reagents used for the kinetic resolution are essentially similar to those used for asymmetric epoxidation except that diisopropyltartrate esters (DIPT, scheme 2.52) are often substituted for the diethylester analogs (DET, scheme 2.52).

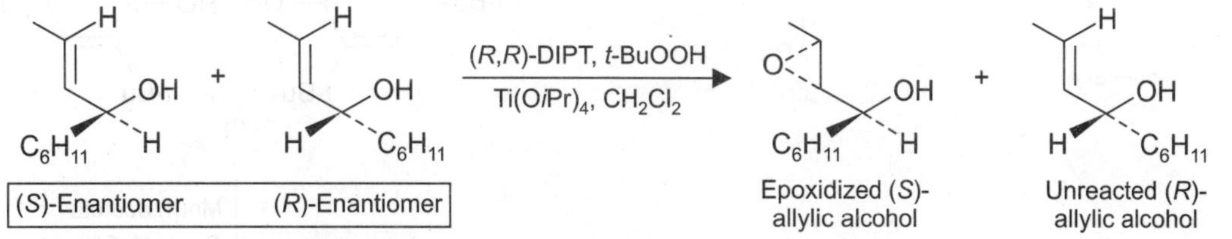

SCHEME 2.52

(*vi*) Asymmetric oxidation of sulfides to sulfoxides

The reagents used for asymmetric epoxidation are used for the asymmetric oxidation of sulfides to sulfoxides, however, in this water is actually needed in order to achieve good selectivity (scheme 2.53, for chiral sulfoxides see scheme 1.6 and 1.7).

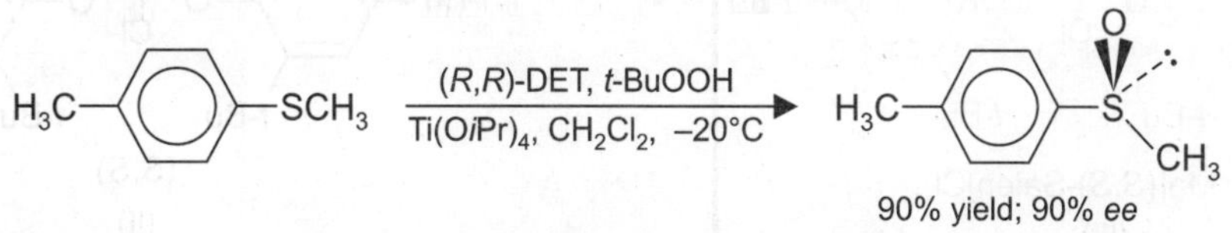

SCHEME 2.53

(*vii*) Enantioselective epoxidation *via* oxygen transfer catalysts

During sharpless epoxidation a substrate coordinates with the metal ion *via* the OH group of the alcohol which brings the double bond into proximity of the peroxy oxygen atom. Reagents that do not require co-ordination prior to reaction are also available. In 1990, Eric Jacobsen reported on a manganese-containing chiral catalyst which epoxidizes olefins with reasonably high *ee* values using low priced cooxidant NaOCl under mild conditions.

The catalyst is made by condensing 1, 2-diaminocyclohexane and a phenolic-aldehyde to generate two carbon-nitrogen double bonds to give a ligand (I, scheme 2.53*a*) which is referred to as "Salen". Salen binds manganese in the +3 oxidation state. This species (II), reacts with NaOCl to form the oxidizing species, represented as a manganese (V) oxo compound, (III, scheme 2.53*a*).

cis- alkenes show high selectivity with Mn (Salen*) (O) Cl reagent to give corresponding epoxides and the catalytic cycle is represented (scheme 2.53*b*). Chiral dioxiranes developed by Professor Y. Shi on the other hand work very well to epoxidize *trans* and trisubstituted alkenes. A dioxirane moiety is generated by the reaction of a keto group with hydrogen persulfate ion (HSO_4^-). The chiral dioxiranes derived from readily available sugar derivatives of D-fructose and L-Sarbose (scheme 2.53*c*) where the pair of hydroxyl groups are blocked by ketal/acetal formation. The enantioselective epoxidation is carried out by treating an alkene with ketone FR* (derived from D fructose or L-sorbose and persulfate ion (scheme 2.53*d*).

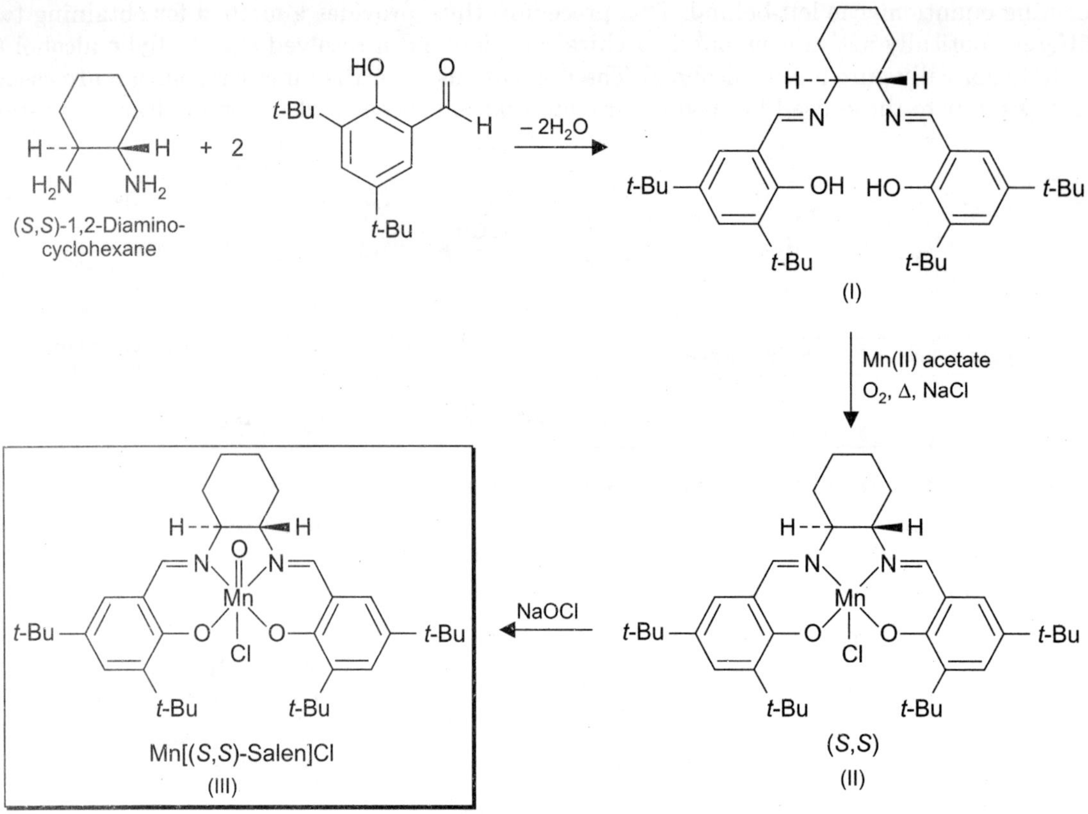

SCHEME 2.53a

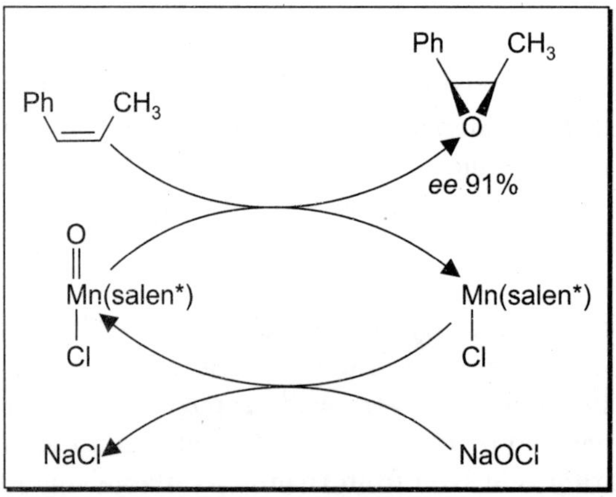

SCHEME 2.53b

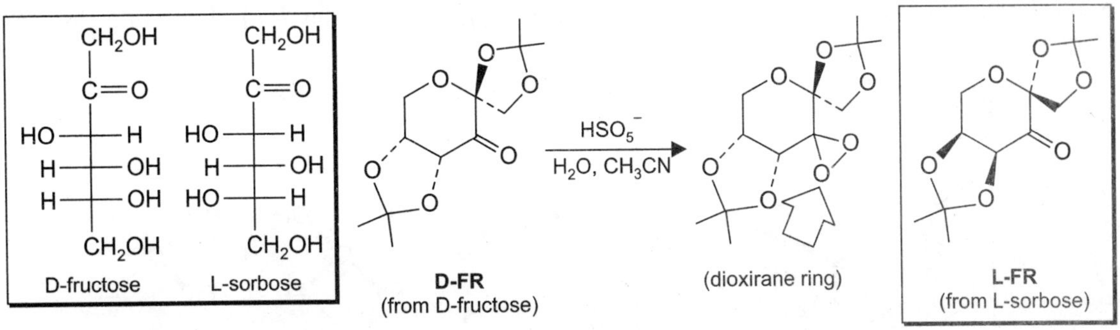

SCHEME 2.53c

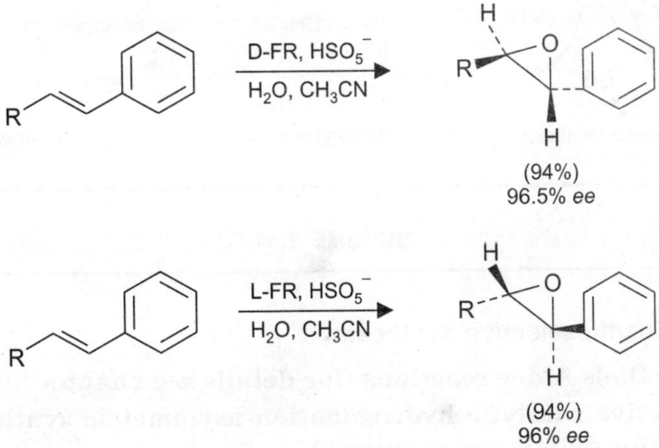

SCHEME 2.53d

(*viii*) Asymmetric reduction using lithium aluminium hydride (reduction via chiral metal hydride complexes)

An unsymmetrical ketone (R^1 COR^2, $R^1 \neq R^2$, with enantiotopic faces) on reduction with lithium aluminium hydride gives a racemic mixture of the secondary alcohol $R^1CH(OH)R^2$. Lithium aluminium hydride is achiral and the approach of this reducing agent to either *Re* or *Si* face of the carbonyl carbon atom is equally facile. However, a chiral hydride donor will transfer an achiral unit $H\bar{:}$ enantioselectively to a prochiral carbonyl group of an enantiotopic ketone (R^1 COR^2) [via two transition states with diastereomeric relationships (of different energies) to give one enantiomer (*R*-) or (*S*-) of the alcohols as the major product].

A commonly used asymmetric reducing agent ('BINAL-H') is derived from optically pure binaphthol, ethanol and lithium aluminium hydride. The tetrahedral (extended) array around aluminium locks effectively the aromatic rings into one configuration and does not allow racemization unless the C—C—C—C—Al—O—ring is disrupted (for the assignment of configuration (see, scheme 1.133*e*). Reduction of variety of ketones (particularly unhindered ketones, acyclic and conjugated enones) proceeds with high selectivity with this reagent (scheme 2.54). Thus *e.g.*, acetophenone on reduction with (*S*)-BINAL-H gives predominantly (*S*)-alcohol (compare this result with scheme 2.46). Asymmetric reduction of ketones like butanone and 4, 4-dimethylpentan-2-one (DIMPO) is carried out (with high enantioselectivity) with (*R, R*)-or

(*S-S*)-dimethylborolane. Generally the *RR*-reducing agent gives the *R*-alcohol while the *SS*-reagent yields the *S*-product.

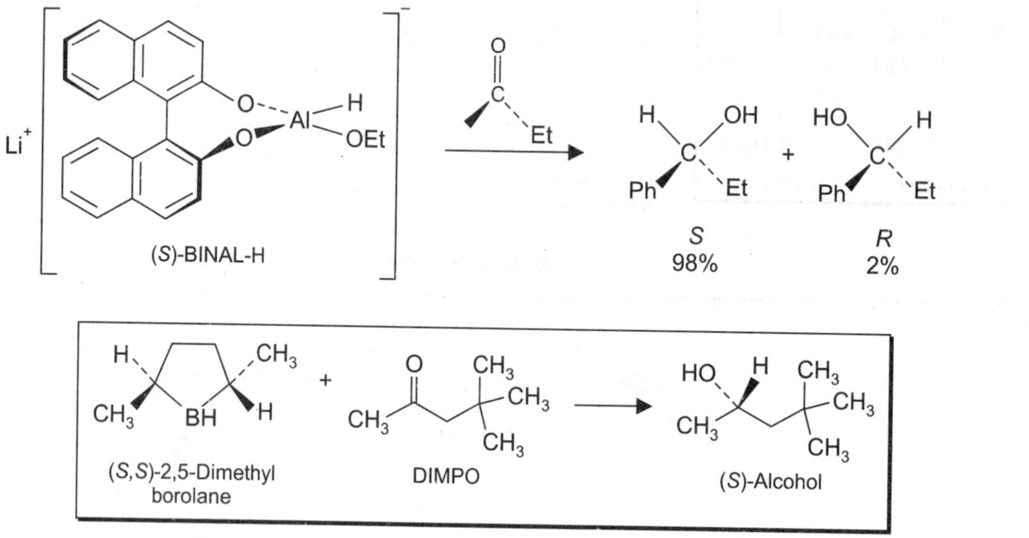

SCHEME 2.54

(e) Miscellaneous Enantioselective Synthesis

(1) **Asymmetric Diels Alder reactions (for details see chapter 8)**

(2) **Enantioselective catalytic hydrogenation asymmetric synthesis of L-alanine and L-dopa (for details see chapter 6)**

(3) **Asymmetric dihydroxylation (for details see chapter 6)**

(f) Phase Transfer Catalysis (Fourth Generation Methods)

This is one of the examples from the fourth generation asymmetric synthesis involving asymmetric catalysis. An efficient enantioselective alkylation of 2-phenylindanone (I, scheme 2.55) has been carried out in a two phase system (toluene-water) catalyzed by quaternary ammonium salt (the benzyl cinchoninium cation II, scheme 2.55). This salt is derived from the alkaloid cinchonine (phase transfer conditions). The high enantioselectivity is believed to be the result of the following factors :

- The quinuclidine ring lies behind the plane of the indanone enolate.
- A highly specific complex is formed between the enolate and the catalyst.
- There is *pi* interaction between the benzyl group of the catalyst and the 2-phenyl group. The 9-OH group of the catalyst provides a directive handle (via H-bonding) and the backside approach of the alkylating agent CH_3Cl is comparatively blocked leading to the formation of (*S*)-(+)–2-methylindanone (III, scheme 2.55).

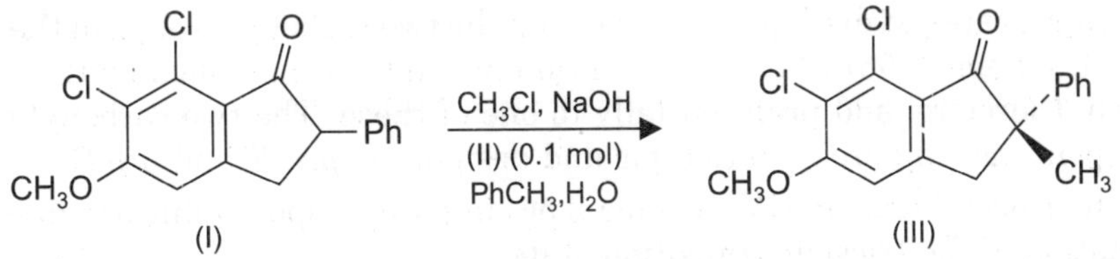

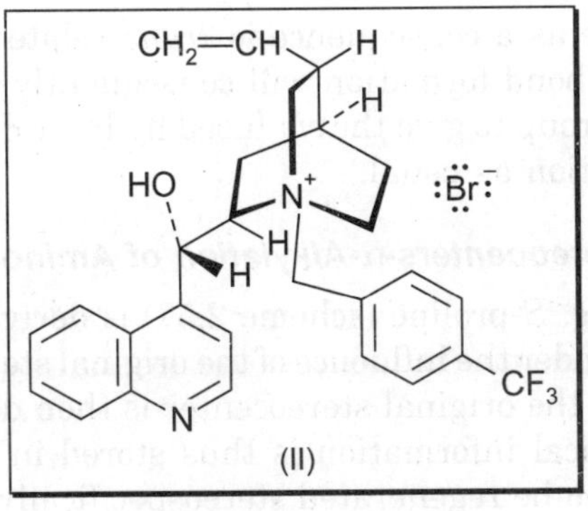

SCHEME 2.55

(g) Enantioselective Intramolecular Aldol Condensation (Carbon-Carbon Bond Forming Reaction–Robinson Annelation Reaction)

The cyclopentan-1, 3-dione (a, scheme 2.56) undergoes an intermolecular aldol condensation under the catalysis of a chiral base (*S*-proline) with high enantioselectivity. When during a Robinson annulation reaction (scheme 2.56, overall process of addition and cyclization), the cyclization step is carried out in the presence of (*S*)-proline, the bicyclic hydroxydiketone (II, scheme 2.56) is obtained in 94% optical purity. The following steps may be noted for this enantioselectivity:

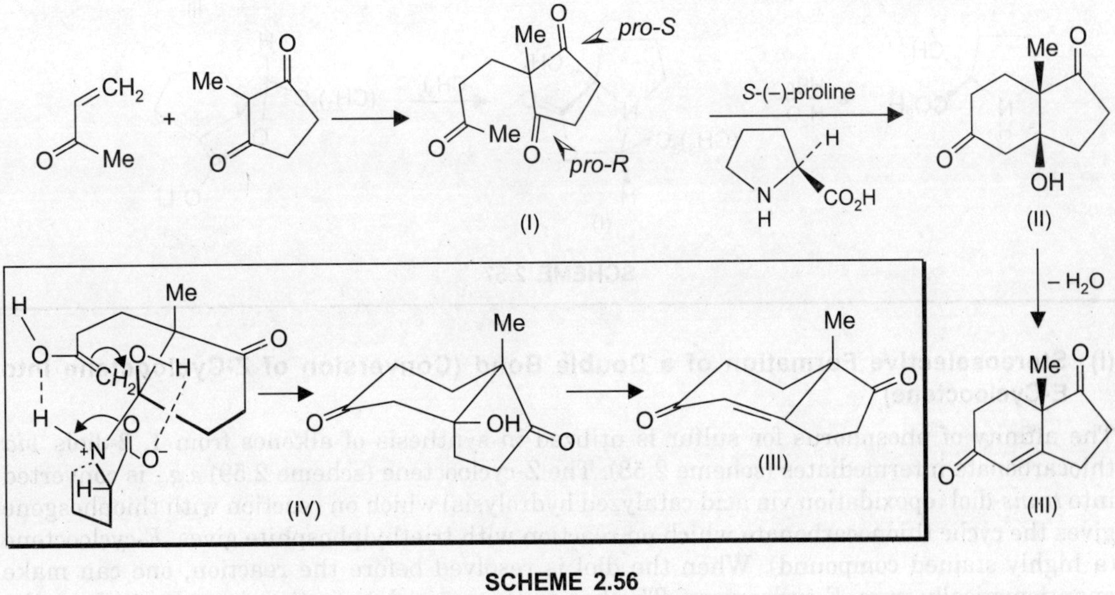

SCHEME 2.56

- The chiral catalyst distinguishes between the two carbonyl groups in the cyclopentane ring (I, scheme 2.56) which are enantiotopic and distinguished by the chiral catalyst which, therefore, add preferentially to one of these. The two carbonyl groups on the cyclopentane ring are enantiotopic and designated *pro R* and *pro S*.
- Two hydrogen bonds in (IV, scheme 2.56) impose a topographical rigidity on the conformation of the tricyclic transition state.
- The proline residue as a consequence is *trans* related to the angular methyl group and carbon-carbon bond formation will consequently take place from the side opposite to the methyl group to give the *cis*-fused hydroxy diketone (II, scheme 2.56) which undergoes dehydration as usual.

(h) Self Regeneration of Stereocenters-α-Alkylation of Amino Acids

A chiral starting material *e.g.* *S*-proline (scheme 2.57) is derivatized in such a way so that a new stereocenter is created under the influence of the original stereocenter as in (I, scheme 2.57). The tetrahedral geometry of the original stereocenter is then destroyed (II, enolate formation with LDA). The stereochemical information is thus stored in the new stereocenter and the original stereocenter can then be regenerated stereospecifically as shown during α-alkylation of amino acids. Although the alkyl group gets attached from the lower face of the molecule on the same side of the bulky *t*-butyl group, this face is in fact less hindered than the upper face. A consideration of the model of the enolate (II, scheme 2.57) shows that the bulky *t*-butyl group occupies a pseudo-equatorial position to make lower face less hindered.

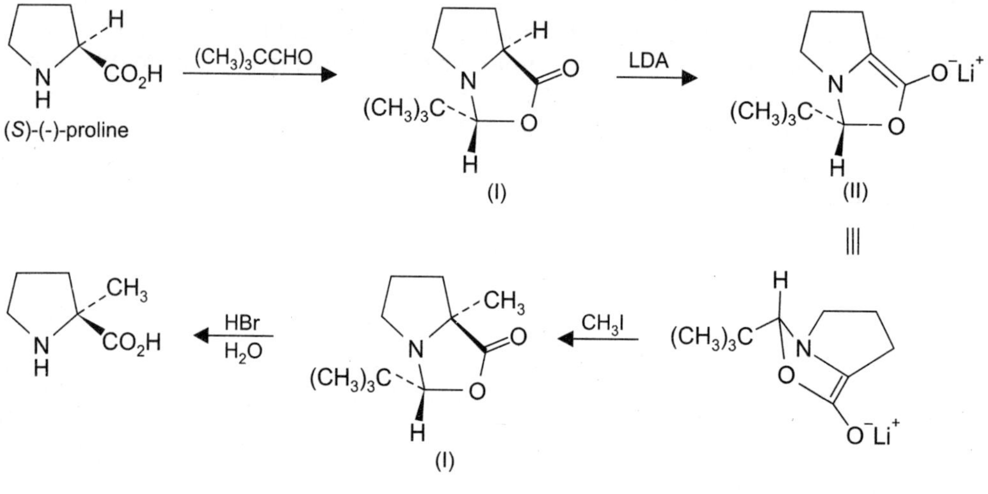

SCHEME 2.57

(i) Stereoselective Formation of a Double Bond (Conversion of Z-Cyclooctene into E-Cyclooctene)

The affinity of phosphorus for sulfur is utilized in synthesis of alkenes from 1, 2-diols *via* thiocarbonate intermediates (scheme 2.58). The Z-cyclooctene (scheme 2.59) *e.g.*, is converted into *trans*-diol (epoxidation via acid catalyzed hydrolysis) which on reaction with thiophosgene gives the cyclic thionocarbonate which on reaction with triethylphosphite gives *E*-cyclooctene (a highly stained compound). When the diol is resolved before the reaction, one can make enantiomerically pure *E*-cyclooctene. The key to this procedure is that in order to form the thionocarbonate the diol has to adopt a conformation (scheme 2.59) which ensures the formation of the *trans*-double bond. One may recall that just as a double bond can be oxidized stereoselectively to *cis*- or *trans*-1, 2-diols, the diols may also be converted into the corresponding olefinic compounds stereoselectively.

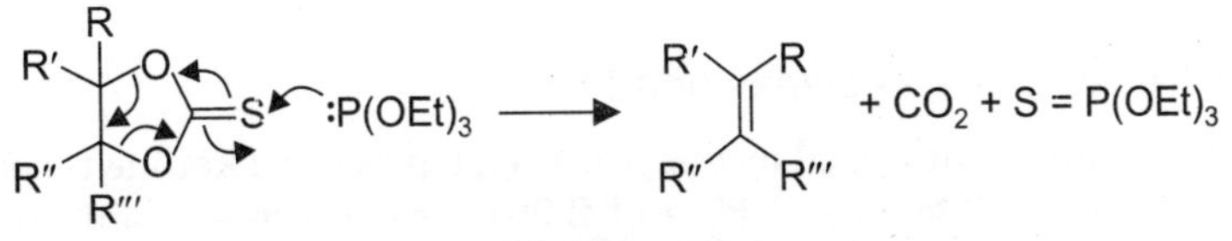

SCHEME 2.58

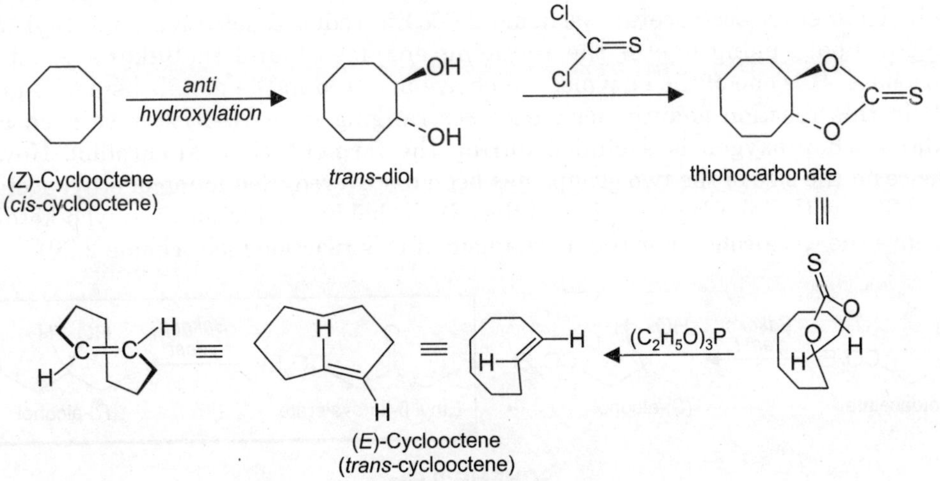

SCHEME 2.59

(j) Asymmetric Synthesis Using Circularly Polarized Light

Absolute chiral synthesis involve the formation of optically active compounds without the incorporation of other optically active compounds. One of the most interesting of some of the known examples of absolute chiral synthesis involves the light induced cyclization–(photoreactions) of 1,2-diarylethylenes to dihydrophenanthrene derivatives (oxidation with iodine and oxygen gives phenanthrene scheme 2.60).

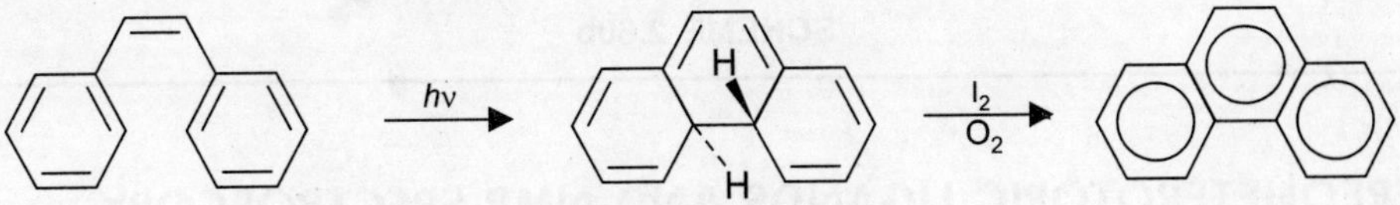

SCHEME 2.60

When 1, 2-diarylethylene is subjected to these reactions in the presence of circularly polarized light to induce cyclization, optically active octahelicene is formed (study the chirality of hexahelicene). With right hand circularly polarized light the (–) enantiomer is formed in 2% optical purity, while with the left hand circularly polarized light the (+) enantiomer is formed to the same extent. The chiral reagent here is the circularly polarized light and the two forms of the product have a nonsuperimposable mirror image relationship to each other. The role of circularly polarized light is reminiscent of an optically active compound in a conventional resolution. It combines with individual enantiomers of the diarylethylene, forming a pair of excited states which are diastereomerically related and, are thus, formed and then decomposed at different rates (scheme 2.60).

(k) Use of Bakers' Yeast, Other Enzymes and Proteins

Recall, that reductions can be catalyzed by enzymes *i.e.*, a prochiral ketone is reduced to optically active secondary alcohol (see, scheme 1.85 and 2.20). The enzyme is only the chiral catalyst which, however, does not provide the hydrogen atom for reduction. The hydrogen atoms, instead are provided by the *relevant coenzyme e.g., NADH*. One may carry out the reduction using whole cells, like bakers yeast where both the enzyme and the coenzyme are provided by the organism. Thus ethyl acetoacetate (scheme 2.60*a*) is reduced selectively to ethyl (*S*)-(+)-3-hydroxybutanoate using one of the reducing enzymes found in Baker's yeast alcohol dehydrogenase. It is one of the enzymes which are involved in the metabolism of D-glucose to ethanol. In this reaction enantiomeric excesses ranging from 70–97% have been achieved particularly when oxygen is excluded during the fermentation (anaerobic). However, a dependence on the size of the two groups has been discovered. Reduction of ethyl acetoacetate (scheme 2.60*a*) with Baker's yeast yielded the (*S*) alcohol but reduction of ethyl β-ketovalerate instead gave the (*R*) alcohol. For the mechanism of this reaction (see scheme 2.20).

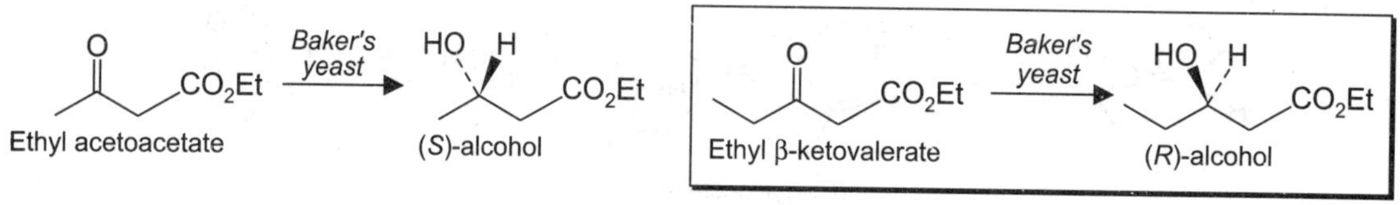

SCHEME 2.60a

Meso compounds undergo enantioselective reactions in the presence of enzymes, horse liver alcohol dehydrogenase (HLADH) selectively oxidises the diol (scheme 2.60*b*) to the lactone.

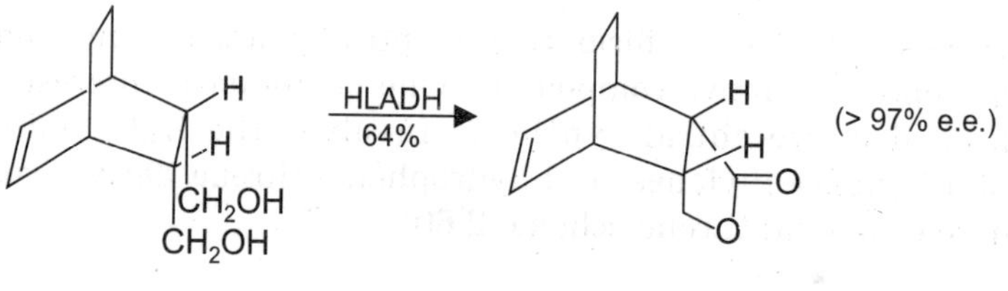

SCHEME 2.60b

2.5 STEREOHETEROTOPIC LIGANDS AND NMR SPECTROSCOPY

(*i*) Introduction

After learning the concepts of prostereoisomerism/prochirality, one can easily classify like atoms or groups of atoms in a molecule according to their symmetry relationships and the following points may be noted:

- The term *equivalent ligands e.g.*, equivalent hydrogens is still widely used for homotopic hydrogens, however, it lacks precision.

- All homotopic hydrogens (also called equivalent) are symmetry equivalent, however, not all symmetry equivalent hydrogens are homotopic (these could well be enantiotopic).
- In achiral conditions symmetry equivalent atoms are chemically equivalent *i.e.*, these undergo reactions at the same rate. These are called chemically equivalent (a term which again lacks precision).
- Under chiral conditions homotopic hydrogens *e.g.*, still cannot be distinguished, however, enantiotopic ligands *e.g.*, hydrogens can be distinguished. Thus in a chiral atmosphere enantiotopic hydrogens are not chemically equivalent (in glycerol *pro-R*-hydroxymethylene group is exclusively phosphorylated with ATP in the presence of an enzyme glycerol kinase (see scheme 2.18*a*).

(*ii*) NMR Identification

(a) Homotopic Ligands

Ligands *e.g.,* hydrogens are homotopic (often termed chemically equivalent) if these can be interchanged by a rotational axis of symmetry. These reside in identical environments and cannot be distinguished by NMR it being an achiral tool. Homotopic protons, therefore, show chemical shift equivalence and are termed *isochronous.*

(b) Enantiotopic Ligands

Enantiotopic nuclei are related by a plane of symmetry (there is no rotational axis of symmetry). Enantiotopic protons are also *isochronous* since their geometrical environments remain identical. The enantiotopic protons, however, become chemical shift non-equivalent when the molecule is placed in a chiral environment and then these can be distinguished. Such an environment is created by using an optically active solvent or by placing, the material in the active site of an enzyme. Under these conditions a diastereotopic relationship is established. For example, the protons of bromochloromethane are enantiotopic (chemically equivalent as there are often called) since these are on a plane of symmetry containing C, Br and Cl (I, scheme 2.61). In a chiral environment (represented by placing a hand a chiral object) to one side, (II scheme 2.61), the protons remain no longer enantiotopic, a diastereotopic relationship is established and the two protons become chemically non-equivalent (The plane of symmetry is lost in the presence of a chiral environment *e.g.*, a hand).

The two methyl groups of dimethylsulfoxide (DMSO, III, scheme 2.61) are enantiotopic and are chemical shift equivalent (*isochronous*) under normal achiral conditions. In the presence of a chiral additive, $PhCHOHCF_3$ with which sulphoxides form some kind of associates, the two methyl groups are seen as a doublet.

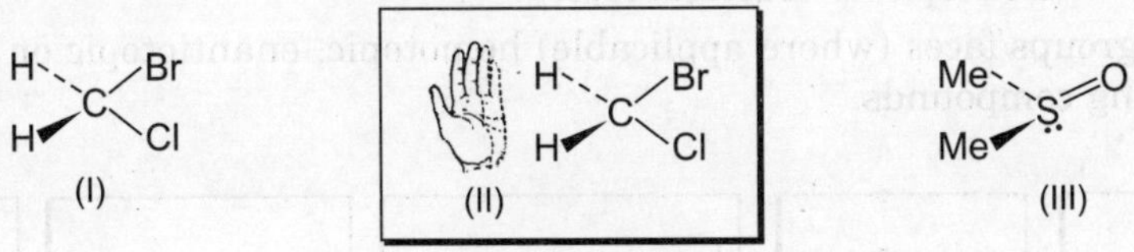

SCHEME 2.61

(c) Diastereotopic Ligands

Diastereotopic ligands are not symmetry equivalent *i.e.*, non-interchangeable by any symmetry operation. Diastereotopic protons *e.g.*, reside in different environments and always in principle display chemical shift differences. As an example, two similar groups, like two hydrogens or two methyl groups adjacent to a stereocenter become diastereotopic (chemically non-equivalent).

The ^{13}CNMR spectrum of S-(+)-3-methyl-2-butanol with two diastereotopic methyl groups (scheme 2.62), as expected displays these as a closely spaced pair of peaks at 17.9 and 18.1 ppm (in all 5 peaks).

One may see that these two diastereotopic methyl groups are always non-equivalent even with free rotation. In any of the fixed staggered conformers (scheme 2.62) one cannot detect a plane of symmetry. Similarly as expected the methylene protons adjacent to a stereocenter are diastereotopic (chemically non-equivalent). Thus, as expected the methyl ester of 2,3-dibromo-2-methylpropionic acid shows an AB quartet (Jab = 10 Hz) for the two diasteretopic protons H_a and H_b.

It may be mentioned that difference in chemical shifts of diastereotopic ligands is often small and may often remain undertectable (*accidental isochrony*).

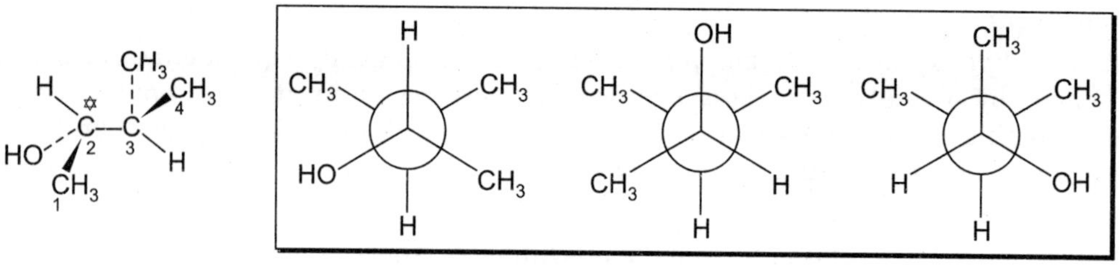

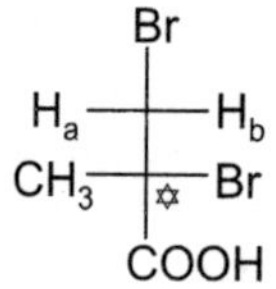

The methylene protons H_a and H_b in 2, 3-dibromo-2-methylpropionic acid being diastereotopic couple with one another and display themselves as an expected AB quartet.

SCHEME 2.62

PROBLEMS

2.1. Using symmetry considerations show if or not the two H's of CH_2Cl_2, CH_2ClF are homotopic or enantiotopic.

2.2. Predict the hydrogen atoms in *cis*-1, 2-dichlorocyclopropane and its *trans*–isomer as homotopic, enantiotopic or diastereotopic.

2.3. Label the groups/faces (where applicable) homotopic, enantiotopic or diastereotopic in the following compounds.

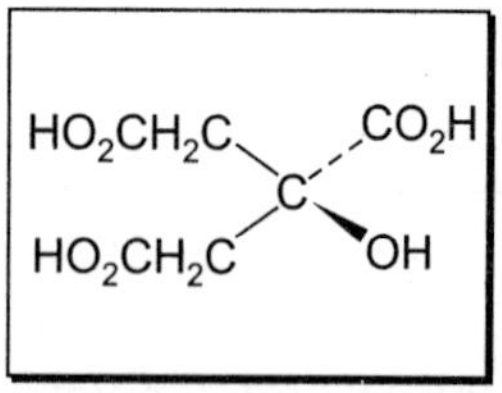

1

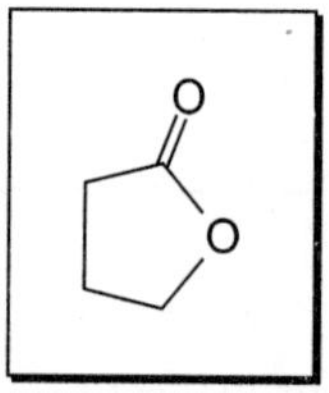

2

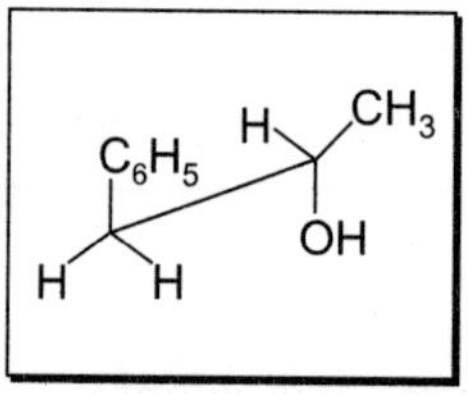

3

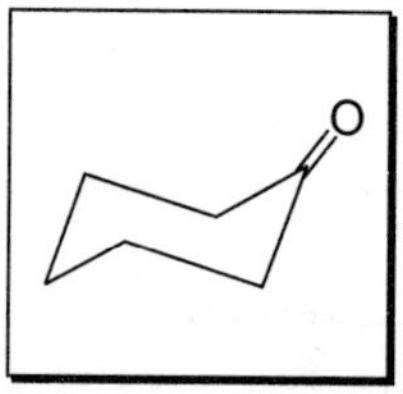

4

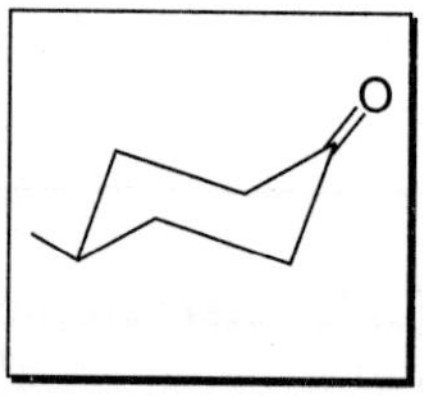

5

2.4. Label the methylene hydrogens of ethanol as *pro-R* and *pro-S* by assigning priorities to the ligands and depicting the path 1 → 2 → 3 as clockwise or anticlockwise.

2.5. Depict the two faces of acetophenone as *Re* or *Si*.

2.6. Show the hydride attack from lithium aluminium hydride from the *Re* face of 2-butanone.

2.7. The enzyme fumarase catalyzed hydration of fumaric acid gives (*S*)-malic acid. Depict the stereochemistry of the reaction.

2.8. Fill in the blanks.

(*i*) The difference between the Cram's rule and Felkin-Ahn model is the ...... of the carbonyl group.

(*ii*) When $R^1 \neq R^2$ the two faces of the carbonyl compound R^1—CO—R^2 are known as ......

(*iii*) Groups or faces which are enantiotopic or diastereotopic are collectively known as ......

(*iv*) Reaction on an allylic alcohol with *tert*-butylhydroperoxide catalyzed by titanium tetraisopropoxide in the presence of (+)- or (–)–diethyltartrate is called ...... epoxidation.

(*v*) *R*-alpine broane is made by the reaction of (+) α-piene and ....

2.9. How the role of chiral enolates determine alkylation of propanoic acid. Discuss the use of (*S*)-proline as a source for chiral enolate formation.

2.10. Using hydroboration–oxidation on the alkene (I), predict the faces of the *pi* bond as enantio- or diastereotopic ?

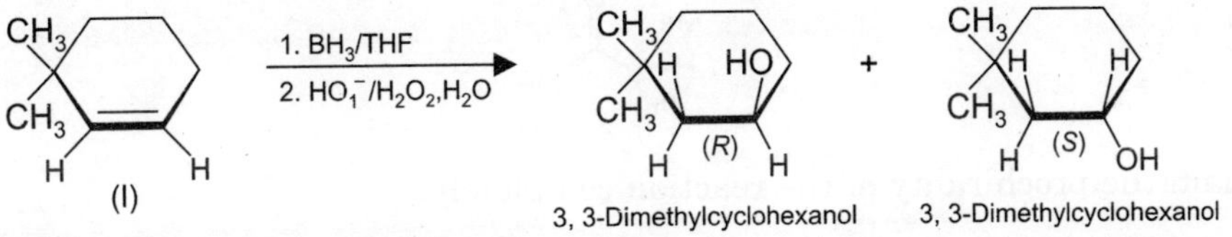

2.11. Label the statements true or false.

(*i*) If the rotation of a molecule about an axis by 360°/*n* followed by reflection in a plane at right angle to that axis gives a superimposable entity, the molecule is said to have a symmetry axis of order *n*.

(*ii*) Faces of a double bond which are not symmetry related are called diastereotopic faces.

(*iii*) Homomorphic ligands are both structurally and configurationally identical when detached.

(*iv*) Homomorphic ligands and faces which are interchangeable by a C_n operation are termed homotopic.

(*v*) A term used to describe the stereoheterotopic ligands or faces when replacement of one such ligand or addition to one such face in an achiral molecule to give chiral products is prochirality and not prostereoisomerism.

2.12. Indicate if the hydrogens marked H^a and H^b are homotopic, enantiotopic or diastereotopic, in the following compounds. Are the methylene protons in (V) enantiotopic and related by σ plane ?

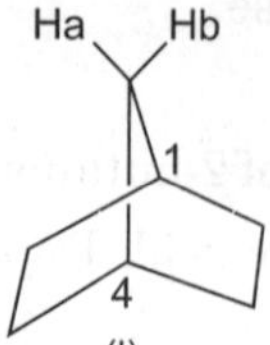
(I)

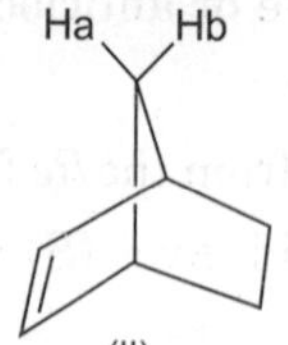
(II)

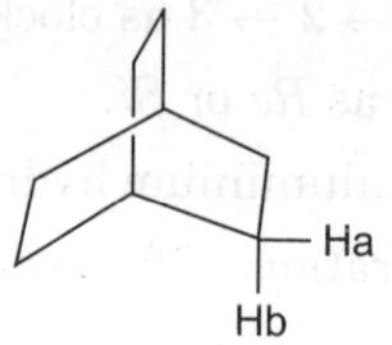
(III)

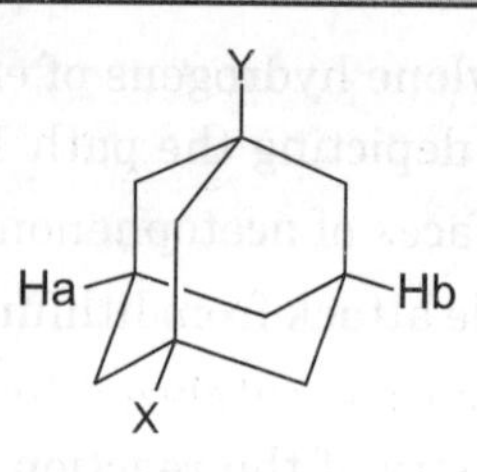
(IV)

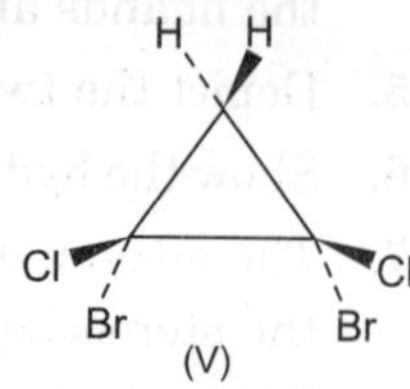
(V)

2.13. Which two assumptions are made in Cram's rule for addition to diastereotopic ketones ?

2.14. Why the faces of carbonyl group in camphor (I) are diastereotopic ? Explain if enantiomers or diastereomers would be formed on reduction of the ketones (II and III) with $NaBH_4$.

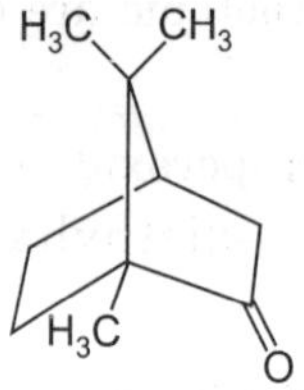
(I)

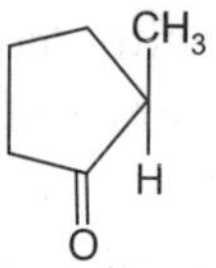
(II)

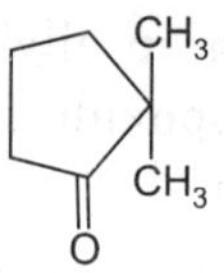
(III)

2.15. Predict the major diastereomer from the lithium aluminium hydride reduction of the following ketone.

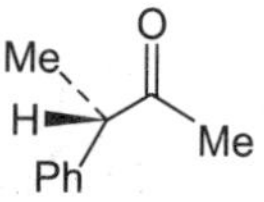

2.16. Explain the prochirality of the reaction completely.

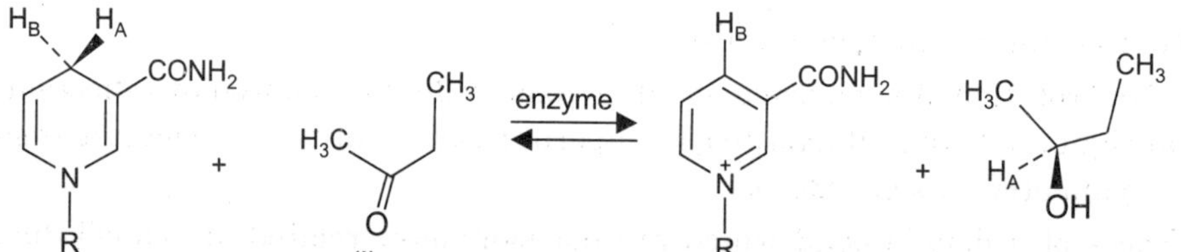
NADH (I) (II)

ANSWERS TO SELECTED PROBLEMS

2.1. 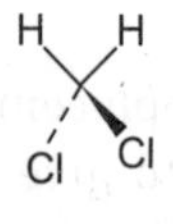 Homotopic protons by C_2 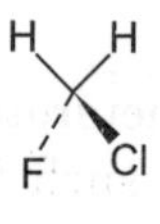 Enantiotopic protons by σ

2.2. The methylene protons in the *cis*-isomer are diastereotopic. These being close to a stereocenter. The other pair of protons is enantiotopic being interchangeable by a plane of symmetry. In the *trans*-isomer methylene protons though close to a stereocenter are, however, homotopic (exception), these being interchangeable by a C_2 axis. The other pair is also homotopic for the same reason (also see, scheme 2.10*a*).

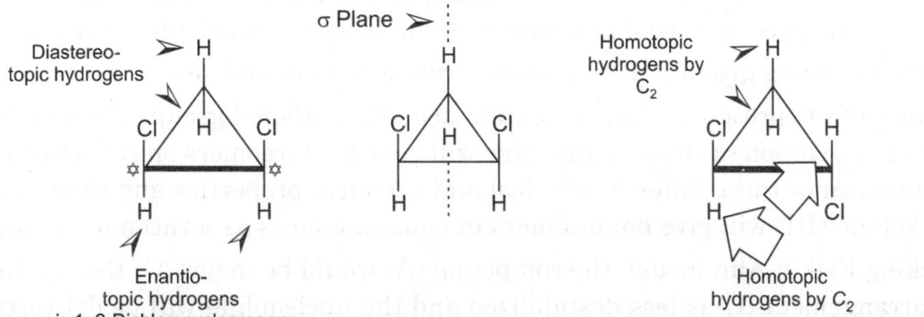

cis-1, 2-Dichlorocyclopropane *trans*-1, 2-Dichlorocyclopropane

2.3. 1. The two-CH_2COOH groups are enantiotopic.

2. The two faces of the carbonyl group in the lactone are enantiotopic, as can be shown by reduction (compare with 2-butanone *i.e.*, an unsymmetrical ketone whose faces are enantiotopic (see, scheme 1.85).
3. In 1-phenylisopropanol, C-2, is a stereocenter and the molecule is chiral C-1 is prochiral since replacement of the stereoheterotopic hydrogens by a group different from other ligands generates diastereoisomeric molecules, because of two centers of chirality. The two C-1 hydrogen atoms, therefore, are diastereotopic.
4. Cyclohexanone has homotopic faces.
5. Monosubstituted derivatives of cyclohexanone have two diastereotopic faces since addition to one face gives a *cis* product while to the other face gives the diastereometric *trans* product.

2.4. See scheme 2.14.

2.5. See scheme 2.15.

2.6. See scheme 2.16.

2.7. See scheme 2.19.

2.8. (*i*) Conformation; (*ii*) enantiotopic; (*iii*) stereoheterotopic, (*iv*) sharpless (*v*) 9-BBN.

2.9. See scheme 2.34.

2.10. The hydroboration–oxidation gives enantiomeric product, the faces are therefore enenatiotopic.

2.11. (*i*) False; (*ii*) true; (*iii*) true; (*iv*) true; (*v*) false, (hint—prostereoisomerism is more general term since in some cases replacement of one or other stereoheterotopic ligands or addition to one or other heterotopic faces may give achiral diastereomers which do contain stereogenic but not chiral elements).

2.12. (I) Homotopic (interchangeable by rotational axis of symmetry); (II) diastereotopic (replacement by a test group would given *syn* and *anti* compounds which are diastereomers, the protons are not interchangeable by a simple axis of symmetry); (III) enantiotopic (gainfully if one makes a model and rotates it to put H[a] and H[b] nearest to the eyes these are on a mirror plane of symmetry); (IV) enantiotopic (the adamantane is an expanded tetrahedron and an adamantane with four different bridgehead substitutents is chiral,

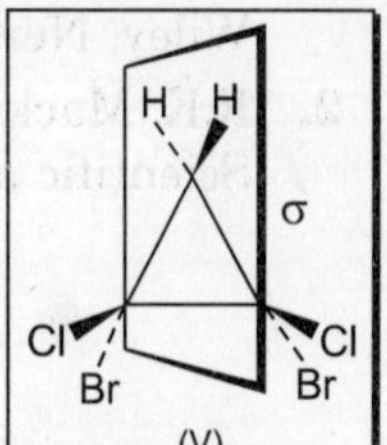

(V)

see schemes 1.5 and 1.110). No, the methylene protons in (V) are instead on the σ plane and diastereotopic (related problem scheme 2.13*a*).

2.13. (*i*) The conformation of the acyclic diastereotopic ketone is frozen with C=O between small and medium sized groups and group L (large) is oriented *trans* to C=O.

(*ii*) The incoming group prefers to enter from less hindered side.

2.14. Because of the presence of stereocenters near the carbonyl group. The ketone (II) has a stereogenic carbon (a stereocenter) this will give diastereomers on reduction (in unequal amounts, since these differ in physical and chemical properties and thus in stabilities). The ketone (III) will give enantiomers in equal amounts as a racemic mixture.

2.15. Invoking Felkin-Ahn model, the compound (A) would be major. Of the two alternatives the arrangement (C) is less destabilized and the nucleophile will be delivered in (C) as shown to give (A) as the major product.

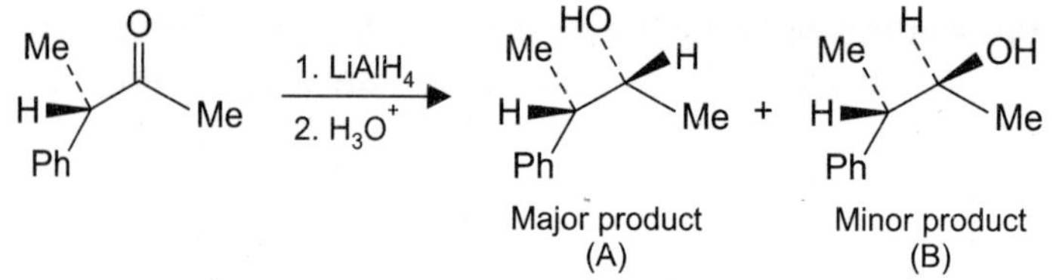

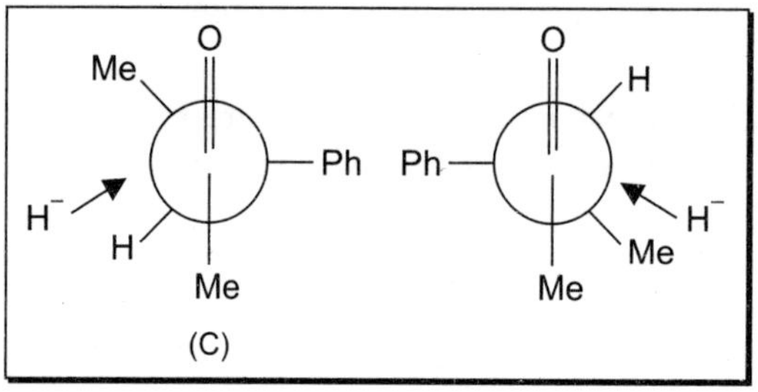

2.16. The ketone (I) has two enantiotopic faces. In the presence of an enzyme the coenzyme (with two diastereotopic hydrogens H_A and H_B) transfers only the *pro R* H_A hydrogen selectively to the rear side of the ketone (*Re* face) to yield a chiral alcohol in which now the methylene hydrogens have become diastereotopic (close to a stereocenter). One may note that the methylene hydrogens in ketone (I) are instead enantiotopic.

REFERENCES AND FURTHER READING

1. Ernest L. Eliel, Samuel H. Wilen, Michael P. Doyle: *Basic Organic Stereochemistry,* Wiley, New York, Chichester, Weinheim, Brisbame, Singapore, Toronto, 2001.
2. R.K. Mackie, D.M. Smith and R.A. Aitken: *Guidebook to Organic Synthesis,* Longman Scientific and Technical, England, 1990.

CHAPTER 3

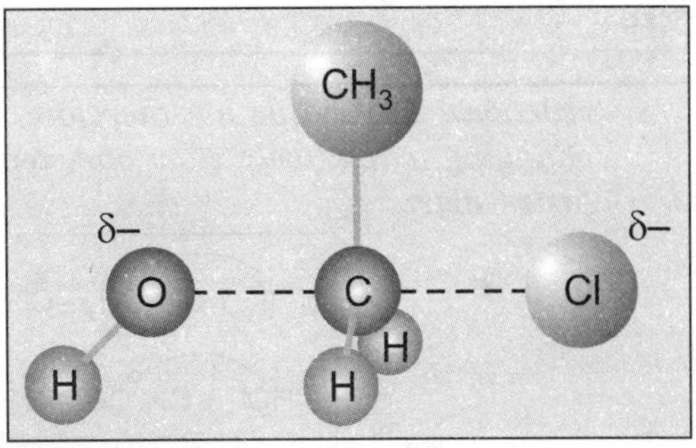

Aliphatic Nucleophilic Substitution

3.1 INTRODUCTION

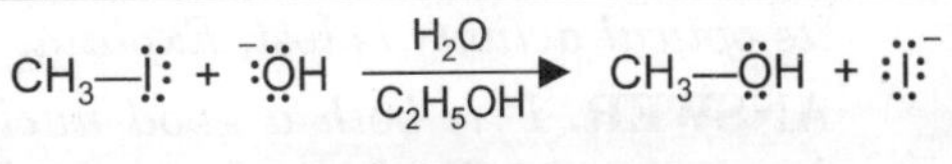

SCHEME 3.1

Nucleophilic substitution is one of the most versatile reactions in organic chemistry. For reaction to occur, the leaving group departs carbon with the pair of bonding electrons. The nucleophile, Nu:⁻, must contain at least one unshared pair of electrons and must be either neutral or negatively charged. One of the most commonly used families of leaving groups is the halides. Iodide is both a good leaving group and a good nucleophile (scheme 3.1). The structure of the alkyl group, the nature of the leaving group and the nucleophile are three major variables that control the mechanism of substitution and the formation of side products. This heterolytic reaction, involves two electron shifts and proceeds mostly by one of the two mechanisms. S_N1 (substitution nucleophilic unimolecular) and S_N2 (substitution nulceophilic bimolecular) which are distinguishable by both kinetic and stereochemical criteria.

EXERCISE 3.1

Why the rate of reaction (I) increases in the presence of very small amount of added NaI as in (II).

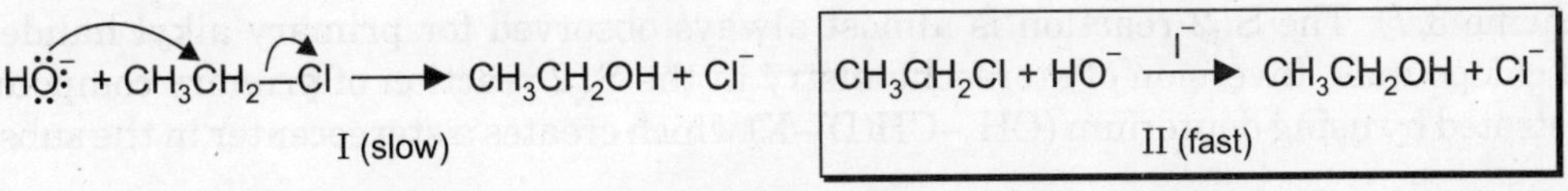

SCHEME 3.1a

ANSWER. *The procedure takes advantage of iodide being an excellent nucleophile and an excellent leaving group. Iodide ion acts as a covalent catalyst (scheme 3.1b). The first S_N2 reaction in the catalyzed reaction (III) is faster than the uncatalyzed reaction (I) since iodide ion is a better nucleophile than hydroxide ion (nucleophile in the uncatalyzed reaction). The second S_N2 reaction in the catalyzed reaction is again faster than the uncatalyzed reaction since iodide ion is a weaker base making it, a better leaving group than chloride ion (the leaving group in the uncatalyzed* *reaction). The iodide ion therefore, increases the rate of formation of ethanol by changing a relatively slow one-step reaction into a reaction with two relatively faster steps.*

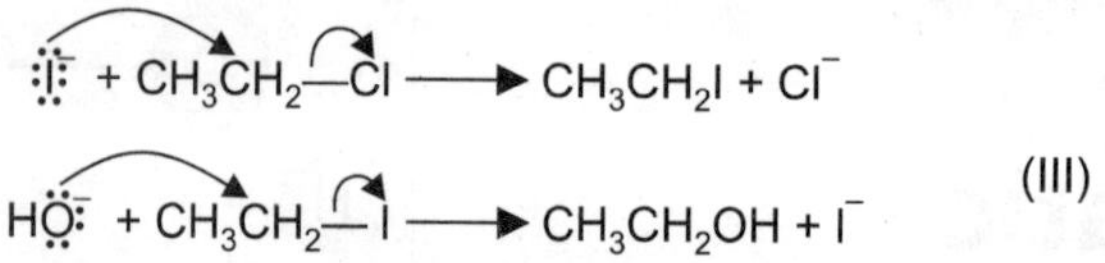

mechanism for the iodide-ion-catalyzed reaction

SCHEME 3.1b

EXERCISE 3.2

On treatment of (S)-2-iodooctane with NaI in solution its optical activity is lost. Explain.

ANSWER. *I^- is both a good nucleophile and a good leaving group. Each displacement is thus attended with inversion at the stereocenter. Since the process is fast it occurs multiple times at every substrate molecule which ultimately leads to racemization.*

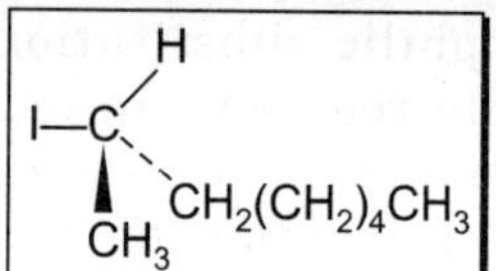

(*S*)-2-iodooctane

SCHEME 3.1c

3.2 S_N2 REACTION MECHANISM AND EVIDENCE

(a) A Mechanism for the S_N2 Reaction—A Stereospecific Reaction

S_N2 reaction at a saturated carbon is a (concerted) one step process without an intermediate (scheme 3.2), that always proceeds with inversion of configuration like the inversion of an umbrella inside out when caught in a wind.

The chemical kinetics of the reaction *e.g.,* between methyl iodide and hydroxide ion has been determined and the rate of reaction is proportional to the concentration of methyl iodide and hydroxide ion. The reaction displays second-order kinetics. The value of the rate constant k_r depends on several factors including temperature and the free energy of the transition state (see scheme 3.7). The S_N2 reaction is almost always observed for primary alkyl halides and related compounds. Inversion of stereochemistry in the S_N2 reaction of primary compounds is demonstrated by using deuterium (CH_3–CH(D)–X) which creates a stereocenter in the substrate.

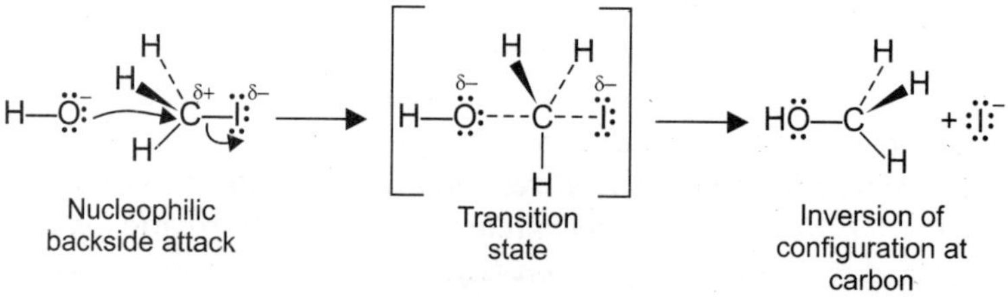

SCHEME 3.2

Most secondary compounds *e.g.* (scheme 3.3) undergo S_N2 reaction with inversion of configuration in which the substituents are alkyl groups. Thus S_N2 reaction is stereospecific and due to inversion of stereochemistry in this reaction, one can design ways to synthesise a desired stereoisomer. When iodide ion is used to displace the bromine in enantiomerically pure reactant (*S*)–(+)-2-bromobutane, it is converted to an enantiomerically pure product (scheme 3.3). The negative iodide (nucleophile) pushes a pair of electrons into the partially positive carbon atom from the back side. The bromine begins to move away with the pair of electrons that had bonded it to the carbon. The transition state shows the formation and breaking of the bonds. The geometry of the transition state appears to be that in which the incoming and leaving groups are both weakly bonded to carbon in a linear fashion. The three remaining bonds of pentacoordinate transition state lie in a plane perpendicular to the two weak bonds. Eventually the bond between the carbon and the leaving group is completely broken and the attacked carbon again becomes tetrahedral with inverted configuration.

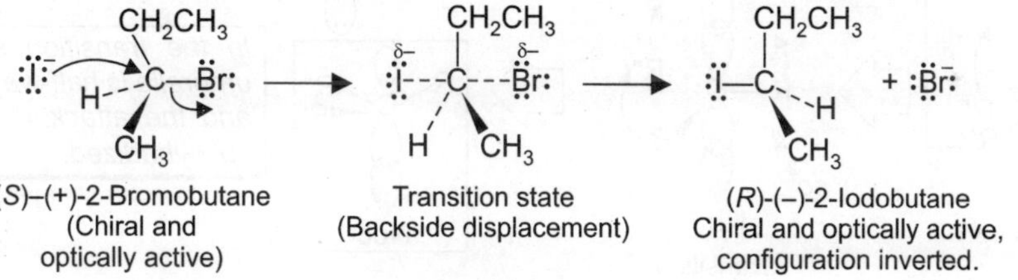

SCHEME 3.3

The potential energy profile for the S_N2 reaction (scheme 3.4) shows that there is only one transition state and no intermediates between the reactants and products. The reactants are slightly higher in energy than the products, since the reaction is known to be exothermic. The transition state is much higher in energy since it involves a five–coordinate carbon atom with two partial bonds. The top of the energy hill corresponds to the transition state of a typical S_N2 reaction. The difference in free energy between the reactants and the transition state is the free energy of activation. The difference in free energy between the reactants and products is the free energy change for a reaction. A reaction which has a low free energy of activation will occur much faster than a reaction with higher one.

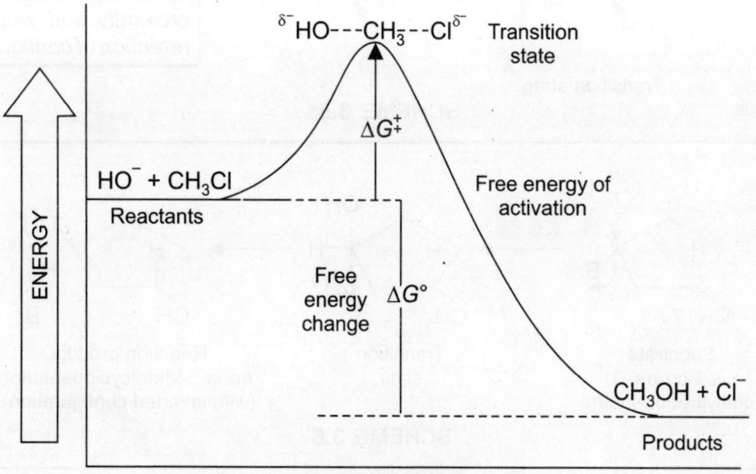

SCHEME 3.4

In the orbital picture of the transition state the carbon undergoing substitution is sp^2-hybridized and has a p orbital perpendicular to the plane of the bonds which partly overlaps with an orbital of the leaving group as well as with that of the incoming nucleophile (scheme 3.5). As the nucleophile approaches the back side and attacks the small back lobe of the sp^3 hybrid orbital used by carbon to bind the halogen atom, the rest of the molecule becomes planar at the transition state by changing the hybridization at carbon to sp^2. The negative charge is no longer located entirely on the nucleophile, it is also partly located on the leaving group. When the reaction proceeds to products, the inversion motion is completed, the carbon returns to the tetrahedral sp^3 configuration, and the leaving group becomes a fully charged anion.

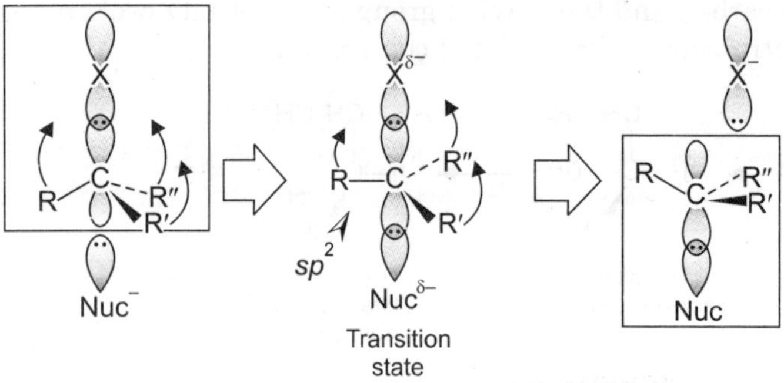

In the transition state, the umbrella is half-way inverted and the attacked carbon is sp^2 hybridized.

SCHEME 3.5

(b) The Stereochemistry of the S_N2 Reaction

In this one step reaction one could imagine that nucleophile could approach the substrate from the same side (front side) of the C—X bond leading to retention of configuration of the attacked carbon (scheme 3.5*a*) or it could attack the carbon of substrate as away as possible from the C—X bond leading to inversion of configuration at the attacked carbon.

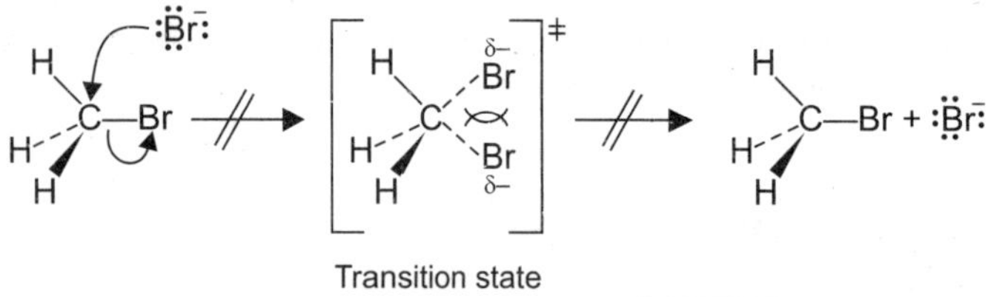

The transition state from front side attack would be too unstable due to two electron rich bromine atoms in close proximity and would lead to retention of configuration

SCHEME 3.5a

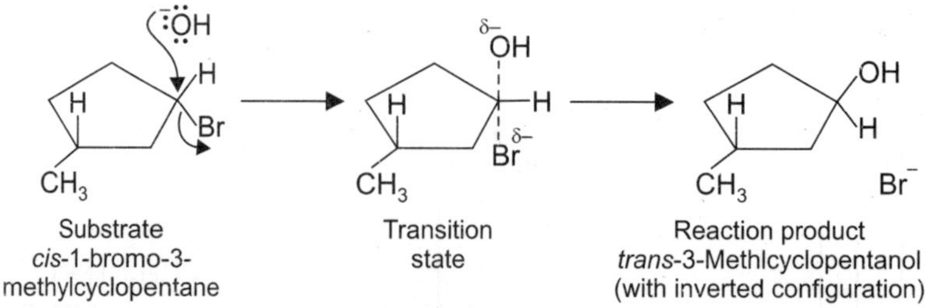

SCHEME 3.6

When the S_N2 reactions occur at a stereocenter, there is always an inversion of configuration at the stereocenter. This has been confirmed with acyclic compounds (see, scheme 3.3). With cyclic compounds the nucleophile ends up being bonded to the opposite side of the ring from which the leaving group departs (a configurational inversion, scheme 3.6). This changes the stereochemical relation between the substituents.

Stereochemical Consequences of S_N2 reactions

- *Optically active compounds give optically active products, provided the nucleophile and the leaving group are not the same and the product is not meso.*
- *In cyclic systems cis, trans relationship may be interconverted.*
- *The guideline of stereochemistry–"diastereomers produce diastereomers" can be well demonstrated (scheme 3.6a).*

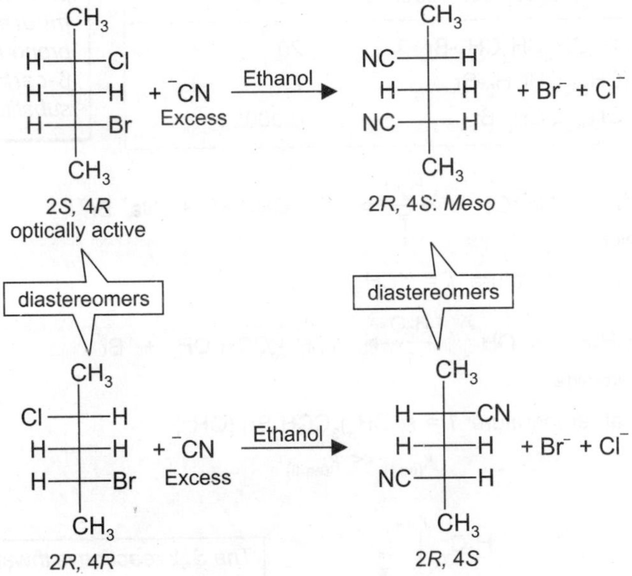

S_N2 Reactions of Molecules with Two Stereocenters
SCHEME 3.6a

(c) The Structure of the Substrate–Requirement of Collinear Transition State

The rate of direct displacement *i.e.*, an S_N2 reaction is very sensitive to the steric bulk of the substituents present on the carbon undergoing such a reaction. This is expected since the degree of coordination (pentacoordinated transition state) increases at the reacting carbon atom. Thus from the steric point of view, the optimum substrate could be CH_3—X. Each replacement of hydrogen by a more bulky alkyl group (at α-carbon) should decrease the rate of reaction. Consequently, the order of reactivity of alkyl groups is expected to be methyl > primary > secondary > tertiary and this is observed. Table 3.1 gives the relative rates of typical S_N2 reactions. Methyl halides react most rapidly and tertiary halides react so slowly as to be unreactive by the S_N2 mechanism.

Neopentyl halides, are primary halides and even then these are unreactive in S_N2 reactions. This situation shows that steric hindrance effects are operative even if the β-carbon is substituted by alkyl groups. An activated complex at the transition state comprises the elements of a molecule of the alkyl halide and a hydroxide ion (eq. I, scheme 3.7).

If the methyl group of ethyl bromide is replaced by a *tert*-butyl group to give neopentyl bromide [2, 2-dimethylbromopropane], *i.e.*, the *primary* nature of the alkyl halide is kept unchanged, the rate constant of the new reaction under the same experimental conditions (eq. II, scheme 3.7) is very much smaller. In neopentyl halides, the *tert*-butyl group blocks the best pathway for rear side displacement of the leaving group (scheme 3.7 and 3.8).

Table 3.1 : Steric hindrance to the S_N2 reaction

Class of halide	Compound	Relative rate
Methyl	CH_3-Br (More reactive)	>1000
Primary (1°)	CH_3CH_2-Br	50
Secondary (2°)	$(CH_3)_2CH$-Br	1
Tertiary (3°)	$(CH_3)_3C$-Br(Less reactive)	<0.001
n-Butyl (1°)	$CH_3CH_2CH_2CH_2$-Br	20
Isobutyl (1°)	$(CH_3)_2CHCH_2$-Br	2
Neopentyl (1°)	$(CH_3)_3CCH_2$-Br	0.0005

When the hydrogens of CH_3Br are successively replaced by $-CH_3$ groups, the rate of reaction with a given nucleophile becomes progressively slower. Steric hindrance is also pronounced even when the β-carbon is increasingly substituted by alkyl gorups.

$$\underset{\text{Ethyl bromide}}{CH_3CH_2Br} + Na^+\,OH^- \xrightarrow[T]{H_2O} CH_3CH_2OH + Na^+\,Br^- \quad ...(I)$$

Rate of reaction at temperature $T = k[CH_3CH_2Br]\,[OH^-]$

$$\underset{\text{Neopentyl bromide}}{(CH_3)_3CCH_2Br} + OH^- \xrightarrow[T]{H_2O} (CH_3)_3CCH_2OH + Br^- \quad ...(II)$$

Rate of reaction at temperature $T = k[(CH_3)_3CCH_2Br]\,[OH^-]$

$$k_{(eq\ I)} << k_{(eq\ II)}$$

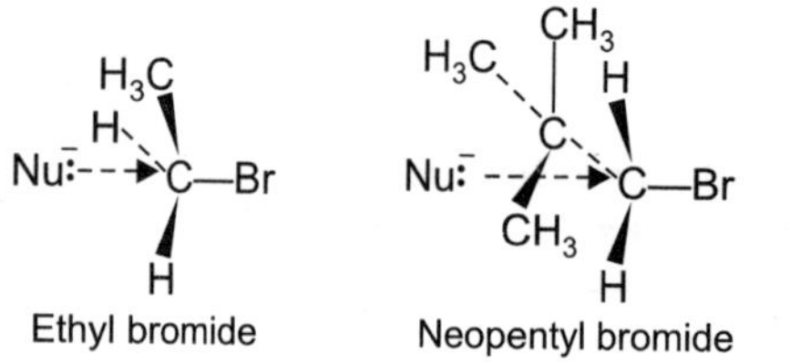

The S_N2 reaction is thwarated in neopentyl bromide due to difficulty of displacement since tert-butyl group shield the rear of C-Br bond.

SCHEME 3.7

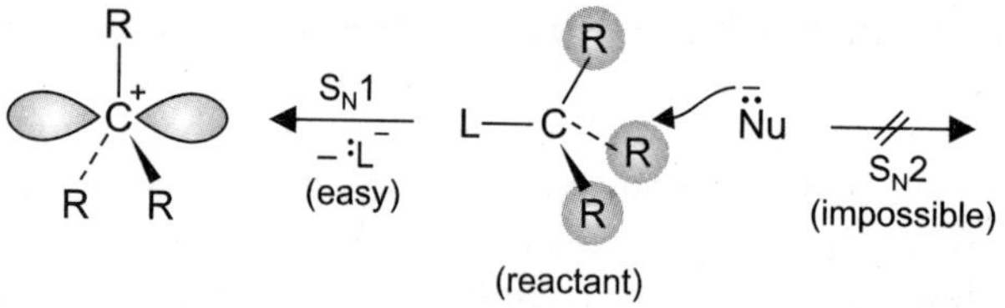

SCHEME 3.8

These examples show the importance of the geometry of the transition state for S_N2 reaction during the access of the nucleophile to the rear of the C—L bond. The nucleophile, the attacked carbon and the leaving group are collinear. In keeping with these requirements of the transition state, in several cyclic systems as well, the S_N2 reaction is either rendered effectively slow or completely stops. This happens (like in acyclic systems) if severe non-bonded interactions are met by the nucleophile as it attempts to gain access to the carbon which is bonded to the leaving group while maintaining the essential collinearity as in the case of neopentyl halides (scheme 3.7). In the case of bicyclo [4.4.0] decane (I, scheme 3.9) the axial angular methyl group hinders the approach of the nucleophilic reagent to the backside of the reaction centre. In the 1-substituted caged compounds (II, III and IV, scheme 3.9), it is almost impossible for the nucleophile to approach the backside of C1 from the "inside" of the molecule. Even if nucleophile could manage to attack from the rear it is not possible to invert the configuration of C-1.

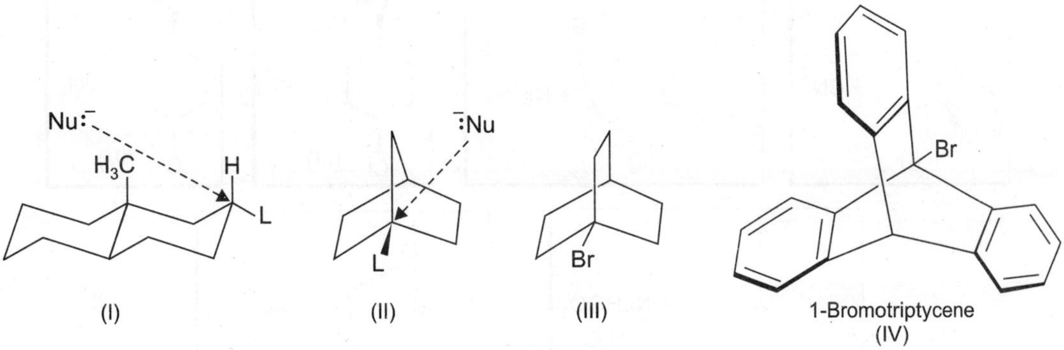

SCHEME 3.9

In an acyclic (noncyclic) haloalkane the hybridization around the halogen-bearing carbon is sp^3 with bond angles around 109°. In the transition state for an S_N2 reaction this carbon get rehybridized to sp^2 and the bond angles involving three unreacting groups around this carbon can easily expand to 120°.

Strong evidence has been provided that the transition state in an S_N2 reaction must be ideally linear. On base treatment of (I, scheme 3.10 removal of α proton), the negatively charged carbon of the ion (II) does not attack the methyl group in an internal S_N2 process to produce (III, scheme 3.10). The kinetic studies have shown that (III) is formed by instead an intermolecular process. The intramolecular (an internal S_N2 reaction) process though with a favourable entropy, does not occur since a linear transition state (IV, scheme 3.10) cannot be

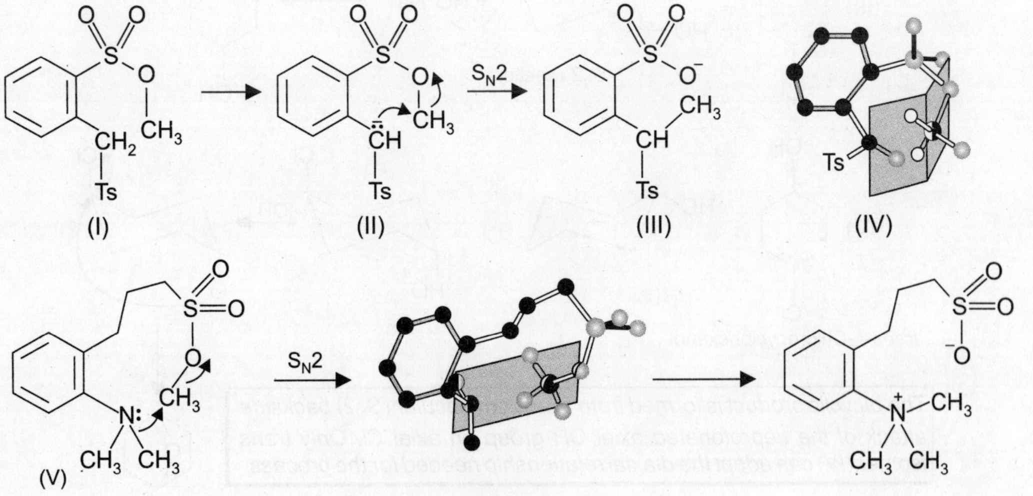

SCHEME 3.10

attained. The bond lengths in a probable six-membered transition state are too short to allow such a geometry. Such a linear arrangement for a successful internal S_N2 process is possible in the nine membered ring of (V, scheme 3.10).

EXERCISE 3.3

Predict the products in the S_N2 reactions (scheme 3.10a).

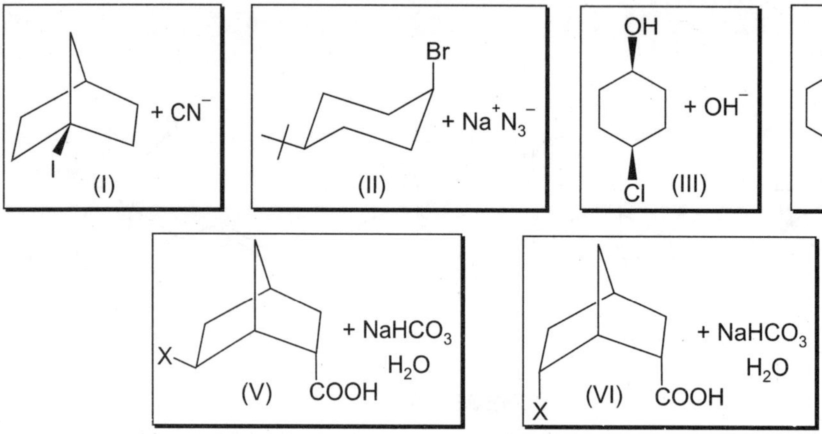

SCHEME 3.10a

ANSWER. *Substrate (I) is tertiary, it cannot undergo S_N2 reaction, the backside of the C–I bond is completely blocked (scheme 3.10b).*

(II) 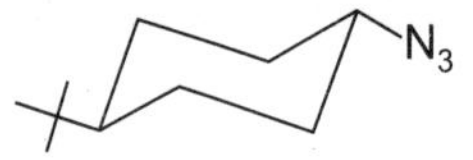

(III) 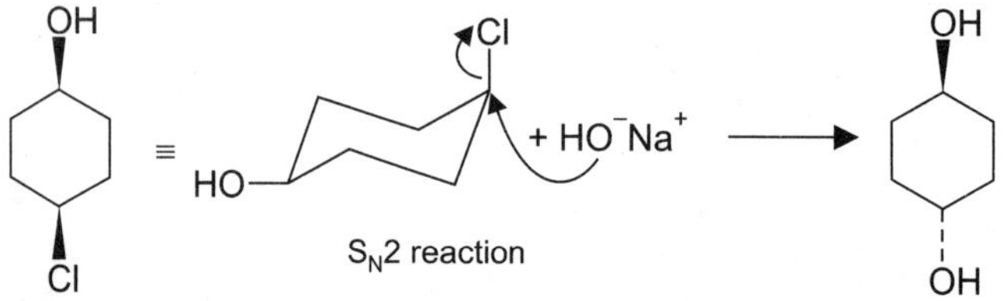

(IV) 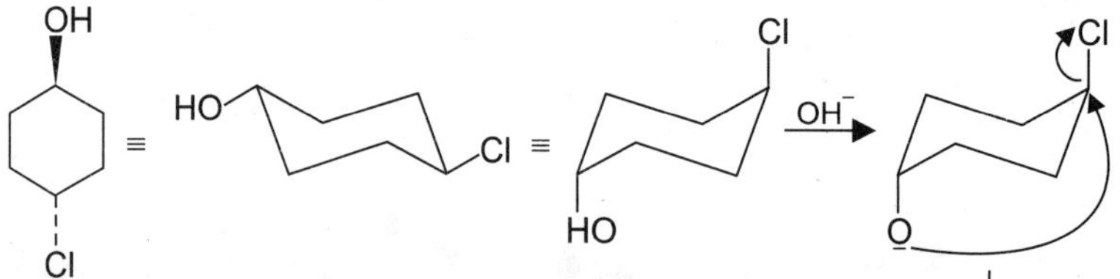

trans-4-Chlorocyclohexanol

The bicyclic product is formed from an intramolecular (S_N2) backside attack of the deprotonated axial OH group on axial Cl. Only trans isomer (IV) can adopt the diaxial relationship needed for the process

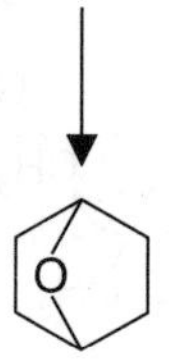

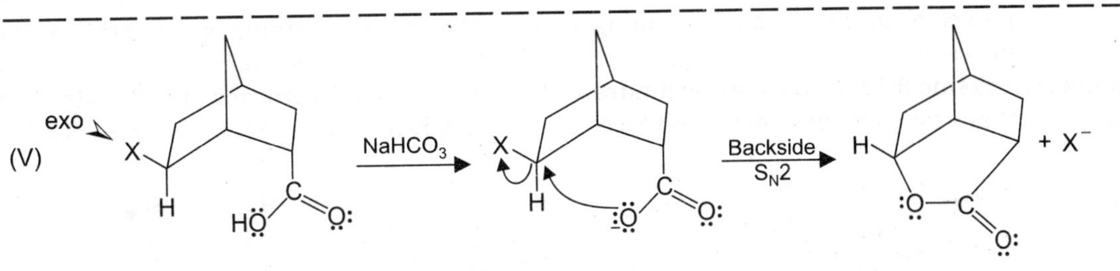

The carboxylate salt is in the correct position for backside S_N2 displacement of X to give a lactone

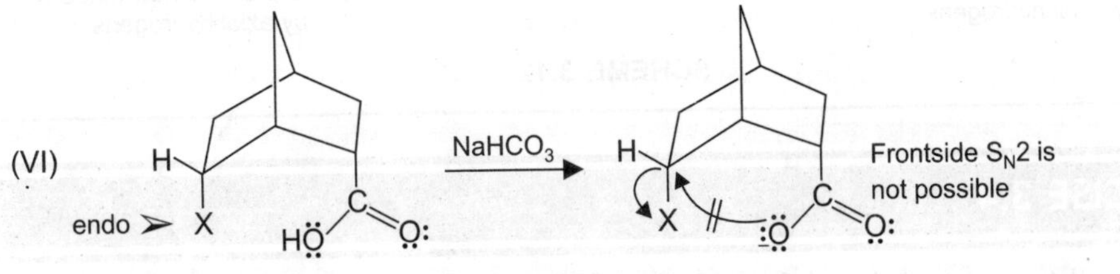

SCHEME 3.10b

Ring compound substrates also undergo normal S_N2 reactions (see scheme 3.6), however, ring size has interesting effects on rate of the reaction (Table 3.2). The answer is in the stability of the transition state. The halocyclopropanes are unreactive to undergo S_N2 reaction due to prohibitive strain both in substrate and in the transition state. The reacting carbon must adopt an sp^2 configuration as the nucleophile replaces the leaving group. The normal bond angle for this hybridization would be 120°, and the cyclopropane ring cannot be distorted much from 60° (scheme 3.11).

Table 3.2 : Relative Reactivities of Cycloalkylbromides in S_N2 reaction with a Nucleophile (Nu^{-}), Iodide Ion

Compound	Relative rate
Cyclopropyl bromide	extremely slow
Cyclobutyl bromide	8×10^{-3}
Cyclopentyl bromide	1.6
Cyclohexyl bromide	1×10^{-2}
Isopropyl bromide (acyclic compound)	1.0

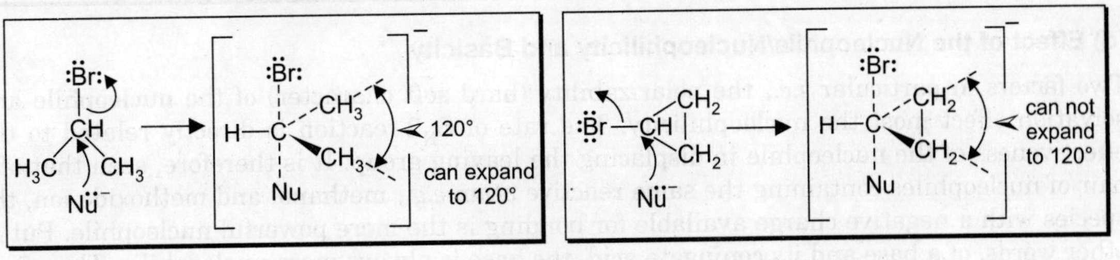

SCHEME 3.11

Halocyclohexanes although seemingly more capable of attaining sp^2 hybridization at the reacting carbon are slower in S_N2 reactions. When a nucleophile approaches an equatorial halide (I, scheme 3.12) it faces steric hindrance by the two axial hydrogens at C-3 and C-5. The leaving axial group on the other hand faces steric hinderance to its exit (II, scheme 3.12).

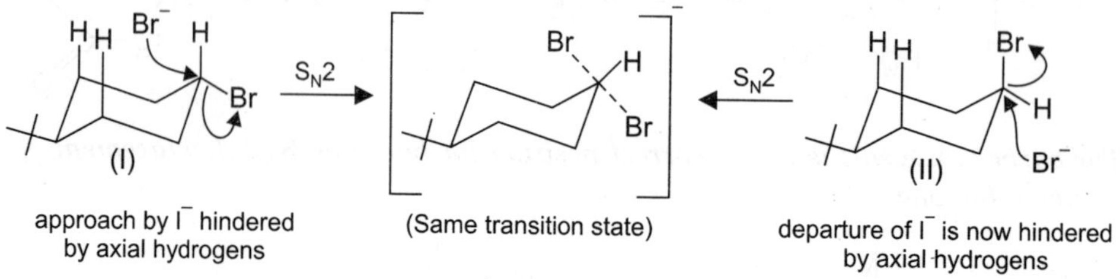

SCHEME 3.12

EXERCISE 3.4

Compared to equatorial cyclohexyl bromide, axial cyclohexyl bromide reacts faster in the S_N2 displacement although the transition state involved is the same (see, scheme 3.12) in both cases. Explain.

ANSWER. *There is no energy difference between the transition states, these being of same energy. The axial iodo conformer (Br), however, increases its energy leading to lowering of the activation energy (scheme 3.12a) for displacement in it compared to the equatorial cyclohexyl bromide.*

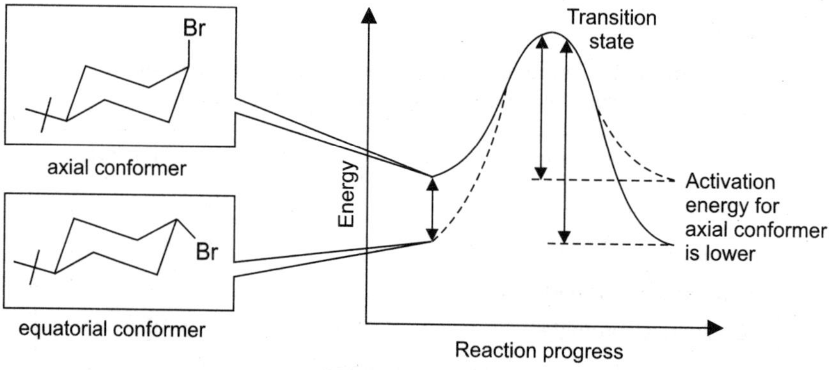

SCHEME 3.12a

(d) Effect of the Nucleophile/Nucleophilicity and Basicity

Two factors in particular *i.e.*, the polarizability (hard soft character) of the nucleophile and solvation effect most the nucleophilicity. The rate of S_N2 reaction is directly related to the effectiveness of the nucleophile in displacing the leaving group. It is therefore, seen that of a pair of nucleophiles containing the same reactive atom *e.g.*, methanol and methoxide ion, the species with a negative charge available for bonding is the more powerful nucleophile. Put in other words, of a base and its conjugate acid, the base is always more nucleophilic. Therefore, OH^- is a stronger nucleophile than H_2O, SH^- a stronger nucleophile than $(CH_3)_2$ SH and NH_2^- is stronger nucleophile than NH_3. This finding is reasonable since a stronger bond between the nucleophilic atom and carbon would mean a more stable transition state and thus a reduced activation energy. Because the S_N2 reaction is concerted the strength of the partially formed new bond will be reflected in the energy of the transition state, the more negative the attacking species, the faster the reaction should be.

There seems to be a good correlation between basicity and nucleophilicity. However, it would be incorrect to explain that methoxide is a much better nucleophile since it is much more basic. Basicity and nucleophilicity are two fundamentally different properties. Basicity (a measure of a thermodynamic phenomenon) is defined by the equilibrium constant for abstracting a proton (scheme 3.13). On the other hand, nucleophilicity (a measure of a kinetic phenomenon) is defined by the rate of attack on an electrophilic carbon. The hydroxide ion (OH^-) is a stronger base than a cyanide ion (CN^-); at equilibrium it has a greater affinity for a proton (the *pKa* of H_2O is ~ 16, whereas the *pKa* of HCN is ~ 10).

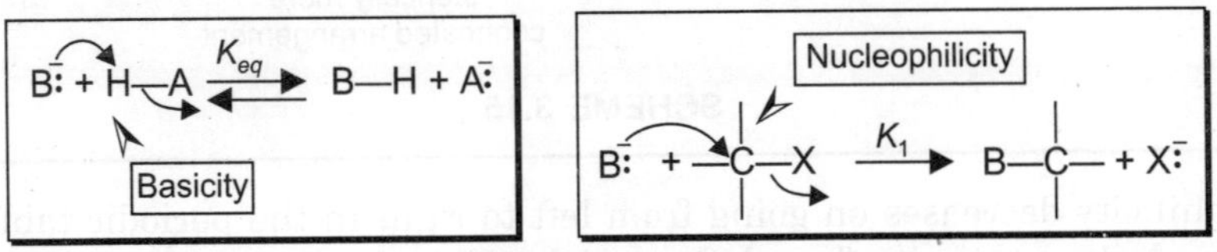

SCHEME 3.13

However, cyanide ion is stronger nucleophile, *i.e.*, it reacts more rapidly with a carbon having a leaving group than a hydroxide ion. In both cases (scheme 3.13), however, a new bond is formed. When the new bond is to a proton, the species has reacted as a base; in case the bond is to a carbon it has reacted as a nucleophile. Moreover, since nucleophilicity is used to describe trends in the kinetic aspects of reactions, the relative nucleophilicity of a given species may differ from substrate to substrate.

A sterically bulky nucleophile is less reactive than a smaller one. As S_N2 reaction is not only sensitive to the bulky groups on the electrophile, but to the bulky groups on the nucleophile as well. This is due to non-bonded repulsions which would develop in the transition state. Thus although *t*-butoxide ion is a stronger base than ethoxide ion, but the bulky *t*-butoxide is a weaker nucleophile (scheme 3.14). Basicity is little effected by steric hindrance, since the attack is on an unhindered proton.

t-Butoxide (hindered)
stronger base, yet weaker nucleophile

Ethoxide (unhindered)
weaker base, yet stronger nucleophile

SCHEME 3.14

It is found that quinuclidine which is a tied-back cage compound reacts faster with ethyl iodide compared to triethylamine (scheme 3.15). In these compounds the nucleophilic nitrogen atom is flanked by three two-carbon chains. In the cage compound, they are tied back and are not free to rotate. In triethylamine, they are free to rotate and thus effectively increase the bulk of the nucleophile. Thus the tied back cage compound quinuclidine reacts much faster with ethyl iodide than does triethylamine.

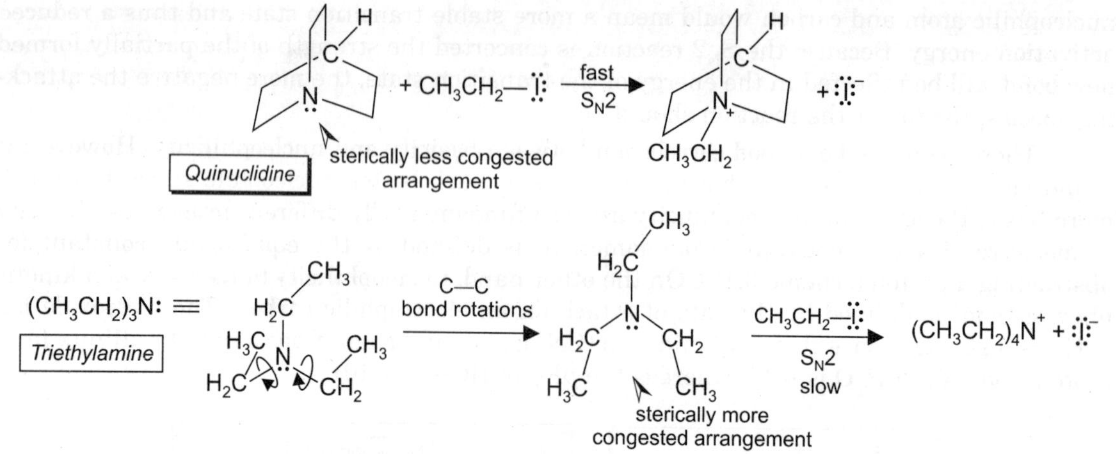

SCHEME 3.15

Nucleophilicity decreases on going from left to right in the periodic table. This follows the increase in electronegativity from left to right. Thus high electronegativity is unfavourable, due to tightly held electrons which are therefore, relatively less available for donation to the substrate for S_N2 process. Thus OH^- is more nucleophilic than F^-; NH_3 is more nucleophilic than H_2O.

EXERCISE 3.5

Devise S_N2 reactions for the synthesis of products (scheme 3.15a).

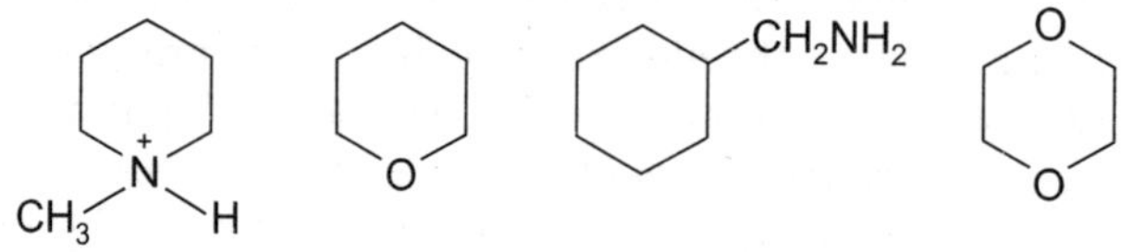

(I) (II) (III) (IV)

SCHEME 3.15a

ANSWER. *(I)*

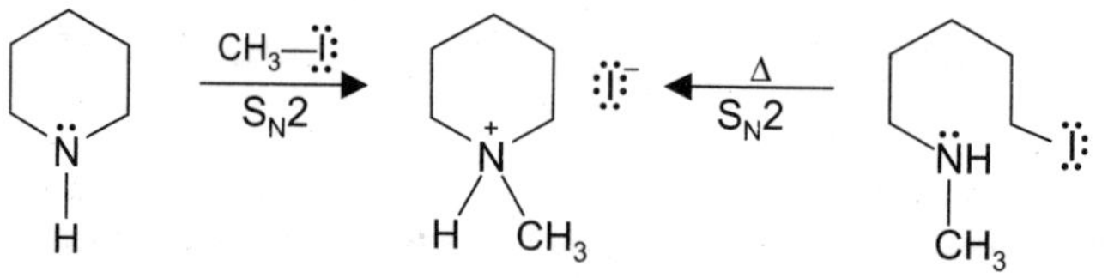

(This can be made by either an S_N2 reaction (intermolecular) using a cyclic amine or via an intermolecular reaction)

(II)

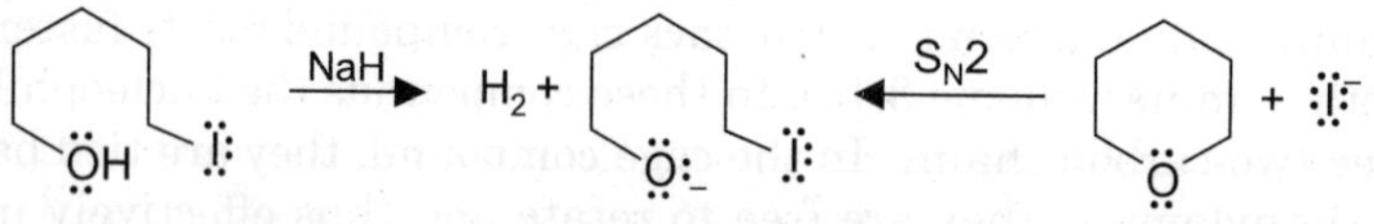

(By an intramolecular reaction by first generating a nucleophile from the OH group by reacting with a base e.g., NaH)

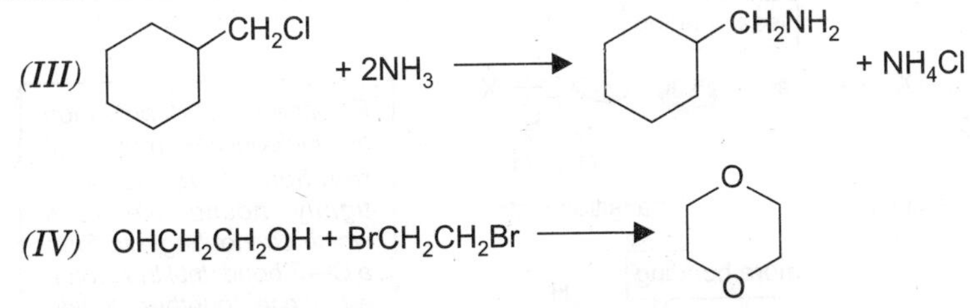

SCHEME 3.15b

Nucleophilicity increases going down the periodic table *i.e.,* Br^- is more nucleophilic than Cl^- and SH^- is more nucleophilic than OH^-. For the halide ions the nucleophilic reactivites follow the order (scheme 3.16).

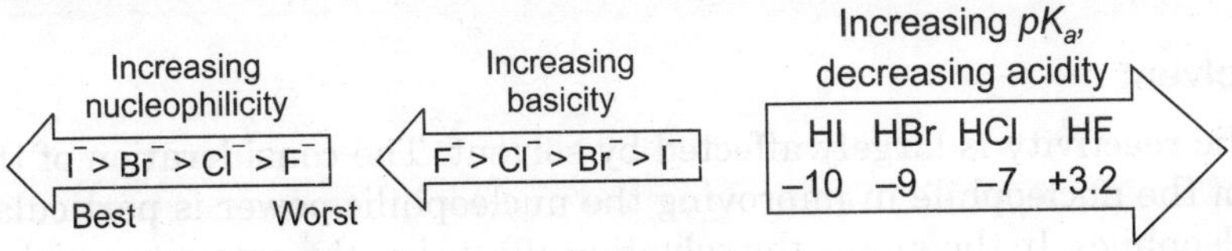

SCHEME 3.16

I^- is the best nucleophile while F^- is poorest. This order is the reverse of the basicity order of the halide ions. Thus one finds that the correlation between basicity and nucleophilicity does not hold for the halides. Going down a column in the periodic table the atoms become larger. One has more electrons at a greater distance from the nucleus. These electrons are more loosely held and the atom with this situation is more polarizable and more nucleophilic. The increase in polarizability (the ease with which the electron cloud on an atom can be distorted is called the polarizability of the atom) leads to an ease of the distortion of the electron could of the attacking atom of the nucleophile. This allows for more effective overlap in the transition state with the back lobe of the slowly rehybridizing sp^3 hybrid used to attain bonding to the leaving group. For these very reasons the basicity of the larger elements is relatively poor than smaller elements, since overlap with the hydrogen 1*s* orbital is poor. Thus, one can use successfully the extent of polarizability to explain as to why I^- is a better nucleophile than F^- (scheme 3.17). It is only with I^- that the partial bond formation begins at relatively large distances as the electron cloud of the nucleophile gets "pulled" to the carbon atom where the S_N2 reaction occurs. Thus the activation energy of the substitution is decreased. This however, is not the case with F^- with has tightly bound electrons that cannot begin to form at C—F bond until the atoms are very close together. One may appreciate that the degree of nucleophilicity increases on going down in the periodic table even for uncharged nucleophiles *i.e.,* $H_2Se > H_2S > H_2O$ and $PH_3 > NH_3$. In these cases particularly, the fact that the increasing polarizability improves nucleophilic power explains the nucleophilicity trend. In these uncharged nucleophiles, the solvent effects must, however, be less pronounced.

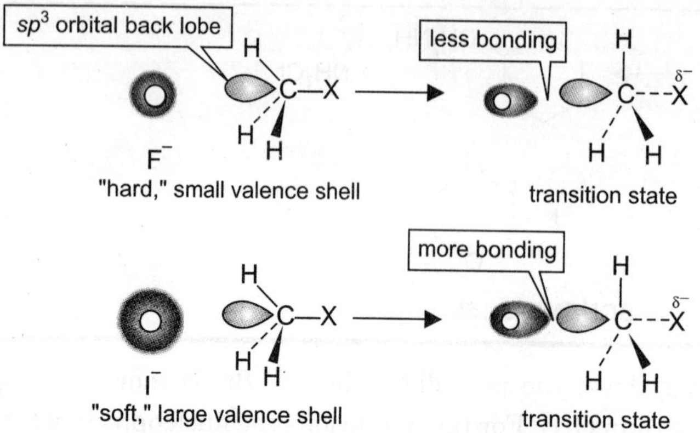

Fluoride ion and iodide ion as nucleophiles in the S_N2 reaction. Fluoride with tightly bound electrons which cannot begin to form a C—F bond until the atoms are close together. Iodide has more loosely bound outer electrons that start bonding earlier in the reaction.

SCHEME 3.17

(e) Effect of Solvent

The nucleophilic reactivity is largely affected by solvent. The consideration of the increase in polarizability of the nucleophile in improving the nucleophilic power is particularly useful for uncharged nucleophiles. In their case, the solvation effects *i.e.,* the extent to which a nucleophilic species is surrounded by solvent molecule, should be much less strong. Nucleophilicity is impeded by solvation. A molecule of a solvent *e.g.,* water or an alcohol (protic solvent) can form hydrogen bonds to an anionic nucleophile (scheme 3.18). Smaller anions (concentrated change) are more tightly solvated than larger ones. This phenomenon, therefore, creates a solvent induced barrier to attack at the substrate. For a solvated anion to act as a nucleophile energy is required to "strip off" some of the solvent molecules. This energy is much larger in the case of a small strongly solvated ion *e.g.* F^- as compared to I^-. Generally, the polarizability of atoms and ions shows an increase with increasing atomic number going down in the periodic table, while solvation of ion (in protic solvents) decreases with increasing atomic number.

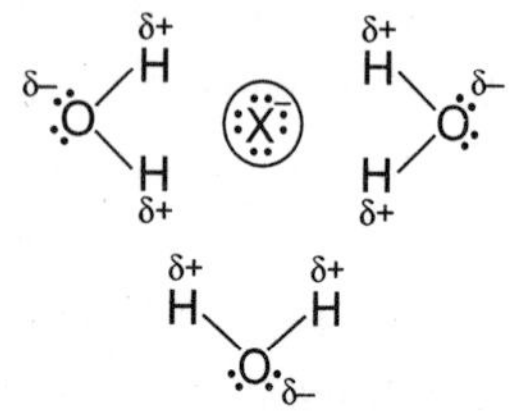

Molecules of the protic solvent e.g., water solvate the halide ion by forming hydrogen bonds to it. A small nucleophile e.g., F^- is more strongly solvated than a larger one e.g., I^-

SCHEME 3.18

Nucleophilic displacement reactions involve ions either as nucleophiles or as products, as a consequence relatively polar solvents are required. A polar solvent has a relatively high dipole moment and dielectric constant. Both hydroxylic (water, methanol, ethanol) as well as polar aprotic solvents dimethyl sulfoxide DMSO, CH_3SOCH_3, dimethyl formamide DMF, $HCON(CH_3)_2$; hexamethylphosphoric triamide HMPT $[(CH_3)_2 N]_3 PO$ are used. Aprotic solvents cannot form hydrogen bonds (no hydroxy groups) thus they do not solvate anions to any appreciable extent. Consequently with anionic nucleophiles, reaction rates are far greater in the polar aprotic solvents. There is lack of stabilization of these nucleophiles by hydrogen bonding, these are therefore, almost "naked" to be highly reactive as nucleophiles. The relative order of nucleophilicity of halide ions in *e.g.* DMSO (scheme 3.20) is the same as their relative basicity. This order is, however, in opposition to their strengths in alcohol or water solution (scheme 3.19).

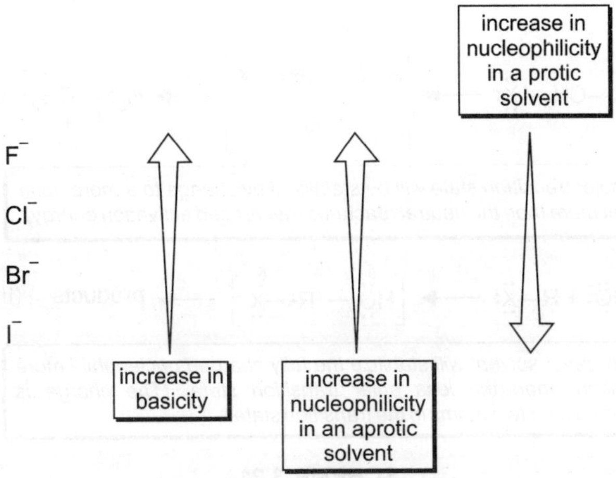

SCHEME 3.19

A polar aprotic solvent dissolves ionic compounds and it solvates cations, the way similar to protic solvents by orienting the negative end of its dipole around the cation (scheme 3.20). It is however, unable to solvate the anion by H-bonding. Moreover, methyl groups in the case of DMSO *e.g.*, shield the S which is the positive end of the dipole, this prevents the solvation of the anion.

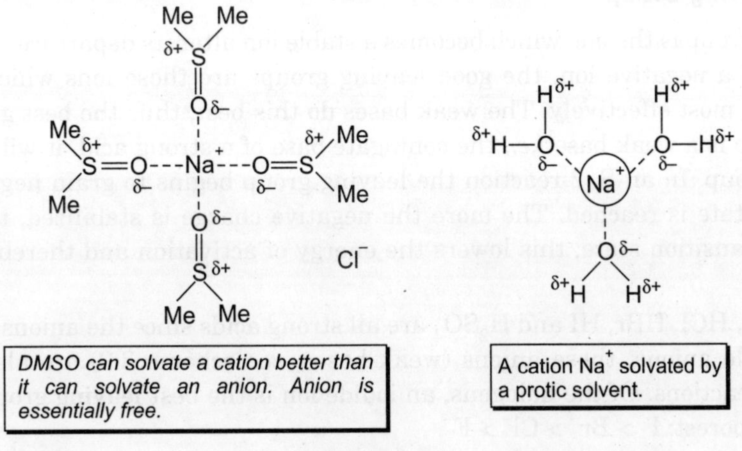

SCHEME 3.20

In summary, the stability of ions depends heavily on solvation. In the gas phase, therefore, fluoride the better base is also the better nucleophile. As the solvent polarity is increased some S_N2 reactions go faster, some slower. Thus *e.g.*, in reaction (I, scheme 3.21) a change to a more polar solvent (from a non-polar hydrocarbon to a much more polar alcohol solvent) results in a faster reaction. Since one is considering rates, the structure of transition state must be considered. A more polar solvent will have little effect on the uncharged starting materials (eq I, scheme 3.21) but will stabilize the charge separated transition state. The result is a decreased activation energy. The reaction thus will go faster in a polar solvent like alcohol (a solvent like alcohol can solvate both positive and negative species, but aprotic solvents can only solvate positive species see scheme 3.20).

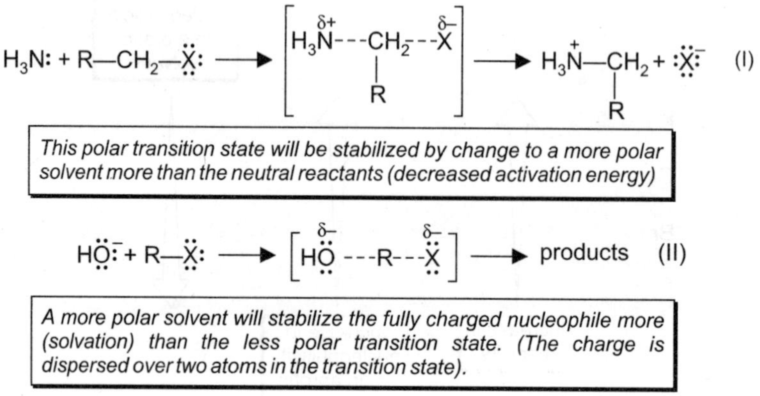

SCHEME 3.21

In the reaction (eq II, scheme 3.21), the nucleophile bears a full negative charge. This charge is dispersed in the transition state over a larger area and therefore, in the transition state the negative charge is less concentrated. Thus in this case a change to a more polar solvent will stabilize the fully charged nucleophile (solvation) more than the transition state, the activation energy increases and the reaction slows.

(f) Effect of Leaving Group

A good leaving group is the one which becomes a stable ion after its departure. As most leaving groups leave as a negative ion, the good leaving groups are those ions which stabilize this negative charge most effectively. The weak bases do this best, thus the best groups are weak bases. If a group is a weak base *i.e.*, the conjugate base of a strong acid, it will generally be a good leaving group. In an S_N2 reaction the leaving group begins to grain negative charge as the transition state is reached. The more the negative charge is stabilized, the lower is the energy of the transition state; this lowers the energy of activation and thereby increases the rate of reaction.

The acids, HCI, HBr, HI and H_2SO_4 are all strong acids since the anions Cl^-, Br^-, I^- and HSO_4^- are stable anions, these anions (weak bases, see scheme 3.9) are also good leaving groups in S_N2 reactions. Of the halogens, an iodide ion is the best leaving group and the fluoride ion is the poorest: $I^- > Br^- > Cl^- > F^-$.

The order of basicity is opposite: $F^- > Cl^- > Br^- > I^-$, the reason that alkyl fluorides are ineffective substrates in S_N2 reactions (scheme 3.22) is related to the relatively low acidity of HF (*pKa* = 3) (Stronger the acid the weaker is its conjugate base, among the halogen acids, HI is the stronger acid see scheme 3.16). Sulfonic acids R SO_2OH are similar to sulfuric acid in acidity and the sulfonate ion RSO_3^- (a weak base) is a very good leaving group. Alkanesulfonate ions, alkyl *p*-toluenesulfonates are therefore, very good substrates in S_N2 reactions (scheme 3.23).

S_N2 reaction				Reactivity
$HO^- + RCH_2I$	$\longrightarrow$	$RCH_2OH + I^-$	(most reactive)	increase in reactivity ↑
$HO^- + RCH_2Br$	$\longrightarrow$	$RCH_2OH + Br^-$		
$HO^- + RCH_2Cl$	$\longrightarrow$	$RCH_2OH + Cl^-$		
$HO^- + RCH_2F$	$\longrightarrow$	$RCH_2OH + F^-$	(least reactive)	

SCHEME 3.22

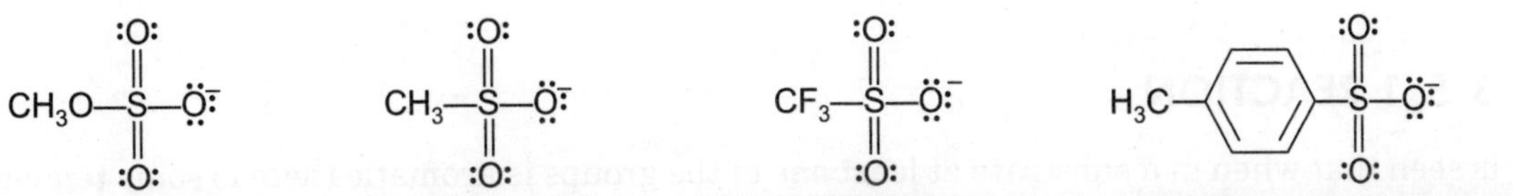

Methyl sulfate ion; Methanesulfonate ion (Mesylate ion); Trifluoromethanesulfonate ion (Triflate ion) (a super leaving group); 4-Methylbenzenesulfonate ion (*p*-Toluenesulfonate ion, tosylate ion)

Sulfate and Sulfonate leaving groups

SCHEME 3.23

The triflate ion ($CF_3SO_3^-$) is one of the best leaving groups known, it is the anion of CF_3SO_3H which is a strong acid much stronger than sulfuric acid (scheme 3.23). Consider the reaction (eq. I, scheme 3.24), the reverse of which though a nucleophilic reactions does not occur *i.e.*, ethyl alcohol and chloride ion do not react. This can be understood if one considers the leaving tendency of Cl^- in the forward reaction and the leaving tendency of OH^- in the reverse reaction in terms of their basicities. For some people it is easy to compare the acid strengths of the conjugate acids. HCl is a much stronger acid than H_2O *i.e.*, Cl^- is a much weaker base than OH^- (since stronger the acid, the weaker its conjugate base). Because it is a weaker base Cl^- is a better leaving group). As a consequence of this OH^- can displace Cl^- in the forward reaction but Cl^- cannot displace OH^- in the reverse reaction.

When the difference between the basicities of the nucleophile and the leaving group is not large, the reaction would be reversible, as in (eq. II, scheme 3.24). The *pKa* values of conjugate acids of the two leaving groups are not very different HBr–9 and HI–10.

$$CH_3CH_2Cl + OH^- \longrightarrow CH_3CH_2OH + Cl^- \quad (I)$$

$$CH_3CH_2Br + I^- \rightleftharpoons CH_3CH_2I + Br^- \quad (II)$$

SCHEME 3.24

(g) Phase Transfer Catalysis–Role of Crown Ethers

Nucleophilic reactivity is increased in the presence of a suitable crown ether. The crown ether 18-crown-6, *e.g.*, coordinates very effectively with potassium ions. Salts like KF, KCN and CH_3COOK which are otherwise insoluble in non-polar solvents like benzene, in the presence of 18-crown-6 dissolve in it. Thus, the situation in the organic phase is the presence of relatively unsolvated anions to carry out nucleophilic substitution on an organic substrate. For example, KF which is both insoluble in benzene and unreactive to organic halides, brings about an efficient nucleophilic displacement on a halide in the presence of 18-crown-6 (scheme 3.25). The role of a crown ether here is of a catalyst which brings the anion into the organic phase, and the process is termed phase transfer catalysis.

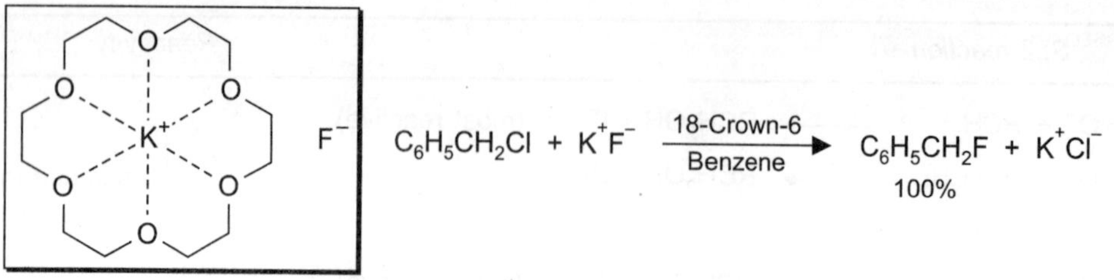

SCHEME 3.25

3.3 S_N1 REACTION

It is seen that when in a substrate at least one of the groups is aromatic there is some tendency for the leaving group in a substrate to leave on its own accord (basis for the S_N1 reaction) rather than it can be expelled by the incoming nucleophile (as is the case in an S_N2 reaction). In summary the substrates which contain good leaving groups and adjacent substituents like a phenyl group which can stabilize a positive charge mesomerically are prone to undergo S_N1 reactions. In these reactions there are two distinct steps, the first of which is the loss of the leaving groups, the second being the introduction of the nucleophile *i.e.,* the departure of the leaving group and the approach of the nucleophile are not linked. The first consequence of S_N1 reaction is that if the substrate is chiral, the product is a racemic mixture (scheme 3.26). Thus when either enantiomer of chloride (scheme 3.26) undergoes an S_N1 reaction the product is almost completely racemized, since the achiral planar intermediate carbocation is attacked both from left and right with equal probability.

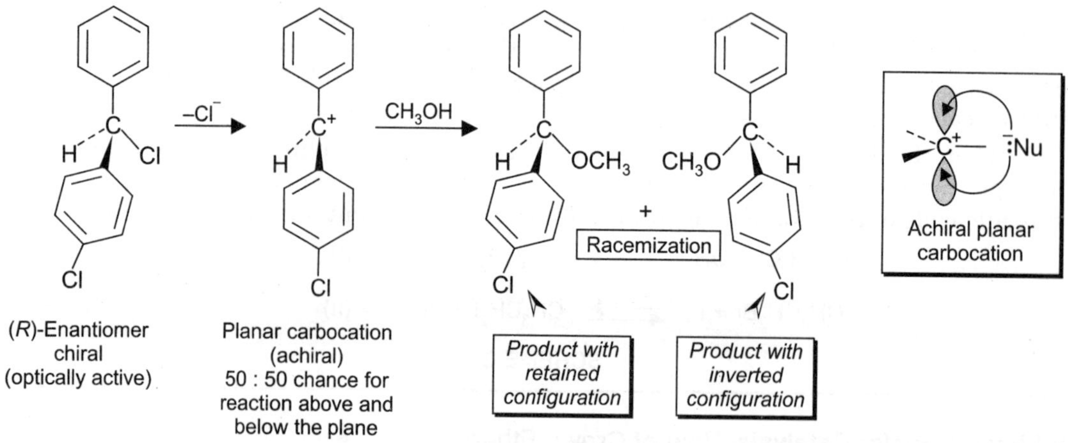

SCHEME 3.26

In the case of *tert.* Substrates *e.g., t*-butyl chloride the crowding prevents an incoming nucleophile from approaching the carbon bearing the leaving group from the rear (S_N2 conditions). Saturated aliphatic carbocations do not have either π bonds or unshared electron pairs and thus resonance effects cannot have much influence on their stabilization. However, the formation of a carbocation in the case of a 3° alkyl halide leads to relief of strain which is associated with the tetrahedral geometry and in the planar carbocation the three methyl groups adopt positions as apart from each other as possible (scheme 3.27) when the cationic carbon atom becomes more highly substituted, additional possibilities for declocalization develop. The *tert*-butyl carbocation has three hyperconjugative interactions (scheme 3.27*a*) with which to stabilize the positive charge, thus it is more stable than the methyl cation which cannot be stabilized by hyperconjugation. Generally, *the more highly substituted the carbocation, the more stable it is*. The *tert*-butyl carbocation has actually been isolated at low temperature and its crystal structure determined.

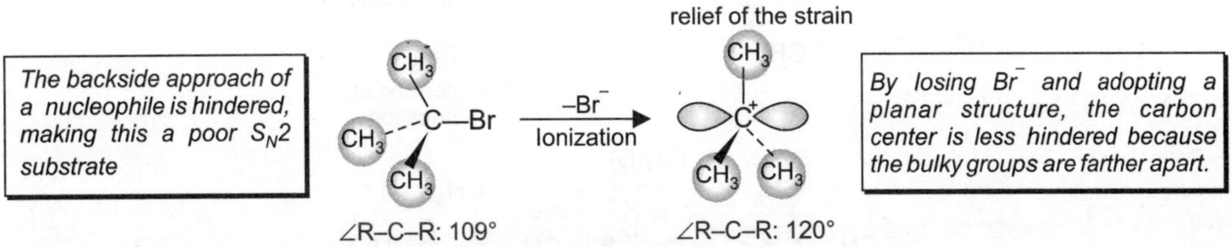

SCHEME 3.27

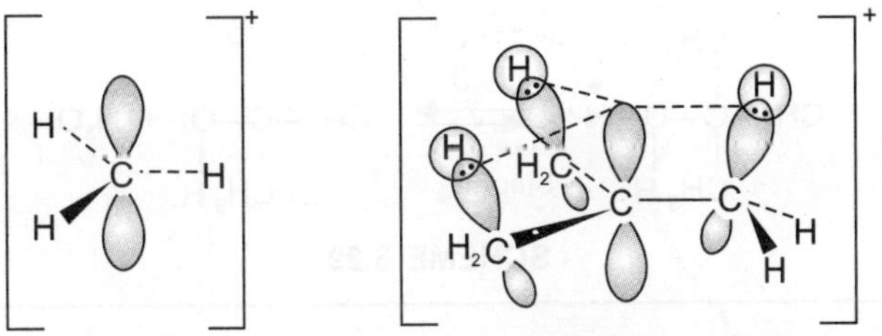

Methyl cation

1, 1-Dimethylethyl cation (*tert*-Butyl cation)

SCHEME 3.27a

(a) The Mechanism of S_N1 Reaction

S_N1 rate = k_r[R—X]

S_N2 rate = k_r [R—X] [Nuc:⁻]

SCHEME 3.28

It is a first order reaction–the rate of reaction depends on the concentration of the alkyl halide but not the concentration of nucleophile (scheme 3.28). A carbocation is formed in the slow step of an S_N1 reaction. Because a tertiary carbocation is more stable and therefore, easier to form than a secondary carbocation which in turn is more stable than a primary carbocation, therefore, on successive replacement of methyl groups of *tert*-butyl bromide by hydrogens, the rate of the S_N1 reaction decreases progressively. This is opposite to the order of reactivity exhibited by alkyl halides in S_N2 reaction.

In step 2, the carbocation *e.g., tert.* butyl cation reacts with the nucleophile, when *e.g.,* H_2O is the nucleophile *i.e.,* during solvolysis in the third step, deprotonation gives the product. The mechanism reflects three different events (*i*) rate determining dissociation of the haloalkane to give a carbocation (scheme 3.29) and a halide ion, nucleophilic attack by water on the carbocation to give an alkyloxonium ion. *Note that the loss of a proton is in fact a separate acid-base reaction that takes place after the substitution process is finished.* Each of the individual steps of the overall S_N1 reaction are in equilibrium, each one is an acid-base reaction, and that the carbocation functions are a Lewis acid in the first and second steps.

An energy diagram tracing the progress of the S_N1 reaction is shown (scheme 3.30). The rate of the reaction reflects the activation energy required to form the carbocation intermediate. The activation energy required for the second step *i.e.,* the addition of the nucleophile to the carbocation is much smaller, thus step 2 is fast. The rate of the second step has no effect on the overall rate of reaction.

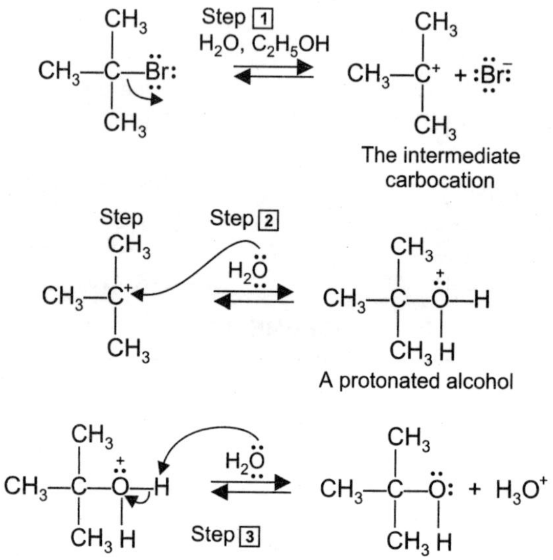

SCHEME 3.29

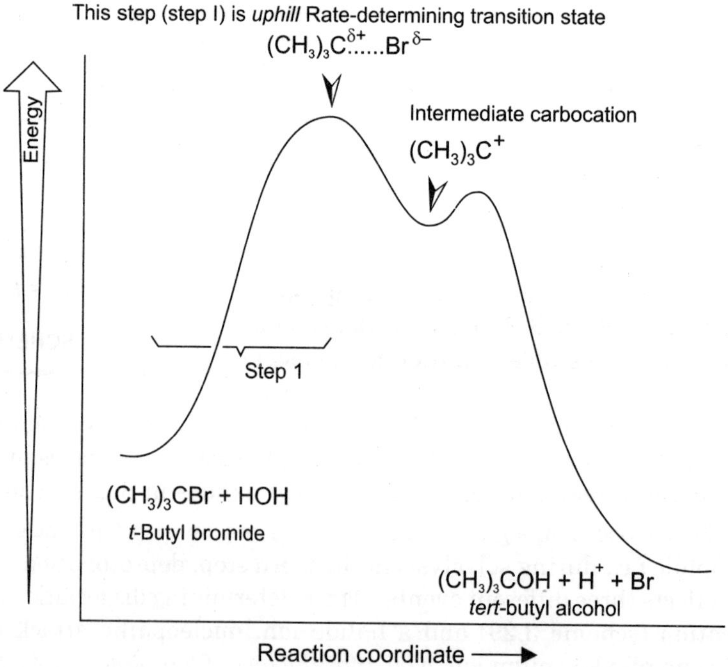

SCHEME 3.30

(b) The Effect of Structure of the Carbon Framework on S_N1 Reactions–Hammond Postulate

Carbon frameworks capable of stabilizing carbocations facilitate the S_N1 reactions, the stabilization generally increasing in the order $CH_3^+ < RCH_2^+ < R_2CH^+ \simeq C{=}C{-}C^+ < R_3C^+ < ArC^+$. Stable carbocations are planar and thus, compounds like *t*-butyl bromide (scheme 3.29) undergo a facile replacement of bromine with the formation of the cation $(CH_3)_3C^+$, since it can assume a planar configuration as required by the sp^2 hybridized central carbon.

For an S_N1 reaction to proceed at a reasonable rate, the formation of relatively stable carbocation is essential since it is then that the free energy of activation for the slow step of the reaction will be low enough for it to proceed. [Recall (scheme 3.30), this step *i.e.*, step 1 is uphill as far as free energy is concerned ($\Delta G°$ for this step is positive) as well as in terms of enthalpy ($\Delta H°$ is also positive). This step is, consequently endothermic. Hammond's postulate states– the transition state for a step which is uphill in energy should show a strong resemblance to the product of that step. Since the product of this step (an intermediate) is a carbocation, any feature that stabilizes it such as dispersal of the positive charge by electron-releasing groups (hyperconjugation) should also stabilize the transition state where the positive charge is generated (scheme 3.31)].

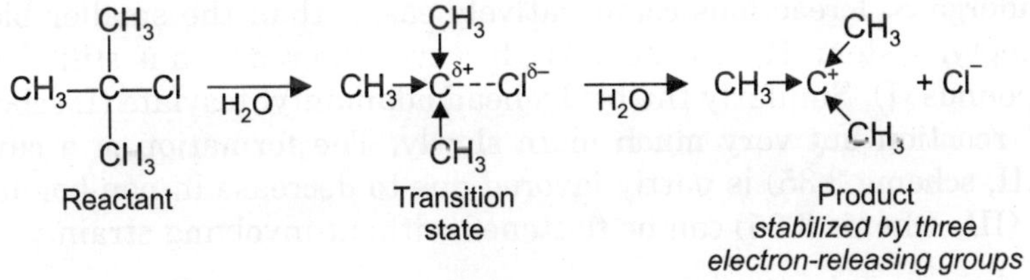

SCHEME 3.31

The carbocation intermediates also get stabilization from delocalization of their positive charge in to adjacent unsaturated groups. Thus benzylic and allylic halides undergo S_N1 reactions more rapidly than the related systems in which such carbocation stabilization is not available.

These arguments explain as to why the reaction of 3-bromocyclohexene (scheme 3.32) with methanol (CH_3OH) is faster than the reaction of bromocyclohexane with methanol. Both the substrates are secondary halides, the reaction with methanol, a neutral nucleophile, will tend to occur by an S_N1 process. Although both carbocations are secondary, the one derived from 3-bromocyclohexene is also a resonance-stabilized allylic carbocation. Its resonance stabilization increases the rate of its formation. No such stabilization occurs in the reaction of bromocyclohexane (scheme 3.32).

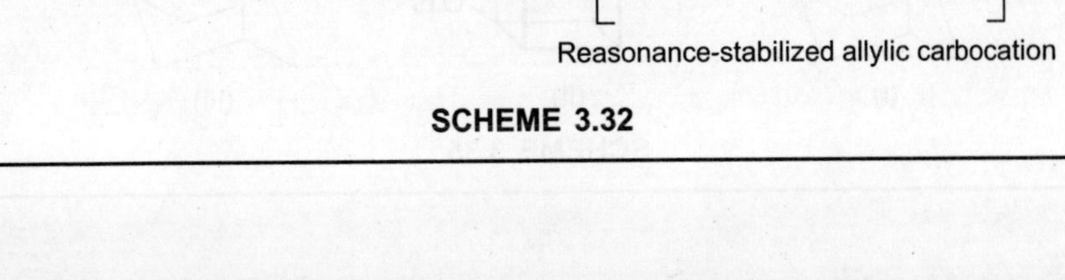

SCHEME 3.32

Consider the acetolyses of *cis*- and *trans*-4-*t*-butyl cyclohexyl tosylates (I and II, scheme 3.33). In the case of *cis* compound I, the tosyl group is in the axial position and thus, suffers from steric 1:3 non-bonded interactions with the axial hydrogens. This leads to a faster reaction than with *trans*-equatorial isomer II by a factor of about 3.4.

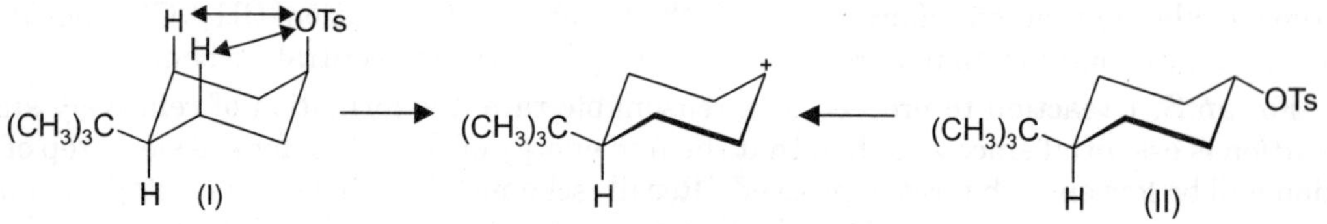

SCHEME 3.33

The S_N1 reactions fail or occur very slowly, if at all, at the bridgehead positions. S_N1 reactions proceed via planar carbocations and bridgehead carbons cannot become the seat of carbocations (S_N2 reactions fail, since the attack of the nucleophile from the rear is hindered). In case however, if the rings are large enough *e.g.,* in bridgehead compounds the strain due to the formation of the carbocation is somewhat relieved. Thus, [2.2.2] bicyclic systems *e.g.* (III, scheme 3.34) undergo S_N1 reactions comparatively easier than the smaller bicyclic systems *e.g.,* the bicycloheptyl system (II, scheme 3.34), however the reaction is still slower than with open chain compounds (I). Similarly the bridgehead adamantyl tosylate (I, scheme 3.35) does undergo an S_N1 reaction but very much more slowly. The formation of a carbocation at a bridgehead in (III, scheme 3.35) is partly favored due to decrease in non-bonded interaction. The carbocation (III, scheme 3.35) can be flattened without involving strain.

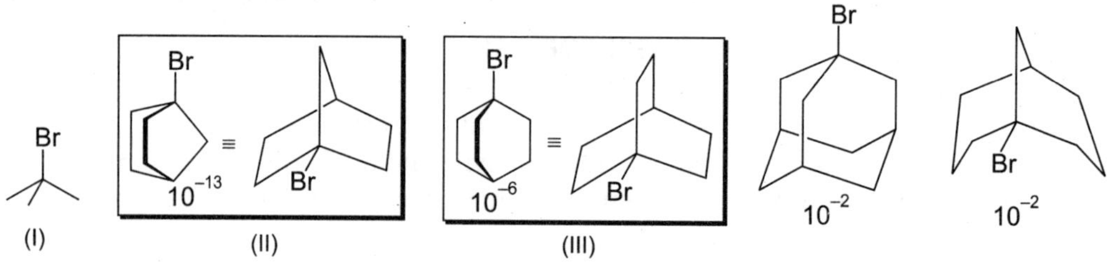

Relative rates of solvolysis of *t*-butyl and bridgehead bromides in H_2O-EtOH

SCHEME 3.34

> *An S_N2 reaction cannot occur and is completely ruled out at a bridgehead position like (I, scheme 3.35, the nucleophile cannot attack from the rear, and the carbon bearing the leaving group cannot be inverted). Very interestingly the cubyl tosylate (II, scheme 3.35) undergoes an S_N1 reaction much faster than expected when compared to the bridgehead adamantyl tosylate and in this compound the CCC bond angles are 90°.*

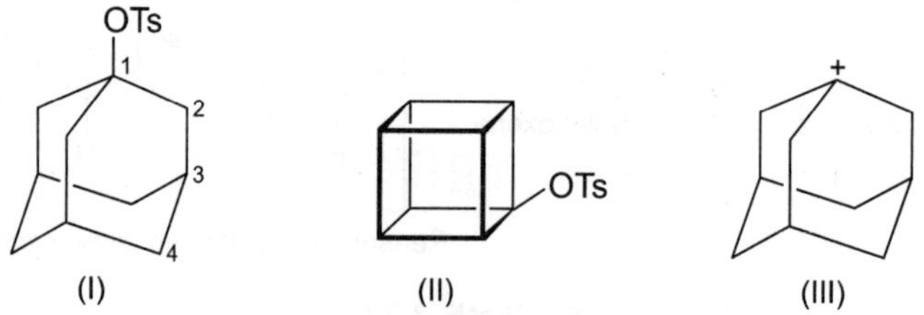

SCHEME 3.35

Interestingly enough, some nucleophilic substitution reactions which normally occur through carbocations do take place at [2.2.1] bridgeheads, provided the leaving group is of the type that it cannot itself function as a nucleophile. In the case of substrate (I, scheme 3.36) chlorobenzene is the nucleophile.

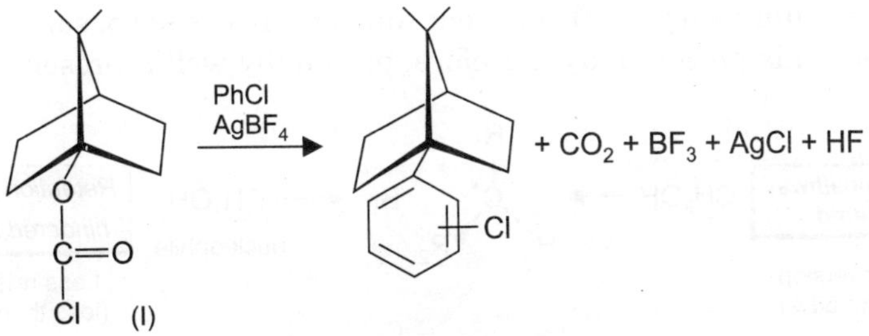

(I)

SCHEME 3.36

(c) Ion Pairs in S_N1 Mechanism

Since the carbocation carbon atom has sp^2 hybridization, it is expected to be planar, and therefore achiral. Equal amounts of enantiomeric products are expected during the reaction of a chiral substrate by attack at either side of the plane with equal probability resulting in complete racemization. Although many S_N1 reactions result in complete racemizations, some others do not. It is found that during many S_N1 reactions there is 5 to 20% inversion and in few cases a small amount of retention of configuration has been observed. These results have been explained by invoking the formation of ion pairs. Thus the S_N1 reactions proceed as shown (scheme 3.37). The following points may be noted.

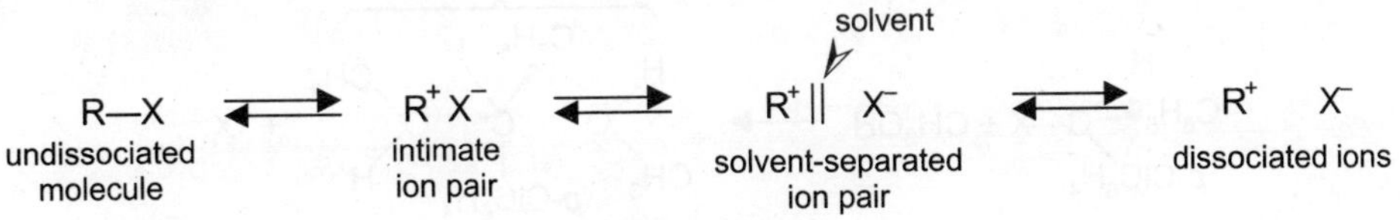

SCHEME 3.37

- In an intimate ion pair (also called contact or tight ion pair) R^+ does not behave like the free cation of the dissociated species (scheme 3.37). There seems to be significant bonding between R^+ and X^- and the asymmetry is well maintained *i.e.,* individual ions retain their original stereochemical configuration. Thus during the course of S_N1

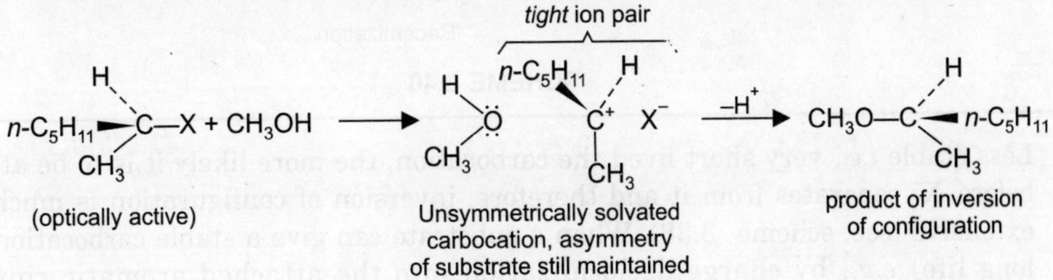

SCHEME 3.38

reaction X^- solvates the cation from the side it departs from, with solvent molecules near the intimate ion pair can solvate it only from the opposite side. The situation is like an unsymmetrically solvated carbocation, where the nucleophilic attack by a solvent molecule (on tight ion pair) gives inverted product. (scheme 3.38).

- In Summary when strong ion pairing exists, one side of the carbocation is shielded from the incoming nucleophile (solvent during solvolysis) so, some enantioselectivity is observed. This process may therefore, be equally well represented (scheme 3.39).

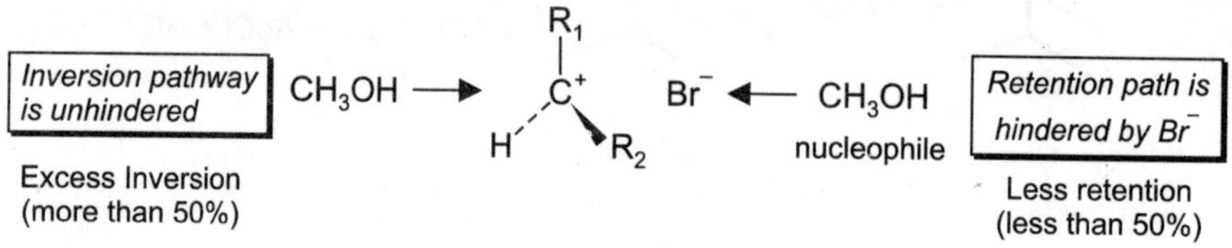

SCHEME 3.39

- If the nucleophile attacks carbocation of the solvent separated ion pair, the stereochemistry of the carbocation is not now maintained as tightly. The leaving groups is still in a position to partially block the approach of the nucleophile to that side of the carbocation. Thus more of the product with inverted configuration will be formed, leading to more racemization (perhaps total).
- If the free R^+ is formed, it is planar and the attack by nucleophile *e.g.*, solvent during solvolysis occurs with equal probability form either face to give complete racemization via symmetrically solvated carbocation (scheme 3.40). One may describe this carbocation as prochiral.

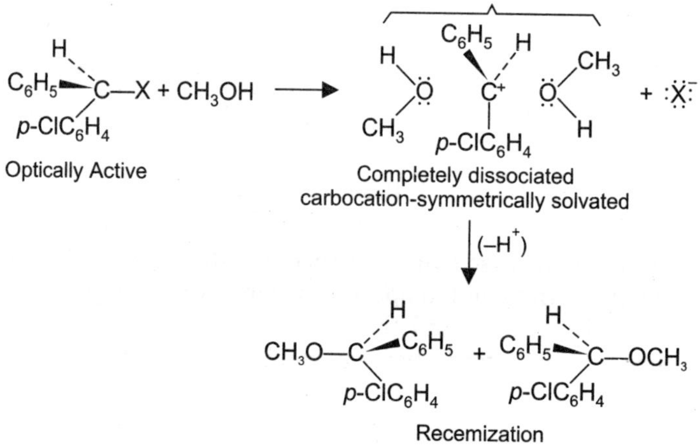

SCHEME 3.40

- Less stable *i.e.*, very short lived the carbocation, the more likely it is to be attacked before X^- separates from it and therefore, inversion of configuration is much more extensive (see, scheme 3.38). When a substrate can give a stable carbocation (with long life) *e.g.*, by charge declocalization into the attached aromatic ring (see, scheme 3.40), the anion hardly makes any contribution towards its stability.

During the hydrolysis of 1-chloro-1-phenylethane (scheme 3.41) the benzene ring stabilizes the intermediate carbocation I and as a consequence the optically active starting material undergoes a predominant (98%) racemization. On replacing the benzene ring by a cyclohexane ring the ion (II, scheme 3.41) formed in the reaction derives most of its stability from the intimate ion pair. Thus, it does not readily ionize to the free ion. Therefore, the product originates mostly by the nucleophilic attack on the ion pair, leading to almost exclusive inversion of configuration.

- The overall balance between racemization and inversion is very sensitive to several factors such as solvent and negative ion and the above two reactions provide extreme situations, most reactions coming between the two.
- In some cases, S_N1 reactions have been found to proceed with partial retention (20 to 50%) of configuration. Ion pairs have been involved to explain for example, the phenolysis of optically active α-phenylethyl chloride, an ether with retention of configuration is obtained and involves a four center mechanism (scheme 3.42).
- It has been suggested that nucleophilic attack on an intimate ion pair can afford inverted product by mechanism that is borderline between S_N1 and S_N2 reaction.

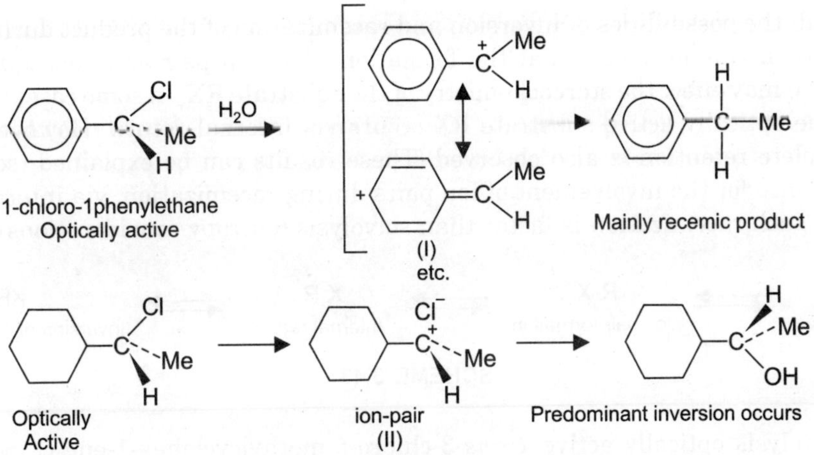

SCHEME 3.41

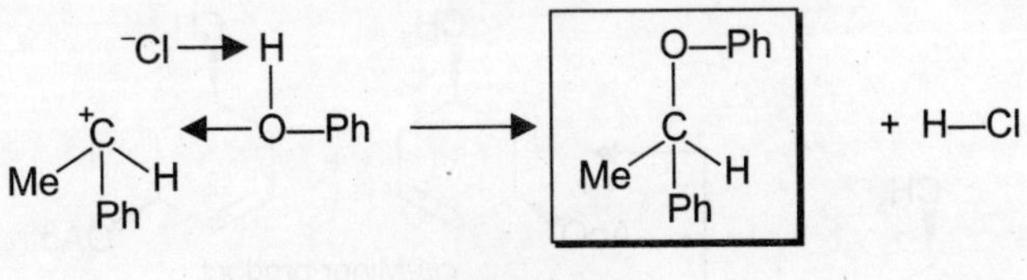

SCHEME 3.42

- The ion associated with another ion of opposite charge is termed counter ion or "gegenion" from the German meaning counter. Thus *e.g.*, in the case of a tight ion pair R^+X^-, R^+ is the counter ion of X^- and vice versa. A reaction in which the intimate ion pair recombines to afford the original substrate is called an internal return.
- *The special salt effect.* The effect produced by addition of an ion, that is generated in a reaction is termed common ion effect which leads to depression in the rate. Thus

e.g., the addition of chloride ion decreases the rate of solvolysis of diphenylmethyl chloride, the first step of the reaction $RCl \rightleftarrows R^+ + Cl^-$ is the rate determining. The presence of Cl^- from another source is subject to mass-law effect on the equilibrium and consequently on the overall rate of reaction.

- A change in the rate of a reaction due to the addition of a salt (electrolyte) that generally does not have an ion in common with reactants or product is termed *special salt effect*.
- The addition of a salt changes, the ionic strength of the solution which in turn affects the free energies of the ions which are being formed or destroyed. For example, addition of $LiClO_4$ (or LiBr) generally increases the rate of an S_N1 reaction (ions being formed $RX \rightleftarrows R^+ + X^-$ in the rate determining step). An increase in the ionic strength of the solution favors their formation. One also looks at these results where the ClO_4^- (or Br^-) traps the solvent separated ion pair to afford $R^+ \parallel ClO_4^-$ and these species being unstable under these conditions give back the product. Thus, the amount of solvent separated ion pair that would have returned to the starting material is reduced leading to an increase in the overall reaction.

  Recall the possibilities of inversion and racemization of the product during solvolysis reaction. One may note that the formation of an ion pair and subsequent internal return may effect the stereochemistry of the substrate RX. In some cases racemization of the optically active substrate RX occurs via internal return. However, partial or complete retention is also observed. These results can be explained (scheme 3.43). Evidence for the involvement of ion pairs during racemization, via internal return is that such racemization is faster than solvolysis (entropy considerations).

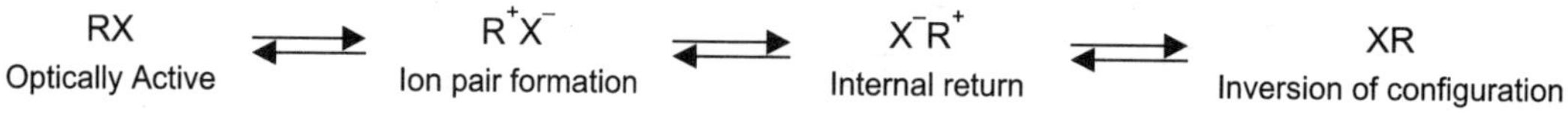

SCHEME 3.43

On acetolysis optically active *trans*-3-chloro-5-methylcyclohex-1-ene (I, scheme 3.44) afforded racemic mixture of *cis*- and *trans*-acetates. This is due to the fact the intermediate allylic cation possesses a plane of symmetry. The rate at which optical activity was lost in this

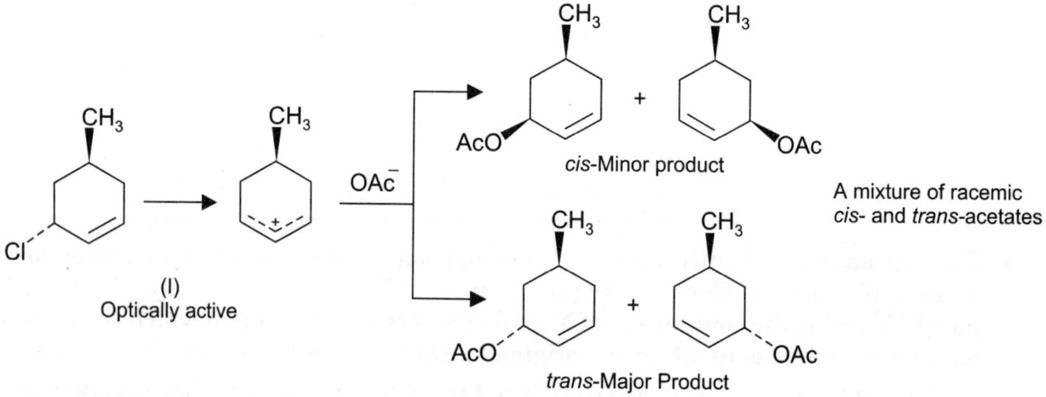

SCHEME 3.44

reaction was found to be faster by a factor of about four. Additionally, it was shown that the chloride racemized without even undergoing solvolysis (internal return) and the loss of its optical activity occurs within an ion pair (scheme 3.45), the product being the racemic *trans*-3-chloro-5-methyl-cyclohex-1-ene. This is so since in the tight ion pair the ions are in close contact and therefore, prevent that movement of the chloride ion from one face of the cyclohexane ring to the other.

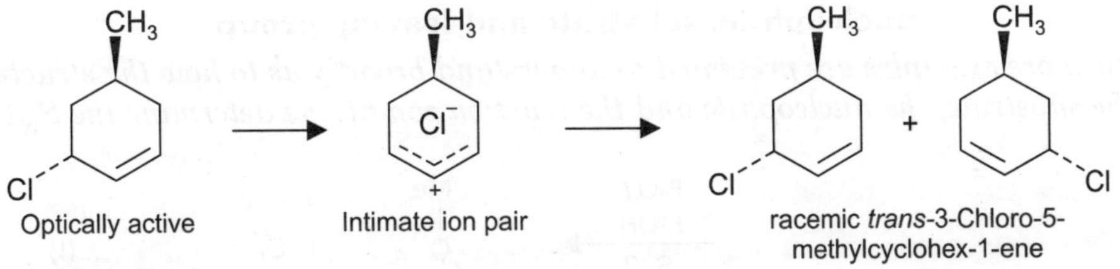

SCHEME 3.45

(d) Role of Solvent

In contrast to the use of aprotic solvents for S_N2 reactions, S_N1 reactions are enhanced by polar and protic solvents, due to the ability of a protic solvent to stabilize both cations and anions. The transition state leading to the intermediate carbocation and halide ion is stabilized more than the reactants (scheme 3.46), thus the free energy of activation is lower. The transition state for this endothermic step is, where separated charges are getting developed and therefore, it assembles the ions that are ultimately formed.

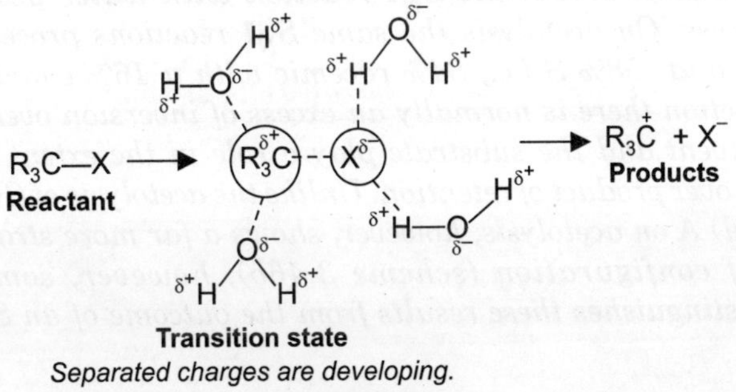

SCHEME 3.46

(e) Effect of Leaving Group and Nucleophile on S_N1 Reactivity (a Comparison with S_N2 Reaction)

In an S_N1 reaction the carbocation formed in the slow step reacts rapidly in the second step with any uncleophile present in the reaction mixture. Thus increasing the rate of the fast step has no effect on the rate of the prior slow carbocation forming step. The rates of S_N1 reactions are unaffected by either the concentration or the identity of the nucleophile. A good nucleophile (HO^-, CH_3O^-) however, favors an S_N2 reaction over an S_N1 reaction.

An S_N1 reaction is favored by a poor nucleophile (H_2O, CH_3OH) not by increasing its rate , but by depressing the competing S_N2 reaction. Thus in summary, an S_N2 reaction is favored by a high concentration of a good nucleophile, while an S_N1 reaction is favored either by low concentration of a good nucleophile or by a poor nucleophile.

A closer look at S_N1 and S_N2 reactions–nature of solvent, nucleophile, substrate and leaving group

A few more examples are presented to understand broadly as to how the structure of the substrate, the nucleophile and the reaction conditions determine the S_N1 or

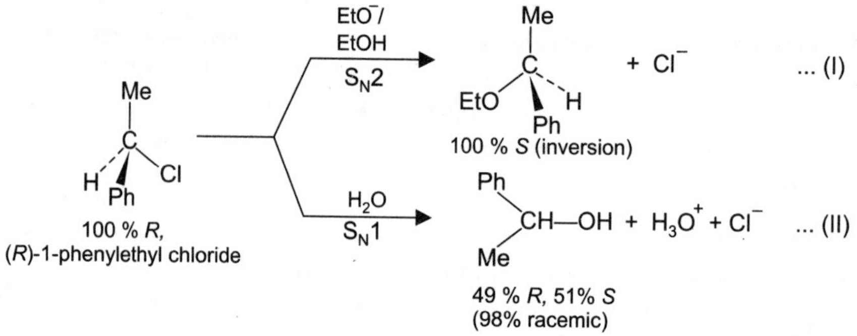

SCHEME 3.46a

S_N2 mechanism for a substitution reaction and how these conditions affect the reactivity within each mechanism. Thus compared to 100% inversion in the S_N2 mechanism (I, scheme 3.46a) the S_N1 reaction with water proceeds with 2% enantiomeric excess. On acetolysis the same S_N1 reactions procedces an acetate mixture (42% R and 58% S i.e., 84% racemic with a 16% enantiomeric excess. Thus in S_N1 reaction there is normally an excess of inversion over retention. The nature of the solvent and the substrate play a role in the extent of formation of inverted product over product of retention. Unlike the acetolysis of (R)-1-phenylethyl chloride, (achiral) A on acetolysis, however, shows a far more stronger preference for inversion of configuration (scheme 3.46b), however, some retention of configuration distinguishes these results from the outcome of an S_N2 reaction.

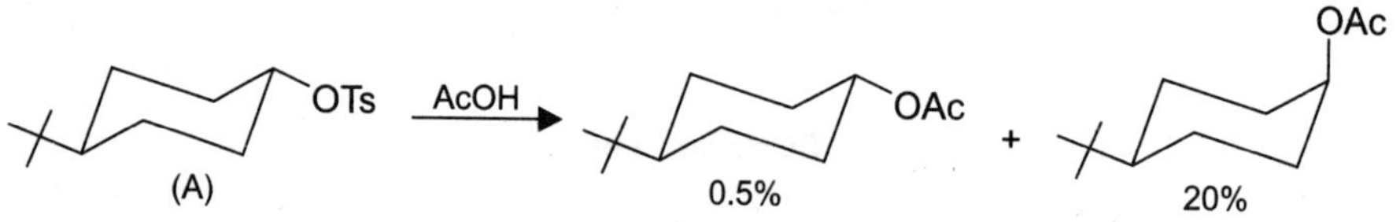

One may also compare the results with (problem 3.14).

SCHEME 3.46 *b*

(f) The Nature of Leaving Group

Both in S_N2 as well as in an S_N1 reaction, the leaving group begins to acquire a negative charge as the transition state is reached. If this developing negative charge at the leaving group is stabilized, this also stabilizes the transition state *i.e.,* lowers its free energy.

Consequently the free energy of activation is lowered and the rate of reaction is increased. The relative abilities of various groups to leave carbon in S_N1 reactions are roughly the same as already discussed for S_N2 reactions. Of the halogens iodide ion is the best leaving group while fluoride ion is the poorest. Similarly *p*-toluenesulfonate ion OTs^- and triflate ion (trifluoromethanesulfonate ion $CF_3SO_3^-$) are very good leaving groups.

(g) Molecular Rearrangements During Nucleophilic Substitutions

A carbocation rearranges provided it becomes more stable as a result of the rearrangement, since carbocations are not involved in S_N2 reactions, therefore, different constitutional isomers as products can be formed from S_N1 and S_N2 reactions. This is seen in the case of neopentyl iodide (scheme 3.47 these systems however, undergo slow S_N1 and S_N2 reactions) which under S_N1 and S_N2 conditions gives different products. The carbocations during an S_N1 reaction can also undergo elimination to give an alkene. Thus S_N1 reaction gives a complicated mixture of products and is thus a synthetic chemists nightmare, while S_N2 reaction is a friendly reaction for synthetic organic chemist.

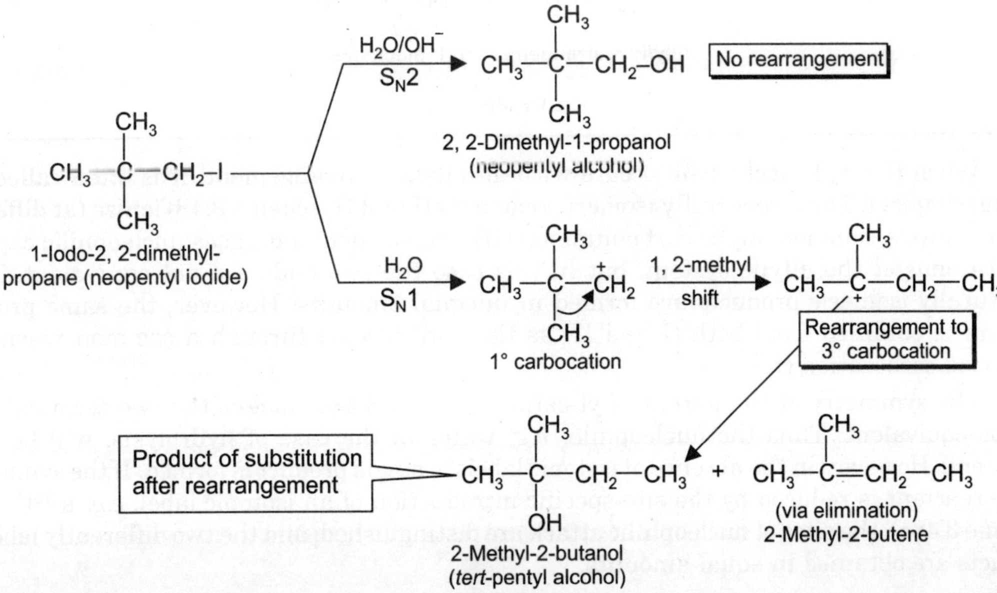

SCHEME 3.47

3.4 NUCLEOPHILIC SUBSTITUTION OF ALLYLIC, SYSTEMS S_N1' AND S_N2' REACTIONS. REARRANGEMENT IN ALLYLIC SYSTEMS

Generally primary alkyl halides undergo substitution by the S_N2 mechanism and do not undergo S_N1 reactions. However, a primary allyl halide is very reactive in an S_N1 reaction *e.g.*, an allyl halide is more than 30 times reactive than an ethyl halide. The reason for the enhanced reactivity lies in the resonance stabilization of the carbocation and the transition state leading to the carbocation as shown for structurally isomeric reactants (scheme 3.48). Carbocations are stabilized by dispersal of the positive charge. Inductive stabilization involves dispersal of the positive charge through sigma bonds, resonance-stabilization involves the dispersal of the positive charge by *pi*-bonds.

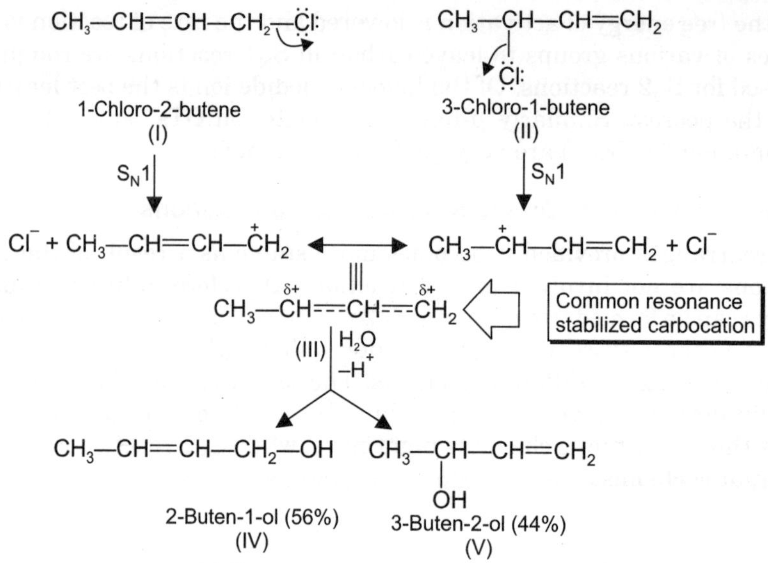

Allylic rearrangement-S_N1' mechanism

SCHEME 3.48

When the S_N1 mechanism occurs with an allylic rearrangement, it is often called the S_N1' mechanism. The structurally isomeric reactants (I and II, scheme 3.48) ionize (at different rates) to give a common allylic carbonium ion (III) which then undergoes nucleophilic capture at both ends of the allylic system, but in this case the two ends are not equivalent. Thus structurally isomeric products are formed in unequal amounts. However, the same product mixture is obtained from both (I and II) as they are formed through a common resonance stabilized carbocation (III).

The symmetry of the parent allyl cation (scheme 3.48*a*) makes the two terminal CH_2 groups equivalent. Thus the nucleophile, *e.g.* water in the case of hydrolysis, will bond to either end. However, in the absence of isotopic labels, a single product is formed. If the symmetry of the reactant is reduced by the site-specific introduction of an isotopic label, *e.g.* a ^{13}C atom (scheme 3.48*a*), the sites of nucleophilic attack are distinguished, and the two differently labelled products are obtained in equal amounts.

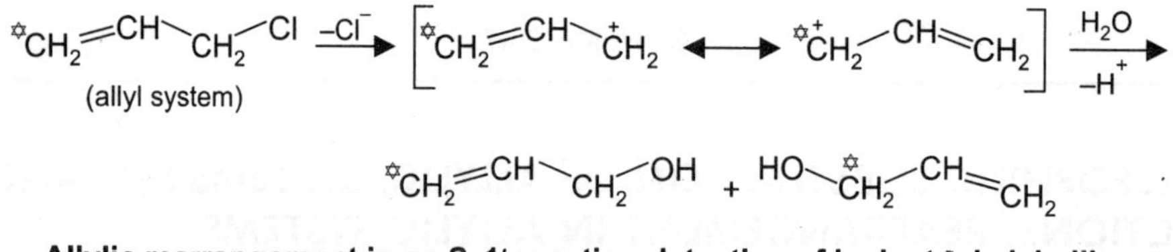

Allylic rearrangement in an S_N1' reaction detection of by isotopic labelling

SCHEME 3.48a

S_N2' is substitution nucleophilic bimolecular with rearrangement, is a concerted nucleophilic displacement where the site of attack is at an atom other than the original point of attachment of the leaving group generally one multiple bond separated from the original point of attachment. This mechanism is kinetically similar with that of the S_N2 reaction and the S_N2' reaction normally occurs exclusively when the S_N2 process is sterically hindered (scheme 3.49).

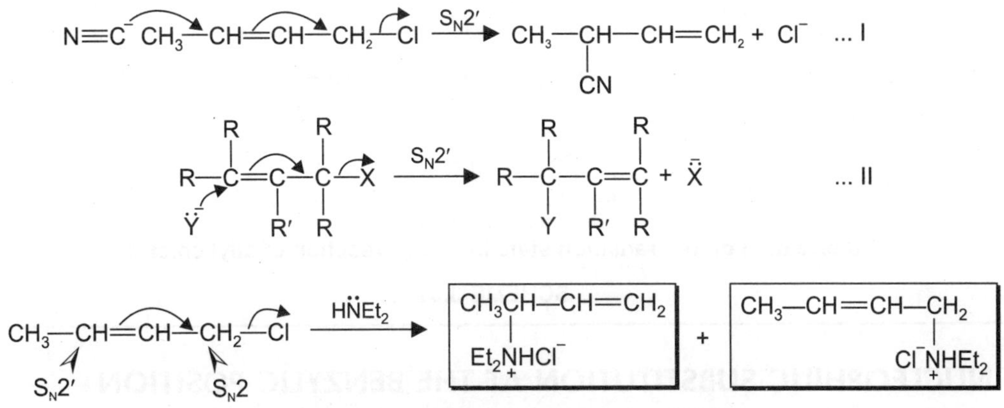

Examples of S_N2 and S_N2' mechanisms

SCHEME 3.49

S_N1' and S_N2' Mechanisms

The S_N1 and S_N1' mechanisms involve via a single common allylic intermediate which is captured by the nuclcophilc at alternative sites. This nucloeophilic capture is regioselective if the two ends of the allylic carbonium ion are not same. However, the S_N2 and the S_N2', are different parallel competing reactions. They occur in relative amounts according to the site selectivity of the given nucleophile for a particular substrate.

Stereochemically, S_N2' reactions are mostly syn, the leaving group and the nucleophile are on the same face of the molecule (scheme 3.50). The cis isomer (I, scheme 3.50) gave cis isomer II with piperidine.

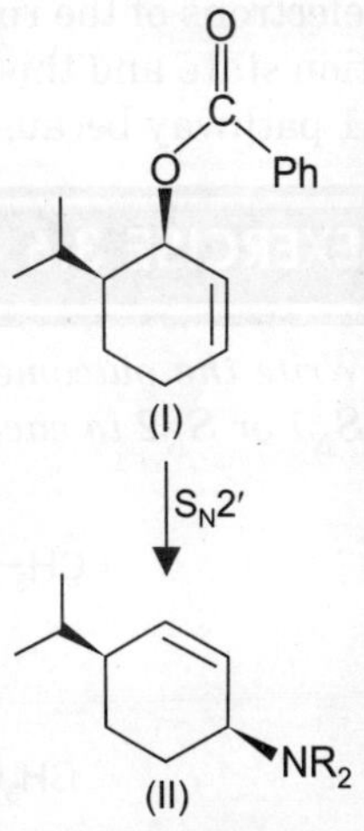

SCHEME 3.50

3.5 NUCLEOPHILIC DISPLACEMENTS AT ALLYLIC HALIDES/TOSYLATES

Allyl halides also undergo S_N2 reaction at faster rates than primary alkyl halides or even methyl halides. The reason for the greater S_N2 reactivity of allyl halides is that the allyl *pi* bond *pi* cloud reduces the energy of the transition state of an S_N2 reaction. In the transition state, the carbon undergoing reaction changes from the sp^3-hybrid state to the sp^2-hybrid state and has a *p*-orbital. This *p*-orbital forms partials bonds with both the incoming nucleophile and the leaving group. The entire grouping of atoms (transition state) have a negative charge. Adjacent *p*-orbitals, as in an allylic group, undergo, partial overlap with the transitional *p*-orbital. Thus, adjacent *p*-orbitals help delocalize the negative charge and thus lower the energy of the transition state (scheme 3.51). Unlike S_N1 reactions on allylic systems which are prone to allylic shift of the double bond (low synthesis utility), allylic halides however, react cleanly (without rearrangement) by way of the S_N2 mechanism.

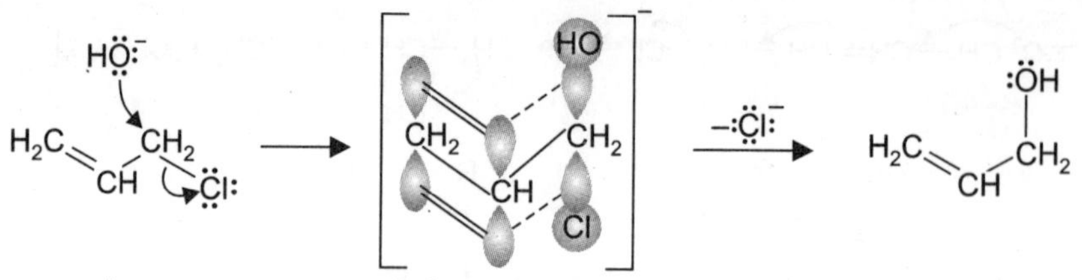

S_N2 *transition state*

Stabilization of the transition state in an S_N2 reaction of allyl chloride

SCHEME 3.51

3.6 NUCLEOPHILIC SUBSTITUTION AT THE BENZYLIC POSITION

Benzylic halides readily undergo S_N2 reactions unless they are tertiary. Tertiary benzylic and allylic halides, like other tertiary halides, are unreactive in S_N2 reactions due to steric hindrance. This high reactivity during S_N2 displacement on the benzylic halide is due to the fact that the *p* orbital that partially bonds with the nucleophile and the leaving group also overlaps with the *pi* electrons of the ring (scheme 3.52). This stabilizing conjugation lowers the energy of the transition state and this increases the reaction rate. Benzylic compounds also react rapidly by the S_N1 pathway because of the relative stability of benzyl cation (scheme 3.52).

EXERCISE 3.6

Write the outcome of the reactions (scheme 3.51a) and predict the reaction to be S_N1 or S_N2 in each case.

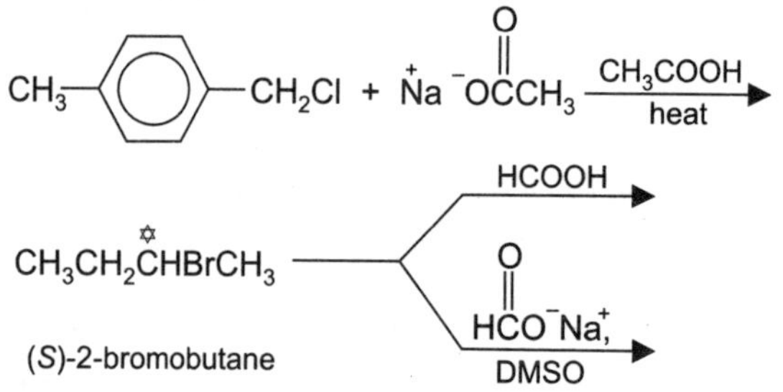

(I)

(II)

(S)-2-bromobutane

(III)

SCHEME 3.51a

ANSWER. *(I) The acetate ion is a weak nucleophile, however benzylic halide is a reactive substrate. The substitution can occur either by an S_N1 or S_N2 path to give*

$CH_3{-}C_6H_4{-}CH_2{-}O\overset{O}{\overset{\|}{C}}CH_3$.

(II) Racemization will occur via an S_N1 reaction, the solvent HCOOH (a carboxylic acid) is highly polar and protic the product being $CH_3CH_2CH(O\overset{O}{\overset{\|}{C}}H)CH_3$.

(III) The product will be the (R) enantiomer $CH_3CH_2CH(O\overset{O}{\overset{\|}{C}}H)\ CH_3$ formed via S_N2 reaction since now the nucleophile is good and the solvent is aprotic.

3.7 NUCLEOPHILIC SUBSTITUTION OF VINYLIC AND ARYL HALIDES

Vinylic halides *i.e.,* haloalkenes in which the halogen is directly attached to the double bond and aryl halides do not show reactivity in S_N1 and S_N2 reactions. S_N1 reactions of these halides are so slow that other kinds of reactions occur instead, such as addition to the multiple bond. The failure to undergo S_N2 reactions by these systems is due to the fact that the approach of the nucleophile to the back of sp^2 carbon is repelled the π electron cloud of double bond or the aromatic ring (scheme 3.53).

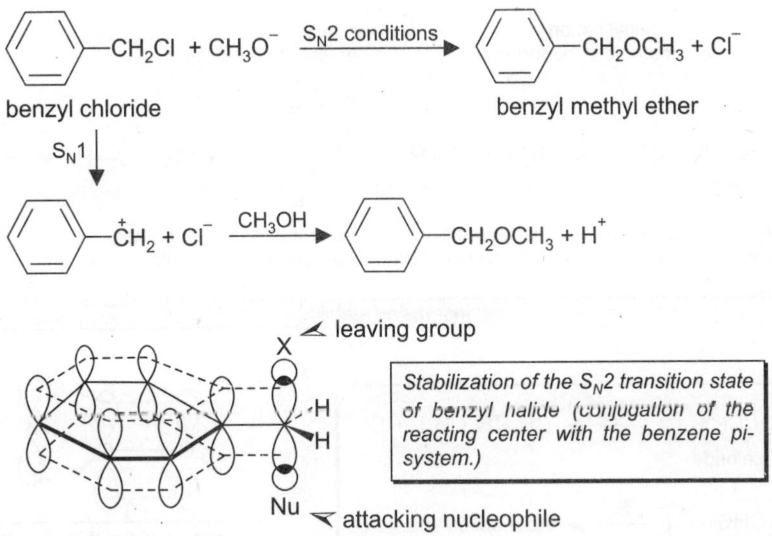

SCHEME 3.52

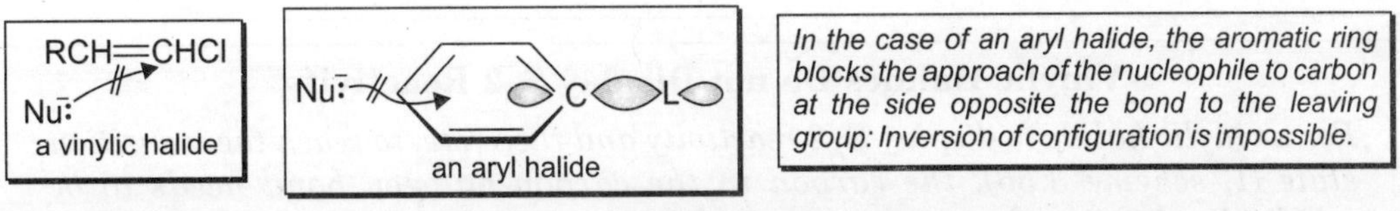

SCHEME 3.53

The difference between the energy needed to form a vinyl cation and that needed for a simple primary carbocation is comparable to the difference between primary and secondary carbocations. Secondary carbocations are common intermediates in many reactions but simple primary carbocations are practically unknown in solution. Primary vinyl cations are similarly not known is S_N1 reactions. The decreased stability of a vinylic cation (scheme 3.54) is due to the positive charge being on a more electronegative carbon. The positive charge in a vinylic cation is on an *sp* carbon, while the positive charge on an alkyl cation is on an sp^2 carbon; *sp* carbons are more electronegative than are sp^2 carbons, and hyperconjugation is less effective in stabilizing the charge on a vinylic cation.

This lack of reactivity is also explained by an increased difficulty in removing an atom with its pair of electrons from a bond to a vinyl orbital with its higher *s*-character than from a simple primary sp^3-orbital. The sp^2 carbon orbital involved in the vinyl halide bond (scheme 3.54) is expected to produce a shorter and stronger bond than the ethyl sp^3-orbital. However, an additional factor leading to a still shorter and stronger bond is π overlap between the *p* orbital of the double bond and a lone-pair orbital of the halogen as shown. This type of π-overlap can also be represented by resonance involving Lewis structures. As a result of this overlap the C—X bond in a vinyl halide attains partial double bond character.

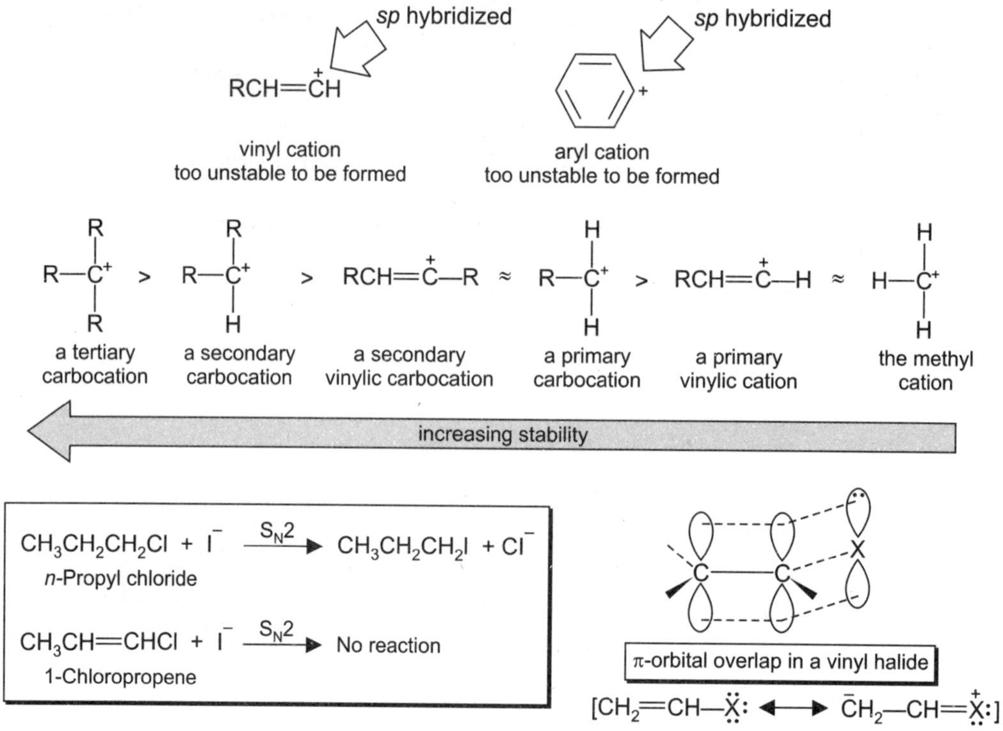

SCHEME 3.54

Vinylic Halides Do not Display S_N2 Reactivity

For a vinylic halide to display S_N2 reactivity and therefore, to reach the transition state (I, scheme 3.55), the carbon in the carbon-halogen bond needs to be rehybridized from sp^2 to sp.

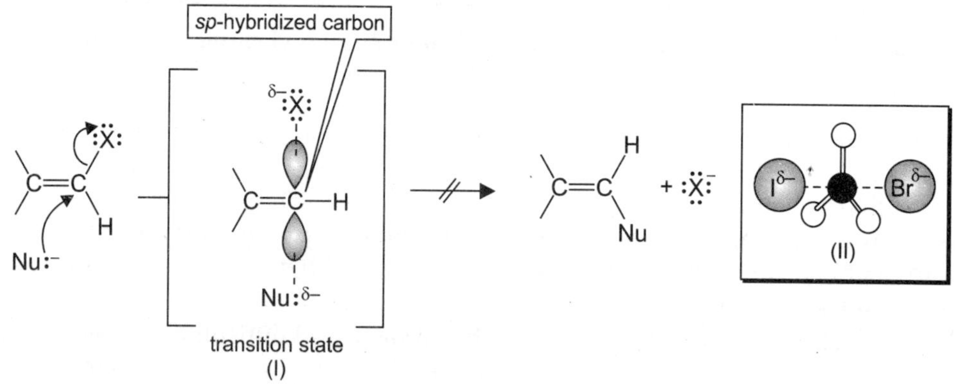

SCHEME 3.55

In an S_N2 reaction of an alkyl halide on the other hand, the hybridization change is sp^3 to sp^2 (II, scheme 3.55). The sp hybridization state has very high energy and conversion of an sp^2-hybridized carbon into an sp-hybridized carbon needs around 21 kJ/mol (5 kcal/mol) more energy. This reduces the rate of S_N2 reaction of vinylic halides.

Secondly the attacking nucleophile (scheme 3.55) would have to approach the vinylic halide at the backside of the halogen-bearing carbon and in the plane of the alkene. This would lead to clash between the nucleophile, leaving group and the alkene substituents R (Scheme 3.55a). This will tend to raise the energy of the transition state, and thus both hybridization and steric effects retard the S_N2 reaction to a point of no reactivity.

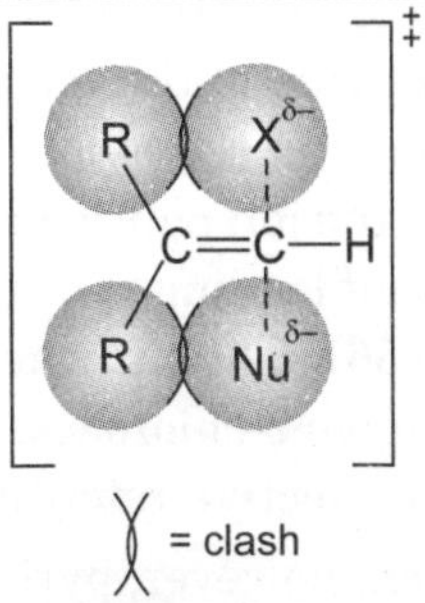

SCHEME 3.55a

3.8 THE S_Ni MECHANISM

Reaction of an alcohol with thionyl chloride converts the OH group into a good leaving group. Moreover, the reagent also provides a nucleophile (halide ion) to replace the leaving group. The oxygen atom of the alcohol replaces one of the chlorine atoms attached to sulfur producing an equivalent of HCl. In a solvent like diethyl ether most of the HCl formed during conversion of the alcohol into chlorosulfite is lost (HCl is a gas) and chloride for the substitution step comes

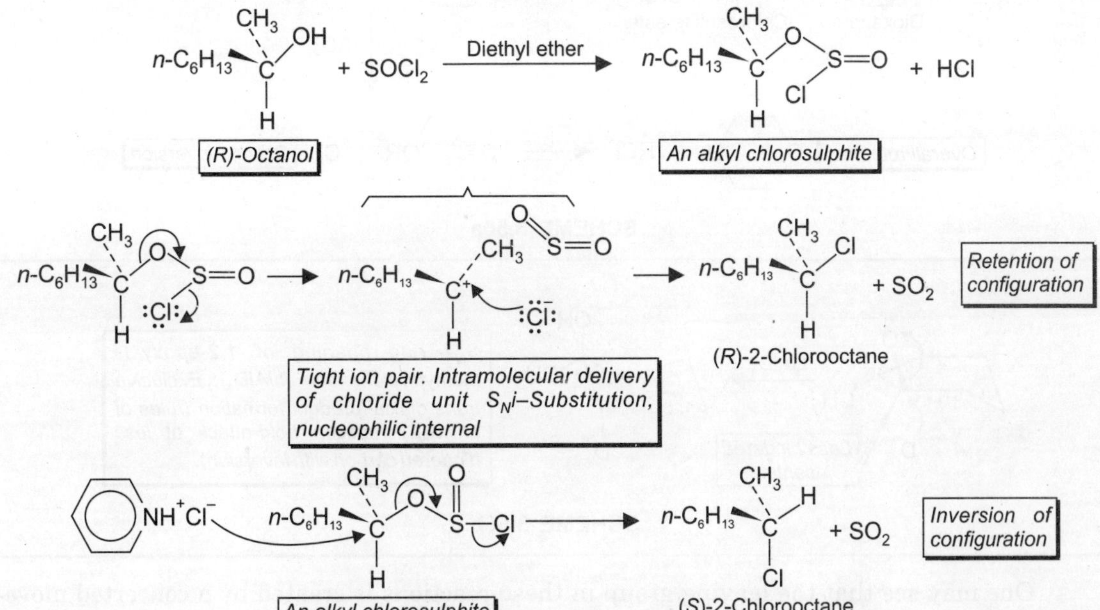

SCHEME 3.56

from the decomposition of the chlorosulfite. As the sulfur dioxide departs, the chloride is oriented at the same side of the reacting carbon atom. In the nonpolar ether, charge separation is also unfavorable and the chloride and carbocation are attracted to each other in a tight ion pair. Collapse of the ion pair leads to the product, when part of the leaving group attacks necessarily from the front leading to retention of configuration in a uncommon process called S_Ni mechanism (scheme 3.56). Thus in the presence of ether optically active 2-octanol reacts with thionyl chloride to give 2-chloroactane with retained configuration. The reaction is then said to proceed with stereoselective retention of configuration.

When, however, pyridine is used as a solvent pyridinium hydrochloride is formed which ensures that chloride ions exist in the reaction medium. The attack by the chloride ion on the electrophilic carbon atom of the alkyl chlorosulfite occurs from the back side by an S_N2 mechanism if the alcohol is 1° or 2° to give the product of inversion of configuration (scheme 3.56). These reactions thus disclose as to how reaction conditions can affect the stereochemical outcome of a reaction.

Compared to ether, tetrahydrofuran or dioxane can solvate a charge better because the ethyl substitutents of diethyl ether provide steric hindrance, making it difficult for the nonbonding electron of the oxygen to approach the compound to be solvated. When the reaction is done in dioxane solution, an oxonium ion with inversion is formed from the solvent and the chlorosulfite ester. The oxonium ion then undergoes substitution by chloride. Two inversions are involved so that the result is overall retention (scheme 3.56*a*).

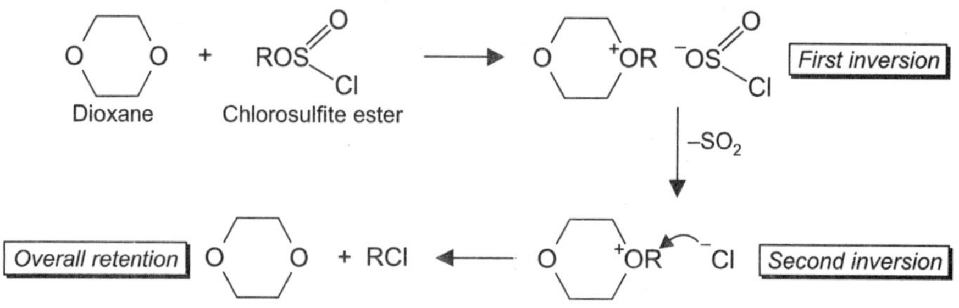

SCHEME 3.56a

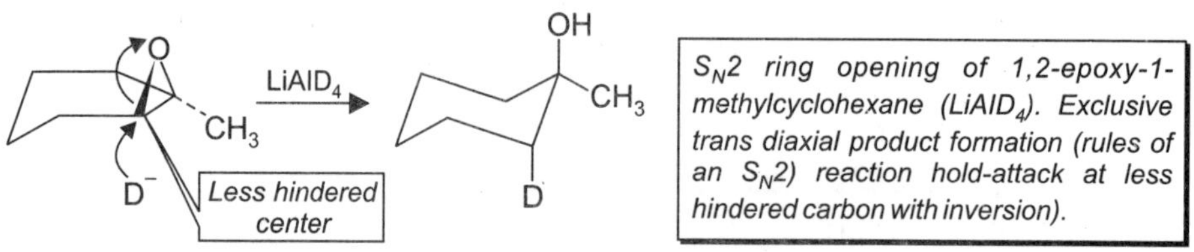

SCHEME 3.56b

One may see that the leaving group in these reactions is created by a concerted movement of electrons which produce the very stable sulfur oxygen double bond of SO_2 and displacement of chloride ion from the ester group. As expected the chlorosulfite esters of 3° alcohols react by an S_N1 mechanism (decomposition of the chlorosulfite ester to R^+, SO_2 and Cl^-) when Cl^- at tacks the carbocation.

3.9 MIXED S_N1 AND S_N2 REACTIONS

A majority of epoxide ring opening reactions occur with inversion of stereochemistry at one end to the ring supporting an S_N2 mechanism under basic or neutral conditions (this indeed is the major pathway). The less substituted carbon is attacked by the nucleophile (scheme 3.56*b*). Under acidic conditions however, it is the protonated epoxide that undergoes the backside displacement *i.e.*, S_N2 type opening. The acid catalysis changes the regioselectivity, the attack is on the more hindered center. Consider the following points:

- The opening of epoxide ring in 1, 2-epoxy-1-methylcyclohexane with HBr, yields one isomer in which the Br and OH groups are *trans* (scheme 3.56*c*) [*In fact when an epoxide ring is fused to a cyclohexane ring, the* S_N2 *opening always gives* 1, 2 *diaxial (trans product) rather than* 1, 2*-diequatorial product (trans product)*]. Thus inversion is observed at the stereocenter, the reaction is S_N2 type.
- Interestingly bromide ion (Br^-) attacks the more hindered epoxy carbon which may suggest an S_N1 mechanism.
- Unlike the carbocation intermediate of a true S_N1 reaction the transition state of this acid catalyzed ring opening is a bridged carbocation in which the oxygen atom is still blocking one face, the reaction therefore, takes place from the other side. As a consequence the stereochemistry is still retained and not lost completely which would have been the case if a free carbocation was involved.

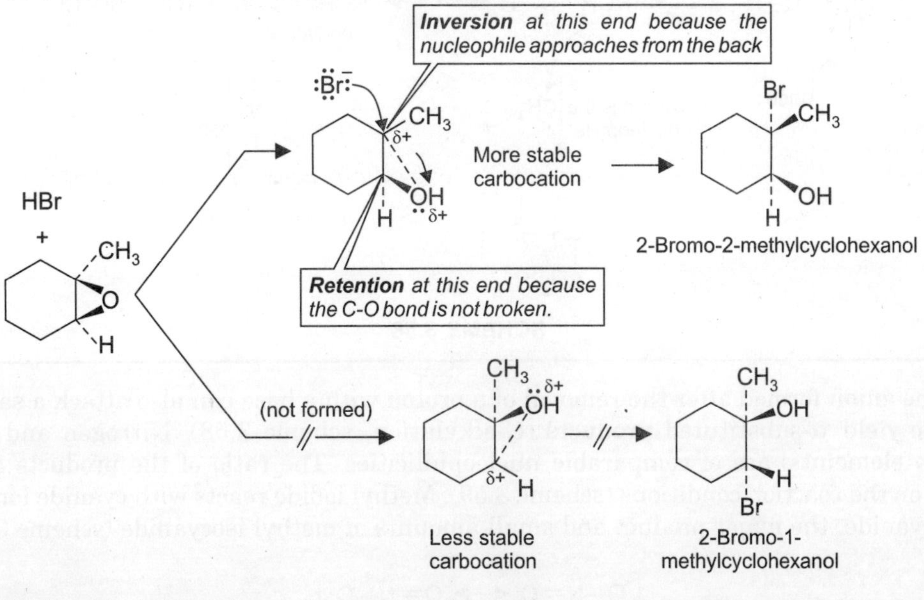

SCHEME 3.56c Acid catalyzed ring opening of 1, 2-epoxy-1-methylcyclohexane. A high degree of S_N1 carbocationic character in the transition state but S_N2 displacement.

- In summary, the stereochemistry of the acid catalyzed ring opening of an unsymmetrical epoxide is S_N2-like since the attack of the nucleophile is from the side opposite the oxonium ion, bridged intermediate. However, the regiochemistry is S_N1 like, due to the partial carbocation character of the transition state—the uneven charge distribution in the transition state counteracts steric hindrance.

3.10 AMBIDENT NUCLEOPHILES–REGIOSELECTIVITY

Ambident nucleophiles (ambi = both, dent "teeth" Latin) are nucleophiles which have two nucleophilic sites ("two teeth") and thus are capable of reacting at two sites. This is shown *e.g.,* for the enolate anion derived from acetone (scheme 3.57), which of these nucleophilic sites will react with an electrophile largely depends on the electrophile and on the reaction conditions. Protonation occurs preferentially on oxygen, since the resonance hybrid resembles more with the enolate since it is the more stable resonance contributor (the negative charge is on a more electronegative atom). Chlorotrialkylsilanes also *e.g.,* react exclusively at the oxygen atom of he enolate in reaction called silylation. Silylation is a nucleophilic substitution at the silicon atom by the oxygen atom of the enolate (scheme 3.58). This preference is due to very strong oxygen-silicon bond compared to carbon-silicon bond. The trimethyl silyl enol ethers can be purified and converted back to the enolate—a process needed for directed aldol condensation. Silyl ethers are also employed for protecting hydroxyl groups.

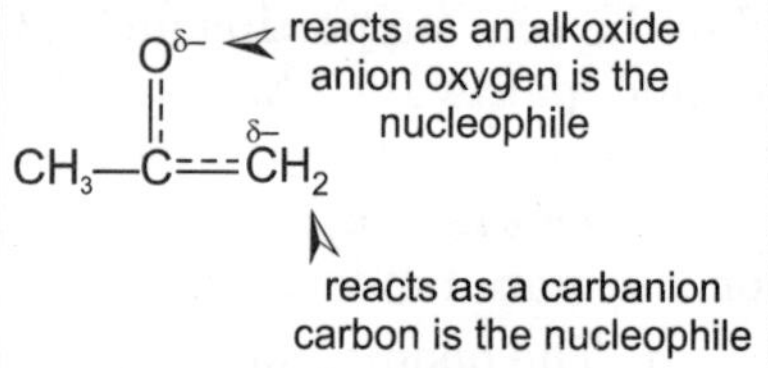

ambident nucleophile

SCHEME 3.57

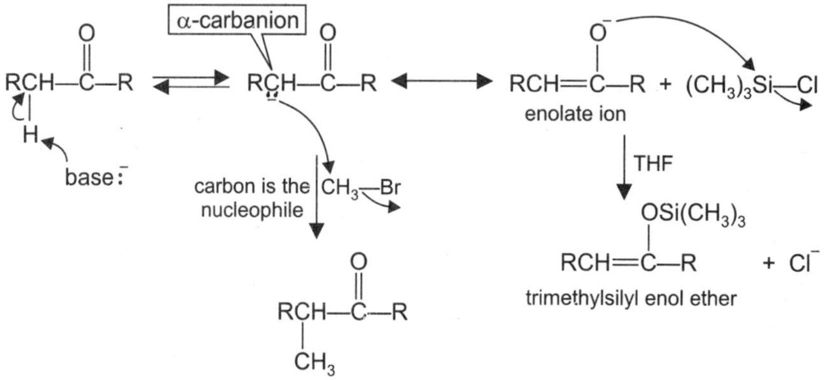

SCHEME 3.58

The anion formed after the removal of a proton with a base can also attack a saturated carbon to yield α-substituted products (*C*-alkylation, scheme 3.58). Nitrogen and oxygen (first-row elememts) are of comparable nucleophilicities. The ratio of the products actually depends on the reaction conditions (scheme 3.59). Methyl iodide reacts with cyanide ion to give methyl cyanide, the major product and small amounts of methyl isocyanide (scheme 3.60).

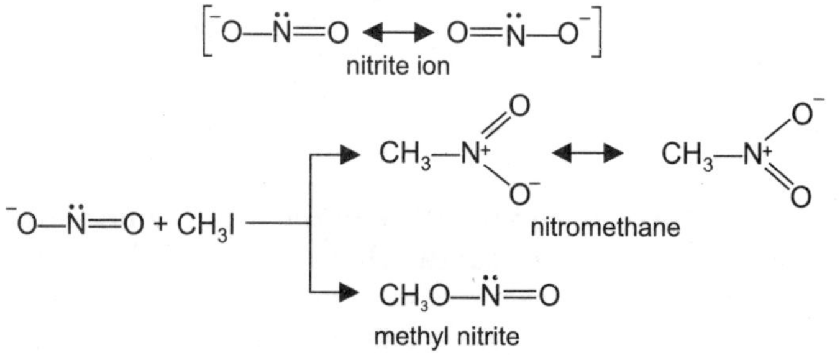

SCHEME 3.59

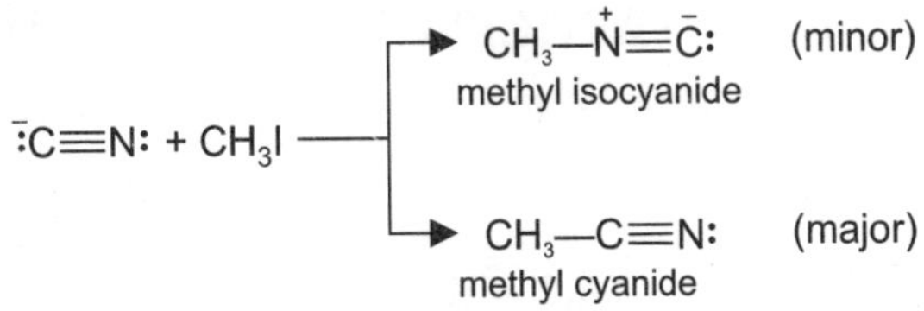

SCHEME 3.60

3.11 SET MECHANISMS

It has been found that in some nucleophilic substitutions, there is an involvement of radicals or radical ions. In such mechanisms, the first step is the transfer of an electron from the nucleophile to the substrate to yield a radical anion (step 1, scheme 3.61). A mechanism which begins this way is termed *SET (single electron transfer mechanism)*. The radical ion then cleaves (step 2) and the radicals formed this way combine to give the product (step 3). Alternately the radical combines with the nucleophilic ion (Y) to give the product.

Racemization to some extent has been observed when SET mechanism is operative. A totally free radical is likely to give totally racemized product, however, inversion is also detected in some SET mechanisms. It is, therefore, suggested that in step I (scheme 3.61), the nucleophile approaches from the back side, although the usual S_N2 mechanism does not operate. The radical R• after formation is confined to a solvent cage with Y• still opposite X^- (scheme 3.62), consequently the steps (scheme 3.61) will lead to inversion and indeed predominant inversion has been found in reactions proceeding with SET mechanism.

In evidence for *SET* mechanism is the involvement of radicals or a radical ion *e.g.*, during the formation of cyclic products when the substrate has a double bond in the 5,6-position (see, schemes 6.75–6.78).

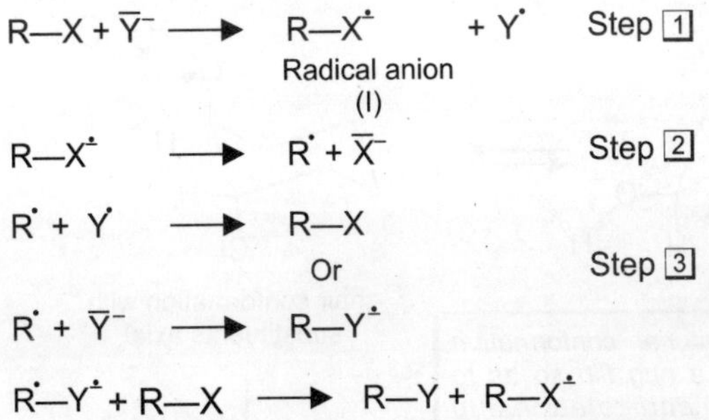

An electron transfer from the nucleophile to the substrate to give a radical anion is the first step of SET (single electron transfer) mechanism. The radical ion cleaves (step 2). The product is formed when radicals combine (step 3) or a radical combines with the original nucleophilic ion (Y) to give a radical anion and then the product.

SCHEME 3.61

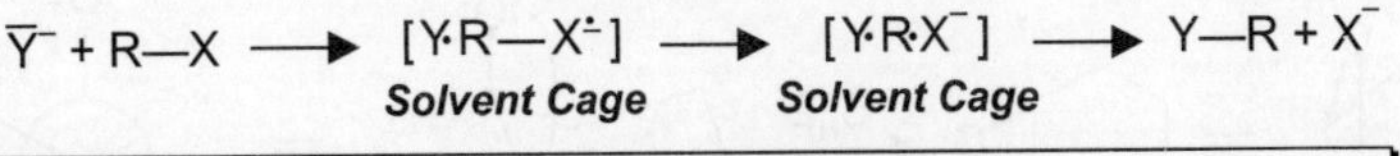

Inversion in some SET mechanisms, nucleophile attacks from back and R• remains in a solvent cage with Y• remaining opposite X⁻.

SCHEME 3.62

3.12 NEIGHBORING GROUP PARTICIPATION—ANCHIMERIC ASSISTANCE

(a) Introduction

The following points may be noted:

- S_N2 reactions at chiral carbons give products with *inverted* configuration. In S_N1 reactions as well, which normally yield at least partially racemized products, it is usually easiest for the nucleophile to attack from the direction opposite to that from which the leaving group is departing. Thus the S_N1 products formed with inversion of configuration should always be more compared with retention.
- While most nucleophilic substitution reactions, whether they proceed by S_N1 or S_N2 mechanisms, result predominantly in inversion of configuration, that is not necessarily the case if the β-carbon-the carbon bonded to the carbon bearing the leaving group–has a substituent with unshared pair of electrons or an aromatic ring *e.g.,* which can also act as a nucleophile. In these cases, nucleophilic substitutions frequently proceed mostly with retention of configuration and with enhanced rates *i.e.,* both the kinetics and the stereochemical outcome are strongly effected.
- Whether the substrate is an open chain compound or a cyclic molecule, the neighboring group effects are the most evident when the leaving group and the neighboring group can easily assume an (S_N2 type geometry) *trans* coplanar arrangement. The stereochemistry of epoxide formation from halohydrins with alkali is related to neighboring group participation. In the case of cyclohexane derivative it proceeds best and at a fast rate if the groups involved are both axial. This is so, as the axial alkoxide is favorably located for a near attack (S_N2 type) on the carbon bearing the axial halogen (scheme 3.62*a*). Halohydrins that cannot achieve a *trans*-coplanar *i.e.,* an *anti* orientation between the nucleophile and the leaving group infact do not give epoxides.

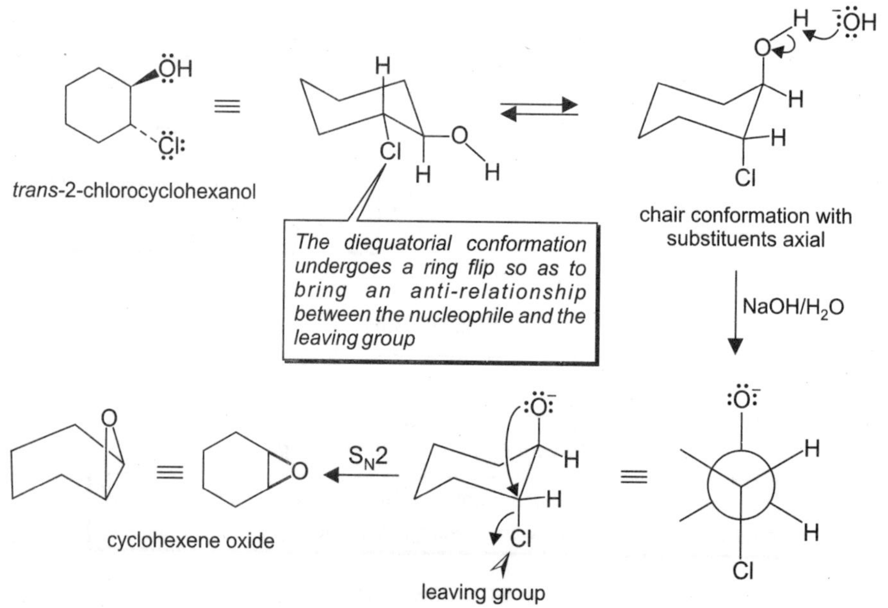

SCHEME 3.62a

- The neighboring group participation infact is a two step process. In the first step the neighboring group (acting as an internal nucleophile) attacks carbon at the reaction center (S_N2 attack) and the leaving group is lost to give a bridged intermediate. This is then attacked in the second step by an external nucleophile (another S_N2 attack) and the internal nucleophile goes back to where it came from, two consecutive S_N2 reactions lead to retention of configuration at the reacting carbon.
- The term bridged ion is synonymous with a nonclassical carbocation. A bromonium or a phenonium ion are infact classical ions, but are often (ambiguously) termed bridged ions. None of these involves a three-center two electron bonding.

(b) Examples of Neighboring Group Participation

(i) An Acyclic (Open Chain) System of 3-Bromo-2-Butanol-Neighboring Group Participation by a Halogen Atom

The following points may be noted:

- The molecule of 3-bromo-2-butanol has two stereocenters, thus it can have four stereoisomers *i.e.,* a pair each of *erythro*- and *threo*-enantiomers (scheme 3.63, also see scheme 1.33, one may see that there can be several ways of writing the structure of a chiral compound. Here each Fischer projection has an equivalent projection (180° rotation in the plane of paper) drawn under each.
- Fischer projections are eclipsed conformations and to study neighboring group participation the concerned groups are to be *trans* and coplanar.
- Firstly one considers the pure (optically active) *threo*-enantiomer and its reaction with HBr to study the neighboring group participation by the halogen atom (bromine).
- The Fischer projection of optically active *threo*-enantiomer (3, scheme 3.63) is first transformed into sawhorse formula (eclipsed form 3*a*, scheme 3.63) following the procedure already explained (see scheme 1.38). The carbon atoms C-2 and C-3 in 3*a* are then rotated appropriately so as to bring Br and OH in *anti* (*trans* coplanar) conformation (3*b*, scheme 3.63).
- Reaction of optically active (3*b*, scheme 3.64) with HBr lead to loss of optical activity to give racemic mixture (2*R*, 3*R* and 2*S*, 3*S* dibromobutane, compare this to the *anti*-addition of bromine to alkenes (see, scheme 1.100).
- Thus in (3*c*, scheme 3.64) the bromine atom acts as a neighboring group displacing the leaving group (resembling intramolecular S_N2 reactions) to give a cyclic bromonium ion with inversion. The brominium ion is then opened by another S_N2 reaction by bromide ion (attack a, scheme 3.64) with inversion. The net result is the formation of (5, scheme 3.64) with overall retention of configuration (*i.e.,* due to two successive displacements with inversion).
- Due to reactions with achiral reagents each end of the (symmetrical, achiral) cyclic bromonium ion (4, scheme 3.64) is indistinguishable from the other end, therefore, exactly 50 per cent of each ring opening reaction occurs at each end. This results in the formation of enantiomers (5 and 6, scheme 3.64) in exactly equal amounts. One may note that the bromonium ion (4, scheme 3.64) has a plane of symmetry thus ends (C2 and C3) are enantiotopic.

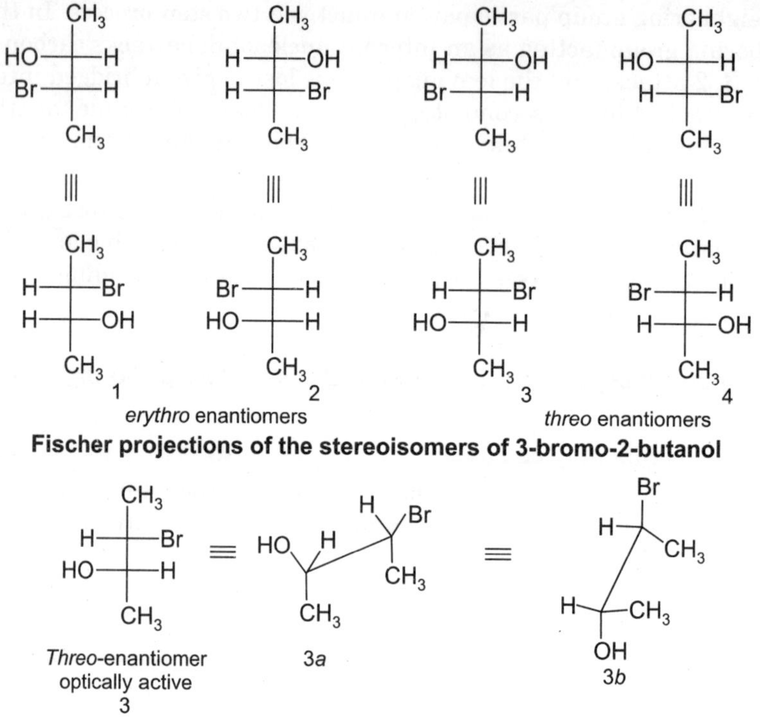

SCHEME 3.63

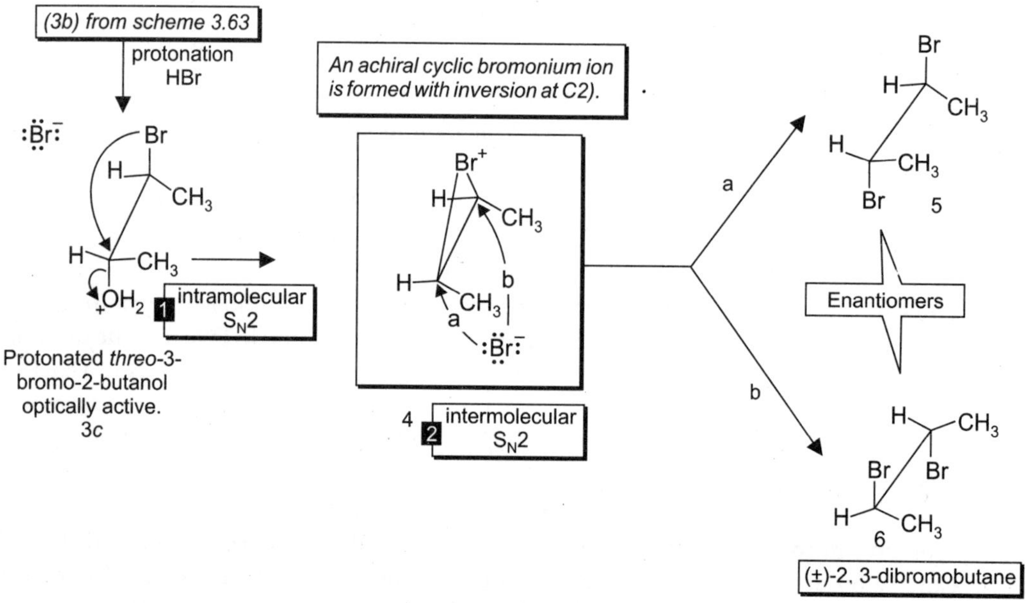

SCHEME 3.64

- Similar arguments will explain the formation of *meso*-2-3-dibromobutane (*R, S*-2,3-dibromobutane) from optically active *erythro*-isomer (1 or 2, scheme 3.63, compare with addition of bromine to *E*-2 butene scheme 1.100).

(ii) Effect of Neighboring Phenyl Group—π Systems of Aromatic Rings

As the aromatic ring also has nucleophilic properties to a certain extent, one may expect that their presence in the β-position under certain conditions will influence the substitution by the formation of a cyclic ion (phenonium ion). The first to prove the validity of this assumption was Cram in the course of the reaction of stereoisomeric toluene-*p*-sulphonates of 3-phenyl-2-butanols (I and II, scheme 3.65) with acetic acid. While reacting the optically active *erythro*-isomer or *threo*-isomer, he obtained in both cases 3-phenyl-2-butyl acetates of the same configuration as the original alcohol. The difference was only in the fact that the optically active *threo*-isomer (I), yielded racemic product only and the *erythro* isomer (II) yielded optically active product. The absence of the *erythro*-isomer in the reaction products when one starts from the *threo*-isomer and *vice versa* shows that in the course of the formation of the phenonium ion a change of the configuration at C-1 takes place, and when it opens, the configuration changes either at C-1 or at C-2. The loss of optical activity in the case of the *threo*-isomer is caused by the formation of the absolutely symmetrical ion (III, scheme 3.65). To enhance understanding, one may show these results by drawing the structures in other stereo-fashion (scheme 3.66). In either case cyclic phenonium ion is formed. This ion from (I, scheme 3.66) has a plane of symmetry and gives equal amount of enantiomers, while the phenoniun ion from II has no plane of symmetry but has a C_2 axis and thus attack at either end of three membered ring gives the same acetate (optically active) with the same configuration as the starting tosylate.

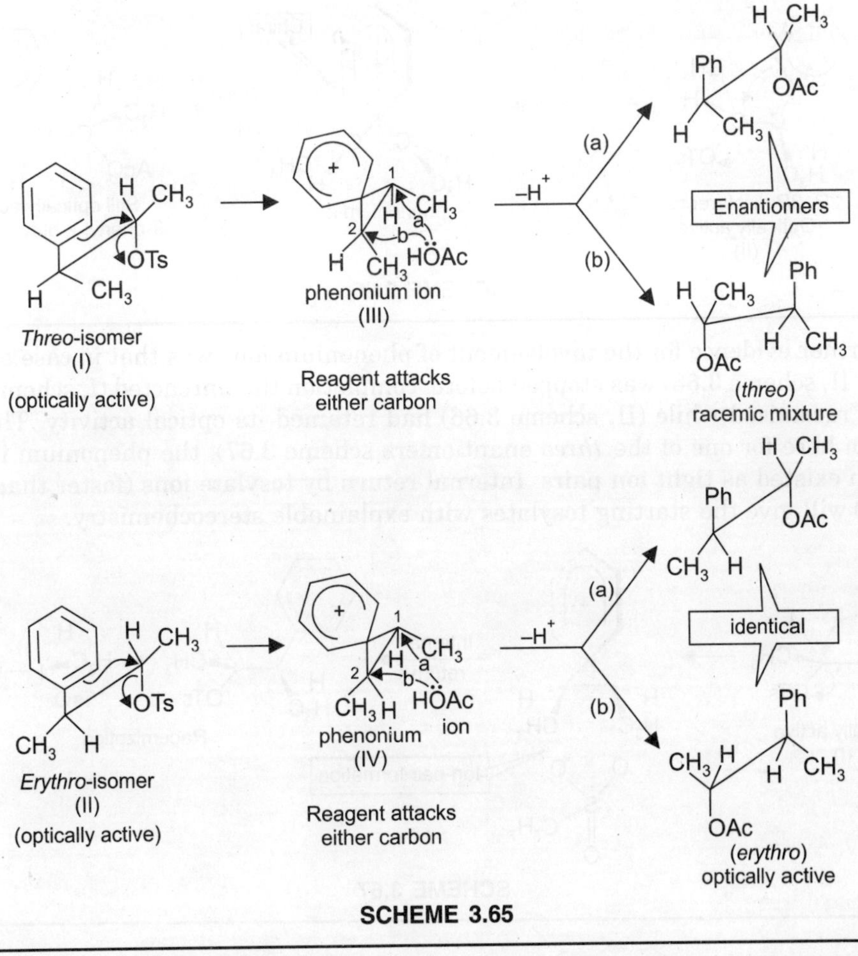

SCHEME 3.65

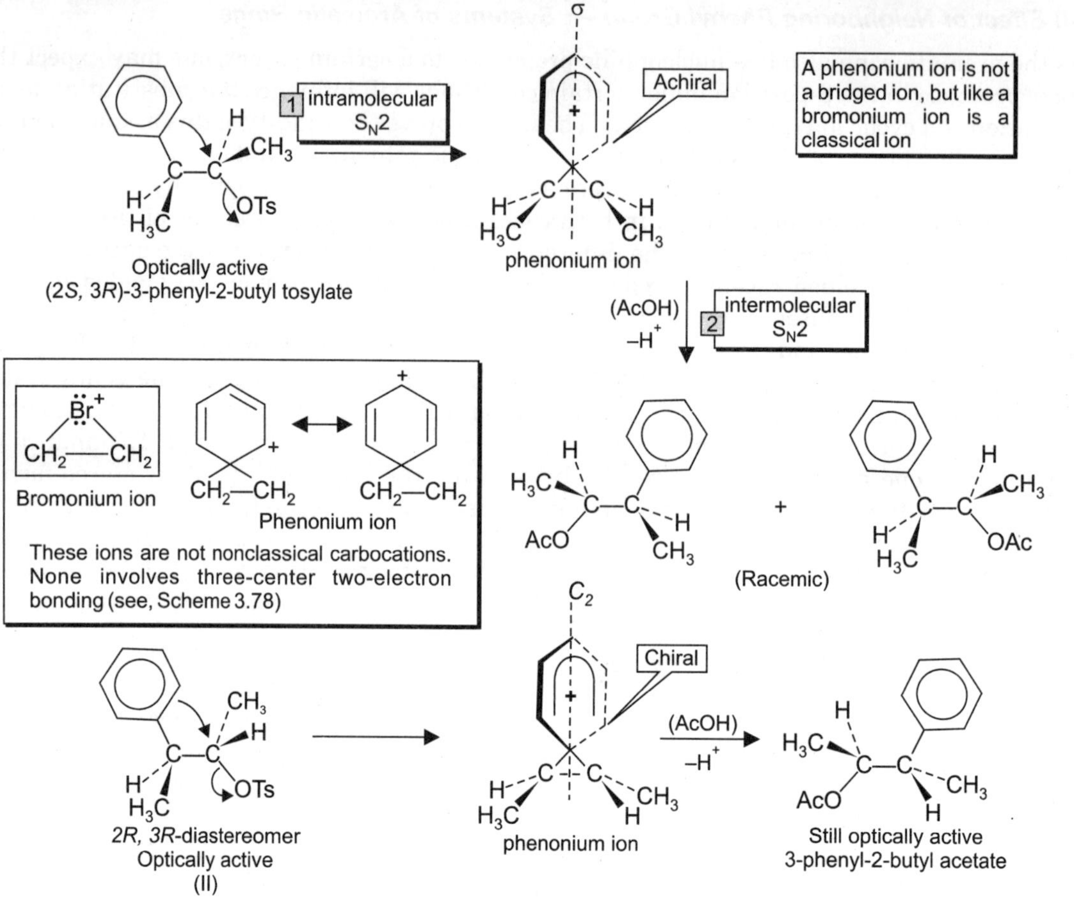

SCHEME 3.66

A further evidence for the involvement of phenonium ions was that in case acetolysis of both (I and II, scheme 3.66) was stopped before completion the unreacted (I, scheme 3.66) had completely racemized while (II, scheme 3.66) had retained its optical activity. Thus in each case (shown here for one of the *threo* enantiomers scheme 3.67), the phenonium ion and the tosylate ion existed as tight ion pairs. Internal return by tosylate ions (faster than attack by acetic acid) will give the starting tosylates with explainable stereochemistry.

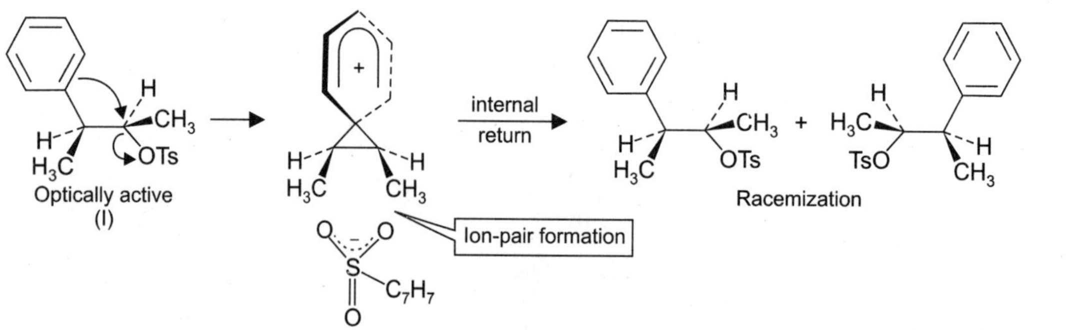

SCHEME 3.67

(iii) Neighboring Group Participation in a Cyclic System—Neighboring Group Participation by Acetoxyl Group

The following points may be noted:

- The rate of solvolysis (acetolysis) of the *cis*- and *trans*-isomers of 2-acetoxycyclohexyl tosylates differ by a factor of about 700, the *trans*-compound being more reactive.
- The diacetate obtained from the *cis*-isomer (1, scheme 3.68) is the *trans* compound (with inverted stereochemistry) which is formed by the usual S_N2 attack as shown in (2, scheme 3.68).
- The *trans*-isomer (3, scheme 3.68) also gives *trans* product but with retention of configuration. Displacement of tosylate by the neighboring *trans*-placed acetoxy group gives a cyclic five membered acetoxonium ion with a resonance hybrid structure (4, scheme 3.68). The acetate ions can then attack either of the two equivalent cyclohexyl ring carbons of this symmetrical achiral acetoxonium ion to give the double inversion product (5 or its enantiomer not shown here, scheme 3.68). Thus when optically active *trans* 2-acetoxycyclohexyl tosylate is solvolyzed, the product is racemic *trans*-diacetate (since the acetoxonium ion is achiral, it can give only racemic product).

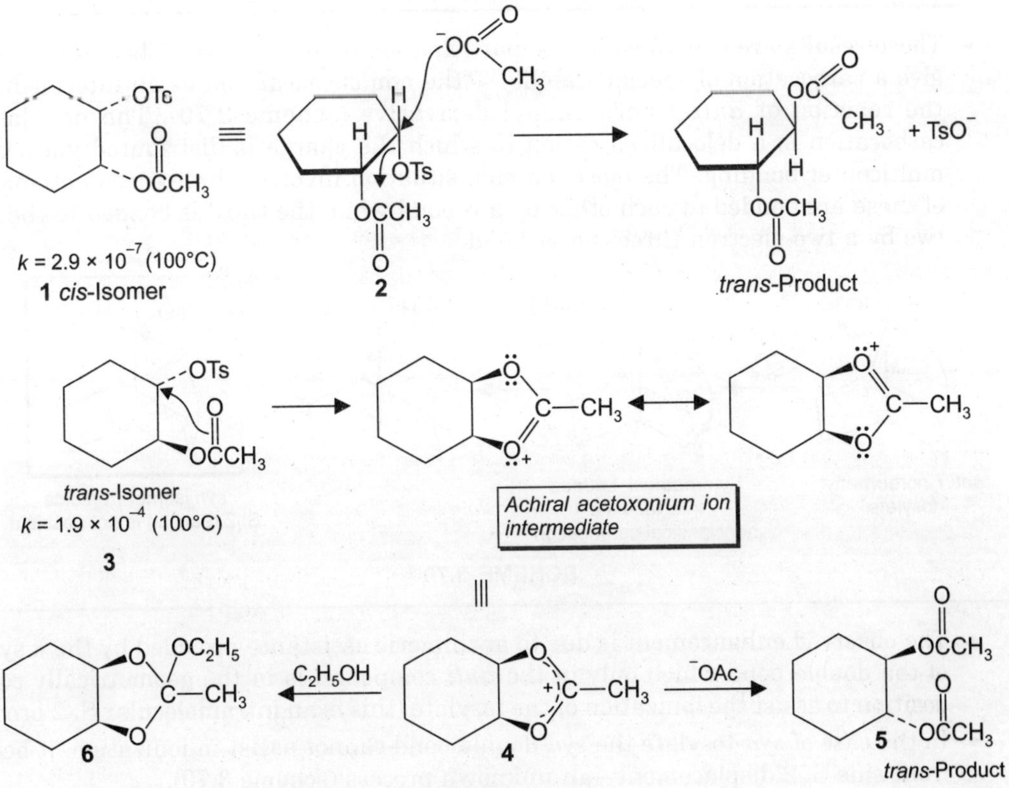

SCHEME 3.68

- That acetoxonium ion is indeed the intermediate is confirmed by capturing it as a cyclic *ortho*-ester (6, scheme 3.68) during solvolysis in the presence of ethanol. Ethanol adds to the "carboxyl carbon" of (4, scheme 3.68) to give the *ortho*-ester.

(iv) Neighboring Group Participation by a Double Bond (Non-Classical Carbocations–Bridged ions)

(a) Acetolysis of anti and syn 7-norbornenyl tosylate and the saturated compound

A graphic example of largest rate enhancement due to a neighboring double bond is observed in the solvolysis of *anti*-7-norbornenyl tosylate (I, scheme 3.69) which reacts 10^7 times as rapidly as its *syn*-isomer (II) and 10^{11} times as rapidly as 7-norbornyl tosylate (III). Moreover, *anti*-norbornenyl tosylate, reacts stereospecifically, with retention of configuration. (A stereo-specific assistance to ionization by π-electrons). The following points may be noted:

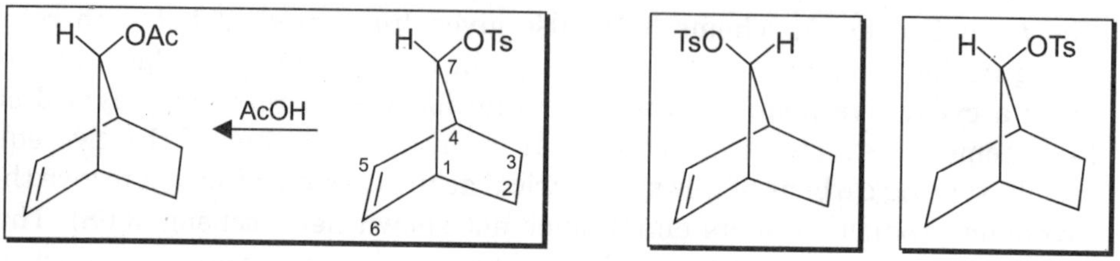

Relative rates (I) 10^{11} (II) 10^4 (III) 1

SCHEME 3.69

- These results are explained by the participation of π electrons of the double bond to give a carbocation of special stability—"the non classical" ion as an intermediate in the reaction of *anti*-7-norbornenyl derivative (scheme 3.70). The nonclassical carbocation is a delocalized cation in which the charge is distributed via a closed multicenter bonding. The most common situation involves three carbon atoms. Two of these are bonded to each other by a σ bond while the third is bonded to the other two by a two-electron three center-bond.

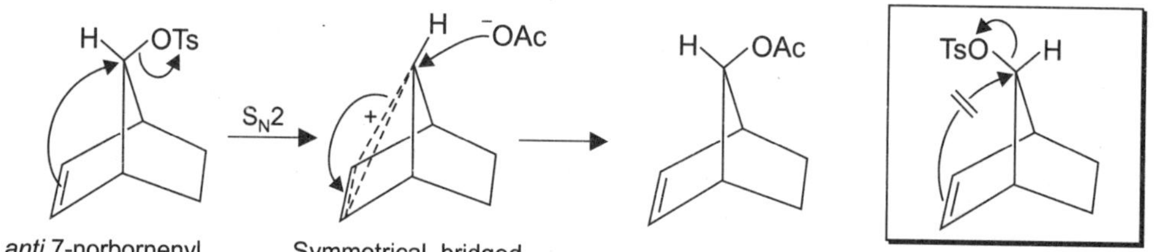

anti 7-norbornenyl tosylate — Symmetrical bridged intermediate nonclassical carbocation — *syn* isomer frontside S_N2 reaction is not possible

SCHEME 3.70

- The observed enhancement is due to anchimeric assistance provided by the π system of the double bond which only in the *anti* compound is in the geometrically correct position to assist the ionization of the tosylate (this is an intramolecular S_N2 process).
- In the case of *syn*-tosylate the *syn* double bond cannot assist in ionization, it being a front side S_N2 displacement—an unknown process (scheme 3.70).
- The anchimeric assistance (neighboring group participation) by the double bond leading to rate enhancement in the case of *anti*-tosylate is faster than the possible direct intramolecular reaction (*i.e.,* S_N2) and is thus the dominant process.
- The attack by acetic acid ($^-$OAc) to the symmetrical bridged intermediate is again another S_N2 process (now intermolecular S_N2 process) and must take place from the

rear of the departing leaving group, here the two partial bonds to C-7 in the bridged intermediate (scheme 3.70) act as a leaving group with respect to the incoming acetic acid as nucleophile.

- The end result of acetolysis of *anti*-tosylate to give *anti*-acetate with retention of configuration is therefore, due to two S_N2 reactions, one intramolecular process to give a bridged ion and one intermolecular process leading to opening of the bridge.

(b) Reasons for remarkably rapid acetolysis of anti-7-norbornenyl tosylate

The following factors may be responsible for this repaid acetolysis reaction :

- The developing cation at C-7 (leading to the formation of a symmetrical bridged intermediate, scheme 3.70) is in a perfect position to interact with the center of the double bond–a region where electron density is highest.
- In comparison, in a homoallylic cation, the cationic carbon can interact with only one end of the double bond.
- In the norbornenyl system the five membered rings are distorted from planarity to large extent, thus C-7 is brought is close proximity to the double bond. Comparing this situation with 4-bromocyclopentene (scheme 3.71) where the ring is almost planar does not display any rate enhancement on heating in water.

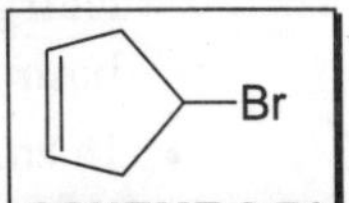

SCHEME 3.71

(c) Evidence–The acetolysis of anti-7-norbornenyl tosylate involves a symmetrical bridged ion intermediate (a highly delocalized non-classical structure) or a rapidly equilibrating mixture of cations.

The most famous controversy of modern times in organic chemistry (Winstein, Cram and H.C. Brown) concerned the question of bridged ions (non-classical carbocations) and unbridged or classical carbocationic intermediates invoked during the participation of neighboring π systems or σ participation. The point of difference had been if the contributing structures are the contributing resonance forms to a single symmetrical bridged ion or these are unsymmetrical

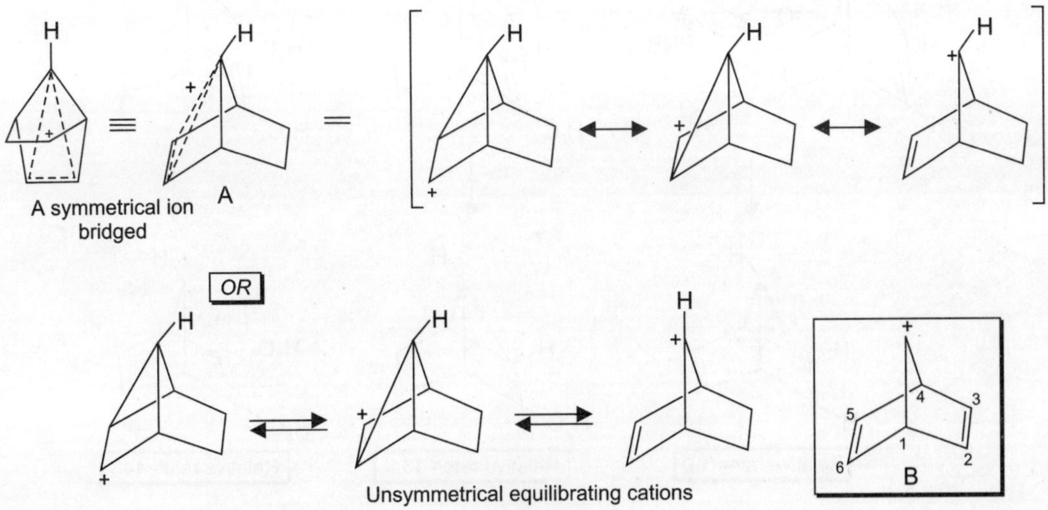

SCHEME 3.72

equilibrating cations each having its own separate existence (scheme 3.72). Evidence for the nonclassical intermediate (A, scheme 3.72) comes from the ^{1}HNHR study. It is shown that in the case of a relatively stable norbornadienyl cation (B, scheme 3.72), the 5 and 6 protons are not equivalent to the 2, 3 protons, to show an interaction between the charged carbon and one double bond which indeed is an evidence for the presence of (A, scheme 3.72).

The following evidence seems to favour the formation of a non classical bridged ion (non-classical carbocation).

- Methyl groups were introduced on the double bond of *anti*-7-norbornenyl derivative–*p* nitrobenzoate of *anti*-bicyclo [2.2.1] hepten-7-ol. The relative rates of acetolysis increased as expected (scheme 3.73) and it was found that the introduction of second mehtyl group on the double bond had a very large effect on solvolysis compared to the first methyl group. These observations showed that the transition state of this solvolysis reaction is the one in which both ends of the double bond share the positive charge at all times *i.e.,* a non classical bridged cation is involved (scheme 3.73). Thus both methyl groups exert their stabilizing effect at the same time.
- In case, however, rapidly equilibrating carbocations were involved (scheme 3.74) and therefore, a positive charge was generated at only one end of the double bond at any given time, a methyl group would assist the formation of the 7-norbornenyl cation. However, the effect of the second methyl group on the other carbon of the double bond should have had little effect (at the most the predicted rate would be only 2 × 13.3 = 26.6 and not that large, 148 as was in fact observed experimentally. Thus in the open (unbridged) cations the two methyl groups can never stabilize the cation at the same time—a fact observed in practice.

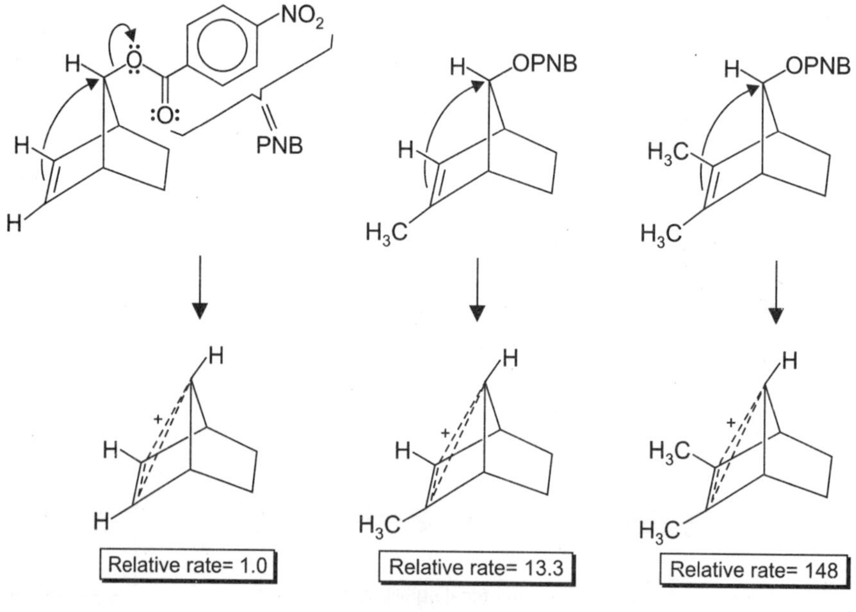

SCHEME 3.73

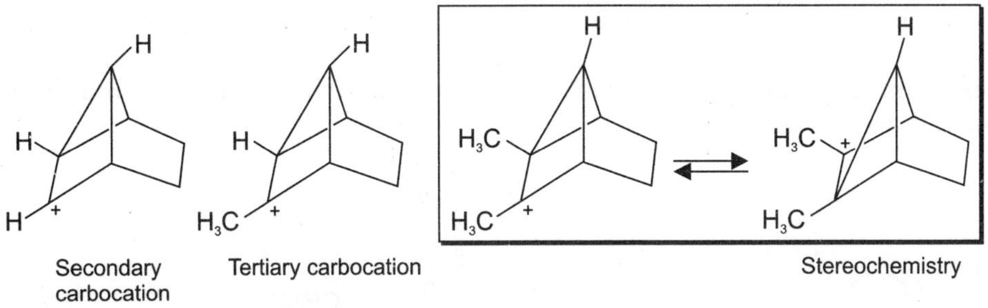

In the pair of equilibrating open cations both the methyl groups cannot exert their stabilizing effect at the same time.

SCHEME 3.74

- The extent of participation of the carbon-carbon double bond in the ionization of *anti*-7-norbornenyl systems is significantly influenced by the substitution at C-7. The placement of an aryl substitutent at C-7 diminishes the rate enhancing effect due to participation of the double bond. When an aryl group is present at C-7, the resulting benzyl-type stabilization of the potential carbocation at C-7 decreases the importance of participation by the double bond. The extent of stabilization further depends on the substitutents on the phenyl ring. For *p*-methoxyphenyl, phenyl and *p*-trifluoromethylphenyl, the rate factor for the unsaturated relative to the saturated system is 3,40 and 3.5×10^4 respectively. The double bond has clearly a much important role to play as a neighboring group with poorly stabilizing *p*-trifluoromethyl substituted system (scheme 3.75).

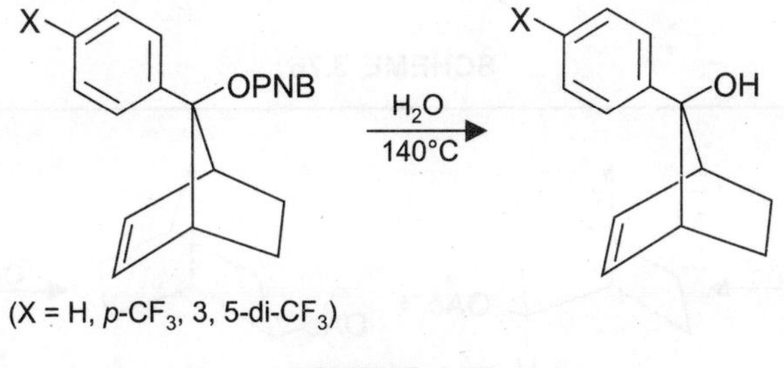

SCHEME 3.75

- Significantly the solvolysis of *anti*-7-norbornenyl derivatives (scheme 3.75) proceeded with retention of configuration to give product with original *anti*-configuration provided the phenyl group at C-7 was either unsubstituted or was substituted with electron withdrawing groups (scheme 3.75). The mixture of *anti* and *syn* isomers was formed when the substitutents were *p*-dimethylamino or *p*-methoxy groups (scheme 3.76).

(v) Neighboring Group Participation by σ Bonds–Anchimeric Assistance by Alkyl Groups the "Non Classical ion" Hypothesis

Evidence has been presented that C=C acts as a neighboring group and that a non classical carbocation (a bridged cation) may be formed. Evidence is also available to show that a suitably located C—C in a substrate can also participate in the departure of the leaving group and non classical carbocations may be involved. (Stereospecific participation of of 1,6 σ-electrons).

The following points may be noted:

- During the acetolysis of *exo*- and *endo*-norbornyl tosylates (scheme 3.77) it is found that the solvolysis of *exo* isomer is 350 times faster than the *endo* isomer. Both the isomers give only the *exo* acetate (retention of configuration). However, optically pure *exo*-tosylate gives 100% racemic product while an optically pure *endo*-tosylate gives 93% racemic product. These observations are in keeping with neighboring group participation and are explained:

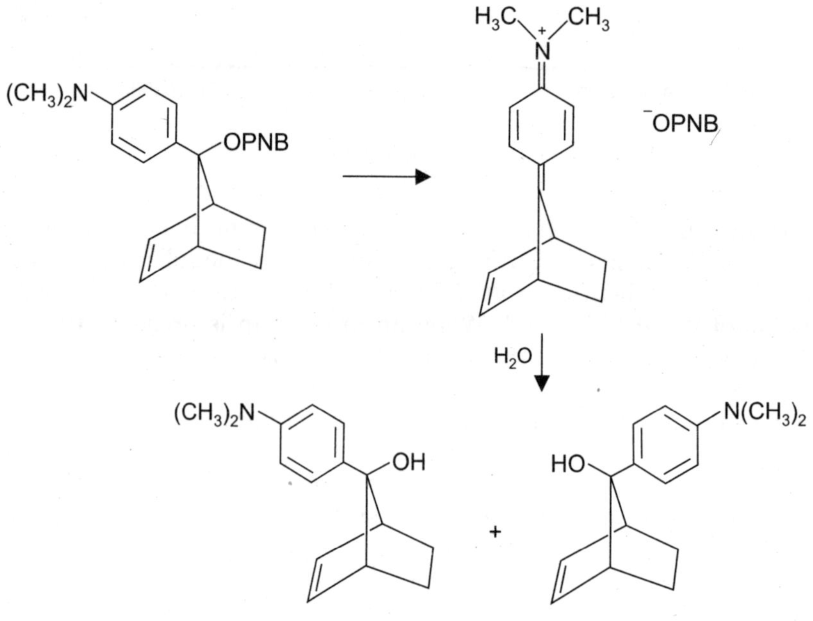

SCHEME 3.76

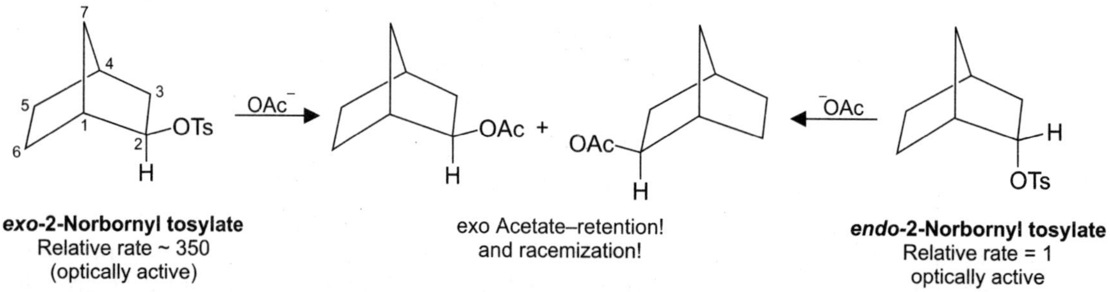

SCHEME 3.77

- In the *exo*-isomer the 1,6 σ-bond is suitably located to act as a neighboring group to lend anchimeric assistance via backside attack to give directly a non classical carbocation (scheme 3.78).
- The *exo*-isomer—a compound with serveral stereocenters can form an intermediate with plane of symmetry which is then expected to give completely racemic products.
- The non-classical carbocation intermediate, (I, scheme 3.78) is achiral having a plane of symmetry passing though C-4, C-5, C-6 and midpoint of the C-1 and C-2 bond (the C-6 has two hydrogens and is pentacoordinate and is the bridging atom in the cation, Thus the attack at both C-1 and C-2 is equally likely these being related by a σ plane (enantiotopic) which gives equal amounts of enantiomeric acetates—a racemic mixture.

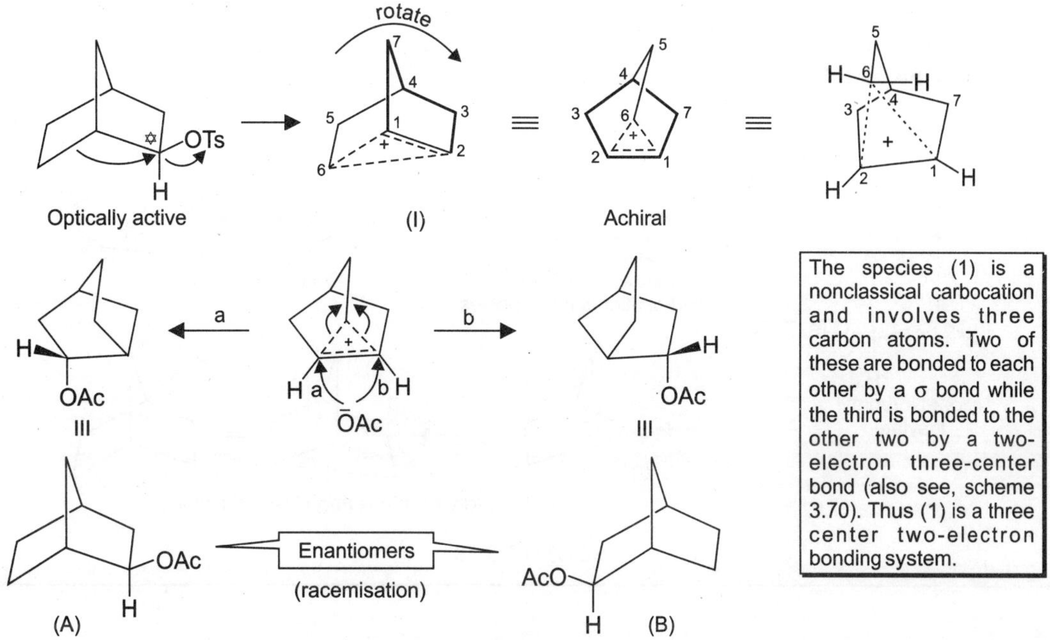

SCHEME 3.78

A Nonclassical Carbocation (a Bridged cation) is a Carbonium ion

The normal coordination number for a positively charged carbon atom without a multicenter bond is 3. When the coordination number of a positively charged carbon is greater than 3 it is considered hypervalent and then it truly represents a carbonium ion. Thus a carbonium ion is a hypervalent carbocation having a valence (coordination number) greater than 3. In nonclassical norbornyl cation (I, scheme 3.78) C6 has a coordination number of 5 and C1 and C2 have a coordination number of 4. Thus nonclassical norbornyl carbocation is an example of a carbonium ion.

- The attack occurs from the *exo*-side due to the cage structure of the non classical carbocation intermediate (I, scheme 3.78). Moreover, the attack must occur from the direction opposite that of bridging interaction and this is *exo*-direction.
- Prof. H.C. Brown, however, put forward an interpretation as an alternative to Winstein's non-classical norbornyl cation (scheme 3.79). The non classical carbocation is stabilized, relative to secondary cation, by C—Cσ bond delocalization (+ve charge is spread). A rapidly equilibrating classical carbocation (scheme 3.79) via 1,2-shift where the inter conversion of classical carbocations was presumed to be repaid relative to capture by the nucleophile (from *exo*-side) would also explain the result. Most of the chemists generally now accent the intermediacy of a non-classical carbocation.
- It was suggested by Winstein that in the case of *endo*-isomer (scheme 3.80) first an open classical carbocation (II, scheme 3.80) is formed which then gives the same non classical carbocation (I, scheme 3.80). Evidence for this interpretation is that the product from solvolysis of the *endo*-brosylate (scheme 3.77) is not completely racemic but contains slightly more of one of the enantiomeric *exo*-acetates (A scheme 3.78)

than (B) corresponding to some inversion. Thus when carbocation (II, scheme 3.80) is formed some of it goes to give (A, scheme 3.78) before it can collapse to the bridged ion (I, scheme 3.80).

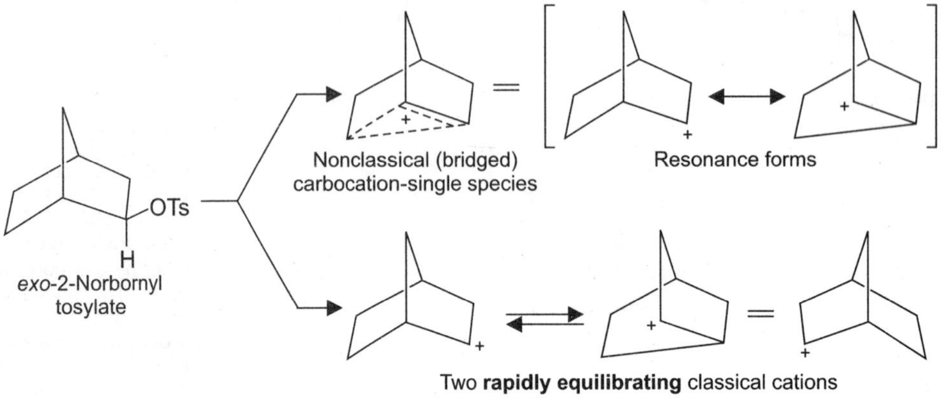

SCHEME 3.79

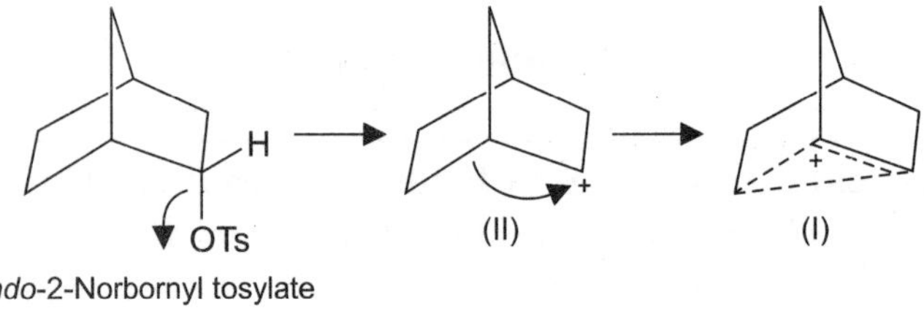

SCHEME 3.80

EXERCISE 3.7

What evidence is against a simple S_N2 displacement on optically active exo-2-norbornyl tosylate (scheme 3.80a) without the σ bond participation ?

ANSWER. *In fact the stereochemical outcome from optically active exo-2-norbornyl tasylate or its epimer (endo-2-norbornyl tosylate) on acetolysis forms the basis for the neighboring group effect and involvement of a bridged non-classical carbocation. In both substrates optical activity is lost and moreover there is retention i.e., formation of only a racemic mixture of exo-tosylates.*

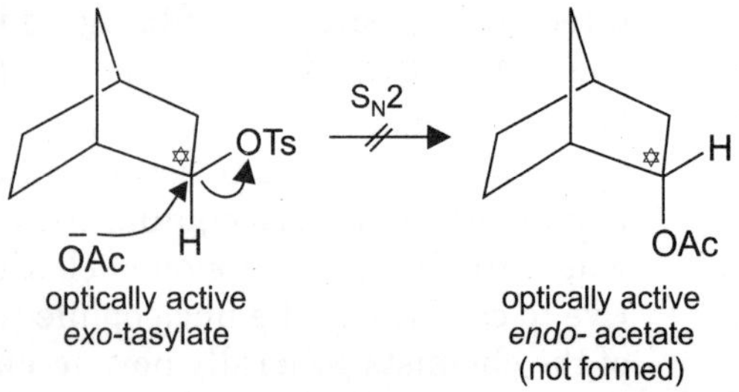

SCHEME 3.80a

Bridging imparts stabilization and nothing more, when other forms of stabilization are available, ions will be then open classical species. ^{13}CNMR spectroscopy distinguishes between equilibrating structures and bridged species. Thus 2-phenylnorbornyl cation (scheme 3.81)

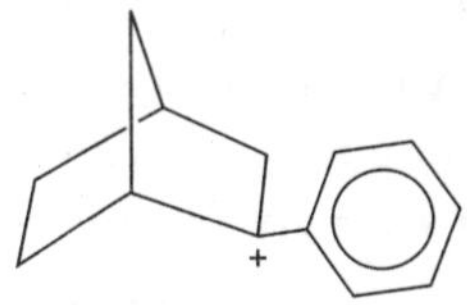

Classical carbocation

SCHEME 3.81

has the classical structure. This benzylic cation gets stabilized by π-electrons of the benzene ring thus bridging has no added advantages. Firstly consider the resonance–stabilized carbocation (scheme 3.82) the two carbons (shown by dots) which share the positive charges as expected are almost equivalent. In equilibrating ionic structures (two independent ions) such carbons differ by about 100 ppm.

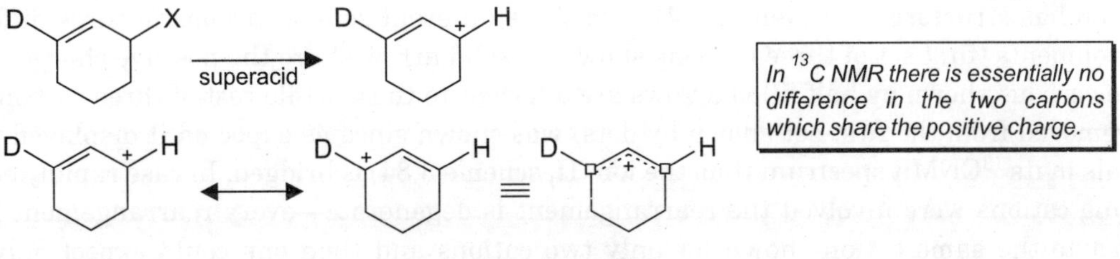

SCHEME 3.82

That 2-norbornyl cation is indeed bridged has been shown by detecting it in a highly polar but non-nucleophilic solvent (super acid media, SbF_5-SO_2). The ^{13}CNMR showed very similar signals for the deuterated and undeuterated positively charged carbons (scheme 3.83).

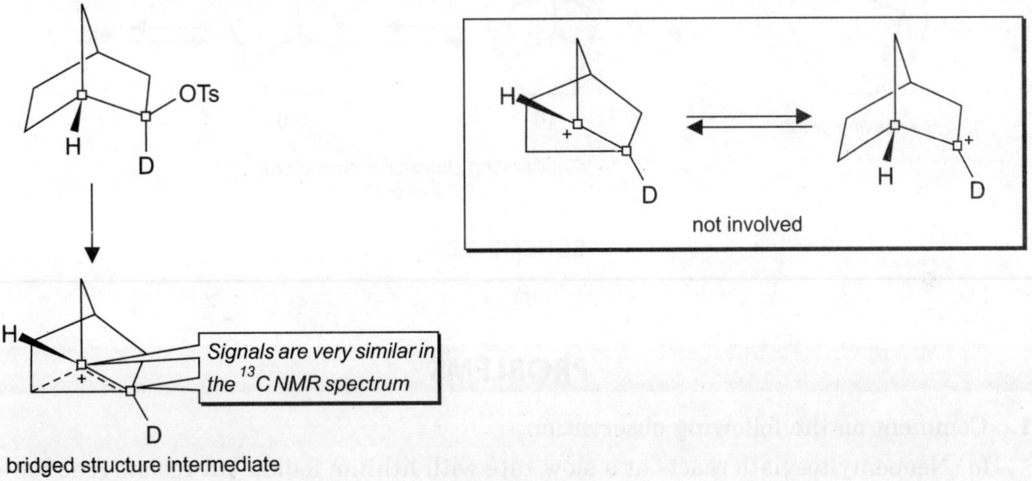

SCHEME 3.83

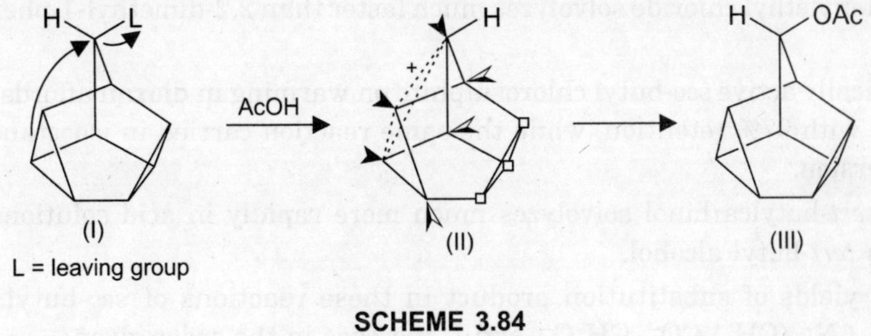

SCHEME 3.84

This evidence excludes the formation of an equilibrating pair of cations, in which case the two dotted carbons should have displayed widely separated signals in ^{13}CNMR (Use of this technique is complicated due to hydride shifts in this case however, the weight of evidence seems to strongly favour the bridged structure for norbornyl cation). [M. Saunders, Yale University

Lastly mention may be made of ion (II, named after its discoverer R.M. Coates, University of Illinois, scheme 3.84) studied at low temperature in super acid which is formed during the acetolysis of (I, scheme 3.84) via the participation of cyclopropane σ bond to give (III). That this ion has structure (II, scheme 3.84) with three different sets of carbons in three different environments (first set of three carbons shown by solid arrow share the positive charge, other three carbons shown by half filled arrows are adjacent to these while rest of three carbons are far removed from the first-set shown by dots) was shown since as expected it displayed three signals in its ^{13}CNMR spectrum thus the ion (II, scheme 3.84) is bridged. In case rapidly equilibrating cations were involved the rearrangement is degenerate—every rearrangement leads in fact to the same cation shown for only two cations and then one could expect only one ^{13}CNMR signal (scheme 3.85 I and I*a* are equivalent).

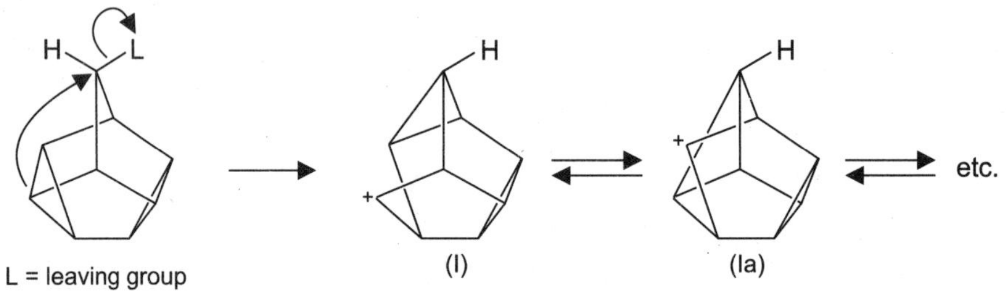

Equilibrating classical carbocations (equivalent)

SCHEME 3.85

PROBLEMS

3.1. Comment on the following observation:

(*a*) Neopentyltosylate reacts at a slow rate with lithium iodide but resists reaction with lithium chloride while *n*-butyl tosylate reacts with either of the nucleophiles at reasonable rate.

(*b*) I-Phenylethyl chloride solvolyzes much faster than 2,2-dimethyl-1-phenylpropyl chloride.

(*c*) Optically active *sec*-butyl chlorosulphite on warming in dioxan affords *sec*-butyl chloride with 97% retention, while the same reaction carried in isooctane leads to 43% inversion.

(*d*) Tri-*tert*-butylcarbinol solvolyzes much more rapidly in acid solutions as compared with *tert*-butyl alcohol.

(*e*) The yields of substitution product in these reactions of *sec*-butyl chloride with $(CH_3)_2N^-$, $(CH_3)_3CO^-$, CH_3O^-, and I^- increase in the order given.

3.2. During displacement of alkyl halides (R—X), the ratio of rates with two amines (Rate ratio = quinuclidine : triethylamine) is given below; with triethylamine it being always slower, how do you account for this rate difference and for its increase in the given series:

CH_3X	57
C_2H_5X	252
$(CH_3)_2CHX$	706

3.3. Name some ionizing and non-ionizing solvents.

3.4. Show the products with mechanism of following reactions :

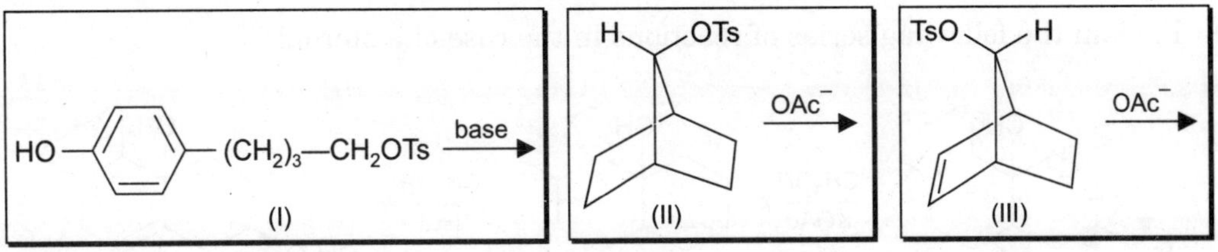

3.5. One of the compounds (I or II, reacts with Cl⁻ to give (III) which one is that. Give the mechanism of its formation.

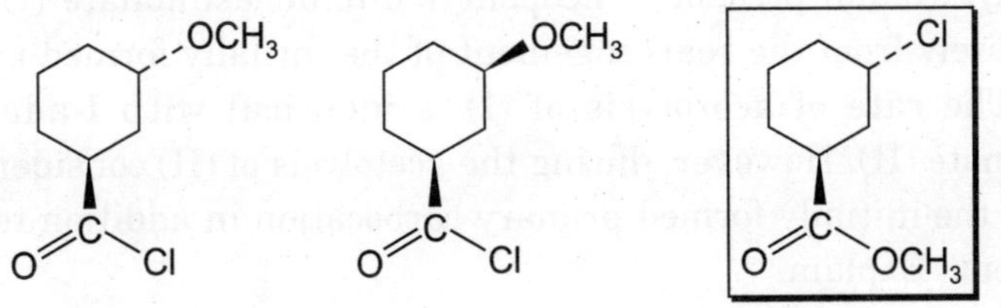

(I) (II) (III)

3.6. What conclusions do you draw from the fact the solvolysis of 1 in the presence of hydride ions ($NaBH_4$) affords small amounts of 2 and 3? (3 is a cyclopropyl derivative).

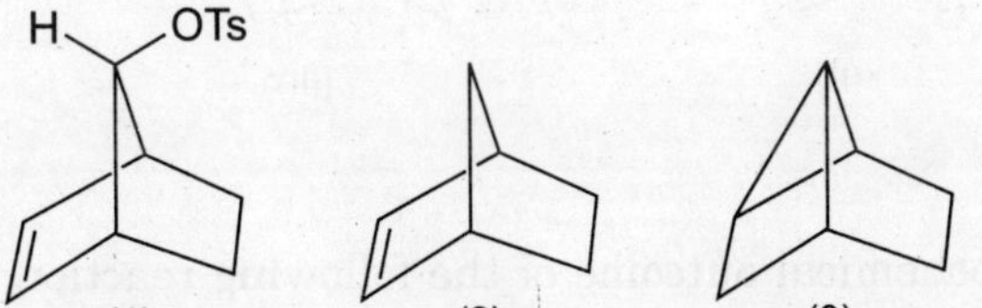

(1) (2) (3)

3.7. Primary halides of the type, $ROCH_2X$ apparently display S_N1 reactivity while most primary halides do not. Propose a resonance based explanation.

3.8. How can you explain the relative rates of solvolysis in ethanol of the following chlorides?

$C_6H_5CH_2Cl$	$C_6H_5CHClCH_3$	$(C_6H_5)_2CHCl$	$(C_6H_5)_3CCl$
(0.08)	(1)	(300)	(3×10^4)

3.9. Which compound in the each pair is expected to undergo a more rapid solvolysis on heating under reflux in ethanol.

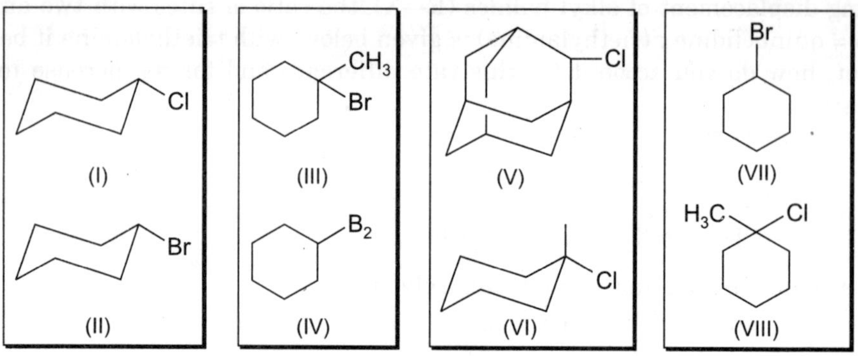

3.10. Explain the following series of reactions in the case of a steroid.

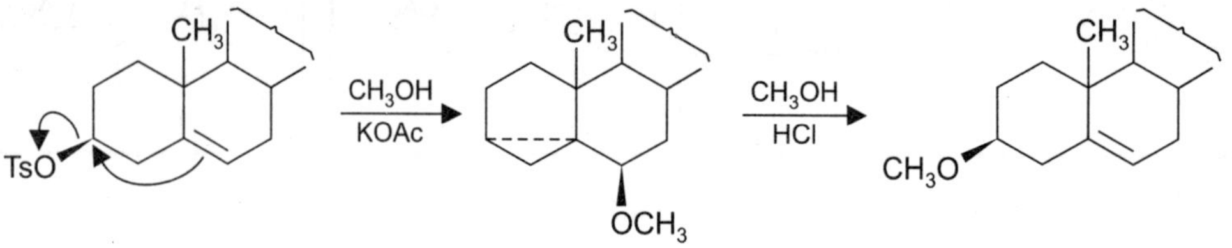

3.11. The neopentyl system present is neopentyl *p*-toluenesulfonate (I) on acetolysis gives products entirely from the rearrangement of the initially formed primary carbocations (Sec. 3.3). The rate of acetolysis of (I) is identical with I-adamantylcarbinyl *p*-toluenesulfonate (II). However, during the acetolysis of (II) considerable product (III) is formed from the initially formed primary carbocation in addition to the product (IV) of rearrangement. Explain.

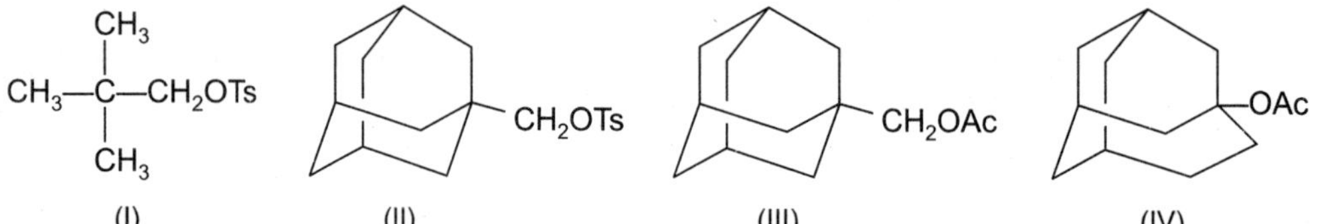

(I) Neopentyl system (II) (III) (IV)

3.12. Discuss the stereochemical outcome of the following reaction.

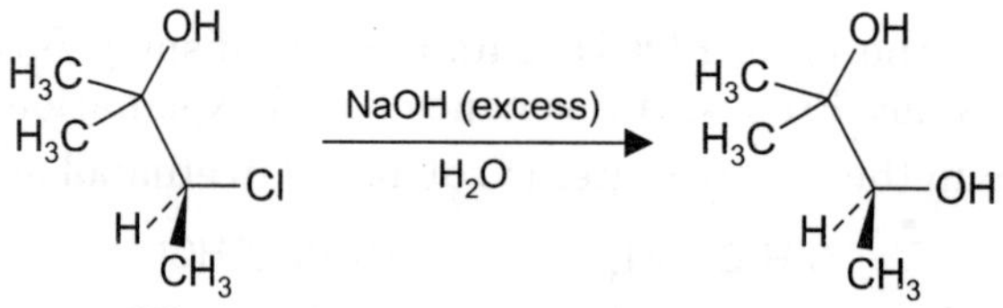

(*S*)- (*S*)-

3.13. What is neighboring group participation? Explain giving an example from a cyclohexane derivative.

3.14. Explain the following solvolysis of optically pure compounds.

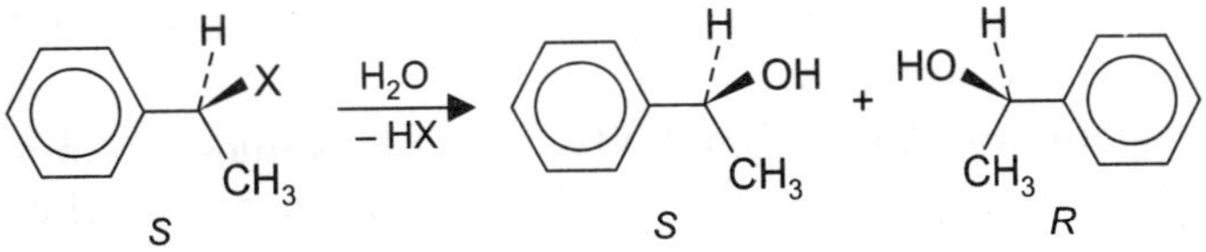

When X is Br: Racemic

When X is Cl: Excess of Inversion

3.15. Explain as to why in the following *endo* tosylate the C-1, C-7 bond fails to provide anchimeric assistance to give the bridged ion?

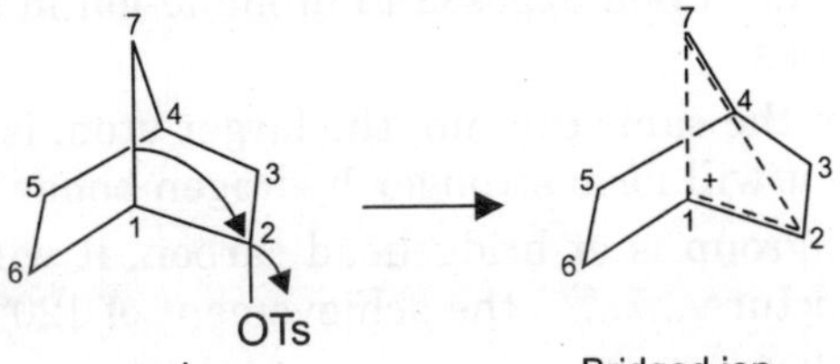

endo Bridged ion

3.16. Why α bromopropionate on treatment with methanol does not undergo the normal S_N2 reaction?

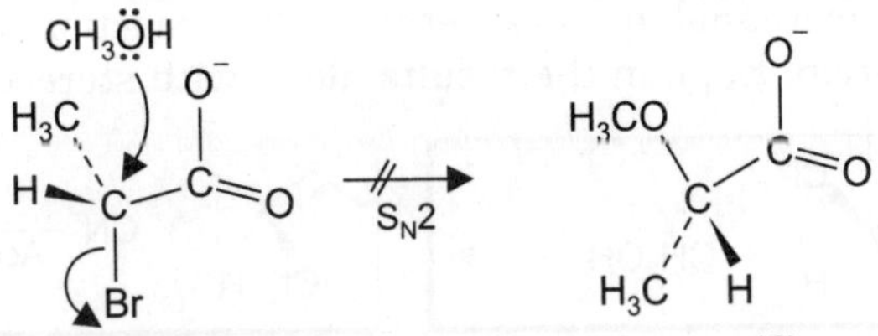

(*R*) (*S*)

3.17. Predict whether the following reactions will proceed by S_N1 or S_N2 reaction pathway.

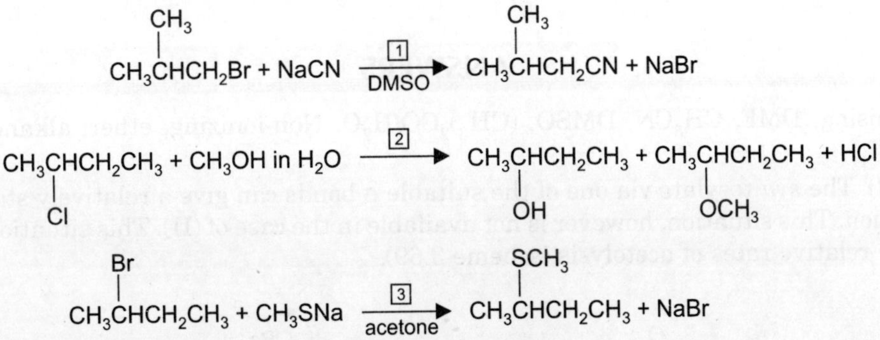

3.18. Indicate if true or false, the following statements;

(*a*) An S_N1 reaction is not affected by the strength of the nucelophile, however, a weak nucleophile favors an S_N1 reaction by disfavoring an S_N2 reaction.

(*b*) A strong nucleophile does not favor an S_N2 reaction

(*c*) A high concentration of a nucleophile favors an S_N2 reaction

(*d*) When the atoms are in the same row the stronger base is the stronger nucleophile.

(*e*) CH_2S^- is a better nucleophile than CH_3O^- since S is larger and more polarizable.

(*f*) If during the rate of S_N2 reaction of CH_3I with CH_3O^- the concentrations of both reactants are doubled with no change in other variables, the reaction would proceed four times as fast.

(*g*) Both N and F are period 2 elements thus the nucleophilicities of F^- and NH_2^- should parallel their base strengths.

Their conjugate acids are HF and NH_3 with pK_a values of ~ 5 and 40 respectively. One therefore, predicts that NH_2^- is the stronger base and thus more nucleophilic than F^-.

(*h*) Both N and O are second period elements and since NH_3 is a stronger base than H_2O, NH_3 is a better nucleophile.

3.19. Fill in the blanks :

(*a*) (*S*)-(+)-2 bromobutane when exposed to bromide ion in acetone is racemized due to .... S_N2 reactions

(*b*) If the atoms are in the same column, the larger atom is the stronger uncleophile in methanol because it will form stronger hydrogen bonds with .... atom

(*c*) When the leaving group is at bridgehead carbon, it will not undergo S_N1 reaction since the cage structure .... the achievement of 120° bond angle required for an sp^2 hybridized carbon.

(*d*) A leaving group at bridgehead position will not undergo an S_N2 reaction due to steric hindrance to .... attack.

3.20. Solvolysis of (*a*) in methanol gives two ethers while reaction of (*b*) gives a product with inverted configuration. Explain the results along with stereochemical outcome.

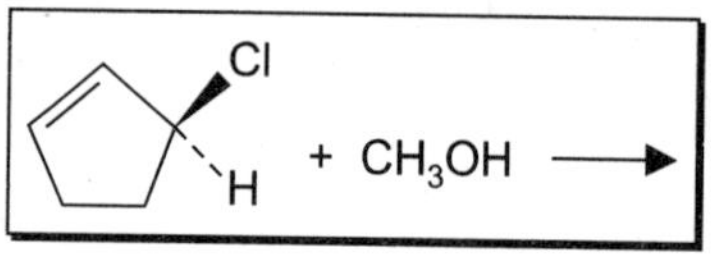

(a)

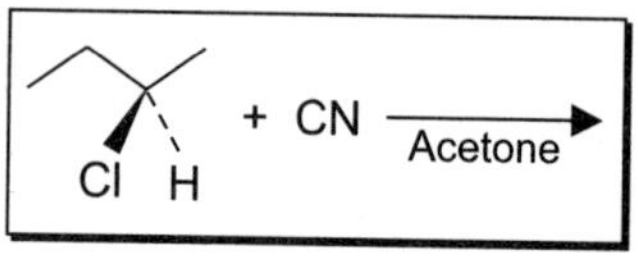

(b)

3.21. What is a cyclic phenonium ion? How one can demonstrate its existence?

ANSWERS

3.3. Ionising, DMF, CH_3CN, DMSO, $(CH_3)_2CO/H_2O$. Non-ionizing, ether, alkanes, $CHCl_3$, CCl_4.

3.4. (III). The *syn* tosylate via one of the suitable σ bonds can give a relatively stable allylic cation. This situation, however is not available in the case of (II). This situation explains the relative rates of acetolysis (scheme 3.69).

(I) 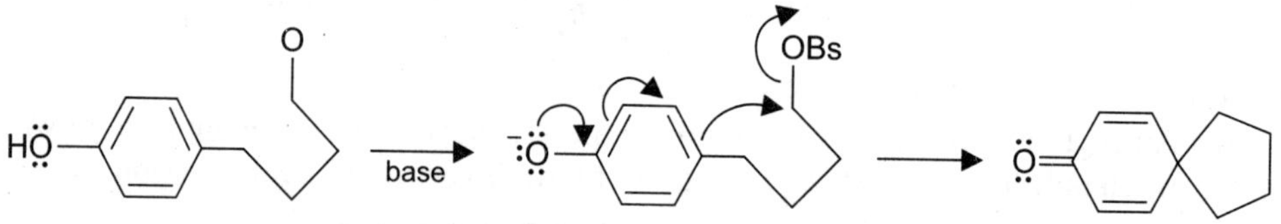

(II) 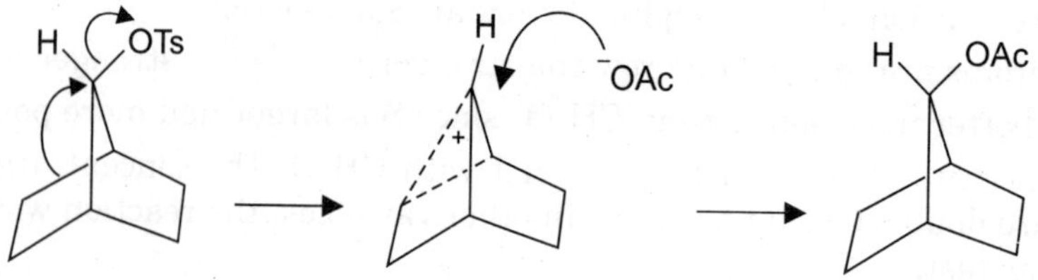

(III) 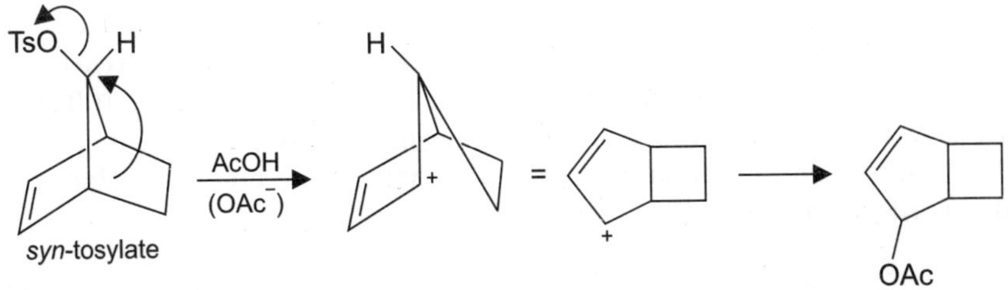

3.5. Only the *cis* compound (II) will react via neighboring group participation. The mechanism is unique since it involves the conversion of a *cis* 1, 3-cyclohexane derivative in to a *trans*-product as well as the change of places between methoxy and chlorine groups.

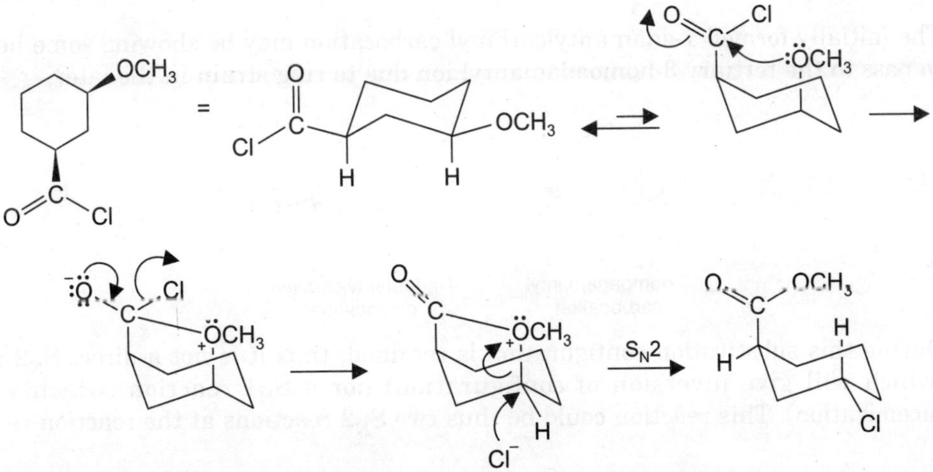

3.6. Neighboring group participation by the suitably located *pi* electrons in the *anti* tosylate involving the non classical intermediate, 1, some resonating structures have "cyclopropane-like" characters.

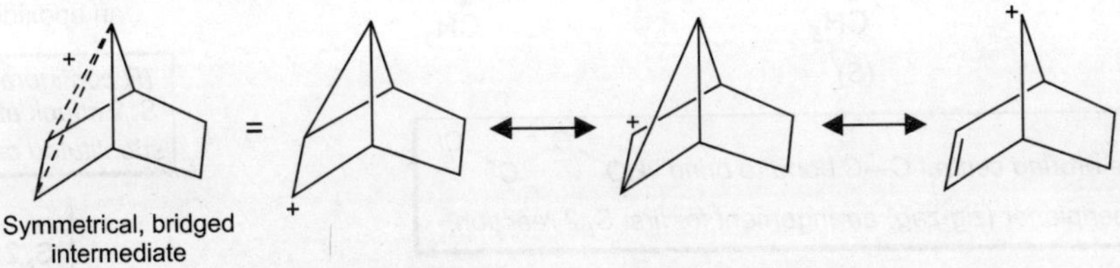

3.7. This is due to carbocation stabilization by a neighboring hetero atom.

$$CH_3CH_2{-}O{-}CH_2Cl \longrightarrow Cl^- + \left[\begin{array}{c} CH_3CH_2{-}\ddot{O}{-}CH_2^+ \\ \updownarrow \\ CH_3CH_2{-}\overset{+}{\ddot{O}}{=}CH_2 \end{array}\right] \xrightarrow[C_2H_5OH]{-H^+} CH_3CH_2{-}O{-}CH_2OC_2H_5$$

3.9. Compound (II) since bromide is a better leaving group than chloride ; compound (III) since it will generate a more stable 3° carbocation; compound (VI); compound (VIII).

3.10. The cylopropyl ring is formed by the participation of the double bond which displaces the leaving group with inversion at the stereocenter. The methoxy group then enters from the face of the starting reactant from which the leaving group departed. The reaction is known as the ***iso*-cholesterol** (or ***i*-cholesterol**) **rearrangement.** On adding acid, the *iso*-cholesteryl ether is converted back into a cholesteryl derivative. It has been shown that displacement reactions of cholesteryl tosylate are about 100 times as fast as reactions of cyclohexyl tosylate. The double bond participates in displacement of the tosylate group to form a homoallylic cation, in which the charge is shared between the two ends of the cation.

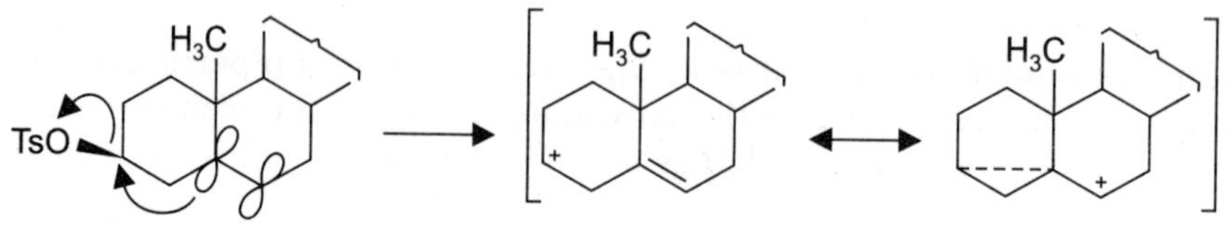

3.11. The initially formed 1-adamantylcarbinyl carbocation may be showing some hesitation to pass to the tertiary 3-homoadamantyl ion due to ring strain in the later.

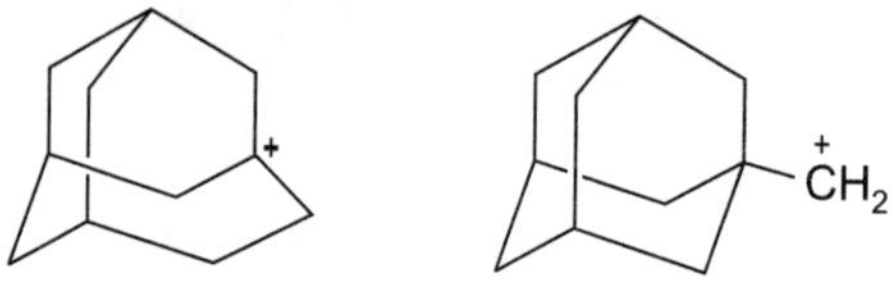

3-homoadamantyl carbocation

1-adamantylcarbinyl carbocation

3.12. During this substitution configuration is retained, thus it is not a direct S_N2 reaction (which will give inversion of configuration) nor a S_N1 reaction (which will give racemization). This reaction could be thus two S_N2 reactions at the reaction center.

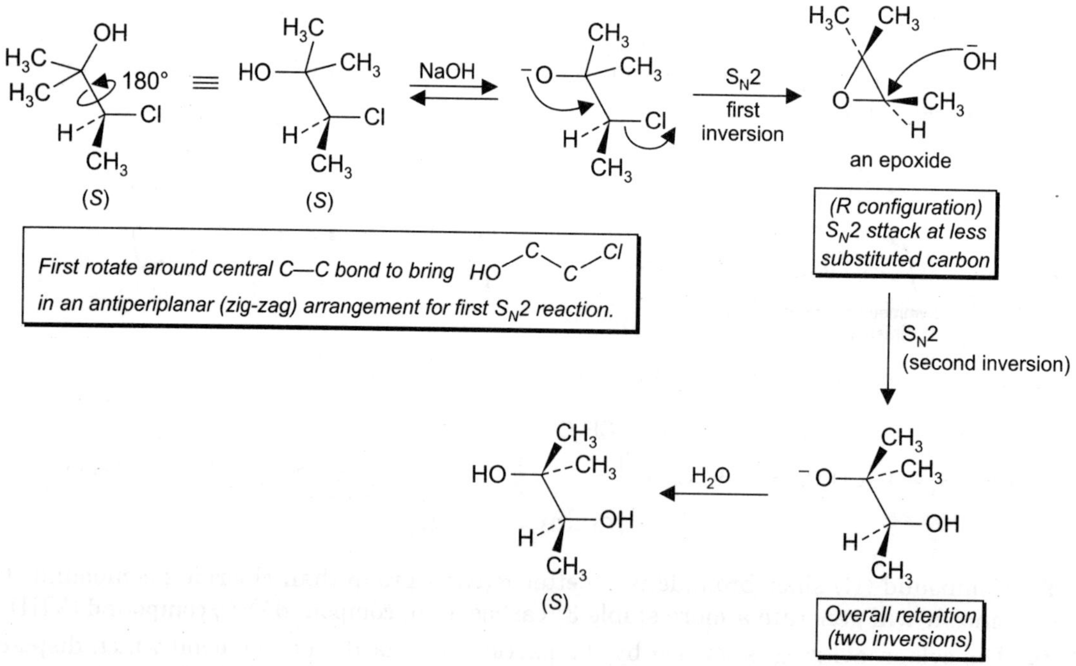

3.14. Chloride is a good but not the best leaving group (see scheme 3.16), its exit to give the carbocation is somewhat sluggish. It thusremains in the vicinity of the carbocationic carbon to shield one face, the attack of the nucleophile thus for sometime occurs entirely from opposite face to give more of inversion. On the other hand bromine is a far better leaving group so it clearly departs without shielding the face from which it departs, in this case the racemization is complete.

3.15. The formation of the bridged ion would involve the contraction of one of the five membered rings of the norbornyl system to a four membered ring to cause strain.

3.16. Due to neighboring group participation by the carboxylate ion to give an α lactone (intramolecular S_N2 reaction). α-Lactones are smallest cyclic esters, which cannot be isolated (reactive intermediates). In the second step methyl alcohol opens the α-lactone ring by a second S_N2 displacement. The overall result of this mechanism is net retention of configuration.

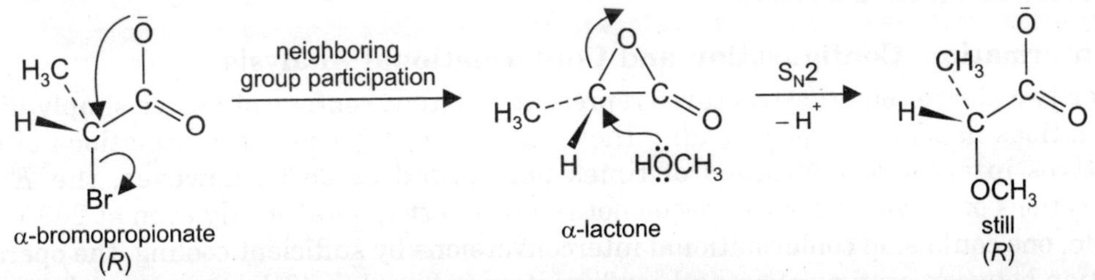

3.17. *Reaction 1.* S_N2 pathway, since the substrate is primary and cyanide ion is a good nucleophile. Dimethyl sulfoxide (DMSO), a polar aprotic solvent, is a particularly good solvent to bring about nucleophile assisted substitution reactions because its ability to solvate cations (in this case, Na^+) is good, while its ability to solvate anions (in this case, CN^-) is poor.

Reaction 2. S_N1 pathway since 2-chlorobutane is a secondary alkyl halide that can ionize to form a fairly stable secondary carbocation. The mixture of methanol and water is a good ionizing solvent. Both water and methanol are poor nucleophiles.

Reaction 3. S_N2 pathway. Bromide ion is a good leaving group on a secondary carbon. The sulfide ion is a strong nucleophile. Acetone, a polar aprotic solvent, is a good medium for S_N2 reaction.

3.18. (*a*) True; (*b*) false; (*c*) true; (*d*) true; (*e*) true; (*f*) true; (*g*) true; (*h*) true.

3.19. (*a*) Reversible; (*b*) smaller; (*c*) prevents; (*d*) backside.

3.20. Compound (*a*) is chiral and (*S*)-enantiomer, ionization of chlorine gives a resonance stabilized secondary allylic carbocation which is achiral and is expected to give almost a racemic mixture of ethers. Compound (*b*) is (*R*)-enantiomer and with a strong nucleophile and aprotic solvent gives S_N2 displacement with inverted configuration of the substrate.

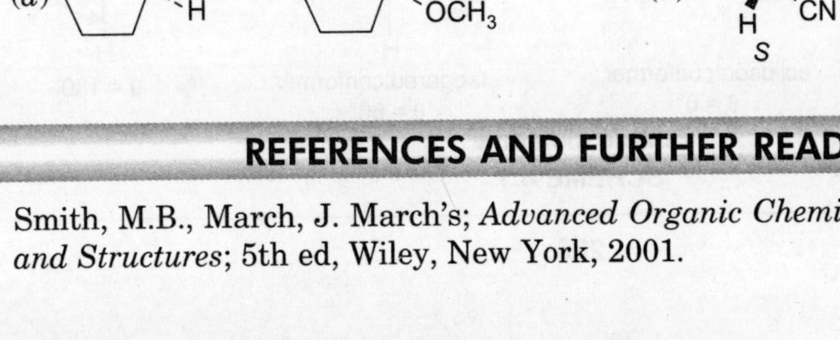

REFERENCES AND FURTHER READING

1. Smith, M.B., March, J. March's; *Advanced Organic Chemistry: Reactions, Mechanisms and Structures*; 5th ed, Wiley, New York, 2001.

CHAPTER 4

Conformations and Stereoisomerism of Acyclic and Cyclic Systems

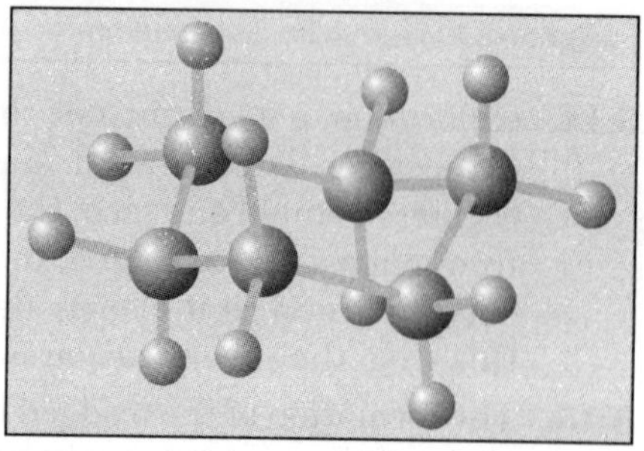

4.1 RESTRICTED ROTATION ABOUT SINGLE BONDS—CONFORMATIONS OF ETHANE AND BUTANE

(*a*) Conformation, Configuration and Conformational Analysis

Whether two stereoisomeric structures represent different configurations or simply different conformations depends on how readily they interconvert. Various conformations of ethane derivatives interconvert millions of times per second at 25°C, however, the *E* and *Z* configurations of ethylene derivatives do not interconvert spontaneously even at 200°C. As, in principle, one could stop conformational interconversions by sufficient cooling, the operational distinction between configuration and conformation is based on if the structures interconvert at room temperature, 25°C. The various shapes that a molecule can adopt by rotation about single bonds are called conformations. Repulsive interaction between bonds (bonding electrons) on adjacent atoms lead to restricted rotation (torsional strain), while the steric strain in a molecule increases as the size of groups increases. A study of various conformations of a compound and their relative stabilities is called conformational analysis.

Conformation and Configuration

- *Stereoisomers separated by high energy barrier (> 100 kJ mol^{-1}) are isolable at room temperature and are called configurational isomers. Stereoisomers separated by relatively low energy barrier (< 60 kJ mol^{-1}) are easily interconvertible at ambient temperature and are called conformational isomers or conformers. Unless a conformation is held rigid by a small ring or double bonds, a molecule can have an infinite number of conformations, but only one configuration.*

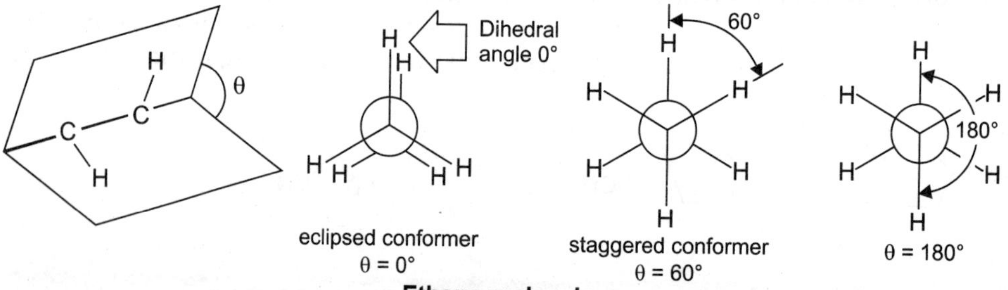

Ethane molecule

SCHEME 4.1

(*b*) Dihedral Angle

The dihedral angle is an important stereochemical parameter, much more than bond length and bond angle which are structural parameters. When an ethane molecule rotates about its carbon—carbon bond, two extreme conformations can result: the staggered conformation and the eclipsed conformation (scheme 4.1). An infinite number of conformations between these two extreme (skewed conformations) are also possible. The potential energy relationships between conformations, is conveniently discussed in terms of dihedral angle. A dihedral angle, (Greek letter θ), is the angle generated by two inter-secting planes. In the Newman projection of the eclipsed conformation of ethane (scheme 4.1), two H—C—C planes are shown, the angle at which these planes intersect (the dihedral angle) is 0° while in a staggered conformation the dihedral angle of the two H—C—C planes is 60°, or 180°.

(*c*) Conformations of Ethane (Conformational Analysis)

Two of the remarkable conformations of ethane are eclipsed conformation and staggered conformation.

The staggered conformations of ethane are the low energy forms while the eclipsed conformations represent transition states. The energy difference between the two conformers, which infact represents the barrier of interconversion between two staggered conformers is about 3 kcal/mol (12 kJ/mol) (scheme 4.1*a*). Thus in an alkane in general, each pair of eclipsed bonds that it contains leads to an energy cost of about 1 kcal/mol (4.2 kJ/mol). This torsional strain is primarily due to electron pair repulsions when the bonds are eclipsed. The activation energy for rotation about the C—C bond in ethane is small, the thermal energy from the surroundings at the room temperature (20 kcal/mol, 83.7 kJ/mol) is sufficient to cause staggered conformations of ethane to interconvert millions of times each second at room temperature. Thus the internal rotation though not completely free but represents a rapid process. The barriers for rotations about most single bonds are in the (3–5 kcal/mol, 12–21 kJ/mol) range to allow fast rotations about most single bonds at room temperature.

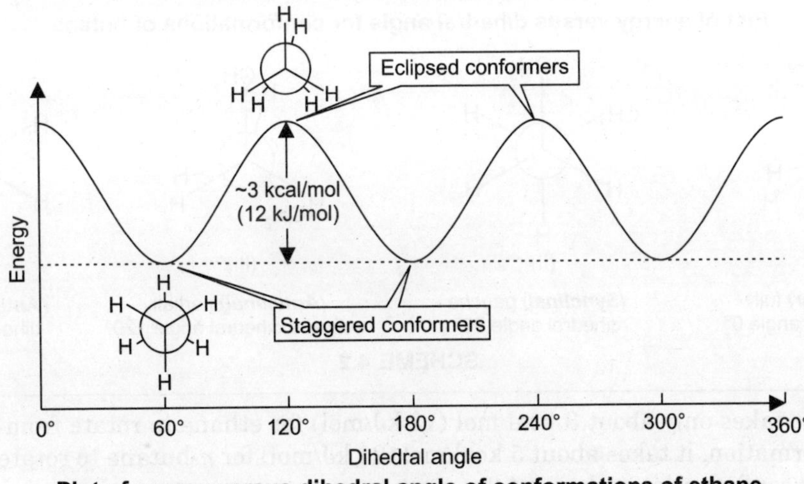

Plot of energy versus dihedral angle of conformations of ethane

SCHEME 4.1a

(*d*) Conformations of Butane (Conformational Analysis)

For a compound of the type A—CH_2—CH_2—B (a 1,2-disubstituted ethane) *e.g.*, *n*-butane there are four extremes. A fully eclipsed conformation *syn*-periplanar) and partly eclipsed conformation *anticlinal*). A fully staggered conformation (*anti*periplanar) and another staggered conformation (*gauche* or *syn*-clinal). In ethane all the staggered conformations are equivalent, in *n*-butane on the other hand there are two different staggered conformations. In one of these (*synclinal*) there is a gauche relationship between methyl groups while in other conformation the relationship is *anti* (*anti*periplanar). Two groups are said to be *gauche* when the dihedral angle between them is 60° and in an *anti* conformation the groups are maximum distance apart, the dihedral angle between them being 180°. These staggered conformations (scheme 4.2) are free of torsional strain however, the *anti*periplanar conformer is the most stable, because the methyl groups are far apart. The methyl groups in the *gauche* conformations are close-enough to each other, so that the van der Waals forces between them are *repulsive*. This repulsion causes the *gauche* conformations to have approximately 0.9 kcal/mol (3.8 kJ/mol) more energy than the *anti* conformation. The eclipsed conformations represent energy maxima in the potential energy diagram. Eclipsed conformations not only have torsional strain, they have additional van der Waals repulsions arising from either eclipsed methyl groups and hydrogen atoms or between the eclipsed methyl groups.

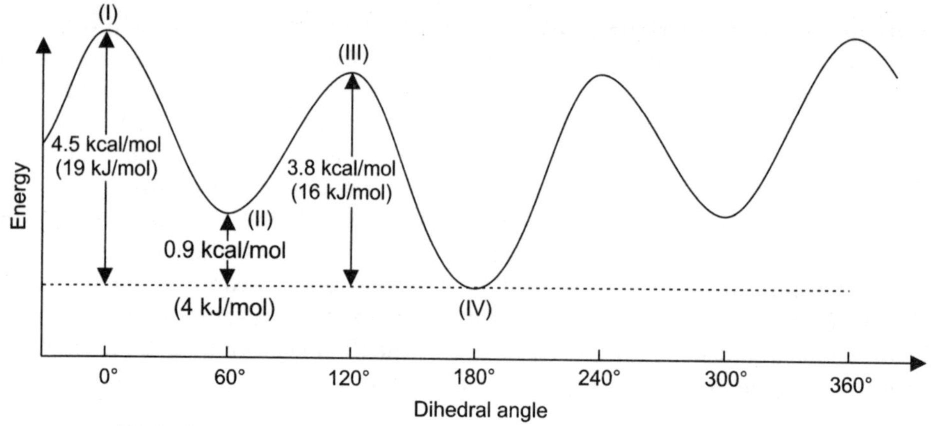

Plot of energy versus dihedral angle for conformations of butane

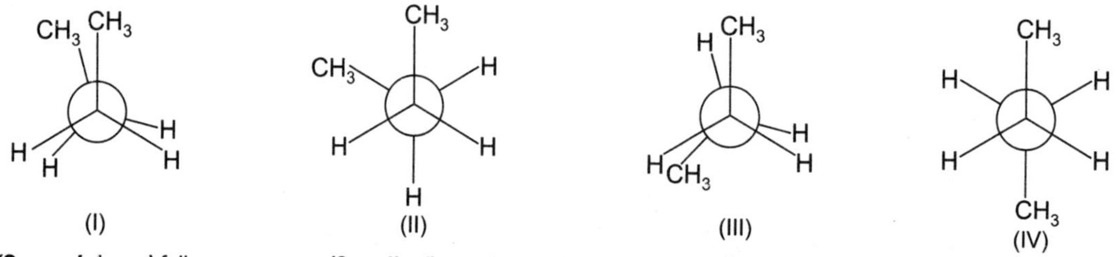

(I) ***(Synperiplanar)*** fully eclipsed dihedral angle 0°

(II) ***(Synclinal)*** *gauche* dihedral angle 60°

(III) ***(Anticlinal)*** partially eclipsed dihedral angle 120°

(IV) ***(Antiperiplanar)*** *anti* dihedral angle 180°

SCHEME 4.2

While it takes only about 3 kcal/mol (12 kJ/mol) for ethane to rotate from staggered to eclipsed conformation, it takes about 5 kcal/mol (21 kJ/mol) for *n*-butane to rotate from *anti* to *syn*periplanar conformation (scheme 4.2).

For butane and most of the molecules of the type A—CH_2—CH_2—A and A—CH_2—CH_2—B, the *anti* conformation is the most stable. In one group of compounds, containing small electronegative atoms like fluorine and oxygen *gauche* form is predominant. There is yet no explanation for this and the presence of intramolecular H bonding *e.g.*, in the case of 2-fluoroethanol as the reason has been ruled out. In one case the two conformational isomers (I and II, scheme 4.3) of a single aliphatic hydrocarbon have been isolated. The isomers have been separately crystallized and are stable at room temperature. The dihedral angles are, however, distorted from 60°, due to steric hindrance between bulky groups [for defining conformations (I and II, scheme 4.3, from the dihedral angle consult scheme 4.4].

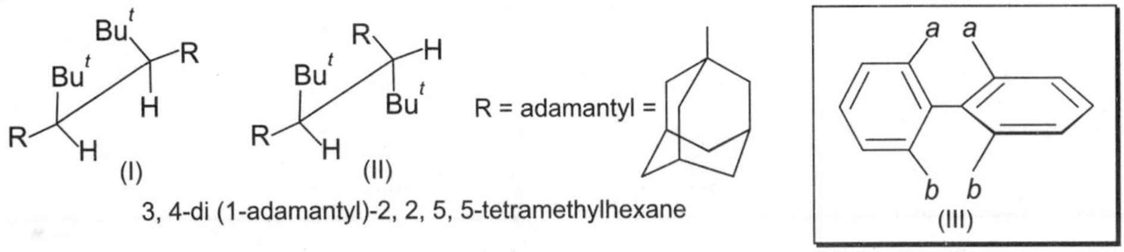

SCHEME 4.3

Atropisomers represent two conformations of a molecule whose interconversion is sufficiently slow under a given set of conditions to allow separation and isolation. Thus, the two enantiomeric conformations of the *ortho*–tetrasubstituted biphenyl (III, scheme 4.3) display hindered rotation about the bonds joining the two rings.

(*e*) Conformations of Unsymmetrically Substituted Alkanes

Unlike ethane and *n*-butane (where various conformations are easily recognized from the dihedral angle of atoms or groups these being symmetrical molecules) when the substituents on the two carbon are different, the following considerations are applied to define the conformations from the dihedral angle.

- In 2-bromopropanamide, all the ligands on the two carbons are different. The dihedral angle is then measured from one substituent on each carbon which is preferred by sequence rules *i.e.,* from oxygen and bromine as in eclipsed and *anti* conformations (scheme 4.4).
- In 3-chloropentan-2 ol a hydrogen atom is common to both carbons, in such cases when an atom or a group is common to both carbons dihedral angle is measured between these two ignoring the higher priority of remaining substituents (scheme 4.4).

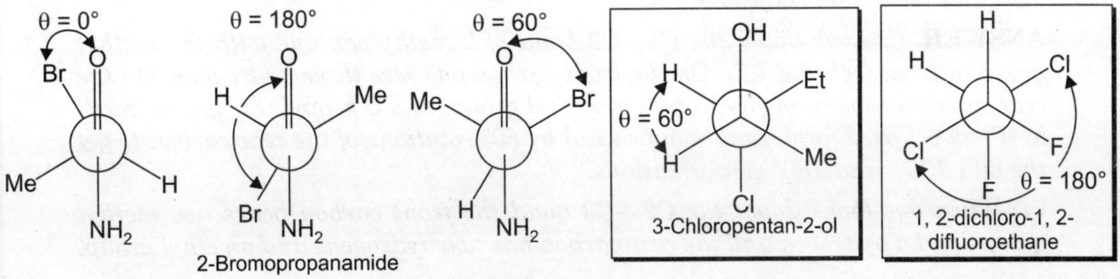

SCHEME 4.4

- When two atoms or groups are common to both the carbons, the pair of highest priority defines the dihedral angle as in 1, 2-dichloro-1, 2-difluoroethane (scheme 4.4)

EXERCISE 4.1

Predict if the following pair of structures (scheme 4.5) are constitutional isomers, different conformations of the same compound or stereoisomers which cannot be interconverted via rotation about single bonds.

(Contd...)

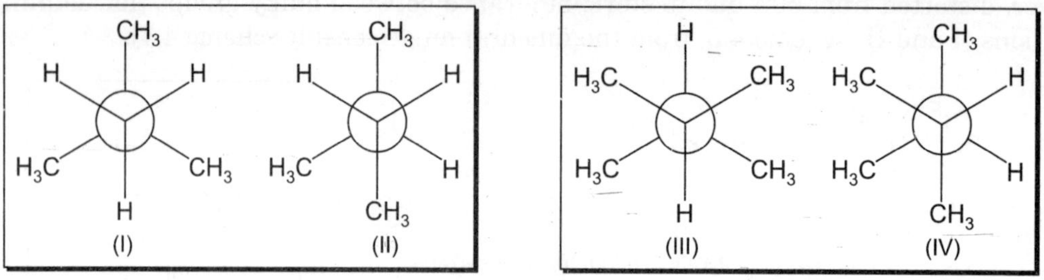

SCHEME 4.5

ANSWER. *The trick is to write the structures in the Lewis form to show their atomic connections (scheme 4.6) to decide the problem. Thus (I) and (II) are constitutional isomers while (III) and (IV) are the different staggered conformations of the same molecule.*

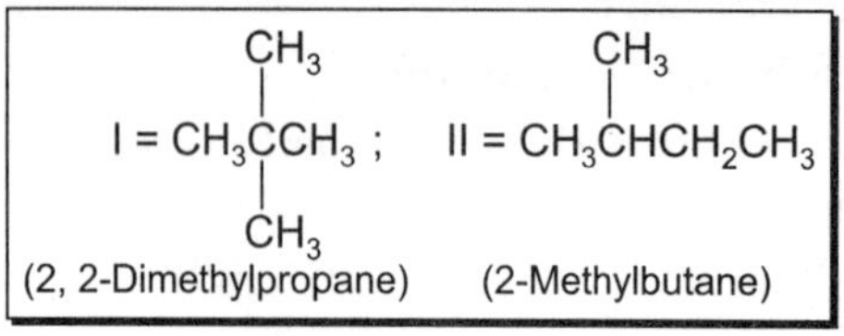

(III) = (IV) = $(CH_3)_2CHCH(CH_3)_2$
(2, 3-dimethyl butane)

SCHEME 4.6

EXERCISE 4.2

Draw one eclipsed conformation in Newman projections (i) by looking down C1 and C2 bond and (ii) C2 and C3 bond in 2-methylpentane. In the case (i) draw 60°, 120° and 180° conformations.

ANSWER. (*i*) *Look down the C1—C2 bond of 2-methylpentane with the methyl group in front (scheme 4.7). On the front carbon one sees three hydrogens. On the rear carbon one sees one hydrogen, a methyl group, and a propyl (Pr) group. Start at 0°, an eclipsed form, and then proceed by 60° rotations of the rear carbon to get the 60°, 120° and 180° conformations.*

(*ii*) *When one looks down the C2—C3 bond the front carbon bears two methyl groups and a hydrogen and the rear carbon has two hydrogens and an ethyl group.*

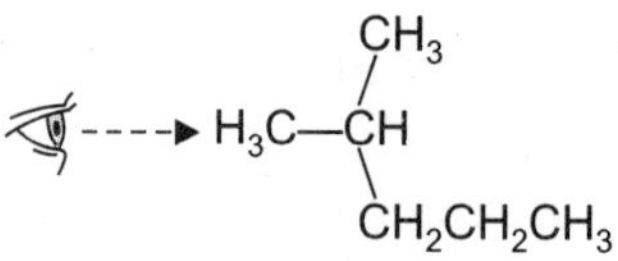

(*i*)

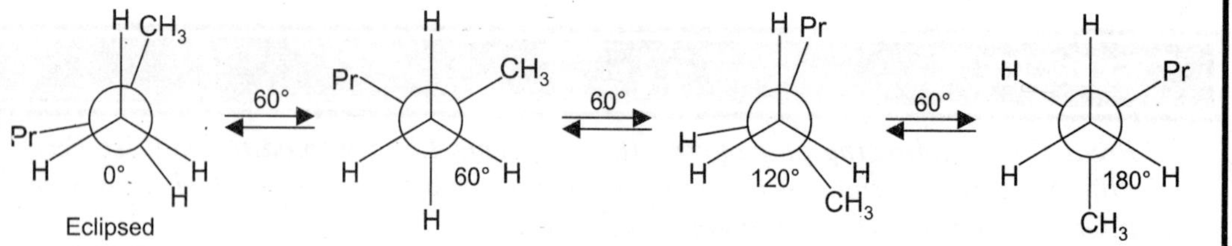

(Contd...)

(*ii*)

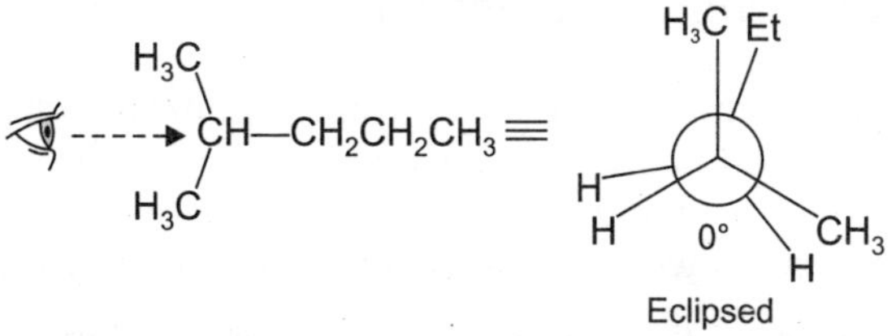

SCHEME 4.7

Dihedral Angle and Torsion Angle

Strictly speaking a torsion angle is not the same as a dihedral angle. Thus while dihedral angle is the (unsigned) angle between two planes defined by A-X-Y and X-Y-B in molecules of the type A-X-Y-B, the torsion angle is the angle subtended by A and B across the bond X-Y. Unlike a dihedral angle, torison angle is directional, (+) when measured in a clockwise direction and (–) when measured in an anticlockwise direction starting from the front substituent A and ending at the back substituent B. One may measure it from 0° to 360° continuously following a clockwise direction and is generally designated by the smaller angle prefixed by (+) or (–) (scheme 4.7a). Thus one says that the lowest potential energy for internal rotation in ethane corresponds to the staggered (θ = ± 60°, 180°) and the highest to the eclipsed conformations (θ – 0°, ± 120°).

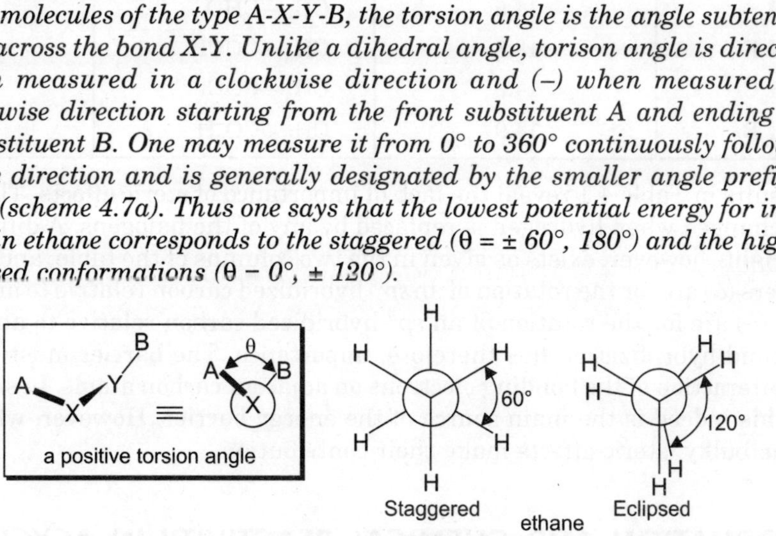

SCHEME 4.7a

In the molecules (scheme 4.7b) the preferred torsion angle (θ) for a pair of bonded carbon atoms depends on the number of substituents on the C atoms, or it may be fixed due to rigid structure.

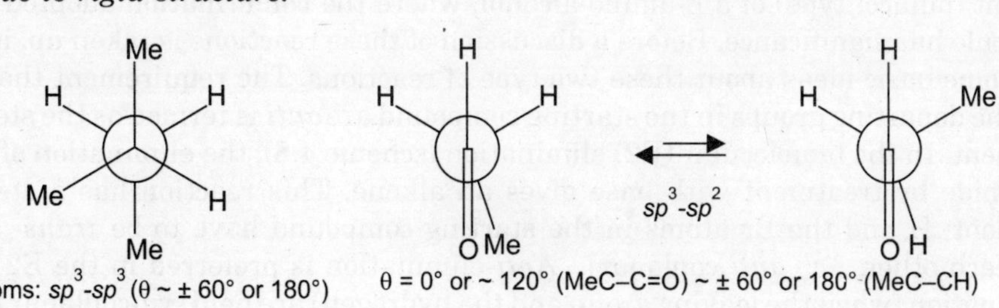

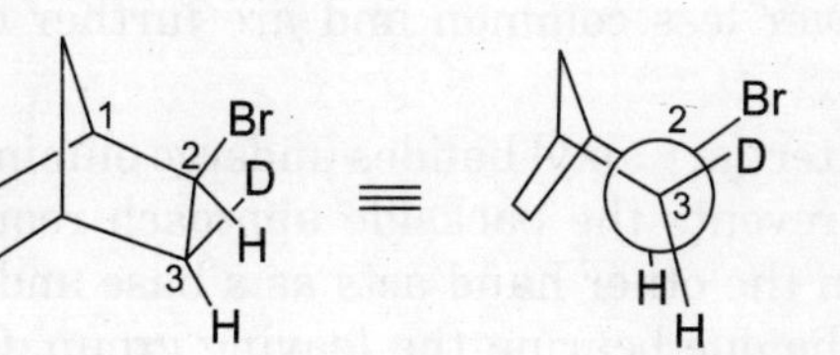

The hydrogen on C3 and C2 bromine have a dihedral angle of 120° between them while it is 0° between C3 D and C2 Br. These angles are fixed and have consequences in e.g., E2 eliminations.

SCHEME 4.7b

4.2 ORIGIN OF CONFORMATIONAL ENERGY

An inspection of the model of ethane shows that the hydrogen atoms on adjacent carbon atoms are far apart making steric interaction negligible. The basic cause of the energy barrier could be studied only after accurate measurements of the energy difference between the maxima and minima for a number of molecules were made with the help of microwave spectroscopy.

Table 4.1 Conformational Energy Barriers of Various Compounds

Compound	*Energy barrier (kJ mol⁻¹)*	*Compound*	*Energy barrier (kJ mol⁻¹)*
	(*a*)		(*b*)
CH_3—CH_3	11.5	CH_3—CHO	4.8
CH_3—CH_2F	13.8	CH_3—$COCH_3$	3.6
CH_3—CH_2Cl	14.9	CH_3—COCl	5.7
CH_3—CH_2Br	14.9	CH_3—CO_2H	2.0

The results in Table 4.1 reveal the lack of importance of steric effects. The barrier is not significantly changed when hydrogen is replaced by any of the halogens. A difference between the barrier heights however, exists as given in the two columns of the table, and it is significant that the barriers (*a*) are for the rotation of an sp^3-hybridized carbon relative to an sp^3-hybridized carbon, while (*b*) are for the rotation of an sp^3-hybridized carbon relative to an sp^2-hybridized carbon. The bond hybridization has therefore, importance. The barrier must therefore, arise mainly from interaction of the bonding electrons on adjacent carbon atoms. In simple molecules like ethane, this indeed is the main source of the energy barrier. However, when substituent groups become bulky, steric effects make their contribution.

4.3 CONFORMATION AND CHEMICAL REACTIVITY IN ACYCLIC SYSTEMS (E2 ELIMINATION AND PINACOL REARRANGEMENT)

Under this section two reactions are discussed *i.e.,* E2 elimination and the molecular rearrangement (pinacol type) of a β-amino-alcohol, where the conformation adopted by the reacting molecule has significance. Before a discussion of these reactions is taken up, it is necessary to have some basic ideas about these two type of reactions. The requirement that the migrating and the departing groups in the starting compound are *anti* is termed as the stereoelectronic requirement. In the bimolecular (E2) elimination (scheme 4.8), the elimination of HBr from an alkylbromide by treatment with base gives an alkene. This reaction has a stereoelectronic requirement; H and the Br atoms in the starting compound have to be *trans-* and coplanar (*anti* to each other, *i.e., anti* coplanar). *Anti*-elimination is preferred in the E2 reaction, the *syn*-elimination (when the leaving group and the hydrogen) are held *syn*-coplanar in an eclipsed conformation of a rigid compound are, however less common and are further discussed in chapter 5.

On treatment with a strong nucleophile, tertiary alkyl halides undergo elimination rather than substitution because steric hindrance prevents the backside approach required by the S_N2 reaction (scheme 4.8). The nucleophile, on the other hand acts as a base and approaches the protons on the carbon atom adjacent to the one bearing the leaving group. One of these hydrogen atoms is removed by the base, the electrons in the C—H bond move to generate the π bond and expel the leaving group, a concerted process E2 elimination (since in the dehydrohalogenation presented here, the rate of reaction depends on the concentration of both substrate and base).

For an E2 reaction, the leaving groups are almost always *anti* in the transition state, which requires this conformation to be accessible if a reaction is to occur readily.

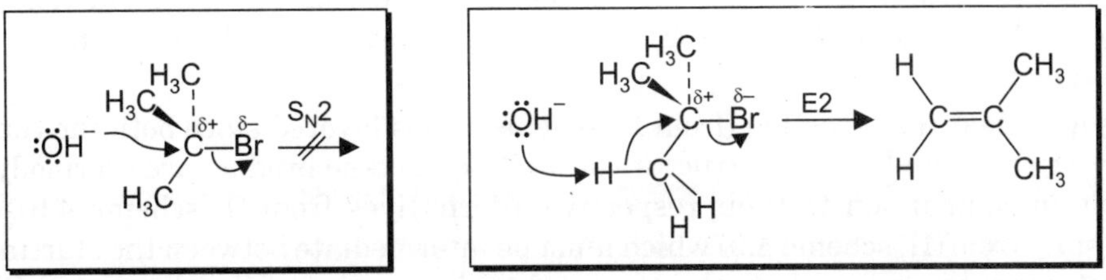

SCHEME 4.8

(*a*) Elimination from 2-Bromobutane

Consider an alkyl bromide, 2-bromobutane. Here there is a choice of H atoms which can line up in an *anti*-periplanar geometry with the Br atom. Elimination of HBr could give but-1-ene or but-2-ene. The reaction is, however, regioselective in favor of but-2-ene (more substituted alkene).

The stereoelectronic requirement of this elimination reaction must involve only those (staggered) conformations of bromobutane molecule in which the ligands to be eliminated attain

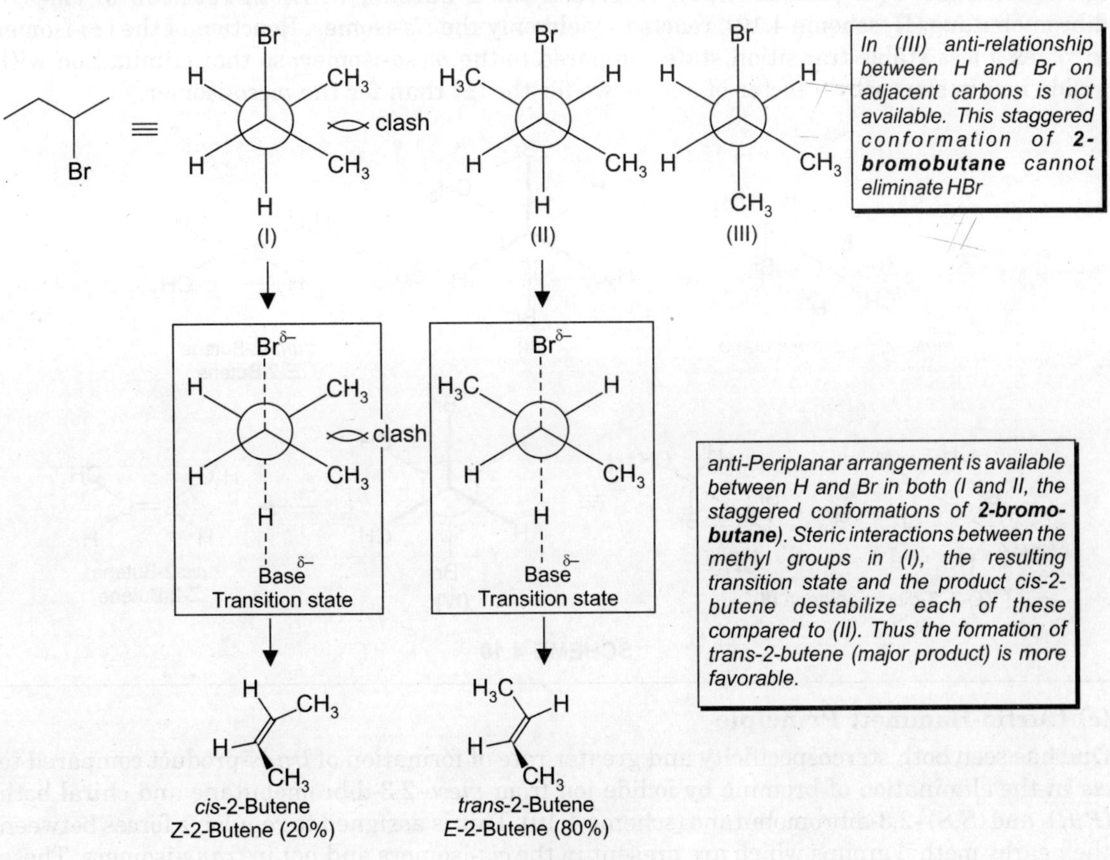

SCHEME 4.9

an antiperiplanar arrangement. There are three possible staggered conformations of 2-bromobutane (I-III, scheme 4.9). Only the conformations (I and II scheme 4.9) satisfy the coplanarity condition of departing groups. Thus, I gives rise to the *cis* olefin while II to its *trans* isomer. It has been found experimentally that the reaction gives mainly the *trans* alkene to prove that the elimination involving rotamer (II) proceeds more readily than via (I). The three arrangements (I-III scheme 4.9) of bromobutane are however, just three of infinite number of possible rotamers.

It is now clear that reaction (from II, scheme 4.9) is favored since both the conformation of the starting compound and the structure of the *trans*-alkene product are thermodynamically more stable in comparison to their respective alternatives from (I, scheme 4.9). Thus, the transition state from (II, scheme 4.9) which must be intermediate between the starting material and product, must be preferable to the one involved in reaction of (I, scheme 4.9) Elimination from 2-bromobutane has been discussed in more detail in Chapter 5 (see, schemes 5.33 and 5.35*a*).

(*b*) Elimination from Stereoisomers of 2-3-Dibromobutane

These stereoelectronic requirements also apply to the bimolecular elimination (E2) of a molecule of bromine from a dibromide, either with iodide ions or by the action of a metal, such as zinc. Elimination can occur only *via* a conformation of the starting compound which places the two bromine atoms in a *anti*-periplanar arrangement regardless of the fact if or not this is the most stable conformation (scheme 4.10), *meso*, 2,3-Dibromobutane can eliminate bromine only from the conformation (I, scheme 4.10), to give *trans*-2 butene, while in reaction of (±)-2,3-dibromobutane (II, scheme 4.10), reacts to yield only the *cis*-isomer. Reaction of the (±)-isomer involves a less stable transition state compared to the *meso*-isomer, so that elimination with iodide ions is slower by a factor of about two for the (±) than for the *meso*-isomer.

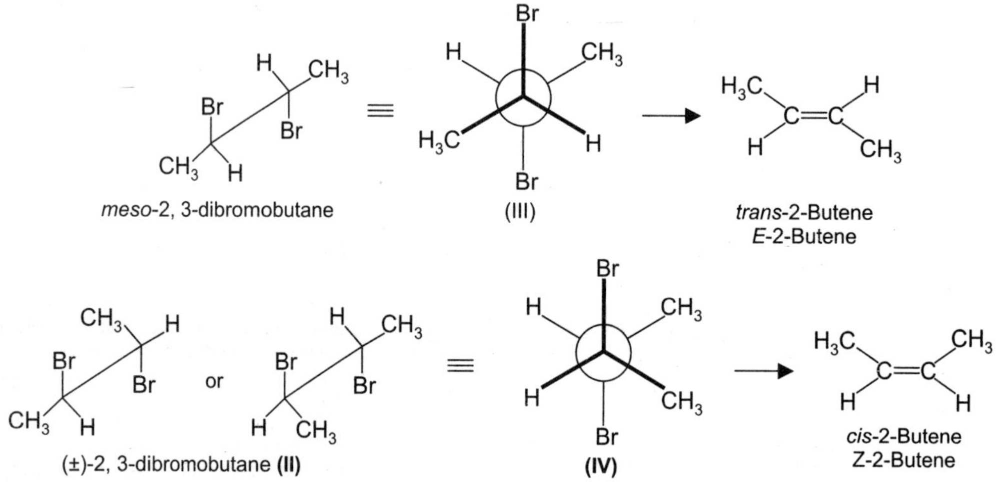

SCHEME 4.10

(*c*) Curtin-Hammett Principle

One has seen both stereospecificity and greater rate of formation of *trans*-product compared to *cis* in the elimination of bromine by iodide ion from *meso*-2,3-dibromobutane and chiral both (*R*,*R*)- and (*S*,*S*) -2,3-dibromobutane (scheme 4.10). This is assigned to repulsive forces between the nearby methyl groups which are present in the *cis*-isomers and not in *trans*-isomers. Those repulsive forces are also present to some degree in the transition states in which the *cis*-product is being developed (for another example of a stereospecific reaction see scheme 5.19).

> *When the anti-conformation, either cannot be attained for geometrical reasons or if attained it would be highly sterically congested, a slower syn pathway for elimination is followed. (syn-elimination is observed in quaternary ammonium ions.*

A compound, however, may undergo a stereoselective elimination (as in scheme 4.9). When 2-bromobutane undergoes base induced elimination of hydrogen bromide, the reaction can occur through either of the two *anti*-conformers (I and II, scheme 4.10*a*) to give *cis* and *trans*-2-butene respectively, the latter predominates. The following points may be noted:

- The formation of *trans* product in larger proportion may be due to two reasons: The conformation of the reactant leading to it may be heavily populated than that which gives the minor (*cis*) product; the transition state which leads to the major (*trans*) product may be of lower energy than the one leading to minor product.
- Both the above factors or either of these may be responsible for the formation of *trans* product.

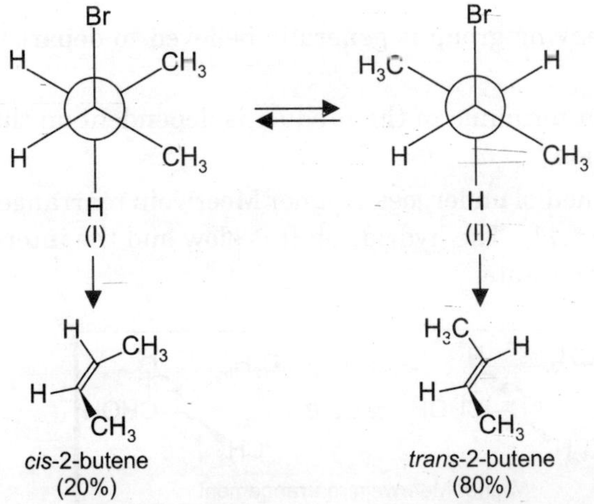

> *Cutrin-Hammett principle relates the product distribution to relative energies of the transition states from e.g., the conformers of 1-bromobutane (conformationally heterogeneous starting material) rather than the relative populations of the conformers. [The products must be non-equilibrating and the conformers be rapidly inter-convertable relative to the rate of product formation]*

SCHEME 4.10a

- The activation energy of the elimination is larger than the rotational energy barrier which separates the two *anti* conformations (scheme 4.10*a*).
- Under such reaction conditions the Curtin-Hammett principle correlates the product distribution with the energies of the transition states for two different pathways (see, scheme 4.3) followed by two different conformers of a substrate which give non-equilibrating products.
- The Curtin-Hammett principle implies that in a chemical reaction that gives one product from one conformer and a different product from another conformer (provided the products do not interconvert while the two conformers are rapidly interconverting relative to the rate of product formation). The product composition is not determined by the relative populations of the ground state populations but depend almost entirely on the relative energies of the representative transition states involved.

- During the dehalogenation of 2-bromobutane, *trans*-2-butene predominates over *cis*-2-butene (scheme 4.9) since the transition state which leads to it is of lower energy *i.e.*, no clash repulsive forces between the two methyl groups in the transition state.

[***One may refer to schemes 5.33 and 5.36b to see the effect of transition state energies on product distribution***].

(*d*) Stereochemistry of Wagner—Meerwein Rearrangement

(i) Introduction

The following points may be noted.

- A reaction proceeding through or intermediate carbocation can give products of substitution, addition of elimination with rearranged carbon skeleton.
- Stereochemical factor involved in molecular rearrangements depends on steric or conformational factors and the timing of the bond-breaking and bond-making processes. When the process is concerted, configuration will be retained in the migrating group and inverted at the migration terminus. When the migration origin becomes a carbocation, as in a Wagner type rearrangement the stereochemical identity is generally lost.
- In most Wagner type shifts the leaving group is generally believed to depart before the migration takes place.
- The configuration at the migration terminus in the product is dependent on the life-time of the intermediate carbocation.
- Optically active 2-methyl-1,2-butanediol undergoes Wagner Meerwein rearrangement to give a racemic product (scheme 4.11). The hydride shift is slow and the intermediate carbocation has a fair chance to rotate.

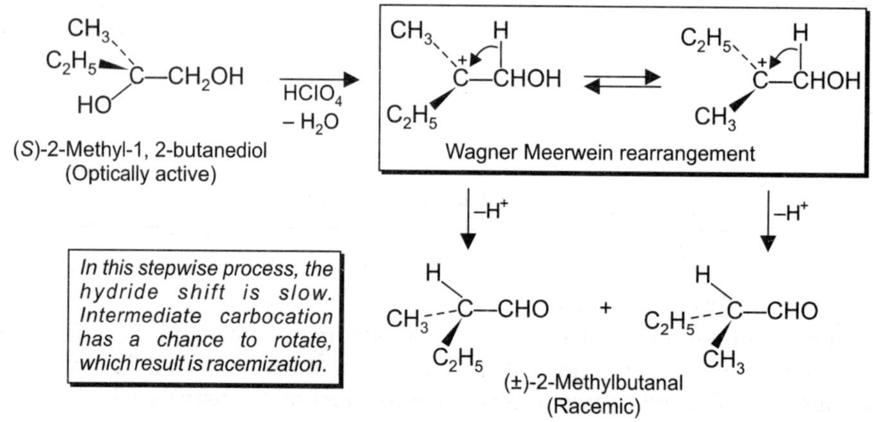

SCHEME 4.11

- In a related deamination process (scheme 4.12), the reaction is very rapid (not much opportunity for the carbocation to rotate) and ends up in a product with significant inversion of configuration at the migration terminus.

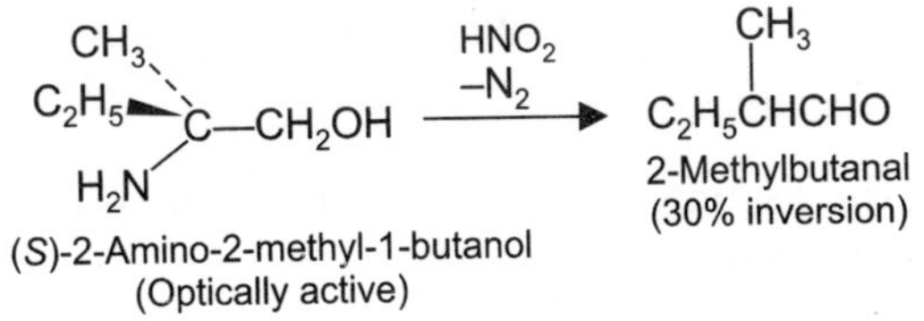

Fast deamination. The intermediate carbocation is not that free to rotate. In the product there is some inversion of configuration at the migration terminus.

SCHEME 4.12

(ii) Pinacolic Deamination—the Role of the Most Stable Conformation

Deaminative rearrangement of one optically active isomer of 2-amino-1, 1-diphenyl-1-propanol proceeded with 90 per cent inversion. When one specific phenyl group was labelled with ^{14}C, the labelled group migrated to give the product of inversion at the migration terminus and the unlabelled group migrated so as to give retention (scheme 4.13). The following points may be noted:

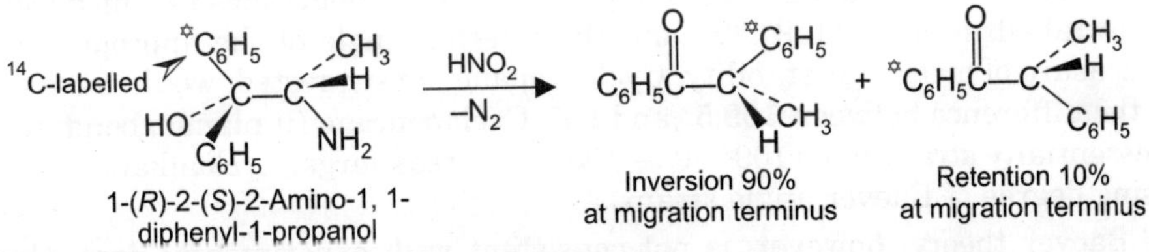

SCHEME 4.13

- The reaction proceeds involving the most stable staggered conformer where the two phenyl groups are adjacent to the smallest atom hydrogen (scheme 4.14).

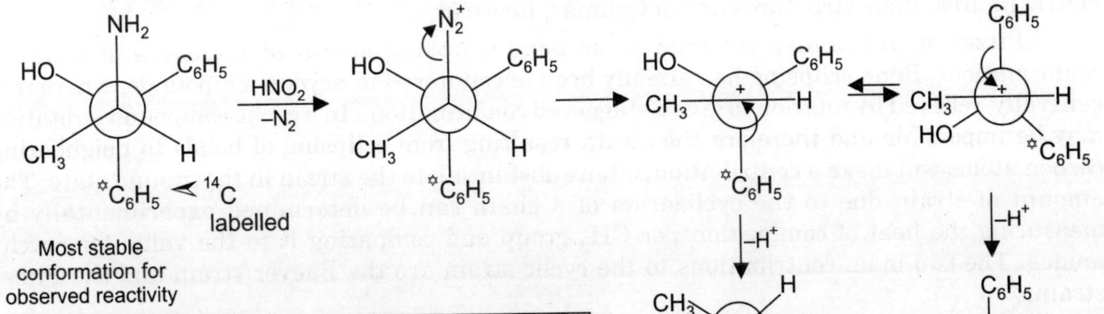

Pinacolic deamination involves the most stable conformer-the two Ph groups adjacent to small H atom. The labelled Ph (being suitably located) migrates to the carbocation to give 90% inversion. A rotation of the carbocation (60°) then initiates Ph migration with retention.*

SCHEME 4.14

- The carbocation is produced by the loss of nitrogen gas and in the compound with ^{14}C labelled phenyl, the migration of the labelled phenyl group gives the major product of inversion (The labelled phenyl group is suitably located in the plane of vacant, *p* orbital of the carbocation).

- If the intermediate carbocation rotates by an angle of 60° (scheme 4.14) a conformation is attained, in which now the unlabelled phenyl migration gives the product of retention of configuration.
- In conclusion the results support the assumption that the observed reactivity is the result of one predominant conformation of the substrate.

4.4 ANGLE AND PITZER STRAIN

Angle strain is also referred to as Baeyer strain. The normal angle between the carbon bonds in an alkane is about 111°; when the carbon atom forms part of a ring, the angle will be controlled by the geometric requirements of the ring. Three-membered ring system in cyclo-propane must be flat (since, mathematically, three points define a plane) so that the bond angles within the ring are forced to be 60° thereby causing strain. The amount by which the bond angle deviates from normal is a measure of the bond-angle strain.

Baeyer (1885) proposed a theory of angle strain for cycloalkanes in which the difference between a tetrahedral angle (109°.30′) and the internal angle of the appropriate polygon is used as a measure of molecular stability. Cyclopropane, he suggested, would have angle strain related to the difference between 109.5° and 60°. Cyclopentane (if planar, bond angle = 108°) would be essentially strain-free (109°.30′ – 108°), whereas larger cycloalkanes would possess an increasing degree of Baeyer angle strain.

The Baeyer theory, however, is not consistent with experimental data. Cycloalkanes larger than cyclopentane show only a small increase in strain energy, and very large cyclic molecules actually become almost strain-free. The problem lies in Baeyer's simple assumption that the rings are flat polygons (geometrical figures). C_3 and C_4 rings have large angle-strain, while the medium-size rings ($C_8 - C_{12}$) possess moderate strain, and large rings show only small strain energies. Though one often draws cyclic compounds as if they were planar geometric figures, their structures are not planar, however.

Pitzer strain is also referred to as bond-eclipsing strain of *gauche* and eclipsed conformations. Bond eclipsing has already been encountered in acyclic compounds, where it is generally relieved by rotation to give a staggered conformation,. In a cyclic compound a rotation may be impossible and therefore the strain resulting from eclipsing of bonds to neighboring carbon atoms will make a contribution, often substantial, to the strain in the ground state. The amount of strain due to the cyclization of a chain can be determined experimentally by measuring the heat of combustion per CH_2 group and comparing it to the value for acyclic analog. The two main contributions to the cyclic strain are the Baeyer strain and the Pitzer strain.

In small rings, the Baeyer strain is particularly significant, while it is less important or negligible in common rings (normal), in medium rings (8-to11-membered), and in large rings (12- and higher membered). In all cycloalkanes the Pitzer strain is operative and tends to be relieved by deviation from planarity of the carbon skeleton. In large rings, transannular interactions account for the relatively marked strain.

4.5 CYCLOALKANE RINGS OTHER THAN CYCLOHEXANE

(*a*) Conformation of Cyclopentane

If one considers a planar pentagonal structure of cyclopentane (scheme 4.15), one would have C—C—C bond angles of 108°, a value so close to the normal tetrahedral angle of 109°.30′ that no significant strain effect would be expected. However, in such a structure all of the hydrogens are completely eclipsed and it would have considerable torsional strain. The molecule finds it energetically favourable to distort largely from a planar conformation to relieve this torsional strain even though it increases the angle strain. The actual structure is of "envelope" shape. The additional bond-angle strain involved in this arrangement is more than compensated by a decrease in eclipsed hydrogens. The out-of-plane methylene group is approximately staggered with respect to its neighbors.

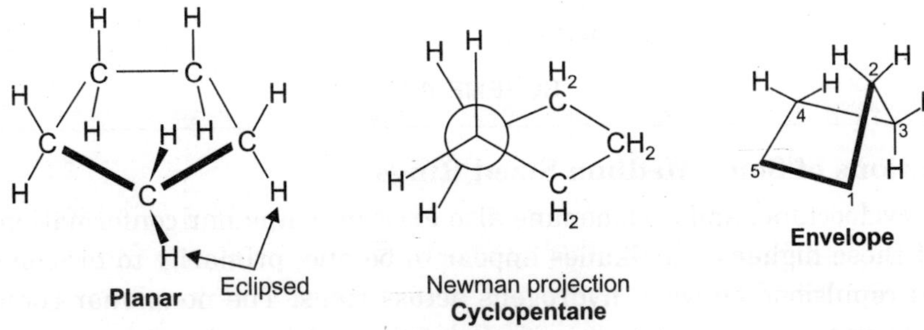

SCHEME 4.15

The envelope conformation of cyclopentane is dynamic. By twisting about the various C—C bonds, successive conformations are realised in which four carbons are in a plane and the fifth is out of plane.

(*b*) Conformation of Cyclopropane and Cyclobutane

In cyclopropane, one wonders as to how the three carbon atoms are connected in a ring without seriously violation the concept of bond angles. However, it is important to know that the highest electron density of the C—C bonds does not lie along the lines connecting the carbon atoms. Bonding electrons lie principally outside the triangular internuclear lines resulting in bent bonds (scheme 4.16).

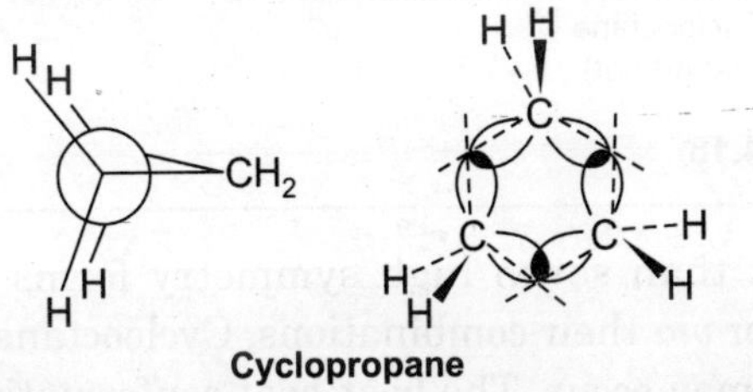

__Cyclopropane__ suffers from both angle and torsional strain. Each C–C bond is eclipsed due to rigidity of the molecule. Three carbons form an equilateral triangle (bond angles 60°) Hybridization at each carbon is sp^3. Angle between sp^3 hybrid AO's is 109.5°.

SCHEME 4.16

In cyclobutane, the internuclear angles of 90° are not as small as in cyclopropane. The C—C bonds are not so bent, and there is less strain per bond. However, there are four strained bonds rather than three and there are eight pairs of eclipsed hydrogen rather than six.

As three points define a plane, the carbon framework of cyclopropane must have a planar structure. However, cyclobutane can exist in a nonplanar conformation. Spectroscopic studies show that cyclobutane and many of its derivatives possess nonplanar structures in which one methylene group is bent at an angle of about 25° from the plane of the other three ring carbons (scheme 4.17). In this conformation, some increase in bond angle strain is compensated by the reduction in the eclipsed hydrogen interactions. The strain that is present in the C7 to C10 compounds is not the result of angle strain (the molecules being puckered), but due to interfering hydrogen atoms.

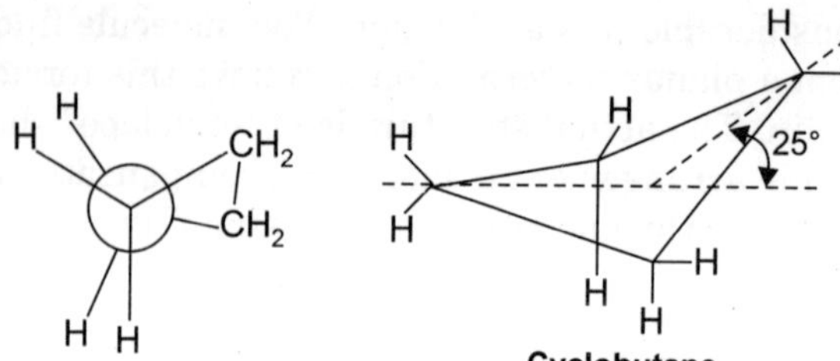

Cyclobutane

Unlike cyclopropane, cyclobutane molecule is not planar but puckered (bending angle 25°). The molecule flips rapidly from one puckered conformation to other

SCHEME 4.17

(*c*) Conformations of Some Medium Sized Rings

Cycloheptane, cyclooctane, and cyclononane also exist in nonplanar conformations. The small instabilities of these higher cycloalkanes appear to be due, primarily to torsional strain and van der Waals repulsions between hydrogens across rings. The nonplanar conformations of these rings, however, are essentially free of angle strain. Although not known with certainty, the most stable conformations of cycloheptane and cyclooctane appear to be those shown in (scheme 4.18).

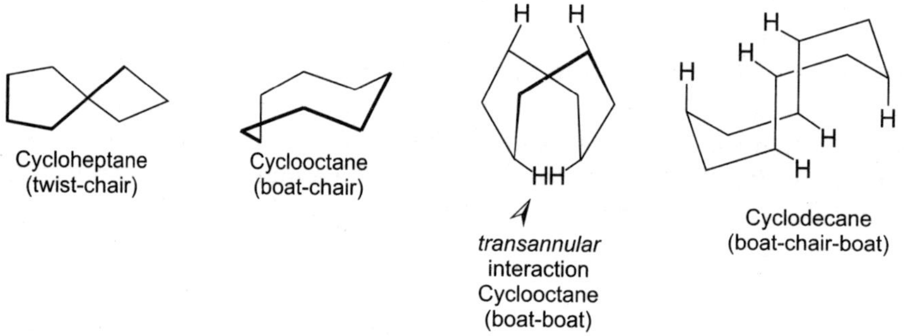

SCHEME 4.18

Puckered cyclooctane can assume no less than seven high symmetry forms which interconvert through inversions, pseudorotations or *via* their combinations. Cyclooctane is the smallest ring in which transannular interactions may occur. The boat-boat conformation has the carbon skeleton which corresponds to a fragment of the diamond lattice (compare with adamantane). A model of this conformation reveals that two hydrogen atoms compete for the same region of space. This is therefore, high-energy conformation. This type of transannular interactions can be quite severe in cyclooctane and other medium ring compounds.

The conformational aspects of cycloalkanes with ten carbon atoms and more are somewhat complicated. An interesting guiding principle is that the lowest energy conformation(*s*) would correspond to the one(*s*) which could be traced in the diamond lattice (scheme 4.19).

X-ray cystallographic studies on cyclodecane derivatives have shown that it adopts the boat-chair boat conformation (scheme 4.19), in the solid state and the relationship of this conformation with diamond lattice is clear. The electron diffraction measurements have shown that in the gas phase as well, that boat-chair boat conformation is dominant for cyclodecane.

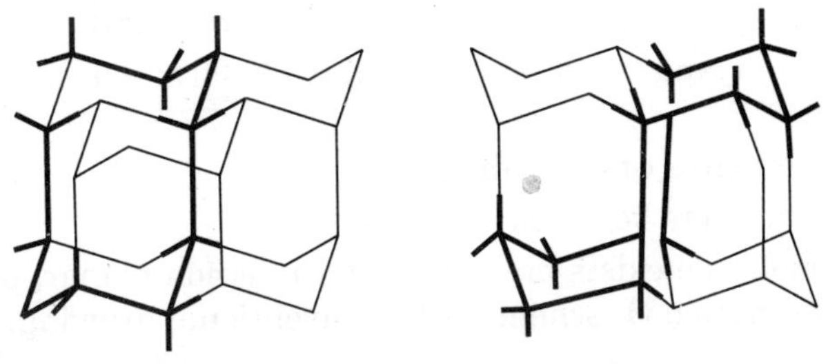

Equivalent diamond-lattice conformations of cyclodecane (boat-chair boat).

SCHEME 4.19

A distinctive feature of medium sized rings is that they display transannular reactions. These reaction do no involve the neighboring atoms; and occur among atoms on opposite sides of the ring. An example of this type of anomalous reaction is *e.g.*, the formolysis of epoxy-cyclodecane to give *trans*-cyclodecane-1,6-diol by a transannular hydride shift (scheme 4.20). Normally such a reaction from a cyclic or an acyclic epoxide affords a 1,2-diol. *Trans*-cyclooctene displays this reaction during the addition of chlorine, when 1,4-dichlorocyclooctane is obtained (10%) involving a 1,5-hydride shift (scheme 4.20) in the ion in addition to the normal product.

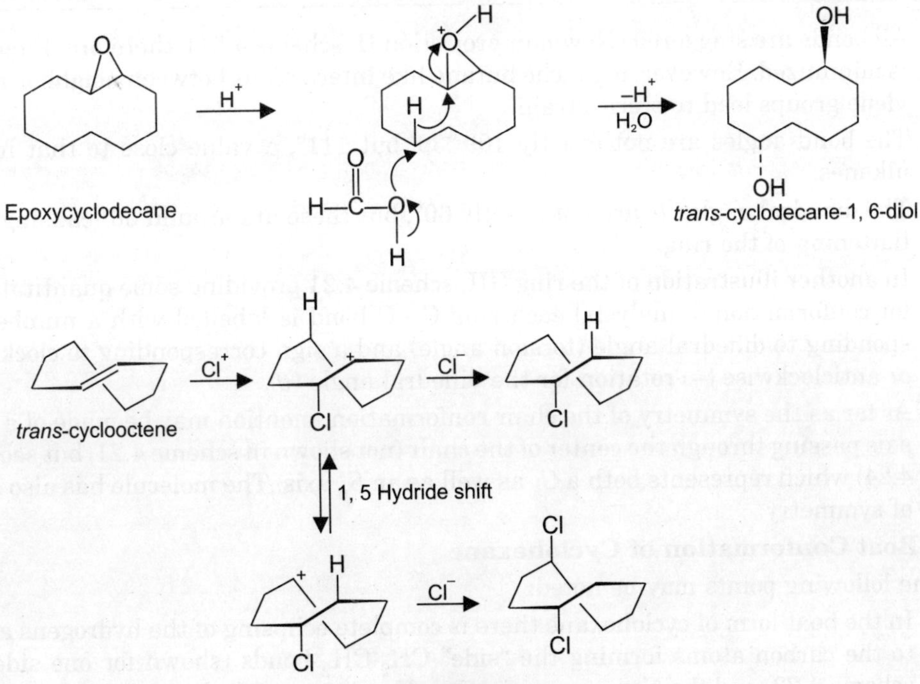

SCHEME 4.20

4.6 CONFORMATIONS OF CYCLOHEXANE (CONFORMATIONAL ANALYSIS)

The importance of Baeyer strain was first appreciated in 1890 by Sasche, who pointed out that two non-planar models of cyclohexane could be constructed in which all the bond angles can be

109° – 28′, so that the systems were free of Baeyer strain. One of these was a fairly rigid form, shaped roughly like a chair and the other was a flexible form whose most symmetrical form was shaped like a boat.

(*a*) The Chair Conformation of Cyclohexane

The following points may be noted:

- Cyclohexane molecule exists predominantly in a non-planar, puckered conformation the chair conformation (I, scheme 4.21, conventional drawing).

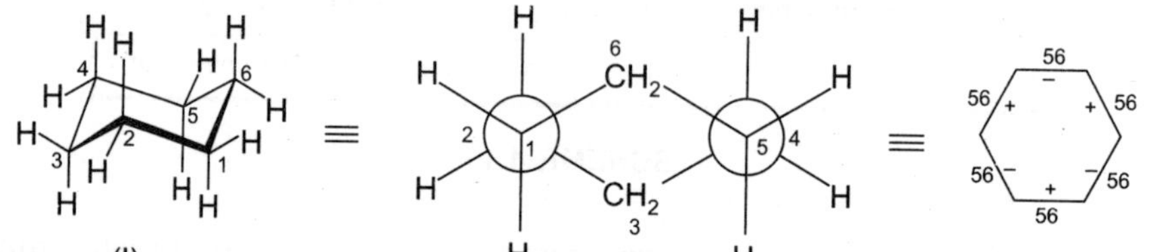

(I) Chair conformation (has both a C_3 and S_6 axis and a center of symmetry)

(II) Newman projection (all bonds are staggered) The dihedral angle is not 60° but close to 56°

(III)

Cyclohexane chair

SCHEME 4.21

- All bonds are staggered (Newman projection II, scheme 4.21), therefore, Pitzer strain is minimized. However, a gauche butane like interactions between neighboring methylene groups lead to steric strain.
- The bond angles are not exactly 109°.30′ but 111°, a value close to that in acyclic alkanes.
- The dihedral angles θ are not exactly 60°, but these are around 56° causing a slight flattening of the ring.
- In another illustration of the ring (III, scheme 4.21 providing some quantitative data for conformational analysis) each ring C—C bond is labelled with a number corresponding to dihedral angle (torsion angle) and a sign corresponding to clockwise (+) or anticlockwise (–) rotation for the dihedral angle (θ).
- As far as the symmetry of the chair conformation, mention may be made of a vertical axis passing through the center of the chair (not shown in scheme 4.21, but see scheme 4.24) which represents both a C_3 as well as an S_6 axis. The molecule has also a center of symmetry.

(*b*) The Boat Conformation of Cyclohexane

The following points may be noted:

- In the boat form of cyclohexane there is complete eclipsing of the hydrogens attached to the carbon atoms forming the "side" CH_2-CH_2 bonds (shown for one side in III, scheme 4.22) and the Newman projection (II, scheme 4.22) shows two eclipsed butane units. Thus the two bonds labeled 0, 0 in (IV, scheme 4.22) are completely eclipsed.
- The "inside" hydrogens on C1 and C4 ("flagpole hydrogens") interfere with each other sterically in transannular interaction (V, scheme 4.22) and there is a second factor which leads to instability for boat form in comparison to chair. The "flagpole" hydrogens lie only 1.83 Å apart than their sum of van der Waals radii 2.5 Å.

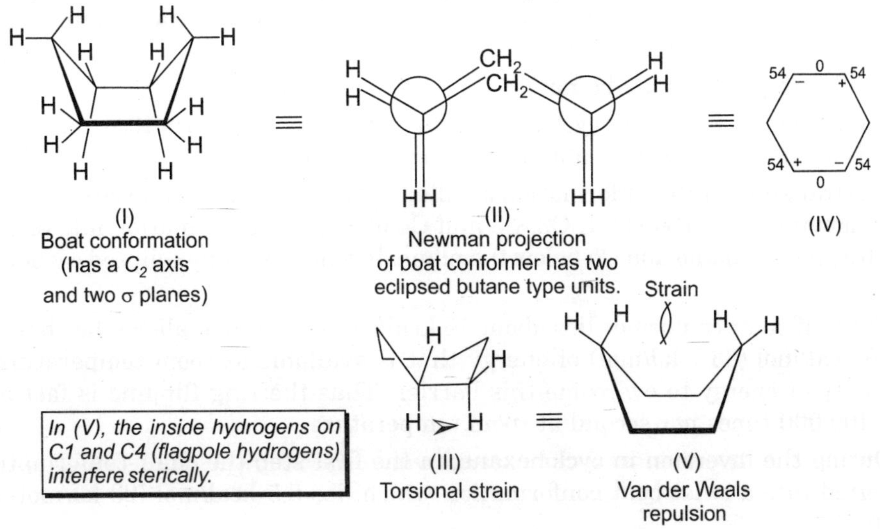

Cyclohexane boat

SCHEME 4.22

- The conventional boat conformation, (I, scheme 4.22) has a C_2 axis and two σ planes (see scheme 1.70)

(*c*) The Twist Boat Conformation of Cyclohexane

One has already seen that three forms of strain reflect largely on the energy of a conformation in a cycloalkane these are bond angle, eclipsing (torsional) and transannular (across the ring) strain. The chair conformation is at the minimum energy possible for cyclohexane system being free from bond angle and eclipsing strain. The only steric strain is due to gauche-butane like interactions between neighboring methylene groups.

The boat conformation is a higher-energy conformation due to eclipsing on its "side", carbons and due to transannular interaction between flagpole hydrogens.

The boat conformation is somewhat stabilized by moving apart the "flagpole" hydrogens. The new conformation called "twist boat" which is realized has minimized transannular interactions between "flagpole" hydrogens and at the same time torsional strain between the pair of side carbons of the initial boat are also relieved (scheme 4.23). The symmetry of the twist boat is reduced and the molecule is chiral.

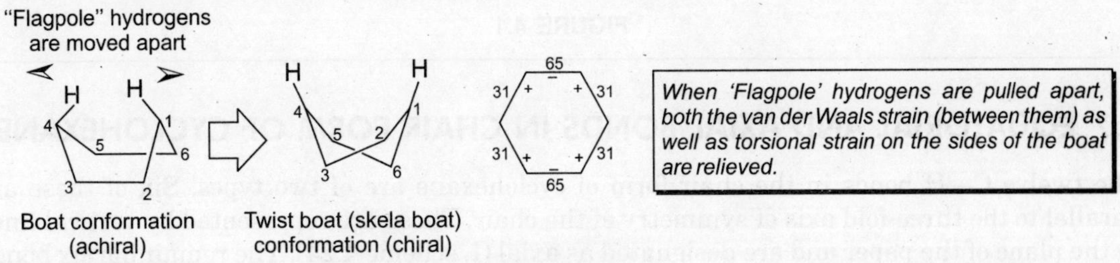

SCHEME 4.23

(*d*) Conformational Analysis of Cyclohexane

One chair conformation of cyclohexane can be easily converted into an alternate chair conformation. The details of this transformation and an estimate of the energies of different intermediates and transition states along the path of one chair to the alternate chair provides another example of conformational analysis.

The following points may be noted:

- The activation energy for cyclohexane ring inversion (Fig 4.1) is 10 kcal/mol (42 kJ/mol). It is a rapid process with a half life of around 10^{-5} s at 25°C.
- The transition state conformation is called the half-chair. The half chair can be realized on moving the carbons C1, C2, C3 and C4 of a chair conformation into one plane and C6 above the plane and C5 below the plane. It contains many eclipsed carbon-hydrogen bond.
- The half chair conformer lies about 10 kcal/mol (42 kJ/mol) above the chair form. The 20 kcal/mol (83.7 kJ/mol) of energy that is available at room temperature provides plenty of energy to overcome this barrier. Thus the ring flipping is fast and occurs ~ 100,000 times per second at room temperature.
- During the inversion in cyclohexane, in the first step the chair conformation is converted into a twist boat conformation which lies 5.5 kcal/mol (23 kJ/mol) above the chair.
- The full boat lies atop of the barrier separating two twist forms and like the half chair it is also a transition state. The twist boat arrangement can pass through a second half chair to give the other cyclohexane.

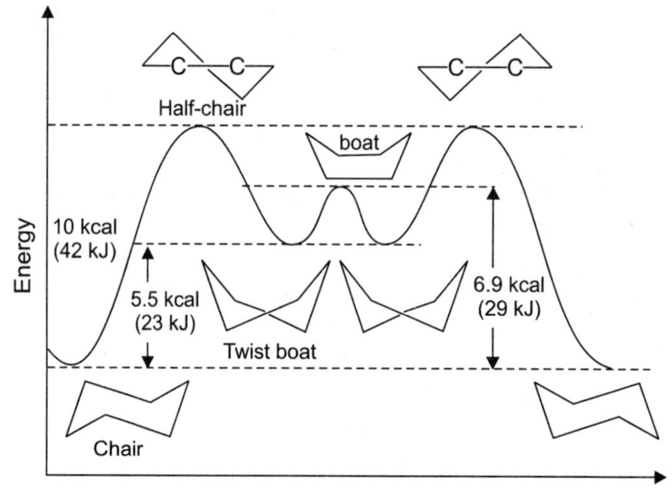

FIGURE 4.1

4.7 EQUATORIAL AND AXIAL BONDS IN CHAIR FORM OF CYCLOHEXANE

The twelve C—H bonds in the chair form of cyclohexane are of two types. Six of these are parallel to the three-fold axis of symmetry of the chair. These are represented by vertical lines in the plane of the paper and are designated as axial (I, scheme 4.24). The remaining six bonds are inclined at an angle of 109°.28′ to the three-fold axis and are described as equatorial. The following points may be noted.

- Of the six axial bonds three are above (up) relative to the mean plane of the ring while three are down (below). Thus, the molecular model of cyclohexane can stand firmly on the table by the three upper or the three lower hydrogens (I, scheme 4.24).

- On each face of chair conformer of cyclohexane there are three axial hydrogens and their orientation (up or down) alternates from one carbon atom to the next (I, scheme 4.24).

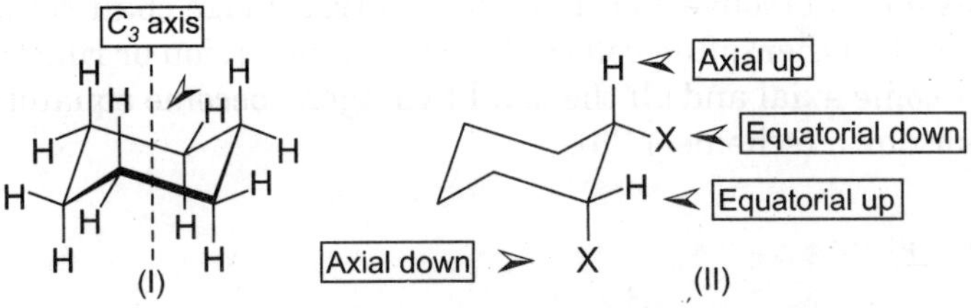

An equatorial bond is recognised (it makes a letter of "Z" with a ring residue). An equatorial bond can be labelled up or down by notionally extending it. It will clearly either go up or go down

SCHEME 4.24

- Similar situation obtains with the six equatorial bonds, three are up the mean plane of the ring while three are down.
- On each carbon one bond is axial while the other equatorial, of these one is up while the other is down.
- Consider a 1,2-disubstituted cyclohexane when the following points become clear.
- If is comparatively easy to recognize the axial bonds on each carbon and whether each is up or down. Thus in (II, scheme 4.24) the relationship between an axial hydrogen and the axial substituent X on adjacent carbons is *trans*-since these are on opposite sides of the ring (in fact their relationship is *trans* and coplanar, the dihedral angle between these being 180° this relationship is also called *anti*).

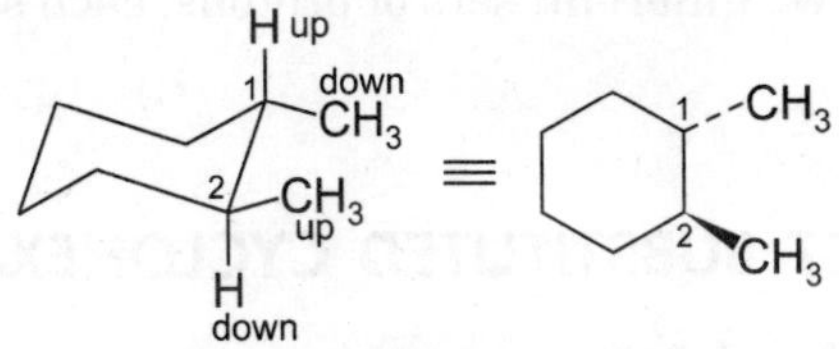

trans-1, 2-dimethylcyclohexane

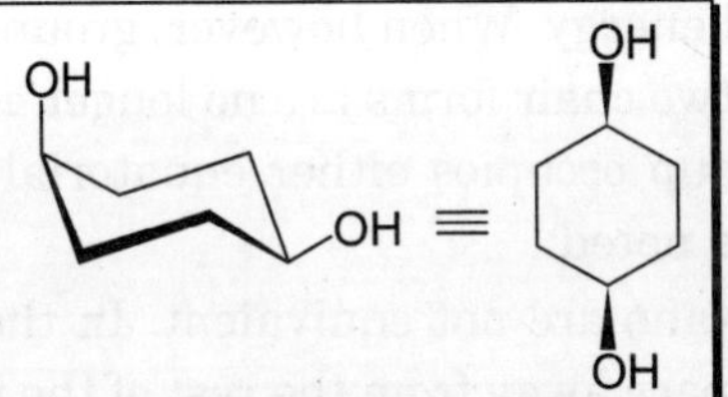

cis-Cyclohexane-1, 4-diol

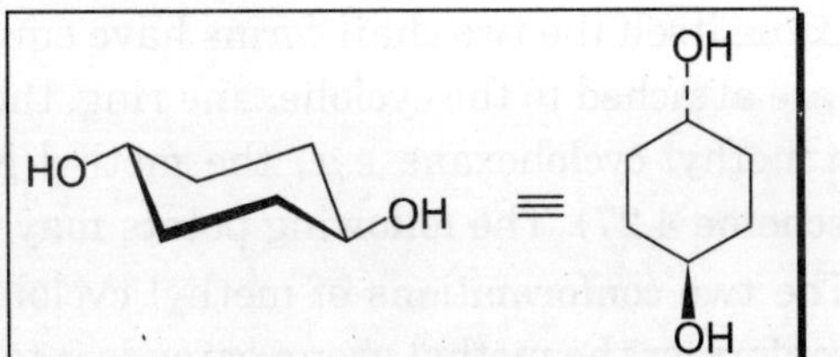

trans-Cyclohexane-1, 4-diol

SCHEME 4.25

- The directionality, is however, not so obvious for equatorial bonds. The two hydrogens (scheme 4.25) are *cis* (axial hydrogen is up, notionally extending the C—H equatorial bond one finds that it as well goes up).
- The two X groups are *cis* as well, but now both are down (notionally extending the C—X equatorial bond one finds that it goes down)
- Both C—H and C—X equatorial bonds represent *trans*-relationship—one goes up while the other down (on notionally extending the two). These 1,2-diequatorial bonds are *trans*-but however not coplanar.

- Based on these argument, one can, therefore transform the chair conformation (scheme 4.25) of *trans*-1, 2-dimethylcyclohexane into a planar regular hexagon and can know if the groups are present on the same or opposite side of the C_6 ring.

At room temperature cyclohexane interconverts rapidly to a mirror image chair conformation. As one chair form converts to the other, often referred to as ring inversion or ring flip, all the equatorial hydrogen atoms become axial and all the axial hydrogens become equatorial without losing their identity as up or down (scheme 4.26).

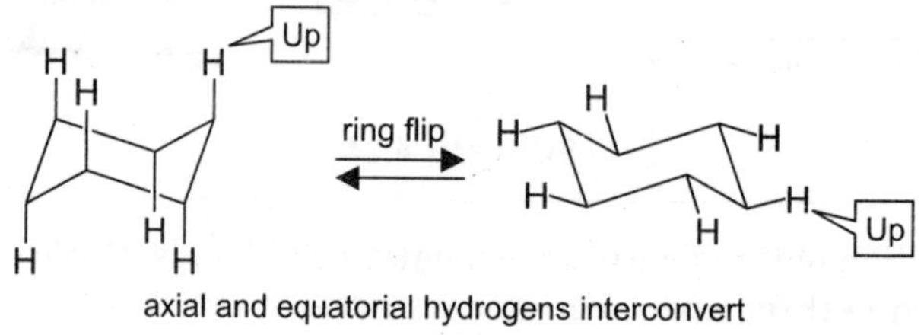

SCHEME 4.26

> *The equatorial-axial interconversion is so rapid at room temperature that all hydrogen atoms on cyclohexane can be considered equivalent. The energy barrier between the alternate chair conformations is only about 10 kcal/mol.*

Ring flipping is slowed down below – 80°C and therefore, axial and equatorial hydrogen atoms can be differentiated as two different, sets of protons, each set giving a separate ^{1}HNMR signal.

4.8 CONFORMATIONS OF SUBSTITUTED CYCLOHEXANES

(*i*) Conformations of Methylcyclohexane

In cyclohexane itself the two chair forms have equal energy. When however, groups other than hydrogen are attached to the cyclohexane ring, the two chair forms are no longer equivalent in energy. In methyl cyclohexane *e.g.,* the methyl group occupies either equatorial or an axial position (scheme 4.27). The following points may be noted:

- The two conformations of methyl cyclohexane are not equivalent. In the equatorial conformer the methyl group extends into space away from the rest of the molecule. In the axial conformer, the methyl substituent is close to other two axial hydrogens on the same side of the molecule *syn* axial hydrogens).
- The first destabilizing influence arises due to 1, 3-diaxial interactions due to steric repulsions between an axial substituent, the methyl group in the present case and another axial substituent (including hydrogen atoms) located two carbon atoms away (scheme 4.27)

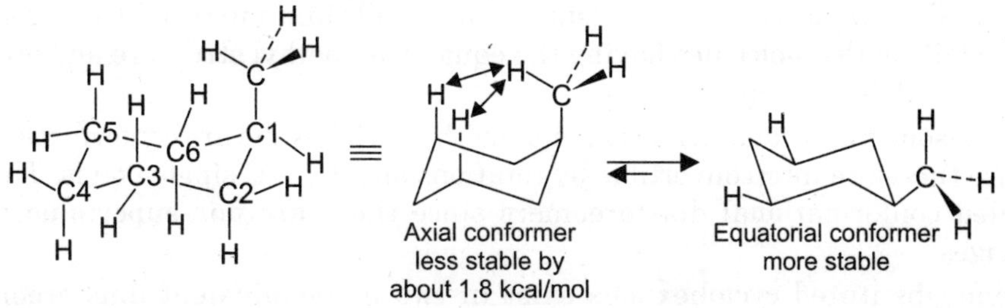

1, 3-Diaxial interactions are caused by steric repulsion between substituents two carbon atoms away.

SCHEME 4.27

- Two *gauche* butane interactions between the axial methyl group and the two ring C—C bonds destabilize the axial conformer while no such interactions exist when the methyl is equatorial. This is the second factor which destabilizes any conformation with an axial substituent (scheme 4.27*a*).

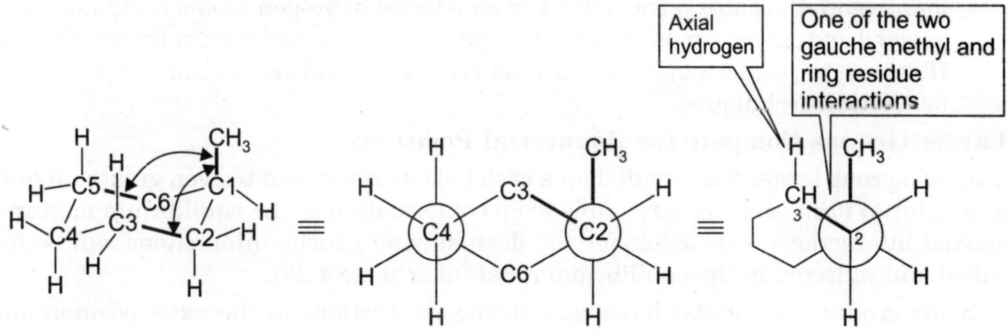

(I) (II) (III)

Two gauche interactions involving CH_3—C1—C2—C3 and CH_3—C1—C6—C5 are clearly seen in chair conformation (I). One can however, see only one of the gauche interactions in the Newman projections (II and III), the other can be seen in viewing these projections along 1–6 bond).

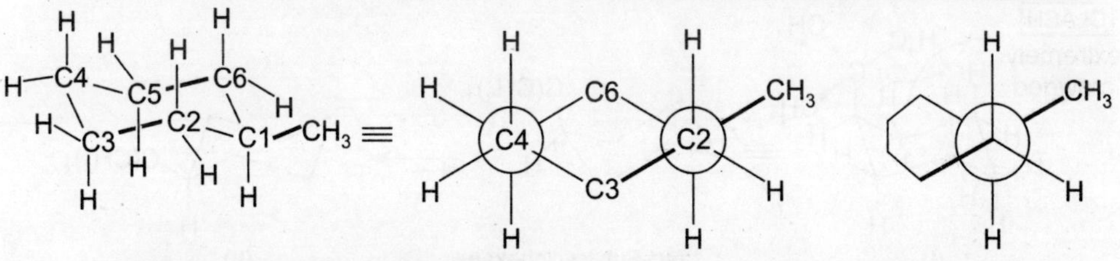

In the equatorial conformer the units CH_3—C1—C2—C3 and CH_3—C1—C6—C5 (shown by thick lines) have now anti orientation

SCHEME 4.27a

- A gauche interaction in butane (see, scheme 4.2) destabilizes that conformation by 0.9 kcal/mole, thus the destabilization caused by the axial methyl in axial methylcyclohexane with its two gauche interactions is twice this value. The equatorial conformation of methylcyclohexane is expected to be about 1.7 kcal/mol (7.1 kJ/mol) more stable than the axial conformation.

- At 25°C methylcyclohexane exists as an equilibrium mixture of two conformations with 95% of the molecules having the equatorial methyl structure and 5% having the axial methyl structure.
- Axial isomer and equatorial isomer of methylcyclohexane are conformational isomers, since these are interconvertible by rotations about C—C single bonds. These are also called conformational diastereomers since these are non-superimposable mirror images.
- Monosubstituted cyclohexanes exist in two non-equivalent diastereomeric chair conformations, one with the substituent in the axial position and the other with the substituent in the equatorial position. Moreover, it would be now clear as to why a planar structure for methylcyclohexane shown with methyl group above this plane cannot be depended on strongly. It not only hides the two conformations for the molecule, but the other rich stereochemical details as well.
- A reason why such nonplanar structures of cyclohexane did not become acceptable until relatively recently was that there should be chemical consequences of having two distinct sets of hydrogen atoms. Given the analytical techniques of the early and mid twentieth century, the axial and equatorial hydrogen atoms could not be distinguished and only one monosubstituted cyclohexane seemed to exist for any substituent. However, the conformational and other stereochemical details can now be established by ^{1}HNMR techniques.

(*ii*) Larger Groups Compete for Equatorial Positions

Presence of a group larger than methyl on a cyclohexane could lead to even greater dominance of the equatorial form over the less stable axial conformation in the equilibrium mixture. The 1, 3-diaxial interactions responsible for the destabilizing gauche interactions will be further magnified and reflected in the equilibrium constant scheme 4.28).

Bulky groups *e.g., t*-butyl have very strong interactions in the axial position and the energy difference in the two forms is particularly pronounced (around 5 kcal/mol) so that 99.9% of *tert*-butylcyclohexane is maintained in the equatorial form (II, scheme 4.28). The *tert*-butyl group is thus used to "lock" the conformation. The better way to put it is that the *tert*-butyl group distorts the equilibrium far toward the much more stable equatorial form (II, scheme 4.28).

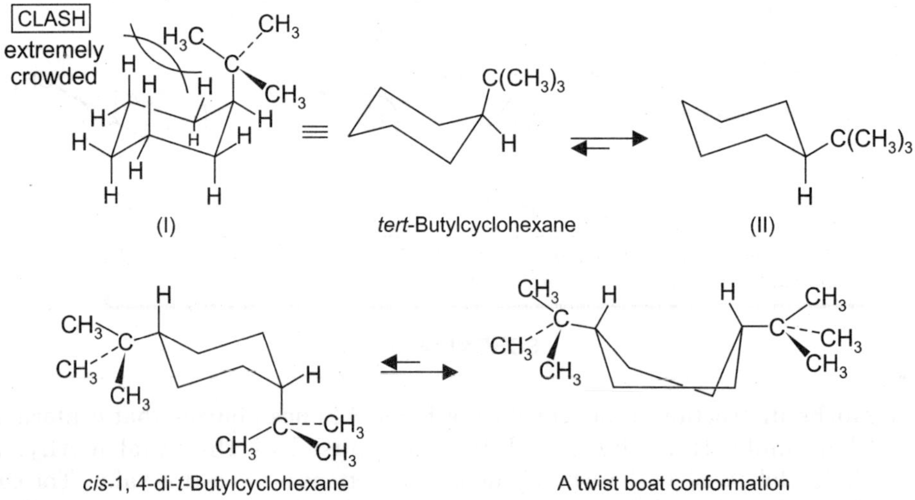

SCHEME 4.28

When the two bulky *t*-butyl groups are present on a chair cyclohexane, these are much less hindered when both occupy equatorial positions. In a case of *cis*-1, 4-di-*t*-butylcyclohexane one group must be axial and other equatorial and none of the chair conformer allows both bulky groups to be equatorial. The molecule, therefore, adopts a twist boat conformation (scheme 4.28).

Actual energy differences for various substituents expressed as $\Delta G°$ value are given in Table 4.2. A direct relationship exists between this difference in energy, called the free energy (symbolized by $\Delta G°$) and the equilibrium constant (K_{eq}) associated with a given equilibrium in solution:

$$\Delta G° = \textit{difference in free energy} = -RT \ \text{In} \ K_{eq}$$

where R is the gas constant (0.00199 kcal/mol °K) and T is the absolute temperature at which the equilibrium is measured (°K). The product of these two numbers and the natural logarithm of K_{eq} gives $\Delta G°$, the *free-energy difference* between the two conformers in kcal/mole.

Table 4.2: Energy Differences showing Preference of Equatorial over Axial Positions in Monosubstituted Cyclohexanes (25°C)

Substituent	*$-\Delta G°$, Free-difference between the two conformers, kcal / mol*	*kJ / mol*
– H	0.0	0.0
– CH_3	1.7	7.1
– CH_2CH_3	1.8	7.5
– $CH(CH_3)_2$	2.1	8.6
– C_6H_5	2.9	12.1
– $C(CH_3)_3$	5.5 Very large	22.8
– F	0.24	1.0
– Cl, – Br, – I	0.5	2.1

The energy and equilibria data in Table 4.2 provides a quantitative picture for structure-conformation relationships.

> *The greater the free energy difference between the two conformers, the larger the K_{eq}, associated with the axial equatorial equilibrium and the greater the preference of a particular group for occupying the equatorial position.*

Generally, substituents larger than hydrogen prefer to be equatorial on a cyclohexane ring to avoid 1, 3-diaxial interactions. The conformational free energies – $\Delta G°$ for the ethyl group (1.8 kcal/mol) (7.5 kj/mol) and isopropyl group (2.1 kcal/mol) (8.6 kJ/mol) are only slightly larger than that for the methyl group (1.7 kcal/mol), (7.1 kJ/mol) while that for the *tert*-butyl group (5.5 kcal/mol) (22.5 kJ/mol) is much larger. The ethyl and propyl groups can be rotated so that a hydrogen is pointed back over the ring to interact with the axial hydrogens, so their effective steric bulk is not much different from that of a methyl group. In contrast, the *tert*-butyl group is forced to have one of its methyl groups pointed over the ring, causing much more severe 1, 3-diaxial interactions (scheme 4.28*a*).

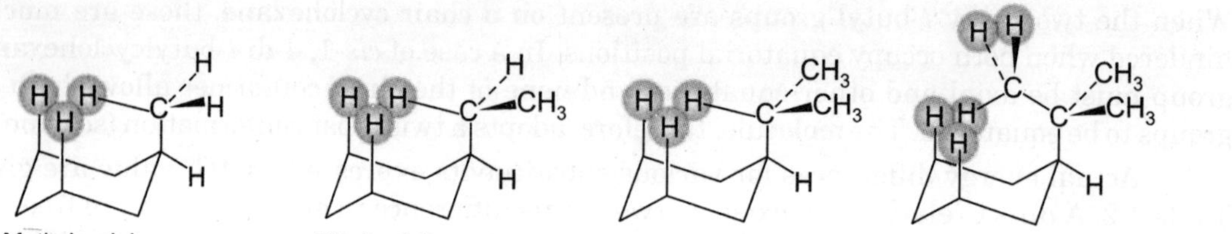

Methylcyclohexane Ethylcyclohexane Isopropylcyclohexane *tert*-butylcyclohexane

SCHEME 4.28a

In iodocydohexane, *e.g.*, iodine has little preference for being axial or equatorial. Although iodine is a large atom, the C—I bond in much longer than C—H bond and therefore, 1, 3-diaxial interactions involving iodine are not significant.

EXERCISE 4.3

(i) Draw the most stable conformation of cis-1-tert-butyl-4-methyl cyclohexane.

(ii) How one can explain the preference for an equatorial orientation in the pairs (scheme 4.28 b)?

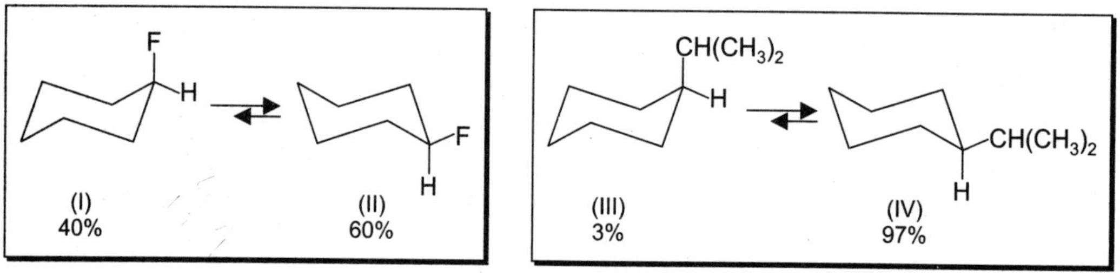

SCHEME 4.28b

ANSWER. *(i) The tert-butyl group must adopt an equatorial orientation, consequantly the methyl group is fixed in the axial position (scheme 4.28c).*

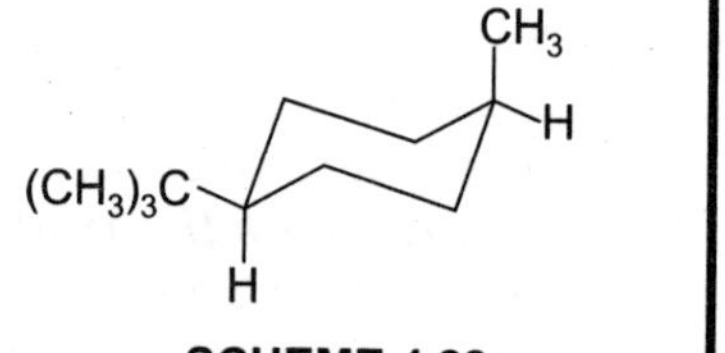

SCHEME 4.28c

(ii) A halogen atom does not require much space, moreover, F and H are nearly similar in size thus the preference of F to take up an equatorial orientation is less. The halogens F, Cl, Br, and I do not differ much in their preference for the equatorial position. As the size of the halogen increases in the order F < Cl < Br < I, so does the carbon-halogen bond distance, and the two effects tend to cancel. The large isopropyl group distorts the equilibrium far towards the more stable equatorial form (IV).

EXERCISE 4.4

What is the relationship between the pairs (scheme 4.28d)-different conformations of the same compound or stereoisomers which cannot be interconverted via rotation around single bonds?

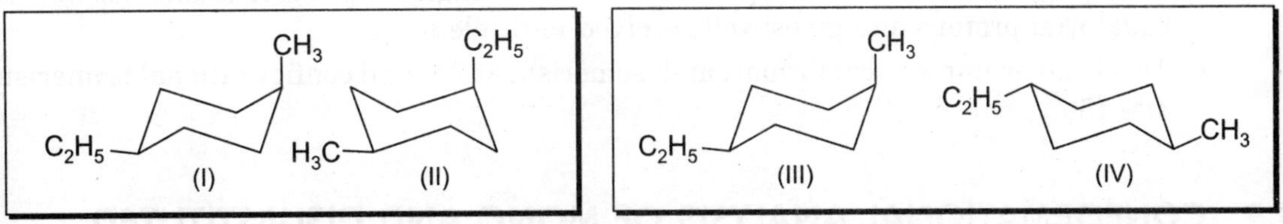

SCHEME 4.28d

ANSWER. *Both (I and II) represent cis-1-ethyl-4-methylcyclohexane (both methyl and ethyl groups are "up"). In (I), the methyl is axial and the ethyl equatorial. The orientations are opposite in (II). The two structures are ring-flipped forms of each other i.e., different conformations of the same compound.*

One may be deceived on the first look since the rings (III and IV) look like ring-flipped forms. However, in both structures ethyl group is equatorial (chair–chair interconversion converts all the equatorial bonds to axial and vice versa). In (III) the two substituents are cis while in (IV) these are trans, thus these are stereoisomers (diastereomers).

(iii) Conformations of Chlorocyclohexane (Isolation of Conformers)

Jensen in 1969 isolated the diastereomeric chlorocyclohexanes working at – 150°. The following points may be noted:

- The methine proton of chlorocyclohexane (which shows only a averaged ^1H NMR signal at room temperature) resolves into two sets of signals at – 115° C (Fig. 4.2).

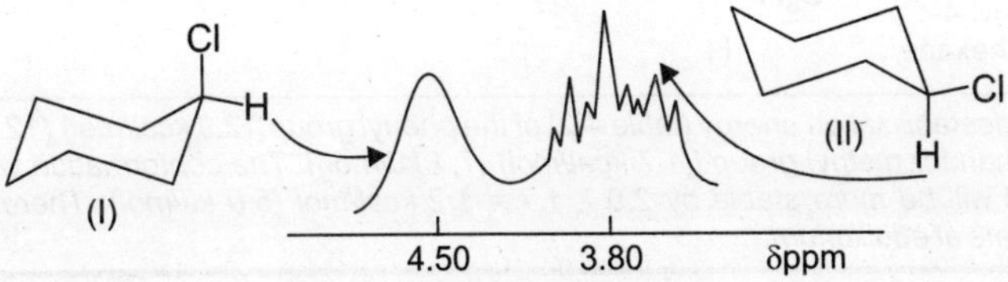

Fig 4.2

- Diastereomeric chlorocyclohexanes are quite stable at – 150°C and can be isolated in their pure forms via fractional crystallization.
- In their pure forms the equatorial conformer displays only one signal (δ 4.50) while the axial conformer displays the other signal (δ 3.80).
- Each conformer gives its respective single spectrum (*i.e.,* only one signal either at δ 3.8 or 4.5 was observed) showing no impurity of the other diastereomer.
- On warming the solution to room temperature, the equilibrium mixture of axial and equatorial conformations of chlorocyclohexanes was reformed from each single isomer.

- Interestingly (unlike at room temperature) the conformers of chlorocyclohexane can be studied independently at – 115° by ^{1}H NMR which displays both the equatorial methine proton and the axial methine proton at different chemical shift values (Fig. 4.2). The broad singlet at δ 4.50 (broad singlet is due to equatorial proton while multiplets at δ 3.80 are due to axial proton. The equatorial proton is coupled with neighboring protons which are all gauche with lower value of coupling constant. The axial proton, however, is coupled to both axial (high coupling constant) as well as equatorial protons and gives well resolved multiplets.
- This is an example of conformational isomerism at 25° and configurational isomerism at – 150°.

4.9 CONFORMATIONAL ANALYSIS OF MONO AND DISUBSTITUTED CYCLOHEXANES —RELATIVE STABILITY OF CONFORMERS

The two forms of chair methylcyclohexane are in equilibrium with the equatorial conformer favored by a ratio of 95 : 5. By using the expression $\Delta G° = -1.36K$ (at 25°C) the difference in energy between the two conformers is 1.7 kcal/mol (7.1 kJ/mol). The energy difference, $\Delta G°$ between the axial and the equatorial isomers of several monosubstituted cyclohexanes has been measured (Table 4.2).

The data (Table 4.2) can be used to determine the relative stability of many substituted cyclohexane conformers. Thus one can analyze different conformations of 1-methyl-4-phenyl cyclohexane (scheme 4.29) for relative stability. The conformation with both groups equatorial is much more stable than the other. Overall, the *trans*-isomer is more stable than the *cis*-isomer by the amount of strain due to the axial methyl group in the *cis*-isomer *i.e.,* by 1.7 kcal/mol (7.1 kJ/mol).

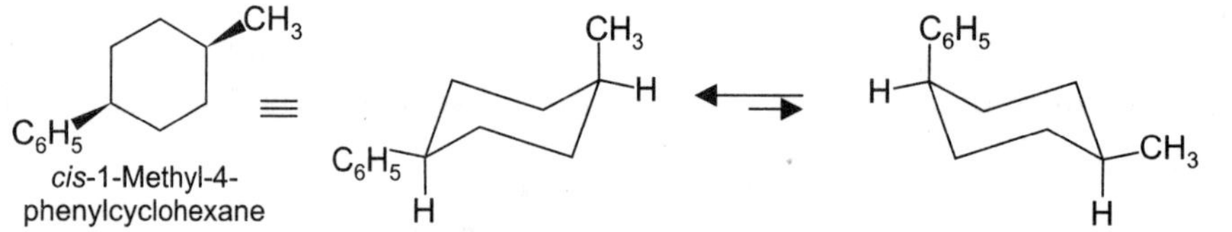

cis-1-Methyl-4-phenylcyclohexane

The axial destabilization energy (table 4.2) of the phenyl group [(2.9 kcal/mol) (12.1 kJ/mol)] is larger than for methyl group [(1.7 kcal/mol) (7.1 kJ/mol)]. The conformation with phenyl equatorial will be more stable by 2.9 – 1.7 = 1.2 kcal/mol (5.0 kJ/mol). Therefore it will predominate at equilibrium.

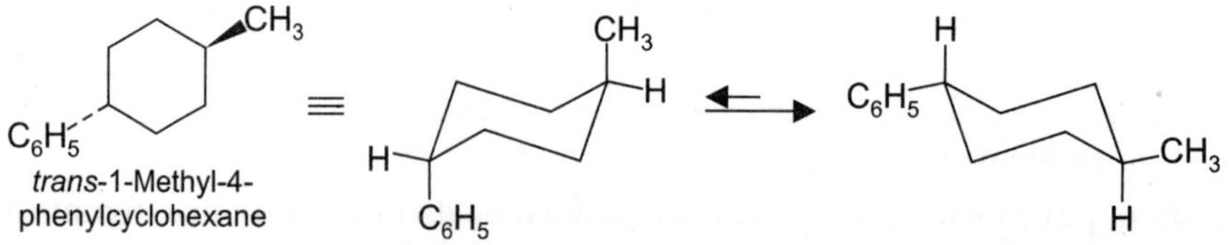

trans-1-Methyl-4-phenylcyclohexane

Total torsion strain energy 1.7 + 2.9 = 4.6 kcal/mol (19.2 kJ/mol)

No strain energy (no group is axial)

SCHEME 4.29

From the conformational analysis one can, therefore, predict the relative stability of its various conformers. In general a substituent (except for hydrogen) in the equatorial position produces the more stable (the preferred) conformation. For this reason β-D-glucose with five equatorial groups is more stable (scheme 4.44) than α-D-glucose with four equatorial group and one axial group.

4.10 STEREOISOMERISM OF DISUBSTITUTED CYCLOHEXANES (*cis*-AND *trans*-ISOMERS AND CHIRALITY)

(A) Introduction (Summary)

Stereoisomerism of alicylic compounds (3-5-membered rings) has already been discusses (scheme 1.65-1.67).

The following points may be noted for cyclohexane systems:

- When two or more substituents are introduced on a ring (infact any size ring) stereoisomerism becomes possible, as already seen for 3-5 membered rings.
- It is often easy to know the number of stereoisomers of a cyclic compound by considering the ring to be flat (even though it may exist in a chair or other conformations).
- The existence of stereoisomers is independent of conformations. Different stereoisomers will have different conformations.
- The first step is therefore, to know the number of stereoisomers (from the flat structures), identification of all the stereoisomers and finally the conformations of each of the stereoisomers.
- To examine the chirality via the plane of symmetry, it is always better to examine both the planar (flat) structure as well as the conformational formula.

(B) Cyclohexane Derivatives—Stereoisomerism through Prostereoisomerism

Single substituent in a cyclohexane ring invariable prefers the equatorial position, but when a second substituent introduced one must consider if it is *cis*- or *trans*- to the first substituent, and whether it is on C_2, C_3 or C_4. In disubstituted derivatives of cyclohexane, the conformational preference will be for a chair containing both substituents equatorial, or, when this is not possible, for the bulkier of the substituents to be equatorial.

One can study the stereoisomerism in cyclohexane derivatives by considering *e.g.*, the prostereoisomerism of a mono-substituted cyclohexane derivative cyclohexanol (scheme 4.30). Cyclohexanol is achiral (see scheme 1.68) however, it contains five centers of prochirality (carbons 1, 2, 3, 5 and 6). The two enantiotopic ligands of C-1 are the two edges of the ring and are designated *pro-R* and *pro-S* (scheme 4.30). The other four prochiral centers carry diastereotopic hydrogens, while carbon-4 has two diastereotopic hydrogen atoms and it itself is proachiral (replacement of either of the hydrogens leads to achiral diastereomeric 1, 4-disubstituted cyclohexane derivatives). Thus, for disubstituted cyclohexanes, stereoisomerism is possible.

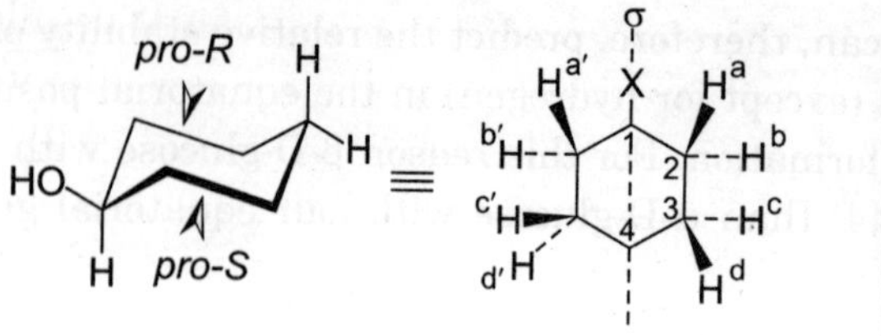

H^a and $H^{a'}$ and also H^b and $H^{b'}$ are mirror images of each other and are enantiotopic, their replacement gives enantiomeric products. Similar is the case with H^c, $HC^{c'}$ and H^d, $H^{d'}$. The CH_2 groups exhibit non-equivalent geminal hydrogens which are diastereotopic

X = OH

SCHEME 4.30

(C) 1, 2-Disubstituted Cyclohexanes (1, 2-Dimethylcyclohexane)

1,2-disubstituted cyclohexanes are probably the most challenging of the disubstituted cyclohexanes. Consider the following points:

- A consideration of the planar hexagon structures of *cis* or *trans* isomers shows that 1, 2-dimethylcyclohexane has two stereocenters, thus it can have four stereoisomers (for the planar structure of *cis*-1, 2-dimethylcyclohexane and its study see, scheme 1.67*c*).

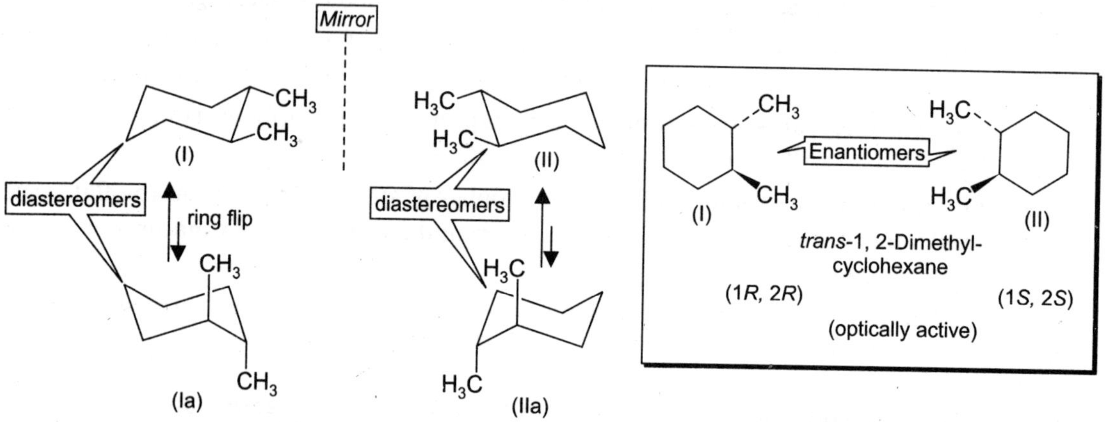

***trans*-1, 2-Dimethylcyclohexane**

SCHEME 4.31

- There are indeed four stereoisomers but one can isolate only three. The *trans* isomer (has no plane of symmetry) and exists as a pair of enantiomers (I and II scheme 4.31) where the diequatorial conformer is the most stable.
- In the case of *trans*-1, 2-dimethylcyclohexane (scheme 4.31 I and II) are enantiomers. On ring flip *trans*–1, 2-dimethyl cyclohexane converts the diequatorial stereoisomer (I) into the diaxial form (I*a*) and these two molecules are diastereomers (conformational diastereomers). Similarly (II and II*a*) are also conformational diastereomers.

A ring flip in trans-1, 2-dimethylcyclohexane will convert both the methyl groups into the less stable diaxial conformation. Remember a 1, 2-disubstituted cyclohexane will be cis if one group is axial and the other equatorial while it will be trans if either both the groups are equatorial or both axial.

- The *cis*-isomer of 1, 2-dimethylcyclohexane has no plane of symmetry when the chair form is considered (see, scheme 1.64). However, the following arguments are necessary to grasp.

- A planar structure of cyclohexane may be helpful when conformational details are not studied, moreover a given cyclic compound cannot be optically active if its planar structure has a plane of symmetry. Whether it is *meso* or racemic may be studies via its conformers.
- The two mirror image conformational structures (I and II, scheme 4.32) are not identical and are non superimposable. Therefore, both are chiral molecules and represent a pair of enantiomers.
- Both (I and II, scheme 4.32) are however, interconvertible by ring flip. The interconversion of enantiomers by ring flip can be best shown by models. Alternatively as shown (I and II, scheme 4.33), have a nonsuperimposable mirror image relationship, however if (I) undergoes a ring flip and the arrangement is rotated through an axis by 90° it becomes identical with its mirror image. Thus ring flip in *cis*-1 , 2-dimethylcyclohexane converts one enantiomer into the other (a process termed enantiomerization) *i.e.,* equilibration between the two conformations produces a racemic mixture.

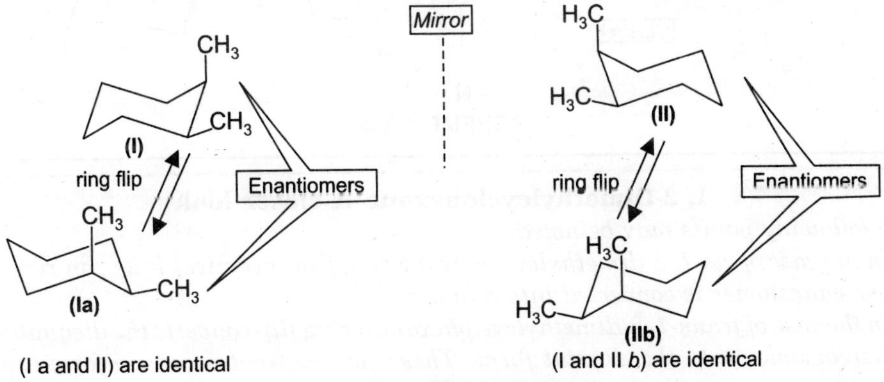

***cis*-1, 2-Dimethylcyclohexane**

SCHEME 4.32

Thus (I and II, scheme 4.32 and 4.33) represent conformational enantiomers a situation similar to that found in gauche butane (see, scheme 1.1*c*). Notice that the configurations of the stereocenters remain unchanged during the ring-flip as expected, since no bonds are broken during this conformational change.

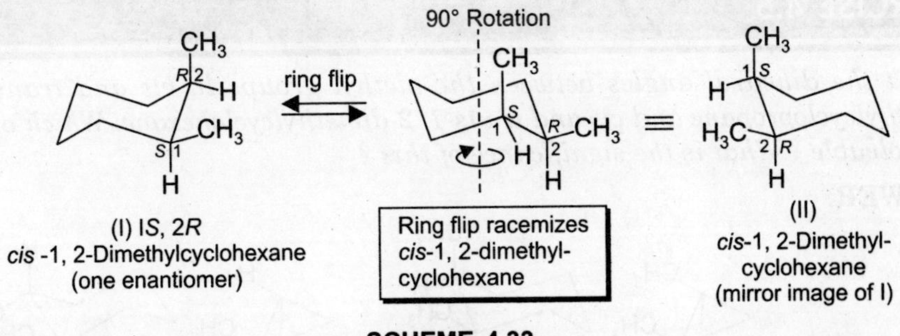

SCHEME 4.33

- At normal temperatures therefore, only three isolable stereoisomers of 1, 2-dimethylcyclohexane exist, *trans*-isomer as a pair of enantiomers and *cis*-isomer as a racemic mixture.

- The *cis*-1, 2-dimethylcyclohexane has a strain energy of 1.7 kcal/mol (7.1 kJ/mol) due to axial methyl group, plus 0.8 kCal/mol (3.3 kJ/mol) due to a gauche interaction between the methyl groups for a total of 2.5 kcal/mol (10.4 kJ/mol). The two conformers (enantiomers) have identical strain energies since one methyl group is axial and other equatorial in both (The equilibrium constant for the ring-flipping process is 1.0).
- The strain energy of *trans*-diaxial-1, 2-dimethylcyclohexane is 1.7 × 2 = 3.4 kcal/mol (14.2 kJ/mol) for two axial methyl groups. The only strain energy in the more stable diequatorial form is 0.8 kcal/mol (3.3 kJ/mol) due to gauche interaction between the two methyl groups (scheme 4.33*a*). Thus the *cis* diequatorial 1, 2-dimethyl conformer is more stable by about 2.6 kcal/mol (10.9 kJ/mol) and the equilibrium is greatly in favor of this conformer (> 99%). The *trans*-diequatorial 1, 2-dimethylcyclohexane is more stable than the *cis*-isomer by about 1.7 kcal/mol (7.1 kJ/mol).

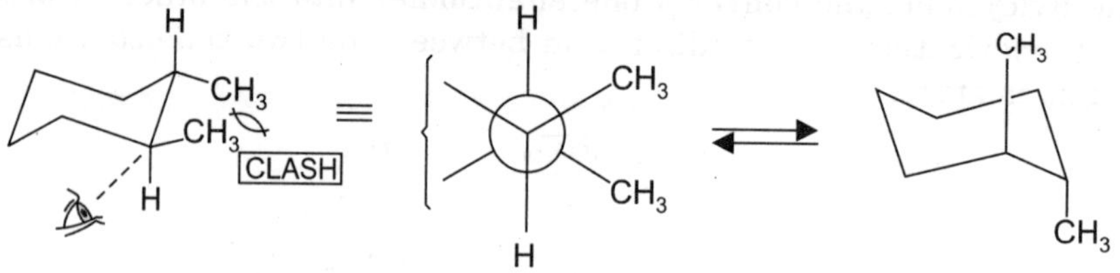

SCHEME 4.33a

1, 2-Dimethylcyclohexane–A closer look

The following points may be noted:

- *In the case of cis-1, 2-dimethylcyclohexane ring flip racemizes it i.e., on ring flip one enantiomer is converted into another.*
- *In the case of trans-1, 2-dimethylcyclohexane a ring flip converts the diequatorial stereoisomer into the diaxial form. These two molecules are conformational diastereomers and not enantiomers. There is no racemization during its ring flip.*
- *In principle in the case of trans-1, 2-dimethylcyclohaxane both diaxial as well as diequatorial forms should be resolvable, both being chiral. However, the trans-diaxial-1, 2-dimethylcyclohexane cannot be isolated, since it simply flips into the more stable diequatorial form.*

EXERCISE 4.5

Depict the dihedral angles between the methyl groups in cis and trans-1, 2-dimethylcyclopropane and cis and trans-1, 2-dimethylcyclohexane. Which of these is resolvable ? What is the significance of this ?

ANSWER.

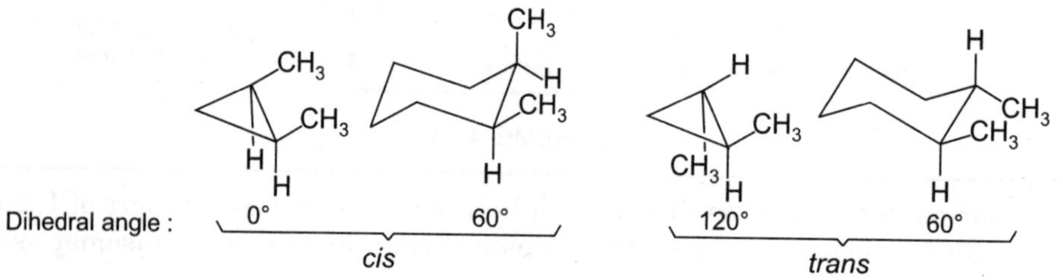

SCHEME 4.33b

Neither cis-1, 2-dimethylcyclopropane nor cis-1, 2-dimethylcyclohexane can be resolved, cyclopropane derivative being meso and the cyclohexane derivative flips into its mirror image. Their counterparts can however, be resolved. This has significance since planar cyclopropanes and nonplanar cyclohexanes behave similarly. Thus in deciding the problems of stereochemistry cyclohexane can be treated in their planar forms.

There is a further interesting and relevant point regarding cis-1, 2 dimethylcyclohexane and like molecules. The configuration of C1 and C2 (IS, 2R) in both (I and II, scheme 4.33) remains the same and these configurations stay with these carbons whatever the conformation recall that a compound can have different conformations but same configuration sec. 1.2, 9). The substituents at the two stereocenters are the same, thus cis-1, 2-dimethylcyclohexane may be thought to represent a meso-compound. In fact, if a particular conformer of a molecule is chiral, the molecule is not necessarily chiral. A molecule cannot be chiral if it is

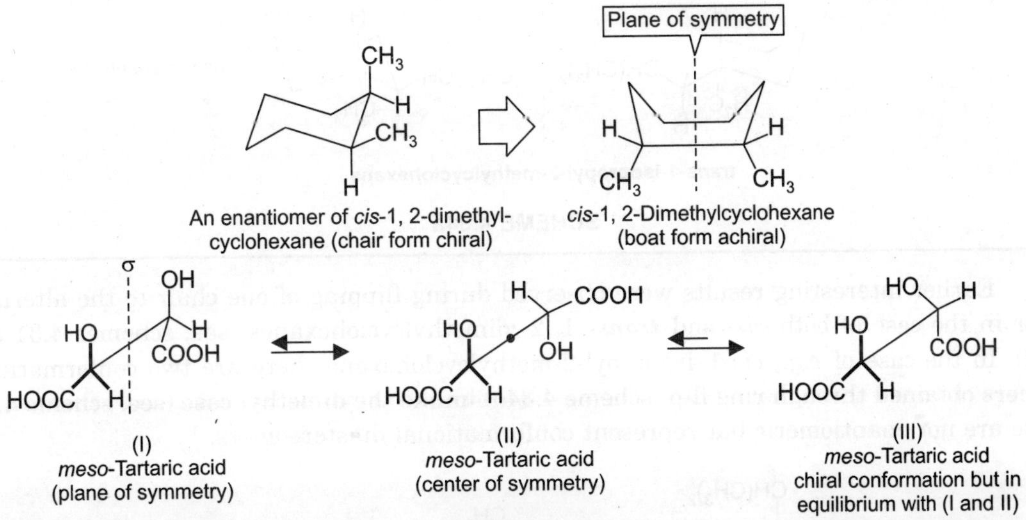

SCHEME 4.34

in equilibrium with a structure (or a conformation) that is achiral. In case there is no achiral conformation then the molecule is chiral. Moreover a molecule cannot be chiral if any two of its conformers are enantiomers. The boat conformation of cis-1, 2-dimethylcyclohexane (scheme 4.34) easily obtainable from either (I and II, scheme 4.34) has a plane of symmetry passing between the stereocenters.

Meso-Tartaric acid can adopt several conformations (I-III scheme 4.34), some chiral and some achiral but since the chiral conformations is in equilibrium with the achiral ones, the molecule overall cannot be chiral.

(D) 1, 2-Disubstituted Cyclohexanes with Two Different Substituents

In case the two substituents are different in a 1, 2-disubstituted cyclohexane, then both *cis* as well as *trans* isomers are resolvable. The *trans* isomer exists in a diequatorial conformation, while the *cis* isomer in equatorial axial conformation with the bulkier of the groups predominantly in the equatorial position.

One may consider the example of 1-isopropyl-2-methylcyclohexanes where *cis* as well as *trans* isomers exist and considering the stable conformers, one can write the conformations of four stereoisomers (scheme 4.34*a*) as a pair of enantiomers.

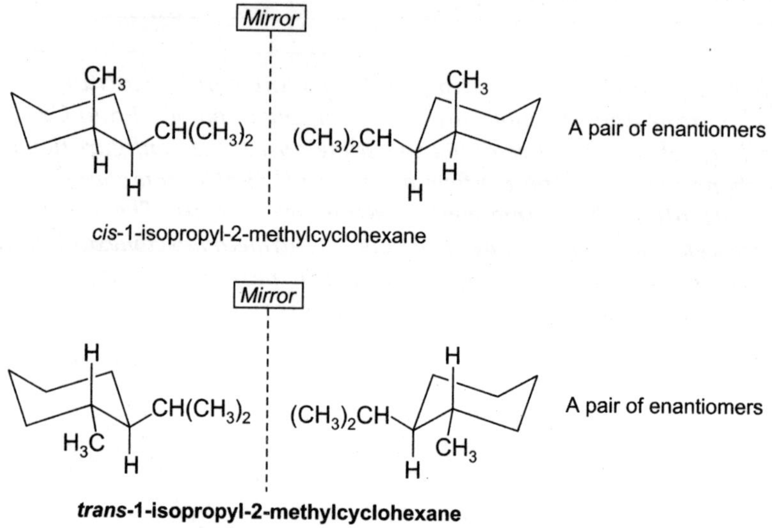

SCHEME 4.34a

Earlier interesting results were observed during flipping of one chair to the alternate chair in the case of both *cis*- and *trans*- 1, 2- dimethylcyclohexanes (see, schemes 4.31 and 4.32). In the case of *e.g.*, *cis*-1–isopropyl-2-methylcyclohexane there are two conformational isomers obtained through ring flip (scheme 4.34*b*) unlike the dimethyl case (see, scheme 4.32) these are not enantiomeric but represent conformational diastereomers.

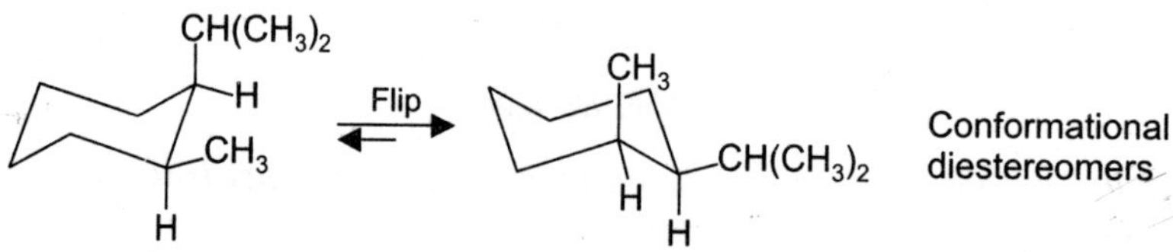

SCHEME 4.34b

(E) 1, 3-Disubstituted Cyclohexanes (1, 3-Dimethylcyclohexane)

Consider the following point-wise discussion:

- 1, 3-Dimethylcycloxane has two stereocenters, one can, therefore, expect ($2^2 = 4$) stereoisomers.
- In reality, however, only three exist, *cis*-1, 3-dimethylcyclohexane has a plane of symmetry both seen in a planar structure and either of the chair conformations (scheme 4.35). Thus it is a *meso* compound.

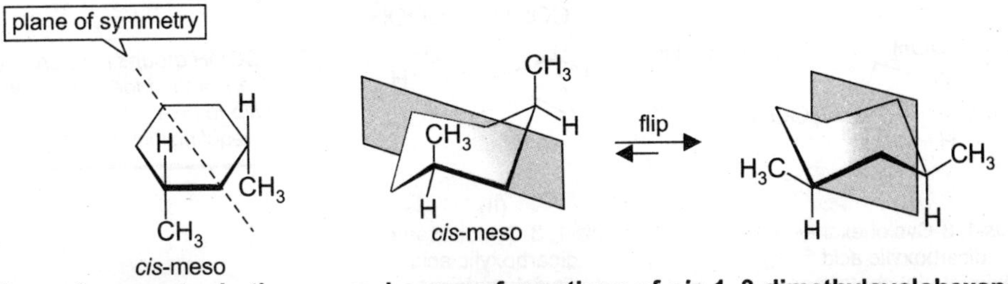

Plane of symmetry in the a, a and e, e conformations of *cis*-1, 3-dimethylcyclohexane

SCHEME 4.35

- *trans*-1, 3-Dimethylcyclohexane, however, does not have a plane of symmetry and exists as a pair of enantiomers (scheme 4.36).

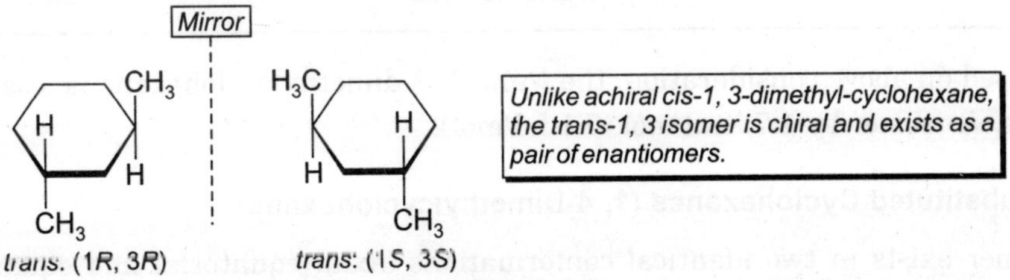

SCHEME 4.36

- In *trans*-1, 3-dimethylcyclohexane (like *cis*-1, 2-dimethylcyclohexane one group is axial and the other equatorial. One has already seen that in *cis*-1, 2-dimethylcyclohexane one enantiomer on ring flip changes into other *i.e.*, ring flip racemizes the compound.
- The ring flip in *trans*-1, 3-dimethylcyclohexane does not racemize this compound but regenerates the same enantiomer *i.e.*, ring inversion here (unlike in *cis*-1, 2-dimethylcyclohexane) only converts (+) into (+) and (–) into (–) enantiomers (scheme 4.37)
- The two substituents in the preferred diequatorial *cis*-1, 3 dimethylcyclohexane are far removed. A preferred diequatorial 1, 3-*cis* isomer can adopt a diaxial conformation by ring flip if a situation so demands. This happens during *e.g.*, anhydride formation from *cis*-1, 3-cyclohexane dicarboxylic acid (scheme 4.38) or in the formation of an intramolecular hydrogen bond formation in *cis*-1, 3, cyclohexanediol.
- In 1, 3-disubstituted cyclohexanes when the two substituents are different both the *cis* as well as *trans* isomers can be resolved, the former exists in the preferred diequatorial conformation while in the latter in an axial equatorial conformer with the bulkier substituent predominantly in the equatorial position.

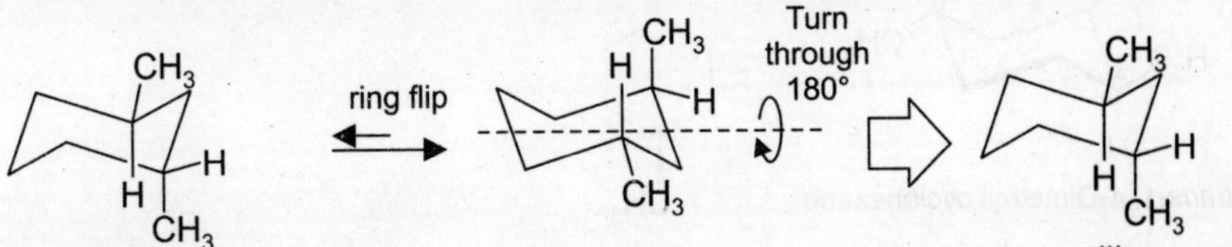

(I)

trans-1, 3-Dimethylcyclohexane

(one enantiomer)

one group is axial and other equatorial

(I)

Ring flip regenerates **the same enantiomer**

SCHEME 4.37

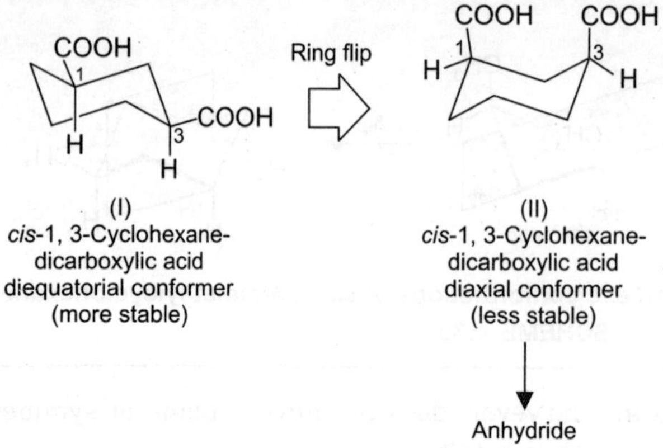

The -COOH groups in (I) are far apart for anhydride formation. Ring flip to give (II) brings them within reacting distance.

SCHEME 4.38

- Based on above consideration, the *trans*-1, 3-dimethylcyclohexane is less stable than the *cis* isomer by 1.7 kcal/mol (7.1 kJ/mol).

(F) 1, 4-Disubstituted Cyclohexanes (1, 4-Dimethylcyclohexane)

The *cis* isomer exists in two identical conformations (axial equatorial and equatorial axial) while the *trans* in two non equivalent (equatorial, equatorial and axial, axial) conformations (scheeme 4.39, only one conformer in each case is shown). The vertical plane passing through C1 and C4 is a mirror plane of symmetry so all the conformers are achiral (even if the two substituents are different). The *cis* and *trans* isomers represent diastereomers.

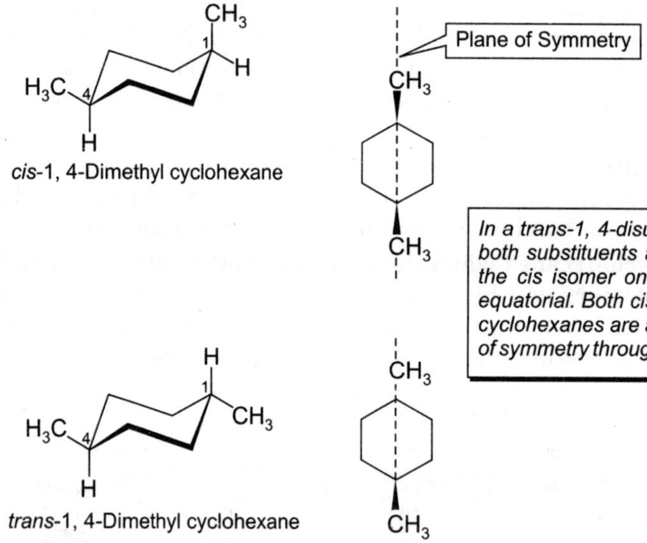

In a trans-1, 4-disubstituted cyclohexane either both substituents are diaxial or diequatorial. In the cis isomer one group is axial while other equatorial. Both cis and trans 1, 4-disubstituted cyclohexanes are achiral diastereomers, (plane of symmetry through C1 and C4)

SCHEME 4.39

(G) Summary

The chirality and nomenclature of different disubstituted cyclohexanes (with the more stable conformer) can be quickly reviewed from (schemes 4.40 and 4,41). When one considers the planar structures of *cis*-1, 3-dimethylcyclohexane and *cis*-1, 2-dimethylcyclohexane (I and II respectively, scheme 4,41) one finds that both have a plane of symmetry and both must be optically inactive. In compound (I) both the conformations (see scheme 4.35) also have a plane of symmetry, thus these are achiral and *meso.* On the other hand (II) consists of two chiral conformations which are enantiomeric. Thus both (I and II, scheme 4.41) have a plane of symmetry in their flat structures however, (I) represents a *meso* compound while (II) a racemic mixture of two separable enantiomers (low temperature). This difference is not clear from the planar structures which therefore, is one of the disadvantages of representing cyclohexanes in their flat form.

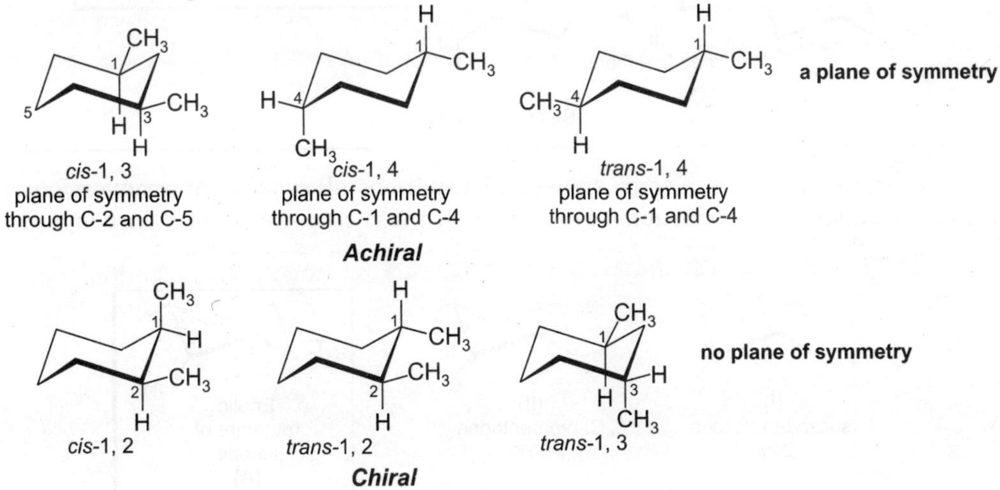

SCHEME 4.40

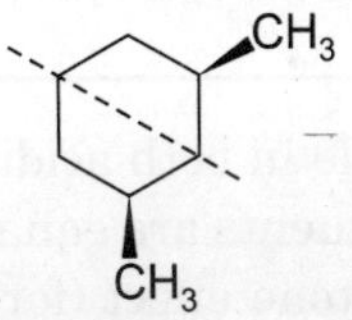

(I)

***cis*-1, 3-Dimethylcyclohexane**

Optically inactive, its conformers (diequatorial and diaxial) have a plane of symmetry. It represents a *meso* compound

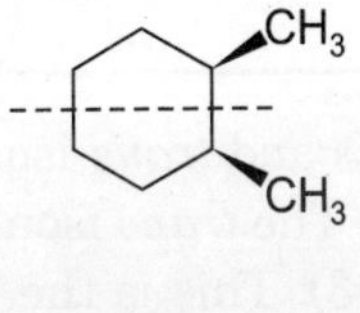

(II)

***cis*-1, 2-Dimethylcyclohexane**

Optically inactive, its conformer has no plane of symmetry, it exists as an enantiomeric pair (*racemic* mixture).

SCHEME 4.41

4.11 EQUILIBRIA OF DISUBSTITUTED CYCLOHEXANES AND RELATED SYSTEMS

Monosubstituted cyclohexanes occur as equilibrium system in which axially and equatorially substituted compounds interconvert rapidly through the flipping of the ring, although one or the other conformational isomer usually predominates.

Disubstituted and more complex cyclohexanes on the other hand can exist in stereoisomeric forms which are not normally interconvertible, whether or not the ring is able to flip. Thus, *cis*- and *trans*-menthane exist (scheme 4,42) as separate compounds although ring-flipping occurs in each. Similarly, *cis*- and *trans*-decalins exist as separate compounds, because ring-flipping cannot interconvert the two.

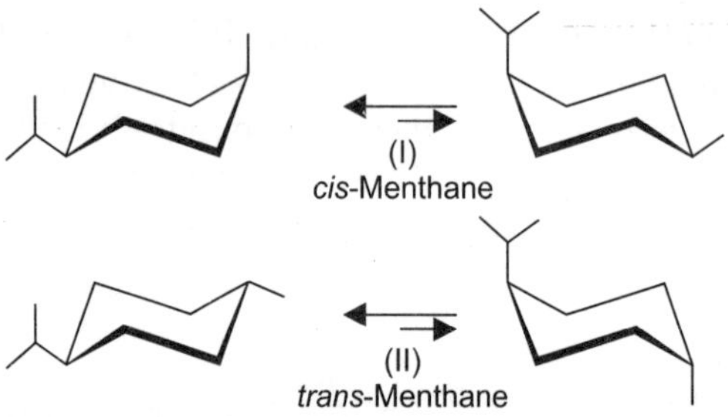

A cis-disubstituted cyclohexane remains cis even after ring flip and same is the case with the trans isomer.

SCHEME 4.42

However, the presence of some structural situations allows the interconversion of such stereoisomers under some conditions. For example, the incorporation of a carbonyl group in

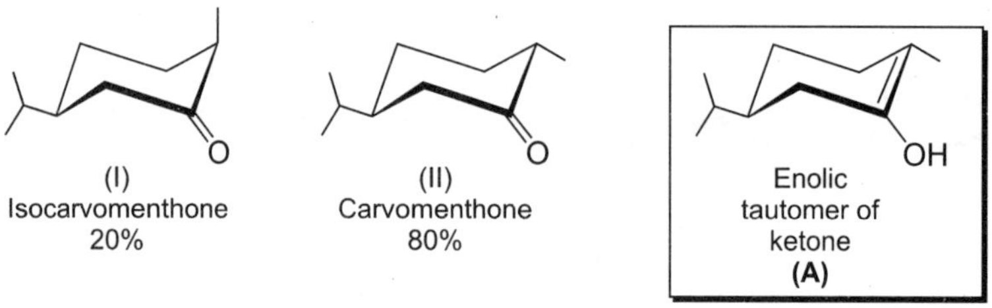

***cis*-Disubstituted cyclohexane can be converted to a *trans*-isomer via an enol.**

SCHEME 4.43

menthane gives *cis* and *trans*-isomers which are interconvertible in both acidic and basic conditions *via* enol A. The *trans* isomer, in which both alkyl substituents are equatorial, predominates (scheme 4.43). This is the basis of 2-alkyl and 3-alkylketone effect (for details see sec. 4.18).

Interconversion occurs, in acidic conditions, through the enolic tautomer of the ketone (scheme 4.43*a*) and, in basic conditions, through the enolate anion *i.e.,* carbanion.

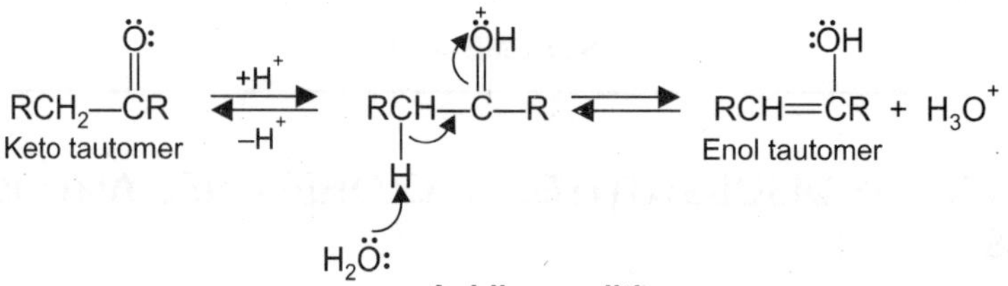

Acidic conditions

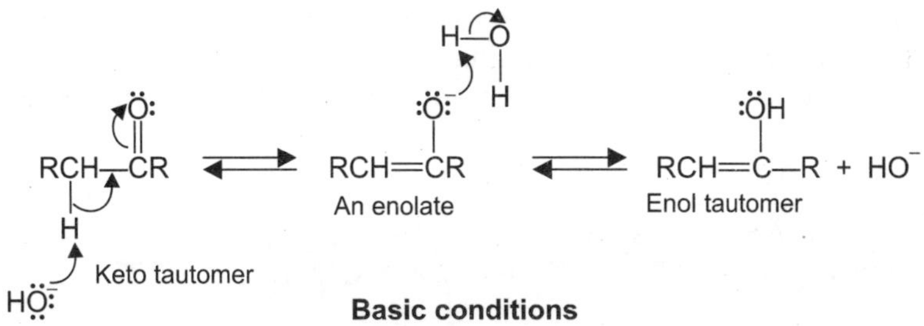

SCHEME 4.43a

Interconversion can also be brought about provided the ring system readily undergoes ring opening and ring-closure. The best known example is glucose. α-D-Glucose and β-D-glucose are interconverted fairly rapidly in solution through the open chain tautomer (scheme 4.44), the β-form predominates, as all the substituents are in equatorial positions (also see schemes 1.108 and 1.109).

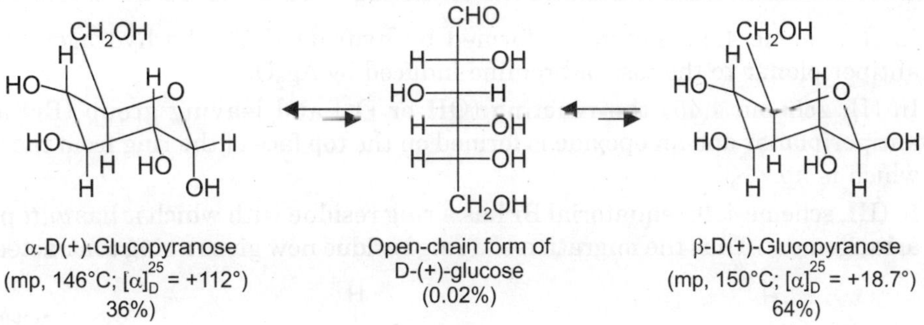

SCHEME 4.44

When crystalline α-D-glucose is dissolved in water, the initial specific rotation of the solution (+ 112°) falls gradually to an equilibrium value of 52°.30′; the β-form has specific rotation + 18°.7′. The phenomenon is known as mutarotation.

4.12 EFFECT OF CONFORMATION ON REACTIVITY—CYCLIC SYSTEMS—INTRODUCTION

One has already learnt about the effect of conformation on reactivity in acyclic systeme *e.g.*, during eliminatoin and Wagner type shifts (schemes 4.9–4.14). Further examples are found in Chapter 5 where again it is seen that for reactivity a particular conformation is adopted (see scheme 5.20).

Several examples have also been discussed in cyclic systems, where to satisfy a particular mechanism a suitable conformation (may be of higher energy) is adopted as in (see schemes 4.38, 5.21 and 5.22, 5.40 and 5.41). Here one will learn of more examples dealing with stereo-electronic effects and that, ring formation, rearrangement and neighboring group participation are interrelated reactions with the stereoelectronic requirement that the groups involved must be *anti*periplanar. This situation normally involves a diaxial orientation, however, a ring residue may also be *anti*periplanar to an equatorial leaving group. Consider the four diastereomeric

2-bromo-4-phenyl-cyclohexanols (schemes 4.45-4.47) the phenyl group is used to anchor the chair conformation and prevent it from flipping. In each case the stereoelectronic requirement for the groups involved in the reaction to be antiperiplanar is obvious. The following points may be noted:

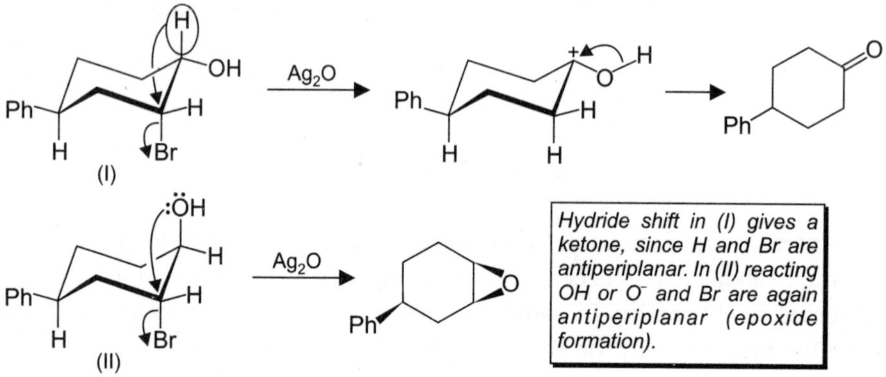

Hydride shift in (I) gives a ketone, since H and Br are antiperiplanar. In (II) reacting OH or O⁻ and Br are again antiperiplanar (epoxide formation).

SCHEME 4.45

- In (I, scheme 4.45) ketone is formed by hydride shift, the hydrogen involved is antiperiplanar to the loss of bromine induced by Ag_2O.
- In (II, scheme 4.45) the reacting (OH or O^-) and leaving group (Br) are again antiperiplanar and an epoxide is formed on the top face of the ring from the OH group which is up.
- In (III, scheme 4.46) equatorial Br has a ring residue with which it has *anti* periplanar arrangement. Thus the migration of a ring residue now gives a ring contracted product.

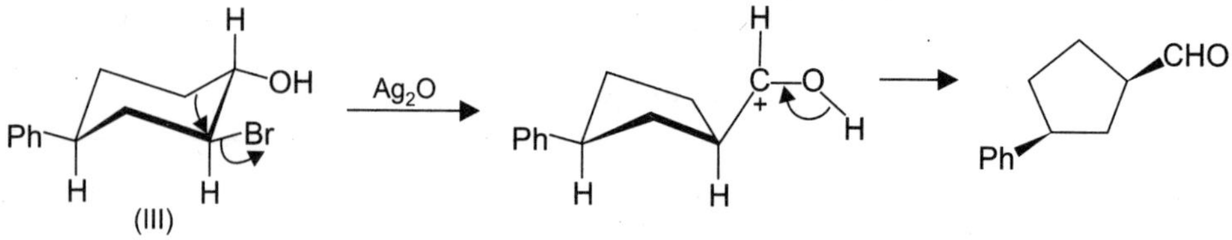

In (III) Ag_2O induces loss of Br (equatorial) and only a ring residue is antiperiplanar to this leaving Br which migrates to give ring contraction.

SCHEME 4.46

- In (IV, scheme 4.47) proper stereoelectronic arrangement is not available though it is the stable starting material. The molecule thus undergoes a ring flip to less stable conformation (IV*a*, 4.47, with *syn*-axial phenyl and bromine). The product is a ketone, the one also obtained from (I, scheme 4.45).

4.13 SOME TYPICAL CYCLOHEXANES

Because of its bulk, the *t*-butyl group has effective preference for an equatorial position. When excessive strain is involved, a distortion of the cyclohexane ring occurs. For example, phenyl and *t*-butyl are both bulky groups, a crystal structure analysis of a compound that has a *cis*-4 phenyl-1-*t*-butyl cyclohexane structure shows that the ring has been stretched out somewhat but still has essentially a chair conformation with axial-phenyl and equatorial-*t*-butyl groups.

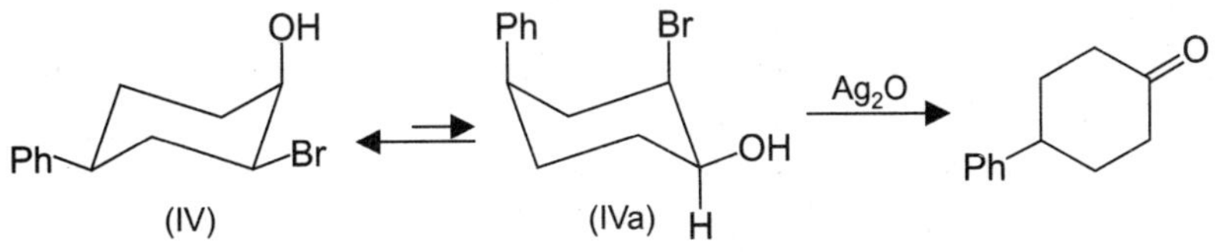

In (IV) proper antiperiplanar arrangement is not available, it reacts via alternate high energy conformation (IVa, obtained by ring flip) which provides this requirement. The cyclohexanone formed is the same as from (I, scheme 4.45)

SCHEME 4.47

In *trans*-1, 3-di-*t*-butylcyclohexane (I, scheme 4.48) a chair-cyclohexane ring would require one *t*-butyl group to be axial, in this compound the cyclohexane ring is twisted in order to avoid placing the *t*-butyl group is an axial position to attain a so-called "twist-form" or "skew-boat" structure (scheme 4.48). This skew-boat form occurs in several compounds containing bulky groups but is not however, a significant conformation for cyclohexane itself.

Similarly in the case of *r*-1-*cis*-*di*-*t*-butyl-*cis*-2-*cis*-5-dihydroxycyclohexane (II, scheme 4,48) an equilibrium exists between a chair and twist boat conformations (The twist boat conformation gets some stabilization from hydrogen bonding).

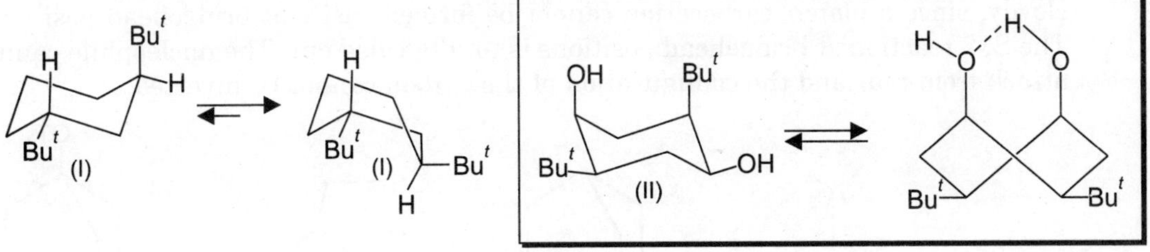

SCHEME 4.48

In (1990) it was reported that every isopropyl group in (I, scheme 4.49) is axial. Here a consideration of space filling models will show that now 1, 2-steric interactions when two isopropyl groups are equatorial overweigh the usual well known, 1, 3-diaxial interactions.

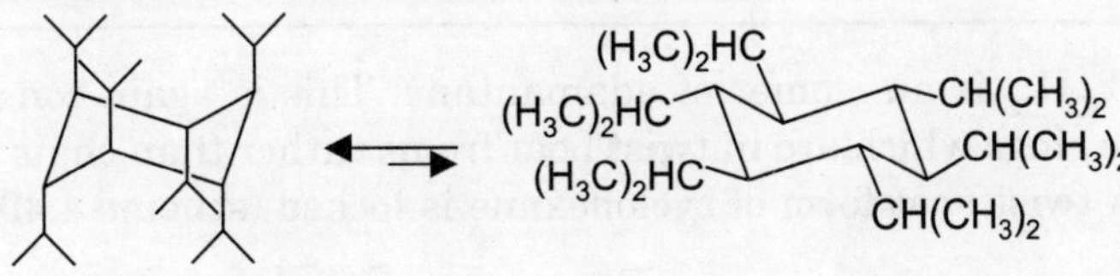

(I)

SCHEME 4.49

4.14 LOCKING OF FLEXIBLE CONFORMATIONS

The following points may be noted:

- The chair form of cyclohexane is normally the most stable conformer.

- A chair form can be anchored in its conformation with an equatorial *t*-butyl group and its flip can be largely arrested (scheme 4.49*a*).
- A cyclohexane ring can be locked into its boat conformation as in norbornane system of camphor.
- The chair form is fixed in the molecule of adamantane $C_{10}H_{16}$ first isolated from mineral oil. It is the smallest unit of the diamond lattice.
- The four six membered rings of adamantane are strain free and are identically arranged in chair forms. Although the model of admantane looks strain free, it has about 7.6 kcal/mol strain which is more than four times the strain in cyclohexane. In adamantane the carbon framework forces the bond angles to be perfectly tetrahedral. However, one knows that in C—CH_2—C system the bond angle does not want to be exactly tetrahedral, the optimum in propane is 112.5° and in cyclohexane it is 111.5°. It is this further reduction to 109.5° in adamantane, that causes strain in the molecule.
- Adamantane has high symmetry and is almost spherical. It has four C_3 axes, three C_2 axes and six σ planes (same as in methane). In fact it is like an expanded tetrahedron and when the four bridgehead positions carry different groups, the compound becomes chiral.
- The introduction of a double bond in the ring system of adamantane is sterically impossible (Bredt's rule) and Bredt's rule is applicable to every carbon in adamantane.
- The bridgehead adamantly tosylate does undergoes an S_N1 reaction but extremely slowly, since a planar carbocation cannot be formed easily at bridgehead positions. The S_N2 reaction at bridgehead positions is totally ruled out. The nucleophile cannot attack from rear and the configuration of the carbon cannot be inverted.

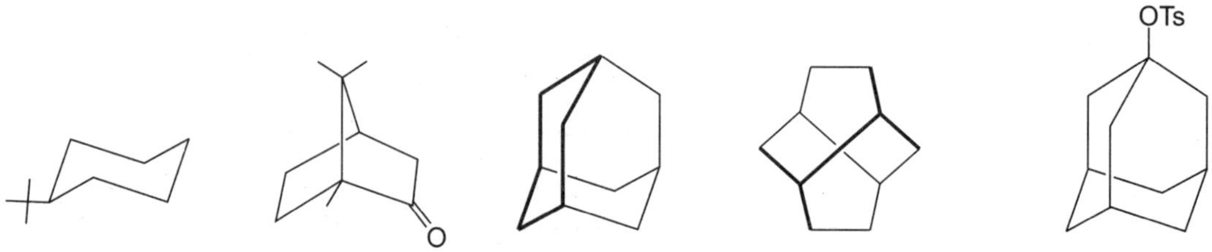

ring flip is arrested | Comphor with cyclohexane ring locked in boat form | Adamantane with each cyclohexane ring in chair form | Twistane with each cyclohexane ring in twist boat form | Adamantyl tosylate

SCHEME 4.49a

- Twistane $C_{10}H_{16}$ is an isomer of adamantane. This is again constructced from four cyclohexane rings which are in twist boat froms rather than chair forms. In twistane, therefore, a twist boat form of cyclohexane is locked (scheme 4.49*a*).

4.15 CONFORMATION OF HETEROCYCLES

Pyrrolidine has been shown, from spectroscopic data and X-ray analysis, to be a puckered ring similar to cyclopentane and also, like cyclopentane, undergoes *pseudo-rotation* (scheme 4.50). Piperidine exists in the chair form. For *N*-ethylpiperidine the ethyl substituent prefers the equatorial position while the lone pair occupies the axial position. Because of intramolecular hydrogen bonding, cyclohexane 1, 4-diol exists in the boat form (or twist-boat form); in some highly substituted 4-hydroxypiperidines boat form is present for the same reason.

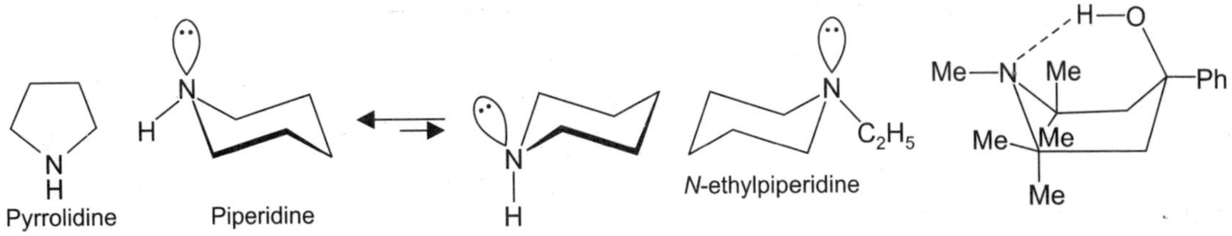

SCHEME 4.50

Like piperidine the six membered heterocycles of oxygen also closely resemble the chair conformation of cyclohexane. In heterocyclic rings the steric repulsions for axial substituents are reduced due to the replacement of methylene groups of cyclohexane by oxygen or nitrogen.

A six membered ring containing an oxygen atom (tetrahydropyran) assumes a chair conformation. At all positions generally alkyl substituents prefer an equatorial arrangement (scheme 4.51), however, an α-alkoxy substituent shows increased stability in the axial position. This effect is called the anomeric effect (see scheme 4.53)

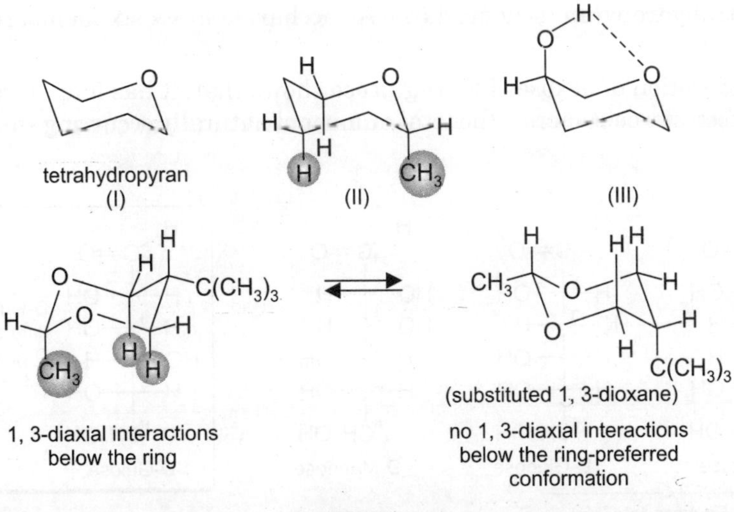

SCHEME 4.51

In (II, scheme 4.51) substitution next to the oxygen atom leads to more severe 1, 3-diaxial interactions compared with cyclohexane because C—O bonds are shorter than C—C bonds and bring the axial groups closer. In the case of situations like (III, scheme 4.51) the presence of an oxygen atom in the ring allows hydrogen bonding that can stabilize hydroxyl groups in the axial position. This is particularly important for sugars because they have several OH groups.

In *cis*-2-methyl-5-*tert*-butyl-1, 3-dioxane (scheme 4.51) in which the preferred conformation has the bulky tert-butyl group axial and the methyl group equatorial. Since the divalent oxygen has no substituents, therefore, the 1, 3-diaxial interactions which are the main unfavorable interactions for axial substituents in cyclohexanes are absent.

It is consistenly seen that 5-alkyl substituents in 1, 3-dioxanes show a smaller equatorial preference than they do in cyclohexane. This decreased preference is due to decreased van der Waals repulsions in the axial orientation, since there are now no hydrogens that are *syn*-axial to the 5-alkyl substituent. A 2-alkyl substituent, however, has a greater preference for the equatorial orientation in a 1, 3-dioxane than in cyclohexane, presumably because the decreased C—O bond length 1.43 Å (relative to C—C bond length 1.54 Å) brings an axial 2-alkyl group closer to the *syn*-axial hydrogens at C(4) and C(6) to result in an increased van der Waals repulsion. Similarly, an axial 4-alkyl substituent in a 1, 3-dioxane suffers a greater van der Waals repulsion with the axial hydrogen at C(2) that it does in cyclohexane.

4.16 CONFORMATION OF SUGARS AND ANOMERIC EFFECT

(A) Conformation of Common Sugars

One has already learnt drawing the Fischer, Haworth and "zig-zag" structures of D-glucose (see, schemes 1.20*g* and 1.33*c*). The conversion of Haworth formula of α-D-glucose to chair conformation has been presented (scheme 1.20*g*). Mechanism of mutarotation is given (scheme 1.109). The following points may be noted regarding conformations of some sugars.

- The six membered ring which includes an oxygen atom is related to the heterocyclic ring tetrahydropyran (scheme 4.52). A saccharide in its six membered form is called a pyranose.
- A consideration of hexoses like D-glucose shows that, it has four stereocenters. There are sixteen stereoisomers—the total number of naturally occurring stereoisomers being eight.

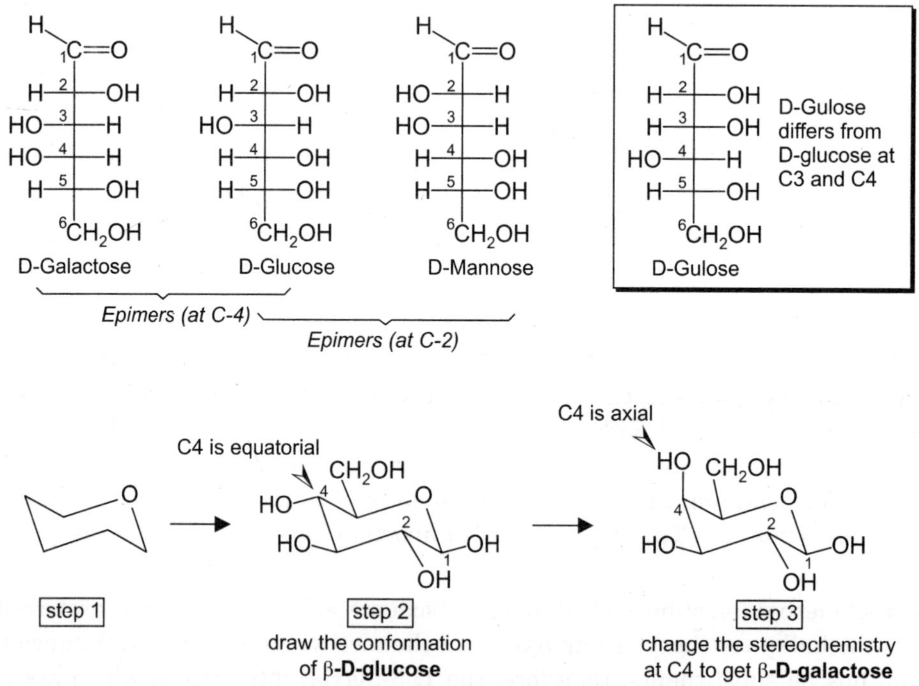

writing the conformation of β-D-galactose

SCHEME 4.52

- Only the three of these namely D-glucose, D-mannose and D-galactose are most widely occurring (α-D-mannose tastes sweet, while its β-form is bitter).
- As the starting point draw the chair conformation of tetrahydropyran.
- Secondly draw the conformation of β-D-glucose which is very easy to draw since all the OH groups are in equatorial position.
- From β-D-glucose it is easy to draw the chair conformation of any sugar provided one knows the epimeric relationship it has with β-D-glucose (β-D-mannose is its C2 epimer while β-D galactose is its C4 epimer). Thus *e.g.*, for drawing the conformation of β-D-galactose change the stereochemistry at C4 from equatorial to axial position.
- In the corresponding α-form of a sugar the OH group at C1 (anomeric carbon) will be down and axial.

EXERCISE 4.6

Draw the conformation of α-D-mannose.

ANSWER.

Step 1. *Draw the chair form of tetrahydropyran.*

Step 2. *Draw β-D-glucose by adding all OH groups in the equatorial position.*

Step 3. *Since D-glucose and D-mannose are C2 epimers change the stereochemistry at C2 (from equatorial to axial to get the conformation of β-D-mannose.*

Step 4. *In an α-anomer the OH group is axial at this carbon thus change the stereochemistry of OH from equatorial to axial at the anomeric carbon to get α-D-mannose (scheme 4.52a).*

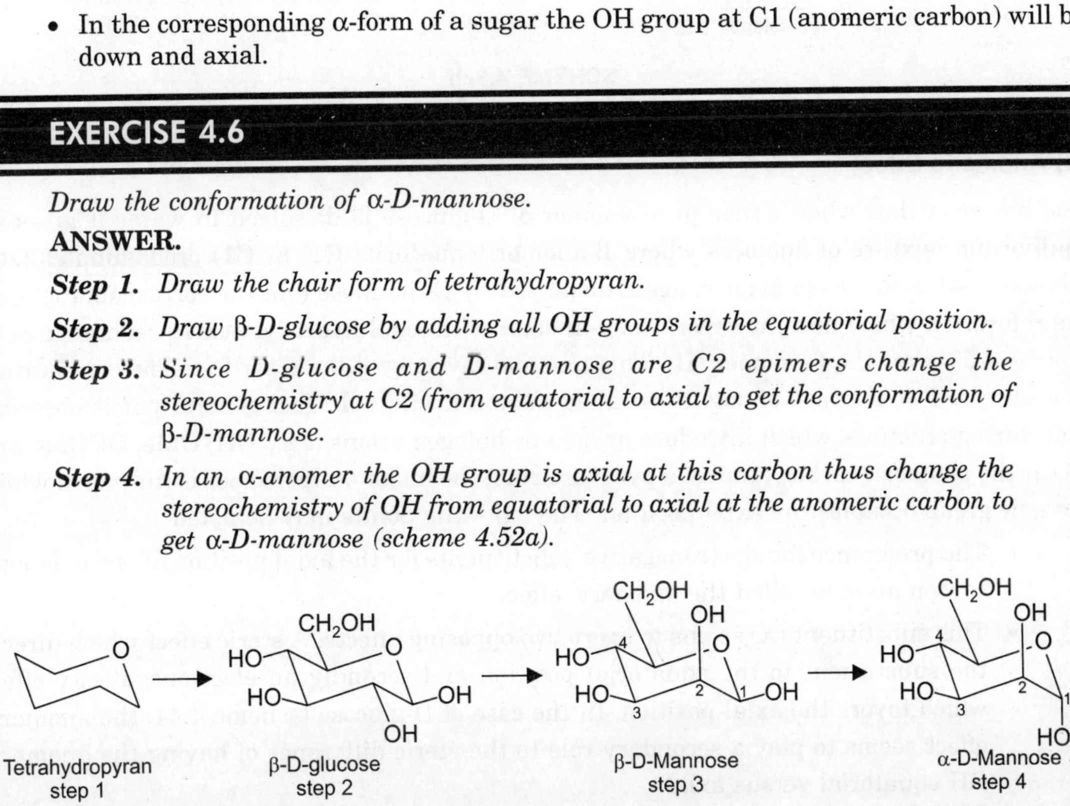

SCHEME 4.52a

(B) Conformation of an L-Pyranose (β-L-Gulose)

To draw the conformation of an L-pyranose, first draw the D-pyranose as explained above. The conformation of *e.g.*, β-L-glucose can be drawn by first drawing β-D-glucose, and then drawing its mirror image.

Step 1. Draw the chair conformation of tetrahydropyran (I, scheme 4.52 *b*).

Step 2. Derive the conformation of β-D-glucose by adding all the OH groups in the equatorial positions.

Step 3. D-gulose differs from D-glucose at C3 and C4 positions (see scheme 4.52) thus in β-D-gulose the OH groups at C3 and C4 must be axial.

Step 4. Finally draw the mirror image of β-D-gulose to get β-L-gulose (scheme 4.52 *b*).

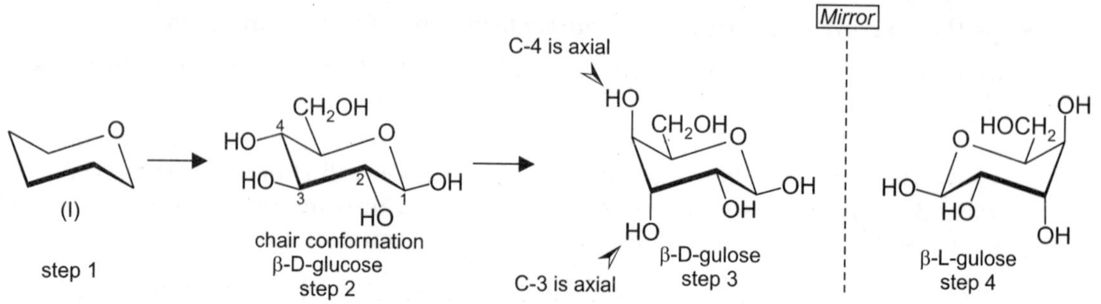

SCHEME 4.52b

(C) Anomeric Effect

One has seen that when either pure anomer of D-glucose is dissolved in water it gives an equilibrium mixture of anomers where β-anomer (equatorial OH at C1) predominates (see, scheme 4.44). Similar behavior is again displayed by D-mannose (via the corresponding open chain form) in which now, however, the α-anomer with axial hydroxyl at C1 predominates (I, scheme 4.53) over the β-anomer (II). This unusual favoring of the α (or the axial) orientation first observed in the case of sugars is called anomeric effect. In fact generally it is observed that during reactions which introduce oxygen or halogen atoms (*e.g.*, OH, OMe, OCOMe and C1) at the anomeric carbon atom of a pyranose ring, the major stereoisomer is the one in which the new group occupies the axial position. The following points may be noted:

- The preference for electronegative substituents for the axial position at the anomeric carbon atom is called the anomeric effect.
- This substituent (X) seems to exert two opposing effects. A steric effect which directs the substituent to the equatorial position and secondly an electronegativity effect which favors the axial position. In the case of D-glucose (scheme 4.44) the anomeric effect seems to play a secondary role to the steric difference of having the anomeric OH equatorial versus axial.
- The methyl mannosides (II and III, scheme 4.53) in 1% methanolic HCl undergo equilibration to a ratio of α : β = 94 : 6 (scheme 4.53). This is an operation of anomeric effect. In α-D-Mannose (or its mannosides, partial structures III and IV, scheme 4.53), the non-bonding electrons on the oxygen of the ring and the nonbonding electrons of the substituent repel eah other when it is in the equatorial position (IV, scheme 4.53). These are farther apart when the substituent is in the axial position.

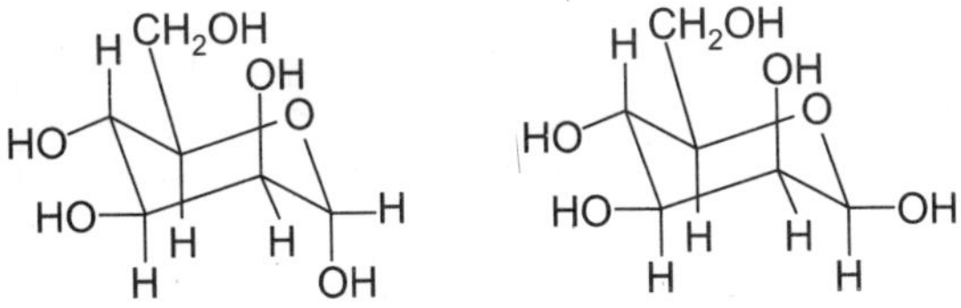

(I) α-D-Mannose (69%)

(II) β-D-Mannose (31%)

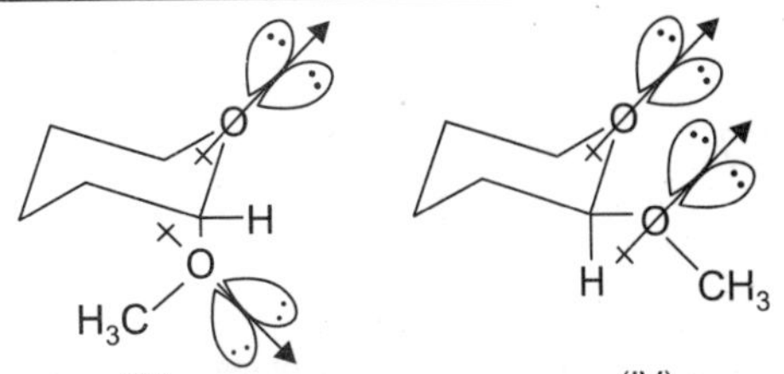

(III) α = 94%

(IV) β = 6%

Methyl mannosides

> *In D-mannose at equilibrium in water more of α-axial anomer is present than β. This discrimination is further enhanced in methyl mannosides. In the Newman projections when the electronegative substituent (X = OH, OCH_3, Cl etc.) is axial it is gauche to one but anti to the other lobe with non-bonding electrons. When the substituent X is equatorial it is gauche to both lobes.*

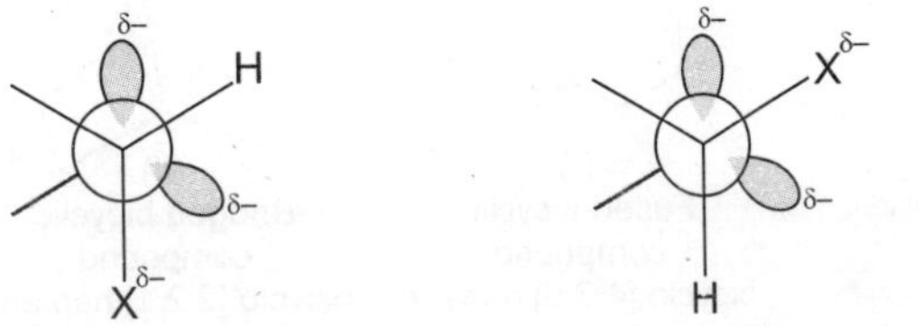

electronegative group gauche to one but anti to another lobe bearing nonbonding electrons

electronegative group gauche to both lobes bearing nonbonding electrons

SCHEME 4.53

4.17 POLYCYCLIC COMPOUNDS

(A) Bicyclo [2.2.1] Heptane A Bridged system–Norbornanes (*Exo, Endo, Syn* and Anti Nomenclature)

Bicyclo [2.2.1] heptane is the most commonly studied system and its trivial name norbornane reflects its relation with natural terpene bornane. In these systems 'nor' signifies the absence of one or more methyl groups. The following points may be noted:

- In norbornane system C1 and C4 of cyclohexane are joined by a methylene bridge thus forcing it to adopt a boat conformation (scheme 4.54). Camphor is an example of a naturally occurring compound with [2.2.1]–heptane system.
- The atoms common to both rings in a bicyclic system are called bridgehead carbons. Bicyclic compounds are further classified according to the relationship of the bridgehead carbons. When the bridgehead carbons of a bicyclic compound are adjacent, the compound is classified as a fused bicyclic compound (I, scheme 4.54). However, when the bridgehead carbons are not adjacent the compound is called a bridged bicyclic compound (II, scheme 4.54).

> *Bridgehead carbons are the carbons joined to each other by three different chains of carbon.*

- The compound (II, scheme 4.54) is named as a bicycloheptane because it is a bicyclic compound containing a total of seven carbon atoms. The numbers in breackets represent the number of carbon atoms in the respective bridges, in order of decreasing size. The sum of the numbers inside the bracket is two less than the total number of carbons since the two bridgehead carbons C1 and C4 are not included.

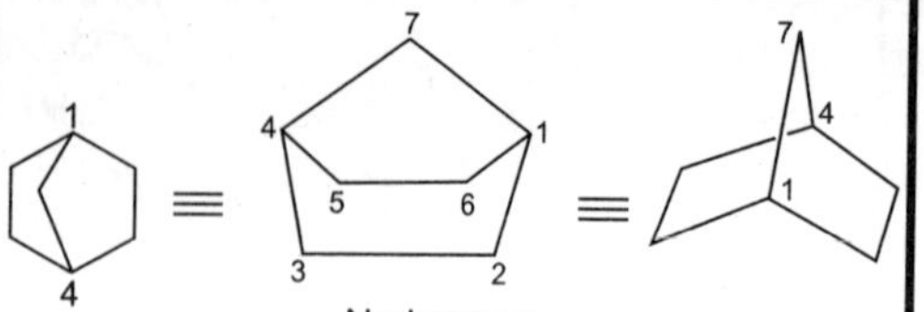

Norbornane
Bicyclo [2.2.1] heptane

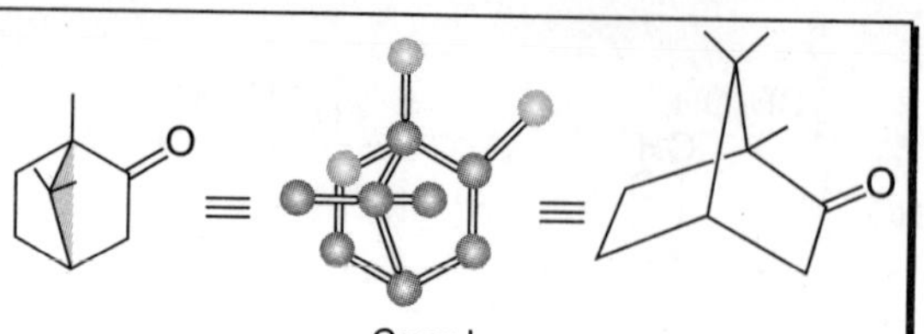

Camphor
contains a bicyclo [2.2.1] norbornane system

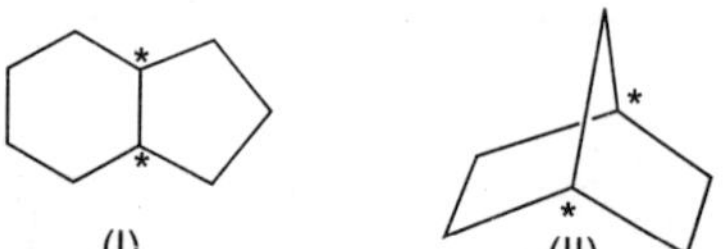

(I)
Fused bicyclic compound
bicyclo [4.3.0] nonane

(II)
Bridged bicyclic compound
bicyclo [2.2.1] heptane

* designate bridgehead carbons

(I) is a fused bicyclic system, since bridgehead carbons are adjacent

(II) is a bridged bicyclic compound since bridgehead carbons are not adjacent.

SCHEME 4.54

- The numbering of carbons beings from one bridgehead carbon and follows the largest bridge first (I, scheme 4.54*a*).

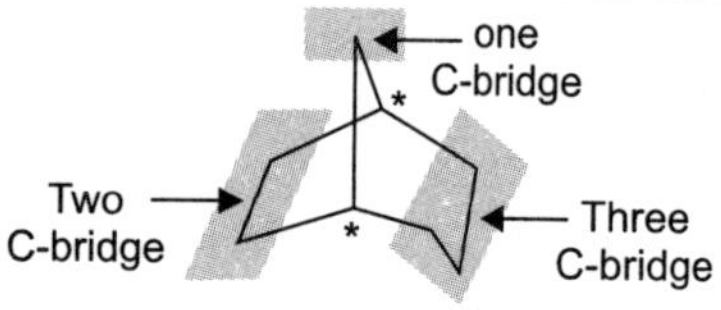

bicyclo [3.2.1] octane
(* = bridgehead carbons)

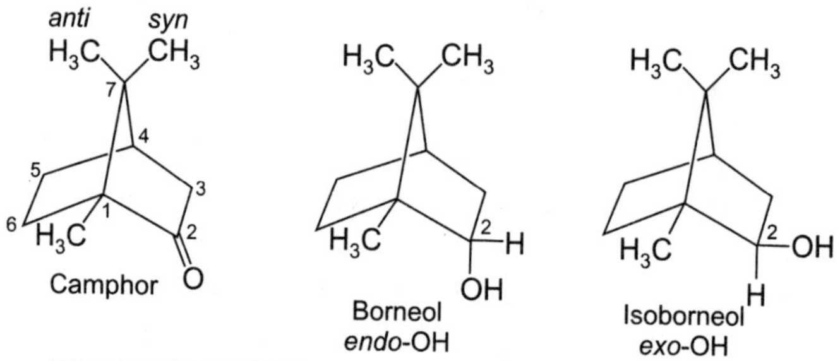

Camphor

Borneol
endo-OH

Isoborneol
exo-OH

In borneol and isoborneol the configuration at only one of the three stereocenters differs (epimeric pair). The configuration at C2 is designated by exo and endo nomenclature which is more convenient than R/S system.

SCHEME 4.54a

In camphor (scheme 4.54*a*) there is a choice as two bridges are the same. In this case numbering is fixed by the C = O group in one of the bridges. The numbering thus requires the lowest possible number to the carbon containing the functional group and this is C2.

- The terms *exo* and *endo* are used for the substituents on a bridge which connects the bridgehead carbons. An *exo* substituent is the one which is on the same side of the smaller bridge and *endo* if it is opposite to the smaller bridge (scheme 4.54*a*). This usage is clear from borneol and isoborneol, the reduction products of camphor. For details of reduction (see scheme 7.8*e*). The configuration at C2 in borneols can be specified by *R* and *S* system of nomenclature, however, an easy system uses the terms *exo* and *endo*.

- When a substituent on the smallest bridge *e.g.*, at C7 in a norbornane system is directed towards the next smallest bridge it is called *syn*, however, when it points to the largest bridge it is called *anti*. When the bridges are of the same size and now if the substituent on the smallest bridge points towards the bridge that itself carries substituents it is called *syn* as in camphor. When the substituent on the smallest bridge *e.g.*, in camphor points towards the unsubstituted bridge it is called *anti*.
- To an extent, a configurational restriction is associated with norbornane and related systems. The molecule of camphor has two stereocenters, the bridgehead carbons C1 and C4 and yet only one (±)-pair is known exist. The bridge can only be *cis* and, therefore, the configurations of the bridgehead carbons are not independent. Thus, for these steric reasons, C–1 and C–4 behave as a single element of chirality. Therefore, the number of stereoisomers is always one half of what would be expected if this restriction is not there. Thus, norbornane monosubstituted in the 2-position has three stereocenters, however, only four stereoisomers exist for such derivatives, *i.e.*, (+)- and (–)-*endo* and (+)- and (–)-*exo* stereoisomers.

(B) Prochirality of Bicyclo [2.2.2] Octane System

This is another interesting bridged bicyclic system where 1, 4 positions of the cyclohexane boat are joined by an ethylene bridge. As is the case with bicyclo- [2.2.1]–heptane system, no cis-trans isomers can exist. It is only possible to fuse a two-carbon bridge to carbons 1 and 4 of cyclohexane ring if the two-carbon bridge is fused in a *cis* fashion. (scheme 4.54*b*).

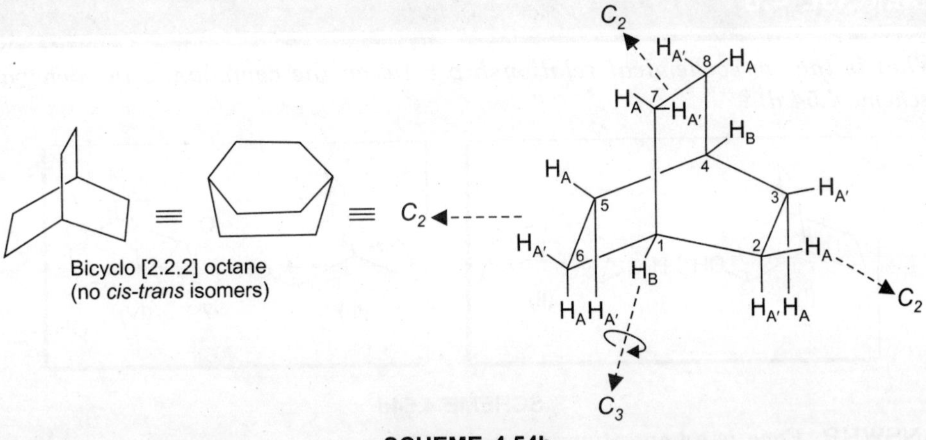

SCHEME 4.54b

In its idealised form, the system has all the adjacent methylene protons eclipsed. It has one C_3 axis, three C_2 axes, and four σ planes (3 vertical and 1 horizontal) and belongs to point group D_{3h}.

The following points may be noted:

- Bicyclo [2.2.2] octane system has three type of H atoms. H_A , $H_{A'}$ and H_B.
- All H_A atoms are homotopic.
- All $H_{A'}$ atoms are also homotopic.
- These atoms are homotopic since these can be exchanged by the C_3 axis or by the three C_2 axes.
- H_A and $H_{A'}$ are however, related as enantiotopic.
- There are no *exo* and *endo* forms here as the two sides of the molecule are identical.

- Substitution at any methylene carbon yields only two enantiomers.
- The two bridgehead hydrogen atoms H_B are homotopic.

EXERCISE 4.7

How many structural isomers can exist for methylbicyclo [2.2.2] octane, write their structures ? Can one of these exist as a diastereomeric pair ?

ANSWER. *Only two structural isomers exist (scheme 4.54c). No, there are no exo and endo forms since all the bridges contain the same number of carbons.*

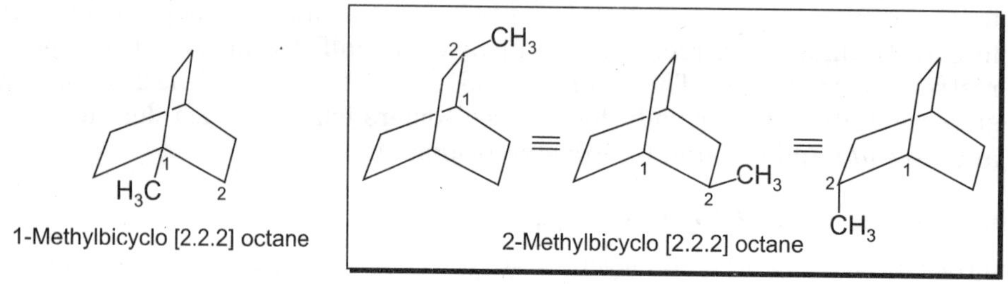

SCHEME 4.54c

EXERCISE 4.8

What is the stereochemical relationship between the compounds in each pair (scheme 4.54 d) ?

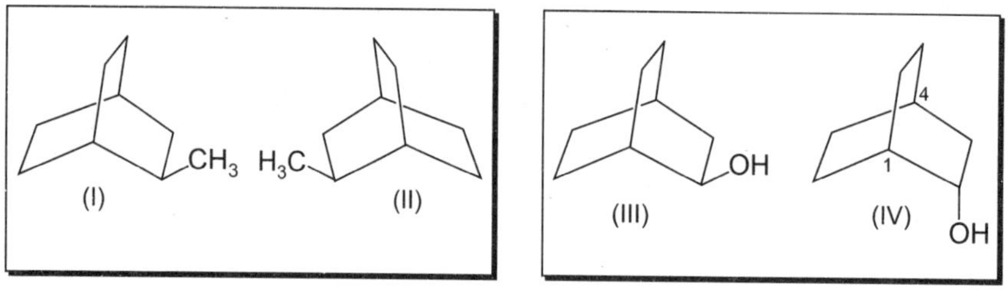

SCHEME 4.54d

ANSWER. *Each is an enantiomeric pair.*

(C) Prochirality of Bicyclo [2.2.1] Heptane System

The following points may be noted:

- The H_B and $H_{B'}$ (*exo* hydrogens) are enantiotopic, and four endo hydrogens H_C and $H_{C'}$ are also enantiotopic.
- The *exo* and *endo* hydrogens are diastereotopic.
- The H_A and H_D pairs are homotopic (norbornane has a C_2 axis).

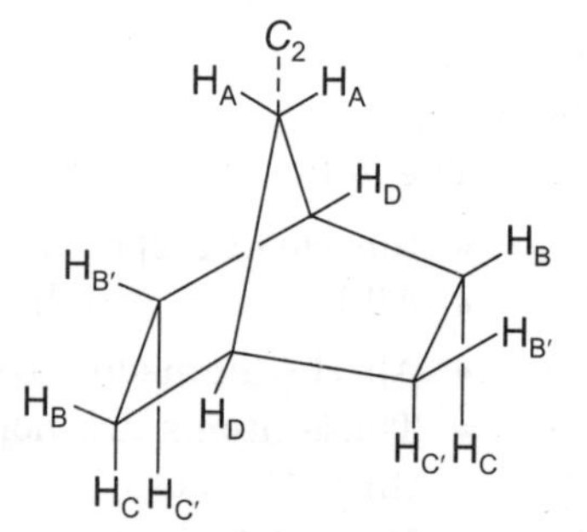

SCHEME 4.54e

EXERCISE 4.9

How many different carbons are present in bicyclo [2.2.1] heptane ? How many structures one can write for methyl bicyclo [2.2.1] heptane ? How many stereoisomers can exist for this compound.

ANSWER. *Three, i.e. C1, C2 and C7. At C1 and C7 one can attach a group only in one way, however at C2 the group could be either exo or endo. The stereoisomers are four (±) exo and (±) endo since monosubstitution at any of the methylene carbons C2, C3, C4 and C5 gives a total of four stereoisomers (scheme 4.54e).*

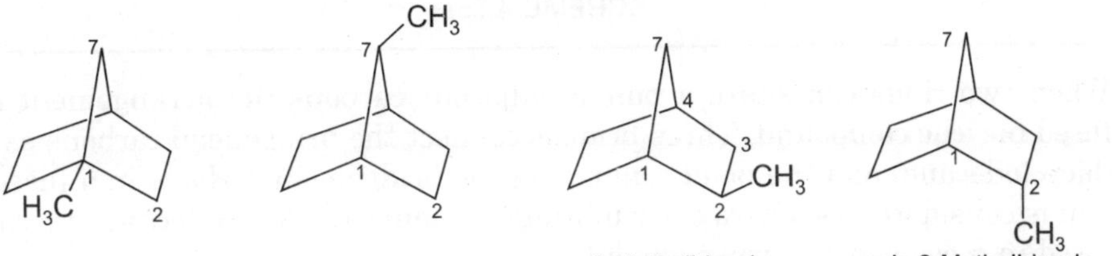

exo-2-Methylbicyclo-[2.2.1] heptane

endo-2-Methylbicyclo-[2.2.1] heptane

SCHEME 4.54e

Cyclic systems with two or more rings joined in the "fused" or "bridged' fashion may lead to some constraints on the number of stereoisomers and structural possibilities for a given system. The bridged-ring compound norbornane is a rigid molecule where the cyclohexane ring is forced in the boat conformation. Norbornane has fewer isomers than predicted, for it may be thought that a bicyclo [2.2.1] heptane structure can exist in which the bridging carbon is linked to the topside of the cyclohexane ring at C-1 and the bottomside at C-4. This system however, would be too strained to be isolated and only a single bicyclo [2.2.1] heptane is known. Similarly less strained bicyclo [2.2.2.] octane (scheme 4.55) is only known in the *cis* fused form. On the other hand, bicyclo [4.4.1] undecane (scheme 4.55) constitutes a system with a methylene group bridging a considerably larger and more flexible ring than cyclohexane. It exists in two isomeric forms. A consideration of molecular models shows this difference rather clearly, and reveals that the *trans* isomer of bicyclo [2.2.1] heptane cannot be constructed while the *trans* isomer of bicyclo [4.4.1] undecane is easily built (scheme 4.55).

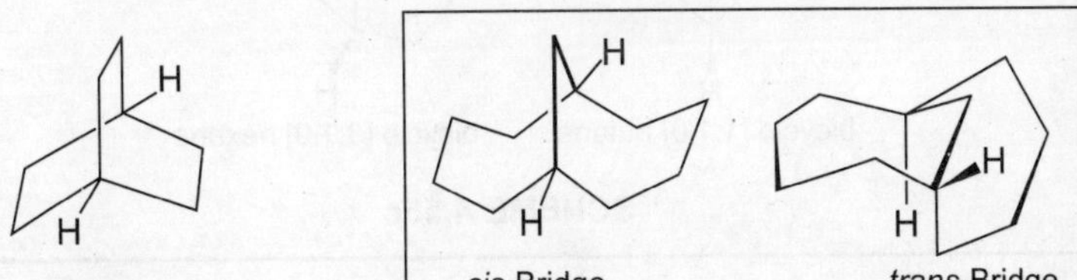

cis-Bridge
Bicyclo [2.2.2]-octane

cis-Bridge *trans*-Bridge
Bicyclo [4.4.1] undecane

SCHEME 4.55

(D) Other Polycyclic Compounds–An Introduction

One may recall the difference between a fused bicyclic compound and a bridged bicyclic compound (see, scheme 4.54). The following points may be noted.

- Two rings may be joined together in three general ways by sharing carbons. When one carbon is shared the mode of attachement is called spiro. In a spiro compound the two rings must lie in different perpendicular planes since the centeral carbon is approximately tetrahedral (scheme 4.55*a*).

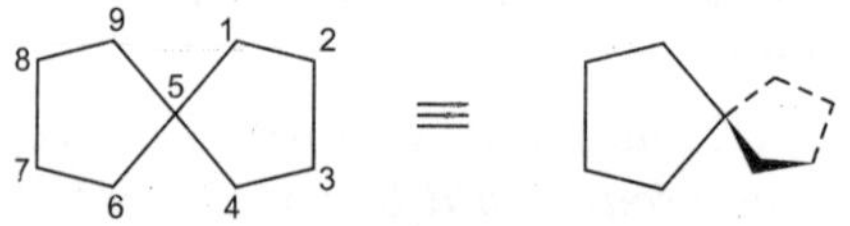

spiro [4.4] nonane
(a spirocyclic compound)

SCHEME 4.55a

- When two rings can share a pair of adjacent carbons the arrangement is called a fused bicyclic compound. Three bridges connect the bridgehead carbons as in decalin (bicyclodecane) and the bond connecting the bridgehead carbons in a fused ring system is considered as a zero carbon bridge (scheme 4.55*b*). In decalin the rings can be fused in a *cis* or *trans* arrangement.

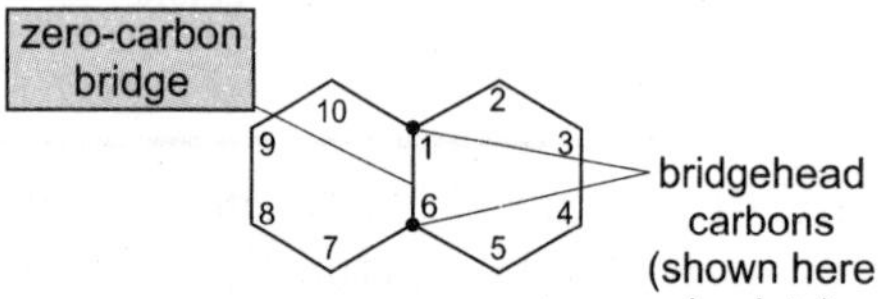

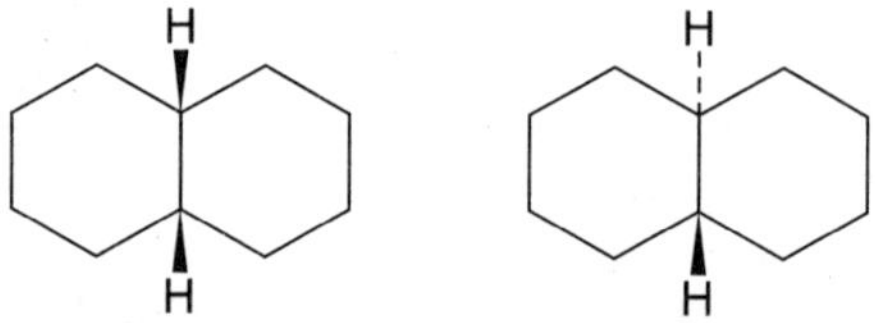

Bicyclo [4.4.0] decane (Decalin)   *cis*-decalin   *trans*-decalin

SCHEME 4.55b

- *Trans*-decalin is found to be more stable than *cis*- decalin (see scheme 4.59). However, when the rings are small only *cis*- fusion is observed since *trans* fusion will introduce too much ring strain. Thus only *cis*- isomers of the compounds (scheme 4.55*c*) are known.

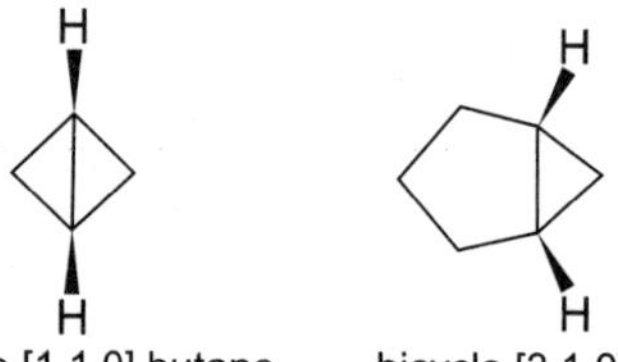

bicyclo [1.1.0] butane   bicyclo [3.1.0] hexane

SCHEME 4.55c

- Recall that in a small bridged compound (*e.g.*, in bicyclo [2.2.1] heptane the two rings share three carbons). The bridge can only be *cis*, but when the ring containing the bridge is large the bridge could as well be *trans* (see scheme 4.55).

(E) Strained Carbocycles—Bredt's Rule

The linear geometry of a triple bond has put a limit to the preparation of stable cycloalkynes to rings of eight or more atoms (I, scheme 4.56). Double bonds with the Z (*cis*) configuration are accommodated by a monocycle of any size. Cyclobutene is a stable compound (bp 2°C) while cyclopropene has been isolated only at temperatures below – 80°C. Cyclohexene is a strain-free compound with an estimated strain energy of 1 kcal/mol (4 kJ/mol). Only its *Z* geometrical isomer is known. Cyclooctene is the smallest cycloalkene of which *E* isomer has been isolated (II, scheme 4.56, also see, scheme 1.64).

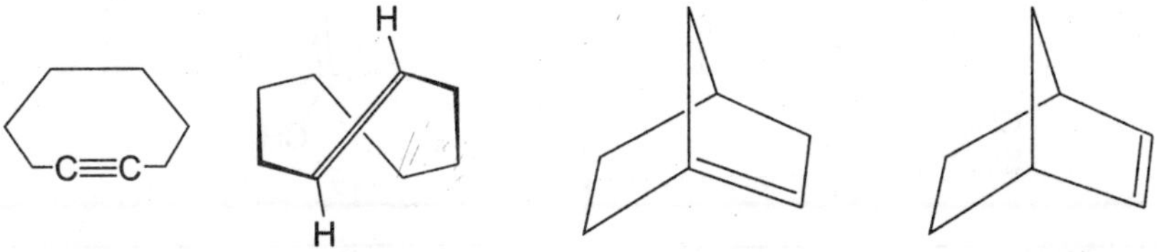

(I) Cyclooctyne (II) *E*-Cyclooctene (one enantiomer) (III) Bicyclo [2.2.1] hept-1-ene (unknown-violation of Bredt's rule) (IV) norbornene a stable compound

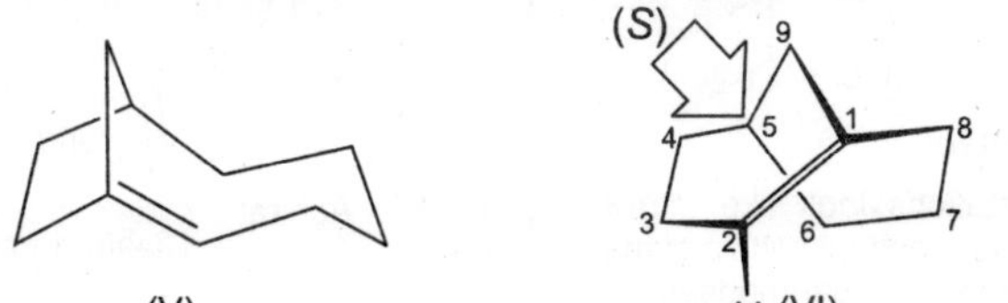

(V) Bicyclo [5.2.1] dec-1-ene (VI) (*S*)-Bicyclo [3.3.1] non-1, (2)-ene

SCHEME 4.56

The Bredt's rule states that a bridged bicyclic compound cannot have a double bond at a bridgehead position unless one of the rings contains atleast eight carbon atoms. Norbornene is a stable compound, since the double bond is away from the bridgehead carbon, however, its isomer (III, scheme 4.56) with a bridgehead double bond is unknown. When however, the number of ring atoms is larger in which the ring strain (the cause of non-formation of bridgehead double bonds) is lower, the presence of a bridgehead double bond is possible.

It may suffice to explain that in (III, scheme 4.56) the *trans* double bond is contained in a six-membered ring to make it incapable of existence. In the case of (V, scheme 4.56), however, the *trans* double bond is instead contained in a nine-membered ring and thus, the compound has been prepared and isolated. A further interesting example of a compound with a bridgehead double bond is of (VI, scheme 4,.56), the compound has a stereocenter (C5), is chiral and rèsembles *trans*-cyclooctene.

During β-eliminations in bicyclic systems, the bridgehead atoms are not involved, provided the ring bearing the incipient "*trans*" double bond has at least eight atoms (schme 4.57).

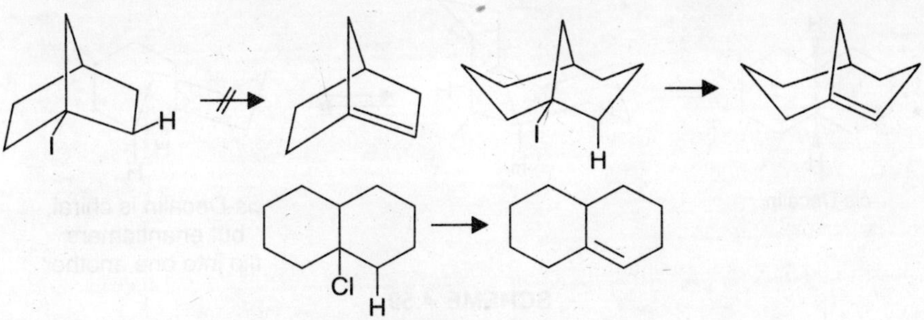

SCHEME 4.57

A simple mental Exercise in stereochemistry

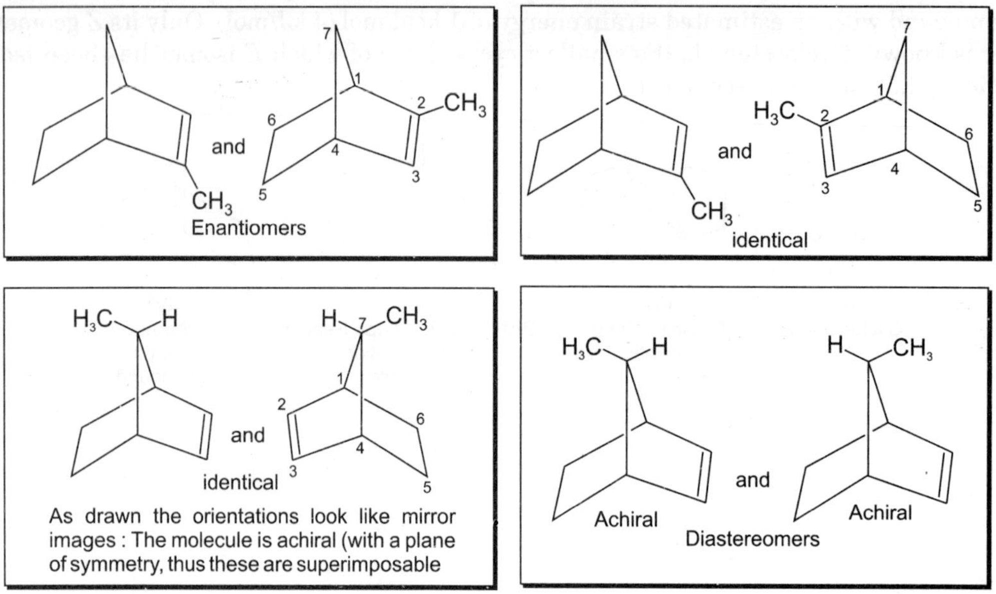

SCHEME 4.58

(F) Fused-Rings: The Decalins

The compound bicyclo [4,4,0] decane, better known as decalin, exists in two diastereoisomeric forms depending on the way in which the two cyclohexane rings are fused together (scheme 4.59). It is also readily determined from the models that *cis* and *trans* decalins are diastereomers as they cannot be interconverted by bond rotations.

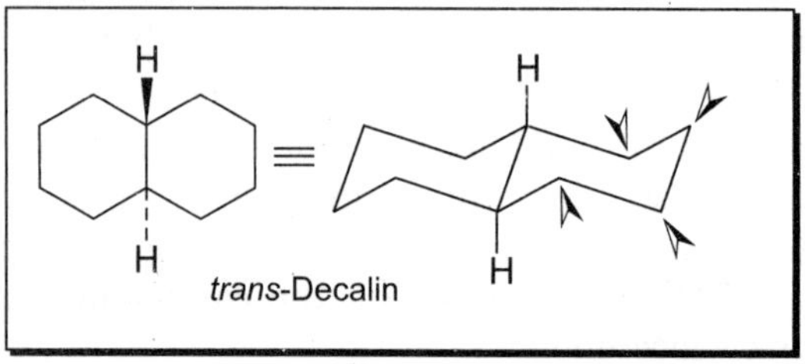

trans-Decalin is rigid, (achiral) with a center of symmetry. Ring, flip is not possible.

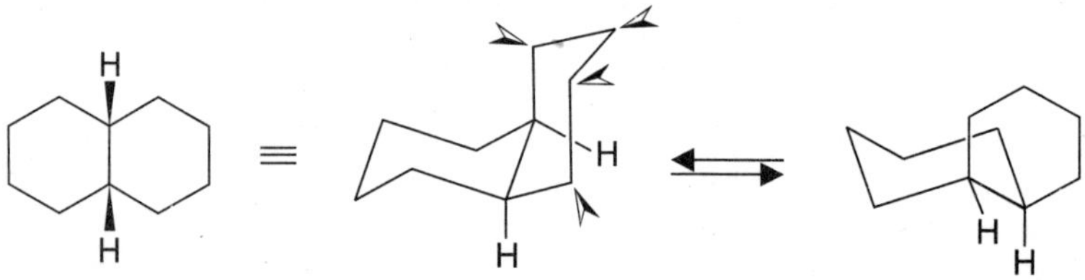

cis-Decalin

cis-Decalin is chiral, but enantiomers flip into one another

SCHEME 4.59

Decalin may be considered as a fusion of a four carbon chain *i.e.*, —CH_2 CH_2 CH_2CH_2— shown by arrows to the chair form of cyclohexane. *Trans*-decalin is fused with this four-carbon chain (equivalent to largest groups) in equatorial positions while the *cis*-decalin this fusion involves equatorial-axial positions (scheme 4.59). As a result in *cis*-decalin the two hydrogens attached at the point of fusion between two rings lie on the same side of the ring while in *trans*-decalin, they are on opposite sides.

In fact, *cis*-decalin exists as an equilibrium between two enantiomeric all-chair conformations which are interconvertible as a result of conformational flipping which is typical of monocyclic cycloheanes. As a result, any substituent attached to the *cis*-decalin system is more or less free to adopt the equatorial orientation. *cis*-Decalin is disymmetric in both conformations which are non-superimposable mirro images of each other. Because of rapid ring flipping between these forms the compound is a non-resolvable *dl*-pair.

On the other hand, *trans*-decalin has a unique and rigid conformation, inversion of which is not possible, as it would otherwise afford a highly strained system with one ring attached to the other by two axial bonds. As a result, a substituent is constrained to remain in a particular conformation which is dependent on its configuration. The rigid conformation of *trans*-decalin has been used in the same manner as *t*-butyl groups to study the relative rates of equatorial and axial substitution. The hindered approach to axial positions has been demonstrated *trans*-Decalin has a centre of symmetry and is therefore, optically inactive, It may be further noted that substituents located at the point of fusion between two rings (angular positions) in *cis* decalins are axial with respect to one ring while equatorial with repect to the other. However in the case of *trans*-decalins these are axial with respect to both the rings. It may be remembered that simple rotation of groups about carbon—carbon bonds cannot bring about inter-conversion of *cis*-and *trans*-decalins. Therefore, in this respect decalins resemble the *cis*-and *trans* 1, 2-disubstituted cyclohexanes and as has already been said, in decalins the 1, 2-substituents are the two ends of four carbon chain.*trans*-Decalin as a result is stabler by about 2.7 kcal/mole (11.3 kJ/mole) as compared with *cis*-form.

The barrier to interconversion of *cis*- and *trans*- decalin can be decreased by heteroatom substitution in some cases *e.g.,* in azadecalin (scheme 4.60). If a nitrogen atom occupies a bridgehead position *cis* and *trans* forms cannot be isolated. The rapid interconversion of the *cis* and *trans* forms in this case is the result of an easy inversion of configuration at nitrogen which eliminates the need for a *trans*-diaxial transition state.

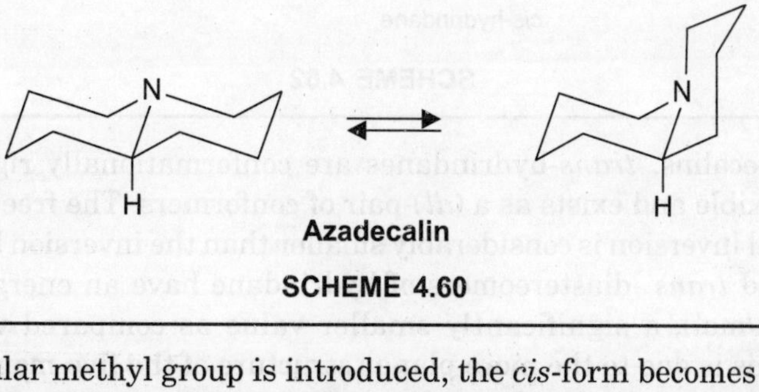

Azadecalin

SCHEME 4.60

When an angular methyl group is introduced, the *cis*-form becomes slightly more stable than the *trans*-form. Steroids and several other natural products contain the 9-methyl decalin system. The introduction of an angular methyl group reduces the differences in energy between the decalins, because in the *trans*-compound (the methyl group is sterically compressed by four axial hydrogen atoms) and by only two (C_2 and C_4) in the case of*cis*-isomer (scheme 4.61).

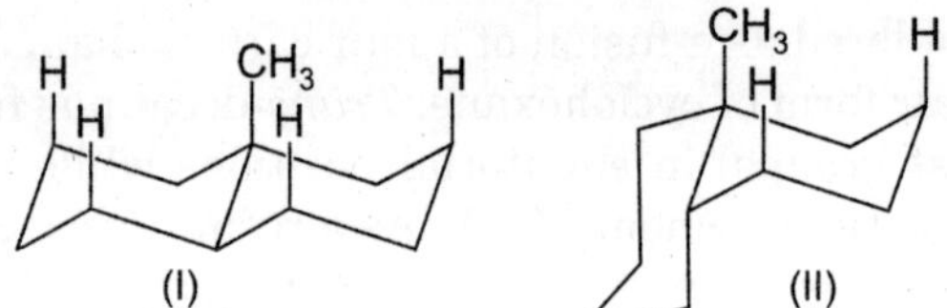

(I)
trans

(II)
cis

9-Methyldecalin

SCHEME 4.61

(G) Fused Rings: The Hydrindanes (Bicyclo [4.3.0] Nonane)

In hydrindane molecule, rings of different sizes are fused and these are represented as the rings C and D of steroids. The dissimilarly of the two rings makes the ring-fused carbon atoms stereocenters. A consideration of planar structures clearly shows the presence of two equivalent stereocenters and predicts the *trans*-isomer as a (±)-pair *i.e.,* two stable *trans*-isomers (scheme 4.62) while *cis*-isomer as a *meso*-compound. However, *cis* isomer in which one axial and one equatorial bond of cyclohexane is used to fuse the cyclopentane ring is chiral and just like *cis*-decalin exists as a (±)-pair, the two enantiomers being interconvertible by ring inversion.

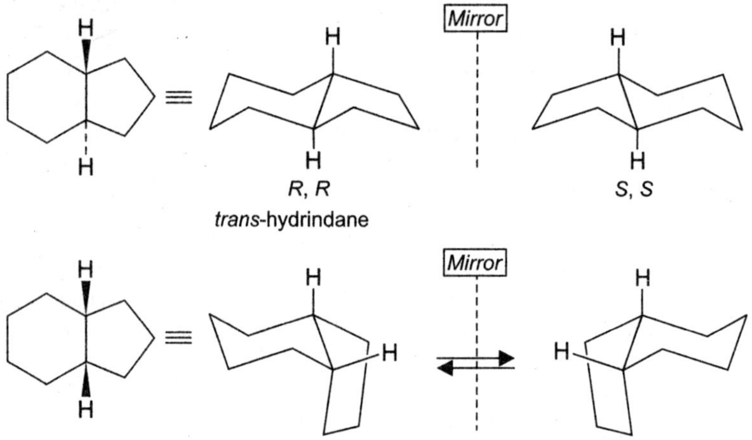

trans-hydrindane

cis-hydrindane

SCHEME 4.62

As with the decalins, *trans*-hydrindanes are conformationally rigid molecules, while *cis*-hydrindane is flexible and exists as a (*dl*)-pair of conformers. The free energy of activation of this conformational inversion is considerably smaller than the inversion barrier of *cis*-decalin. Moreover, the *cis* and *trans* -diastereomers of hydrindane have an energy difference around 2.9 kJ/mol (0.7 kcal/mol), a significantly smaller value as compared with that of decalin diastereoisomers. This is due to the more planar structure of the five-membered ring and as a result there is less strain in the fused system.

(H) Fused Rings: The Steroid Nucleus, Bridged Alkaloids

The tetracyclic ring system of perhydrocyclopentanophenanthrene is found in steroids and is composed of three fused 6-membered rings (perhydrophenanthrene itself exists in ten stereoisomeric forms; four enantiomeric pairs and two *meso* isomers). The prefixes *cis*- and *trans*-refer to the stereochemistry at the bonds of fusion between rings while *syn*- and *anti* (scheme 4.63) have reference to the orientation of terminal rings with respect to each other. The most stable of these is the *trans-anti-trans* form, an all-chair conformation. The *trans-syn-trans* form is the least stable as this structure can be realized only when the middle ring has the energetically unfavorable boat conformation.

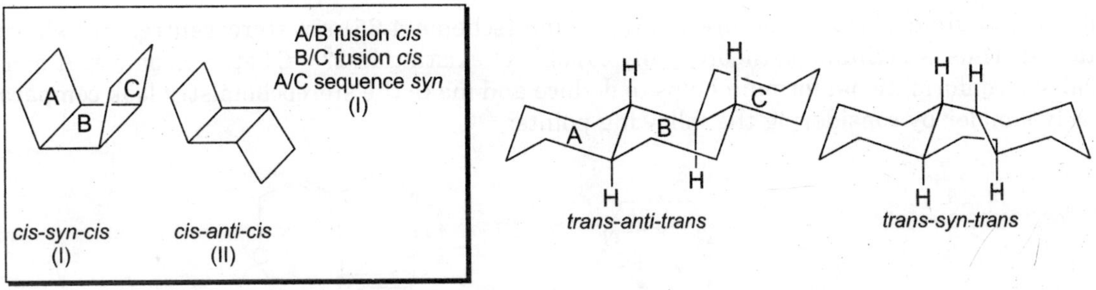

SCHEME 4.63

(*i*) Steroids

The following points may be noted:

- Steroids can broadly belong to one of the two families that resemble the decalins at the A and B rings. The cholestane family has an A/B *trans*-decalin system and the coprostane series has a A/B *cis*-decalin system.
- In fact the steroid nucleus contains three ring junctures (A/B, B/C and C/D). In cholestane, all the A/B, B/C and C/D ring junctions are more stable *trans* (scheme 4.64).

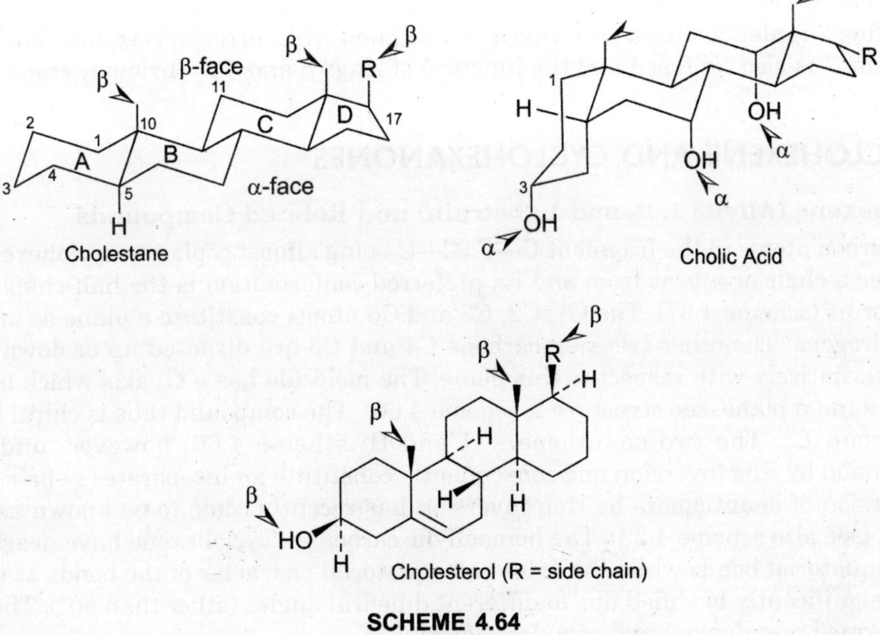

SCHEME 4.64

- The two methyl groups at ring junctions are called angular methyl groups.
- A group below the plane of the steriod ring system *trans*-to the angular methyl groups is called an α-group (corresponding face is called α) while the one above the plane *cis*-to the angular methyl groups is called the β-group (corresponding face is called β face).
- Cholic acid has the A/B *cis*-ring junction.

***(ii)* Alkaloids**

In the two-dimensional structure of strychnine (scheme 4.65) six stereocentres are shown starred. Thus, strychnine is theoretically capable of existing in $2^6 = 64$ stereoisomers. A more realistic conformational formula helps to deduce and make the stereochemistry look comparatively simpler by considering the following points:

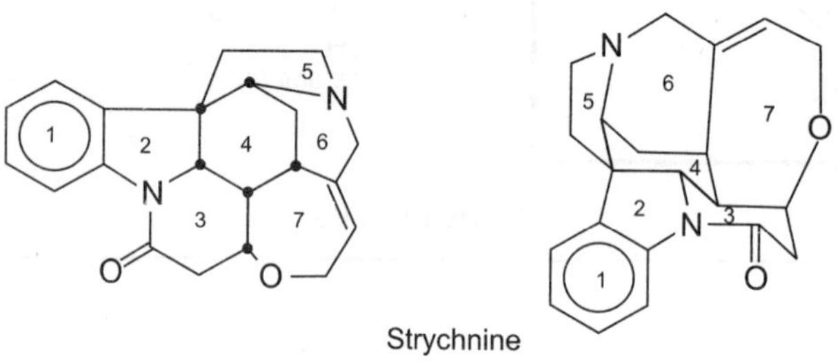

Strychnine

SCHEME 4.65

- Ring 4 is in chair conformation thus, forcing the bridging of ring 6 with it *via* and *cis*-diaxial bonds.
- This makes the fusion of ring 5 with 6 *via* one axial and one equatorial bond, *i.e.*, *cis* fusion as it is geometrically impossible to bridge 1, 2-diaxial *trans* positions with a ring of ordinary size.
- Ring 2 is also *cis* fused with ring 4 *i.e.*, its bond with nitrogen is equatorial. Similarly, ring 7 is also *cis* fused and the juncture of rings 3 and 4 is obviously *trans*.

4.18 CYCLOHEXENE AND CYCLOHEXANONES

***(i)* Cyclohexene (Allylic 1, 2- and 1, 3-strain) and Related Compounds**

The four carbon atoms in the fragment C—C=C—C being almost coplanar, cyclohexene cannot adopt either a chair or a boat from and its preferred conformation is the half-chair shown in different forms (scheme 4.67). The C1, C2, C3 and C6 atoms constitute a plane as are the two alkene hydrogens. The other two ring carbons C4 and C5 are disposed up or down or (down and up) alternatively with respect to this plane. The molecule has a C_2 axis which bisects the double bond (no σ plane, see structure I, scheme 4.66). The compound thus is chiral belonging to point group C_2. The two enantiomers (I and II, scheme 4.66) however, undergo fast interconversion by ring inversion and consequently constitute an inseparate (±)-pair. This fast interconversion of enantiomers by ring inversion has recently come to be known as enantiomerization (see also scheme 4.33). The homoallylic carbons of cyclohexene have nearly normal axial and equatorial bonds while the axial and equatorial character of the bonds at the allylic carbons is significantly modified due to different dihedral angles (other than 60°). These bonds are thus termed pseudoaxial and pseudoequatorial.

A cyclohexene with endocyclic double bond of the type (III, scheme 4.66) suffers from allylic 1, 2-strain also called $A^{1,2}$ -strain when *R* and *R′* are bulky. They interfere to an extent that the axial conformer will predominate. This is so *e.g.,* when in (III, scheme 4.66, *R* = Ph and *R′* = *t*-butyl), the conformer (IV) is predominant.

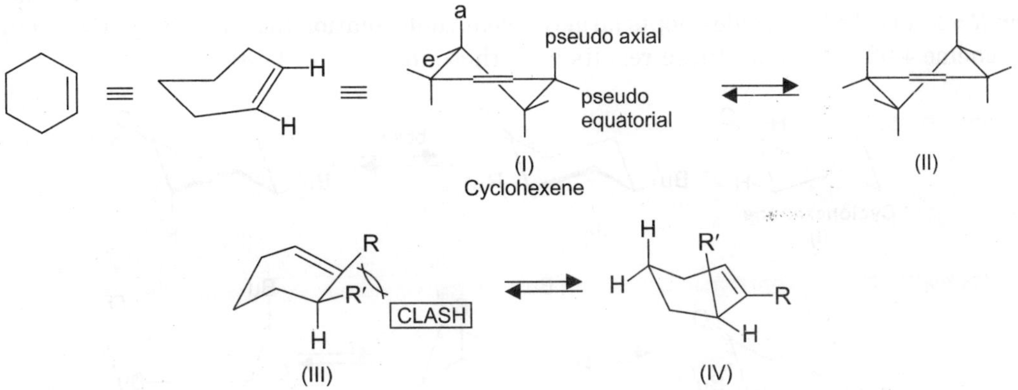

SCHEME 4.66

***(ii)* Alkylidenecyclohexanes**

In these compounds with exocyclic double bond (I, scheme 4.67). The allylic segment $R'—C_\alpha = C_\beta—C_\gamma—R$ is almost coplanar and thus *R* and *R′* are almost eclipsed (unfavourable interaction) and this interaction becomes further unfavourable due to shorter C = C in between. This interaction is called allylic 1, 3-strain or $A^{1,3}$ -strain as the groups involved are at 1 and 3 positions of an allylic system. As a result axial conformer (II, scheme 4.67) may predominate particularly if the groups *R* and *R′* are bulky.

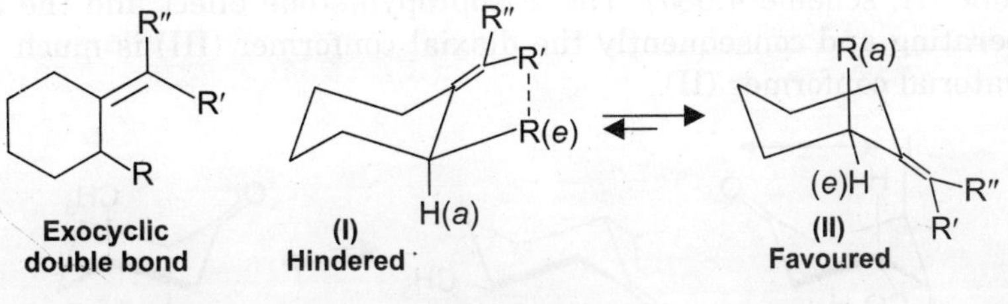

SCHEME 4.67

***(iii)* Cyclohexanone Ring—Alkyl Ketone Effect**

Unlike cyclohexane, the ring in cyclohexanone is buckled slightly from the chair structure in order to accommodate the trigonal carbon (optimal CCC angle 120°). Thus the equatorial hydrogens on the α-carbon (C2) are almost eclipsed with the carbonyl oxygen (I, scheme 4.68). An equatorial substituent may be destabilized to some extent due to steric repulsions. This decreases the energy difference between the axial and the equatorial conformers when compared to those in cyclohexane. This decrease is termed as 2-alkylketone effect which is measured by the difference in – ΔG°R in cyclohexane and – ΔG°R in cyclohexanone. Equilibrium data of 2-alkyl-4-*t*-butylcyclohexanones (II and III, scheme 4.68) provides data of these values. When

$R = CH_3$ this value is, however, negligible, the CH_3 group in equatorial position is indeed far, to experience any appreciable steric repulsion with C=O, moreover, eclipsing of CH_3 with C=O is electronically favourable as is so in the case of propanal in which the preferred conformation has C=O and CH_3 groups eclipsed. With the increase in size of $R(C_2H_5)$ a value of around 3 kJ/mol is obtained for the 2-alkylketone effect when $R = i - Pr$ the value increases to 7 kJ/mol. When $R = t$-butyl the molecule adopts largely a boat conformation to avoid t-butyl/C=O eclipsing (III, scheme 4.68). Compare these results with those in (scheme 4.43).

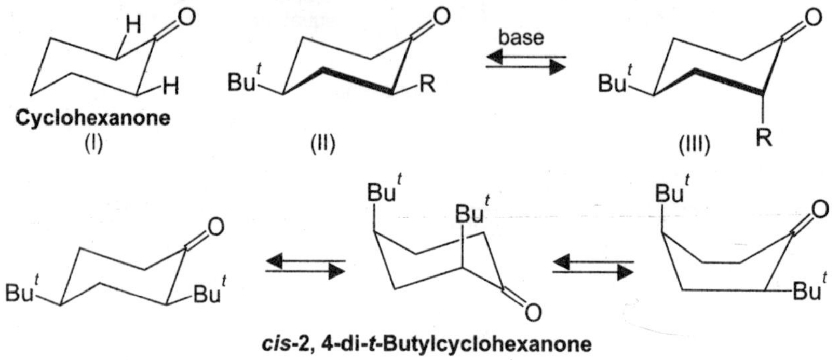

SCHEME 4.68

Based on these arguments 2-bromocyclohexanone will predominantly exist as axial conformer which will have greater stability over the equatorial conformer. For these reasons *cis*-2, 4-di-*tert*-butylcyclohexanone adopts largely a boat conformation (scheme 4.68).

In the case of a 3-alkylcyclohexanone (1, scheme 4.68*a*) one 1, 3-interaction between the axial *R* group and axial *H* is missing. This is equivalent to one butane gauche interaction (3.75 kJ/mol) when $R = CH_3$. This decrease in $-\Delta G°$ value of *R* is termed 3-alkylketone effect. In the case of menthone (II, scheme 4.68*a*). The 2-isopropylketone effect and the 3-methylketone effect are cooperating and consequently the diaxial conformer (III) is much more stabilized than the diequatorial conformer (II).

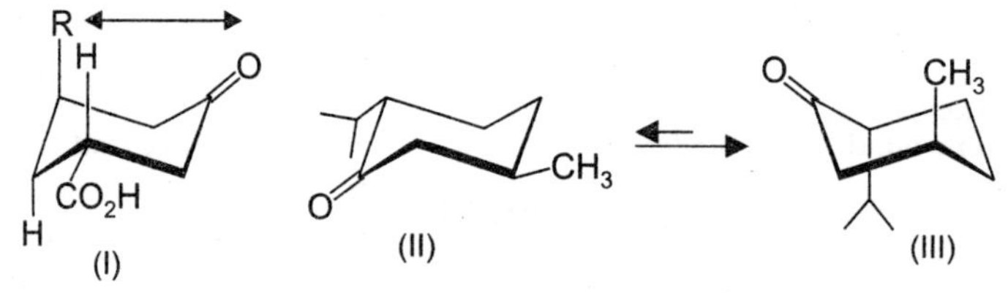

SCHEME 4.68a

(*iv*) Reactivity in Cyclohexanones a Comparison with Cyclopentanones

Cyclohexanone is reduced with hydride ion from borohydride to give cyclohexanol much faster than cyclopentanone. This difference in reactivity shows that unlike cyclopentanone cyclohexanone suffers from eclipsing strain (torsional strain) due to eclipsing of its oxygen atom by two α-hydrogens which are equatorial (see I, scheme 4.68). This strain is relieved on the formation of cyclohexanol (scheme 4.68*b*). However, on the formation of cyclopentanol there is an increase in the torsional strain.

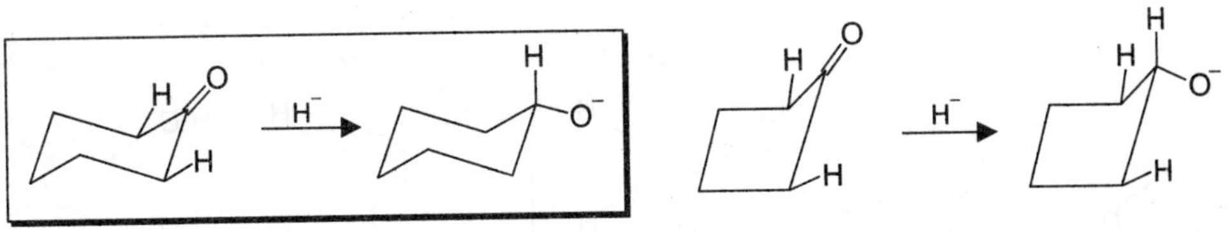

SCHEME 4.68b

The results may be compared with *e.g.*, S_N1 reactions where the reactivity is less in cyclohexyl derivatives compared to their cyclopentyl analogues. The formation of transition state in non-planar cyclohexane leads to increase in strain, while this is not so in cyclopentane which is almost flat leading to decrease in strain in the transition state (scheme 4.68*c*).

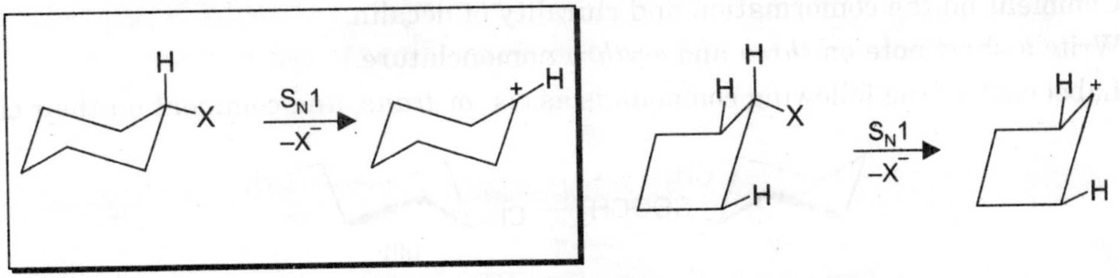

SCHEME 4.68c

More on Cyclopentane Ring

The geometry of cyclopentane ring has been discussed as a planar and an envelope form (see, scheme 4.15). The additional half chair form may have axial, equatorial or isoclinal bonds (i). In a monosubstituted cyclopentane a substituent is equatorial at the top of the ring (scheme 4.68d).

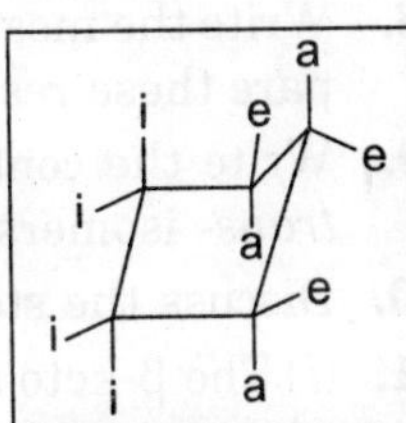

SCHEME 4.68d

EXERCISES AND PROBLEMS

4.1. Comment on the stereochemical and the identity aspects of the following Newman projections of tartaric acid.

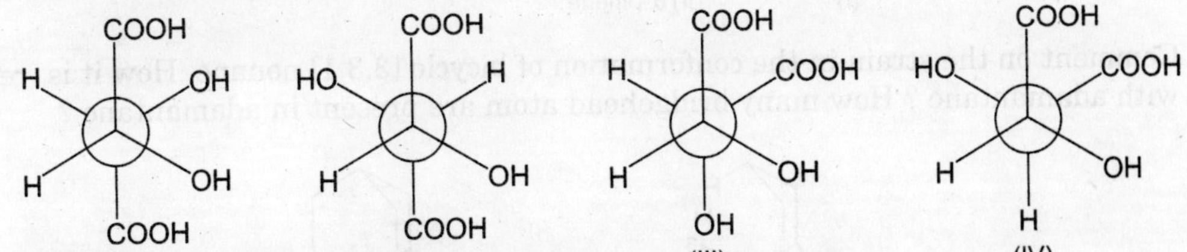

(I) (II) (III) (IV)

4.2. (*i*) Comment on the relationship among the following Newman projections of 2, 3-dichlorobutane, which is the *meso*-compound ?

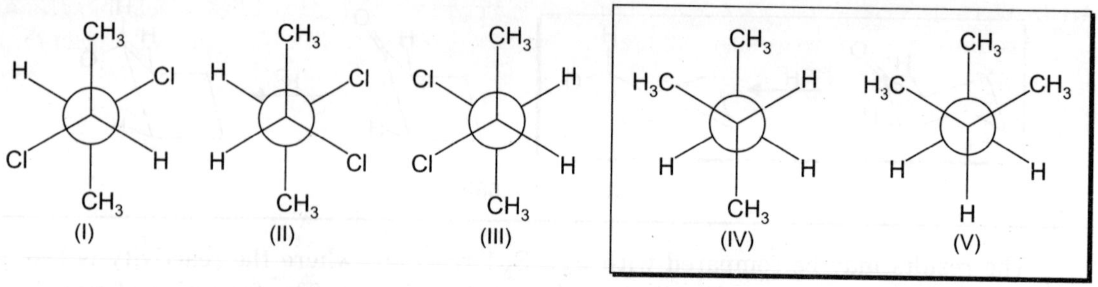

(*ii*) Of the two staggered conformations (IV and V) of 2-methylbutane, one is more stable than the other. Explain.

4.3. Give quantitative data and conformational analysis for cyclopentane.

4.4. Comment on the conformation and chirality of decalin.

4.5. Write a short note on *threo* and *erythro* nomenclature.

4.6. Label each of the following compounds as *cis-* or *trans*, and comment on their chirality.

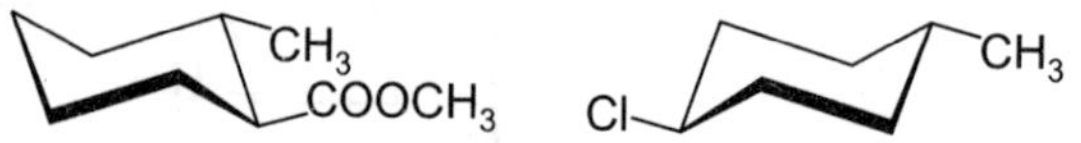

4.7. Comment on the configurational and conformational aspects of disubstituted cyclohexane derivatives. How their chirality depends on the nature of substituents *R* and *R′* ?

4.8. Write the most stable conformation of *trans*-1, 2-dimethylcyclohexane, is it chiral ? Compare these results with its *cis*-isomer.

4.9. Write the conformations of 1, 3-dimethylcyclohexane. Discuss the chirality of *cis* and *trans-* isomers.

4.10. Discuss the stereoisomerism of *cis-* and *trans*-1-*tert-* butyl-2-methylcyclohexane.

4.11. (*i*) The β-keto acid (I) is highly resistant to decarboxylation, explain.

(*ii*) Draw bicyclo [3.1.1] heptane. What system is present in (II).

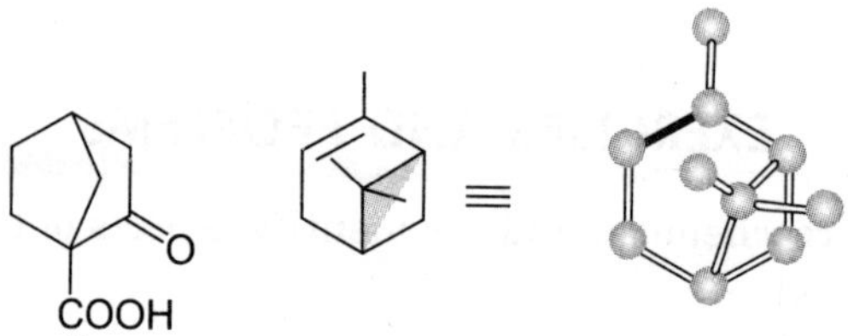

4.12. Comment on the strain in the conformation of bicyclo [3.3.1] nonane. How it is related with adamantane ? How many bridgehead atom are present in adamantane ?

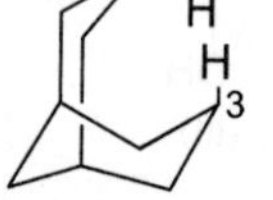

Bicyclo [3.3.1] nonane chair-chair form

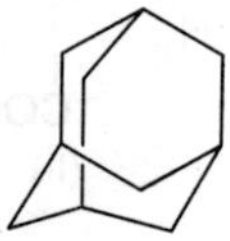

Adamantane

4.13. Write the conformations of *cis* and *trans* isomers of 1-bromo-4-butylcyclo hexane and comment on the favored conformation in each case by calculating their relative energies (consult Table 4.2).

4.14. Compared to rotation about single bonds *e.g.,* in ethane (sp^3-sp^3 carbon-carbon bonds), what influences you expect on the rotational isomerism about sp^2-sp^3 hybridized carbon-carbon single bonds as for example in CH_3—CHO ?

4.15. When cyclopentane is reacted with excess chlorine at a high temperature, seven dichlorocyclopentanes are formed. Comment on their geometrical isomerism and chirality.

4.16. (*a*) Discuss the geometrical isomerism of 1-bromo-2-chlorocyclobutane. Give stereodescriptors to different stereocenters.

(*b*) Assign stereodescriptors to the two stereocenters in the compound (I).

(*c*) Assign nomenclature to compound (II).

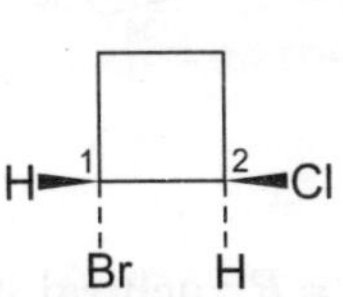

(I)

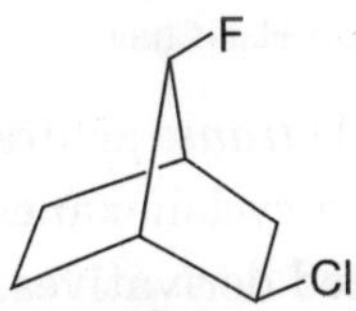

(II)

4.17. (*R*)-Camphor is reduced with lithium aluminium hydride to give a pair of alcohols. Give nomenclature to each stereocenter in these. How these alcohols are stereochemically related ?

4.18. Write the products when each of the following compounds is treated with HONO/H$^+$.

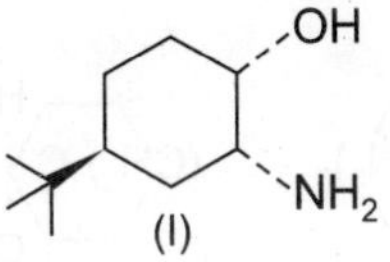

(I)

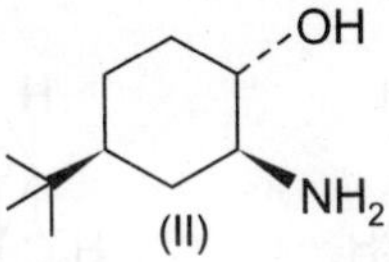

(II)

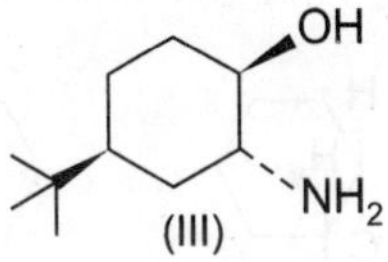

(III)

ANSWERS TO SELECTED PROBLEMS

4.1. These are the staggered conformations as these are drawn keeping in mind:

(*a*) That a compound cannot be chiral if any rotamer is non-disymmetric or if each chiral conformation is in equilibrium with its enantiomer.

(*b*) Rotation about single bonds has to be considered to prove or disprove identity.

(*c*) While comparing stereochemical relationship of two compounds it is not sufficient to compare only one conformation of each compound. Rotation about single bonds is necessary to find the closest correspondence Newman projections (I and II) are enanatiomers while *e.g.*, (III) is a *meso*-compound; (I and IV) are identical.

4.2. (*ii*) The conformation (IV) with one *gauche* (CH_3,CH_3) and one *anti* (CH_3, CH_3) relationship is more stable than (V) which has two (CH_3, CH_3) *gauche* interactions.

4.3. In the envelope from, one atom projects out of the plane of the four other atoms. It has a plane of symmetry σ). The other flexible form is the half-chair form wherein three neighboring carbon atoms are coplanar, while the other two are above and below the plane, respectively, and equidistant from it. This conformer has C_2 symmetry. The envelope and half-chair forms interconvert through intermediate conformations with no symmetry.

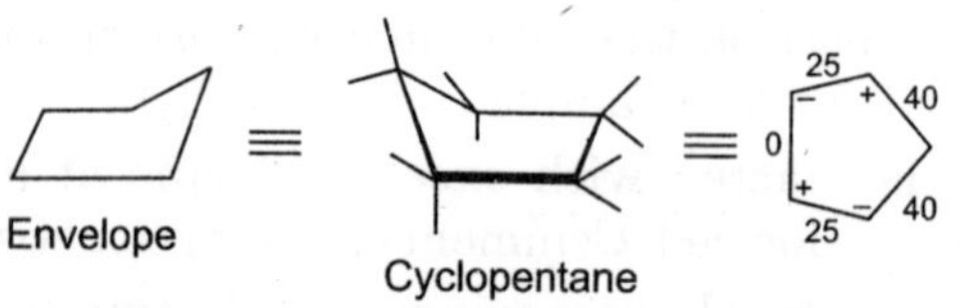

Envelope Cyclopentane

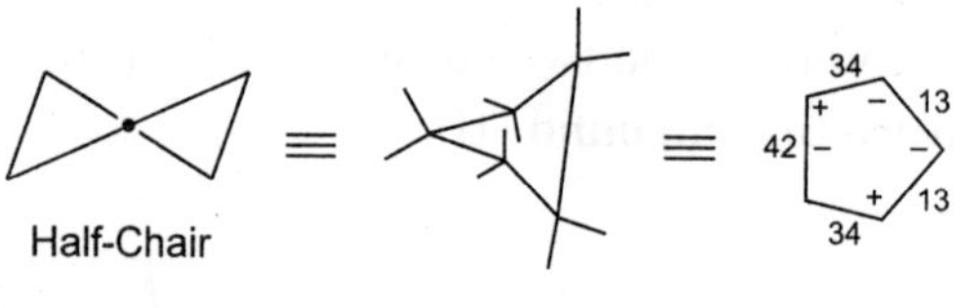

Half-Chair

4.6. (I) *trans*, Chiral; (II) *trans,* achiral.

4.7. (*i*) 1,1-Disubstituted cyclohexanes are achiral.

(*ii*) 1, 2-Disubstituted derivatives, *cis* chiral if R ≠ *R*′; achiral if R = *R*′, *trans* chiral.

(*iii*) 1, 3-Disubstituted derivatives, *cis* chiral if R ≠ *R*′; achiral if R = *R*′, *trans* chiral.

(*iv*) 1, 4-Disubstituted derivatives, *cis* achiral; *trans* achiral.

4.10. The compound has two stereocenters and exists as four stereoisomers. In the *cis*-compound, one group must be axial and other equatorial (the larger will be equatorial). In the *trans*-compound the diequatorial conformer is more stable. The four stereoisomers are written) compare (scheme 4.32–4.34).

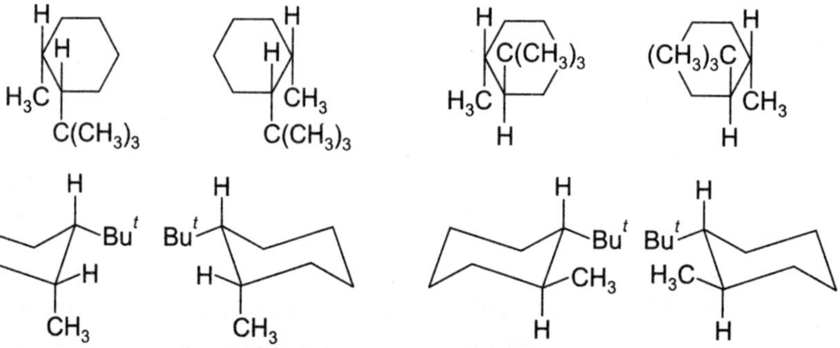

cis-1-*tert*-butyl-2-Methylcyclohexane enantiomers (stable conformers) *trans*-1-*tert*-butyl-2-Methylcyclohexane enantiomers (stable conformers)

4.11. (*i*) The mechanism of decarboxylation of β-keto acids involves a cyclic six-centre transition state to give an enol which quickly tautomerizes to the ketone. In this case of bridgehead bicyclic keto acid the product would be highly strained bridgehead-enol (violation of Bredt's rule).

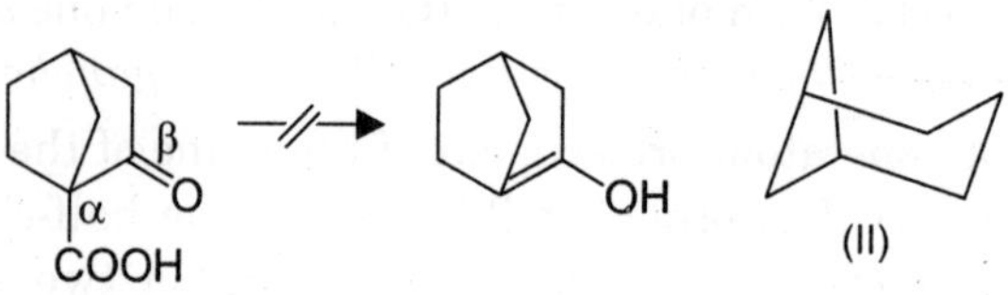

(II)

(*ii*) [3.1.1] Heptane system.

4.12. Bicyclo [3.3.1] nonane constitutes a 1, 3 fusion of two cyclohexane chairs which are free of angle strain. However, the molecule in the conformation suffers from serious trans-annular interaction between the axial hydrogen at C-3 and C-7. The more attractive conformation is the one in which both chairs and somewhat stretched out to reduce this strain.

Adamantane is the compound obtained by bridging the conflucting methylene groups, C-3 and C-7 in the chair-chair form. It has four bridgehead atoms.

4.13.

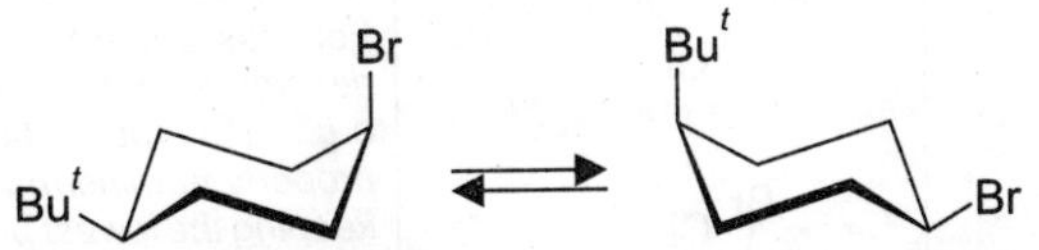

Relative energy : 0 + 0.5 = 0.5 kcal/mole ~ 5 + 0 = ~ 5 kcal/mole

$\Delta G° \cong$ 4.5 kcal/mole

***cis*-1-Bromo-4-*t*-butycyclohexane**

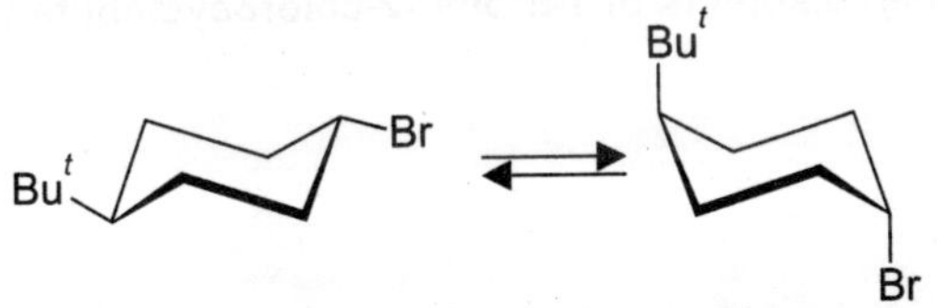

Relative energy : 0 + 0 = 0 ~ 5 + 0.5 = ~ 5.5 kcal/mole

$\Delta G° \cong$ 5.5 kcal/mole

***trans*-1-Bromo-4-*t*-butycyclohexane**

4.14. In saturated compounds generally the staggered conformation is preferred and the one with the groups at 180° *i.e.*, farthest from each other is the best staggered conformation. When however, a single bond is connected to a double bond, the double bond at one end of the single bond eclipses a hydrogen at the other end in the preferred conformation which for acetaldehyde thus, represents a slightly eclipsed form (by 9°).

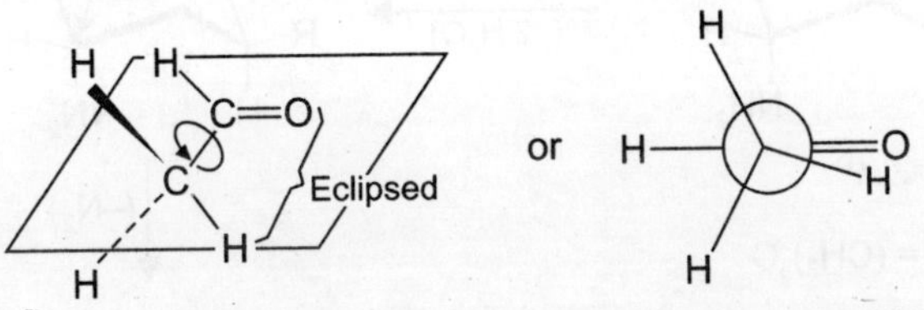

4.15. In common with other alicyclic rings the chirality can be discussed. The 1, 1-dichloro-derivative is achiral. The 1, 2 and 1, 3-compounds have two stereocenters. In each case *cis*-isomer is *meso* and the *trans* isomer exists as a pair of enantiomers.

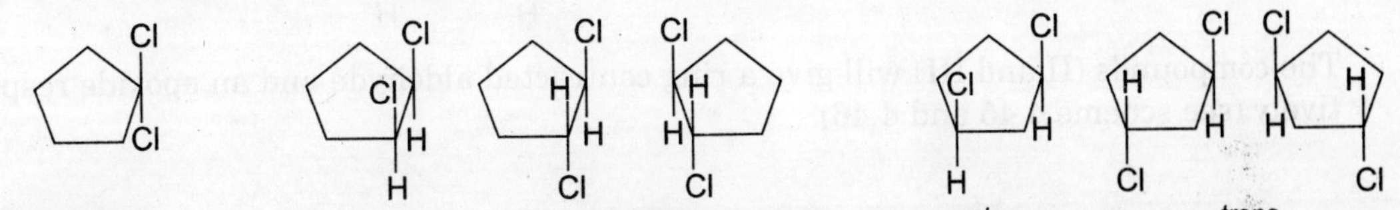

1, 1-Dichloro-cyclopentane (achiral)

cis (*meso*) *trans* (±) pair
1, 2-Dichlorocyclopentane

cis (*meso*) *trans* (±) pair
1, 3-Dichlorocyclopentane

4.16. (*a*) The compound has two stereocenters and therefore, has four stereoisomers (there is no possibility of finding a *meso* isomer—the substituents on two stereocenters being different) one enantiomeric pair has *cis* geometry and the other *trans*-. These *cis* and *trans* isomers are diastereomers (recall from *cis* and *trans*-alkenes).

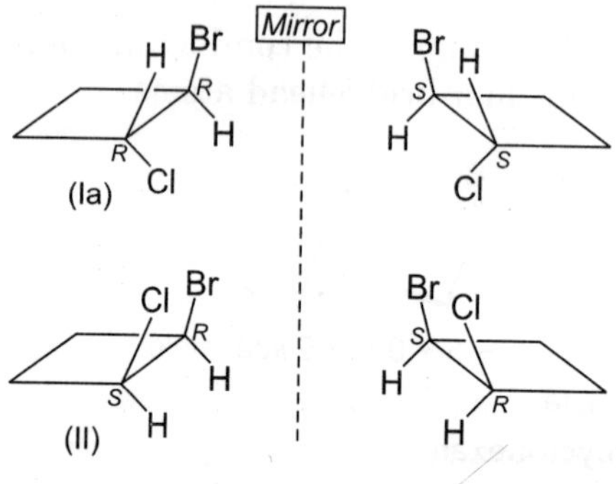

For assignment of R and S descriptors a geometrical figure e.g., (II) must be oriented properly in mind/make a model keeping the lowest priority group away.

Stereoisomers of 1-bromo-2-chlorocyclobutane

(*b*)

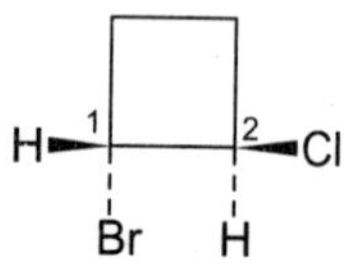

C1 = *R*; C2 = *R*

(I)

(*c*) II, 2-*exo*-chloro-7-*syn*-fluorobicyclo [2.2.1] heptane.

4.18. [*Hint.*] Write each compound as its chair conformation *e.g.*, (I) gives a cyclohexanone derivative by the loss of N_2 and hydride shift.

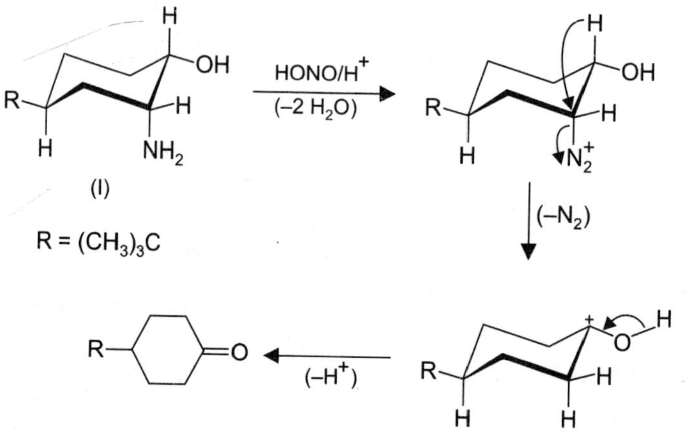

The compounds (II and III) will give a ring contracted aldehyde and an epoxide respectively (see scheme 4.45 and 4.46).

REFERENCES AND FURTHER READING

1. Ernest L. Eliel, Samuel H. Wilen, Michael P. Doyle: *Basic Organic Stereochemistry*, Wiley, New York, Chichester, Weinheim, Brisbane, Singapore, Toronto, 2001.

CHAPTER 5

Stereochemistry of Elimination Reactions

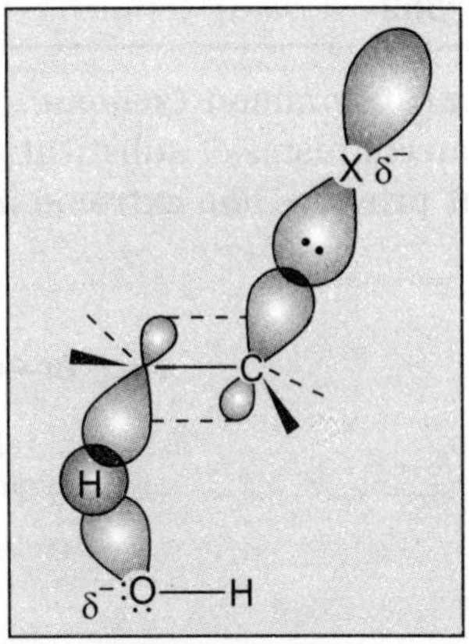

5.1 INTRODUCTION

The term elimination normally refers to the loss of two atoms or groups from a molecule and in the commonly known elimination reactions multiple bonds are formed through the loss of groups bonded to adjacent atoms. The process is usually termed 1, 2-elimination or β-elimination because the two departing ligands are on adjacent atoms. Often E1 and E2 reactions are referred to as β-eliminations (scheme 5.1).

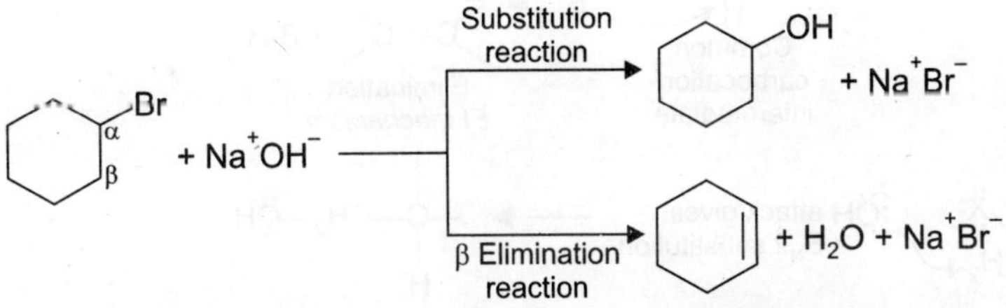

Generally in S_N1 and S_N2 reactions, substitution is accompanied by some elimination of H—L, where L is a leaving group to give an alkene.

SCHEME 5.1

Other elimination reactions may also occur in organic compounds, with the departing groups located at 1, 3 or more remote sites in the molecule to yield cyclic products. This type of elimination reaction is actually an intramolecular substitution (scheme 5.2).

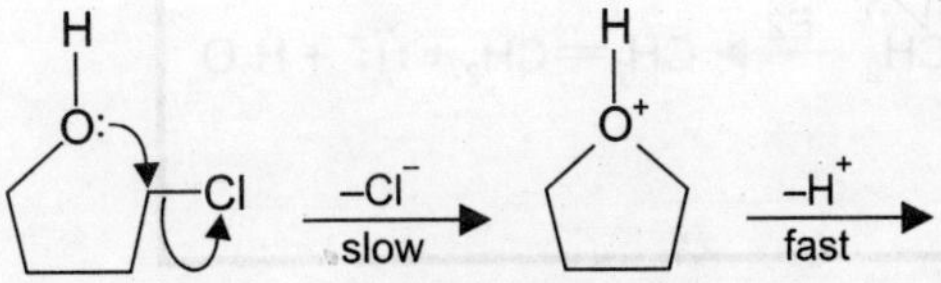

SCHEME 5.2

Still another mode of elimination may involve two groups departing from the same atom. Such 1,1-or α-eliminations are used for producing the very reactive species known as carbenes (scheme 5.3).

Elimination reactions always face a competition from substitution reactions, since all nucleophiles are potential bases as all bases are potential nucleophiles.

Elimination reactions 1,2 can occur by a variety of mechanisms and three mechanistic pathways are normally distinct routes for β-elimination reactions. Of these E2 and E1 are the most common (scheme 5.4). These two processes are closely related to the S_N2 and S_N1 mechanisms of substitution. A third mechanism is designated as ElcB, although less common, it provides one extreme in the possible elimination pathways (see scheme 5.7).

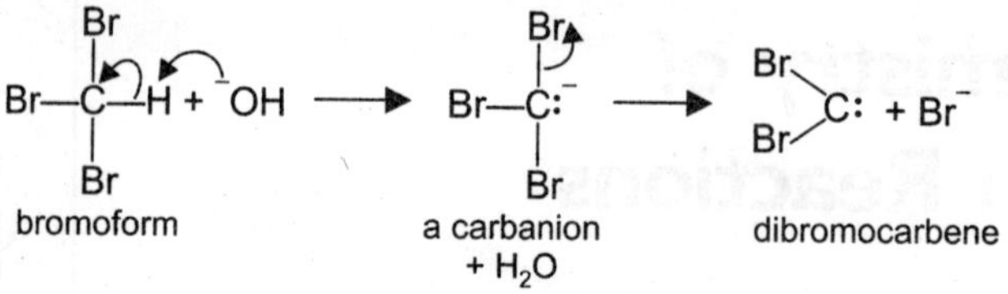

SCHEME 5.3

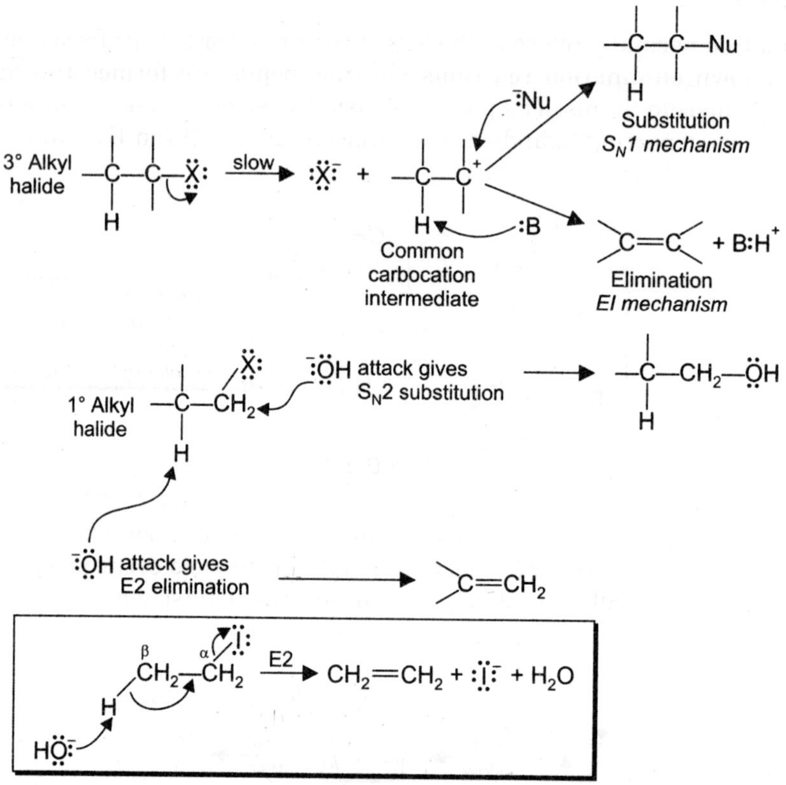

SCHEME 5.4

5.2 MECHANISMS E1, E2 AND ElcB

(A) Carbocation Mechanism (E1 Mechanism)

In these elimination reactions the initial step involves cleavage of the bond between the leaving group and its carbon with the generation of fully dissociated carbocation. These elimination reactions are called, E1 reactions (Elimination unimolecular).

E1 reactions almost always compete with S_N1 reactions as the initial step for both the reactions is similar. When *tert*-butyl bromide is treated with water, the reaction produces the substitution product as well as the elimination product (2-methyl-propene) (scheme 5.5).

The initial step for both of these reactions is the formation of a *tert*-butyl cation (or an ion pair) and represents the rate-limiting step for both the reactions (scheme 5.5).

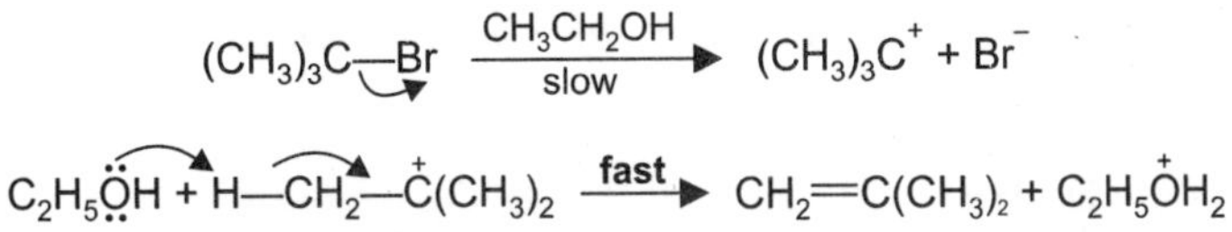

Rate = k_1 [$(CH_3)_3CBr$]

E1 Mechnism: Carbocation Intermediates

SCHEME 5.5

If a solvent molecule reacts as a nucleophile at the positive centre of the *tert*-butyl cation the product is *tert*-butyl alcohol (S_N1), however, if a solvent molecule acting as a base, accepts one of the β-protons, the product is 2-methylpropene (E1).

(B) The Concerted Reaction (E2 Mechanism)

Both the substrate and the nucleophile participate in the single step. Reaction kinetics are typically second-order. The nucleophile which is commonly a base abstracts a proton from the β-atom to the leaving group (scheme 5.6).

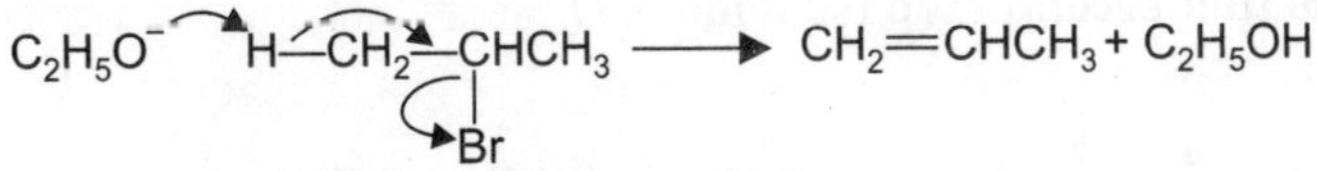

Rate = $k_2[(CH_3)_2CHBr]\,[C_2H_5O^-]$

E2 Mechanism: Synchronous Elimination

SCHEME 5.6

The following evidences support the existence of these general mechanistic representations :

- The rate expressions for the reactions are proper second order.
- The reactions display considerable hydrogen-deuterium isotope effect. When eliminations are carried out by replacing leaving hydrogen with deuterium, the reaction rates get slow. Breaking a carbon-deuterium bond is slower than cleaving a similar carbon-hydrogen bond (kinetic isotope effect). Therefore, these results are consistent with a mechanism which involves carbon-hydrogen bond cleavage in the transition state.
- The reactions also show a substantial element effect. The relative reactivity of 2-phenylethyl halides towards ethoxide ion is in the order:
  $$C_6H_5CH_2CH_2I > C_6H_5CH_2CH_2Br > C_6H_5CH_2CH_2Cl > C_6H_5CH_2CH_2F$$
  An element effect of this kind is again in keeping with a transition state involving carbon-halogen bond cleavage. Thus, 2-phenylethyl iodide reacts fastest because the carbon-iodine bond is weakest; 2-phenylethyl fluoride reacts slowest as the carbon-fluorine bond is strongest.
- E2 reactions, generally, are not accompanied by rearrangements. This is consistent with a mechanism which does not involve the formation of carbocations.
- The most compelling evidence for E2 mechanism is, however, the stereochemical outcome. This mechanism is stereospecific as illustrated (scheme 5.20).

EXERCISE 5.1

Explain why the major product from the elimination (scheme 5.6a) contains deuterium?

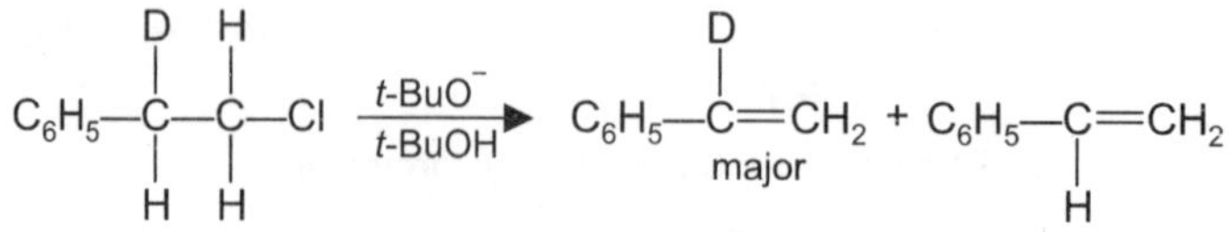

SCHEME 5.6a

ANSWER. *The rate determining step requires the breaking of a C—H or C—D bond and C—H bond is easier to break.*

(C) The ElcB Mechanism

In the E1 mechanism the bond to the leaving group breaks during the first (slow) step to give the carbocation and the bond to the hydrogen breaks in a second step. In the E2 mechanism, both of these bonds break in a single step. There is a third possible mechanism in which the bond to the hydrogen breaks during the first step to give a carbanion and the bond to the leaving group breaks in a second step (scheme 5.7).

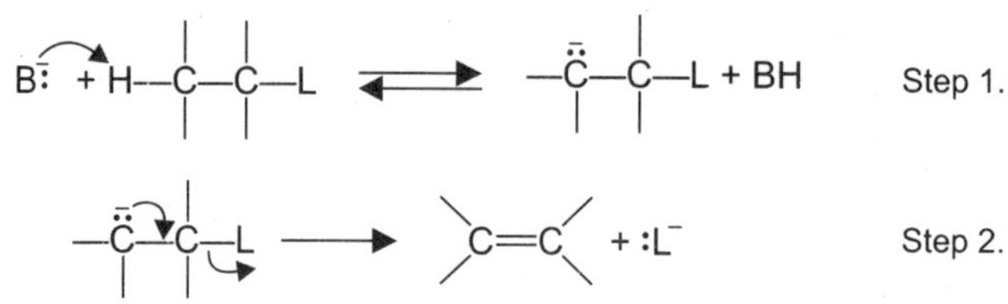

ElcB Mechanism: Carbanion Intermediates

SCHEME 5.7

This third mechanism where base abstracts a β-hydrogen atom to form a carbanion in the first step, followed by the departure of the leaving group to give an alkene in the second step is called ElcB mechanism. The second step often determines the rate, since this step involves a unimolecular reaction of the conjugate base of the initial reactant, the mechanism is designated as elimination, unimolecular, conjugate base-or ElcB. This mechanism dominates by the presence of substituents that stabilize the intermediate carbanion. A poorer leaving group, which makes the second step less likely to be concerted with the first, also favors this mechanism.

When a good carbanion-stabilizing group is present on the carbon from which the proton is lost, the ElcB mechanism becomes a favorable process for elimination (scheme 5.8). In such cases *e.g.*, dehydration of the aldol product occurs by an ElcB process (scheme 5.8) where even a poor leaving group (hydroxide ion) is eliminated. The leaving group (hydroxide ion) is a strong base which is not the normal leaving group in an E1 or E2 process. The elimination pathway is very common and the aldol condensation is often difficult to stop at the β-hydroxy-aldehyde stage, particularly when the reaction is carried out above room temperature (The carbonyl group plays two critical roles, it stabilizes the intermediate carbanion and also it provides an additional driving force for elimination to provide stability to the conjugated product). Thus the dehydration of alcohols provides examples of all the three mechanistic pathway *i.e.*, E1, E2 and ElcB mechanisms. Secondary and tertiary alcohols eliminate water under acidic conditions (to convert a poor leaving hydroxyl group in to good leaving group H_2O) generally via E1 mechanism involving carbocations. Primary alcohols, instead dehydrate (acidic conditions) via E2 mechanism since unstable primary carbocation would be formed via E1 pathway. Dehydration is particularly easy when a conjugated double bond can be formed (scheme 5.9). Example of dehydration of an alcohol under basic conditions via an ElcB mechanism is during aldol condensation (scheme 5.8).

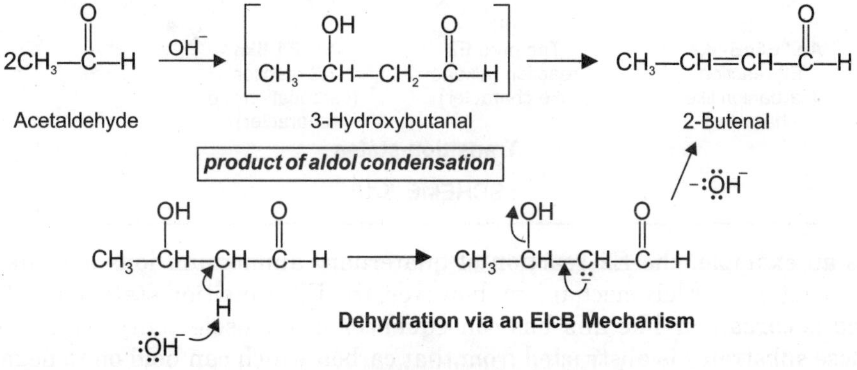

SCHEME 5.8

In summary, the requirements for the ElcB reaction are a poor leaving group and an easily lost proton.

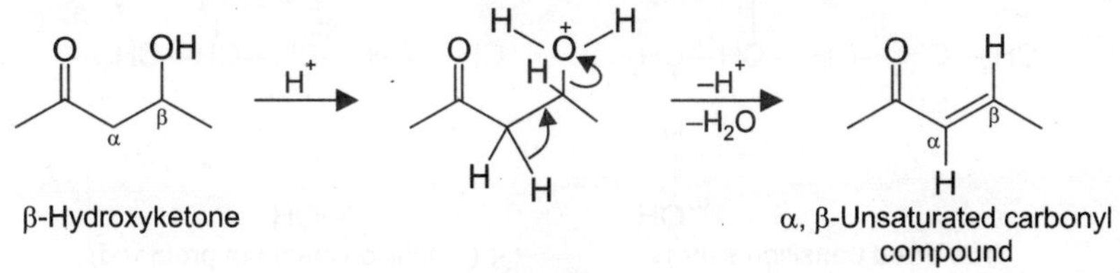

SCHEME 5.9

5.3 E1, E2 AND ElcB VARIABLES—THEIR MECHANISTIC SPECTRUM

The three most common elimination mechanism, E1, E2 and ElcB have been studied (scheme 5.5-5.7). These mechanisms are largely similar than different in each case, the leaving group departs with its pair of electrons while the hydrogen come off without them. These, however, differ in the order of steps, there is in fact a spectrum of mechanisms which range from one extreme (pure E1) to other extreme (pure ElcB). The following points may be considered:

- The pure E2 mechanism is somewhere in the middle of this spectrum when in the transition state the bonds to hydrogen and leaving group (X) are both breaking (II, scheme 5.10).
- In the E2 reaction, however, the transition state may not be exactly in the middle of this spectrum but somewhere to one side or other. The C—H bond may begin to cleave ahead of the C—X bond as in the transition state (I scheme 5.10). The reaction can

still be bimolecular, however, the transition state will be "carbanion–like" and resemble the ElcB reaction.

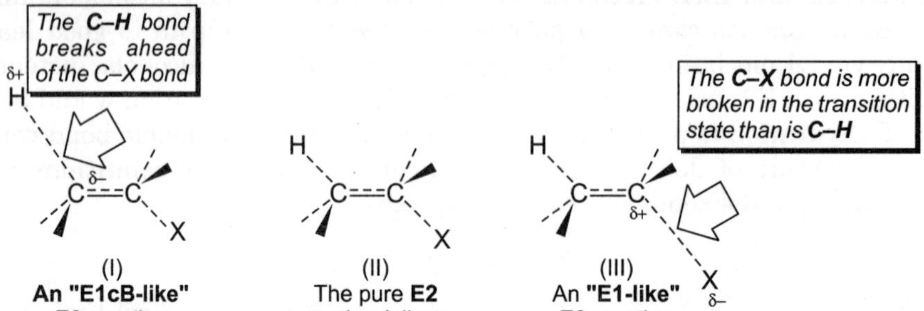

(I) An "E1cB-like" E2 reaction (carbanion like character)
(II) The pure E2 reaction (alkene like character)
(III) An "E1-like" E2 reaction (carbocation like character)

Transition states

SCHEME 5.10

- As an example, the E2 reaction of quaternary ammonium ions and alkyl fluorides does not go by ElcB mechanism, however, the E2 transition state is itself "ElcB-like" and is carbanion–like and has consequences in regioselectivity. Thus the proton in these substrates is abstracted from that carbon which can hold on to negative charge the best *i.e.*, a carbon with least alkyl groups.

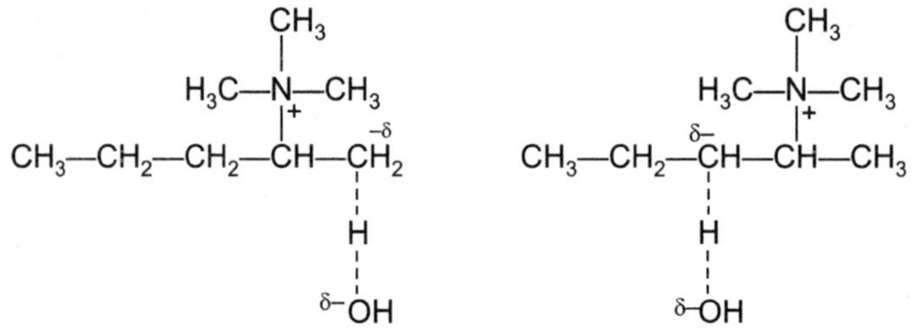

(Preferred transition state) (Transition state not preferred)

SCHEME 5.11

- Similarly the leaving group (X also called the nucleofuge) may depart just before the proton *i.e.*, the C—X cleavage is ahead of C—H cleavage. The situation is then expressed as an E2 reaction with a small amount of E1 character (See III, Scheme 5.10). In fact such a transition state leads to a carbocation intermediate *i.e.*, E1 elimination.

5.4 COMPETITION BETWEEN ELIMINATION AND SUBSTITUTION

Often one has to face a competition among four different mechanisms while carrying out an elimination reaction and these are S_N1, S_N2, E1 and E2. Thus one has to decide if the combination is going to be S_N2/E1 or E2 or S_N1/E1. Several factors influence this outcome and these are the structure of the substrate, the identity of the base or nucleophile, the solvent and the leaving group.

A particular substrate generally responds to changes in reaction conditions so as to favour one pathway over the other. Changes in the base, nucleophile generally have a marked effect on the elimination-substitution pathways. Smaller, but significant effect is also exerted by the reaction medium and temperature.

Significantly in the elimination-substitution competition, substitution is generally the predominant reaction. Substitution involves less bond reorganization (less bonds are broken and formed) and is therefore, generally more favourable energetically.

(a) Effect of Structure of the Substrate and Basicity Versus Nucleophilicity

(*i*) S_N2/E2 Condition : Substitution is Slowed by Steric Hindrance (Crowding) while Elimination is not.

The following points may be considered:

- When the reactant is a primary alkyl halide, there is no possibility to form carbocations. Primary alkyl halides undergo only S_N2/E2 reactions (a high concentration of a good nucleophile/strong base). Thus primary substrates due to low steric hindrance are excellent for S_N2 reactions with almost any nucleophile (eq. I, Scheme 5.12).
- When either the primary alkyl halide or the nucleophile/base is sterically hindered, the nucleophile will face difficulty to attack the back side of the α-carbon, thus elimination product will predominate (eqs. II and III, scheme 5.12). Potassium *tert*-butoxide is a strong base and not a very good nucleophile due to its steric bulk, and is therefore, used for this purpose.

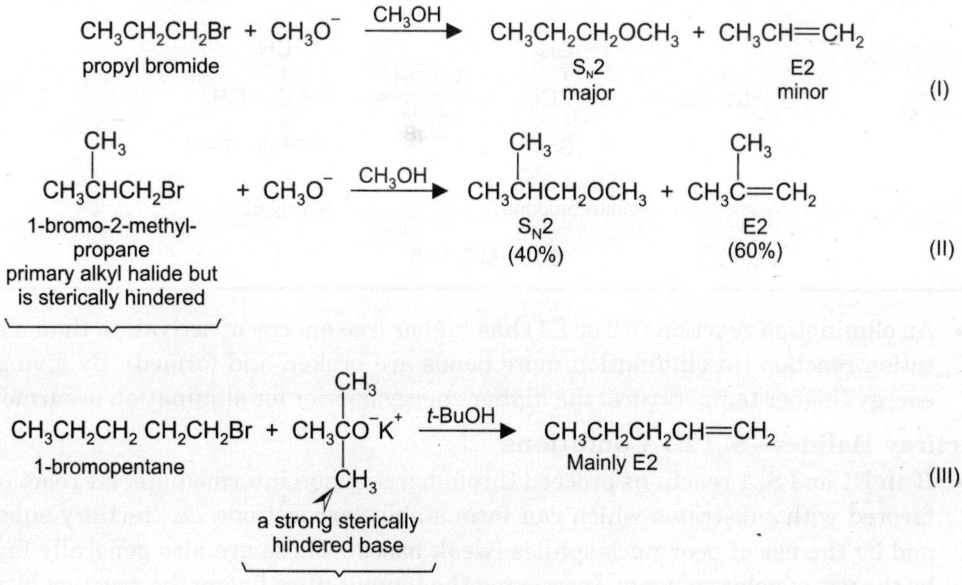

SCHEME 5.12

- When the reactant is a secondary halide, the elimination reaction is highly favored under S_N2/E2 conditions (strong base) *e.g.*, with the use of OH^- or OR^- (particularly hindered one) as shown (scheme 5.13). Thus the elimination is the main reaction from 2-bromopropane.

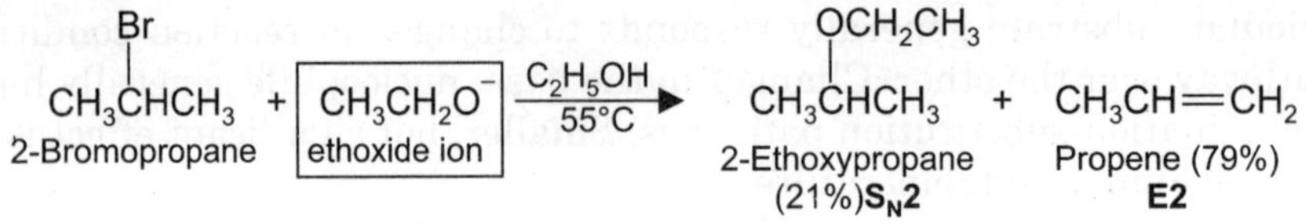

SCHEME 5.13

- Compared with the use of strongly basic ethoxide ion (scheme 5.13), with weakly basic acetate ion only S_N2 substitution is observed (scheme 5.14, acetic acid is a stronger acid pKa = 4.76 than ethanol (pKa = 15.9) to show that acetate ion is a weaker base than ethoxide ion).

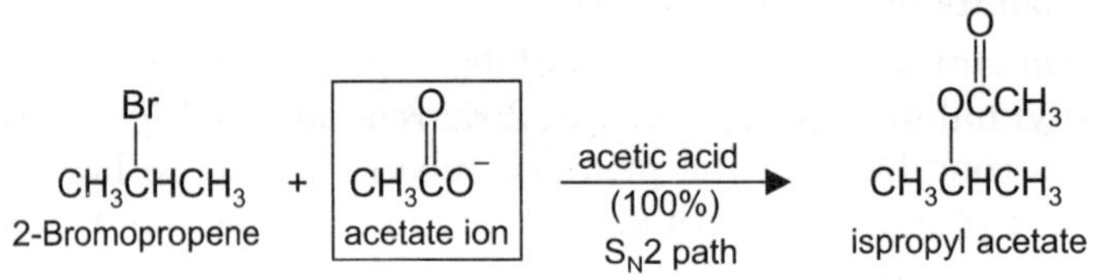

SCHEME 5.14

- With tertiary alkyl halides an S_N2 reaction cannot take place and therefore, elimination reaction is highly favoured (scheme 5.15) under S_N2/E2 conditions (particularly at higher temperatures little substitution whatsoever may occur, is through S_N1 process).

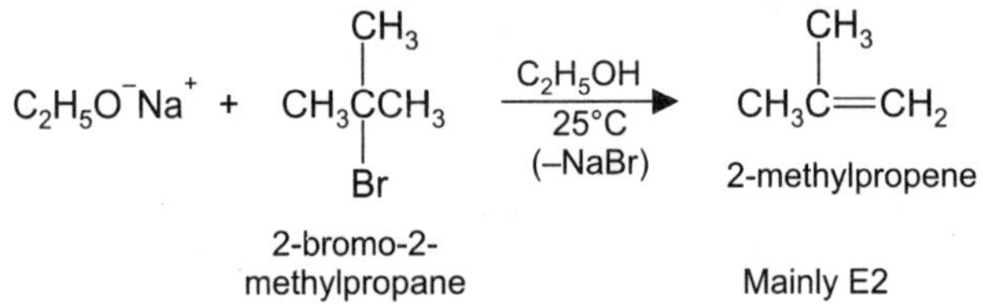

SCHEME 5.15

- An elimination reaction (E2 or E1) has higher free energy of activation than a substitution reaction (in elimination more bonds are broken and formed). By giving more energy (higher temperature) the higher energy barrier for elimination is surmounted.

(*ii*) Tertiray Halides—S_N1/E1 Conditions

- Both E1 and S_N1 reactions proceed through a common intermediate. E1 reactions are favored with substrates which can form stable carbocations *i.e.*, tertiary substrates and by the use of poor uncleophiles (weak bases). These are also generally favoured by the use of polar solvents. Increasing the temperature favors the reaction by the E1 mechanism.
- High yields of elimination product can be obtained by forcing E2 mechanism instead on a tertiary substrate *i.e.*, the use of a strong base (OH^- or OR^- see, scheme 5.15).

(*iii*) Summary of the Products Expected in Substitution/Elimination Reactions

These are given in Table 5.1.

Table 5.1

RCH_2L 1°	R_2CHL 2°	R_3CL 3°
Only $S_N2/E2$		*$S_N1/E1$ or E2*
Give mainly S_N2 products, except with a hindered strong base [*e.g.*, $(CH_3)_3CH^-$] when mainly E2 pathway is followed.	Give mainly S_N2 products with weak bases (*e.g.*, I^-, CN^-, RCO_2^-) and mainly E2 with strong bases (*e.g.* RO^-).	No S_N2 reactivity. In solvolysis gives S_N1/E1 at lower temperature S_N1 is favoured. When a strong base (*e.g.*, RO^-) is used E2 pathway predominates.

(b) Effect of Solvent

Primary alkyl halides undergo only E2 elimination reactions. E1 reactions are not possible because of the difficulty encountered in forming primary carbocations. *Secondary and tertiary* alkyl halides can undergo both E2 and E1 reactions. For these alkyl halides which can undergo both E2 and E1 reactions, the E2 reaction is favoured by the same factors that favour an S_N2 reaction and similarly the E1 reaction is again favoured by the same factors that favour an S_N1 reaction. *An E2 reaction is favoured by a high concentration of a strong base and an aprotic polar solvent* (DMSO, DMF). An E1 reaction is favoured by a weak base and a protic polar solvent (H_2O, ROH).

Further details are similar as already discussed (See, schemes 3.18–3.21).

5.5 STEREOCHEMISTRY OF E2–*anti*-ELIMINATION REACTIONS

In reactions like E2 elimination (it is a concerted reaction, the two substituents are eliminated in the same step) several bonds are made and broken simultaneously. This electron reorganization has strict requirements for the stereochemical relationship of these bonds (stereoelectronic requirement) as the reaction proceeds. These stereochemical factors occur since the bonds to the eliminated substituents (H and L) must be in the same plane since the sp^3 orbital

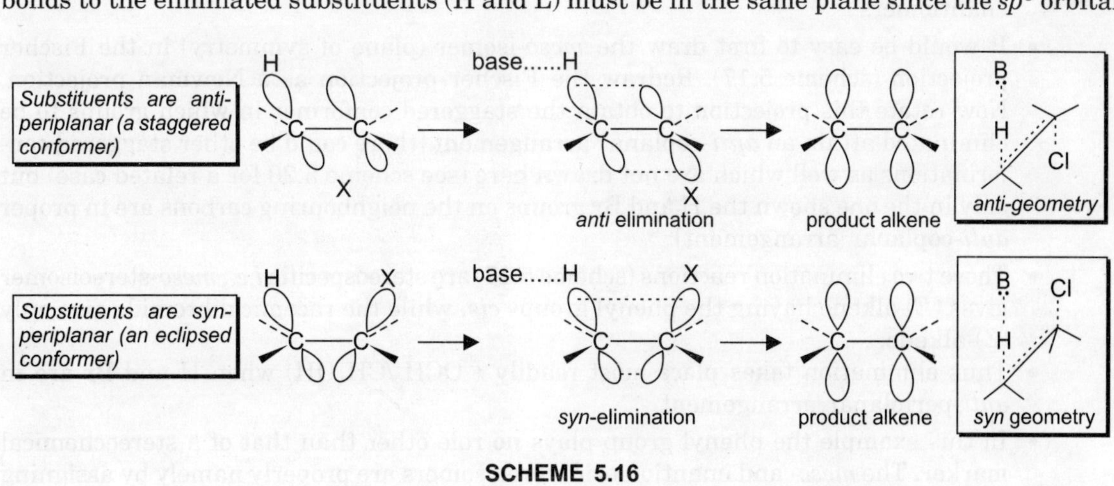

SCHEME 5.16

of the carbon bonded to H and the sp^3 orbital of the carbon bonded to leaving group L become overlapping *p* orbitals in the alkene. Thus the orbitals must overlap in the transition state (this overlap increases from the start of the reaction and provides significant stabilization in the transition state to help offset the energy cost of breaking other bonds). The overlap can only occur provided the orbitals are parallel (scheme 5.16) and for to be parallel, these must lie in the same plane (*i.e.*, coplanar).

To be in the same plane the C—H and C—L bonds can either be parallel to one another on the same side of the molecule (*syn*-coplanar) or on the opposite side of the molecule (*anti*-coplanar). Thus when the eliminated groups are on the same side of the molecule, the reaction is termed *syn*-elimination and if these are eliminated from opposite sides, the reaction is called *anti*-elimination (scheme 5.16). The following points may be noted:

- The *anti*-coplanar conformation is more stable (scheme 5.16, the dihedral angle between H and L is 180°) since it is staggered when compared to *syn*-coplanar conformation in which all bonds are eclipsed.
- *Anti*-elimination is preferred in E2 reactions.
- *Syn*-elimination is however, less common but it does occur in substrates where structure is rigid and as a consequence, the hydrogen and leaving group are held *syn*-coplanar in an eclipsed or nearby eclipsed conformation (see, scheme 5.26).
- Moreover, *anti*-elimination occurs more rapidly since the attacking electron rich base avoids the repulsion with electron-rich halogen on the same side of the molecule.
- The preference for *anti*-elimination has been proved always when the acyclic or cyclic substrate is able to achieve *anti*-periplanar geometry.

The following examples illustrate the above points. One must study the stereostructure of the substrate by considering the number of stereoisomers which may be imagined/drawn initially as Fischer projections and then translated into Newman or sawhorse projections or perspective drawings as already learnt (see, schemes 1.38, 1.44 and 1.77) to reach the results of elimination.

Example 1. *E2-elimination from 1, 2-dibromo-1, 2-diphenylethane (sodium ethoxide-ethanol)*

- The compound has two stereocenters, since the group on each of the stereocenters are the same one expects only three stereoisomers, a *meso*-compound and a pair of enantiomers.
- It would be easy to first draw the *meso*-isomer (plane of symmetry) in the Fischer projection (scheme 5.17). Redraw the Fischer projection as a Newman projection. Now rotate this projection to obtain the staggered conformer in which groups to be eliminated attain an *anti*-coplanar arrangement [there could be other staggered conformations as well which are not drawn here (see scheme 5.20 for a related case) but only in the one shown the H and Br groups on the neighbouring carbons are in proper *anti*-coplanar arrangement].
- These two elimination reactions (scheme 5.17) are stereospecific *i.e.*, *meso*-stereoisomer gives (*E*)-alkene having the phenyl groups *cis*, while the racemic dibromide give only (*Z*)-alkene.
- Thus elimination takes place most readily ($^-OCH_3/CH_3OH$) when H and Br are in *anti*-periplanar arrangement.
- In this example the phenyl group plays no role other than that of a stereochemical marker. The *meso*- and enantiomeric stereoisomers are properly namely by assigning *R* and *S* descriptors.

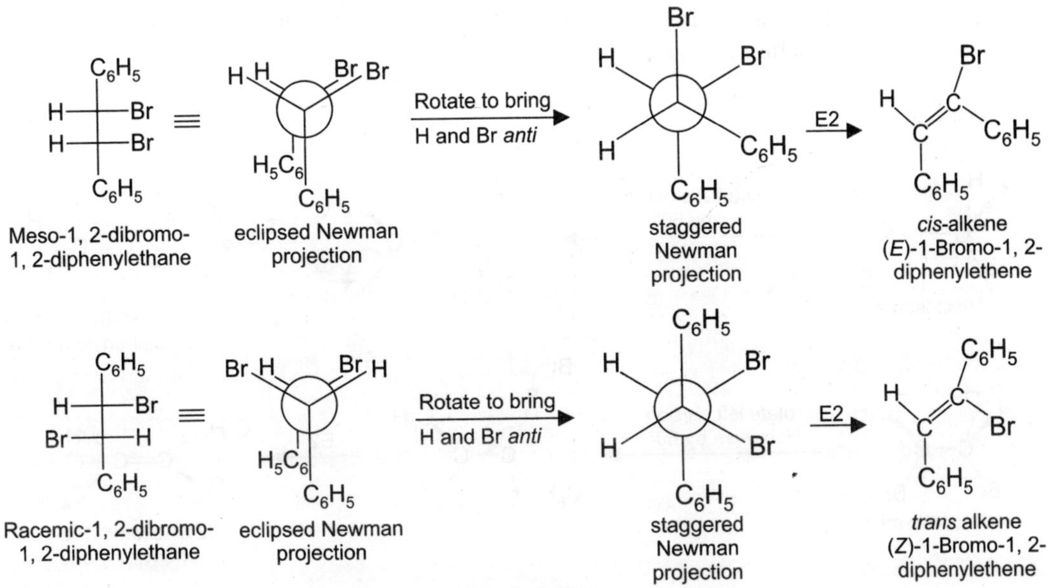

SCHEME 5.17

- The same result can be seen by drawing sawhorse projections (scheme 5.17*a* in the case of racemic mixture the eclipsed forms are not drawn).

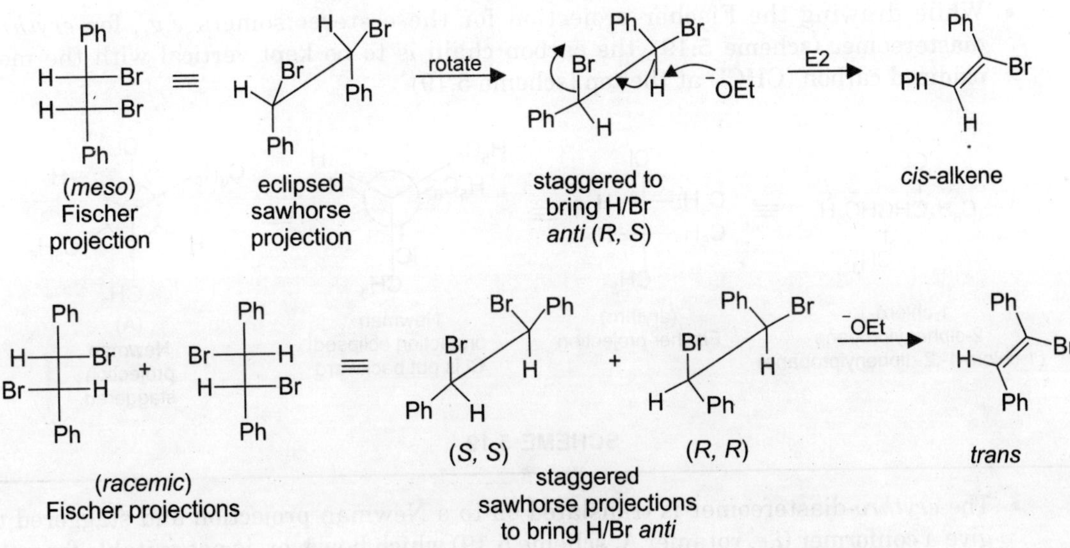

SCHEME 5.17a

- Alternatively one may also redraw the Fischer projection (scheme 5.17) on a wedge and dash structure, or start afresh by first drawing the *meso*-isomer on a wedge and dash structure to show the plane of symmetry (scheme 5.18). The left carbon has to be rotated (only 60) to bring H and Br into *anti*-coplanar arrangement. This rotation now brings H—C—C—Br bonds of the molecule in the plane of the paper shown with continuous lines and consequently the dihedral angle is now 180° (For a related situation see IV, scheme 1.41).

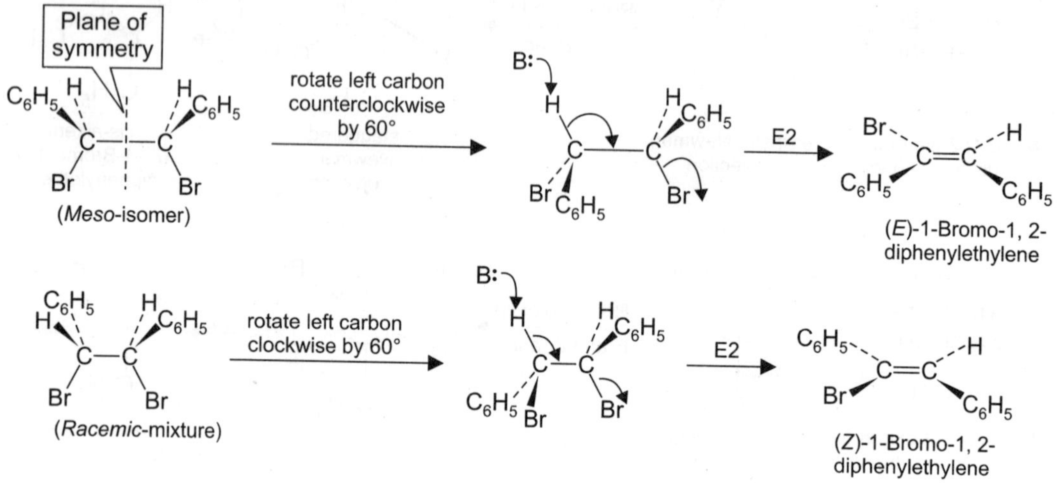

SCHEME 5.18

Example 2. ***E2 elimination from 1-chloro-1, 2-diphenylpropane***

- This compound has two stereogenic carbons (stereocenters), thus it would exist in four stereoisomeric forms (as pairs of enantiomers).
- While drawing the Fischer projection for these stereoisomers *e.g.*, for *erythro*-diastereomer (scheme 5.19), the carbon chain is to be kept vertical with the most oxidized carbon (CHCl) at the top (scheme 5.19).

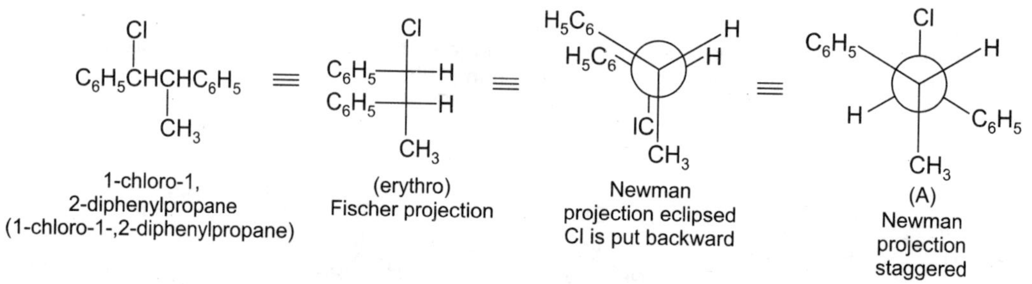

SCHEME 5.19

- The *erythro*-diastereomer is translated on to a Newman projection and staggered to give a conformer (*i.e.* rotamer A, scheme 5.19) which however, is not suitable for *anti*-elimination.
- In this rotamer, the front carbon is rotated further appropriately so as to bring H and C1 in *anti*-periplanar relationship (I, scheme 5.20). This undergoes E2 elimination (NaOEt/EtOH) to give *Z*-alkene. Thus, out of three possible staggered rotamers of *erythro*-diastereomer, only rotamer (I, scheme 5.20) is suitable for *anti*-elimination and that too to give *Z*-alkene.
- Similarly in the *threo*-diastereomer the rotamer (II) gives *E*-alkene.

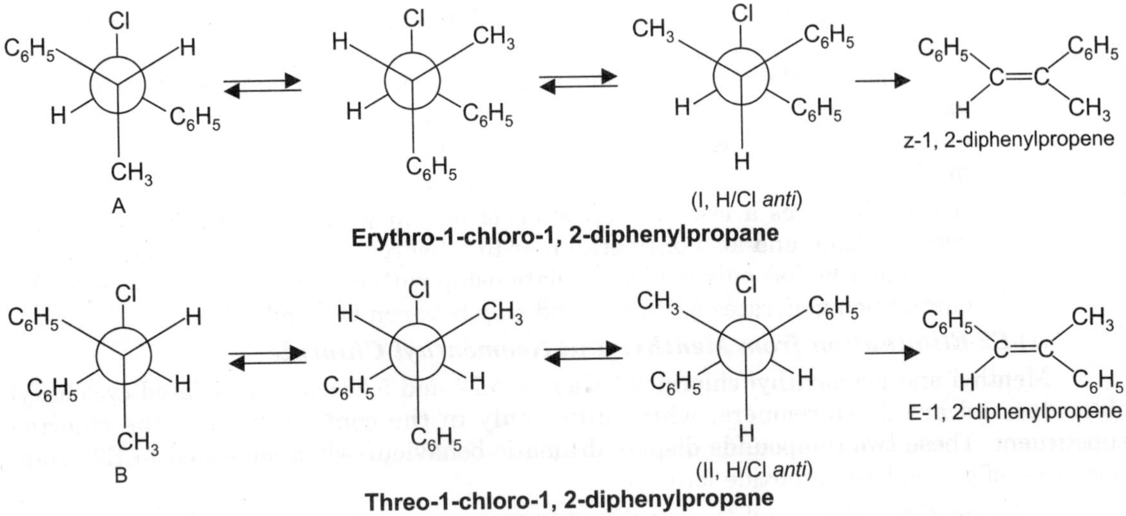

SCHEME 5.20

- The elimination reaction (E2) of each diastereomer proceeds in a stereospecific *anti*-manner *i.e.*, when the H atom and Cl are in the *anti*-periplanar arrangement.
- Recall that *erythro* diastereomer A can be converted into B by interchanging groups around one of the stereocenters *e.g.*, H and C_6H_5 on the rear carbon.

Example 3. ***E2-elimination in cyclohexane rings—Conformation and reactivity.***

In open chain compounds a molecule can easily adopt a conformation where the leaving groups are *anti*periplanar. In cyclic compounds, this is not always the case which thus effects the outcome of the reaction. In a six membered ring, the two leaving groups must be *trans*-diaxial so as to assume the requisite *anti*coplanar arrangement prior concerted E2 elimination.

(i) E2-Elimination from Trans-1-Chloro-2-Methycyclohexane

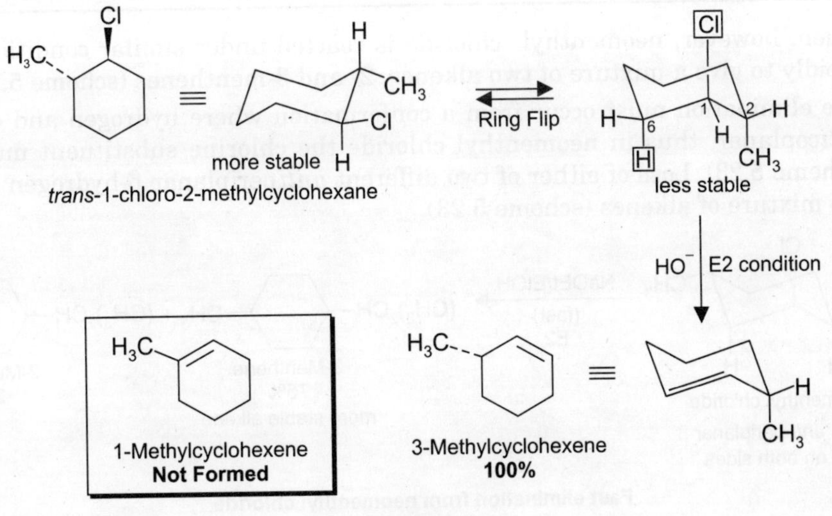

SCHEME 5.21

- In the more stable conformation of the compound (scheme 5.21) the chlorine being in equatorial position, cannot have any *anti*periplanar relationship with any C—H bond of the ring (The chlorine atom in the equatorial position, however, is *anti*periplanar to the C—C bonds of the ring system which is not involved in elimination). Thus E2-elimination cannot occur from this more stable conformation.
- A ring flip gives a less stable conformation, in which now, the chlorine atom becomes axial and also *anti*periplanar to a hydrogen at C-6. However, since the axial chlorine has only a gauche relationship with hydrogen at C-2, concerted E2 elimination generates a double bond only between C-1 and C-6.

(ii) E2-Elimination from Menthyl and Neomenthyl Chloride

Menthyl and neomenthyl chlorides (schemes 5.22 and 5.23) are substituted cyclohexyl chlorides and are diastereomers, which differ only in the configuration of the chlorine substituent. These two compounds display dramatic behaviour when subjected to E2 conditions (use of strong base, ethoxide ion).

- In menthyl chloride all the substituents are equatorial. When it is treated with sodium ethoxide in ethanol, elimination occurs very slowly to produce the alkene 2-menthene as the only product (scheme 5.22).

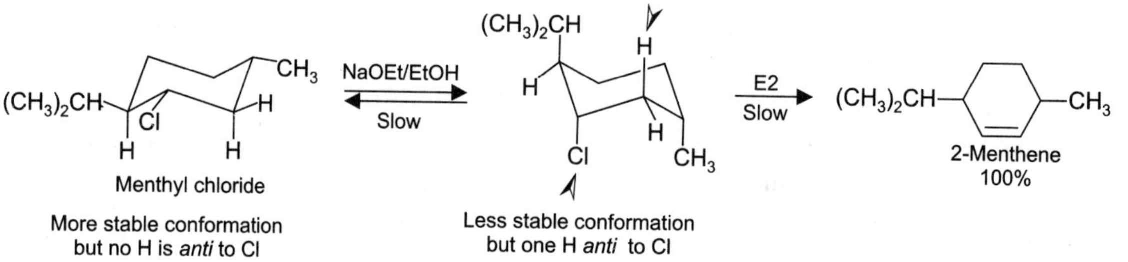

Slow E2 elimination but from less stable conformation of menthyl chloride

SCHEME 5.22

- When, however, neomenthyl chloride is reacted under similar conditions it reacts rapidly to give a mixture of two alkenes, 2- and 3-menthenes (scheme 5.23).
- The elimination must occur from a conformation where hydrogen and chlorine are *anti*coplanar, thus in neomenthyl chloride the chlorine substituent must be axial (scheme 5.23). Loss of either of two different *anti*periplanar β-hydrogen atoms gives the mixture of alkenes (scheme 5.23).

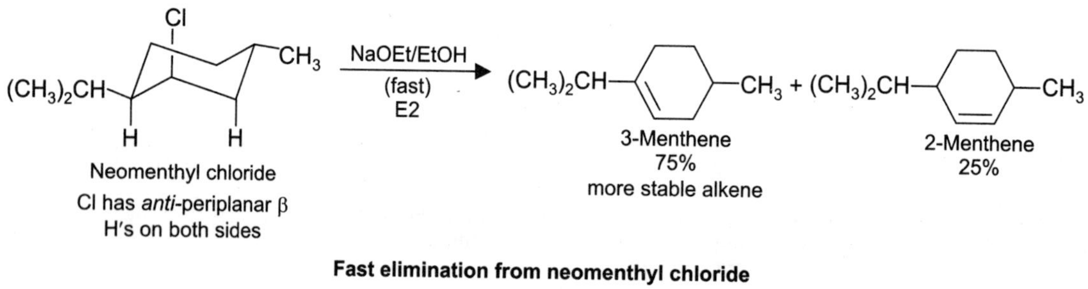

Fast elimination from neomenthyl chloride

SCHEME 5.23

- In menthyl chloride, the chlorine atom must be equatorial. The elimination, therefore, occurs from a less favourable conformation (see, scheme 5.22) obtained by ring flip. At any one time only a small percentage of the molecules have this conformation. A low population of this reactive conformation leads to a slow elimination reaction.

EXERCISE 5.2

On attempted E2 reaction, (I) reacted much faster than (II) while (III) reacted much faster than (IV, scheme 5.24). Write the product of reaction in each case.

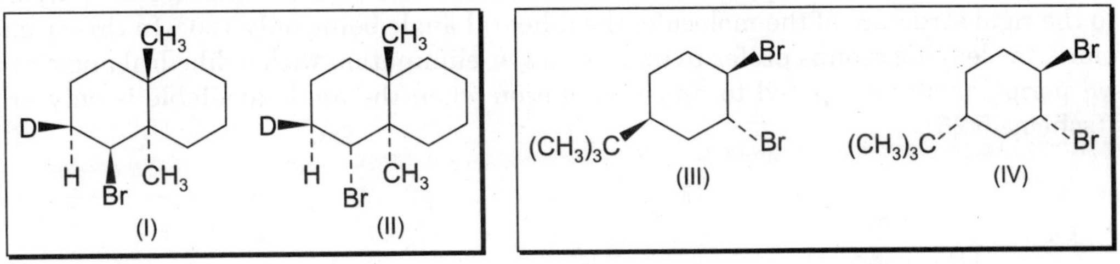

SCHEME 5.24

ANSWER. *The conformations of trans-decalin systems in (I and II) are in (V and VI) respectively. Conformation of (III) is in (VIII) which can undergo E2 reaction while conformation of (IV) depicted in (IX) cannot do E2 unless it adopts a high energy boat conformation which could give the same product as from (VIII) via a slow E2 reaction. A ring flip in (IX, scheme 5.25) is highly unlikely since, bulky t-butyl group is most stable in equatorial orientation.*

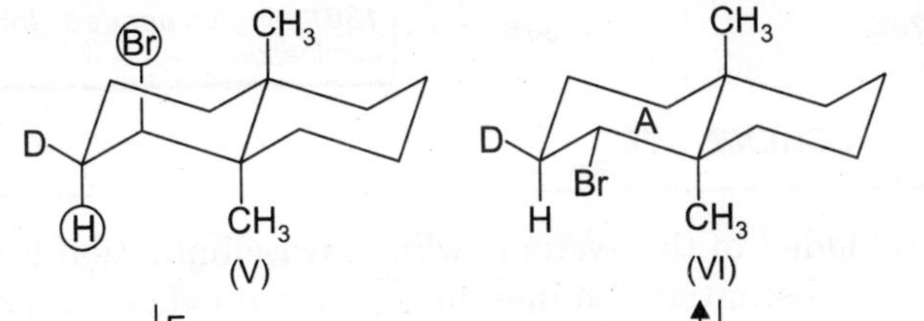

In (VI) ring flip is not possible (trans fusion). Attainment of a boat conformation in ring (A) brings D/Br anticoplanar for E2 elimination. Since this attainment of boat form requires energy, elimination in (VI) is very slow.

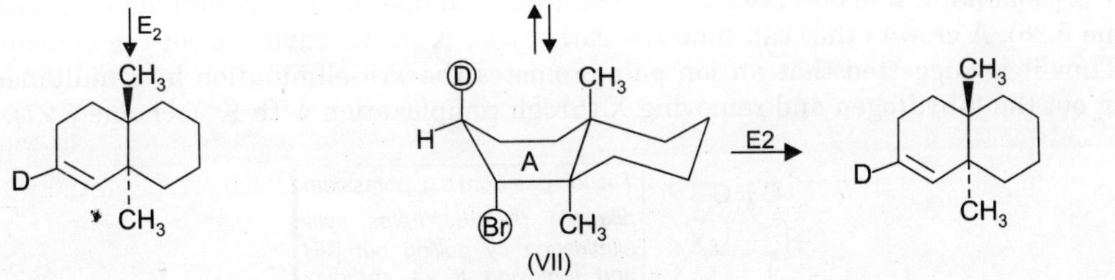

SCHEME 5.24

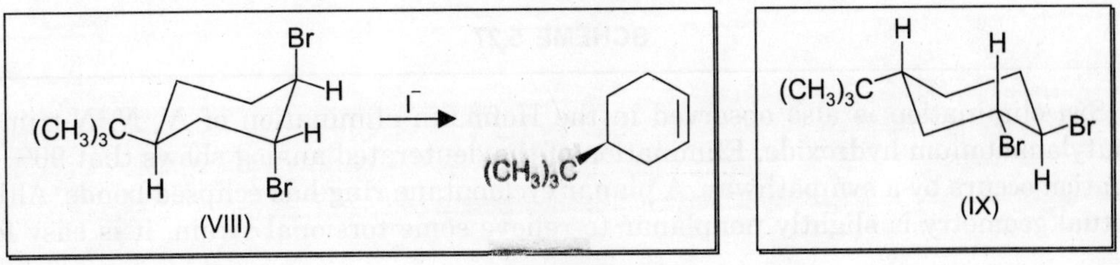

SCHEME 5.25

5.6 E2-*syn*-ELIMINATION

Some rigid molecules are not able to attain the favourable *anti*-periplanar conformation of the departing groups. Elimination is, thus, inhibited and may proceed by another mechanistic or stereochemical pathway.

It has been shown that arrangements in which the departing groups are coplanar on the same side of the molecule may undergo concerted elimination. The groups are termed to be *syn*-periplanar, and the process represents a *syn*-elimination. The deuterated norbornyl system (scheme 5.26), gives almost 100% of the product with no deuterium. In this case the *exo* tasylate group cannot achieve *anti*-periplanarity *i.e.*, a dihedral angle of 180° with an *endo* β-hydrogen due to the rigid structure of the molecule, the dihedral angle being only 120°. In this situation, therefore, the leaving groups prefer to undergo a *syn*-elimination with a dihedral angle around 0° (*syn*-periplanarity) compared to *anti*-elimination when the angle available is only around 120° (scheme 5.26).

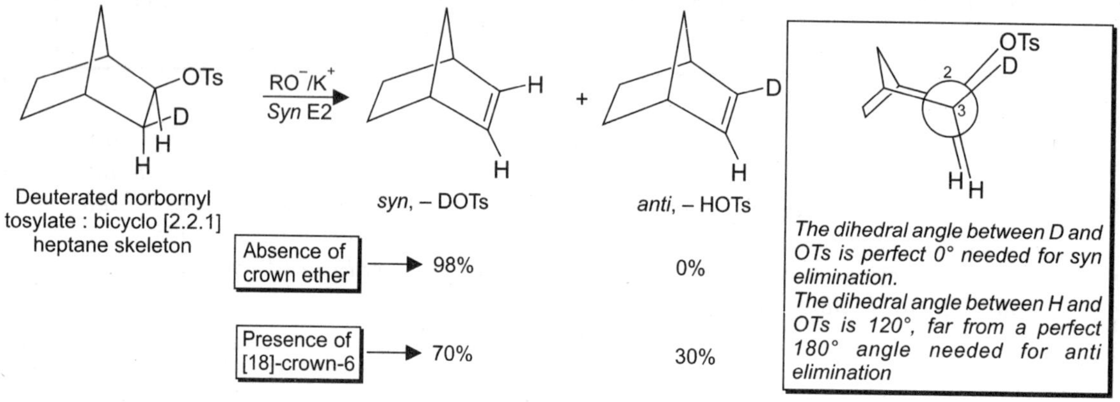

SCHEME 5.26

When suitable crown ethers are added to the systems where *syn*-elimination is exclusive or predominant, it is observed that substantial elimination then occurs via *anti* pathway (scheme 5.26). A crown ether can bind the cation *e.g.*, K^+ in its cavity to set the alkoxide ion free. Thus it is suggested that an ion pair promotes the *syn*-elimination by simultaneously pulling out the β-hydrogen and removing X throgh complexation with K^+ (scheme 5.27).

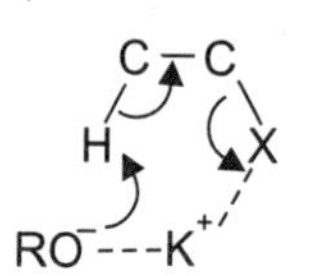

The ion-pair from e.g. potassium alkoxide ($R\bar{O}\,K^+$) helps syn-elimination by pulling out β-H and removing X via complex formation with K^+

SCHEME 5.27

Syn elimination is also observed in the Hofmann elimination of *N, N, N*-trimethylcyclobutylammonium hydroxide. Elimination of the deuterated analog shows that 90% of the elimination occurs by a *syn* pathway. A planar cyclobutane ring has eclipsed bonds. Although the actual geometry is slightly nonplanar to relieve some torsional strain, it is easy for the cyclobutane ring to attain the eclipsed geometry needed for *syn* elimination than to attain a nonplanar geometry to have the nitrogen and a hydrogen *anti*planar (scheme 5.28).

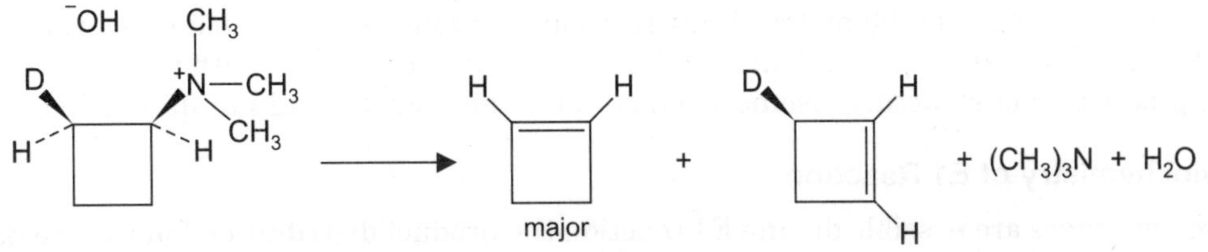

SCHEME 5.28

In E2-eliminations the significant condition for the reactions is that the departing groups should be coplanar no matter *anti* or *syn*. In *anti*-eliminations the incoming and outgoing bonding electrons are kept as far apart as possible. In *syn*-eliminations, however, the eclipsing of the groups in the reaction leads to a less favorable situation, thus, *syn*-eliminations are considerably less favorable, occuring at a slower rate than *anti*-eliminations.

Two points may be noted. Firstly *anti*-elimination requires a dihedral angle of 180°. If this angle cannot be achieved, *anti*-elimination is slowed or entirely prevented. Secondly, for the simple systems *syn*-elimination does not occur to any significant extent unless *anti*-elimination is greatly prevented by failure to achieve the 180° dihedral angle.

Six-membered rings singularly, from among rings of four to thirteen members present systems where strain-free *anti*-periplanar conformations can be achieved. It is, therefore, not surprising that *syn*-elimination is rare in six-membered rings. Cycloalkyltrimethylamonium hydroxides on elimination give the following precentages of *syn*-elimination products with ring sizes: four-membered 90%; five-membered 46%; Six-membered 4%; Seven-membered 31 to 37%. It may be noted that the $^+NMe_3$ group has a greater tendency for *syn*-elimination than other common leaving-groups such as OTs, Cl and Br.

The *cis*-and *trans*-isomers of a cyclopentane derivative (scheme 5.29) yielded the same olefin by E2-elimination, thus, the *trans* isomer undergoes *syn*-elimination while the *cis*-isomer an *anti*-elimination. Moreover *anti*-elimination has been shown to proceed slightly faster than *syn*-elimination. This is understandable, since in cyclopentane, only small twisting of the fairly flexible ring is needed to attain a *syn*-periplanar arrangement of the groups *cis* to each other. As attainment of the *anti*-periplanar relationship of the groups *trans* to each other is somewhat difficult as the ring shall have to be bent well away from planarity.

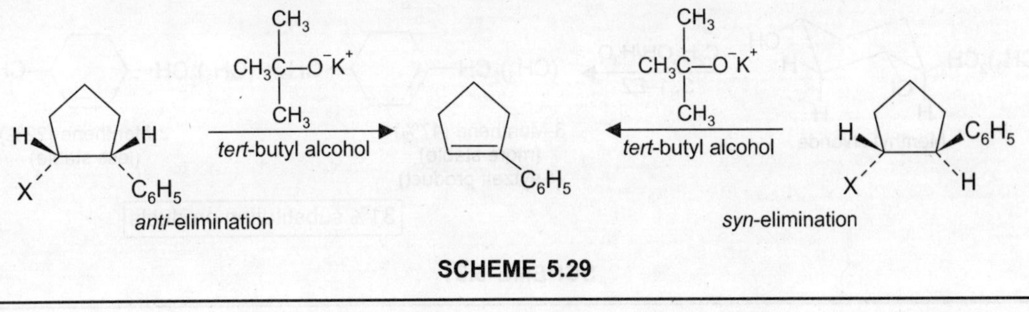

SCHEME 5.29

5.7 ORIENTATION OF THE DOUBLE BOND—REGIOCHEMISTRY OF THE ELIMINATION REACTIONS

While discussing the stereochemistry of substitution reactions, the concern has been stereogenic carbon bearing the leaving group. During elimination one is concerned with the regiochemistry *i.e.*, the production of structural isomers which have the double bond in different positions.

(a) Regiochemistry of E1 Reaction

When regioisomers are possible during E1 reaction the product distribution follows the Saytzeff rule—the double bond goes mainly toward the most highly substituted carbon. Put differently the major product is the one with more alkyl groups on the carbons of the double bond. This is so since in E1 mechanism the leaving group is gone to give a carbocation before the choice is made as to which direction the generated double bond will go. Thus the direction is determined almost exclusively by the relative stabilities of the olefins produced via elimination. From the heat of combustion data, it is known that olefin stability increases with alkyl substitution. Thus E1 reactions occur under thermodynamic control. The more stable product is formed preferentially. The eliminations which occur under thermodynamic control are said to give the Saytzeff product (scheme 5.30).

For E1 elimination, therefore, Saytzeff rule governs the orientation whether the leaving group is neutral or positive (this is so, since leaving group is not present when the choice for the direction of double bond is made).

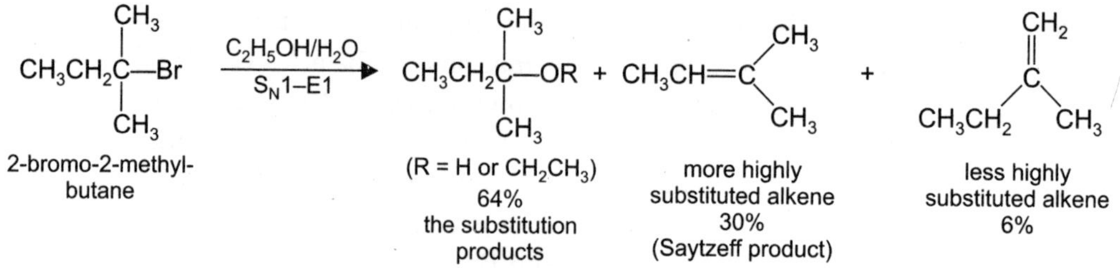

SCHEME 5.30

(b) A Comparison of Regiochemistry of E1 with E2 Reaction

Most E2 reactions also follow Saytzeff rule as seen in the case of neomenthyl chloride (see scheme 5.23). For the *anti*-E2 reactions a *trans* β proton is necessary (an *anti*-coplanar arrangement) and in case this is available only in one direction, that is the way the double bond will form. Thus menthyl chloride (see, scheme 5.22) gives less substituted 2-menthene in

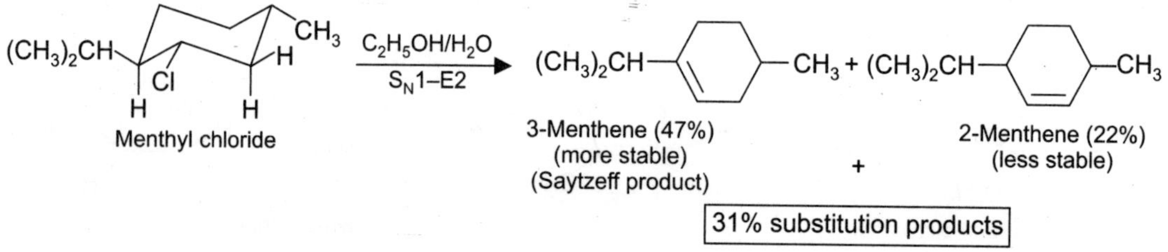

SCHEME 5.31

100% yield. Interestingly in E1 reactions only a carbocation is formed and only at this point a C—H bond must be aligned parallel to the empty *p* orbital of the carbocation for the *pi* overlap to occur as the C—H bond beings to break. Thus *anti*-periplanar requirement is no longer required for E1 reactions in cyclohexane rings and under E1 conditions (scheme 5.31) menthyl chloride gave Saytzeff product as the major alkene along with the less substituted alkene.

(c) Regiochemistry in E2 Reactions (Orientation of Double bond)

As already stated most E2 reactions follow Saytzeff rule. Another example is provided during the E2 elimination of 2-bromobutane (*sec*-butyl bromide, scheme 5.32) where more substituted alkene is formed as the major product (elimination E2 from 2-bromobutane has been discussed in detail in Scheme 4.9). The regioselectivity of the E2 reaction is determined by the transition state which is stabilized by the same factors which stabilize the alkene (scheme 5.33). The transition state for the formation of 2-butene (more substituted) by the loss of H from C3 resembles it and is thus of lower energy than the transition state which gives less substituted 1-butene by the loss of proton from C1 (scheme 5.33). Thus E2 reaction of 2-bromobutane follows two distinct pathways to give different regioisomers. The base can approach a proton at either C1 or C3.

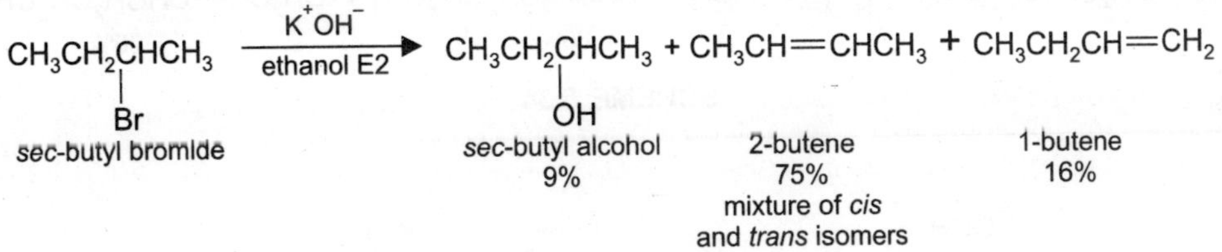

SCHEME 5.32

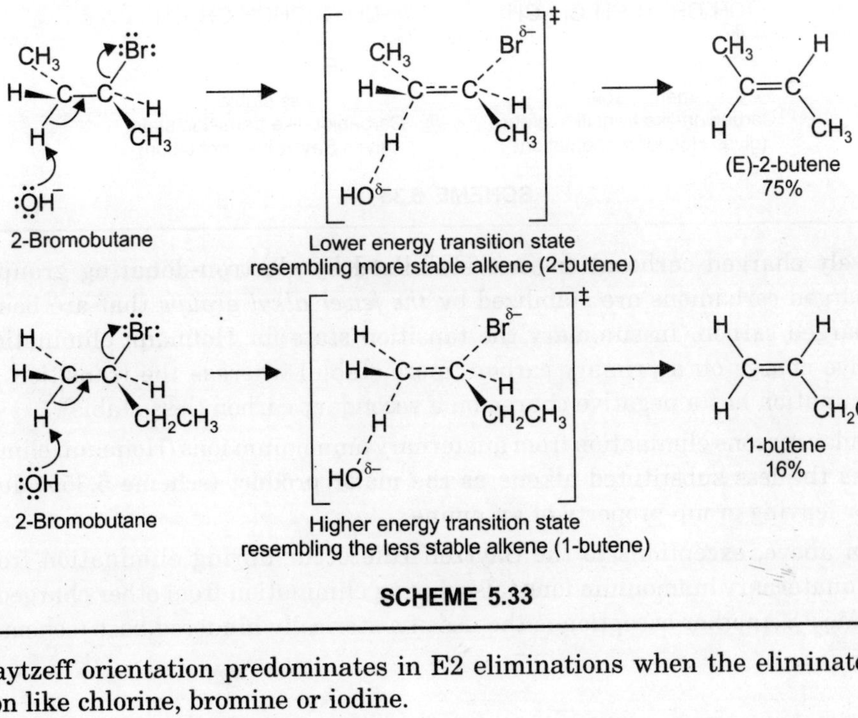

SCHEME 5.33

Saytzeff orientation predominates in E2 eliminations when the eliminated group is a halide ion like chlorine, bromine or iodine.

Thus compared to the major product of E2 dehydrohalogenation of alkyl chlorides, alkyl bromides, and alkyl iodides (normally the most substituted alkene) the major product of the E2 dehydrohalogenation of alkyl fluorides is however, the least substituted alkene (scheme 5.34). When a hydrogen and a chlorine, bromine, or iodine are eliminated from an alkyl halide, the halogen starts to leave as soon as the base begins to remove the proton (scheme 5.33). Loss of the halogen and its bonding electrons prevents the buildup of negative charge on the carbon that is losing a proton, giving the transition state alkene character rather (See II, scheme 5.10) than carbanion character. Of the halogen ions, the fluoride ion is the strongest base and therefore the poorest leaving group. Thus when a base begins to remove a proton from an alkyl fluoride, the fluoride ion has less tendency to leave compared to other halide ions. Consequently negative charge develops on the carbon that is losing the proton, giving the transition state a carbanion character rather than an alkene character. Normally, carbanion transition states are unstable, however, in this case the carbanion transition state is stabilized by the strongly electron-withdrawing fluorine (scheme 5.35).

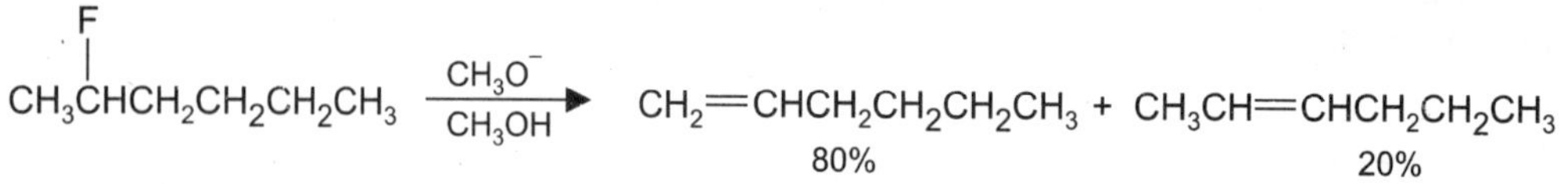

SCHEME 5.34

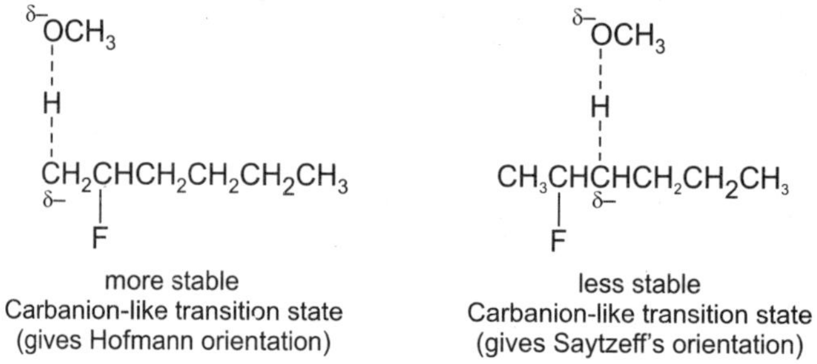

SCHEME 5.35

Positively charged carbocations are stabilized by electron-donating groups, while negatively charged carbanions are stabilized by *the fewer alkyl groups* that are bonded to a negatively charged carbon. In summary the tansition state for Hofmann elimination has a partial negative charge on a primary carbon (more stable) whereas the transition state for Saytzeff's elimination has a negative charge on a secondary carbon (less stable).

For similar reasons elimination from quaternary ammonium ions (Hofmann elimination), normally gives the less substituted alkene as the major product (scheme 5.35*a*) due to the relatively poor leaving group property of an amine.

As seen above, exceptions to the Saytzeff rule occur during elimination from alkyl fluorides and quaternary ammonium ions (also during elimination from other charged leaving groups *e.g.*, SMe_2^+). Another exception is the use of a sterically hindered base (scheme 5.35*b*).

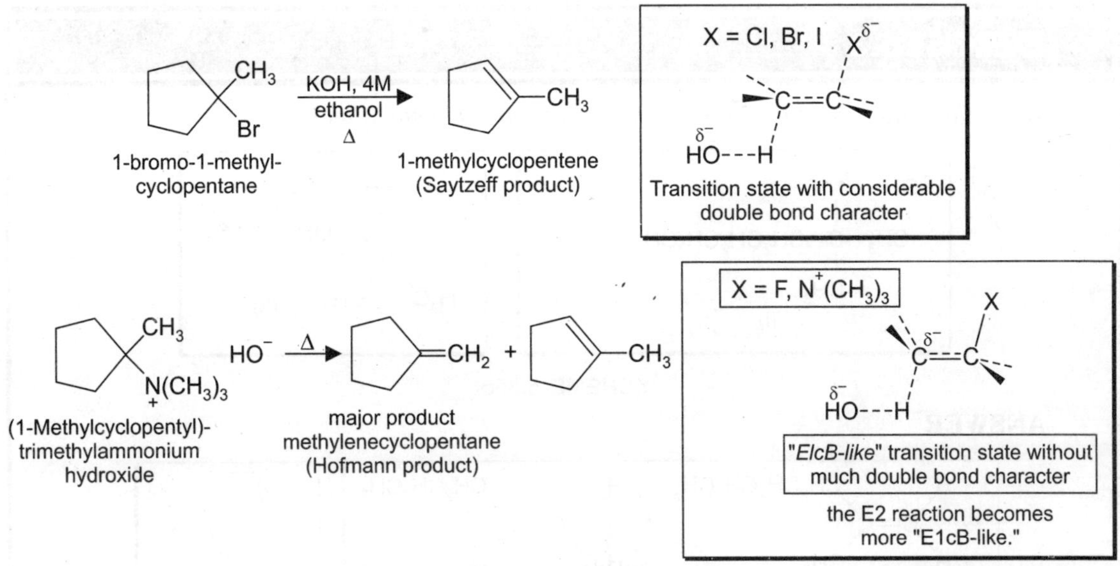

SCHEME 5.35a

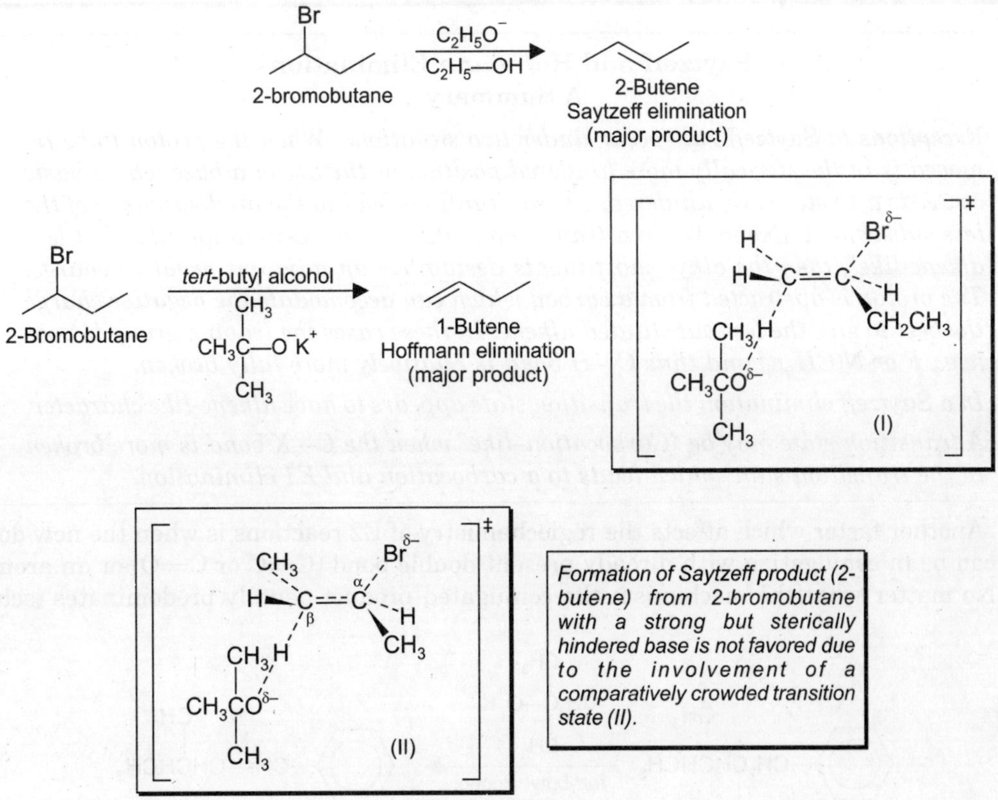

SCHEME 5.35b

EXERCISE 5.3

Write the major product from the eliminations (scheme 5.35c).

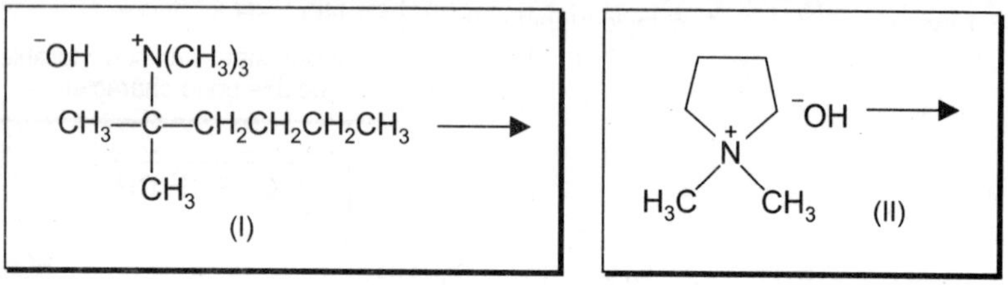

SCHEME 5.35c

ANSWER

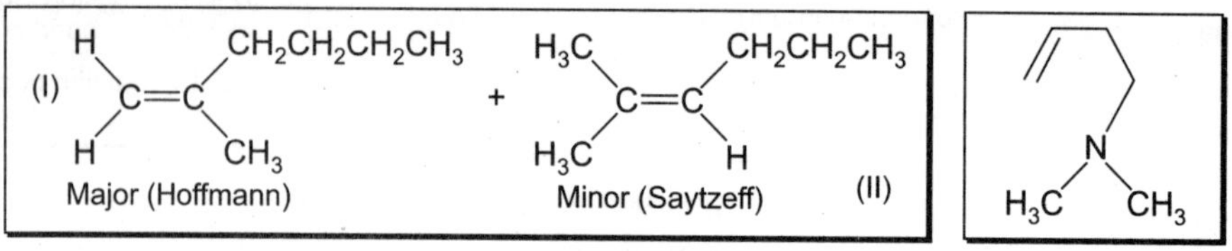

SCHEME 5.35d

Saytzeff and Hoffmann Eliminations
A Summary

Exceptions to Saytzeff rule occur under two situations. When the proton to be removed is in the sterically more hindered position or the use of a base whose basic center is also sterically hindered. These situations lead to the predominance of the less substituted alkene. When a transition state is more "Carbanion-like" but less alkene-like", then the alkyl substituents destabilize an adjacent negative charge. The proton is abstracted from a carbon which can accomodate the negative charge the best to give the less substituted alkene. In these cases the leaving group is poor [e.g., F or $N(CH_3)_3$] and thus C—H bond is relatively more fully broken.

In a Saytzeff elimination the transition state appears to have alkene-like character.

A transition state may be "Carbocation–like" when the C—X bond is more broken in the transition state which leads to a carbocation and E1 elimination.

Another factor which affects the regiochemistry of E2 reactions is when the new double bond can be in conjugation with already present double bond (C=C or C=O) or an aromatic ring. No matter what the mechanism the conjugated product usually predominates (scheme 5.36).

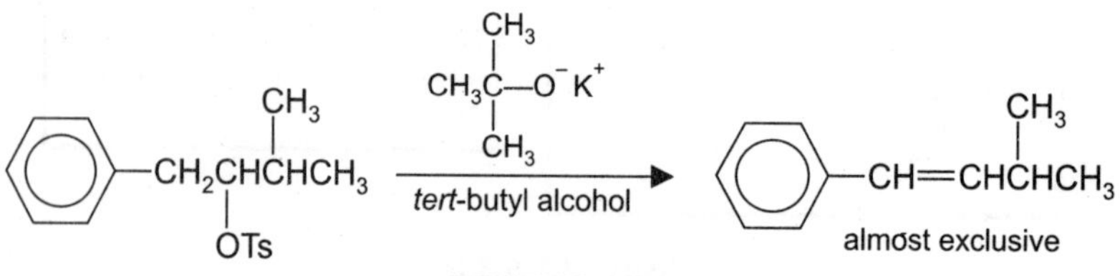

SCHEME 5.36

No matter what the mechanism is, a double bond never goes to the bridgehead carbon in a small ring substrate *e.g.*, [2.2.1] heptane system scheme 5.37 (violation of Bredt's rule).

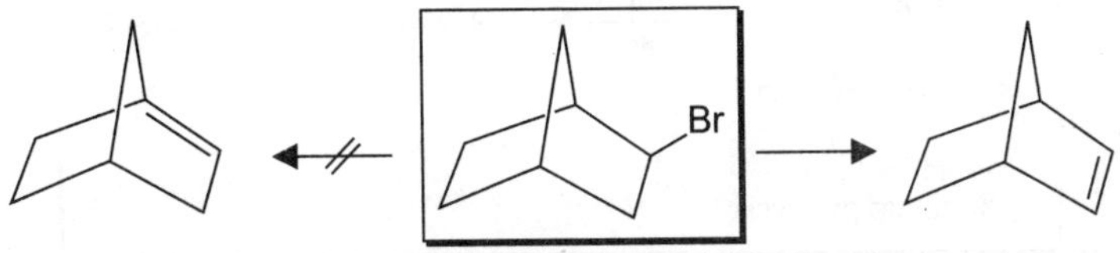

SCHEME 5.37

Lastly a completely different explanation is available (H.C. Brown) for Hoffmann orientation. Field effects have been ignored and the orientation is thought to be a consequence of steric effect. The larger charged groups (compared to neutral smaller groups) block access to the hydrogens (scheme 5.37*a*).

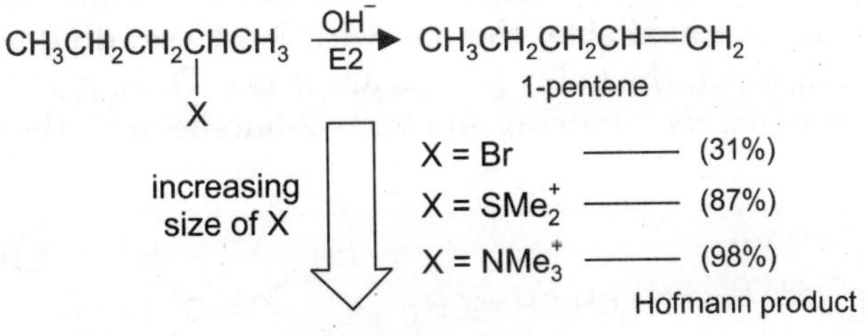

SCHEME 5.37a

However, this steric explanation is not in keeping with the results from 2-halopentanes (see, scheme 5.34) on E2 elimination to give 1-pentene. In this case 2-fluoropentane with the smallest of the halogens gave 1–pentene with predominant Hofmann orientation (80%) followed by Cl 37%, Br 25% and I 20%. Recall that the "Carbanion-like" transition state (see, scheme 5.35) explains the results of Hoffmann orientation involving substrates with poor leaving groups (F, $N(CH_3)_3$ etc.).

5.8 PYROLYTIC ELIMINATIONS

Pyrolytic elimination reactions of acetates, benzoates and xanthates take place through cyclic six-membered transition states which require a *syn* (*cis*) arrangement and proceed by a concerted pathway (scheme 5.38). The order of ease of decomposition among commonly used esters is xanthate > benzoate > acetate.

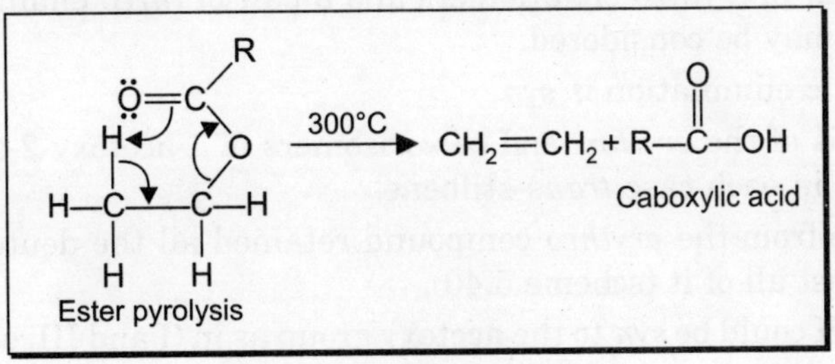

SCHEME 5.38 (Continued)

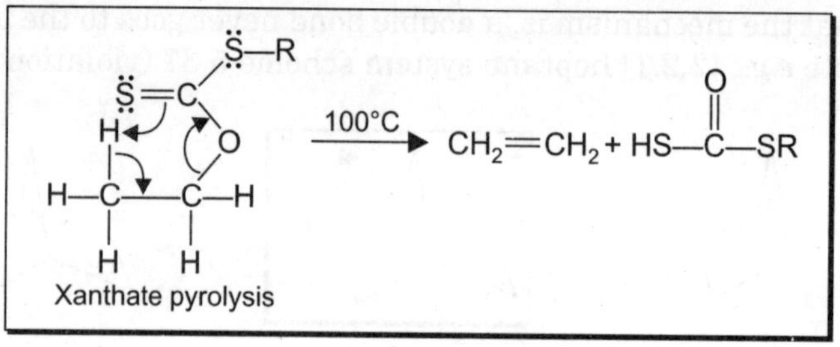

SCHEME 5.38

(*i*) Acyclic Systems

Like *syn*-elimination during reactions is solution, the pyrolytic reactions also require that the groups to be eliminated depart from the same side of the molecule (*syn*-stereochemistry) in the transition state, even if they are not eclipsed in the starting material. If two or more *syn* periplanar arrangements are available, the one with least crowding is preferred, generally leading to the product in which the bulky groups are *trans*. Thus, pyrolysis of 2-butyl acetate affords a mixture of 1-butene, *cis*-2-butene, and *trans*-2-butene, with the *trans*/*cis* ratio around 2 (scheme 5.39).

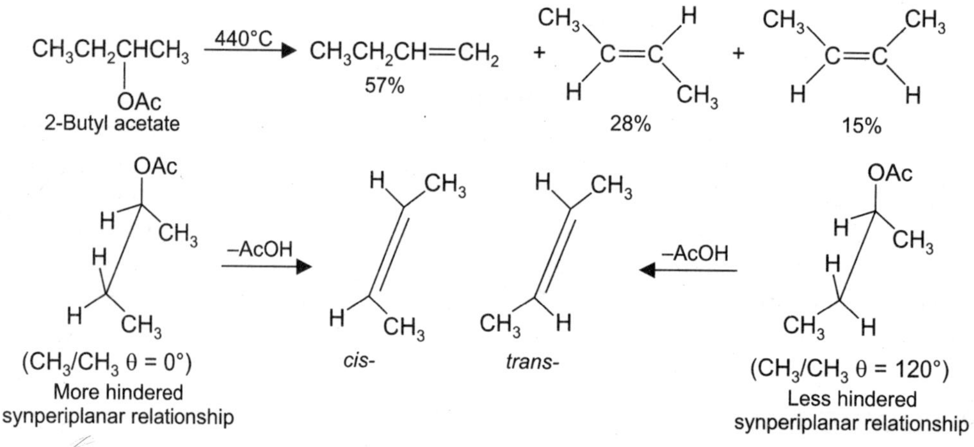

SCHEME 5.39

That the pyrolytic eliminations have a *syn* character is also evident in the pyrolysis of the *erythro* and *threo* isomer of 1-acetoxy-2-deutero-1, 2-diphenylethane (2 stereocenters, 4 stereoisomers as a pair of *erythro* enantiomers and a pair of *threo* enantiomers, scheme 5.40). The following points may be considered:

- This pyrolytic elimination is *syn*.
- The pyrolysis of the *erythro* and *threo* isomers of 1-acetoxy-2-deutero-1, 2-diphenylethane gave in each case *trans*-stilbene.
- The stilbene from the *erythro* compound retained all the deuterium while the *threo* compound lost all of it (scheme 5.40).
- Either D or H could be *syn* to the acetoxy group as in (I and III, scheme 5.40), however, these conformations are most unsatisfactory since the phenyl groups are eclipsed.

- Preferred conformations for *syn* elimination are those in which the phenyl groups are as far removed as possible from each other. These conformations are (II and IV scheme 5.40).
- This pyrolytic elimination may also be studied by considering the preferred conformations of the compounds drawn in wedge formulas (scheme 5.41)

> **Pyrolytic syn Elimination compared with E2 Anti Elimination**
>
> *During pyrolytic elimination from the erythro product one gets the trans isomer of the alkene, since the reaction is via cyclic transition state so, the opposite stereochemistry is obtained than was achieved in the E2 mechanism. The cyclic pyrolytic elimination is labelled E_i mechanism (intramolecular or internal elimination).*

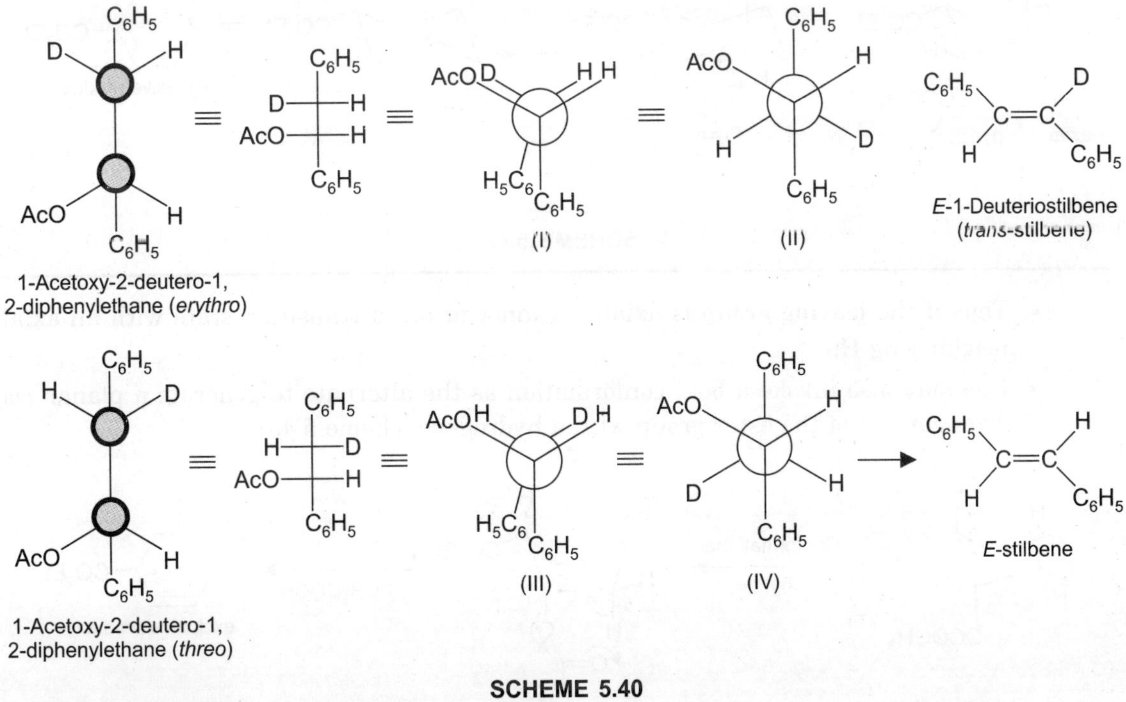

SCHEME 5.40

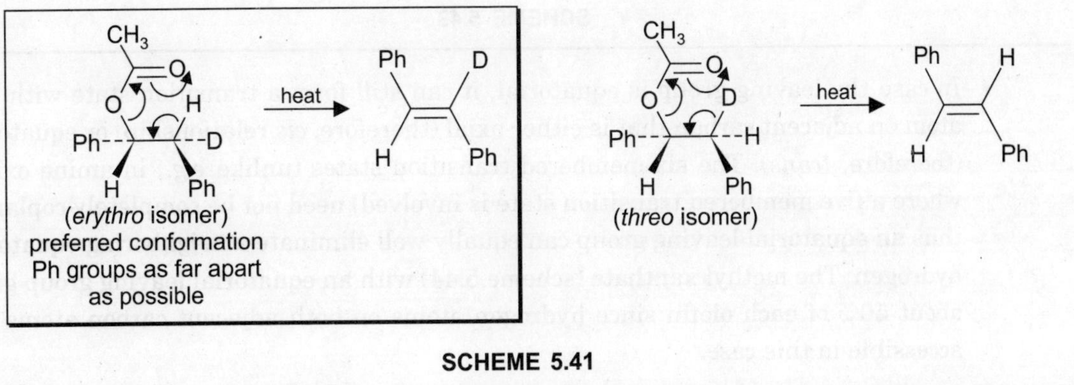

SCHEME 5.41

(*ii*) Alicyclic Systems

In these systems some restrictions are imposed by the conformations of the leaving groups and the necessity to form the cyclic intermediate. The following points may be considered:

- In a cyclohexane system if the leaving group is axial then the hydrogen on the adjacent carbon must be equatorial *i.e.* a *cis* relationship between the leaving groups. Thus the cyclohexylacetate (I, scheme 5.42) with the leaving group in axial position eliminates ethanoic acid to form only one alkene. The double bond in the direction of carbethoxyl group does not form even though it would be conjugated, since there is no equatorial hydrogen on that side.

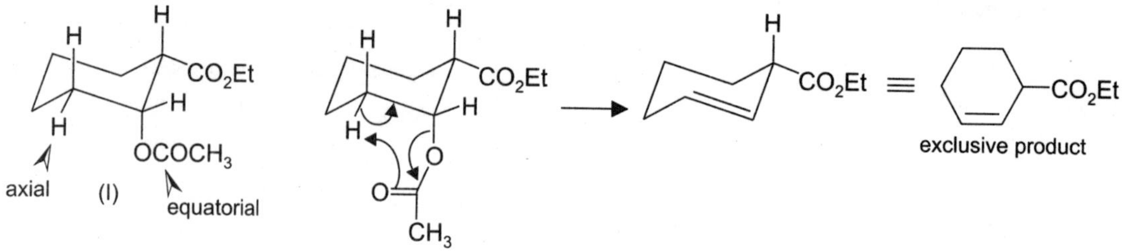

SCHEME 5.42

- Thus if the leaving group is axial, it cannot attain a transition state with an axial neighboring H.
- One may also invoke a boat conformation as the alternate to generate a planar *cis* arrangement of the ester group with a hydrogen (scheme 5.43).

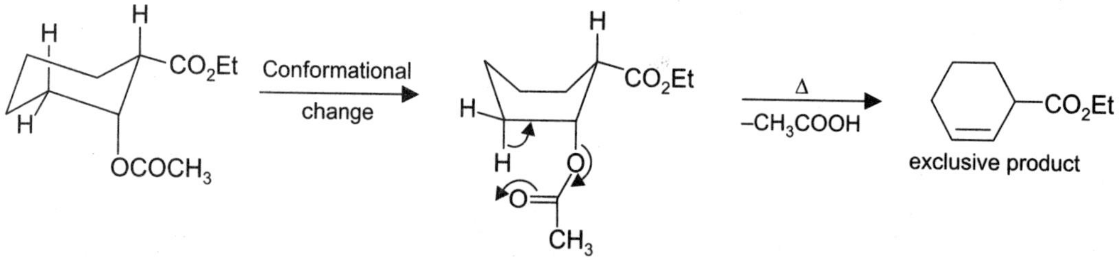

SCHEME 5.43

- In case the leaving group is equatorial, it can still form a transition state with a H atom on adjacent carbon that is either axial (therefore, *cis* relationship) or equatorial (therefore, *trans*). The six-membered transition states (unlike *e.g.*, in amine oxides where a five membered transition state is involved) need not be completely coplanar, thus an equatorial leaving group can equally well eliminate a neighboring equatorial hydrogen. The methyl xanthate (scheme 5.44) with an equatorial leaving group gives about 50% of each olefin since hydrogen atoms on both adjacent carbon atoms are accessible in this case.

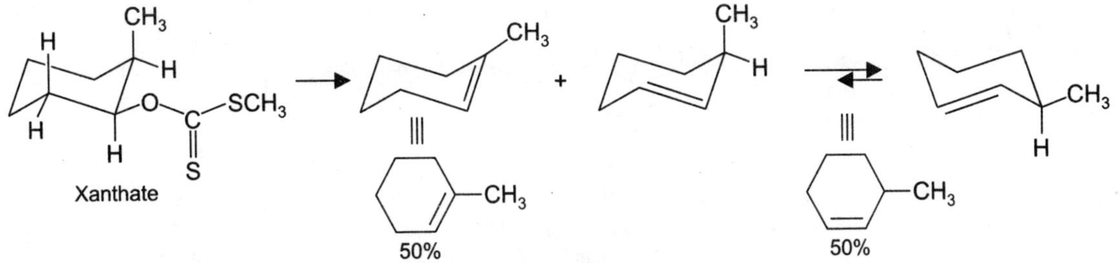

SCHEME 5.44

- More explanation with an equatorial leaving group eliminating an equatorial hydrogen (*trans*-arrangement) is the involvement of stable *trans*-decalin type transition state (scheme 5.44*a*).

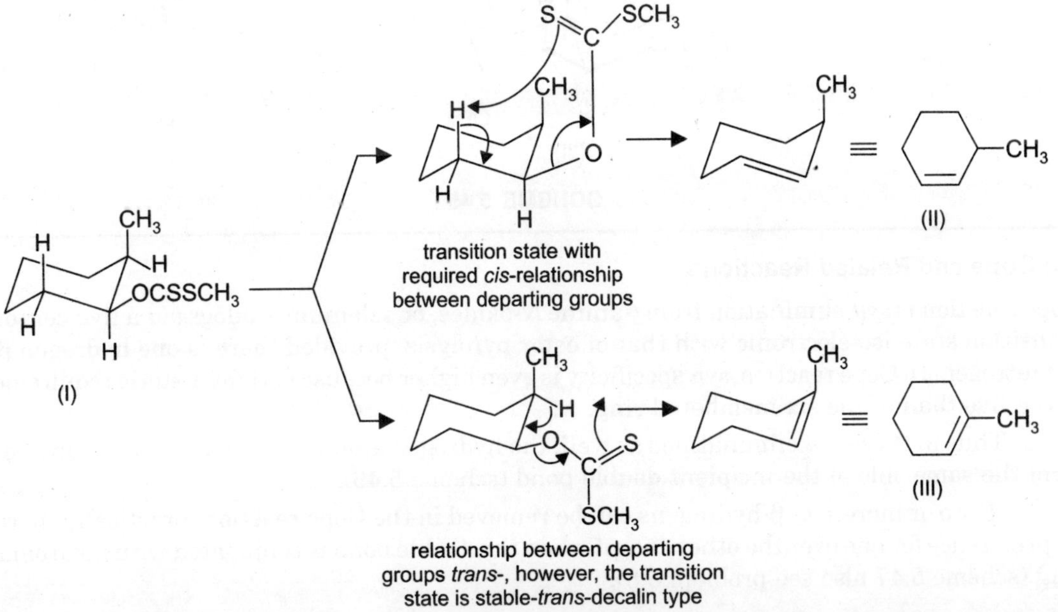

SCHEME 5.44a

- The differing stereochemical preferences of E2 and pyrolytic eliminations make them complementary methods for creating specific olefinic centers. Either the alcohol is made to undergo E2 elimination or is first esterified and then subjected to pyrolysis. The 4-hydroxy compound in 5α-chloestane series on E2 elimination eliminates C5 hydrogen to give more substituted olefin rather than C3 hydrogen (eq. 1, scheme 5.45). When the C4 hydroxy compound (I, scheme 5.45) is first esterified (benzoate, OBz) and then heated the C4 ester group (axial leaving group) can only eliminate the available equatorial hydrogen at C3 (eq. 1, scheme 5.45). In (II, scheme 5.45) hydrogen atom is stereochemically suitably located on both sides to the departing group (equatorial leaving group, OBz) at C6 and C8. The preference is therefore, for the formation of a highly substituted olefin (eq. 2, scheme 5.45). In case of (II, scheme 5.45) the leaving axial group has only one choice to eliminate C6 equatorial hydrogen.

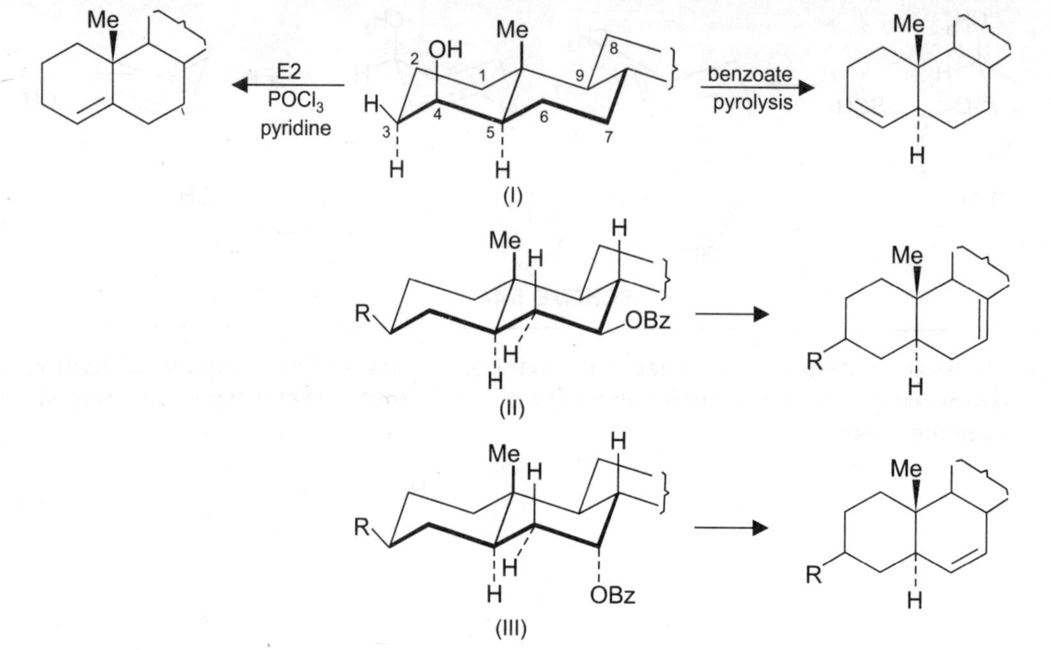

1

2

3

SCHEME 5.45

(*iii*) Cope and Related Reactions

Cope reaction is *syn* elimination from *t*-amine *N*-oxides, or selenium analogs *via* a five-centered transition state iso-electronic with that of ester pyrolysis, provided there is one hydrogen β to the nitrogen. In Cope reaction, *syn* specificity is even higher because of tighter steric requirement of the five than of the six membered ring.

Thus in these *syn* eliminations as well, the hydrogen atoms and the leaving group depart form the same side of the incipient double bond (scheme 5.46).

If two or more *syn*, β-hydrogens can be removed in the Cope reaction, practically there is no preference for one over the other except when the double bond is conjugated with an aromatic ring (scheme 5.47 also see problem 5.8).

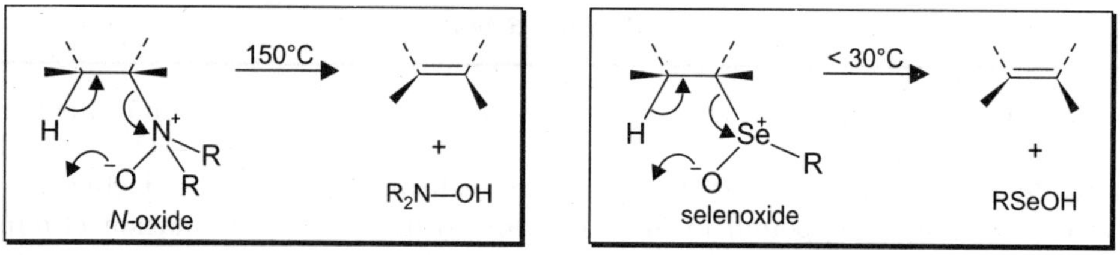

SCHEME 5.46

As already discussed in cyclohexane systems an axial equatorial alignment of the eliminating groups is required. Thus, (I, scheme 5.47) gives the exclusive formation of (II) since only one axial H (circled) is available for elimination. In the case of (III) two axial protons are available for elimination, however the major product is formed by the exclusive loss of one of these protons to give a double bond in conjugation with the aromatic ring.

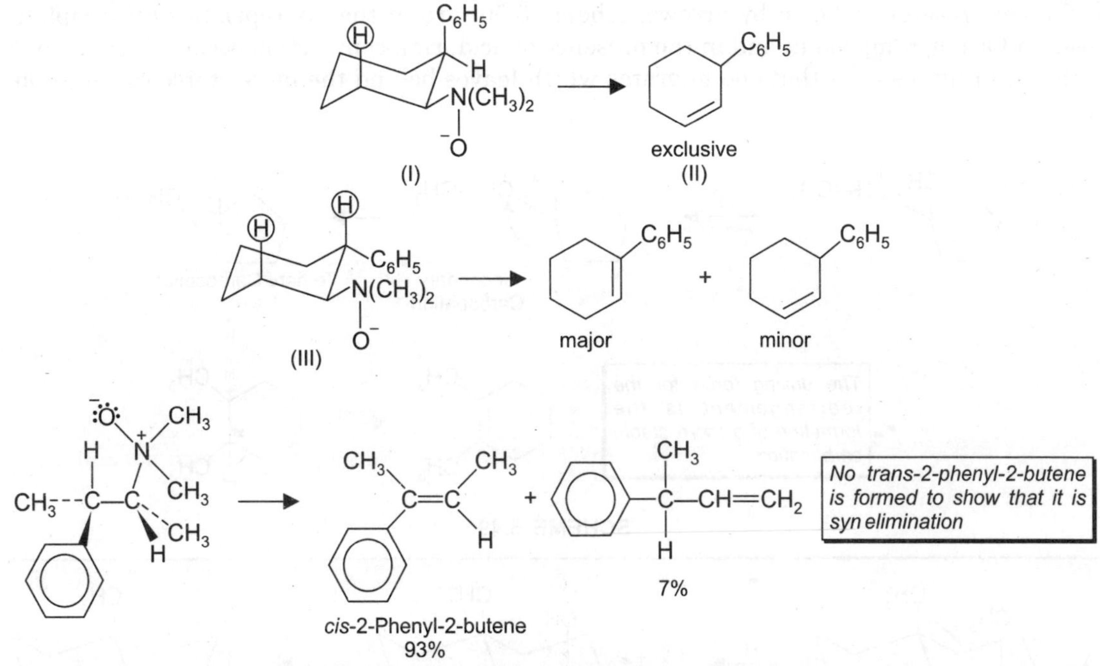

SCHEME 5.47

Moreover, as expected, less hindered elimination would be expected to give the major alkene (scheme 5.48).

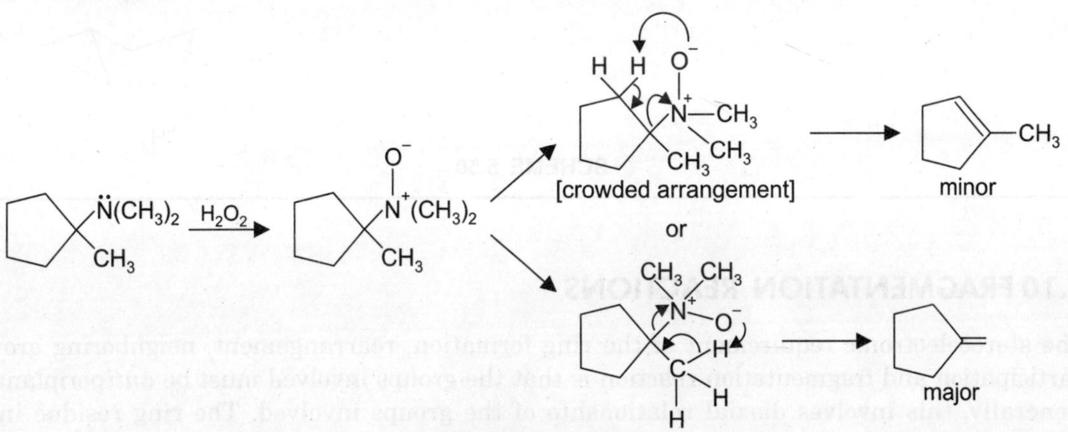

SCHEME 5.48

5.9 MOLECULAR REARRANGEMENT DURING ELIMINATION

During the dehydration of *e.g.*, *sec*-alcohols (formation of carbocations) a rearrangement may take place. In (scheme 5.49) a rearrangement of carbocation leads to change in ring size.

When the achievement of the *anti* conformation in cyclohexane systems is not possible, molecular rearrangement may precede elimination. Thus, in the *trans* decalin derivative, the hydroxyl group is held in the equatorial position because the ring system cannot flip, but two of the ring residues (shown by arrows, scheme 5.50) are in the appropriate *anti*-periplanar position for rearrangement, and in the presence of acid, ring-contraction occurs. Interestingly, of the two ring residues that one migrates which leaves behind the more stable carbocation.

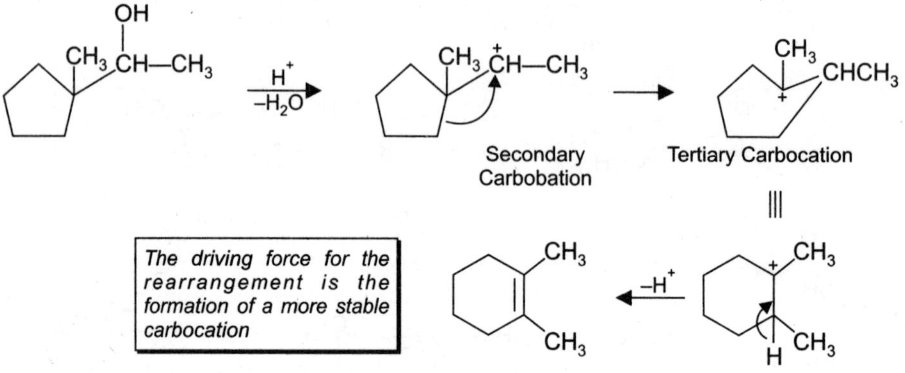

SCHEME 5.49

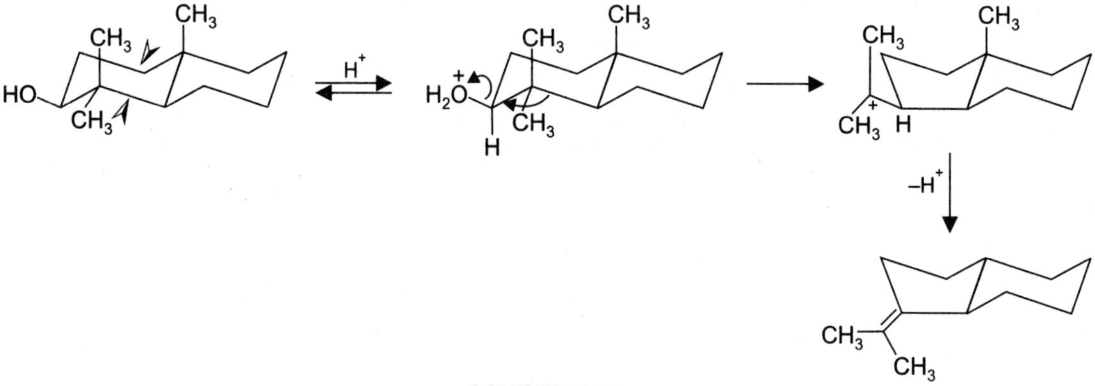

SCHEME 5.50

5.10 FRAGMENTATION REACTIONS

The stereoelectronic requirement of the ring formation, rearrangement, neighboring group participation and fragmentation reaction is that the groups involved must be *anti*periplanar. Generally, this involves diaxial relationship of the groups involved. The ring residue in a cyclohexane also has an antiperiplanar, relationship to an equatorial leaving group. This situation obtains in several fragmentation reactions (scheme 5.51). Thus one may note that eliminations are not limited to the formation of a single bond in a reaction, stereospecific multiple eliminations (fragmentations) may take place if coplanar bonds are suitably aligned with leaving groups.

1, 3-Disubstituted cyclohexanes which have a hydroxyl substituent and a good leaving group like a tosylate (toluene-*p*- sulfonate) undergo a base-induced fragmentation provided that the leaving group can attain an equatorial position. The process allows the formation of medium sized cyclic alkenes.

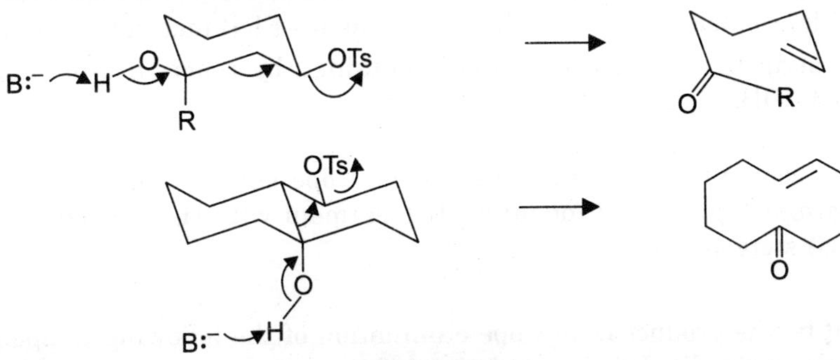

SCHEME 5.51

PROBLEMS

5.1. Explain the results in the following E2 reactions:

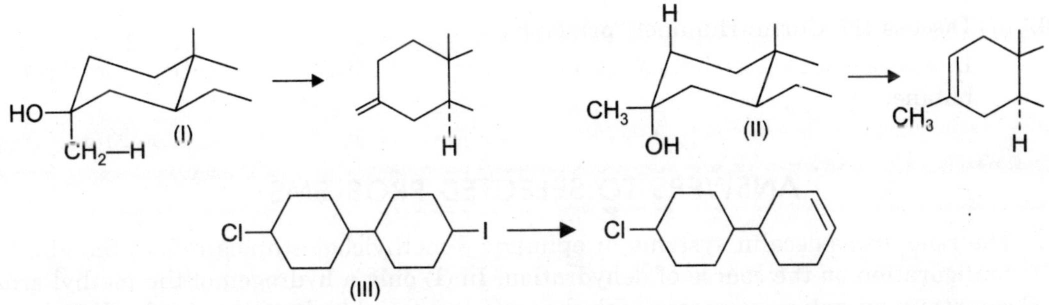

5.2. *Anti*-elimination occurs readily for *cis*-but nor for *trans*-1-bromo-4-(1, 1, dimethylethyl) cyclohexane. Suggest a reason for this difference.

5.3. (*a*) In the rigid *trans*-declain system (I, shown partly) of Δ8-lanosterol, the C-3 hydroxyl is equatorial, the compound, however, undergoes an easy E2 elimination. Explain. (*b*) the norbornane *i.e.* [2.2.1] heptane system of compound (II) undergoes an E2 elimination giving only the product with methylenic double bond. Explain. (*c*) The isomer (III) of 1, 2, 3, 4, 5, 6-hexachlorocyclohexane undergoes E2 elimination 7000 times more slowly than any of its stereoisomers. Explain.

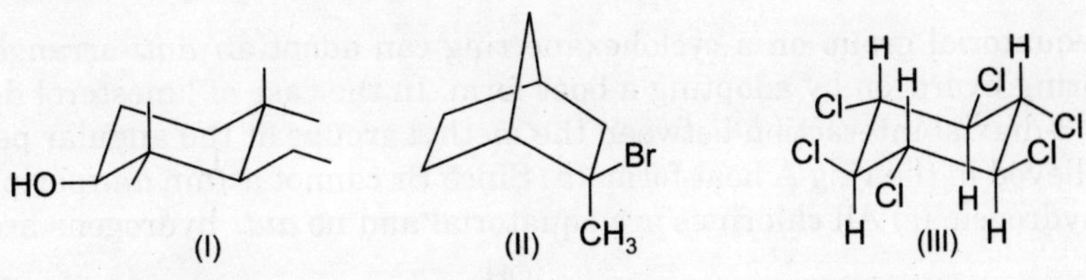

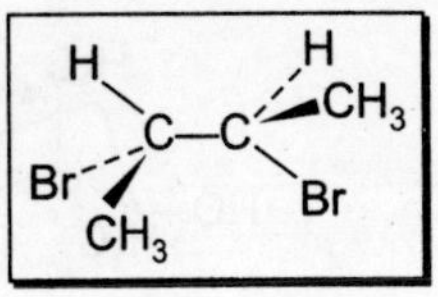

5.4. Write the configuration of the alkene formed by the elimination of one molar equivalent of HBr from the compound.

5.5. Write the mechanism of ElcB reaction. What are the structural features in the substrate which are essential for the operation of this mechanism ?

5.6. What is aldol condensation ? Which step resembles ElcB mechanism in this reaction ?

5.7. Draw the structure of all the products obtained by reaction 2-bromopentane with $NaOCH_3/CH_3OH$.

5.8. Predict the configuration of alkane from the Cope elimination of 2-dimethyl-amino 3-phenylbutane (after its treatment with H_2O_2) from the (2*R*, 3*S*) stereoisomer.

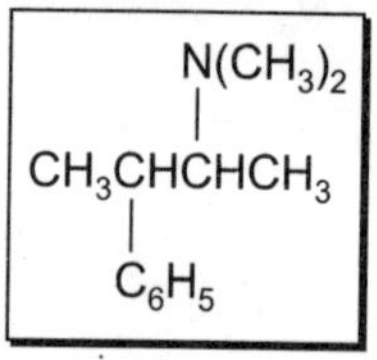

5.9. Which will be the product from Cope elimination of the following compound ? Explain the stereochemistry by drawing it chair conformation.

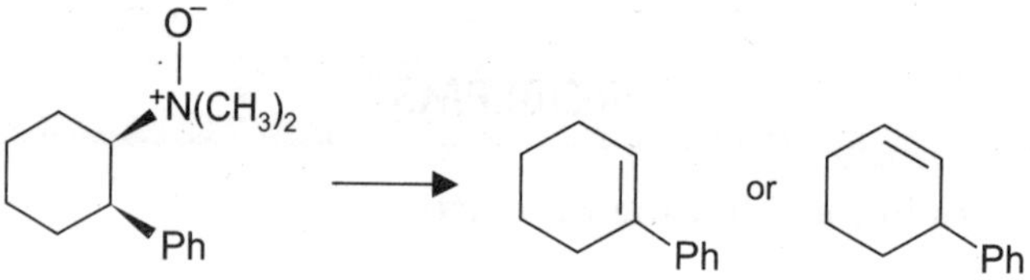

5.10. (*a*) Discuss the Curtin-Hammett principle.

(*b*) Discuss the regioselectivity and stereoselectivity during E2 elimination from 2-bromo-butane.

ANSWERS TO SELECTED PROBLEMS

5.1. The rigid *trans*-decalin systems in epimeric 3-methylcholestanols reflect the effect of configuration on the course of dehydration. In (I) only a hydrogen of the methyl group can attain an *anti* relationship with the ring equatorial hydroxyl group. In (III) since I^- is a better leaving group, selective elimination of HI is allowed.

5.2. The ring is anchored by the bulky *t*-butyl group which occupies equatorial position. Of the two stereoisomers only in the *cis*-isomer *anti*-hydrogens are available.

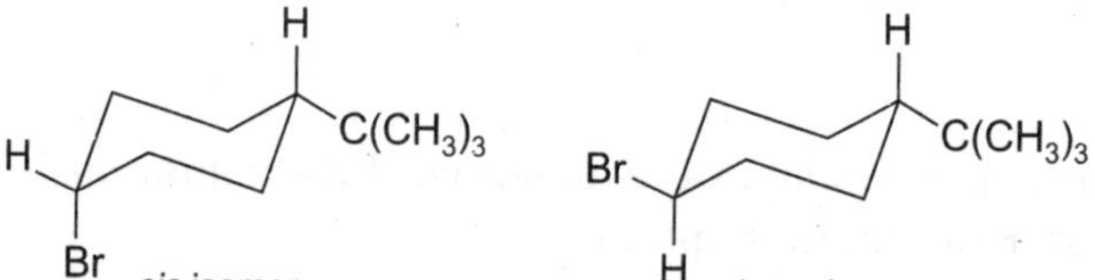

cis isomer *trans* isomer

5.3. (*a*) An equatorial group on a cyclohexane ring can adopt an *anti*-arrangement with a neighboring hydrogen by adopting a boat form. In the case of lanosterol derivative, the serve 1, 3-diaxial interaction between the methyl groups at the angular position and at C4 is relieved in the ring A boat form. (*b*) Since Br cannot attain *anti*periplanarity with a ring hydrogen. (*c*) All chlorines are equatorial and no *anti* hydrogens are available.

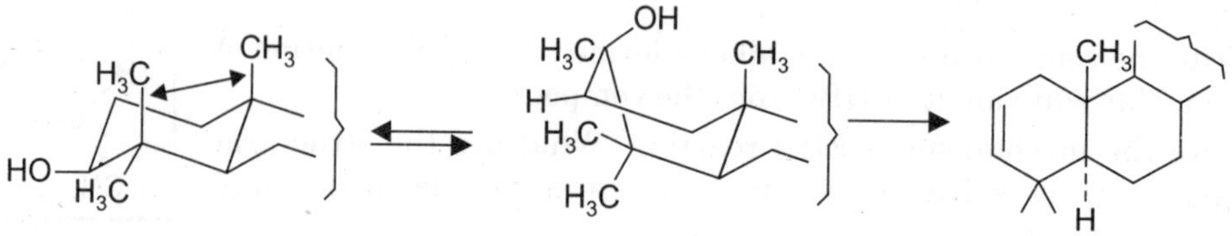

5.4. The alkene formed is (*E*)-2-bromo-2-butene.

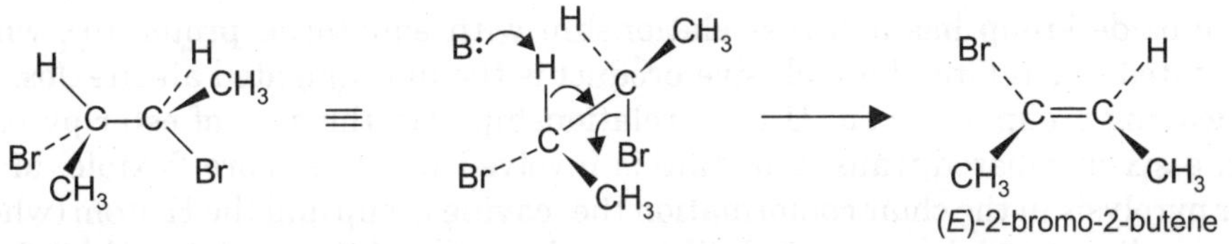

5.7. With the secondary substrate with little steric hindrance at C2, S_N2 pathway is operative. Elimination (E2) gives three alkenes by the loss of a proton from C3 to give 2-butene and the loss of a proton from C1 gives 1-butene. The choice between the geometric isomers of 2-butene depends as to which prochiral proton is removed.

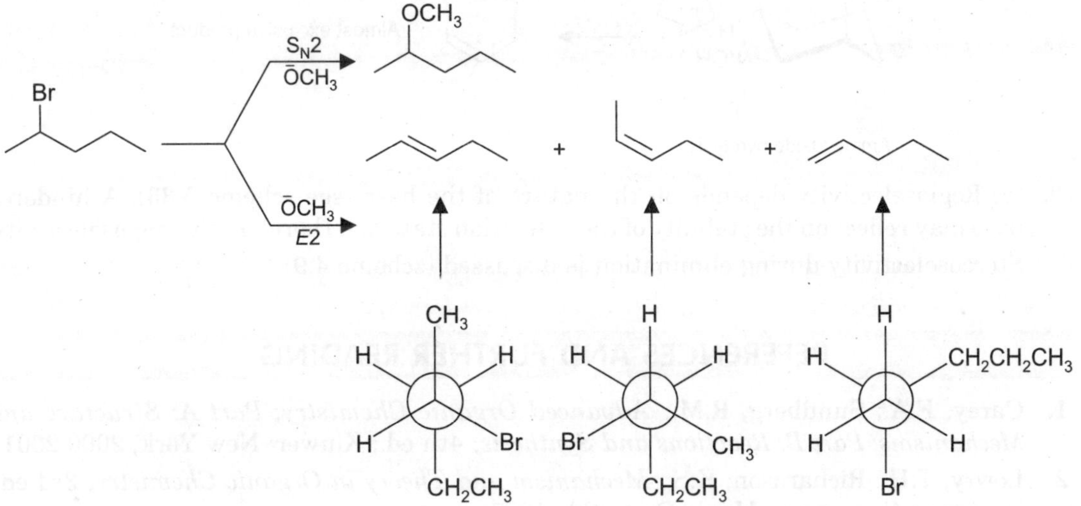

5.8. In this case the compound has two stereocenters (*erythro* enantiomeric pair and *threo* enantiomeric pair *i.e.*, in all four stereoisomers). The stereoisomer (2*R*, 3*S*) is drawn as a solid and hatched wedge stereodrawing (flying wedge representation—a sort of "zig-zag") depicting a *syn* relationship of the dimethylamino group (which will eliminate as its oxide after H_2O_2 treatment) and the β-hydrogen.

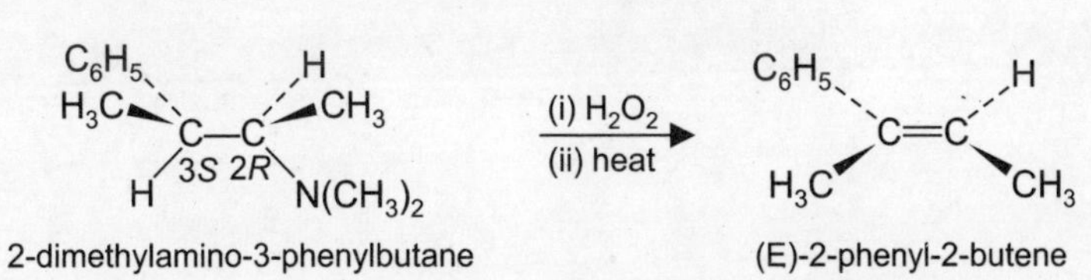

This stereoisomer on Cope elimination gives the *E*-alkene [its enantiomer (2*S*, 3*R*) stereoisomer will also give the same *E*-alkene. The diastereomer of (2*R*, 3*S*) is (2*S*, 3*S*) *i.e.* different stereochemistry on only one of the stereocenters, this third stereoisomer or its enantiomer (2*R*, 3*R*) will give the same *Z*-alkene, whether reacted separated or as a racemate].

5.9. Cope reaction is a pyrolytic *syn*-elimination which involves a concerted five membered cyclic transition state (the, hydrogen atom and the leaving group depart from the same side of the incipient double bond). For this, five membered transition state the five atoms making up the ring must be coplanar (Thus an equatorial group must eliminate an axial H) and therefore, deviations from planarity are not found here. The equatorial amine oxide group has a *trans*-relationship with equatorial proton (*b*), which is not eliminated. Therefore, the exclusive product is the unconjugated alkene (loss of equatorial leaving group and axial H a *cis*-relationship). (In the case of ester pyrolysis, however, a six membered transition state is involved which is more flexible. Moreover, in ester pyrolysis in the chair conformation the leaving group and the H atom (when present as *trans*-diequatorial groups) indeed come close enough (*syn*-relationship) for the reaction to occur. Thus a gauche interaction is sufficient in ester pyrolysis for the reaction to occur.

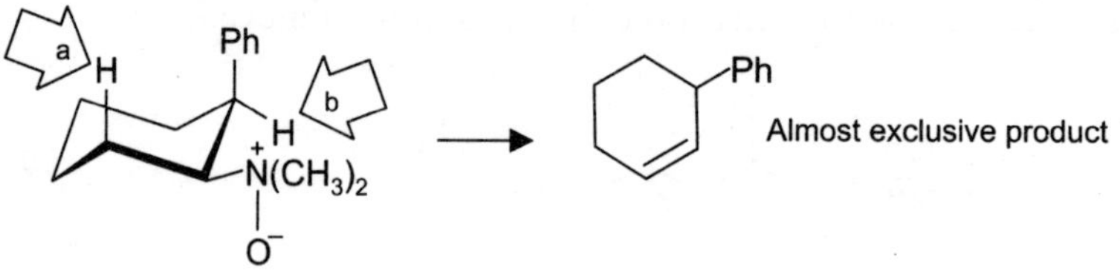

Amine Oxide pyrolysis

5.10. (*b*) Regioselectivity depends on the nature of the base (see scheme 5.33). A hindered base may reflect on the stability of the transition state and therefore the regioselectivity. Stereoselectivity during elimination is discussed (scheme 4.9).

REFERENCES AND FURTHER READING

1. Carey, F.A.; Sundberg, R.M.; *Advanced Organic Chemistry; Part A: Structure and Mechanisms; Part B: Reactions and Synthesis;* 4th ed.; Kluwer: New York, 2000/2001.
2. Lowry, T.H.; Richardson, K.S.; *Mechanism and Theory in Organic Chemistry;* 3rd ed.; Benjamin-Cummings: Menlo Park, CA, 1997.

CHAPTER 6

Stereochemistry and Mechanism of Some Additions to Carbon–Carbon Multiple Bonds

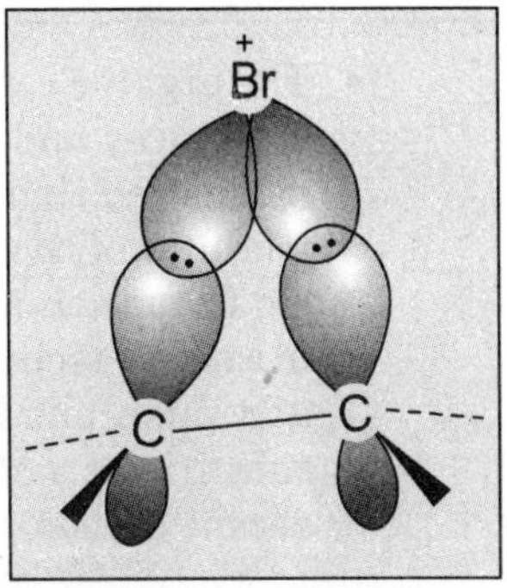

Addition to a Carbon–Carbon double or triple bond can take place in four different ways :

- By the initial attack of an electrophile (*e.g.*, bromination of a double bond).
- By the initial attack of a nucleophile (*e.g.*, addition of a polar reagent to an α, β-unsaturated carbonyl compound).
- By the initial attack of a free radical.
- In the fourth type the mechanism involves the attack at the two carbons of the double or triple bond simultaneously (*e.g.*, epoxidation of alkenes).

6.1 ADDITION REACTIONS OF ALKENES INVOLVING ELECTROPHILES

In an alkene the electron density of a *pi* bond is maximum below and above the plane of the double bond. The most favourable way for the attack of the electrophile based on stereoelectronic considerations is along these electron rich regions *i.e.*, perpendicular to the plane of the double bond. The attacking total reagent might be divided into an electrophilic and a nucleophilic portion (E-Nu). Each part of this reagent may add to the double bond of the reactant molecule from the same side (stereospecific *syn* addition) or from opposite side (stereospecific *anti* addition), or the reaction may be nonstereospecific.

(A) Addition of Bromine to an Alkene-Stereochemistry of Bromine Addition (*Anti* Addition)

Bromination of an alkene was discussed in some detail during the study of stereoselective and stereospecific reactions (See, schemes 1.100 and 1.105). Here a review is given and the following points may be noted:

- This mechanism involves the addition of electrophilic reagents in a two step process. An organic cation is formed in the first step which reacts with the nucleophile in the second. The organic cation may have a bridged structure or a carbocationic nature which may be viewed as in (scheme 6.1).

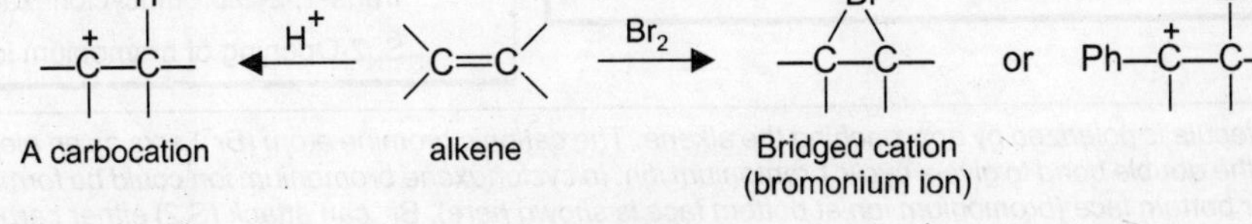

SCHEME 6.1

- The organic cations are not simple cations but have bridged or otherwise complexed structures and the best understood of these is the bromonium ion which is a three-membered ring and is formed by donation of bromine lone-pair electrons to the vacant *p* orbital of the neighbouring carbocation. Although (scheme 6.2) depicts bromonium ion formation as stepwise, this is done only for clarity. The bromonium ion is formed in a single step by interaction of the alkene with Br^+. Alkene π electrons attack bromine and push out bromide ion to give a bromo carbocation. The neighboring bromo substituent stabilizes the positive charge by using two of its electrons to overlap the vacant carbon *p* orbital, to give a three-membered-ring bromonium ion.

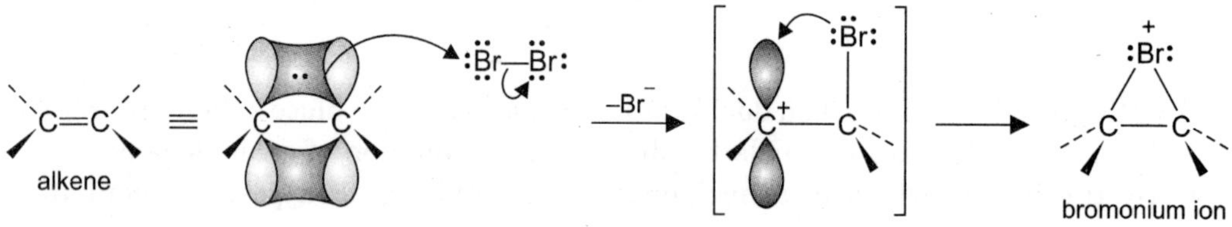

SCHEME 6.2

- The bromonium ion is often represented as a hybrid (scheme 6.3) in which the structure (A), a tertiary carbocation is more important contributor to the hybrid compared to (B) a secondary carbocation.

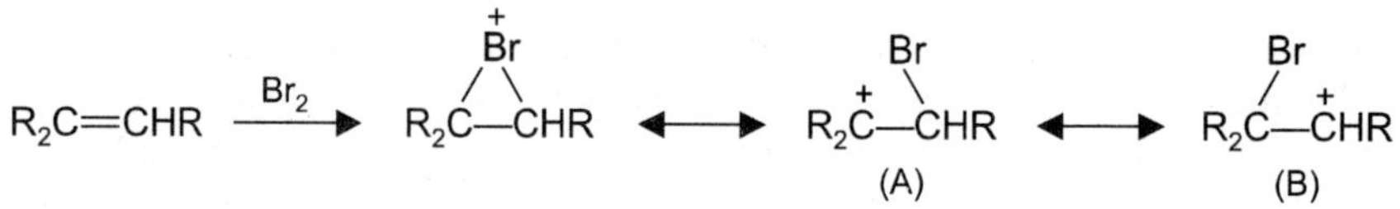

SCHEME 6.3

- When an alkene *e.g.*, cyclohexene reacts with bromine, *anti* addition occurs. Bromonium ion is formed in the first step (scheme 6.4) which is then attacked by the bromide ion from the opposite side of the bromonium ion in an S_N2 type reaction to result in *anti* addition to give a *trans* product.

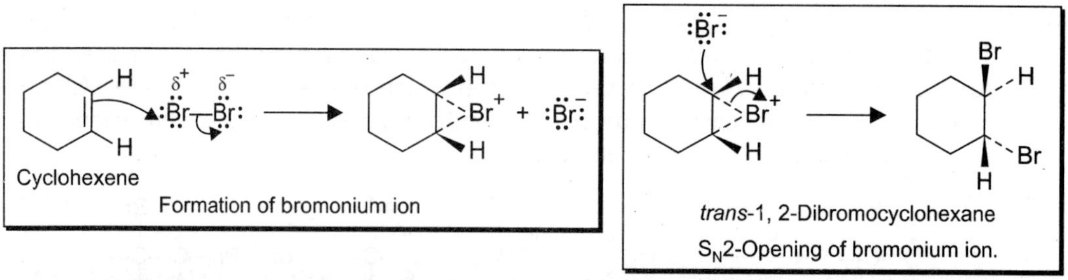

Bromine molecule is polarized by approaching the alkene. The cationic bromine atom (Br^+) acts as an electrophile and adds to the double bond to give a cyclic bromonium ion. In cyclohexene bromonium ion could be formed either at top face or bottom face (bromonium ion at bottom face is shown here). Br^- can attack (S_N2) either carbon of the bromonium ion. Attack at one carbon gives trans-1,2-dibromocyclohexane (one enantiomer). The attack at other carbon will give its enantiomer. (not shown here).

SCHEME 6.4

- Addition of bromine to cyclohexene is a stereoselective reaction (*cis*-1,2-dibromocyclohexene is not formed).
- The nucleophilic attack by the bromide ion (S_N2) reaction on the bromonium ion causes an inversion of configuration at the carbon being attacked. This inversion of configuration at one carbon of the ring gives one enantiomer of *trans* 1, 2-dibromocyclohexane. The other enantiomer results from the equally probable attack of the bromide ion at the other carbon of bromonium ion.

The trans product formed from cyclohexene is drawn as a single enantiomer. Opening of the bromonium ion must take place in two equivalent ways to produce a racemic mixture i.e. both the enantiomers are formed in equal amounts. In such cases, however, it is the practice not to draw both the enantiomers, unless needed for clarity. Thus for clarity the anti bromination of cyclohexene may be shown to give a racemic product. (Scheme 6.5).

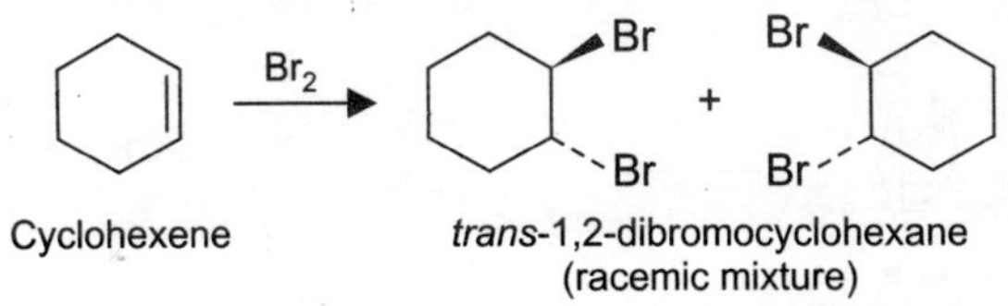

anti-Bromination of cyclohexene shown to give trans-1,2-dibromocyclohexane as a racemic mixture (The hydrogens are omitted for clarity).

SCHEME 6.5

- If the addition of bromine to cyclohexene had involved a nucleophilic attack of the double bond by bromine with the resulting formation of a planar carbocation as an intermediate, some *cis*-1, 2-dibromo compound would be expected as a product because in the second step the attack of bromide ion can take place from either side of the planar carbocation (scheme 6.5*a*).

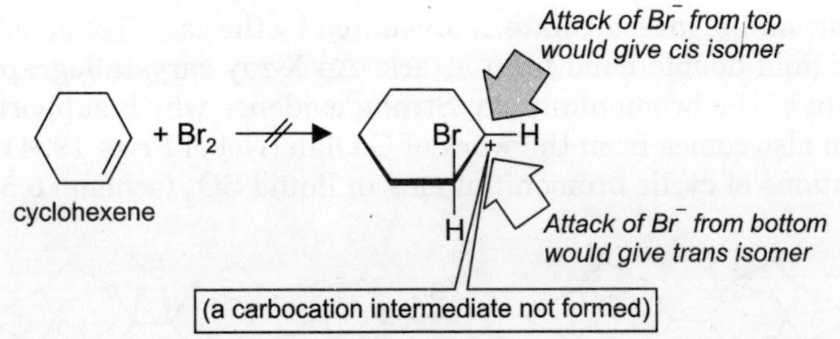

SCHEME 6.5a

- During the addition of chlorine, the intermediate chloronium ion is believed to be less strongly bridged *i.e.*, the cyclic structure (see, scheme 6.3) is of considerably less important in the hybrid.

Strong bridging ensures that the anti product is likely to be the major if not exclusive. The bridged species formed by the attack of sulfur and selenium electrophiles have strong bridging features (scheme 6.5b). Thus Se atom in CH_3SeCl (methylselenyl chloride) is electrophilic and has lone pairs with $CH_3Se^{\delta+}$ as the electrophile similarly in CH_3SCl, $CH_3S^{\delta+}$ is the electrophile. A strong bridged ion behave like an epoxide in an S_N2 reaction and the strength of bridged species *therefore, determines both the regioselectivity and stereoselectivity of the addition reaction as seen (scheme 6.5b). Thus the nucleophile attacks preferentially the less substituted carbon atom from the opposite side of the bridge. With these strongly bridged species, even if the approach of the nucleophile from the opposite side is hindered, anti addition is still the preferred pathway as seen in addition to norbornene (scheme 6.5b).*

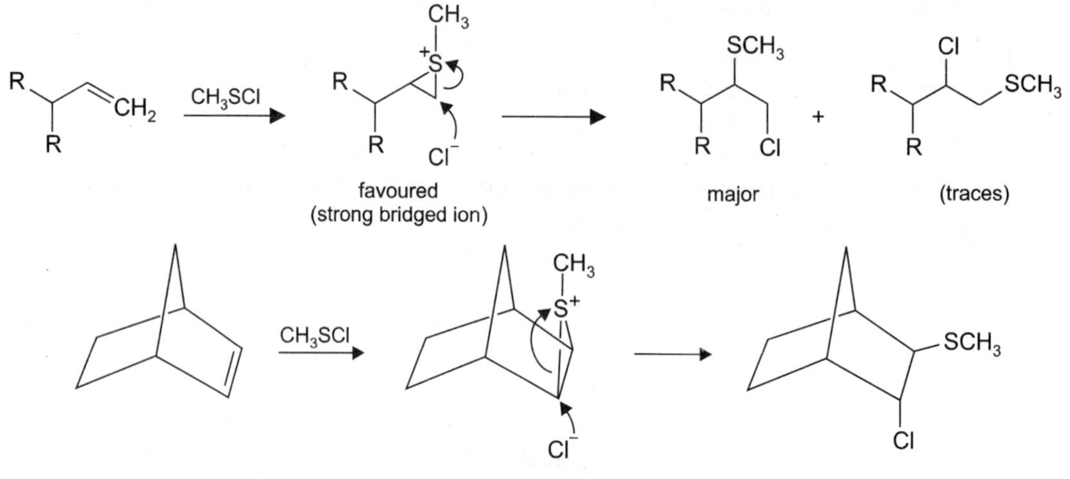

SCHEME 6.5b

(B) Evidence for the Existance of Bridged Bromonium Ion

One case where a bromonium ion is stable to further addition of nucleophiles is (II, scheme 6.5c). This bromonium ion is formed by the addition of bromine to hindered alkene (I, scheme 6.5c). The bromonium ion (II) formed initially does not go to second stage to give a dibromo compound. Attack by Br^- on the intermediate is prevented by the cage like structure which shields each end of the original double bond from attack. An X-ray carystallographic study has confirmed the structure of the bromonium ion. Strong evidence which supports the involvement of a bromonium ion also comes from the work of G.Olah (Nobel Prize 1994) who prepared and studied stable solutions of cyclic bromonium ions in liquid SO_2 (scheme 6.5c).

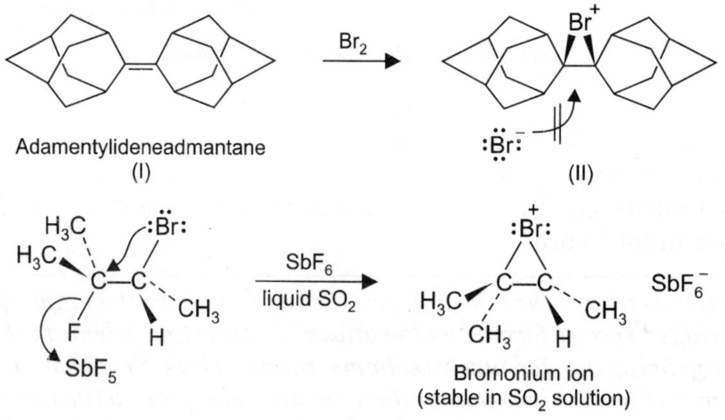

SCHEME 6.5c

- A symmetric alkene like cyclohexene can be approached by the electrophile either from top face (as in scheme 1.105) or from bottom face (as in scheme 6.4). However, the presence of substituents may hinder the approach of the electrophile from the side on which these are present. Thus (I, scheme 6.6) will form the bromonium ion on the side opposite to the methyl group. In conformationally rigid systems the *trans* product formed is diaxial rather than diequatorial as shown in rigid-decalin structure (I, scheme 6.6). In case ring flipping is possible, the diaxial compound formed initially converts into the diequatorial compound. This is so in cyclohexene (scheme 6.6). The second example is of cholesterol a naturally occuring compound, whose top face (β-face) is hindered by the presence of two bulky angular methyl groups and the side chain at C17 on that face. Thus cholesterol exclusively forms a bromonium ion from the bottom (α-face), which is subsequently attacked by Br^- at C6 (S_N2 conditions) to form 5α, 6β-diaxial dibromide (scheme 6.6*a*). The 5α, 6β-diaxial bromide undergoes a change when kept in chloroform solution to more stable diequatorial isomer. This is an example of a slow stereoisomerization and not a conformational change.

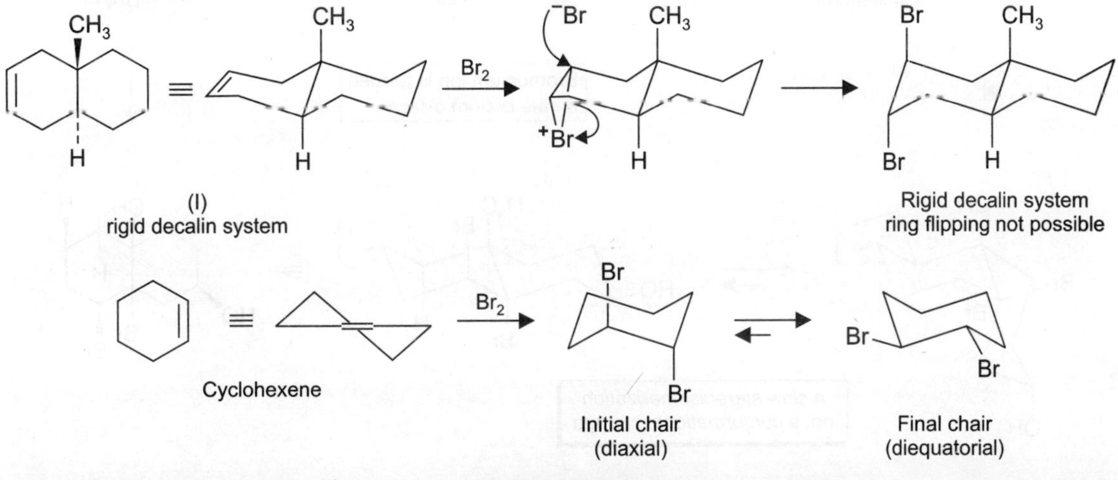

SCHEME 6.6

EXERCISE 6.1

Write the conformation of the product of addition of bromine to 4-tert-butylcyclohexene.

ANSWER. *tert-Butyl group holds the molecule overwhelmingly in a conformation in which the tert-butyl group is equatorial. The favoured conformation of the product thus holds the bromines in the axial positions (scheme 6.6a).*

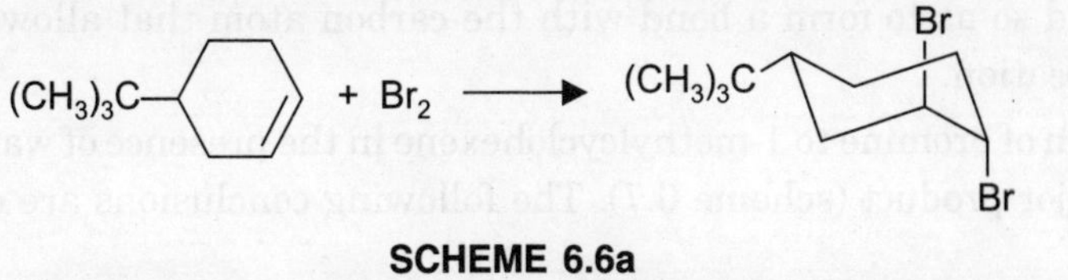

SCHEME 6.6a

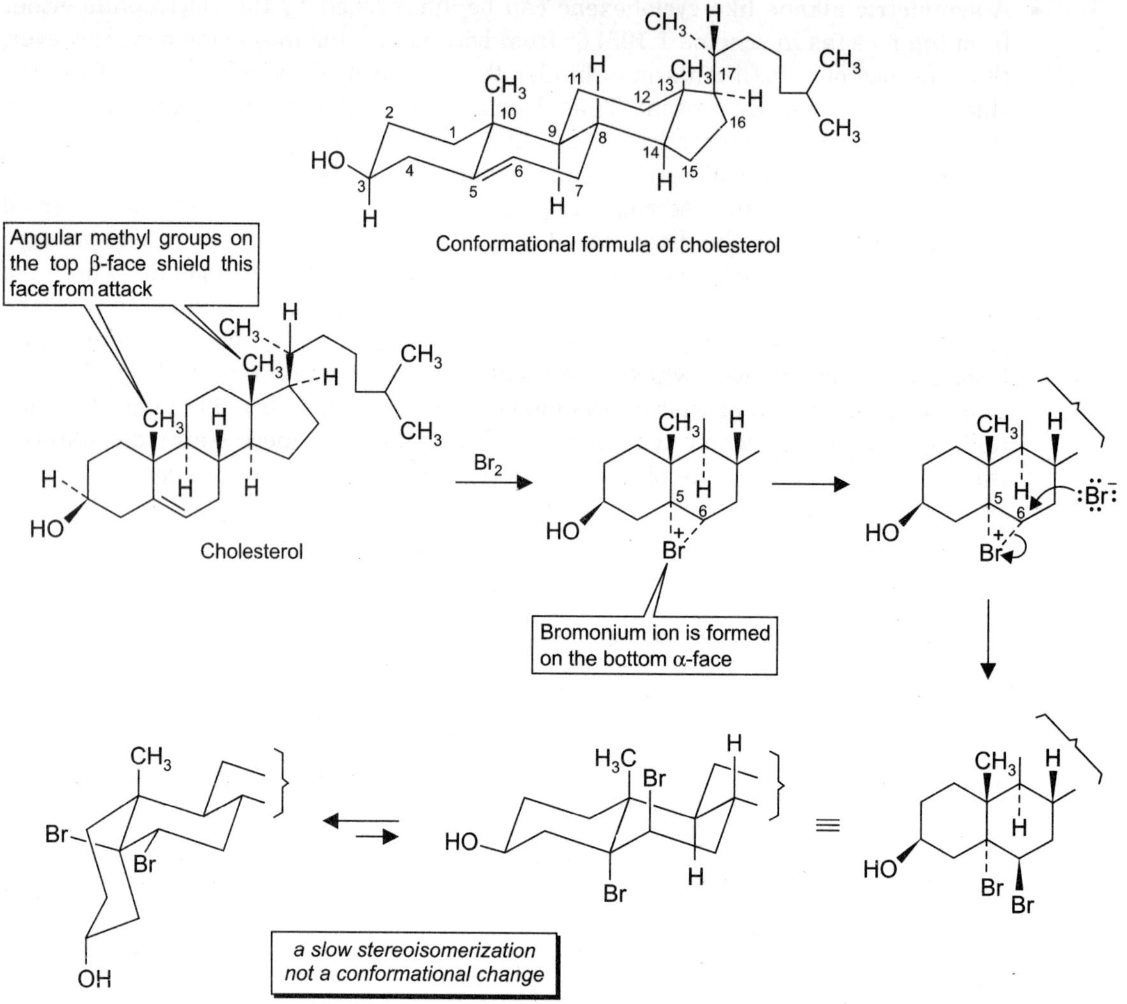

SCHEME 6.6b

(C) Addition of Bromine in the Presence of Water—Halohydrin Formation—Markovnikovs Regiochemistry

When a nucleophile reacts with a bromonium ion, the addition takes place with Markovnikovs regiochemistry. In the formation of bromohydrins, bromine adds to the least substituted carbon and the hydroxyl group (from nucleophilic attack by water) bonds to the more substituted carbon (*i.e.*, the carbon which can accommodate more of the positive charge in the bromonium ion.) Markovnikovs rule states: In the ionic addition of an unsymmetrical reagent an electrophile adds to the π bond so as to form a bond with the carbon atom that allows the formation of a most stable carbocation.

The addition of bromine to 1-methylcyclohexene in the presence of water gives the tertiary alcohol as the major product (scheme 6.7). The following conclusions are drawn:

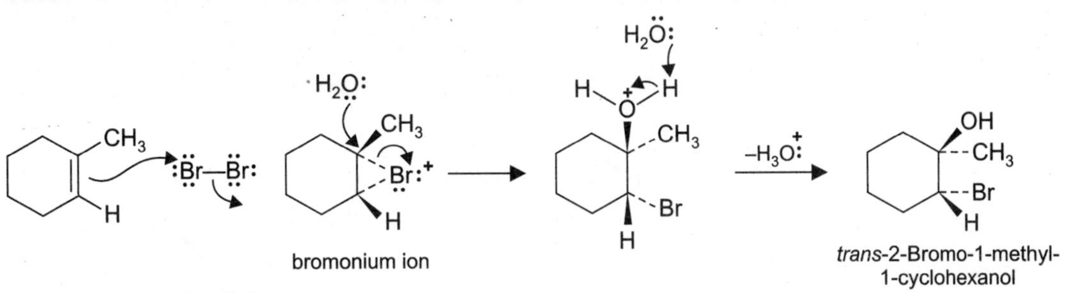

SCHEME 6.7

- The nucleophile (H_2O) attacks the more highly substituted carbon of the ring since this is more positively polarized than the other. The bromonium ion is below the ring and thus H_2O attacks from top of the ring.
- Since, tertiary alcohol is the major product, the reaction is regioselective. Nucleophile (H_2O) attacks at the more substituted carbon of bromonium ion since it has greater δ^+.
- Stereochemically the orientation of OH and Br is *trans* (*anti*-addition), thus the intermediate is a cyclic bromonium ion which is attacked by the nucleophile at the opposite face.
- A racemic mixture is obtained since there is no chiral influence in the reaction.
- In the first step of halohydrin formation, Br^+ is the only electrophile in the reaction mixture. The second step is fast and of the two nucleophiles Br^- and H_2O, H_2O is the solvent and is therefore, available in large concentration. The bromonium ion is thus more likely to collide with a molecule of water (than with Br^-) to give protonated bromohydrin, which immediately loses a proton.
- A strongly bridged bromonium ion shows reactivity like an epoxide in an S_N2 reaction (see schemes 3.56*b* and 3.56*c*).

(D) The addition to acyclic alkenes follows the same mechanistic and stereochemical pathways. The stereospecific bromination of *cis* and *trans*-2-butene has been discussed (see, scheme 1.100). The following points is the quick review.

- Consider the formation of bromonium ion from (*Z*)-but-2-ene, the bromonium ion could be formed above or below the plane of the already existing double bond (scheme 6.8). The bromonium ions (I≡II) represent the same *meso* species.
- The bromonium ion (I) can be attacked by the nucleophile from the opposite side of the bridge [note that in (I) there is mirror plane of symmetry and the two positions of attack path (1) and path (2) are enantiotopic]. Similar outcome is seen if one considers the attack of the nucleophile on (II, scheme 6.8). The bromination of the diastereomeric (*E*)-but-2-ene gives an enantiomeric pair of bromonium ions (III and IV) with C_2 axis. The positions of attack by Br^- are thus related by a C_2 axis, and attack on the either carbon of the bridged intermediate (which are homotopic) gives the same product.

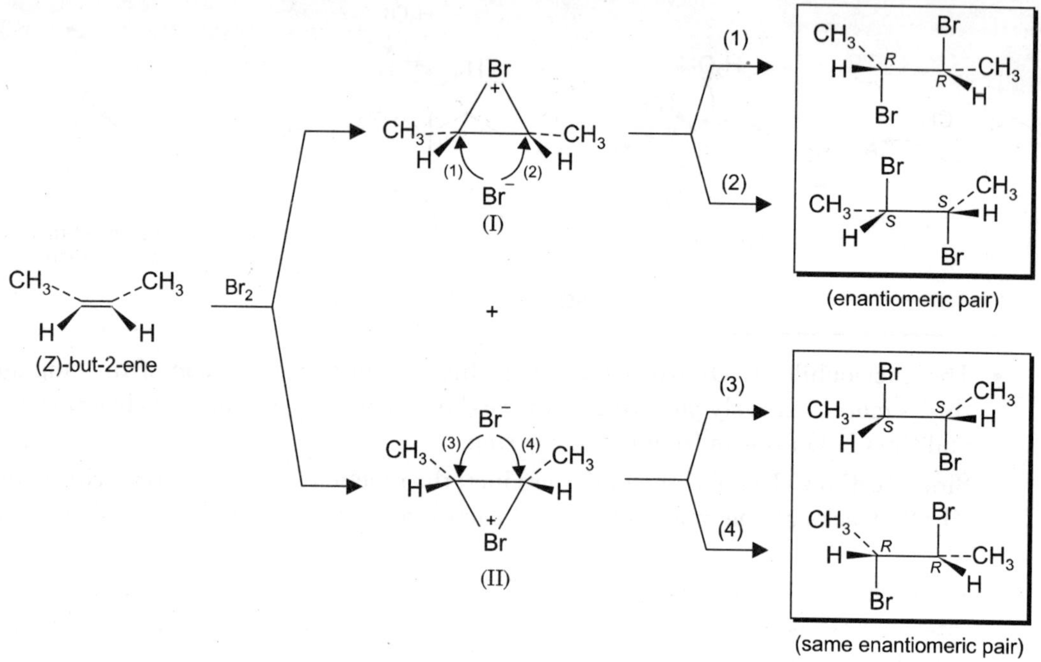

SCHEME 6.8

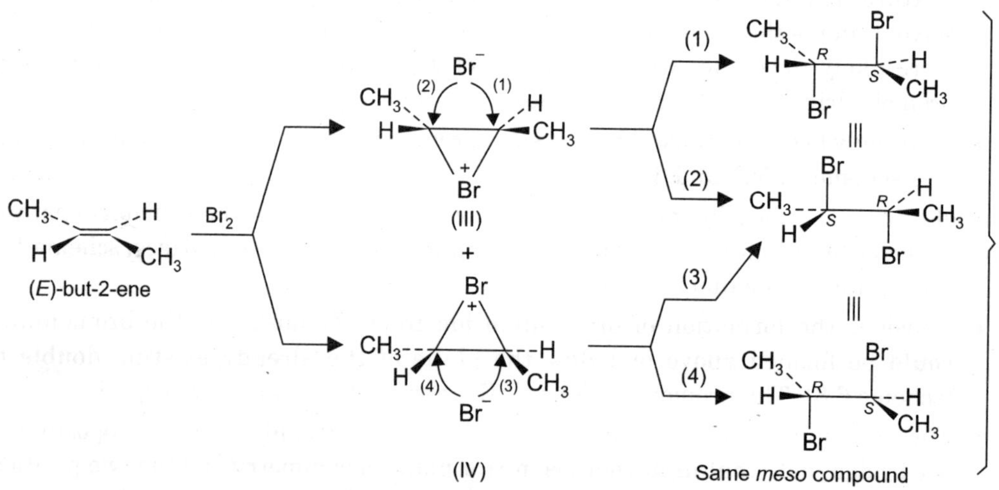

SCHEME 6.8a

EXERCISE 6.2

Give the product of addition of bromine to trans-2-butene both via syn and anti pathway. Which of these pathways is consistant with the accepted mechanism ?

ANSWER.

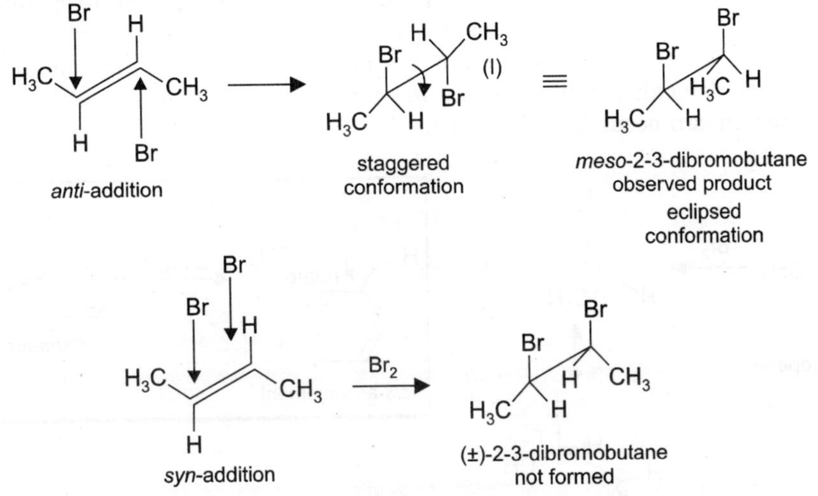

SCHEME 6.8b

(E) Stereochemistry of *Syn*-Addition of Bromine to an Alkene

An addition reaction which proceeds through a bromonium ion intermediate results in *anti*-addition (opening of the bromonium ion in S_N2 fashion) to give *trans*-products. However, a reaction which proceeds through a carbocation intermediate should result in a mixture of *syn*- and *anti*-addition since the planar carbocation can be attacked from either side by the incoming nucleophile. In these situations the alkene contains atleast one aryl group attached to the double bond. The stabilization of the positive charge in the intermediate (after electrophilic attack) is by delocalization on to the aromatic ring. As a result the effectiveness of bridging (formation of the bromonium ion) is reduced.

Acenaphthylene (scheme 6.9) on reaction with bromine gives large amounts of the usual *trans* dibromide (*anti* addition) alongwith a large amount of *cis* dibromide (*syn* addition). This is due to the formation of a resonance stabilized open carbocation which competes favourably with the formation of a bromonium ion, S_N2 reaction on cyclic bromonium ion will give only the *trans* product, while the open cation will give both *cis* and *trans* products (scheme 6.9).

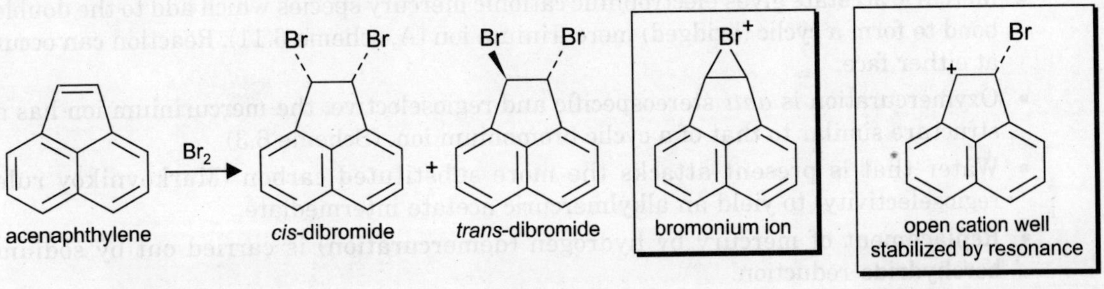

SCHEME 6.9

Similarly 1-phenylpropene $C_6H_5CH = CHCH_3$ (*cis* or *trans* isomer, scheme 6.10) displays a nonstereospecific course (both *syn* and *anti* addition) during the addition of bromine. This may well be compared with the stereospecific addition of bromine to *E*- and *Z*-2 butene isomers (scheme 1.100). Thus in alkenes in which atleast one aryl group is attached to the double bond, the selectivity is less since now the positive charged is instead stabilized in the intermediate by its delocalization on to the aromatic ring. As a result, effective bridging is reduced and rotation can occur (scheme 6.10).

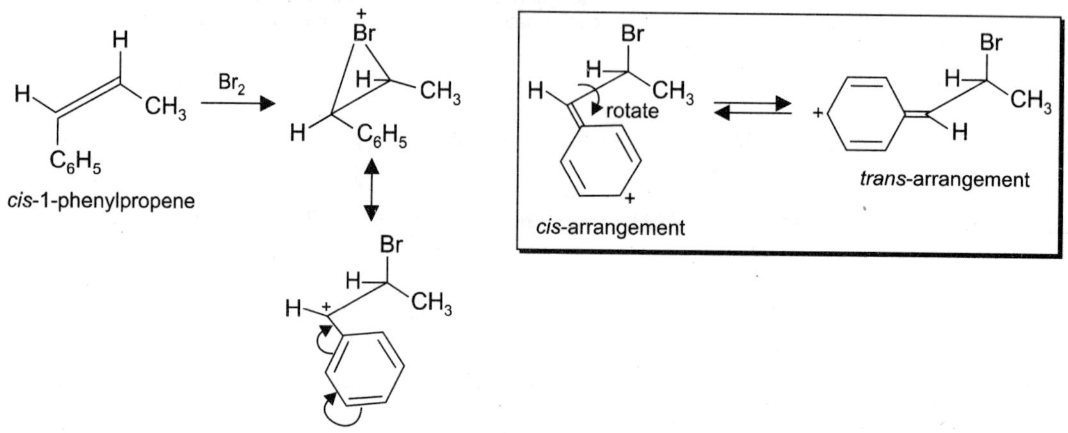

SCHEME 6.10

Additionally, increase in the polarity and ion-solvating ability of the solvent also stabilizes a carbocation than the bromium ion intermediate which as excepted leads to a decrease in *anti* stereoselectivity. This is seen in the addition of bromine to stilbene to give 95% *anti* addition in solvents of low dielectric constants but is only 50% *anti* in a solvent with $\varepsilon = 35$.

However, irrespective of the fact, whether the intermediate is an open carbocation or a cyclic bromonium ion, the mechanism is called Ad_E2 (electrophilic addition bimolecular).

(F) Oxymercuration/Reduction of an Alkene (Markovnikovs Hydration of an Alkene). A Metal Acting as an Electrophile

This is a two step process, an alkene is first treated with mercuric acetate and the cyclic mercurinium ion is then reduced with sodium borohydride. The following points may be noted :

- Mercuric acetate gives electrophilic cationic mercury species which add to the double bond to form a cyclic (bridged) mercurinium ion (A, scheme 6.11). Reaction can occur at either face.
- Oxymercuration is *anti* stereospecific and regioselective, the mercurinium ion has a structure similar to that of a cyclic bromonium ion. (Scheme 6.3)
- Water that is present attacks the more substituted carbon (Markovnikov rule regioselectivity) to yield an alkylmercuric acetate intermediate.
- Replacement of mercury by hydrogen (demercuration) is carried out by sodium borohydride reduction.
- The alcohol obtained on demercuration is similar to the product of Markovnikov hydration of the substrate. It is however, a valuable alternative to acid-catalyzed hydration, since no carbocations are involved and thus no rearrangements are observed.

$$Hg(OAc)_2 \rightleftharpoons {}^{+}Hg(OAc)\ AcO^{-}$$

Mercuric acetate

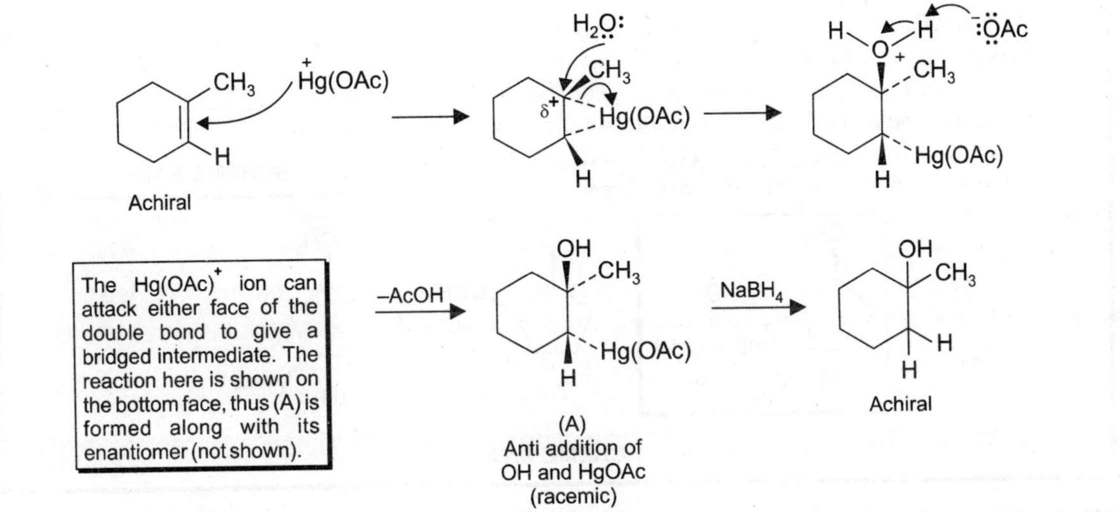

SCHEME 6.11

When mercuration is carried out in an alcohol solvent (solvomercuration) the product contains an alkoxy group (– OR) and the method converts alkenes to ethers *via* the addition of an alcohol across the double bond of the alkene (scheme 6.12). Again the solvent attacks the mercurinium ion at the more highly substituted end of the double bond.

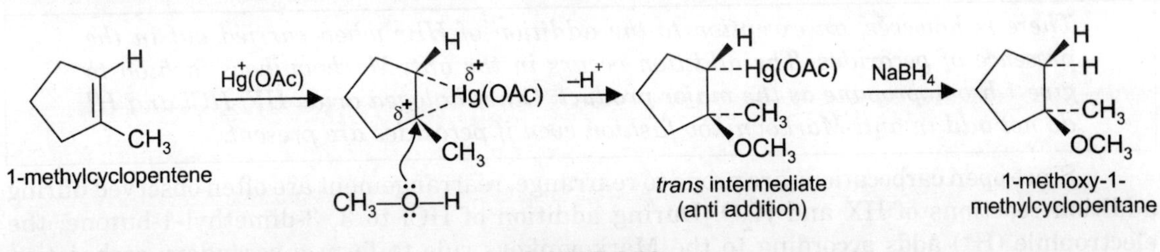

SCHEME 6.12

When a hydroxyl group is part of the reactant and geometrically suitably placed an intramolecular bond formation can take place (scheme 6.12*a*).

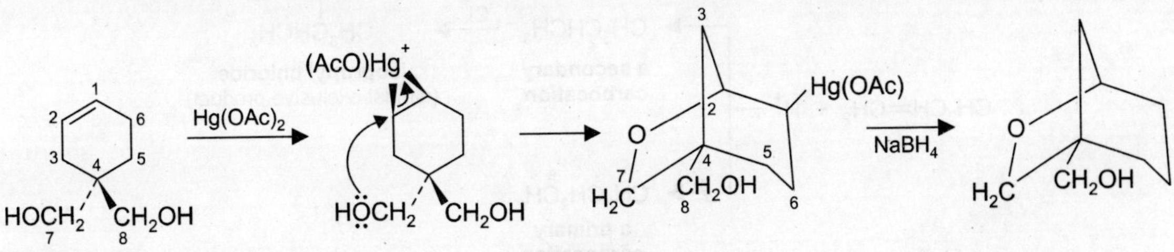

SCHEME 6.12a

EXERCISE 6.3

What products you expect from the reaction (scheme 6.12b).

ANSWER. *In this case the methyl group at top blocks this position, (thus the electrophilic cationic mercury species will add from the bottom face of the molecule to give (I, scheme 6.12c) as the major intermediate. Reduction with $NaBH_4$ in CH_3OH will give (II and III scheme 6.12c) as racemic mixtures.*

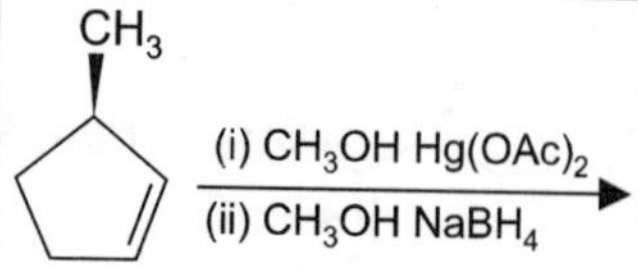

SCHEME 6.12b

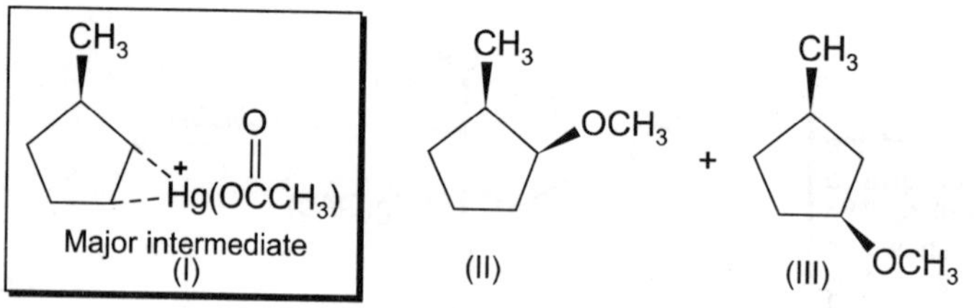

SCHEME 6.12c

(G) Addition of HX-Hydrohalogenation and Hydration of Alkenes Markovnikovs Additions

The halogen acids *e.g.*, HCl add to alkenes (unsymmetrical) via the addition of a proton (electrophile) to give a carbocation. The orientation of addition follows the Markovnikov rule—The electrophile adds so as to form the most stable carbocation where the nucleophile then adds. In the addition of HCl to propene (scheme 6.13), the main product (almost exclusive) is isopropyl chloride (2-chloropropane) and therefore, the reaction is regioselective.

There is however, an exception to the addition of HBr when carried out in the presence of peroxides. The addition occurs in the anti-Markovnikovs fashion to give 1-bromopropane as the major product. Other halogen acids HF, HCl and HI do not add in anti-Markovnikov fashion even if peroxides are present.

Since open carbocations are prone to rearrange, rearrangement are often observed during addition reactions of HX and H_2O. During addition of HCl to 3, 3-dimethyl-1-butene, the electrophile (H^+) adds according to the Markovnikovs rule to form a secondary carbocation (scheme 6.14). A 1, 2-methyl from the neighboring carbon shifts with its pair of electrons to the positively charged carbon to give a more stable tertiary carbocation and the major product of addition is derived from this carbocation.

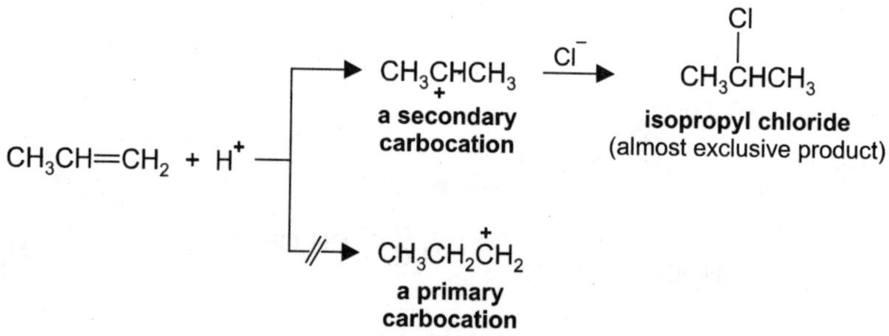

SCHEME 6.13

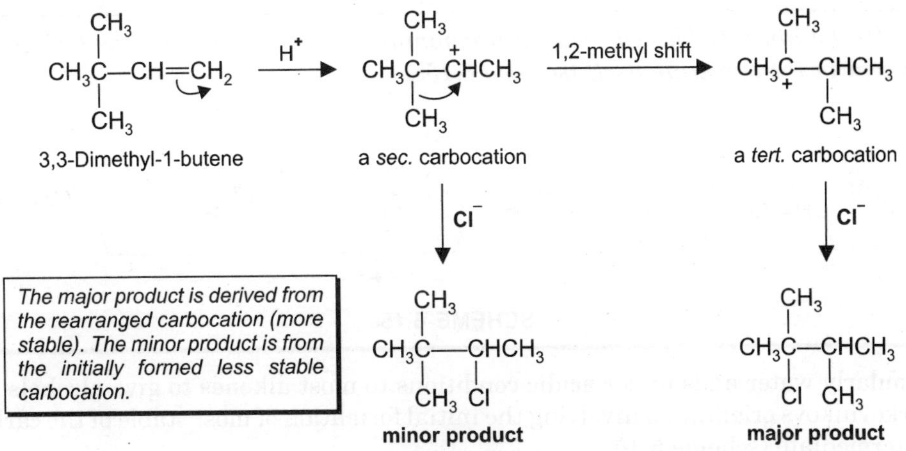

SCHEME 6.14

In the second example the ring residue shifts (scheme 6.15) to form a more stable tertiary carbocation, then the initially formed secondary carbocation. Here the 1, 2-shift has the added advantage because the five membered ring formed during ring expansion is more stable with less angle strain.

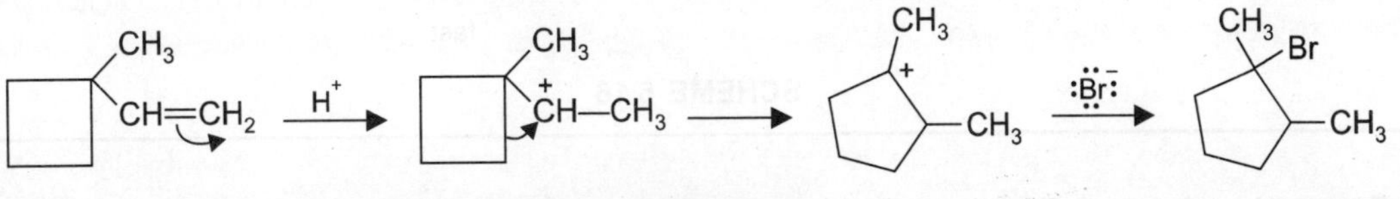

SCHEME 6.15

EXERCISE 6.4

Predict the products from reactions (I and II, scheme 6.15a).

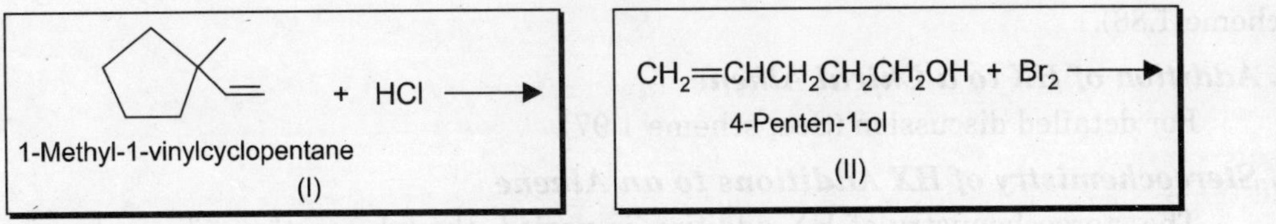

SCHEME 6.15a

ANSWER. *In (I) the initially formed carbocation (secondary) rearranges to give a more stable carbocation (tertiary, scheme 6.15b).*

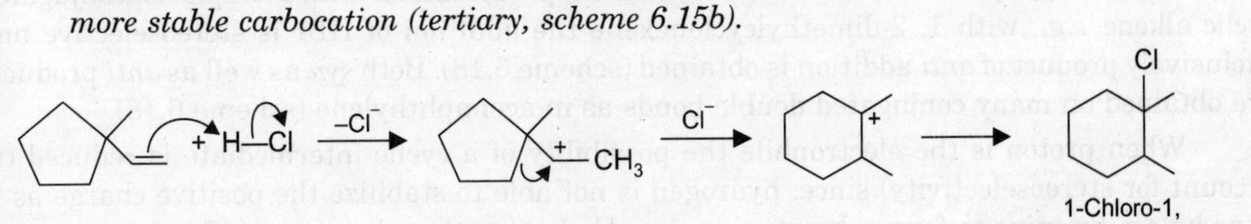

SCHEME 6.15b

In (II) the initially formed bridged bromonium ion is attacked by the OH group present in the substrate itself (scheme 6.15c).

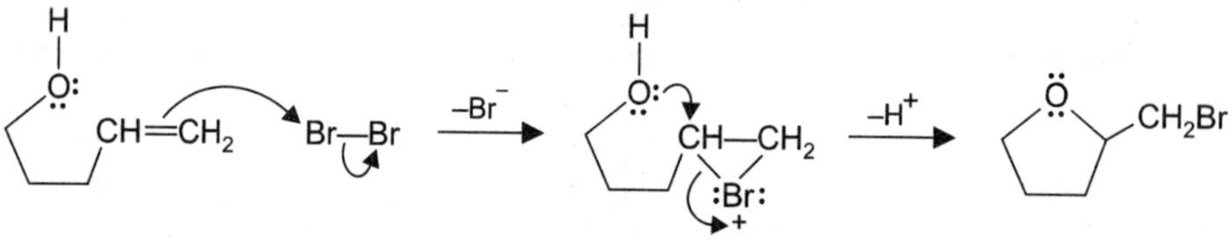

SCHEME 6.15c

Similarly water adds under acidic conditions to most alkenes to give alcohols resulting from Markovnikovs orientation involving the initial formation of most stable of the carbocations as the intermediate (scheme 6.16).

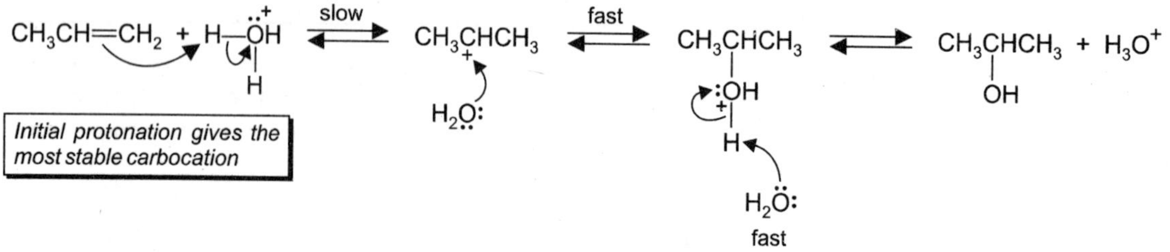

SCHEME 6.16

(H) Stereochemistry of Addition of HX and H_2O to Alkenes

1. *Addition to an Achiral Alkene*

Consider the addition of HBr to 1-butene which produces 2-bromobutane with a stereocenter. The product thus would exist as a pair of enantiomers. The product is formed as a racemate since the achiral trigonal planar carbocation formed after the electrophilic addition by proton is attacked equally from either face (scheme 6.17). Further detailed discussion is given (See, scheme 1.88).

2. *Addition of HX to a Chiral Alkene*

For detailed discussion (See, scheme 1.97).

3. *Stereochemistry of HX Additions to an Alkene*

The stereochemistry of HX addition is varied, though mostly *anti* product is formed [Examples are known of predominant *syn, anti* and non-stereoselective addition as in the case of *e.g.*, 1,2-dimethylcyclohexene]. In fact *anti* addition predominates with a simple nonconjugated cyclic alkene *e.g.*, with 1, 2-dimethylcyclohexene the addition of HBr is stereoselective and exclusively product of *anti* addition is obtained (scheme 6.18). Both *syn* as well as *anti* products are obtained on many conjugated double bonds as in acenaphthylene (scheme 6.18).

When proton is the electrophile the possibility of a cyclic intermediate is reduced (to account for stereoselectivity) since, hydrogen is not able to stabilize the positive charge as is done by *e.g.* bromine to form a bromonium ion. Hydration though regiospecific often proceeds by random *anti* and *syn* stereochemical pathway (scheme 6.18*a*). Thus addition of water to 1, 2-dimethylcyclohexene gives almost equal amount of products of *syn* and *anti* addition.

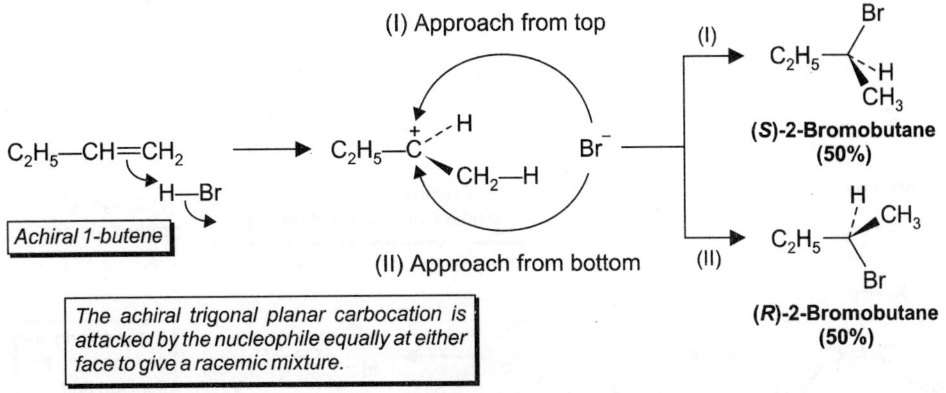

SCHEME 6.17

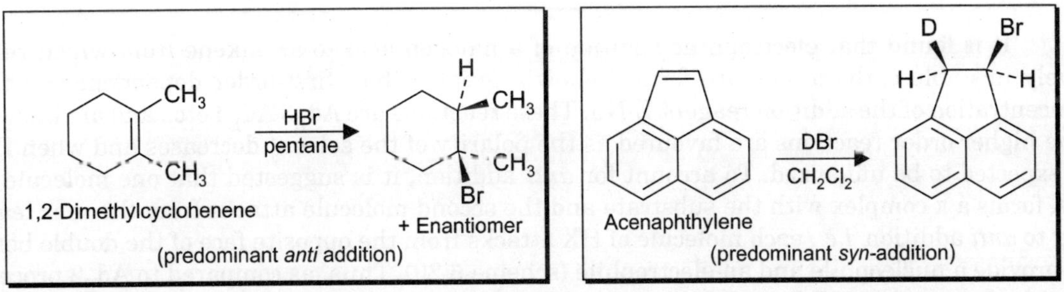

SCHEME 6.18

Consider the addition of DCl (deuterium chloride) to *Z*-1-phenyl-propene (scheme 6.19). This is Ad_E2 addition and results in the formation of *erythro* product in greater amount (*Syn* addition) than the *threo* isomer (*anti* addition). These results are clearly in contradiction to the formation of a free carbocation. In this case one may explain the predominant *syn* addition by invoking the formation of a "tight ion-pair" (scheme 6.19) in which the nucleophile is very closely associated with the carbocation that it attaches itself to the cationic centre before the bond can rotate to give a product of *syn* addition. However, interestingly in other additions which involve carbocation intermediates, *anti* addition takes place to a greater extent and this fact further puts to doubt the true nature of the cation.

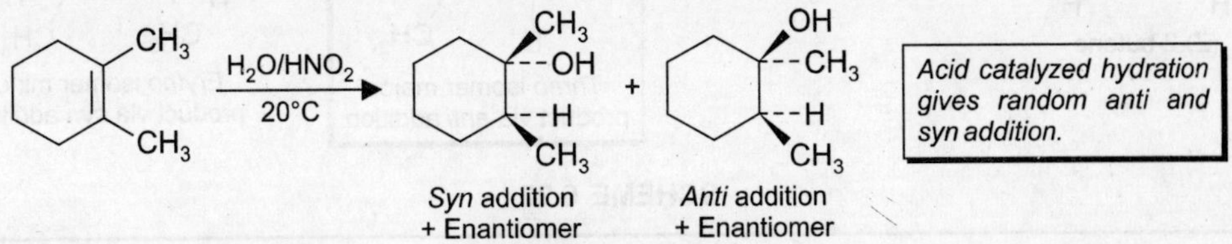

SCHEME 6.18a

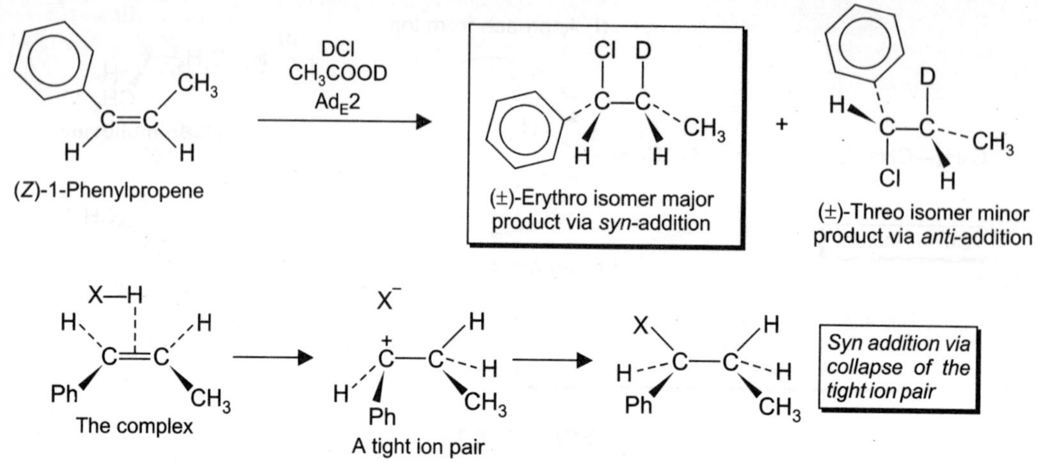

SCHEME 6.19

It is found that electrophilic addition of a halogen acid to an alkene from which relatively unstable carbocations are formed show a greater than first-order dependence on the concentration of the addition reagent E-Nu. These reactions are Ad_E3 Ad_E4 etc., and are known. The higher order reactions are favoured as the polarity of the solvent decreases and when HX is expected to be unionized. To account for *anti* addition, it is suggested that one molecule of HX forms a π complex with the substrate and the second molecule attacks from the rear leading to *anti* addition, *i.e.*, each molecule of HX attacks from the opposite face of the double bond to provide a nucleophile and an electrophile (scheme 6.20). Thus, as compared to Ad_E2 process during the reaction of DCl to *Z*-1-phenylpropene (scheme 6.19) which leads to *erythro* product in larger amount, a similar addition to *cis*-2 butene follows the Ad_E3 process leading instead to the formation of threo product in larger amount (scheme 6.20).

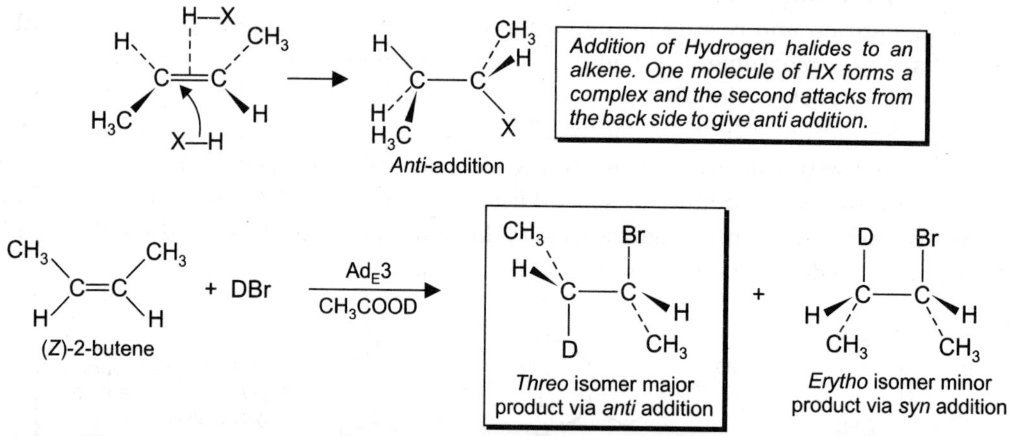

SCHEME 6.20

An interesting case of electrophilic *syn* addition is also observed with *trans*-cyclooctene (scheme 6.21). This compound gives *trans*-1, 2-disubstituted cyclooctanes. The preference for *syn* addition is due to the inaccessibility of the rear face of the double bond which is shielded by the chain of methylene groups of the eight membered ring.

Addition of HX to norbornene presents an interesting example where steric factors and molecular rearrangement are involved. Addition of DBr to norbornene shows preference for *syn* addition *i.e.*, *syn-exo* since *endo* side is sterically hindered. The *syn* addition leading to (I, scheme 6.21) is via the collapse of ion pair. The presence of deuteriam at C3 and C7 position also indicate the involvement of a bridged ion (II, scheme 6.21) in addition to ion pair formation.

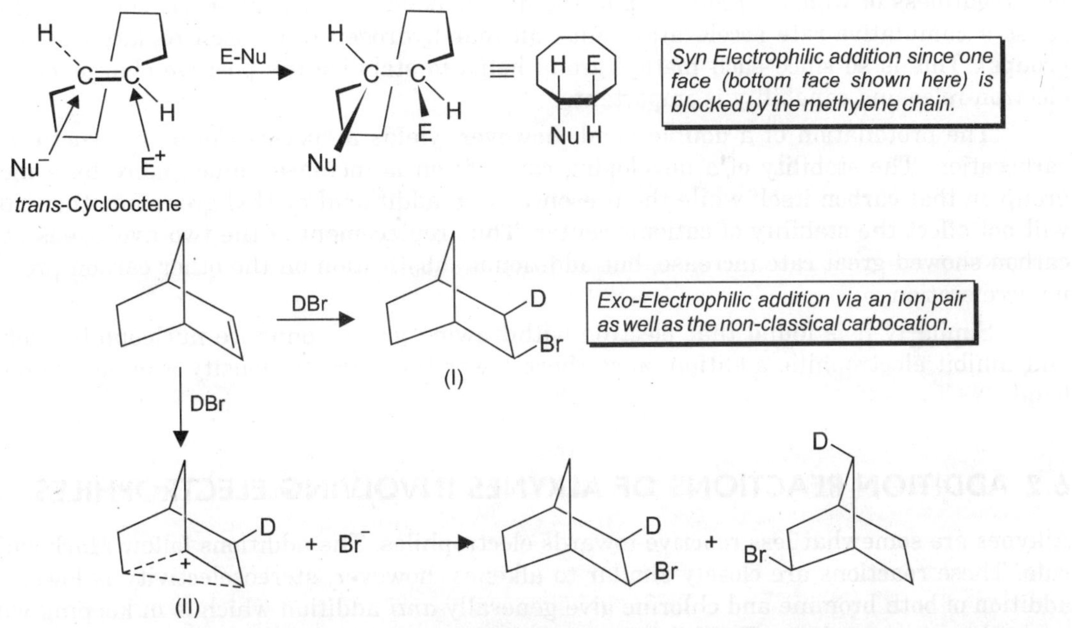

SCHEME 6.21

EXERCISE 6.5

What product one expects from reaction (scheme 6.21a) ? Depict the stereochemical outcome at every step.

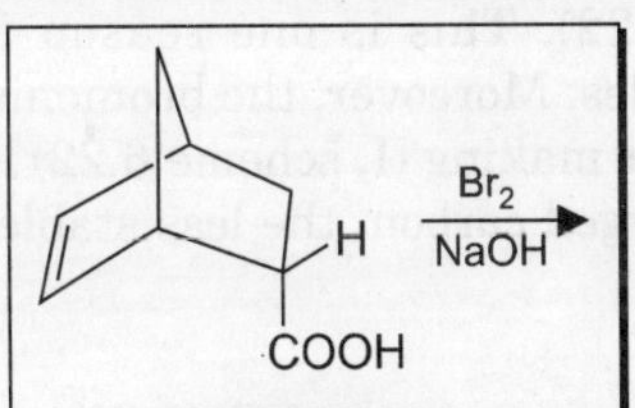

SCHEME 6.21a

ANSWER. *It is a typical reaction involving the formation of a cyclic bromonium ion on the less hindered exo-face of the molecule followed by intramolecular attack by COOH group (scheme 6.21b)*

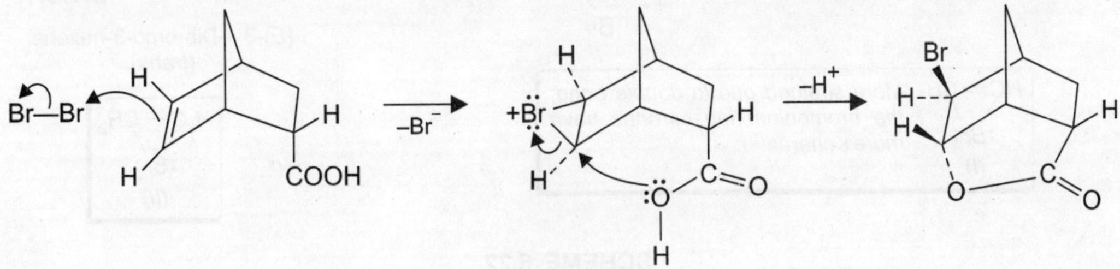

SCHEME 6.21b

(I) Reactivity—Effect of Substituents

Presence of alkyl substituents on an alkene effects the rate of electrophilic substitution and helps in deciding on the nature of the intermediate cation *e.g.*, if or not a cyclic species like bromonium ions is involved. Thus *e.g.*, if a bromonium ion is involved, an increase in the substitution of methyl groups on CH_2=CH_2 increases, the rate of addition of bromine (in methanol) regardless of which carbon atom of the double bond is methylated (*i.e.* the alkyl groups cause a cumulative rate acceleration until all four hydrogen have been replaced by methyl groups). This is so since each methyl group helps to stabilize the positive charge *i.e.*, total electron-releasing capability is important.

The protonation of a double bond, however, yields a species closer in structure to a carbocation. The stability of a developing carbocation is increased much more by a methyl group on that carbon itself while the presence of an additional methyl group on other carbon will not effect the stability of cationic center. Thus replacement of the two hydrogens at one carbon showed great rate increase, but additional substitution on the other carbon produced no acceleration.

Similarly it is found that electron-withdrawing groups enhance nucleophilic addition and inhibit electrophilic addition since these lower the electron density around the double bond.

6.2 ADDITION REACTIONS OF ALKYNES INVOLVING ELECTROPHILES

Alkynes are somewhat less reactive towards electrophiles. The additions follow Markovnikov rule. These reactions are closely similar to alkenes, however, stereoselectivity is lower. The addition of both bromine and chlorine give generally *anti* addition which is in keeping with a bridged ion intermediate. The following two reactions prove these points.

(*i*) Bromination of Alkynes

A bromonium like intermediate is invoked to explain the *anti*-stereoselectivity (scheme 6.22). The three membered ring bromonium ion derived from an alkyne (I, scheme 6.22) has a full double bond and is thus strained as compared to a bromonium ion from an alkene (II, scheme 6.22). This is one reason for alkynes to be less reactive than alkenes towards electrophiles. Moreover, the bromonium ion carbons of (I) have more *s* character than those of (II) further making (I, scheme 6.22) less stable than (II), the more the *s* character in the positively charged carbon, the less stable is the carbocation.

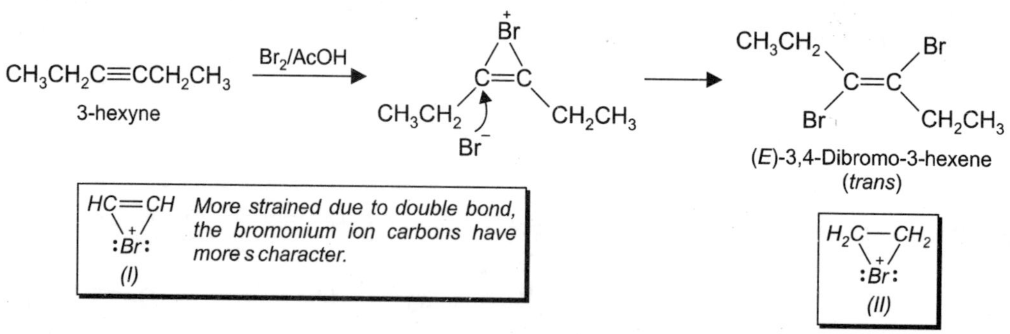

SCHEME 6.22

(ii) Hydration of Alkynes

In the presence of sulfuric acid and Hg (II) salts as catalysts, alkynes add water. The terminal alkynes add water according to Markovnikov rule—The hydrogen adds to the carbon atom of the triple bond bearing and hydrogen atom (scheme 6.23). Mechanistically Hg^{2+} attacks the triple bond to form a bridged mercurinium ion intermediate. Water then attacks the most substituted carbon of the cyclic intermediate from the side opposite the bridge. The proton is then transferred to solvent to give a mercuric enol. It rearranges to mercuric ketone and cleavage of carbon mercury bond by water is followed by keto-enol tautomerism.

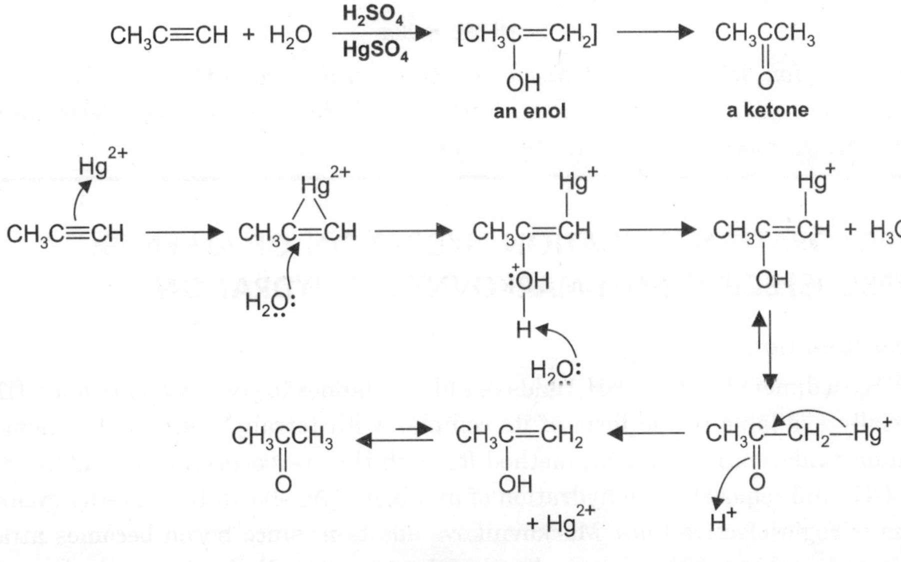

SCHEME 6.23

EXERCISE 6.6

Explain why phenylacetylene gives a mixture of cis and trans dibromides on reaction with bromine?

ANSWER. *Mostly additions of bromine to acetylenes give only trans dibromides, however, there is a competition between the formation of a carbocation and a cyclic*

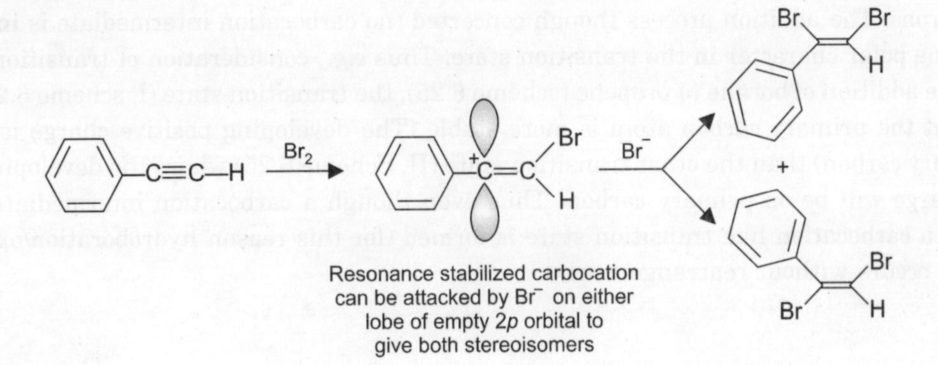

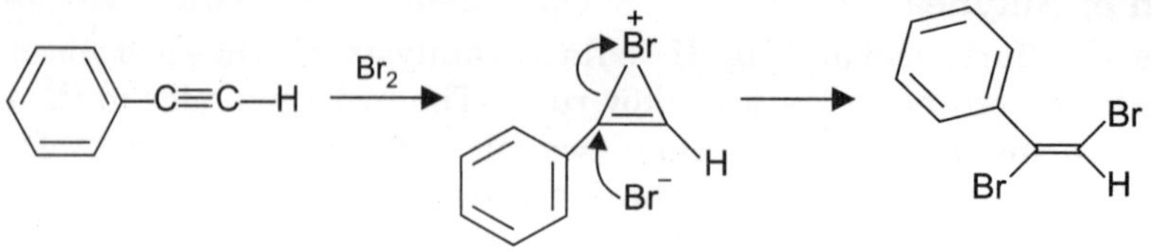

a bromonium ion
intermediate would have
given only *trans* dibromide

SCHEME 6.23a

bromonium ion. In the present case a resonance stabilized carbocation is formed as the intermediate (scheme 6.23a). With a cyclic bromonium ion intermediate only trans dibromide would have been formed.

6.3 HYDROBORATION/OXIDATION HYDRATION OF ALKENES: A STEREOSPECIFIC ANTI-MARKOVNIKOV HYDRATION

(*a*) General Reaction

Diborane B_2H_6 (a dimer of borane, BH_3) adds readily to alkenes to give organoboranes. [Diborane is commercially available in the form of its complex with tetrahydrofuran (I, scheme 6.24)]. Hydroboration oxidation is an useful method for both the stereospecific (*syn* addition of both —H and —OH) and regioselective hydration of an alkene. As shown for 1-methylcyclopentene the reaction is regioselective (*anti* Markovnikovs addition) since boron becomes attached to less substituted carbon of the double bond (scheme 6.24). Following hydroboration, the organoborane is oxidized by treatment with hydrogen peroxide in aqueous base and the organoborane is converted to an alcohol. Thus hydroboration–oxidation leads to the overall hydration of an alkene. The method was developed by H.C. Brown (Nobel Prize 1979). Because only *syn* addition occurs hydroboration–oxidation is stereospecific, *syn* addition and leads to the formation of only the pair of enantiomers which has the added groups on the same side of the ring.

(*b*) Regiochemistry of Addition and Mechanism

Borane, the monomer of diborane is very reactive electrophile since it has only a sextet of electrons. The addition process though concerted (no carbocation intermediate is involved) has some polar character in the transition state. Thus *e.g.*, consideration of transition states from the addition of borane to propene (scheme 6.25), the transition state (I, scheme 6.25) with boron at the primary carbon atom is more stable (The developing positive charge is on the secondary carbon) than the other transition state (II, Scheme 6.25) where the developing positive charge will be on primary carbon. Thus even though a carbocation intermediate is not formed a carbocation like transition state is formed (for this reason hydroboration/oxidation process occurs without rearrangement.

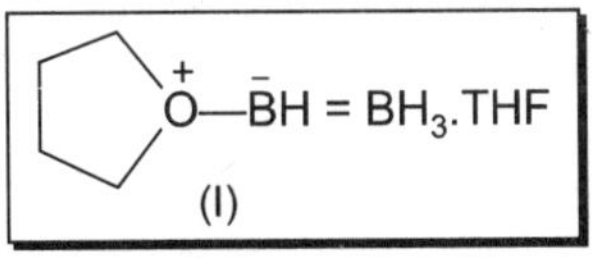

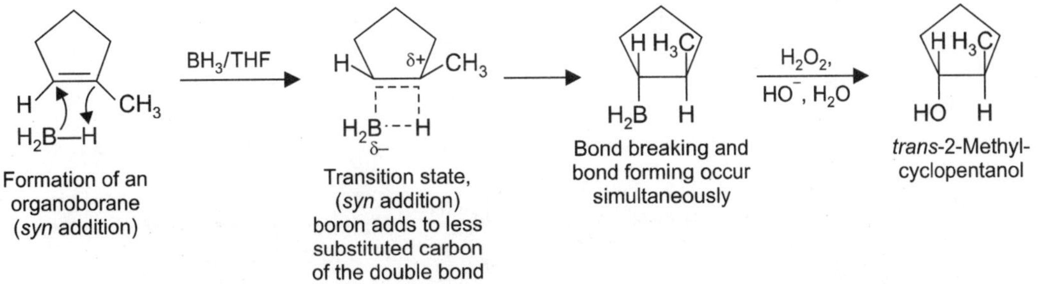

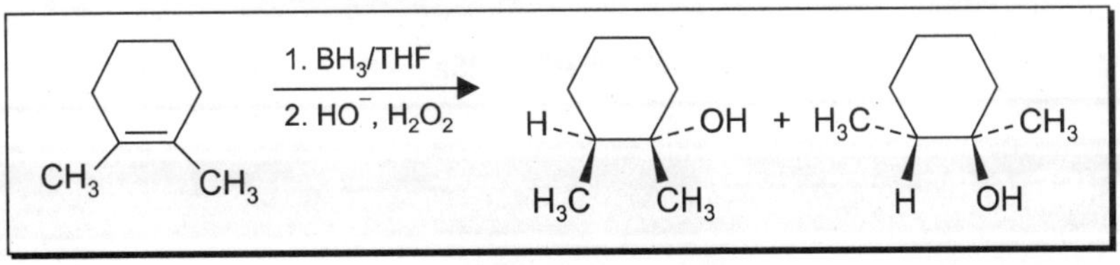

SCHEME 6.24

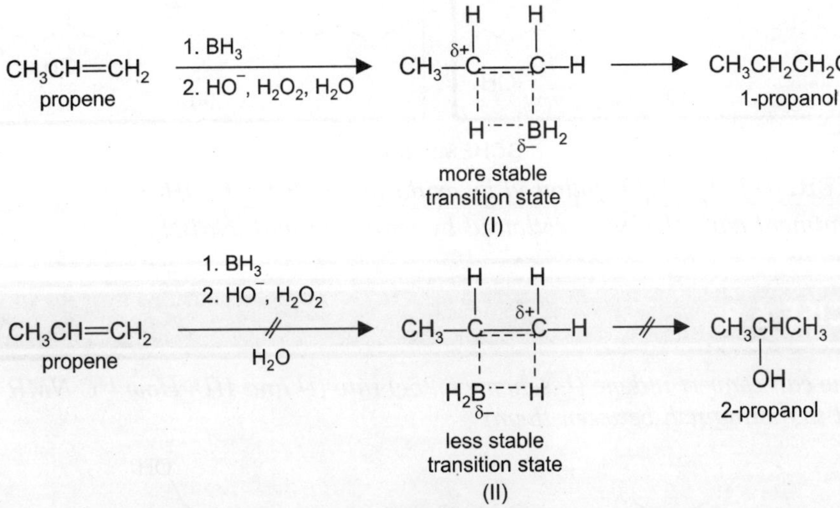

SCHEME 6.25

EXERCISE 6.7

Show the products of Markovnikov and anti Markovnikov addition of H_2O to the double bond of 1-methylcyclohexene and reaction conditions for each pathway.

ANSWER. *These are shown (scheme 6.25a).*

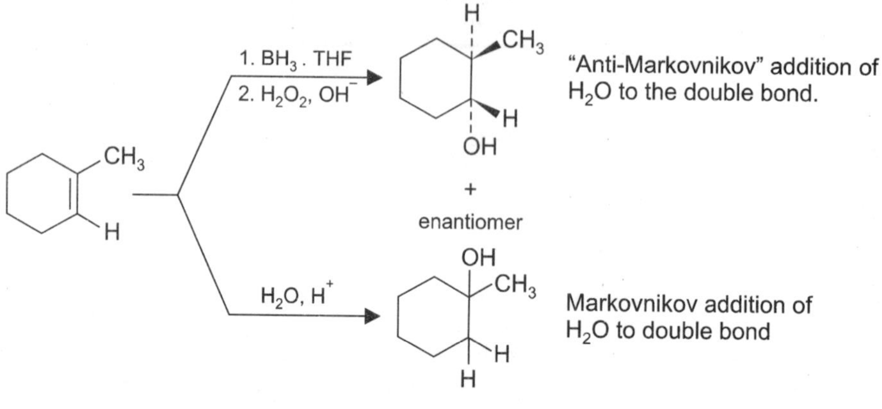

SCHEME 6.25a

EXERCISE 6.8

What reagents are needed to bring about the transformations (scheme 6.25b) ?

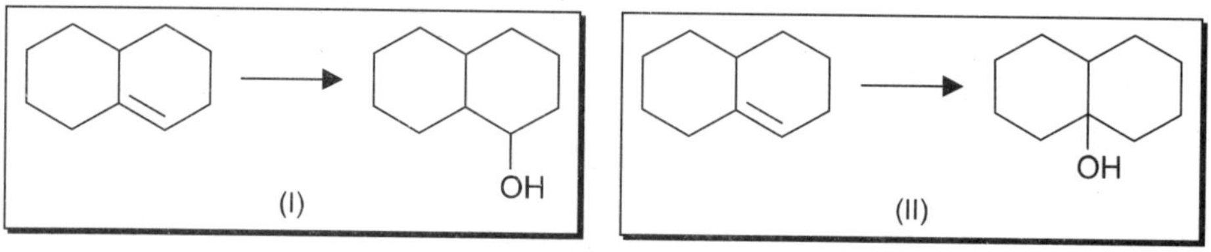

SCHEME 6.25b

ANSWER. *(i) BH_3 . THF, followed by oxidation with H_2O_2/OH^-*
(ii) Treatment with $Hg(OAc)_2$ followed by reduction with $NaBH_4$.

EXERCISE 6.9

How one can convert indene (I, scheme 6.25c) into (II and III). How ^{13}C NMR can be used to distinguish between them?

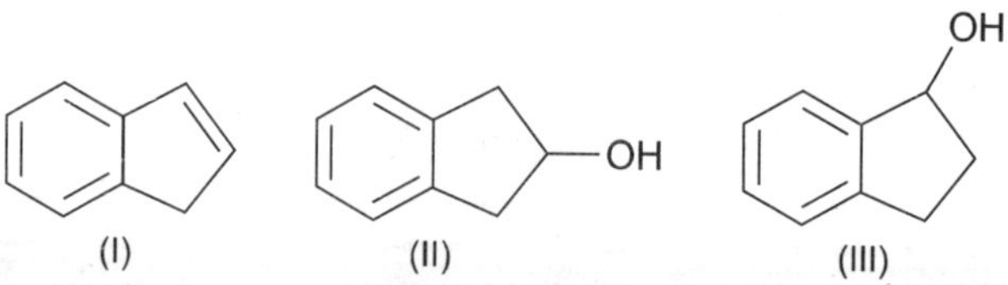

SCHEME 6.25c

ANSWER. *Compound (III) can be obtained by Markovnikov hydration (scheme 6.25d) compound (II) can be made by hydroboration-oxidation which is an anti-Markovnikov addition (scheme 6.25e). Compound (II, plane of symmetry) shows five ^{13}CNMR signals while compound (III) shows nine.*

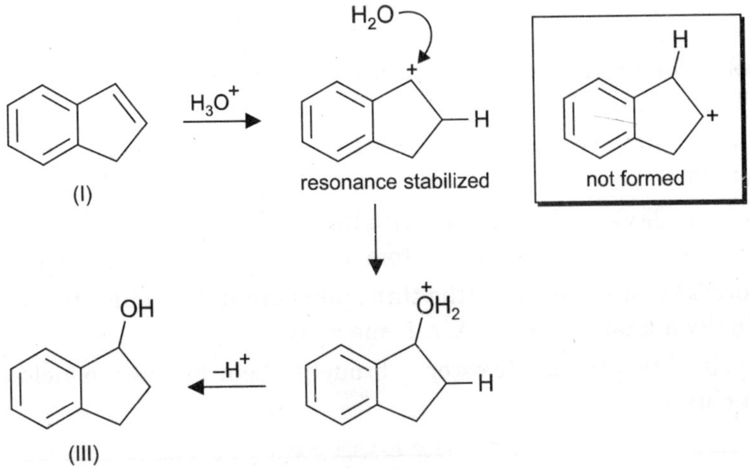

SCHEME 6.25d

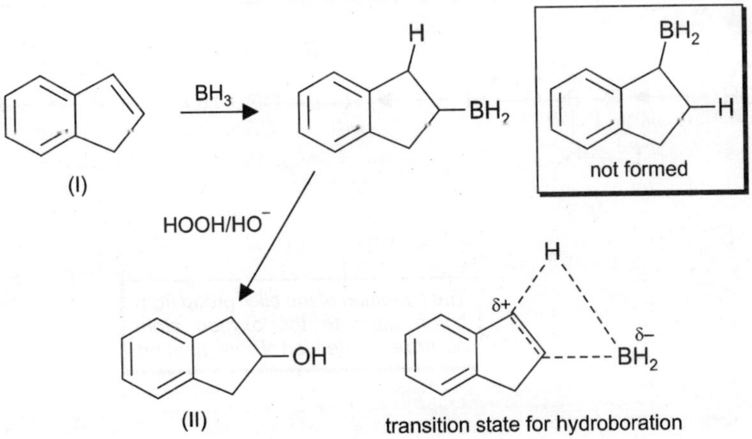

SCHEME 6.25e

The second tactic to explain the regiochemistry of hydroboration is steric effect. The borane containing group (BH_2) is much bulkier than hydrogen, thus boron attaches with the less hindered end (with more hydrogens) of the double bond.

(*c*) Hydroboration–Oxidation Reaction

An organoborane RBH_2 formed initially can react further with the alkene in stages to finally form a trialkylborane R_3B, *i.e.*, until all three hydrogens of borane are replaced by alkyl groups (scheme 6.26).

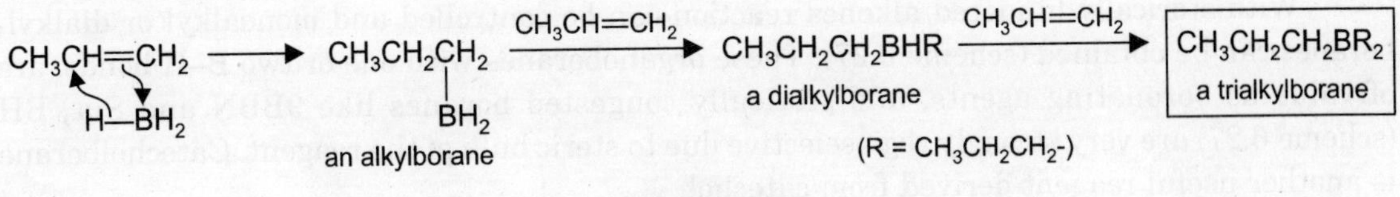

SCHEME 6.26

Trialkylboranes can be oxidized with basic aqueous hydrogen peroxide to give alcohols in which the hydroxy function has replaced the boron atom. The net result of this two step sequence *hydroboration-oxidation* is the addition of elements of water to a double bond (I, scheme 6.26*a*). The mechanism involves the following steps:

- The nucleophilic hydroperoxide ion attacks the electron—poor boron atom (step 1, scheme 6.26*a*).
- In step 2, an alkyl group migrates with its electron pair and with retention of configuration to the neighboring oxygen atom and a hydroxide ion is expelled.
- The process is repeated till all the three alkyl groups get migrated to oxygen atoms to yield finally a trialkylborate (A, scheme 6.26*a*).
- Hydrolysis of this inorganic ester is brought about by base to yield and alcohol and sodium borate.

$$3CH_3CH{=}CH_2 + BH_3 \xrightarrow{THF} (CH_3CH_2CH_2)_3B \xrightarrow[H_2O]{HO^-,\ H_2O_2} 3CH_3CH_2CH_2OH + BO^{3-} \quad (I)$$

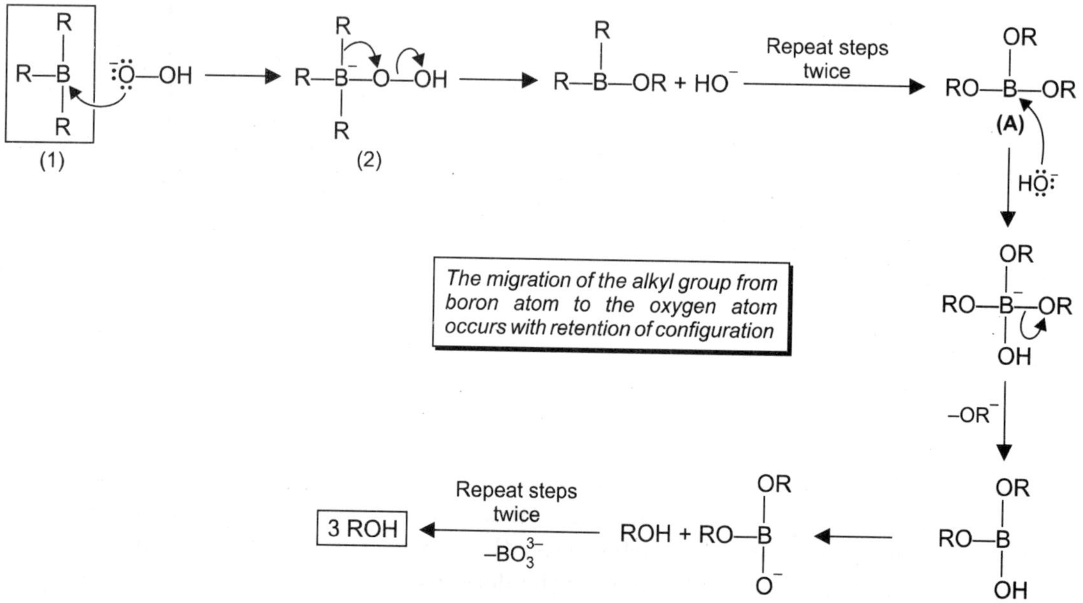

The oxidation of an alkyl borane

SCHEME 6.26a

(*d*) Sterically Congested Boranes

With sterically hindered alkenes reaction can be controlled and monoalkyl or dialkylboranes can be obtained (scheme 6.27). These organoboranes with one or two B–H bonds, are of value as boronating agents, and sterically congested boranes like 9BBN and Sia_2 BH (scheme 6.27) are very strongly regioselective due to steric bulk of the reagent. Catecholborane is another useful reagent derived from catechol.

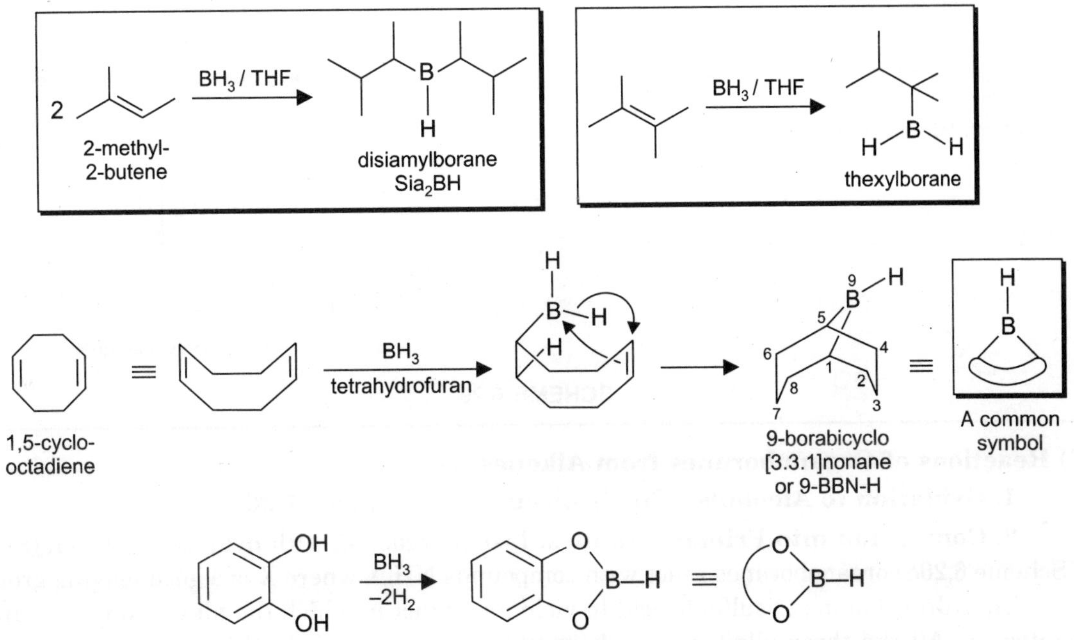

SCHEME 6.27

The boron atom adds predominantly to the less substituted carbon atom. However, with 1, 2-disubstituted alkenes this selectivity is lost or is less pronounced (scheme 6.28). By using a bulkyl reagent *e.g.* 9-BBN the boron atom almost exclusively adds to that carbon atom which has the less bulky of the two substituents.

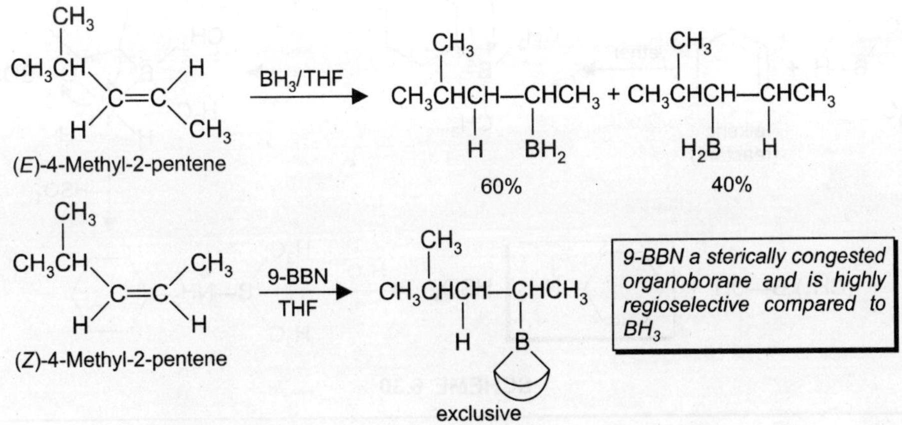

SCHEME 6.28

(e) Isomerization of Organoboranes

On heating organoboranes to about 150°C an isomerization occurs. Alkenes are formed and reconverted to organoboranes till the least substituted organoborane is formed (scheme 6.29). By using this method a readily available internal alkene can be converted *e.g.*, into a primary alcohol.

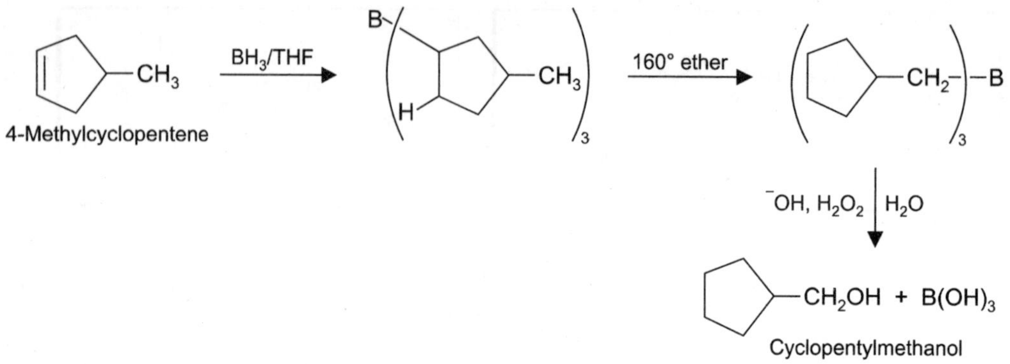

SCHEME 6.29

(*f*) Reactions of Organoboranes from Alkenes

1. Oxidation to Alcohols (This is discussed in scheme 6.26a)

2. Conversion into Primary Amines. Like its reaction with hydroperoxide ion (Step 2, Scheme 6.26*a*) organoboranes react with compounds NH_2X where X is a good leaving group *e.g.*, with hydroxylamine-*O*-sulfonic acid to yield an adduct in which the alkyl group migrates to nitrogen. All the three alkyl groups, however, do not migrate in this case in contrast to hydroperoxide ion. Methyl groups attached to boron resist migration. Thus dimethylborane is used in the first step and only the alkyl group derived from the alkene reactant migrates in the reaction (scheme 6.30, a good leaving group HSO_4^- initiates the migration of the alkyl group to nitrogen). The amine-borane salt is hydrolyzed with water to give the amine salt with is treated with dilute sodium hydroxide to neutralize the acid to give free amine.

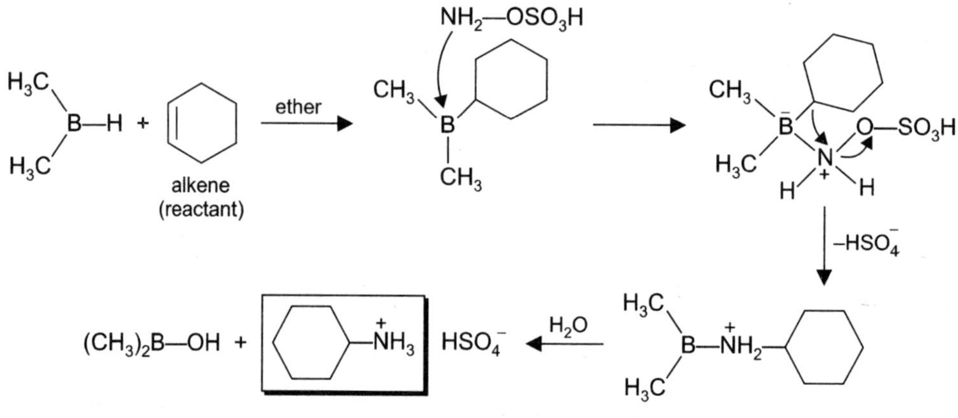

SCHEME 6.30

3. Conversion into Hydrocarbons (Protonolysis). Carboxylic acids readily cleave the carbon-boron bond. The process which replaces a hetero atom with hydrogen is termed protonolysis. The carboxylic acid firstly forms an adduct by reacting with electron deficient boron atom. A six membered transition state is involved in the reduction process (scheme 6.31) which also accounts for complete retention of configuration at the site of reduction *i.e.*, the carbon-carbon bond.

The protonolysis of carbon-carbon bond is used to introduce deuterium at a specific position in an organic compound (*e.g.* I, scheme 6.31) and also to synthesize a *cis* alkene from an alkyne.

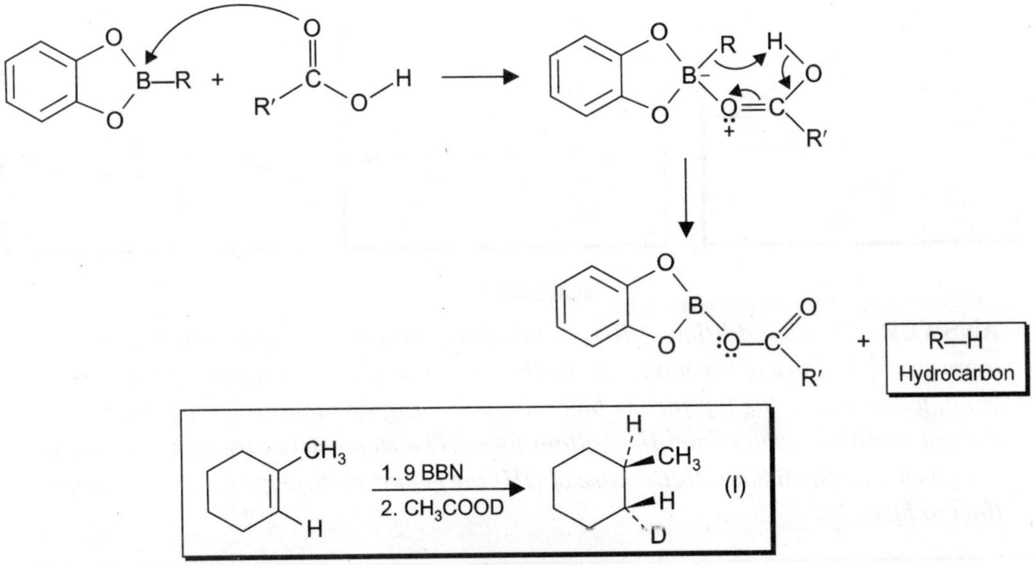

SCHEME 6.31

Stereospecific and Stereoselective Reactions Revisited

One may recall the definitions of stereospecific and stereoselective reactions. If stereoisomers of a substrate react under identical reaction conditions to give products that are stereoisomers of each other, the reaction is called stereospecific. When the reactant is non-stereoisomeric but leads to stereoisomeric products in unequal amounts the reaction is said to be stereoselective.

More precisely stereospecific reactions have a mechanism which demands a specific stereochemical outcome. Stereoselective reactions offer alternative pathways so that the reaction proceeds via. (i) the most favourable pathway (kinetic control) or (ii) via a pathway which gives the most stable stereoisomer as the major product (thermodynamic control). Thus hydroboration-oxidation of α-pinene is stereoselective. Addition occurs at the less hindered face of the double bond and only one alcohol is formed in high yield (scheme 6.32).

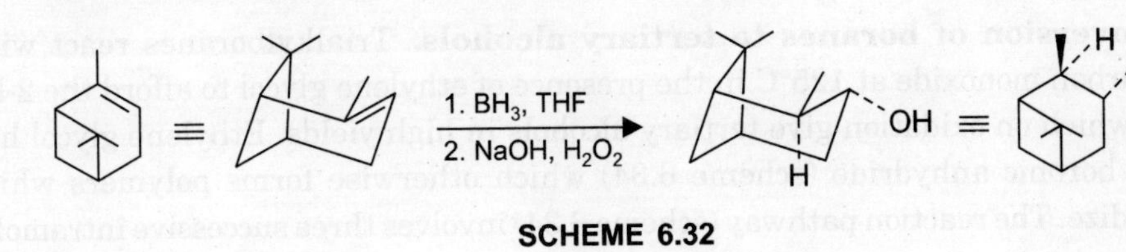

SCHEME 6.32

EXERCISE 6.10

Write the stereostructures of the products from reactions (scheme 6.33).

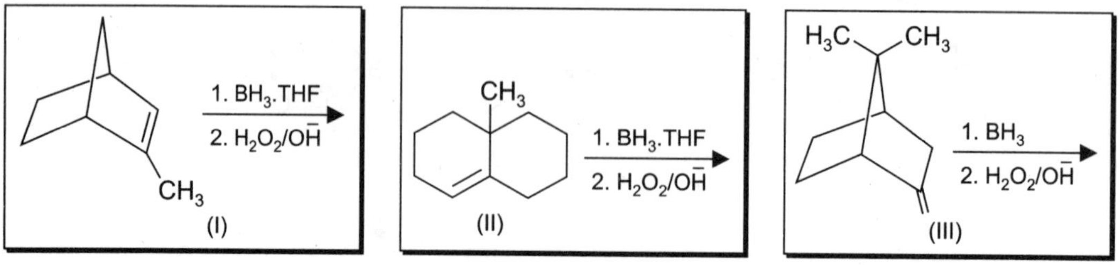

SCHEME 6.33

ANSWER. *The syn addition of BH_3 across the double bond occurs mostly from the more accessible face of the molecule. In the case of (I) it is the exo face of the double bond, while in (II) the top face is blocked by the angular methyl group, therefore, the syn addition occurs from the bottom faces. The stereostructure of the products are given (scheme 6.33a). In the case of (III) the gem dimethyl group at C7 hinders the exo face.*

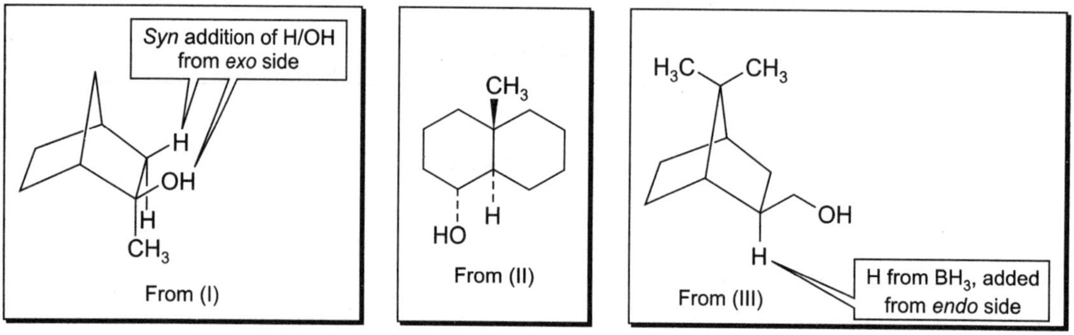

SCHEME 6.33a

4. Carbonylation of Organoboranes. Carbonmonoxide forms Lewis acid-base complexes with organoboranes, and these adducts undergo boron-to-carbon migration of the boron substituents. The reaction can be controlled and therefore, directed in the migration of one, two or all the three substituents on boron. Conditions can be defined so that carbonylation of organoboranes can lead to primary, secondary and tertiary alcohols, aldehydes and open chain, cyclic and polycyclic ketones.

***(i)* Conversion of boranes to tertiary alcohols.** Trialkylboranes react with one molecule of carbon monoxide at 125°C in the presence of ethylene glycol to afford the 2-bora-1, 3-dioxolanes which on oxidation give tertiary alcohols in high yields. Ethylene glycol helps to intercept the boronic anhydride (scheme 6.34) which otherwise forms polymers which are difficult to oxidize. The reaction pathway (scheme 6.34) involves three successive intramolecular migrations.

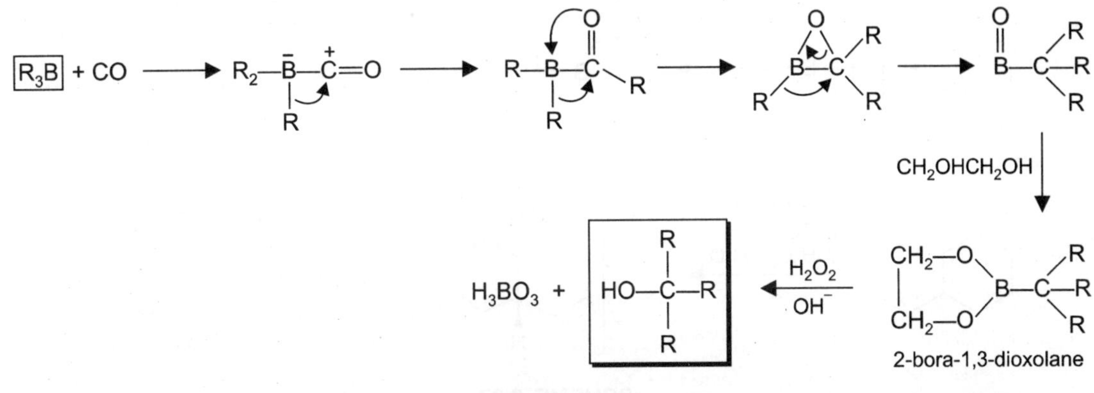

SCHEME 6.34

***(ii)* Conversion of boranes to ketones and secondary alcohols.** When the carbonylation of a trialkylborane is conducted in the presence of water, the migration of the third alkyl group (step 4, Scheme 6.34) is intercepted. The hydrate is formed (scheme 6.35), this can be oxidized (OOH^-) to a ketone or hydrolyzed to a secondary alcohol.

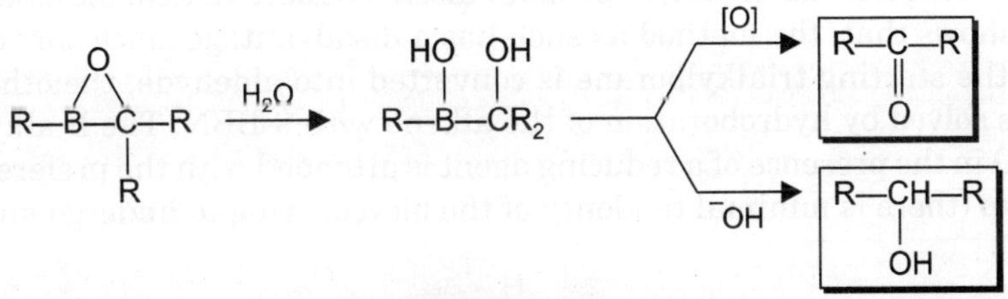

SCHEME 6.35

The reaction can be used to prepare unsymmetrical ketones by using 'mixed' organoboranes prepared from thexylborane (scheme 6.36). The success of this method depends upon the thexyl group being noncompetitive with the other groups in migration steps. Thus the sequential introduction of two alkenes (scheme 6.36) leads to the synthesis of unsymmetrical ketones.

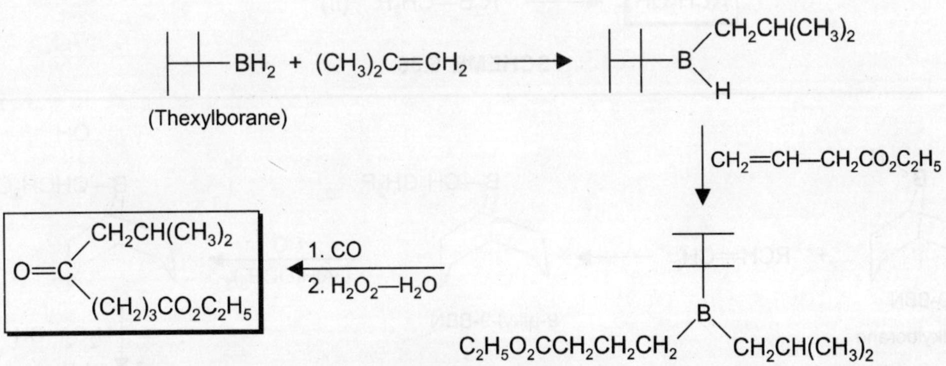

SCHEME 6.36

By working with appropriate dines one can end up with cyclic or bicyclic ketones (scheme 6.37).

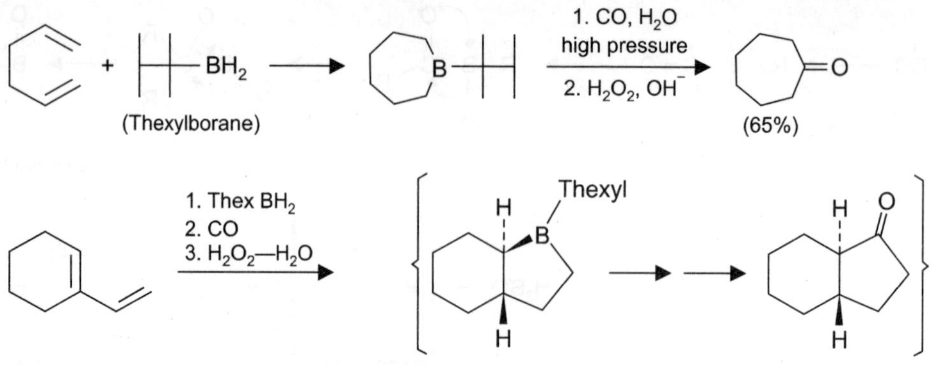

SCHEME 6.37

(*iii*) Conversion of boranes to aldehydes and primary alcohols

When the carbonylation of the trialkylborane is done in the presence of a reducing agent *e.g.* lithium trimethoxyaluminium hydride, the reducing agent intercepts the intermediate after only one boron-to-carbon migration has taken place (see, scheme 6.34). This on oxidation (route I, scheme 6.38) gives aldehydes, or primary alcohols (route II, Scheme 6.38). An inspection of scheme 6.38 shows that, the method as such has a disadvantage since only one of the three alkyl groups of the starting trialkylborane is converted into aldehyde, the others are wasted. This difficulty is solved by hydroboration of the alkene with 9-BBN. The B-alkyl derivative on reaction with CO in the presence of a reducing agent is attended with the preferential migration of the alkyl group (there is minimal tendency of the bicyclic ring to undergo migration) to give

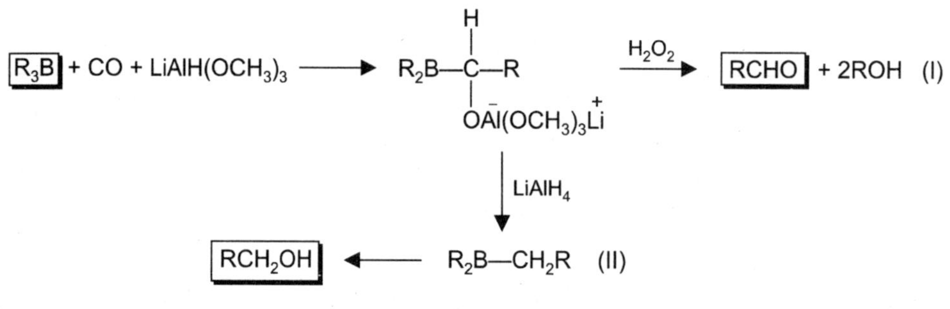

SCHEME 6.38

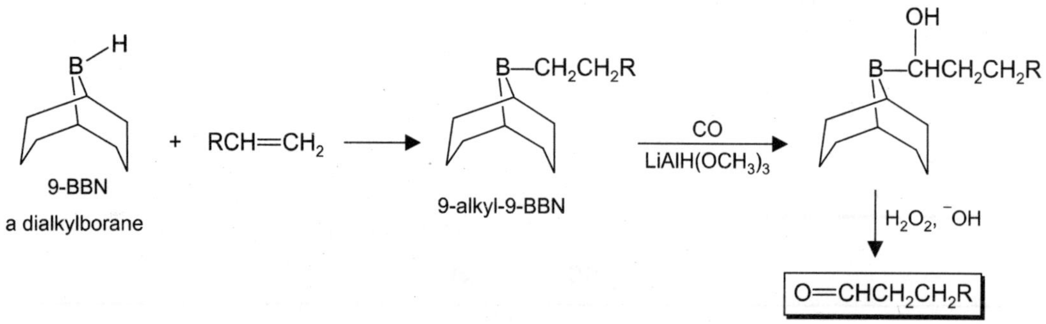

SCHEME 6.39

high yields of the aldehyde. While working with B-alkyl-9-BBN, since only the 9-alkyl group migrates, this method, thus converts (high yields) an alkene to a primary alcohol or an aldehyde containing one more carbon (scheme 6.39).

5. Cyanidation of Organoboranes. Cyanide ion is isoelectronic with carbon monoxide and a borane reacts similarly to initially form an adduct with cyanide ion, on reaction with sodium cyanide (scheme 6.40). Thus this is another reagent besides carbon monoxide which serves as the electrophilic migration terminus. The nitrogen atom in the initial adduct is not sufficiently electron attracting to induce migration. This ability is increased by acylation of the cyano group with trifluoroacetic anhydride. Two alkyl groups are transferred (scheme 6.40) at low temperature to give ketones by oxidation. The method has a merit over carbonylation due to low temperature conditions of this reaction. Moreover, as in the carbonylation reaction, one can avoid the wastage of the alkyl groups by working with thexylborane. Also, the thexyl group does not migrate and unsymmetrical ketones can be easily obtained.

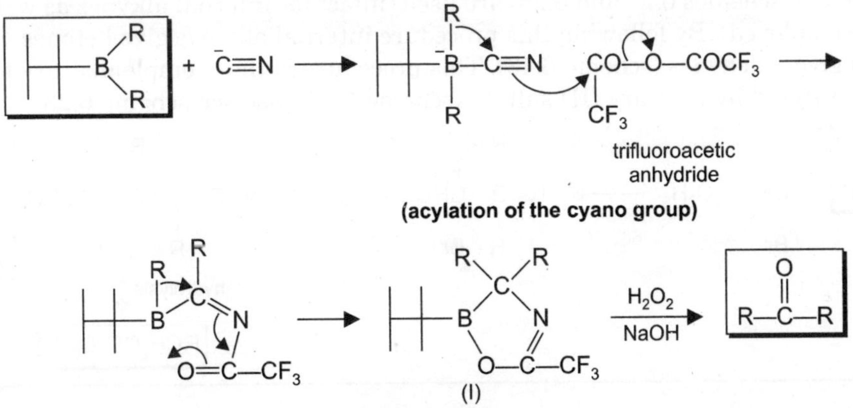

SCHEME 6.40

One can induce the migration of the third group also on (I, scheme. 6.40) by using an excess of trifluoroacetic anhydride, to afford tertiary alcohols on oxidation (scheme 6.41).

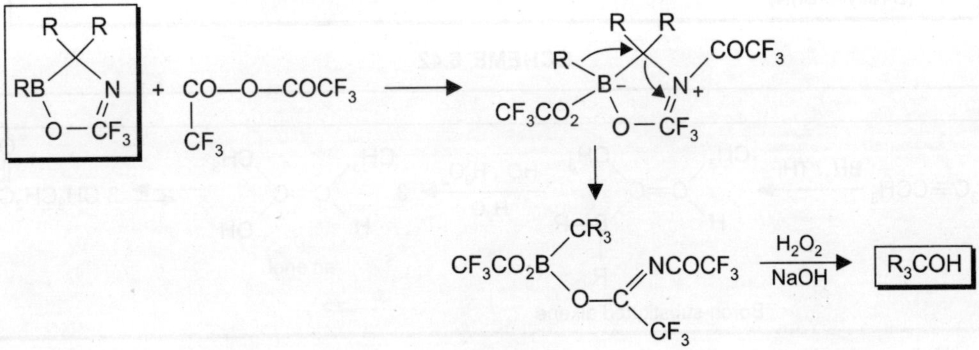

SCHEME 6.41

6. Synthesis of Esters. Excellent yields of esters can be obtained by reacting trialkylboranes with *e.g.*, enthylbromoacetate, (an α-haloester) in the presence of a base (eq. I, scheme 6.42). The enolate adds to the organoborane to give a tetracoordinate boron intermediate on which the migration of an alkyl group occurs with displacement of bromide ion. Hydrolysis of the rearranged products gives an ester. Similar reactions are displayed by α-haloketones and α-halonitriles. The key step in this reaction is again the migration of an alkyl group from boron to adjacent carbon atom. Only one of the three alkyl groups of the trialkylborane is used in this reaction. This problem is solved as usual by using an alkyl derivative of 9-BBN instead of trialkylborane (eq. II, scheme 6.42). By following this method not only the alkyl group is fully utilized, but 9-BBN takes no part in alkylation. When β-alkenyl derivatives of 9-BBN are employed, one ends up with β-γ-unsaturated esters.

7. Hydroboration of Alkynes and Reaction of Derived Organoboranes. Borane adds to alkynes in a way similar to alkenes *i.e.,* one mole of BH_3 reacts with three moles of alkyne to form one mole of boron substituted alkene. On oxidation boron is replaced by an OH group and the enol rearranges to ketone (scheme 6.43). For this synthesis of ketones, the reaction must stop at the alkene stage. For internal alkynes, the second addition is prevented by the bulky groups on the boron substituted alkene. In the case of a terminal alkyne, it is difficult to stop the addition at the alkene stage. Thus hindered boranes like, disiamylborane or 9-BBN-H (See schemes 6.27 and 6.28) are used (infact for internal alkynes as well, hindered boranes are employed). By following this procedure internal alkynes give ketones while terminal alkynes give aldehydes (scheme 6.43). This procedure is thus complementary to the direct hydration catalyzed by mercury (II) sulfate (scheme 6.44, also see scheme 6.23).

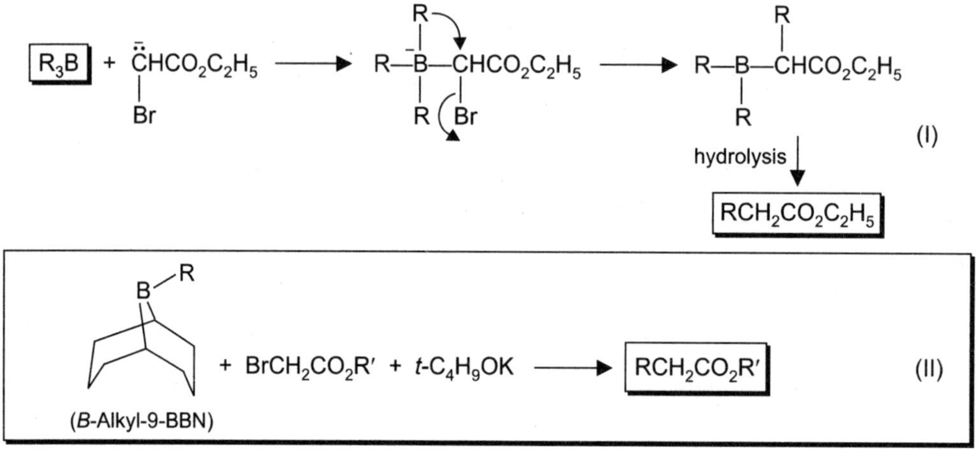

SCHEME 6.42

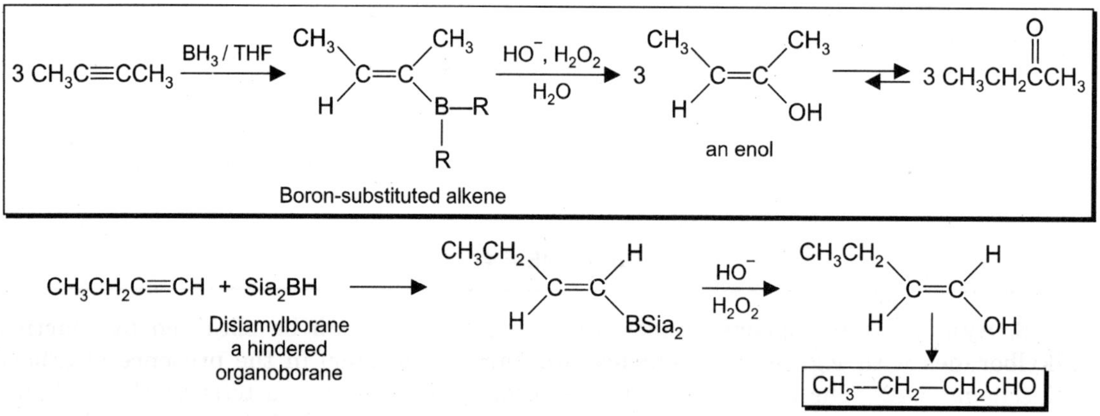

SCHEME 6.43

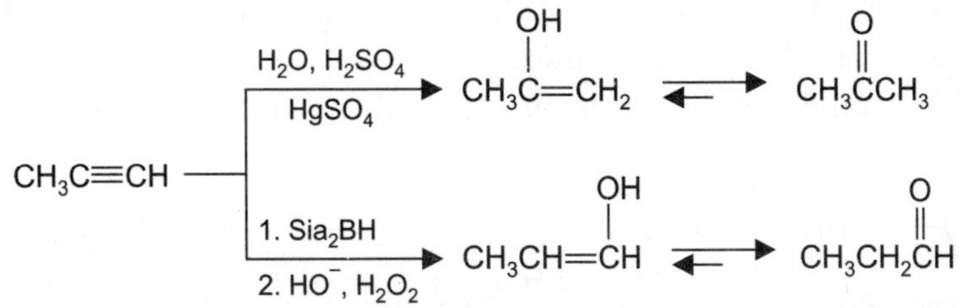

SCHEME 6.44

***(i)* Synthesis of Z-alkenes**

Vinylboranes, prepared by the monohydroboration of triple bonds (scheme 6.43) on reaction with iodine undergo an alkyl migration from boron to carbon, within an iodonium ion (scheme 6.45). Z-alkene is formed *via anti* elimination after the alkyl group migration from an *anti*-periplanar transition state (I, scheme 6.45). Thus this method, provides a pathway for the synthesis of a Z-alkene from a monosubstituted alkyne.

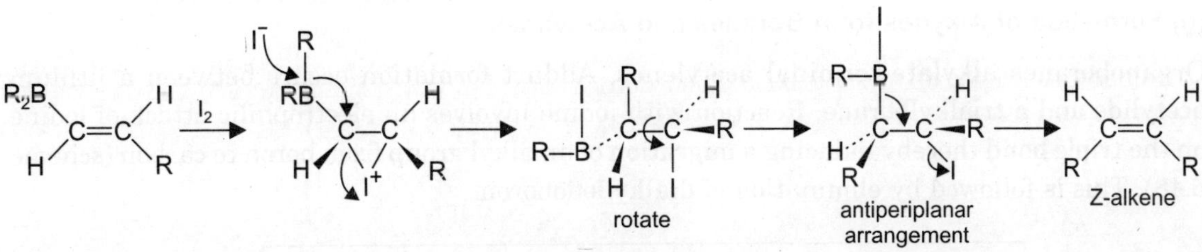

SCHEME 6.45

***(ii)* Synthesis of E-alkenes**

For the synthesis of E-alkenes a 1-haloalkyne is used. The addition of a dialkylborane gives an α-halovinylborane (*i.e.* α-haloalkenylborane). On treatment with methoxide ion, this intermediate undergoes boron to carbon migration (scheme 6.46) to afford an alkylated alkenylborane. Protonolysis gives an *E*-alkene.

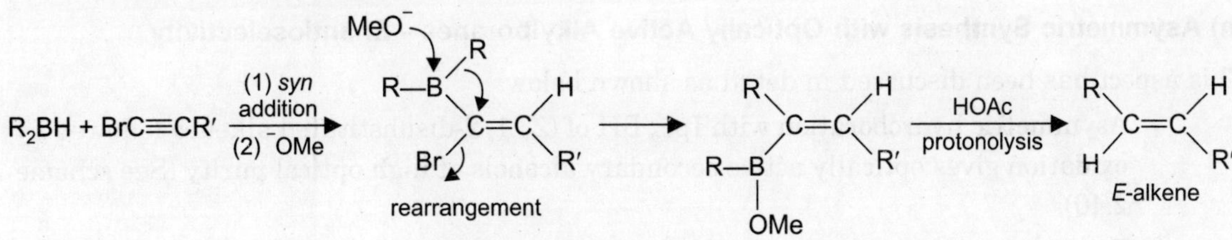

SCHEME 6.46

***(iii)* Synthesis of conjugated dienes**

The principles used in the synthesis of *E*- and *Z*-alkenes can be applied for an appropriate geometry of the double bond(s) in a conjugated diene. Thus *e.g.*, *E*, *E*-dienes can be made as shown (scheme 6.47). Hydroboration of a 1-haloalkyne with thexylborane affords a thexyl-1-chloroalkyenylborane. This reacts with another alkyne to give thexyldialkenylborane. Reaction with methoxide ion inducts a rearrangement of alkenyl and protonolysis affords the diene.

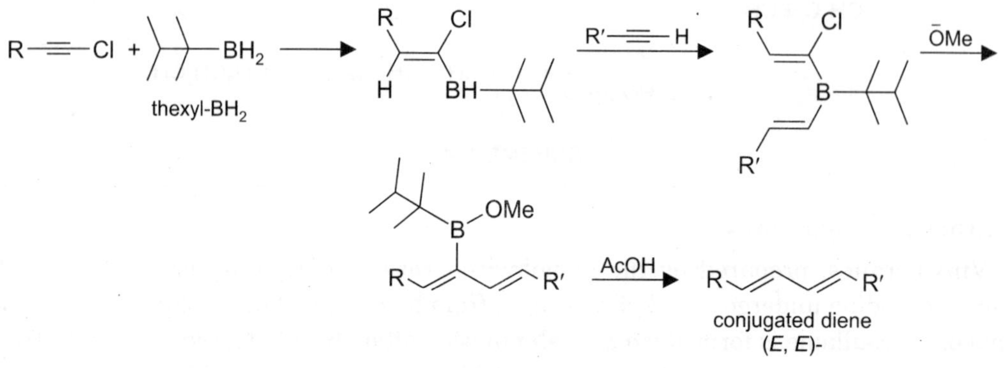

SCHEME 6.47

(g) Formation of Alkynes from Boranes and Acetylides

Organoboranes alkylate terminal acetylenes. Adduct formation occurs between a lithium acetylide and a trialkylborane. Reaction with iodine involves an electrophilic attack of iodine on the triple bond thereby inducing a migration of an alkyl group from boron to carbon (scheme 6.48). This is followed by elimination of dialkyliodoboron.

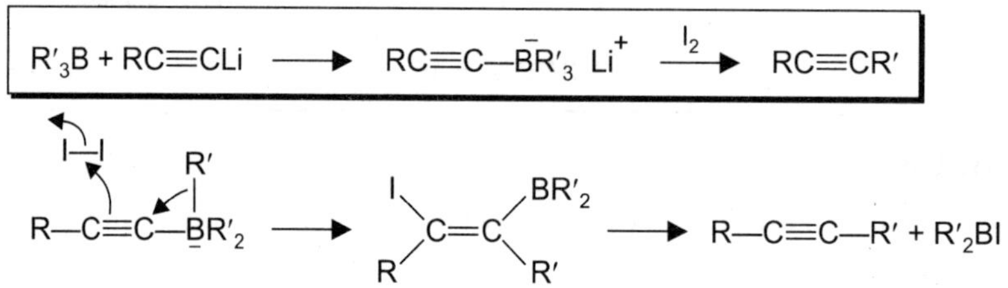

SCHEME 6.48

(h) Asymmetric Synthesis with Optically Active Alkylboranes—Enantioselectivity

This aspect has been discussed in detail as shown below:

- Asymmetric hydroboration with Ipc_2 BH of (*Z*) 1, 2-disubstituted alkenes followed by oxidation gives optically active secondary alcohols of high optical purity (See scheme 2.40).
- For enantioselective reduction of carbonyl compounds with alpine borane (See scheme 2.44).

(i) Approach of the Borane to the Alkene

Finally an alkene reacts at the less hindered face during hydroboration *i.e.*, *exo*-face in the case of norbornene (See scheme 6.33) and α-face in the case of cholesterol (scheme 6.49).

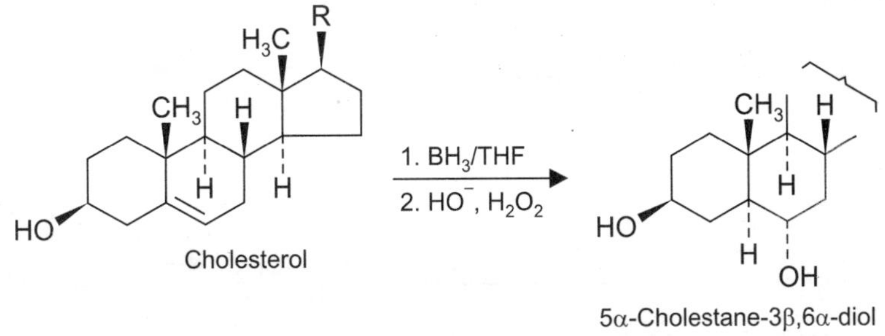

SCHEME 6.49

Approach of a Borane to the Alkene

Significantly, hydroboration oxidation is a syn-addition of H/OH to the double bond. The stability of the product is not important since the thermodynamically more stable product is not always formed. An example is of β-pinene which gives the less stable cis-myrtanol (scheme 6.49a, also see exercise 6.7). Thus, the reaction pathway is syn-addition, irrespective of direction of attack.

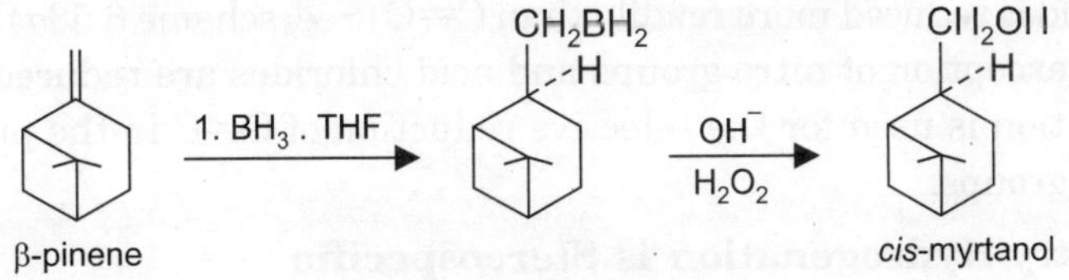

SCHEME 6.49a

6.4 HYDROGENATION OF ALKENES AND ALKYNES

(A) Catalytic and Metal Ammonia Reductions

Hydrogenation of an olefin in the presence of a catalyst (*e.g.*, Ni, Pt, Pd) takes place under varying conditions of heat and pressure. The catalyst is believed to provide a surface to form a complex with the olefin and at the same time promote the activation of the adsorbed hydrogen molecule. Under the reaction conditions, the molecular hydrogen splits into hydrogen atoms which then add in a step wise way to the olefin and form two new C—H bonds on the same side of the ring. This type of addition is called *syn* addition and a *cis*-disubstituted cycloalkane will be obtained if the reactant is a 1, 2-disubstituted cycloalkene (scheme 6.50). Catalytic hydrogenation is therefore, *stereospecific*. The two hydrogen atoms are added to the same face of the double bond **(*syn* addition).** Thus, 1-ethyl-2-methylcyclohexane is hydrogenated over platinum to give specifically *cis*-1-ethyl-2-methylcyclohexane. Hydrogen can add from above or from below the molecular plane with equal probability. Therefore, each stereocenter is generated as both image and mirror image, and the product is racemic.

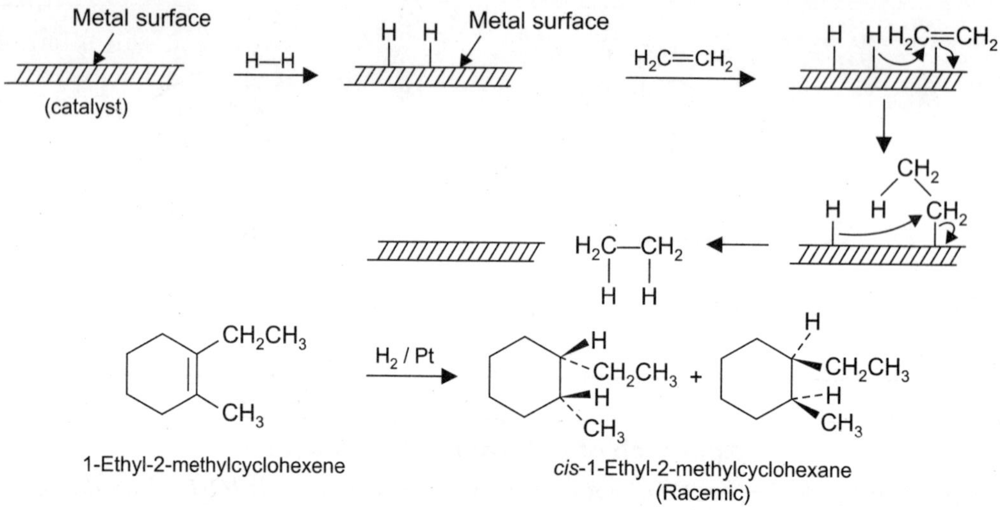

Two hydrogen atoms add to the same face of the *pi* bond (*syn* stereochemistry)

SCHEME 6.50

(*i*) Selectivity

The C≡C bond is reduced more readily than C=C (See, scheme 6.53*a*). Other unsaturated groupings with the exception of nitro groups and acid chlorides are reduced less readily. Thus catalytic hydrogenation is used for the selective reduction of C=C in the presence of aromatic rings and carbonyl groups.

(*ii*) Stereochemistry-Hydrogenation is Stereospecific

When addition of hydrogen to an alkene creates two stereocenters (scheme 6.50), only two of the four stereoisomers are formed since only *syn* addition can take place (The other two stereoisomers would arise *via anti* addition). Which two stereoisomers would be formed will depend on the geometry of the reacting alkene *i.e.*, whether it is *cis* or *trans*. *cis*-Alkene on *syn* addition will give *erythro*-pair of enantiomers (scheme 6.51, when the groups attached to the two stereocenters are the same (R=R′), the *meso*-stereoisomer will be formed instead of *erythro*

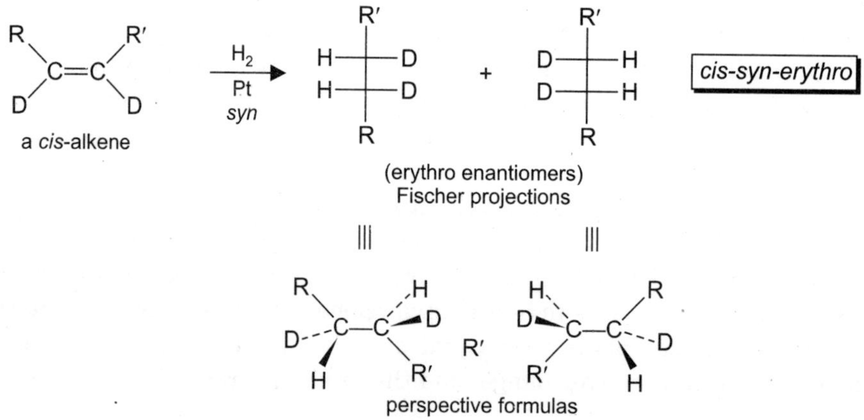

SCHEME 6.51

enantiomers. When the alkene is *trans*, the *syn* addition of hydrogen will afford *threo* enantiomers. This shows that hydrogenation is a stereospecific reaction (scheme 6.52).

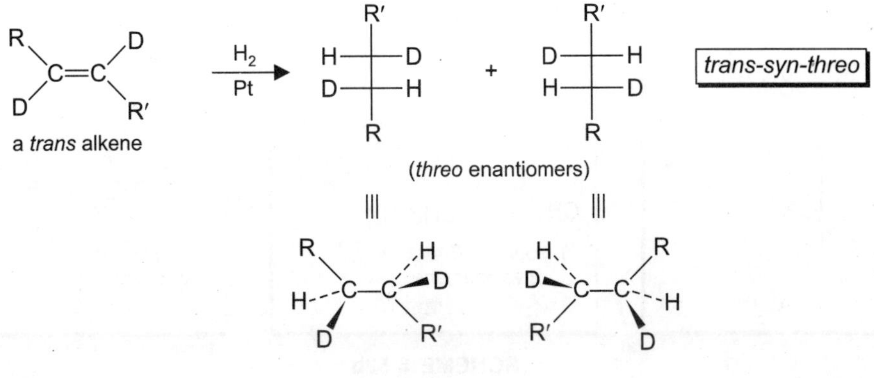

SCHEME 6.52

Problem Solving Hint

One can predict the outcome of reactions which generate two stereocenters by remembering the system cis, syn, erythro where every terms means to the same side. The sterechemical outcome can be predicted by changing any two of the terms in these systems at one time in a given reaction :

- *cis, syn, erythro (e.g., scheme 6.51)*
- *trans, syn, threo (e.g., scheme 6.52)*
- *cis, anti, threo (e.g., scheme 6.20)*

In rigid cyclic systems *e.g.*, cholesterol, the addition of H_2 is from the less hindered α side, thus cholesterol gives mainly *trans* ring-junction on reduction, since the angular methyl group hinders the fit of the catalyst on the β side of the molecule. However, when the C-3 OH group is axial, the fit to the catalyst is hindered from the α side of the molecule as well and reduction gives a mixture of *cis* and *trans* decalins in equal amounts (scheme 6.52*a*).

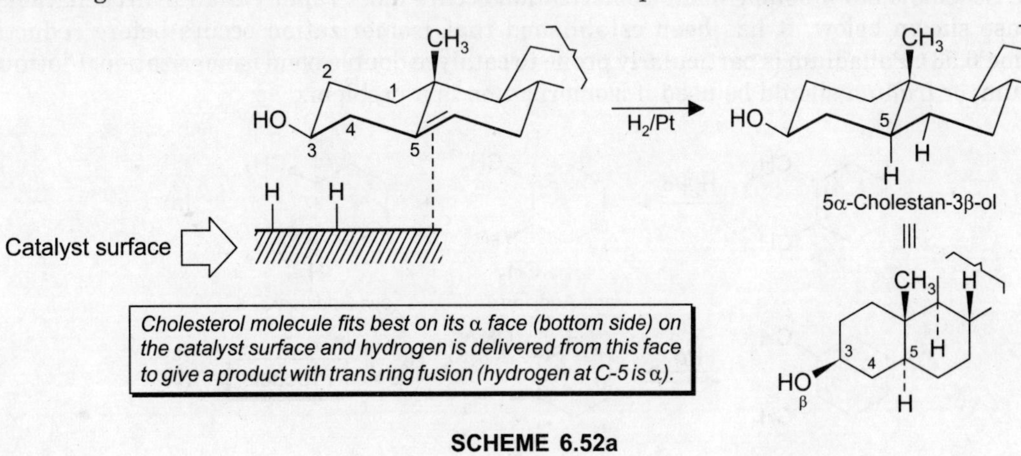

SCHEME 6.52a

EXERCISE 6.11

Predict the products along with stereochemical outcome of the reactions (scheme 6.52b).

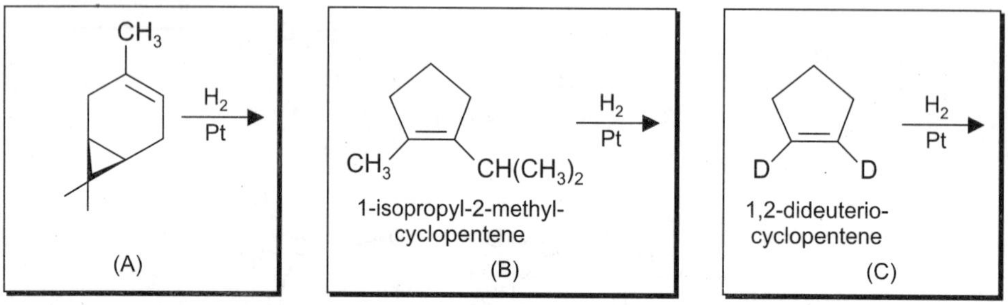

SCHEME 6.52b

ANSWER. *(A) Steric hindrance completely inhibits hydrogenation on one side of the ring, thus hydrogen is added on the side opposite to the three membered ring to give (IV, scheme 6.52c). (B) syn Addition of H_2 to a cyclic compound gives cis stereoisomers (V). (C) The product formed will be meso since the addition from either top or bottom face of the double bond gives only the same meso compound (VI).*

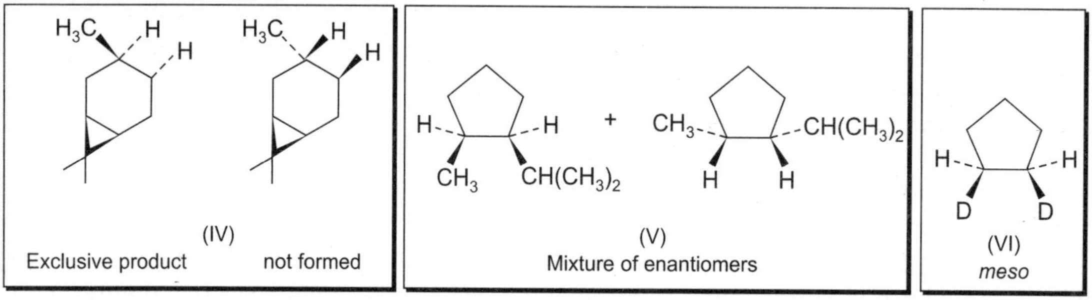

SCHEME 6.52c

Although *syn* addition of H_2 is generally observed, *anti* addition is sometimes observed, (eq. a, Scheme 6.53) if double bond isomerization occurs more rapidly than hydrogenation. In the case shown below, it has been established that isomerization occurs before reduction (scheme 6.53). Palladium is particularly prone to catalyze double bond isomerization. Platinum, rhodium, or iridium should be used if isomerization is a problem.

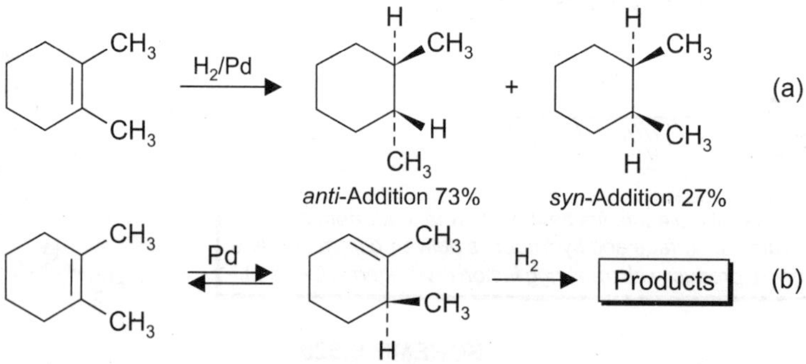

SCHEME 6.53

The catalytic hydrogenation of alkynes also leads to *syn* addition of two H atoms and proceeds with *syn* stereochemistry (scheme 6.53*a*). The hydrogenation of an alkyne can be stopped at the alkene stage by using a less reactive ("poisoned" catalyst). Lindlar's catalyst is one such poisoned palladium catalyst using barium sulphate poisoned with quinolene.

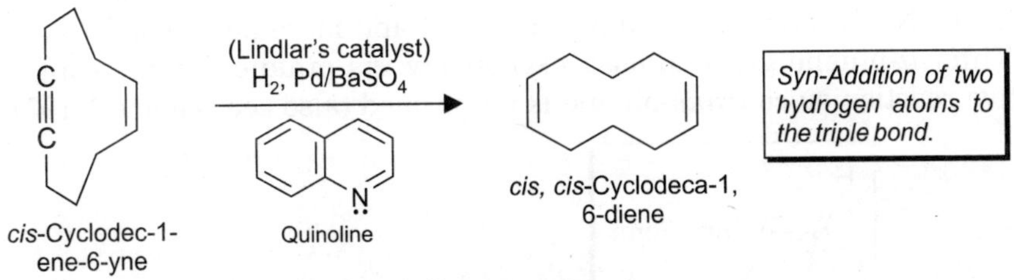

SCHEME 6.53a

Sodium metal in liquid ammonia reduces alkynes with *anti* stereochemistry to give *trans*-alkenes. Electrons are transferred more readily to acetylenes as compared to olefins (greater reactivity of acetylenes towards nucleophiles). The reagents of choice are the metal-amine and metal-ammonia systems. The reduction is selectively *anti*, since the vinyl radical formed during reduction is more stable in the *trans* configuration (scheme 6.54). The method is complementary to the stereospecifically *syn* addition of hydrogen during catalytic methods. Thus, both *cis*- and *trans*-cyclooctene can be obtained from cyclooctyne.

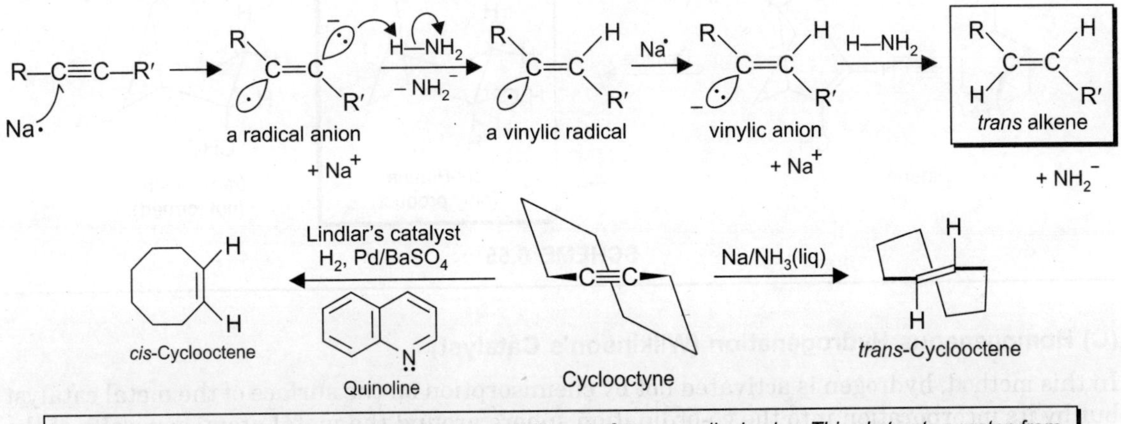

Sodium atom donates an electron (addition) to the alkyne to form an radical anion. This abstracts a proton from ammonia to give a vinylic radical. The transfer of another electron gives a vinylic anion, which is more stable in trans form. It removes a proton from ammonia to give trans alkene.

SCHEME 6.54

(B) Diimide Reductions

An interesting nonmetallic compound often used for the hydrogenation of carbon-carbon multiple bonds is diimide which is formed by oxidation of hydrazine (NH_2NH_2). Diimide is generated in the reaction mixture as needed.

Diimide is an unstable compound, it decompose to nitrogen and hydrogen (in the absence of additives), but when it is generated in the presence of an alkene, rapid reduction occurs. The driving force being the great stability of the nitrogen molecule compound to the —N=N-system (scheme 6.55). Although both the *syn* and *anti* forms of diimide are produced by the oxidation of hydrazine, only the *syn* form reduces the double bonds by a cyclic mechanism. The addition is therefore, stereochemically *syn*.

Diimide is a highly selective reagent employed to reduce readily the symmetrical bonds like C≡C, C=C, N=N, and not the unsymmetrical and more polar bonds like C≡N, N=O, S=O, etc. Thus α-pinene gives almost exclusively *cis*-pinane which is an example of a stereoselective reaction since *trans*-pinane is not formed (also see scheme 1.107).

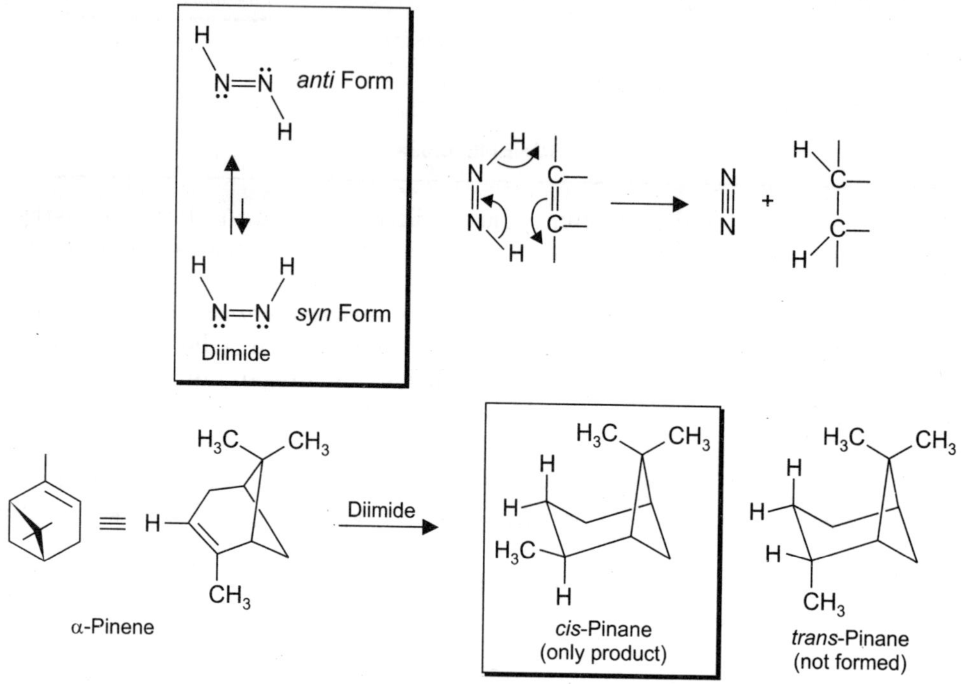

SCHEME 6.55

(C) Homogeneous Hydrogenation (Wilkinson's Catalyst)

In this method, hydrogen is activated not by chemisorption on the surface of the metal catalyst but by its incorporation into the co-ordination sphere around the metal atom, generally of the group VIII *e.g.*, rhodium, ruthenium. In the heterogeneous hydrogenation reactions two phases are involved, the solid phase of the catalyst (Pd, Pt, Ni etc.) containing the adsorbed hydrogen, while the liquid phase of the solution has the unsaturated reactant. When however, one uses a transition metal complex *e.g.*, tris (triphenylphosphine) chlororhodium, $Rh[(C_6H_5)_3P]_3Cl$ (called Wilkinson's catalyst), hydrogenation occurs in a single phase-in solution.

The Wilkinson's catalyst in solution probably exchanges reversibly one Ph_3P group for a loosely held solvent molecule (A, scheme 6.56). The catalyst reacts with hydrogen to form a metal hydride (B, the dihydrido complex). The H—H bond is cleaved and each hydrogen is separately bonded to rhodium, and thus the metal uses one of its electron pairs and is oxidized to the rhodium (III) state. In the next step alkene reacts with the complex and replaces the solvent molecule. The alkene-metal bond formation is a result of the overlap of an empty orbital of the metal with π cloud of the alkene. Two hydrogen atoms are then transferred to the alkene one at a time and the net result is a *syn* addition.

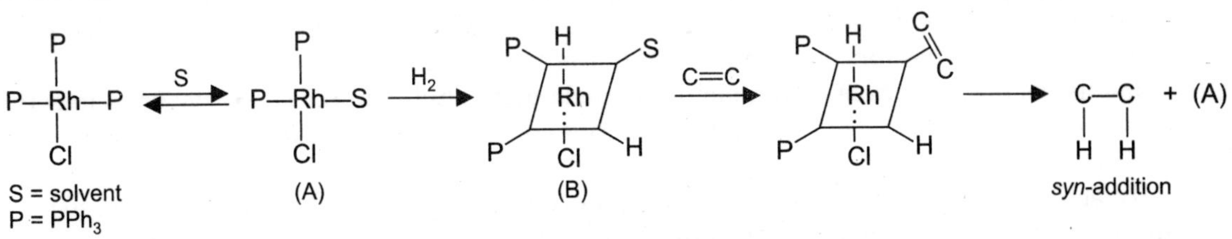

SCHEME 6.56

When *e.g.*, maleic acid is hydrogenated using D_2 in place of H_2 with Wilkinson's catalyst two stereocenters are generated in the product (2,3-D_2) butanedioic acid. This acid can exist as a *meso* compound, or as a pair of enantiomers *i.e.*, only three stereoisomers can exist. However, maleic acid gives only the *meso* acid to show, that hydrogenation with Wilkinson's catalyst involves *syn*-addition (scheme 6.57). If one starts with fumaric acid only racemic acid is formed.

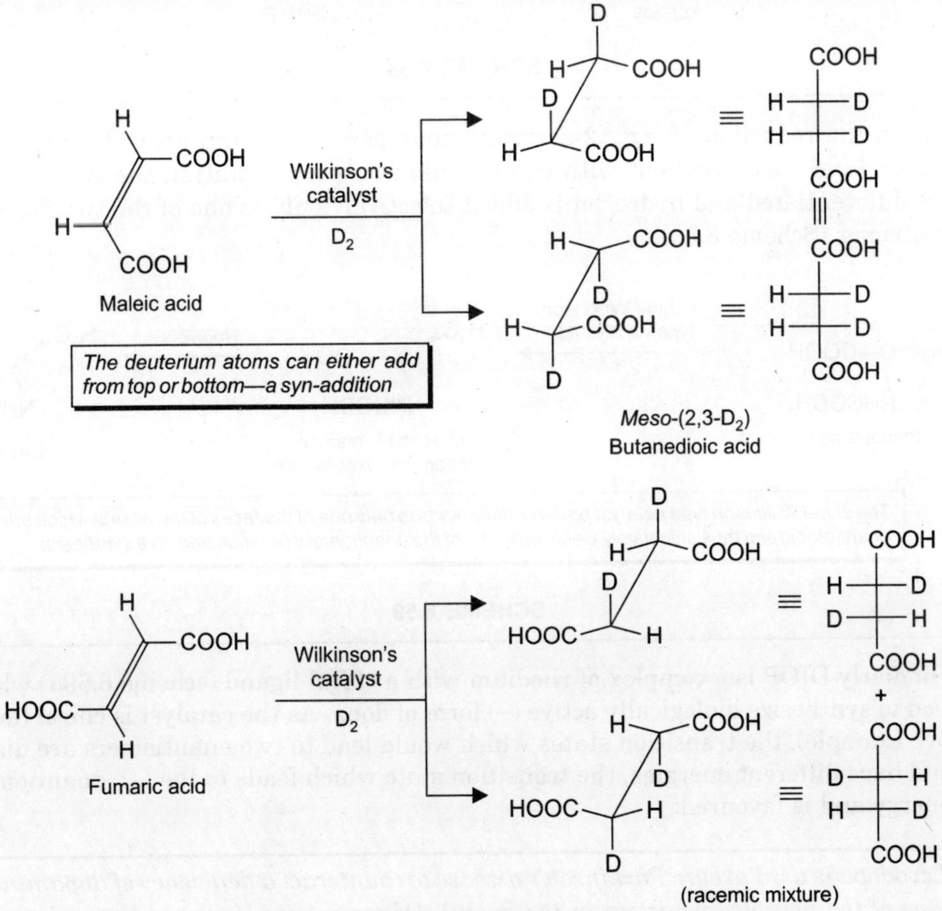

SCHEME 6.57

(D) Enantioselective Catalytic Hydrogenation with Chiral Wilkinson–Type Catalysts

The hydrogenation of an achiral starting material with *e.g.*, optically inactive Wilkinson catalyst will always produce a racemic modification. Thus, 2-acetylaminopropenoic acid (with enantiotopic faces) on hydrogenation with Wilkinson's catalyst give racemic acetylalanine since there is equal probability for hydrogen to add to the either face. Thus, here the situation is just the same as seen in the case of maleic or fumaric acids (Scheme 6.57). However, when the Wilkinson's catalyst is made chiral (optically active) by attaching optically active ligands, then the hydrogenation occurs in a chiral medium and can give optically active products. Several optically active ligands have been developed where the chirality is due to either stereogenic carbon or stereogenic phosphorus (scheme 6.58).

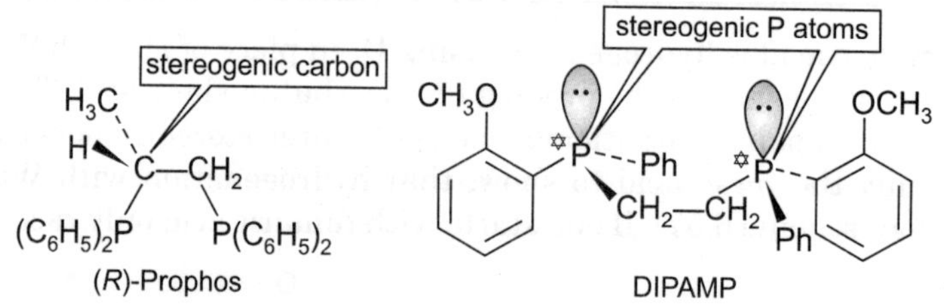

SCHEME 6.58

When the reduction of *e.g.*, 2-acetylaminopropenoic acid (an achiral compound with enantiotopic faces) is carried out with chiral Wilkinson's type catalyst, the two enantiotopic faces are differentiated and hydrogen is added selectively only to one of the two faces to give one enantiomer (Scheme 6.59).

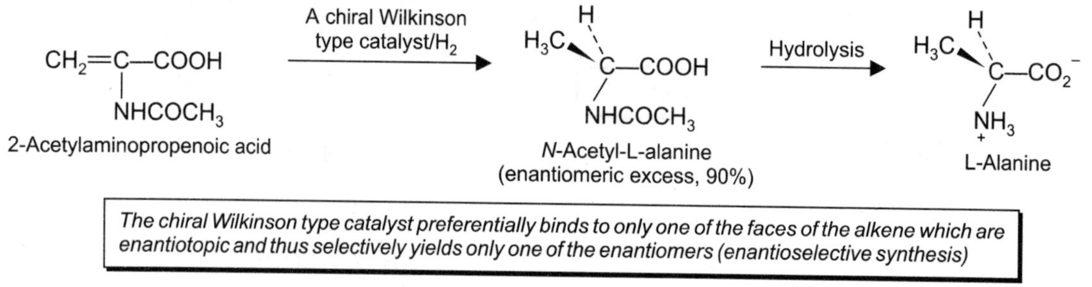

SCHEME 6.59

Similarly DIOP is a complex of rhodium with a chiral ligand (scheme 6.59*a*) which has been used to synthesize biologically active (–) form of dopa. As the catalyst is chiral (and as in the above example), the transition states which would lead to two enantiomers are diastereotopic, and have different energies, the transition state which leads to the (–) -enantiomer is of lower energy and is favoured.

> *Levodopa is used to cure Parkinson's disease to counteract a deficiency of dopamine (one of the neurotransmitters in the brain). Dopamine itself cannot be used as a drug, because it cannot cross the "blood-brain barrier".*

(–)-DOPA is an amino acid derived from tyrosine is capable of crossing the blood-brain-barrier to reach the site of action and is converted by brain enzymes into dopamine. Significantly only the (–) enantiomer of dopa is converted into dopamine while the other enantiomer is toxic to the patient.

Rhodium (DIOP)Cl_2—a chiral catalyst for enantioselective reduction =

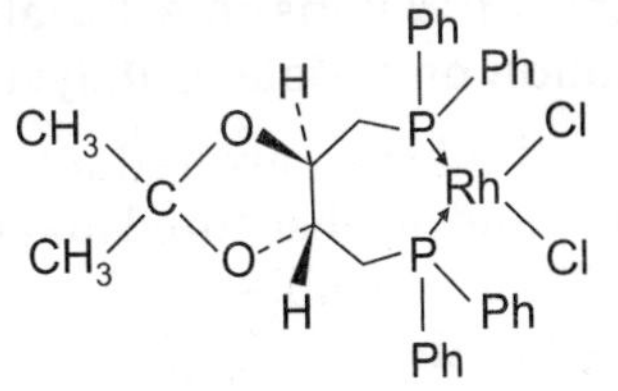

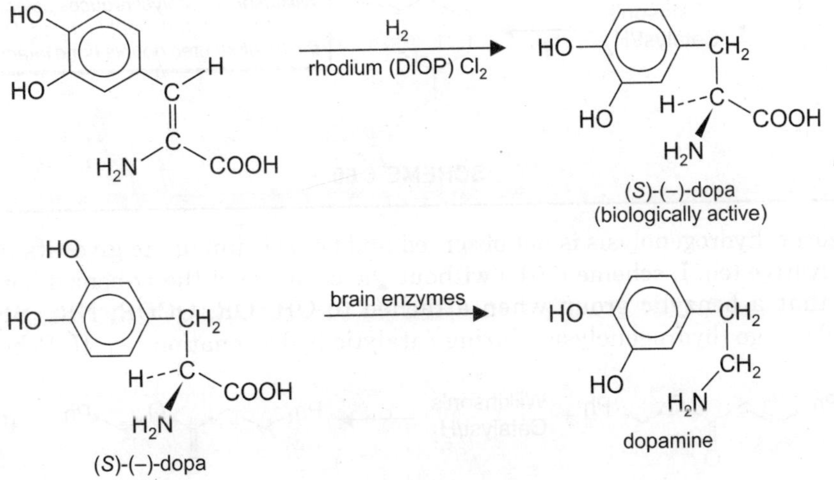

SCHEME 6.59a

Similarly Ru(BINAP)Cl_2 is a chiral ruthenium phosphine catalyst where the chirality is derived from 1, 1′-binaphthyl system. It is used as homogeneous hydrogenation catalyst for asymmetric reduction (Scheme 6.59*b*) *i.e.*, creation of a new stereocenter with the exclusive formation of one enantiomer. Thus, when the ligands in a Wilkinson's catalyst [$(Ph_3P)_3RhCl$] are chiral, optical activity is induced in the product of hydrogenation. An example of these ligands is binaphthyl-phosphine BINAP.

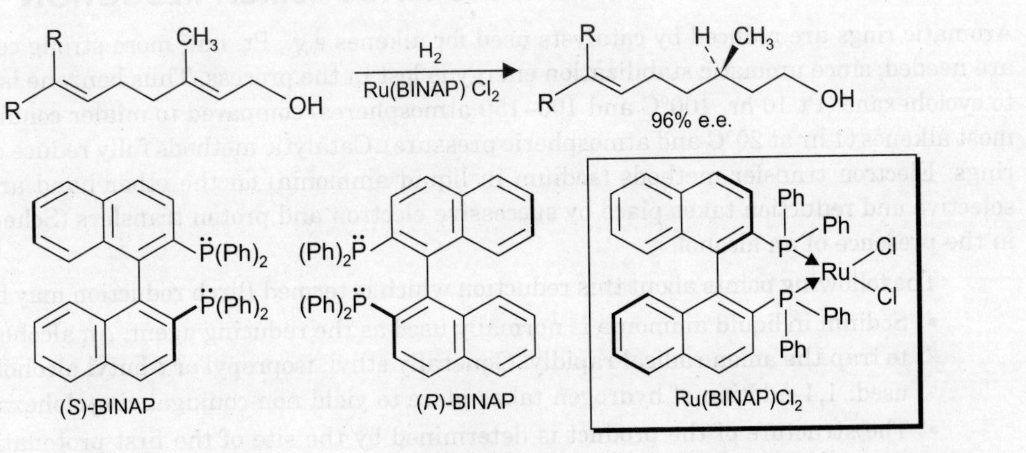

SCHEME 6.59b

(E) Advantages of Wilkinson's Catalyst (Homogeneous Catalytic Hydrogenation) over Hydrogenation on a Metal Catalyst

Only alkenes and alkynes are reduced, and groups like C=O, C≡N and NO_2 remain intact. Moreover, mono- and disubstituted alkenes are far more reactive than tri- and tetrasubstituted double bonds. Thus a double bond may be selectively reduced as in carvone (Scheme 6.60) which is reduced to the dihydroderivative with trisubstituted double bond intact.

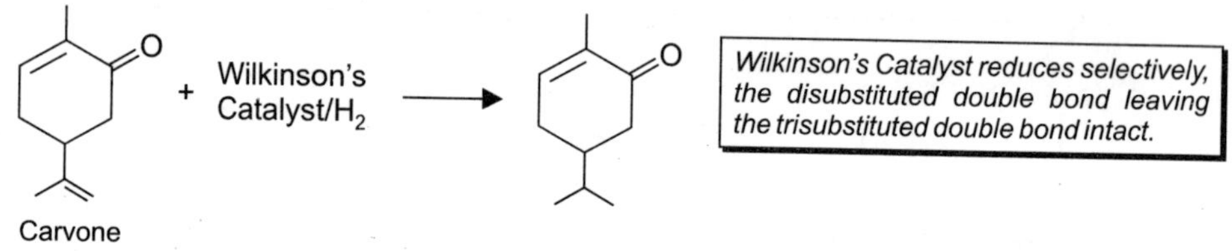

SCHEME 6.60

Moreover, hydrogenolysis is not observed and benzyl cinnamate gives its corresponding dihydro derivative (eq. I, scheme 6.61) without the cleavage of the *O*-benzyl bond. It may be mentioned that a benzylic group when attached to OH, OR, OCOR, NR_2, SR or halogen undergoes cleavage (hydrogenolysis) during catalytic hydrogenation (eq. II, Scheme 6.61).

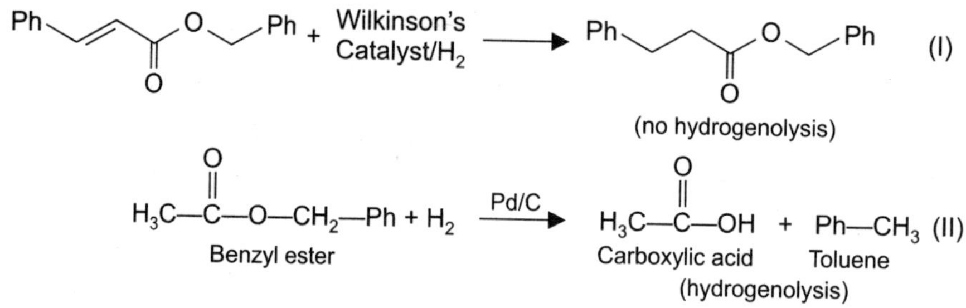

SCHEME 6.61

6.5 HYDROGENATION OF AROMATIC RINGS—BIRCH REDUCTION

Aromatic rings are reduced by catalysts used for alkenes *e.g.*, Pt, but, more strong conditions are needed, since aromatic stabilization energy is lost in the process. Thus benzene is reduced to cyclohexane (Pt 10 hr, 100°C and 100–150 atmospheres) compared to milder conditions for most alkenes (1 hr at 20°C and atmospheric pressure). Catalytic methods fully reduce aromatic rings. Electron transfer methods (sodium in liquid ammonia) on the other hand are highly selective and reduction takes place by successive electron and proton transfers (Scheme 6.62) in the presence of an alcohol.

The following points about this reduction which is termed Birch reduction may be noted:

- Sodium in liquid ammonia is normally used as the reducing agent. An alcohol is used to trap the anion radical rapidly. Generally ethyl, isopropyl or *t*-butyl alcohol may be used. 1,4-Addtion of hydrogen takes place to yield non-conjugated cyclohexadienes.
- The structure of the product is determined by the site of the first protonation. The anion thus formed reacts at the central carbon of the delocalized system to complete

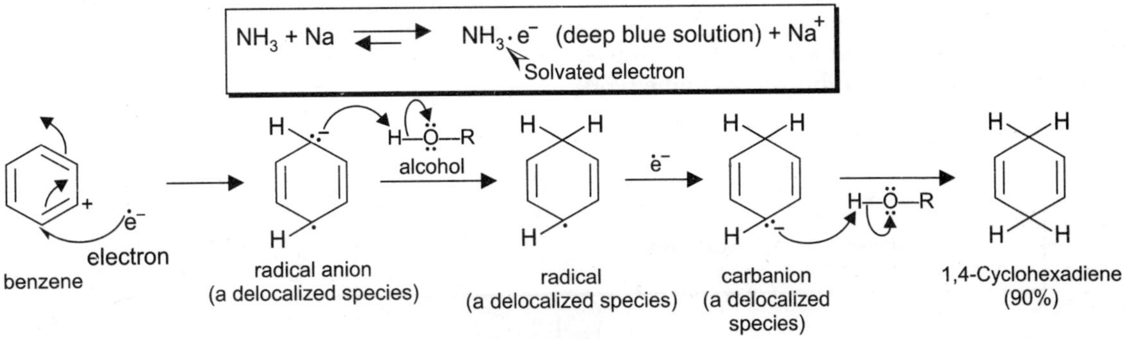

Birch Reduction

SCHEME 6.62

the 1,4-addition. Electron-releasing substituents direct the first protonation to the *ortho* position (scheme 6.63) as in the Birch reduction of methoxybenzene (anisole) which gives 1-methoxy-1,4-cyclohexadiene, which can be hydrolyzed by dilute acid to 2-cyclohexenone, and this method provides a useful synthesis of 2-cyclohexenones.

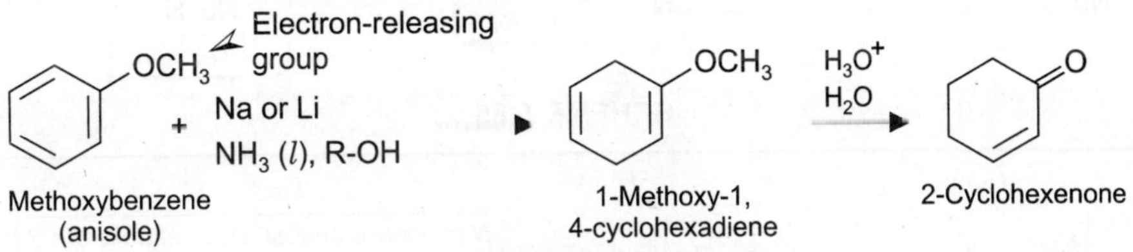

SCHEME 6.63

In contrast, the $-CO_2^-$ substituent directs protonation to the *para* position (scheme 6.64).

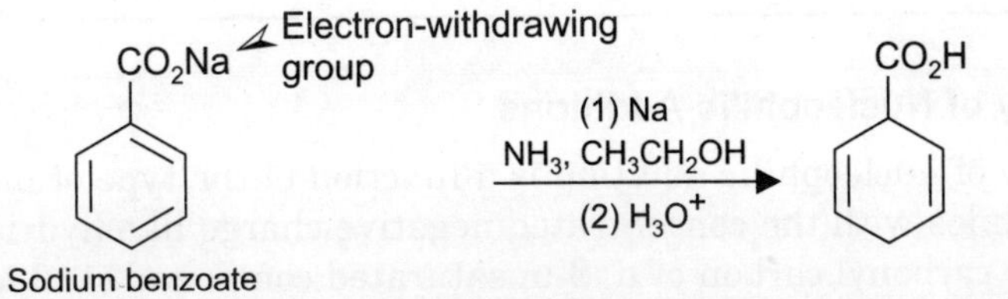

SCHEME 6.64

6.6 ADDITION OF NUCLEOPHILES TO CARBON-CARBON DOUBLE BOND

Alkenes are not attacked by nucleophiles unless the carbon-carbon double bond is conjugated to a group of $-M$ type. Because of conjugation, α, β-unsaturated carbonyl compounds have two electrophilic sites. The carbonyl carbon (eq. I, scheme 6.65) and the β-carbon (eq. II, scheme 6.65) and both can form a bond with a nucleophile as seen during the addition of a polar reagent H-Nu to an α, β-unsaturated carbonyl compound. Nucleophilic addition to the carbonyl carbon is called 1, 2-addition and to the β-carbon is called 1,4-addition or conjugate addition since addition occurs to a double bond conjugated with the carbonyl group. The initially formed product of the conjugate addition is an enol which tautomerizes to a ketone or an aldehyde (Scheme 6.65). The overall reaction amounts to the addition to the carbon-carbon double bond with the nucleophile adding to the β-carbon and a proton adding to the α-carbon. Significantly nucleophile almost never attacks at the 3 position (*i.e.* α carbon) since the resulting carbanion will not be resonance stabilized (scheme 6.66).

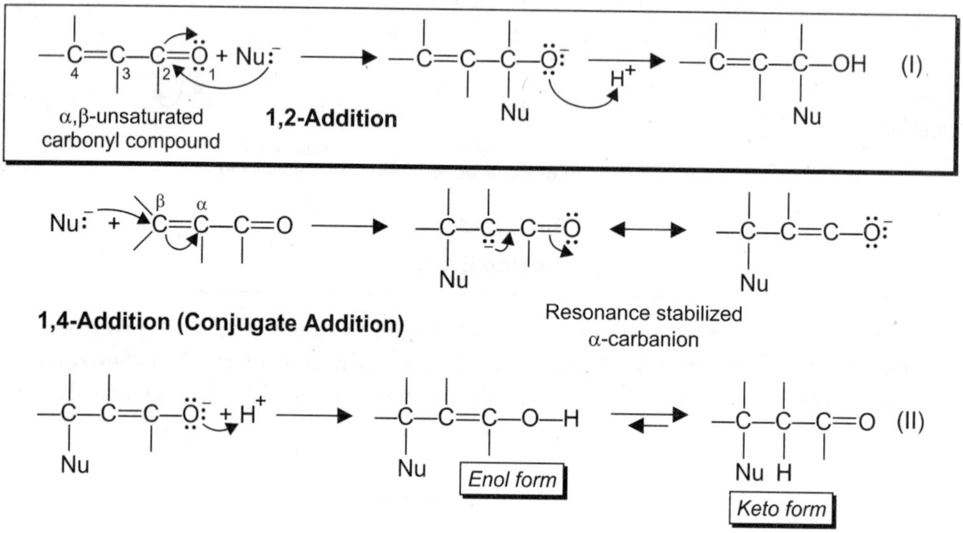

SCHEME 6.65

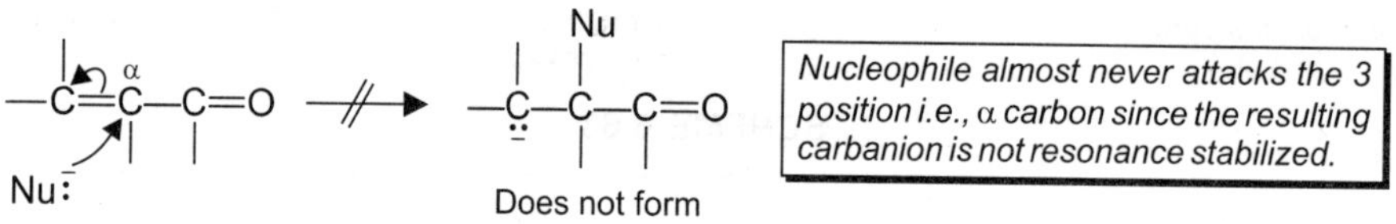

SCHEME 6.66

(a) Regioselectivity of Nucleophilic Additions

The regioselectivity of nucleophilic addition is a function of the type of nucleophile employed. Generally, nucleophiles with the concentrated negative charge like hydride ion and Grignard reagents add to the carbonyl carbon of α, β-unsaturated compound (1, 2-addition) while those with diffused delocalized charge *e.g.*, enolate anions add to the β carbon to give conjugate addition. Put in other words strong bases add to the carbonyl group while weak bases to the double bond. The following points may be noted:

- The 1, 2-addition occurs more rapidly than conjugate addition, therefore, the product from 1, 2-addition is formed first and, if addition is irreversible, it will be the final product of the reaction.
- In case 1, 2-addition is reversible the initially formed 1, 2-addition product goes back to the starting material, allowing the slow conjugate addition to occur. Since conjugate addition is irreversible, the conjugate addition product will accumulate and consequently will be the major product.
- A carbon-oxygen double bond is stronger than a carbon-carbon double bond. Thus a 1, 2-addition product is less stable than the 1,4-addition product (scheme 6.67). These additions are therefore, under kinetic control versus thermodynamic control of product formation.

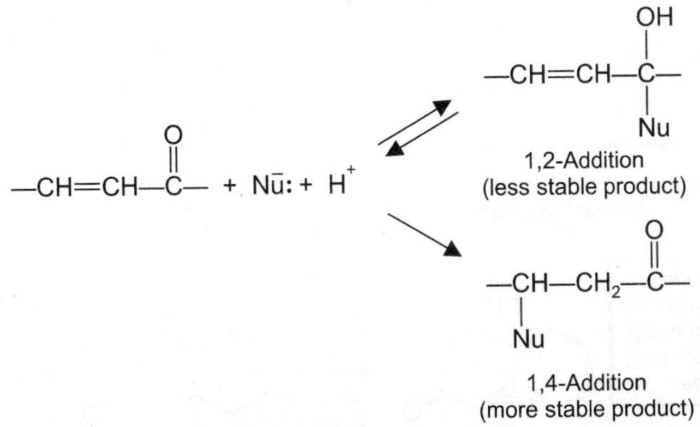

SCHEME 6.67

- The 1, 2-addition will be irreversible when the nucleophile is a strong base, since strong bases are poor leaving groups and cannot be eliminated from the 1, 2-addition product. 1, 2-addition will be reversible if the nucleophile is a weak base and, therefore, can be eliminated from the 1, 2-addition product. When 1, 2-addition is reversible, the conjugate addition product will be the major product of the reaction even though it is more slowly formed.

(b) Michael Addition

Many different nucleophiles can add to the α, β-carbonyl compounds. When the anion is resonance stabilized enolate ion (*e.g.*, an anion of a diketone) and is used as a nucleophile in a conjugate addition, the reaction is termed a Michael addition. In fact the term Michael addition is often used rather loosely to include all conjugate addition reactions.

The Michael reaction takes place with a wide variety of α, β-unsaturated carbonyl compounds as well as α, β-unsaturated nitrites and nitro compounds. Effective nucleophiles may be derived from a β-diketone, β-keto-ester, β-ketonitrite or even an enamine.

(i) Michael Reaction During Robinson Ring Annulation

The process fuses a cyclohexanone ring on the starting cyclic ketone and involves a Michael addition as well as an aldol condensation (scheme 6.68). The reaction begins by the addition of the enolate anion of the ketone to methyl vinyl ketone in a conjugate sense to the β-carbon of the enone (Michael reaction) and a new carbon-carbon bond is formed. This is followed by another carbon-carbon bond forming reaction-Aldol condensation. In the last step, the aldol formed undergoes dehydration on heating to give the enone. (The reaction is made enantioselective under chiral influence, scheme 2.56).

(ii) Michael Addition Using Enamines

Enamines are used as nucleophiles to carry out Michael additions (scheme 6.69) and then the reaction is specifically named Stork enamine reaction.

(iii) Conjugate Addition with Gilman Reagents

Lithium diorganocopper reagents (Gilman reagents) add to α, β-unsaturated aldehydes and ketones to give products of 1,4-(*i.e.*, conjugate) addition (scheme 6.70). The reaction is closely related with Michael reaction.

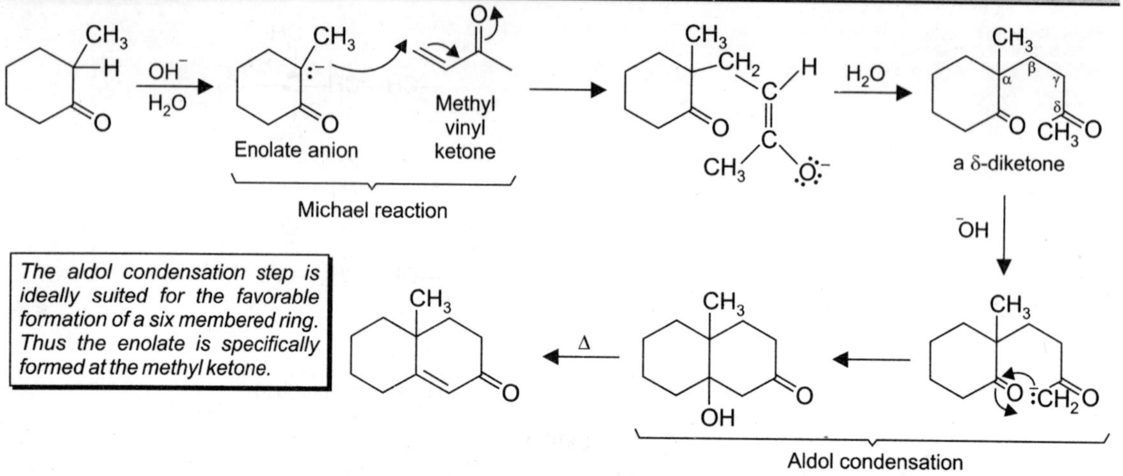

The aldol condensation step is ideally suited for the favorable formation of a six membered ring. Thus the enolate is specifically formed at the methyl ketone.

SCHEME 6.68

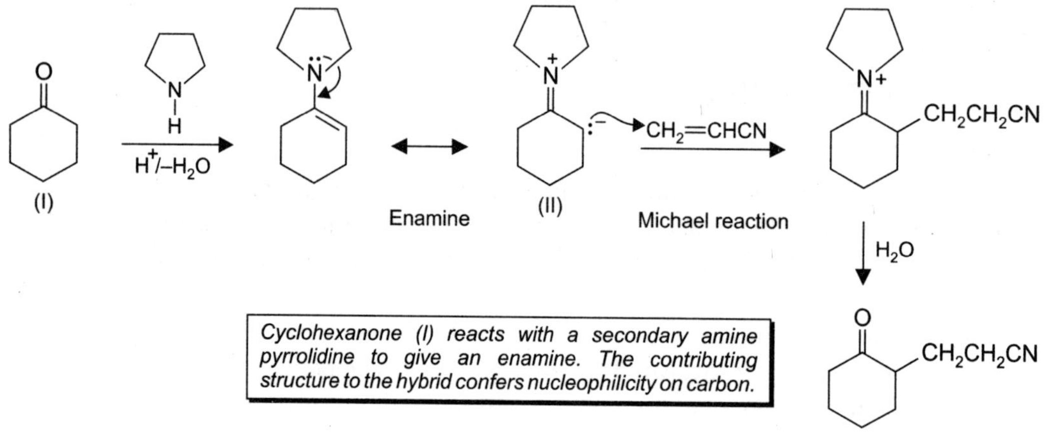

Cyclohexanone (I) reacts with a secondary amine pyrrolidine to give an enamine. The contributing structure to the hybrid confers nucleophilicity on carbon.

SCHEME 6.69

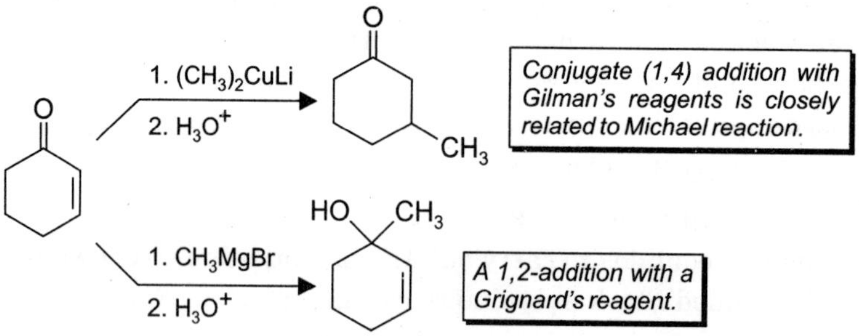

Conjugate (1,4) addition with Gilman's reagents is closely related to Michael reaction.

A 1,2-addition with a Grignard's reagent.

SCHEME 6.70

6.7 ADDITION OF FREE RADICALS TO π-SYSTEMS

The free radicals add readily to an alkene to give a new bond and a new free radical (scheme 6.71). In the process, the π-bond is broken—the carbon–carbon σ bond which is formed is about 30 kcal stronger than the π-bond which is broken. The reactions are initiated by a peroxide or azoisobisbutyronitrile (AIBN) which on heating or irradiation give reactive radicals R• (scheme 6.71).

The free radical reactions used in synthesis are chain reactions and a chain is terminated by the reaction of radicals among themselves.

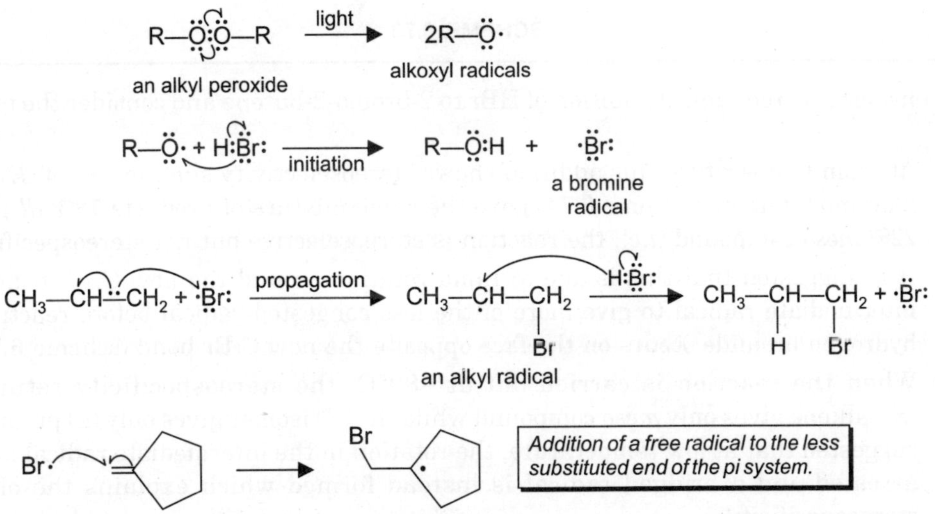

SCHEME 6.71

(i) Regioselectivity

A free radical addition occurs readily at the less substituted carbon atom of an unsymmetrical alkene like propene (scheme 6.71). All substrates $CH_2{=}CHX$ are preferentially attacked at the CH_2 regardless of the nature of X or of the attacking radical and the major reason is therefore, steric. When one considers the free radical addition of HBr to an unsymmetrical alkene, it comes out that the addition is *anti*-Markovnikov. This compared with the observed orientation of HBr (Markovikov electrophilic) shows that in both situations the formation of the more stable secondary intermediate is involved (scheme 6.72).

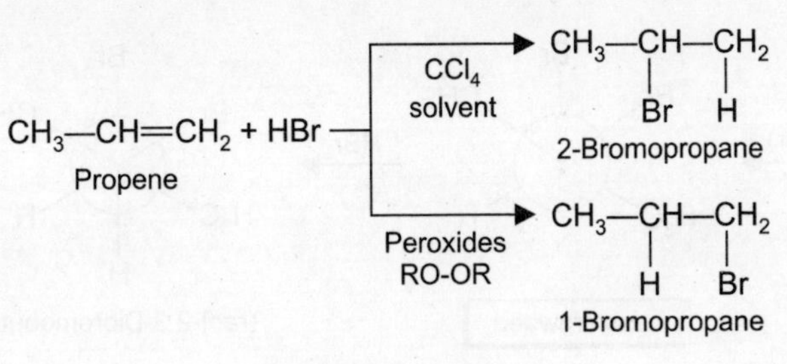

SCHEME 6.72

(ii) Stereochemistry

Many free-radical additions are stereoselective, while others are not. Thus the addition of HBr to cyclohexenes gives *anti*-addition to give *trans*-diaxial addition products (scheme 6.73) and is best explained on the basis of a bromine bridged intermediate (scheme 6.73).

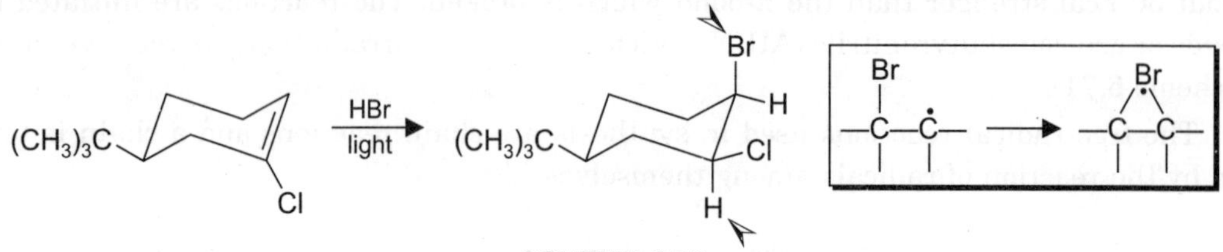

SCHEME 6.73

Consider the free radical addition of HBr to 2-bromo-2-butene and consider the following points:

- At room temperature this addition shows stereoselectivity since either of (*E*)- or (*Z*)-2-bromo-2-butene (scheme 6.74) gave the same mixture of products 78% *dl* pair and 22% *meso* compound (*i.e.*, the reaction is stereoselective but not stereospecific).
- It is suggested that this is due to rapid rotation around the new C—C bond in the intermediate radical to give more of the less congested radical before reaction with hydrogen bromide occurs on the face opposite the new C-Br bond (scheme 6.74).
- When the reaction is carried out at –80°C, the stereospecificity returns–the (*E*)–alkene gives only *meso* compound while the (*Z*) isomer gives only (±) product. It is suggested that at low temperature, the rotation in the intermediate radical is almost arrested and a bridged radical is instead formed which explains the observed stereospecificity.

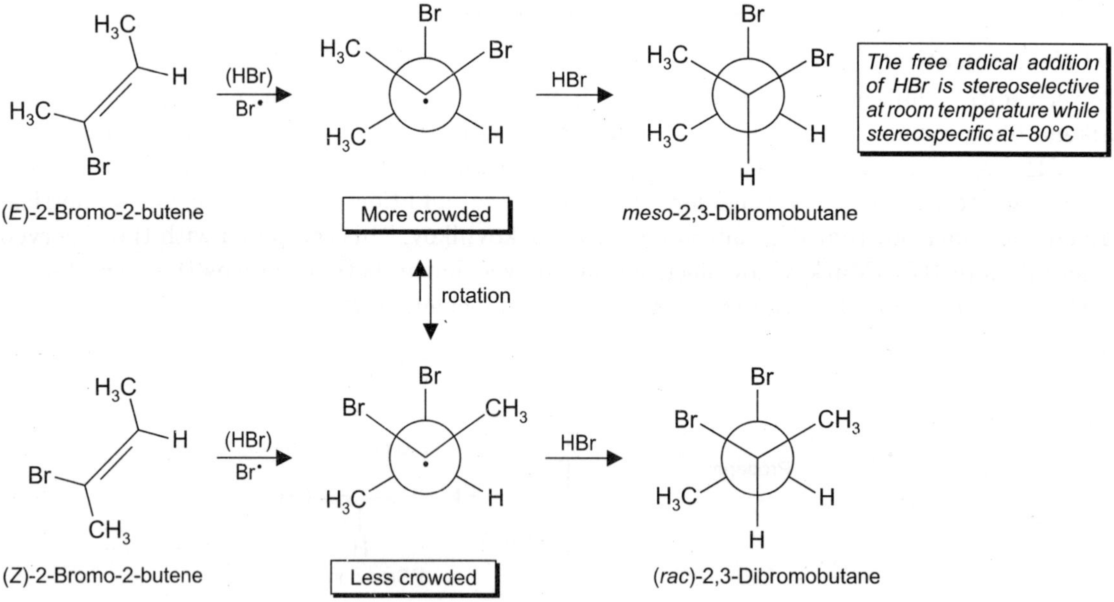

SCHEME 6.74

(iii) Regioselectivity of Intramolecular Addition of a Free Radical to a Double Bond—Free Radical Cyclization

Intramolecular olefin additions produce rings and these reactions are much faster than other reaction pathways. Intramolecular addition of radicals generated at a carbon having a 5, 6-double bond can give either a five membered ring or a six membered ring, but mostly a five membered ring is greatly preferred kinetically (scheme 6.75, methylcyclopentane from 6-bromo-1-hexene). The mechanism and explanation for the formation of these products is as under:

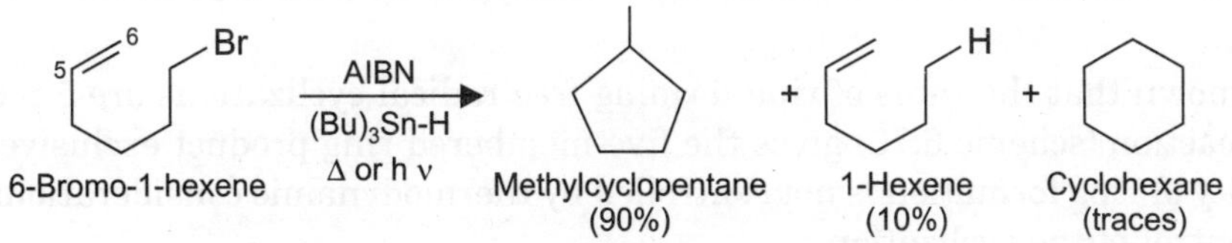

SCHEME 6.75

- Tributyltin hydride is a versatile reagent for the generation of a free radical at a specific carbon atom in a molecule. Tributyltin radical is generated by the abstraction of a hydrogen atom from tributyltin hydride with AIBN (scheme 6.76) which is the initiation step. This radical abstracts efficiently the halogens (except fluorine), PhS and PhSe. Since the Sn-H bond is weak, it does not abstract hydrogen from an alkyl C—H group.
- The method thus provides a convenient way to convert an alkyl halide to the corresponding hydrocarbon (R—X→R—H) no matter what the structure of substrate.
- When this method is applied to an unsaturated alkyl halide, two pathways are available to the free radical. The 5-hexenyl radical (scheme 6.77) can react in two ways. Its reaction with tributyltin hydride gives 1-hexene and tributyltin radical is generated (eq. II, scheme 6.77). This process leads to the substitution of the halogen atom by hydrogen (reduction process).

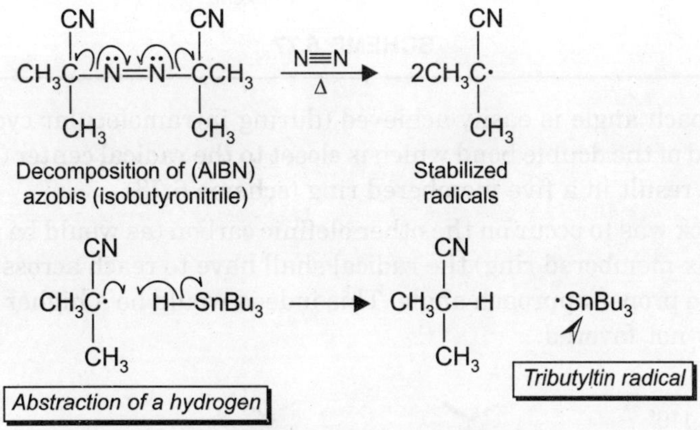

$$(Bu)_3Sn\cdot + \boxed{R\text{-}Br} \longrightarrow (Bu)_3SnBr + R\cdot$$

$$R\cdot + (Bu)_3Sn\text{-}H \longrightarrow \boxed{RH} + (Bu)_3Sn\cdot$$

SCHEME 6.76

- In the second pathway (eq. I, scheme 6.77) cyclization takes place by the addition of the alkyl radical to the π-bond generating a different alkyl radical. This cyclized radical then reacts with tributyltin hybride to give the saturated hydrocarbon and regenerating tributyltin radical to propogate the chain reaction.

These two pathways *i.e.* substitution (reduction) and addition compete with each other. The course of reaction can be changed by changing the concentration of tributyltin hydride. Since the substitution reaction is a bimolecular process high concentration of tributyltin hydride will favour it, while a lower concentration of tributyltin hydride will favour the cyclization process.

It is known that the rates of ring-forming free radical cyclizations are 5 > 6 > 7. It was found that reaction (scheme 6.75) gives the five-membered ring product exclusively. Thus the rigioselectivity of ring formation is not controlled by thermodynamic considerations, but rather by kinetic control of the cyclization.

- It is found that bond formation between the radical and the alkene stereoelectronically requires an approach angle of about 110° between the free radical center and the olefinic plane. This is due to the fact that free radical addition results from donation of the unpaired electron on the radical into the π antibonding orbital of the olefin, which coincidentally makes an angle of about 110° with the olefinic plane (scheme 6.78).

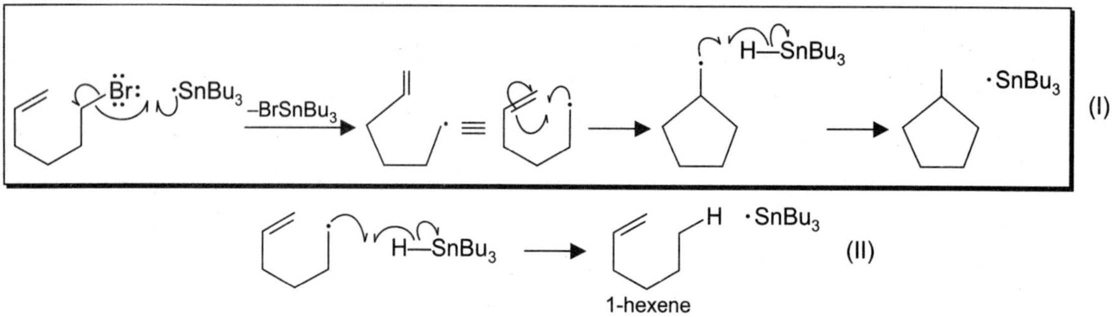

SCHEME 6.77

- This approach angle is easily achieved (during intramolecular cyclization) by attack on that end of the double bond which is closet to the radical center (favourable entropy factors) to result in a five membered ring (scheme 6.78).
- If the attack was to occur on the other olefinic carbon (as would be required for formation of a six-membered ring) the radical shall have to reach across the double bond to achieve the proper approach angle. This indeed would be a higher energy path and is kinetically not favored.

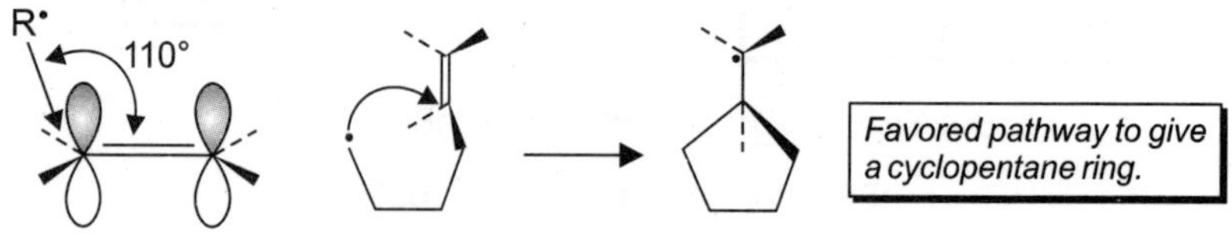

SCHEME 6.78

One may appreciate that while Diels Alder reaction is one of the methods to form fused six-membered rings, radical cyclization is superior to synthesize a fused ring system containing a five membered carbocycle. The power of the method for the synthesis of fused ring system with a five membered carbocycle is presented (scheme 6.79). In this case the reaction is induced at a bridgehead position where radical reactions are normally difficult since the radical cannot achieve planar geometry.

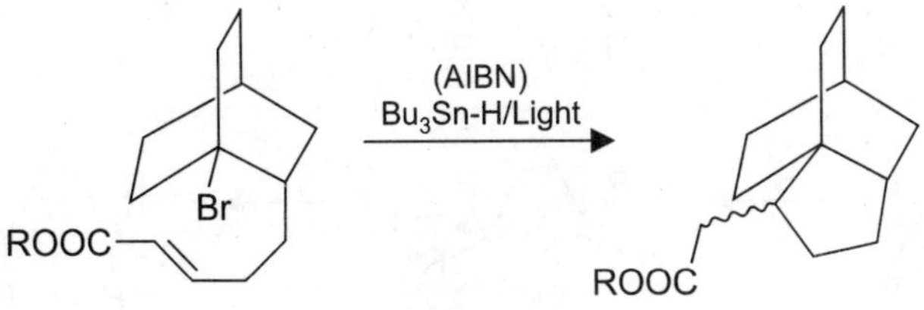

Scheme 6.79

6.8 ADDITIONS TO CYCLOPROPANE RING SYSTEMS

Cyclopropane is an alkane, however, it reacts like a double bond due to strain in its small ring. It is hydrogenated like an alkene and undergoes electrophilic additions (scheme 6.80). In appropriate cases involving substituted cyclopropanes these additions mostly follow Markovnikov rule (exceptions are however, known, and then the regioselectivity is found to be less). 1,1,2-Trimethylcyclopropane adds HX as shown (6.81). Thus H^+ the electrophile adds to the carbon with larger number of hydrogens (this is the CH_2 shown by an arrow) while the nucleophile adds to the carbon that can best stabilize the carbocation *i.e.* the tertiary rather than the secondary carbocation (scheme 6.81).

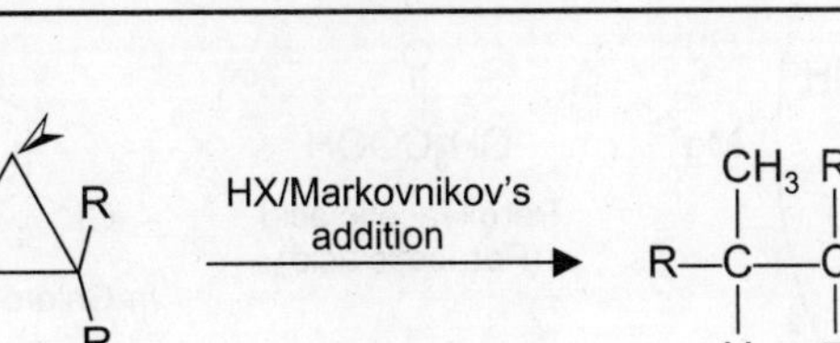

SCHEME 6.80

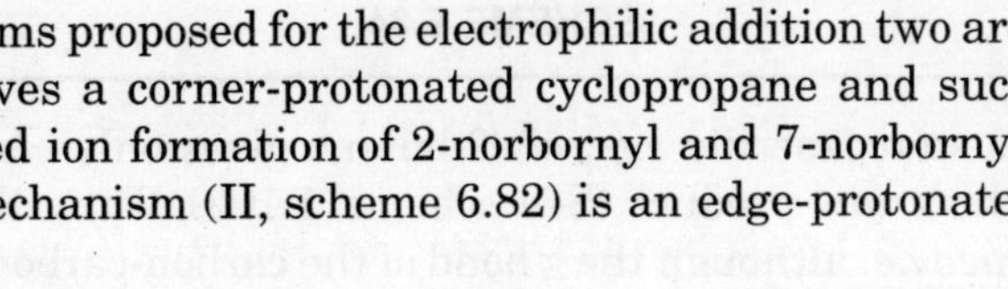

SCHEME 6.81

Of the mechanisms proposed for the electrophilic addition two are presented. Mechanism (I, scheme 6.82) involves a corner-protonated cyclopropane and such mechanisms are also involved during bridged ion formation of 2-norbornyl and 7-norbornyl systems (see, schemes 3.70 and 3.78). The mechanism (II, scheme 6.82) is an edge-protonated cyclopropane.

Free radicals also add to a cyclopropane system *e.g.*, bromine adds by a free radical mechanism in the presence of uv light. During the addition (scheme 6.83), the initial radical attacks the least substituted carbon and the second going to the most substituted carbon.

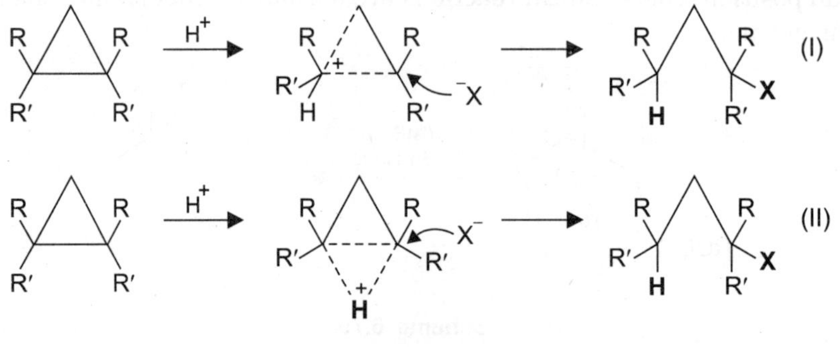

SCHEME 6.82

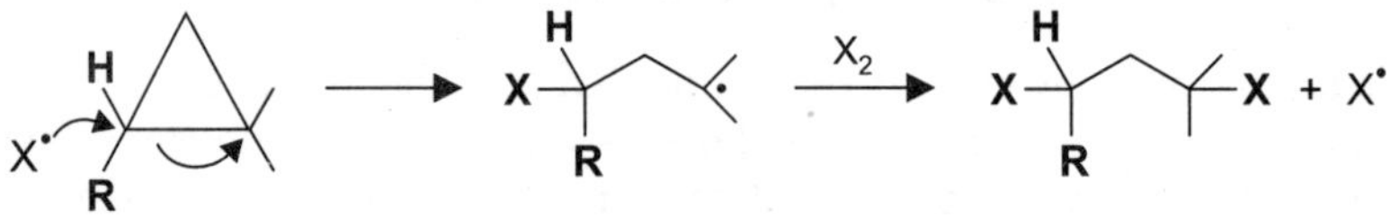

SCHEME 6.83

6.9 EPOXIDATION OF ALKENES

(a) Epoxidation with Peracids

Alkenes react with peroxyacids to give epoxides. The commonly used peracid is 3-chloroperbenzoic acid (MCPBA), however, it being shock sensitive, recently megnesium monoperoxyphthalate (MMPP) has been introduced as a safer substitute (Scheme 6.84). The following points may be noted:

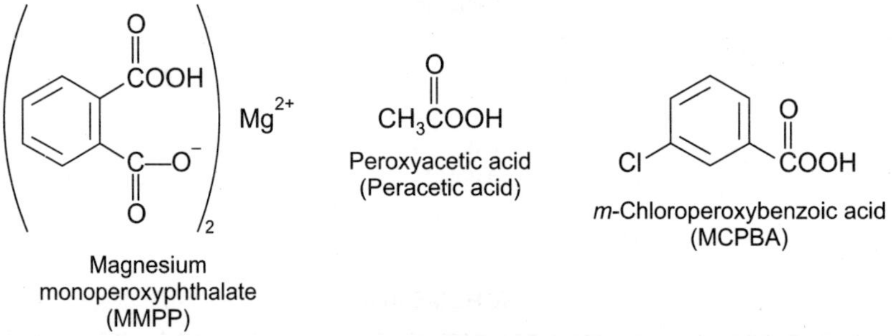

SCHEME 6.84

- The reaction is stereospecific *e.g.*, *cis*-2-butene gives only *cis*-epoxide while *trans*-2-butene gives the *trans*-product (See scheme 1.102). Thus the configuration of the alkene is retained *i.e.* although the π bond of the carbon-carbon double bond is broken, at no time, there is free rotation around the remaining σ bond. Thus the bond making and bond breaking steps are concerted—epoxidation is a single step transfer of the peroxy oxygen atom to the double bond (scheme 6.85).

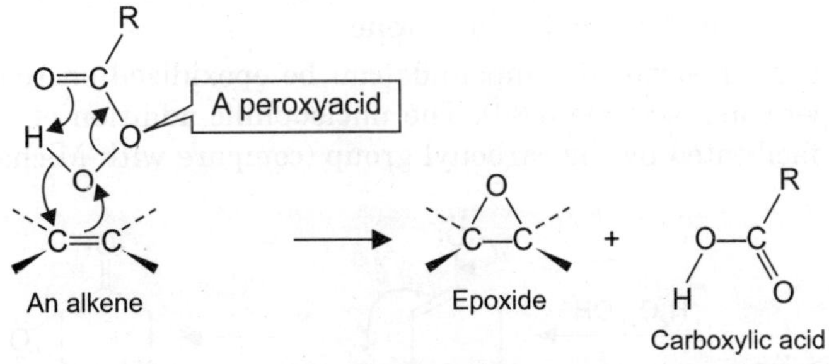

SCHEME 6.85

- Epoxidation is a regioselective reaction, since the more electron rich double bond reacts when more than one double bond is present. This also shows that the peroxy acid behaves as an electrophile, thus 1,2-dimethyl 1,4-cyclohexadiene reacts selectively at the more highly substituted of the two double bonds (scheme 6.86).

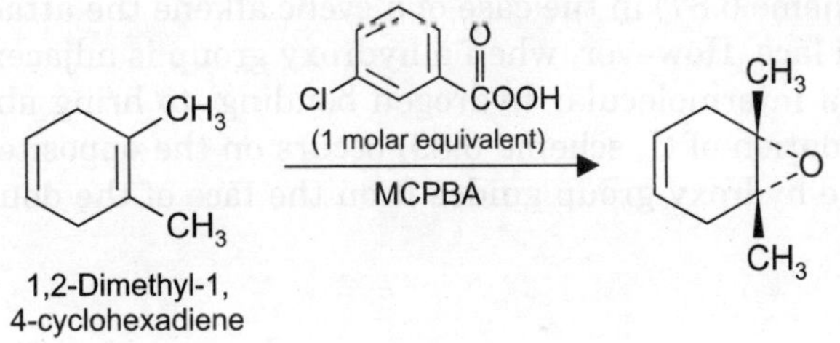

SCHEME 6.86

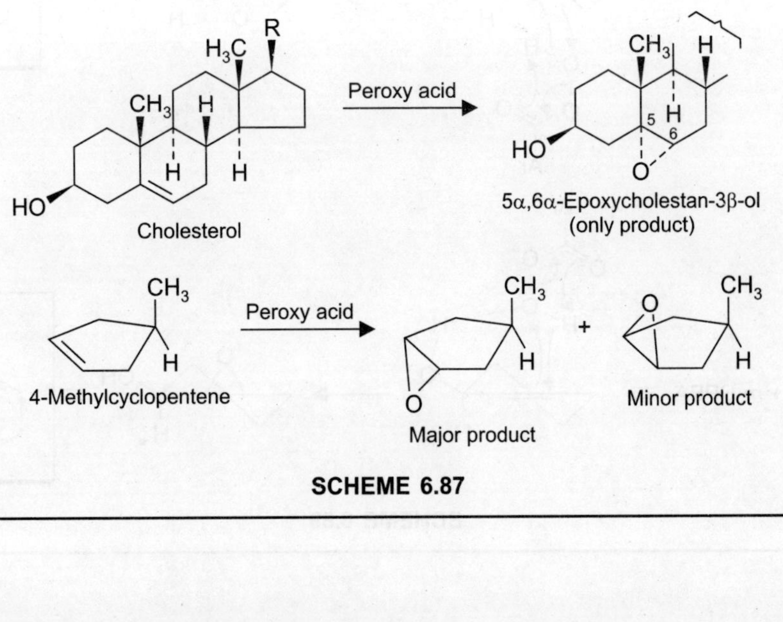

SCHEME 6.87

- Cyclic alkenes are attacked by the peracid predominantly from the less hindered side. Cholosterol molecule *e.g.*, presents extensive steric hindrance from its top face (β-face) by the angular methyl groups. Thus many reagents like a peroxyacid react preferentially at the relatively unhindered α-face (scheme 6.87). Similar is the situation with epoxidation of 4-methylcyclopentene.
- α, β-Unsaturated carbonyl compounds can be epoxidized on reaction with alkaline hydrogen peroxide (scheme 6.88). The nucleophilic addition of hydroperoxide anion to C=C is facilitated by the carbonyl group (compare with Michael reaction).

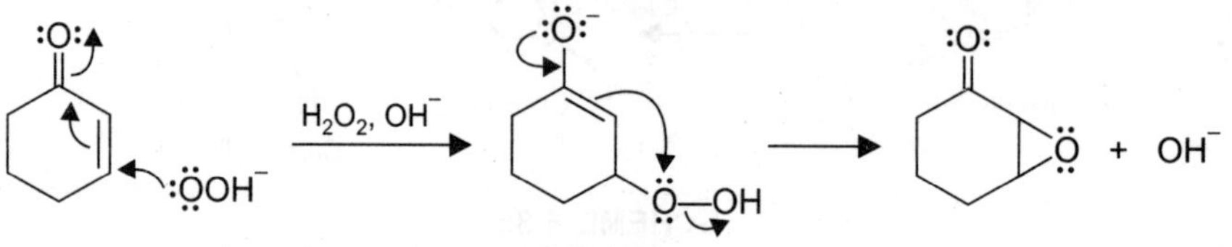

SCHEME 6.88

(b) Enantioselective Epoxidation—Sharpless Reaction

(i) Introduction

As seen above (See, scheme 6.87) in the case of a cyclic alkene the attack of the peracid occurs from the less hindered face. However, when a hydroxy group is adjacent to the double bond it directs the peracid (via intermolecular hydrogen bonding) to bring about epoxidation on the same face. Thus epoxidation of (I, scheme 6.89) occurs on the opposite face of methoxy group for steric reasons while hydroxy group guides it on the face of the double bond occupied by it (II, scheme 6.89).

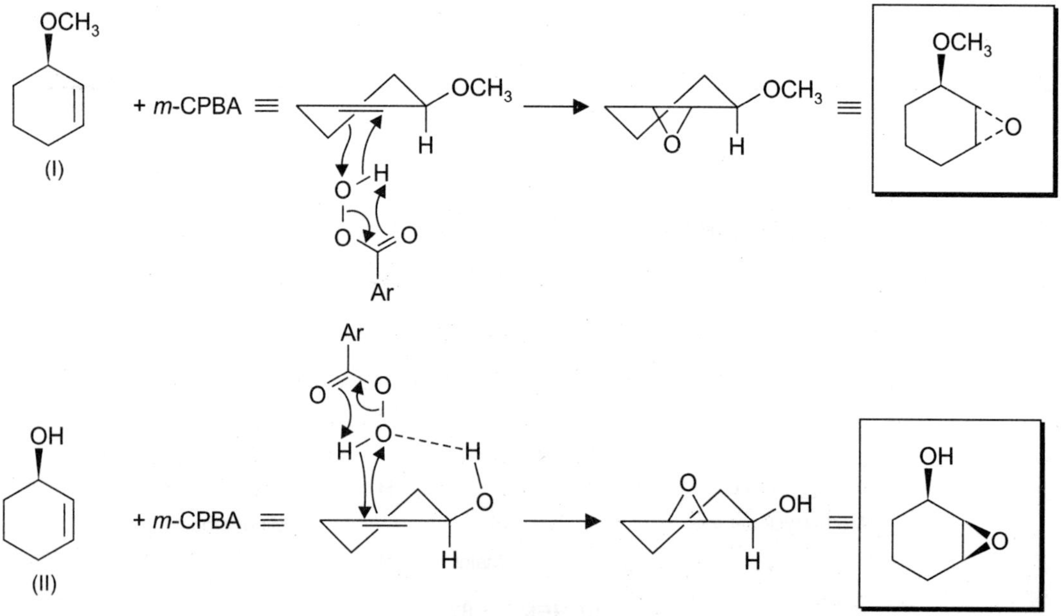

SCHEME 6.89

Like peroxyacids, alkylhydroperoxides (ROOH) *e.g.*, *tert*-butylhydroperoxide bring about epoxidation of an alkene under catalysis by a transition metal (scheme 6.90) like molybdenum, vanadium or titanium. Mechanistically the reaction involves the initial coordination of the metal to both the allylic hydroxyl group and *tert*-butylhydroperoxide. The double bond of the alkene unit then displaces the peroxy group and the said epoxidation process, therefore, involves the displacement of two alkoxy ligands, a process similar to Sharpless epoxidation.

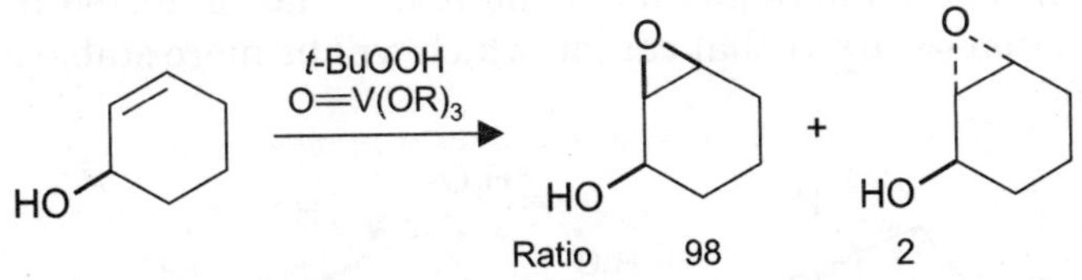

SCHEME 6.90

Significantly the diastereoselective epoxidation (scheme 6.90) becomes highly enantioselective by adding a chiral material. This elegant modern method of asymmetric epoxidation developed by Sharpless uses as a chiral molecule (+)-diethyltartrate (+)-DET or its enantiomer (scheme 6.91). Thus the reaction of a primary allylic alcohol with *tert*-butylhydroperoxide catalyzed by titanium (IV) tetraisopropoxide in the presence of (+) or (–)-DET gives specifically one enantiomer of the pair of epoxides. Further mechanistic and other details are discussed in detail (See schemes 2.48–2.51).

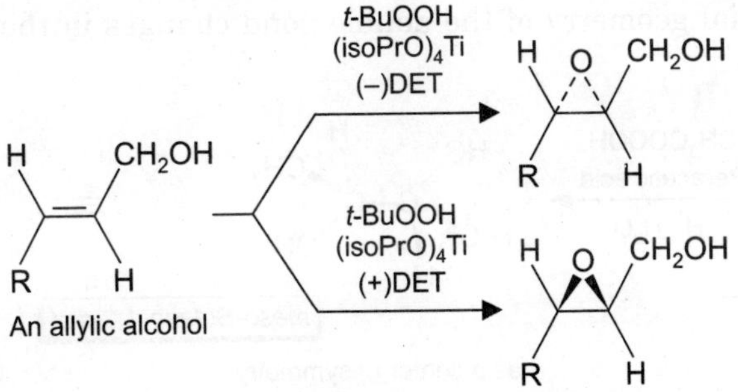

SCHEME 6.91

(c) Epoxides from Halohydrins

These are discussed (see scheme 3.61*a*).

(d) Some Reactions of Epoxides

(i) Hydrolysis (Acidic or Basic Conditions) Gives 1,2-Diols

Acid or base catalyzed hydrolysis of an epoxide gives glycols with *anti*-stereochemistry (scheme 6.92). The reactivity to the weak nucleophile (water) is enhanced by protonation of the ring oxygen. The protonated epoxide is then attacked from the side opposite the epoxy ring (S_N2 conditions) to give a *trans*-product (scheme 6.92) as a racemic mixture, the mirror image structure is however, not drawn (scheme 6.92). The following points may be noted:

- Other weak nucleophiles like alcohols and bromide ion are also effective when the reactivity of the epoxide is increased by protonation (See, schemes 3.56*b* and 3.56*c*).

- In the case of an unsymmetrical epoxide different products are obtained under acid-catalyzed and neutral conditions (See, scheme 3.56*b* and 3.56*c*).
- Under neutral (or basic) conditions S_N2 opening is observed *i.e.*, nucleophile attacks at the less hindered carbon of the epoxy ring.
- Under acid catalyzed epoxide ring opening the stereochemistry is predicted as from an S_N2 reaction *i.e.*, nucleophile approaches from the side opposite the leaving oxygen. The regiochemistry where possible is however, that predicted from an S_N1 reaction, the substitution occurs at that carbon which will be more stable as a carbocation (See scheme 3.56*c*).

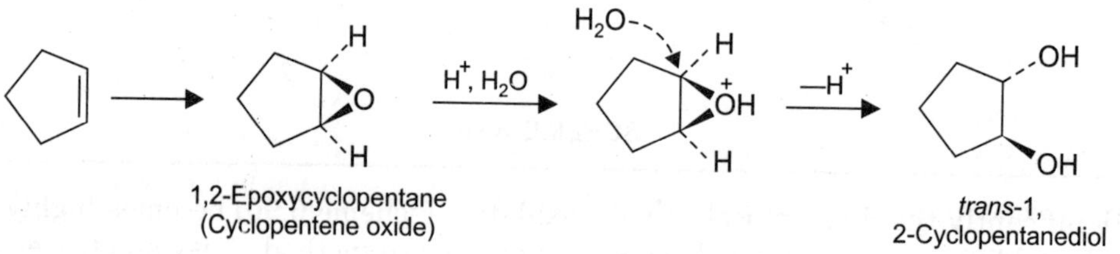

SCHEME 6.92

(ii) Direct Hydroxylation

When a stronger peracid, like peracetic or performic acid is used, the epoxide cannot be isolated (scheme 6.93). It directly reacts to give *anti* hydroxylation (via a hydroxy acetate or hydroxy formate). Thus *trans* 2-butene gives *meso* product, since the reaction involves one inversion of configuration, the initial geometry of the double bond changes in the product (compare with problem 6.2).

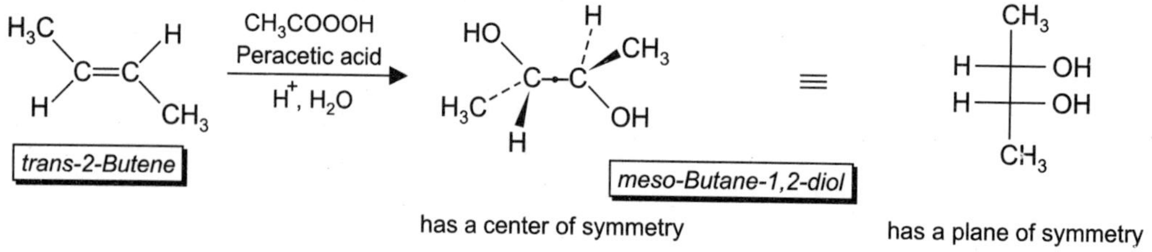

SCHEME 6.93

(iii) Formation of Trans-Diaxial Products

The ring opening of epoxides with base from a rigid cyclohexene devivative gives *trans* diaxial products (scheme 6.94), the attack of OH^- at other carbon would have given diequatorial product which is not observed (also see problem 6.19).

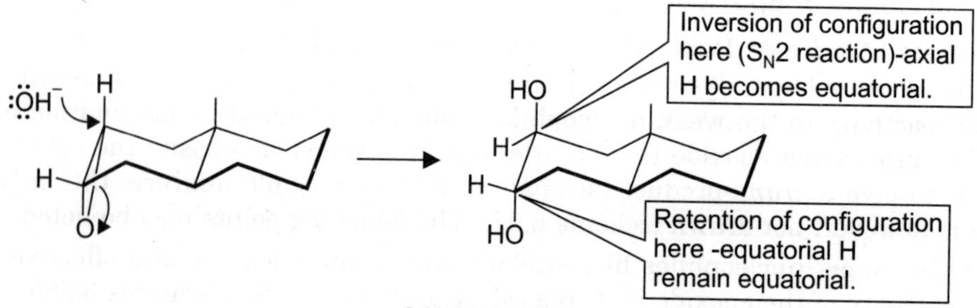

SCHEME 6.94

(iv) Reduction with Lithium Aluminium Hydride

Epoxides are reduced to alcohols with lithium aluminium hydride and as epoxides are derived from alkenes, the overall process is the hydration of an alkene. The method is complementary to the hydroboration method, since the hydride selectively attacks the less substituted carbon (S_N2 mechanism) of the epoxide to give more highly substituted alcohol (scheme 6.95). In rigid cyclohexane systems an axial alcohol is formed (scheme 6.95, also see scheme 3.56*b*).

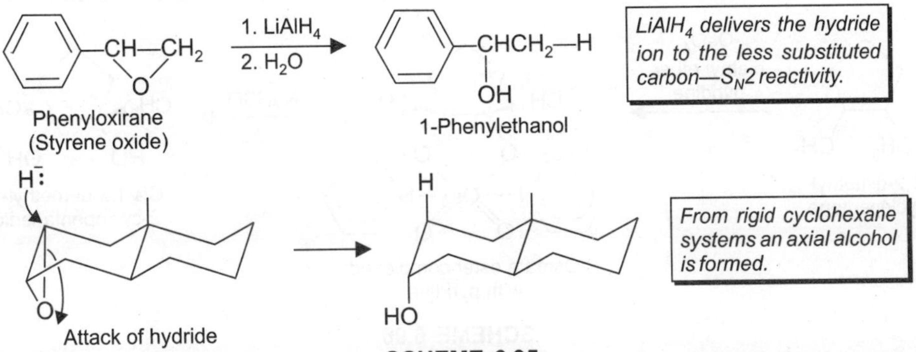

SCHEME 6.95

EXERCISE 6.12

Write the stereostructure of the product from the reduction of cis-4-tert-butylcyclohexene oxide with $LiAlD_4$.

ANSWER. *The opening of the epoxide will give the trans-diaxial product (scheme 6.95a).*

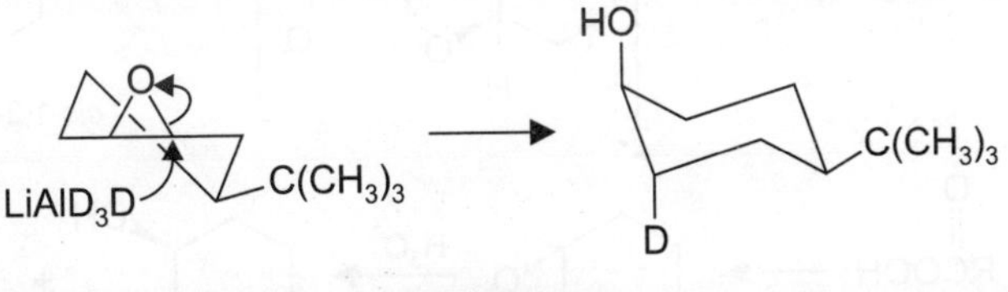

SCHEME 6.95a

6.10 HYDROXYLATION OF ALKENES—DIOL FORMATION

(i) Syn—Hydroxylation and Woodward Method

Conversion of an epoxide into a *trans* 1, 2-diol has already been discussed. The process involves the delivery of oxygen *syn* to the double bond to give an epoxide and its subsequent acid catalysed hydrolysis (scheme 6.92).

An alkene can be converted into a *cis*-1, 2-diol on reaction with OsO_4 or $KMnO_4$. With OsO_4 *e.g.*, a cyclic osmate ester is formed which is then hydrolyzed with sodium sulphite to give the addition product *i.e.*, a diol by the net addition of two hydroxyl groups to the same face of the alkene *i.e. syn* addition (scheme 6.96). Pyridine complexes with osmium atom and is often added as a catalyst. An alternative method is to react the alkene with a catalytic amount of OsO_4 along with an oxidizing agent such as H_2O_2 or an amine oxide to get the diol. The following points may be noted:

- Thus stereoselective formation of *trans*-1,2-diol from the oxirane of cyclopentene with acid in water is complementary to the oxidation of cyclopentene with potassium permanganate or osmium tetroxide which gives *cis*-1,2-cyclopentanediol. The stereochemistry of *cis*-diol is determined from the cyclic intermediate (scheme 6.96 and 6.96*a*) as shown for cyclohexane. If like oxirane of cyclopentene, cyclohexene is first oxidized with a peroxyacid and the epoxide thus obtained is opened in aqueous acid, *trans*-1,2-cyclohexanediol is formed as a racemic mixture (scheme 6.96*a*).

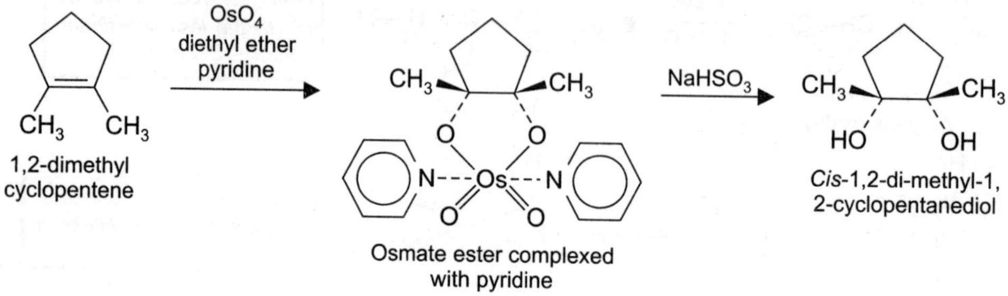

SCHEME 6.96

> *The chirality of cis- and trans-1,2-cyclohexane diols is to be understood on the lines of cis- and trans-1,2-dimethylcyclohexanes (see, schemes 4.31–4.33).*

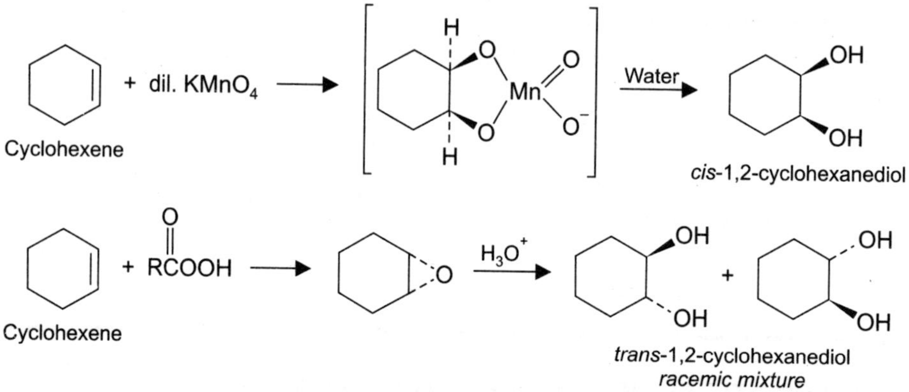

SCHEME 6.96a

- In rigid cyclic systems, OsO_4 attacks from the less hindered side, and the net result is the formation of the more stable of the two possible *cis* diols (scheme 6.97), compare with the diol in scheme 6.98).

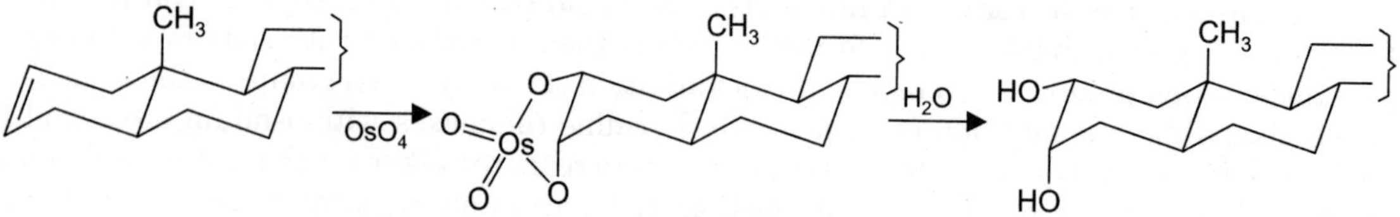

SCHEME 6.97

- In Woodward hydroxylation (iodine silver acetate wet *syn*-hydroxylation), an alkene on reaction with iodine, aqueous acetic acid and silver acetate also gives *cis* diols (scheme 6.98). Iodine reacts with the double bond to give an iodonium ion which undergoes displacement by acetate in the S_N2 displacement, giving a *trans*-iodoacetate (I, scheme 6.98). Anchimeric assistance by the acetate group, together with the powerful bonding capacity of silver ion for iodide gives a cyclic acetoxonium ion which reacts with water to give a *cis*-hydroxyacetate. Hydrolysis gives the *cis*-diol (scheme 6.98).

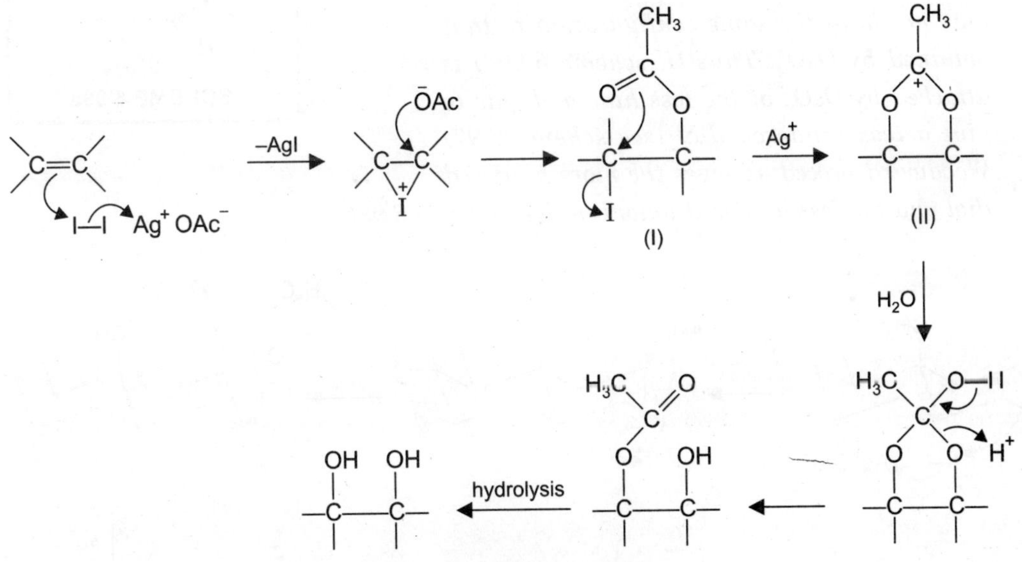

Woodward hydroxylation

SCHEME 6.98

(ii) Anti Hydroxylation (Prevost Method)

The *trans*-diols are formed from acetoxonium ion by reaction with nucleophilic acetic acid (scheme 6.99), under dry conditions, followed by hydrolysis of the diacetate. The cation (II, scheme 6.98) is resonance stabilized and under dry conditions the attack on the cation by acetate ion in a bimolecular process and gives the *trans*- diacyl compound (scheme 6.99).

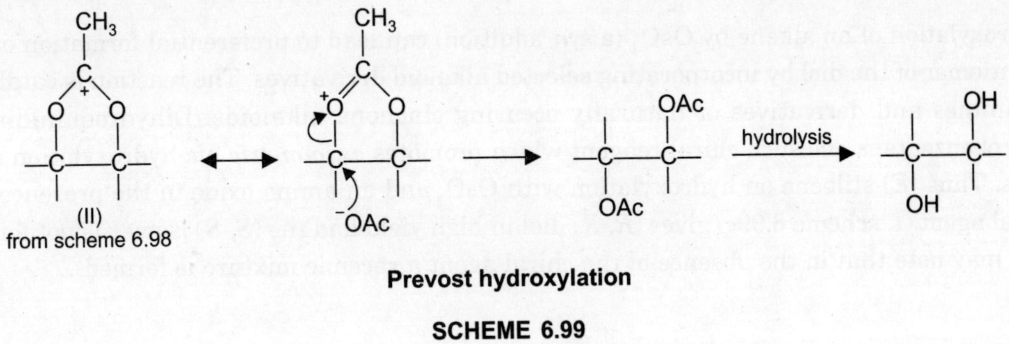

Prevost hydroxylation

SCHEME 6.99

EXERCISE 6.13

What products will be formed from the steroid (scheme 6.99a) on reaction with (i) OsO_4 and (ii) with $I_2/CH_3COOAg/CH_3COOH/H_2O$?

ANSWER. *With conformationally rigid molecules the cis diol formed by Woodward method may not have the same configuration to that obtained by OsO_4. Thus (I, scheme 6.99a) is attacked by OsO_4 at the less hindered side to give a less hindered diol (see scheme 6.97) Woodward procedure gives the more hindered diol (via the less hindered iodonium ion, scheme 6.99b).*

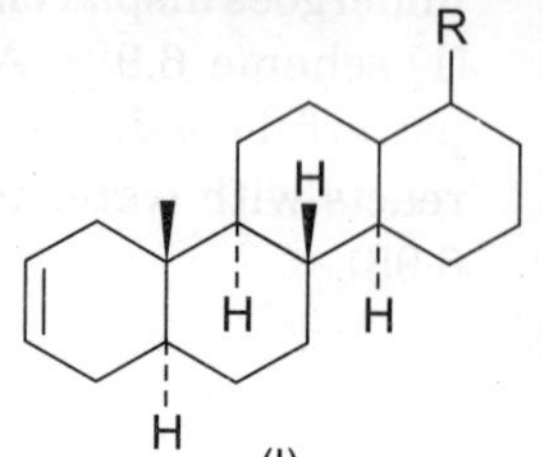

(I)

SCHEME 6.99a

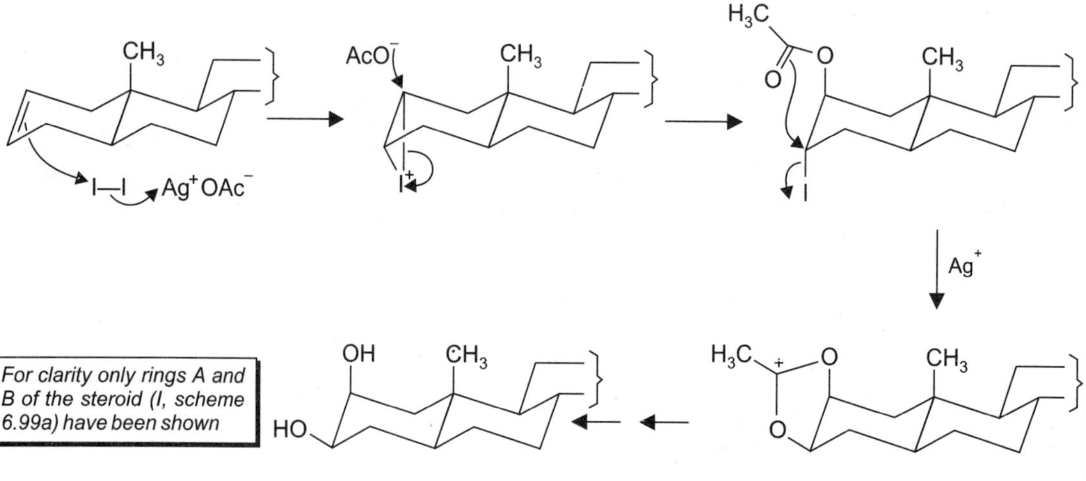

For clarity only rings A and B of the steroid (I, scheme 6.99a) have been shown

SCHEME 6.99b

(iii) Asymmetric Dihydroxylation of Alkenes with OsO_4 and Chiral Amine Ligands—Enantioselective Addition

Hydroxylation of an alkene by OsO_4 (a *syn* addition) can lead to preferential formation of one enantiomer of the diol by incorporating selected alkaloid derivatives. The reaction is catalyzed by amines and derivatives of naturally occuring cinchona alkaloids. Dihydroquinidine-4-chlorobenzoate is one such chiral reagent which promotes *asymmetric bis*-hydroxylation reactions. Thus (*E*)-stilbene on hydroxylation with OsO_4 and an amine oxide in the presence of a chiral agent (I, scheme 6.99c) gives (*R*, *R*) diol in high yield and the (*S*, *S*) isomer is not formed (one may note that in the absence of the chiral agent a racemic mixture is formed).

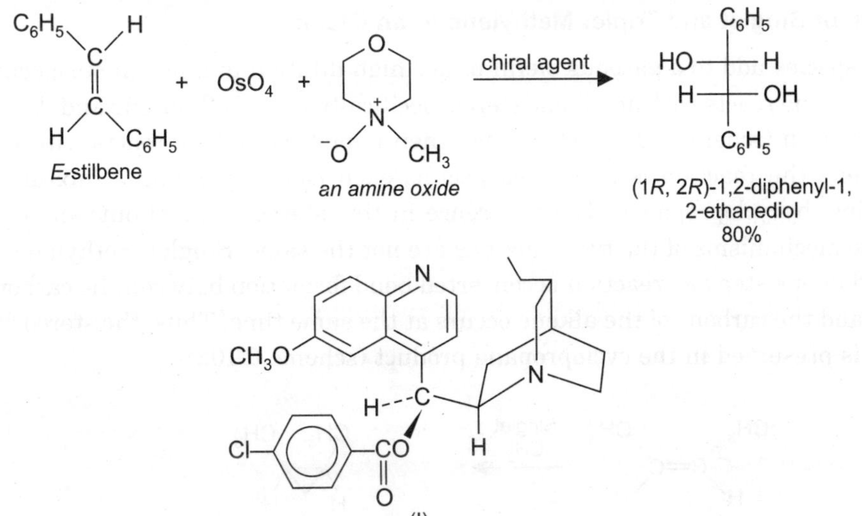

(I)
dihydroquinidine 4-chlorobenzoate
a chiral agent

SCHEME 6.99c

6.11 ADDITION OF CARBENES (METHYLENE) TO ALKENES

(A) Spin Multiplicity of Divalent Carbenes

Carbenes are highly reactive, electrophilic divalent carbon species with two unshared electrons. If both unshared electrons occupy the same orbital, the spins of two electrons must be paired, and the carbene is said to be in the singlet state. On the other hand, if the two unshared electrons occupy two different orbitals; their spins are unpaired and the carbene is said to be in the triplet state. It is generally believed that carbenes in the triplet state are more stable than those in the singlet state.

(B) Generation of Divalent Carbon Species

The generation of carbenes may be accomplished in a number of ways (scheme 6.100) but it is usually carried out in the presence of the intended olefinic reactant during the synthesis of cyclopropane derivatives.

It is generally believed that carbenes generated by photolysis are initially formed as singlet species and will add stereospecifically to liquid olefins. However, in the gaseous phase with low olefin concentration, such species may lose energy to become the more stable triplet species through collision with the container wall or with inert gas molecules (*e.g.*, N_2).

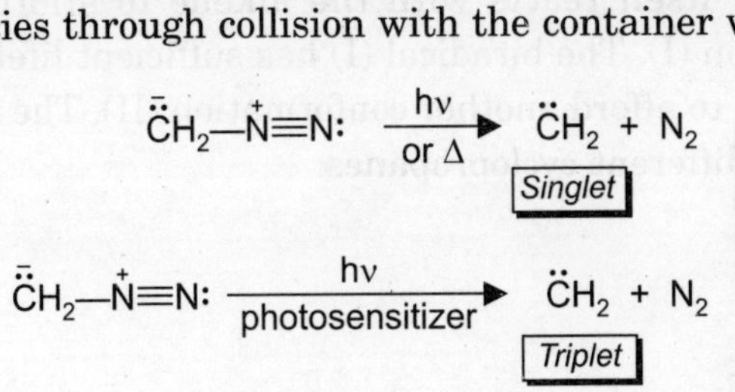

Carbenes (singlet species) can be generated by the decomposition of diazomethane (1,1-elimination) by heat or light. The triplet state is obtained by a special photochemical method using a photosensitizer.

SCHEME 6.100

(C) Addition of Singlet and Triplet Methylene to an Olefin

The triplet species add to a gaseous olefin under high dilution in a nonstereospecific manner. Singlet methylene reacts with an alkene stereospecifically, *e.g.*, *cis*-2-butene and its *trans* isomer react as shown in (scheme 6.101). However, when triplet methylene reacts with either *cis*-or *trans*-2-butene the reaction is nonstereospecific, each isomer yielding a mixture of *cis*-and *trans*-1,2-dimethycyclopropanes. The difference in this stereochemical outcome is due to the fact that the mechanisms of the two reactions are not the same. Singlet methylene adds to the double bond in one step *i.e.*, reaction is concerted bond formation between the carbon of singlet methylene and the carbons of the alkene occurs at the same time. Thus, the stereochemistry of the alkene is preserved in the cyclopropane product (scheme 6.102).

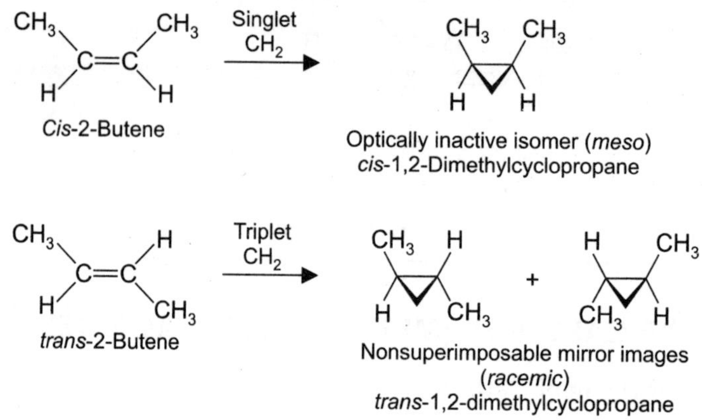

SCHEME 6.101

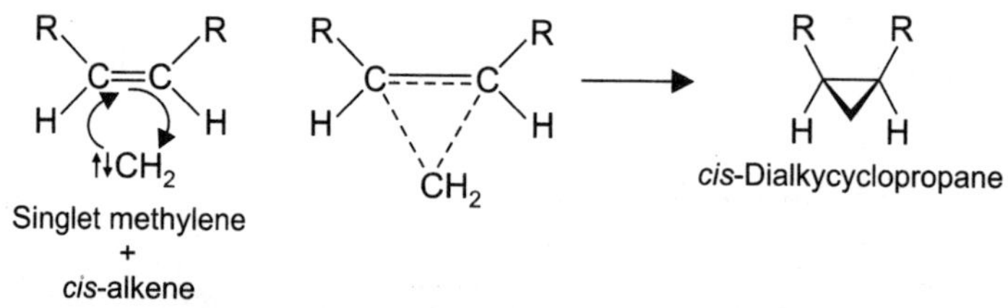

SCHEME 6.102

The electrons in triplet methylene are not paired, therefore, it reacts in a stepwise process (scheme 6.103). Triplet methylene, a diradical itself reacts with the alkene to afford an intermediate biradical in the form of conformation (I). The biradical (I) has sufficient lifetime to allow rotation of groups joined by single bonds to afford another conformation (II). The ring closure on (I) and (II) yields diastereomerically different cyclopropanes.

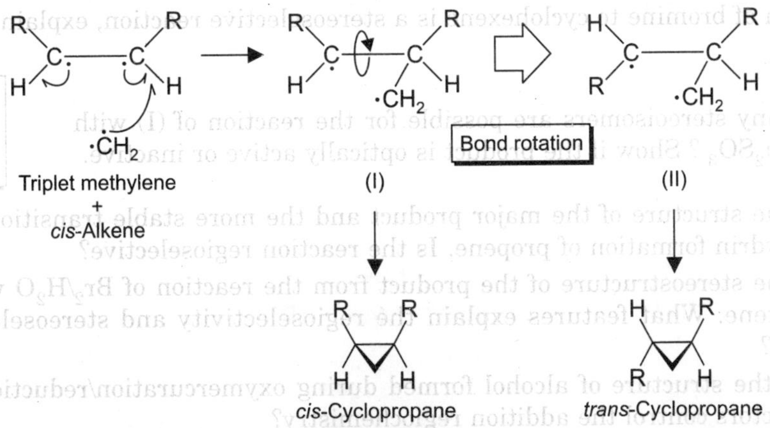

SCHEME 6.103

Stereospecific addition has also been observed in reactions involving complex carbenes which are not free species, the carbenoids (scheme 6.104). In this addition diiodomethane and zinc-copper couple are stirred together with an alkene. The diiodomethane and zinc react to yield a carbene like species called a carbenoid.

$$CH_2I_2 + Zn(Cu) \longrightarrow \underset{\text{A carbenoid}}{ICH_2ZnI}$$

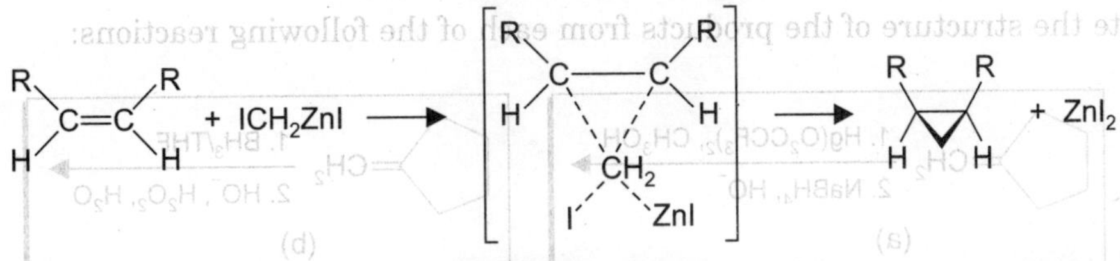

SCHEME 6.104

PROBLEMS

6.1. *meso*-Butane-1, 2-diol can be made from two diastereomeric alkenes. Give the mechanism of the two pathways and the geometry of the diastereomeric alkenes.

6.2. Predict the stereochemistry of the products of hydroxylation with osmium tetroxide followed by hydrolysis from *cis* and *trans*-crotonic acids. What product will be formed from crotonic acid on treatment with performic acid ?

6.3. Designate the two faces of 1-butene. Discuss the stereochemical consequences of its reaction with bromine.

6.4. Predict the outcome of *anti*-addition of bromine of *cis*-cyclodecene and its diastereomer. From the stereochemistry of the products formed in each case, is the reaction stereoselective or stereospecific?

6.5. Write the conformation of the major product formed on bromination of 4-*tert*-butylcyclohex-1-ene.

6.6. Explain giving mechanism why the addition of bromine to 2-butene is a stereospecific reaction.

6.7. Addition of bromine to cyclohexene is a stereoselective reaction, explain giving mechanism.

6.8. How many stereoisomers are possible for the reaction of (I) with OsO_4/Na_2SO_3 ? Show if the product is optically active or inactive.

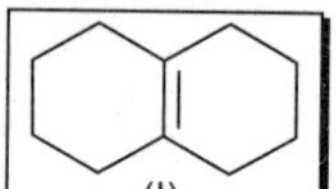
(I)

6.9. Write the structure of the major product and the more stable transition state during bromohydrin formation of propene. Is the reaction regioselective?

6.10. Write the stereostructure of the product from the reaction of Br_2/H_2O with 1-methylcyclohexene. What features explain the regioselectivity and stereoselectivity of the reaction?

6.11. Predict the structure of alcohol formed during oxymercuration/reduction of propene. What factors control the addition regiochemistry?

6.12. Depict the addition or bromine to *E*-2 butene. Detect the plane or center of symmetry in the product by writing suitable stereostructures.

6.13. Why the addition of bromine to 2-butene is termed as a stereospecific reaction ? Explain briefly by drawing stereostructure of the products.

6.14. Why the bromonium ion from (*Z*)-2-butene is achiral while from (*E*)-2-butene is chiral?

6.15. Why *trans* cyclooctene displays electrophilic *syn* addition? Is cyclooctene chiral, if so, why?

6.16. Norbornene on electrophilic addition of DBr gives only the *exo-syn* product. How this product can arise?

6.17. Write the structure of the products from each of the following reactions:

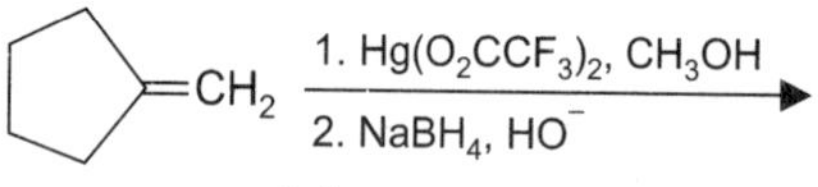

(a)

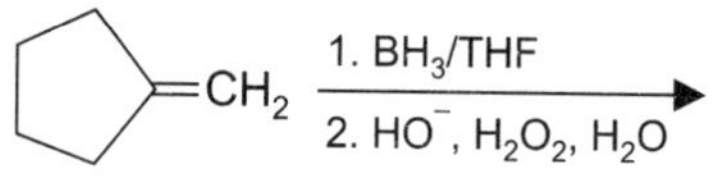

(b)

6.18. *trans*-2-Butene is treated with MCPBA and the isolated epoxide is subjected to acid catalyzed hydrolysis. Predict the stereochemistry of the product.

6.19. The following epoxide on acid catalyzed hydrolysis as expected gives a *trans* diol which could have either structure (I) or (II). Predict the correct product with reasoning.

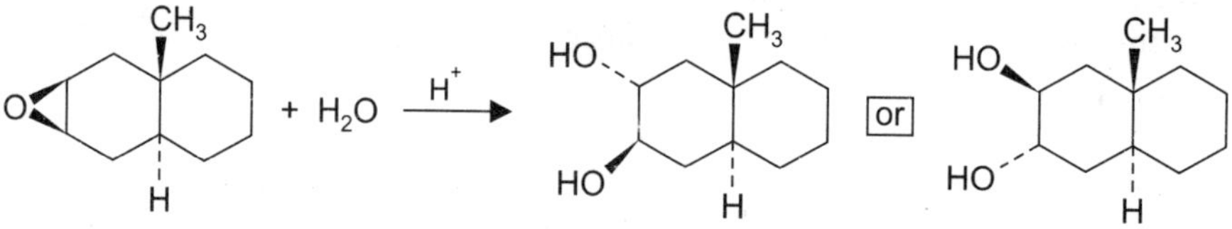

(I) not formed (II)

6.20. Predict the stereochemical outcome of hydroxylation of maleic and fumaric acids with osmium tetroxide.

6.21. Write the stereochemistry of the product of opening the following epoxide (1-phenylcyclohexene oxide) with hydroxide.

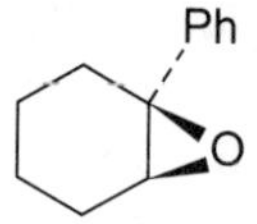

6.22. Opening of an epoxide in a rigid cyclohexane system leads to *trans*-diaxial products. An exception to this rule is the opening of the following 2β, 3β-epoxylanost-8-ene with HBr. Explain.

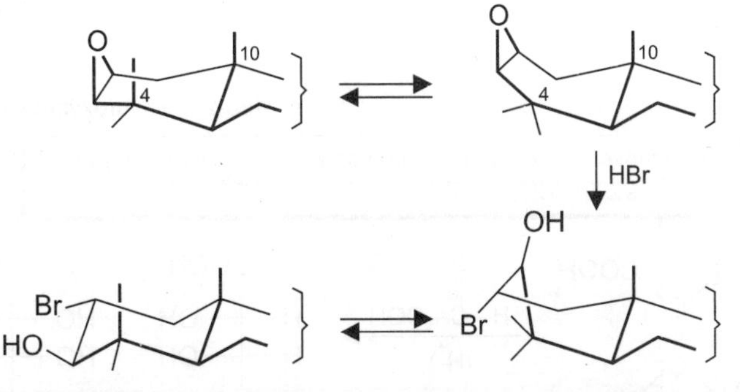

6.23. Write the Fischer projections of the products of bromination of *cis*- and *trans*-2-pentene. Write the perspective formulas of these products.

6.24. Write the Fischer projection of the product of addition to *trans*-2-butene, and transform it into its perspective drawing.

6.25. Predict the products from the reactions

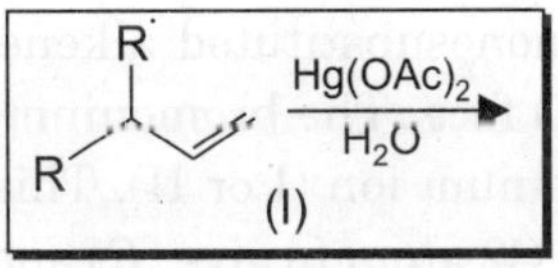

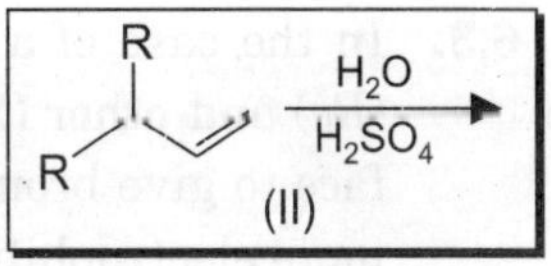

ANSWERS TO SELECTED PROBLEMS

6.1. *vicinal*-Dihydroxy compounds can be made from alkenes by two methods (i) epoxidation and S_N2 ring opening of the epoxide (*anti* addition) and (ii) via osmate ester formation and ring opening by treatment with $NaHSO_3$ (*syn* addition). The pathway in (I) is similar to bromination of an alkene and makes *trans*-2-butene the starting material (the inversion of configuration in the product forming step as *e.g.* in scheme 6.92 will give the *meso* product as in scheme 6.93). The pathway (II) does not involved inversion of configuration and the same *meso* product will be obtained from *cis*-2-butene.

6.2. Consult answer of problem 6.1. During hydroxylation with osmium tetroxide no inversion of configuration is involved at any stage. During hydroxylation with performic acid, the acid catalyzed ring opening of epoxide is attended with inversion of configuration.

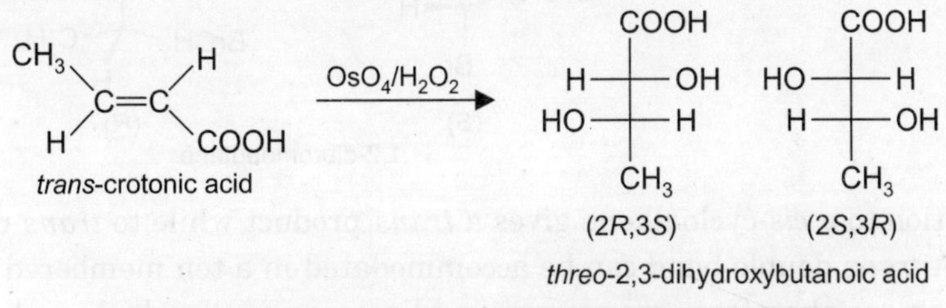

No inversion of configuration at any stage of hydroxylation. The geometry of substrate is maintained in the product.

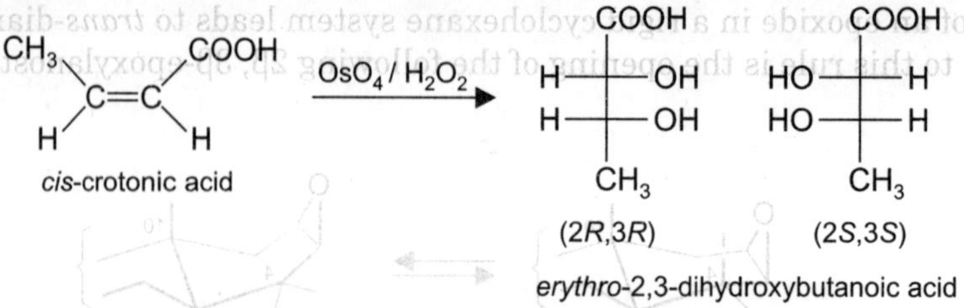

No inversion of configuration at any stage of hydroxylation. The geometry of substrate is maintained in the product.

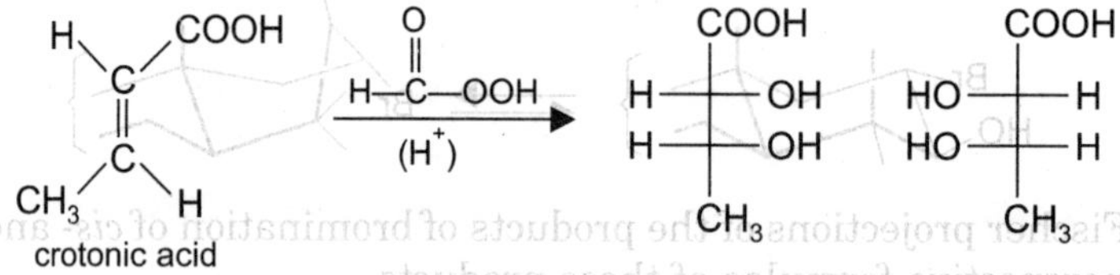

One inversion of configuration during hydroxylation. The initial geometry of the substrate changes in the product.

6.3. In the case of a monosubstituted alkene there are always two enantiotopic faces one (*Re*) and other (*Si*) face. The bromonium ion is equally likely to be formed from either face to give bromonium ion (I or II). This is then attacked by Br^- (S_N2 fashion) on the more electrophilic C2 atom to give (*R*) and (*S*) 1, 2-dibromobutane as a racemic mixture. Any alternative attack at C1 will also be the same for both the bromonium ions and will still give the same racemic mixture.

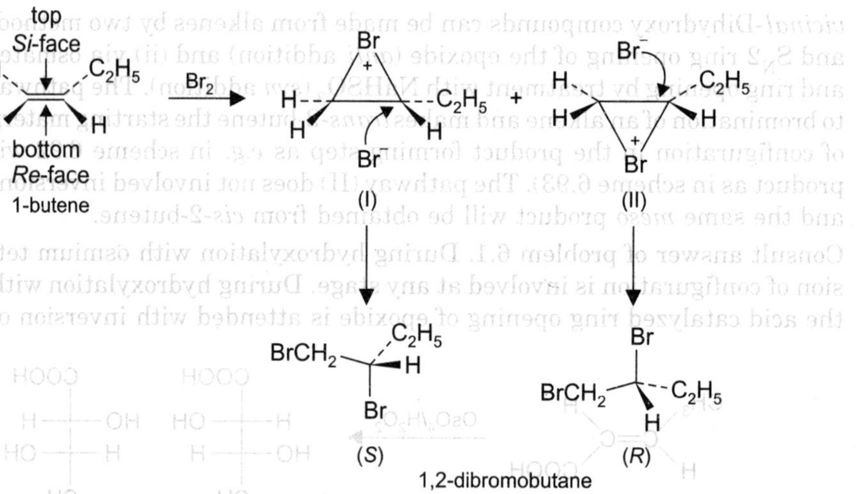

6.4. *anti*-Addition to a *cis*-cycloalkene gives a *trans* product while to *trans* cycloalkene a *cis* product. A *trans* double bond can be accommodated in a ten membered ring. The reacting alkenes are therefore, stereoisomers (diastereomers), which on bromination give different stereoisomers. The reaction is therefore, stereospecific (Note, *cis*-1,2-dibromocyclodecane has a plane of symmetry).

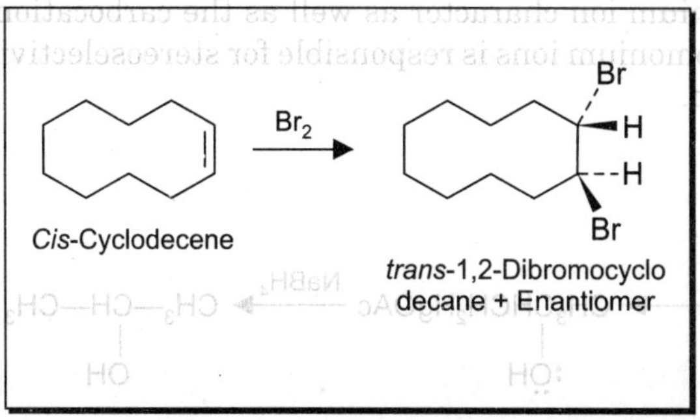

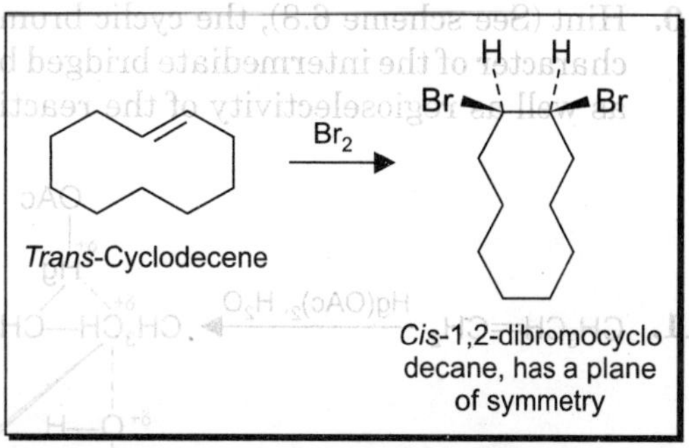

6.5. The bulky *tert*-butyl group locks the cyclohexene and preferentially occupies the equatorial position. Bromonium ion can be formed either from the top face or bottom face with equal facility (here the bromonium ion at the bottom face is shown). Attack at the appropriate carbon of the bromonium ion then takes place to give the diaxial product (also see schemes 6.5 and 6.6).

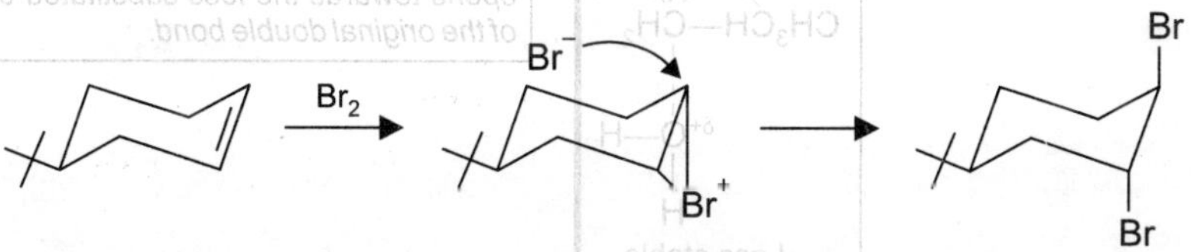

6.8. Due to symmetry in the molecule there are no stereocenters in the product, however, the product can still display *cis-trans* isomerism. Due to *syn* addition only the *cis* isomer will be formed which is achiral.

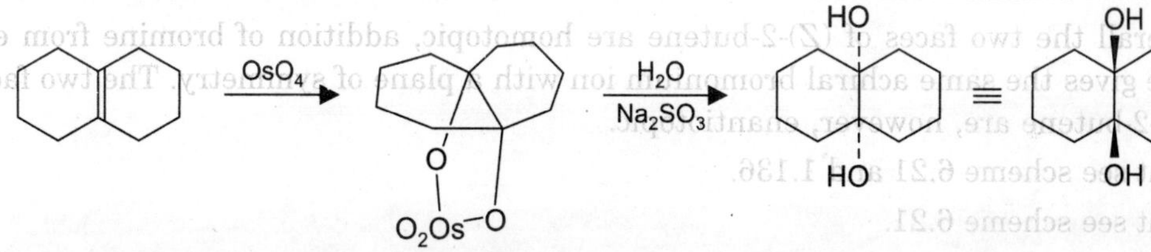

6.9. Yes the reaction is regioselective.

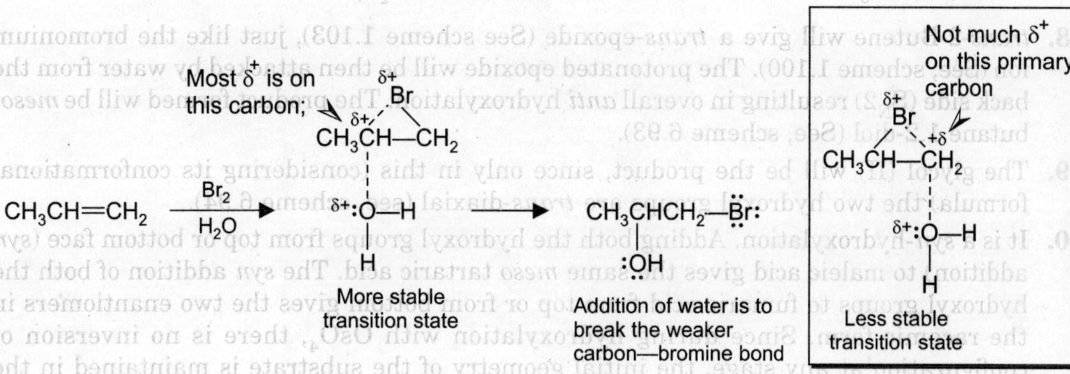

6.10. Hint (See scheme 6.8), the cyclic bromonium ion character as well as the carbocationic character of the intermediate bridged bromonium ions is responsible for stereoselectivity as well as regioselectivity of the reaction.

6.11. 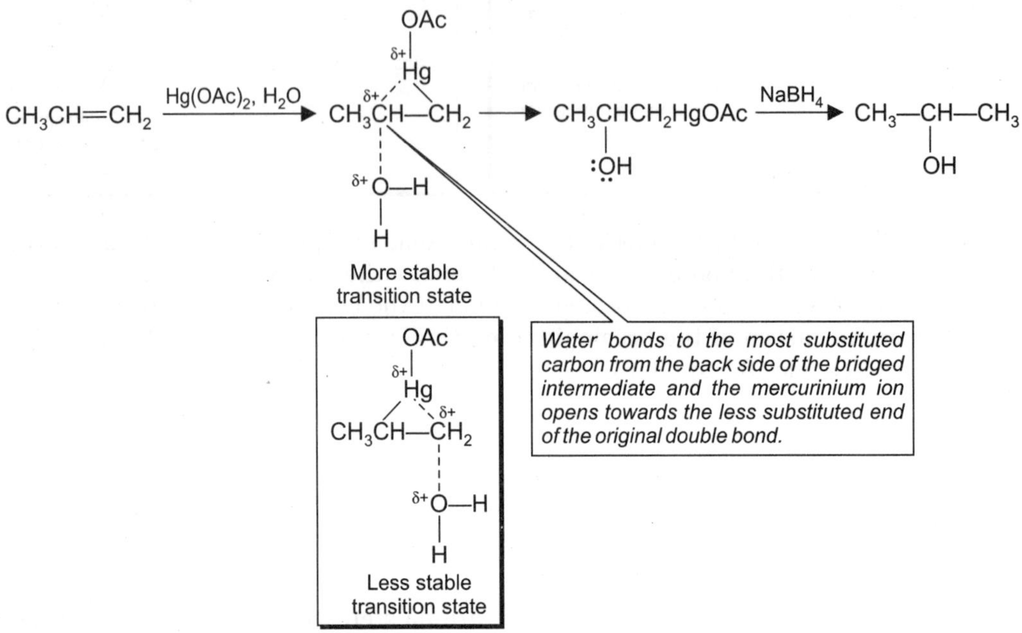

6.12. Hint (See scheme 1.100).

6.13. Hint see scheme 1.100 and 6.4.

6.14. Overall the two faces of (*Z*)-2-butene are homotopic, addition of bromine from either face gives the same achiral bromonium ion with a plane of symmetry. The two faces of (*E*)-2-butene are, however, enantiotopic.

6.15. Hint see scheme 6.21 and 1.136.

6.16. Hint see scheme 6.21.

6.17. (*a*) 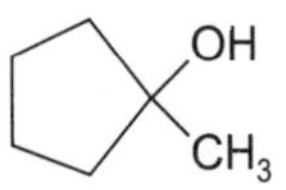

(*b*) 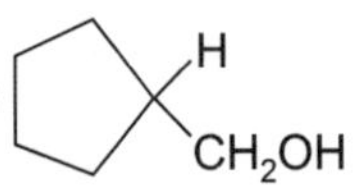

6.18. *trans*-2-Butene will give a *trans*-epoxide (See scheme 1.103), just like the bromonium ion (See, scheme 1.100). The protonated epoxide will be then attacked by water from the back side (S_N2) resulting in overall *anti* hydroxylation. The product formed will be *meso*-butane 1,2-diol (See, scheme 6.93).

6.19. The glycol (II) will be the product, since only in this (considering its conformational formula) the two hydroxyl groups are *trans*-diaxial (see, scheme 6.94).

6.20. It is a *syn*-hydroxylation. Adding both the hydroxyl groups from top or bottom face (*syn* addition) to maleic acid gives the same *meso* tartaric acid. The *syn* addition of both the hydroxyl groups to fumaric acid from top or from bottom gives the two enantiomers in the racemic form. Since during hydroxylation with OsO_4, there is no inversion of configuration at any stage, the initial geometry of the substrate is maintained in the product.

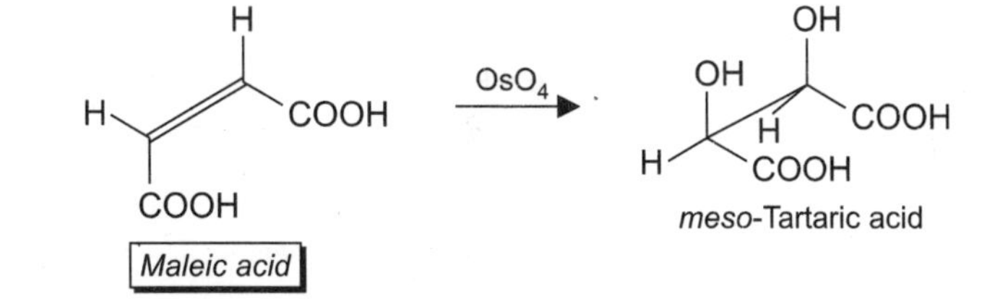

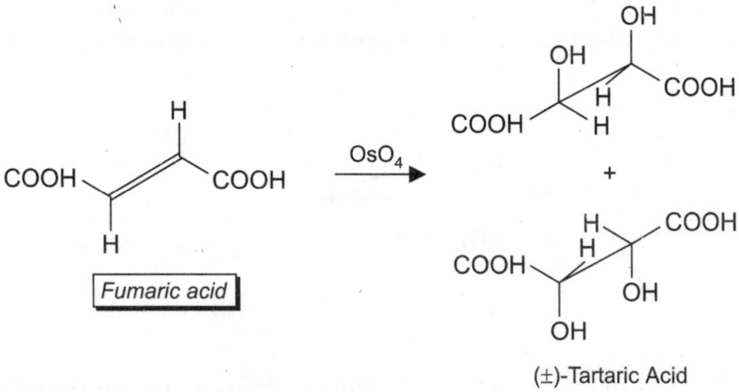

6.21. Opening of the epoxide with base proceeds via an S_N2—like transition state to give a single diastereomer.

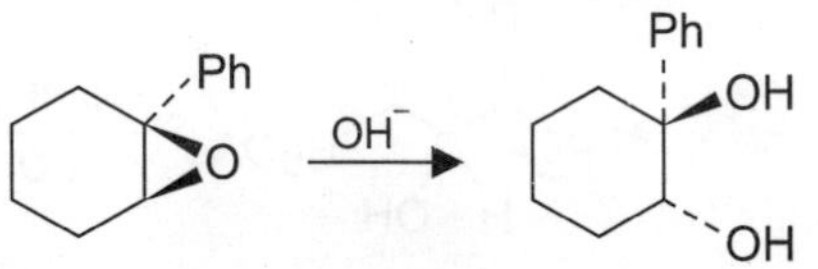

6.22. The chair conformation of ring A is destabilized by 1 : 3 type of non bonded interactions among the epoxy oxygen and the two β oriented methyl groups at C4 and C10. Therefore, ring A attains a boat conformation in which the diaxial addition takes place and a reversion to chair form leads to the diequatorial product.

6.23. When two stereocenters are generated via a bromonium ion intermediate, only one pair of enantiomers would be formed—addition of bromine is stereospecific (see, Scheme 1.100). Remember the stereochemical sequence *cis-syn-erythro* (all terms mean to the same side). One can change any two of the terms at a given situation to predict stereochemical outcome. Thus *cis*-2-pentene on *anti*-bromonation will give *threo*-enantiomers and *trans*-2-pentene on *anti* addition will give *erythro* enantiomers. Stereochemically, since after the formation of a bromonium ion there is one inversion of configuration during its opening, the initial geometry of the reactant changes in the substrate-*cis*-alkene gives *threo* product while *trans*-alkene gives *erythro*-product

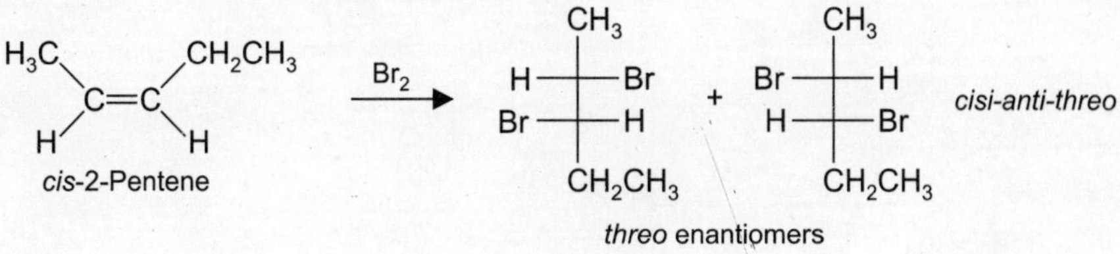

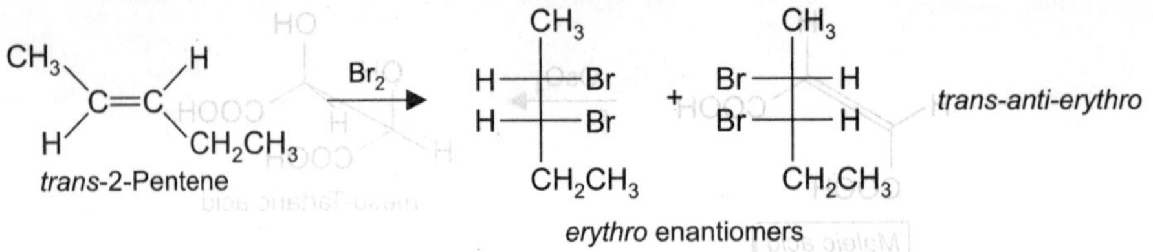

6.24. The two stereocenters in the reaction product have the same substituents, thus the *erythro* isomers are identical and represent a *meso* compound (compare with scheme 6.4 and 1.100).

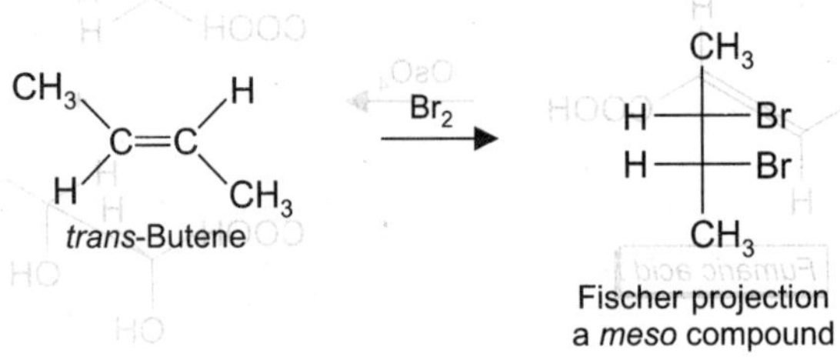

6.25. Both are hydration reactions. Oxymercuration-demercuration process does not involve the formation of a carbocation, thus in this case rearrangements are not common which, however is the case under acidic conditions.

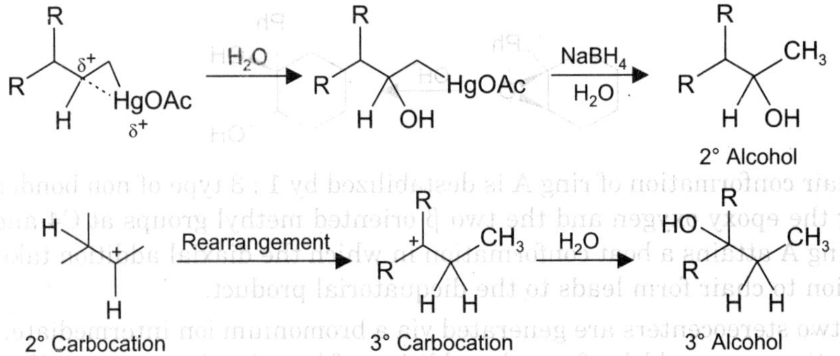

REFERENCES AND FURTHER READING

1. M.B. Smith, J. March, *March's Advanced Organic Chemistry; Reactions, Mechanisms and Structures;* 5th ed.; Wiley, New York, 2001.
2. F.A. Carey, R.M. Sundberg, *Advanced Organic Chemistry;* 5th ed.; Springer, 2007.

CHAPTER 7

Addition to Carbon Hetero Multiple Bonds—Stereochemistry and Mechanism

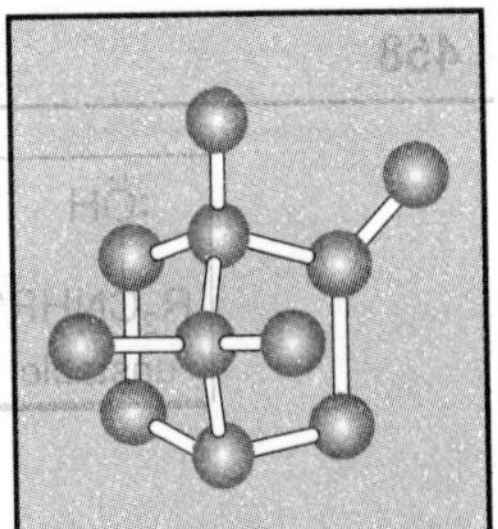

Camphor

7.1 INTRODUCTION

One type of a reaction typical of the carbonyl group in aldehydes and ketones is the addition of the reagent H-Nu across the carbon—oxygen double bond. The proton ends up on the more electronegative oxygen atom, while the nucleophile becomes attached to the electron deficient carbonyl atom (eq. I, scheme 7.1). The following points may be noted:

- A good nucleophile reacts with the partially positive sp^2 hybridized carbon atom of the trigonal carbonyl group, which results in the formation of the tetrahedral sp^3 hybridized carbon atom. In this process, oxygen acquires a formal negative charge in the tetrahedral intermediate. The oxygen anion (an alkoxide ion) then picks up a proton either from the solvent or from added acid, and an alcohol is produced (eq. I, Scheme 7.1). This happens when Nu is not electronegative and in that case the tetrahedral intermediate is stable. Thus aldehydes and ketones undergo nucleophilic

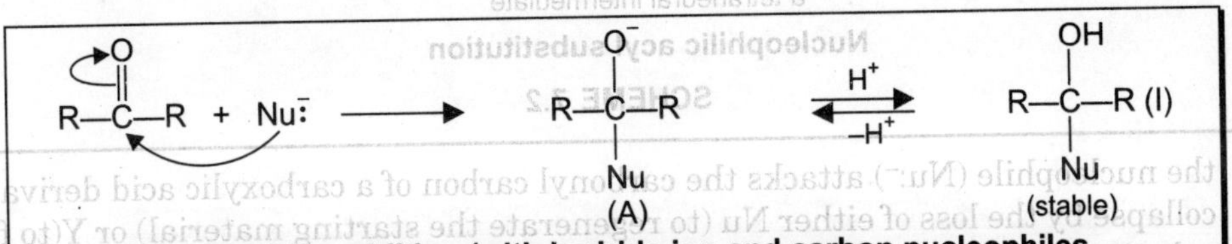

Nucleophilic addition (with hydride ion and carbon nucleophiles *i.e.*, when Nu is not electronegative)

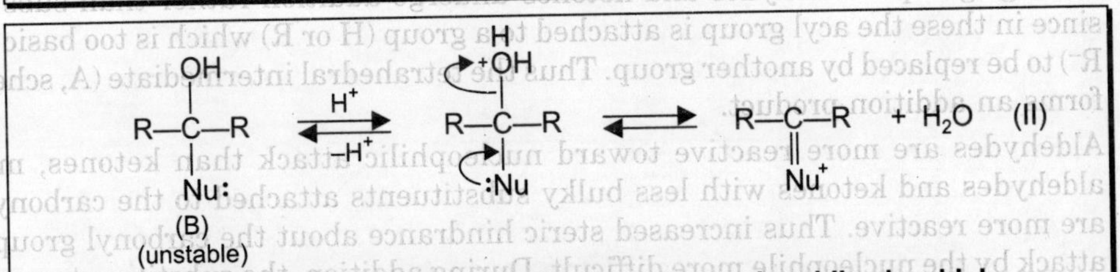

Nucleophilic addition-elimination with nucleophiles in which Z is electronegative (e.g., an amine)

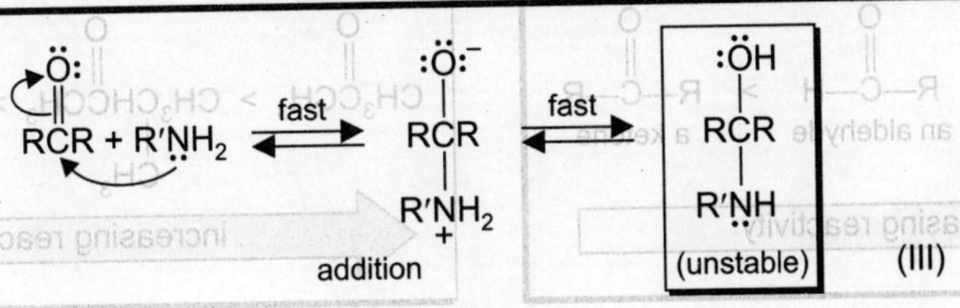

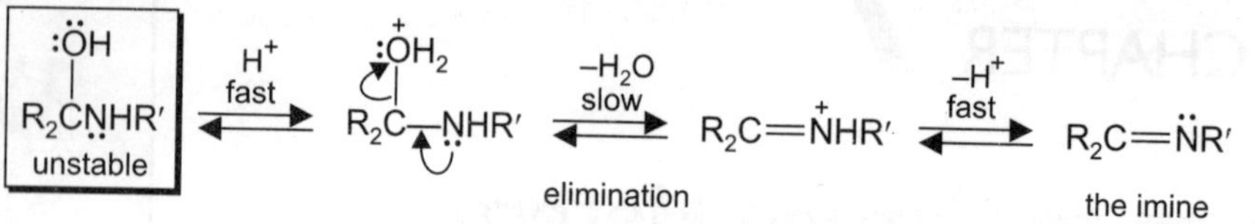

SCHEME 7.1

addition reactions. When, however, the adding nucleophile is one in which Nu is electronegative (Nu = O or N nucleophile) the initial tetrahedral product will not be stable (B, scheme 7.1). Elimination of water will then give a product with a double bond. Thus when the attacking atom of the nucleophile has a pair of nonbonding electrons in the addition product, water will be eliminated. This is called a nucleophilic addition—elimination reaction and imine formation is an example (eq. III, scheme 7.1).

- The acyl group of a carboxylic acid or its derivative is attached to a group (Y) which can be replaced by another group, these compounds therefore, react with nucleophiles to give acyl substitution products (scheme 7.2). The acyl group of aldehydes and ketones is attached to a group (H or alkyl) which are too basic to be replaced by another group, thus aldehydes and ketones give only addition products on reaction with nucleophiles (scheme 7.1). The tetrahedral intermediate (scheme 7.2) formed when

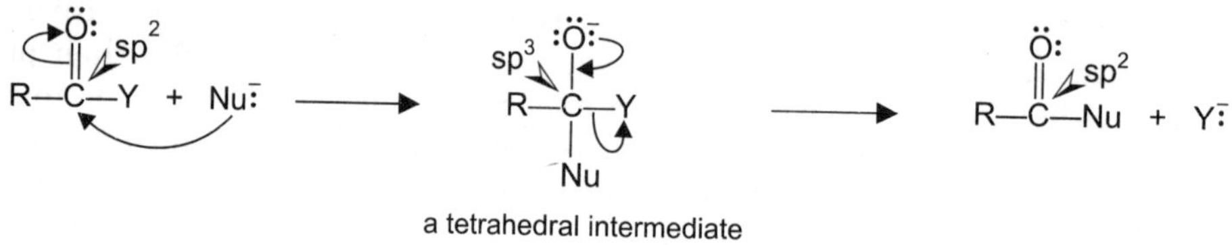

Nucleophilic acyl substitution

SCHEME 7.2

the nucleophile (Nu:⁻) attacks the carbonyl carbon of a carboxylic acid derivative can collapse by the loss of either Nu (to regenerate the starting material) or Y(to form the substitution product). This depends on the principle, the weaker bases are better leaving groups. Aldehydes and ketones undergo addition rather than substitution, since in these the acyl group is attached to a group (H or R) which is too basic (H^- and R^-) to be replaced by another group. Thus the tetrahedral intermediate (A, scheme 7.1) forms an addition product.

- Aldehydes are more reactive toward nucleophilic attack than ketones, moreover aldehydes and ketones with less bulky substituents attached to the carbonyl group are more reactive. Thus increased steric hindrance about the carbonyl group makes attack by the nucleophile more difficult. During addition, the substituents are pushed

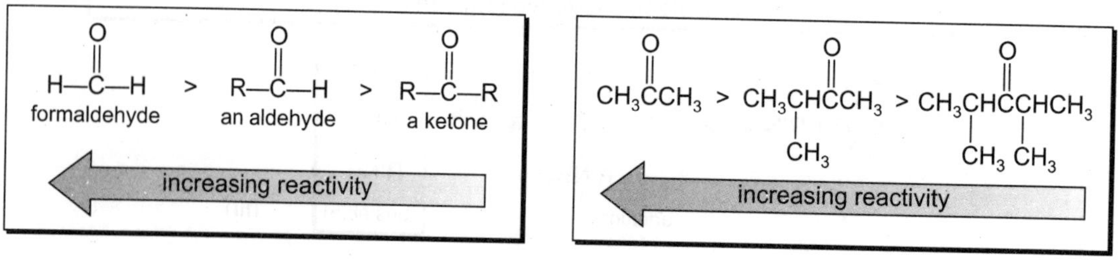

SCHEME 7.2a

closer to one another and the bond angle about carbon changes from 120° in the carbonyl group to 109.5° in the product. The larger the substituents, the more difficult it is to push them closer to one another (scheme 7.2*a*).

- Cyclic ketones are more reactive over open-chain ketones. There is less crowding in the product, and nucleophilic attack on carbon is easier in cyclic compounds in which the carbonyl group is held rigidly in place by the ring while the substituents are held out of its way (scheme 7.3).

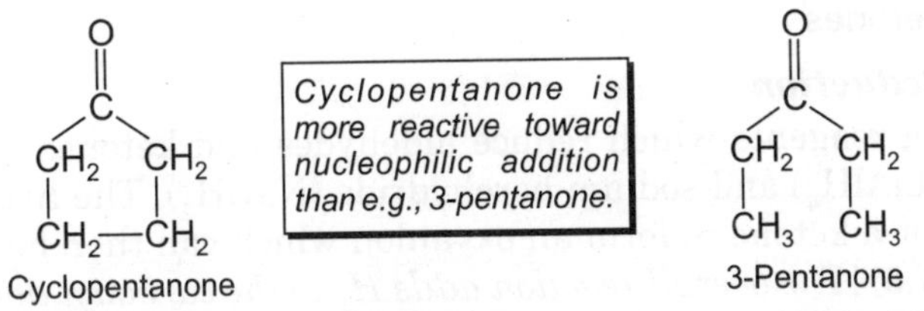

SCHEME 7.3

- The alkyl groups release electrons, so attachment of alkyl groups to the carbonyl carbon should decrease its positive character and make it less susceptible to nucleophilic attack. The ketones should be and are less reactive than aldehydes in addition reactions (scheme 7.4).

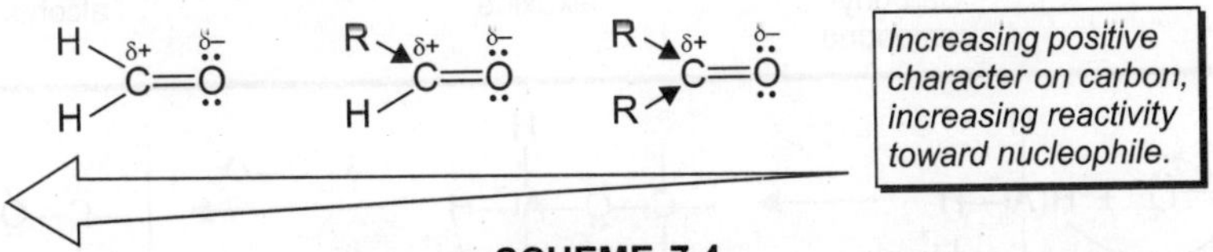

SCHEME 7.4

- Aromatic aldehydes and ketones are less reactive than aliphatic compounds toward nucleophilic addition since the resonance structures show that the carbonyl carbon is deactivated toward addition (scheme 7.4*a*).

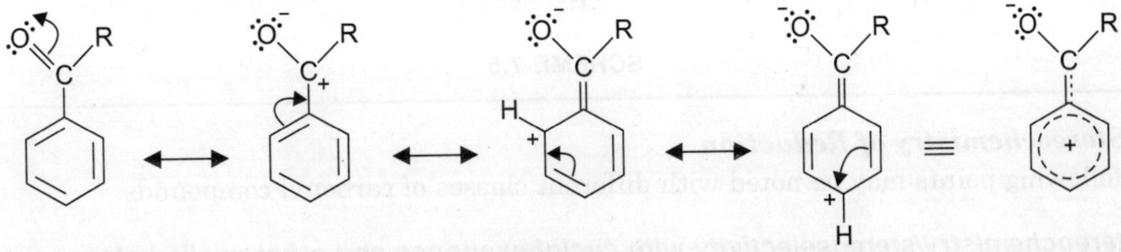

SCHEME 7.4a

7.2 SOME GENERAL AND STEREOCHEMICAL ASPECTS OF ADDITION TO CARBONYL COMPOUNDS

The following points may be noted:

- The faces of aldehydes other than (formaldehye) and unsymmetrical ketones R—CO—R′ (R ≠ R′) are enantiotopic. In such carbonyl compounds, on nucleophilic addition a new stereocenter is generated. The product however, is optically inactive since equal mixture of enantiomers is formed (see scheme 2.7*a*).

- **When, however, there is chirality in R or R′ in a ketone R—CO—R′ (or in an aldehyde), the faces become diastereotopic. The racemic mixture will not be formed on nucleophilic addition and Cram's rule predicts the stereochemistry of the diastereomeric product (see scheme 2.24—2.28).**

7.3 MECHANISM AND STEREOCHEMISTRY OF METAL HYDRIDE REDUCTION

(a) Aldehydes and Ketones

(i) Mechanism of Reduction

Two hydride donating reagents which reduce aldehydes and ketones to alcohols are lithium aluminium hydride ($LiAlH_4$) and sodium borohydride ($NaBH_4$). The hydride ion (nucleophile) adds to an aldehyde or a ketone to form an oxyanion which can then be protonated by adding water (eq. I, scheme 7.5) [*The overall reaction adds H_2 to the carbonyl double bond. Each of the four hydrogen atoms in lithium aluminium hydride is available for transfer to the carbonyl group, and the complete reduction mechanism is represented (eq. II, scheme 7.5)*].

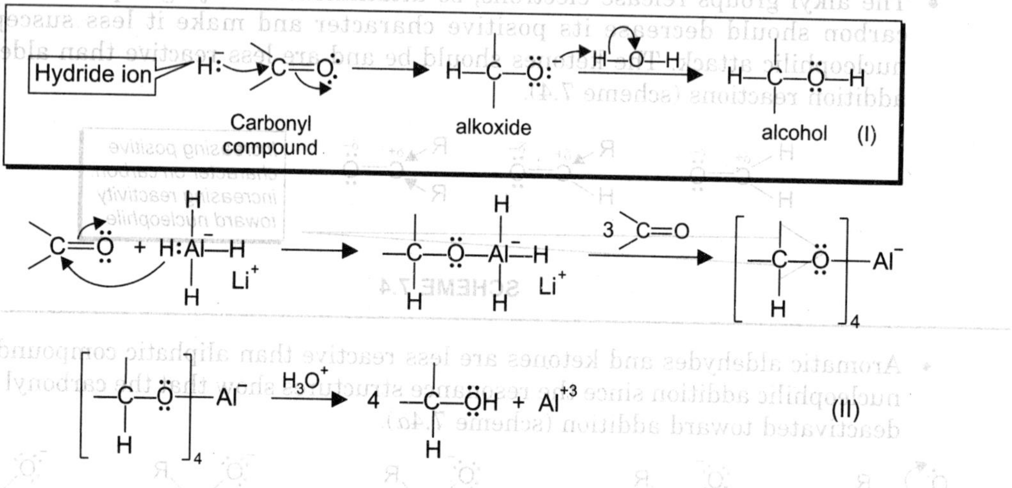

SCHEME 7.5

(ii) Stereochemistry of Reduction

The following points may be noted with different classes of carbonyl compounds.

A. Stereochemistry/stereoselectivity with cyclohexanones and other cyclic ketones

During the addition of nucleophiles to cyclohexanones, the nucleophile can approach from either the axial or the equatorial side, the carbonyl oxygen then moves into an equatorial or axial position respectively. With comparatively small hydride donors like AlH_4^- and BH_4^- ions the more stable equatorial alcohol generally predominates. The nucleophile *i.e.*, the hydride ion is delivered axially (scheme 7.6), and the reaction is highly stereoselective.

According to Felkin's model, in cyclohexanones the outcome of reduction (and addition of other nucleophiles) is based on the balance of both steric and torsional strain. Formation of an equatorial alcohol (*e.g.*, by an axial attack of hydride) requires a staggered transition state (I, scheme 7.7) which could suffer from steric strain between the nucleophile and the β-axial substituents (substituents at C-3). Thus with less sterically demanding nucleophile *i.e.*, AlH_4^-

and with no axial substituent at C-3 *i.e.*, with only a tiny hydrogen at C-3, the energy of (I) is lower and an equatorial alcohol predominates. Formation of an axial alcohol *i.e.*, attack by nucleophile *e.g.*, a hydride at equatorial position requires a partially eclipsed transition state (II, scheme 7.7) with torsional strain between the nucleophile and the axial α-hydrogen substituents.

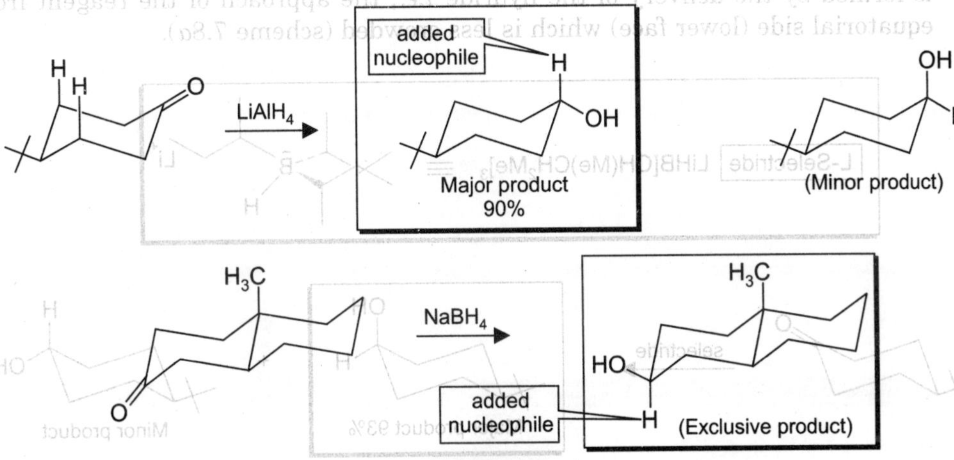

SCHEME 7.6

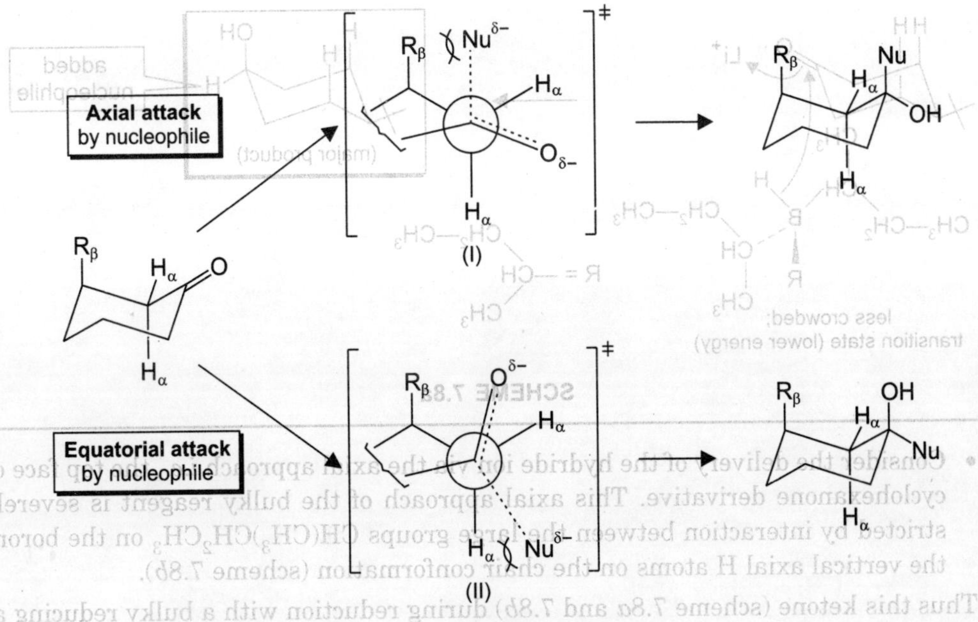

SCHEME 7.7

When the nucleophile is sterically demanding like a bulky hydride "selectride" (scheme 7.8) then an axial alcohol formation predominates *i.e.*, now the nucleophile ($H:^-$) is delivered from the equatorial side. Here one has a situation opposite to the one in (scheme 7.6) *i.e.*, selectride is a nucleophilic hydride ion source like $NaBH_4$ but it is much bulkier than $NaBH_4$. The following points may be noted:

- This reduction with selectride like $NaBH_4$ reduction is not reversible and thus is under kinetic control.
- Approach path and stability of the transition state are important. The major product is formed by the delivery of the hydride *i.e.*, the approach of the reagent from the equatorial side (lower face) which is less crowded (scheme 7.8*a*).

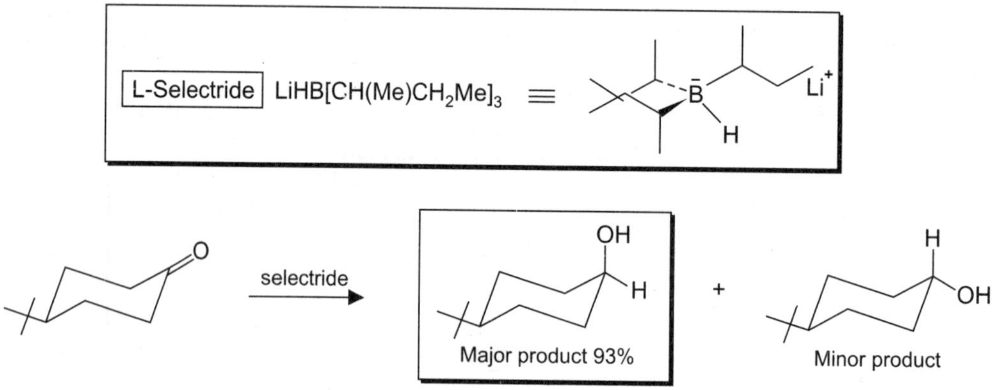

SCHEME 7.8

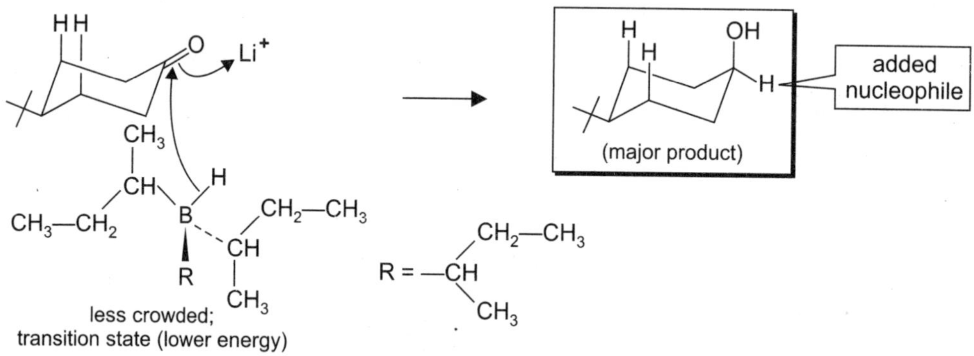

SCHEME 7.8a

- Consider the delivery of the hydride ion via the axial approach *i.e.*, the top face of the cyclohexanone derivative. This axial approach of the bulky reagent is severely restricted by interaction between the large groups $CH(CH_3)CH_2CH_3$ on the boron and the vertical axial H atoms on the chair conformation (scheme 7.8*b*).

Thus this ketone (scheme 7.8*a* and 7.8*b*) during reduction with a bulky reducing agent is under kinetic control and the selectride is facially selective. The major product being formed by the delivery of the hydride via equatorial face which involves a stable and less crowded transition state as compared to the one involved via axial approach by the bulky reducing agent.

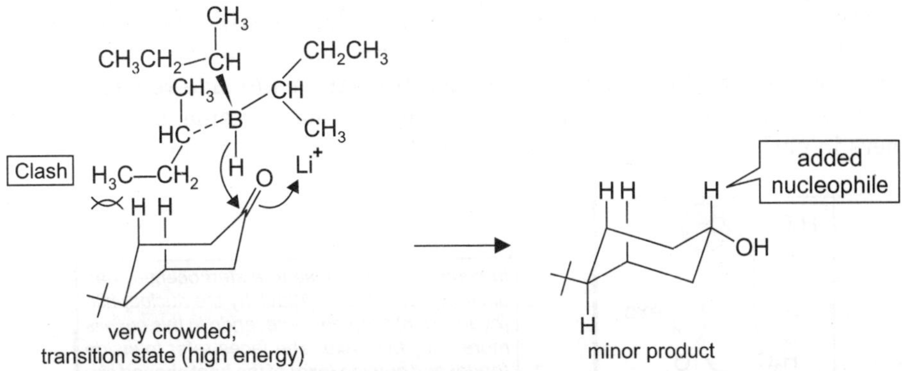

SCHEME 7.8b

EXERCISE 7.1

Predict the outcome of the products from reactions (scheme 7.8c).

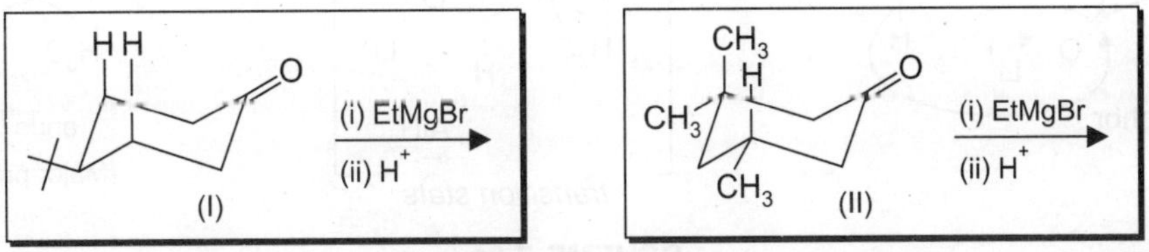

SCHEME 7.8c

ANSWER. *In both reactions the nucleophile is comparatively bulky and thus will be delivered from the equatorial side. With the presence of an axial substituent at C3 position in the cyclohexanone substrate (II, scheme 7.8c) the nucleophile is exclusively delivered from the equatorial side and the stereoselectivity becomes more pronounced. The products are in scheme 7.8d. Recall that axial approach by the nucleophile dominates with small nucleophiles like* $LiAlH_4$.

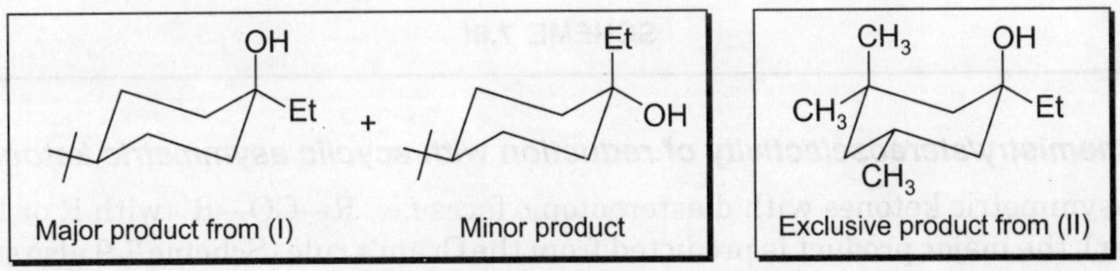

SCHEME 7.8d

B. Stereochemistry (stereoselectivity) of addition to norbornane systems

Consider the reduction of camphor. The two faces of the carbonyl group of camphor (*i.e., exo* and *endo* are diastereotopic) have different accessibility to nucleophilic reagents. The approach to the bottom (*endo*) face is hindered by the U-shaped cavity of the molecule, but the top (*exo*) face is strongly hindered by the overhanging C-7 methyl group. Thus reduction with $LiAlH_4$ occurs largely by approach of the nucleophile *e.g.*, AlH_4^- from, the *endo* side (scheme 7.8*e*). A more bulky reducing agent *e.g.*, $LiAl(OBu^t)_3H$ is more selective still and gives almost exclusive delivery of hydride to the *endo* face. In the case of norcamphor, however, there is no hindrance from the *exo* side, but the hindrance to approach from the *endo* side remains. Thus, hydride reduction occurs by attack of the AlH_4^- now from the *exo* side (scheme 7.8*f*). For the number of stereoisomers in each case (see scheme 4.54*a*) and relevant discussion.

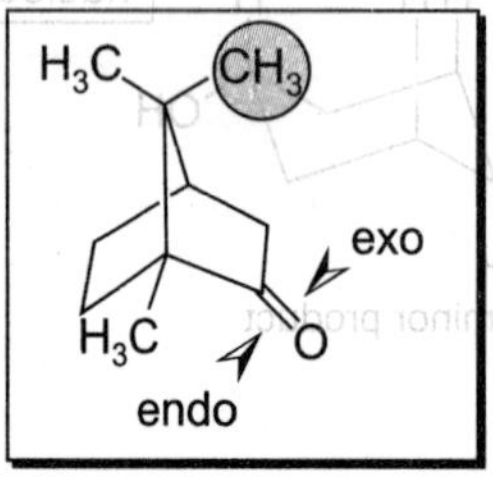

In camphor C=O (close to a stereocenter) has diastereotopic faces. Attack by the nucleophile (hydride ion) from one face, endo in this case is more easy than exo. The faces refer to inside (endo) and outside (exo) of the boat shaped six-membered ring.

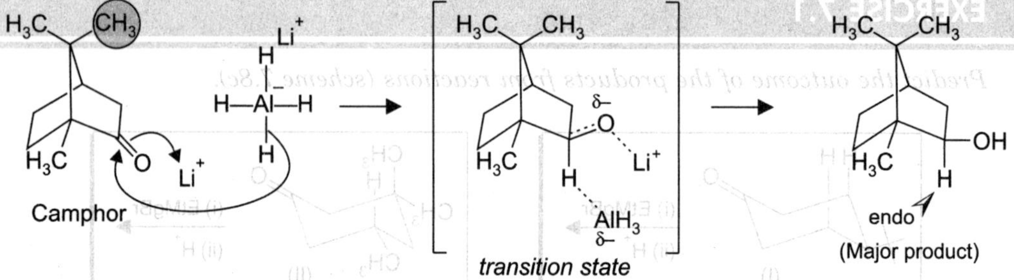

SCHEME 7.8e

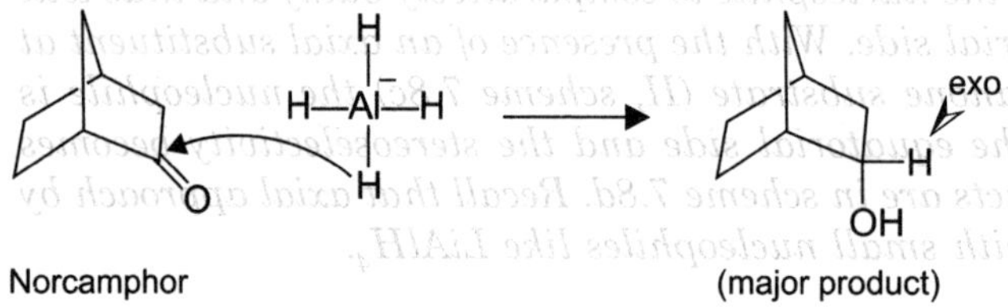

In norcamphor as well, the faces of the C=O are diastereotopic, now the attack by the nucleophile is from the exo side.

SCHEME 7.8f

C. Stereochemistry/stereoselectivity of reduction with acyclic asymmetric ketones

In acyclic asymmetric ketones with diastereotopic faces *i.e.*, R—CO—R′ (with R or R′ having a stereocenter), the major product is predicted from the Cram's rule (Scheme 7.9 also see schemes 2.24—2.28).

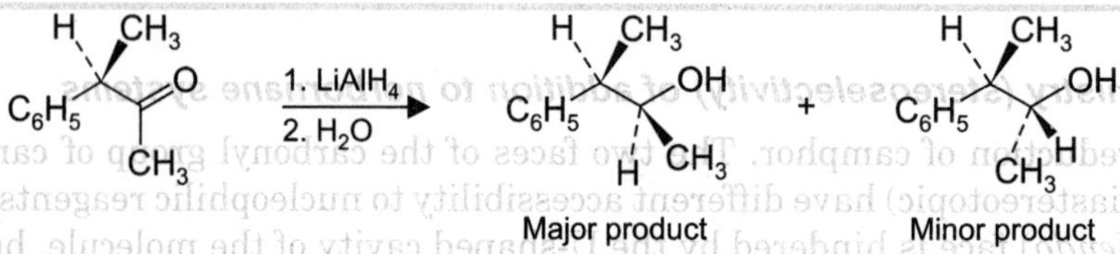

SCHEME 7.9

Nucleophilic additions to Conformationally locked Cyclohexanones

The nucleophilic attack on top or bottom face of the ketone takes place at an angle of 109° to the carbonyl group (Bürgi-Dunitz angle, scheme 7.9a). This trajectory helps to explain the outcome of the nucleophilic additions. This angle is easily attained during the selectride reduction of 4-tert-butylcyclohexanone (see, scheme 7.8a) when the regent approaches the lower face (equatorial approach) without much steric strain in the transition state. Similarly during the reduction ($LiAlH_4$) of camphor this ideal approach angle is not attainable via attack from exo face which is hindered by the geminal methyl groups (see, scheme 7.8e). The hydride reductions are not reversible and thus under kinetic control. Recall that in the case of irreversible reactions, the stabilities of the transition states to give the products have to be considered. In the case of a reversible reaction, however, the most stable compound will predominate at equilibrium i.e., the reaction will be under thermodynamic control (an example of thermodynamic control is in Meerwein Ponnodorf reduction (see scheme 7.16c).

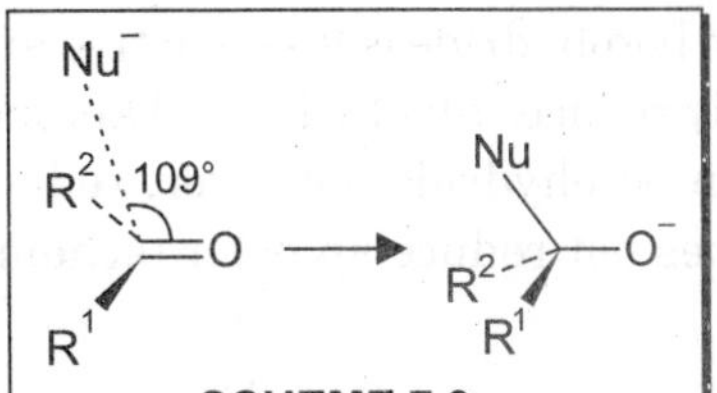

SCHEME 7.9a

D. Stereoselectivity/stereochemistry of reduction with prochiral ketones

A chiral, optically active reducing agent, *e.g.*, Ipc.BBN (alpine-borane, scheme 7.10 brings about asymmetric reduction of unsymmetrical (prochiral ketones R—CO—R′ R ≠ R′) to optically active secondary alcohols with a very high degree of enantioselectivity even when the ketone itself is achiral. The chiral reagent is prepared from 9-BBN and α-pinene (scheme 7.10 see

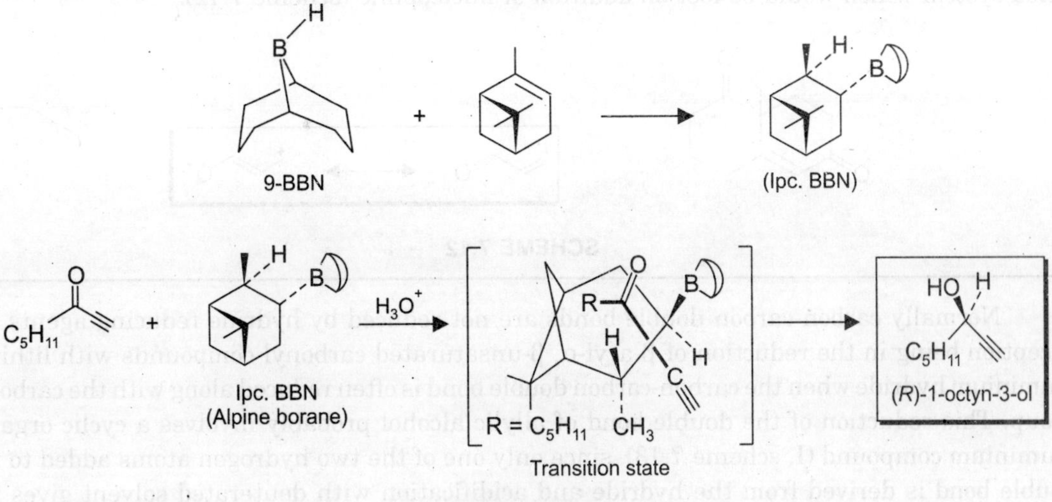

SCHEME 7.10

also schemes 2.42—2.44). Thus *e.g.*, 1-octyn-3-one gives *R*-1-octyn-3-ol with almost 95% optical purity. It is suggested that transition state has the pentyl group lying over the ring of the reducing agent to give the product with observed stereochemistry.

(iii) Chemoselectivity

Although both sodium borohydride and lithium aluminium hydride are sources of hydride ion, sodium borohydride is less reactive and both safer and easier to use. Lithium aluminium hydride not only reduces aldehydes and ketones, but also acids, esters and many other functional groups. Sodium borohydride, however, reduces only aldehydes, ketones, acid chlorides and imines. It *e.g.,* does not reduce an ester (scheme 7.11).

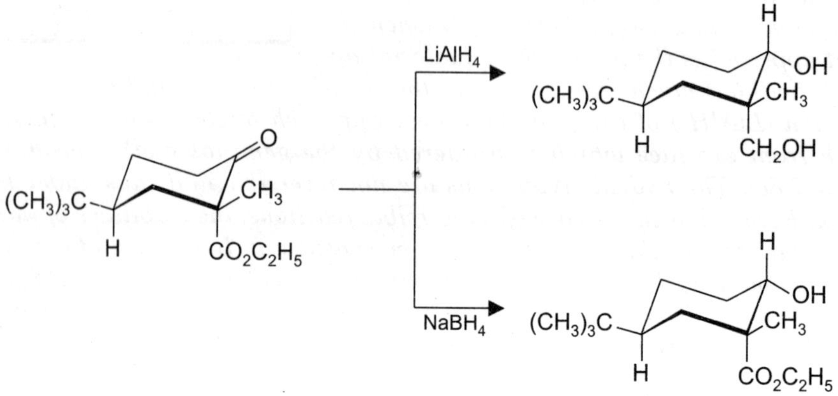

SCHEME 7.11

Sodium borohydride also selectively reduces carbonyl groups in different environments. Thus when one equivalent of sodium borohydride is used, only the non-conjugated carbonyl group is reduced, the other being less reactive due to the delocalization energy in the conjugated system which would be lost on addition of nucleophile (scheme 7.12).

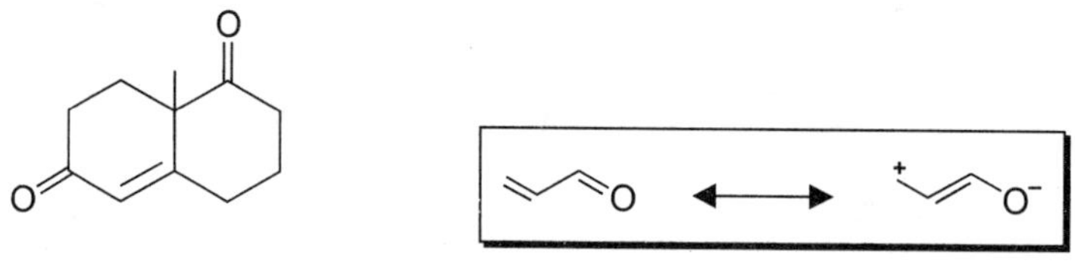

SCHEME 7.12

Normally carbon-carbon double bonds are not reduced by hydride reducing agents, an exception being in the reduction of β-aryl-α, β-unsaturated carbonyl compounds with lithium aluminium hydride when the carbon-carbon double bond is often reduced along with the carbonyl group. This reduction of the double bond of allylic alcohol probably involves a cyclic organo-aluminium compound (I, scheme 7.13), since only one of the two hydrogen atoms added to the double bond is derived from the hydride and acidification with deuterated solvent gives the deuterated alcohol (scheme 7.13).

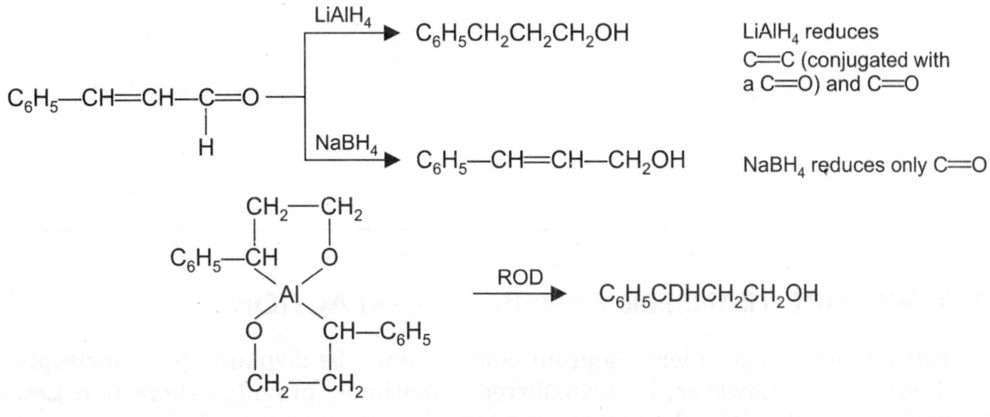

SCHEME 7.13

(b) Carboxylic Acids

Carboxylic acids are carbonyl compounds which have a leaving group (step 3, scheme 7.14). In the first step of the reduction of carboxylic acid with lithium aluminium hydride, the hydride ion reacts with the acidic hydrogen of carboxylic acid to form H_2 gas (This reaction destroys some of the expensive reagent, thus acids are first converted into esters before reduction see, scheme 7.15). Nucleophilic addition of hydride ion occurs in step 2 with the loss of the $^{-}OAlH_2$ from the intermediate (I, scheme 7.14) to give an aldehyde. Another addition of hydride ion (step 4) to the intermediate aldehyde completes the reduction to primary alcohol. Thus, in summary the reduction of a carboxylic acid to the primary alcohol involves two successive additions of hydride ions with the loss of $OAlH_2^-$.

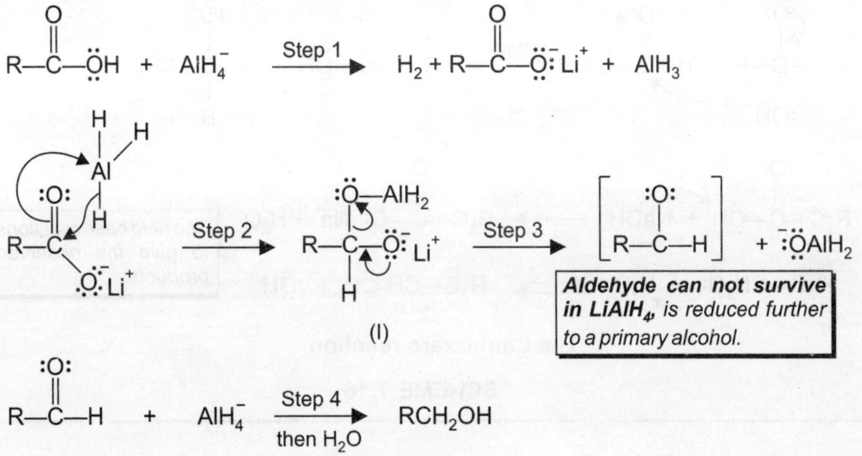

SCHEME 7.14

(c) Esters

When an ester reacts with hydride ion, the first reaction is a replacement of alkoxide ion by way of the tetrahedral intermediate to give an aldehyde. The aldehyde (which is not isolated) reacts with another hydride ion to give a primary alcohol (scheme 7.15): With the use of less reactive reducing agents like DIBAH (diisobutylaluminium hydride, i-$(C_4H_9)_2AlH$) the intermediate aldehyde can be isolated.

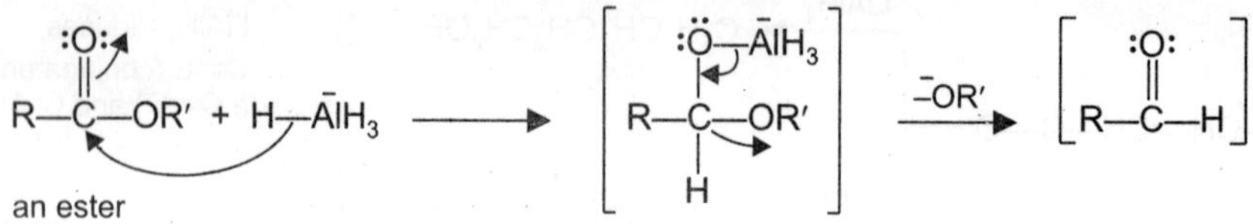

SCHEME 7.15

7.4 THE HYDRIDE TRANSFER CANNIZZARO REACTION

The hydride ion is not a good leaving group and it cannot be displaced by a nucleophile (see, scheme 7.5*a*). It can, however, be transferred sometimes, provided there is a Lewis acid positioned in exactly the right place to accept it (entropy requirements). In Cannizzaro reaction, the two partners have not to find each other, an aldehyde with no α-hydrogens undergoes a nucleophilic addition of hydroxide ion to the carbonyl group and is followed by hydride transfer to another molecule of the aldehyde. Protonation gives a molecule of an alcohol and a carboxylic acid related to the original aldehyde (scheme 7.16). [*One may compare this situation, when, however, a powerful base like OH^- can be lost during aldol condensation, (see scheme 5.8)*].

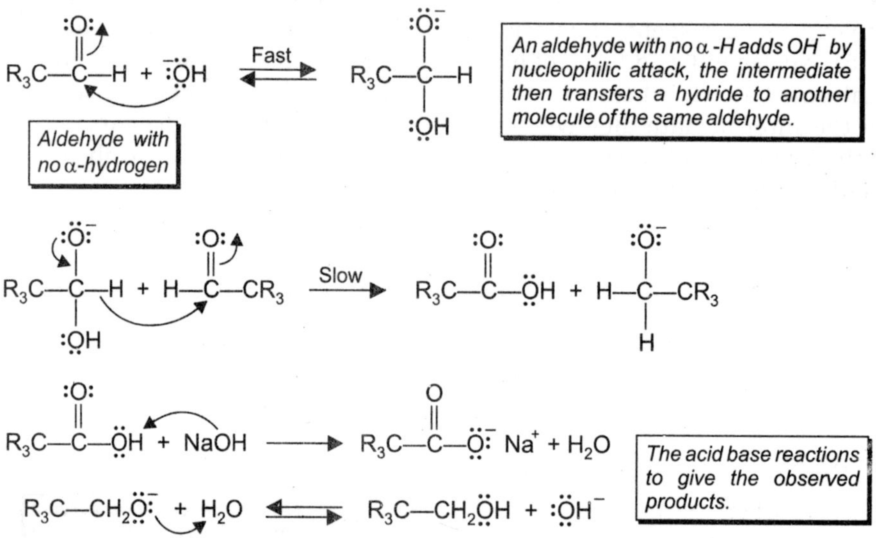

The Cannizzaro reaction

SCHEME 7.16

7.5 MEERWEIN—PONNDORF REDUCTION THE HYDRIDE TRANSFER REACTION

In the Meerwein–Ponndorf–Verley reaction, equilibrium is established between the carbonyl group to be reduced and isopropanol on the one hand and also between the product alcohol and acetone in the presence of aluminium isopropoxide. Since acetone is the lowest boiling component of the mixture, it is regularly distilled off so that the equilibrium is displaced to the right (scheme 7.16*a*).

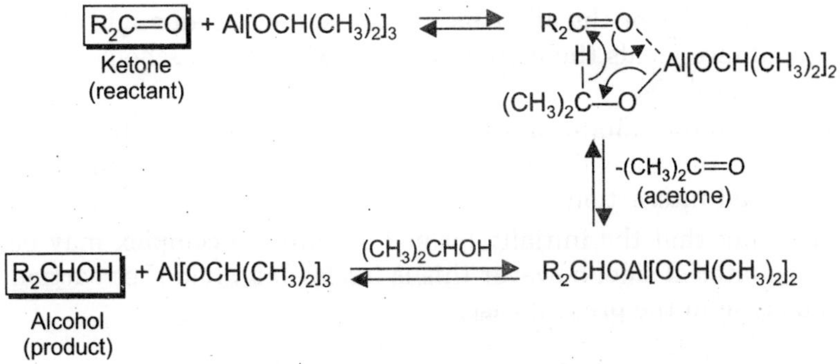

SCHEME 7.16a

Consider the Meerwein–Ponndorf reduction of 4-*tert*-butylcyclohexanone (scheme 7.16*b*) and compare these results with $NaBH_4$, $LiAlH_4$ and bulky hydride reduction of the same cyclohexanone (scheme 7.6, 7.8*a* and 7.8*b*). Based on the accepted mechanism (scheme 7.16*a*), aluminium alkoxide coordinates with the carbonyl oxygen of ketone and a hydride ion is transferred via a six membered transition state. The following points may be noted :

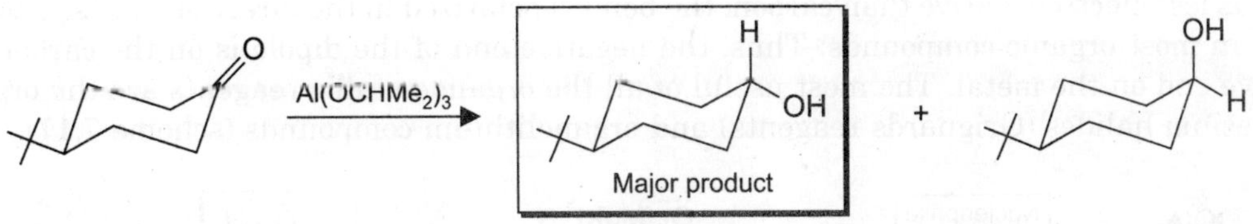

SCHEME 7.16b

- This is a reversible reaction and is under thermodynamic control and therefore, the most stable product (A, scheme 7.16*c*) will be the major product at equilibrium (the major product will thus not be the one that is easily formed *i.e.*, under kinetic control).

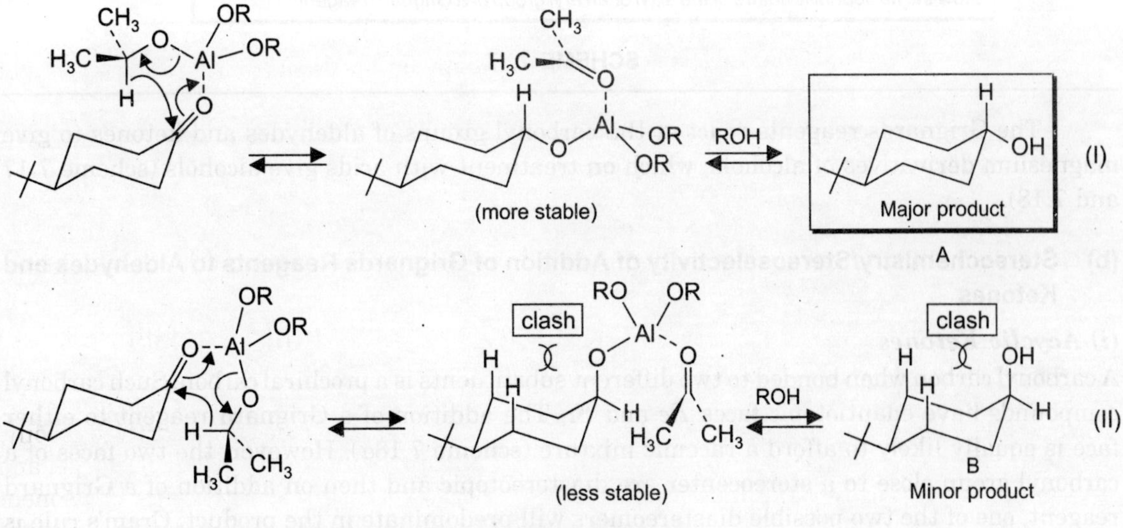

SCHEME 7.16c

- The attack is at one or the other face of the ketone at an angle of 109°. Attack from the top axial face yields the more stable alkoxide (eq. I, scheme 7.16*c*) and the most stable of the alcohols.
- The attack from the equatorial side (eq. II) yields the less stable alkoxide and alcohol (B), since both the alkoxide and the corresponding alcohol suffer from unfavourable 1, 3-nonbonded interactions.
- One may think that the initially formed aluminium complex may be formed easily from the equatorial side, however this is irrelevant when the reaction is under equilibrium as is so in the present case.

7.6 ADDITION OF ORGANOMETALLIC COMPOUNDS TO CARBON—HETERO MULTIPLE BONDS

(a) Addition of Organomagnesium Compounds (Grignard Reagents) to Aldehydes and Ketones

Although the carbon metal bond is covalent, many organometallic compounds react as though they are carbanions and are useful as carbon nucleophiles (scheme 7.17). This is so since, the metal is less electronegative than carbon, the bond is polarized in the direction opposite to that found in most organic compounds. Thus, the negative end of the dipole is on the carbon and positive end on the metal. The most useful of all the organometallic reagents are the organomagnesium halides (Grignards reagents) and organolithium compounds (scheme 7.17).

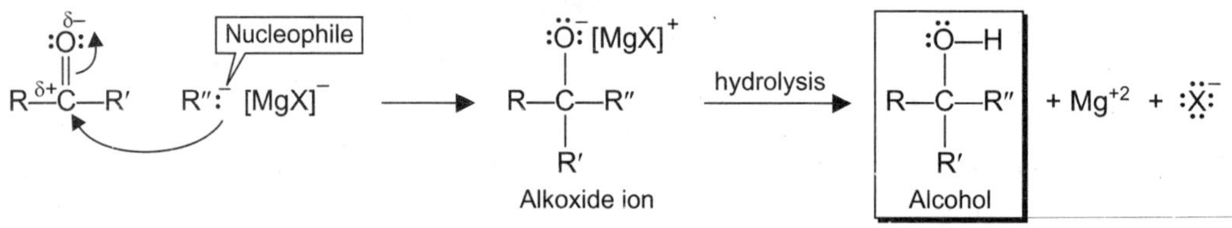

An organometallic compound often reacts like a carbanion R:⁻ (a nucleophile/base) to show the nucleophilic nature of the alkyl or an aryl group of a Grignard reagent.

SCHEME 7.17

The Grignards reagents react at the carbonyl groups of aldehydes and ketones to give magnesium derivatives of alcohols, which on treatment with acids give alcohols (scheme 7.17 and 7.18).

(b) Stereochemistry/Stereoselectivity of Addition of Grignards Reagents to Aldehydes and Ketones

(i) Acyclic Ketones

A carbonyl carbon when bonded to two different substituents is a prochiral carbon. Such carbonyl compounds have enantiotopic faces *Re* and *Si*. The addition of a Grignard reagent to either face is equally likely to afford a racemic mixture (scheme 7.18*a*). However, the two faces of a carbonyl group close to a stereocenter are diastereotopic and then on addition of a Grignard reagent, one of the two possible diastereomers will predominate in the product. Cram's rule is used to predict the stereochemistry of the major product as in the case or α-phenylpropionaldehyde, where *erythro*-compound is formed as the major product (scheme 7.18*b*, compare this with scheme 2.25, in scheme 7.18*b*. This compound is considered as Felkin-Ahn model, the end result being the same).

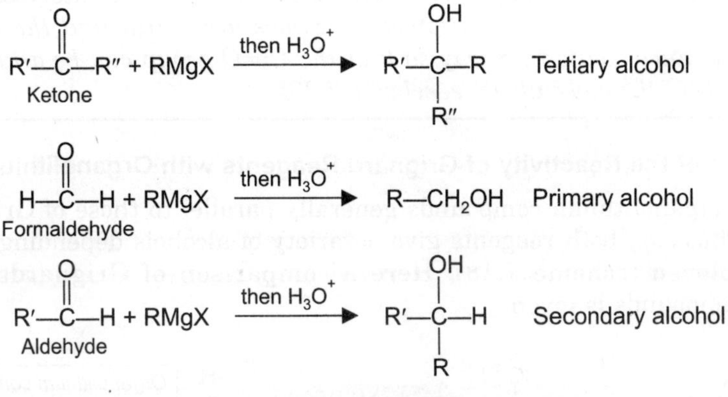

SCHEME 7.18

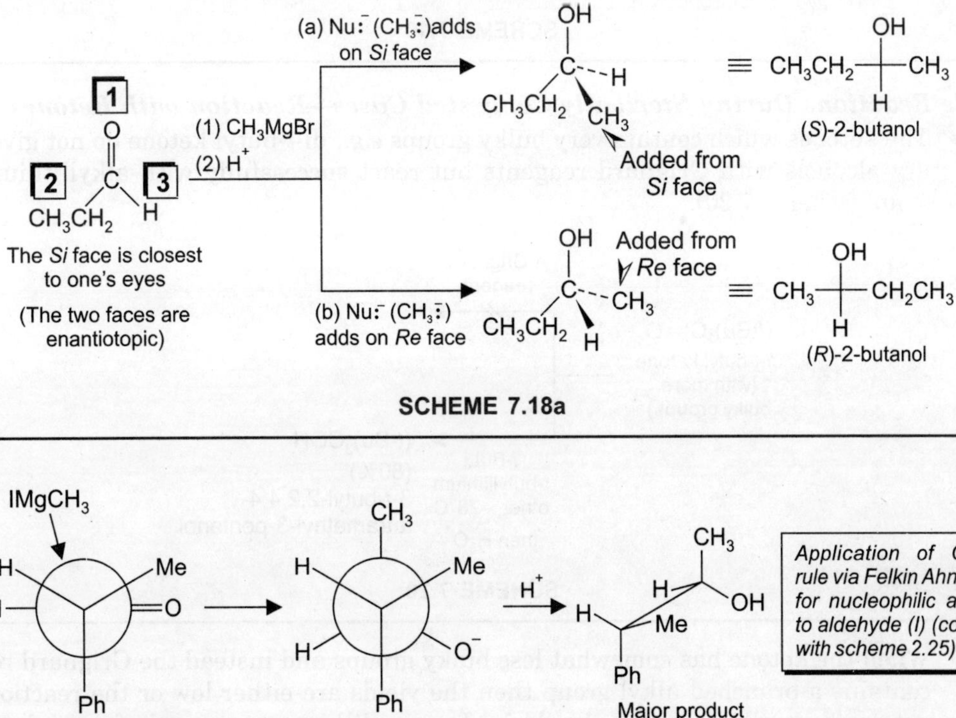

SCHEME 7.18a

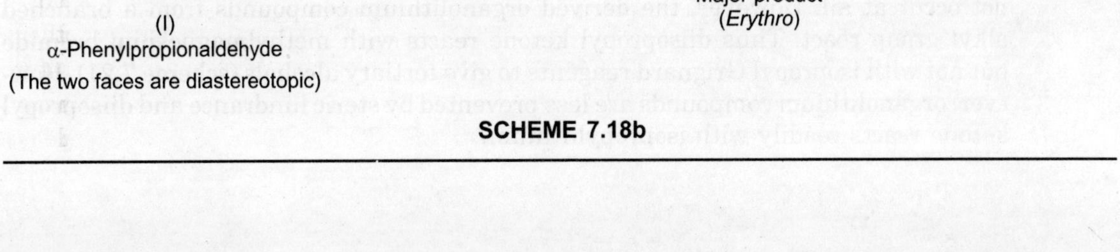

SCHEME 7.18b

Enantioselective Synthesis

Much effort has been spent to develop ways to perform the enantioselective reduction of ketones. Several drugs are chiral and have the alcohol functionality. In Chapter 2, methods are presented for making enantiomerically pure alcohols e.g., via hydroboration of an alkene (see scheme 2.41). The enzyme catalyzed addition to a ketone gives only one enantiomer. Enzymes are chiral and these e.g., can position, the nucleophile in a way so that the C=O group can be attacked from only one side of the molecule (See, scheme 2.20)

(c) A Comparison of the Reactivity of Grignard Reagents with Organolithium Compounds

The reactions of organolithium compounds generally parallel to those of Grignards reagents (scheme 7.19). Thus *e.g.*, both reagents give a variety of alcohols depending on the carbonyl component employed (scheme 7.18). Here a comparison of Grignards reagents with organolithium compounds is given.

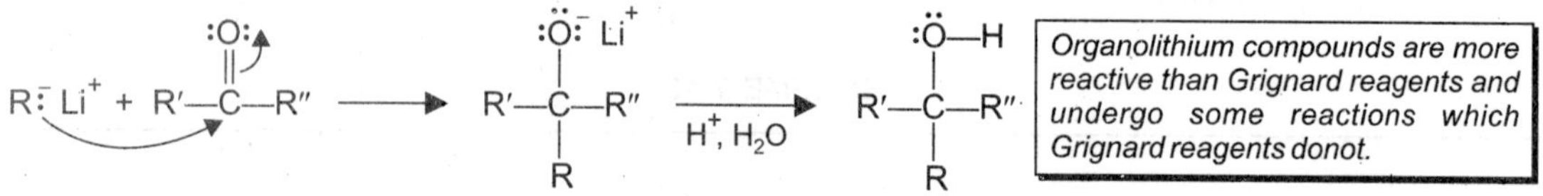

SCHEME 7.19

(i) Side Reactions During Sterically Congested Cases—Reaction with Ketones

- The ketones which contain very bulky groups *e.g.*, di-*t*-butyl ketone do not give tertiary alcohols with Grignard reagents but react successfully with alkyllithium reagents (scheme 7.20).

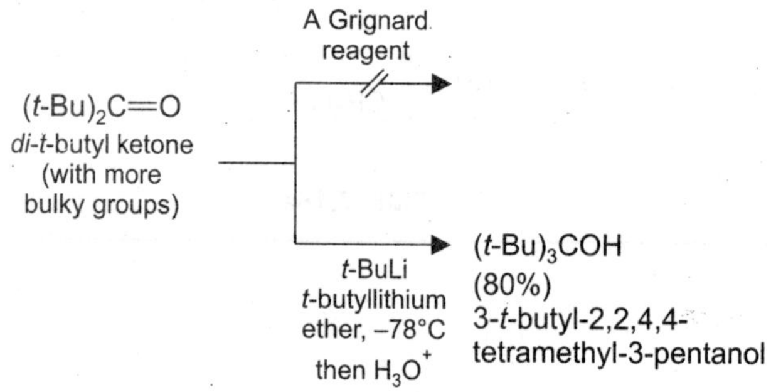

SCHEME 7.20

- When the ketone has somewhat less bulky groups and instead the Grignard reagent contains a branched alkyl group then the yields are either low or the reaction does not occur at all. However, the derived organolithium compounds from a branched alkyl group react. Thus diisopropyl ketone reacts with methylmagnesium bromide but not with isopropyl Grignard reagents to give tertiary alcohols (scheme 7.21). However, organolithium compounds are less prevented by steric hindrance and diisopropyl ketone reacts readily with isopropyllithium.

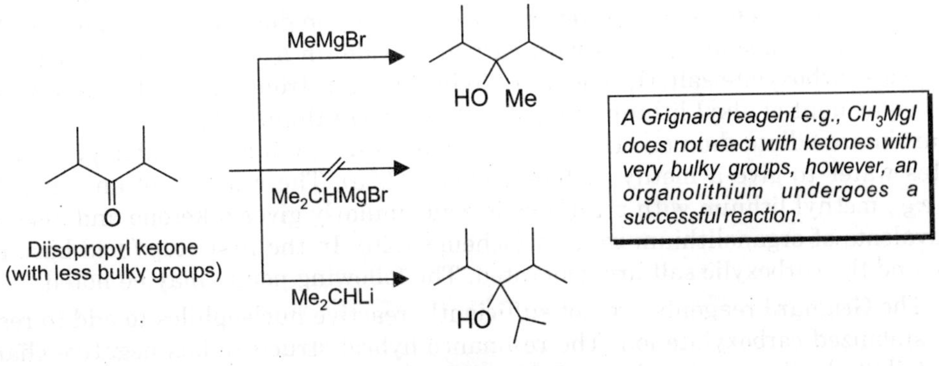

SCHEME 7.21

- When steric hindrance is the problem, in the case of an attempted Grignard reaction enolization and reduction may take place. During enolization the organomagnesium compound acts as a base and not as a nucleophile and abstracts an α-proton (activated proton) from the ketone to give the enolate. When water is added, the enolate is hydrolyzed and tautomerizes to regenerate the starting ketone (scheme 7.22).

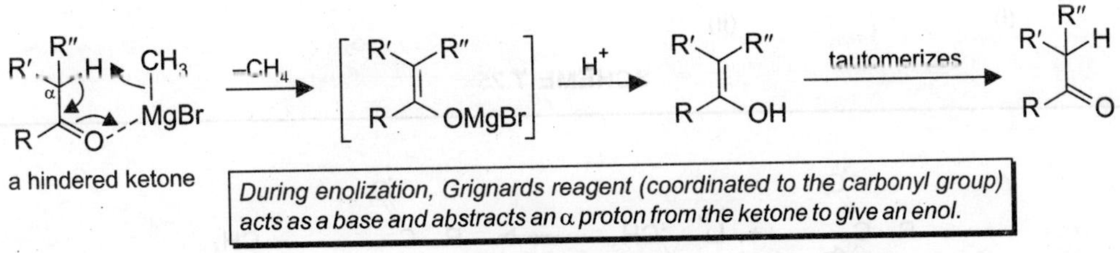

SCHEME 7.22

- Another important side reaction in hindered cases is reduction when the Grignard reagent has a β-hydrogen, the carbonyl group is reduced by hydride ion transfer. The process occurs in a six membered cyclic transition state (with Grignard reagent coordinated to the carbonyl group) and an alkene is formed (scheme 7.23).

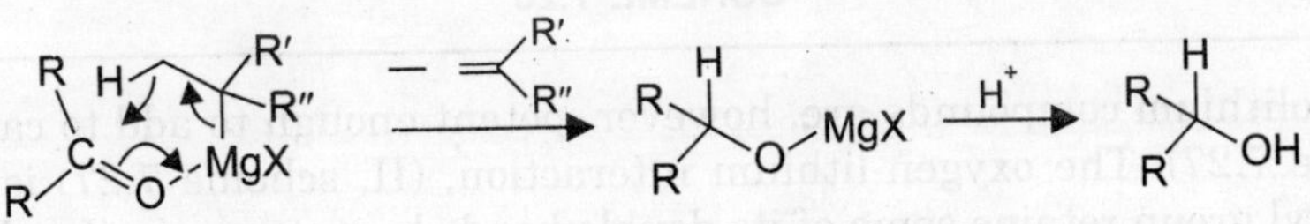

SCHEME 7.23

(ii) Reaction with Carbondioxide

The Grignard reagents add to carbon dioxide to afford salts of carboxylic acids from which free acids are formed on treatment with mineral acid (scheme 7.24).

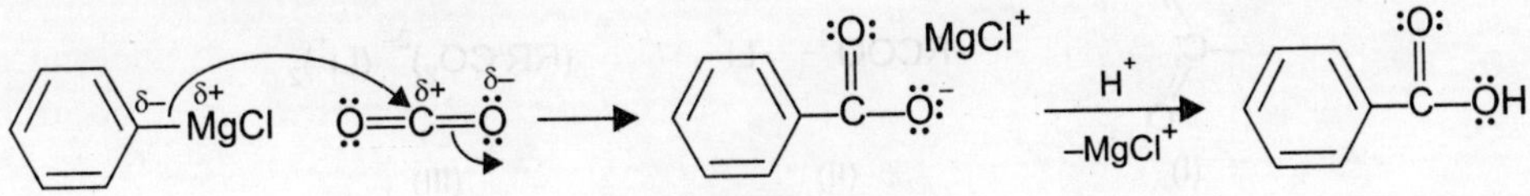

SCHEME 7.24

Organolithium compounds however, react with carbon dioxide to give ketones. Initially an organolithium compounds reacts with carbon dioxide (in a way similar to Grignard reagent) to give a carboxylate salt, (I, scheme 7.25) involving a strong oxygen-lithium interaction. A second equivalent of alkyl lithium reagent *e.g.*, methyl lithium adds to this carboxylate ion to give an intermediate dianion (II) which on hydrolysis gives an unstable gem diol which eliminates water to give a methyl ketone (scheme 7.25). The reaction of an organolithium reagent *e.g.*, methyl lithium with a carboxylic acid similarly gives a ketone and also requires two equivalents of organolithium reagent (scheme 7.26). In the first step (acid base process) methane and the carboxylic salt are generated. The following points may be noted:

- The Grignard reagents are not sufficiently reactive nucleophiles to add to resonance stabilized carboxylate ion. The resonance hybrid structure has negative charge distributed all over (I, scheme 7.27) making a reaction with a negatively charged nucleophile unlikely.

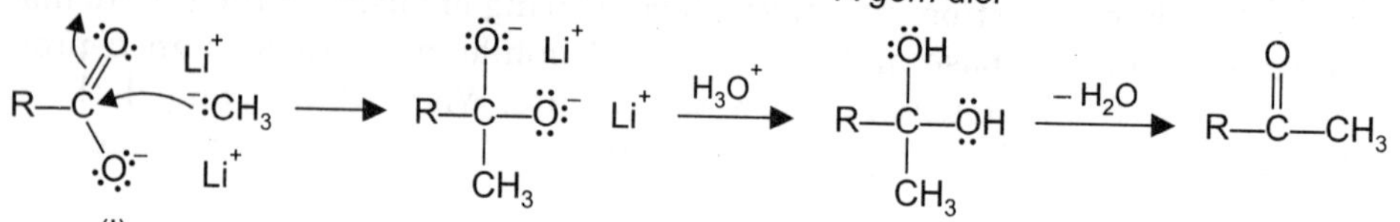

SCHEME 7.25

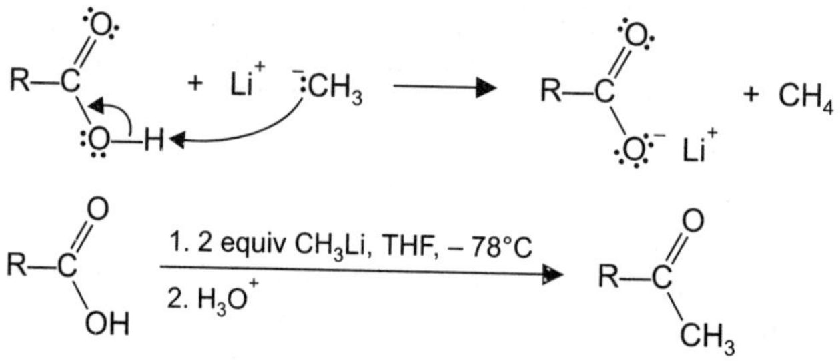

SCHEME 7.26

- Organolithium compounds are, however, potent enough to add to carboxylate ion (I, scheme 7.27). The oxygen-lithium interaction, (II, scheme 7.27) is strong, so that carbonyl group retains some of its double bond character as in (I, scheme 7.25) so as to react like a carbonyl group of a ketone. Moreover, the dianion (III, scheme 7.27) is also very stable due to strong interactions between the anionic oxygen atoms and the lithium cations. Temperatures below –75°C are needed to prevent the decomposition of this lithium salt (also see, Scheme 8.15).

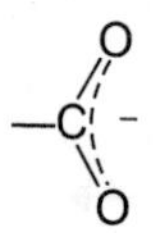

(I)

$RCOO^- \text{---} Li^+$

(II)

$(RR'CO_2)^{2-}(Li^+)_2$

(III)

SCHEME 7.27

(iii) Reaction with α, β-*Unsaturated Ketones*

The Grignard reagents generally react with α, β-unsaturated ketones to give 1, 4-adducts by addition to the C═C described as 1, 4-addition (*i.e.,* conjugate addition). There is however, an increasing competition from addition to C═O (*i.e.,* 1, 2-addition). The 1, 4-addition is predominant when there is a large substituent(s) attached to the C═O group (scheme 7.28). Organolithium compounds show much greater tendency to react by 1,2-addition. [*It is now known, that absolutely pure Grignards reagent (free from transition metal impurities like copper) brings about only 1, 2-addition, 1, 4-addition is promoted by the addition of catalytic amount of copper (I) salts. The precise mechanism is however, not known*].

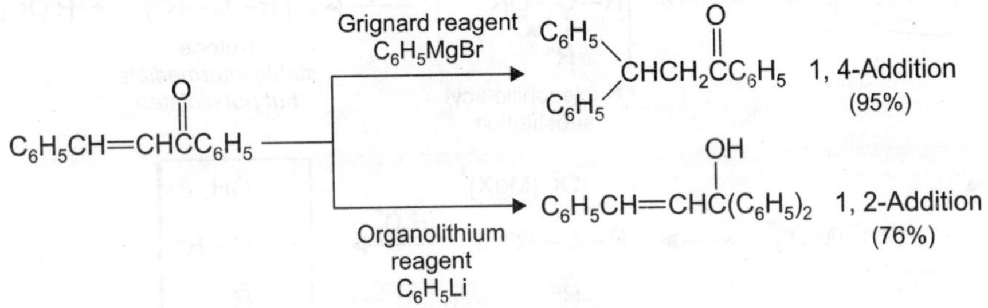

SCHEME 7.28

(d) Addition of Grignard Reagents to Nitriles-Synthesis of Ketones

Grignard reagents add to the triple bond of nitrites to afford magnesium derivatives which do not undergo further addition. These on hydrolysis give ketones via the unstable ketimines (scheme 7.29).

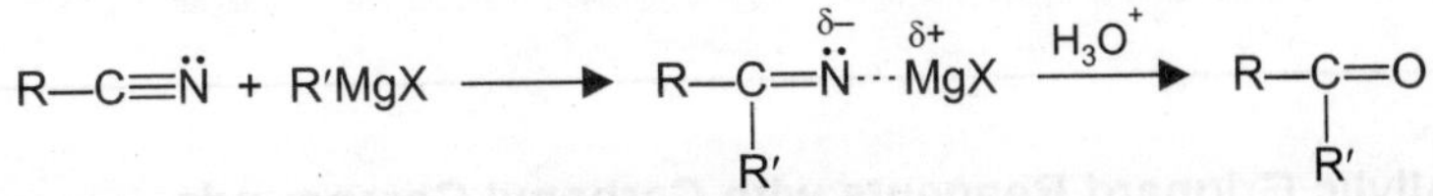

SCHEME 7.29

(e) Reaction of Grignard Reagents with Esters and Acid Chlorides

In the previous sections, one has seen a characteristic reaction of aldehydes and ketones. (*e.g.,* reduction with $LiAlH_4$) which is a nucleophilic addition to the carbon—oxygen double bond. During some of these reductions *e.g.,* of acids and esters firstly a nucleophilic addition at the carbonyl carbon atom occurs. The intermediate thus formed then eliminates a leaving group.

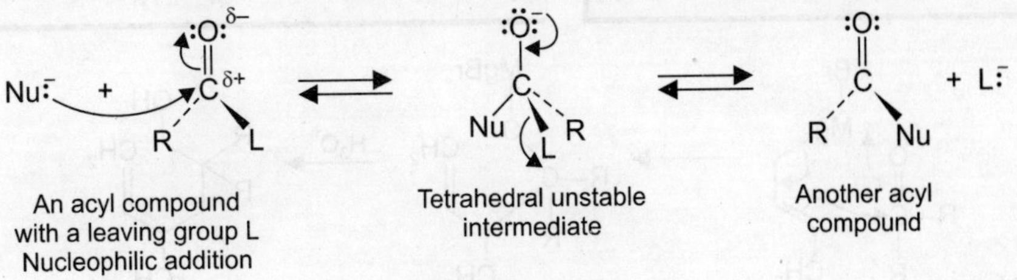

SCHEME 7.30

Thus the overall process in the case of an acyl compound is the nucleophilic addition, followed by elimination to give another acyl compound. The process is called nucleophilic substitution (Scheme 7.30), which is generally the pathway for carboxylic acids and their derivatives that takes place at their acyl (carbonyl) carbon atoms.

Thus an ester (L = OR, scheme 7.30) reacts with a Grignard reagent to give a ketone which is finally reduced to a tertiary alcohol (scheme 7.31). Similarly an acid chloride (L = Cl, scheme 7.30) gives a ketone, which is subsequently reduced to an alcohol.

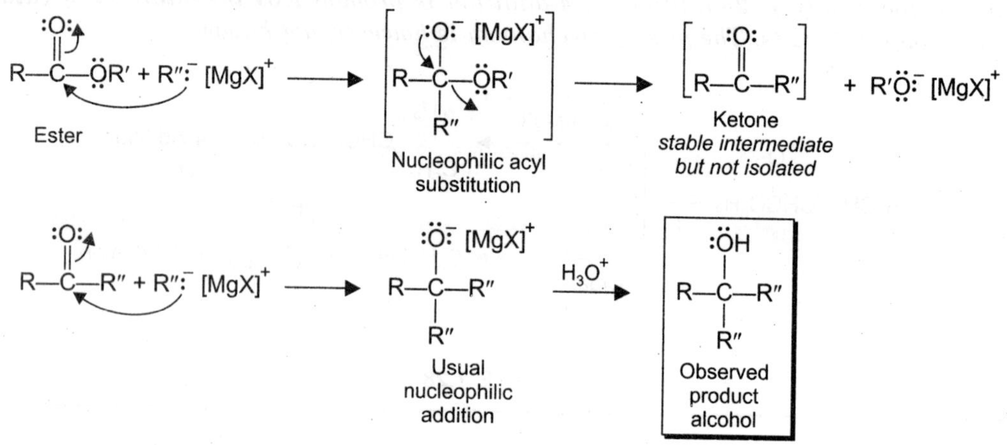

SCHEME 7.31

> *For a successful nucleophilic acyl substitution on a carboxylic acid derivative, the incoming nucleophile must not be a much weaker base than the group that is to be replaced.*

(f) Reaction of Allylic Grignard Reagents with Carbonyl Compounds

Reaction of an allylic Grignard reagent with a carbonyl compound is attended with an allylic shift (shift of the double bond) and occurs through a six membered transition state (scheme 7.32). One may note that allyl Grignard reagents are in rapid equilibrium, the major product of their reaction with an electrophile *e.g.*, carbon atom of C═O has less substituted double bond due to double bond shift.

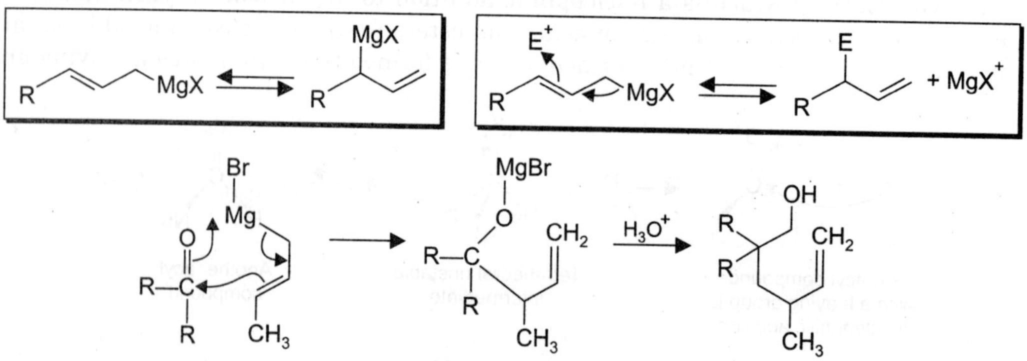

SCHEME 7.32

7.7 CONJUGATE ADDITION OF ORGANOCOPPER REAGENTS TO α, β-UNSATURATED CARBONYL COMPOUNDS–GILMAN REAGENTS

Lithium dialkylcuparates react with α, β-unsaturated compounds to give conjugate addition where these regents act as carbon nucleophiles. This reaction is the most prominent feature of these reagents in organic synthesis. These are made when a copper (I) halide reacts with two equivalents of an alkyllithium in diethyl ether or THF, the first equivalent forms an alkyl copper compound.

The driving force for this reaction is due to the preference of lithium, the more electropositive metal to exist as an ionic compound ($Li^+ I^-$). The copper of an alkyl copper reagent is a Lewis acid and therefore, the "alkyl anion" from the second equivalent of the organolithium reagent reacts with it to give lithium dialkyl cuparate (scheme 7.32*a*).

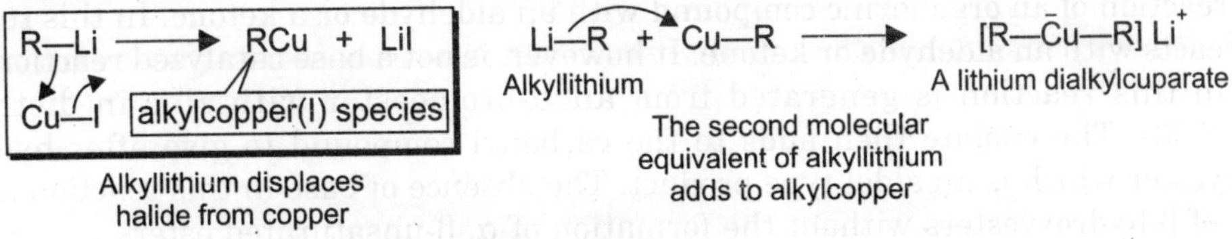

SCHEME 7.32a

Lithium dialkylcuparates react some what like Grignard or lithium reagents; however since the "alkyl anion" is complexed by copper, a less electropositive element than lithium, the alkyl-copper bond has more covalent character (I, scheme 7.32*b*) and these reagents are less reactive. Their reaction with acid chlorides and aldehydes is typical, and these react very slowly with ketones, and not at all with esters. The reaction of lithium dialkylcuparates with acid chlorides gives ketones in excellent yield.

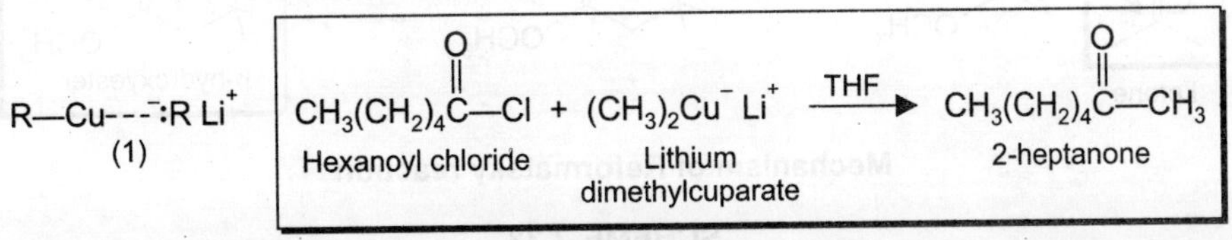

SCHEME 7.32b

The reaction of lithium dialkyl-cuparates with α, β-unsaturated carbonyl compounds gives excellent yields of 1, 4-addition products (scheme 7.32*c*). One can introduce two different alkyl groups in an enone. Initial product of nucleophilic attack at the β-carbon gives an enolate anion which can be further alkylated by an alkyl halide (scheme 7.32*d*).

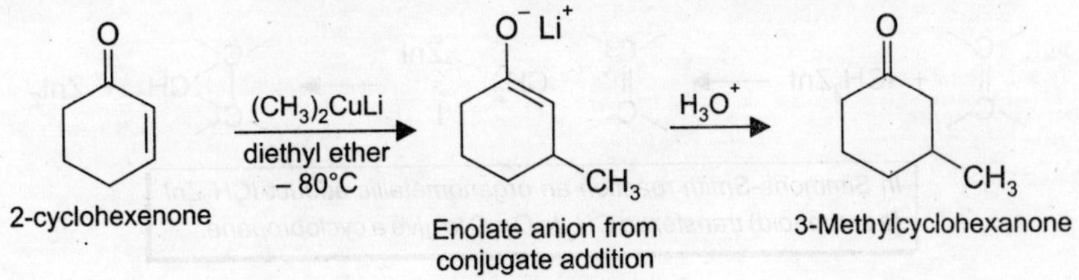

SCHEME 7.32c

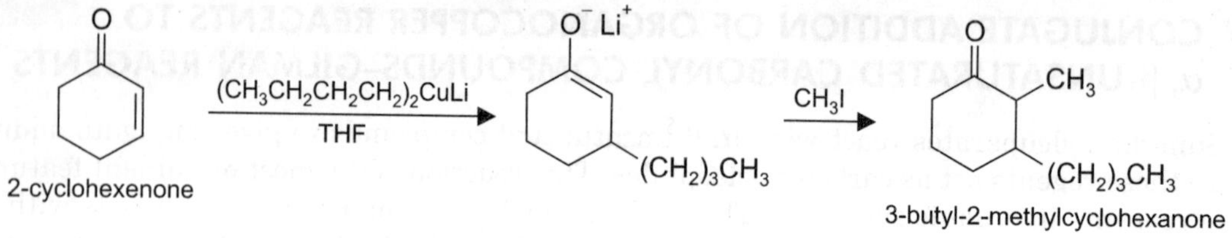

SCHEME 7.32d

7.8 ORGANOZINC COMPOUNDS

(i) The Reformatsky Reaction

It is the reaction of an organozinc compound with an aldehyde or a ketone. In this reaction an enolate reacts with an aldehyde or ketone. It however, is not a base catalyzed reaction, and the enolate in this reaction is generated from an α-bromoester with zinc in diethyl ether (Scheme 7.33). The enolate then adds to the carbonyl compound to give after hydrolysis a β-hydroxyester which is an aldol type product. The absence of base in this reaction allows the isolation of β-hydroxyesters without the formation of α, β-unsaturated esters.

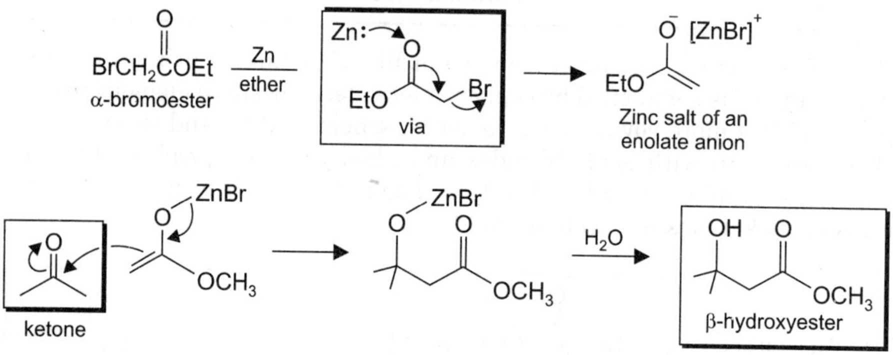

Mechanism of Reformatsky reaction

SCHEME 7.33

(ii) The Simmons Smith Reaction

The Simmons Smith procedure involves reaction of a double bond compound with diiodomethane (CH_2I_2) and a Zn-Cu couple, the attacking species is an organozinc intermediate (ICH_2ZnI), a carbene like species called a carbenoid (scheme 7.34). Overall in this reaction a CH_2 group is transferred to an alkene to form a cyclopropane.

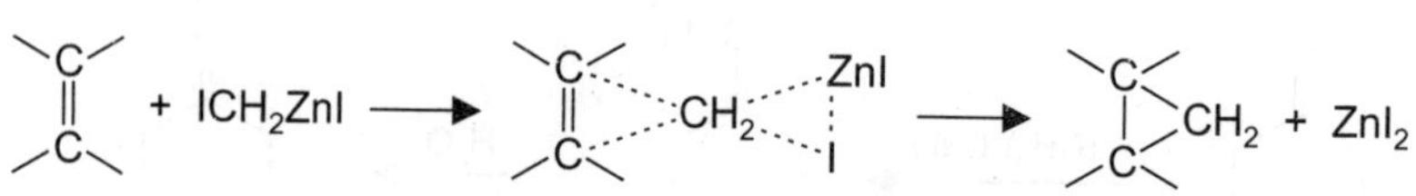

In Simmons-Smith reaction an organometallic adduct-ICH_2ZnI (a carbenoid) transfers a CH_2 to C = C to give a cyclopropane.

SCHEME 7.34

7.9 THE WITTIG REACTION (USE OF CARBANIONS IN ORGANIC SYNTHESIS)

A. Introduction/Mechanism

Elimination reactions give a mixture of olefins. The Wittig reaction is an alternative regiospecific conversion of a carbonyl group to an alkene, when the $>C{=}O$ bond of an aldehyde or a ketone is converted to a $>C{=}C<$ bond (scheme 7.35). Unlike Wittig reaction, reaction of *e.g.*, a ketone with Grignard reaction and dehydration of the tertiary alcohol thus formed instead gives an isomeric mixture of olefins.

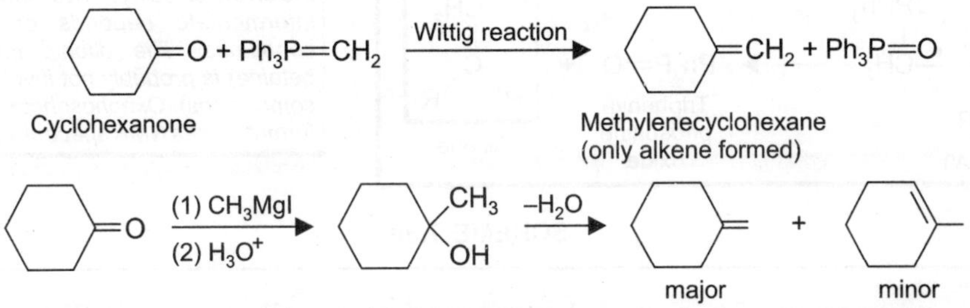

SCHEME 7.35

- The Wittig reaction is carried out by the reaction of a carbonyl compound with ylides. Ylides are special class of anions (Scheme 7.36).

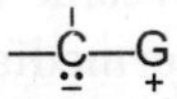

Ylide

Ylides are carbanions, where G is appropriately substituted N, S or P. Wittig reaction involves phosphorus ylides—the negative charge on carbon is balanced by the positive charge on adjacent phosphorus.

SCHEME 7.36

- Ylides are prepared from triphenyl phosphine and an appropriate alkyl halide (S_N2 displacement) to give an alkyltriphenylphosphonium salt. This on deprotonation with a base gives a carbanion (Scheme 7.37).

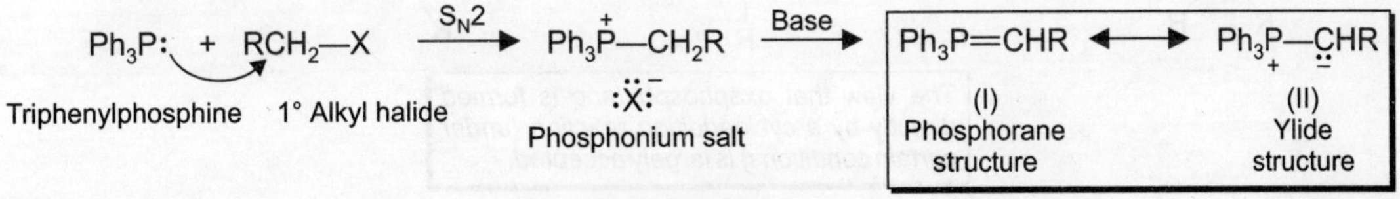

SCHEME 7.37

- The negative charge on carbon is balanced by the positive charge on the adjacent phosphorus and such zwitterions are often called ylides (II, Scheme 7.37). The other resonance structure is called (Phosphorane structure I). The ylides react as strong nucleophiles.
- Traditional mechanism of Wittig reaction is the nucleophilic addition of the ylide to an aldehyde or a ketone to give a diion (A, Scheme 7.38 called a *betaine*) as an intermediate. Its cyclization to the oxaphosphetane (B) and its elimination to triphenylphosphine oxide and an alkene completes the mechanism.

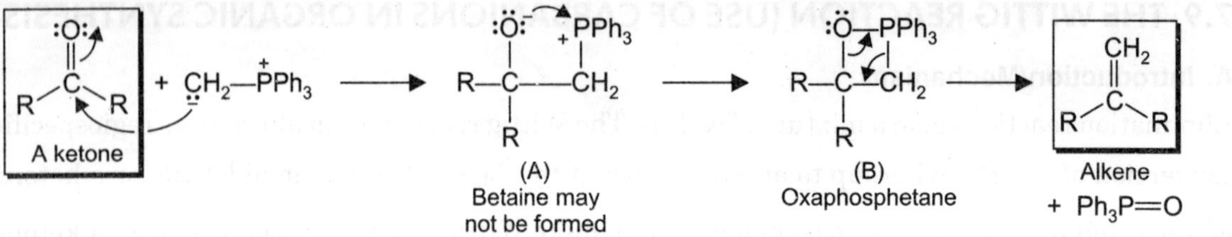

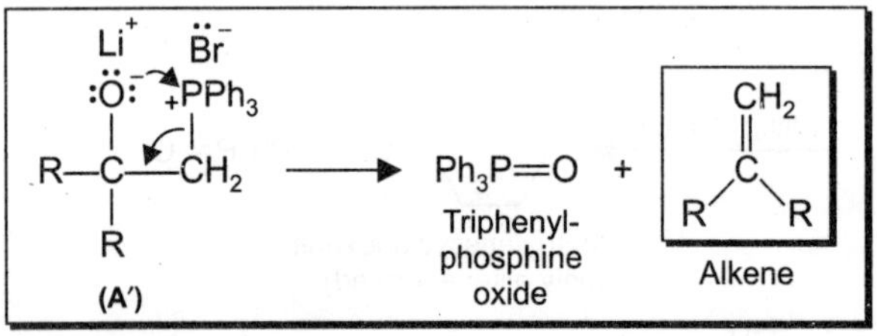

Mechanistically, the nature of the intermediate depends on the reaction conditions. The diionic intermediate (a betaine) is probably not involved (at least in some cases). Oxaphosphetane intermediate formation has NMR spectral support.

SCHEME 7.38

- The conversion of a carbonyl compound to an alkene using Wittig reaction is regiospecific unlike the use of Grignard reaction, followed by dehydration of the resultant tertiary alcohol which will yield mixture of isomeric alkenes.
- Much evidence including NMR, however, shows that the diion (A, Scheme 7.38) may not be an intermediate, and the oxaphosphetane intermediate may indeed be involved. The diion (betaine) intermediates have, however, been isolated in some Wittig reactions in the presence of salts like lithium bromide when the ionic intermediate is stabilized by interaction with lithium and bromide ions. Bond reorganization in either intermediate (A or A′ or B, Scheme 7.38) leads to the products.
- However, evidence has accumulated that oxaphosphetane is formed directly by a [2+2] cycloaddition reaction [$\pi^{2s} + \pi^{2a}$] as shown (Scheme 7.39).

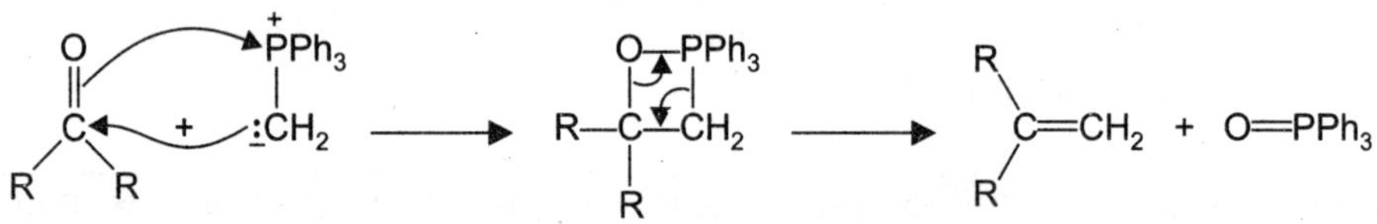

The view that oxaphosphetane is formed directly by a cycloaddition reaction (under certain conditions) is largely accepted.

SCHEME 7.39

(B) Stereochemistry/Stereoselectivity—Wittig Reaction though Regiospecific is not Stereoselective

This largely depends on reaction conditions and stability of the ylides.

(i) Non-Stabilized Ylides/Stereoselectivity

When the groups attached to the original alkyl halide are hydrogens or simple alkyl groups (See scheme 7.37). The α-hydrogen of the phosphonium salt is only weakly acidic and a very strong base *e.g.,* butyllithium is needed to give the ylide. This ylide is very reactive, unstable and highly nucleophilic. It gives the oxaphosphetane (irreversibly) which affords the alkene spontaneously. A mixture of both *E* and *Z* isomers is then obtained (Scheme 7.40).

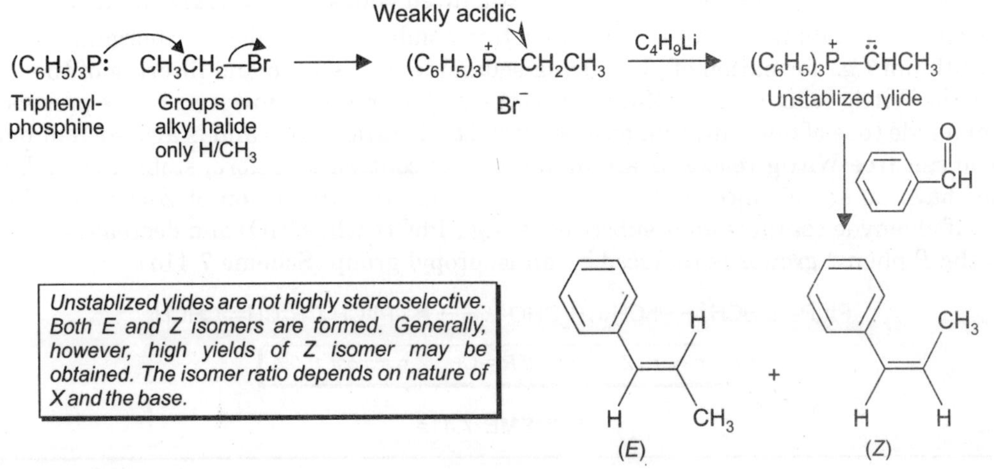

SCHEME 7.40

(ii) Stabilized Ylides

When in the original alkyl halide a group of *-M* type *e.g.,* an ester is adjacent to an anionic carbon, the ylide can be formed using less strong bases. This type of a ylide is stable and can be isolated. The stability allows it to react *reversibly* with the carbonyl compound and at the end of the reaction the (*E*) isomer usually predominates (Scheme 7.41).

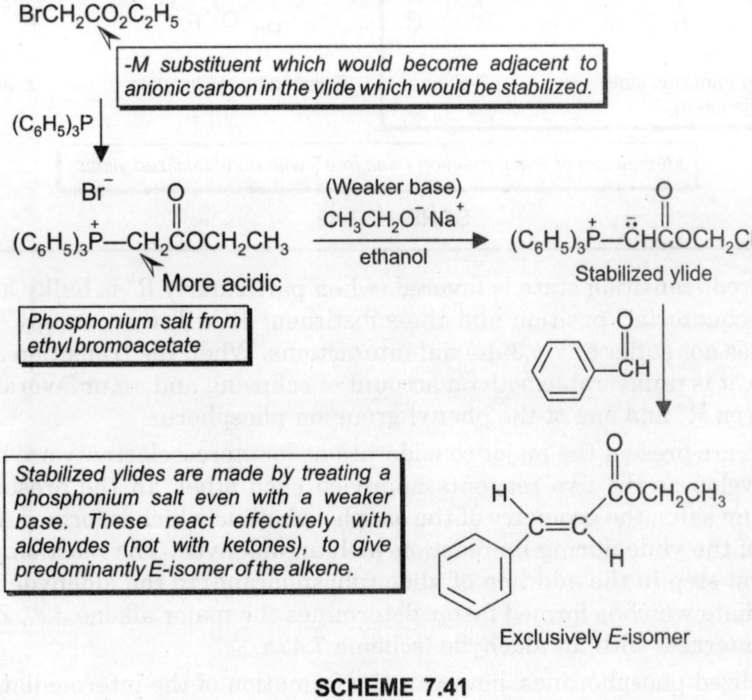

SCHEME 7.41

(iii) Wittig Reaction "Salt Free" or in the Presence of Salts and Steric Consideration—Finer Aspects of Mechanism

Wittig reactions using non-stabilized ylides are less stereoselective and lead to both *E*- as well as *Z*-isomers (often *Z*-isomer predominating), stabilized ylides, however, give mainly *E*-alkenes.

With non-stabilized ylides under salt free conditions (absence of inorganic salts like that of lithium *e.g.*, formation of ylide using sodamide as the base and filtering of the sodium halide), the Wittig reaction gives Z-alkene as the major product. However, in the presence of a lithium halide (use of butyl lithium as a base) the *E* : *Z* ratio varies and is also solvent dependent. The salt free Wittig reactions are under kinetic control, therefore, stabilities of *E* and *Z* alkenes have no significance. Two steric factors effect the proportion of *Z*-isomer, which increases if aldehyde contains an α-substituent (*e.g.*, $PhCH_2CR_2$ CHO) and decreases *e.g.*, when one of the *P*-phenyl groups is replaced by an isopropyl group (Scheme 7.41*a*).

$$Ph_3\overset{+}{P}—\overset{-}{C}HCH_3 + PhCH_2CR_2CHO \longrightarrow PhCH_2CR_2CH{=}CHCH_3$$

If R = H, E: Z = 6 : 94; If R = CH_3, Z = almost 100%

SCHEME 7.41a

The "salt free" Wittig reactions of nonstabilized ylides follow the cycloaddition mechanism (scheme 7.39). For *Z*-alkene formation the spatial orientations are as in puckered four-membered transition state (I, scheme 7.42) while the *E*-alkene formation involves an almost planar transition state (II, scheme 7.42).

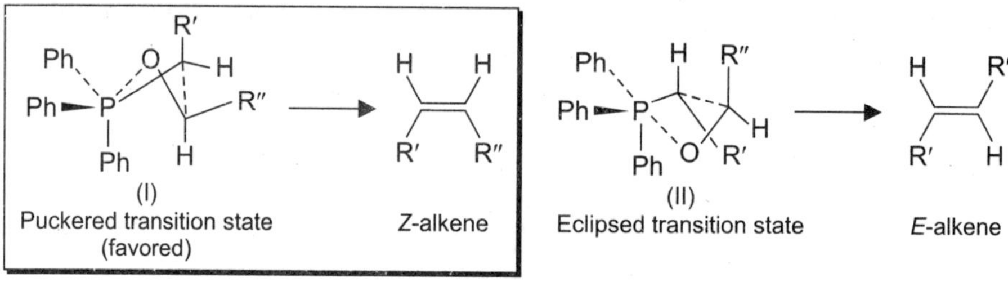

Mechanism of Wittig reaction ("salt free") with non-stabilized ylides.

SCHEME 7.42

The puckered transition state is favored (when particularly R″ is bulky and then it can adopt a "pseudo-equatorial" position and the substituent R′ will then occupy "pseudo-axial" position but it does not suffer any 1,3-diaxial interactions. When the transition state is planar (II, Scheme 7.42), it is unfavorable both on account of eclipsing and an unfavorable 1,3-diaxial interaction between R″ and one of the phenyl group on phosphorus.

When salts are present the major considerations for stereoselectivity are based on steric effects which develop as the two reagents approach each other. In the presence of a polar solvent and lithium salts, the geometry of the oxaphosphetane which is formed depends on the steric approach of the ylide during its reaction with an aldehyde. The reaction yields more of *Z*-alkene. The first step in the addition of ylide (phosphorane) to the aldehyde is irreversible and the intermediate which is formed faster determines the major alkene, *i.e.*, *Z*-alkene when a reactive ylide interacts with an aldehyde (scheme 7.42a)

With stabilized phosphoranes, however, the formation of the intermediate is reversible (Scheme 7.42*a*) and this allows the formation of the more stable stereoisomeric form of oxaphosphetane. Thus an interconversion to the more stable and thus more abundant isomeric form and a *syn*-elimination affords the *E*-alkene.

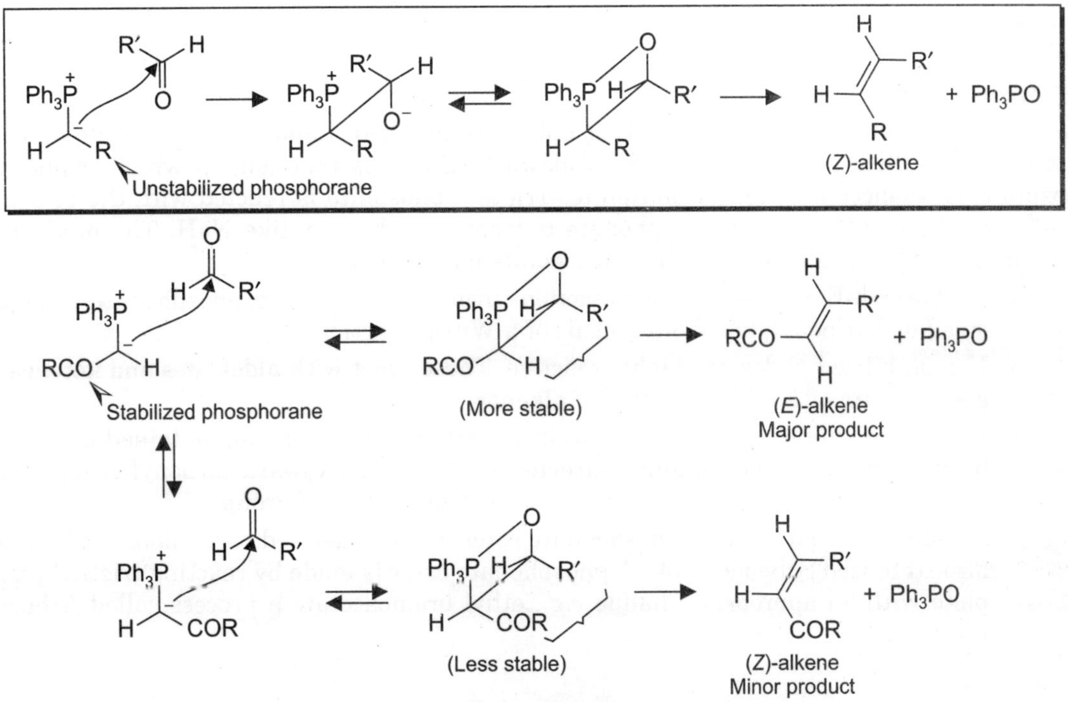

SCHEME 7.42a

Thus in summary when lithium salts are present in the Wittig reaction both with stabilized and non-stabilized ylides alike, the stepwise mechanisms may be operating and the adducts are diastereomeric diionic species (I and II, Scheme 7.43) which eliminate to the *Z*- and *E*-alkenes respectively.

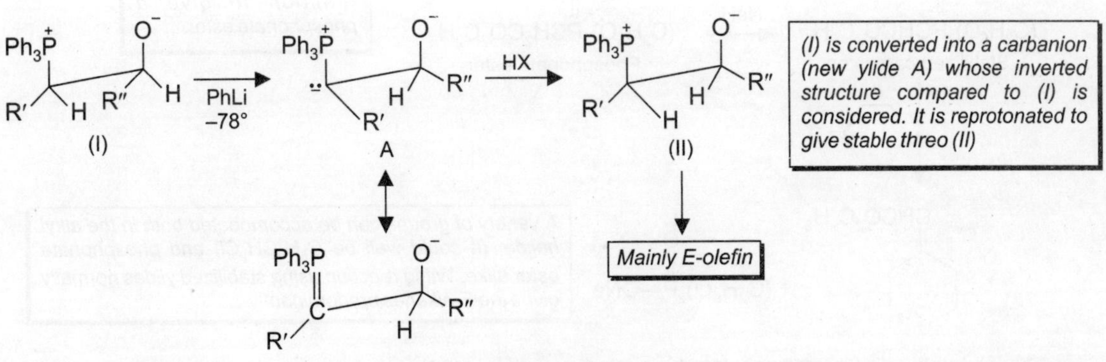

SCHEME 7.43

It is however, possible to form mainly *E*-alkenes from the non-stabilized ylides by the Schlosser modification of the Wittig reaction. The ylide is generated as a lithium bromide complex and the reaction with an aldehyde is carried out at very low temperature–78° so that the diionic species remain stable and do not undergo elimination step. Then a second molar equivalent of phenyl-lithium is added giving a new ylide (*A*, Scheme 7.43) which is protonated. The inverted structure of the ylide (carbanion is considered) to give almost exclusively the more stable *threo*- diion, its decomposition gives mainly *E*-alkene (also see answer to problem 7.3).

(iv) Wadsworth-Emmons Reaction

The stabilized ylides, though useful, react only with aldehydes and these do not do so with ketones. A modification is introduced (Wadsworth-Emmons reaction) in which triphenylphosphine is replaced by triethyl phosphite. Triethyl phosphite is reacted with the relevant halide and the resulting alkyl phosphonate is treated with base, like NaH. The product of reaction is mainly (*E*) alkene. The following points may be noted:

- Wadsworth-Emmons—Horner reaction involves the use of a phosphonate ester instead of a tripenylphosphonium salt of a Wittig reaction.
- Non-stabilized ylides are highly reactive. These react with aldehydes and ketones to give mixture of *E* and *Z* isomers of alkenes.
- A stabilized ylide (—*M* group adjacent to carbanionic carbon are obtained using less basic conditions. Nucleophilic character and reactivity toward carbonyl group is decreased, since the negative charge is delocalized into —*M* group.
- A useful alternative uses phosphonate ester and the derived carbanions, (which are also stabilized Scheme, 7.44). A phosphonate ester is made by reacting triethyl phosphite with an appropriate halide *e.g.*, ethyl bromoacetate a process called Arbuzov reaction.

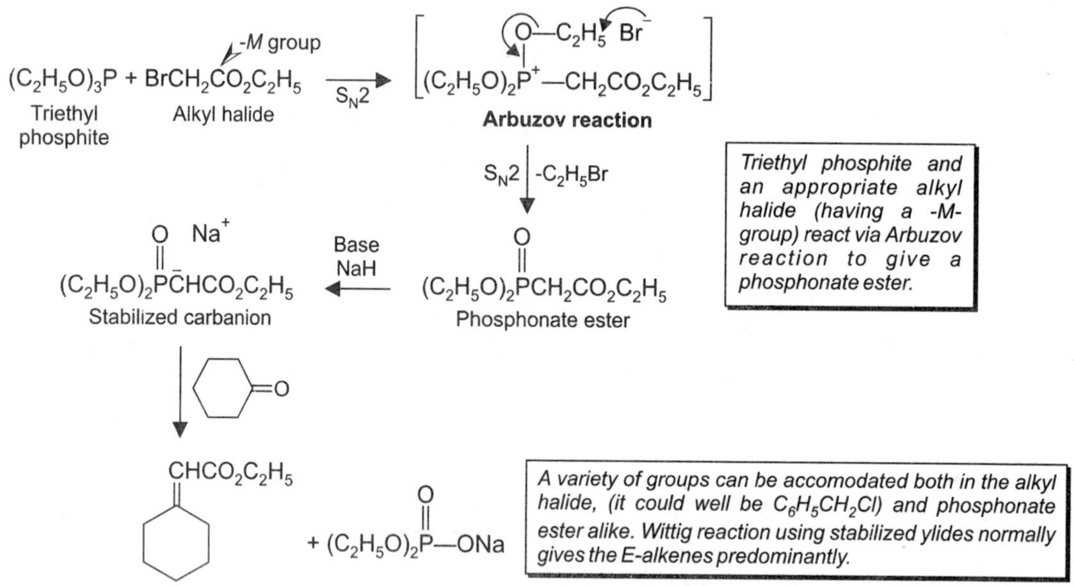

Wadsworth-Emmons reaction

SCHEME 7.44

- Of the stabilized ylides, the one derived from a phosphonate ester is more nucleophilic (scheme 7.45).

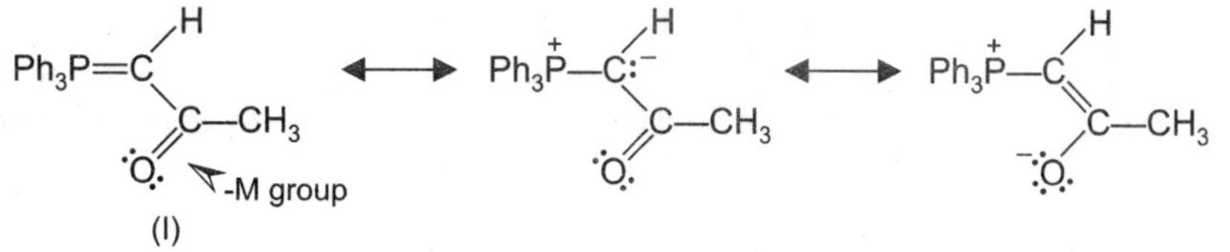

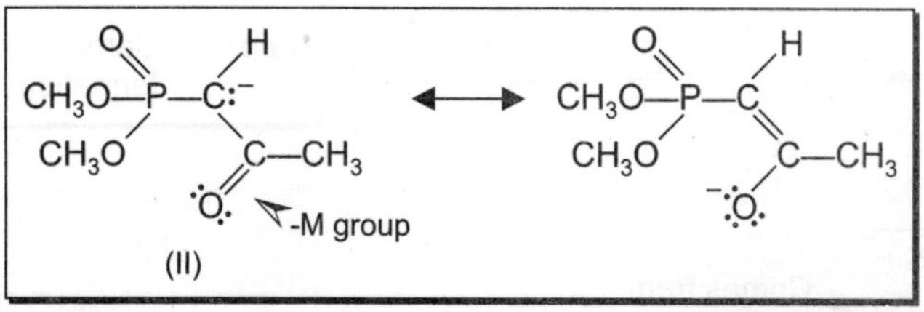

*A familiar stabilized ylide (I, for **Wittig reaction**) is less nucleophilic due to resonance. The stabilized ylide (II) from phosphonate ester (**Wadsworth-Emmons reaction**) is more nucleophilic (oxo and alkoxy groups on P ensure that negative charge is more on carbon)*

SCHEME 7.45

- The elimination step from Wadsworth and Emmons reaction is highly stereospecific and is thought to involve the formation of oxyanions of β-hydroxyphosphonates (Scheme 7.46). The reactions proceed under thermodynamic control, the intermediate oxyanion (I, Scheme 7.46) with *threo*-configuration is more stable thermodynamically and gives *E*-alkene as the main product.

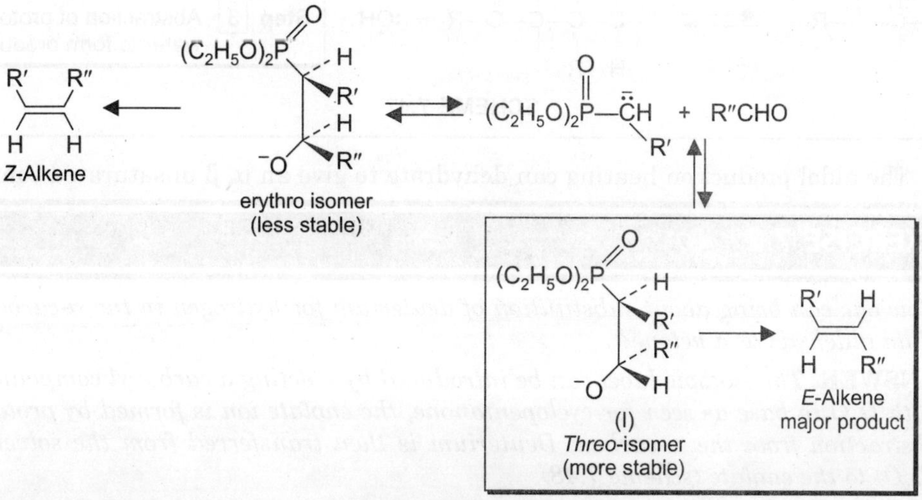

SCHEME 7.46

7.10 MECHANISM OF SOME CONDENSATION REACTIONS INVOLVING ENOLATES

1. Aldol Condensation

A. *Introduction*

The stereochemical details of the reaction are discussed in detail in Chapter 2, here an outline of the reaction is given and following points may be considered:

- Two important aspects of aldol addition of carbonyl compounds are the acidity of α-hydrogens and tendency of their carbonyl groups to undergo nucleophilic substitution under basic conditions.

- Thus the carbanion obtained by the removal of a proton from the α carbon of the molecule (This carbanion is resonance stabilized enolate anion) acts as a nucleophile to attack the carbonyl group of the second molecule to give an alkoxide anion (scheme 7.47).

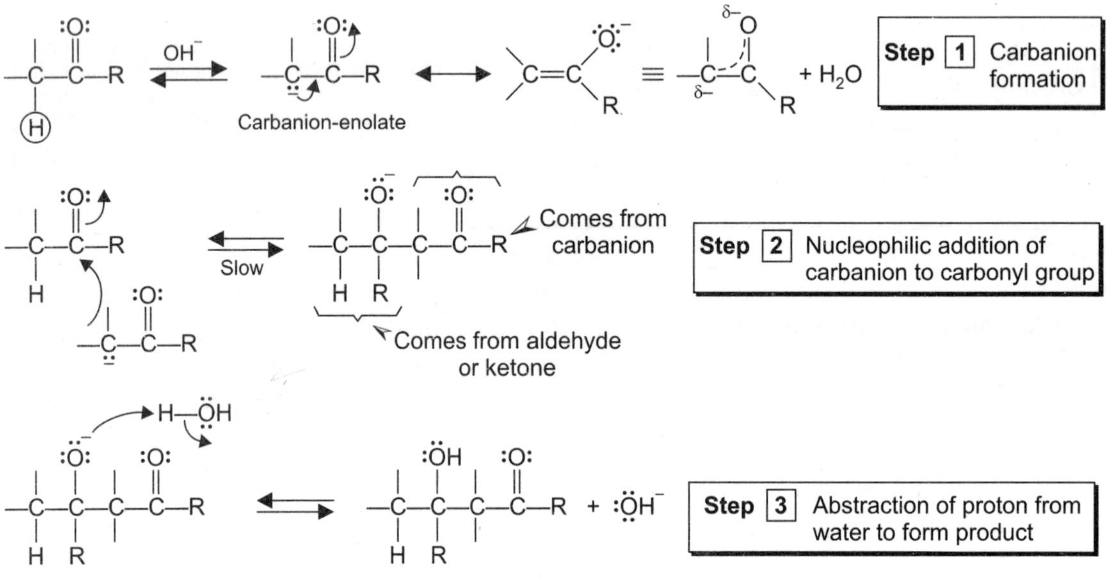

SCHEME 7.47

- The aldol product on heating can dehydrate to give an α, β unsaturated compound.

EXERCISE 7.2

How one can bring about substitution of deuterium for hydrogen in the α-carbon of an aldehyde or a ketone?

ANSWER. *The isotopic label can be introduced by reacting a carbonyl compound with D_2O in base as seen for cyclopentanone, the enolate ion is formed by proton abstraction from the α-carbon. Deuterium is then transferred from the solvent (D_2O) to the enolate (scheme 7.48).*

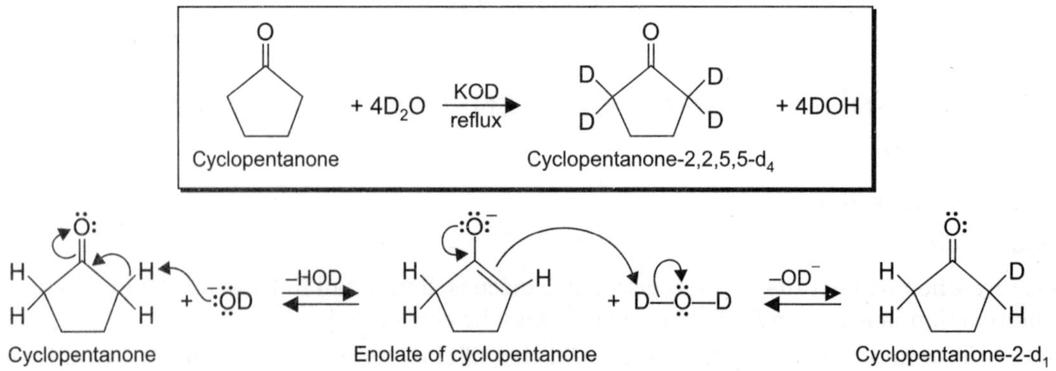

SCHEME 7.48

EXERCISE 7.3

What are the stereochemical consequences of enolization on (R)-sec-butyl phenyl ketone?

ANSWER. *The proton abstraction from the α-carbon converts the chiral substrate into an achiral enol (or enolate ion). Thus the process leads to racemization (scheme 7.48a).*

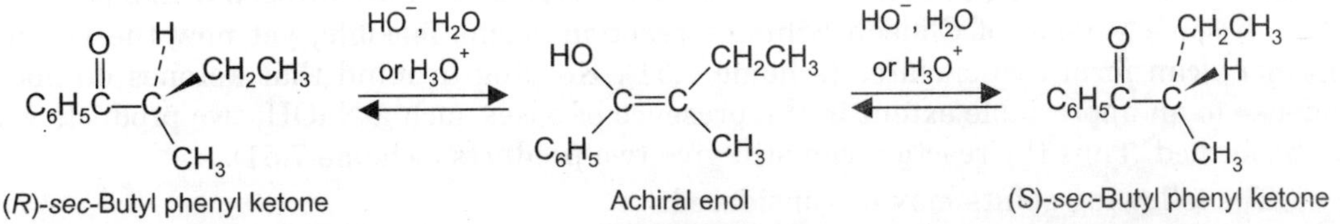

SCHEME 7.48a

- In the real world of organic synthesis, one rarely does a simple aldol condensation between two identical aldehydes or two identical ketones, since this leads to a mixture of products. The aldol reaction which starts from two different carbonyl compounds is called a *crossed aldol reaction.*
- A usual simple crossed aldol reaction is of no synthetic importance, since if both reactants have α-hydrogens a complex mixture of products is formed. Thus at least four products would be formed via an aldol reaction between CH_3CHO and CH_3CH_2CHO with sodium hydroxide.
- Crossed aldol reactions (using NaOH) can be of practical value if one component does not have an α-hydrogen. Thus this component will not be able to undergo self condensation since no enolate anion can be formed from it. If benzaldehyde is placed in base and propanal is added slowly at 10°, reaction (scheme 7.49) is reasonably successful.

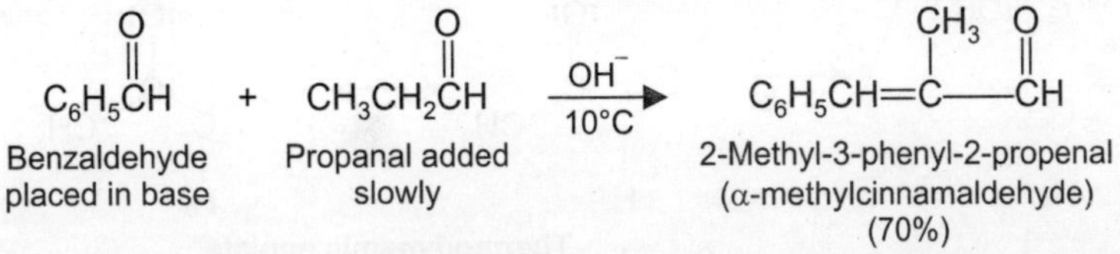

SCHEME 7.49

- A crossed aldol reaction in which one component is a ketone is called a Claisen-Schmidt reaction (scheme 7.50).

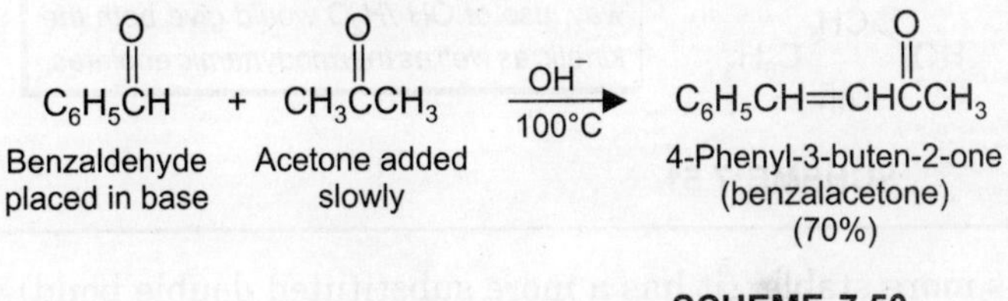

The ketones under these conditions do not self condense appreciably. A crossed aldol reaction in which one component is a ketone is called a ***Claisen-Schmidt reaction.***

SCHEME 7.50

- There are even difficulties in carrying out crossed aldol condensations. Thus *e.g.*, in the reaction (scheme 7.50) the only enolate possible from the ketone can add to either ketone itself or aldehyde to generate two products (only one product is shown; although ketones also undergo base catalyzed aldol condensation but for them equilibrium is unfavorable and the aldol addition is reversible).

B. Directed Aldol Condensations

In (scheme 7.50) during aldol type condensation only one enolate can be formed. Other difficulties may be involved *e.g.*, when one pairs benzaldehyde with no α-hydrogen with 2-pentanone, although some version of Claisen Schmidt reaction seems feasible, yet now the ketone 2-pentanone can form two enolates (scheme 7.51). Keeping in mind that ketones do not self condense to an appreciable extent in the presence of bases such a NaOH, two products would still be formed. Thus the reaction can still give two products (scheme 7.51).

The following points may be considered:

- Under ordinary conditions with bases like NaOH ketones *e.g.*, 2-pentanone can give two enolates and thus in principle two products would be formed with an aldehyde like benzaldehyde (I and II, scheme 7.51).

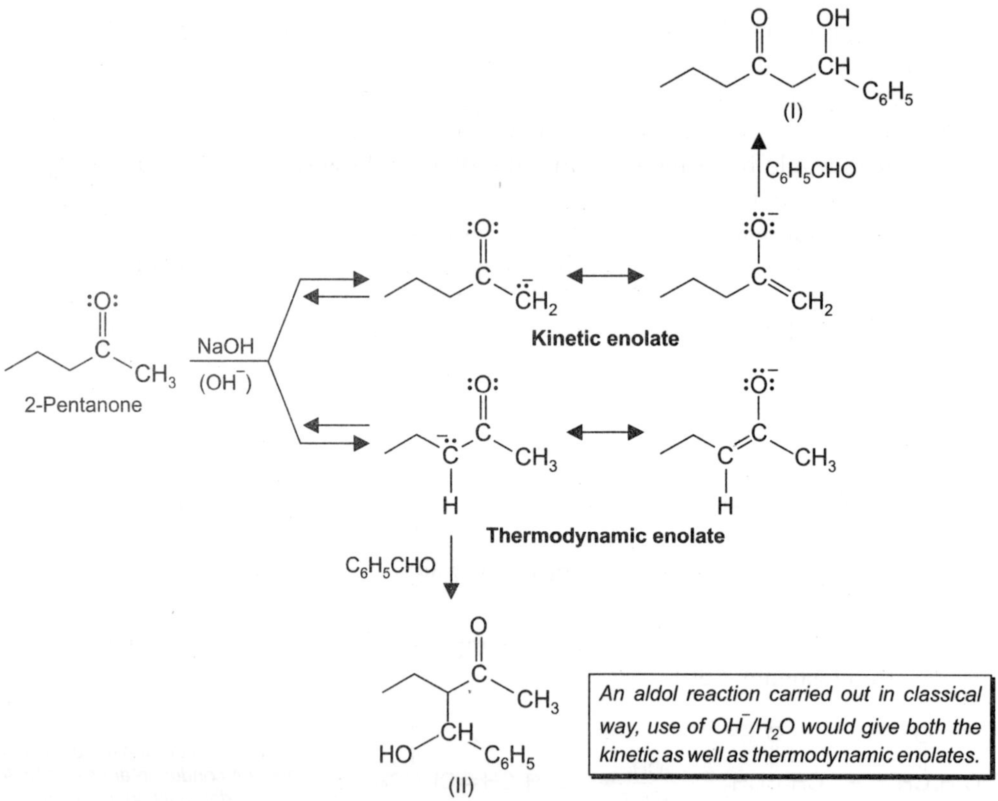

SCHEME 7.51

- Thermodynamic enolate is more stable, (it has a more substituted double bond) with a weaker base like OH^-/H_2O, this enolate will predominate at equilibrium (scheme 7.51).

- Kinetic enolate is less stable and recently methods have been developed to cleanly make the kinetic enolate as the major product. The kinetic enolate is formed easily and faster when a strong and very bulky base like LDA is employed. LDA has a relative easy access to the less hindered α-hydrogen and has difficulty to gain access to the more hindered parts of the carbonyl molecule (scheme 7.52).

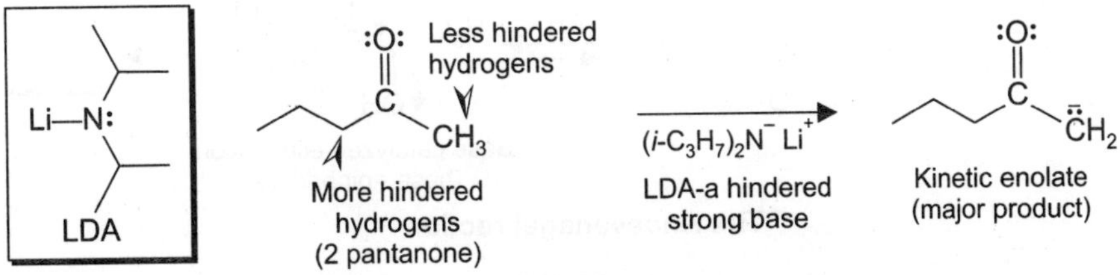

SCHEME 7.52

- One can make the aldol condensation regioselective with unsymmetrical ketones like 2-pentanone (see, scheme 7.51) by first generating the kinetic enolate using LDA. Its reaction with benzaldehyde will give only one condensation product (*i.e.*, I, Scheme 7.51). Such aldol reactions are termed directed aldol condensations.
- One can create conditions to generate exclusively *Z*-enolates as well (See scheme 2.32).

Stereochemistry of Directed Aldol Condensation

Aldol condensation of two different carbonyl compounds can give a molecule with two stereocenters (two stereogenic carbons). Two diastereomers and their enantiomers are possible to give in all four stereoisomers. This reaction can be made both diastereoselective as well as enantioselective. The details of this state-of the-art synthesis are given in Chapter 2.

2. The Knoevenagel Reaction

The condensation of aldehydes and ketones usually not containing an α-hydrogen with compounds having an active methylene group (*i.e.*, a methylene bonded to two groups of *–M* type) like malonic ester can take place even with a weaker base, which is enough to provide a sufficient concentration of the enolate ion. In those reactions where amines like piperidine are used, such condensations are termed Knoevenagel condensation scheme 7.53). The initially formed condensation product, then undergoes a base catalyzed elimination. When malonic acid is employed, one of the carboxyl group gets eliminated. Thus using benzaldehyde and malonic acid, one ends up with cinnamic acid in high yield (scheme 7.54). In some cases it is possible for the second molecule of the active methylene compound to add to the C=C bond of the product of Knoevenagel reaction scheme 7.55) via Michael reaction.

The Knoevenagel reaction has more synthetic value with aromatic aldehydes than aliphatic aldehydes. The addition to the C=C of the product from an aromatic aldehyde should be less likely due to loss of conjugation to the aromatic system.

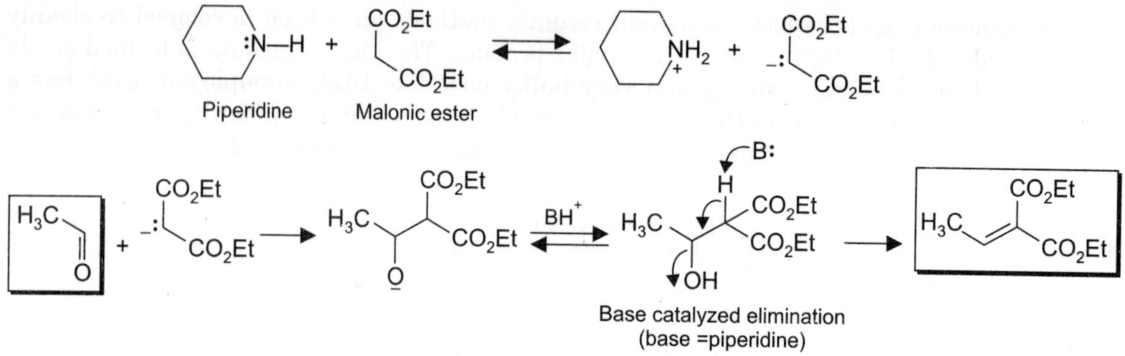

The Knoevenagel rection

SCHEME 7.53

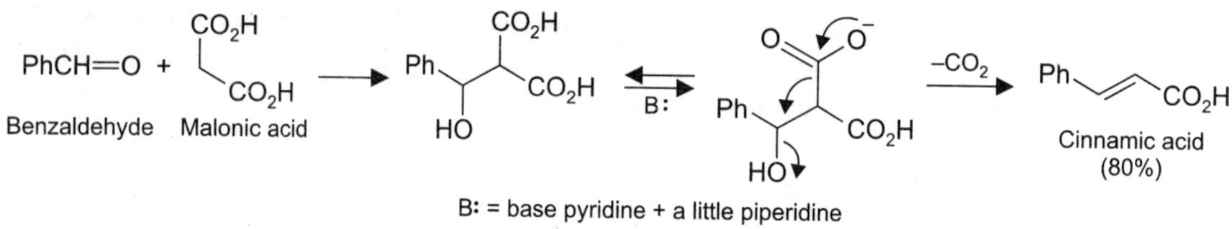

SCHEME 7.54

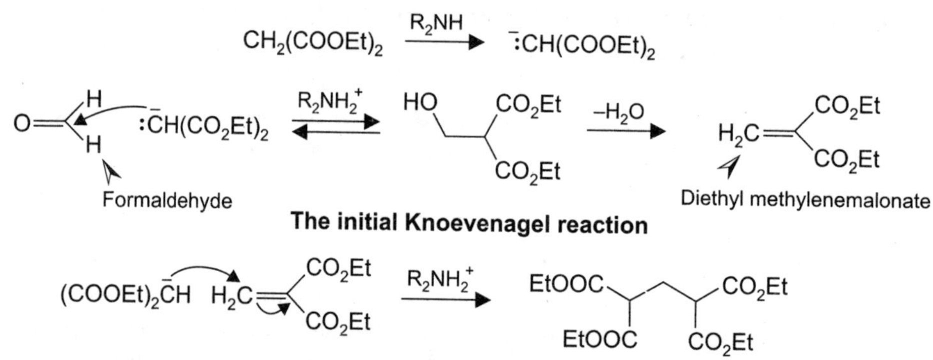

Followed by Michael addition

SCHEME 7.55

3. The Claisen Reaction

The Claisen reaction (different from Claisen ester condensation) is a base catalyzed reaction between an aldehyde which does not have an active hydrogen and an ester which contains an active hydrogen. Thus *e.g.*, benzaldehyde condenses with ethyl acetate in the presence of sodium ethoxide to give ethyl cinnamate in high yield (scheme 7.56). One may note that carbonyl in an aldehyde is more reactive towards nucleophiles when compared to the carbonyl group of an ester. The aldehydes with α-hydrogens atoms are not suitable partners for the reaction, since these prefer to under self condensation.

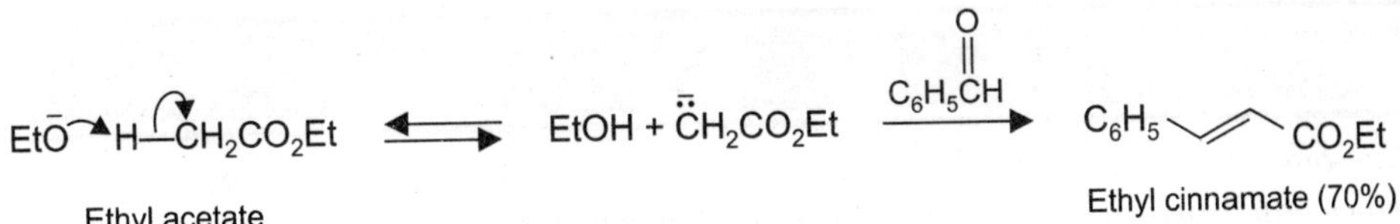

The Claisen reaction

SCHEME 7.56

4. The Claisen (Ester) Condensation

On treatment with strong bases like sodium ethoxide, esters which contain an α-hydrogen undergo self condensation termed Claisen (ester) condensation (scheme 7.57). The mechanism involves the conversion of one molecule of ester to a nucleophile (ester enolate) by the base and the second molecule serves as a substrate (scheme 7.57). The Claisen condensation reaction involves a series of equilibria *i.e.*, each step in the condensation is reversible. One may note that the formation of the new carbon-carbon bond is not thermodynamically favorable. The formation of the α-carbanion(ester enolate anion *pKa* ~ 25 scheme 7.57) is not a favorable equilibrium reaction with ethoxide, (*pKa* of ethanol = 15.9) and consequently only a low concentration of enolate forms. Attack of the α-carbanion on the second molecule of ester (acylation step), however, gives the product. This reaction is predicted to have an equilibrium constant around 1 and is reversible. In the final step, β-keto ester reacts with the alkoxide generated from substitution to give the enolate anion of the product. The equilibrium reactions are shifted toward this anion due to its stability since two carbonyl groups stabilize the common α-carbanion. The neutral β-keto-ester product is isolated via acidification of the reaction mixture.

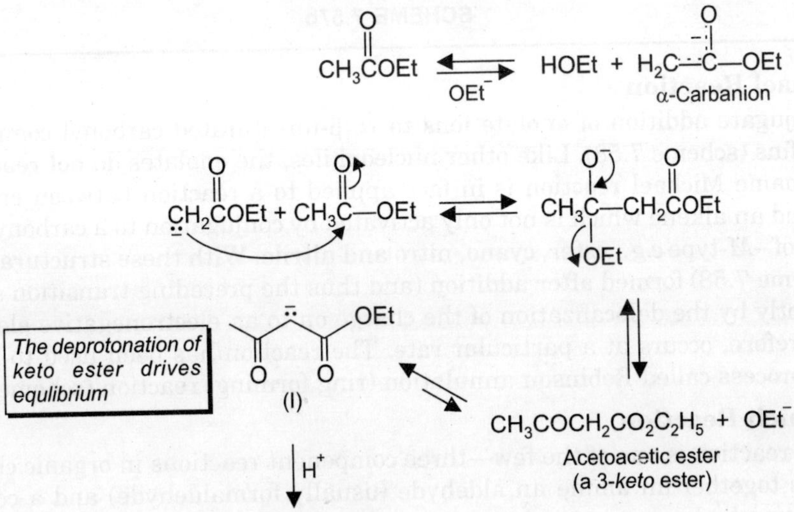

The Claisen condensation

SCHEME 7.57

EXERCISE 7.4

How one can prove that the deprotonation of the ketoester is the key step which drives equilibrium during the Claisen ester condensation ?

ANSWER. *In cases where the β-keto-ester product does not contain a C—H group adjacent to both keto and ester group. Thus the marked acidic character is absent, the equilibria are unfavourable for a successful reaction with ethoxide as the base. Ethylisobutyrate does not give the expected product (scheme 7.57a) with OEt⁻ as the base.*

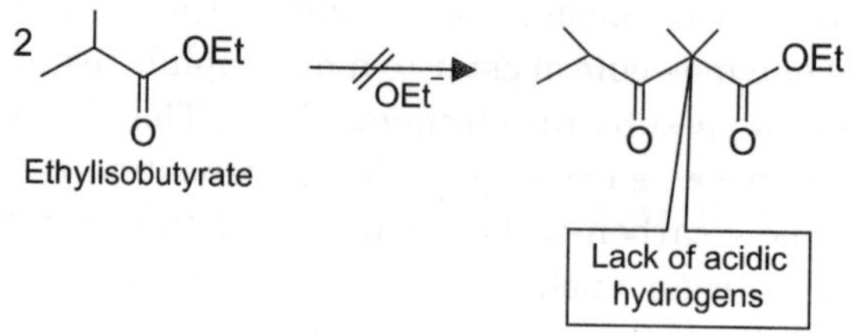

SCHEME 7.57a

This problem can be solved by using a far more stronger base than ethoxide ion *e.g.*, triphenylmethide ion (added as sodium triphenylmethyl). This completely removes ethanol formed in the reaction (scheme 7.57*b*)

$$EtOH + Ph_3C^- \rightleftharpoons EtO^- + Ph_3CH$$

SCHEME 7.57b

5. The Michael Reaction

This is a conjugate addition of enolate ions to α, β-unsaturated carbonyl compounds *i.e.*, to activated olefins (scheme 7.58). Like other nucleophiles, the enolates do not react with simple olefins. The name Michael reaction is in fact applied to a reaction between enolate forming component and an alkene which is not only activated by conjugation to a carbonyl group but to other groups of –*M*-type *e.g.*, ester, cyano, nitro and nitrile. With these structural features, the anion (I, scheme 7.58) formed after addition (and thus the preceding transition state) is stabilized sufficiently by the delocalization of the charge on to an electronegative element and the addition, therefore, occurs at a particular rate. The reaction has been used to build alicyclic rings in the process called Robinson annulation (ring forming) reaction (scheme 7.59).

6. The Mannich Reaction

The Mannich reaction is one of the few—three component reactions in organic chemistry. The reaction links together an amine an aldehyde (usually formaldehyde) and a compound containing an active hydrogen *e.g.*, a ketone with an α-hydrogen (scheme 7.60). Thus the net result of Mannich reaction among CH_2O, RNH_2 and $R'COCH_3$ is the extension of the —CH_3 of the ketone by —CH_2NHR. The products of the Mannich reaction are known as Mannich bases and many are useful as intermediates in synthesis.

The mechanism of Mannich reaction involves the following steps:

- An aldehyde *e.g.*, formaldehyde reacts with an amine (in an alcohol solvent containing HCl) to give an iminium salt (Scheme 7.60).

- The ketone undergoes enolization (eq. I, Scheme 7.61) (the activation of the α-carbon of the ketone as a nucleophile) and the enol undergoes nucleophilic attack on the electrophilic iminium carbon (eq. II, Scheme 7.61).

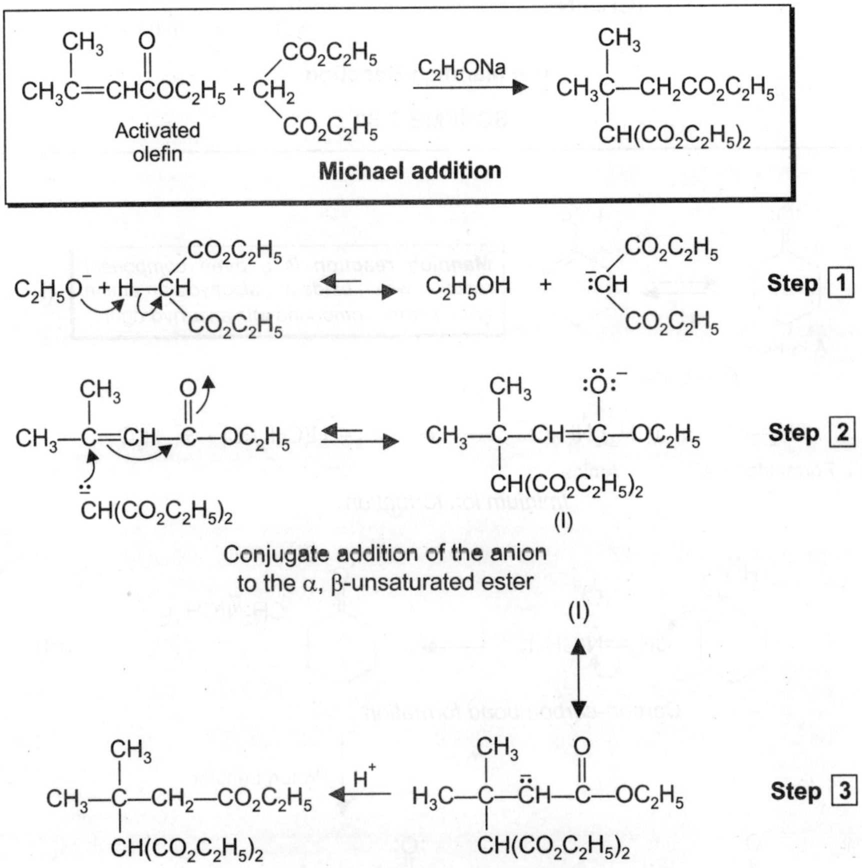

Mechanism of Michael addition

SCHEME 7.58

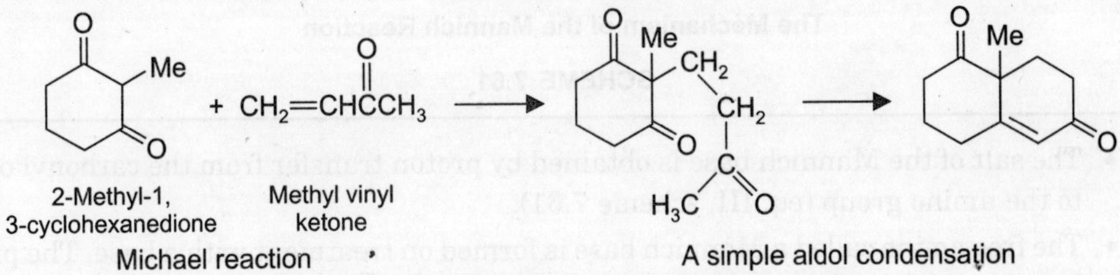

The Robinson annulation (ring forming) reaction

SCHEME 7.59

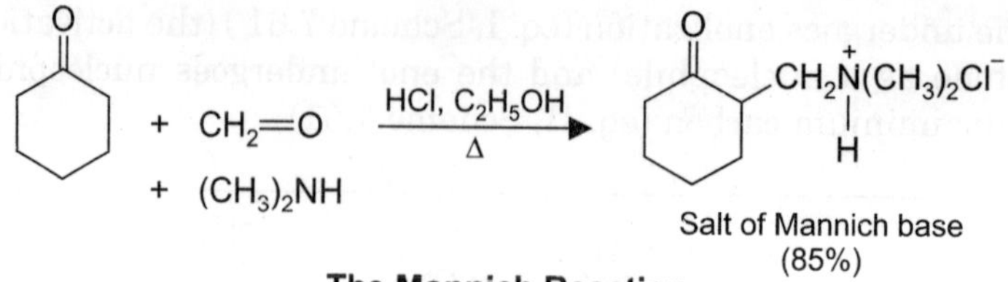

The Mannich Reaction

SCHEME 7.60

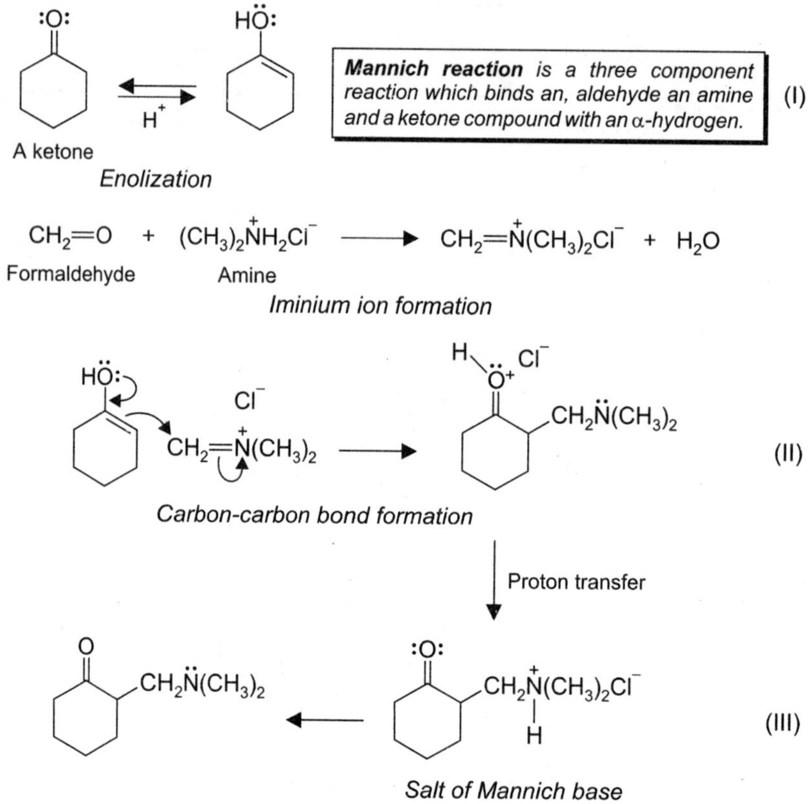

The Mechanism of the Mannich Reaction

SCHEME 7.61

- The salt of the Mannich base is obtained by proton transfer from the carbonyl oxygen to the amino group (eq. III, scheme 7.61).
- The free amine called a Mannich base is formed on treatment with a base. The process is very useful for the synthesis of natural products, tropinone skeleton was made in a single reaction by a double Mannich condensation. Succindialdehyde (I, scheme 7.62) is condensed with methylamine and acetone in the presence of an acid catalyst.

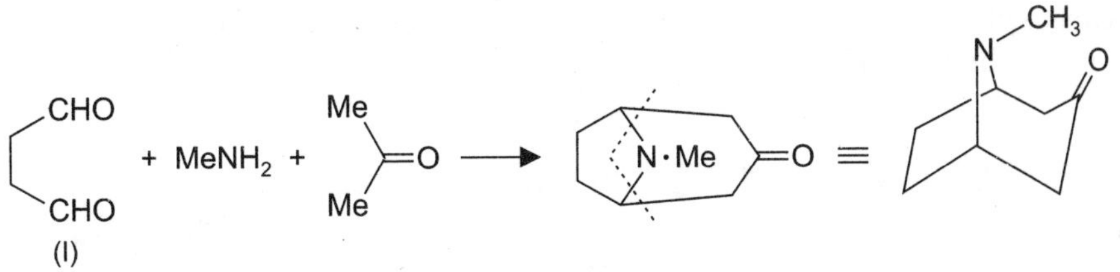

SCHEME 7.62

7. The Benzoin Condensation

This reaction is the cyanide ion catalyzed intermolecular condensation of an aromatic aldehyde to give an acyloin. The condensation of benzaldehyde gives benzoin (scheme 7.63). The aromatic aldehydes do not form cyanohydrins, but the cyanide addition product. Base abstraction of a proton gives a carbanion which reacts with a second molecule of the aldehyde.

$$\underset{\text{Benzaldehyde}}{2C_6H_5CHO} \xrightarrow{KCN} \underset{\text{Benzoin}}{C_6H_5CHOHCOC_6H_5}$$

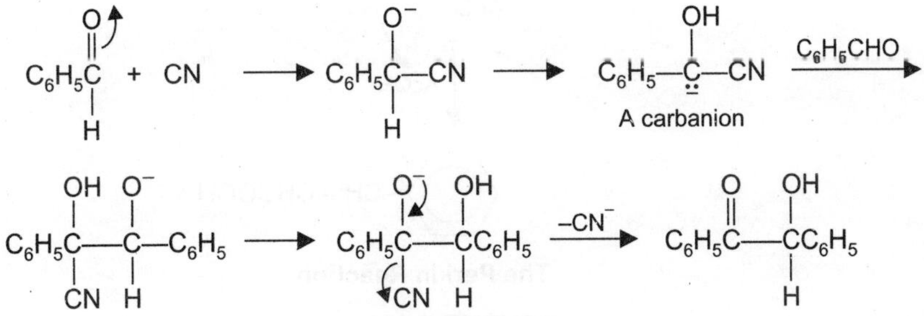

The benzoin reaction

SCHEME 7.63

The success of the reaction is due to the cyanide ion. Firstly it is a reactive nucleophile and secondly it has the capacity to delocalize the negative charge on the carbanion. Thus the carbanion formation is assisted. Moreover, with aromatic aldehydes (unlike aliphatic aldehydes) the negative charge of the carbanion further delocalizes on the aromatic ring (scheme 7.64) and this factor provides the extra driving force for the reaction.

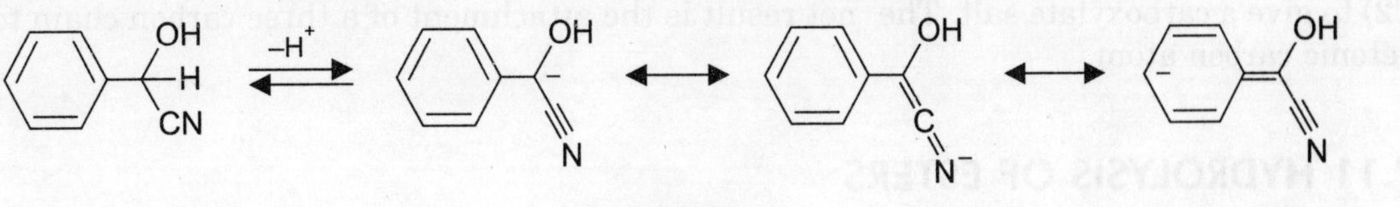

SCHEME 7.64

8. The Perkin Reaction

This reaction is used for the synthesis of α, β-unsaturated acids, and is an aldol type condensation between an aromatic aldehyde (ArCHO) with no α hydrogens and an enolate of an acid anhydride (carboxylic acid anhydride $(RCO)_2O$) catalyzed by a carboxylate ion (the potassium salt of the carboxylic acid, RCOOK). The anhydride gives the enolate by reacting with basic carboxylate ion (scheme 7.65).

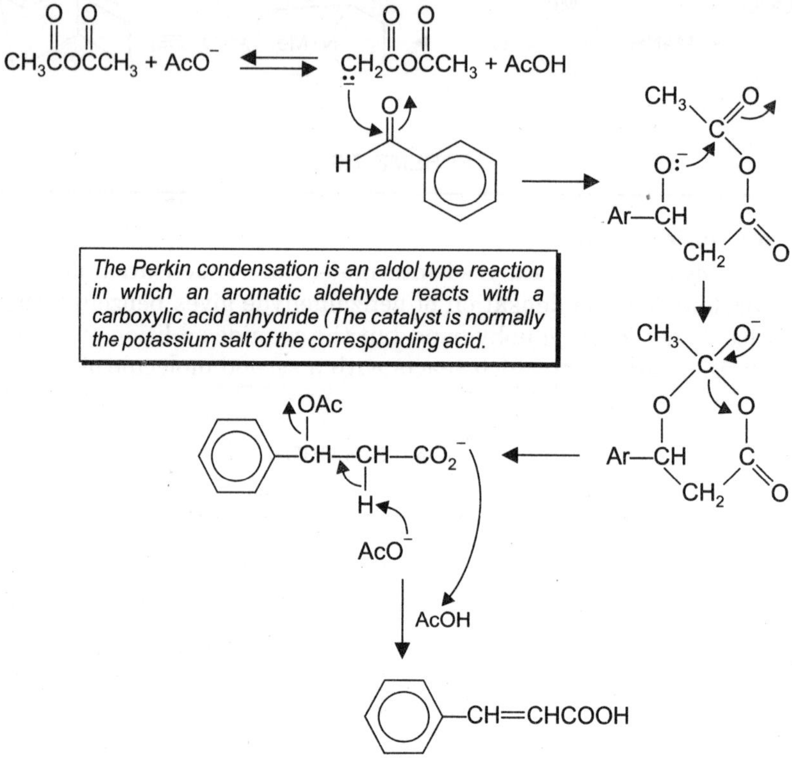

The Perkin condensation is an aldol type reaction in which an aromatic aldehyde reacts with a carboxylic acid anhydride (The catalyst is normally the potassium salt of the corresponding acid.

The Perkin Reaction

SCHEME 7.65

9. The Stobbe Condensation

This is a condensation between dialkyl succinates and non enolizable aldehydes and ketones in the presence of bases like NaOEt. Mechanistically the enolate from the ester adds to the carbonyl group of the ketone (scheme 7.66). One may compare this situation with that in Claisen ester condensation where it is presence of a base, the enolate from the ketone displaces alkoxide ion from the ester. In the Stobbe condensation the product (I, scheme 7.66) undergoes cyclization to give a lactone intermediate, which then undergoes elimination (E1 or E2) to give a carboxylate salt. The net result is the attachment of a three carbon chain to the ketonic carbon atom.

7.11 HYDROLYSIS OF ESTERS

A. Ester Hydrolysis and Esterification

The conversion of an ester into its acid and alcohol moieties (eq. I, Scheme 7.67) is termed ester hydrolysis. This hydrolysis can involve cleavage either at the acyl oxygen or alkyl oxygen bond (eq. II, scheme 7.67). A variety of mechanisms may be involved both depending on the nature of R and R′ and the conditions of the reactions. The mechanisms of esterification are based on the principle of microscopic reversibility.

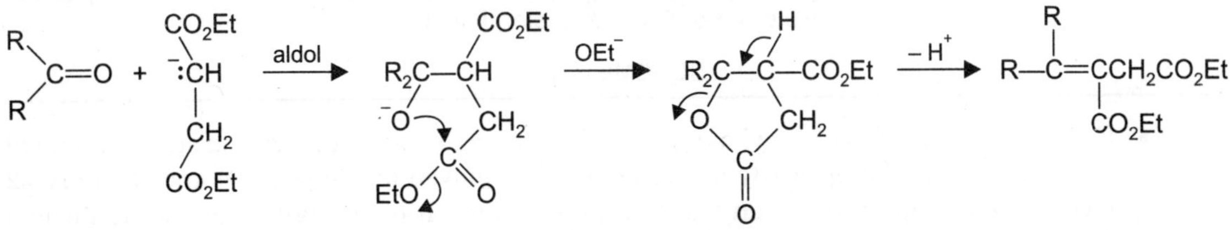

The Stobbe Condensation

SCHEME 7.66

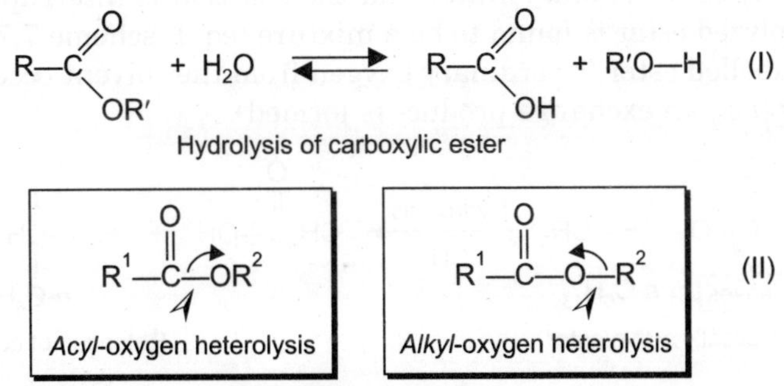

SCHEME 7.67

B. Mechanisms for Ester Hydrolysis

1. B_{AC} 2 Mechanism (Ester Hydrolysis) [Base-Promoted (B), Bimolecular (2) Hydrolysis with Acyl-Oxygen (AC) Cleavage]

This is the most common mechanism for base promoted ester hydrolysis and involves nucleophilic attack by base on the ester to give an unstable tetrahedral intermediate which then decomposes to the product *i.e.*, an acid and alcohol. The following points prove this mechanism.

- The use of ^{18}O as an isotopic tracer has shown that the base promoted reaction generally proceeds by acyl oxygen cleavage and this is not dependent on the structure of the R and R′ groups (scheme 7.68). Thus the alcohol produced (*e.g.*, ethanol if ethylpropionate labelled with ^{18}O, C_2H_5-CO-^{18}O-C_2H_5 is used) was found to be enriched in ^{18}O.

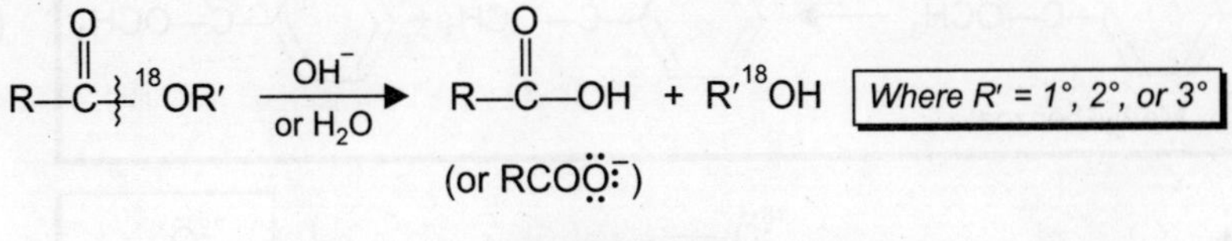

SCHEME 7.68

- Chemical kinetics of the alkaline hydrolysis of ester is second order, with the rate depending on the concentration of both ester and the base (scheme 7.69). Thus, the reaction involves the attack on ester by the base.

Rate of reaction = k_2 [ester] [OH^-]

SCHEME 7.69

- Further evidence for the acyl oxygen fission during basic hydrolysis is stereochemical in nature. When an optically active ester is employed (scheme 7.70) and if the $B_{AC}2$ mechanism is operative then the complete retention of configuration of the alcohol formed is indicative of the formation of the ion RO^- (acyl-oxygen fission) *i.e.*, the bond R—O is never broken thus the mechanism (scheme 7.71) involving an unstable tetrahedral intermediate is operative. This mechanism is consistent with the following evidence. When the carboxyl labelled methyl benzoate (scheme 7.72), undergoes alkaline hydrolysis in ordinary water and the reaction is interrupted after some time the unhydrolyzed ester is found to be a mixture (eq. I, scheme 7.72). The exchange of ^{18}O of the labelled ester for ordinary oxygen from the solvent occurs to give the unlabelled ester (*i.e.*, an exchange product is formed).

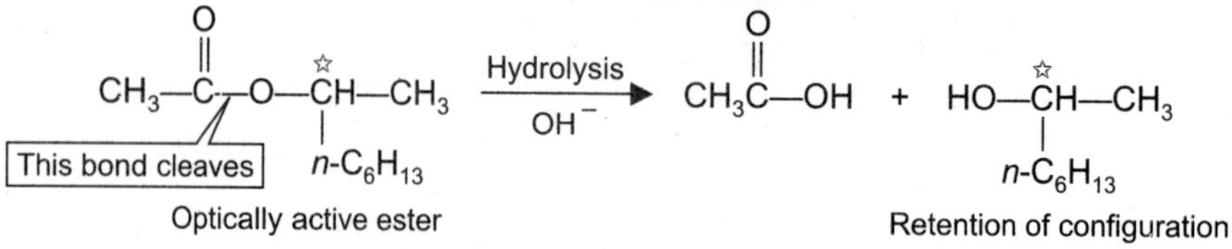

SCHEME 7.70

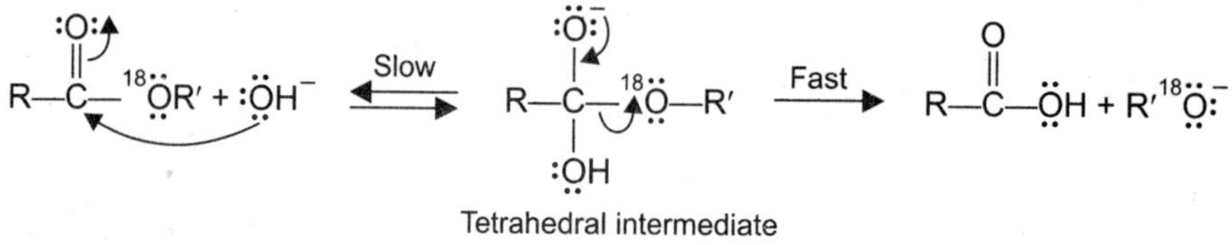

SCHEME 7.71

The initially formed intermediate (A, scheme 7.72) via the addition of hydroxide ion to the starting ester is the alkoxide ion, a strong base. This intermediate reacts with water to give the intermediate (B, scheme 7.72) having two chemically equivalent hydroxyl groups. Now the hydroxide ion can attack either of these two equivalent OH groups (scheme 7.73).

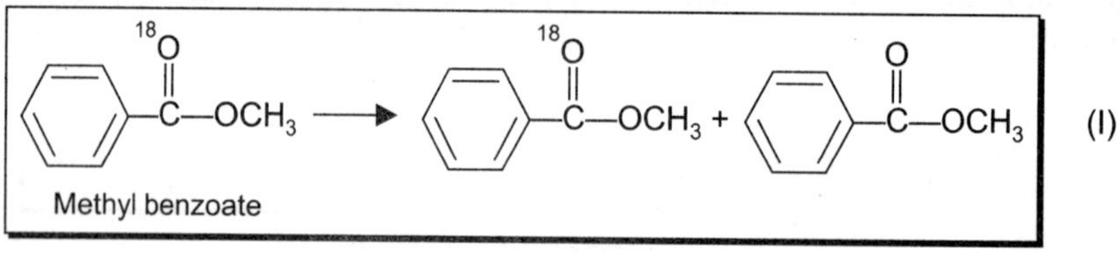

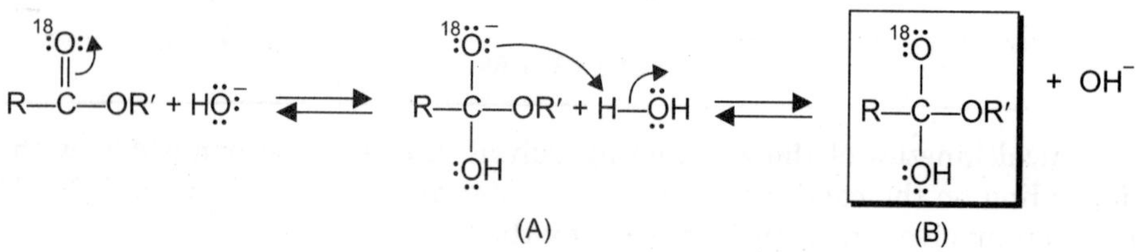

SCHEME 7.72

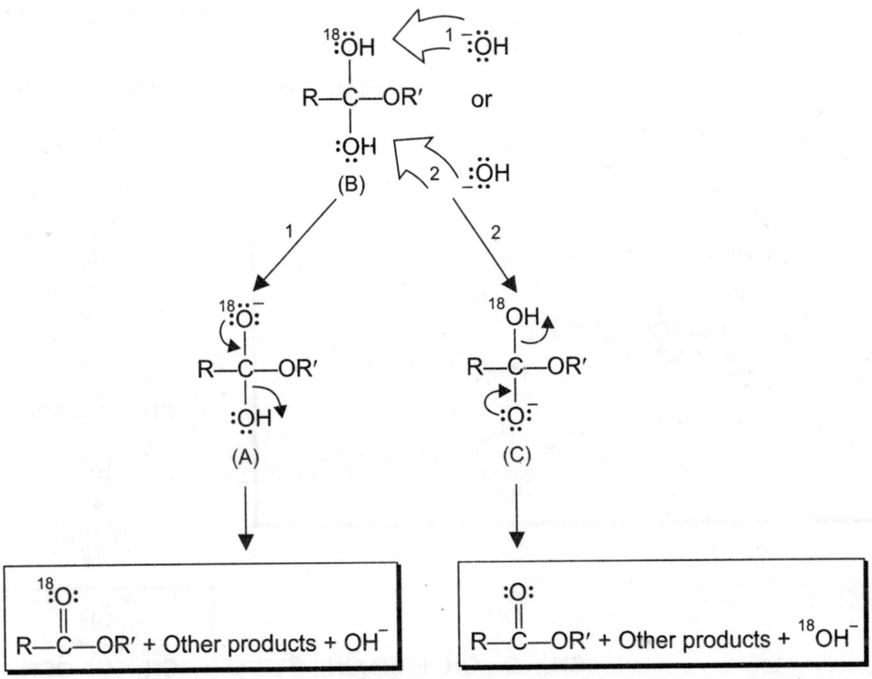

SCHEME 7.73

In case the base removes the proton from ^{18}OH group the original intermediate (A, scheme 7.72) is reformed which then gives the starting ^{18}O labelled ester (or the hydrolysis products). In case the base removes the proton from the OH group of the intermediate (B, scheme 7.73), then a new intermediate (C, scheme 7.73) is formed which can either give the ester with no ^{18}O or could decompose to give hydrolysis products.

2. A_{AC}2 [Acid-Catalyzed (A) Biomolecular (2) Hydrolysis with Acyl Oxygen (AC) Fission]

This is the most common acid catalyzed hydrolysis mechanism and involves nucleophilic attack by water on the protonated ester to give a tetrahedral intermediate (I, scheme 7.74) Recall that the OH and OR groups have approximately the same basicity and both tetrahedral intermediates (I and II) are formed. When tetrahedral intermediate I collapses, it eliminates H_2O in preference to CH_3O^- (H_2O is a weaker base), to reform the ester. When tetrahedral intermediate II collapses it eliminates CH_3OH and not HO^- since CH_3OH is a weaker base, thereby forming the carboxylic acid.

Esters in Acidic and Basic Solution

In acidic solution, protonation of carbonyl oxygen initiates, nucleophilic attack by a weak nucleophile like H_2O. In alkaline solution, the carbonyl carbon of an ester can be attacked by a good nucleophile e.g., OH^- without prior protonation. The tetrahedral intermediate formed in each case then undergoes rapid collapse to the observed products.

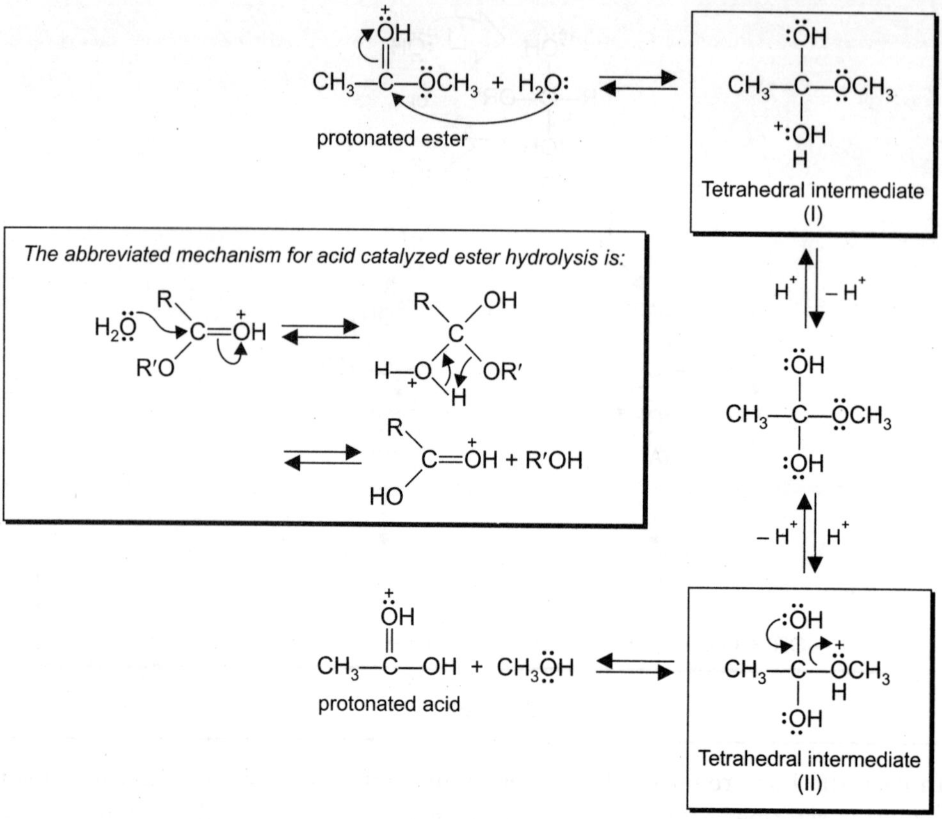

SCHEME 7.74

The oxygen O–18 studies show that the acid catalyzed hydrolysis is sensitive to the structure of alkyl group. For primary and most secondary alkyl groups the acyl oxygen bond is broken (scheme 7.74*a*).

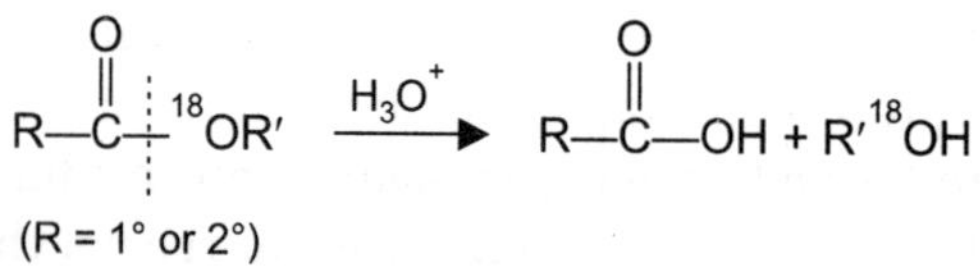

SCHEME 7.74a

As H_2O and CH_3OH have nearly the same basicity it will be equally easy for tetrahedral intermediate I to collapse to reform the ester and for the tetrahedral intermediate II to collapse to form the carboxylic acid (scheme 7.74). Thus, when the reaction attains equilibrium, both ester and carboxylic acid will be obtained (scheme 7.74*b*).

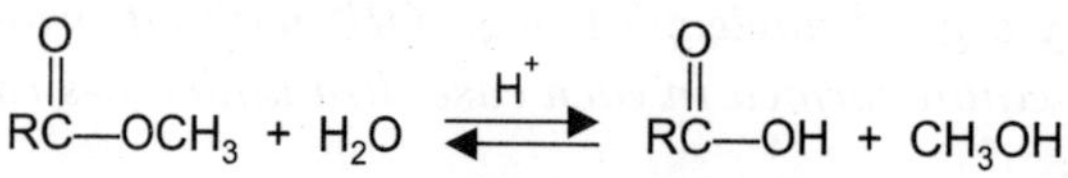

Both ester and carboxylic acid will be formed when the reaction attains equilibrium

SCHEME 7.74b

When water is in excess the equilibrium will be forced to the right (scheme 7.74*c*).

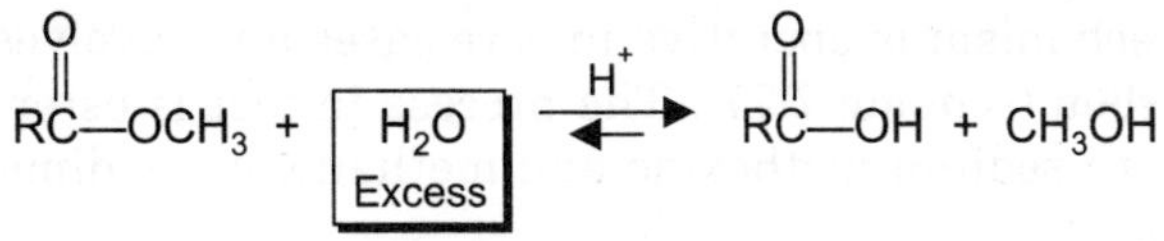

SCHEME 7.74c

The mechanism for the acid-catalyzed reaction of a carboxylic acid and an alcohol to yield an ester and water is the exact reverse of the mechanism for the acid-catalyzed hydrolysis of an ester to give a carboxylic acid and an alcohol. To get an ester product the reaction is carried out under conditions that will drive the equilibrium to the left–using excess alcohol or removal of water on its formation (scheme 7.74*d*).

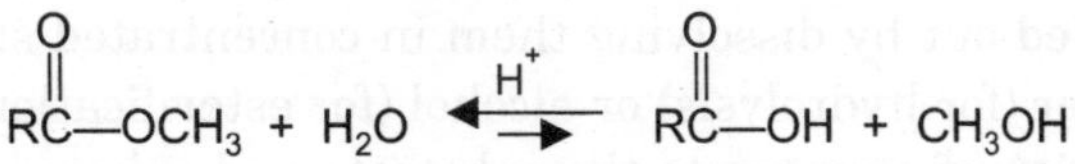

SCHEME 7.74d

3. $A_{AL}1$*[Acid-Catalyzed (A) Unimolecular (1) Hydrolysis with Alkyl Oxygen (AL) Cleavage]*

This hydrolysis involves the fragmentation of a protonated ester to give a carboxylic acid and a carbocation, which is rapidly trapped by water. This mechanism (S_N1 type) is operative only in those cases where R^+ is a relatively stable carbocation. It is observed that for most tertiary alkyl groups, considerable ^{18}O ends up in the carboxylic acid, to indicate alkyl oxygen fission (scheme 7.75). Thus for esters derived from tertiary alcohols the reaction begins with the protonation of the ester (scheme 7.76) and the leaving group is a stable tertiary carbocation. The carbocation subsequently reacts with water to give an alcohol. In keeping with this mechanism (scheme 7.76) when the acid catalyzed hydrolysis of *t*-butyl acetate is carried out in water enriched with ^{18}O, *t*-butanol containing ^{18}O was isolated.

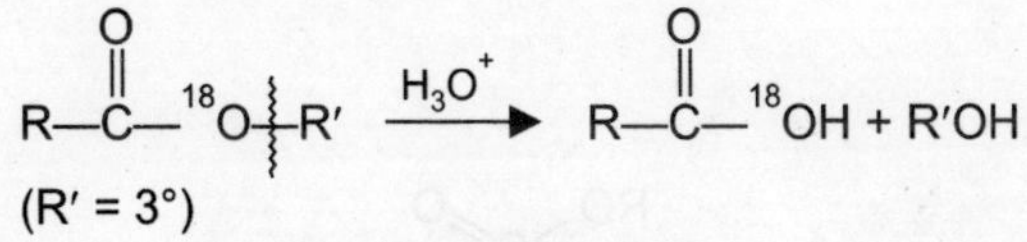

SCHEME 7.75

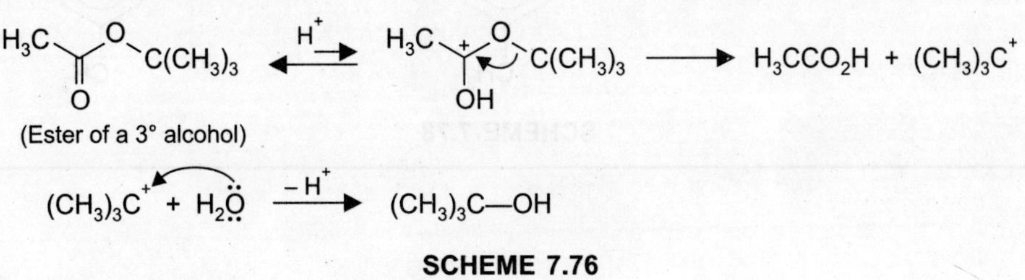

SCHEME 7.76

4. B_{AL} 2 [Base Promoted (B) Bimolecular (2) Ester Hydrolysis with Alkyl Oxygen (AL) Cleavage]

The ester hydrolysis mechanism is operative in rare cases and involves nucleophilic displacement at the alcohol carbon (scheme 7.77). The process in fact is ester cleavage, thus methyl benzoate on reaction with sodium methoxide and methanol gives dimethyl ether and sodium benzoate (scheme 7.77).

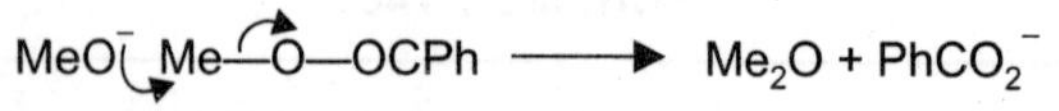

SCHEME 7.77

5. A_{AC}1 Ester Hydrolysis [Acid-Promoted (A) Unimolecular (1) Ester Hydrolysis with Acyl-Oxygen (AC) Cleavage and Esterification]

Esterification and hydrolysis of sterically hindered substrates (esters and acids *e.g.*, mesitoic acid or its ester) is carried out by dissolving them in concentrated sulphuric acid and subsequent reaction with water (for hydrolysis) or alcohol (for esterification). The mechanism operates *via* the initial addition of a proton to the substrate and subsequent heterolytic fission in the rate controlling step to give an acylium ion (scheme 7.78) which then reacts further. The formation of the acylium ion is due to the delocalization of its charge on to the aromatic ring and relief in the steric congestion.

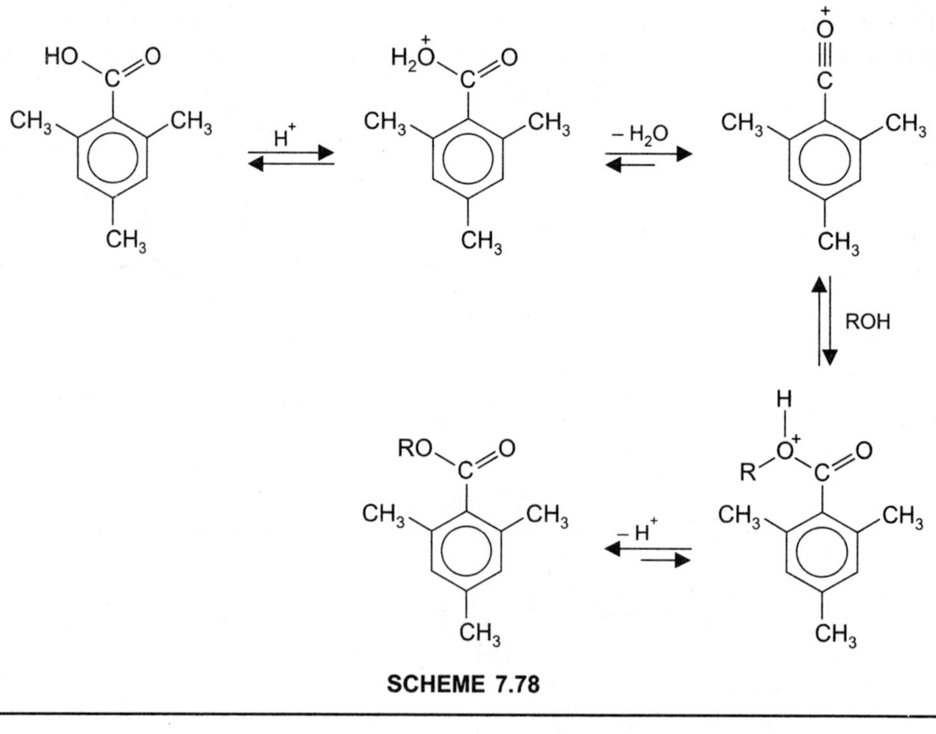

SCHEME 7.78

7.12 HYDROLYSIS OF AMIDES

1. Hydrolysis of an Amide, in Aqueous Acid

The following points may be considered:

- The acid protonates the carbonyl oxygen increasing the susceptibility of the carbonyl carbon to nucleophilic attack by water (scheme 7.79) to give a tetrahedral intermediate.
- Proton is transferred and an amine is expelled. This amine then reacts with H^+ to give an amine salt.
- The acid hydrolysis of amides follows the mechanism in analogy with ester hydrolysis. Under acidic conditions, the formation of the tetrahedral intermediate is promoted (attack by poor nucleophile water becomes feasible). The acid also changes the relative leaving abilities of the two groups on the tetrahedral intermediate. CH_3NH_2 is lost being a weaker base than OH^-.
- The formation of the ammonium ion at the end shows that H^+ is a reactant and not a catalyst. The factor that drives the reaction to completion and excludes the reverse reaction is that although the amine is a nucleophile, the ammonium ion is not.

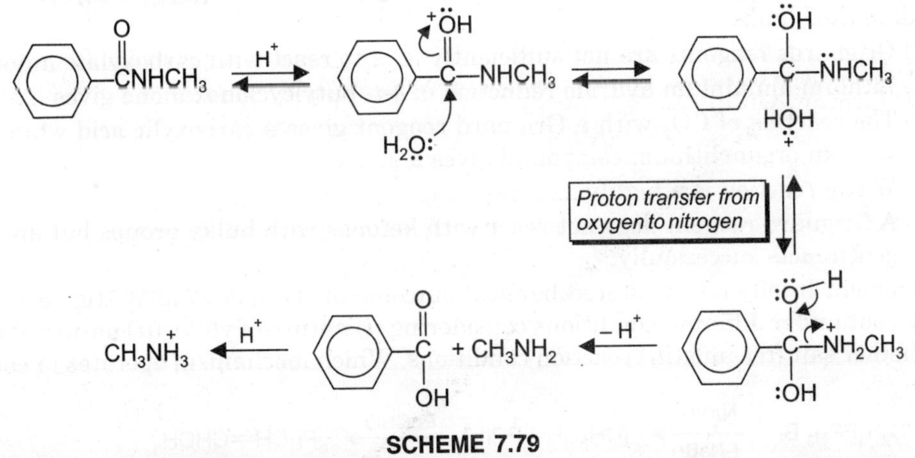

SCHEME 7.79

2. Hydrolysis of an Amide in Aqueous Base

The following points may be noted:

- In the hydroxide-ion-promoted hydrolysis of an amide; OH^- rather than water is the nucleophile. Since OH^- is a better nucleophile than water it is better to form the tetrahedral intermediate (Scheme 7.80).

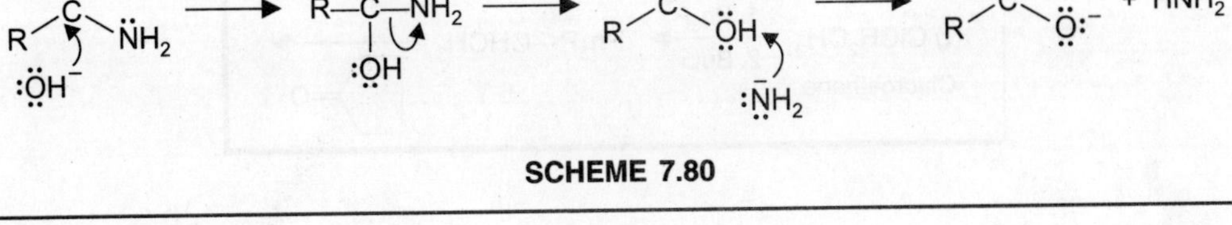

SCHEME 7.80

- The two leaving groups on the tetrahedral intermediate are OH^- and NH_2^- (since OH^- is the weaker base it is more easily eliminated, however, occasionally and more often NH_2^- is ejected).
- When, NH_2^- is ejected, the carboxylic acid thus formed immediately loses a proton. This step is irreversible (The carboxylate ion cannot be attacked by nucleophiles) and the reaction is driven towards products.
- Since hydroxide ion is consumed in the reaction, it is a reagent and not a catalyst.

PROBLEMS

7.1. Point out if the following statements are true or false.

(*i*) The stabilized phosphorane ylides in the Wittig reaction react only slowly.

(*ii*) The carbanions derived from the phosphonates are more nucleophilic in Wittig type reactions.

(*iii*) The Wittig reactions use phosphonate esters.

(*iv*) A betaine is a true intermediate in all Wittig reactions.

(*v*) An organolithium compound is less nucleophilic than Grignard reagent.

7.2. Fill in the blanks

(*i*) Grignards reagents are not sufficiently .... to react with carboxylate anions.

(*ii*) Lithium aluminium hydride reduction of 4-*t*- butylcyclohexanone gives .... alcohol.

(*iii*) The reaction of CO_2 with a Grignard reagent gives a carboxylic acid while reaction with an organolithium compound gives a ....

(*iv*) Wittig reaction is a highly .... reaction.

(*v*) A Grignard reagent does not react with ketones with bulky groups but an .... reagent reacts successfully.

7.3. Comment briefly on the stereochemical outcome of the following Wittig reactions carried out under different conditions considering (*i*) nature of ylide, (*ii*) nature of the base, whether salt free and (*iii*) reaction conditions. Which mechanism operates in each case ?

$$\underset{\text{(I)}}{C_2H_5\overset{+}{P}Ph_3\,\overset{-}{Br}} \xrightarrow[\text{(-NaBr)}]{NaNH_2} [CH_3\overset{-}{C}H{-}\overset{+}{P}Ph_3] \xrightarrow{PhCHO} \underset{\text{Mainly Z-alkene}}{PhCH{=}CHCH_3}$$

$$\underset{\text{(II)}}{C_2H_5\overset{+}{P}Ph_3\,\overset{-}{Br}} \xrightarrow{PhLi} CH_3\overset{-}{C}H\overset{+}{P}Ph_3 \xrightarrow[\text{(ii) PhLi; (iii) } H^+ (Me_3COH)]{\text{(i) PhCHO } (-78°)} \underset{\text{Mainly E-alkene}}{PhCH{=}CHCH_3}$$

$$\underset{\text{(III)}}{Ph_3\overset{+}{P}CH_2CO_2C_2H_5\,\overset{-}{Br}} \xrightarrow{NaOH} Ph_3\overset{+}{P}{-}\overset{-}{C}HCO_2C_2H_5 \longrightarrow \underset{\text{Mainly } E\text{-alkene}}{PhCH{=}CHCO_2C_2H_5}$$

7.4. Complete the following equations by writing the products formed.

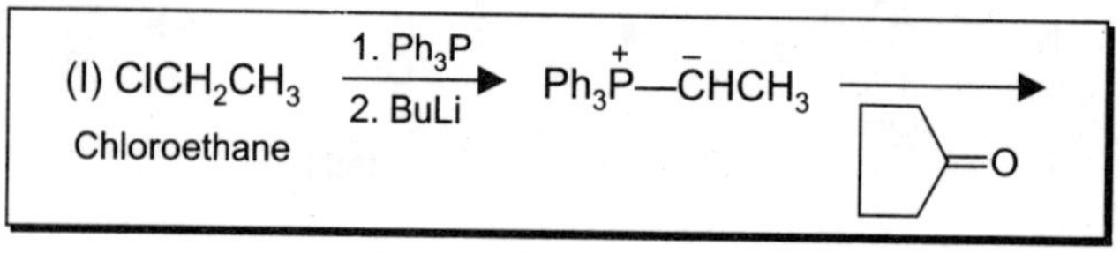

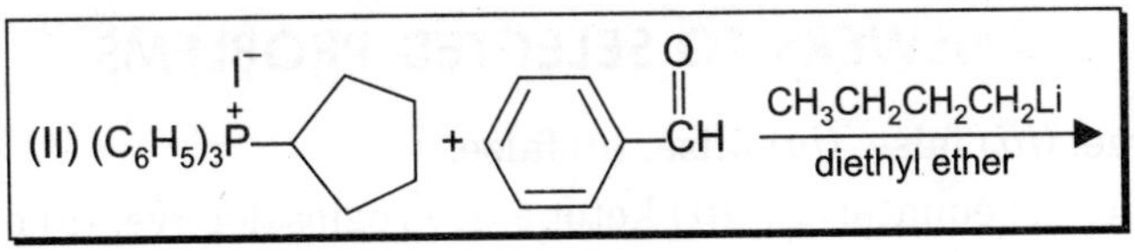

7.5. What reactions/reagents are needed in the Wittig reaction to synthesize the following compounds ?

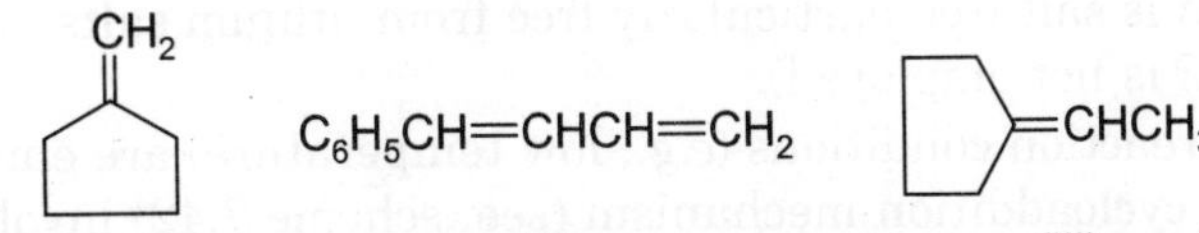

(I) (II) (III)

7.6. An organic compound (A) on reaction with C_6H_5MgBr followed by acidification gives a compound (B). The compound (B) on mild oxidation gives a compound C. Its mass spectrum $M^{+\bullet}$ $m/z = 106$, shows a base peak at $m/z = 105$ and another strong peak at $m/z = 51$. Assign a structure to A.

7.7. An E2 elimination is not completely regioselective and can form a mixture of alkenes *e.g.*, from 3-bromohexane. How can 2-hexene be synthesized completely regioselectively ?

$$CH_3CH_2CH_2CH(Br)CH_2CH_3 \xrightarrow{HO^-} CH_3CH_2CH{=}CHCH_2CH_3 \ (\text{3-Hexene}) + CH_3CH_2CH_2CH{=}CHCH_3 \ (\text{2-Hexene})$$

***E2 elimination** may not be completely regioselective.*

7.8. Write the structure of the product from the following reaction. What is the name of this reaction ? Give a mechanism.

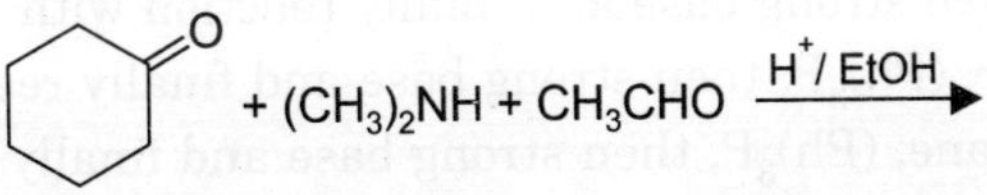

7.9. What is a directed aldol condensation ? How one can control the regioselectivity during this reaction using a ketone which could give two different enolates ?

7.10. How an aldol reaction can be made regioselective, diastereoselective and enantioselective ? Explain taking suitable compound(s) and only the reagents and conditions in each case.

7.11. The relative nucleophilic ability to interact with the carboxylate anion explains the difference in reactivity toward carbon dioxide of Grignard reagent and organolithium compounds. Explain.

7.12. 4-*t*-Butylcyclohexanone on reduction with lithium aluminium hydride gives equatorial alcohol as the major product while with a Grignard reagent (C_2H_5MgBr), the axial alcohol is the major product. Explain.

7.13. Write the mechanism of Reformatsky reaction. How is it related with aldol condensation?

7.14. Benzaldehyde can be made to react under certain conditions to give cinnamic acid. Name two such reactions.

7.15. LDA is a very strong base but a poor nucleophile being hindered. Give an example.

ANSWERS TO SELECTED PROBLEMS

7.1. (*i*) True (*ii*) true, (*iii*) false, (*iv*) false, (*v*) false.

7.2. (*i*) Nucleophilic, (*ii*) equatorial, (*iii*) ketone (*iv*) regioselective, (*v*) organolithium.

7.3. I (*i*) The reaction uses nonstabilized ylide prepared from CH_3CH_2Br, triphenylphosphine followed by deprotonation.

(*ii*) The base $NaNH_2$ followed by filtering of NaBr formed (before reaction) shows that the reaction is salt free particularly free from lithium salts (since a base like PhLi or *n*-C_4H_9Li is not employed).

(*iii*) No special reaction conditions (*e.g.*, low temperature) are employed. This reaction follows the cycloaddition mechanism (see, scheme 7.42) involving a favored puckered transition state to give mainly *Z*-alkene.

II (*i*) The reaction again uses only non-stabilized ylide prepared from CH_3CH_2Br.

(*ii*) The base is PhLi–lithium salts are present.

(*iii*) Special reaction temperature conditions – 78° are used during adding of carbonyl compound. Thus adduct(s) though formed do not eliminate at this low temperature. A second molar addition of base leads to the formation of a new ylide which changes into a more stable *threo* diionic structure (see scheme 7.43) which eliminates to give mainly *E*-alkene.

III (*i*) The alkyl halide $BrCH_2COOC_2H_5$ contains a–*M* group, the ylide would be obtained with a comparatively weaker base. The ylide is stabilized and is likely to react reversibly with the carbonyl compound to give *E*-alkene as the major product.

7.4. 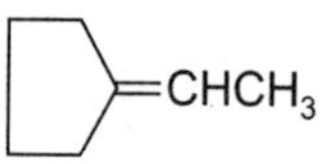

(I)

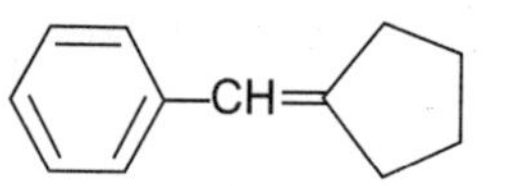

(II)

7.5. (*i*) CH_3I, $(Ph)_3P$, then strong base and finally reaction with cyclopentanone.

(*ii*) $CH_2{=}CHCH_2Br$, $(Ph)_3P$, then strong base and finally reaction with benzaldehyde.

(*iii*) Chlorocyclopentane, $(Ph)_3P$, then strong base and finally with acetaldehyde.

7.6. The mass spectrum of compound (C) is typical of an aromatic aldehyde (benzaldehyde). Phenylmagnesium bromide must react with formaldehyde to give a primary alcohol $C_6H_5CH_2OH$ which on mild oxidation gives C_6H_5CHO. The compound A is therefore formaldehyde HCHO.

7.7. A Wittig reaction is completely regioselective, the product can have double bond only in one location. 2-Hexene can be made, as shown below.

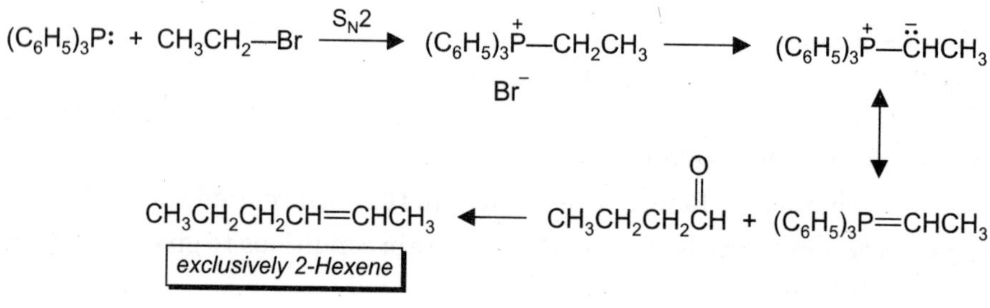

7.8. It is a Mannich reaction.

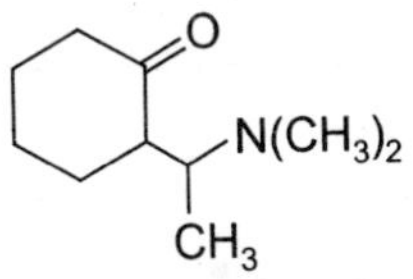

7.10. (see chapter 2).

7.13. One may use Knoevenagel or Perkin reaction.

REFERENCES

1. J.A. March, *Advanced Organic Chemistry, Reactions Mechanism, and Structure,* 4th ed., Wiley Interscience, New York, 1992; M.B. Smith and J.A. March, *Advanced Organic Chemistry Reactions, Mechanism, and Structure,* 5th ed., Wiley, New York, 2001.
2. B. Mundy and J. Ellerd, *Name Reactions and Reagents in Organic Synthesis,* Wiley, New York, 1988.
3. M.B. Smith, *Organic Synthesis,* McGraw-Hill, New York, 1994, Chapter 9, pp. 857-964.
4. R.C. Larock, *Comprehensive Organic Transformations*, 2nd ed., Wiley VCH Publishers, 1999.
5. D.G. Morris, *Stereochemistry*, RSC Tutorial Chemistry Series, Royal Society of Chemistry, Cambridge, 2001.
6. J.R. Hanson, *Functional Group Chemistry*, Royal Society of Chemistry, Cambridge, 2001.

CHAPTER 8

Pericyclic Reactions

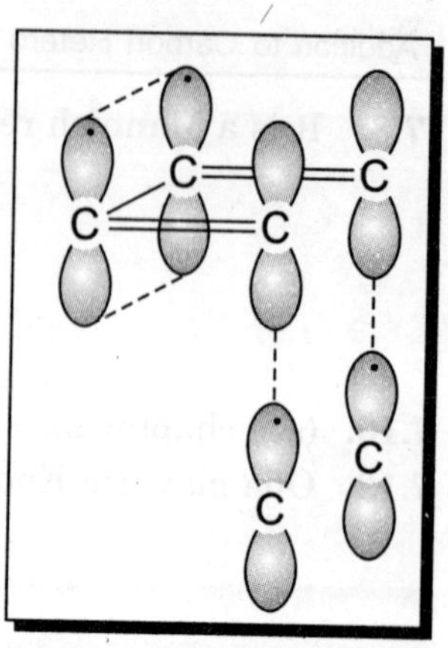

In a pericyclic reaction, there is a concerted bond reorganization and the essential bonding changes occur within a cyclic array of the participating atomic centers. These reactions do not involve the intermediate formation of either ions or radicals. Pericyclic reactions are also largely unaffected by polar reagents, solvent changes, radical initiators etc. These can however, be influenced only thermally *i.e.*, reactants are in their ground state or photochemically, *i.e.*, the excited state of a reactant is involved in the reaction.

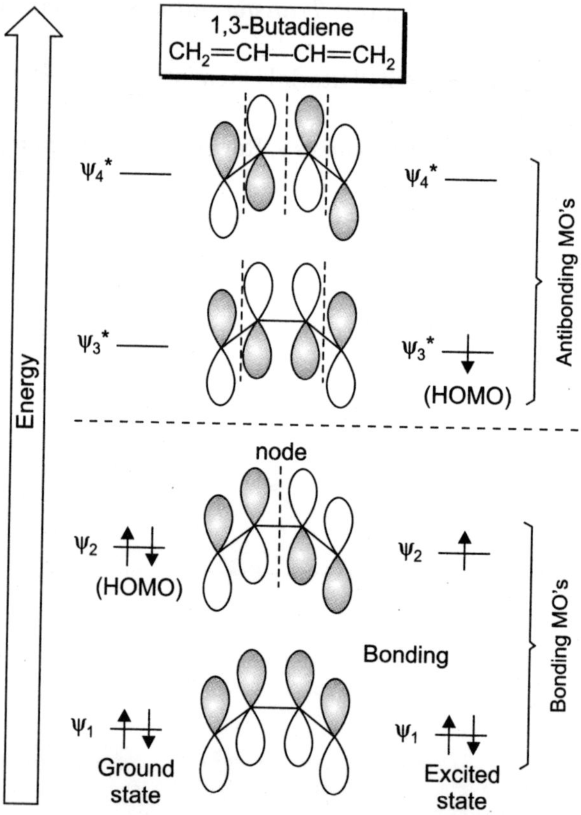

The four π molecular orbitals of 1,3-butadiene (asterisk denotes an antibonding orbital)

SCHEME 8.1

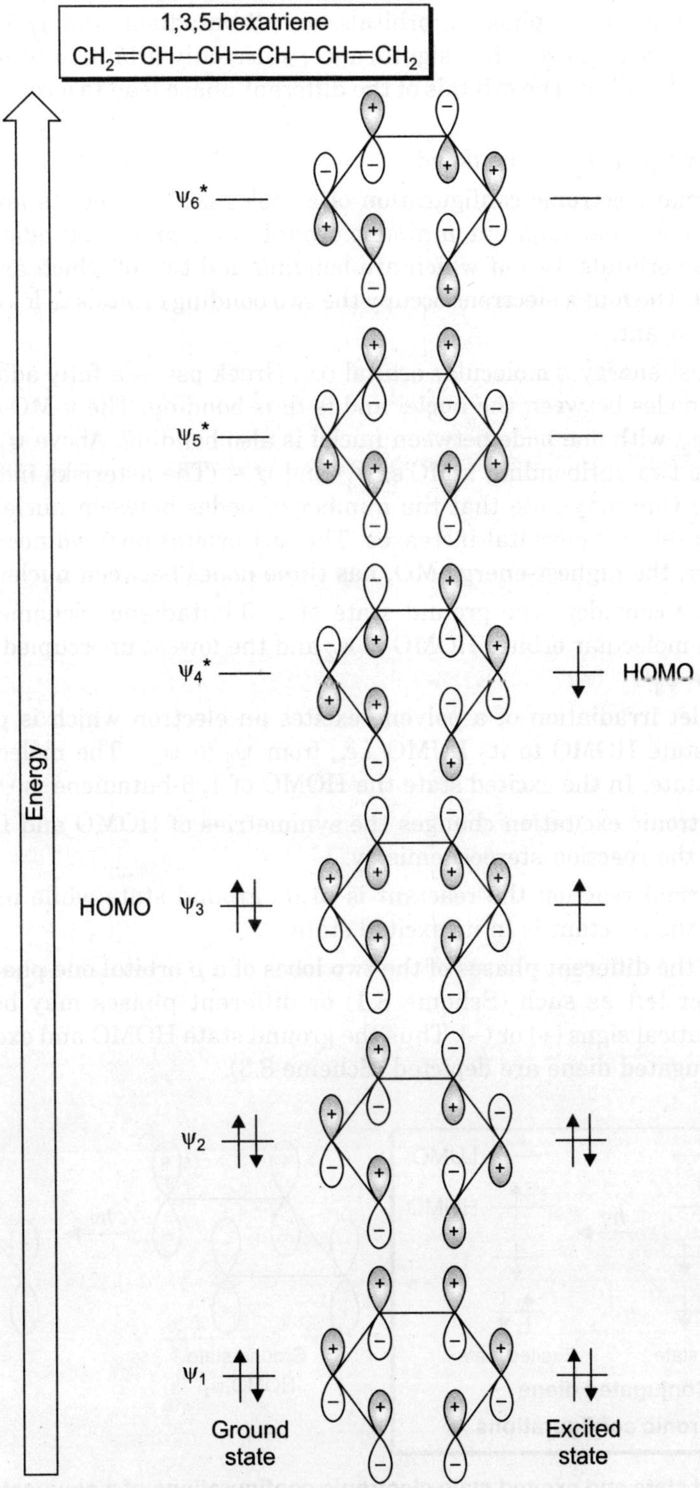

The six π molecular orbitals of 1,3,5-hexatriene

SCHEME 8.2

A consideration of the phase of orbitals of 1, 3-butadiene and 1, 3, 5-hexatriene are presented (Schemes 8.1 and 8.2) has significance, since only orbitals of the same phase will overlap to result in bonding. The orbitals of the different phase lead to a repulsive anti-bonding situation.

The following points may be noted:

- The normal electronic configuration of a molecule is called its ground state. When one combines *four* adjacent *p* atomic orbitals *e.g.,* in 1, 3-butadiene a set of four π molecular orbitals, two of which are bonding and two of which are antibonding are obtained. The four π electrons occupy the two bonding orbitals to leave the antibonding orbitals vacant.
- The lowest-energy π molecular orbital (ψ_1, Greek psi) is a fully additive combination with no nodes between the nuclei and is thus bonding. The π MO of the next lowest energy ψ_2, with one node between nuclei is also bonding. Above ψ_1 and ψ_2 in energy there are two antibonding π MO's, ψ_3* and ψ_4*. (The asterisks indicate antibonding orbitals.) One may note that the number of nodes between nuclei increases as the energy level of the orbital increases. The ψ_3* orbital has two nodes between nuclei while ψ_4*, the highest-energy MO, has three nodes between nuclei.
- When one considers the ground state of 1, 3-butadiene (Scheme 8.1) the highest occupied molecular orbital HOMO is ψ_2 and the lowest unoccupied molecular orbital LUMO is ψ_3*.
- Ultraviolet irradiation of a polyene excites an electron which is promoted from its ground state HOMO to its LUMO *i.e.,* from ψ_2 to ψ_3*. The molecule is then in an excited state. In the excited state the HOMO of 1, 3-butadiene is ψ_3*.
- The electronic excitation changes the symmetries of HOMO and LUMO and it also changes the reaction stereochemistry.
- In a thermal reaction the reactant is in its ground state while in a photochemical reaction the reactant is in its excited state.
- To show the different phases of the two lobes of a *p* orbital one phase may be shaded and other left as such (Scheme 8.1) or different phases may be represented by mathematical signs (+) or (–). Thus the ground state HOMO and excited state HOMO of a conjugated diene are depicted (Scheme 8.3).

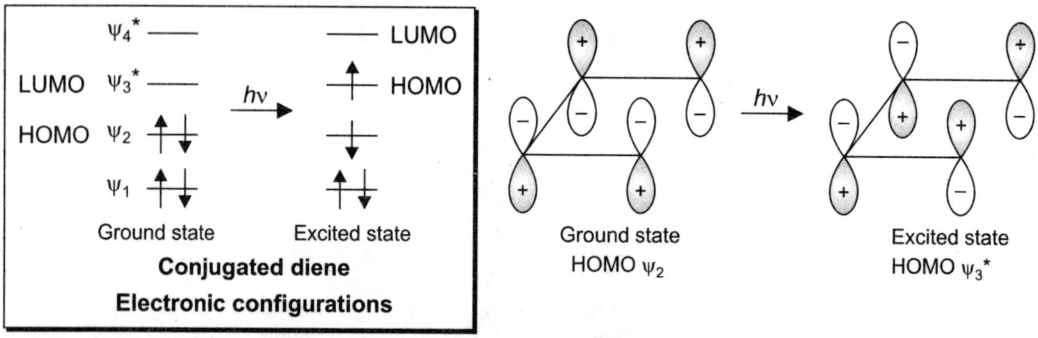

Ground state and excited state electronic configurations of a conjugated diene

SCHEME 8.3

A consideration of the six π molecular orbitals of 1, 3, 5-hexatriene (Scheme 8.2) shows that ψ_3 is the HOMO in the ground state and ψ_4* is the LUMO. In the excited state of 1, 3, 5-hexatriene ψ_4* is the HOMO and ψ_5* is the LUMO.

8.1 CONSERVATION OF MOLECULAR ORBITAL SYMMETRY

The resulting stereochemistry of the concerted pericyclic reactions depends on whether the reaction is thermal or photochemical. Woodward and Hofmann in 1965 pointed out that symmetry of the molecular orbitals which participate in the chemical reaction determines the course of the reaction and they proposed the principle of the conservation of orbital symmetry in concerted reactions. A pericyclic reaction can take place only provided the symmetry of all reactant molecular orbitals is the same as the symmetry of the product molecular orbitals (symmetry allowed reaction). In other words the lobes of the reactant molecular orbitals must be of correct algebraic sign for bonding overlap to take place in the transition state to give the product.

In a concerted reaction, the symmetry present in the reactants is maintained during the course of the reaction and is present in the product as well (principle of conservation of orbital symmetry). For example, in the Diels-Alder reaction, the reactants, the diene and the dienophile each has a plane of symmetry which is maintained in the *transition* state as well as in the product cyclohexene (Scheme 8.4).

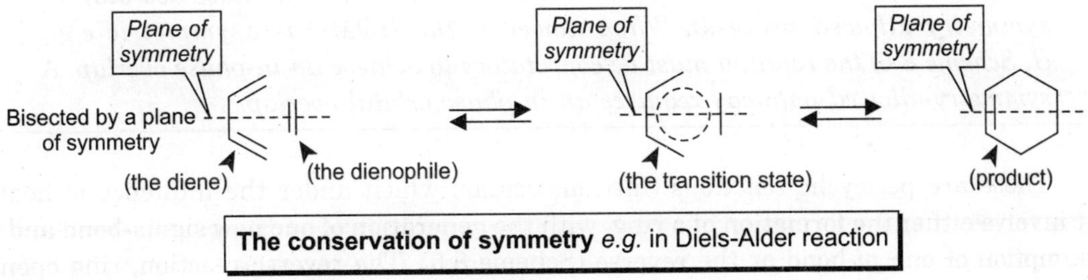

SCHEME 8.4

8.2 METHODS TO EXPLAIN PERICYCLIC REACTIONS

Three approaches have been employed to explain the pericyclic reactions and these are:

1. Woodward Hoffman Rules—Correlation Diagrams

These rules require the smooth passage of the participating molecular orbitals of the reactants into the molecular orbitals of the product. The process is described by a correlation diagram. In case the conversion of the reactant orbitals into the product orbitals is favoured in terms of energy and if orbital symmetry is conserved in the process, the reaction is called symmetry allowed (The symmetries of the orbitals to be maintained are around the mirror plane and the two fold axis of rotation as a symmetry element). In case either of these conditions (energy and orbital symmetry) is not met the reaction is called symmetry-forbidden.

2. HOMO-LUMO Method-FMO Approach

The Fuki's method concentrates on the so called frontier molecular orbitals, the HOMO (highest occupied molecular orbital) and the LUMO (lowest unoccupied molecular orbital). In the ground state 1,3-butadiene ψ_2 is HOMO and ψ_3^* is the LUMO (Scheme 8.1). The Fuki's FMO approach examines as to how the orbitals of HOMO or in some cases, the orbitals of HOMO of one component and the LUMO of other overlap to form new bonds. If the overlaps are favourable (bonding overlaps) then the reaction is allowed and if not favourable (antibonding overlaps) then the reaction is forbidden.

3. Möbius-Hückel Analysis (PMO Method)

The idea behind this method is that a pericyclic reaction which proceeds through a *transition* state which has aromatic characteristics (electron interactions is energetically favourable) is allowed process. For system of $4n + 2$ electrons Hückel transition states are aromatic; for systems of $4n$ electrons Möbius transition states are aromatic.

8.3 ELECTROCYCLIC REACTIONS (FMO-APPROACH)

Electrocyclic reactions are brought about by heat or light and are concerted and stereospecific. A symmetry-allowed pathway requires in-phase orbital overlap. When the HOMO is symmetric which has the end orbitals identical e.g., (II, Scheme 8.6) the rotation will have to be disrotatory to achieve the in-phase overalap (a symmetry allowed process). When however, the HOMO is asymmetric e.g., (I, Scheme 8.6) the rotation must be conrotatory to achieve an in-phase overlap. A symmetry-allowed pathway requires an in-phase orbital overlap.

These are pericyclic reactions (intramolecular) which under the influence of heat or light involve either the formation of a ring, with the generation of one new sigma-bond and the consumption of one *pi*-bond or the reverse (Scheme 8.5). The reverse reaction, ring opening proceeds by the same mechanism, but in reverse.

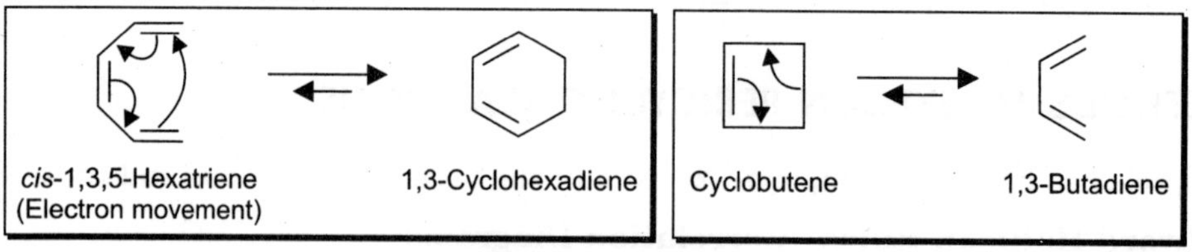

Electrocyclic reactions

SCHEME 8.5

The stereochemistry of electrocyclic reactions can be studied by using suitably substituted molecules. An intriguing feature of electrocyclic reaction is the stereochemistry of the product which depends on whether the reaction is thermally induced or photo-induced.

Regarding Electron Movement

In electrocyclic reactions conjugated polyenes close to give rings or rings open to polyenes shown by having electrons "chase each other's tails" around in a circle. These are among the conceptually simplest organic reactions. (It doesn't matter in which direction the curved arrows depicting electron motions are drawn in electrocyclic reactions, as long as they all move in the same direction (Scheme 8.5).

The following points may be noted:

- All the electrocyclic reactions are accounted for by orbital-symmetry arguments (FMO approach) by looking only at the symmetries of the two outermost lobes of the polyene. Thus the inner lobes of a reactant may not be shown (Scheme 8.6) and if shown, these may not be labelled +ve or –ve. The lobes of like sign can be either on the opposite side or on same side of the molecule (I and II respectively; Scheme 8.6). For bond formation the outermost lobes must rotate—a positive lobe overlapping a positive lobe or a negative lobe overlapping a negative lobe. When the two lobes of like sign are on the same side of the molecule (symmetric arrangement) the two orbitals (on the ends of the *pi* system) must rotate in different direction (clockwise and counterclockwise) and this motion is termed DISROTATORY (Scheme 8.7). When, however, the lobes of like sign are on opposite side of the molecule (asymmetric arrangement) both orbitals must rotate in the same direction (both clockwise or both counter-clockwise) and this motion is termed CONROTATORY.

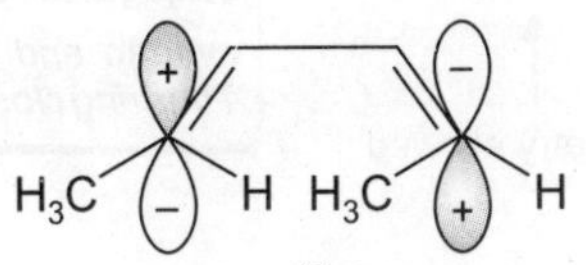

(I)
Ground state HOMO ψ_2 of a substituted butadiene (2*E*, 4*Z*-hexadiene)

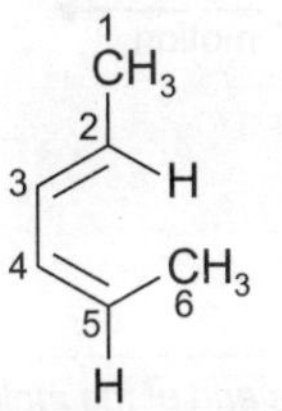

(2*E*,4*Z*)-hexadiene

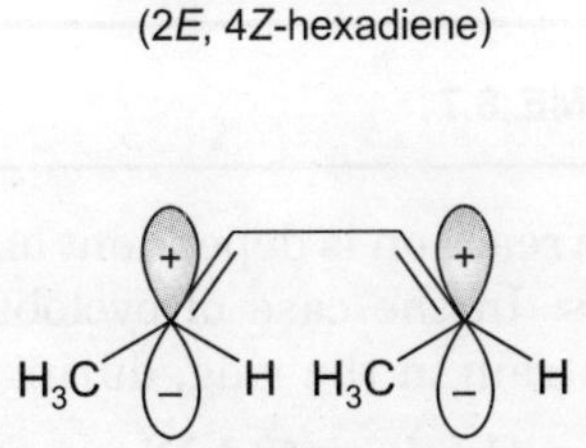

(II)
Excited state HOMO ψ_3* of a substituted butadiene (2*E*, 4*Z*-hexadiene)

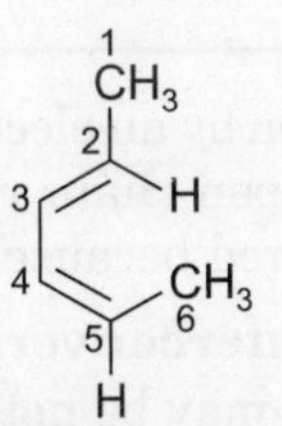

(2*E*,4*Z*)-hexadiene

SCHEME 8.6

- The stereochemical outcome of an electrocyclic reaction depends on the number of double bonds in a polyene and on whether the reaction is thermal or photochemical. A thermal electrocyclic reaction involving 4*n* pi electrons (*n* = 1, 2, 3,...) proceeds with conrotatory motion (*i.e.*, a motion in which the bonds rotate in the same direction) while the photochemical reaction involves disrotatory motion (a motion, in which the bonds rotate in opposite directions).

A thermal reaction involving $(4n + 2)$ *pi* electrons (where $n = 0, 1, 2,...$) proceeds with disrotatory motion while the photochemical reaction proceeds with conrotatory motion.

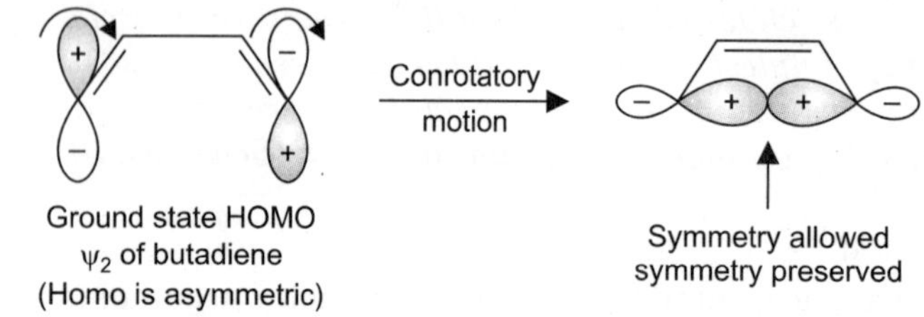

A bonding σ-molecular orbital between C1 and C4 is formed when the p-orbitals rotate via conrotatory motion i.e., rotation in the same direction either both clockwise or both anticlockwise.

The p orbitals at the end of the molecular orbital are not identical, in one, (+) lobe is at top while in other it is below. The molecular orbital is asymmetric. The ring closure is conrotatory.

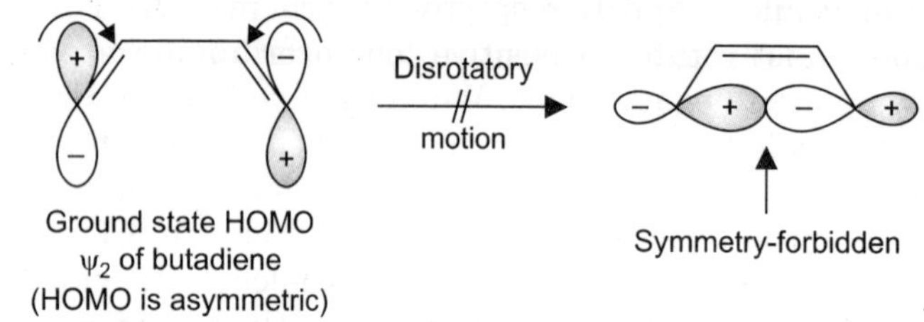

With the two end p-orbitals non-identical i.e., when the molecular orbital is asymmetric a disrotatory motion is symmetry forbidden.

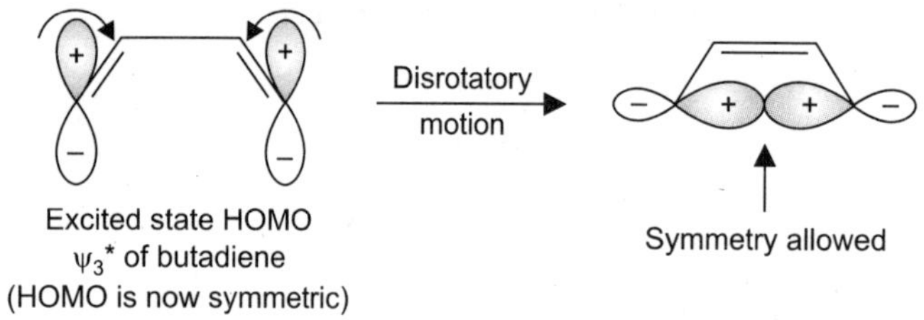

The new σ-bond is formed by rotation of the p orbitals at the end of conjugated system (head to head overlap and rehybridization to sp^3). Thus ring closure must be disrotatory.

The p orbitals at the end of the molecular orbital are identical i.e., both have (+) lobes on the top and (–) lobes at the bottom thus the HOMO is symmetric. The ring closure in disrotatory.

SCHEME 8.7

- The direction taken by an electrocyclic reaction is dependent on the relative stabilities of the ring and open-chain reactants. In the case of cyclobutanes the open chain structure is favoured because of the strain in the ring, during the thermal reaction.

A. Thermally Induced Interconversion of a Conjugated Diene and a Cyclobutene

The following points may be noted:

- The symmetry of the HOMO of a conjugated polyene undergoing ring closure determines the outcome of the electrocyclic reaction. All conjugated polyenes with asymmetric HOMO's undergo conrotatory ring closure. The ground state HOMO of a conjugated compound with an even number of double bonds *e.g.*, in a conjugated diene is asymmetric and should undergo conrotatory ring closure (Scheme 8.8). Moreover the reaction is remarkably stereospecific (Scheme 8.8). One may recall that there are always two conrotatory modes clockwise and anticlockwise and both are equally probable. It is obvious that neither of the two possible disrotatory modes can be a favourable process each being an antibonding situation.

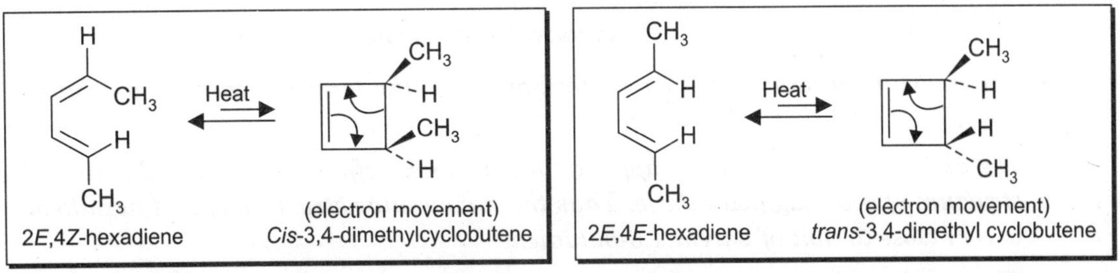

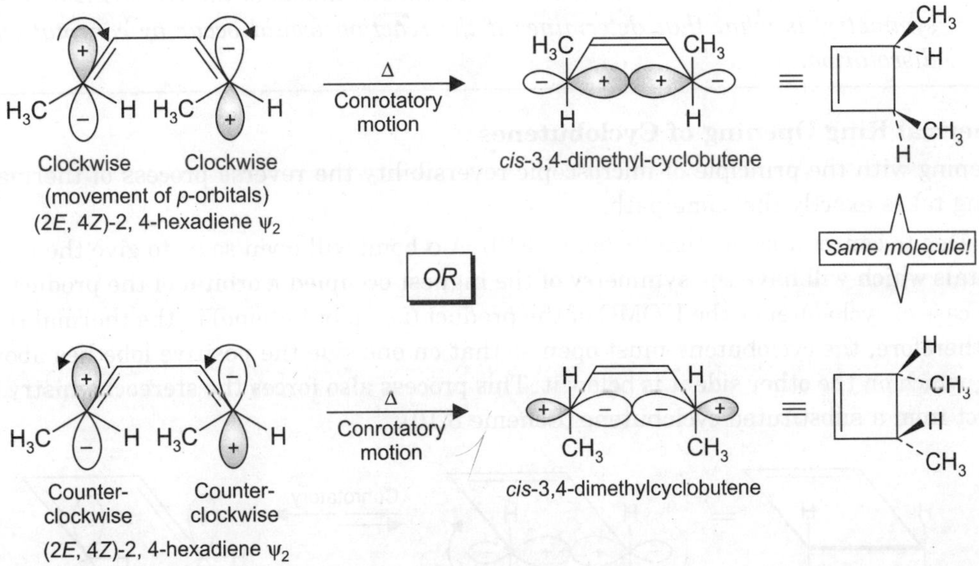

SCHEME 8.8

EXERCISE 8.1

Depict the conrotatory ring clasure of 2E, 4E-hexadiene and predict the stereochemistry of the product.

ANSWER. *Conrotatory ring closure is depicted in the counterclockwise fashion, the product is trans-3, 4-dimethylcyclobutene (Scheme 8.9). Same product would result by conrotation via clockwise motion.*

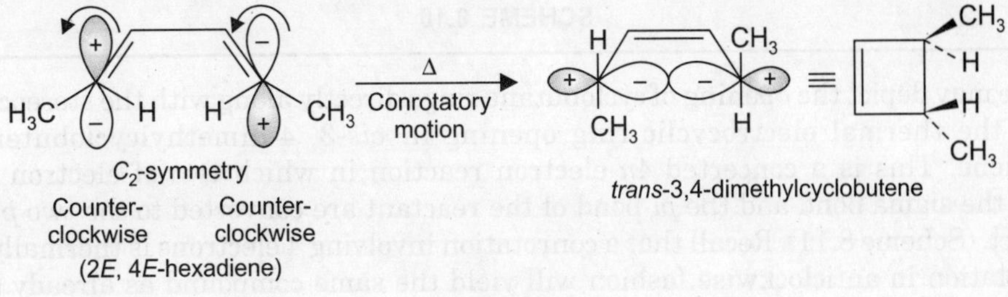

SCHEME 8.9

More on Electrocyclic Reactions

- *The diene must assume a s-cis-conformation for the terminal carbons p orbitals overlap.*
- *The alkyl substituents (as an approximation) do not effect the π molecular-orbital structure of a conjugated alkene. Thus the π molecular structure of 2, 4-hexadiene is very close to that of parent 1,3-butadiene.*
- *The relative orbital phase at the terminal carbon atoms of the HOMO (the orbital symmetry) is what that determines if the reaction would occur by conrotation or disrotation.*

B. Thermal Ring Opening of Cyclobutenes

In keeping with the principle of microscopic reversibility the reverse process of thermal ring opening takes exactly the same path.

Due to conrotatory motion (Scheme 8.10). A σ bond will open so as to give the resulting *p* orbitals which will have the symmetry of the highest occupied π orbital of the product. Since in the case of cyclobutenes the HOMO of the product (*i.e.*, a butadiene) in the thermal reaction is ψ_2 therefore, the cyclobutene must open so that on one side the positive lobe lies above the plane, which on the other side it is below it. This process also forces the stereochemistry in the product from a substituted cyclobutene (Scheme 8.10).

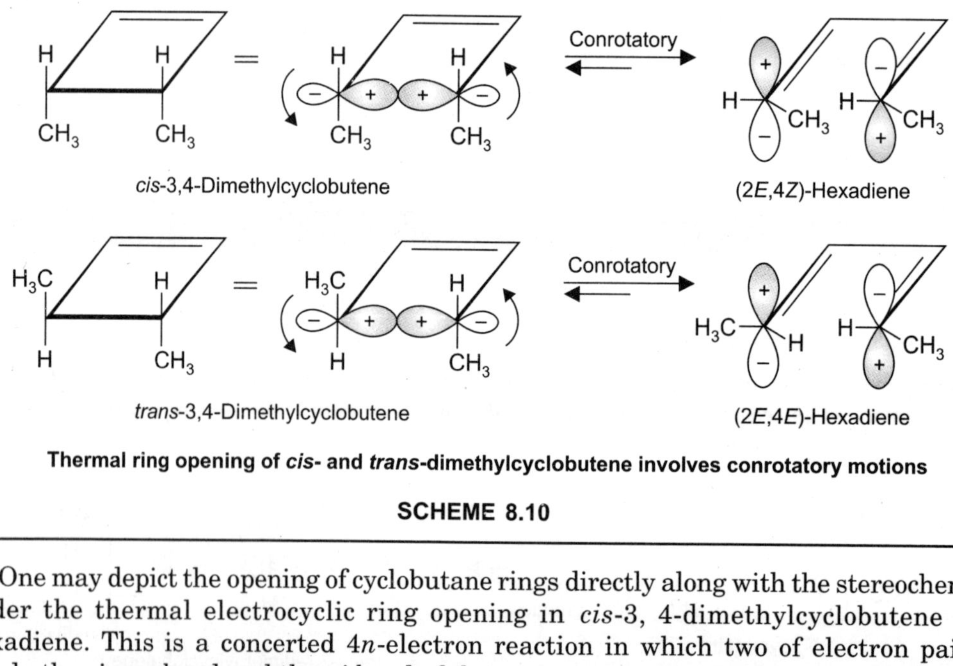

Thermal ring opening of *cis*- and *trans*-dimethylcyclobutene involves conrotatory motions

SCHEME 8.10

One may depict the opening of cyclobutane rings directly along with the stereochemistry. Consider the thermal electrocyclic ring opening in *cis*-3, 4-dimethylcyclobutene to 2*E*, 4*Z*-hexadiene. This is a concerted 4*n*-electron reaction in which two of electron pairs are involved ; the sigma bond and the *pi* bond of the reactant are converted to the two *pi* bonds of the product. (Scheme 8.11). Recall that a conrotation involving 4 electrons is thermally allowed. The conrotation in anticlockwise fashion will yield the same compound as already seen (see, Scheme 8.8).

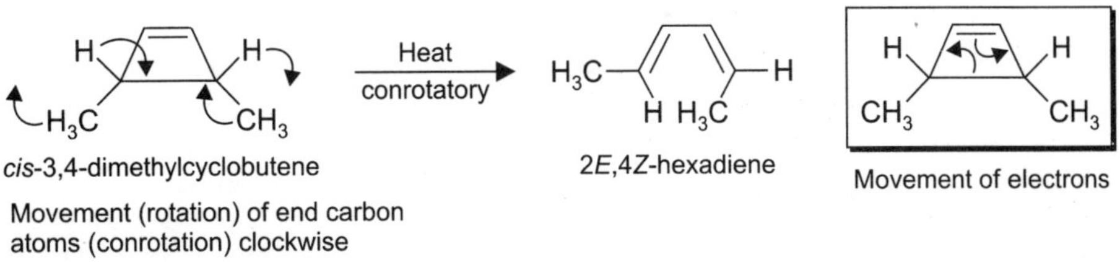

SCHEME 8.11

EXERCISE 8.2

Depict two possible conrotatory (thermal) modes of ring opening in trans-3, 4-dimethylcyclobutene.

ANSWER. *The two conrotatory modes lead to different products. (The two disrotatory modes on cis- and trans-disubstituted cyclobutenes are in Scheme 8.20a).*

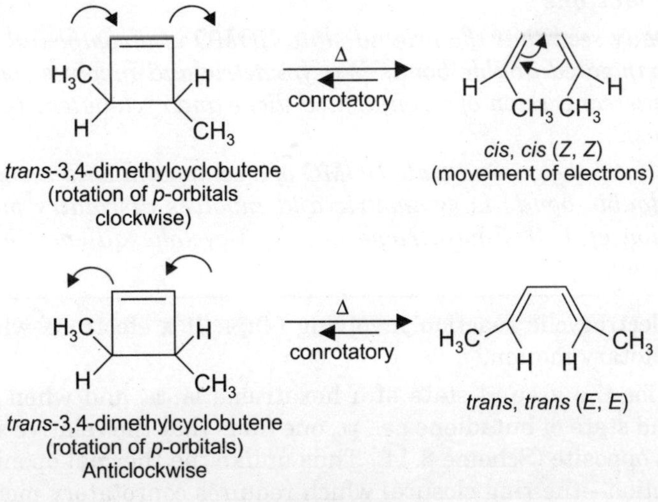

SCHEME 8.12

Thus one can also easily depict these conrotations (and also disrotations) directly without drawing molecular orbitals (Scheme 8.13). When a polyene undergoes an electrocyclic ring closure to give a cycloalkene, the terminal carbons of the polyene chain must rotate about 90° to convert the *p* orbitals on these carbons into the sp^3 orbitals forming the new σ bond. The substituents on these carbons are to be rotated into a plane which is approximately at right angles to the newly formed ring. Conversely, during the ring opening of a cycloalkene, the substituents on the atoms forming the bond undergoing cleavage will rotate into the plane of the new double bonds.

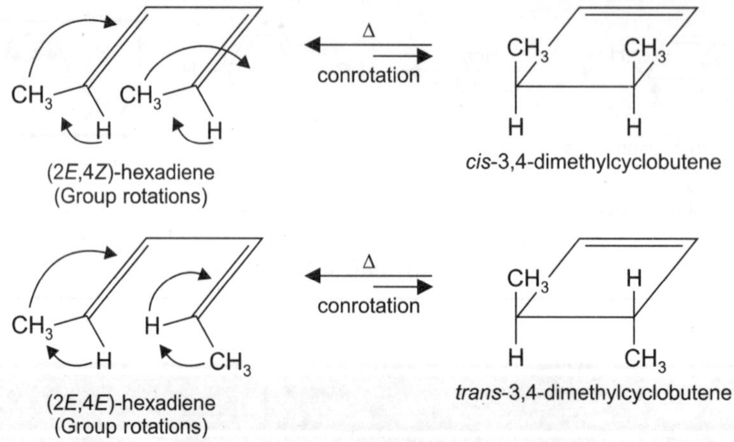

SCHEME 8.13

C. Thermal Electrocyclic Interconversion of 1,3,5-Hexatriene and 1,3-Cyclohexadiene

> ***Thermal Electrocyclic Ring Closure or Ring Opening–Conrotatory or Disrotatory Motions***
>
> *One has already seen that the ground state HOMO of a compound with an even number of conjugated double bonds is asymmetric and involves conrotatory motion e.g., in interconversion of a conjugated diene and cyclobutene (involvement of 4n pi electrons).*
>
> *On the other hand the ground state HOMO of a compound with an odd number of conjugated double bonds is symmetric and involves disrotatry motion e.g., in interconversion of 1, 3, 5-hexatriene and 1, 3-cyclohexadiene (involvement of 4n + 2 electrons).*

A thermal electrocyclic reaction involving $(4n + 2)$ π electrons where $(n = 0, 1, 2,...)$ proceeds with disrototary motion.

The HOMO for the ground state of a hexatriene is ψ_3 and when compared with the HOMO of the ground state of butadiene *i.e.*, ψ_2 one finds that the relative symmetry about the terminal carbons is opposite (Scheme 8.14). Thus unlike the thermal opening of a cyclobutene (or the reverse reaction—the ring closure) which requires conrotatory motion, in the thermal opening of a 1,3-cyclohexadiene and likewise the ring closure requires a disrotatory motion. Based on these arguments $2E$, $4Z$, $6E$-octatriene gives specifically *cis*-5,6-dimethyl-cyclohexadiene (Scheme 8.15).

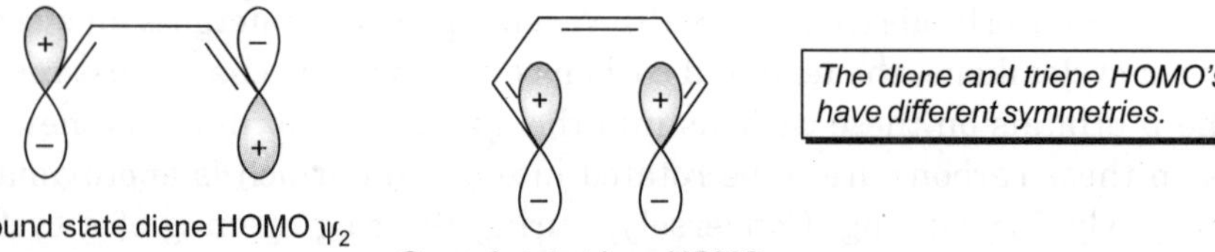

Ground state diene HOMO ψ_2

Ground state triene HOMO ψ_3

A conjugated diene and a conjugated triene react in opposite stereochemical senses. The diene opens and closes by conrotatory motion while a triene opens and closes by disrotatory motion.

SCHEME 8.14

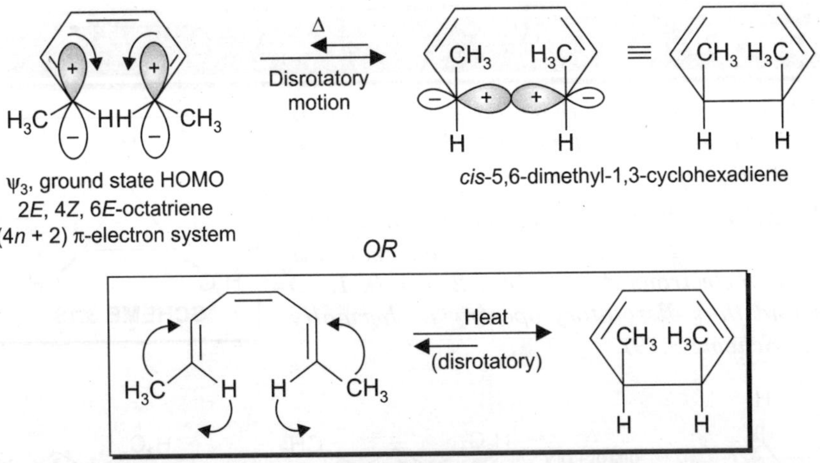

SCHEME 8.15

Similarly one finds that in the thermal cyclization of (2*E*, 4*Z*, 6*Z*)-octatriene as well the methyl groups rotate in the disrotatory fashion (Scheme 8.16).

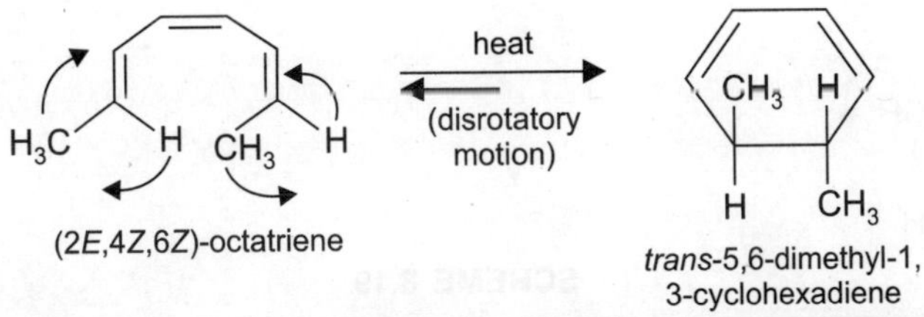

SCHEME 8.16

A thermal ring opening *e.g.*, in the case of *cis*-5, 6-dimethyl-1, 3-cyclohexadine (see Scheme 8.15) must also be disrotatory. The ground state HOMO of the derived triene is to be ψ_3 (Scheme 8.17).

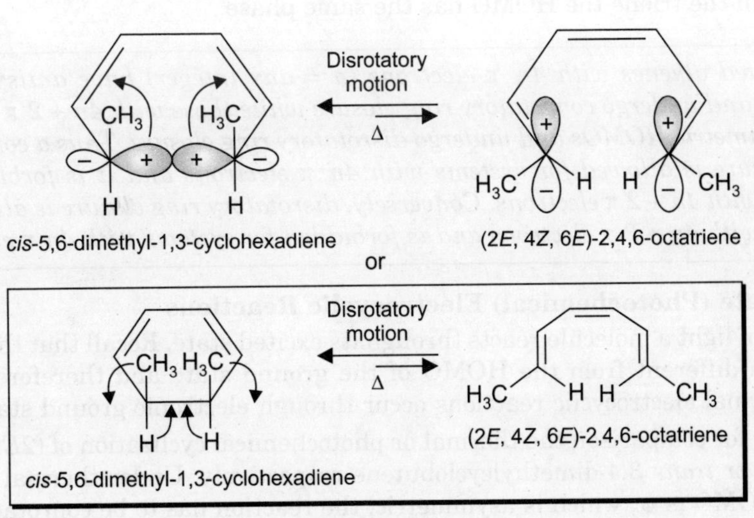

SCHEME 8.17

EXERCISE 8.3

What stereochemistry of the allyl carbocation is expected under thermal conditions from cis-dimethylcyclopropyl carbocation (Scheme 8.18)?

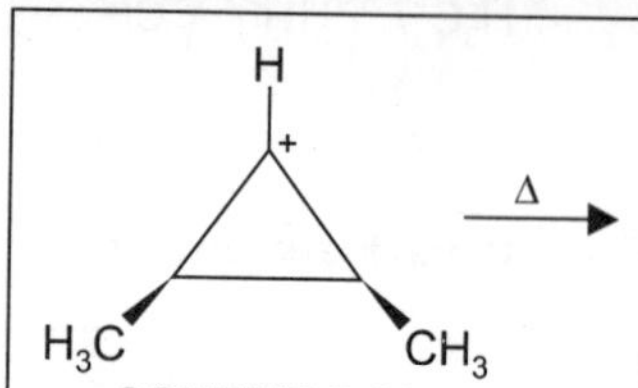

SCHEME 8.18

ANSWER. *The allyl carbocation will be formed by electrocyclic ring opening. Since the reactant involves two electrons it is a (4n + 2, n = 0, 1, ...) system and thus disrotatory opening is thermally allowed (Scheme 8.19).*

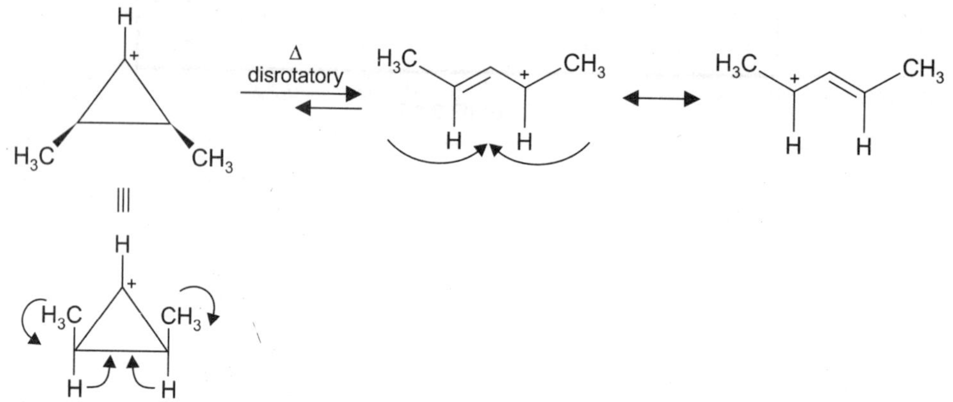

SCHEME 8.19

D. Summary—Electrocyclic Thermal Reactions

Electrocyclic closure of a conjugated diene is conrotatory, while that of a conjugated triene is disrotatory. This is due to the difference is the phase relationships within the HOMO at the terminal carbons of these π systems. In the diene the HOMO has opposite phase at these two carbons while in the triene the HOMO has the same phase.

> *Conjugated alkenes with 4n π electrons (n = any integer) have antisymmetric HOMOs and undergo conrotatory ring closure while those with 4n + 2 π electrons have symmetric HOMOs and undergo disrotatory ring closure. Thus a conrotatory ring closure is allowed for systems with 4n π electrons and it is forbidden for systems with 4n + 2 π electrons. Conversely, disrotatory ring closure is allowed for systems with 4n + 2 π electrons and is forbidden for systems with 4n π electrons.*

E. Excited State (Photochemical) Electrocyclic Reactions

On absorption of light a molecule reacts through its excited state. Recall that the HOMO of the excited state is different from the HOMO of the ground state and therefore, has different symmetry (thermal electrocyclic reactions occur through electronic ground states).

Consider for comparison the thermal or photochemical cyclization of (2*E*, 4*Z*) hexadiene which gives *cis* or *trans* 3,4-dimethylcyclobutene respectively. Under thermal cyclization the ground state HOMO is ψ_2 which is asymmetric, the reaction has to be conrotatory in order to achieve in-phase overlap (Scheme 8.20).

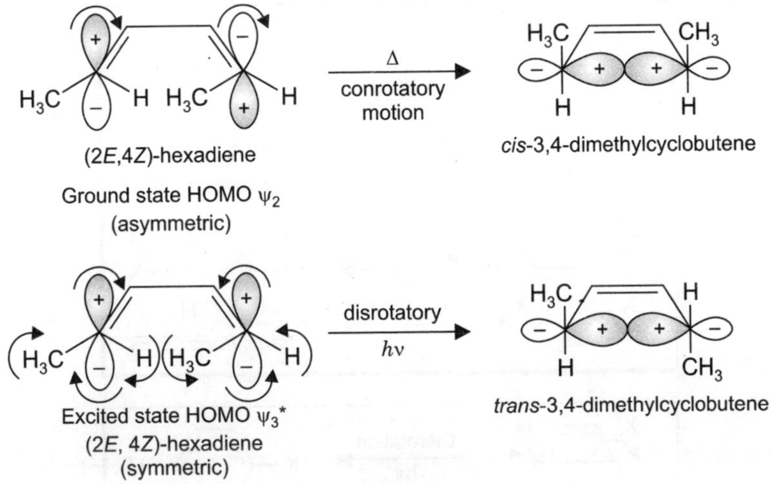

SCHEME 8.20

Similarly when (2*E*, 4*E*)-hexadiene on heating cyclizes to form *trans*-3, 4-dimethylcyclobutene, none of the *cis*-isomer is formed. In the reverse reaction the cyclobutene opens to produce only the (*E*, *E*)-isomer of the hexadiene. When the same isomer is photolyzed rather than heated, only *cis*-3, 4-dimethylcyclobutene is produced as the only product (Scheme 8.20*a*). Thus the reaction is completely stereospecific and involves conrotation under thermal conditions and disrotation on photochemical cyclization.

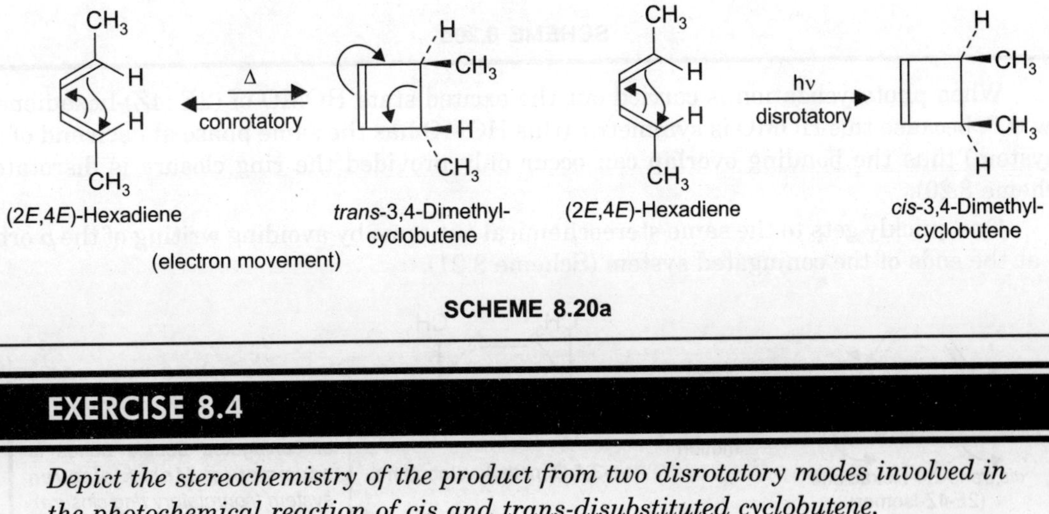

SCHEME 8.20a

EXERCISE 8.4

Depict the stereochemistry of the product from two disrotatory modes involved in the photochemical reaction of cis and trans-disubstituted cyclobutene.

ANSWER. *The products will be formed by disrotation, a cis-disubstituted cyclobutene opens to give cis, cis and trans, trans isomers of butadiene (Scheme 8.20b), while a trans-disubstituted cyclobutene gives the same product (in which there is one cis and one trans double bond).*

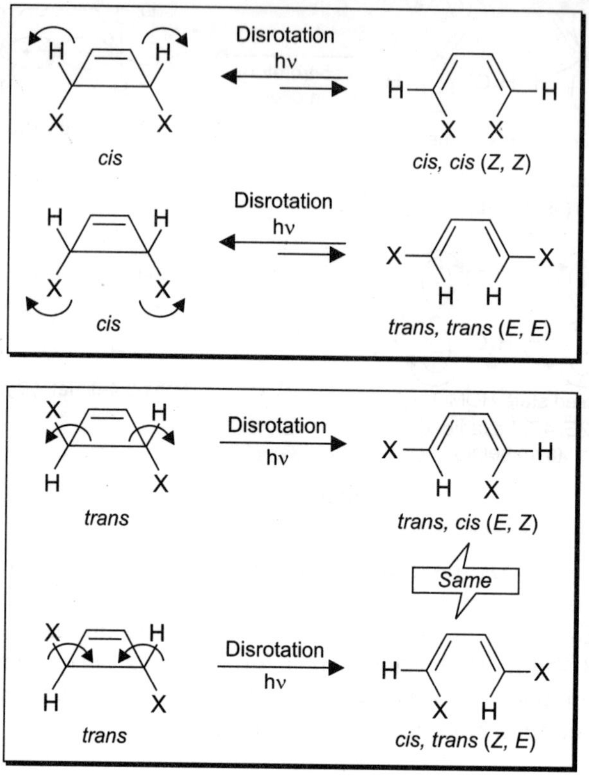

SCHEME 8.20b

When photocyclization is carried out the excited state HOMO of (2*E*, 4*Z*)-hexadiene is now ψ_3^*, because this HOMO is symmetric (this HOMO has the same phase at each end of the π system) thus the bonding overlap can occur only provided the ring closure is disrotatory (Scheme 8.20).

One quickly gets to the same stereochemical out come by avoiding writing of the *p* orbitals at the ends of the conjugated system (Scheme 8.21).

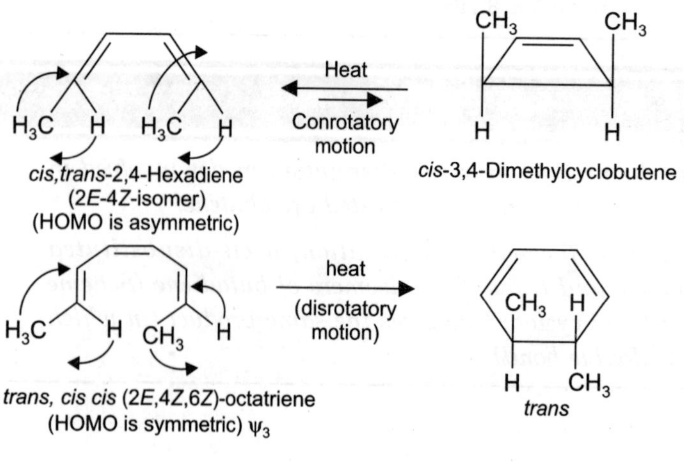

The ground state HOMO of a compound with an even number of conjugated double bonds is asymmetric (4n) π-electron system (conrotatory ring closure). The ground state HOMO of a compound with odd number of conjugated double bonds (4n + 2) π-electron system is symmetric (disrotatory ring closure).

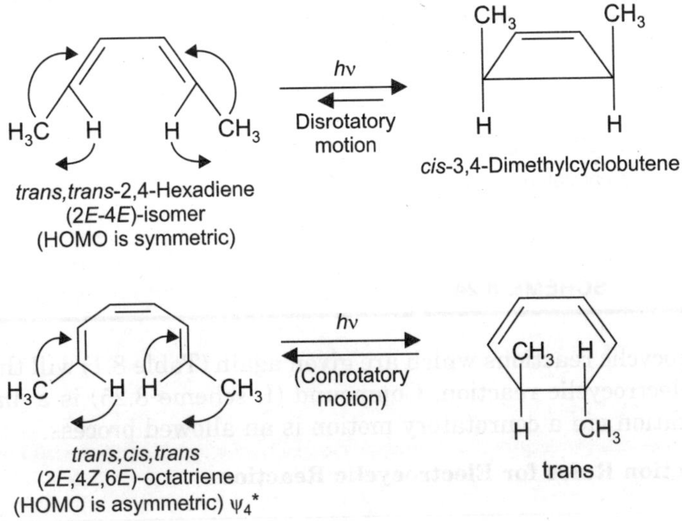

> *In reactions under photochemical conditions every thing is reversed. The ground state and excited state HOMO's have opposite symmetries. If the ground state HOMO is symmetric the excited state HOMO is asymmetric.*

SCHEME 8.21

EXERCISE 8.5

Predict the stereochemical outcome from the electrocyclic reactions (Scheme 8.22)?

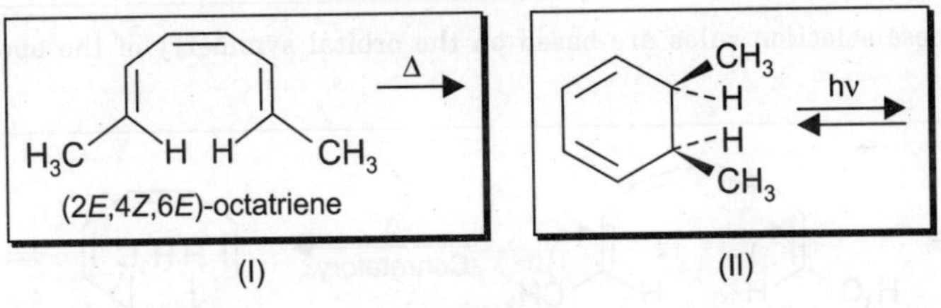

SCHEME 8.22

ANSWER. (I) *Reactant has odd number of double bonds $(4n + 2)\ \pi$ electron system, ground state HOMO of the triene (ψ_3) is symmetric. Ring closure is disrotatory.*

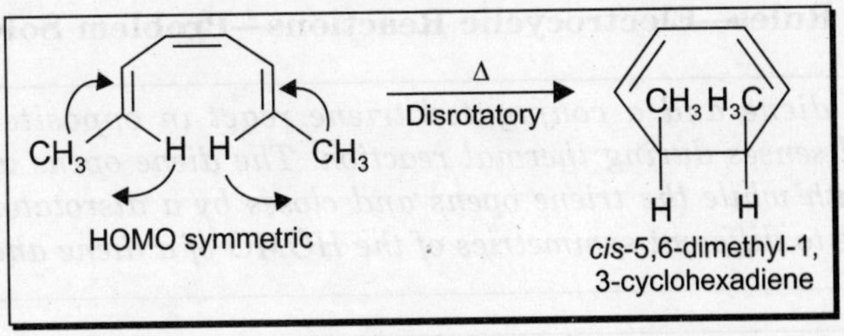

SCHEME 8.23

(II) Reactant is a (4n + 2) π electron system, photochemically it involves a conrotatory motion.

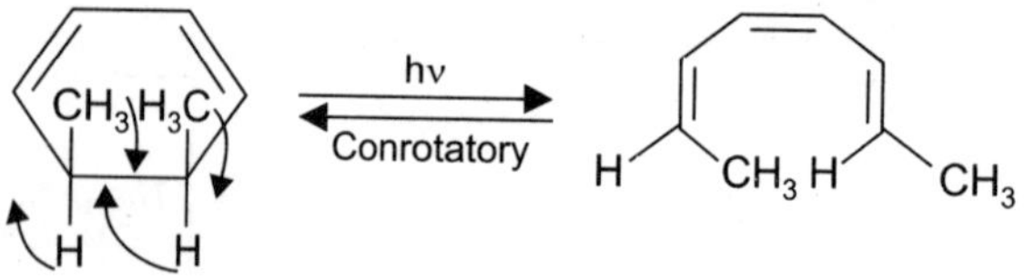

SCHEME 8.24

The selection rules for electrocyclic reactions which are given again (Table 8.1) will thus help to know the outcome of an electrocyclic reaction. Compound (I, scheme 8.25) is a $4n$ π system, therefore, a thermal cyclization *via* a conrotatory motion is an allowed process.

Table 8.1. Selection Rules for Electrocyclic Reactions

Number of electrons*	Mode of activation	Allowed stereochemistry
$4n$	thermal photochemical	conrotatory disrotatory
$4n + 2$	thermal photochemical	disrotatory conrotatory

*n = an integer. These selection rules are based on the orbital symmetry of the open-chain (conjugated alkene) reactant.

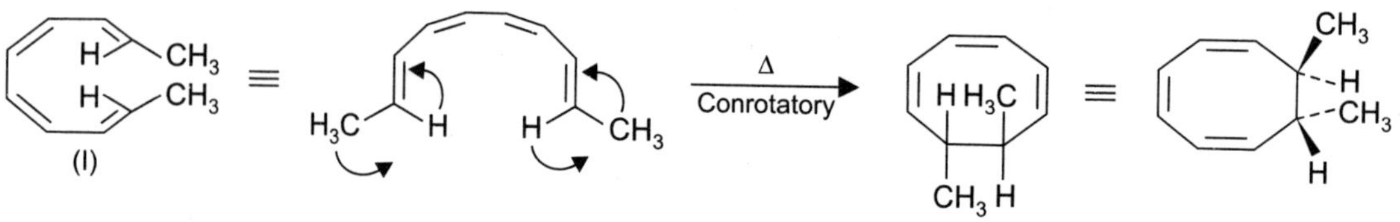

SCHEME 8.25

F. Stereochemical Rules—Electrocyclic Reactions—Problem Solving Hint

A conjugated diene and a conjugated triene react in opposite (alternating) stereochemical senses during thermal reaction. The diene opens and closes by a conrotatory path while the triene opens and closes by a disrotatory path. These results are due to different symmetries of the HOMO of a diene and a triene.

On electronic excitation (photochemical reactions) the symmetries of HOMO and LUMO change and with it changes the reaction stereochemistry (which is reversed) and stereochemical rules for electrocyclic reactions are given (Table 8.1).

Problem Solving Hint 1

If in a reactant the bonds to the substituents point in opposite directions then these substituents will have cis stereochemistry in the product provided the motion is disrotatory (Scheme 8.25a)

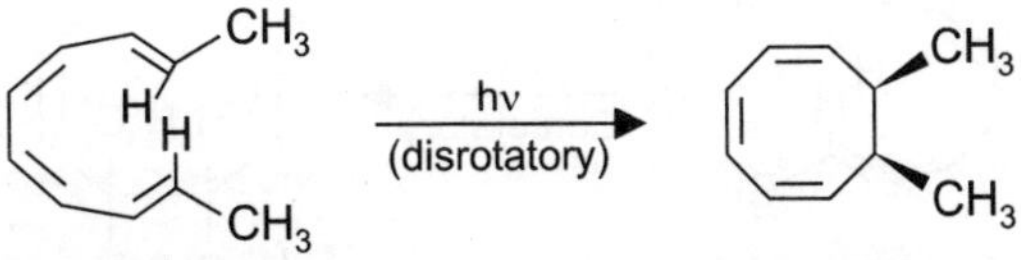

(4*n*) π-electron system substituents point in opposite directions

cis-stereochemistry

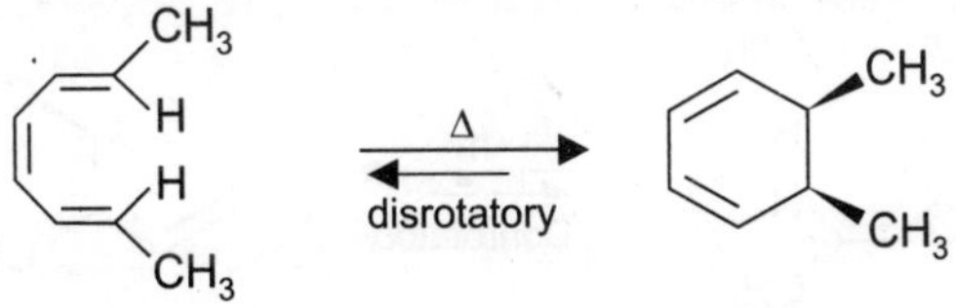

(4*n* + 2) π-electron system substituents point in opposite directions

cis-stereochemistry

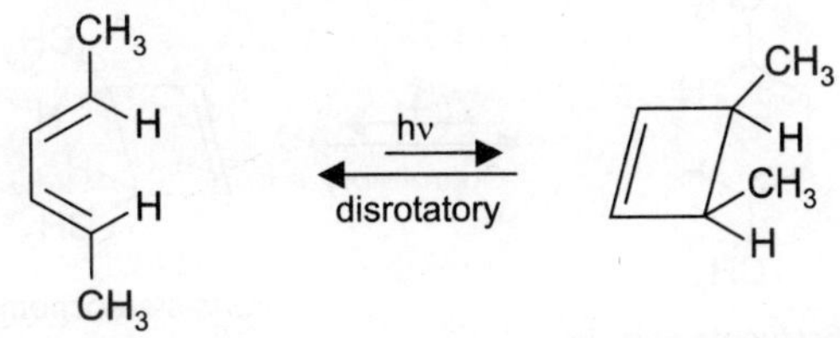

(4*n*) π-electron system substituents point in opposite directions

cis-stereochemistry

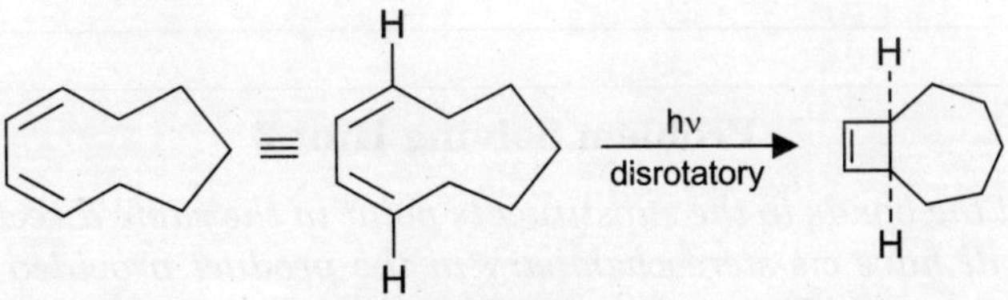

4*n* and (Z, Z)
Substituents point in opposite directions

SCHEME 8.25a

Problem Solving Hint 2

If in a reactant the bonds to the substitutents point in opposite directions then these substituents will have trans stereochemistry in the product provided the motion is conrotatory (Scheme 8.26).

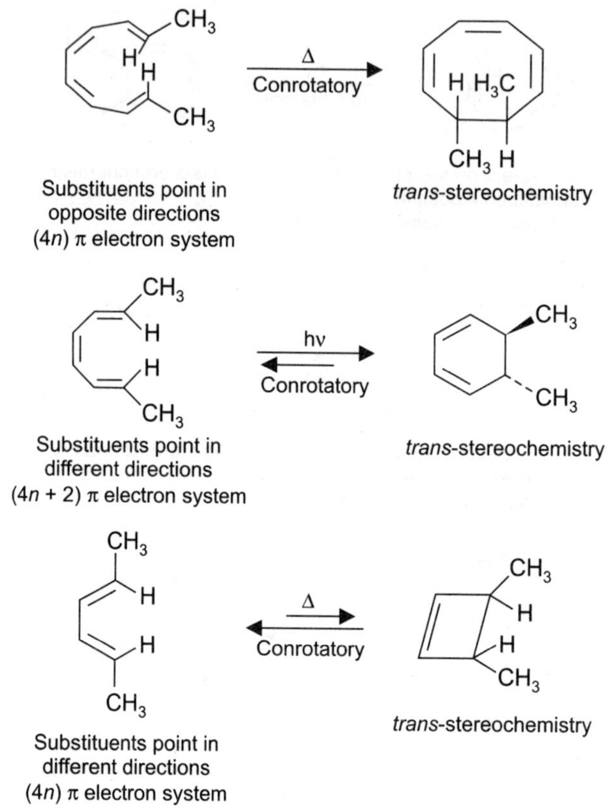

SCHEME 8.26

Problem Solving Hint 3

If in a reactant the bonds to the substituents point in the same direction then these substituents will have cis-stereochemistry in the product provided the motion is conrotatory (Scheme 8.27).

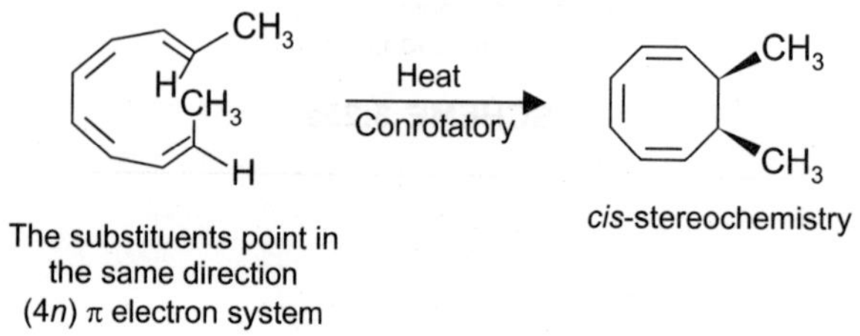

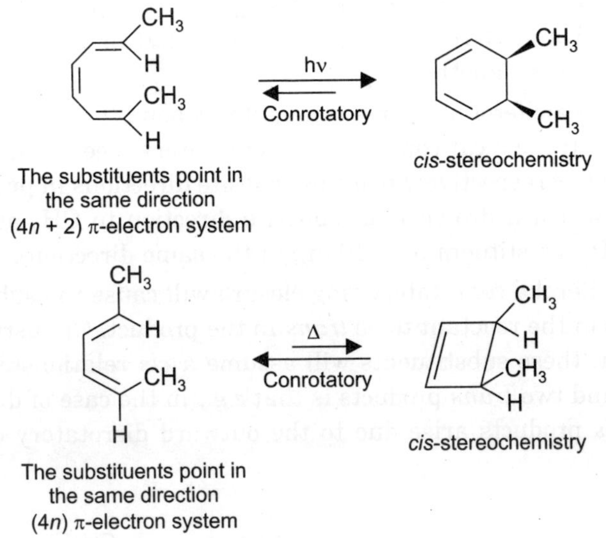

SCHEME 8.27

Problem Solving Hint 4

If in a reactant the bonds to substituents point in the same direction then the substituents will have trans-stereochemistry in the product provided the motion is disrotatory (Scheme 8.28).

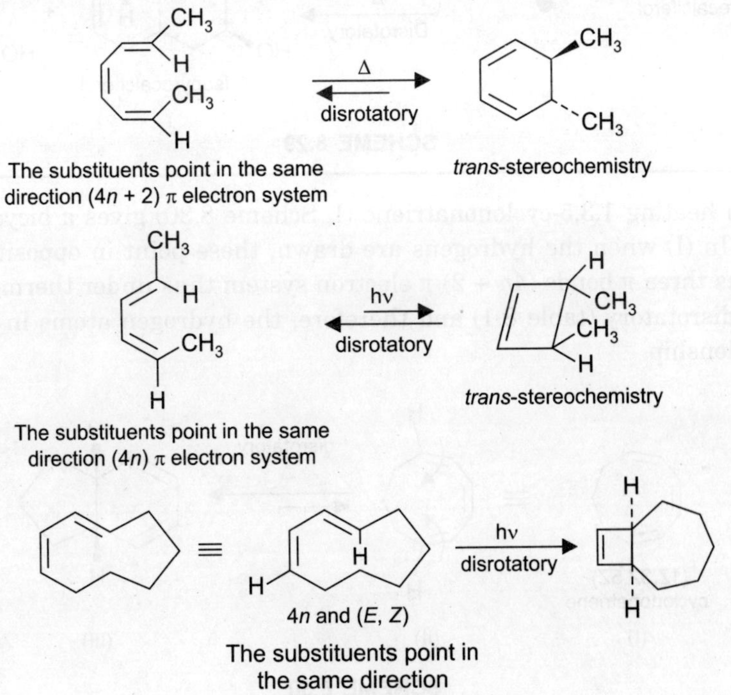

SCHEME 8.28

Example 1: Consider, firstly the photochemical ring closure of precalciferol which gives ergosterol and lumisterol (I and II respectively; Scheme 8.29) in which hydrogen and methyl substituents are *trans* to one another.

The reactant precalciferol has three conjugated π bonds $(4n + 2)$ π electron system thus ring closure under photochemical conditions is conrotatory (see Table 8.1). The methyl and hydrogen at C-10 and C-9 respectively point in opposite directions in precalciferol (the H atom at C-9 is not shown which if drawn is in opposite direction to CH_3 at C-10, note that ring residue at C-9 and CH_3 substituent at C-10 are in the same direction).

Thus in precalciferol a conrotatory ring closure will cause the substituents which point in opposite directions in the reactant to be *trans* in the product. On disrotatory ring closure of precalciferol, however, these substituents will assume a *cis* relationship. The reason for the formation of two *cis* and two *trans* products is that *e.g.*, in the case of disrotatory ring closure (Scheme 8.29) two *cis* products arise due to the outward disrotatory or inward disrotatory motion.

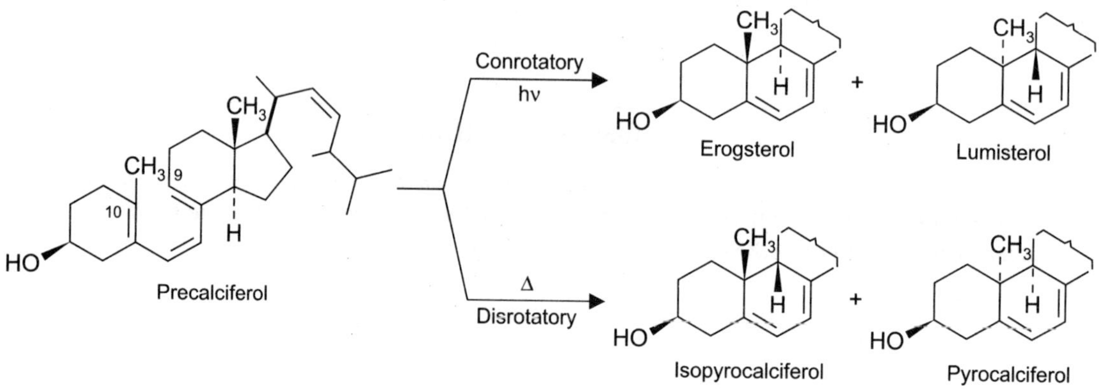

SCHEME 8.29

Example 2: On heating 1,3,5-cyclononatriene (I, Scheme 8.30) gives a bicyclic product with *cis*-ring fusion. In (I) when the hydrogens are drawn, these point in opposite directions (II). The reactant has three π bonds $(4n + 2)$ π electron system thus under thermal conditions the ring closure is disrotatory (table 8.1) and therefore, the hydrogen atoms in the product will have a *cis* relationship.

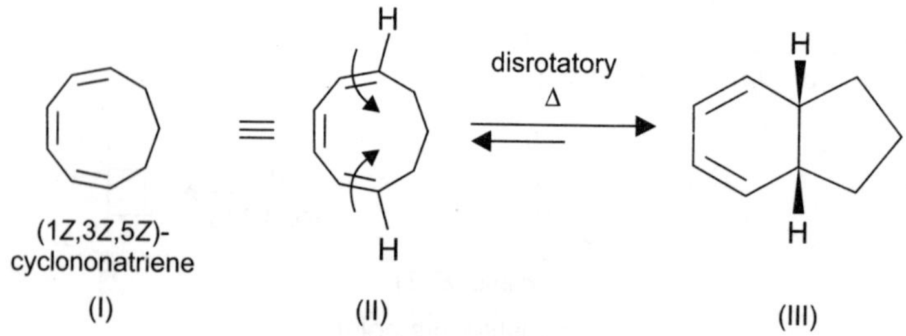

SCHEME 8.30

Example 3: Consider the photochemical ring opening electrocyclic reaction of (I, scheme 8.31). The product (II) undergoing ring closure has three conjugated double bonds and thus under photochemical conditions (see Table 8.1) ring closure or ring opening is conrotatory. To get a product with *cis* hydrogens in (I, Scheme 8.31) the substituent hydrogens have to point in the same direction (II). Thus compared to 1,3,5,-cyclononatriene (I, Scheme 8.30) in which all the three double bonds have *Z* geometry in (II, Scheme 8.31) one of the double bonds has *E* configuration.

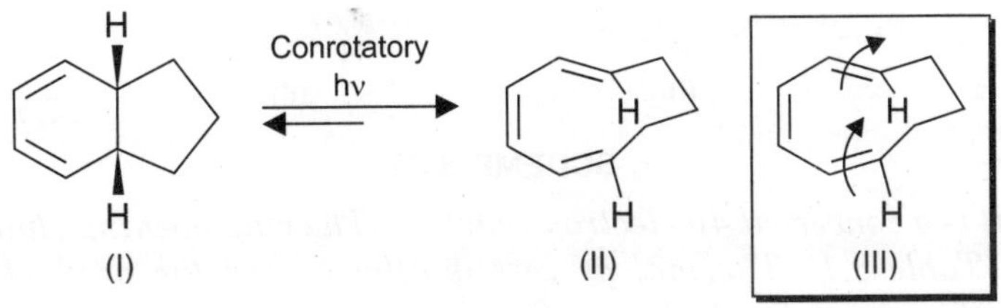

SCHEME 8.31

Example 4: 1,3-Cycloheptadiene (I, Scheme 8.32) closes to the cyclobutene by a disrotatory motion under photochemical conditions with expected *cis* ring fusion. A controtatory opening of cyclobutene (II) is thermally allowed, since the hydrogens in (II, Scheme 8.32) are *cis*, these must point in the same direction in the ring opened product (III) *i.e.,* one of the double bonds in III must be *cis* and the other must be *trans*. However, a *trans*-double bond cannot be accommodated in a seven membered ring and thus (II, Scheme 8.32) is stable under thermal conditions. Similar arguments prove that compound (IV, Scheme 8.32) does not undergo ring opening under thermal conditions while (V) does.

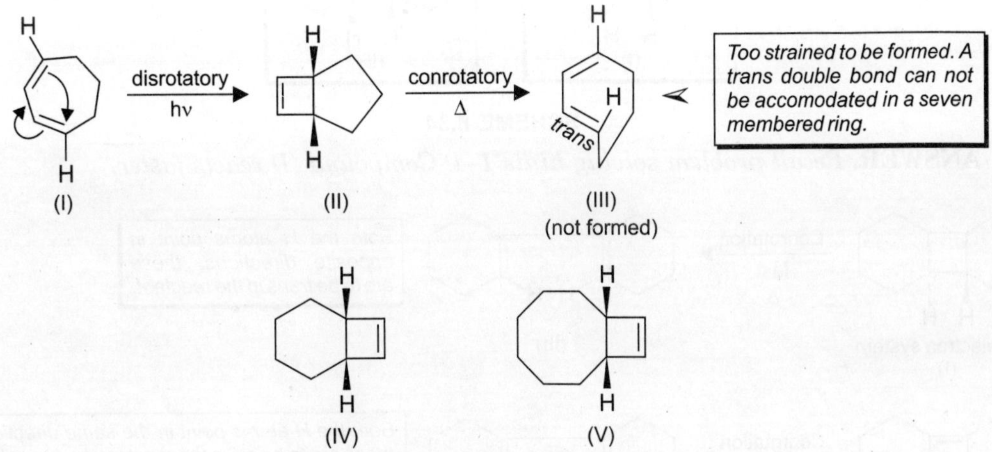

SCHEME 8.32

EXERCISE 8.6

Predict if the conversion shown (Scheme 8.33) is allowed or forbidden?

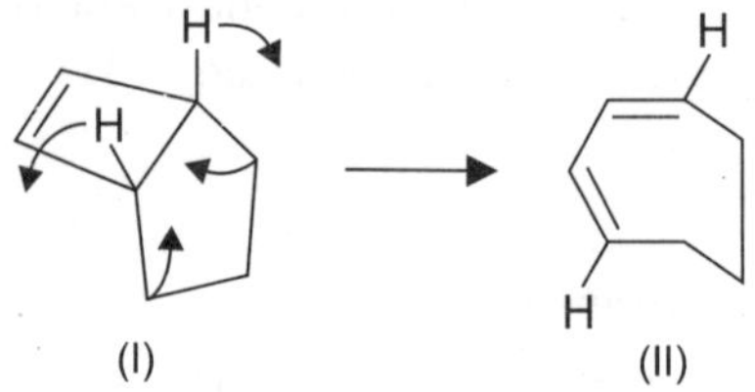

SCHEME 8.33

ANSWER. *It is a concerted 4n electron reaction. The ring opening shall have to be conrotatory (Table 8.1). Thermal process is allowed but the product is strained and is therefore, not formed (see Scheme 8.32). However, the conversion (Scheme 8.33) to unstrained all cis-diene is a disrotatory process as shown and is forbidden by the selection rules (Table 8.1).*

EXERCISE 8.7

One of the cyclobutenes (Scheme 8.34) on heating reacts very fast while the other reacts at an extremely slow rate and at much higher temperature. Explain.

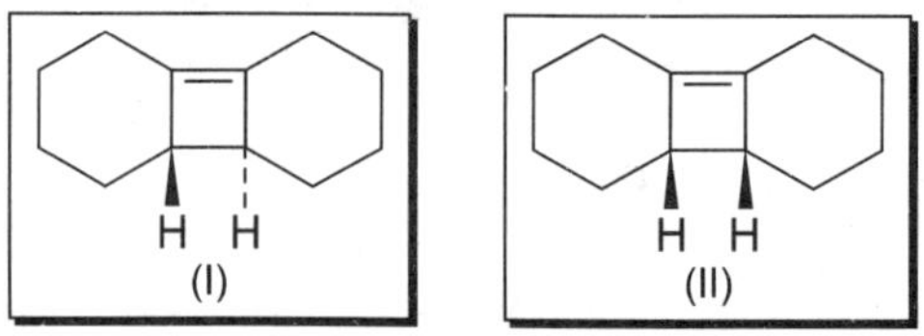

SCHEME 8.34

ANSWER. *Recall problem solving hints 1–4. Compound (I) reacts faster.*

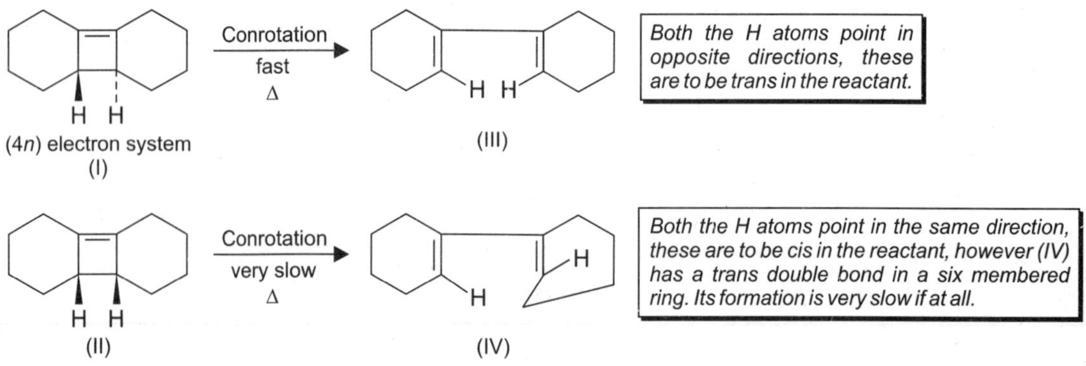

SCHEME 8.35

The compound (II) can however, undergo an easy photochemical opening to the butadiene (III see Problem 8.9), involving dextrorotation. One may recall that e.g., on ring closure a butadiene under thermal or photochemical conditions gives cyclobutenes with opposite configurations. (see, Scheme 8.20a)

EXERCISE 8.8

How one can convert trans-5, 6-dimethyl-1, 3-cyclohexadiene into its cis isomer?

ANSWER. *This may be achieved by electrocyclic ring opening followed by electrocyclic ring closure under appropriate conditions (Scheme 8.36).*

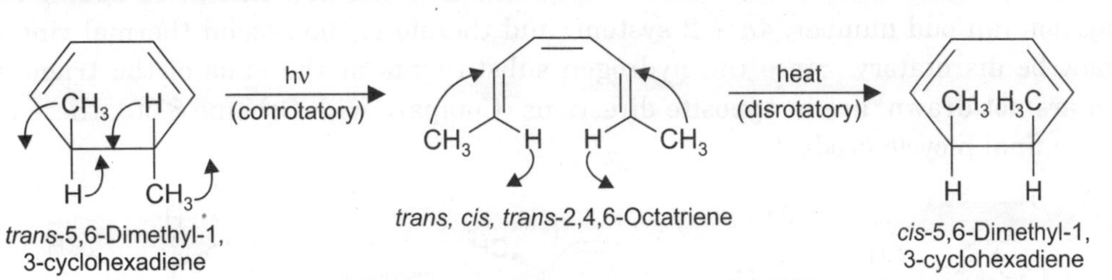

SCHEME 8.36

Example 5: **Dewar benzene has been synthesized and is stable at room temperature (at 25°C, the half life for its conversion to benzene is two days). It is much less stable than benzene due to angle strain and no stabilization due to aromaticity. Dewar benzene could therefore, easily isomerize to benzene. However, this conversion is an electrocyclic reaction (Scheme 8.37), it involves two pair of electrons one pair of π electrons and one pair of sigma electrons of the Dewar benzene, the third pair of electrons is located in exactly the same place both in the reactant and the product and is thus not involved in the reaction).**

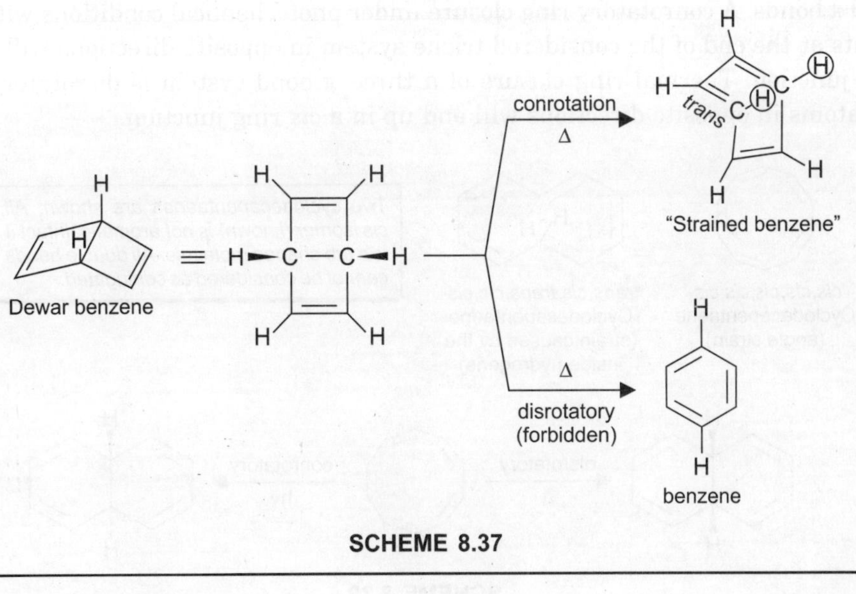

SCHEME 8.37

A thermally allowed electrocyclic reaction with two pairs of electrons (4*n*) π electron system must be conrotatory and the opening of the cyclobutene ring in Dewar-benezene (Scheme 8.37) would result in strained isomer of benzene ("strained benzene") with a *trans* double bond in six membered ring. (In strained benzene two hydrogens on the newly created diene moiety which point to the same side are circled. Thus the otherwise thermally allowed conversion of Dewar benzene into benzene is geometrically impossible since low energy pathway for this conversion is not available.

Example 6: The tetraene (Scheme 8.38) has an even number of π bonds (4*n*) π electron system and therefore, under thermal conditions (– 10°C) it will undergo conrotatory ring closure. Since the two methyl substituents on the tetraene point to opposite directions these will be *trans* in the ring closed cycloctatriene. In cycloctatriene one now has three double bonds in conjugation (an odd number, 4*n* + 2 system) and therefore, the second thermal ring closure will now be disrotatory. Since the hydrogen substituents at the ends of the triene system (which are not drawn) are in opposite directions (Compare with Scheme 8.30), these must be *cis* in the final bicycle product.

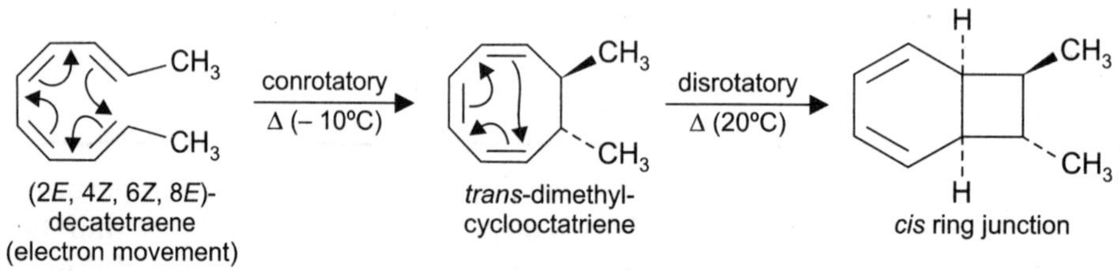

SCHEME 3.38

Example 7: A remarkable distinction between photochemical and thermal reactions is displayed by all *cis*-cyclodecapentaene (Scheme 8.39). One may consider its reacting system with three conjugated π bonds. A conrotatory ring closure under photochemical conditions with hydrogen substituents at the end of the considered triene system in opposite directions will end up in a *trans* ring junction. Thermal ring closure of a three π bond system is disrotatory, and with hydrogen atoms in opposite directions will end up in a *cis* ring junction.

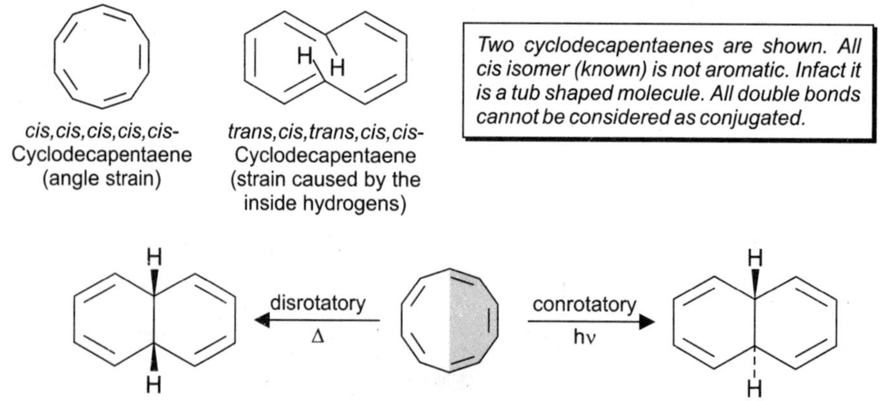

SCHEME 8.39

8.4 CYCLOADDITIONS (FMO-APPROACH)

The reactions of alkenes (the dienophiles) and polyenes (conjugated dienes) in which two molecules react to form a cyclic product, with π electrons being used to form two new σ bonds are called cycloaddition reactions. These reactions are classified on the basis of π electrons involves, in each component, the [4 + 2] cycloaddition being the well known Diels-Alder reaction (Scheme 8.40). The reaction of two alkenes to form a cyclobutane derivative is termed a [2 + 2] cycloaddition reaction (Scheme 8.40). A cycloaddition reaction requires only heat or light for initiation, radical and ionic intermediates are not involved.

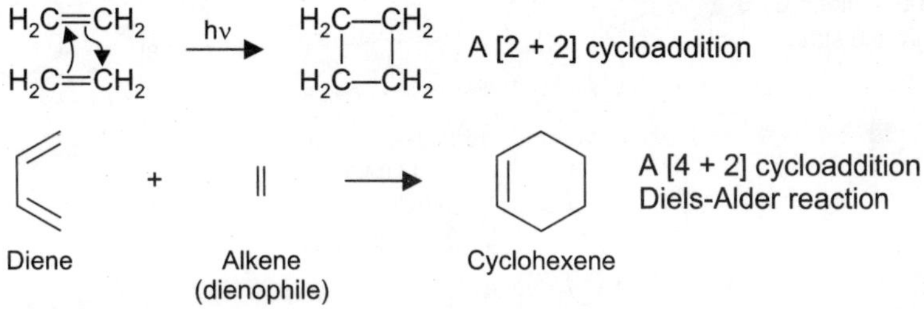

SCHEME 8.40

Molecular Orbitals of Ethylene

On heating ethylene its π electrons are not promoted, but remain in the ground state ψ_1. Irradiation with UV light excites an electron from ψ_1, the ground-state HOMO, to ψ_2^*, which becomes the excited-state HOMO. Interaction between the excited-state HOMO of one alkene and the LUMO of the second alkene indicates that a photochemical [2 + 2] cycloaddition reaction can occur by a suprafacial pathway (Scheme 8.41).

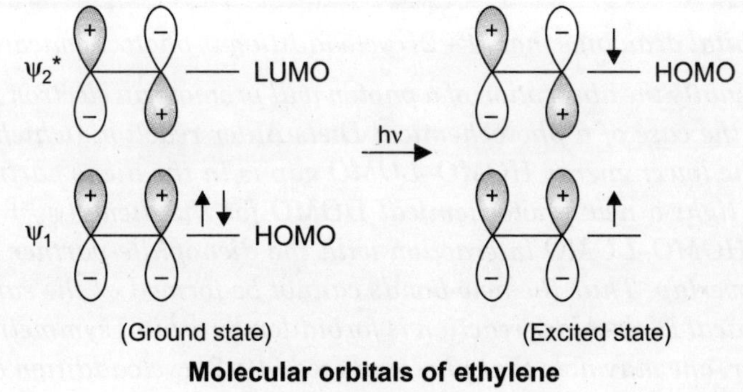

Molecular orbitals of ethylene

SCHEME 8.41

A. Diels-Alder Reaction—[4 + 2] Cycloadditions

These are concerted, thermal [4 + 2] cycloadditions. A consideration of orbital interactions (two combinations) accounts for this (Scheme 8.42), *i.e.*, the overlap can take place between the HOMO of one component and the LUMO of the other and *vice versa*.

As with electrocyclic reactions in cycloadditions as well, one is only concerned wih the terminal lobes. The simplest [4 + 2] system involves the cycloaddition of 1, 3-butadiene (the diene) and ethylene (the dienophile) which is a thermally induced reaction. This thermally allowed reaction involves *e.g.*, the HOMO of 1, 3-butadiene (ψ_2) with the LUMO of ethylene ψ_2* (one could equally well use the diene LUMO and the alkene HOMO).

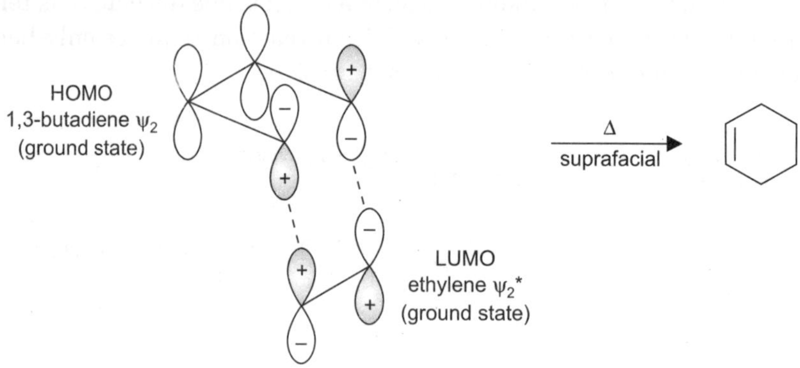

Symmetry-allowed thermal [4 + 2] cycloaddition : 1,3-butadiene and ethylene

SCHEME 8.42

In either case, the overlap brings together lobes of the same phase. Addition to the lobes on the same side of a π system is called suprafacial addition, while addition to lobes on opposite sides of a π system is termed antarafacial addition (for an example of antarafacial addition see Scheme 8.57). These modes of addition are identified by the symbols *s* and *a* respectively. Thus cycloaddition of two π bonds each reacting suprafacially would be called [$\pi^2 s + \pi^2 s$] reaction.

EXERCISE 8.9

Explain by orbital drawings that [4 + 2] cycloaddition is photochemically forbidden.

ANSWER. *Usually the absorption of a photon will promote an electron from HOMO to LUMO. In the case of a photochemical Diels-Alder reaction (which is the most uncommon) the lower energy HOMO–LUMO gap is in the diene partner. Thus on absorption of light a new photochemical HOMO for the diene (ψ_3*) is generated and now the HOMO–LUMO interaction with the dienophile partner involves one antibonding overlap. Thus the new bonds cannot be formed at the same time and the photochemical Diels-Alder reaction is forbidden by orbital symmetry (I, Scheme 8.43). However, one may note that photoinduced [4 + 2] cycloaddition cannot occur if either the diene or the dienophile is excited (II, Scheme 8.43).*

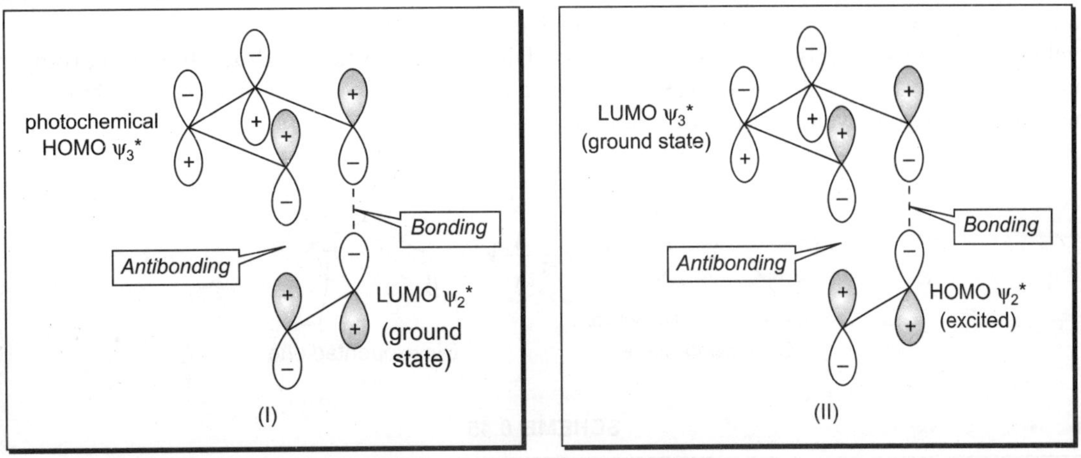

SCHEME 8.43

When in the dienophile there is conjugation to a group of – *M* type *e.g.*, carbonyl, nitro etc. The reaction occurs under milder conditions and gives good yields. The substituent lowers the energy of the LUMO of the dienophile so as to bring it closer in energy to the HOMO of the diene. Consequently the bonding interaction in the transition state increases. As expected, the reactivity is also increased by an electron releasing group in the diene (Scheme 8.16*a*). Conversely, when the diene contains an electron-withdrawing substituent the dienophile requires an electron-releasing substituent for ready reaction. In this situation the interaction is between diene's LUMO and the dienophile's HOMO. Thus, the bonding at the transition state is more effective when the HOMO of one reactant and the LUMO of other are more closely matched in energy. The following points may be noted:

- *s-cis Conformation of the Diene.* As correctly shown (Scheme 8.40) the diene reacts in the *s-cis* conformation, which allows the ends of the conjugated system to reach the doubly bonded carbons of the dienophile. That the *s-cis* geometry of the diene is essential is shown by the unreactive nature of the fixed *transoid* dienes (I and II, Scheme 8.44). Moreover, as expected the substituents in the diene may also effect the cycloaddition sterically. The substituents effect the equilibrium proportion of the diene in the required *cisoid* form (Scheme 8.44). Consequently *Z* alkyl or aryl substituents in the 1-position (III, Scheme 8.44 of the diene slow down the reaction by sterically hindering formation of the *cisoid* conformation, while bulky 2-substituents (IV) make it fast.

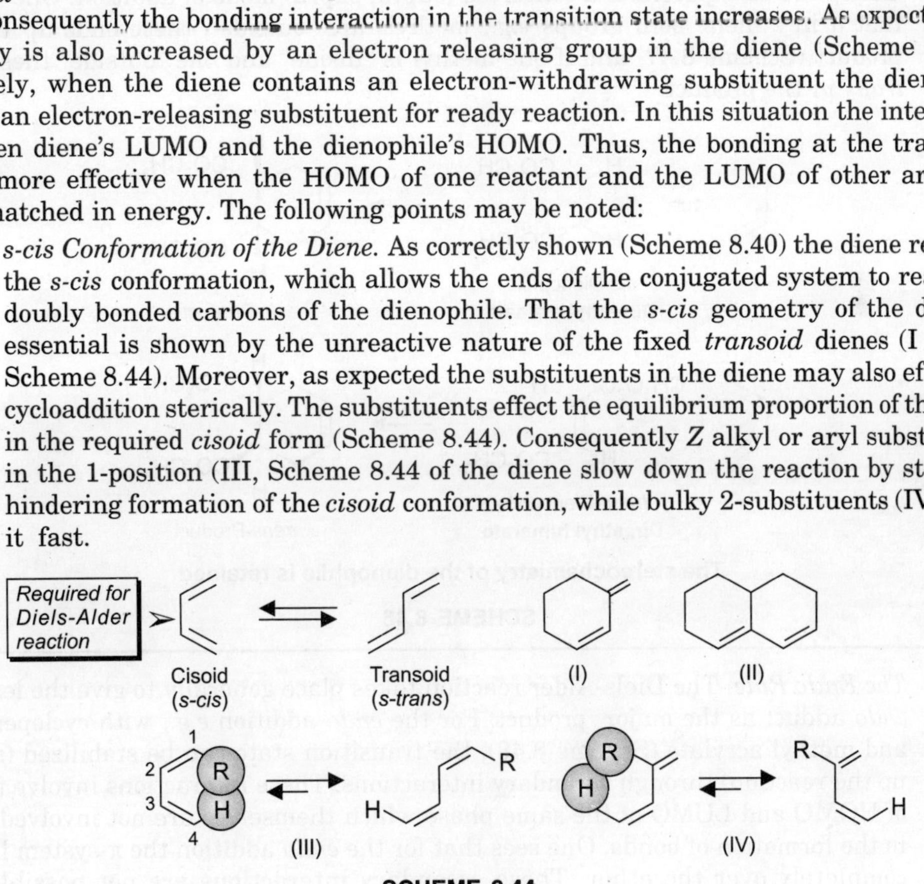

SCHEME 8.44

Cyclic dienes are among the useful dienes and particularly reactive in Diels-Alder reactions as the two double bonds are held in the *s-cis* conformation in five or six membered rings. Cyclopentadiene is highly reactive and forms a Diels-Alder adduct with itself. On heating, the commercially available dimer undergoes a *retro-Diels-Alder* reaction (The term *retro* means the reverse) to give cyclopentadiene (Scheme 8.45).

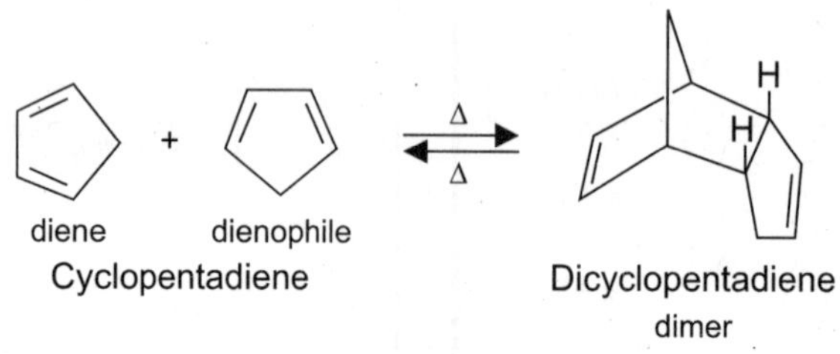

SCHEME 8.45

- *syn-Stereochemistry.* That the Diels-Alder reaction is concerted (both the new bonds are formed in the same transition state) is shown by the fact, that it proceeds with retention of configuration of both the diene and the dienophile (Schemes 8.46 and 8.47) *i.e.*, it proceeds stereoselectively *syn* with respect to both the diene and the dienophile as expected of a concerted (supra, supra) mode of addition. One may note that if in a diene both groups *e.g.*, methyls are "outside", these ends up *cis* in the product (Scheme 8.47) and if one methyl is "inside" and one "outside" these end up *trans* in the product.

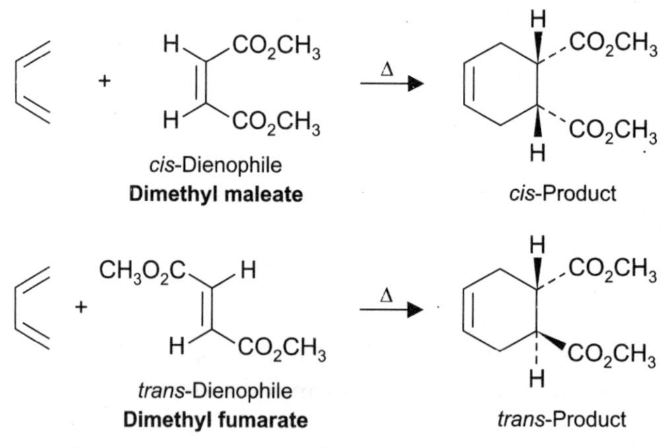

The stereochemistry of the dienophile is retained

SCHEME 8.46

- *The Endo Rule.* The Diels-Alder reaction takes place generally to give the less stable *endo* adduct as the major product. For the *endo* addition *e.g.*, with cyclopentadiene and methyl acrylate (Scheme 8.48), the transition state can be stabilized (speeding up the reaction) through secondary interactions. These interactions involve the lobes of HOMO and LUMO of the same phase which themselves are not involved directly in the formation of bonds. One sees that for the *endo* addition the π-system lies more completely over the other. These secondary interactions are not possible in the transition state for *exo* addition since the relevant set of centers in the diene and the

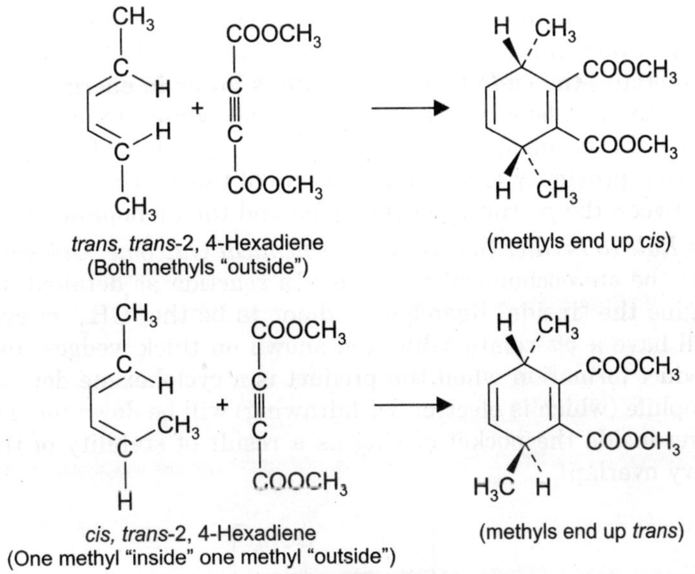

The stereochemistry of the diene is retained

SCHEME 8.47

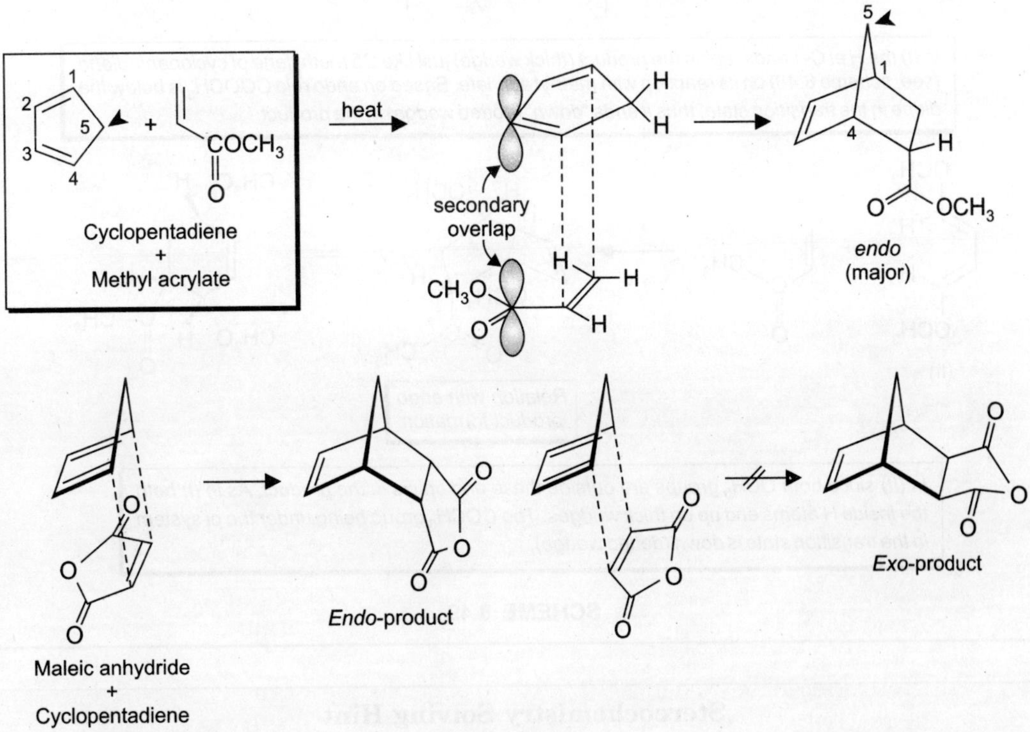

SCHEME 8.48

dienophile are now too far apart, from each other. Thus the preference for *endo* selectivity (which infact is due to several steric and electronic influences on the transition state—the *endo* transition state is lower in energy) is observed when the dienophile has a π bond in its electron withdrawing group *e.g.*, CN or C=O. The *p* orbitals of this group approach the central carbon atoms C2 and C3 of the diene and the resulting proximity leads to an overlap of the *p* orbitals and secondary overlap effects between the *p* orbitals of the diene and the dienophile (Scheme 8.48).

- *The endo Rule to Predict General Stereochemical Outcome.* One can use the *endo* rule to predict the stereochemical outcome of a reaction as detailed (Scheme 8.49). One can imagine the "inside" ligands of a diene to be the "CH_2" of cyclopentadiene and these will have a *cis*-relationship and shown on thick wedges. In keeping with the *endo* product formation when the product is a cyclohexene derivative the group on the dienophile (which is electron withdrawing) will be down (on a dotted wedge) due to it being inside the pocket of ring as a result of stability of the transition state (secondary overlap).

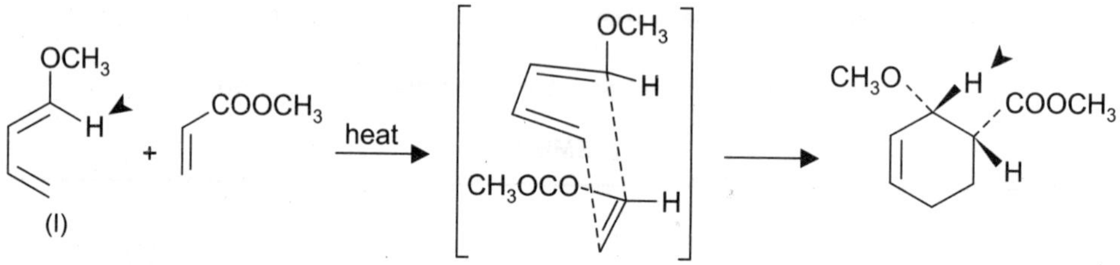

In (I) the H at C-1 ends "up" in the product (thick wedge) just like C5 methylene of cyclopentadiene (see, scheme 8.48) on its reaction with methyl acrylate. Based on endo rule $COOCH_3$ is below the diene in the transition state, thus it ends "down" (dotted wedge) in the product.

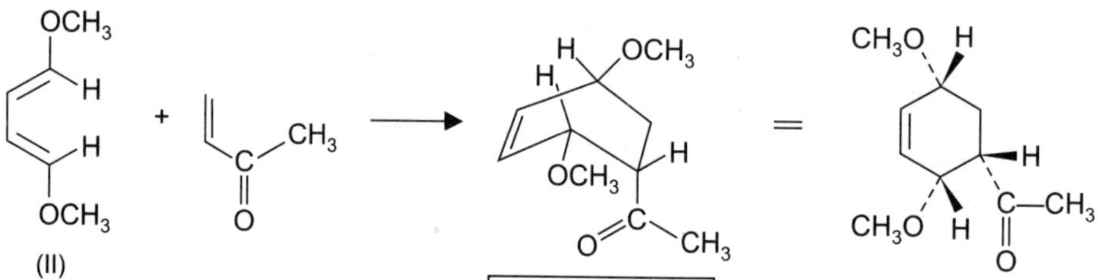

Relation with endo product formation

In (II) since both OCH_3 groups are outside these end up cis in the product. As in (I) both the inside H atoms end up on thick wedges. The $COCH_3$ group being under the pi system in the transition state is down (dotted wedge).

SCHEME 8.49

Stereochemistry Solving Hint

Considering schemes 8.47 and 8.49 one can evolve an useful stereochemistry solving hint of syn addition of concerted Diels-Alder reaction.

Two inner substituents always end up with a cis relationship in the product, and same is the case with two outer substituents. An inner substituent on one carbon ends up always trans to an outer substituent on the other.

When a ring junction is created then as expected the stereochemistry at the ring junction must be cis for a syn addition. Thus when maleic anhydride adds to trans, trans-2, 4-haxadine two diastereomeric syn addition products are possible (Scheme 8.49a).

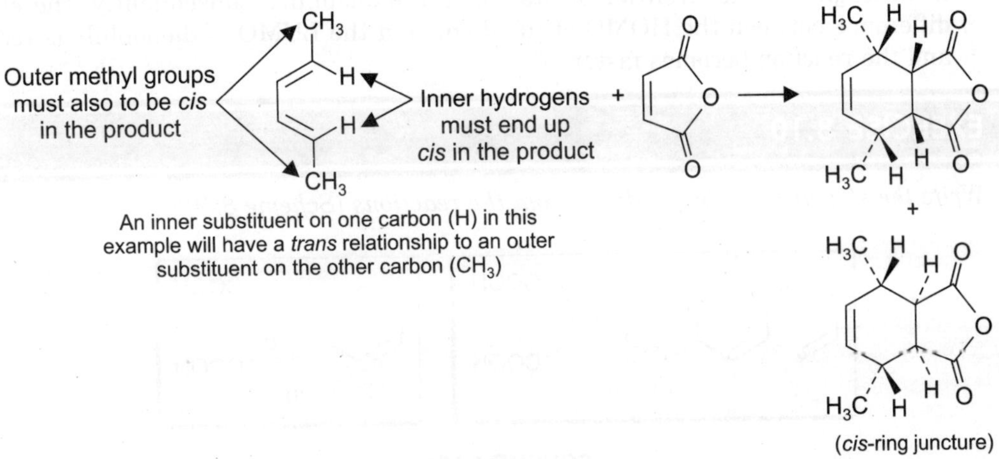

SCHEME 8.49a

- *Regioselectivity.* Cycloaddition of an unsymmetrically substituted diene and an unsymmetrically substituted dienophile can lead to regioisomers (Scheme 8.50).

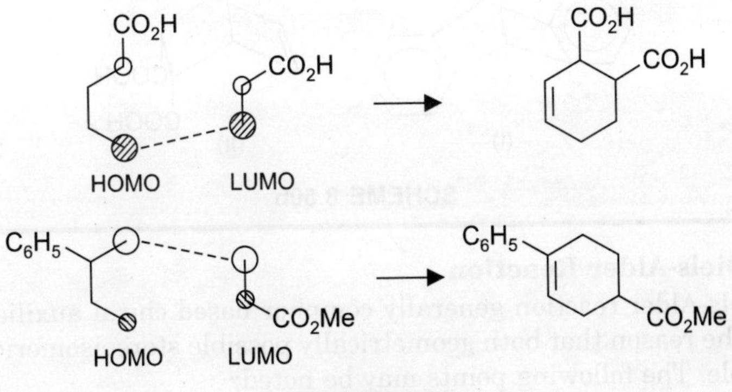

SCHEME 8.50

One has already seen that in a normal Diels-Alder reaction *i.e.,* between an electron rich diene and electron-deficient dienophile, the main interaction is between the HOMO of the diene and LUMO of the dienophile (In this situation these orbitals are more closely matched in energy, the better is the overlap and thus the reaction occurs more readily). However, the orientation of the product from an unsymmetrical diene and an unsymmetrical dienophile depends mostly on the atomic orbital coefficients at the reacting terminii. The atoms with the larger terminal coefficients on each reactant, bond preferentially in the transition state, because of better orbital overlap. Consequently with 1-substituted butadienes the major product is 1,2 ("*ortho*") adduct while with 2-substituted butadienes, the major adduct is 1,4 ("*para*").

In the case of butadiene-1 carboxylic acid and acrylic acid the frontier orbitals are polarized as shown (Scheme 8.50). The size of the circles as shown is roughly proportional to the size of the coefficients and an allowed reaction leads to 1,2-adduct. Similarly, now with 2-phenyl-butadiene and methyl acrylate the major product formed would be 1,4.

- *Lewis Acid Catalysts.* Some Diels-Alder reactions are catalyzed by Lewis acid catalysts. These catalysts form complexes with the polar groups on the dienophile which lower the energies of the frontier orbitals of the dienophile. Consequently, the energy difference between the HOMO of the diene and the LUMO of dienophile is reduced and the reaction becomes faster.

EXERCISE 8.10

Write the structure of the products from the reactions (Scheme 8.50a).

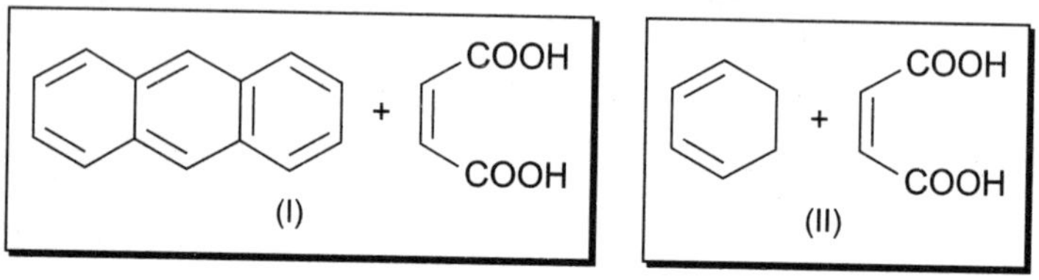

SCHEME 8.50a

ANSWER.

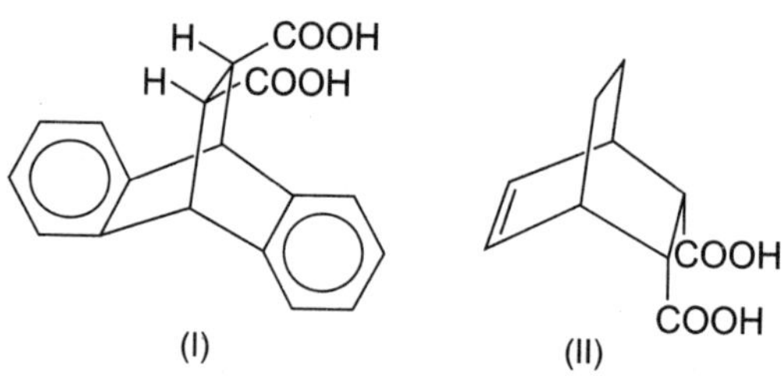

SCHEME 8.50b

B. Asymmetric Diels-Alder Reaction

In asymmetric Diels-Alder reaction generally comphor based chiral auxiliaries are employed and this is due to the reason that both geometrically possible stereoisomeric forms of camphor are readily available. The following points may be noted:

- In the case of a monosubstituted alkene there are two enantiotopic faces *Re* or *Si*.
- A diene *e.g.*, buta-1,3-diene thus can add either to *Re* or *Si* face of a dienophile like acrylic acid to give a cyclohexene in which the stereocenter generated via this Diels-Alder reaction could be either *S* (*Re* approach of the diene) or *R* (*Si* approach of the diene, Scheme 8.51).
- Diels-Alder reaction, therefore, creates stereocenters and when both reactants are achiral and no other chiral influence is there, racemic mixtures are obtained.

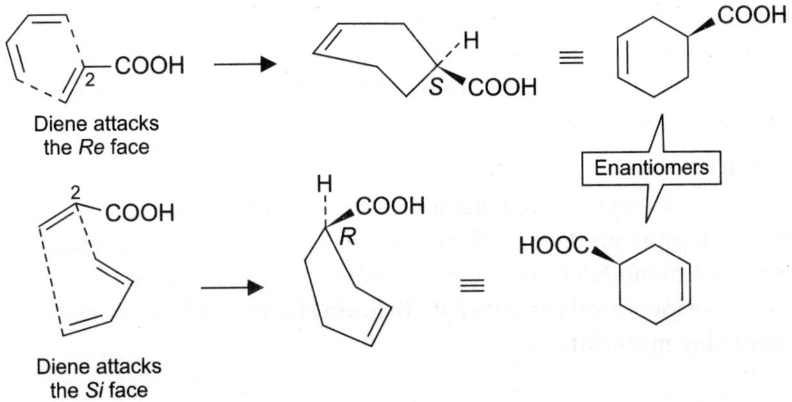

SCHEME 8.51

- When there is chiral influence, *e.g.*, in the dienophile as is so in chiral acrylate (optically active, Scheme 8.52) derived from acrylic acid and optically active alcohol (I, Scheme 8.52) derived from camphor, one of the faces of the double bond in the dienophile gets hindered.

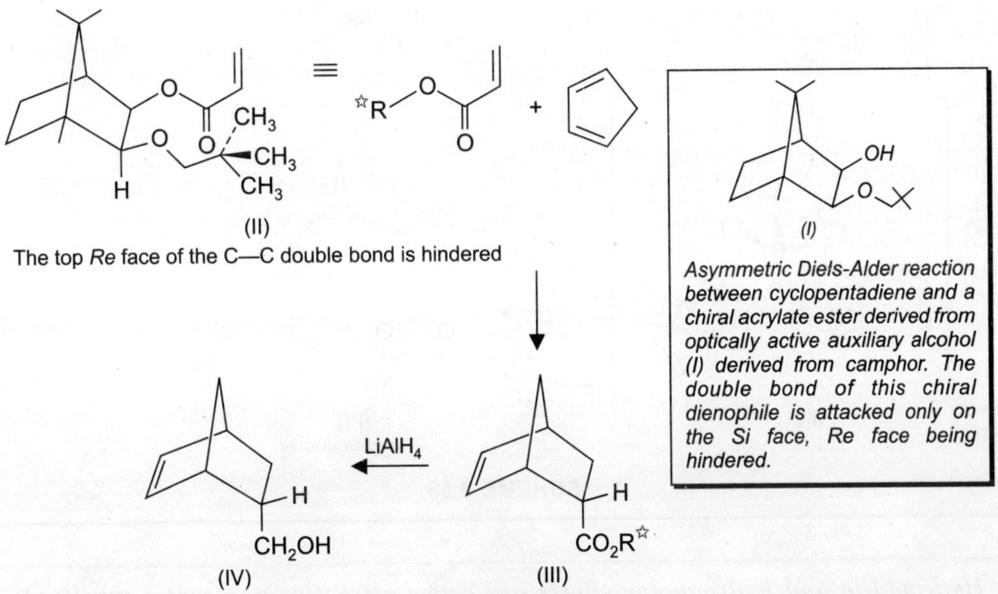

SCHEME 8.52

- The top *Re* face of the carbon-carbon double bond of the dienophile is hindered by the *t*-butyl group of neopentyl unity. This forces the addition to occur preferentially from the *Si*-face (back of the double bond to give (III) almost exclusively, to give only one of the enantiomers.
- One knows that Diels-Alder cycloadditions follow the *endo*-rule.
- Reduction of the ester with lithium aluminium hydride gives the product (IV) in an optically pure form and regenerates the camphor derived chiral auxiliary.

Another example is in (Scheme 8.53). The dienophile (E)-methyl crotonate becomes chiral when optically active alkyldichloroborane (I) complexes with it to yield (II). Now the approach of diene from the rear face of the dienophile is blocked (naphthyl group). Attack occurs from front to give (IV) in optically pure state (*trans*-geometry of dienophile is preserved in IV).

C. Hydrophobic Effects

It has been shown that some intermolecular Diels-Alder reactions are accelerated under hydrophobic effects in aqueous media. This was the case when cyclopentadiene reacted with methylvinyl ketone (Scheme 8.54) and it was observed that any additive which increased the hydrophobic effect also increased the rate *e.g.*, lithium chloride increases the hydrophobic effect by salting out nonpolar material.

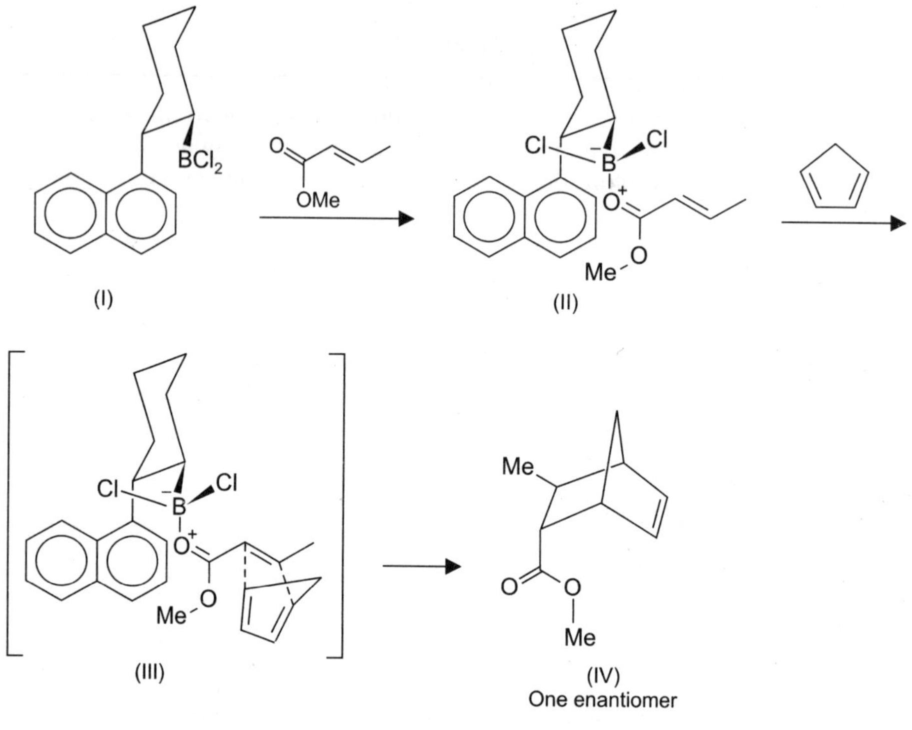

SCHEME 8.53

> *Hydrophilic and hydrophobic effects are water attractive and water repellant respectively. Soaps e.g., sodium oleate have a hydrophilic site (COO^-) and a hydrophobic site (the hydrocarbon chain).*

β-Cyclodextrin has a hydrophobic cavity and if the system of a particular Diels-Alder combination can fit within the cavity a significant rate enhancement is observed. This is found in the case of methyl vinyl ketone and cyclopentadiene system in aqueous medium with β-cyclodextrin as additive. α-Cyclodextrin, however, has a smaller cavity which is not able to accommodate the reactive species and the rate is significantly diminished in its presence compared to that in β-cyclodextrin.

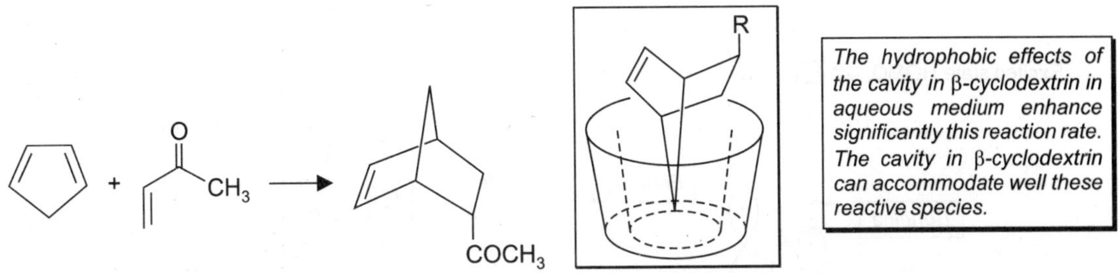

SCHEME 8.54

D. [2+2] Cycloadditions

In the dimerization of ethylene, a thermal [2+2] cyclization would involve overlap of HOMO, of one molecule with the LUMO, of the other (see, Scheme 8.41). If in this concerted reaction both bonds to a component are formed on the same face *i.e.,* the process is suprafacial, the lobes of opposite phase would approach each other (Scheme 8.55). This interaction which is suprafacial with respect to both components [π^{2s} + π^{2s}] is therefore, antibonding and repulsive and the concerted reaction, does not take place (symmetry forbidden process).

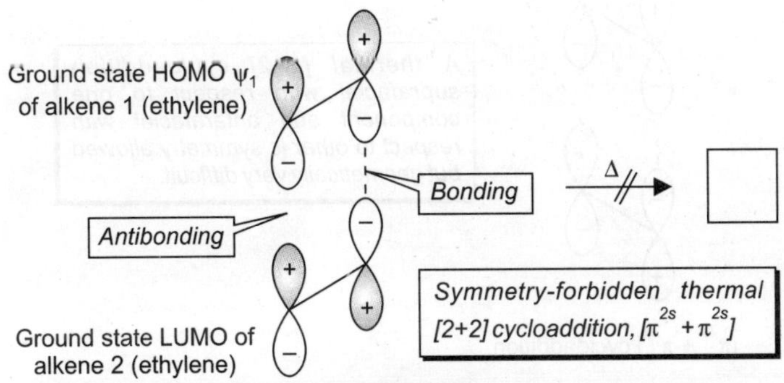

SCHEME 8.55

The photochemical [2+2] cycloadditions which are suprafacial with respect to both the components [π^{2s} + π^{2s}] will, however, permit a previously forbidden reaction to become a symmetry allowed process. During [2 + 2] cycloadditions, irradiation of an alkene with UV light excites an electron from ψ_1, the ground state HOMO to ψ_2* which now becomes the excited state HOMO. The interaction between the excited state HOMO of one alkene and LUMO of the second alkene is now a symmetry allowed process (Scheme 8.56).

The stereochemistry of the Diels-Alder reaction reveals that these are also π^{4s} + π^{2s} processes. However, a thermal [2+2] cycloaddition could occur provided it is suprafacial with respect to one component and antarafacial with respect to the other *i.e.,* it is π^{2s} + π^{2a} (Scheme 8.57). This process, through symmetry allowed is geometrically very difficult.

Thus the photochemical [2+2] cycloaddition reaction occurs smoothly and represents one of the best techniques to synthezise cyclobutane rings and cage compounds (Scheme 8.58). The CO double bond of an aldehyde or a ketone can act as one component in [2+2] cycloaddition with an alkene to form an oxetane (Scheme 8.58). The reaction can occur inter-or intramolecularly.

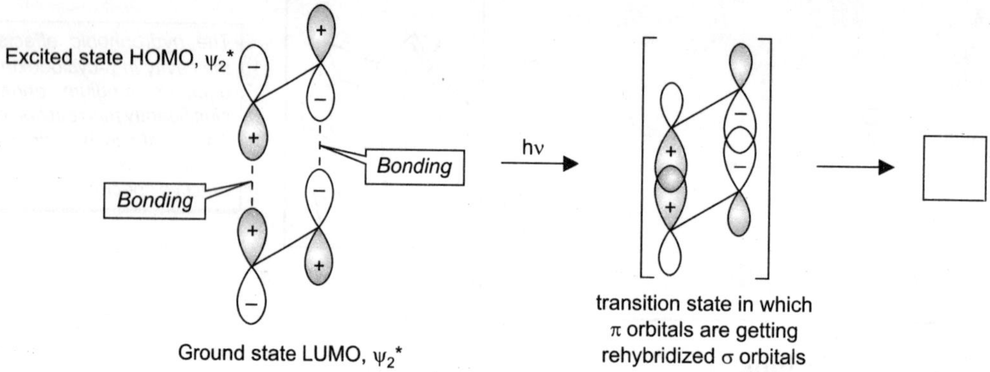

Symmetry-allowed photochemical [2+2] cycloaddition $[\pi^{2s} + \pi^{2s}]$

SCHEME 8.56

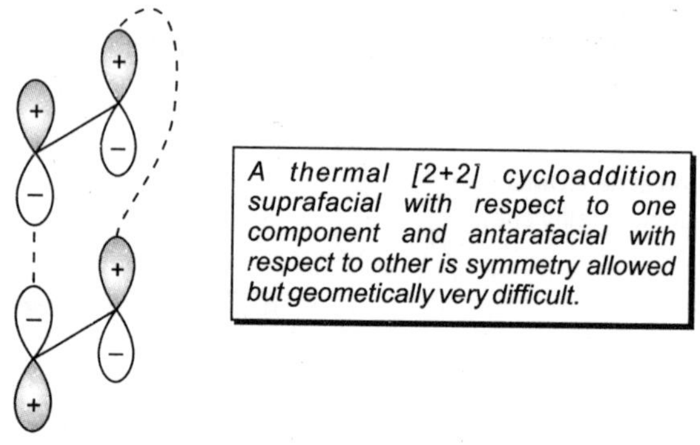

SCHEME 8.57

Both thermal as well as photochemical cycloaddition reactions take place by opposite stereochemical pathways. As with electrocyclic reactions one can categorize cycloadditions according to the total number of electron pairs (double bonds) taking part in the rearrangement. Thus, a Diels-Alder [4+2] reaction between a diene and a dienophile involves an odd number (three) of electron pairs and takes place by a ground state (thermal) suprafacial pathway. A [2+2] thermal reaction between two alkenes involves an even number (two) of electron

Table 8.2 Cycloaddition Reactions Stereochemical Rules

Electron pairs (double bonds)	Thermal reaction	Photochemical reaction
Even number	Antarafacial	Suprafacial
Odd number	Suprafacial	Antarafacial

pairs and must takes place by an antarafacial pathway. However, it may be impressed that both suprafacial and antarafacial cycloaddition pathways are symmetry allowed. Only the geometric constraints inherent in twisting a conjugated *pi* electron system out of planarity make antarafacial reaction geometrically difficult in most of the cases. One may note that preferences for cycloadditions may be summarized further (Table 8.3) to quickly known the success of a particular cycloaddition and provides a rule of thumb.

Table 8.3 Cycloaddition Reactions

Number of Electron Pairs	Allowed Cycloaddition
Odd	Thermal
Even	Photochemical

As a last example of cycloaddition, cyclopentadiene reacts with cycloheptatriene system to give a product—which is [6+4] suprafacial process (Scheme 8.59, this would include an aromatic transition state with 10 electrons).

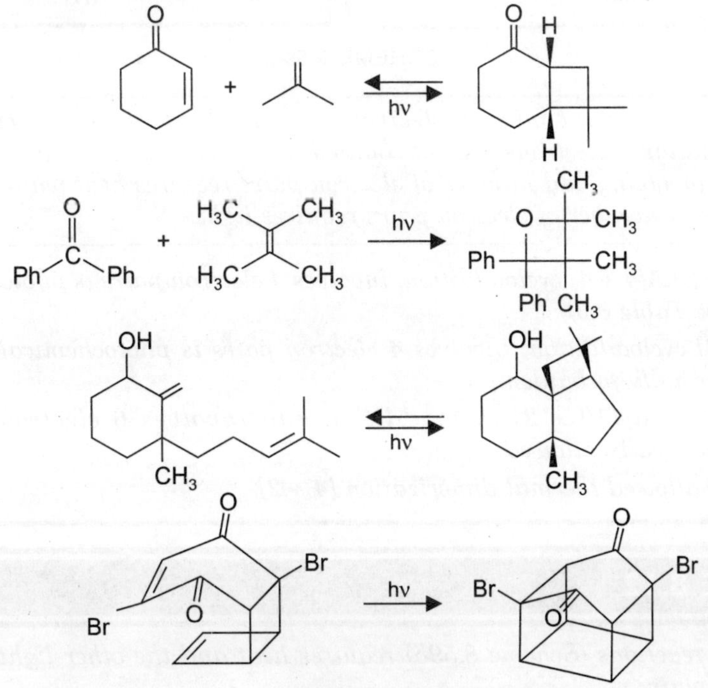

SCHEME 8.58

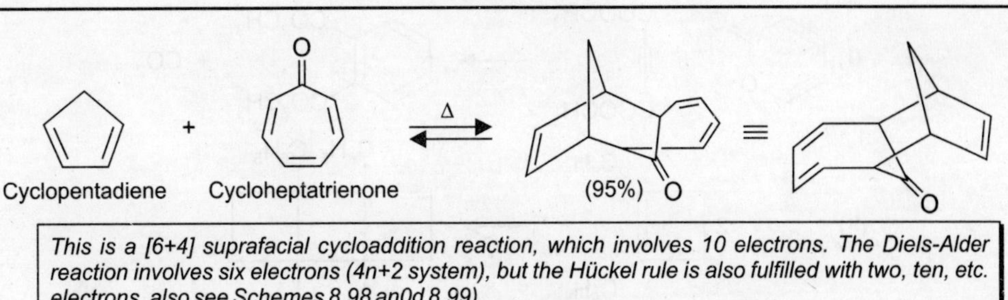

This is a [6+4] suprafacial cycloaddition reaction, which involves 10 electrons. The Diels-Alder reaction involves six electrons (4n+2 system), but the Hückel rule is also fulfilled with two, ten, etc. electrons, also see Schemes 8.98 an0d 8.99)

SCHEME 8.59

EXERCISE 8.11

Indicate if the following reactions (Scheme 8.59a) are allowed or forbidden.

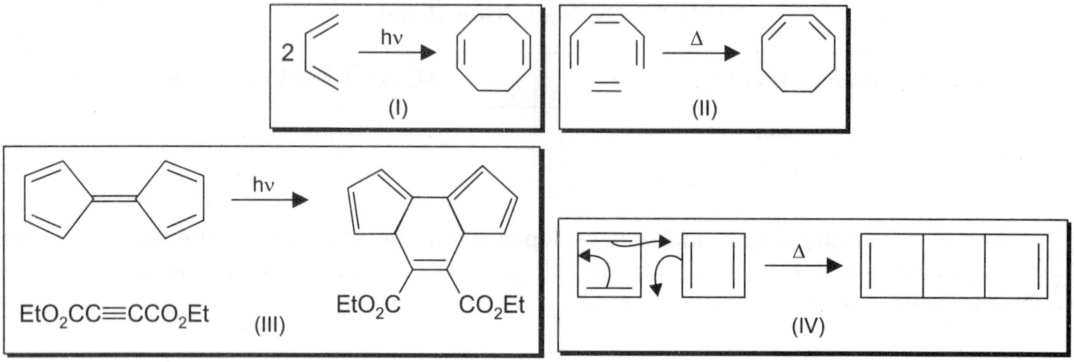

SCHEME 8.59a

> **Hint.** *If in a reactant more π electrons are involved in the cyclization, the nonparticipating π electrons are not counted for the classification (see Table 8.3). A reaction involving odd number of electron pairs requires heat while a reaction involving even number of electron pairs requires light.*

ANSWER. *(I) A[4 + 4] cycloaddition, involves 4 electron pairs is photochemically allowed (see Table 8.3).*

(II) A[6 + 2] cycloaddition, involves 4 electron pairs is photochemically allowed and thermally forbidden.

(III) Since it is a [10 + 2] cycloaddition and involves 6 electron pairs, is photochemically allowed.

(IV) It is the allowed thermal dimerization [4 + 2].

EXERCISE 8.12

One of the reactions (Scheme 8.59b) requires heat and the other light. Which is which ? Explain.

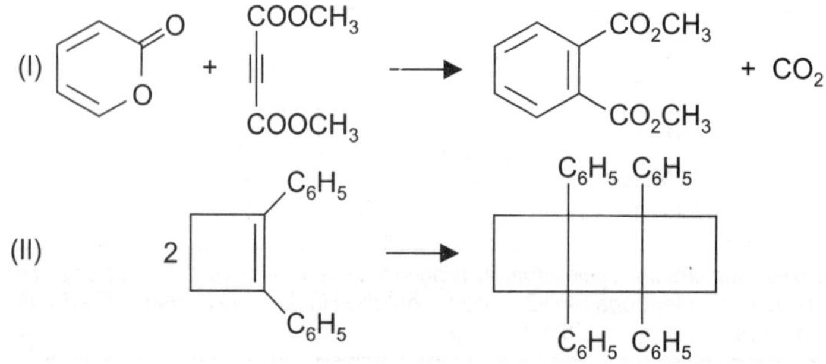

SCHEME 8.59b

ANSWER. *(I) This reaction is thermally allowed ; it being a [4 + 2] cycloaddition. The initially formed intermediate then undergoes a thermally allowed reverse Diels Alder reaction (Scheme 8.59c).*

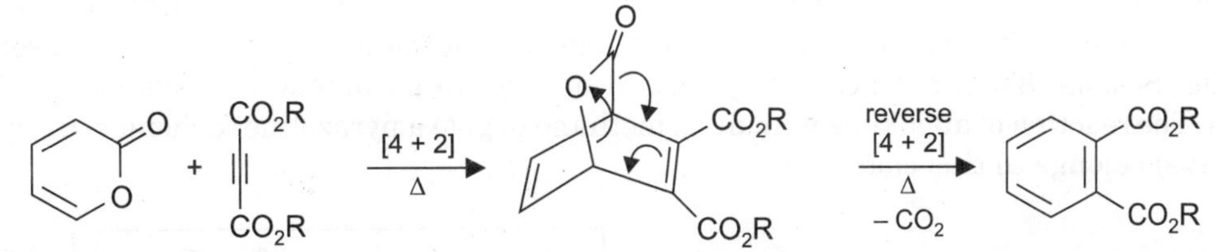

SCHEME 8.59c

(II) This is a photochemical [2 + 2] dimerization with the stereochemistry as dictated from end-to-end overlap of p-orbital components (Scheme 8.59d).

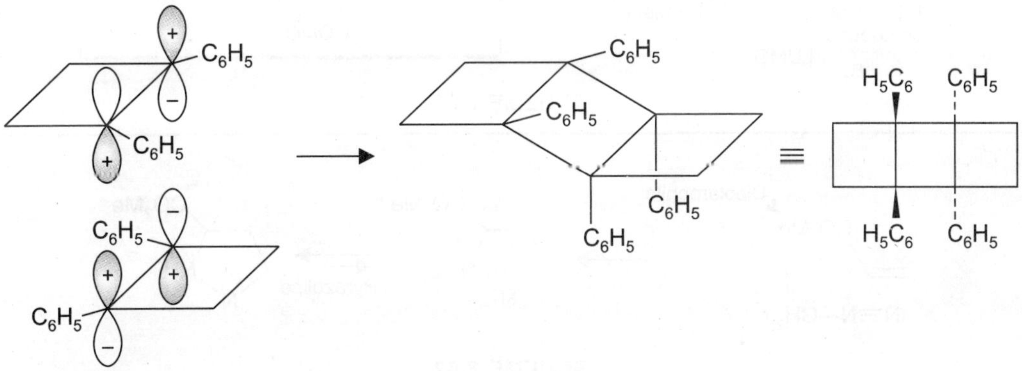

Orbitals of the same phase overlap

SCHEME 8.59d

8.5 1,3-DIPOLAR CYCLOADDITIONS

These cycloadditions are analogous to the Diels-Alder reaction in that they are concerted [$\pi^{4s} + \pi^{2s}$] reactions. The 1,3-dipolar components are compounds whose representation requires ionic structures which include ones with charges on atoms bearing 1,3-relationship, as in diazomethane (Scheme 8.60). These type of molecules which are called, 1,3-dipoles are isoelectronic with allyl anion. These have four π electron and each has at least one charge separated responance structure with opposite charges in a 1,3 relationship. The other reactant (dipolarophile) in a dipolar cycloaddition has unsaturated bonds like, C=C, C≡C, C=O and C≡N. The 1,3-dipolar cycloadditions form useful reactions for the synthesis of five membered heterocyclic rings.

'1,3-dipolar' compound
diazomethane

SCHEME 8.60

Mechanistically the transition state for 1, 3-dipolar cycloaddition is not very polar and the reaction rate is not strongly sensitive to solvent polarity. The loss of charge separation which is implied, is more apparent rather than real, since most 1, 3-dipolar compounds are not highly polar. The polarity associated with a single structure is balanced by other contributing structures.

A 1, 3-dipole represents a structural variant of the diene component in the Diels-Alder reaction; in the dipolar compound, four π-electrons are distributed over three atoms instead of the four in a diene. Moreover, the HOMO and LUMO of a 1, 3-dipole are similar in symmetry to that in a diene with respect to the two-fold axis and to the mirror plane which bisects the molecule (Scheme 8.61), a concerted cycloaddition *e.g.,* to an alkene is a symmetry allowed process. The reaction of an alkene with diazomethane to give a pyrazoline (Scheme 8.62 pyrazole derivative) belongs to this class.

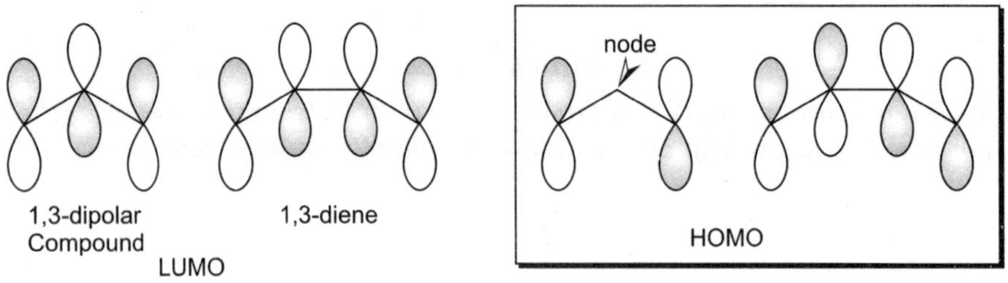

SCHEME 8.61

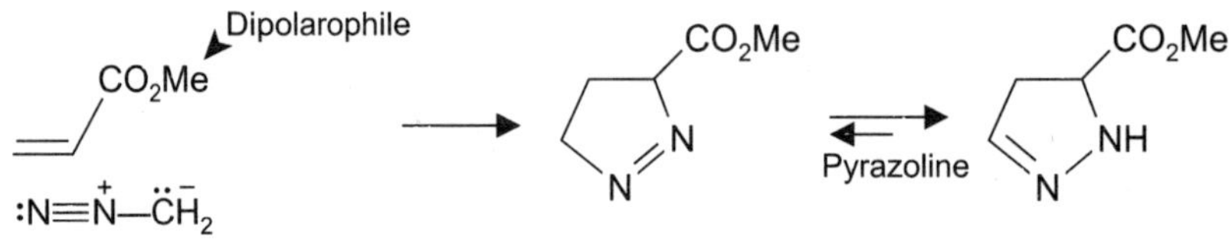

SCHEME 8.62

8.6 CHELETROPIC REACTIONS

In a cheletropic reaction two σ bonds that terminate at a single atom are made or broken during a concerted reaction (Scheme 8.63). In the case of molecules, sulfur dioxide or carbon monoxide the HOMO is that which has a lone pair of electrons in the plane having the atoms, while the LUMO represents the *p* orbital perpendicular to this plane (Scheme 8.64).

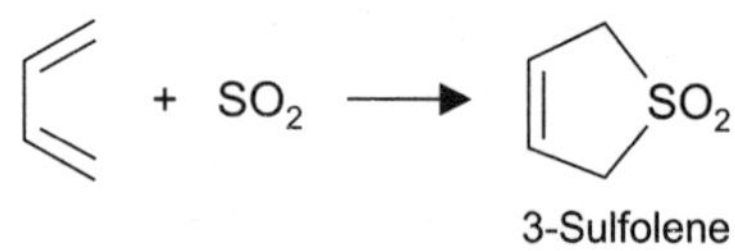

(a cheletropic reaction)

SCHEME 8.63

For a symmetry allowed cycloaddition of *e.g.,* SO_2 to a diene, the molecule of SO_2 must lie in a plane which bisects the *s-cis* conformation of the diene (Scheme 8.64). The interaction is suprafacial for diene and SO_2. In the transition state, the terminal carbon atoms of the diene must move in the disrotatory manner in order that the HOMO of SO_2 can interact with the LUMO of the diene, or the LUMO of SO_2 with the HOMO of the diene.

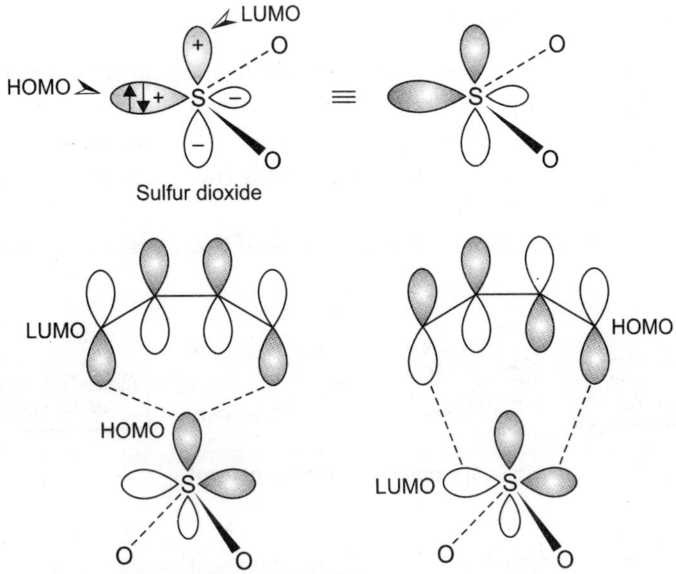

SCHEME 8.64

In keeping with these arguments the *trans, trans*-1,4-disubstituted dienes give specifically more crowded *cis*-substituted 3-sulphones (Scheme 8.65). By similar arguments *cis, trans*-disubstituted dienes, on the other hand afford *trans*-substituted-3-sulphones. As with electrocyclic reactions, the opposite stereochemistry is observed when the reaction is photochemical rather than thermochemical (Scheme 8.65).

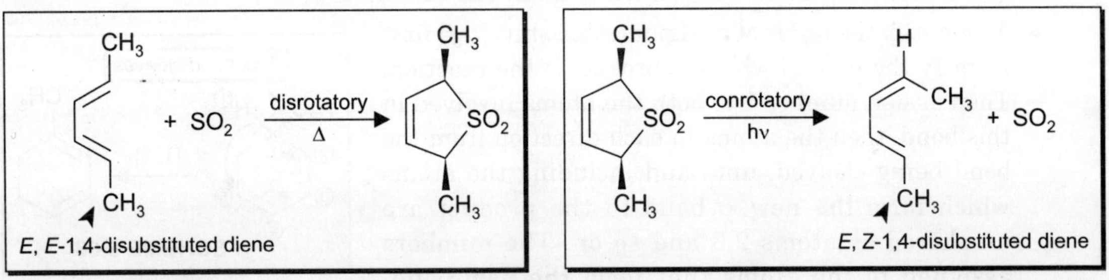

SCHEME 8.65

3-Sulpholene, a solid, is a convenient substitute for geseous butadiene. Butadiene is generated at high temperatures from 3-sulpholene in a reverse reaction (Scheme 8.66) and when a dienophile is present it is trapped in a Diels-Alder reaction. The Diels-Alder reaction in itself is usually reversible and has been used to protect double bonds (see Exercise 8.11).

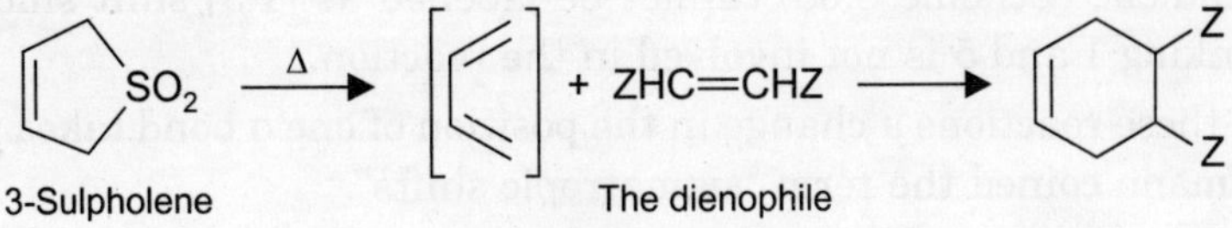

SCHEME 8.66

8.7 SIGMATROPIC REARRANGEMENTS

A sigmatropic rearrangement is a concerted intramolecular shift of an atom or a group of atoms. During this arrangement at σ bond is broken in the reactant and a new σ bond is formed in the product and the π bonds rearrange. The following points may be noted:

- The number of π bonds does not change, both the reactant and the product contain the same number of π bonds.
- The σ bond that cleaves can be in the middle of the π system or at end of the π system (Scheme 8.67).

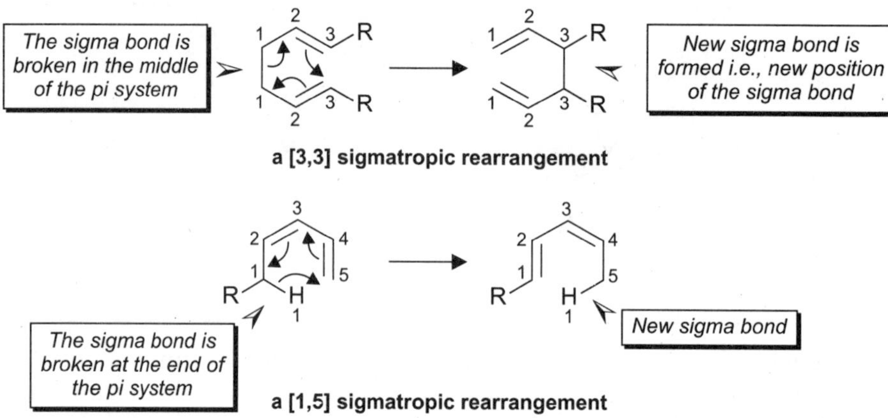

SCHEME 8.67

- The σ bond that breaks is bonded to an allylic carbon.
- To identify the order of a sigmatropic shift [*i*, *j*] first identity the σ bond which is broken in the reaction. Then assign number 1 to both the atoms involved in this bond, then the atoms in each direction from the bond being cleaved, upto and including the atoms which form the new σ bond in the product are numbered as atoms 2,3 and so on. The numbers assigned to the atoms that form the new bond, separated by commas are put within the brackets to show the reaction order (Scheme 8.67). Similarly the migration of hydrogen (Scheme 8.68) is another example of [1,5] sigmatropic shift. The order [1,5] is not due to the fact that hydrogen migrates from C1 to C5 but since the hydrogen (one of the two atoms given the number 1) forms part of the new σ bond and had also formed part of the old σ bond. Only all the atoms taking part in the reaction have to be counted. Thus the rearrangement of cyclohexadiene (Scheme 8.68) cannot be labelled as [1,3] shift since the methylene group linking 1 and 5 is not involved in the reaction.

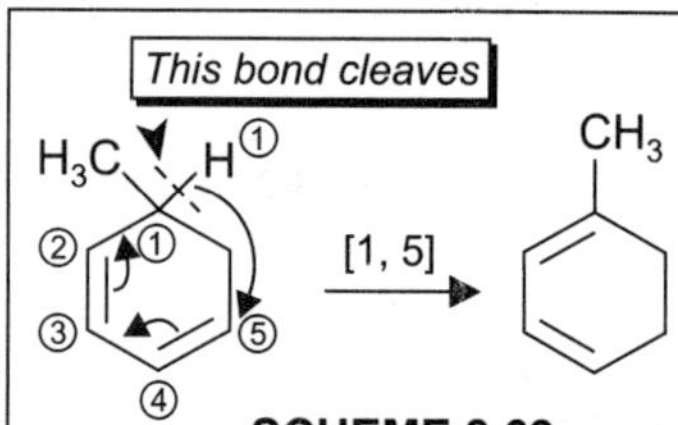

SCHEME 8.68

- Since in these reactions a change in the position of one σ bond takes place, Woodward and Hofmann coined the term "sigmatropic shifts".
- A [3,3] sigmatropic rearrangement of a 1,5-diene (when the six atoms involved are all carbons) is known as the Cope rearrangement (Scheme 8.69).

- The oxygen analog of the Cope rearrangement is called the Claisen rearrangement. Often one of the π bonds is part of an aromatic ring (II, Scheme 8.69). Allyl vinyl ethers also undergo Claisen rearrangement.

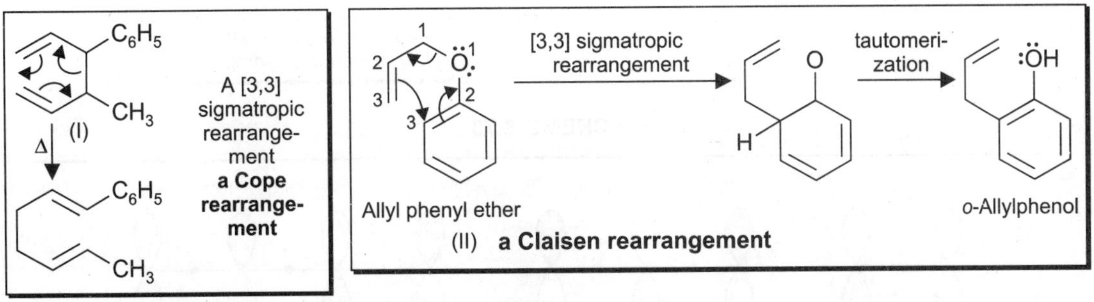

SCHEME 8.69

- More examples of sigmatropic rearrangements are in (Scheme 8.69*a*).

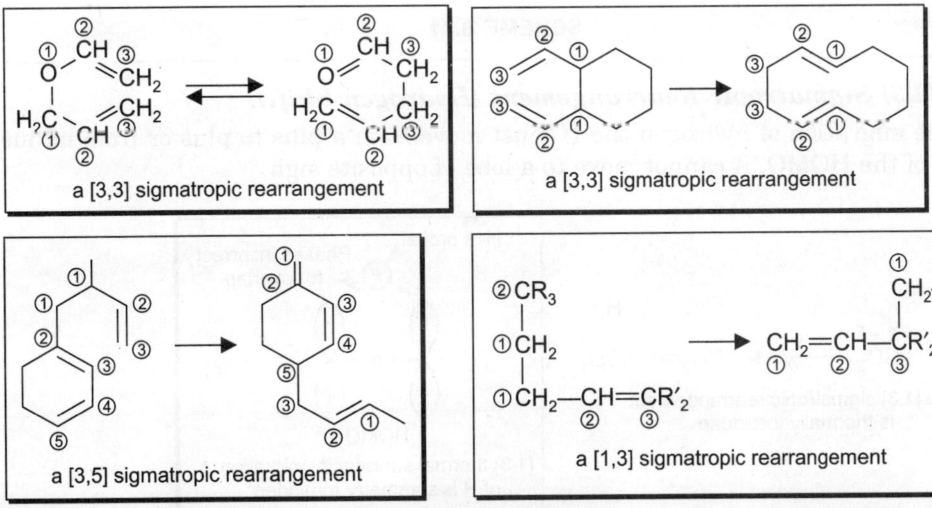

SCHEME 8.69a

A. Sigmatropic Migration of Hydrogen

(i) Introduction.

A hydrogen atom is reported to migrate from one end of a system of π bonds to the other, under thermal or photochemical rearrangements. In the transition state the hydrogen must be in contact with both ends of the chain at the same time. There are two distinct processes by which a sigmatropic migration can occur. If the hydrogen moves along the top or bottom face of the π-system *i.e.,* migrating group remains associated with same face of the conjugated system throughout the process, the migration is termed *suprafacial.* When the hydrogen moves across the π system either from top to bottom or vice versa *i.e.,* the migrating group moves to the opposite face of the π-system during the course of migration then it is called *antarafacial.*

In a given sigmatropic rearrangement, the migrating group is bonded to both the migration source and the migration terminii in the transition state. It is imagined that the migrating H atom breaks away from the rest of the system which is treated as a free radical. Thus in a simplest case involving a [1,3] shift of hydrogen (Scheme 8.70), the frontier orbital analysis treats this system as a hydrogen atom interacting with an allyl radical. The electron of the hydrogen atom is in a 1*s* orbital which has only one lobe. The HOMO of an allylic free radical depends on the number of carbons in the π-framework (Scheme 8.71).

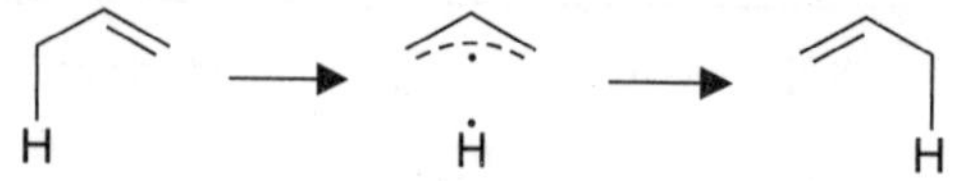

Imaginary transition state for a [1,3] sigmatropic hydrogen shift.

SCHEME 8.70

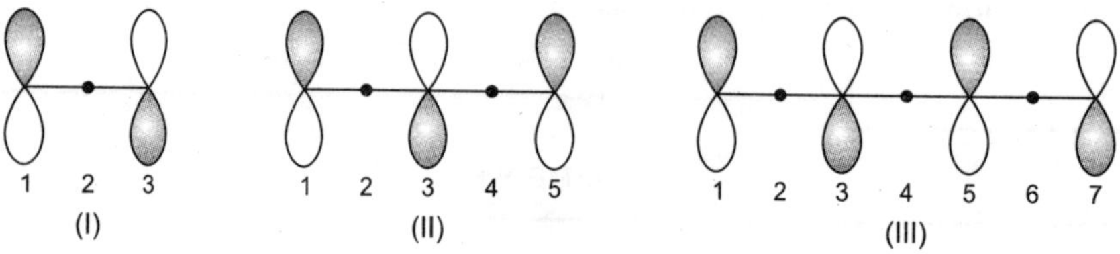

The HOMO of the *allylic* radicals

SCHEME 8.71

(ii) [1,3] Sigmatropic Rearrangement (Hydrogen Shift).

In the migration of hydrogen the H must move from a plus to plus or from minus to a minus lobe of the HOMO, it cannot move to a lobe of opposite sign.

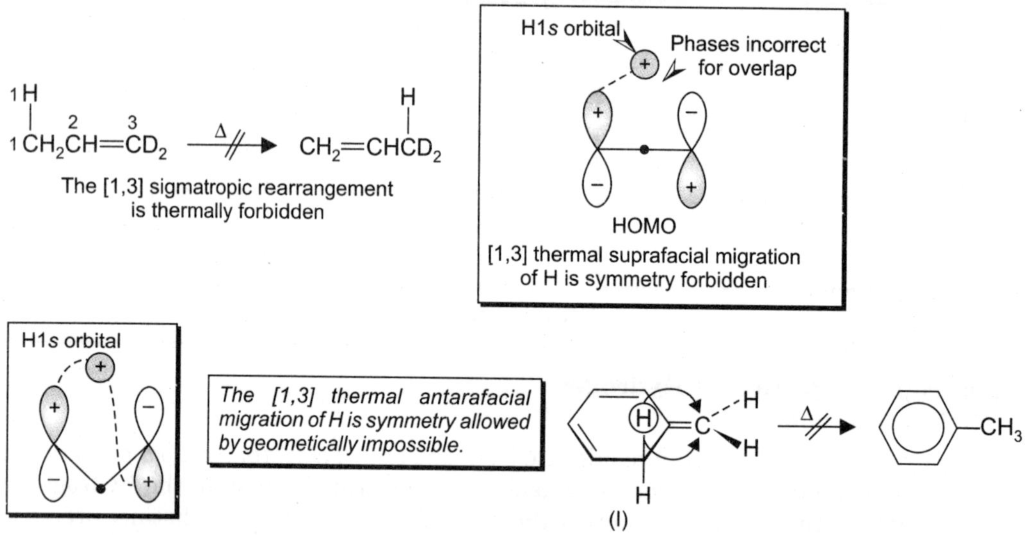

SCHEME 8.72

During a thermal [1,3] sigmatropic migration of a hydrogen, the overlap of the hydrogen 1*s* orbital with the HOMO of the allyl radical (I, Scheme 8.71, asymmetric) is bonding at one end and antibonding at the other end for the suprafacial migration (Scheme 8.72). Thus [1,3] sigmatropic suprafacial migration of hydrogen (under thermal conditions) is symmetry-forbidden (Scheme 8.72). However, in the antarafacial process (Scheme 8.72) the hydrogen atom shall have to cross over the *pi* system to the other face to form a four membered ring transition state, a geometrically very difficult situation. Thus over all thermal [1,3] sigmatropic rearrangements are rare. The stability of the triene (I, Scheme 8.72) which is not thermally isomerized to toluene, which is thermodynamically more stable is due to a symmetry-forbidden process (suprafacial H migration is symmetry forbidden, antarafacial H migration though symmetry allowed but sterically forbidden).

In a photochemical reaction promotion of an electron means that now (I, Scheme 8.73) becomes the HOMO. Suprafacial pathway for [1,3] shift now becomes an allowed process and antarafacial pathway forbidden. Thus, the compound (II, Scheme 8.73) displays a [1,3] hydrogen shift under photochemical conditions.

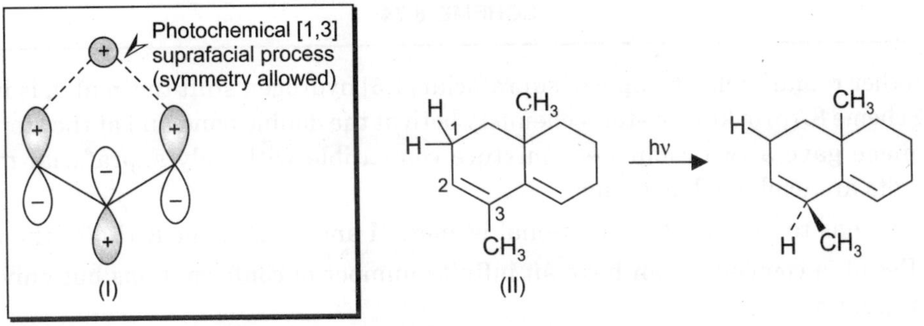

The [1,3] sigmatropic rearrangement is photochemically allowed

SCHEME 8.73

A [1,3] sigmatropic rearrangement involves a π bond and a pair of σ electrons so in all two pairs electrons are involved similarly a [1,5] sigmatorpic rearrangement involves three pairs of electrons. Woodward Hoffmann rules for sigmatropic rearrangements are given in Table. 8.4.

Table 8.4 Woodward-Hoffmann Rules for Sigmatropic Rearrangements

Number of pairs of electrons in the reacting system	Reaction condition	Allowed mode
Even number	Thermal	Antarafacial
	Photochemical	Suprafacial
Odd number	Thermal	Suprafacial
	Photochemical	Antarafacial

Thus, since a [1,3] sigmatropic migration involves two pairs of electrons, an antrafacial rearrangement for a 1,3-shift under thermal conditions does not take place due to geometrical constraints. 1,3-Shifts do take place photochemically [Table 8.4, moreover, since under photochemical conditions HOMO becomes symmetric (see, Scheme 8.73) hydrogen can migrate by suprafacial pathway].

(iii) [1,5] Sigmatropic Hydrogen Shift.

The [1,5] sigmatropic shift of hydrogen or deuterium atoms is well known. These involve three pairs of electrons, thus these occurs via a suprafacial pathway under thermal conditions (see Table 8.4). These shifts can be analyzed by examining a hydrogen atom and a pentadienyl radical whose HOMO (see III, Scheme 8.71) in bonding at both the migration origin and the migration terminus. Thus the migration maintains orbital symmetry when the migrating group remains on the same side of the conjugated system (suprafacial process, Scheme 8.74).

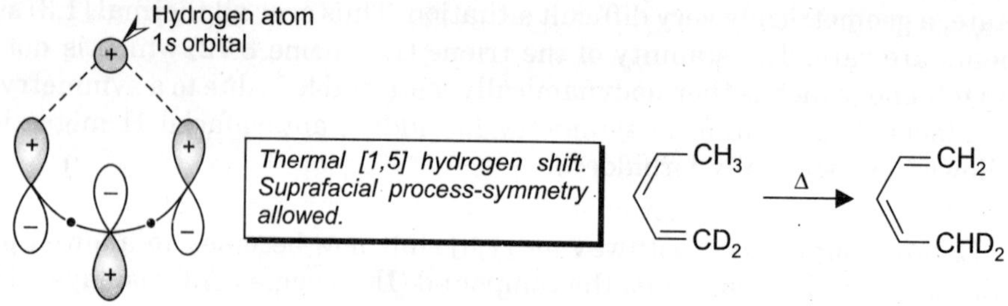

The [1,5] sigmatropic rearrangement is thermally allowed

SCHEME 8.74

Another remarkable example of suprafacial [1,5] hydrogen shift thermally, is in the 1,3-diene (I, Scheme 8.75) of known stereochemisty both at the double bond and at the stereocenter. This 1,3-diene gave a two component mixture compatible with only suprafacial migration. These results are explained as under:

- One has to consider two rotational isomers (I and I*a*, Scheme 8.75) for the reaction.
- Recall, a compound can have an infinite number of conformations but only one configuration.
- In (I) the methyl group is directed toward C4-C5 double bond while in (I*a*) it is now ethyl group that is directed toward C4-C5 bond.

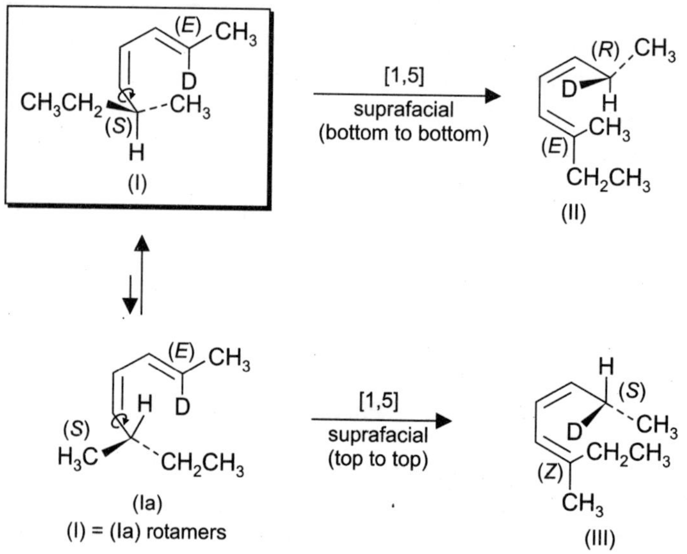

SCHEME 8.75

- There are two suprafacial [1,5] pathways for the hydrogen in these two conformations (I and I*a* Scheme 8.75) "top to top" as in (I) or "bottom to bottom" as in (Ia).
- Each of these suprafacial pathways gives a product with specific stereochemistry and both are formed.

- Considering the two stereogenic units in (I, Scheme 8.75) 4 stereoisomers could be considered. Two (II and III) are formed during suprafacial migration by the symmetry allowed pathway.
- If one considers, the antarafacial pathway the remaining two stereoisomers (as a diastereomeric pair) would have been formed (Scheme 8.76) which however, is not the case.

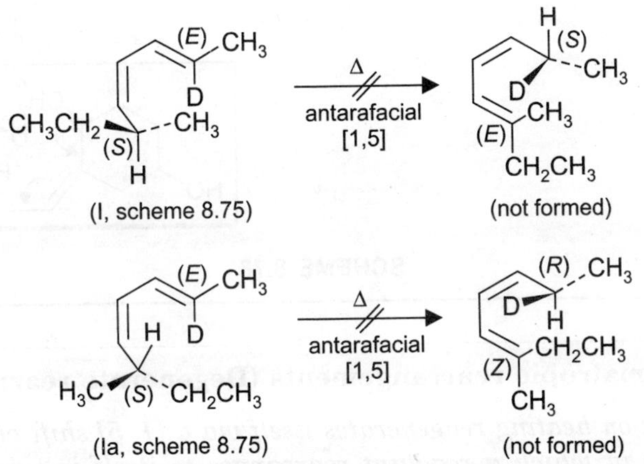

SCHEME 8.76

Heating of indene (Scheme 8.77) causes the scrambling of the label to all the three non-aromatic positions. It is only via [1,5] shift of H or D (by including the *p*-orbitals of the benzene ring) that one can account for the results.

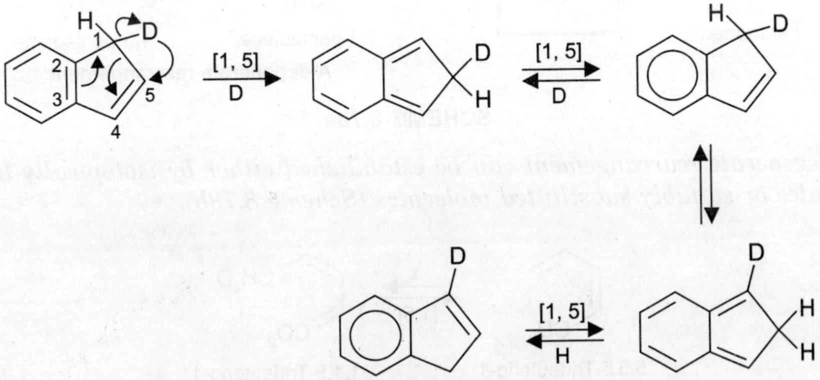

SCHEME 8.77

(iv) [1,7] Sigmatropic Hydrogen Shift

In the case of [1,7] hydrogen shifts, in a triene, the orbital symmetry rules (see III, Scheme 8.71) predict that the transfer of hydrogen must be antarafacial compared to [1,3] shift, the transition state is not much strained and the shift is sterically feasible. This is seen in the thermal interconversion of vitamin D series (Scheme 8.78).

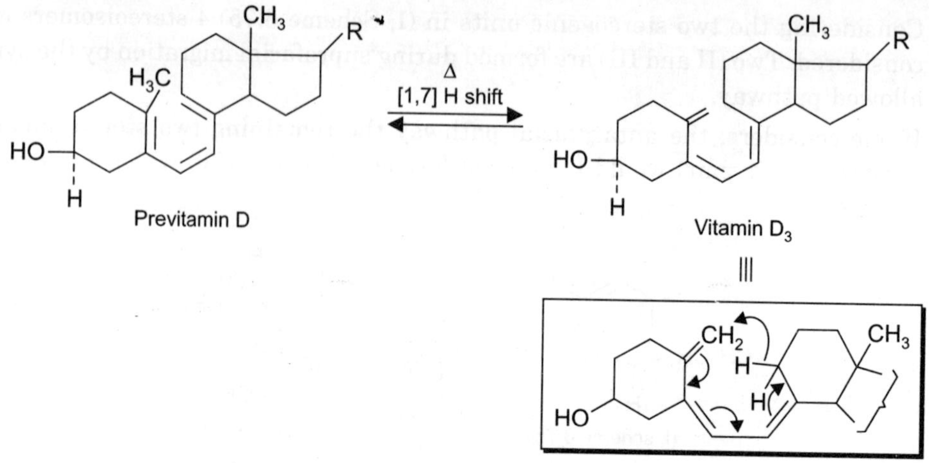

SCHEME 8.78

More on sigmatropic rearrangements (Degenerate rearrangements)

1, 3–Pentadiene on heating reagenerates itself via a [1, 5] shift of hydrogen, this kind of process in which a reactant rearranges to itself is termed degenerate rearrangement (Scheme 8.78a). Significantly as expected the [1, 3] shift is symmetry farbidden.

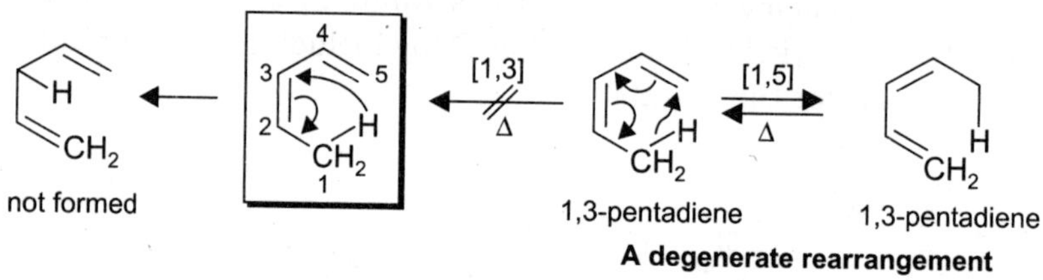

SCHEME 8.78a

The degenerate rearrangement can be estabilished either by isotopically labeled molecules or suitably substituted molecules (Scheme 8.78b).

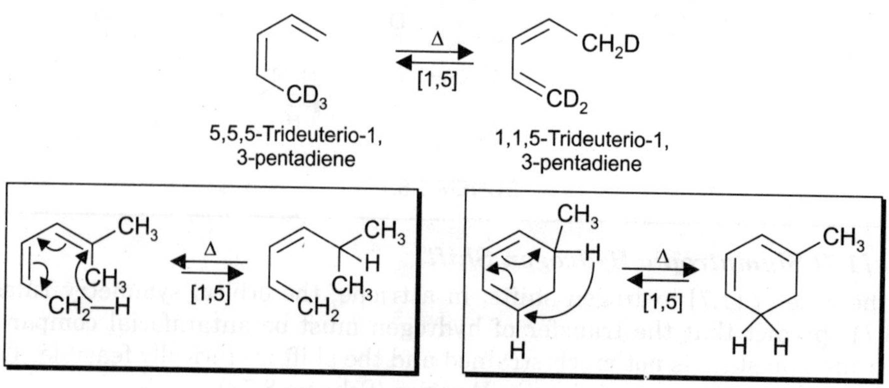

SCHEME 8.78b

Three membered rings can often play the role of a double bond and a [1, 5] H shift can take place just like in 1, 3-pentadiene and involves the opening of the cyclopropane (Scheme 8.78c).

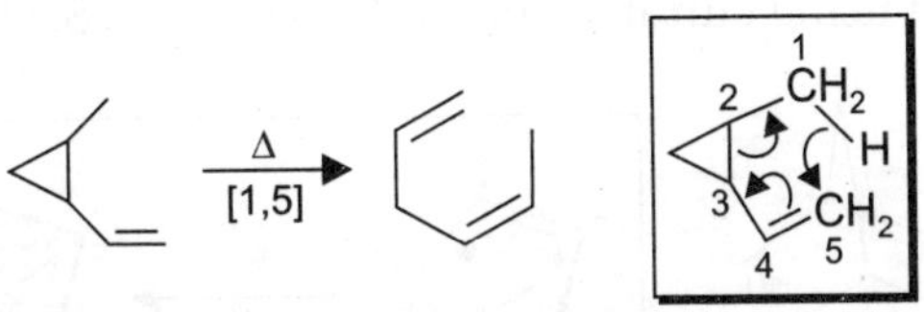

SCHEME 8.78c

B. Sigmatropic Migrations of Carbon

As compared to a hydrogen atom which has its electrons in a 1*s* orbital that has only one lobe, a carbon free radical (free imaginary transition state) has its odd electron in a *p* orbital which has two lobes of opposite sign. Recall that a [1,3] sigmatropic suprafacial migration of hydrogen (thermally) is symmetry forbidden while an antarafacial reaction though allowed is geometrically improbable (see Schemes 8.72–8.74). Interestingly an additional possibility would exist if an alkyl group (carbon) rather than a hydrogen was potential migrator. A [1,3] shift can now be suprafacial migration (I, Scheme 8.79) if the migrating group does so antarafacially *i.e.*, it would result in inversion of configuration of the migrating group. Thus carbon can simultaneously interact with the migration source and the migration terminus using either one of its lobes or both of its lobes (Scheme 8.79). Considering suprafacial rearrangement, carbon will migrate using one of its lobes if the HOMO is symmetric (II, Scheme 8.79). This happens during a thermal suprafacial [1,5] process. When carbon migrates with only one of its lobes interacting with migration source and migration terminus, the migrating group retains its configuration since bonding is always to the same lobe. When the carbon migrates using both of its lobes. (asymmetric HOMO, I Scheme 8.79), a [1,3] thermal suprafacial migration would involve opposite lobes. Thus, if the migrating carbon was originally bonded *via* its positive lobe, it must now use its negative lobe to form the new C—C bond. The stereochemical outcome of such a process is the inversion of configuration in the migrating group.

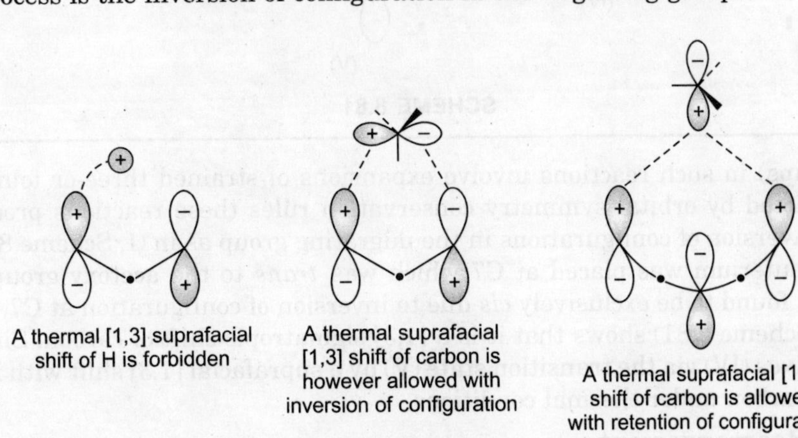

SCHEME 8.79

In summary, a suprafacial [1,5] thermal rearrangement proceeds with retention of configuration at the migrating carbon, while the related [1,3] suprafacial process proceeds with inversion. In the thermal conversion of (I, Scheme 8.80) to (II) a carbon atom migrates across an allyl system to leave C-1 and ending up at C-3 via a [1,3] shift. The inversion of configuration is observed using suitable substrates (Scheme 8.81). The [1,3] shifts of carbon

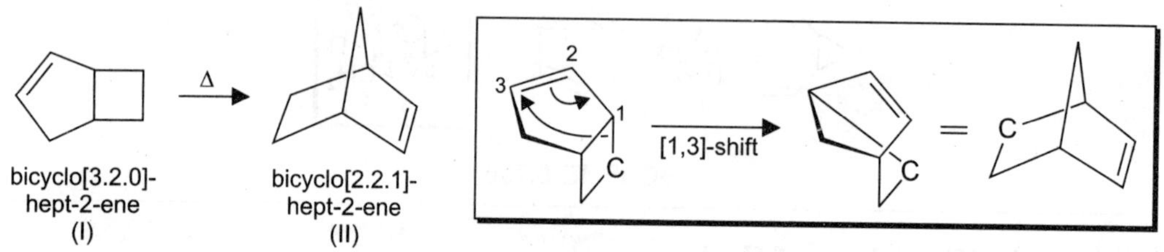

SCHEME 8.80

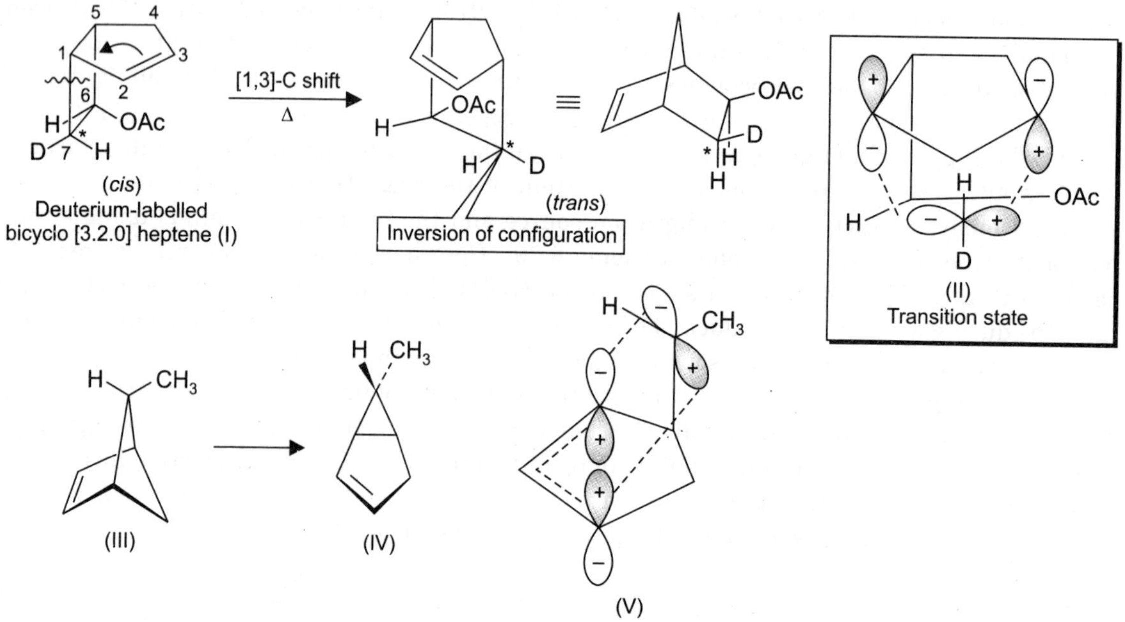

SCHEME 8.81

(*i.e.*, alkyl groups) in such reactions involve expansions of strained three-or four-membered rings. As predicted by orbital symmetry conservation rules these reactions proceed almost entirely with inversion of configurations in the migrating group as in (I, Scheme 8.81). In this case, a label deuterium was placed at C7 which was *trans* to the acetoxy group. After the reaction, it was found to be exclusively *cis* due to inversion of configuration at C7. The transition state (II, Scheme 8.81) shows that it is a [1,3] sigmatropic shift of carbon. Similarly (III, Scheme 8.81) gives (IV) via the transition state (V) by a suprafacial [1,3] shift with inversion at the migrating carbon under thermal conditions.

C. The Cope Rearrangement

A 1,5-diene on heating is rearranged to another 1,5-diene by concerted formation of a 1, 6-bond, breaking of the 3,4-bond and migration of both double bonds in a [3,3] sigmatropic rearrangement known as Cope rearrangement (see, Scheme 8.69). The compound rearranges by a [3,3] sigmatropic pathway and is also hypothetically pictured as split into two allyl radicals (Scheme 8.82). Interaction of the HOMO's of these allyl radicals is bonding at both ends, so the reaction is thermally allowed. The stereochemical outcome of this rearrangement is in keeping with their occurrence generally through the chair-shaped transition states (Scheme 8.83). *Meso* 3,4-dimethyl-1, 5-hexadiene gives *cis*, *trans*-2, 6-octadiene (in the starting compound the two methyl groups are having *cis*-relationship, in the chair form of a cyclohexane only 1, 2-axial, equatorial relationship is *cis*) while a boat shaped transition state would give *cis*, *cis*-product or *trans*, trans-product (Scheme 8.84).

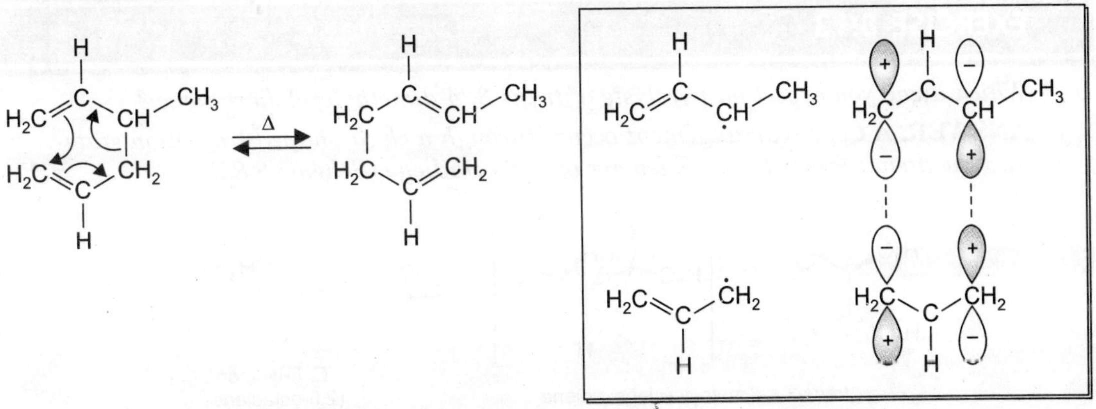

The [3,3] sigmatropic rearrangement is thermally allowed

SCHEME 8.82

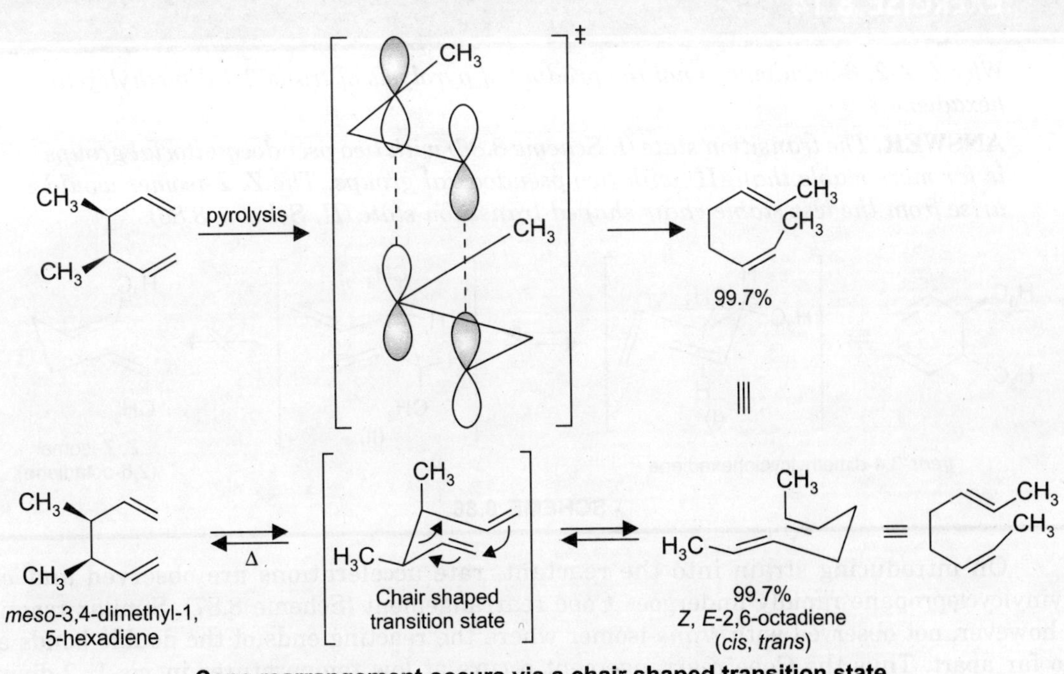

Cope rearrangement occurs via a chair shaped transition state

SCHEME 8.83

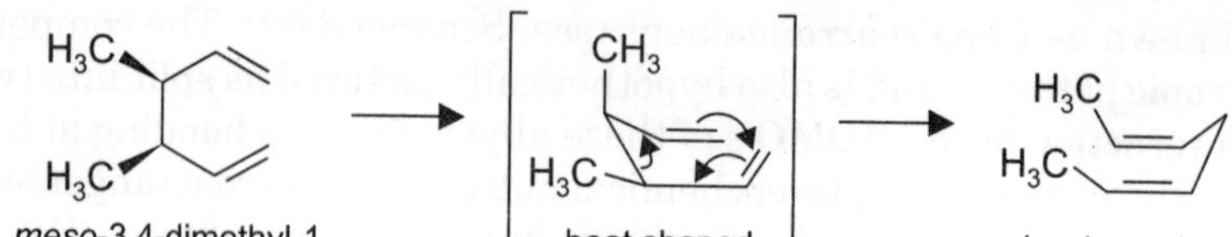

meso-3,4-dimethyl-1, 5-hexadiene — boat shaped transition state — *cis, cis*-product not formed

SCHEME 8.84

EXERCISE 8.13

Which diene you expect on pyrolysis of trans-3, 4-dimethylcyclohexadiene ?

ANSWER. *A Cope rearrangement occurs through a chair shaped transition state and the diene expected is E, E isomer of 2, 6-octadiene (Scheme 8.85).*

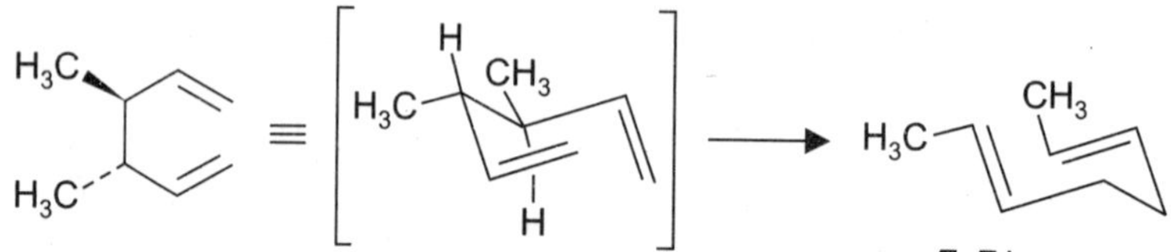

trans-3,4-dimethylcyclohexadiene

E, E-isomer (2,6-octadiene)

SCHEME 8.85

EXERCISE 8.14

Why Z, Z-2, 6-octadiene is not the product of pyrolysis of trans-3,4-dimethylcyclohexadiene ?

ANSWER. *The transition state (I, Scheme 8.86) with two pseudoequatorial groups is far more stable than (II) with two pseudoaxial groups. The Z, Z-isomer would arise from the less stable chair shaped transition state (II, Scheme 8.86).*

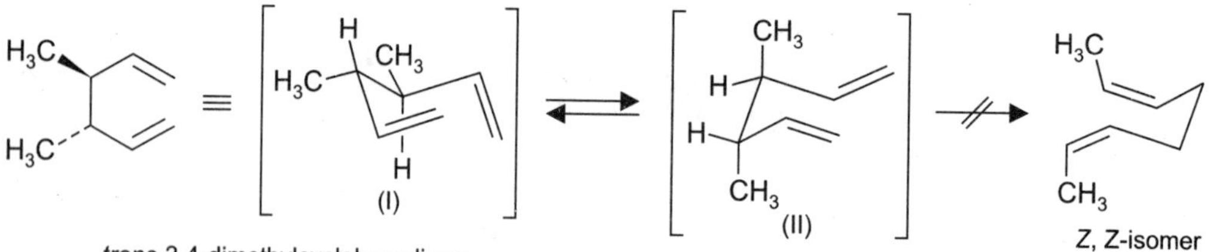

trans-3,4-dimethylcyclohexadiene

Z, Z-isomer (2,6-octadiene)

SCHEME 8.86

On introducing strain into the reactant, rate accelerations are observed and *cis*-divinylcyclopropane rapidly undergoes Cope rearrangement (Scheme 8.87). Similar reaction is however, not observed with *trans*-isomer where the reacting ends of the double bonds are too far apart. Thus the Cope rearrangement occurs at low temperatures in *cis*-1, 2-divinyl cyclopropane compared to Cope rearrangement of 1, 5-hexadiene itself which requires temperatures in the range of 200–300°C.

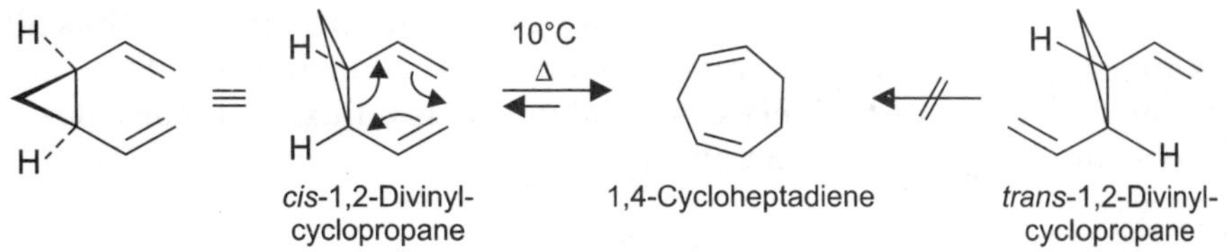

SCHEME 8.87

EXERCISE 8.15

Which two conformations of cis-1, 2-divinylcyclopropane can be considered for a possible Cope rearrangement ?

Which of these conformations is capable of undergoing this rearrangement ?

ANSWER. *See Scheme 8.87a. Two conformations can be adopted (I and II, Scheme 8.87a), (I) is less stable due to steric strain between double bond and a H atom and only this conformation can undergo Cope rearrangement. The conformation (II) will be unreactive since then the product will have two trans double bonds in a seven membered ring (an impossible geometrical situation).*

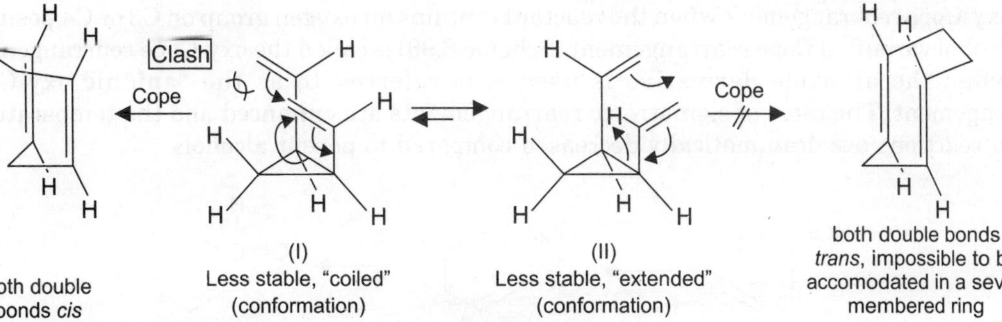

SCHEME 8.87a

D. Fluxional Molecules—A Degenerate Rearrangement

Divinylcyclopropane rearrangements can proceed even with more ease in case the entropy of activation is made still negative by incorporating both vinyl groups into a ring. The system then becomes homotropilidene (Scheme 8.88) which undergoes a degenerate Cope rearrangement

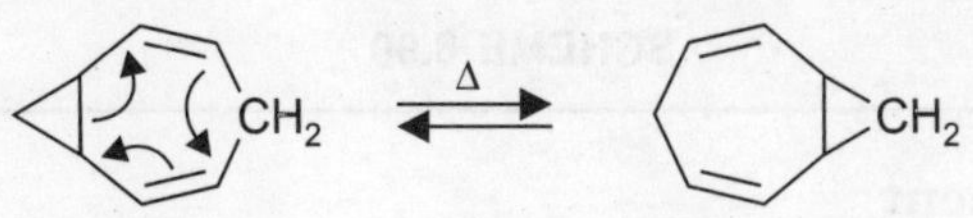

The *degenerate* Cope rearrangement of homotropilidene

SCHEME 8.88

A degenerate rearrangement leads to a product which is indistinguishable from the reactant. By bridging the two methylene groups in homotroplidene one gets a molecule of bullvalene (Scheme 8.89). This is converted into itself at 25°C. At 100°C the ^{1}HNMR spectrum of bullvalene shows a single peak at 4.22 ppm. Bullvalene has a three fold rotational axis; thus all the three double bonds are equivalent. The Cope rearrangement can occur in each of the three faces of the molecule and is degenerate in every case (Bullvalene is a fluxional molecule—a molecule which undergoes rapid degenerate rearrangement).

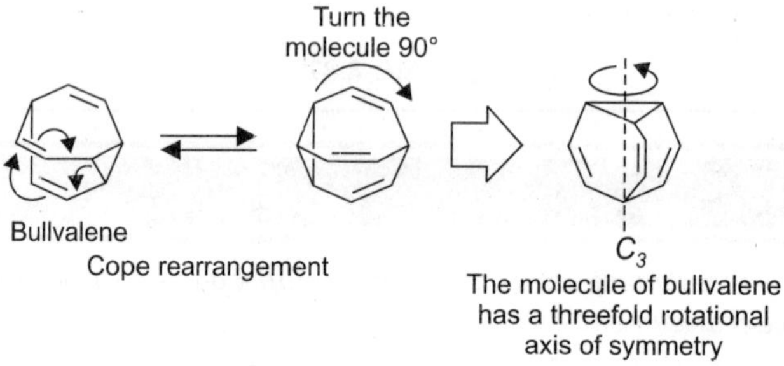

SCHEME 8.89

E. Oxy-Cope Rearrangement

As seen in other pericyclic reactions Cope rearrangement is also reversible and the position of equilibrium depends on the relative stability of the isomers. This problem can be checked and the forward reaction can be made to predominate provided the product reacts further. This is so in oxy-Cope rearrangement when the reactant contains an oxygen group on C3 or C4 position. The alcohol variant of Cope rearrangement (Scheme 8.90) is called the oxy-Cope rearrangement and when the alkoxide derivative is used it is referred to as the "anionic oxy-Cope rearrangement. The rates of sigmatropic rearrangements are enhanced and the temperatures for the reactions are drammatically decreased compared to parent alcohols.

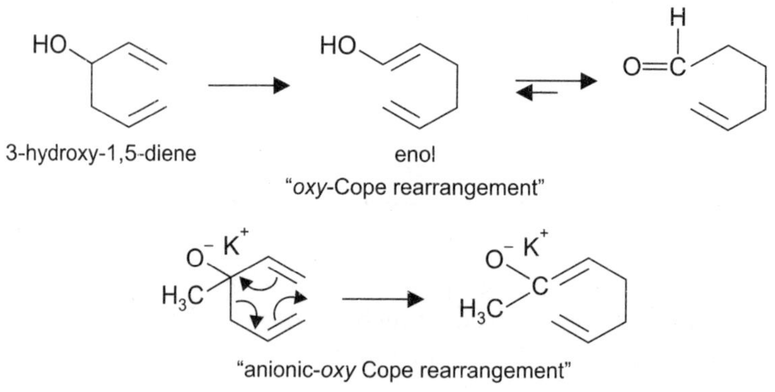

SCHEME 8.90

F. Aza-Cope Rearrangement

It is well known that the presence of oxygen atom adjacent to the π bond accelerates the Cope rearrangement. Similarly a nitrogen usually as an iminium salt fragement in the diene also induces an aza-Cope rearrangement. Thus the reactant (II, Scheme 8.91) derived from (I) underwent a fast aza-Cope rearrangement at low temperature.

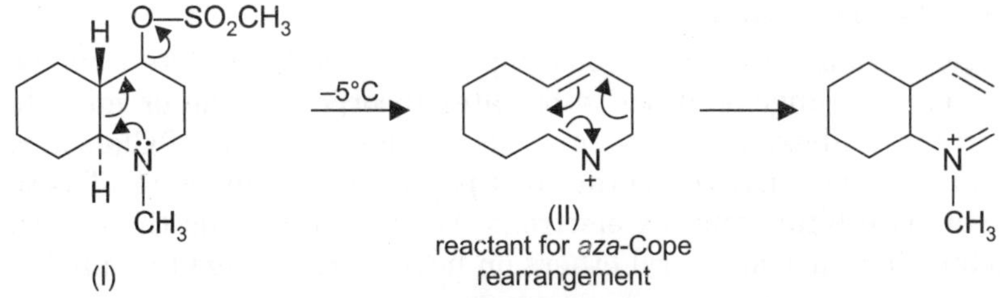

***Aza*-Cope rearrangement**

SCHEME 8.91

EXERCISE 8.16

Write the structure of products from the reactions (Scheme 8.92).

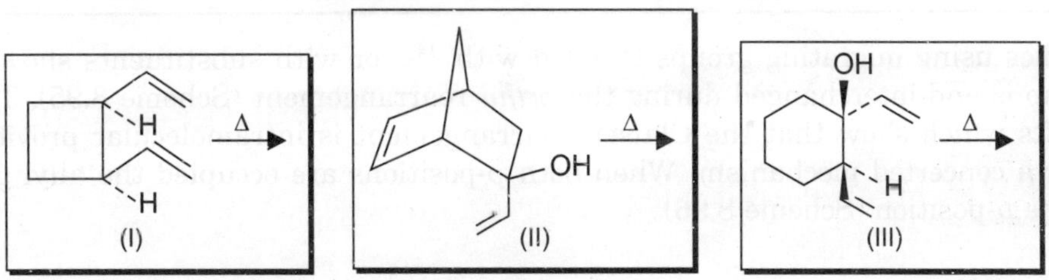

SCHEME 8.92

ANSWER. *Always look for the presence of a 1, 5-diene unit which will hint towards a Cope rearrangement (draw the arrow to form a bond between C1 and C6 and breaking a bond between C3 and C4, Scheme 8.93).*

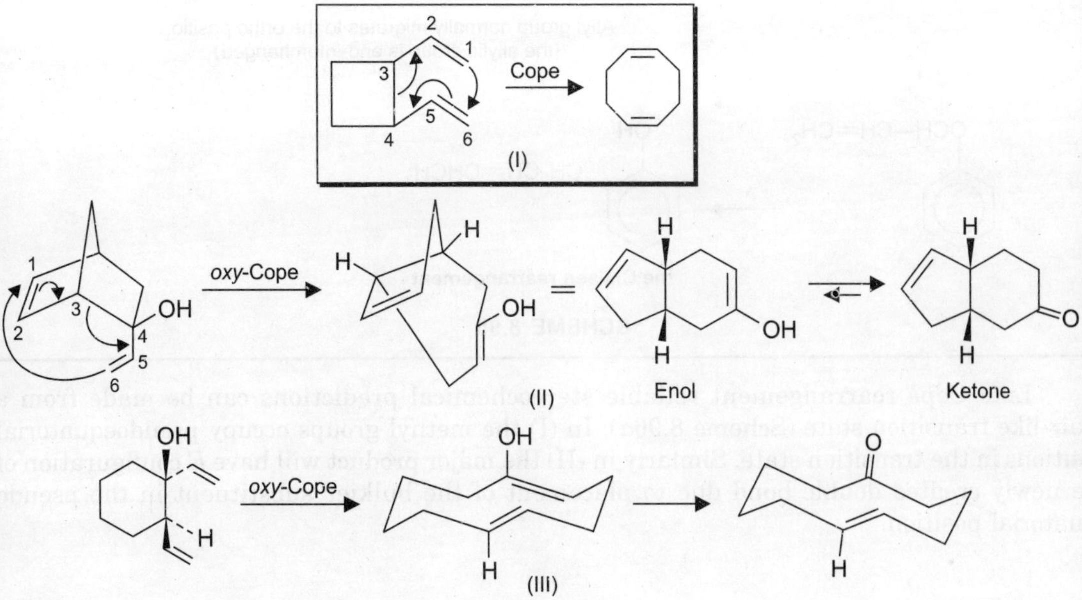

SCHEME 8.93

G. The Claisen Rearrangement

Claisen rearrangement also involves a [3,3] sigmatropic pathway like Cope rearrangement, however, in Claisen rearrangement the substrates incorporate one or more heteroatoms in place of carbon in the 1,5-hexadiene system. The simplest example of Claisen rearrangement is the thermal conversion of allyl vinyl ether to 4-pentenal (Scheme 8.94). The transition state involves a cycle of six orbitals and six electrons. With six electrons the transition state has aromatic character. Similarly allyl aryl ethers on heating rearrange to *o*-allyl phenols.

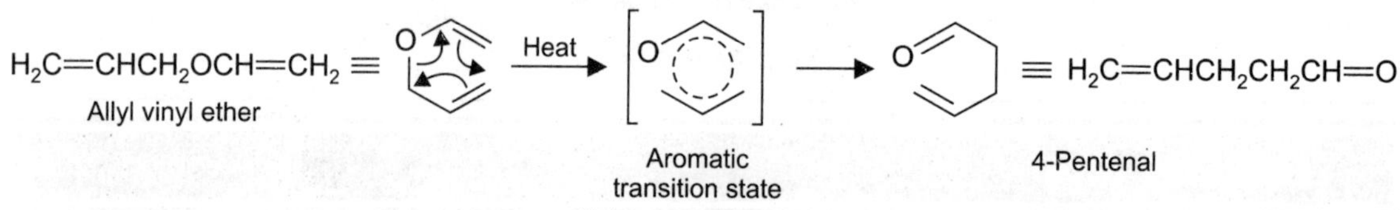

Claisen rearrangement

SCHEME 8.94

Studies using migrating groups labelled with ^{14}C or with substituents show that the allylic group is end-interchanged during the *ortho* rearrangement (Scheme 8.95). These and other results which show that the Claisen rearrangement is intramolecular provide strong support for a concerted mechanism. When both *o*-positions are occupied the allyl group migrates to the *p*-position (Scheme 8.96).

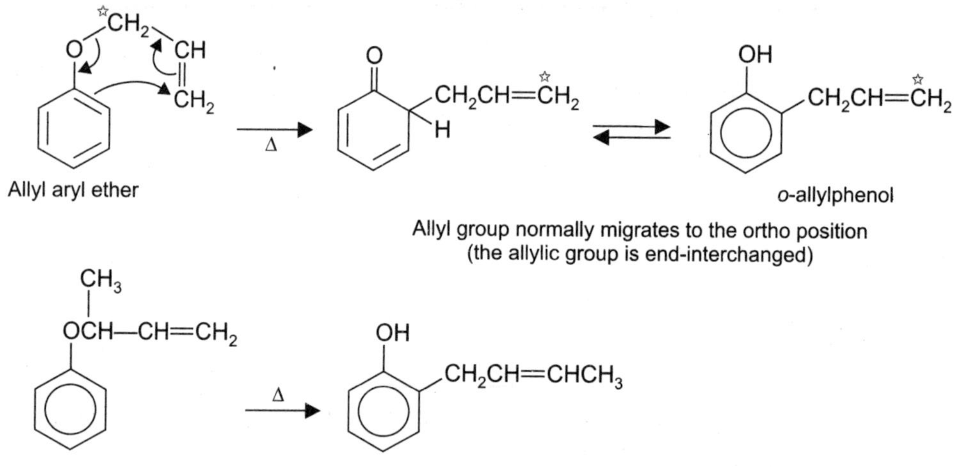

the Claisen rearrangement

SCHEME 8.95

Like Cope rearrangement reliable stereochemical predictions can be made from a chair-like transition state (Scheme 8.96*a*). In (I) the methyl groups occupy pseudoequatorial positions in the transition state. Simlarly in (II) the major product will have *E* configuration of the newly created double bond due to placement of the bulkier substituent in the pseudo equatorial position.

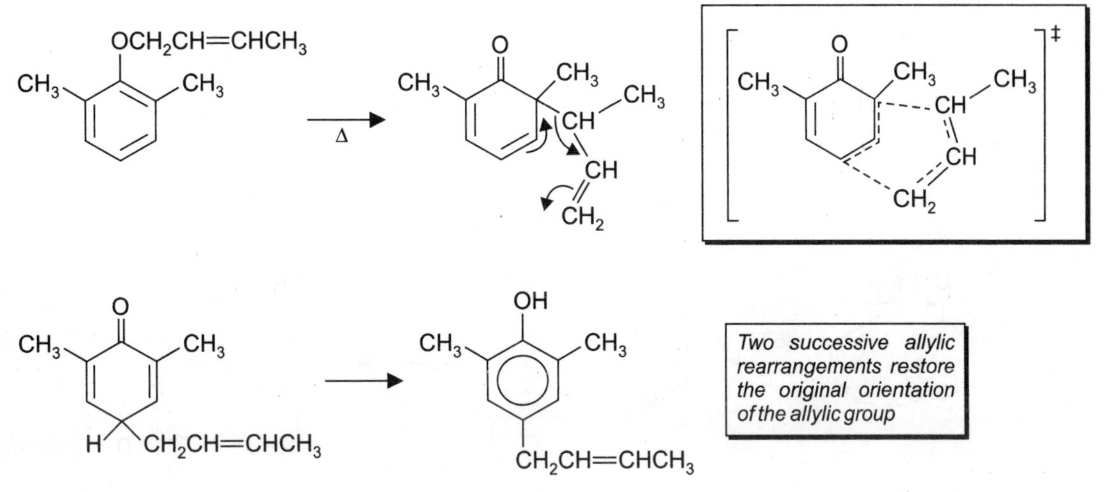

SCHEME 8.96

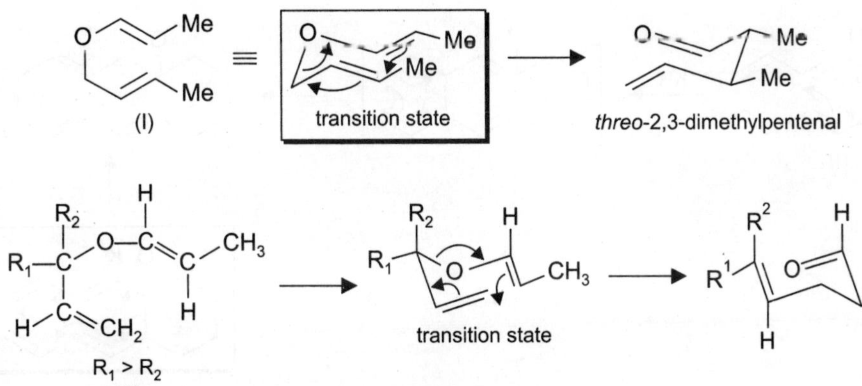

Claisen rearrangement

SCHEME 8.96a

H. [5, 5] Sigmatropic Shifts

Thermal [5, 5] shifts are facile, however the compounds undergoing such rearrangement are not common. One type of substrates are pentadienyl ethers which give 4-pentadienylphenols as the major products along with minor products arising from *ortho*-Claisen rearrangement (Scheme 8.96*b*). With the help of deuterium labeling it has been shown that major products arise from direct [5, 5] sigmatropic shifts and not by two consecutive [3, 3] shifts.

It is proved that [5, 5] shifts occur very fast in negatively charged compounds. An interesting reaction is oxy-Cope rearrangement of (I, Scheme 8.96*c*) the arrangement does not proceed by a sequence of consecutive [3, 3] shifts, however it is indeed a result of [5, 5] shifts, (Scheme 8.96*c*).

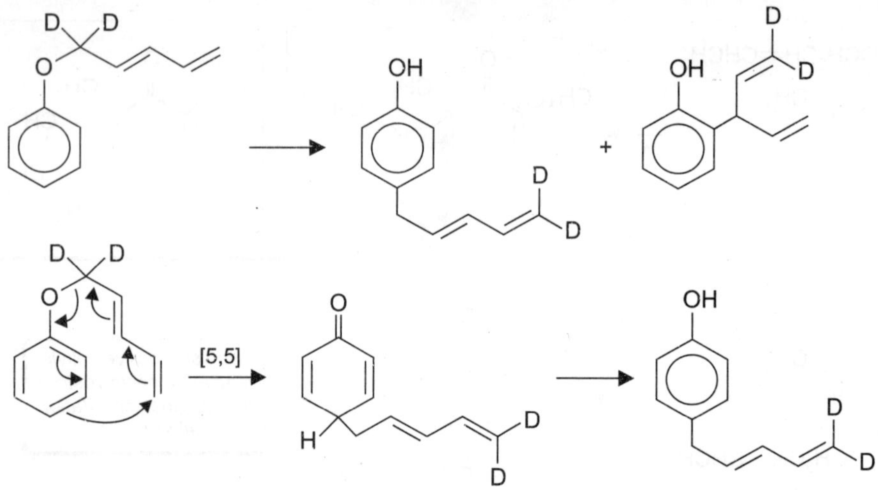

[5,5] Sigmatropic shifts

SCHEME 8.96b

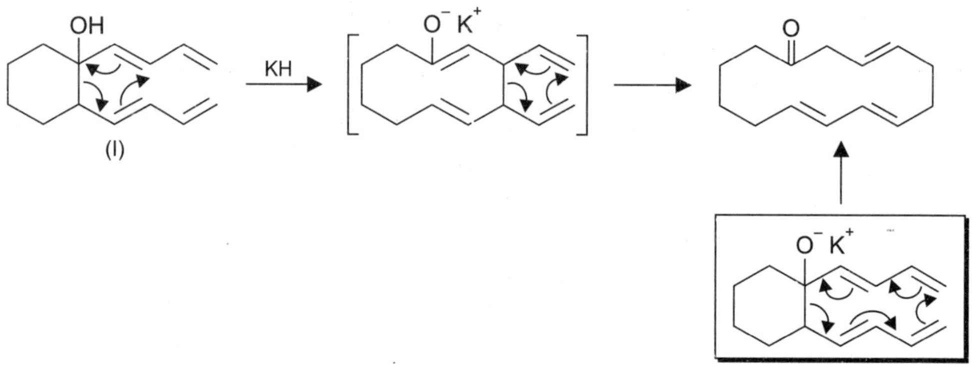

SCHEME 8.96c

8.8 THE ENE REACTION

In this reaction an alkene having an allylic hydrogen atom reacts thermally with a dienophile (C=C, C=O, N=N etc., called enophile) with the formation of a new σ bond to the terminal carbon of the allyl group. This is followed by the 1,5-migration of the allylic hydrogen and subsequent change in the position of allylic double bond. The reaction thus resembles both cycloaddition and a [1,5]-sigmatropic shift of hydrogen.

Mechanistically, the reaction is a concerted process, there being little charge development in the transition state. It shows a primary kinetic isotope effect to show C—H bond breaks in the rate determining step (the reverse process occurs in the pyrolysis of esters). The interaction of a hydrogen atom with the HOMO of the allyl radical and the LUMO of the enophile (Scheme 8.97) is a symmetry allowed process. A good example of ene reaction is found during allylic oxidation of alkenes with selenium dioxide.

The reaction shows a primary kinetic isotope effect of C—H bond breaking in the rate determining step. The ene reaction of β-pinene with maleic anhydride (Scheme 8.97) gives the product without skeletal rearrangement of the strained four membered ring in the β-pinene, to show the concerted nature of the reaction (rather than *e.g.*, the formation of a cationic intermediate). The ene reaction, however requires higher temperature than in Diels-Alder reaction, but occurs faster with conjugated enones with Lewis acid catalysts. Coordination of the Lewis acids with the enophile lowers the energy of LUMO.

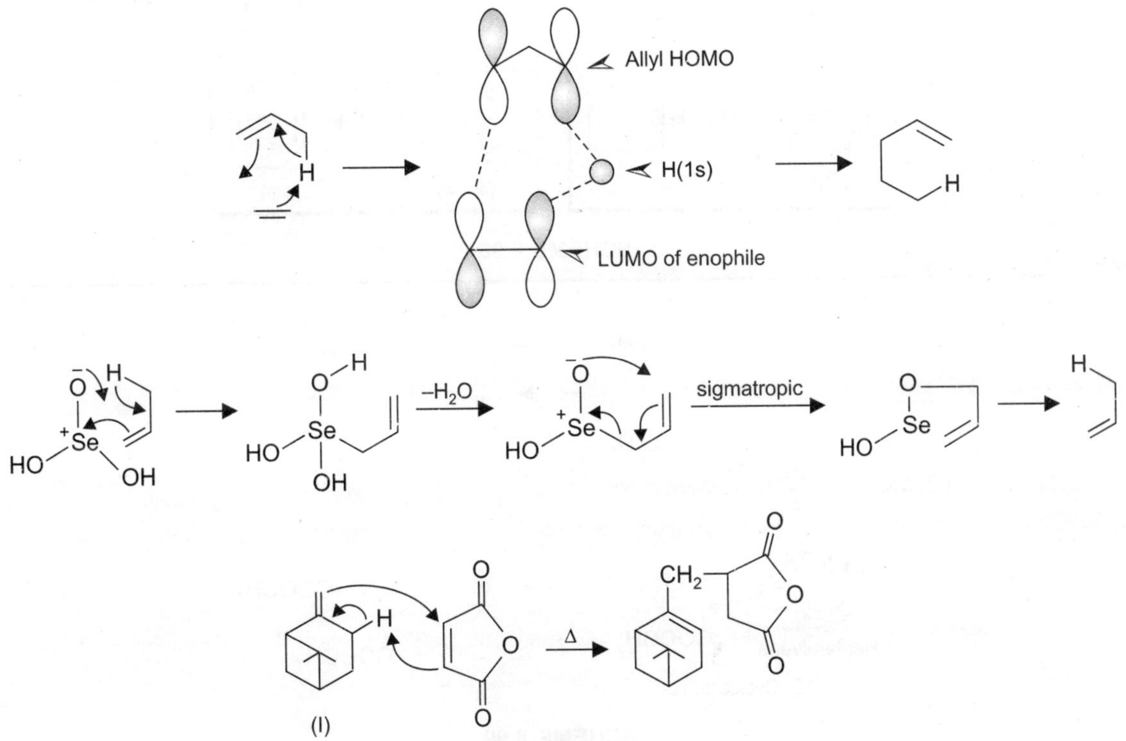

Ene reaction

SCHEME 8.97

8.9 AROMATIC TRANSITION STATES

The reactions with aromatic transition states (2, 6, 10, 14 ... 4*n* + 2) delocalized electrons are permitted, while the *anti*-aromatic systems (4, 8, 12 ... 4*n*) delocalized electrons are the forbidden ones. It is clear that the Diels-Alder reaction has a 6π aromatic transition state isoelectronic with benzene while the forbidden cyclobutane formation has the unfavourable 4π transition state, isoelectronic with cyclobutadiene. This is the reason that although [4+2] reactions are common and general, the analogous concerted [2+2] and [4+4] thermal cycloaddition reactions generally do not occur, since the corresponding transition states involve 4 and 8 electrons respectively (Scheme 8.98 also see Table 8.3). However, the other hand, cycloaddition reactions which involve 6 or 10 electrons occur readily (Scheme 8.99). It may be mentioned that *e.g.*, a [4+2] cycloaddition involves four π electrons of one system and two on another.

Most pericyclic reactions involve six electrons, but the Huckel rule is also obeyed with two, ten etc. electrons. An important pericyclic reaction with two electrons is the rearrangement of carbocations. The pericylic nature of such a transition state is shown (Scheme 8.100). The cation with a total of two electrons has a filled shell and relative stability. The corresponding transition state for a carbanion involves four electrons and an unfilled shell. Accordingly, carbocation rearrangements are common, while carbanion rearrangements are not. Moreover, Wagner-Meerwien and related rearrangement order [1,2], occur in carbocation because of the allowed *s, s* pathway, but not in carbanions which would requires *s,a*. The migrating group retains its chirality. By contrast, the [1,3] shift (see, Scheme 8.79) is a accompanied with inversion of configuration.

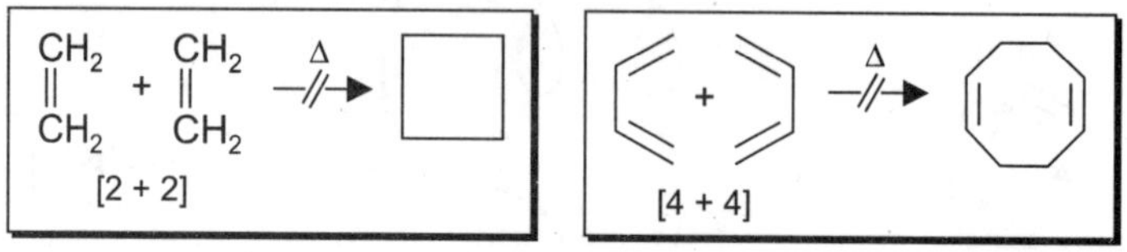

SCHEME 8.98

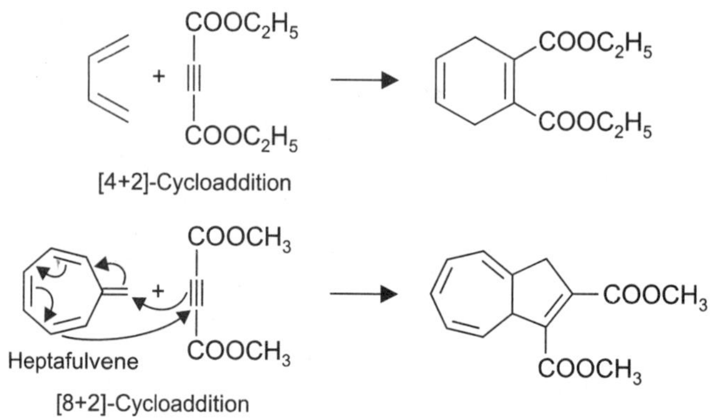

SCHEME 8.99

8.10 MÖBIUS-HÜCKEL ANALYSIS (PMO) APPROACH

The concerted reactions are analyzed by the classification of transition states as aromatic or antiaromatic. These predictions yield the same results as by other methods. In the previous section a mention was made of aromatic transition states. Hückel's rule of aromaticity states that a monocyclic planar conjugated system with $(4n + 2)\pi$ electrons is *aromatic* and therefore, stable in the ground state. On the other hand a system with $(4n)$ π electrons is unstable and is called antiaromatic. Further it has been shown that these rules are reversed by the presence of a node (a phase dislocation) in the array of atomic orbitals. The Möbius-Hückel concept is used to analyze the pericylic reactions without using the actual molecular orbitals. The following points may be noted:

- Each atom of the interacting system is assigned a *p* orbital with one lobe black and one while (or some other designation). A hydrogen atom is represented by a circle of one color representing an *s* orbital.
- One draws each reactant with the black lobes on one side and the white on the other. Then one considers the transition state of a particular reaction, counts the number of

electrons and the nodes in the array to reach the conclusion if that reaction is symmetry allowed or forbidden.

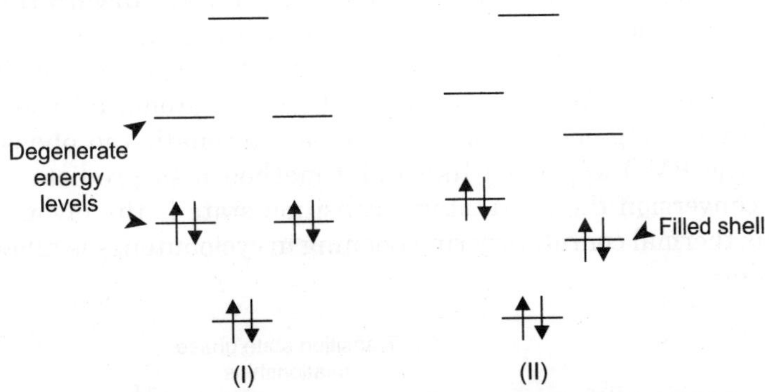

MO energy levels for (I) benzene and (II) a six-electron transition state

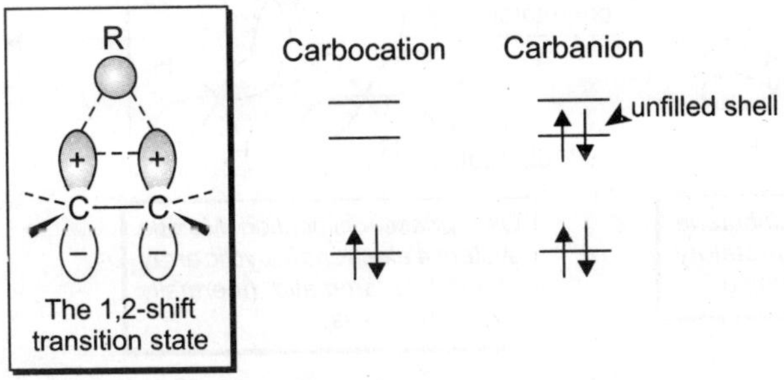

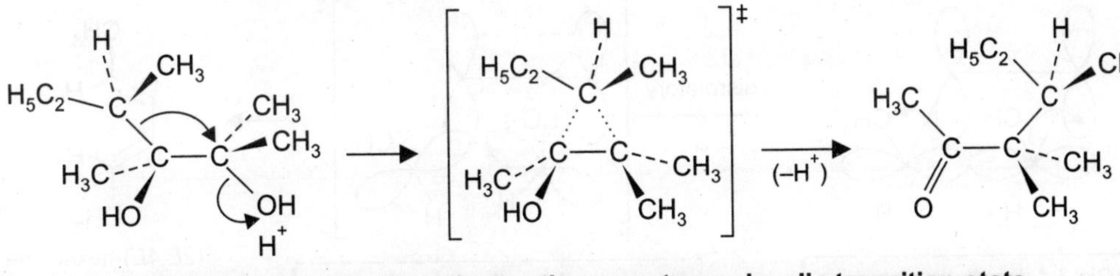

1,2-Rearrangement involves a three-center pericyclic transition state

SCHEME 8.100

- A Hückel system has zero (or any even number) of nodes (phase changes) around the orbital array. A Hückel system with $4n + 2$ electrons is aromatic and with $4n$ electrons is antiaromatic. An array with an odd number of phase dislocations is called an *anti* Hückel system (Mobius system). An *anti*-Hückel system with $4n$ electrons is aromatic and with $4n + 2$ electrons is antiaromatic.
- The condition for aromaticity in *anti*-Hückel system is opposite to that for Hückel system.

1. Electrocyclic Reactions

(a) Thermal Ring Opening of Cyclobutenes—4n Systems

On the basis of FMO approach under thermal conditions the observed stereochemistry of the products indicates a conrotatory motion. Consider the thermal ring opening in *cis* 3,4-dimethylcyclobutene which on the basis of FMO method gives $2E, 4Z$-hexadiene (Schemes 8.6 and 8.10).

The Hückel-Möbius approach also predicts conrotatory motion under thermal conditions and predicts the formation of same diene. Consider the basis set for the starting butadiene (Scheme 8.101), the tilt at C-1 and C-4 as the butadiene system rotates toward the transition state is different for conrotatory and disrotatory modes. The transition state for conrotatory ring opening has one sign inversion (phase dislocation) and with four electrons it is aromatic. The conrotatory transition state for cyclobutene ring opening is therefore, aromatic. The transition state for disrotatory cyclobutene ring opening however, is anti-aromatic (no phase dislocation with 4 electrons). Thus the PMO approach like FMO method also predicts that for butadiene-cyclobutene interconversion the conrotatory transition state is the favoured aromatic transition state and thus thermal conrotatory ring opening in cyclobutenes is allowed and disrotatory opening is forbidden.

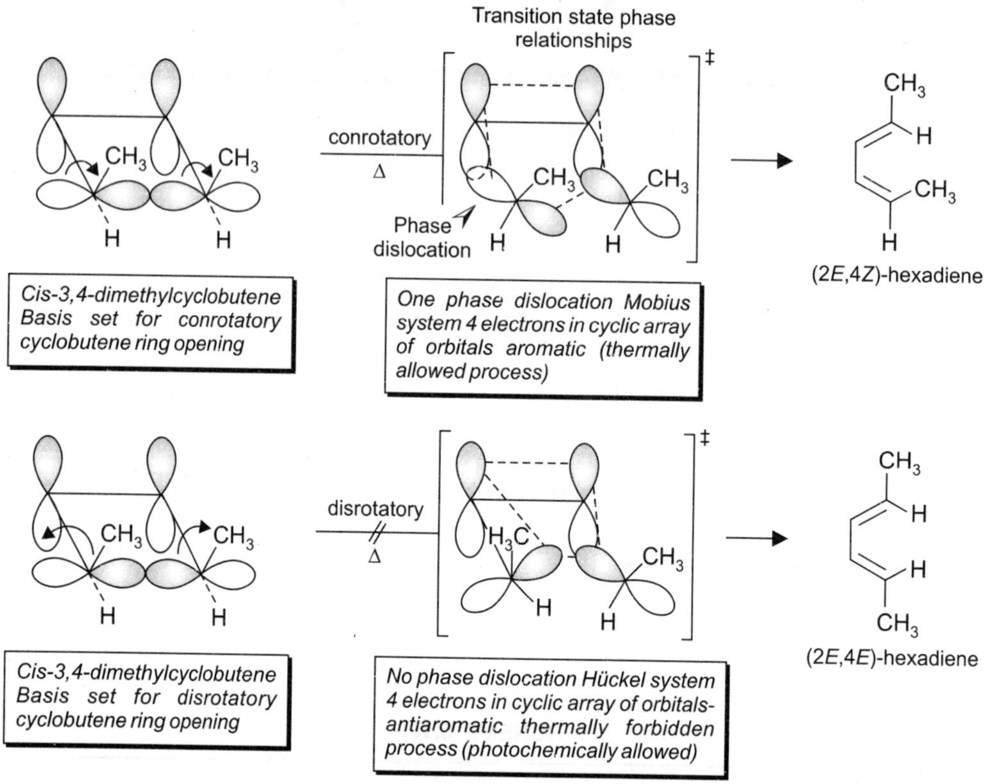

SCHEME 8.101

(b) Thermal Ring Closure of an Octatetraene—4n Systems

In this case also a conrotatory ring closure (Scheme 8.102) is allowed since in the transition state basis set there is one phase dislocation at the forming σ bond. The transition state is therefore, aromatic (8 electrons Möbius system). Disrotatory ring closure for the reaction (Scheme 8.102) will be however, antiaromatic eight electron Hückel system.

As with FMO approach the general statement is that thermal electrocyclic reactions in the conrotatory mode are allowed for $4n$ electron transition states.

The stereochemistry and other aspects are just the same as already discussed during FMO approach. Moreover, since Woodward-Hoffmann rules for photochemical reactions are always the reverse for thermal reactions, one can always predict the outcome of a reaction under photochemical conditions (Schemes 8.101 and 8.102).

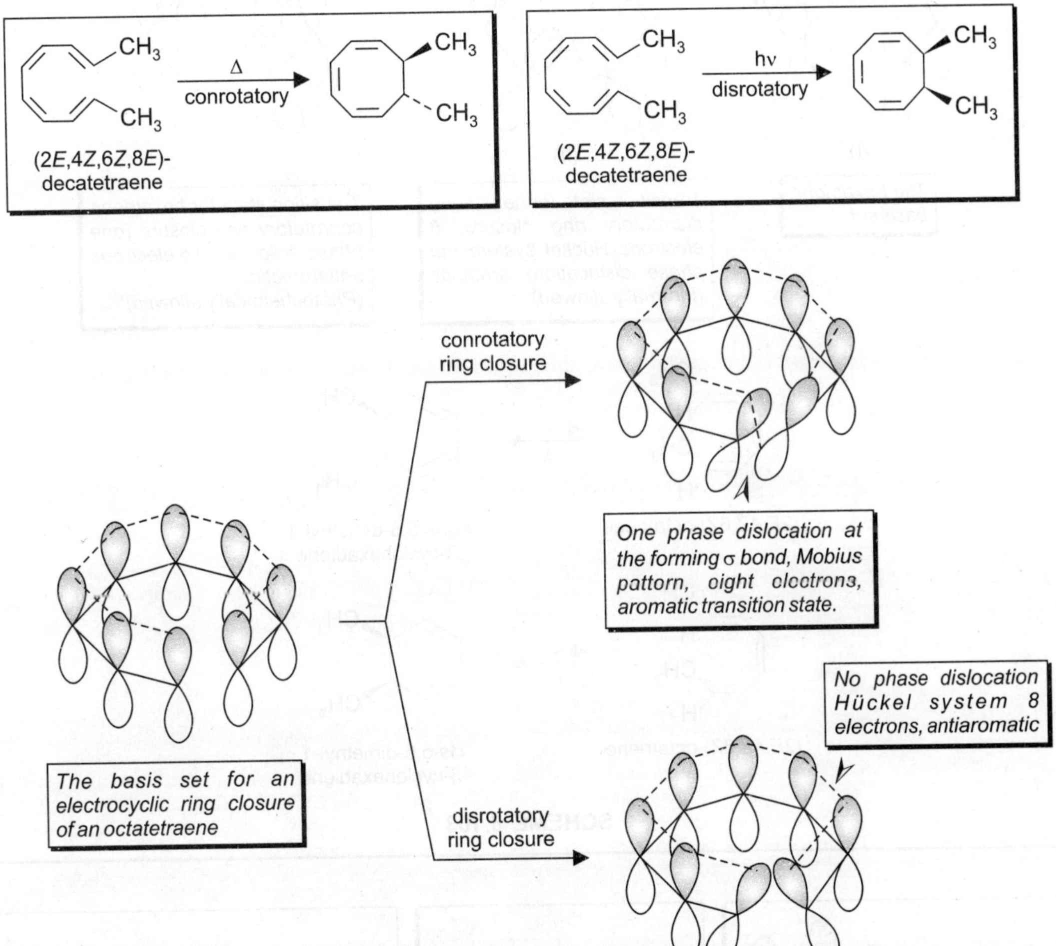

SCHEME 8.102

(c) Thermal Hexatriene Ring Closure—4n + 2 Systems

In FMO approach for $4n + 2$ electron transition states, in electrocyclic reactions, the disrotatory mode is allowed for thermal reactions and the conrotatory mode for photochemical reactions. In PMO method as well, this is found to be true. Starting from hexatriene basis set (I, Scheme 8.103), the transition state for disrotatory ring closure is a six-electron Hückel system and thus aromatic (II). The conrotatory ring closure proceeds through an antiaromatic anti-Hückel transition state. It is, therefore, correctly predicted that thermal ring closure of substituted hexatrienes should be disrotatory and the photochemical reaction should proceed via the opposite conrotatory path.

2. Diels-Alder Reaction

The selection rules may also be derived by the consideration of the aromaticity of the transition state of a Diels-Alder reaction. Same conclusions (as in the case of FMO approach) are again reached and these are summarized in Scheme 8.104).

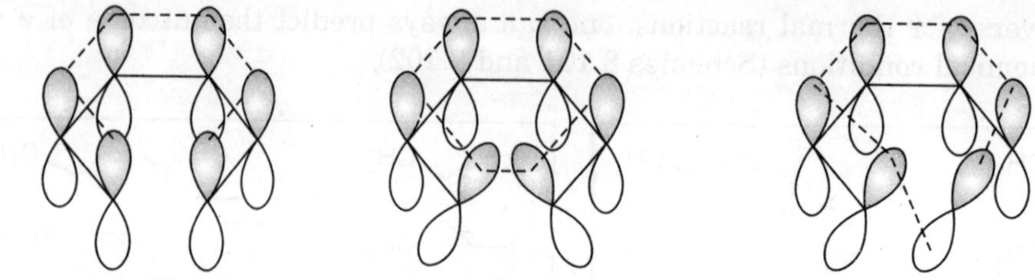

(I) The hexatriene basis set

(II) Transition state for hexatriene disrotatory ring closure, 6 electrons Hückel system (no phase dislocation) aromatic (thermally allowed)

(III) Transition state for hexatriene conrotatory ring closure (one phase dislocation) 6 electrons antiaromatic. (Photochemically allowed)

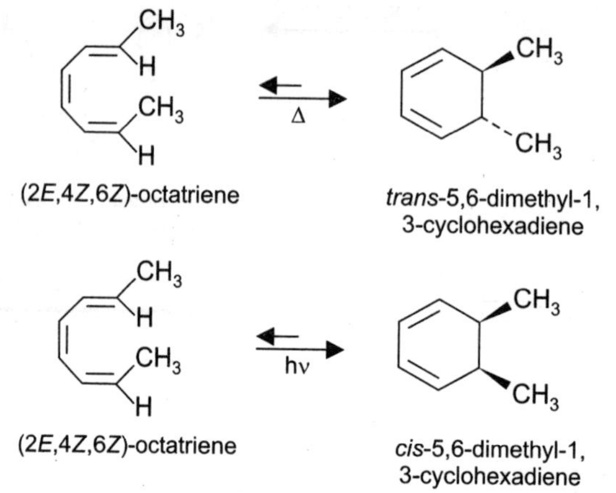

SCHEME 8.103

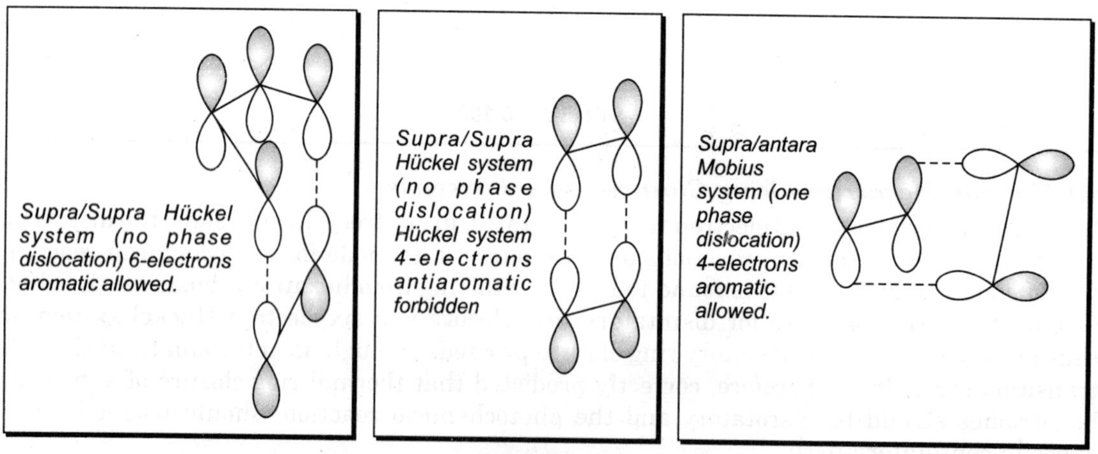

SCHEME 8.104

Antarafacial-Suprafacial cycloaddition is highly sterically hindered and is, therefore, less common. In case two ethylene molecules are brought together in such a way that a Mobius activated complex can be realized (see Scheme 8.104) the process become a suprafacial/ antarafacial addition. This process should be allowed since it is a Mobius system with four electron and a node. For this process to be realized the ethylene molecules have to approach each other in a perpendicular geometry. The completion of this addition involves distortion of the carbon framework. The process, therefore, is difficult although allowed. For this reason simple alkenes do not display this addition. The highly strained triene (Scheme 8.105) however, spontaneously dimerizes thermally and represents [π^2a + π^2s] transition state. Reaction of heptafulvalene with tetracynoethylene is a remarkable example of a [$\pi^{14}a$ + π^2s] thermal cycloaddition leading to a product of *anti* addition. The transition state involves a negative overlap which corresponds to a Mobius cyclic electronic system, a favourable transition state for a 16-electron (4*n*) cyclic system (Scheme 8.105).

As already discussed many known cycloadditions [*p* + *q*] involve pericyclic electrons equal to 6, 10, 14 etc., and involve Hückel aromatic transition states. The [14 + 2] cycloaddition, however, does not fit the Hückel rule.

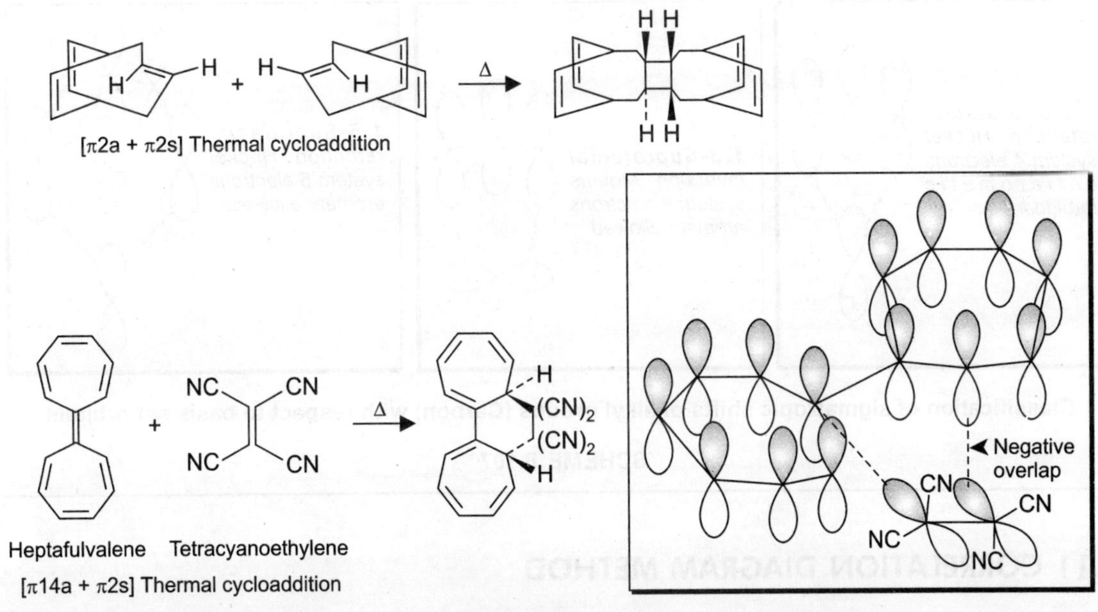

SCHEME 8.105

3. Sigmatropic Shifts

Consider the simplest case of 1,3-sigmatropic shift of a hydrogen. In the FMO approach the hydrogen 1*s* orbital interacts with an allyl radicals HOMO. A thermal [1,3] suprafacial shift is symmetry forbidden, the antarafacial is symmetry allowed, but energetically very unfavourable (see, Scheme 8.72). A consideration of basis set atomic orbitals and their classification as aromatic or antiaromatic reaches the same conclusions (Scheme 8.106). The 1,3-suprafacial shift of hydrogen is forbidden, but the suprafacial 1,5-shift is allowed. The 1,7-shifts should be antarafacial, when an alkyl group (carbon) migrates, an additional stereochemical feature has to be considered. Again in agreement with FMO approach, the allowed processes include, the suprafacial 1,3-shift with inversion and the suprafacial 1,5-shift with retention (Scheme 8.107).

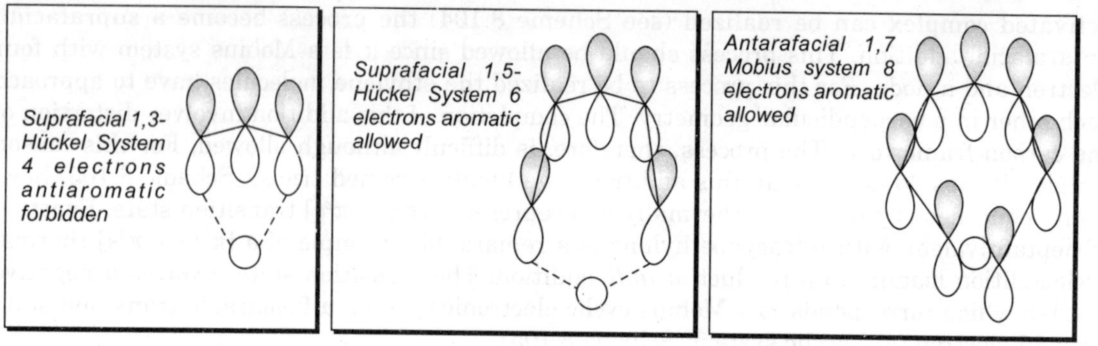

Classification of sigmatropic shifts of hydrogen atom with respect to basis set orbitals

SCHEME 8.106

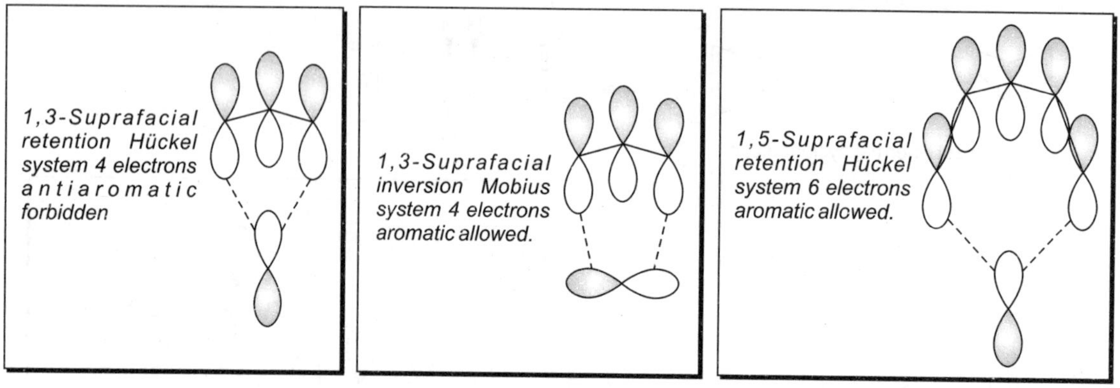

Classification of sigmatropic shifts of alkyl groups (Carbon) with respect to basis set orbitals

SCHEME 8.107

8.11 CORRELATION DIAGRAM METHOD

In this method a correlation of the geometrical symmetry of the orbitals between reactants and the products is involved. Based on this, a correlation diagram is developed which compares the symmetry characteristics from this comparison. A reaction can be easily predicted to be symmetry allowed or symmetry forbidden. The following points may be noted:

- Molecular orbitals (of the reactant and the product) are either symmetric or antisymmetric around the mirror plane or around a two fold axis of rotation (C_2). Thus π orbital of ethylene in the ground state is symmetric (S) with respect to mirror plane (I, Scheme 8.108) while it is antisymmetric (A) with respect to rotational axis C_2 (II, Scheme 8.108). A two fold axis of rotation may be regarded as a pin at right angles passing in the middle. If a molecular orbital is spun 180° (*i.e.*, C_2 axis 360°/2), around this axis it would either yield an identical orbital *i.e.*, symmetric or an orbital with all signs the opposite of what these were originally *i.e.*, antisymmetric (A). Thus when the π orbital of ethylene is rotated it is found that the symmetry around the axis of rotation is not maintained. One would see that after the rotation the shaded lobes would be instead on the bottom rather than at the top (as was so in the original). Thus π orbital of ethylene is antisymmetric (A) in relation to its axis of rotation (II Scheme 8.108).

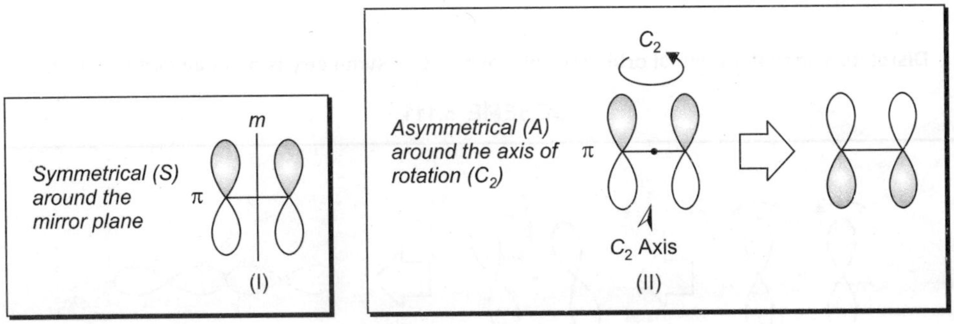

SCHEME 8.108

- On the other hand the same operations show that the antibonding π^* orbital of ethylene is antisymmetric around the mirror plane but symmetric around the C_2 axis of rotation (Scheme 8.109).

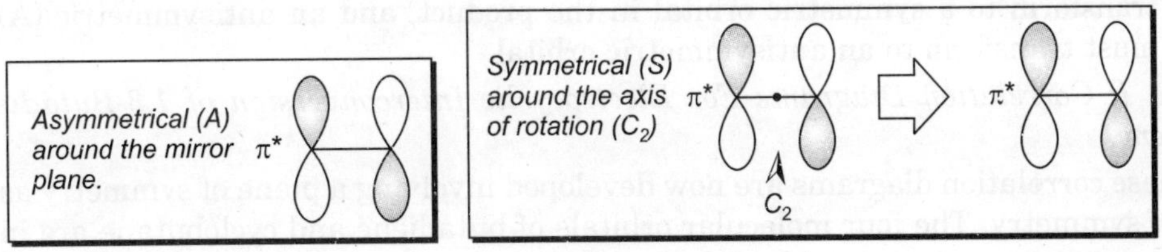

SCHEME 8.109

- Similarly the orbitals of the reactant and the product can be labelled. Thus the sigma orbital of a C-C covalent bond has a mirror plane of symmetry as well as C_2 symmetry, a σ^* orbital is antisymmetric both with respect to plane of symmetry as well as rotational axis of symmetry (Scheme 8.110).

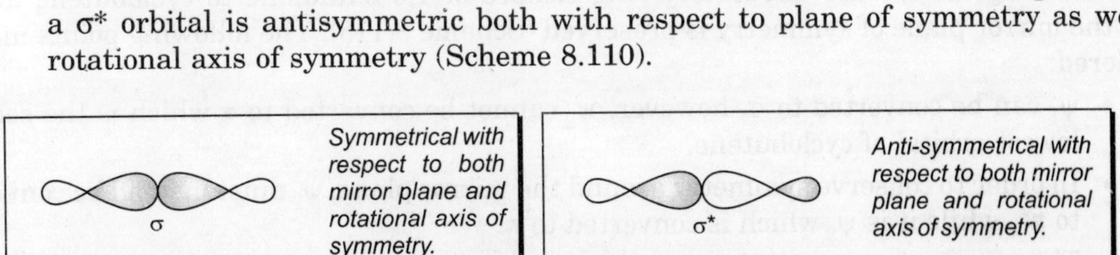

SCHEME 8.110

- During a disrotatory electrocyclic conversion, a plane of symmetry is maintained throughout (Scheme 8.111). If the reaction proceeds by a conrotatory motion a two fold axis of Symmetry (C_2) is preserved throughout (Scheme 8.112).

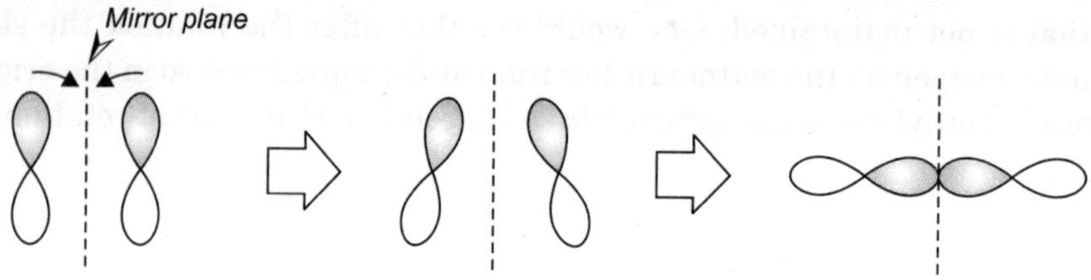

Disrotatory motion (terminal orbitals) a mirror plane of symmetry is maintained throughout

SCHEME 8.111

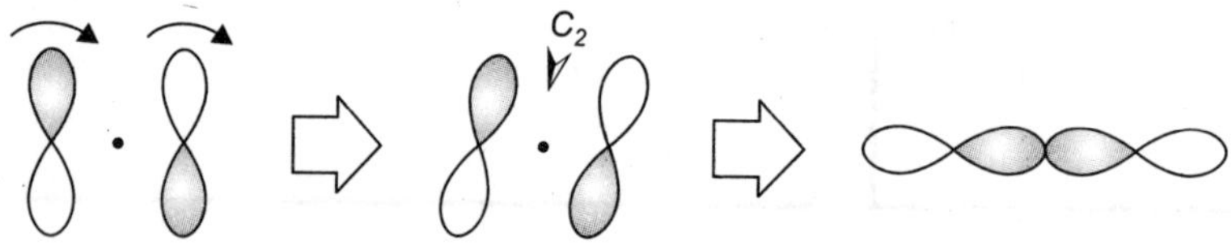

Conrotatory motion (terminal orbitals) a C_2 axis of symmetry is preserved throughout

SCHEME 8.112

- The most stable transition state is the one that conserves the symmetry of the reactant orbitals in passing to product orbitals—a symmetric (S) orbital in the reactant must transform to a symmetric orbital in the product, and an antisymmetric (A) orbital must transform to an antisymmetric orbital.

Example 1: *Correlation Diagrams For Electrocyclic Interconversion of 1,3-Butadiene and Cyclobutene.*

These correlation diagrams are now developed involving a plane of symmetry as well as an axis of symmetry. The four molecular orbitals of butadiene and cyclobutane are inspected for the two symmetry elements. [*In Scheme 8.113, the symmetric properties (plane of symmetry) of molecular orbitals of butadiene and cyclobutene along with the correlation diagram are depicted together. However, these symmetry properties can be translated onto a correlation diagram (as shown on the bottom of Scheme 8.113) for its study.*] Firstly one considers the correlation diagram for the disrotatory ring closure of 1,3-butadiene to cyclobutene during which the mirror plane of symmetry is preserved (Scheme 8.113). The following points may be considered:

- ψ_1 can be converted to σ, however, ψ_2 cannot be converted to π which is the second-lowest orbital of cyclobutene.
- In order to conserve symmetry around the mirror plane, ψ_2 must instead be converted to π^*, while it is ψ_3 which is converted to π.
- This symmetry correlation requires crossover between bonding and antibonding orbitals (Scheme 8.113). This is thermally an unfavourable energetic process (*i.e.*, symmetries of the molecular orbitals with respect to mirror plane do not show ground state correlation) and thus the disrotatory process is forbidden.
- When one considers the correlation diagram for the thermal conrotatory ring closure of 1,3-butadiene (Scheme 8.114) considering the symmetries of the orbitals in relation to the axis of rotation the following results arise.

- ψ_1 is now antisymmetric in relation to its axis of rotation (if the ψ_1 orbital is rotated 180° around the axis, the shaded lobes would come on the bottom instead of on top) similar is the case with π (Scheme 8.114).

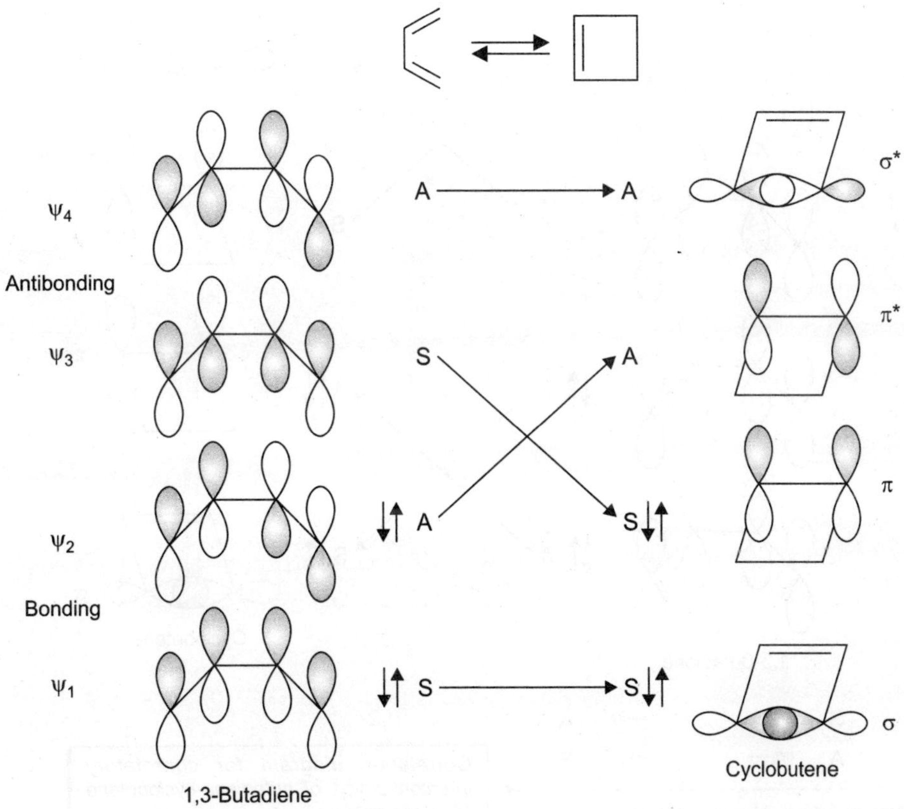

Symmetry properties of molecular orbitals and correlation diagram for disrotatory interconversion of butadiene-cyclobutene

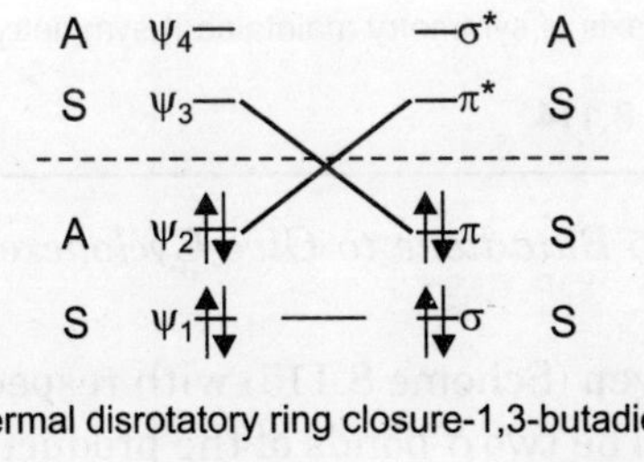

Correlation diagram for disrotatory interconversion of butadiene-cyclobutene (Mirror plane of symmetry is preserved)

Thermal disrotatory ring closure-1,3-butadiene (mirror symmetry maintained) symmetry forbidden

SCHEME 8.113

- However, ψ_2 and π^* are symmetric as in both the cases a 180° rotation would bring shaded lobes back to the top left and bottom right of the orbital.
- Similar operations reveal that σ is symmetric while σ^* is antisymmetric around the axis of rotation.
- These data show that correlation exists between the ground state bonding orbitals, therefore, a thermal conrotatory motion is symmetry allowed process.

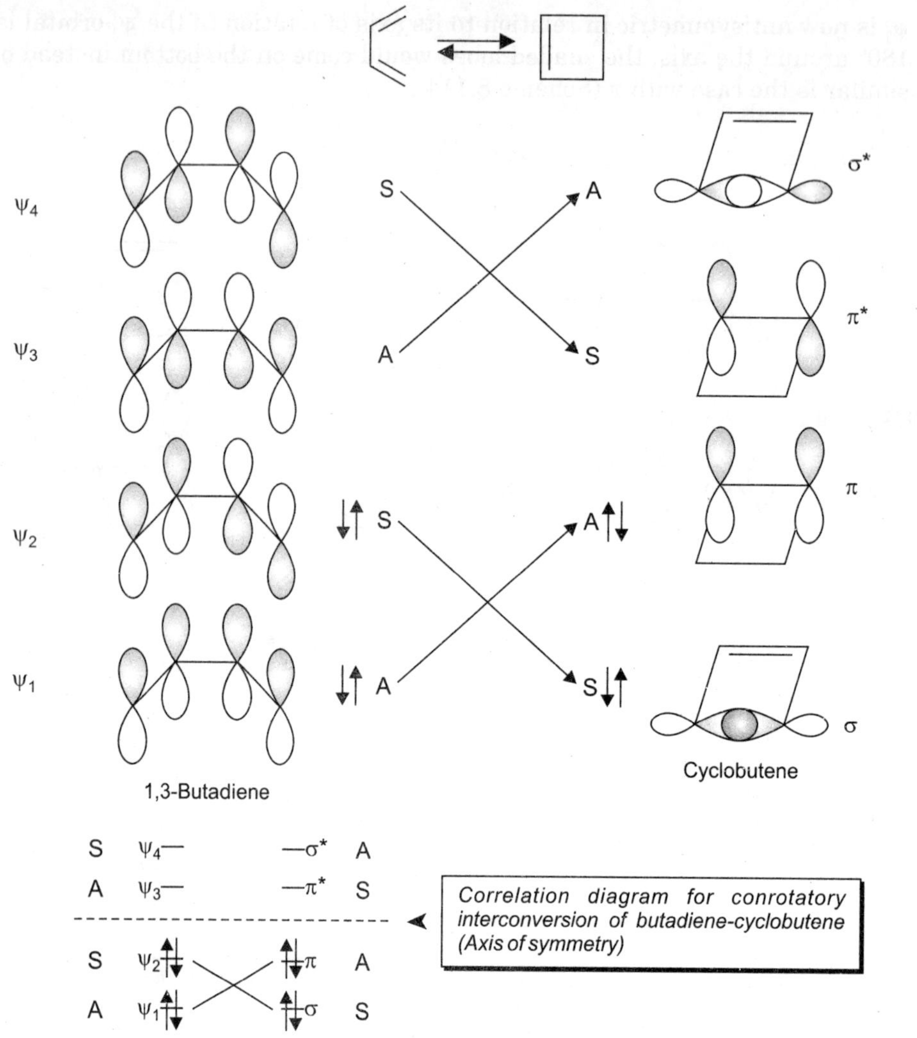

Thermal conrotatory ring closure-1, 3-butadiene (C_2 axis of symmetry maintained) symmetry allowed

SCHEME 8.114

Example 2: *[4 + 2] Cycloaddition of Ethylene to Butadiene to Give Cyclohexene Suprafacial Suprafacial Thermal Cycloaddition.*

The orbital symmetry relationships are given (Scheme 8.115) with respect to the mirror plane of symmetry of the whole reacting system. The two σ-bonds of the product are considered as a symmetric and antisymmetric combination. After the classification of the orbitals with respect to symmetry these are arranged according to energy and the correlation lines can be drawn. (Scheme 8.115). It is found that all bonding levels of the reactants correlate with product ground state orbitals (orbital symmetry is conserved within the bonding orbitals and also within the antibonding set and no cross over between the two sets occurs). This therefore, is an allowed reaction.

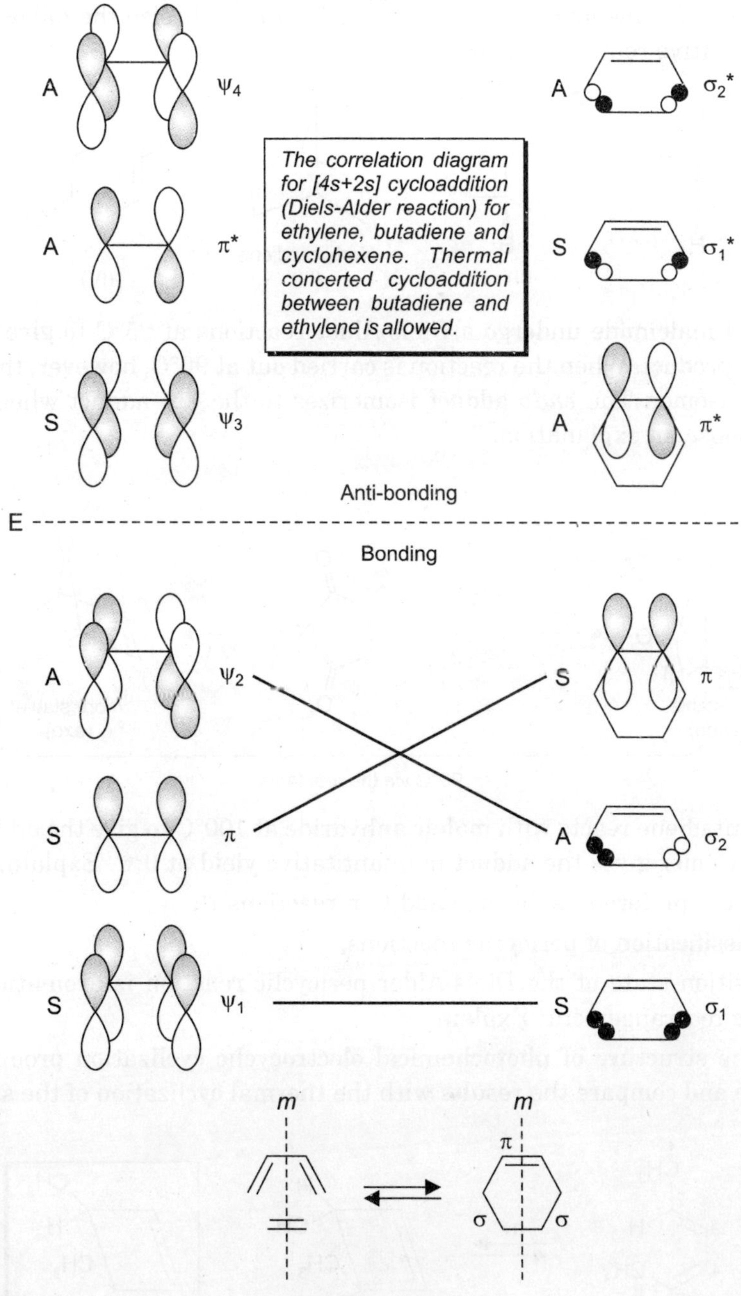

The correlation diagram for [4s+2s] cycloaddition (Diels-Alder reaction) for ethylene, butadiene and cyclohexene. Thermal concerted cycloaddition between butadiene and ethylene is allowed.

SCHEME 8.115

PROBLEMS

8.1. Write the stereostructure of the compound obtained by the Diels-Alder reaction of dimethyl maleate with butadiene.

8.2. Which diene and dienophile one would employ to synthesize the following compounds? Give alternative route for one of these.

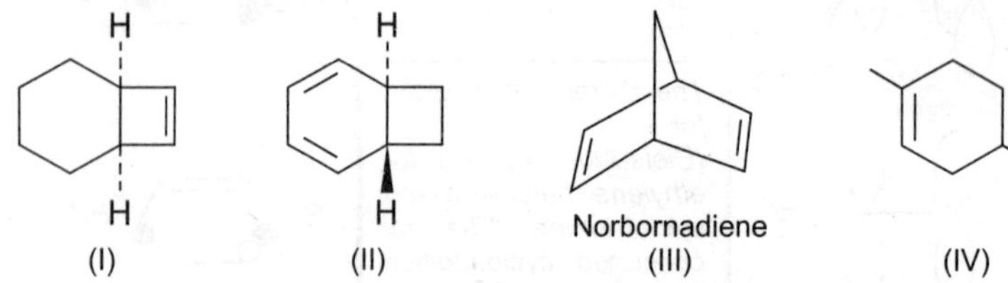

8.3. Furan and maleimide undergo a Diels-Alder reactions at 25°C to give *endo* adduct as the major product. When the reaction is carried out at 90°C, however, the major product is the *exo* isomer. The *endo* adduct isomerizes to the *exo* adduct when it is heated to 90°C. Propose an explanation.

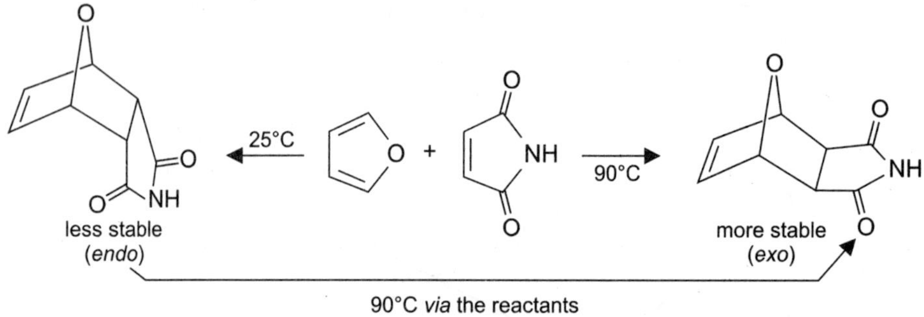

8.4. (*Z*)-1,3-pentadiene reacts with maleic anhydride at 100°C to give the adduct in 4% yield, while (*E*)-isomer gives the adduct in quantitative yield at 0°C. Explain.

8.5. What are the preferences for cycloaddition reactions ?

8.6. Give a classification of pericyclic reactions.

8.7. The transition state of the Diels-Alder pericyclic reaction is aromatic and compares with Cope rearrangement. Explain.

8.8. Predict the structure of photochemical electrocyclic cyclization product of (2*E*, 4*Z*)-hexadiene and compare the results with the thermal cyclization of the same compound.

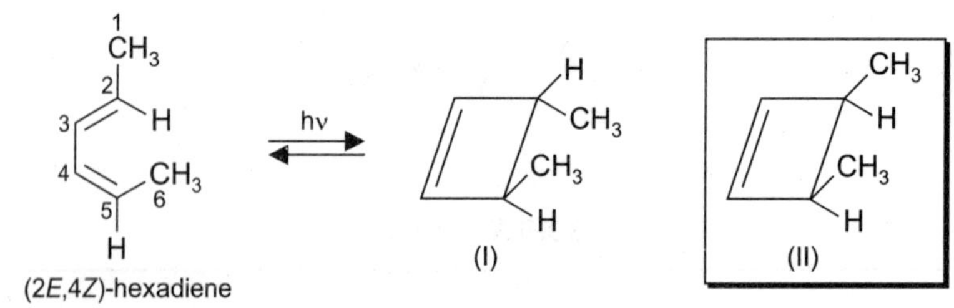

8.9. Give the stereostructure of the products from the following electrocyclic reactions of (I and II) carried out under photochemical reactions. Discuss if each reaction takes place in a conrotatory or disrotatory fashion.

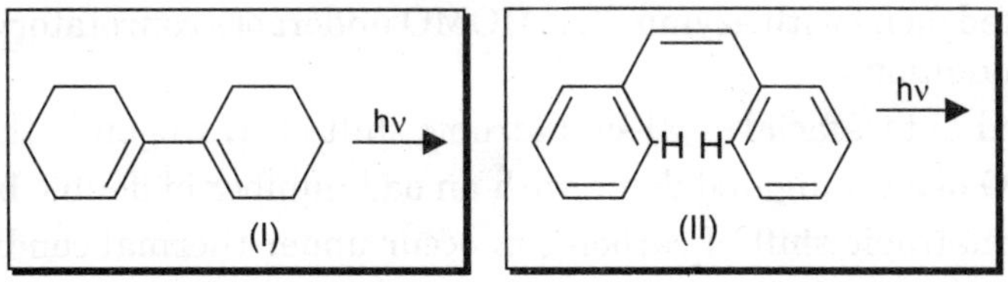

8.10. On thermal ring opening *cis* 3,4-dimethylcyclobutene gives two dienes (I and II). One of these is formed almost exclusively which is this diene and how it is formed?

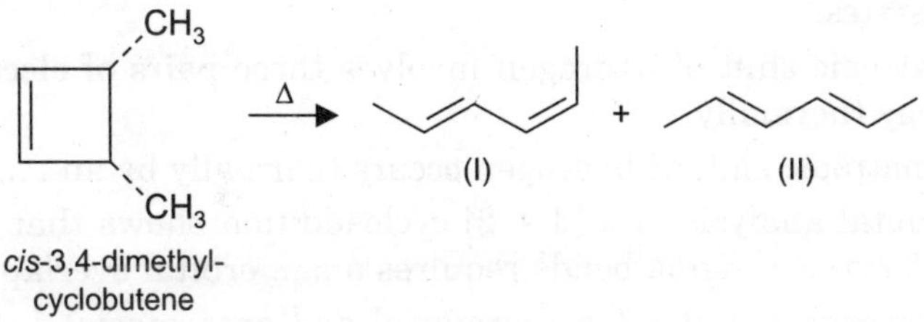

8.11. Predict whether conrotatory or disrotatory motion will take place under the conditions mentioned against each compound. Write the structure of the product with stereochemistry in each case.

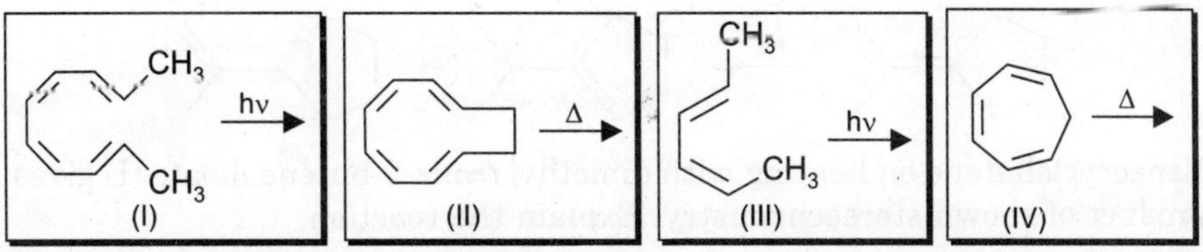

8.12. Explain briefly, taking one common example as to how FMO (frontier molecular orbital), method, PMO method and correlation diagram can be used for analyzing a pericyclic reaction.

8.13. Draw a correlation diagram for disrotatory conversion of butadiene to cyclobutene. Is the process allowed or forbidden? Explain.

8.14. A [3,3] sigmatropic rearrangement is thermally allowed via hypothetically formed allyl radicals. Explain by drawing appropriate bonding interactions.

8.15 (*a*) Explain briefly as to how a conjugated diene under photochemical conditions undergoes cyclization via a disrotatory path?

(*b*) Under which conditions, thermal or photochemical, the following ring closure will take place? Explain the stereochemistry at the ring fusion.

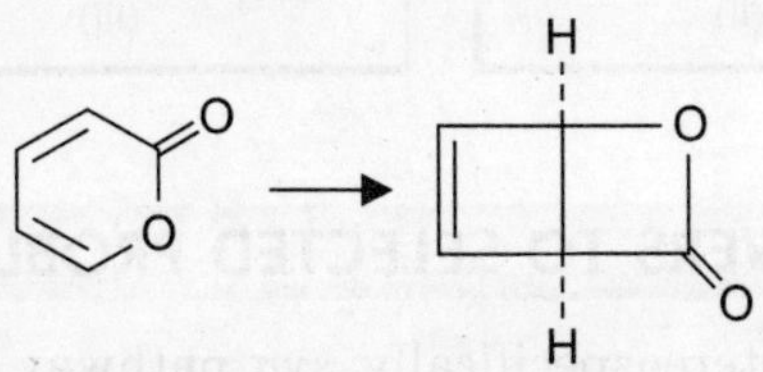

8.16. Which of the following statements are true or false.

(*i*) A conjugated diene with an even number of double bonds undergoes conrotatory ring closure under thermal conditions.

(*ii*) A conjugated diene with asymmetric HOMO undergoes conrotatory ring closure under thermal conditions.

(*iii*) A concerted antarafacial [1,3]-sigmatropic shift of hydrogen is thermally allowed.

(*iv*) The HOMO of a conjugated diene with an odd number of double bonds is symmetric.

(*v*) A [1,3] sigmatropic shift of carbon can occur under thermal conditions.

8.17. Fill in the blanks:

(*i*) A 1, 3-migration of carbon can take place thermally with ...... of configuration.

(*ii*) Pericyclic reactions are concerted, unaffected by catalysts or solvents and have ...... transition states.

(*iii*) [1,5] Sigmatropic shift of hydrogen involves three pairs of electrons and occurs by ...... pathway thermally.

(*iv*) A [1,7] sigmatropic shift of hydrogen occurs thermally by an ...... pathway.

(*v*) Frontier orbital analysis of a [4 + 2] cycloaddition shows that overlap of in phase orbitals to form new sigma bonds requires a ...... orbital overlap.

8.18. (*a*) Draw the transition states for suprafacial and antarafacial 1, 3 hydrogen shift by drawing the phase interactions in the basis sets. Show which process is thermally forbidden and which thermally allowed?

(*b*) Explain the results of the following photochemical reaction.

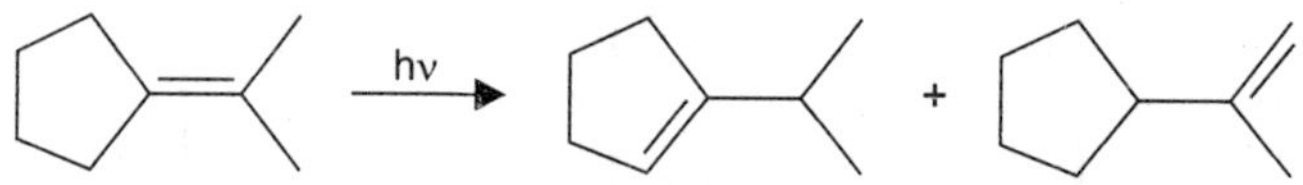

8.19. (*a*) Benzocyclobutene on heating with dimethyl *trans*-2-butene dioate (I) gives a bicyclic product of shown stereochemistry. Explain the reaction.

(*b*) Write the product with stereochemistry of the Diels-Alder reactions (II and III).

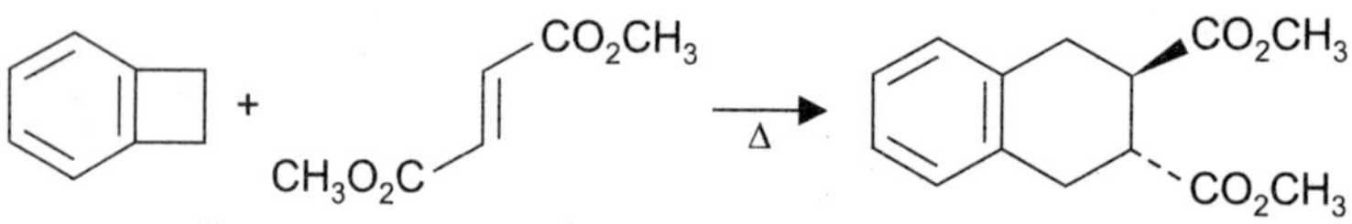

(I)

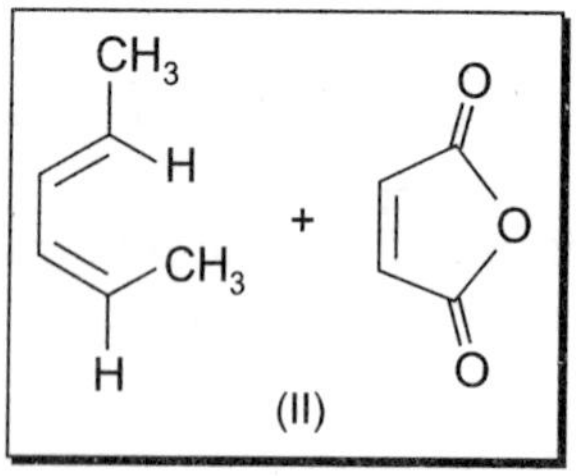

(II)

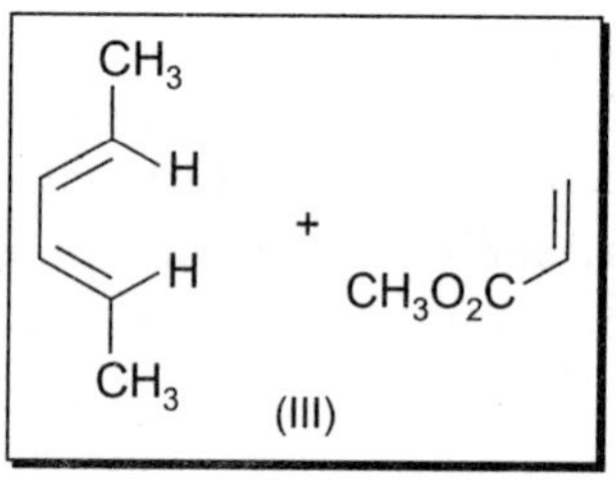

(III)

ANSWERS TO SELECTED PROBLEMS

8.1. The reaction follows a stereospecifically *syn* pathway. The product is *cis*-dimethyl cyclohexene-4, 5-dicarboxylate. Therefore, the groups which are *cis* in the olefin also occupy *cis*-positions in the cyclohexene ring.

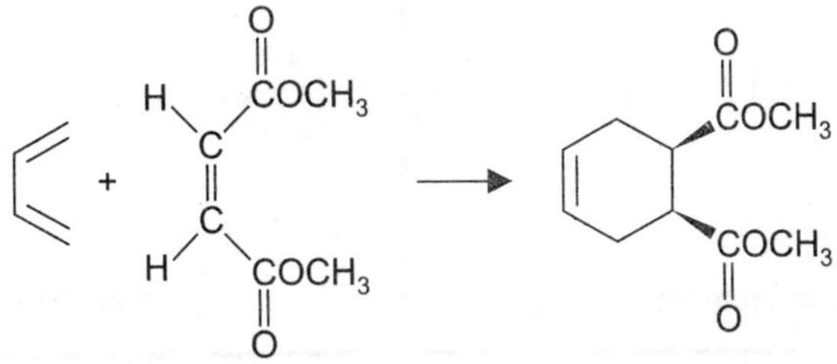

8.2. Cyclopentadiene and acetylene.

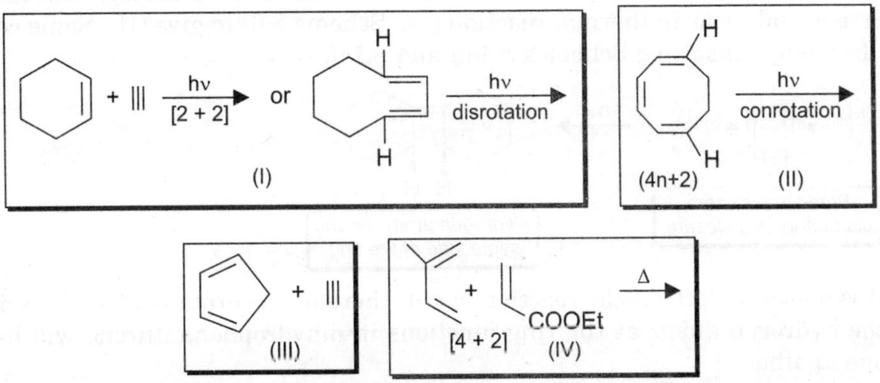

8.3. The *exo* product is thermodynamically more stable. The less stable *endo* isomer (kinetically favoured adduct) is formed faster and predominates at 25°C, the reaction is effectively irreversible. At 90°C this product is in rapid equilibrium with the reactants, consequently, the less rapidly formed but more stable *exo* isomer accumulates with time.

8.4. The bulky 1-*cis* (*Z*-) methyl substituent, sterically hinders formation of the *cisoid* conformation with a hydrogen at C-4. In the *E*-isomer the *cisoid* conformation is attained easily due to only tiny H, H interaction.

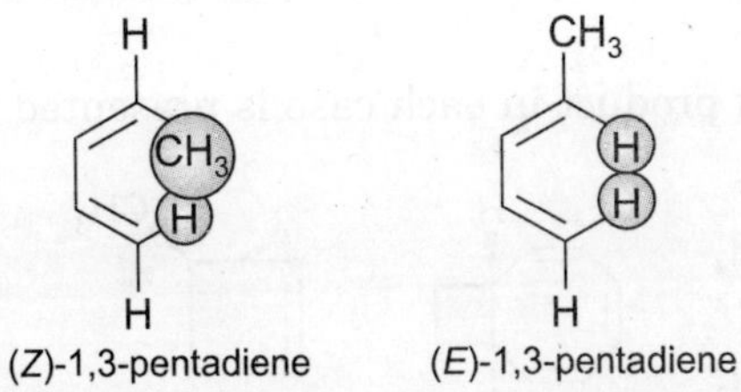

8.5. The rule of thumb is that when the reactants involve odd number of electron pairs, the cycloaddition is allowed thermally, while with even number of electron pairs, the cycloaddition is allowed photochemically.

8.6. Electrocyclic reactions are stereochemically classified as conrotatory and disrotatory, cycloadditions and sigmatropic rearrangements are classified as suprafacial or antarafacial.

8.7. In both the cases, the transition states involve six orbitals and six electrons.

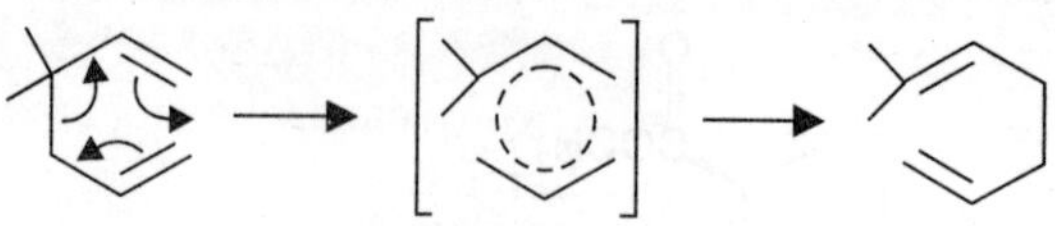

Aromatic transition state

the **Cope rearrangement**

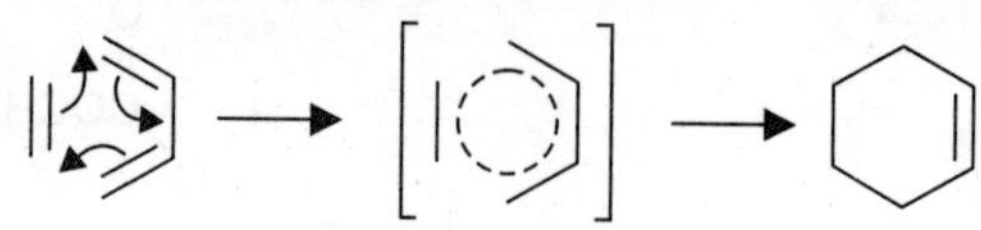

Aromatic transition state

the Diels-Alder reaction

8.8. In the photochemical cyclization disrotatory motion is required for bond formation, one methyl rotates up while the other down to give trans-3,4-dimethylecyclohexene (I). The reverse would occur in thermal reaction (see Scheme 8.6) to give (II). Same conclusions are drawn by consulting Schemes 8.14*a* and 8.14*b*.

8.9. (I)

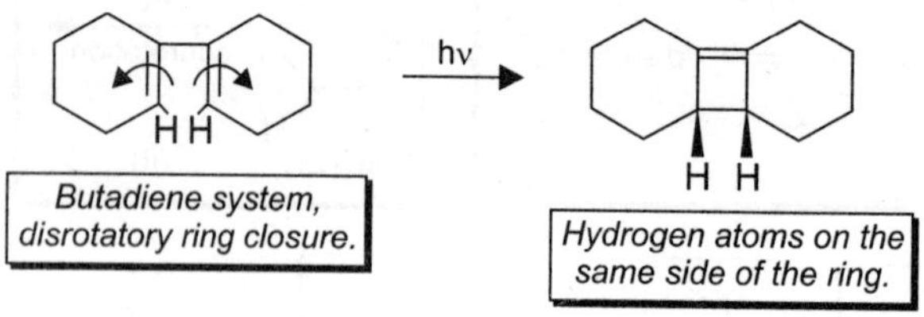

(II) It is a $4n + 2$ electrocyclic reaction; $n = 1$, therefore, conrotatory in the excited state, the hydrogen atoms at the ring junctions in dihydrophenanthrene will be *trans* to one another.

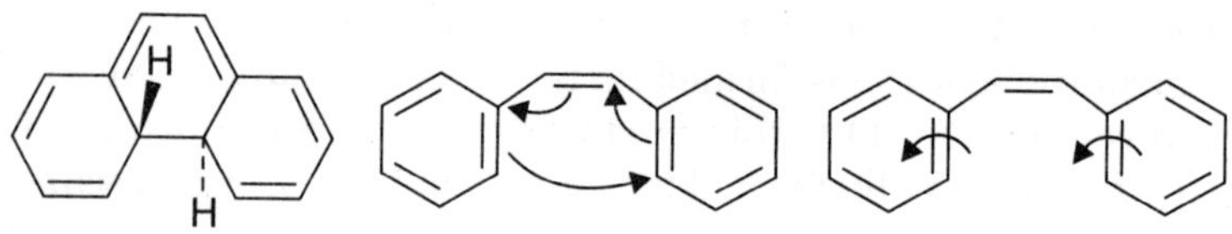

8.10. (Hint see eq. III, Scheme 8.4). It is diene. Thermally a compound with two π bonds undergoes conrotatory ring closure. Since in the product the two methyls are *cis* placed, these point in the same direction in the reactant. In diene (I) the methyls point in the same direction.

8.11. The stereostructures of product in each case is presented (refer to Table 8.1).

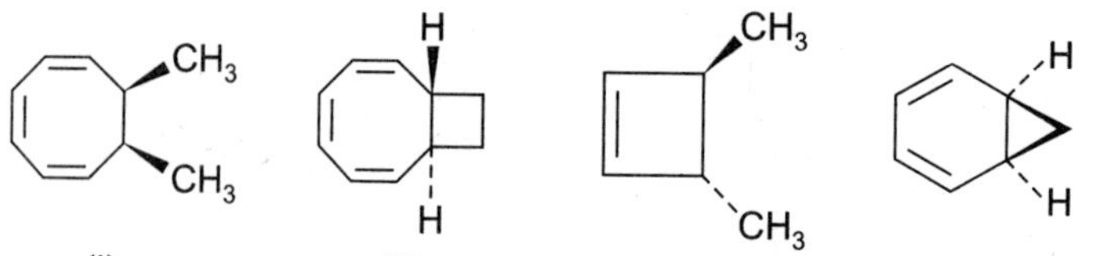

(I) (II) (III) (IV)

(I) The reactant being a ($4n$) π electron system will undergo disrotatory ring closure under photochemical conditions. Since the two methyl groups point in opposite directions these will be *cis* in the product.

(III) The reactant is (2*E*, 4*Z*) hexadiene ($4n$) π electron system. Photochemically it will undergo ring closure by disrotation. Since the methyl substituents point in the same direction, the product will have these in *trans* relationship.

(IV) The reactant 1,3,5-cycloheptatriene is a ($4n + 2$) π electron system. It will underdo disrotatory ring closure under thermal conditions, since in the triene, the hydrogen substituents (not shown) point in opposite directions (compare with Scheme (8.14*d*) these will be *cis* in the product norcaradiene. However, due to the strain, norcaradiene cannot be isolated and it reverts back to the starting material. In case the ring size of the starting triene is large as in 1,3,5-cyclononatriene (Scheme 8.14*d*) the product is stable and is isolated as the exclusive product.

8.12. One may discuss the Diels-Alder reaction between ethylene and butadiene using all the three techniques (Schemes 8.16, 8.52 and 8.62).

8.15. (*b*) It is a 2,3-butadiene system. Even number of double bonds shows that the electrocyclization will be under photochemical conditions (see table 8.1). Since in the reactant α-pyrone, the hydrogens on the diene system are pointing in opposite directions (compare Scheme 8.14*a*, 8.14*f*) these will be *cis* in the product on disrotatory motion.

8.16. (*i*) True; (*ii*) true; (*iii*) true; (*iv*) true; (*v*) true.

8.17. (*i*) Inversion; (*ii*) cyclic; (*iii*) suprafacial; (*iv*) antarafacial; (*v*) suprafacial.

8.18. (*b*) This is a photochemical 1,3-hydrogen shift (see, Scheme 8.34). Two products are expected since two different allylic hydrogens can undergo this shift.

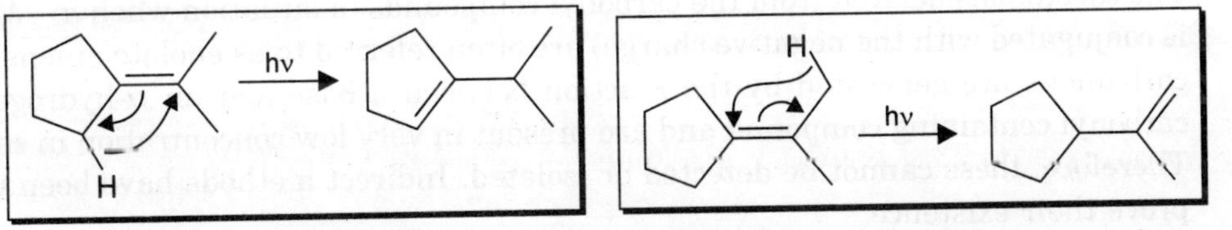

8.19 (*a*) It is a combination of electrocyclic reaction (to give IV) followed by Diels-Alder reaction.

(*b*) The reaction (II) gives the product (V) with the stereochemistry at centres other than ring junction as shown.

This is deriveable from the discussion presented (Scheme 8.19*c*). Maleic anhydride is a *cis*-alkene, since the Diels-Alder reaction is a *syn* addition, the stereochemistry at the ring junction must be *cis*.

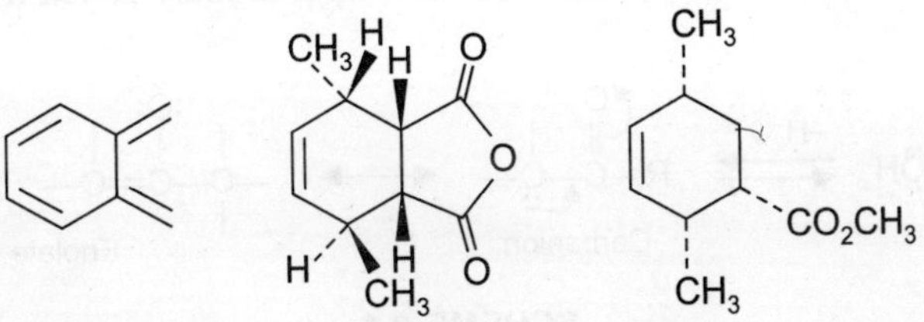

(IV) (V) (VI)

REFERENCES

1. M.B. Smith and J. March, *March's Advanced Organic Chemistry*, Wiley, New york, 2001.
2. Fleming, *Frontier Orbitals and Organic Chemical Reactions*, Wiley, London, 1976.
3. R.B. Woodward and R. Hoffmann, *The Conservation of Orbital Symmetry*, Verlag Chemie, Weinheim, 1971.

CHAPTER 9

Aliphatic Electrophilic Substitution

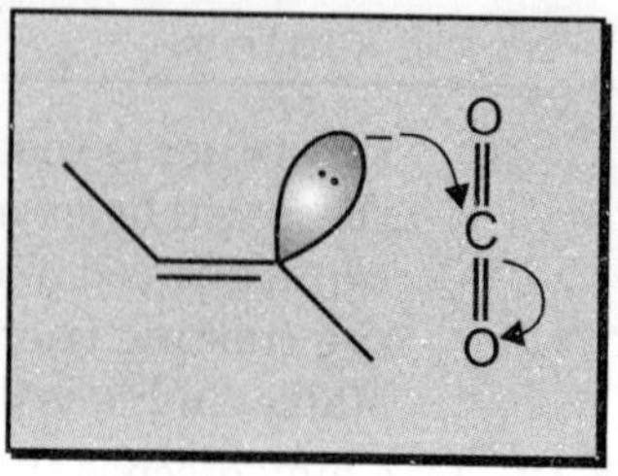

9.1 INTRODUCTION

When a positive species is removed from a carbon atom a carbanion is formed leading to aliphatic electrophilic substitution (Displacement of one electrophile by another). Thus, before a systematic discussion on aliphatic electrophilic substitution is takes up, it is necessary to learn about carbanions. The following points may be noted:

- The carbanions derived from the carbonyl compounds (a situation when $a - M$ group is conjugated with the negative charge) are often referred to as enolate anions. These carbanions are generated by the reaction between a base and an α-hydrogen in a carbonyl containing compound and are present in very low concentration in solution. Therefore, these cannot be detected or isolated. Indirect methods have been used to prove their existence.
- Isotope exchange studies is one method to prove their existence. The α-hydrogen from an aldehyde or a ketone is abstracted with a base like sodium hydroxide, however, the equilibrium lies far to the left and favors the neutral carbonyl compound. The stability of an α-carbanion is due to resonance (Scheme 9.1). When an aldehyde or a ketone is dissolved in deuterium oxide containing sodium deuteroxide as the base the α-hydrogen is quickly substituted by deuterium (Scheme 9.2). This is an example of aliphatic electrophilic substitution of H by D via a carbanion.

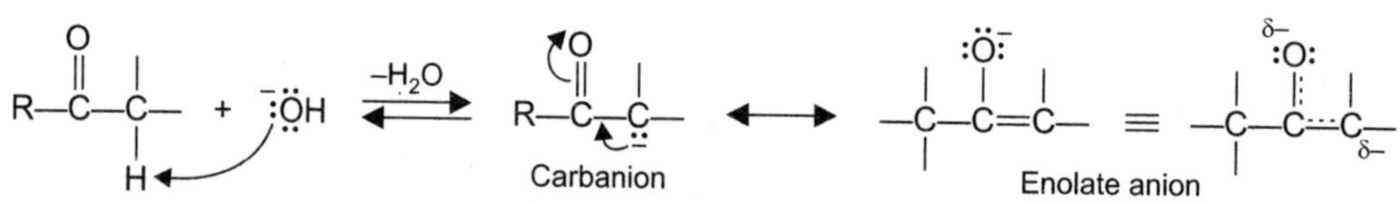

SCHEME 9.1

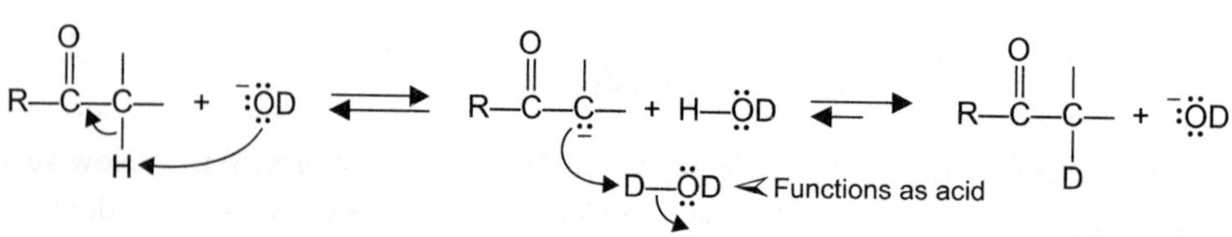

SCHEME 9.2

- The carbanions which are stabilized by resonance (*i.e.*, enolate anions) are planar. When the carbanion is planar, racemization occurs during electrophilic substitution [an example of this has already been discussed (see, Scheme 1.91)].
- If the carbanion is pyramidal and can hold its structure, then retention of configuration should be expected. Largely however, a pyramidal carbanion even undergoes inversion (umbrella effect see, Scheme 1.8) like an acyclic chiral amine and is configurationally unstable. An example of non stabilized carbanion is the Grignard reagent prepared from optically active *sec.* butyl bromide (Scheme 9.3) which reacts with acetone to give optically inactive product. That the Grignard reagent itself becomes optically inactive has been proved with other optically active Grignard reagents and their carbanions.

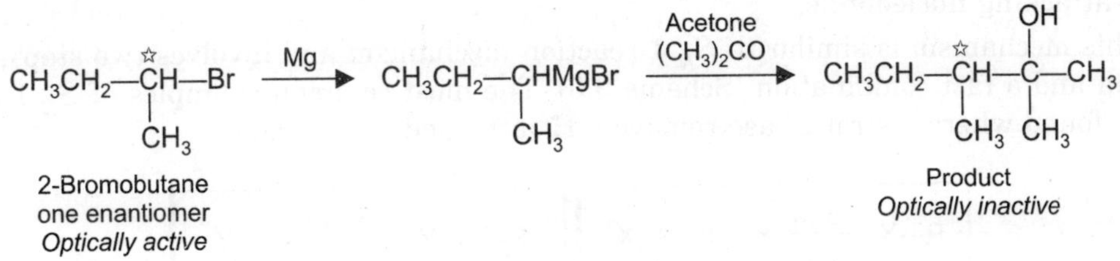

SCHEME 9.3

- The leaving group in an aliphatic electrophilic substitution reaction is an electron deficient species. In an aliphatic system the commonest type of leaving group is a metal ion as it can easily bear a positive charge. Thus aliphatic electrophilic substitution reaction are often encountered in organometallic pathways.
- Weakly acidic protons of an optically active alkane can undergo deuterium exchange (strong base is needed, aliphatic electrophilic substitution) and show that carbanions are involved since racemization occur (Scheme 9.4). When three different groups are attached to the carbanion, one pyramidal structure is the nonsuperposable mirror image of the other and the two undergo a rapid interconversion.

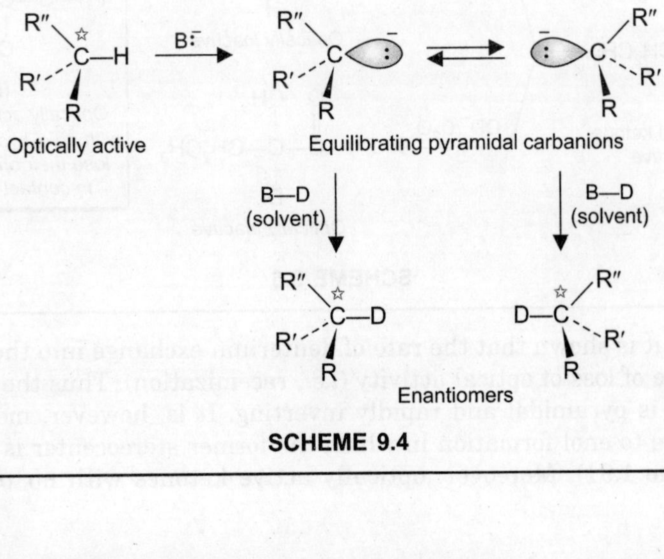

SCHEME 9.4

> *The equilibration of the pyramidal carbanion appears to be faster than protonation i.e., deuteration in this case. In case the carbanion reacts with a proton or deuteron from solvent as soon as it is generated and before, it has an opportunity to undergo equilibration optically active product may be formed.*

9.2 DIFFERENT MECHANISMS FOR ALIPHATIC ELECTROPHILIC SUBSTITUTION (DISPLACEMENT OF ONE ELECTROPHILE BY ANOTHER)

(a) S_E1 Mechanism (Substitution Electrophilic Unimolecular)

The S_E1 reaction mechanism is a multistep electrophile interchange with bond breaking (The rate limiting step) followed by bond making. The reaction is first order in substrate and zero order in attacking nucleophile.

This mechanism is similar to S_N1 reaction mechanism and involves two steps, a slow ionization and a fast combination (Scheme 9.5). The most common examples of S_E1 mechanism are found where a strong base remove a H$^+$ attached to a carbon.

$$\text{R—X} \xrightarrow[\text{Step 1}]{\text{Slow}} \text{R:}^- + \text{X}^+ \qquad \text{R:}^- + \text{Y}^+ \xrightarrow{\text{Step 2}} \text{RY}$$

S_E1 Mechanism-substitution electrophilic unimolecular

SCHEME 9.5

An evidence of S_E1 mechanism is during base catalyzed tautomerization of optically active phenyl *sec*-butyl ketone (Scheme 9.6) with NaOH which is attended with racemization. Isotopic exchange shows the formation of α-carbanion (See, Scheme 9.2) and the incorporation of deuterium into the α-position in place of hydrogen and racemisation still occurs.

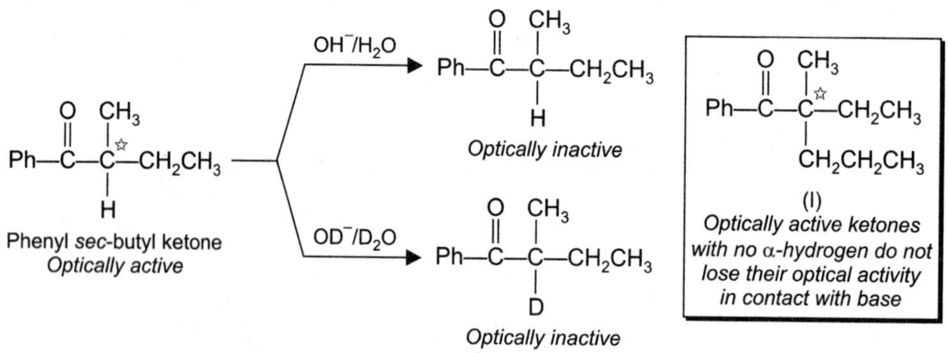

SCHEME 9.6

Significantly it is shown that the rate of deuterium exchange into the ketone is exactly the same as the rate of loss of optical activity (*i.e.*, racemization). Thus the carbanion formed as an intermediate is pyramidal and rapidly inverting. It is, however, more likely that the carbanion is flat due to enol formation in which the former stereocenter is planar and hence achiral (see, Scheme 1.91). Moreover, optically active ketones with no α-hydrogen (*e.g.*, I

Scheme 9.6) do not lose any optical activity on standing with base. Thus when in an optically active compound in which the chirality is due to a stereogenic carbon α to the carbonyl group is reacted with a base racemization results.

In case there is another stereocenter in the molecule, the less stable epimer is converted to the more stable epimer and this equilibration (unlike racemization) has preparative value. *cis*-Decalone undergoes epimerization in basic solution and when equilibrium is reached, the solution is found contain 90% *trans* 1-decalone and about 5% *cis*-1-decalone (Scheme 9.7). These results may be compared with first and second order asymmetric transformations (Chapter 1). *cis* and *trans* Decalones are also drawn with each ring in the chair conformation (Scheme 9.7, however, in *cis* decalin both the ring juncture hydrogens are drawn as down). There is more on equilibration in Scheme 4.43. Keto-enol tautomerism is both acid (see, Scheme 4.43*a*) as well as base catalyzed.

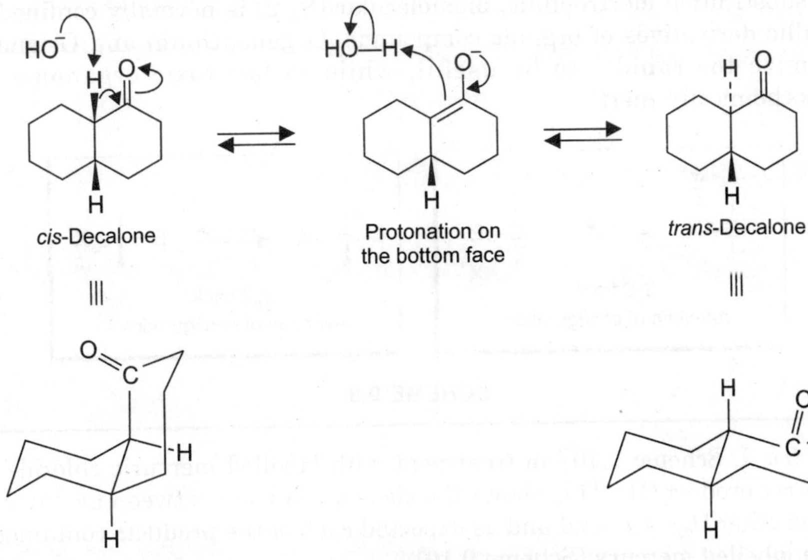

SCHEME 9.7

(*b*) Retention of Configuration During S_E1 Mechanism

As already discussed S_E1 reactions whether they involve planar carbanions or oscillating pyramidal species, racemization is generally observed. Vinyl carbanions (I, Scheme 9.8) can, however, hold their configuration and S_E1 reactions with a vinyl system proceeds with retention of configuration. Thus *cis*-2-bromo-but-2-ene on reaction with lithium and then with carbon dioxide gives a carboxylic acid with retention of its configuration (angelic acid, the *trans* isomer tiglic acid is formed to the extent of only 5%).

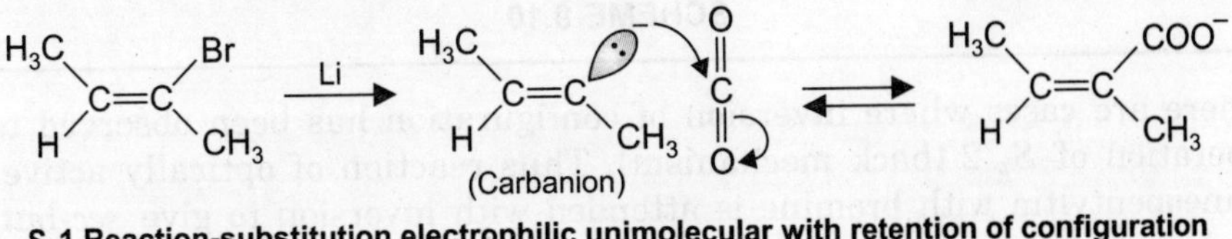

S_E1 Reaction-substitution electrophilic unimolecular with retention of configuration

SCHEME 9.8

(c) $S_E 2$ Mechanism (Substitution Electrophilic Biomolecular)

The $S_E 2$ reaction mechanism is the concerted displacement of one electrophile by another. The biomolecular mechanism for electrophilic aliphatic substitution is similar to $S_N 2$ mechanism since a new bond forms and the old bold breaks. The following points may be noted:

- In $S_N 2$ nucleophilic mechanism, the incoming group, the nucleophile has a pair of electrons and thus must attack from backside at a position 180° (front side attack not feasible since electron clouds repel) and the net result is inversion of configuration.
- In $S_E 2$ aliphatic electrophilic substitution, the attacking species is an electrophile with a vacant orbital, thus it could attack from front ($S_E 2$ front retention of configuration) or from back ($S_E 2$ back inversion of configuration Scheme 9.9).
- Mostly in second order electrophilic substitutions retention of configuration has been observed to show frontside attack and thus the operation of $S_E 2$ (front) mechanism.
- The substitution electrophilic, biomolecular ($S_E 2$) is normally confined to organometallic derivatives of organic compounds. Organolithium and Grignard reagents racemize too rapidly to be useful, while rather toxic organomercurials are stereochemically inert.

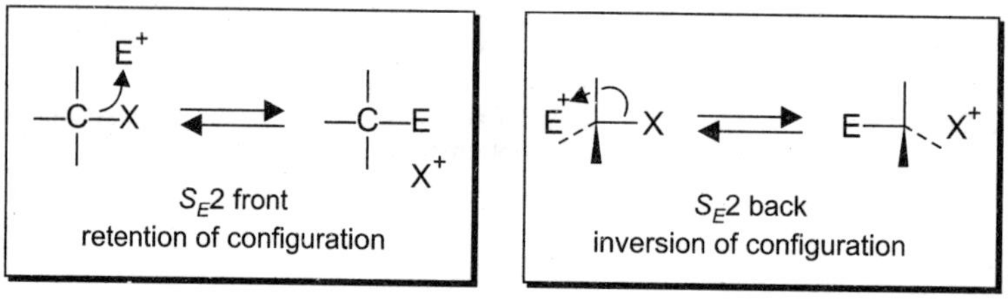

SCHEME 9.9

- Thus (*cis* I, Scheme 9.10) on treatment with labelled mercuric chloride gave exclusively *cis* product (II). This shows the cleavage of bond between mercury and ring as well as other Hg—Cl bond and as expected each of the products contained about half of the labelled mercury (Scheme 9.10).

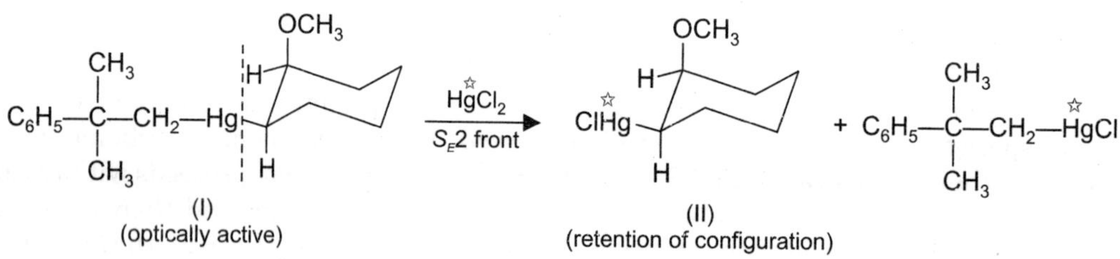

S_E2 Reaction-substitution electrophilic bimolecular with retention of configuration

SCHEME 9.10

- There are cases where inversion of configuration has been observed to show the operation of $S_E 2$ (back mechanism). Thus reaction of optically active *sec*-butyltrineopentyltin with bromine is attended with inversion to give *sec*-butyl bromide (Scheme 9.11).

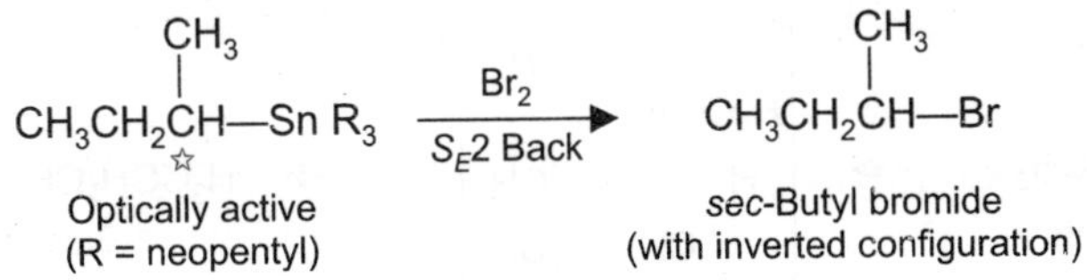

SCHEME 9.11

(c) $S_E i$ Mechanism (Substitution Electrophilic Internal)

There is another mechanism $S_E i$ which is a concerted four center electrophilic interchange kinetically and stereochemically identical with $S_E 2$ front. The two mechanisms differ in the transition states. In the case of $S_E i$ mechanism, it is suggested that the electrophile can attack from the front and a portion of the electrophile may assist the exist of the leaving group to form a bond with it as the new bond is made (Scheme 9.12). As already said mostly in second order electrophilic substitutions retention of configuration has been observed thus like $S_E 2$ (front) $S_E i$ proceeds with retention.

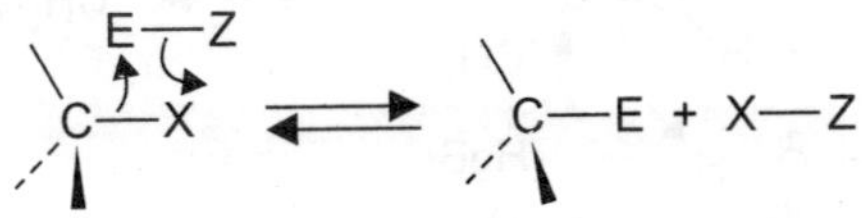

($S_E i$ mechanism)
(retention of configuration)

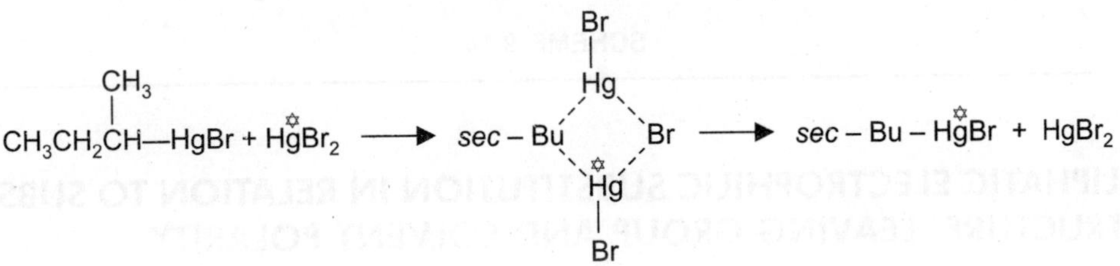

(transition state)
$S_E i$ Mechanism-substitution electrophilic internal

SCHEME 9.12

9.3 ELECTROPHILIC SUBSTITUTION ACCOMPANIED BY DOUBLE BOND SHIFT

(a) $S_E i'$—Substitution Electrophilic with Rearrangement

This is a concerted electrophile interchange which is kinetically identical with $S_E i$ but involves a multicenter transition state and formation of a rearranged product. Crotylmercuric bromide reacted with HCl 10^7 times faster than n-butylmercuric bromide and the product was exclusively 1-butene formed by double bond shift. This double bond shift is assigned to $S_E i'$ mechanism (Scheme 9.13).

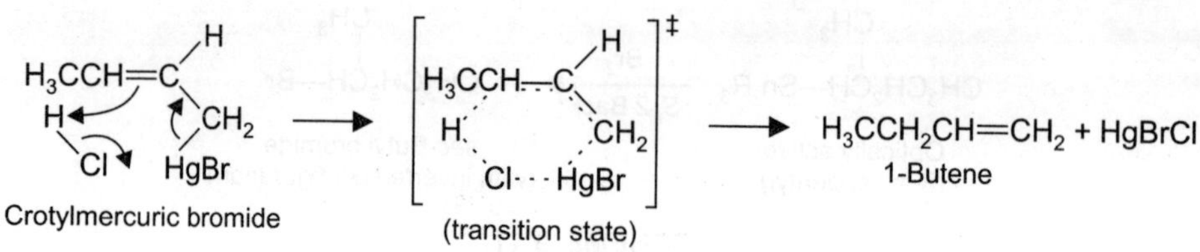

S_Ei' substitution electrophilic with rearrangement

SCHEME 9.13

(b) S_E 2′-Substitution Electrophilic Bimolecular with Rearrangement

This is a mechanism which is kinetically identical to the S_E 2 process but involves the formation of a rearranged product. Thus crotylmercuric bromide on reaction with acetic acid in perchloric acid follows the S_E 2′ pathway to give 1-butene (Scheme 9.14).

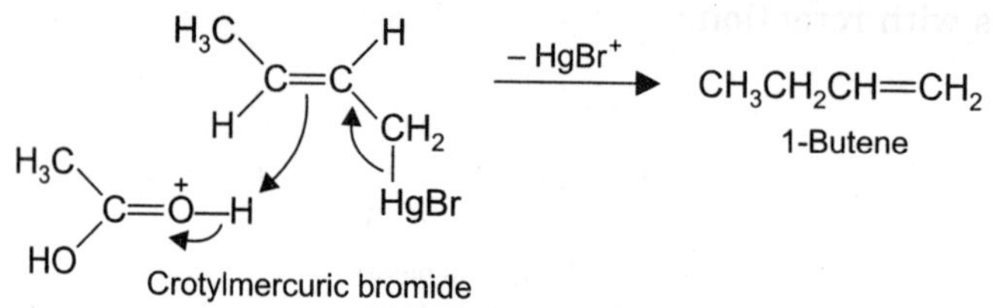

S_E2'-Substitution electrophilic bimolecular with rearrangement

SCHEME 9.14

9.4 ALIPHATIC ELECTROPHILIC SUBSTITUTION IN RELATION TO SUBSTRATE STRUCTURE, LEAVING GROUP AND SOLVENT POLARITY

(A) Effect of Substrate

The following points may be noted:

- Electron donating groups decrease rates of S_E 1 reaction while electron withdrawing groups increases the rates. This is in line with a reaction in which the rate determining step is similar to the cleavage of a proton from an acid.
- S_E 2 (front) reactions take place easily at bridgehead carbons. S_E 2 (front) reactions also are easy with neopentyl systems (Scheme 9.15) the reactions are fast (only slightly slower than ethyl S_N 2). As comparison S_N 2 reactions at bridgehead carbons do not take place while S_N 2 reactions with neopentyl systems are extremely slow:

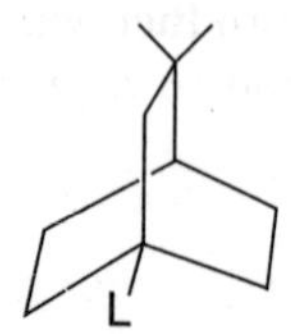

[2.2.2] Bridge head system

With leaving group (L) at bridge head S_E2 reactions occur easily (S_N2 reactions are unsuccessful.

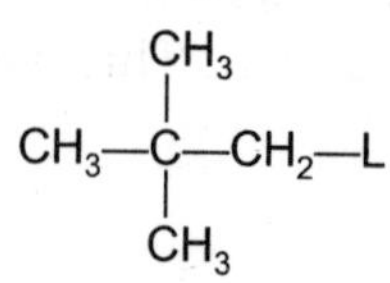

Neopentyl system

S_N2 reactivity of neopentyl systems is very slow, while S_E2 is fast.

SCHEME 9.15

- As expected in S_E2 (back) the reactivity of various alkyl groups is similar with S_N2, Me > Et > Pr > *iso*-Pr > neopentyl.

(B) Effect of Leaving Group

S_E1 mechanism is normally operative with carbon leaving groups, while in the case metallic leaving groups the mechanism is almost always S_E2 or S_Ei.

(C) Effect of Solvent

Solvents play a role on the preferred mechanism of electrophilic substitution. As expended an increase in the solvent polarity increases the probability of the mechanism becoming ionizing S_E1 rather than S_E2. An S_E2 mechanism does not involve ions and is a concerted process. Solvents/added salt, help to distinguish S_E2 (front or back) from S_Ei. An increase in the polarity has little effect on S_Ei process while in comparison S_E2 reactions are not effected. When neutral starting materials in a reaction acquire charges in the transition state (as would be the case in S_E2 mechanism) the addition of added ions aids such reactions.

9.5 MORE REACTIONS INVOLVING ALIPHATIC ELECTROPHILIC SUBSTITUTION

(A) Isomerization of Double Bonds

When an unsaturated compound is treated with a strong base, double bond shifts to give the thermodynamically stable compound. The following points may be noted:

- The double bond isomerizes as so to be in conjugation with a double bond, a carybonyl group or an aromatic system if it is already present.
- An excocyclic double bond in a six membered ring goes to the endocylic position.
- But for the situations described above the Zaitzeff's rule is followed (The double bond goes to the carbon with the least number of hydrogens).
- A terminal double bond in a chain moves to the internal position (Scheme 9.16).

$$\text{R—CH}_2\text{—CH=CH}_2 \xrightarrow[\text{Me}_2\text{SO}]{\text{KNH}_2} \text{R—CH=CH—CH}_3$$

Prototropic rearrangement *involves electrophilic substitution with allylic rearrangement with strong base.*

SCHEME 9.16

The mechanism of the reaction (a prototropic shift) is an electrophilic substitution with allylic rearrangement (compare this with nucleophilic allylic shifts in Scheme 3.48 and 3.52, the double bond shifts also occur photochemically and during electrocyclic and sigmatropic rearrangements).

Mechanism of double bond shifts with strong base involves the abstraction of a proton to afford a resonance stabilized carbanion, which recombines with a proton at the position so as to give the more stable alkene (Scheme 9.17).

$$R\text{—}CH_2\text{—}CH{=}CH_2 + \overset{\text{base}}{B} \longrightarrow R\text{—}\ddot{C}\bar{H}\text{—}CH{=}CH_2 \longleftrightarrow R\text{—}CH{=}CH\text{—}\ddot{C}\bar{H}_2 + BH^+ \quad \textbf{Step 1}$$

$$R\text{—}\ddot{C}\bar{H}\text{—}CH{=}CH_2 \longleftrightarrow R\text{—}CH{=}CH\text{—}\ddot{C}\bar{H}_2 \xrightarrow{BH^+} R\text{—}CH{=}CH\text{—}CH_3 + B \quad \textbf{Step 2}$$

SCHEME 9.17

When alkenes are treated with acids, the double bonds are found to shift now by first a proton gain followed by proton loss (Scheme 9.18). However, an acid catalyzed double bond shift is not a useful preparative method, since the intermediate carbocation is likely to give a mixture of products.

$$CH_3\text{—}CH_2\text{—}CH{=}CH_2 + H^+ \longrightarrow CH_3\text{—}CH_2\text{—}\overset{+}{C}H\text{—}CH_3 \quad \textbf{Step 1}$$

$$CH_3\text{—}CH_2\text{—}\overset{+}{C}H\text{—}CH_3 \longrightarrow CH_3\text{—}CH{=}CH\text{—}CH_3 + H^+ \quad \textbf{Step 2}$$

SCHEME 9.18

(B) Halogenation of Aldehydes and Ketones

In some of the earlier situations, one has seen examples where hydrogen acts as the electrophile *i.e.*, during base catalyzed double bond shift and keto-enol tautomerization. Halogen act as an electrophile during halogenation of aldehydes and ketones in the α-position. For example, acetone is brominated (Scheme 9.19) the following points may be noted:

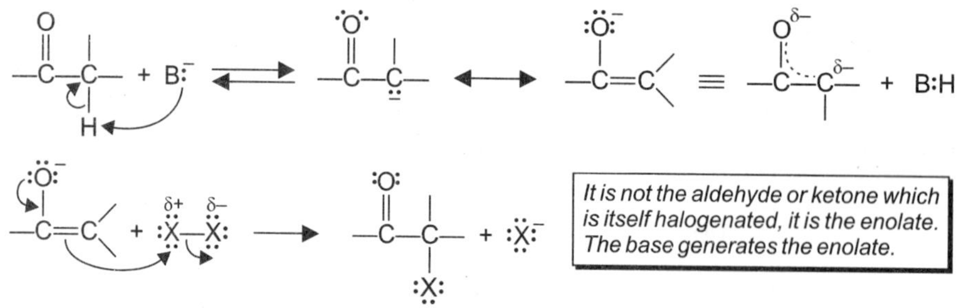

SCHEME 9.19

- α-Bromoacetone gives only α, α-dibromoacetone and not α, α′-dibromoacetone ($BrCH_2COCH_2Br$).
- The rate of halogenation depends on the concentrations of base and of aldehyde. Thus these two are involved in the slow rate determining step. It is independent of the halogen concentration and the nature of halogen.
- Mechanistically, the reaction involves the halogenation of the initially formed enolate ion (removal of acidic α-hydrogen by base). The purpose of the catalyst is to provide the enolate.
- The halogenation can be carried further and the second or the third halogen is added to the same carbon as the first, rather than on the other side of the carbonyl group. This is due to fact that *e.g.*, in CH_3COCH_2Br the protons of the methylenic group are

more acidic. These being on a carbon which is attached to two electron withdrawing groups.

- Halogenation of carboxylic acids and acyl halides (Hell-Volhhard-Zelinsky reaction) and the haloform reaction of methyl ketones follow similar aliphatic electrophilic substitutions.

Lastly mention may be made of Stork enamine reaction which can replace a hydrogen in the α-position of a ketone with a variety of groups (see, Scheme 6.69).

Decarboxylation of carboxylic acid with a variety of groups at β-position *e.g.*, β-keto acids involve a cyclic six center mechanism. However, in this case itself it is the enol which tautomerizes to the product (Scheme 9.20). This mechanism is in keeping with the fact that bridgehead bicyclic, keto acids are resistant to decarboxylation since the formation of transition state will be a violation of Bredt's rule (Scheme 9.20, also see Scheme 4.56).

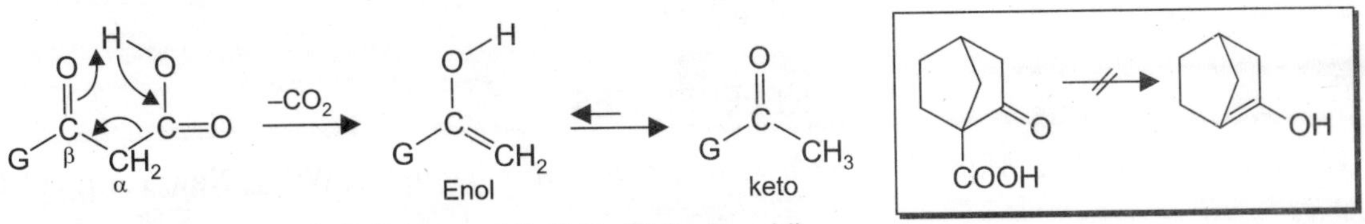

where G = OH (diacid); R (β keto acid); H (β aldehyde acid)

SCHEME 9.20

PROBLEMS

9.1. (*a*) Write the mechanism of S_E1 reaction. How does it compare with S_N1 reaction ?

(*b*) Why generally an S_E1 reaction leads to racemization ?

9.2. How a ketone can be monobrominated in the α-position ? What is the evidence for the axial or the equatorial orientation of the halogen in the α-position of a rigid cyclohexanone ?

9.3. S_E2 reactions are known to occur easily at bridgehead positions while S_N2 reactions are prohibited. Explain.

9.4. Point out if the following statements are true or false.

(*i*) The S_E2 reaction is the electrophile analog of S_N2 mechanism.

(*ii*) The S_E2 reactions involve concerted displacement of one electrophile by another.

(*iii*) The S_Ei mechanism is both kinetically and stereochemically indistinguishable from S_E2.

(*iv*) The S_E1 reaction leads to epimerization (or racemization) if the reaction site is a stereogenic carbon.

(*v*) Carbanions have pyramidal geometry, but are planar when stabilized by resonance.

9.5. Write the mechanism of S_Ei reaction. How does it compare with S_E2 mechanism.

9.6. Each of the following chiral (optically active) ketones is separately treated with base to study the base catalyzed racemization. Explain the results in each case. Name the compound (including *R* and *S*) which will undergo racemization.

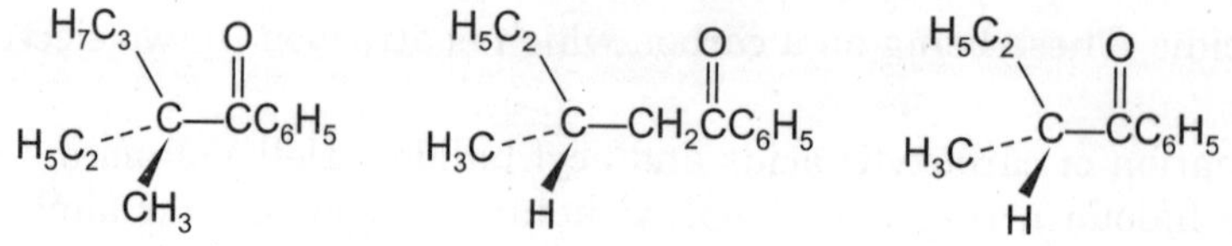

(I) (II) (III)

ANSWERS TO SELECTED PROBLEMS

9.4. (*i*) true, (*ii*) true, (*iii*) false, (*iv*) true, (*v*) true.

9.6. Only ketone III will undergo racemization. In (I) there is no α-hydrogen while in (II), the stereocenter is not involved in enol formation. Compound III is (*R*)-*sec*-butyl phenyl ketone.

REFERENCES

1. M.B. Smith and *J. March, March's Avanced Organic Chemistry*, Wiley, New York, 2001.

Index

acetophenone, 208
achiral, 17
acyl iron complexes, 202
adamantane, 129, 139, 326
addition:
  boron reagents, 404
  bromine (*anti*), 385
  bromine (*syn*), 393
  bromonium ion, 386
  camphor/norcamphor, 464
  carbenes, 447
  carbonyl compounds, 459
  congested boranes, 408
  conjugate, 477
  cyclohexanones, 460
  cyclopropanes, 437
  diastereotopic faces, 180
  Gilman reagents, 431
  metal hydride reduction, 460
  nucleophiles, 429, 457
  organometallic reagents, 470
  prochiral ketones, 465
  radicals, 433
  regioselectivity, 390
  stereochemistry, 385
  unsymmetrical reagents, 396
AIBN, 435
aldol reaction, 197, 485
  diastereoselectivity, 197
  enantioselectivity, 201
alkaloid, 340
alkenes
  bromination 385
  carbene addition, 448
  epoxidation, 438
  faces, 186
  free radical additions, 433
  hydroboration/oxidation, 404
  hydrogenation, 419
    diastereoselective, 121
    enantioselective, 425
    homogeneous, 424
    stereospecific, 420
  hydroxylation
    asymmetric, 446
    (*anti*), 445
    (syn), 443
  oxymercuration, 394
aliphatic electrophilic substitution, 586
alkyl ketone effect, 343
alkynes,
  bromination, 402
  conjugated dienes from, 418
  hydration, 403
  hydroboration, 416, 418
  hydrogenation, 419
  synthesis from boranes, 442
allenes, 25, 131, 140
allylic rearrangement, 4, 254
alpine borane, 465
alternating axis of symmetry, 91
ambident nucleophile, 262
amides, 503
amino acids, 47
ammonium salts, 23, 160
anchimeric assistance, 264, 273
angle strain, 289
annulenes, 151
anomer, 12
anomeric effect, 328
antarafacial, 551
anti, 56
aromatic transition states, 567

asymmetric:
- aldol condensation, 197, 485
- alkylation, 200, 202
- asymmetric transformations, 104
- center, 17
- compounds, 94
- epoxidation, 208
- hydroboration, 192, 204
- oxidation, 211
- reduction, 213
- synthesis, 190
- transformation, 126

aza cope rearrangement, 562
aza-decalin, 339
aza-enolates, 202
atropisomers, 133
axial chirality, 129
axis of symmetry C_n, 88
axial bonds, 304

bakers' yeast, 187, 218
basicity, 234
9BBN, 206, 409
BBN (ipc) 465
benzoin condensatoin, 495
betaine, 480
bicyclo [2.2.2] octane, 333
BINAL-H, 214
BINAP, 427
binaphthyls, 138
biphenyls, 6, 133, 142
Birch reduction, 428
boat conformation, 302
boranes, 404
bridged rings, 331
Bredt's rule, 336
bromination of
- butene, 116, 392
- cyclohexene, 119, 387
- brucine, 100

bromonium ion, 386, 388
Burgi-Dunitz angle, 465
1-butene
- addition of HBr, 399
- epoxidation, 120

2-butene
- bromination of, 116, 392
- epoxidation of, 118

t-butyl cyclohexane, 326

camphor, 39, 88, 326, 332, 464
Cannizzaro reaction, 468
carbene, 447
carbenoids, 449
carbohydrates, 12, 37, 328
carbocation, 25
carbonyl compounds, 449
- disstereotopic faces, 181
- enatiotopic face, 178

carvone, 86
catenanes, 139
center of symmetry, 20, 90
chair conformation, 302
chemoselectivity, 466,
chiral, 16
- amines, 2, 13, 151
- auxiliaries, 199
- axes, 25, 131
- azaenolates, 202
- boranes, 204
- catalysts, 204
- center, more than one, 51
- crown ethers, 104
- cyclohexanes, 314
- enolates, 200
- hydrazones, 203
- metal complexes, 27, 151, 213
- molecules, 19
- organogermanium compounds, 23
- organolithium compounds, 470
- organosilicon compounds, 23
- oxozaborolidine, 207
- phosphorus compounds, 24, 151
- plane, 26, 146
- point groups, 96
- reagents, 204
- substrates, 201
- sulfones, 23
- sulfonium salts, 24, 153
- sulfoxides, 24

chirality, 1, 19
- axes, 25
- plane, 26
- restricted rotaton, 139

chiral compounds, 7
chirotopic, 58
chelotropic reactions, 548
cholesterol, 87
Cis-trans isomers, 28, 72, 78
Claisen reaction, 490

Claisen arrangement, 564
configuration, 14, 48, 286
  absolute, 15
  and configurational descriptions, 47
  inverted, 15
  relative, 15
configurational descriptors
  D, *L*, 31, 47
  l, *u*, 70
  M, *P*, 145, 149
  R, *S*, 32, 70, 140, 143
conformation, 14, 286
  alkanes, 289
  boat, 302
  butane, 287
  chair, 302
  cheletropic reactions, 548
  chemical reactivity, 292, 322
  cyclobutane, 299
  cyclopentane, 298
  cyclopropane, 299
  ethane, 287
  medium sized rings, 300
  preferred, 310
  twist boat, 303
conformational analysis, 287
  alkanes (substituted), 289
  butane, 287
  cyclohexane, 301, 303
  energy, 292
  ethane, 289
  substituted cyclohexanes, 306, 312, 324
  sugars, 328
conformational isomers, 2, 301
conformational diastereomers, 25, 314
conformers
  heterocycles, 326
  isolation, 311
  locking of, 325
  stability, 312
  sugars, 328
conglomerates, 99
conrotation, 513
conservation of molecular arbital symmetry, 511
constitutional isomers, 2
Cope reaction, 378
Cope rearrangement, 558
correlation diagram, 511, 574
Cram's rule, 192
18-crown,-6, 242
chromatography, 104
crystallization induced asymmetric
  transfermation, 127
Curtin-Hammett principle, 294
cumulene, 132
cycloalkanes, 84, 298
cycloalkenes, 76
cycloaddition, 533
  [4 + 2], 533
  [2 + 2], 543
  1, 3-dipolar, 547
cyclobutane, 79
cyclodextrins, 105
cyclohexane, 81
  polysubstituted, 82, 84, 314, 325
cyclohexanone, 342
cyclooctatetraene, 27
cis-cyclooctene, 213
trans-cyclooctene, 2, 148, 217, 401
  (resolution), 102
cyclopentane, 80
cyclopropane, 78

d/*l* nomenclature, 31
D, L-notation, 31
decalins, 338
degenerate rearrangement, 556
Dewar benzene, 531
dextrorotatory, 107
diastereoisomers, 2, 10, 12
  formation of, 113
diastereoselective, 115, 191
diastereotopic ligands and faces, 180
diaxial 1, 3-interaction, 308
DIBAL, 462
Diels-Alder reaction, 533
  asymmetric, 540
dihedral angle, 287, 291
diimide reduction, 423
DIMPO, 214
DIOP, 451
DIPAMP, 152, 426
dissymmetric, 94
disrotatory, 513
DMSO, 219, 239
DOPA, 192, 426

E (*entgengen*), 72
electrocyclic reactions, 512
elements of symmetry, 19

elimination reactions, 351
  Cope, 378
  crown ethers (role), 366
  examples, 357, 360
  pyrolytic, 373
  rearrangements, 379
  regiochemistry, 368
  spectrum of mechanisms, 355
  stereochemistry, 359
  substitution/elimination, 356
  syn (E2), 366
  E1, 352, 355
  E2, 352, 355
  ElcB, 352, 355
enamine, 432
enantiomers, 5, 9
  formation of, 108
enantiomeric excess, 124, 157
enantiomerism, 5, 16
enantiomerization, 126
enantioselective, 202, 472
enantiotopic ligands and faces, 177
endo face, 331
endo isomer, 331
ene reaction, 566
endo rule, 536
enolates, 110, 197, 485
  boron, 198
  chiral, 200
  lithium, 198
enzymes, 9, 102, 187, 215
epimers, 125
epoxides, 80, 118, 438, 441
equatorial bonds, 304
equilibria (cyclohexanes), 321
erythro, 51, 62
D-erythrose, 62
ester hydrolysis, 496–502
exo-endo nomencalture, 331
exo face, 351
E/*Z* nomenclature, 72

Felkin-Anh, 192
flexible conformations, 325
fluxional molecules, 4, 561
flying wedge formula, 28
Fischer projection, 29
  disadvantages, 53
  R, S configuration, 44
  interconversion, 49, 60
fragmentation reactions, 380
formamides, 73
fumarase catalyzed hydration, 62, 187
fumaric acid, 110
fused rings, 338

geometric isomerism, 71
  monocyclic compounds, 76
  nomenclature, 72
  oximes, 73
geraniol, 210
Gilman reagent, 431, 477
glucose, 3, 7, 37, 125, 126
  D, 30
  L, 30
Grignard reagent, 432, 463, 470
  (reaction with)
    acid chlorides, 475
    aldehydes and ketones, 470
    carbon dioxide, 473
    Cram's rule, 471
    comparison with organolithium compounds, 472
    cyclohexanones, 463
    esters, 475
    stereochemistry of, 470
    sterically congested ketones, 472
    α, β-unsaturated ketones, 475
Grigards reagents allylic, 476
(*R*)-glyceraldehyde, 7, 30, 50
halohydrin, 390
Hammond postulate, 245
Haworth projection, 38
helical structures, 26, 149
heterotopic ligands, faces, 174
hexahelicene, 27, 101, 150
Hofmann rule, 369, 372
HOMO, 510
homochiral molecules, 14
homomers, 13
homomorphic ligands, 174
homotopic ligands and faces, 175
hydration, 396
hydrindanes, 340
hydroboration, 122, 404
  (asymmetric), 207, 418
hydrogenation
  alkenes, 419
  alkynes, 423
  aromatic rings, 428

hydrophobic effects, 566
hydroxylation, 443

inclusion compounds, 104
Invertomers, 13
ion pairs, 247
Ipc BBN, 465
isomerization, 593
isomers, 2

keto-enol tautomerism, 322
ketones
  diastereotopic faces, 181
  enantiotopic faces, 178
  Re/*Si* nomenclature, 184
  re, *si* nomenclature, 189
kinetic resolution, 210
Knoevengal reaction, 489

l-configuration, 70
laevorotatory, 107
lithium aluminium hydride, 460
  (reduction of)
    acyclic asymmetric ketones, 464
    aldehydes, ketones, 460
    bulky hydride, 462
    camphor/noncamphor, 464
    carboxylic acids, 467
    esters, 467
    prochiral ketones, 465
    α, β-unsaturated ketones, 467
lithium dialkyl cuparates, 477
lithium diisopropylamide (LDA), 489

Mannich reaction, 492
Markovnikov rule, 390, 396
Masamune, 56
Meerwein Ponndorf reduction, 468
meso compounds, 13
Michael addition, 431, 492
Mobius-Hackel analysis, 568
MTPA, 157
mutarotation, 322

neighboring group participation, 264
  acetoxyl group, 269
  double bonds, 270
  halogens, 265
  sigma bonds, 273
Newman projection, 54, 67

NMR spectroscopy, 153, 218
NOE, 155
non-classical carbocations, 270, 273
norbornane (systems), 270, 332, 334
stereoheterotopic ligands, 174
nucleophilicity, 234
nucleophilic acyl sustitution, 458
nucleophilic additions, 457

optical activity, 107
optical purity, 124
organoboranes, 404
  alcohols (from), 405, 412, 413, 414
  aldehydes (from), 414
  E-alkenes (from), 417
  Z-alkenes (from), 417
  alkynes (from), 418
  amines (from), 410
  asymmetric synthesis (with), 418
  carbonylation, 412
  congested, 408
  cyanidation, 415
  dienes (from), 417
  esters (from), 415
  hydrocarbons (from), 410
  isomerization, 409
  ketones (from), 413
  oxidation, 407
  protonolysis, 410
organometallic compounds, 470, 478
osmium tetraoxide, 446
oxazaborolidine, 207
oxaphosphetane, 480
oximes, 73
oxy-Cope rearrangement, 562
oxymercuration, 394

Pericyclic reactions, 508
  correlation diagram, 511, 574
  HOMO-LUMO, 510
  PMO method, 512, 568
  Perkin reaction, 495
pseudo asymmetric centre, 59
phase transfer catalysis, 241
photochemical electrocyclic reactions, 50
pinacolic deamination, 297
α-pinene, 121
Pitzer strain, 298
planar chirality, 146
plane of symmetry, 20, 89

PMO method, 568
point groups, 95
polarimeter, 107
polycyclic compounds, 331
Prevost hydroxylation, 445
prochirality, 173, 333
prostereoisomerism, 173
prototropy, 3
pro *R*, 183
pro *S*, 183
(*S*)-prolinol, 200
(*S*)-(—)-proline, 200
propanoic acid, 173
(asymmetric α-substitution), 200
protein, 9
pseudochirality center, 57
pyrolytic elimination, 373

R/*S* notation, 31, 40, 47, 56
r/*s* notation, 59, 71, 85
racemate, 97
racemisation, 107
radicals, 108, 433–37
RAMP, 203
Re/*Si* nomenclature, 184
re, si nomenclature, 189
Reformatsky reaction, 478
residual isomers, 14
resolution, 97, 125
cis-retinal, 76
Robinson annulation, 215, 431, 493
rotation reflection axis (*Sn*), 19, 91

Salen, 211
SAMP, 203
sawhorse projection, 54
Saytzeff rule, 369, 372
s-*cis*/s-*trans*, 74
S_E1, 588
S_E2, 590
S_Ei, 591
S_Ei', 591
selectride, 461
sequence rules, 32
SET mechanism, 263
Sharpless (asymmetric epoxidation), 209, 440
shift reagents, 156
Si face, 184
Si/*Re* nomenclature, 184
sigmatropic rearrangements, 550
carbon migration, 557
hydrogen migration, 551
[1, 3,), 552
[1, 5], 553
[1, 7], 555
[5, 5], 565
Simmons Smith cyclopropanation, 478
S_N1, 113, 242
S_N1', 253
S_N2, 112, 226
S_N2', 253
S_Ni, 259
sodium borohydride, 460
specific rotation, 107
spiranes, 25, 131
stereoaxis, 25, 129, 145
stereocenter, 16
stereogenic carbon, 16, 21
stereogenic centers other than carbon, 22
stereogenic centre, 18
stereogenic units, 27
stereoheterotopic ligands, 134, 218
stereoisomers, 2, 4, 86
stereoplane, 26, 146
stereospecific reactions, 116, 119
steroid, 340
stereoselective reactions, 116, 119, 184, 190
Stobbe condensation, 496
strained carbocycles, 336
strain (allylic), 342
s-*trans*, 74
substitution
aliphatic electrophylic, 586
nucleophilic, 225
sugars, 328
sulfone, 23
sulfoxide, 23
suprafacial, 551
symmetry elements, 19, 88, 93
syn, 56
syn addition, 393
*syn*clinal, 288
syn elimination, 366, 375
syn periplanar, 288

TAPA, 101
tartaric acid, 2, 18, 30, 61, 65, 66, 95
tautomers, 3
thalidomide, 122

three dimensional formulas, 28
  interconversions, 29, 49, 60
threo, 51, 62
threose, 51, 62
torsion angle, 291
torsional chirality, 71
TOT, 105
transannular interaction, 301
transition states,
  anti aromatic, 567
  aromatic, 567
tributyltin hydride, 435
twistane, 87, 326
twist boat, 303

u-configuration, 70

valence isomers, 3
vision process, 75
vision process, 75
vitamin A, 76

Wadsworth-Emmons reaction, 484
Wagner-Meerwein rearrangement, 296
wedge formula, 28, 49
Wilkinson's catalyst, 424, 426
Wittig reaction, 479, 482
Woodward-Hofmann rules, 511
Woodward hydroxylation, 445

X-ray diffraction, 15

Ylide
  resonance stabilised, 479, 480
  non-stabilised, 480
  Z (*zusammen*), 72
  Zig-Zag structures, 55